BIOGEOCHEMISTRY OF INLAND WATERS

A DERIVATIVE OF ENCYCLOPEDIA OF INLAND WATERS

BIOGEOCHEMISTRY OF INLAND WATERS

A DERIVATIVE OF ENCYCLOPEDIA OF INLAND WATERS

EDITOR

PROFESSOR GENE E. LIKENS
Cary Institute of Ecosystem Studies
Millbrook, NY, USA

ELSEVIER

Amsterdam • Boston • Heidelberg • London • New York • Oxford
Paris • San Diego • San Francisco • Singapore • Sydney • Tokyo
Academic Press is an imprint of Elsevier

ACADEMIC PRESS

Academic Press is an imprint of Elsevier
525 B Street, Suite 1900, San Diego, CA 92101-4495, USA
30Corporate Drive, Suite 400, Burlington, MA01803, USA
32 Jamestown Road, London, NW1 7BY, UK
Radarweg 29, PO Box 211, 1000 AE Amsterdam, The Netherlands

British Library Cataloguing in Publication Data
A catalogue record for this book is available from the British Library

Library of Congress Cataloging-in-Publication Data
Biogeochemistry of inland waters : a derivative of Encyclopedia of Inland Waters / editor, Gene E. Likens.
p. cm.
Includes bibliographical references and index.
ISBN 978-0-12-810224-4 (alk. paper)
1. Fresh water. 2. Biogeochemistry. I. Likens, Gene E., 1935-
QH541.5.F7.B556 2010
577.6'14–dc22
2010010993

ISBN: 978-0-12-810224-4

For information on all Academic Press publications
visit our website at Elsevierdirect.com

10 11 12 10 9 8 7 6 5 4 3 2 1

EDITOR

Professor Gene E. Likens is an ecologist best known for his discovery, with colleagues, of acid rain in North America, for co-founding the internationally renowned Hubbard Brook Ecosystem Study, and for founding the Institute of Ecosystem Studies, a leading international ecological research and education center. Professor Likens is an educator and advisor at state, national, and international levels. He has been an advisor to two governors in New York State and one in New Hampshire, as well as one U.S. President. He holds faculty positions at Yale, Cornell, Rutgers Universities, State University of New York at Albany, and the University of Connecticut, and has been awarded nine Honorary doctoral Degrees. In addition to being elected a member of the prestigious National Academy of Sciences and the American Philosophical Society, Dr. Likens has been elected to membership in the American Academy of Arts and Sciences, the Royal Swedish Academy of Sciences, Royal Danish Academy of Sciences and Letters, Austrian Academy of Sciences, and an Honorary Member of the British Ecological Society. In June 2002, Professor Likens was awarded the 2001 National Medal of Science, presented at The White House by President G. W. Bush; and in 2003 he was awarded the Blue Planet Prize (with F. H. Bormann) from the Asahi Glass Foundation. Among other awards, in 1993 Professor Likens, with F. H. Bormann, was awarded the Tyler Prize, The World Prize for Environmental Achievement, and in 1994, he was the sole recipient of the Australia Prize for Science and Technology. In 2004, Professor Likens was honored to be in Melbourne, Australia with a Miegunyah Fellowship. He was awarded the first G.E. Hutchinson Medal for excellence in research from The American Society of Limnology and Oceanography in 1982, and the Naumann-Thienemann Medal from the Societas Internationalis Limnologiae, and the Ecological Society of America's Eminent Ecologist Award in 1995. Professor Likens recently stepped down as President of the International Association of Theoretical and Applied Limnology, and is also a past president of the American Institute of Biological Sciences, the Ecological Society of America, and the American Society of Limnology and Oceanography. He is the author, co-author or editor of 20 books and more than 500 scientific papers.

Professor Likens is currently in Australia on a Commonwealth Environment Research Facilities (CERF) Fellowship at the Australian National University.

CONTRIBUTORS

J H Aldstadt III
University of Wisconsin-Milwaukee, Milwaukee, WI, USA

W M Alley
U.S. Geological Survey, San Diego, CA, USA

J L Ammerman
SEAL Analytical, Inc., Mequon Technology Center, Mequon, WI, USA

J P Antenucci
University of Western Australia, Nedlands, WA, Australia

J F Atkinson
University of Buffalo, Buffalo, NY, USA

D L Bade
Kent State University, Kent, OH, USA

M T Barbour
Tetra Tech, Owings Mills, MD, USA

D Bastviken
Stockholm University, Stockholm, Sweden

L Boegman
Queen's University, Kingston, ON, Canada

B Boehrer
UFZ – Helmholtz Centre for Environmental Research, Magdeburg, Germany

H A Bootsma
University of Wisconsin-Milwaukee, Milwaukee, WI, USA

P A Bukaveckas
Virginia Commonwealth University, Richmond, VA, USA

N Caraco
Cary Institute of Ecosystem Studies, Millbrook, NY, USA

M J Coates
Deakin University, Warrnambool, Vic., Australia

J J Cole
Cary Institute of Ecosystem Studies, Millbrook, NY, USA

D J Conley
Lund University, Lund, Sweden

C S Cronan
University of Maine, Orono, ME, USA

E A Dreelin
Michigan State University, East Lansing, MI, USA

K R Echols
US Geological Survey, Columbia, MO, USA

M C Feller
University of British Columbia, Vancouver, BC, Canada

K Fienberg
University of Minnesota, Minneapolis, MN, USA

A M Folkard
Lancaster University, Lancaster, UK

E Foufoula-Georgiou
University of Minnesota, Minneapolis, MN, USA

W Geller
UFZ – Helmholtz Center for Environmental Research, Magdeburg, Germany

A E Giblin
Marine Biological Laboratory, Woods Hole, MA, USA

C Gilmour
Smithsonian Environmental Research Center, Edgewater, MD, USA

D S Glazier
Juniata College, Huntingdon, PA, USA

C R Goldman
University of California, Davis, CA, USA

E M Gross
University of Konstanz, Konstanz, Germany

G Harris
Lancaster University, UK

J Hauxwell
DNR Science Operations Center, Madison, WI, USA

B R Hodges
University of Texas at Austin, Austin, TX, USA

G Hornberger
Vanderbilt University, Nashville, TN, USA

R Howarth
Cornell University, Ithaca, NY, USA

J A Hubbart
University of Missouri, Columbia, MO, USA

J R Jones
University of Missouri, Columbia, MO, USA

N O G Jørgensen
University of Copenhagen, Fredericksberg, Denmark

P Y Julien
Colorado State University, Fort Collins, CO, USA

G Katul
Duke University, Durham, NC, USA

S S Kaushal
University of Maryland Center for Environmental Science, Solomons, MD, USA

R Kipfer
Swiss Federal Institute of Environmental Science and Technology (Eawag), Swiss Federal Institute of Technology (ETH), Ueberlandstr, Duebendorf, Switzerland

S Knight
University of Wisconsin Trout Lake Station and Wisconsin Department of Natural Resources, Boulder Junction, WI, USA

H-P Kozerski
Institute of Freshwater Ecology and Inland Fisheries, Berlin, Germany

J W LaBaugh
U.S. Geological Survey, Reston, VA, USA

W M Lewis
University of Colorado, Boulder, CO, USA

A Lorke
University of Koblenz-Landau, Landau/Pfaly, Germany

S MacIntyre
University of California, Santa Barbara, CA, USA

R W Marino
Cornell University, Ithaca, NY, USA

M D Mattson
Massachusetts Department of Environmental Protection, Worcester, MA, USA

R M McNinch
Michigan State University, East Lansing, MI, USA

J C Meadows
US Geological Survey, Columbia, MO, USA

R Menzel
Humboldt Universität zu Berlin, Berlin, Germany

M Meybeck
Université Pierre et Marie Curie, Paris, France

E Michael Perdue
Georgia Institute of Technology, Atlanta, GA, USA

S G Monismith
Stanford University, Stanford, CA, USA

T N Narasimhan
University of California at Berkeley, CA, USA

H M Nepf
Massachusetts Institute of Technology, Cambridge, MA, USA

J R Nimmo
U.S. Geological Survey, Menlo Park, CA, USA

R H Norris
University of Canberra, Canberra, ACT, Australia

K Novick
Duke University, Durham, NC, USA

Y Olsen
Norwegian University of Science and Technology, Trondheim, Norway

C E Orazio
US Geological Survey, Columbia, MO, USA

F Peeters
Universität Konstanz, Mainaustrasse, Konstanz, Germany

Y T Prairie
Université du Québec à Montréal, Montréal, QC, Canada

E Prepas
Lakehead University, Thunder Bay, ON, Canada

G Putz
Lakehead University, Thunder Bay, ON, Canada

V H Resh
University of California, Berkeley, CA, USA

C S Reynolds
Centre of Ecology and Hydrology and Freshwater Biological Association, Cumbria, UK

B L Rhoads
University of Illinois at Urbana-Champaign, Urbana, IL, USA

G Riedel
Smithsonian Environmental Research Center, Edgewater, MD, USA

J B Rose
Michigan State University, East Lansing, MI, USA

F J Rueda
Universidad de Granada, Granada, Spain

M Schultze
UFZ – Helmholtz Center for Environmental Research, Magdeburg, Germany

N Serediak
Lakehead University, Thunder Bay, ON, Canada

R W Sheibley
Washington Water Science Center, Tacoma, WA, USA

D W Smith
Lakehead University, Thunder Bay, ON, Canada

V H Smith
University of Kansas, Lawrence, KS, USA

M Søndergaard
University of Aarhus, Denmark

C E W Steinberg
Humboldt Universität zu Berlin, Berlin, Germany

R W Sterner
University of Minnesota, St. Paul, MN, USA

K M Stewart
State University of New York, Buffalo, NY, USA

E Struyf
Lund University, Lund, Sweden

A N Sukhodolov
Institute of Freshwater Ecology and Inland Fisheries, Berlin, Germany

M A Timofeyev
Baikalean Research Centre, Irkutsk, Russia

R W Tiner
University of Massachusetts, Amherst, MA, USA

L J Tranvik
Uppsala University, Uppsala, Sweden

H M Valett
Virginia Tech, Blacksburg, VA, USA

A V Vähätalo
University of Helsinki, Helsinki, Finland

J Vidal
Universidad de Granada, Granada, Spain

W F Vincent
Laval University, Quebec City, QC, Canada

E von Wachenfeldt
Uppsala University, Uppsala, Sweden

C J Watras
Wisconsin Department of Natural Resources, Madison, WI, USA

F M Wilhelm
University of Idaho, Moscow, ID, USA

T C Winter
US Geological Survey, Denver, CO, USA

E Wohl
Department of Geosciences, Colorado State University, Ft. Collins, CO, USA

A Wüest
Eawag, Surface Waters – Research and Management, Kastanienbaum, Switzerland

CONTENTS

Editor v

Contributors vii–ix

Introduction to the Biogeochemistry of Inland Waters and Factors Affecting Flux and Cycling of Chemicals xv–xvi

PROPERTIES OF WATER

Chemical Properties of Water *J H Aldstadt III, H A Bootsma, and J L Ammerman* 1

Physical Properties of Water *K M Stewart* 10

Pressure *J F Atkinson* 17

Gas Exchange at the Air–Water Interface *D L Bade* 28

Light, Photolytic Reactivity and Chemical Products *A V Vähätalo* 37

HYDROLOGY

Hydrological Cycle and Water Budgets *T N Narasimhan* 51

Atmospheric Water and Precipitation *K Fienberg and E Foufoula-Georgiou* 58

Snow and Ice *G Hornberger and T C Winter* 68

Evapotranspiration *G Katul and K Novick* 69

Vadose Water *J R Nimmo* 76

Ground Water *W M Alley* 88

Ground Water and Surface Water Interaction *H M Valett and R W Sheibley* 95

Groundwater Chemistry *J W LaBaugh* 107

Fluvial Export *M Meybeck* 118

Fluvial Transport of Suspended Solids *P Y Julien* 131

Streams *E Wohl* 134

Rivers *P A Bukaveckas* 143

Springs *D S Glazier* 155

Wetland Hydrology *R W Tiner* 177

HYDRODYNAMICS AND MIXING IN LAKES, RESERVOIRS, WETLANDS AND RIVERS

Biological-Physical Interactions *C S Reynolds* 189
Density Stratification and Stability *B Boehrer and M Schultze* 196
The Surface Mixed Layer in Lakes and Reservoirs *S G Monismith and S MacIntyre* 207
Small-Scale Turbulence and Mixing: Energy Fluxes in Stratified Lakes *A Wüest and A Lorke* 222
The Benthic Boundary Layer (in Rivers, Lakes, and Reservoirs) *A Lorke and S MacIntyre* 230
Currents in Rivers *A N Sukhodolov, H-P Kozerski, and B L Rhoads* 239
Currents in Stratified Water Bodies 1: Density-Driven Flows *F Peeters and R Kipfer* 247
Currents in Stratified Water Bodies 2: Internal Waves *L Boegman* 256
Currents in Stratified Water Bodies 3: Effects of Rotation *J P Antenucci* 277
Currents in the Upper Mixed Layer and in Unstratified Water Bodies *F J Rueda and J Vidal* 286
Flow in Wetlands and Macrophyte Beds *A M Folkard and M J Coates* 301
Flow Modification by Submerged Vegetation *H M Nepf* 313
Hydrodynamical Modeling *B R Hodges* 320

INORGANIC CHEMICALS: CYCLES AND ECOSYSTEM DYNAMICS

Chemical Fluxes and Dynamics in River and Stream Ecosystems *W M Lewis* 335
Dissolved CO_2 *J J Cole and Y T Prairie* 343
Alkalinity *M D Mattson* 348
Major Cations (Ca, Mg, Na, K, Al) *C S Cronan* 354
Chloride *S S Kaushal* 361
Iron and Manganese *A E Giblin* 368
Micronutrient Elements (Co, Mo, Mn, Zn, Cu) *C R Goldman* 378
Biogeochemistry of Trace Metals and Mettaloids *C Gilmour and G Riedel* 383
Nitrogen Fixation *R W Marino and R Howarth* 392
Nitrogen *R Howarth* 400
Phosphorus *N Caraco* 408
Silica *D J Conley and E Struyf* 414
Salinity *G Harris* 418

ORGANIC COMPOUNDS: CYCLES AND DYNAMICS

Allelochemical Reactions *E M Gross* 425
Carbohydrates *N O G Jørgensen* 437
Carbon, Unifying Currency *Y T Prairie and J J Cole* 453
Dissolved Humic Substances: Interactions with Organisms *C E W Steinberg, M A Timofeyev, and R Menzel* 457

Interactions of Dissolved Organic Matter and Humic Substances *L J Tranvik and E von Wachenfeldt* 464

Lipids *Y Olsen* 471

Methane *D Bastviken* 480

Natural Organic Matter *E Michael Perdue* 503

Organic Nitrogen *N O G Jørgensen* 517

Nutrient Stoichiometry in Aquatic Ecosystems *R W Sterner* 537

Redox Potential *M Søndergaard* 549

POLLUTION AND REMEDIATION

Acidification *W Geller and M Schultze* 557

Aquatic Ecosystems and Human Health *R M McNinch, J B Rose, and E A Dreelin* 569

Bioassessment of Aquatic Ecosystems *R H Norris and M T Barbour* 577

Deforestation and Nutrient Loading to Fresh Waters *M C Feller* 585

Distribution and Abundance of Aquatic Plants – Human Impacts *S Knight and J Hauxwell* 601

Effects of Climate Change on Lakes *W F Vincent* 611

Eutrophication *V H Smith* 617

Fires *E Prepas, N Serediak, G Putz, and D W Smith* 630

Floods *J A Hubbart and J R Jones* 644

Mercury Pollution in Remote Fresh Waters *C J Watras* 648

Pollution of Aquatic Ecosystems I *F M Wilhelm* 658

Pollution of Aquatic Ecosystems II: Hydrocarbons, Synthetic Organics, Radionuclides, Heavy Metals, Acids, and Thermal Pollution *K R Echols, J C Meadows, and C E Orazio* 668

Vector-Borne Diseases of Freshwater Habitats *V H Resh* 677

Index 687

INTRODUCTION TO THE BIOGEOCHEMISTRY OF INLAND WATERS AND FACTORS AFFECTING FLUX AND CYCLING OF CHEMICALS

Biogeochemistry is a multidisciplinary approach to the study of the transformation, flux, and cycling of chemical compounds in aquatic and terrestrial ecosystems (e.g., Likens et al., 1977; Likens and Bormann, 1995; Schlesinger, 1997). The science of biogeochemistry combines aspects of biology, ecology, geology, chemistry, and often hydrology and meteorology. Cycling of elements occurs within the boundaries of an ecosystem, whereas flux refers to the transfer of materials across the boundaries of an ecosystem (Likens, 1992). Conceptual models are informative for thinking about and, in particular, quantifying these complicated transformations, linkages, interactions, and fluxes (**Figure 1**). Materials moving across the boundaries of an ecosystem represent the biogeochemical connections of that particular ecosystem with the remainder of the biosphere and provide critical points for management interventions (Bormann and Likens, 1967). For example, inputs of acidic materials from the atmosphere (acid deposition) can degrade aquatic and terrestrial ecosystems. Federal legislation in the United States (Clean Air Act Amendments of 1990) was passed to reduce these inputs and their impact on the structure and function of recipient ecosystems. Likewise, outputs from the drainage basin (watershed-ecosystem) of waste water or agricultural or industrial chemicals can pollute or eutrophy receiving systems such as rivers flowing into a lake or estuary. Management interventions using this understanding can then be taken to reduce these impacts.

This volume contains seven sections: first, a brief introduction to the biogeochemistry of inland waters and the factors affecting the flux and cycling of chemicals; second, the properties of water that impact on the biogeochemistry of an ecosystem; third, the hydrologic factors that affect biogeochemical flux and cycling; fourth, the hydrodynamics and mixing in lakes, reservoirs, wetlands, and rivers that are important to biogeochemical dynamics; fifth, the cycles and ecosystem dynamics of inorganic chemicals; sixth, the cycles and

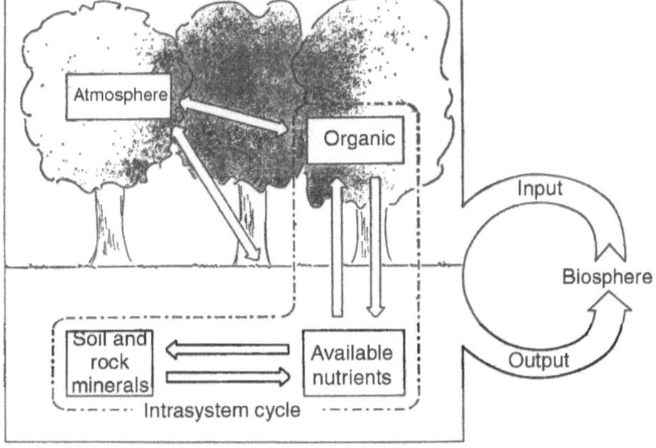

Figure 1 A conceptual model of the major biogeochemical relationships for a terrestrial ecosystem Bormann FH and Likens GE (1967), Reprinted with permission from AAAS.

ecosystem dynamics of organic compounds; and seventh, pollution and remediation of the biogeochemical components of aquatic ecosystems. As such, the contents of this book are broadly drawn to cover a wide variety of topics related to biogeochemistry.

The articles in this volume are reproduced from the Encyclopedia of Inland Waters (Likens, 2009). I thank the authors of the articles in this volume for their excellent and up-to-date coverage of this important topic.

Gene E. Likens
Cary Institute of Ecosystem Studies
Millbrook, NY
December 2009

References Cited/Further Reading:

Bormann FH and Likens GE (1967) Nutrient cycling. *Science* 155(3761): 424–429.
Hutchinson GE (1950) The biogeochemistry of vertebrate excretion. *Bulletin of the American Museum of Natural History* 96: 554.
Likens GE (1992) *The Ecosystem Approach: Its Use and Abuse. Excellence in Ecology*, vol. 3, p. 167. Germany: Ecology Institute, Oldendorf/Luhe.
Likens GE (ed.) (2009) *Encyclopedia of Inland Waters*, 3 vols. Oxford, UK: Elsevier/Academic Press.
Likens GE and Bormann FH (1995) *Biogeochemistry of a Forested Ecosystem*, 2nd edn., p. 159. New York: Springer.
Likens GE, Bormann FH, Pierce RS, Eaton JS, and Johnson NM (1977) *Biogeochemistry of a Forested Ecosystem*, p. 146. New York: Springer.
Schlesinger WH (1997) *Biogeochemistry: An Analysis of Global Change*, 2nd edn., p. 588. London: Academic Press.
Vernadsky WI (1945) The biosphere and the nöösphere. *American Scientist* 33(1): 1–12.

PROPERTIES OF WATER

Contents

Chemical Properties of Water
Physical Properties of Water
Pressure
Gas Exchange at the Air-Water Interface
Light, Photolytic Reactivity and Chemical Products

Chemical Properties of Water

J H Aldstadt III and H A Bootsma, University of Wisconsin-Milwaukee, Milwaukee, WI, USA
J L Ammerman, SEAL Analytical, Inc., Mequon Technology Center, Mequon, WI, USA

Water is H_2O, hydrogen two parts, oxygen one, but there is also a third thing, that makes it water and nobody knows what it is.

—D.H. Lawrence (1929)

Introduction

Water is the most abundant molecule on Earth. In spite of being so common, water is quite unusual – from its high melting and boiling points to its tremendous solvating power, high surface tension, and the largest dielectric constant of any liquid. In this article, we present an overview of the chemical properties of water. The phrase 'chemical property' is context dependent, which we define in general as a description of the way that a substance changes its identity in the formation of other substances. A universally accepted set of chemical properties does not exist in the same way that there is, more or less, a standard set of physical properties for a given substance. Whereas a given substance has intrinsic physical properties (such as melting point), by our definition chemical properties are clearly tied to change. In addition to reactivity, a substance's 'chemical properties' also typically include its electronegativity, ionization potential, preferred oxidation state(s), coordination behavior, and the types of bonding (e.g., ionic, covalent) in which it participates. Because these properties are extensively studied in general chemistry courses, we will not further discuss them here. Rather, we move beyond the basic general chemistry concepts and focus upon water in a limnologic context – particularly, its bulk fluid structure and aspects of its chemical reactivity in the hydrosphere.

In the following pages, we begin by briefly reviewing the molecular structure of water and then discuss models for its structure in 'bulk' solution. We then turn our attention to the hydration of ions and an overview of important reactions that involve water, including acid–base, complexation, precipitation, and electron transfer. We conclude with a look at trends in the chemical composition of freshwater that are fundamental to the field of limnology.

The Structure of Water

Knowledge of the structure of water is the basis for understanding its unique chemical and physical properties. Like the other nonmetallic hydrides of the Group 16 elements, water is a triatomic molecule that forms a nonlinear structure. In terms of group theory, water has two planes of symmetry and a twofold rotation axis and is therefore assigned point-group C_{2v}. The H–O–H angle is 104.5°, formed as a result of the distortion of the O–H bond axes by the two pairs of nonbonding electrons on the oxygen atom. Although water is often described as having four sp^3-hybridized molecular orbitals in a slightly distorted tetrahedral geometry, models based solely upon that configuration fail to accurately predict the properties of liquid water, particularly the extent and influence of hydrogen bonding on the structure of the bulk fluid state. However, a tetrahedral geometry is in fact present in the solid state, giving rise to the sixfold axis of symmetry that is characteristic of ice, and in large part as the basis of the networks that form in the bulk liquid, though in a rapidly fluctuating dynamic state.

Models for the bulk fluid structure of water are a function of the noncovalent van der Waals forces that exist between water molecules. There are five major types of van der Waals forces that occur between neutral molecules and ions in solution:

(1) London (or dispersion) forces, in which transient dipoles form by variations in electron density between neutral molecules; (2) Debye forces, in which the dipole of a molecule induces the formation of a dipole in an adjacent neutral molecule; (3) Keesom forces, which form between neighboring dipoles; (4) Coulombic forces, the electrostatic attraction (and repulsion) of ions; and (5) hydrogen bonds, which involve the electrophilic attraction of a proton to electronegative atoms such as oxygen and nitrogen. All of these forces are present in aqueous solution to varying degrees – hydrogen bonding being the most dominant. The high negative charge density of the oxygen atom relative to the high positive charge density of the hydrogen atom creates a large (1.84 D) electric dipole moment for the water molecule (**Figure 1**). Because of the large dipole moment, the partial positive charge on the H atom is attracted to electron density, while the partial negative charge on the O atom causes the attraction of electrophilic H atoms. In this way, hydrogen bonds are formed, representing the strongest of the van der Waals forces that exist between neutral molecules. While each hydrogen bond is ~20 times weaker than a typical covalent bond, each water molecule can participate in multiple hydrogen bonds – one to each H atom and one (or more) to each nonbonded pair of electrons on the O atom.

The key to understanding the structure of bulk water – and its abnormal properties – is understanding the way that noncovalent hydrogen bonds affect its intermolecular interactions. Although one might expect that the random translational motion of molecules in a liquid results in an amorphous structure, the extensive network of hydrogen-bonded molecules in the liquid state of water gives rise to a surprisingly very high degree of order. Water has considerable short-range order that continues to a distance of at least ~10–15 Å from the 2.75 Å diameter water molecule. Hydrogen bonds are certainly not peculiar to water, but in water they form such elaborate, extensive, and strong networks that they create a 'bulk' structure with significant order, order that is in fact maintained up to its boiling point.

A great deal of research has been devoted to improving our understanding of water's structure in condensed phases – broadly divided into studies of short-range and long-range order, the latter defined as beyond ~15 Å. These research endeavors have been both theoretical and empirical, with theoreticians employing advanced computational tools for molecular modeling, and experimentalists armed with a wide variety of spectroscopic techniques. Models for the structure of water in the solid phase (i.e., in the various ices that can form) generate little controversy because theoretical models can be directly verified by crystallographic and neutron-scattering techniques. Because of the much more limited atomic motion in the solid state, crystallographic methods have provided an accurate picture of the various ices that form as a function of temperature and pressure. The most common type of ice under ambient conditions is hexameric ice, in which six water molecules are hydrogen bonded to form a hexagonal ring, as shown in **Figure 2**. The most stable state for this structure is a so-called 'chair' conformation (analogous to cyclohexane), in which H–O···H bonds alternate around the ring (where '–' is a covalent bond and '···' is a hydrogen bond). Also shown in **Figure 2** is the 'boat' conformation, an energetically less stable conformation than the 'chair' structure. Each O atom has a nearly tetrahedral arrangement of H atoms

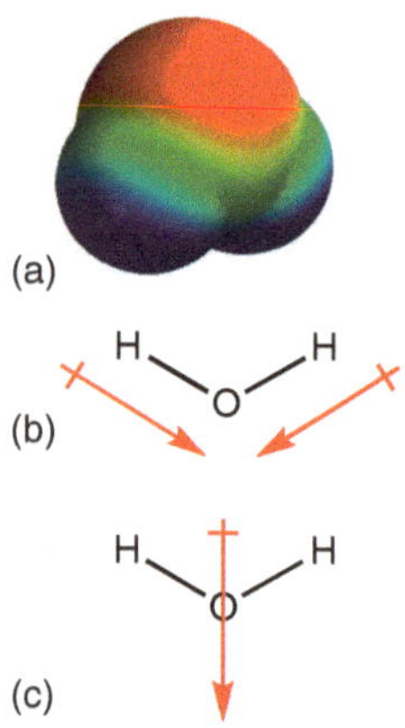

Figure 1 (a) The distribution of electron density in molecular water (red = high, blue = low). Representation of the electric dipolar nature of molecular water, as contributing dipoles along each O–H axis (b) and as a net dipole (c).

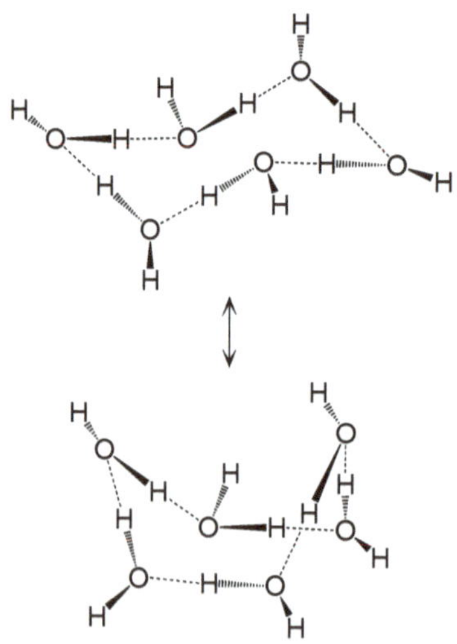

Figure 2 The arrangement of hexagonal water into a 'chair' conformation (top) and less stable 'boat' conformation (bottom).

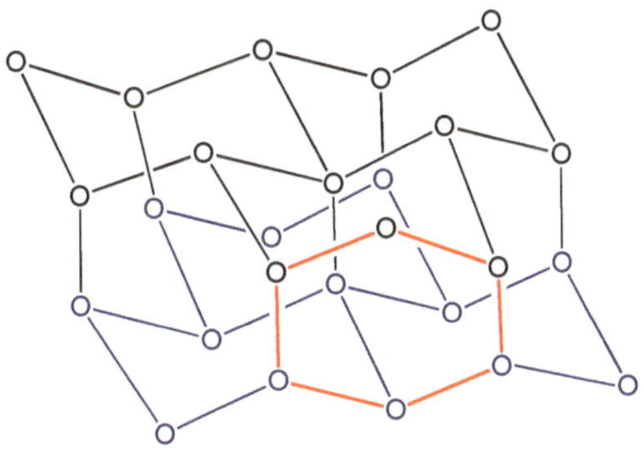

Figure 3 The structure of the most common form of ice (hexagonal ice), an arrangement based upon the HOH 'chair' hexamer. Each oxygen atom is at the approximate center of a tetrahedron formed by four other oxygen atoms. The sixfold axis of symmetry is shown in red for a layer of water 'chairs' (black) overlaying another layer (blue). (Hydrogen atoms are not shown for clarity.)

surrounding it, in which two H atoms are covalently bonded and two noncovalently as hydrogen bonds. The sixfold axis of symmetry found in ice (**Figure 3**) is the result of the building blocks of cyclic hexamers.

Unlike models for ice, much controversy continues to surround models for the structure of liquid water. This may be somewhat surprising given that water is a simple molecule, yet general agreement on a realistic model remains elusive despite the application of powerful computational and experimental approaches. Predicting the precise arrangements of hydrogen-bonded neighboring water molecules is challenging because the structures are in a state of rapid flux (at subpicosecond timescales). Some insight into the structure of bulk water can be gleaned by examining the structural changes that occur upon the melting of ice. When ice melts, the increase in temperature causes a slight disruption of the hydrogen-bonded network, thereby initially causing the ice crystalline lattice to collapse. Whereas the structure of ice is >80% ordered, only an ~10% decrease in order occurs upon transition to the liquid phase. In this way, much if not most of the short-range order is maintained, which in fact continues to persist in part all of the way to the boiling point at 100 °C, where the order is essentially lost completely. The partial collapse of the ordered environment during melting results in slightly more compact hexameric chairs. Consequently, water has the very unusual property of maximal density at a temperature that is higher than its melting point. Above 4 °C, further disruption of the intricate networks of cyclic hexamers by more intensive thermal agitation causes the structures to become more open with a consequent decrease in water's density.

Water forms clusters in the liquid state. The presence of 'ice-like' structures in water, based on not only hexameric but also pentameric and octameric building blocks, along with 'free' swimming water molecules in more amorphous regions, is the generally accepted model (**Figure 4**). However, there have been intriguing studies that suggest that there are regions that are far more complex than the structures analogous to ice. Curiously, one of the earliest is found in Plato's dialogue *Timaeus*, where the ancient Greek's classification of matter – Earth, Fire, Air, and Water – is described in mathematical (geometric) terms. In the Platonic conception of 'substance,' matter is intrinsically composed of triangles. Earth is cubic (i.e., two equilateral triangles each comprising six faces), Fire is tetrahedral (four triangles), and Air is octahedral (eight triangles). In Plato's view, water is the most complex structure, taking the form of an icosahedron. A regular icosahedron has 20 faces, with five equilateral triangles meeting at each of the 12 vertices. Thus, along with the dodecahedron, these regular convex polyhedra comprise the famous 'Platonic Solids.' This ancient conception of water may seem quaint, yet it is strikingly similar in concept to several recent theoretical models of the structure of water in the bulk liquid phase. Clusters based on dodecahedra and icosahedra have been proposed by molecular modeling and supported by experiment to exist in water – though the evidence remains somewhat controversial. Early work by Searcy and Fenn on protonated water clusters by molecular beam mass spectrometry found that a large peak in the spectrum, which corresponded to 21 water molecules (a so-called 'magic number') was present, that is, for a cluster of unusual stability. Speculation arose that the structure of this 'magic' cluster was a dodecahedral complex of 20 water molecules, each vertex occupied by an oxygen atom and a hydronium ion trapped within (e.g., as in clathrates). Recent work by Dougherty and Howard has indeed found evidence for dodecahedral clusters, and Chaplin has proposed a theoretical model for the formation of icosahedral clusters, a model that has been supported by recent neutron scattering experiments.

Solvation by Water

Ions in aqueous solution interact with one another and with other nonelectrolytes, and their presence in water's dipolar electronic field creates relatively strong noncovalent bonds such that the hydrated ion is the form that undergoes further interactions and chemical reactions, and has consequent implications for the rates of these processes. Only in the gas

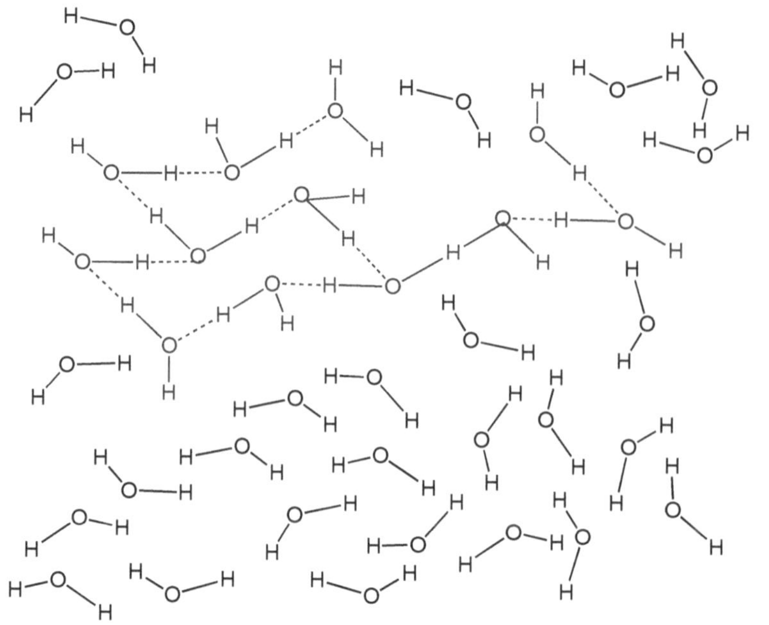

Figure 4 Proposed models for the structure of bulk water. (Top) The ''flickering cluster'' model, with ice-like ordered regions (high-lighted in blue) surrounded by amorphous regions where little short-range order is present. Molecular modeling and some experimental evidence suggests that quite complex structures, such as dodecahedra (bottom left) and icosahedra (bottom right), may also exist.

phase do 'bare' (unsolvated) ions exist; in the liquid phase, all ions are hydrated to some degree.

To appreciate the solvating power of water, the solubility parameter (δ) provides a useful measure, defined as the ratio of the energy required to completely break all intermolecular forces that maintain the liquid state. We represent δ quantitatively as

$$\delta = \sqrt{\left(\frac{\Delta E_V}{V}\right)}$$

where ΔE_V is the total energy required to vaporize a solute. One can think of δ as the 'cohesive energy density' of a substance. Of course δ correlates strongly with polarity, with water not surprisingly having the highest value of δ when compared to other common solvents (**Figure 5**).

Before studying an example of the structure of a hydrated metal ion, we recognize that each water molecule is already 'solvated' to a very high degree of structural complexity. And because of the autoionization reaction of water, which we can represent as a net reaction:

$$H_2O \rightleftharpoons H^+ + OH^-, \qquad [1]$$

protons and hydroxide ions are formed that also become hydrated. Realistic structures of the reaction [1] products continue to be the subject of debate, but much evidence suggests that a more realistic way to describe the autoionization of water is

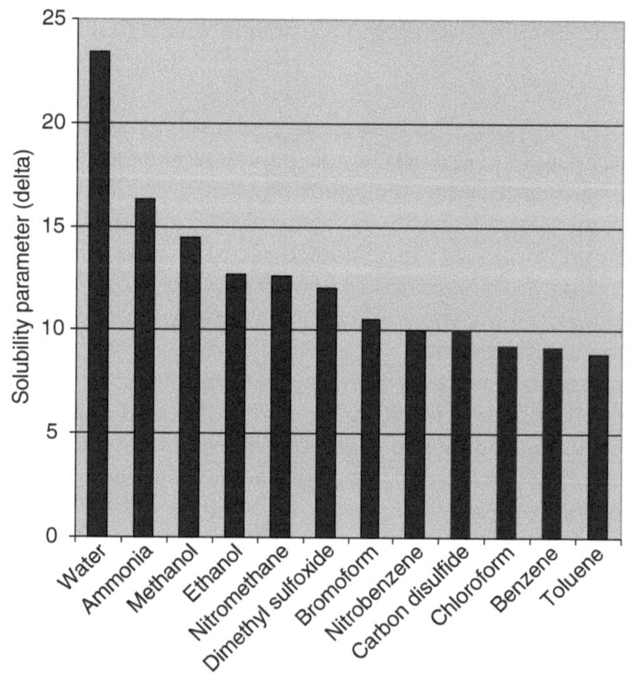

Figure 5 A comparison of Hildebrand's solubility parameter (δ) for various liquids (25 °C).

$$6H_2O \rightleftharpoons (H_2O)_2H^+ + (H_2O)_3OH^- \qquad [2]$$

Proposed structures for these ions are shown in **Figure 6**. For convenience, the simplistic products of

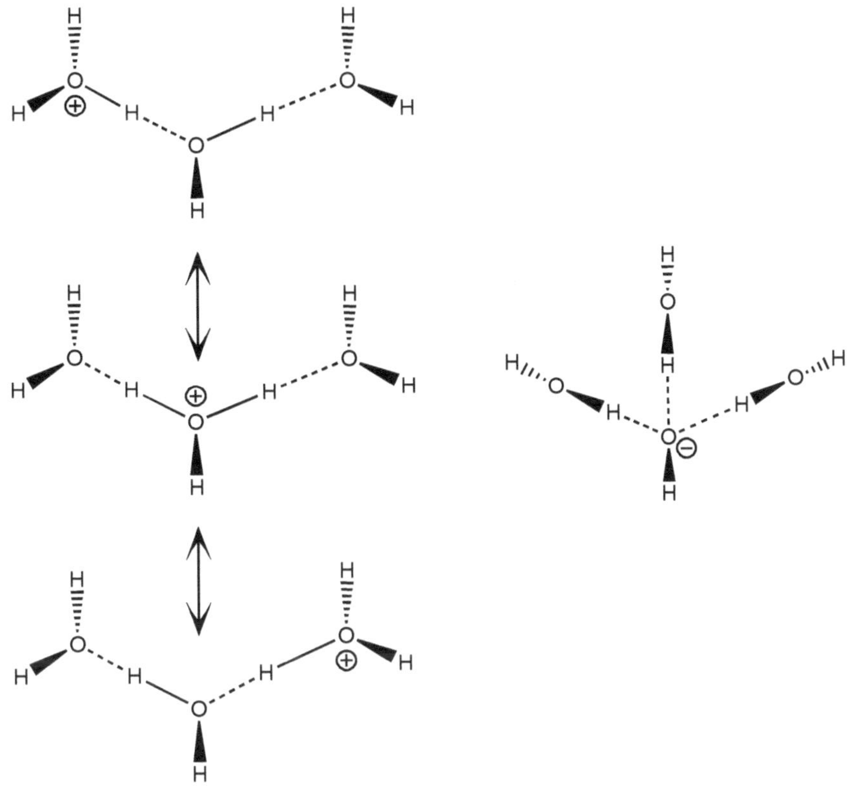

Figure 6 'Proton hopping' among three water molecules which together constitute a more accurate representation of a hydrated proton ($H_5O_2^+$). The center structure is the most energetically stable of the three shown. A more realistic structure for solvated hydroxide ion ($H_7O_4^-$) is also shown (right). Hydrogen bonds are denoted by dashes (---).

reaction [1] are commonly used in the literature. However, more complex structures, such as those depicted in **Figure 6**, are themselves not yet fully accepted as realistic.

For ions in aqueous solution, the structures formed by hydration reactions are driven by geometric and electronic factors. The number of water molecules that coordinate as ligands to an ion typically varies from four to nine, and is a function of factors that include ion size, the number of vacant orbitals present, and the degree of ligand–ligand repulsion. Given the great interest in pollution by toxic metals, our understanding of cation hydration is more extensive than for anions, yet hydration of the latter should not be surprising given the dipolar nature of water as a ligand.

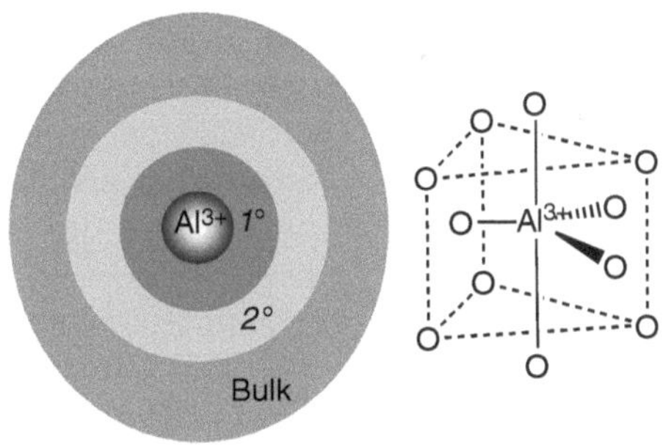

Figure 7 The 'concentric shell' model (left) for the hydration spheres surrounding a cation, showing the primary, secondary, and bulk solution shells. The primary hydration shell of aluminum ion (right), a tricapped trigonal prismatic geometry in which only the O atom positions for the 11 coordinating water molecules are shown.

In **Figure 7**, the 'concentric shell' model for the hydration of an ion is illustrated for aluminum ion, which exists under ambient conditions in the +3 oxidation state. Three regions form the shells – an inner layer, known as the primary (1°) shell, an intermediate layer known as the secondary (2°) shell, and a third region comprised of the bulk fluid. The structure of the 1° shell is highly ordered, as shown in **Figure 7** for the tricapped trigonal prismatic arrangement of 11 water molecules closely surrounding the trivalent cation. In the 2° shell, the influence of the Al(III) ion's high charge density would create a more loosely held though structurally defined layer. The bulk fluid extends beyond the 2° shell where the range of the ion's force field has no apparent effect on the fluid structure. It is important to note that the concentric shell model is simplistic, focusing on the strongest inner layers that are present. That is, the model ignores long-range ordering effects, which, because of their weakness, are inherently difficult to study. For example, molecular modeling (theoretical) studies have suggested that for heavy metal ions in aqueous solution, the surrounding water would be affected by the electronic field of the ion to a distance corresponding to several dozen or more layers of water molecules. Only beyond these layers would the bulk water reflect the 'undisrupted' structural state of a pure solution of water.

The Reactivity of Water

While we may tend to think of water as relatively inert, it is actually a very reactive molecule, with the oxygen atom behaving as a strong electrophile and the protons involved in autoionization reactions. However, water's reactivity is attenuated by its extensive hydrogen bonding. The eightfold ratio between water's single relatively heavy (O) atom and two light (H) atoms, and the charge inequity that exists between them, gives rise to a rapid exchange of protons between adjacent water molecules (proton hopping). In a pure solution of water, proton hopping among water molecules is constantly occurring at a high rate – even at pH 7 where it is slowest, it occurs on the order of 1000× per second (**Figure 5**). In studies of hydrogen bonding and the solvation of ions by water, the exchange of protons is even faster than the millisecond timescale observed for a bulk solution of pure water. Nevertheless, water is treated as a stable molecule because the net structure (H–O–H) is maintained in spite of its intrinsic dynamic state.

Fundamentally, chemical reactions occur as means for a species (atom, molecule, or ion) to increase its thermodynamic stability. We can generally classify chemical reactions into two broad categories: (1) those that involve changes in oxidation state, and (2) those that involve changes in coordination environment. While the former redox processes stand alone, the latter type of reaction can be divided into acid–base, complexation, and precipitation reactions. We can illustrate these three subcategories of coordination reactions by the example of a series of hydrolytic reactions involving the Al(III) ion:

$$Al^{3+}_{(aq)} + H_2O \rightleftharpoons Al(OH)^{2+}_{(aq)} + H^+_{(aq)} \quad pK_1 = 8.2$$

$$Al(OH)^{2+}_{(aq)} + H_2O \rightleftharpoons Al(OH)^{+}_{2(aq)} + H^+_{(aq)} \quad pK_2 = 19.0$$

$$Al(OH)^{+}_{2(aq)} + H_2O \rightleftharpoons Al(OH)_{3(s)} + H^+_{(aq)} \quad pK_3 = 27.0$$

$$Al(OH)_{3(s)} + H_2O \rightleftharpoons Al(OH)^{-}_{4(aq)} + H^+_{(aq)} \quad pK_4 = 31.4$$

(The subscript 'aq' denotes 'in aqueous solution,' a reminder that all of these species are hydrated, the structures of which are not shown.) While none of the reactions above cause changes in oxidation states, all are acid–base reactions because of the generation of a (hydrated) proton. All can be classified as complexation reactions as well because of hydroxide ion acting as a ligand in its coordination to the metal cation, with the formation of complex cations and anions (with the exception of the third reaction). For the third reaction, because of the formation of a solid product, we classify it as a precipitation reaction. Chemical reactions in the environment that involve water as a reactant or product – i.e., each type of reaction illustrated above as well as redox reactions – represent an enormous volume of scholarly work; the interested reader is therefore referred to the 'Further Reading' listed at the end of this article and elsewhere in this Encyclopedia.

Trends and Patterns in Limnology

The chemistry that is mediated by water in natural aquatic systems varies in space and time. Often this variability is expressed in the form of trends and patterns, and by understanding their causes it is possible to gain insight into the mechanisms that control water chemistry. Ultimately, variation in the chemistry of lakes and rivers can be attributed to three controlling factors: (1) physical processes and properties, including lake morphometry, weather, and climate; (2) geologic setting; and (3) biological factors, including the abundance and composition of biota within the water body and its watershed. Each of these factors may in turn be influenced by human activities. A discussion of how these factors influence water chemistry is best facilitated by examining some observed patterns for three important classes: dissolved gases, major ions, and nutrients.

Dissolved Gases

The dissolved gases of primary interest in most aquatic ecosystems are oxygen and carbon dioxide. Both of these molecules are nonpolar, therefore, as they partition at the air–water interface their hydration by water is minimal and consequently their solubility is very low. The only van der Waals forces that act upon them are very weak Debye forces, in which water's strong dipolar field induces a transient dipole in the nonpolar molecule's electronic field. These gases are of primary importance because they both influence and reflect biological processes. As a result, they serve as tracers of electron flow (i.e., energy flow) in an ecosystem. Reactions that convert energy into an organic form will reduce CO_2. In the case of photosynthesis, energy is derived from light and water is the electron donor, with the resultant production of O_2. CO_2 can also be reduced by chemoautotrophic bacteria, using other alternate electron donors, such as ammonium (NH_4^+), methane (CH_4), and hydrogen sulfide (H_2S). In each case, anabolic processes result in a loss of dissolved CO_2. Conversely, the decomposition of organic material results in the production of CO_2 and the loss of O_2, if that gas is available. In general, the balance between carbon reduction and oxidation in lakes and rivers is controlled by light-driven photosynthesis. This, and the physical exchange of gases between water and the atmosphere, results in deep waters having higher CO_2 concentrations and lower dissolved O_2 concentrations than surface waters. In lakes that are chemically or thermally stratified, the combination of decomposition and reduced vertical mixing can result in anoxia in the hypolimnion. In lakes that are well mixed, anoxia will occur in the sediment. Under these conditions, bacteria will use other electron donors in the metabolism of organic carbon. The electron donor used depends on the relative availability and the Gibbs' free energy of reaction resulting from the use of that donor. As a result, a vertical redox gradient is created, in which the various electron acceptors serially decrease with depth.

For lakes of a given size and within a geographic/climatic region, dissolved gas concentrations can vary according to the loading of nutrients and organic carbon. Lakes with high nutrient loads will exhibit large diurnal fluctuations in surface dissolved O_2 and CO_2 concentrations, because of high photosynthetic rates during the day and high respiration rates at night. Lakes with high organic carbon loads may be persistently supersaturated with CO_2 and undersaturated with O_2.

Temperature is a key property that determines the solubility of gases in water (**Figure 8**). This has ramifications both for the distribution of dissolved gases within lakes, and for the relationship between climate and dissolved gases, especially O_2. Within large temperate lakes in which plankton metabolism is generally slow, there is usually sufficient dissolved O_2 at all depths to support aerobic organisms. Smaller lakes that stratify may develop an anoxic hypolimnion, with the probability of anoxia increasing with the duration of stratification and lake productivity. In tropical lakes and rivers, warm temperatures result in lower dissolved O_2 saturation concentrations, and higher decomposition rates, making these systems more prone to anoxia than their temperate counterparts.

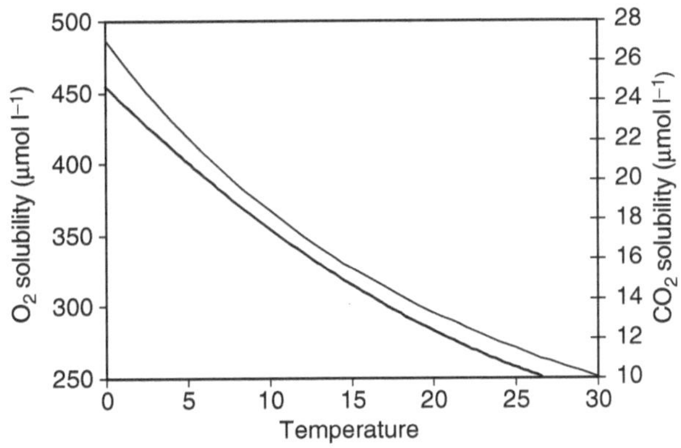

Figure 8 The solubility of oxygen (— —) and carbon dioxide (— —) in fresh water at a pressure of one atmosphere and an atmospheric CO_2 partial pressure of 380 μatm.

Major Ions

Major ions are those that contribute significantly to the salinity of water. Major cations generally include Ca^{2+}, Mg^{2+}, Na^+, and K^+, while major anions may include HCO_3^-, CO_3^{2-}, Cl^-, SO_4^{2-}, and sometimes NO_3^- All of these species are of course solvated by water, and the concentric shells that are formed may extend relatively far into the 'bulk' water. The absolute and relative abundance of the hydrated major ions in rivers and lakes are controlled by three factors: basin geology, rainfall, and evaporation–crystallization processes. Hence geographic variations in major ion composition can be related to one or more of these factors. For example, the relatively low Ca^{2+} concentrations in lakes and rivers of Precambrian Shield regions of North America and northern Europe are because of the dominance of igneous granite in their watersheds, while the high sodium and chloride concentrations of lakes in many dry regions is because of evaporative concentration of these salts. Because the above three factors differentially influence various major ions, the salinity and relative abundance and distribution of ions can be used to infer which of these processes is most significant for a given water body (**Figure 9**). Some exceptions to the pattern shown in **Figure 9** occur, especially in Africa, where a combination of intense weathering, low Ca^{2+} concentrations in rock, and evaporative concentration can result in moderately high salinities that are dominated by Na^+ and HCO_3^-.

Nutrients

Most algae require a minimum of 14 essential nutrients to grow. The nutrient that limits algal growth in a water body depends on the availability of these nutrients relative to algal demand. In most water

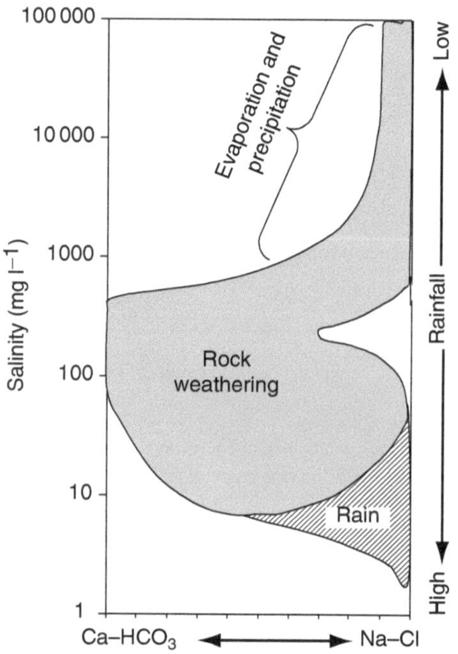

Figure 9 Influence of geology and climate on salinity and major ion composition of inland waters.

bodies, phosphorus or nitrogen is the limiting nutrient, but trace elements such as iron and molybdenum may also be limiting in some systems.

The effects of accelerated nutrient loading to lakes and rivers resulting from human activities, referred to as eutrophication, are well documented in the scientific literature, and are not addressed here. Phosphorus input to lakes and rivers is controlled primarily by rock composition and weathering intensity, but the availability of phosphorus to algae is influenced by the availability of other elements, and by biologically mediated processes. In iron-rich waters, inorganic phosphorus is bound as insoluble ferric phosphate or adsorbed onto ferric oxides and oxyhydroxides, and such systems tend to be unproductive and phosphorus limited. In calcareous regions, including the Laurentian Great Lakes, calcium minerals may serve as a source of phosphorus through weathering, but this phosphorus is often biologically unavailable because of adsorption to minerals such as calcium carbonate and precipitation with calcium to form apatite. The equilibrium between dissolved and particulate phosphorus is influenced by redox potential, with phosphorus dissolution being accelerated under anoxic conditions. Such conditions also promote denitrification, through which biologically available nitrate is ultimately reduced to nitrogen gas, which cannot be assimilated by most algae. Over an annual cycle, water column anoxia is more prevalent in tropical lakes than in temperate lakes, increasing phosphorus availability while promoting nitrogen loss. As a result, nitrogen limitation of algae tends to be more common in the tropics. These patterns can be modified by lake depth. In deep lakes, phosphorus is sequestered more efficiently into sediment, and as a result these lakes tend to have a lower concentration of phosphorus in the water column relative to shallow lakes with similar external phosphorus loads.

Conclusion

Primarily because of an extensive network of hydrogen bonding, water is structurally complex and has very unusual properties. Ions and molecules are solvated by water, and the resulting structures affect their reactivity and hence their toxicity, transport, and fate. Understanding how nutrients and pollutants are transformed by their interaction with water is essential to understanding the dynamics of the Earth's hydrosphere. Furthermore, these chemical transformations affect how these compounds are transported to other environmental compartments (e.g., the lithosphere and biosphere).

Water is said to be the most studied molecule. Yet while theory and experiment have greatly improved our knowledge of water's structure, important questions remain only partially answered. In particular, questions concerning the structure of water in the liquid state, specifically how hydrogen bonding determines long-range ordering effects, continue to intrigue researchers. Given the astonishing properties of such a simple molecule, one might conclude that hydrogen bonding is indeed that 'third thing' to which D.H. Lawrence was alluding.

See also: Dissolved CO_2; Groundwater Chemistry; Major Cations (Ca, Mg, Na, K, Al); Mercury Pollution in Remote Fresh Waters; Physical Properties of Water.

Further Reading

Baird C and Cann M (2005) *Environmental Chemistry*, 3rd edn. New York, NY: W.H. Freeman.

Barrett J (2003) *Inorganic Chemistry in Aqueous Solution*. Cambridge, UK: Royal Society of Chemistry.

Chaplin MF (2000) A proposal for the structuring of water. *Biophysical Chemistry* 83: 211–221.

Cotton FA, Wilkinson G, Murillo CA, and Bochmann M (1999) *Advanced Inorganic Chemistry*, 6th edn. New York, NY: Wiley-Interscience.

Dougherty RC and Howard LN (1998) Equilibrium structural model of liquid water: evidence from heat capacity, spectra, density, and other properties. *Journal of Chemical Physics* 109: 7379–7393.

Kusalik PG and Svishchev IM (1994) The spatial structure in liquid water. *Science* 265: 1219–1221.

Manahan SE (2005) *Environmental Chemistry,* 8th edn. Boca Raton, FL: CRC Press.
Marcus Y (1985) *Ion Solvation.* Chichester, UK: Wiley-Interscience.
Martell AE and Motekaitis RJ (1989) Coordination chemistry and speciation of Al(III) in aqueous solution. In: Lewis TE (ed.) *Environmental Chemistry and Toxicology of Aluminum*, pp. 3–19. Chelsea, MI: Lewis Publishers.
Searcy JQ and Fenn JB (1974) Clustering of water on hydrated protons in a supersonic free jet expansion. *Journal of Chemical Physics* 61: 5282–5288.
Stumm W and Morgan JJ (1996) *Aquatic Chemistry: Chemical Equilibria and Rates in Natural Waters,* 3rd edn. New York, NY: Wiley-Interscience.
VanLoon GW and Duffy SJ (2000) *Environmental Chemistry: A Global Perspective.* New York, NY: Oxford University Press.
Wallqvist A and Mountain RD (1999) Molecular models of water: Derivation and description. *Reviews in Computational Chemistry* 13: 183–247.
Wetzel RG (2001) *Limnology: Lake and River Ecosystems,* 3rd edn. San Diego, CA: Academic Press.
Zwier TS (2004) The structure of protonated water clusters. *Science* 304: 1119–1120.

Relevant Websites

http://www.lsbu.ac.uk/water/ – "Water Structure and Science" by Professor Martin Chaplin, London South Bank University, London, England, UK.
http://witcombe.sbc.edu/water/chemistry.html – "The Chemistry of Water" by Professor Jill Granger, Sweet Briar College, Sweet Briar, Virginia, USA.
http://webbook.nist.gov/chemistry/ – The National Institute of Standards and Technology's "Chemistry WebBook," Gaithersburg, Maryland, USA.

Physical Properties of Water

K M Stewart, State University of New York, Buffalo, NY, USA

Introduction

Water is an indispensable and remarkable substance that makes all forms of life possible. Speculation about possible past or present life on other planets within our solar system, or on any extraterrestrial body somewhere within the universe, is conditioned on the evidence for or against the existence of past or present water or ice. Humans can and did survive and evolve without petroleum products (gas and oil) but cannot survive and evolve without water. Water is the most important natural resource.

By far the greatest volume (~76%) of water on Earth is in the oceans. A smaller fraction (~21%) is found within sediments and sedimentary rocks. A still smaller fraction (~1% of the overall volume) is fresh water, and of that 1%, about 73% is in the form of ice (mostly contained within the Greenland and Antarctic ice caps), and only about 23% of that 1% is liquid fresh water. If we consider further that about one-fifth of the world's liquid fresh water is contained within the five St. Lawrence Great Lakes in North America, and another approximately one-fifth is contained within the deepest freshwater lake on Earth, Lake Baikal, in Russia, we are left with an unevenly distributed resource. It is obvious that if the expanding human populations around the world do not conserve and manage this precious resource very carefully, they put themselves at great peril.

Liquid water can be formed through some hydrogen bonding and electrostatic attraction of two slightly positively charged atoms of the gaseous hydrogen (H) and one slightly negatively charged atom of the gaseous oxygen (O) to form one molecule of water (H_2O). **Figure 1** provides two views of that polar molecule. **Figure 1(a)** and **1(b)** show the somewhat lopsided or asymmetrical arrangement of two smaller hydrogen atoms, separated by an angle of ~105°, and a larger oxygen atom. **Figure 1(a)** is a simple 'ball and spoke' representation whereas **Figure 1(b)** shows the shared electron orbits, positive (+) and negative (−) poles, and the number (eight each) of protons and neutrons in the nucleus of the oxygen atom.

The relative elemental simplicity of water is somewhat deceptive because of the great influence that some of the unusual properties of water have on the physics, chemistry, and biology of the world generally, and on the distribution of life specifically. The following discussion will describe briefly some of these unusual properties and provide examples of how these properties may help us understand the world of inland waters.

Density

Density may be simply defined as the amount of weight or mass contained in a specific volume. If the volumes of all substances could be standardized to one size, e.g., one cubic centimeter (cm^3), then a measure of the weight or mass in that fixed volume gives the density. **Table 1** lists a few comparative densities (rounded to two decimals) of two liquids (water and mercury) and some selected solids.

Density differences in inland waters may be caused by variations in the concentrations of dissolved salts, by changes in the water temperature, and in pressure. For the vast majority of inland lakes, only vertical differences in salt concentrations and temperatures are of significant influence to mixing processes. Fixed or uniform additions of salts to the water tend to cause *linear increases* in the density of water. In contrast, fixed or uniform changes in the temperature (both below and above 4 °C) of water cause *nonlinear changes* in the density of water (see **Table 2**). The density of pure water is maximum at a temperature of 4 °C (3.98 °C to be precise). It is at this temperature that the interatomic and intermolecular motions and intermolecular distances of water molecules are least. One consequence of this reduction is that more molecules of H_2O can fit into a fixed space at 4 °C than at any other temperature. This compaction allows the most mass per unit volume and thus the greatest density. It is especially noteworthy that the temperature at which water has the maximum density is above its freezing point.

Because the differences in densities, within a few degrees above and below 4 °C, are very slight, it takes relatively little wind energy to induce substantial vertical mixing when water temperatures are within those ranges. An example period, for those lakes that become covered with ice in the winter, would be shortly before an ice cover develops and shortly after the ice cover departs. However, it takes much more energy to cause extensive mixing when the density differences are high, such as is common between the usually warm upper waters and colder lower waters of Temperate Zone lakes during summer. The greater the top-to-bottom differences in temperature, the

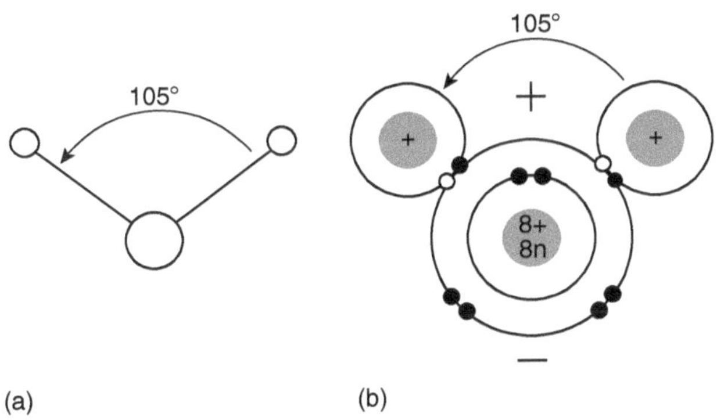

Figure 1 Two schematic representations (a) and (b) of a water molecule. (Modified from various sources.)

Table 1 Some comparative densities of water and other substances or elements

Substance	*Densities (g cm^{-3})*
Wood	
Seasoned balsa	0.11–0.14
Seasoned maple	0.62–0.75
Seasoned ebony	1.11–1.33
Water	1.00
Calcium	1.55
Aluminum	2.70
Iron	7.87
Lead	11.34
Mercury	13.55
Uranium	18.95
Platinum	21.45

Information from multiple sources.

Table 2 Comparative densities of average ocean water (salinity ~35%), freshwater ice, and pure distilled water at different temperatures

Temperatures (°C)	*Densities (g cm^{-3})*
20.0	1.02760, ocean water (salinity 35%)
0.0	0.9168, freshwater ice
0.0	0.99987, pure water (from here on)
2.0	0.99997
3.98–4.00	1.00000
6.0	0.99997
8.0	0.99988
10.0	0.99973
12.0	0.99952
14.0	0.99927
16.0	0.99897
18.0	0.99862
20.0	0.99823
22.0	0.99780
24.0	0.99733
26.0	0.99681
28.0	0.99626
30.0	0.99568
32.0	0.99505

Values from Hutchinson (1957), Pinet (1992), and Weast and Astle (1979).

greater the top-to-bottom differences in density and, consequently, greater are the energies required for wind-induced mixing.

There is an old, but still valid, cliché in the northern hemisphere that '... it is cold up north and warm down south.' Water temperatures in more northerly Temperate Zone lakes tend to average cooler than those of more southerly tropical lakes. Interestingly, although the upper-water summer temperatures in tropical lakes are somewhat higher than those of Temperate Zone lakes, the lower-water temperatures in tropical lakes are substantially higher than those ordinarily found in the lower waters of Temperate Zone lakes. It might there fore seem that there would be an easy top-to-bottom mix of the water in tropical lakes. Indeed some shallow tropical lakes, with only slight top-to-bottom temperature differences, may have this. However, because of the nonlinear increases in water density with temperature, tropical lakes can be surprisingly stable and resistant to much vertical mixing. **Table 2** provides a listing of some comparative densities. Let us consider two hypothetical lakes with just a 2 °C spread between their lower and upper waters. For example, if a Temperate Zone lake in the spring, not long after the ice departed, had lower and upper waters of 4.0 and 6.0 °C, respectively, the density difference would be $1.00000 - 0.99997 = 0.00003\,\mathrm{g\,cm^{-3}}$. In contrast, a warmer tropical lake whose lower and upper water temperatures may be 26.0 and 28.0 °Cs would have density differences that are much greater ($0.99681 - 0.99626 = 0.00055\,\mathrm{g\,cm^{-3}}$). Thus, the top-to-bottom ratio or density difference of these two lakes with a temperature difference of just 2 °C would be 55/3 or ~18 times as great in the tropical lake as in the Temperate Zone lake. The example above is only hypothetical but it shows the nonlinear influence of density changes with temperature, a property of water that influences, to varying degrees, the stratification and mixing of lakes around the world.

Heat Capacity/Specific Heat

Heat is a form of energy and, as such, we can measure changes in the temperature of a given volume of a substance and determine its heat capacity. Water is the common standard used and its heat capacity (arbitrarily defined as the heat needed to increase the temperature of 1 g of water by 1 °C) is comparatively large. When the mass is also considered then the number of calories needed to raise 1 g of a substance by 1 °C is termed its *specific heat*. For water, the value is $1\,\mathrm{cal\,g^{-1}}$. That quantity may not seem like much but, compared to other materials, the heat capacity or specific heat of water ($1.00\,\mathrm{cal\,g^{-1}}$) and ammonia

Table 3 The specific heat ($cal\,g^{-1}$) of selected substances compared to that of ice, pure water, and ammonia

Aluminum	0.215
Copper	0.092
Gold	0.030
Lead	0.030
Silver	0.056
Zinc	0.092
Ethyl alcohol	0.60
Ice (at 0 °C)	0.51
Water	1.000
Ammonia	1.23

Information from multiple sources.

($1.23\,cal\,g^{-1}$) are much greater than that of most other substances (**Table 3**). Consequently, these two liquids are commonly used to exchange heat in refrigerators and air conditioners.

Along with its ever changing and mesmerizing aesthetic qualities, inland waters are of immense importance in the storage and release of heat. In terms of freshwater lakes, the influence of their heat capacity can be seen most easily around very large lakes located in Temperate Zone latitudes and more inner continental areas. It is in these areas that even larger swings in seasonal air temperature would ordinarily occur in the absence of those lakes. Parts of the immediate surrounding areas of Lake Baikal in Russia (this is actually the world's deepest freshwater lake as well as one with the greatest volume of water) and the five St. Lawrence Great Lakes of North America are prime examples of the 'thermal buffering' these large lakes provide to their surroundings because of their large heat capacity.

For humans, this may mean some 'beneficial economic consequences' as portions of a lake's heat capacity are slowly released or 'shed' to down-wind regions as the fall and winter seasons progress. The immense thermal capacity of Lake Baikal is such that the lake and its immediate environments are roughly 10 °C warmer in December and January, and about 7 °C cooler in June and July, than in the cities of Irkutsk (about 50 km to the west of the southern half of Lake Baikal) and Ulan-Ude (about 70 km to the east of the lake). Several coastal and near-coastal regions of the St. Lawrence Great Lakes also provide impressive beneficial evidence of the influence of the Great Lake's heat capacity. There may be reduced costs associated with home and business heating in some coastal regions. An extended or milder autumnal period permits greater production in near-shore plantations of fruit trees and vineyards. Economic benefits may also accrue in some coastal regions of higher terrain during winter, when enhanced snows permit additional winter skiing, snowmobiling, and other winter sports.

However, some influences of a lake's heat capacity have 'detrimental economic consequences'. There are costs involved with snow removal, increased vehicular accidents (because of slippery roads), the corrosion of cars (attributable to road salts), and the potential long-term ecological changes associated with lake and stream salinization. There are also greater heating costs in spring as cooler water bodies extend their cooling influence inland. In late fall and winter, before an ice-cover develops, heavy snows may result when water vapor, being formed by evaporative processes off a relatively warm lake, is buoyed into much colder Arctic air (northerly Temperate Zone) crossing the lake. The rising water vapors may freeze, coalesce to ice crystals, and be carried down wind to shore areas where they fall out as snow. Perhaps the most dramatic of all the detrimental consequences is seen following the sometimes paralyzing effect of occasional, but intense, 'lake-effect' snow storms of mesoscale proportions. The lake-effect snow storms tend to have their greatest impact at the downwind end of the St. Lawrence Great Lakes after very cold Arctic air ($\geq$13 °C colder than the temperature of the lake) has moved across a long axis of the lakes and deposited its snows. These deposits or drops of snow may be in a broader synoptic pattern, but sometimes they are in very narrow bands of thick snow that may bring auto traffic, schools, and businesses to a stop. In the St. Lawrence Great Lakes region of North America, three of the better known areas where unusually heavy deposits of lake-effect snows may occur are (1) portions of the Upper Peninsula of Michigan on the southeastern shore of Lake Superior, (2) the southeasterly and easterly shores of Lake Ontario, especially the Tug Hill Plateau area of New York State, and (3) the easterly end of Lake Erie, around Buffalo, NY. Indeed, the St. Lawrence Great Lakes have been considered 'weather factories' capable of causing twists of climate found in few other parts of the world.

Heat of Fusion/Melting

This is just the amount of heat exchanged during a phase shift from either liquid water to solid ice, or from solid ice to liquid water. One gram of water at 0.0 °C can be converted to ice at 0.0 °C if 80 cal ($79.72\,cal\,g^{-1}$ to be precise) are released in the process. The same quantity, i.e., 80 cal, is required to melt that 1 g of ice back to 1 g of water. No further caloric additions or subtractions are needed to effect the phase shift.

Because of the heat needed to melt ice, researchers might intuitively expect to see a brief but substantial drop in the mean or weighted lake-water temperature when the ice cover of a lake melts in the spring season. For example, assume there is a hypothetical northerly latitude and a 20-m deep lake in late winter (March). Consider that the lake is covered with 50 cm of ice at 0.0 °C. Consider further that the weighted mean temperature of the 1950 cm (essentially 1950 g) water column below the ice is 3.0 °C. The heat content of that water column would be 5850 cal (1950 g × 3 cal g^{-1} = 5850 cal). Assuming that there are no further gains or losses of heat to the lake, the amount of heat required to melt the ice would be 3680 cal (80 cal g^{-1} × 50 cm of ice × 0.92 g cm^{-1}, allowing for density of pure ice rounded to two decimals = 3680 cal). If some of the caloric content of the water column could be used to melt all the ice, the total caloric content would drop to 2170 cal (5850 cal − 3680 cal = 2170 cal). If those 2170 cal were now equally distributed within a 1-cm^2 square and 20-m (2000 cm, essentially 2000 g) deep water column, the mean water temperature would need to drop from 3 to 1.08 °C (2170 cal/2000 cal = 1.08 °C). A drop of about 2 °C during the melting of ice would be large!

As it turns out, the hypothetical example in the above paragraph is not realistic. Some background follows. Many years ago as a graduate student, I took daily measurements of ice thickness and top-to-bottomwater temperatures for two winters and right through the spring ice break up in a Midwestern U.S. lake. From conversations with others, I was told to expect, and did anticipate, a substantial drop in mean water temperature as the ice melted… especially in the last few days of ice cover when the ice thinned rapidly. However, I did not measure any big drops in lake temperature and, in retrospect, should not have anticipated them. The reasons researchers do not see large decreases in lake temperatures with ice loss reflect some interacting physics. For example, there may be somewhat differing weather patterns each spring. The ice generally melts over an extended period of time, from several days to several weeks, not suddenly. Half or more of the total ice thickness may be lost from the top of the ice by melting from warming air temperatures above the ice, not necessarily from waters that are just above freezing below the ice. Because of its *albedo* (*percent of incoming solar radiation that is reflected back into space*) dark or open water generally reflects only a small fraction of the incoming solar radiation, whereas white snow cover on a frozen lake can reflect a large fraction of incident radiation. Indeed, snow cover extending into the spring period can delay the date the ice disappears. However, with increasing amounts of solar radiation, rising air temperatures, melting snows, and darkening ice, the water below the ice may be gaining some heat from solar inputs at the same time it is losing some heat in melting an overlying ice cover. Moral of the story: Do not expect a big drop in mean water temperature as an ice cover melts on a lake.

Heat of Vaporization/Condensation

As was the case for 'Heat of Fusion/Melting,' the heat of vaporization/condensation also represents the amount of heat exchanged during a phase shift. For vaporization, it is the quantity of heat (540 cal g^{-1}) needed to convert 1 g of water to 1 g of water vapor. The same amount of heat is exchanged or released in the phase shift during the condensation of 1 g water vapor to 1 g of water.

Aquatic scientists may be naturally impressed with the large amount of heat exchanged (80 cal g^{-1}) in the phase shift from water to ice, or from ice to water, but the amount of heat exchanged (540 cal g^{-1}) in the phase shift from water to water vapor, or water vapor to water is 6.75 times larger (540/80 = 6.75). Although the importance of this large amount of heat exchange via vaporization or condensation may be underappreciated by humans, it is huge. On a small but critical scale for life, water evaporating off perspiring warm-blooded animals, including humans, helps maintain body temperatures within narrow survivable limits. On a global scale, the seemingly endless phase shifts between liquid water and water vapor in the atmosphere are key determinants in the redistribution of water and heat within the hydrological cycle around the world.

Isotopes

An isotope is one of two or more forms of the same chemical element. Different isotopes of an element have the same number of protons in the nucleus, giving them the same atomic number, but a different number of neutrons giving each elemental isotope a different atomic weight. Isotopes of the same element have different physical properties (melting points, boiling points) and the nuclei of some isotopes are unstable and radioactive. For water (H_2O), the elements hydrogen (atomic number 1) and oxygen (atomic number 16) each have three isotopes: ^1H, ^2H, and ^3H for hydrogen; ^{16}O, ^{17}O, and ^{18}O for oxygen. In nature, the ^1H and ^{16}O (usually just given as O) isotopes are by far the most common. In water, the water molecule may be given as ^1H_2O or hydrogen oxide, ^2H_2O or deuterium oxide, and ^3H_2O or tritium oxide, the radioactive one.

Both of the latter two are sometimes called heavy water because of their increased mass. However, the phrase 'heavy water' gained notoriety primarily because of the association of ^2H_2O or deuterium oxide, also called the deuterated form of water, in the development of nuclear weapons. Many elements have isotopes, but the isotopes of hydrogen and oxygen are of particular interest because fractionation occurs in vapor–liquid–solid phase changes. Heavier molecular 'species' tend to be enriched in the condensation phase and lighter molecular 'species' in the vapor phase. Some isotopes can be used to great advantage as tracers in understanding water movements and exchanges within atmospheric, oceanic, lake, stream, and groundwater systems.

Sublimation

Water is said to be sublimated, sublimed, or undergo sublimation when it passes directly from a solid (ice) stage to a gas (vapor stage) without becoming a liquid in between. The latent heat of sublimation, i.e., the heat required to make the form of water change from ice to a water vapor, is $679\,\text{cal}\,\text{g}^{-1}$. This quantity is larger than the heat required to melt ice ($80\,\text{cal}\,\text{g}^{-1}$) and vaporize water ($540\,\text{cal}\,\text{g}^{-1}$) combined ($80 + 540 = 620\,\text{cal}\,\text{g}^{-1}$). Because there may be multiple heat sources and sinks (e.g., the air above the ice and the water below the ice) associated with changing ice thickness on frozen Temperate Zone lakes, it is a challenge to assess the quantitative role that sublimation may play in those changes.

Some practical effects of sublimation may be visualized by observing a reduction in the volume of some dry ice (solid CO_2) or camphor. In another example, after several weeks of continuing subfreezing temperatures and deep frost, and assuming that no deicing salts were used, sublimation is most likely responsible for the slow disappearance of an ice sheet over the surface of a frozen sidewalk. Sublimation is also the main process by which wet clothes, which were hung out to dry in subfreezing temperatures, may dry. In the latter case, the water on the clothing quickly freezes to ice, but then slowly vaporizes through sublimation, and the clothes dry. In more recent years, freeze-dried vegetables, fruits, and other products (including instant coffee) provide other examples where the practical application of sublimation is utilized to both market and preserve food.

Surface Tension and Cohesiveness

Surface tension may be regarded as the resistance offered by liquid water to forces attempting to deform or break through the surface film of water. It is an interesting property and, for water, the surface tension measured in Newton's per meter ($\text{N}\,\text{m}^{-1}$), is high and shows a slight *increase* as the temperature falls from 100 ($0.0589\,\text{N}\,\text{m}^{-1}$) to 0 °C ($0.0765\,\text{N}\,\text{m}^{-1}$). The molecules of water are strongly attracted to each other through their *cohesiveness* (attraction of like substances). The properties of surface tension and cohesiveness work together in water in shaping the small rounded water droplets seen on a table top or a car windshield. The same properties help to form the slightly flattened to spherically-shaped raindrops as they fall through the air.

The primary force for restoring larger wind-generated surface and internal waves of lakes is gravity, but the primary force for restoring the much smaller capillary waves or ripples on a lake's surface seems to be surface tension of the water itself.

The surface tension of water is sometimes used to advantage in parlor games in which someone claims that he/she can float a more dense (than water) steel needle on less dense water. When the needle is lowered slowly and carefully with its long axis paralleling the surface of the water, it may be possible to 'float the needle' because the high surface tension of the water may prevent the needle from sinking. Do not try this by lowering one of the sharp ends of the needle first because a point application of the needle will exceed the surface tension of the water film, and the needle will sink rapidly.

When responding to a 'fire call' in fire trucks, water is the most common and practical substance used by firemen. Water is cool, it suppresses heat, it puts out fires and sometimes there is much water to spare. However, the *high surface tension of water can reduce its effectiveness in suppressing some fires.* Surfactants are compounds that reduce the surface tension of water. In their response to a 'fire call' firemen often quickly attach hoses to street fire hydrants and spray water from that source on a burning structure. Although the addition of tiny quantities of surfactants to water may help put out fires, it is not practical (or safe) to add surfactants to an entire distribution system of a city. However, the addition of tiny quantities of surfactants to the volume (roughly $1.89\,\text{m}^3$ or 500 gallons in the United States) of water being carried in the fire truck would make that truck water 'wetter.' Some combustibles could be penetrated more easily by this wetter water of reduced surface tension and selected fires could be put out more rapidly.

There is a specialized community of organisms, sometimes called neuston, associated with the surface film. For many observers of nature, it is always fascinating to see small insects such as pond skaters or water-striders (*Gerris* sp., within the insect Order Hemiptera), and whirligig beetles (*Gyrinus* sp. and *Dineutes* sp., within the insect Order Coleoptera), running around on the surface of ponds, sheltered lakes, and some streams. Because of padded ends to the long middle and hind feet of water striders, and the much shortened but paddle-like feet of the whirligig beetles, the high surface tension of the water is such that the insects may dimple, but not break through, the surface film.

One of the easiest ways of getting popcorn into your mouth is by touching your tongue to some popcorn in a container. Here again it is the surface tension of the water on your tongue that lets you 'hold on' to the light popcorn easily.

Viscosity

This property may be thought of as the internal friction or resistance exerted on one substance (gas, liquid, or solid) as that substance tries to flow or move through the same or another liquid. One way of visualizing the influence that liquids or semiliquids of progressively greater viscosities might exert would be to take three glass marbles (same diameter and density) and drop one in each of three similar-sized glasses, one glass containing water, one light oil, and one honey, all at the same temperature. The marble would descend quite rapidly in water, more slowly in the light oil, and very much more slowly in the glass of honey. In this example, honey would obviously exert the most friction or resistance to movement through it and have the greatest viscosity. Viscosity is usually measured in poises ($N\,s\,m^{-2}$) or centipoises (= 0.01 P). Water at 20 °C has a viscosity of 0.01002 P or 1.002 cP.

The rate of passive descent through a liquid reflects the density of the liquid itself as well as the surface area and density of the substance moving through it. Viscosity changes with water temperature in that viscosities decrease as water temperatures rise and increase as water temperatures fall. Many fish are powerful enough, slippery from mucous on their skin, and shaped so they can 'slip through' water relatively easily. In contrast tiny zooplankton, with multiple projections on their body, are ordinarily challenged as they attempt to move in any direction and particularly so when moving in cool waters.

Colligative Properties

These are the four special properties of water that are significantly altered or modified when solutes are added to and dissolve in water. The alterations or modifications of a colligative property (regarded as a binding property) may be predictable in dilute solutions when the number of solute particles is known. It is the number of solute particles, not their chemical nature, that determines the extent to which a property is modified.

The four colligative properties of water are *vapor pressure* (when water is in equilibrium with its own vapor), *osmotic pressure* (the pressure controlling the diffusion of a solvent across a semipermeable membrane), *boiling point* (the temperature at which water undergoes a phase shift to a gas), and *freezing point* (the temperature at which water undergoes a phase shift to a solid). Even at standardized pressures and temperatures, the extent to which a property is modified depends on the number of solute particles added. Generally, if we add a fixed number of solute particles of a sugar or salt to a liter of pure water, there would be some consequences. The vapor pressure would be lowered but the osmotic pressure would rise. The boiling point (also termed *boiling-point elevation*) would be elevated a bit above the usual boiling point of 100.0 °C. In the latter case, a watery mixture with solutes (e.g., a well-mixed soup being heated for a meal) would have to get hotter than the boiling point of pure water before it would boil. The freezing point would be lower than 0.0 °C. A practical application of this (also termed *freezing-point depression*) is easily seen, in parts of the northerly Temperate Zone in winter, following the application of deicing salts to melt the ice and snow on roads and sidewalks. Although not a colligative property as such, a simple increase in physical pressure also lowers the melting point of ice (~0.007 °C/atm) and helps form snowballs (when the snow is not too cold) and form a lubricating layer of water under the blade of an ice skate.

See also: The Surface Mixed Layer in Lakes and Reservoirs.

Further Reading

Eichenlaub VL (1979) *Weather and climate of the Great Lakes Region*, pp. 1–27. Notre Dame, IN: University of Notre Dame Press.

Hutchinson GE (1957) Geography, Physics, and Chemistry. *A Treatise on Limnology*, vol. I. Chap. 3. New York, NY: John Wiley.

Klaff J (2002) *Limnology, Inland Water Ecosystems*, pp. 35–40. Upper Saddle River, NJ: Prentice-Hall.

Kozhov MM (1963) *Lake Baikal and Its Life*, pp. 15–19. The Hague, Netherlands: Dr. W. Junk Publishers.

van der Leeden F, Troise FL, and Todd DK (1990) *The Water Encyclopedia*, 2nd edn., pp. 774–777, Chelsea, MI: Lewis Publishers.

Lemmon EW, McLinden MO, and Friend DG (2003) Thermophysical properties of fluid systems. In: Linstrom PJ and Mallard WG (eds.) *NIST Chemical Webbook, NIST Standard Reference Database Number 69*. Gaithersburg, MD: National Institute of Standards and Technology. http://webbok.nist.gov.
Morgan JJ and Stumm W (1998) Properties. In: Kroschwitz JI and Howe-Grant M (eds.) *Encyclopedia of Chemical Technology*, 4th edn., vol. 25, pp. 382–405. New York, NY: John Wiley.
Mortimer CH (2004) Lake Michigan in motion. Responses of an Inland Sea to Weather, Earth-Spin, and Human Activities, p. 143. Madison, WI: University of Wisconsin Press.
Pinet PR (1992) *Oceanography, An Introduction to the Planet Oceanus*, pp. 120–128. St. Paul, MN: West Publishers.
Scott JT and Ragotzkie RA (1961) The heat budget of an ice-covered inland lake. *Tech. Rep. 6. ONR Contract 1202 (07)*. Madison, WI: Deptartment Meteorology, University of Wisconsin.
Voet D, Voet JG, and Pratt CW (1999) *Fundamentals of Biochemistry*, pp. 23–27. New York, NY: John Wiley.
Wallace RA, Sanders GP, and Ferl RJ (1991) *Biology. The Science of Life*. ch. 2. New York, NY: Harper Collins.
Weast RC and Astle MJ (1979) *CRC Handbook of Chemistry and Physics*, 60th edn. Boca Raton, FL: CRC Press.

Pressure

J F Atkinson, University of Buffalo, Buffalo, NY, USA

Introduction

Pressure is an example of a surface stress (force per unit area), and is important in many problems of fluid mechanics, as well as in a variety of biological responses in natural water bodies. Pressure has an impact on each of the physical, chemical, and biological components of a fluid system. In this article, we deal mostly with the analysis of the physical impacts of pressure, and provide several examples of biological responses to variations in pressure, including algae, fish, and humans.

The study of fluid mechanics is ultimately the study of the response of a fluid to applied forces, subject to certain constraints such as continuity, or mass balance, and constitutive functions such as the relationship between shear stress and rate of strain. Fluid forces can generally be divided into body forces and surface forces. Body forces act on all the fluid within a defined control volume, or fluid element, while surface forces act on the surface that encloses that volume. Furthermore, the surface forces can be separated into those that act either normal to, or in the plane of, the surface of interest (**Figure 1**). Here we consider both forces and stresses. There are many situations in which analysis on the basis of forces is more useful. For example, calculation of forces acting on a pipe bend or forces acting on a sluice gate or canal lock may be needed in the design of a pipe distribution system or for structures in an open channel flow, respectively. For differential analysis of fluid flow, such as would be used in describing details of the velocity distribution, stresses are generally of more direct interest than forces. Stresses acting in the plane of a surface are called shear stresses, and stresses acting perpendicular to a surface are called normal stresses. Normal stress is generally called pressure, and pressure at a point has the same magnitude in all directions. Although by mathematical convention, the normal stress in **Figure 1** is considered positive as shown, pointing outwards from the fluid element, pressure is considered positive as a compressive stress, since fluids cannot withstand tension forces (this is a fundamental difference between the mechanical description of fluids and solids). Thus, pressure acting on a fluid element as a result of the surrounding fluid is the negative of the normal stress and points inward on all faces of the element.

In general, pressure can be divided into two parts, the hydrostatic and dynamic components. Hydrostatic pressure is generated by the force of gravity acting on fluid that lies above a particular point of interest at which the value of pressure is desired, and dynamic pressure is due to the movement of a fluid. As shown below, hydrostatic pressure also can be assumed in many analyses of fluid flow, especially in environmental applications. For so-called Newtonian fluids, which encompass most fluids of practical interest including water, it is assumed that the shear stress is proportional to the rate of strain, which in turn can be approximated by the velocity gradient in the direction normal to the surface on which the shear stress acts. Thus, in a static fluid, or in a fluid that is moving but in which there are no velocity gradients (i.e., no relative motion), the only forces to consider are the body and normal forces.

In the present article, although many of the concepts apply to fluids in general, it will be assumed that the main fluid of interest is water, either fresh or saline. The main property of interest that pertains to water and is useful for the analysis of pressure distributions is that of incompressibility, meaning that the density of the fluid is not a function of pressure. In addition, a normal right-handed Cartesian coordinate system will be assumed.

Hydrostatic Pressure

As the name implies, hydrostatic pressure (here denoted by p_h) is the pressure that is exerted under conditions of no fluid motion. To understand how p_h varies, first consider a small element of fluid within a larger volume of that same fluid (**Figure 2**). The top of this element is at a depth h below the water surface. Similar to **Figure 1**, the element has sides dx, dy, and dz, aligned with each of the respective coordinate directions. The forces acting on the element include its own weight and the forces of the surrounding fluid acting on the surface of the element. Since there is no fluid motion, or more precisely, no relative motion of the fluid element with respect to the surrounding fluid, the only surface force to consider is that of pressure. If the element is to be in equilibrium, and therefore to remain motionless (as a result of Newton's First Law), the forces acting on it in each direction must be in balance.

In **Figure 2**, the sides of the fluid element are numbered for convenience, with sides 1 and 2 referring to faces with perpendiculars (to the faces) along the

x direction, sides 3 and 4 referring to sides with perpendiculars along the y direction, and sides 5 and 6 with perpendiculars along the z direction. Note that the z direction is oriented parallel to the direction of gravity, but points positive upwards. If p_h is considered to be the pressure at the center of the element, then pressures along each of the faces can be expressed in terms of p_h using a truncated Taylor Series expansion, and force (product of stress and area on which it acts) balances in each of the three coordinate directions may be written as

$$(x)\, p_{h1}(dy\,dz) = p_{h2}(dy\,dz) \Rightarrow \left(p_h - \frac{\partial p_h}{\partial x}\frac{dx}{2}\right) = \left(p_h + \frac{\partial p_h}{\partial x}\frac{dx}{2}\right) \qquad [1]$$

$$(y)\, p_{h3}(dx\,dz) = p_{h4}(dx\,dz) \Rightarrow \left(p_h - \frac{\partial p_h}{\partial y}\frac{dy}{2}\right) = \left(p_h + \frac{\partial p_h}{\partial y}\frac{dy}{2}\right) \qquad [2]$$

$$(z)\, p_{h5}(dx\,dy) = p_{h6}(dx\,dy) + g(\rho\, dx\,dy\,dz) \Rightarrow \left(p_h - \frac{\partial p_h}{\partial z}\frac{dz}{2}\right) = \left(p_h + \frac{\partial p_h}{\partial z}\frac{dz}{2}\right) + \rho g\, dz \qquad [3]$$

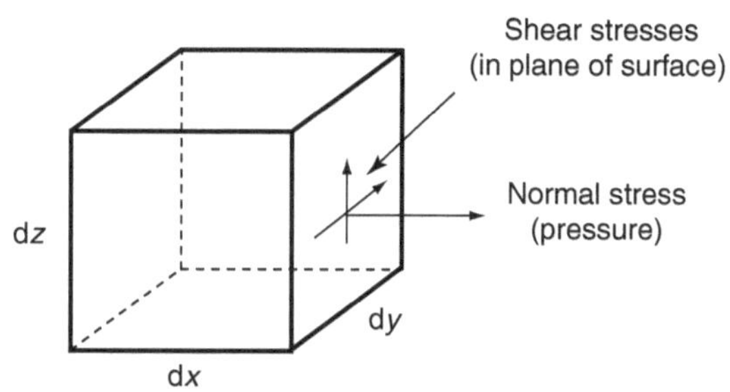

Figure 1 Stresses acting on one surface of a small fluid element.

where p_{hi} is the hydrostatic pressure acting on face i, g is gravitational acceleration, and ρ is the fluid density. Note that dx, dy, and dz are assumed to be small enough that any variations in p_{hi} over the area of the side may be neglected. Equation [3] differs from eqns [1] and [2] only in that the fluid weight is included in the force balance. Simplifying eqns [1]–[3] results in

$$\frac{\partial p_h}{\partial x} = \frac{\partial p_h}{\partial y} = 0;\ \frac{\partial p_h}{\partial z} = -\rho g = -\gamma \qquad [4]$$

where $\gamma = \rho g$ is the specific weight of the fluid.

Keeping the assumption of constant density and letting the fluid be water, the pressure on the top of the fluid element in **Figure 2** is given by integrating the last part of eqn [4], resulting in $p = p_0 + \gamma h$, where p_0 is the pressure at the water surface, and $h = z_0 - z$ is the water depth, where z_0 is the elevation at the water surface and z is the elevation at the top of the element. Also note that the weight (W) of the water column above the fluid element is $W = \gamma h A$, where $A = dx\,dy$ is the area of the element. If $p_0 = 0$ is assumed, then the pressure acting on the top surface of the fluid element in **Figure 2** is simply the weight of water sitting above the element, divided by the area (dx dy). Using similar reasoning, the pressure at the water surface is actually given by the weight of air above the surface, per unit area (note that pressure can be defined in terms of weight, whether or not the density is constant, but if density is not constant the weight must be determined by integration of the density distribution). In fact, Newton long ago deduced that air has mass by determining that water could be pumped up a tube by extracting air at the top only to a maximum height of about 33 ft (10 m), since the weight of a column of water 33 ft high weighs approximately the same as a similar column of air stretching through the entire

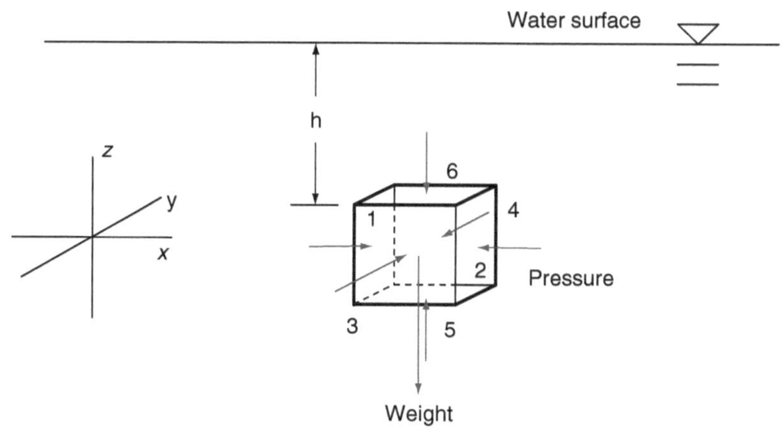

Figure 2 Forces acting on a small fluid element under static conditions (no fluid motion); forces include the weight of the fluid enclosed by the rectangular surface and the pressure acting on each of the faces (six totally) of the element.

atmosphere. Atmospheric pressure changes slightly with weather and with location (elevation), but is usually assumed to be 14.7 psi, 1 atm, or about 101 kPa. For convenience, the pressure at the water surface is often taken as zero, which facilitates the interpretation of pressure in terms of weight of water above a given point of interest. Pressures written with respect to real air pressure as a reference, or boundary condition, are said to be 'absolute' pressures, while pressures written in terms of zero pressure at the water surface are said to be 'gauge' pressures. In absolute terms, pressure at a depth of 33 ft (10 m) is approximately double what it is at the water surface, and pressure increases by 1 atm for every increase in depth of 33 ft. The lowest possible pressure in absolute terms is zero, which is the pressure in deep space, while gauge pressures may take negative values (indicating a vacuum). In the remainder of this article all pressures will be assumed to be in 'gauge' terms.

The relationship between pressure and depth is further illustrated by considering the case where no external forces are applied (except gravity), and in addition it is assumed that density, and therefore specific weight, is constant, as is usually the case with water. Then, a simple integration of eqn [4] gives the general expression for pressure difference between two points at different vertical locations in the fluid,

$$\Delta p_{\mathrm{h}} = p_{\mathrm{h}2} - p_{\mathrm{h}1} = -\gamma \Delta z = \gamma(z_1 - z_2) \qquad [5]$$

where z_1 and z_2 are the two locations at which p_{h} is evaluated (**Figure 3**).

Equation [4] describes the distribution of pressure in a static fluid. In words, pressure in a static fluid is constant on a horizontal plane, and has a gradient in the vertical direction equal to $-\gamma$. In other words, pressure decreases while moving upward in a static fluid, at a rate given by $-\gamma$. Finding the actual

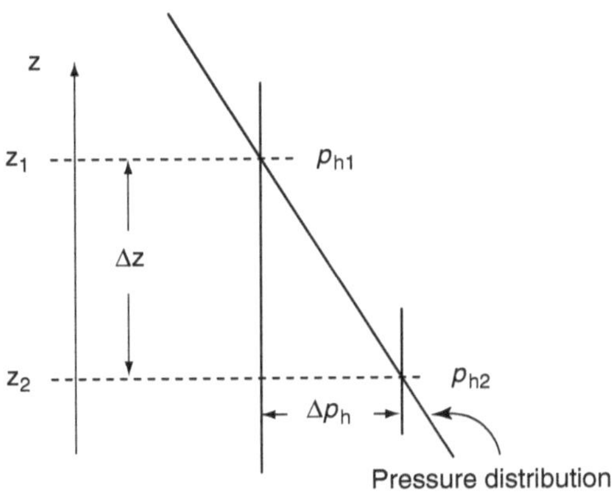

Figure 3 Pressure variations in a static fluid with constant density.

pressure at any point requires integrating the last part of eqn [4] while imposing a boundary condition given by a known pressure at some location, as was done above. The direction of the vertical gradient implies that pressure increases while moving downward in the fluid. However, pressure may be forced to increase upwards, if a fluid were to be subjected to a downward acceleration with a magnitude greater than g. For example, consider a fluid in a container (so there is no relative movement of fluid particles with respect to each other, i.e., no velocity gradients) subjected to a downward acceleration of magnitude α, while still assuming gravity is effective, in which case an extra force would have to be included in eqn [3]. The last part of eqn [4] would then become

$$\frac{\partial p_{\mathrm{h}}}{\partial z} = -\rho(g - \alpha) \qquad [6]$$

and if $\alpha > g$, the pressure would increase upwards in the fluid. Similarly, if the acceleration was upward, α and g would be additive, and the magnitude of the pressure gradient would be greater than γ. For a fluid mass in free-fall, there is no effective weight, since there is no resistance to gravity, and eqn [3] would result in a zero vertical gradient and pressure would be constant everywhere within the falling mass (equal to zero, in gauge pressure terms).

For fluids being accelerated in the x or y directions, the appropriate force, or stress would have to be included in either eqn [1] or [2], and in general the corresponding pressure gradient would no longer be zero. If a force were applied to accelerate a fluid in a container in a horizontal direction, say in the x direction with magnitude α_x, then eqn [1] would have to include this force and the resulting gradient would be

$$\frac{\partial p_{\mathrm{h}}}{\partial x} = -\rho \alpha_x \qquad [7]$$

If the container had a surface open to the atmosphere, the gradient in eqn [7] would be exhibited by a gradient in surface elevation (**Figure 4**), where pressure along a horizontal line would be directly proportional to water depth, since there is still no relative motion of fluid in any part of the container, relative to fluid in any other part of the container. An example of this type of problem would be a tanker truck carrying water with a free surface and accelerating (or decelerating) along a highway. Using the above results, it is possible to calculate the maximum acceleration possible before water would spill over the sides of the tank.

The situation is similar for a fluid in solid-body rotation. Considering a fluid rotating around a vertical axis, applying a similar force balance in the radial direction as was previously done in vertical and horizontal directions (eqns [1]–[3]) results in

$$\frac{\partial p_h}{\partial r} = \rho r \omega^2 \qquad [8]$$

where r is the radial coordinate and ω is the angular velocity. Upon integrating eqn [8], assuming constant density, pressure is found to vary parabolically with radial distance, and then considering a surface open to the atmosphere, the fluid surface will take a parabolic shape as well, since in this case rotation (about a vertical axis) does not affect forces or pressure distribution in the vertical direction. Although of interest in many applications of fluid mechanics, situations in which a fluid is artificially accelerated, either linearly or in rotation, are rare for environmental applications and these situations will not be considered further here.

Density Variations

When density is not constant, its variation with depth must be known in order to integrate the last part of eqn [4] to obtain the pressure distribution. For example, if density increases linearly with depth, $d\rho/dz = -k$, where k is a (positive) constant, then the pressure variation would be quadratic with depth and

$$p_{h2} - p_{h1} = g\left[\frac{k}{2}(z_1^2 - z_2^2) + \rho_0(z_1 - z_2)\right] \qquad [9]$$

where ρ_0 is the density at $z = 0$. As previously noted, when working with water, the incompressibility assumption may be applied, so the density is not affected by pressure (density of water becomes a function of pressure only in the deepest parts of oceans), and for inland waters it may be assumed that density depends on temperature and salinity only. This relationship is expressed through an equation of state,

$$\rho = \rho(T, S) \qquad [10]$$

where T is temperature and S is salinity. For most inland waters, salinity is small enough (or zero) that only temperature variations are important. For fresh water, the temperature of maximum density is 4 °C, and the variation of density with temperature may be approximated by a parabolic function such as

$$\rho = \rho_0\left[1 - 0.00663(T - 4)^2\right] \qquad [11]$$

where $\rho_0 = 999.97\ \mathrm{kg\ m^{-3}}$ is the density at 4 °C and T is in (°C).

Independent of any density stratification, eqn [4] must still hold. For example, it can easily be shown that lines of constant density must be horizontal in a static fluid, since equilibrium would not otherwise be possible. To see this, consider a force balance applied to the fluid element as shown in **Figure 5**. With lines of constant density oriented as shown, ρ increases while moving along a horizontal line towards the right. According to the last part of eqn [4], the pressure would then also increase while moving along a horizontal line to the right (assuming a horizontal water surface), resulting in a non-zero pressure gradient, violating the first part of eqn [4]. This situation is impossible in a static fluid, although it may be possible in a fluid moving in a body of water large enough that the earth's rotation (i.e., Coriolis acceleration) could generate a pressure gradient (tilt in water surface) to counter-balance the pressure gradient associated with the horizontal density gradient.

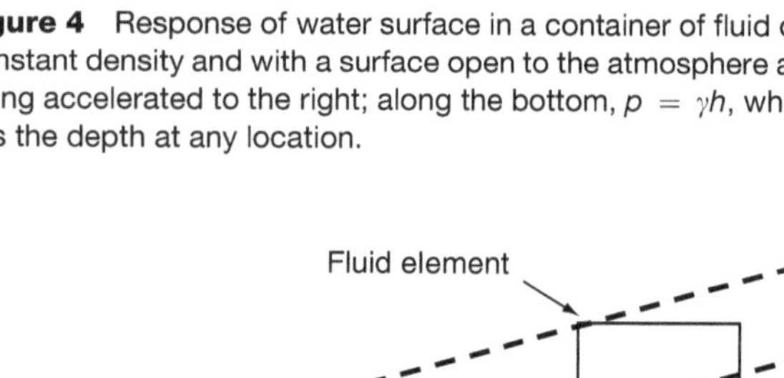

Figure 4 Response of water surface in a container of fluid of constant density and with a surface open to the atmosphere and being accelerated to the right; along the bottom, $p = \gamma h$, where h is the depth at any location.

Hydrostatic Forces on Submerged Surfaces

In addition to development of the fluid equations of motion, understanding of pressure has practical applications in calculating forces acting on submerged objects. In general, both hydrostatic and dynamic

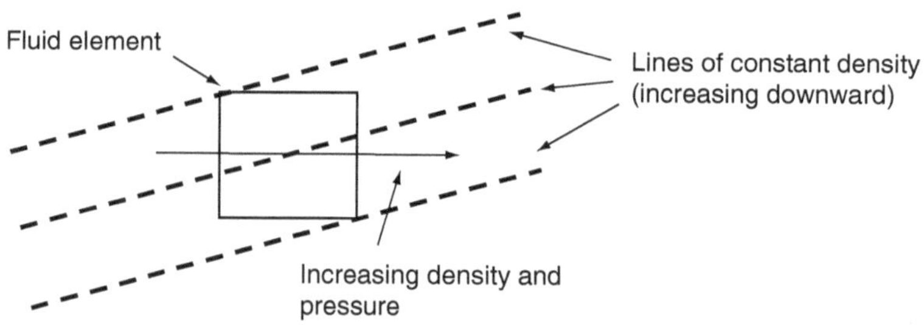

Figure 5 Fluid element in fluid where lines of constant density are tilted.

pressures must be considered, but initially we look at hydrostatic forces only.

To start, consider a submerged rectangular planar surface oriented at an angle with respect to horizontal, in a fluid as shown in **Figure 6** (note that the surface is shown in an oblique view). To simplify the discussion, it is assumed that the fluid has constant density, although the general approach here is easily extended to conditions of varying density. As discussed above, the pressure at any depth h is $p_h = \gamma h$, which acts perpendicularly on the surface, indicated on the projection of the surface to the right in **Figure 6**. The differential force acting on a small area element of the surface dA is $dF = p\,dA = \gamma hB\,dh(\sin\theta)^{-1}$, where B is the width of the surface. Note that dA is oriented horizontally and dh is assumed to be small enough that pressure may be assumed to be approximately constant over the area dA. The total pressure force is then obtained by integrating between the limits h_1 and h_2,

$$F = \frac{\gamma B}{\sin\theta}\frac{1}{2}(h_2^2 - h_1^2) \qquad [12]$$

This force may be decomposed into horizontal and vertical components,

$$F_x = (-)F\sin\theta = -\frac{\gamma B}{2}(h_2^2 - h_1^2);$$
$$F_z = (-)F\cos\theta = -\frac{\gamma B}{2}(h_2^2 - h_1^2)\cos\theta \qquad [13]$$

where the negative signs indicate forces in the negative coordinate directions. It is useful to note that the pressure at the area centroid of the surface is

$$p_{hc} = \gamma\frac{h_1 + h_2}{2} \qquad [14]$$

where p_{hc} is the (hydrostatic) pressure at the area centroid. If this pressure acted on the entire area of the surface, given by $B(h_2 - h_1)(\sin\theta)^{-1}$, the total force would be identical to the result of eqn [12]. In fact, it can be shown that this is a general result, that the total pressure force on a submerged planar surface may be calculated as the product of the area and the pressure evaluated at the area centroid. It also can be shown (see 'Further Reading') that the location of the resultant pressure force acts through the centroid of the pressure prism defined by the pressure distribution. Similarly, using moment balances it can be shown that the resultant vertical force acting on a horizontal surface passes through the centroid of that area.

For submerged curved surfaces, consider forces acting on the control volume designated by the cross-hatched area shown in **Figure 7**. Applying a force balance in the horizontal direction gives

$$\sum F_x = 0 \Rightarrow F_{sx} = \frac{\gamma B}{2}(h_2^2 - h_1^2) \qquad [15]$$

where F_{sx} is the force of the curved surface acting on the control volume in the x direction (pointing to the right), B is the width of the surface into the page, and the right-hand-side of eqn [15] is pressure force acting on the right side of the control volume, given by eqn [12]. In the vertical direction, the force balance gives

$$\sum F_z = 0 \Rightarrow F_{sz} = \gamma h_1 B\Delta x + \gamma V_f \qquad [16]$$

where F_{sz} is the force of the curved surface acting on the control volume in the z direction (assumed to point upwards), Δx is the length of the curved segment projected in the vertical direction, and V_f is the cross-hatched volume. The first term on the right-hand-side of eqn [16] is the pressure force acting on the top surface of the control volume, and the second term is the weight of fluid in the control volume. Equations [15] and [16] show that the forces acting on a curved surface may be calculated by considering the projected areas of the surface in the horizontal and vertical directions. In other words, the shape of

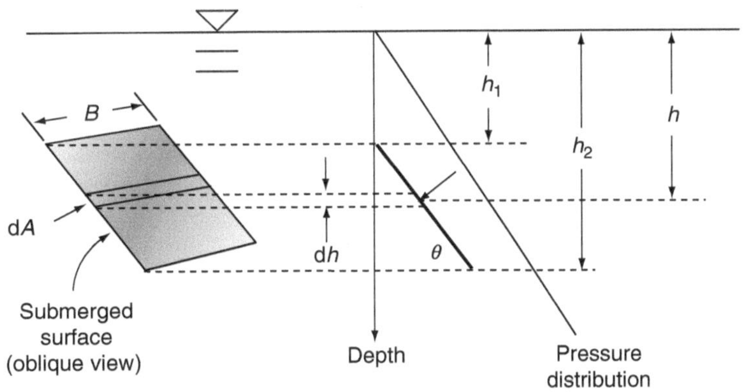

Figure 6 Hydrostatic forces acting on a submerged planar surface.

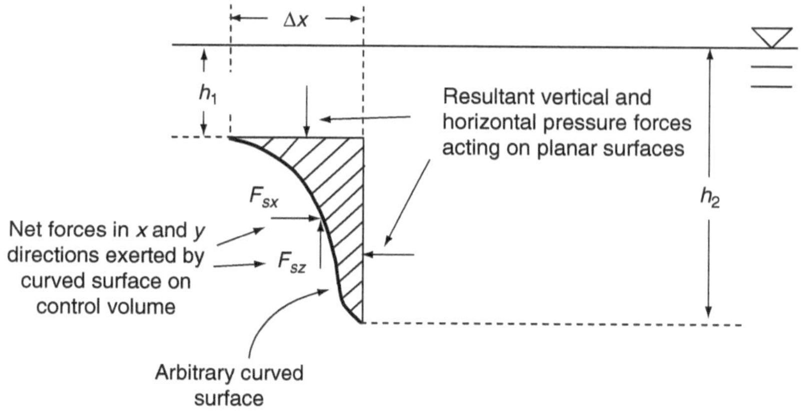

Figure 7 Pressure forces acting on a curved surface.

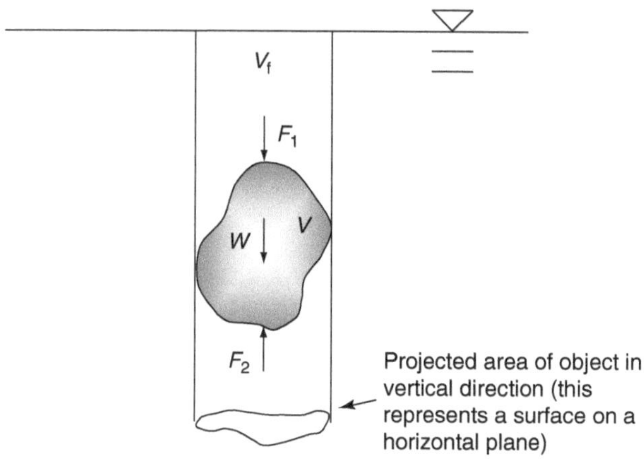

Figure 8 Object submerged in a fluid of constant density.

the surface does not matter, except in so far as it determines the control volume (cross-hatched area in **Figure 7**).

Buoyancy

Buoyancy is directly related to hydrostatic pressure, and may be considered as the net upward vertical force due to pressure acting on a submerged object. This result can be seen by considering an arbitrarily shaped object with volume V submerged in a fluid of constant density ρ (**Figure 8**). A surface is drawn below the object to illustrate a space defined by vertical sides drawn everywhere tangential to the object (this surface is the area of the object projected vertically). A volume of fluid V_f is contained within this surface and above the object. The forces acting on the object are the forces of fluid acting on its upper and lower surfaces, F_1 and F_2, respectively, and the object's weight W. There are no net forces in the horizontal direction because there are no horizontal variations in pressure. It is easily seen (by applying a force balance to the fluid above the object, for example) that the force F_1 is simply the weight of fluid sitting above the object, occupying volume V_f. The force F_2 acting on the lower surface is the same as the weight of fluid that would have occupied the volume $(V + V_f)$ – this can be seen by thinking of a column of fluid with a surface drawn around the lower half of the object and applying the same sort of force balance as above and noting that pressure is the same upwards as downwards. The force balance for the object in the vertical direction is then

$$\rho V_f + W = \rho(V + V_f) \Rightarrow W = \rho V \qquad [17]$$

This result shows that the weight of the object is balanced by the weight of fluid that has been displaced by the object. The displaced weight of fluid is called the buoyancy force. It should be emphasized that the depth at which the object rests is arbitrary (unless the density of the ambient fluid varies), as is its shape. The most important consideration here is the displaced volume. If the average density of the object (total mass divided by V) is equal to the density of the fluid, the object will be in equilibrium at any location in the water column. If the object's density is greater than that of the fluid, it will sink until it hits the bottom. If the density is less than that of the fluid, the object will float at the surface, with the degree of submersion depending on the relative difference in densities and the displaced volume. Sometimes reference is made to the 'submerged weight' of an object, which is simply the actual weight minus the weight of displaced fluid in which the object is placed. Buoyancy is what allows ships, which are typically made of materials much denser than water, to float. The ship simply has to be designed so that the weight of displaced water is greater than the weight of the ship and all its contents.

Applying a force balance to fluid elements in a fluid with density stratification gives rise to the concept of 'relative buoyancy,' where the density of the fluid element is not the same as the density of its surroundings. There is a net reduced effect of gravity since the weight of the fluid element is partially offset by buoyancy. For analysis of these situations, the net effective gravity is referred to as reduced gravity, or relative buoyancy, defined by

$$g' = g\frac{\Delta\rho}{\rho_0} \qquad [18]$$

where $\Delta\rho$ is the difference between the density of the element and its surroundings, and ρ_0 is a reference density value (usually taken as that of the surrounding fluid). Reduced gravity appears in problems associated with density-stratified flows, and may take positive or negative values, depending on the sign of $\Delta\rho$.

Dynamic Pressure

As previously noted, dynamic pressure is associated with fluid motion. The simplest illustration of dynamic pressure is obtained by considering the Bernoulli equation, which for cases of steady flow, constant density, and negligible frictional losses may be written as

$$\frac{v^2}{2g} + \frac{p}{\gamma} + z = H \qquad [19]$$

where v is the fluid velocity and H is a constant known as the Bernoulli constant, or total head, and is given by conditions of the problem (i.e., eqn [19] states that total head is constant for a given set of flow conditions). Each of the additive terms in eqn [19] has units of length. The first term is known as the velocity head, the second is the pressure head, and the third is the elevation head, or simply elevation. Definitions of these terms, as well as concepts of hydraulic grade line (HGL) and energy line (EL) for the case of open channel flow are shown in **Figure 9**. It may be noted that the Bernoulli equation represents a statement of conservation of energy, where H represents the total energy of the flow in units of length, or head. In terms of real energy units, eqn [19] would be multiplied by mass and by g. The EL is a graphical representation of the magnitude of H; so in a system where energy is conserved, the EL, or magnitude of H, is at a constant level when moving from one location to another in the flow. In other words, considering two sections in the flow as in **Figure 9**, the total head should be the same at both sections, $H_1 = H_2$ (note that velocity is assumed to be uniform at each cross section in this example – more detailed discussion is needed when velocity gradients are considered). The Bernoulli equation is developed for comparison of conditions at difference points along a common streamline, or in the case of irrotational flow as is usually assumed for open channel flow, for any two points in the flow field. By considering a case where velocities are zero everywhere, eqn [19] reduces to a statement of hydrostatic pressure, where $(p\gamma^{-1} + z)$ is a constant. This sum is known as piezometric head, and constancy of piezometric head (i.e., constant position of the HGL) in a static fluid is easily seen to be consistent with eqn [7].

In a moving fluid there is a sort of inverse relationship between velocity and pressure, as indicated in eqn [19]. That is, regions of higher velocity generally have lower pressures, and vice-versa. This is the main effect, for example, that produces lift in airfoils and allows aircraft to fly – airfoils are designed so that

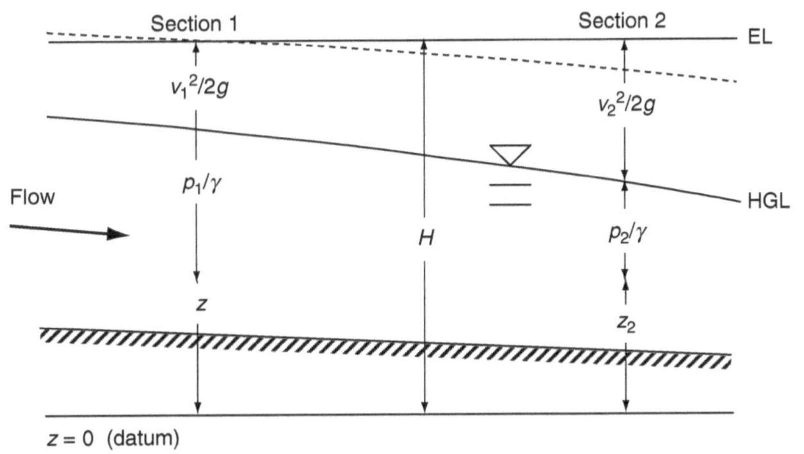

Figure 9 Definition of head terms for Bernoulli equation (eqn [19]). For hydrostatic pressure variations the sum of the elevation and pressure heads is equal to the elevation at the water surface, the position of which is also known as the hydraulic grade line (HGL). The energy line (EL) represents the elevation of total head; in an energy-conserving system the EL is horizontal, but in cases where there is energy loss, as may be induced by friction, the EL slopes downward in the direction of flow (shown as a dashed line).

there is a faster flow of air over the top of the airfoil than over the bottom, resulting in lower pressure on the top than on the bottom, with a net upward force resulting. For applications in water flow, a typical problem might involve calculating the pressure force acting on an object submerged in a flow. A simple situation of this type is illustrated in **Figure 10**, where a flat plate is placed perpendicular to a moving stream of water. At point 1 the velocity is v_1, the pressure is p_1, and the elevation is z_1. At point 2, which is at the surface of the plate, the velocity is (ideally) zero, while the elevation $z_2 = z_1$. Applying the Bernoulli equation [19] then gives

$$p_2 = p_1 + \frac{1}{2}\rho v_1^2 \qquad [20]$$

Point 2 is known as a stagnation point, which is defined anywhere where the velocity is zero, and the second term on the right-hand-side of eqn [20] is the dynamic pressure component. In this case, the dynamic pressure acting at point 2 attains the highest value possible, since point 2 is a stagnation point and $v_2 = 0$ (any velocity $v_2 > 0$ would reduce p_2 by an amount $\rho v_2^2/2$). When comparing pressures at two points in a fluid, any difference due to different velocities comes from a dynamic pressure effect, which depends on the difference in velocities squared (Δv^2).

In general, to calculate the total pressure force acting on a submerged object would require an integration of the pressure distribution on the surface of the object, which as described previously would require detailed knowledge of the velocity distribution. Fortunately, in many cases a simpler approach may be applied, based on general force balance and continuity considerations. For example, consider calculations of force acting on a sluice gate as shown in **Figure 11**. Again assuming constant velocities at each cross section (1 and 2), an integral momentum equation may be applied, along with the continuity and Bernoulli equations, to solve for the net hydraulic (pressure) force acting on the gate. Considering a control volume consisting of the water between sections 1 and 2, and assuming steady flow, continuity states that flow rate of water entering the control section must be the same as that leaving, so

$$v_1 h_1 = v_2 h_2 \qquad [21]$$

where unit width has been assumed (i.e., two-dimensional flow is considered for this problem). Using the channel bed as datum and neglecting head loss, the Bernoulli equation gives

$$\frac{v_1^2}{2g} + h_1 = \frac{v_2^2}{2g} + h_2 \qquad [22]$$

where h is the piezometric head. If the depths h_1 and h_2 are known, eqns [21] and [22] can be used to find

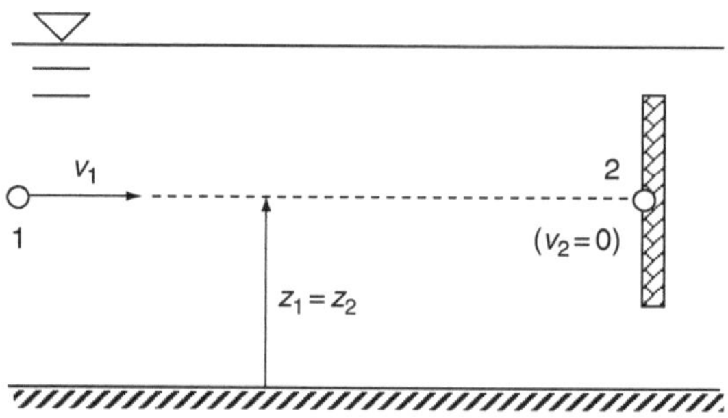

Figure 10 Dynamic pressure force acting on planar surface in moving stream of water; point 2 is a stagnation point.

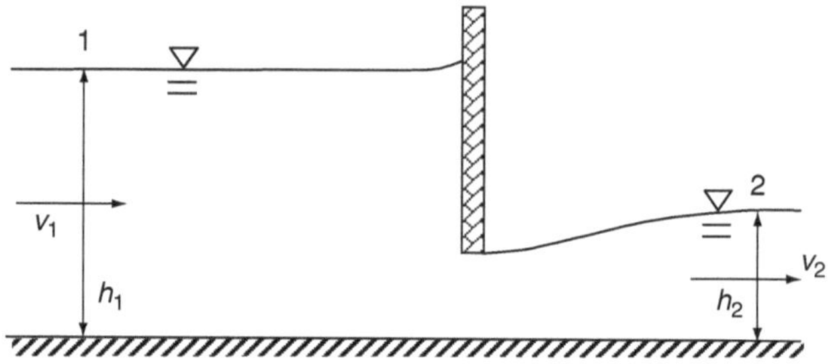

Figure 11 Forces acting on a sluice gate in (two-dimensional) open channel flow.

the velocities at each section, and therefore the flow in the channel. Applying the Reynolds Transport Theorem to evaluate forces on the control volume gives the integral momentum equation

$$p_{c1}h_1 - p_{c2}h_2 - F_g = \rho q(v_2 - v_1) \qquad [23]$$

where p_c is the (hydrostatic) pressure evaluated at the centroid of each cross section (recall previous discussion of forces on submerged surfaces), F_g is the total force exerted by the gate on the fluid in the control volume, assumed to act in the negative x direction, and $q = vh$ is the two-dimensional flow rate, or flow per unit width. The force F_g is the net integrated effect of the pressure distribution on the gate, resulting from both hydrostatic and dynamic components, and is found using eqn [23] without needing to actually calculate the pressures directly. Thus, with a simple measurement of depths upstream and downstream of the gate, the net force on the gate is found, where the force on the gate is in the opposite direction as the force found from eqn [23].

Pressure in the Equations of Motion

In several of the above examples it has been implicitly assumed that the pressure variation was approximately hydrostatic, even in the case where velocity was not zero. This assumption is evaluated in this section, which explores the impact of pressure differences on the equations of motion, as would be used in developing mathematical models of flows and circulation for environmental analyses in lakes and streams. The equations governing fluid flow consist of statements of conservation of mass (continuity), momentum, and energy. Of particular interest here are the momentum equations, or Navier–Stokes equations, which in component form are written as

$$\frac{\partial u}{\partial t} + u\frac{\partial u}{\partial x} + v\frac{\partial u}{\partial y} + w\frac{\partial u}{\partial z} - fv = -\frac{1}{\rho}\frac{\partial p}{\partial x} + \nu\nabla^2 u \qquad [24]$$

$$\frac{\partial v}{\partial t} + u\frac{\partial v}{\partial x} + v\frac{\partial v}{\partial y} + w\frac{\partial v}{\partial z} + fu = -\frac{1}{\rho}\frac{\partial p}{\partial y} + \nu\nabla^2 v \qquad [25]$$

$$\frac{\partial w}{\partial t} + u\frac{\partial w}{\partial x} + v\frac{\partial w}{\partial y} + w\frac{\partial w}{\partial z} = -\frac{1}{\rho}\frac{\partial p}{\partial x} - g + \nu\nabla^2 w \qquad [26]$$

where f is the Coriolis parameter, defined as twice the daily rotation rate of the earth times the sine of the latitude, and ν is kinematic viscosity. For inland waters, except for very large lakes such as the Laurentian Great Lakes of North America, the Coriolis term may be neglected. Also, it is easily seen that in the case of no motion, $u=v=w=0$, eqns [24]–[26] reduce to eqn [4]. The main hydrostatic pressure equation refers to the vertical distribution of pressure. In most cases of natural flows, the motions are predominantly in horizontal directions, so w is small, as are vertical accelerations, so that all terms in eqn [26] are negligible except for the first two terms on the right-hand-side, consistent with hydrostatic pressure variation in the vertical direction. There are certain situations where this is not the case, such as during fall or spring overturns in lakes, but these situations are generally of limited temporal duration. It should be noted that assuming a hydrostatic pressure variation in the vertical direction does not necessarily imply any assumption for horizontal pressure gradients.

For applications in model development for inland waters, it is useful to explore the impact of the pressure term in the Navier–Stokes equations. Here, consider the vector form of eqns [24]–[26]

$$\frac{\partial \vec{v}}{\partial t} + \vec{v}.\nabla\vec{v} + 2\vec{\Omega}\times\vec{v} = \vec{g} - \frac{1}{\rho}\nabla p + v\nabla^2\vec{v} \qquad [27]$$

where $\vec{\Omega}$ is the earth's rotation vector. The pressure term, as discussed previously, may be considered as the sum of hydrostatic and dynamic components, where the hydrostatic part may be written as

$$p_b = p_r - \int_{z_r}^{z} \rho g \mathrm{d}z \qquad [28]$$

where p_r is a reference value at $z=z_r$. Note that eqn [28] is simply the integrated form of the last part of eqn [4]. Letting $p=p_h+p_d$, where p_d is the dynamic pressure, and substituting eqn [28], the pressure gradient term in eqn [27] may be written as

$$\frac{1}{\rho}\nabla p = \frac{1}{\rho}\nabla p_r - \frac{g}{\rho}\nabla\int_{z_r}^{z}\rho \mathrm{d}z + \frac{1}{\rho}\nabla p_d = \frac{1}{\rho}\nabla p_r - \frac{g}{\rho}\int_{z_r}^{z}\nabla\rho \mathrm{d}z - g\nabla z + g\frac{\rho_r}{\rho}\nabla z_r + \frac{1}{\rho}\nabla p_d \qquad [29]$$

Then, substituting eqn [29] into eqn [27] and applying the Boussinesq approximation (neglect density variations except in the buoyancy term), the result is

$$\frac{\partial \vec{v}}{\partial t} + \vec{v}.\nabla\vec{v} + 2\vec{\Omega}\times\vec{v} = -\frac{1}{\rho_0}\nabla(p_r + p_d) + \frac{g}{\rho_0}\int_{z_r}^{z}\nabla\rho \mathrm{d}z - g\nabla\rho z_r + v\nabla^2\vec{v} \qquad [30]$$

where ρ_0 is a reference density value, usually the density at 4 °C in freshwater systems. On the right-hand-side of eqn [30], the gradient of reference pressure, ∇p_r, may usually be neglected. The second term is the effect of density variations, which are important for stratified fluids, and the third term is the effect of reference surface gradients (such as waves). Along with continuity and energy equations, eqn [30] may serve as a general starting point for developing models of fluid motion in natural waters, although in many instances it is possible to neglect some of the terms and use a simplified form of the equation.

Biological Responses

The above discussion focuses on the physical description of pressure, how it varies in a fluid, how forces are manifested on submerged objects, and how it is incorporated in the equations that would form the basis of hydrodynamic and water quality models of inland waters. Other considerations apply to various species that live or play in water, and examples of issues related to algae, fish, and humans are presented briefly here.

For submerged objects buoyancy is the main force of interest. As shown previously, buoyancy is the net result of pressure forces in the vertical direction. Pressures in horizontal directions, or at least in the direction of movement, are of interest in determining drag that must be overcome to maintain such movement. The simplest biological response and movement in the water column is achieved through the process of buoyancy regulation, which is used by certain species of algae to position themselves optimally in regions of preferable temperature, light, and nutrient availability. These algae are mostly of the blue-green type (or cyanobacteria), which also can cause nuisance and even harmful (toxic) blooms. Buoyancy regulation is achieved by increasing or decreasing gas volume in vesicles in the cells. By increasing or decreasing volume, the displaced water and hence the buoyancy force acting on the cell is altered. Increasing volume increases the buoyancy force and causes the cell to rise, and up to five-fold variations in rising or falling rates have been observed. The actual rate of rise depends on the density structure of the water column in which the organism floats, since buoyancy is the weight of the displaced water, which changes as a function of density.

For fish the physical interactions with the water environment are more complicated, due to locomotion. Some fish make use of a gas-filled cavity, called a swim bladder, or gas bladder, to maintain buoyancy and stability. Additional uplift forces can be obtained by swimming, through the same Bernoulli effect noted above, that allows airplanes to fly. However, this dynamic lift is achieved only when there is forward motion. With respect to the swim bladder, in order to maintain a constant buoyancy, the volume of the bladder must remain approximately constant as the fish swims in different depths where pressure changes, and this requires some interesting physiological responses. Near the surface, the pressure in the water is close to atmospheric pressure, but as previously described, pressure increases by about 1 atm for every 10 m of depth. Unlike water, air is compressible and volume decreases as pressure increases, so there is a tendency for reduced buoyancy at greater depths. In order to maintain neutral buoyancy within a water column, an effort must be made by the fish with swim bladders to keep the volume of their swim bladder constant. Methods of maintaining some 'hydrostatic equilibrium' varies among different groups and species of fish. This maintenance is usually done by slight secretions or resorptions of gas within the swim bladder itself, or by release of gas through a duct. 'Physoclists' are fish (e.g., the perch, *Perca fluviatilus*) that have special gas secreting and resorbing sites on the swim bladder wall that let the fish descend or ascend, respectively. 'Physostomes' are fish (e.g., the eel, *Anguilla anguilla*) that have a pneumatic duct that extends from the swim bladder to the esophagus. Such a duct allows these fish to release, or 'burp' some expanding gas as the fish ascends. Still other fish (e.g., castor oil fish, *Ruvettus pretiosus*) are able to maintain neutral buoyancy through alterations in their quantities of lipid storage. Interestingly, fish like tuna (e.g., the little Pacific mackerel tuna, *Euthynmus affinus*) and sharks have no swim bladders. This latter group, as well as species like dolphins, gain hydrodynamic lift by their shape, but they must swim continuously to keep from sinking.

Perhaps of more direct interest for humans is the attention one must pay to pressure when diving. As a diver moves into deeper or shallower water, the pressure changes and affects the balance between concentrations of gases in the dissolved (liquid) and gaseous phases, following Henry's Law. This law expresses the equilibrium between the dissolved phase concentration of a gas and its pressure in the surroundings. In essence, as outside pressure increases, gases are 'pushed' into the dissolved phase. The problem for divers occurs when they ascend too quickly following a deep dive. As pressure is reduced, gases move into the gaseous phase, and if pressure is reduced too rapidly, the gas cannot leave the blood stream quickly enough and gas bubbles, mostly nitrogen, form in the blood. In other words, the re-dissolution process does not have enough time to accommodate the gasses moving out of solution due to the pressure change. This situation leads to 'the bends,' also known as decompression or caisson sickness (the latter definition comes from the situation where workers would work in pressurized caissons, or boxes lowered in streams for construction of structures such as bridge towers – the interior of the caisson was pressurized to equal that of the surrounding water to prevent water from entering the work area, and workers leaving the pressurized area too quickly would suffer the same symptoms as divers who ascended too quickly from a dive; this was a significant problem in the building of the Brooklyn Bridge, for example). Symptoms of 'the bends' include pain in the joints, muscle cramps, sensory system failure, and, in extreme cases, even death.

While the material in this article goes into some depth with regard to the scientific and engineering analysis of pressure, it is important to recognize that different species have different responses to pressure variations. The above examples represent only a small sampling of these reactions, and how pressure and its net vertical force, buoyancy, is important in regulating our physical environment.

Further Reading

Batchelor GK (1967) *An Introduction to Fluid Dynamics*. Cambridge: Cambridge University Press.

Morris HM and Wiggert JM (1972) *Applied Hydraulics in Engineering*. 2nd edn. New York: Wiley.

Munson BR, Young DF, and Okiishi TH (1998) *Fundamentals of Fluid Mechamics*. 3rd edn. New York: Wiley.
Pelster B (1998) Buoyancy. In: Evans DH and Boca R (eds.) *The Physiology of Fishes*, 2nd edn. Boca Raton, FL: CRC Press.
Rubin H and Atkinson J (2001) *Environmental Fluid Mechanics*. New York: Marcel Dekker, Inc.
Shames IH (2003) *Mechanics of Fluids*. 4th edn. New York: McGraw-Hill.
Turner JS (1973) *Buoyancy Effects in Fluids*. Cambridge: Cambridge University Press.

Relevant Websites

http://padi.com – Diving information and diving tables.
http://www.americandivecenter.com/deep/preview/pd04.htm – Diving information and diving tables.
http://hyperphysics.phy-astr.gsu.edu/Hbose/pman.html – General discussion of water pressure.
http://www.atozdiving.co.nz/waterpressure.htm – Calculator for pressure in salt water.

Gas Exchange at the Air–Water Interface

D L Bade, Kent State University, Kent, OH, USA

Introduction

Gas exchange at the air–water interface is a key component of the biogeochemistry of all aquatic ecosystems. Organisms (e.g., algae and bacteria) can directly influence the flux of gases by controlling their concentrations. Aquatic systems can be a source of gases when they accumulate from biochemical reactions. Or, inland water bodies can act as a sink for gases when they are consumed in biochemical reactions. Release of volatile pollutants will also be controlled by the exchange across the air–water interface. The importance of air–water gas exchange is made even more evident in our attempts to understand global carbon cycles and the movement of CO_2 between air and sea. Exchange of CO_2 and certain other gases, such as CH_4, from inland aquatic systems has been shown to be of global significance. Most of the examples in this chapter will be drawn from the knowledge base developed in lakes and oceans. Specifically examined will be the gas exchange that is limited by molecular diffusion directly at the air–water interface in relatively still bodies of water.

Basic Gas Exchange Principles

Gas exchange at the air–water interface is influenced by both chemical and physical parameters. The chemical aspects of gas exchange are dominated by the chemical gradient across the air–water interface. These gradients can be induced by changes in temperature or salinity, which influence the solubility of the gas, or by chemical or biological processes that produce or consume the gas. Physical factors include the turbulent and molecular diffusion of the gas. Turbulent transport moves gases near the aqueous boundary, where they approach the diffusive sub-layer. In this boundary layer near the surface, turbulent forces are attenuated owing to viscous properties of the boundary and diffusive transport dominates (e.g., **Figure 1(a)**). Therefore an understanding of these physical and chemical properties is crucial for understanding gas exchange phenomena. Fick's law of diffusion can be used to describe the dependency of gas exchange on chemical and physical properties. From Fick's law, gas flux F [mol m^{-2} day^{-1}] is given as:

$$F = -D\frac{\partial c}{\partial z} \qquad [1]$$

where D is the molecular diffusion coefficient (m^2 d^{-1}) and $\partial c/\partial z$ is the concentration gradient (mol m^{-4}) across the boundary layer of depth Z (m). Although D can be quantified experimentally, the concentration gradient is difficult to measure because Z is extremely small (on the scale of μm's). Therefore Fick's law is restated so as to account for the bulk properties of air–water gas exchange. Here the gas flux is defined as:

$$F = k\Delta C \qquad [2]$$

where k has several names such as, 'gas transfer velocity' or 'piston velocity' or 'gas exchange coefficient,' and units of m day^{-1} or more commonly cm hour^{-1}. The concentration gradient, ΔC, is the difference between the concentrations on either side of the boundary. The term piston velocity refers to the rate of approach towards equilibrium of the concentration gradient, and can be conceptualized as the depth of water that is equilibrated with the equilibrium air concentration over the specified period of time. We will now explore the physical and chemical details of gas exchange represented by eqn. [2].

Physical Factors

Kinetic energy is transferred to water bodies by several means; most notable for gas exchange in lakes are wind shear and buoyancy flux. In fact, most work on lakes has focused solely on the influence of wind and largely has ignored the possibility of other sources of kinetic energy. This kinetic energy is dissipated as it cascades from large to small turbulent eddies. These eddies cause mixing within the water column, and are responsible for diffusion of mass throughout the mixed-layer of the water column. However, close to the interface between air and water, turbulent eddies and mixing are suppressed owing to viscous forces of the water. In the layer near the interface where viscous forces dominate, turbulent diffusivity is decreased and molecular diffusivity dominates. The same situation of a turbulent layer and diffusive sub-layer can be described for the air above the water. **Figure 1(a)** diagrams the turbulent layers and diffusive sub-layers at a hypothetical air–water interface. An important term, the Schmidt number (Sc), describes the dependence of the diffusive sublayer on kinematic viscosity of the water and molecular diffusivity. The Schmidt number is defined as:

$$\mathrm{Sc} = \frac{\nu}{D} \qquad [3]$$

where ν is the kinematic viscosity (m^2 day^{-1}). Kinematic viscosity is the dynamic viscosity divided by the density.

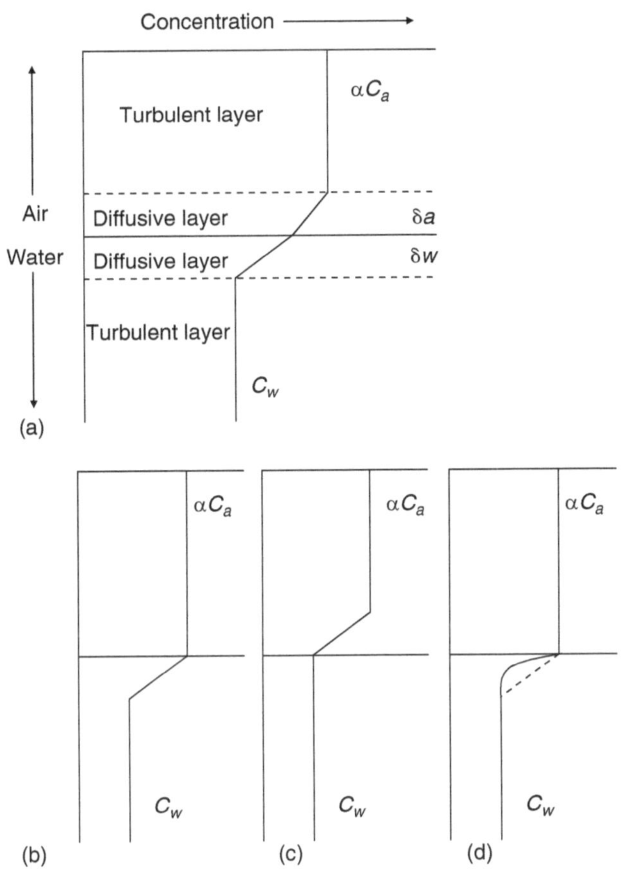

Figure 1 Conceptual diagram of the turbulent layers and diffusive sub-layers at the air–water interface. The depth of the diffusive sublayer in (a) depends on the diffusivity, solubility, and reactivity of the gas in question. The schematic in (b) represents the case for slightly soluble gases that are limited by the aqueous phase boundary. In (c) the gas represented is for a soluble gas and exchange is limited by the air phase. Under conditions chemically enhanced diffusion, the gas can react with water, causing an increased gradient in concentration.

Because the diffusivities of gases are generally much decreased in water compared with their diffusivities in air, their flux across the air–water interface is most often controlled by the aqueous diffusive sublayer (**Figure 1b**). Exceptions occur for those gases that are extremely soluble in water, or react chemically with water (**Figure 2c**). All of the major constituents of the atmosphere are limited in their exchange by the water phase.

Mechanistic Models of Gas Transfer

Several alternative models have been proposed to explain the mixing and diffusion across the boundary layer of the air–water interface. The earliest of these is the 'stagnant film model' that results directly from eqns. [1] and [2] and is captured in **Figure 1**. Here, δ_w (from **Figure 1**), or analogously, Z (from eqn. [1]), is spatially and temporally constant, and

$$k = \frac{D}{Z} \qquad [4]$$

Considering this model, more turbulent environments have decreased values of Z (the diffusive layer becomes smaller) and an increased gas transfer results. Theoretical and experimental evidence, however, suggests that this relationship is incorrect. However, it provides a useful starting point for evaluating some characteristics of gas exchange.

Using this simple model it is possible to see how gas transfer coefficients are related to the Schmidt number, and how gas transfer coefficients can be converted from one gas to another. Combining eqns. [3] and [4] for two different gas results in

$$\frac{Z}{\nu} = \frac{1}{Sc_1 k_1} = \frac{1}{Sc_2 k_2} \qquad [5]$$

More generally

$$\frac{k_1}{k_2} = \left(\frac{Sc_1}{Sc_2}\right)^{-n} \qquad [6]$$

where the coefficient $n = 1$ for the stagnant film model, but n may have other values (usually 1/2 or 2/3) depending on the particular model of the boundary layer.

One model developed, after recognition of shortcomings in the stagnant film model, was termed the 'boundary layer model.' Because a distinct discontinuity between turbulent and diffusive layers does not actually exist, the boundary layer model describes a smooth transition from turbulent to molecular transport. This model contains the assumption of a rigid smooth surface of the water. The boundary layer model describes the gas exchange coefficient as:

$$k = 0.082\left(\frac{\rho_a}{\rho_w}\right)^{0.5} \mathrm{Sc}^{-0.67} u^* \qquad [7]$$

where ρ_a and ρ_w are the densities of air and water, respectively, and u^* is the friction velocity which can be related to wind stress. Note that the Schmidt number exponent for this model is $n = 2/3$. This exponent has been corroborated by empirical evidence under conditions of low wind speed, i.e., the water surface is mostly smooth. Thus, the inherit assumptions of this model make it applicable only under conditions of low wind speed.

Another refinement of the stagnant film model is the 'surface renewal model.' This model describes the boundary layer as being influenced by turbulent eddies that bring water from below and periodically renew the water within the surface boundary layer.

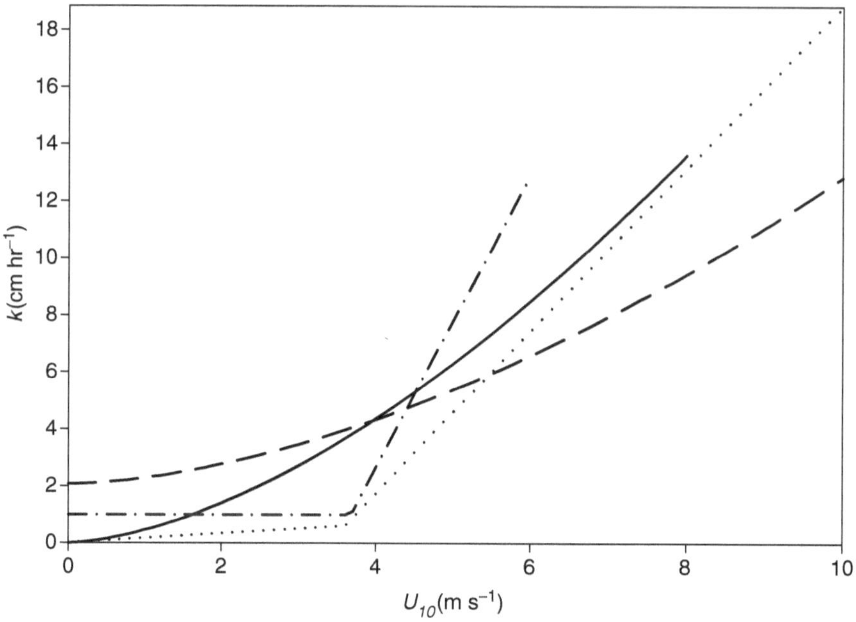

Figure 2 Models of k_{600} related to wind speed at 10 m (U_{10}): solid line, Wanninkhof (1992); dashed line, Cole and Caraco (1998); dotted-dashed line, Crusius and Wanninkhof (2003) (their first model presented in **Table 3**); and dotted line, Liss and Merlivat (1986). See **Table 3** for sources.

The time period (t) required to renew the surface boundary layer is used to determine the gas exchange coefficient by:

$$k = \left(\frac{D}{t}\right)^{0.5} \quad [8]$$

In this model the Schmidt number exponent is $n = 1/2$. Empirical evidence suggests that a Schmidt number of 1/2 is appropriate for conditions where the water surface is not smooth. Many elaborations of the surface renewal model exist, and the model is currently quite popular in the field of gas exchange. The flexibility of this model lies in the fact that many turbulent measurements can be used in quantifying the renewal time. This is an important advancement, since other factors besides just wind stress can lead to turbulence in a water column (e.g., convection currents or turbulence generated at sediment or thermocline interfaces).

In general, the relationship between the gas transfer velocity, the Schmidt number and the turbulent parameters can be established as:

$$k = \mathrm{Sc}^{-n} f(u, l) \quad [10]$$

where the Schmidt number describes the thermodynamic conditions, n depends on the characteristics of the waters surface, and $f(u,l)$ describes k as being a function of the hydrodynamics. Specifically, u is the turbulent velocity scale, and l is the turbulent length scale. The turbulent velocity scale is also known as the root mean squared velocity and is the standard deviation of the velocity fluctuations. The turbulent length scale is based on the distance over which velocity fluctuations are correlated with each other. Several important length scales can be determined for water bodies. For instance, large or small eddies can be considered most important for gas exchange, and parameterization for each situation can be made using the general model mentioned earlier.

Chemical Factors

The concentration gradient across the air–water interface, ΔC, is described in several analogous ways. One way of representing the gradient, as depicted in **Figure 1**, is in terms of concentrations,

$$\Delta C = (C_w - \alpha C_a) \quad [11]$$

where α is the dimensionless Ostwald solubility coefficient defined as the volume of gas (at a given temperature and pressure) dissolved per unit volume of solvent. Subscripts w and a refer to water and air concentrations, respectively. The gradient can also be expressed as

$$\Delta C = (C_w - K_s p_a) \quad [12]$$

where p_a is the partial pressure/fugacity of gas in the air in units of atm, and K_s is the solubility coefficient (mol l^{-1} atm^{-1}). Finally, the gradient is often measured as the partial pressures of the gas in both the air and water phases, and represented as

$$\Delta C = K_s(p_w - p_a) \quad [13]$$

Solubility of Gases

Solubility of the major gases in the atmosphere differs by several orders of magnitude. For instance CO_2 is much more soluble than CH_4. The solubility of most gases is inversely related to temperature. Over the range in temperatures experienced in most water bodies, the solubility changes by a factor of 2, and decreases approximately $-2\%\ ^{\circ}C^{-1}$ for slightly soluble gases. Salinity also influences the solubility of gases. Known as the Setschenow effect, the solubility of gases in seawater are decreased by approximately 20% when compared with fresh water. **Table 1** gives the solubilities of several selected gases. The values are given as a Bunsen coefficient, β, which is the volume of gas at standard temperature and pressure (STP), dissolved in a unit volume of solution at the temperature (T; degrees Kelvin) when the total pressure and fugacity are 1 atm. The Ostwald solubility coefficient (from eqn. [11]) is related to the Bunsen coefficient by:

$$\alpha = \beta\left(\frac{T}{T_0}\right) \quad [14]$$

where $T_0 = 273.15$ K. The solubility coefficient (eqns. [12] and [13]) is

$$K_s = \alpha/RT \quad [15]$$

where R is the gas constant (e.g., 0.082057 l atm K^{-1} mol^{-1}).

Diffusion Coefficient

As described already, the molecular diffusion of gases across the boundary layer is a controlling factor of gas exchange. Like solubility, the diffusion coefficient is also influenced by temperature and salinity. Over the typical range of temperatures of inland waters, diffusion coefficients increase by a factor of three with temperature. In saltwater, diffusion is decreased by about 5%.

Schmidt numbers are a useful means of describing diffusion coefficients because they also relate gas exchange coefficients of different gases (eqn. [6]). Some selected Schmidt numbers can be calculated from **Table 2**. Because diffusion coefficients are temperature dependent and ultimately influence gas exchange coefficients, there is a need to normalize gas transfer coefficients for empirical comparison. Gas exchange coefficients can be normalized to a particular Schmidt number. Often, exchange coefficients are normalized to a Sc = 600, which is the Schmidt number of CO_2 at 20 °C in fresh water or to a Sc = 660, the Schmidt number of CO_2 at 20 °C in seawater. The gas exchange coefficient of a particular gas, x (k_x), at a Schmidt number of 600 (k_{600}) is

$$k_{600} = k_x\left(\frac{600}{Sc_x}\right)^{-n} \quad [16]$$

Chemical Enhancement

Chemical reaction of gases with water can lead to deviations from expected gas transfer relationships. Perhaps the most notable case is for CO_2, which undergoes hydration when dissolved in water via two reactions

$$CO_2 + H_2O = H_2CO_3 \quad [17]$$

and

$$CO_2 + OH^- = HCO_3^- \quad [18]$$

The first reaction is relatively slow and does not influence gas transfer rates significantly. However, the reaction with hydroxide ions is fast, and causes the

Table 1 Temperature and salinity dependence of the Bunsen solubility coefficient (β) using the fit to the equation $\ln\beta = A_1 + A_2(100/T) + A_3\ln(T/100) + S[B_1 + B_2(T/100) + B_3(T/100)^2]$, where T is temperature in degrees Kelvin, and S is salinity in per mille (See the text for conversion between β, α and K_s)

Gas	A_1	A_2	A_3	B_1	B_2	B_3
O_2	−58.3877	85.8079	23.8439	−0.034892	0.015568	−0.0019387
CH_4	−68.8862	101.4956	28.7314	−0.076146	0.04397	−0.0068672
CO_2[a]	−60.2409	93.4517	23.3585	0.023517	−0.023656	0.0047036
N_2	−59.6274	85.7661	24.3696	−0.05158	0.026329	−0.0037252
SF_6	−520.606	250.6	(75.701)[b]	−0.0117		

Abstracted from Wanninkhof R (1992) Relationship between wind speed and gas exchange over the ocean. *Journal of Geophysical Research* 97: 7373–7382. The reader is urged to consult and cite the original publication and references therein.

[a]The solubility of CO_2 is expressed as K_s instead of a Bunsen solubility coefficient.

[b]For SF_6 the term for A_3 in the Bunsen coefficient equation is $A_3\ln(T)$, rather than $A_3\ln(T/100)$.

Table 2 Schmidt number relationship with temperature (0–30 °C) for fresh water and seawater (35%): $Sc = A - Bt + Ct^2 - Dt^3$, with t in degrees Celsius

Gas	*A*	*B*	*C*	*D*
O_2 (fresh)	1800.6	120.1	3.7818	0.047608
(seawater)	1953.4	128	3.9918	0.050091
CH_4	1897.8	114.28	3.2902	0.039061
	2039.2	120.31	3.4209	0.040437
CO_2	1911.1	118.11	3.4527	0.04132
	2073.1	125.62	3.6276	0.043219
N_2	1970.7	131.45	4.139	0.052106
	2206.1	144.86	4.5413	0.056988
SF_6	3255.3	217.13	6.837	0.08607
	3531.6	231.4	7.2168	0.090558

The top line for each gas is the freshwater value. Abstracted from Wanninkhof R (1992) Relationship between wind speed and gas exchange over the ocean. *Journal of Geophysical Research* 97: 7373–7382. The reader is urged to consult and cite the original publication and references therein.

concentration gradient in the boundary layer to increase (**Figure 1d**), thus enhancing gas transfer. The reaction with hydroxide becomes more important at high pH. One example where chemical enhancement can be important is in eutrophic systems. The high productivity causes a decrease in CO_2 which leads to an increase in pH. Chemical enhancement is also most pronounced at low wind speeds. At higher wind speeds and increased turbulence, the chemical reaction no longer becomes rate limiting and physical processes dominate the rate of gas transfer. The chemical enhancement of CO_2 also has important implications when considering stable isotope dynamics, as a large isotopic fractionation takes place under conditions of chemical enhancement.

Other Factors

Meteorological Conditions

Thermal conditions at the air–water interface influence gas transfer in several ways. Heat can be lost from the surface of the water by several mechanisms including, evaporation, sensible heat loss and long wave radiation. As this heat is lost, the surface layer becomes cooler and convection occurs, due to the buoyant forces of the warmer, less dense water below. This penetrative convection contributes another term to the turbulence in the water column. Including this component of the turbulence regime in lakes aids in the understanding of why gas exchange rates may remain relatively high, even at low wind speeds (see Empirical models of gas exchange, below). Also related to heat loss from the surface is the presence of a 'cool skin' in the top ~1 mm of water. This cool skin alters the solubility of gas in the boundary layer and results in a different concentration gradient than would be estimated from a measurement some distance below. The temperature difference is small, and this factor is probably most important when considering fluxes over large areas, such that small bias could lead to large errors.

Surfactants

Organic compounds are found on the surface of all natural waters. Generally, surface slicks reduce gas exchange, sometimes significantly. Occasionally they can be found in amounts that require a third layer to be considered when examining gas exchange. Under most conditions, however, they act to change the hydrodynamic conditions of the air–water interface. The presence of surfactants has been shown to reduce wind stress, and also to dampen waves. The role of surfactants is most pronounced at low wind speeds, since at high speeds wave action will distribute the organic compounds throughout the water column, reducing their impact on surface properties.

Bubbles

The formation of bubbles, usually caused by the breaking of waves, can have a significant impact on gas transfer by enhancing gas exchange. A positive relationship exists between the gas exchange coefficient and the fractional area of bubble plumes caused by whitecapping waves. The most obvious impact of bubbles might be an increased surface area of gas exchange, yet this has been shown to account for only a small percentage of the enhanced gas exchange. Also important is the change in internal pressure of a bubble due to the vertical displacement it experiences. This increased pressure effectively alters the ΔC from what would be experienced at the surface.

Methods for Estimating Gas Exchange Rates

Purposefully Released Tracers

Several types of volatile tracers can be used to examine gas exchange. The key aspect of the tracer is that it can be measured with high precision at low concentrations, and remains relatively conservative with regards to other processes besides atmospheric exchange. Perhaps the most useful tracer employed to this point has been sulfur hexafluoride, SF_6. The background concentration of this human-made tracer is low and the detection limit is several-fold lower than ambient concentrations. In addition, SF_6 is inert. Other easily manageable volatile tracers included propane, ethane, and methyl chloride. However, these tracers are not as stable as SF_6. Studies using any of these types of tracers can constrain gas exchange on scales ranging from 0.5 days to several days. Greater precision is gained by measuring the concentration change over longer time periods. But, the power to mechanistically explain the drivers of gas exchange is diminished at longer time scales.

By injecting a volatile tracer into a water body and following the change in concentration gradient through time it is possible to estimate the gas exchange coefficient. In eqn. [2], F can be replaced by:

$$F = \frac{\partial M}{\partial t}\frac{1}{A} = \frac{\partial C_w}{\partial t}\bar{h} \qquad [19]$$

where M is the mass of the tracer, A is the area of the flux and $\bar{h}$ is the average depth of the volume (V) of water that is exchanging with the atmosphere ($\bar{h} = V/A$). Equation 2, then yields:

$$k = \frac{\partial C_w}{\partial t}\frac{\bar{h}}{\Delta C}, \qquad [20]$$

which after rearranging and integration from the initial to final time (subscripts i and f, respectively) results in,

$$k = \ln\left(\frac{\Delta C_i}{\Delta C_f}\right)\frac{\bar{h}}{(t_f - t_i)} \qquad [21]$$

For a tracer such as SF_6, where $\alpha C_a << C_w$, the concentration gradient can be approximated by $\Delta C = C_w$.

The technique of deliberate volatile tracer additions is particularly suited for small lakes. For larger bodies of water, or where advection is of concern, the use of dual tracers is required. With the dual tracer method, two tracers are added, one being the volatile tracer. The second tracer can either be a nonvolatile tracer or another volatile tracer with differing diffusivities. Using a second volatile tracer also requires that the relative rates of exchange between the two tracers be known. With the dual tracers, concentration changes due to dilution can be factored out to ultimately estimate gas exchange. Some nonvolatile tracers include salts, dyes, and bacterial spores.

Chambers and Enclosures

Floating chambers can be deployed on the surface of the water, and changing headspace concentrations within the chamber used to estimate gas exchange. Naturally present or tracer gases can be measured, and depending on the rate of concentration change, periods as short as several hours can be examined. Thus, it is possible to directly measure the flux of a particular gas, as opposed to estimating it from another tracer gas and Schmidt number relationship. The analysis of the time series of concentrations in the headspace is done in the same way as the tracer analysis and the equation for gas exchange is:

$$k = \left(\frac{\bar{h}_H}{\alpha(t_f - t_i)}\right)\ln\left(\frac{\Delta C_i}{\Delta C_f}\right) \qquad [22]$$

where $\bar{h}_H$ is the average height of the chamber above the water.

One noted drawback of the chamber method is that the chamber itself may alter the ambient conditions at the air–water interface. The chamber may reduce turbulence (e.g., reducing wind stress), alter the pressure (e.g., bobbing motions), or cause an unnatural atmosphere (e.g., increased water vapor). However, when a relative comparison of exchange among two or more gases is of interest, the chambers may prove to be very useful. One example is calculating the enhancement of CO_2 flux under conditions of high pH. By measuring CO_2 and a nonreactive gas such as methane, and examining the ratios of their exchange coefficients (an indicator of enhancement), most biases imparted by the chamber should be eliminated.

Enclosures within the water column can also be employed to estimate gas exchange. Gas concentration can be altered or tracers added to examine the response time of the concentrations to return to equilibrium. Alternately for biologically active gases, chambers excluded from the atmosphere can be monitored congruently with the bulk water concentrations. The difference between the two concentration time-series can be attributed to gas exchange with the atmosphere. As with the chamber method, the influence of the apparatus on the results needs to be considered.

Eddy Correlation Methods

Eddy correlation (covariance) methods directly estimate the flux of gas above water surfaces by measuring the fluctuations of the vertical wind speed and gas concentrations at a fixed point. The velocity fluctuations are represented by $v' = v - \bar{v}$, where v is the instantaneous vertical velocity and $\bar{v}$ is the mean. Concentration fluctuations are represented likewise. For demonstration, if CO_2 is evading the surface of a lake, there will be a concentration profile of CO_2 going from high near the lakes surface to low some distance above the lake surface. An upward fluctuation from the mean vertical velocity (positive v') will result in a positive concentration fluctuation (positive C') at the point of measurement, and the instantaneous flux from lake to air will be positive $F = v'C'$. Inversely, a downward fluctuation in vertical velocity (negative v') will result in a negative concentration fluctuation (negative C'), also resulting in a positive flux. Therefore, the overall flux is obtained as the correlation of the vertical velocity fluctuations and the concentration fluctuations:

$$F = \overline{v'C'} \quad [23]$$

This method is used extensively in terrestrial ecosystems, and appears that it may have utility in aquatic ecosystem also, but application to this point is sparse. Eddy flux measurements require basic meteorological instruments, the most important being a sonic anemometer which can resolve wind speeds in three dimensions. In addition, some means of measuring gas concentrations at high frequency is required. Often an infra-red gas analyzer is used for appropriate gases.

Empirical Models of Gas Exchange

Most studies which have derived models for estimating gas exchange coefficients have used tracer injections of SF_6, and have relied on wind velocity measurements as the predictor variable. Wind velocity has dominated in the study of gas exchange because it is probably one of the more important factors leading to turbulence at the air–water interface. Also, wind is relatively easy to measure. The convention for reporting wind speed measurements is for wind speeds at a 10 m height above the water surface. In practice, wind speeds are rarely measured at this height, and instead are calculated based on a logarithmic wind profile. For the measurement of wind speed at a height of x meters (U_x), the wind speed at a height of 10 m (U_{10}) is

$$U_{10} = U_x\left[1 + \frac{(C_d)^{0.5}}{\kappa}\ln\left(\frac{10}{x}\right)\right] \quad [24]$$

where κ is the Van Karman constant (0.4) and C_d is the drag coefficient at a height of 10 m. The value for C_d typically ranges from 1.1×10^{-3} to 1.5×10^{-3}.

Table 3 lists several gas exchange coefficient relationships, which have been derived from experiments carried out on lakes of different sizes. Considerable uncertainty in actual rates of gas exchange and appropriate representation of wind velocities (average vs. instantaneous) has led to disparity among the models (**Figure 2**). At low wind velocity there seems to be little dependence of gas exchange on wind speeds. At higher velocities, gas exchange rates increase rapidly as a function of wind speed. A notable difference between these models is the intercept at zero wind speed. Because other sources of energy can cause turbulence it seems unlikely that gas exchange rates should decrease to zero. Therefore, other models which take into account other important sources of

Table 3 Models of k_{600} as a function of wind speed. The reader is urged to consult the original publications for descriptions of the wind speed over which the relationships are valid

K_{600} (cm hour^{-1})	*Wind speed (U_{10})*	*Source*
$k = 0.45U_{10}^{1.64}$		1
$k = 2.07 + 0.215U_{10}^{1.7}$		2
$k = 0.72U_{10}$	$U_{10} < 3.7\ \mathrm{m\ s^{-1}}$	3
$k = 4.33U_{10} - 13.3$	$U_{10} \geq 3.7\ \mathrm{m\ s^{-1}}$	
$k = 0.168 + 0.228U_{10}^{2.2}$		3
$k = 1$	$U_{10} < 3.7\ \mathrm{m\ s^{-1}}$	3
$k = 5.14U_{10} - 17.9$	$U_{10} \geq 3.7\ \mathrm{m\ s^{-1}}$	
$k = 0.17U_{10}$	$U_{10} \leq 3.6\ \mathrm{m\ s^{-1}}$	4
$k = 2.85U_{10} - 9.65$	$3.6\ \mathrm{m\ s^{-1}} < U_{10} \leq 13\ \mathrm{m\ s^{-1}}$	
$K = 5.9U_{10} - 49.3$	$U_{10} > 13\ \mathrm{m\ s^{-1}}$	

Sources

1. Wanninkhof R (1992) Relationship between wind speed and gas exchange over the ocean. *Journal of Geophysical Research* 97: 7373–7382.
2. Cole JJ and Caraco NF (1998) Atmospheric exchange of carbon dioxide in a low-wind oligotrophic lake measured by the addition of SF_6. *Limnology and Oceanography* 43: 647–656.
3. Crusius J and Wanninkhof R (2003) Gas transfer velocities measured at low wind speed over a lake. *Limnology and Oceanography* 48: 1010–1017.
4. Liss PS and Merlivat L (1986) Air–sea gas exchange rates: Introduction and synthesis. In: Buat-Menard PE (ed.) *The role of air–sea exchange in geochemical cycling*. pp 113–127. Boston, MA: Reidel.

turbulence, such as penetrative convection, should be considered essential in future research. This is why parameterizing the 'surface renewal model' has become a recent focus.

Other Important Aspects of Gas Exchange

Three additional aspects of gas exchange in aquatic environments that have received less attention than gas exchange in lakes include gas exchange in streams, rivers and estuaries, macrophyte-mediated gas transfer, and ebullition of gases from sediments. Excellent work has been conducted in these additional disciplines; however, gas exchange remains less constrained in these systems. This lack of constraint is in a large part due to the variability that can exist within each of these types of environments. Because of the variability, it is often difficult to apply previously derived models from other systems, and parameterizing gas exchange for individual studies is often required.

Gas exchange in lotic systems is given as a reaeration rate, which is the gas transfer velocity divided by the mean depth of the stream, and these are normalized to oxygen exchange at 20 °C. As in lakes, turbulence is key component of gas exchange. Turbulence in lotic systems is complicated by the interaction of the water column with the bottom surface. Factors that are often considered important for gas exchange in flowing waters include: slope, discharge, velocity, width, depth, energy dissipation, roughness, shear velocity, Froude number and Reynolds number. Models usually include a subset of these parameters, and not all the parameters listed are independent of one another. The simplest models typically estimate reaeration as:

$$k_{O_2} = aU_w^b H_w^c \qquad [25]$$

where U_w is the water velocity, H_w is the mean depth and a, b, and c are fitted parameters. Models that give satisfactory prediction among systems have been difficult to create. Even within a system extrapolation may be difficult. For instance in a stream, gas exchange will vary considerably between pool and riffle sections. In high gradient streams, bubble entrainment can significantly alter gas exchange and will vary depending on discharge.

Emergent macrophytes can exhibit significant control on the flux of gases in aquatic ecosystems. The movement of gas is of significant importance for aquatic plants, since their roots are often located in sediments that are depleted in oxygen. Therefore significant amounts of gases are moved between the atmosphere and the rooting zone. Oxygen is supplied to the roots from the atmosphere, and in exchange, other gases can be transported from the sediment to the atmosphere. Emergent macrophytes typically have hollow or pithy spaces that facilitate gas movement. Gas can move through these areas by passive diffusion or they can be moved actively by several mechanisms. As O_2 is consumed by roots, CO_2 is produced and subsequently dissolved in the water. The dissolution in pore waters creates a pressure gradient that causes gases to flow towards the roots. However, sediments are often rich in CO_2 and this mechanism is not always important, and may even cause gas to flow from the roots to the atmosphere. Emergent macrophytes can also create a pressurized system through thermal osmosis and the evaporation of water. Macrophytes significantly influence the transport of CH_4 and S containing gases. Chambers are the most highly used method for quantifying the role of macrophytes in gas exchange. Care must be exercised to ensure that the chamber does not significantly alter the thermal or light regime of the plants, since these factors are key components to the plants ability to transport gases.

Ebullition (bubble transport) is especially important in sediments that generate significant quantities of CH_4. Methane is only sparingly soluble in water. If the production rate of CH_4 is low, most of it will escape the sediments via diffusion. However, if production rates exceed the diffusion rate, concentrations of CH_4 will increase to the point that bubbles will form. Bubbles do not instantly release from the sediments, but often require some sort of disturbance. The disturbance can be a change in atmospheric or hydrostatic pressure, temperature, or some other sort of physical disturbance. When bubbles rise through the sediments, they can strip other gases as well. Ebullition is usually measured by using inverted funnel-shaped traps, which capture bubbles so that they can be sampled for volume and chemistry. Since physical disturbance is important for ebullition, it is critical that the design of the trap allow for sampling that will not actually be the cause of bubble release.

Conclusion

From the earlier-mentioned text, one should be able to make rough estimates of gas exchange for many lake settings. A general idea about wind speeds allow for an estimate of k_{600}. Applying the Schmidt number dependence permits the calculation of k for the gas of interest. Chemical gradients could be estimated from previous studies or be measured manually. Some temporal variation can be expected for both the gas exchange coefficient and chemical gradients. Nonetheless, a general idea about the magnitude of the flux could be deduced quite easily. Methods of direct measurement are most important in systems (e.g., streams and wetlands) in which empirically derived predictive models are not always applicable.

See also: Chemical Properties of Water; Density Stratification and Stability; Dissolved CO_2; Gas Exchange at the Air-Water Interface; Methane; Nitrogen; Physical Properties of Water; Pollution of Aquatic Ecosystems I; Pollution of Aquatic Ecosystems II: Hydrocarbons, Synthetic Organics, Radionuclides, Heavy Metals, Acids, and Thermal Pollution; Redox Potential; Small-Scale Turbulence and Mixing: Energy Fluxes in Stratified Lakes.

Further Reading

Anderson DA, Striegl RG, Stannard DI, *et al.* (1999) Estimating lake-atmosphere CO_2 exchange. *Limnology and Oceanography* 44: 988–1001.

Bade DL and Cole JJ (2006) Impact of chemically enhanced diffusion on dissolved inorganic carbon stable isotopes in a fertilized lake. *Journal of Geophysical Research-Oceans* 111: C01014.

Brustaert W and Jirka GH (eds.) (1984) *Gas transfer at water surfaces.* Dordrecht: Reidel.

Chanton JP and Whiting GJ (1995) Trace gas exchange in freshwater and coastal marine environments: ebullition and transport by plants. In: Matson PA and Harris RC (eds.) *Biogenic Trace Gases: Measuring Emissions from Soil and Water,* pp. 98–125. Oxford, UK: Blackwell.

Donelan MA (1990) Air–sea interaction. In: LeMehaute B and Hanes D (eds.) *The Sea: Ocean Engineering Science, no. 9,* pp. 239–292. New York: Wiley.

Donelan MA, Drennan WM, Saltzman EM, and Wanninkhof R (eds.) (2002) *Gas Transfer at Water Surfaces.* (Geophysical Monograph 127) Washington, DC: American Geophysical Union.

Frost T and Upstill-Goddard RC (1999) Air–sea gas exchange into the millennium: Progress and uncertainties. *Oceanography and Marine Biology: An Annual Review* 37: 12–45.

Jähne B and Haußecker H (1998) Air–water gas exchange. *Annual Review of Fluid Mechanics* 30: 443–468.

Jähne B, Münnich KO, Bösinger R, *et al.* (1987) On the parameters influencing air–water gas exchange. *Journal of Geophysical Research* 92: 1937–1949.

MacIntyre S, Wanninkhof R, and Chanton JP (1995) Trace gas exchange across the air–water interface in freshwaters and coastal marine environments. In: Matson PA and Harris RC (eds.) *Biogenic Trace Gases: Measuring Emissions from Soil and Water,* pp. 52–97. Oxford, UK: Blackwell.

Melching CS and Flores HE (1999) Reaeration equations derived from U.S. Geological Survey database. *Journal of Environmental Engineering* 125: 407–414.

Wanninkhof R (1992) Relationship between wind speed and gas exchange over the ocean. *Journal of Geophysical Research* 97: 7373–7382.

Light, Photolytic Reactivity and Chemical Products

A V Vähätalo, University of Helsinki, Helsinki, Finland

Photochemistry Starts from the Absorption of Solar Radiation

Photochemical reactions of organic matter can take place in surface waters and any irradiated surfaces of organic or inorganic substrata. The prerequisite for photochemical reactions is the absorption of radiation. The photolytic ultraviolet (UV) and short wavelength visible radiation (~290–500 nm) is primarily responsible for abiotic photochemical reactions. In many surface waters, chromophoric dissolved organic matter (CDOM) dominates the absorption of photolytic solar radiation. In strict sense, CDOM is an optical definition for dissolved organic matter responsible for the absorption of solar radiation, but in this article CDOM is also treated as a representative organic matter having characteristics similar to humic substances.

Primary Photophysics and Subsequent Secondary Chemical Reactions

During absorption, a photon ($h\nu$) delivers its energy (250–400 kJ mol^{-1} at 280–500 nm) to the absorber molecule and excites the molecule from its ground state (S_0) to the first excited electronic singlet state (S_1); $CDOM(S_0) + h\nu \rightarrow CDOM(S_1)$. Excited CDOM releases its excitation energy mostly as heat (with 0.95–0.98 probability) but also as fluorescence (probability of ~0.01; $CDOM(S_1) \rightarrow CDOM(S_0) + h\nu$) at the nanosecond time scale. The excited singlet state of CDOM may be converted into an excited triplet state (T_1, probability of 0.01–0.04; $CDOM(S_1) \rightarrow CDOM(T_1)$). The excited triplet state of CDOM has greater than microsecond-lifetime and may undergo secondary (thermal) chemical reactions, but it can also transfer its excitation energy to another compound or release it as heat or a photon (phosphorescence).

Numerous reaction pathways are available for excited states (e.g., homolysis of bonds, heterolysis, ionization, intermolecular oligomerization, intramolecular rearrangements, and photoisomerization). Because the energy of photon can be larger than the dissociation energy of chemical bonds, the excited molecules may dissociate (probability of $\leq 10^{-5}$) or ionize CDOM (e.g., $CDOM + h\nu \rightarrow CDOM^+ + e^-_{(aq)}$; probability of 10^{-3}–10^{-4}). The fragmentation may take place via homolysis or heterolysis of δ- and π-bonds, e.g., next to the carbonyl functional groups of CDOM (e.g., type I and II Norrish reactions). The fragmentation of excited molecules results in radical species (such as organoperoxy radicals), which further can lead to the expulsion of carbon monoxide, carbon dioxide, or other small molecules.

The excited CDOM (triplet states in particular) can transfer their excitation energy to other molecules (often called a quencher). Dissolved dioxygen is a common acceptor of this energy. Energy transfer to O_2 generates reactive oxygen species (singlet oxygen, $^1O_2$; superoxide, O_2^-; hydroxyl radicals. ·OH and hydrogen peroxide, H_2O_2) and is seen as the photochemical consumption of O_2. Hydroxyl radical is the most reactive oxygen species and can react effectively with organic and inorganic matter as well as cell constituents of organisms. In surface waters, the lifetime of H_2O_2 can be greater than several hours, while those of $^1O_2$, O_2^-, and ·OH are much lower than a second. The photochemical formation of reactive oxygen species can proceed also via photolysis of nitrate and nitrite as well as via photo-Fenton reaction involving iron.

The photochemical reactions concerning the primary absorber are frequently called direct photochemical reactions. A direct photoreaction takes place, e.g., when CDOM absorbs solar radiation and becomes photochemically transformed. An example is the photobleaching of CDOM, when photochemistry decreases the absorption of CDOM. If photochemical reactions concern other molecules than the primary absorber, they are called indirect photochemical reactions and the primary absorber is called a sensitizer. The indirect photochemistry may concern molecules that cannot absorb solar radiation themselves but react with, e.g., the photochemically generated reactive oxygen species. For example, algal-derived DOM is largely transparent to solar radiation and cannot be directly photochemically transformed alone. In the presence of CDOM sensitizer, solar radiation can transform algal DOM into a biologically recalcitrant form.

Photochemistry of Organic Carbon

In many fresh waters, the majority (50–90%) of DOM consists of heterogeneous high molecular weight (>1000 g mol^{-1}) colored humic substances (CDOM), which are sensitive to photochemical decomposition sensitive to direct photochemical decomposition. Large molecular mass, heterogeneous structure, and nonhydrolyzable bonds make CDOM recalcitrant against biological metabolism in the water column.

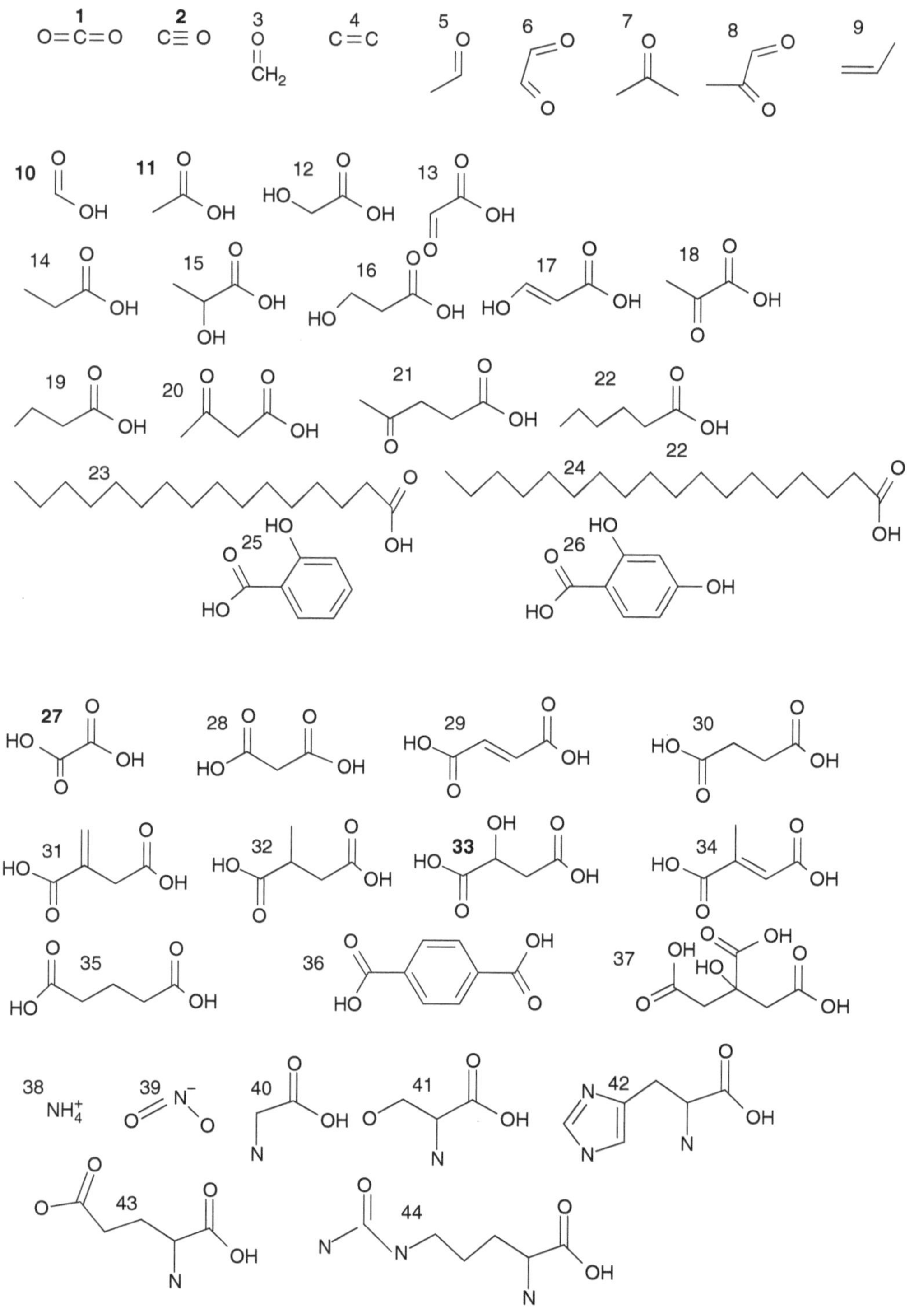

Figure 1 Identified photoproducts of irradiated natural dissolved organic matter. The most abundant products are marked with bold. (1) carbon dioxide; (2) carbon monoxide; (3) formaldehyde; (4) ethene; (5) acetaldehyde; (6) glyoxal; (7) acetone; (8) methyl glyoxal; (9) propene. Monocarboxylic acids: (10) formic acid; (11) acetic acid; (12) glycolic acid; (13) glyoxylic acid; (14) propionic acid; (15) lactic acid; (16) 3-hydroxy propanoic acid; (17) 3-hydroxy 2-propenoic acid; (18) pyruvic acid; (19) butanoic acid; (20) 3-oxobutanoic acid; (21) levulinic acid; (22) hexanoic acid; (23) hexadecanoic acid; (24) octadecanoic acid; (25) 2-hydroxy benzoic acid; (26) 2,4-dihydroxy benzoic acid. Di-tricarboxylic acids: (27) oxalic acid; (28) malonic acid; (29) fumaric acid; (30) succinic acid; (31) methylene succinic acid; (32) methyl succinic acid; (33) malic acid; (34) methyl malic acid; (35) pentane dioic acid; (36) terephtalic acid; (37) citric acid. N containing products: (38) ammonium; (39) nitrite; (40) glycine; (41) serine; (42) histidine; (43) glutamic acid; (44) citrulline. Kieber RJ, Zhou X, and Mopper K (1990) Formation of carbonyl compounds from UV-induced photodegradation of humic substances in natural waters: Fate of riverine carbon in the sea. *Limnology and Oceanography* 35: 1503–1515. Mopper K, Zhou X, Kieber RJ, *et al.* (1991).

Because the biologically recalcitrant CDOM is photochemically reactive, the photochemical transformation of CDOM and photooxidation of nonhydrolysable bonds of CDOM are major photoreactions in lakes.

The most abundant identified photochemical product of natural CDOM is carbon dioxide (CO_2) or, more correctly, inorganic carbon existing as CO_2, H_2CO_3, HCO_3^-, or CO_3^{2-} depending on pH (**Figure 1**). Photochemical reactions produce low molecular mass organic compounds from CDOM (**Figure 1**). Identified products include carboxylic acids (33 of those in **Figure 1**), few carbonyl compounds, volatile alkenes, and carbon monoxide. When the photochemical production of CO and CO_2 has been determined from the same waters, the photoproduction rate of CO has been 5–7% of that of CO_2. The magnitude of four photoproduced carboxylic acids (formic, acetic, oxalic, and malonic acid) has corresponded to 30–50% (mean 34%) of the photoproduced CO_2 in 38 Swedish lakes and also in other fresh waters.

Most of the identified organic photoproducts of CDOM (**Figure 1**) are directly bioavailable for bacterioplankton, and are called biologically available photoproducts (BAPs). The total magnitude of BAPs is quantified by measuring bacterial carbon consumption, which is caused by the irradiation of CDOM. For example, bacteria assimilated BAPs at a magnitude similar to the photoproduction of CO_2 during the irradiation of humic lake water (Lake Valkea-Kotinen, Finland) when cultured in the irradiated water after the irradiation. Solar radiation mineralized 27 μmol carbon per liter of Parker River DOM (USA) into CO_2 but produced 54 μmol carbon per liter BAPs. According to those studies, the amount of BAPs can be nearly similar to or twofold larger than the photoproduction of CO_2. Thus, in addition to the direct photochemical mineralization of CDOM, a combined action of photochemistry and microbial respiration based on BAPs leads to the mineralization of organic carbon.

Photochemical reactions decompose the aromatic moieties of CDOM and typically decrease the molecular mass of CDOM. The decrease in aromaticity, in molecular mass and changes in the functional groups change the chemical and physical properties of CDOM. Photochemical reactions increase the water solubility of humic acids, decrease the light absorption of CDOM, decrease the sorption of CDOM towards surfaces or hydrophobic organic chemicals, and change the composition of functional groups (ligands) responsible for the chelating of metals. Photochemical reactions can also increase molecular mass of CDOM and even precipitate it. However, further irradiation of photochemically formed high molecular mass CDOM eventually result in low molecular weight compounds.

Solar radiation decomposes the fluorophores and light absorbing moieties of CDOM in a process called photobleaching. An extensive photobleaching requires the presence of O_2. The photochemical loss of carbonyl chromophores into low molecular weight organic compounds, and the destruction of aromatic and quinone structures of CDOM are likely responsible for the photobleaching. Under extensive exposures to solar radiation, photochemical reactions can destroy completely the chromophores and fluorophores of organic matter. The photobleaching of CDOM takes place faster than the mineralization of dissolved organic carbon (DOC), and after complete photochemical loss there will be some nonchromophoric DOC left. Experimental and modeling studies suggest that solar radiation-induced photobleaching is primarily responsible for the total decomposition of CDOM in surface waters, and contributes remarkably to the observed in-lake losses of the imported CDOM. Because CDOM is largely responsible for the absorption of UV radiation and low wavelength visible radiation, photobleaching increases penetration of solar radiation into the water column, and thus interferes with the UV radiation exposure of organisms, the light availability of photosynthetic organisms, and the thermal stratification of the water column.

Photochemical degradation of dissolved organic carbon and its impact on the oceanic carbon cycle. *Nature* 33: 60–62. Backlund P (1992). Degradation of aquatic humic material by ultraviolet light. *Chemosphere* 35: 1869–1878. Allard B, Borén H, Pettersson C, and Zhang G (1994) Degradation of humic substances by UV irradiation. *Environment International* 20: 97–101. Wetzel RG, Hatcher PG, and Bianchi TS (1995) Natural photolysis by ultraviolet irradiance of recalcitrant dissolved organic matter to simple substrates for rapid bacterial metabolism. *Limnology and Oceanography* 40: 1369–1380. Corin N, Backlund P, and Kulovaara M (1996) Degradation products formed during UV-irradiation of humic waters. *Chemosphere* 33: 245–255. Kulovaara M, Corin N, Backlund P, and Tervo J (1996) Impact of UV254-radiation on aquatic humic substances. *Chemosphere* 33: 783–790. Bertilsson S and Tranvik L (1998) Photochemically produced carboxylic acids as substrates for freshwater bacterioplankton. *Limnology and Oceanography* 43: 885–895. Riemer DD, Milne PJ, Zika RG, and Pos WH (2000) Photoproduction of nonmethane hydrocarbons (NMHCs) in seawater. *Marine Chemistry* 71: 177–198. Tarr MA, Wang W, Bianchi TS, and Engelhapt E (2001) Mechanisms of ammonia and amino acid photoproduction from aquatic humic and colloidal matter. *Water Research* 35: 3688–3696. Kieber RJ, Li A, and Seaton PJ (1999) Production of nitrite from the photodegradation of dissolved organic matter in natural waters. *Environmental Science and Technology* 33: 993–998. Shiller AM, Duan S, van Erp P, and Bianchi TS (2006) Photo-oxidation of dissolved organic matter in river water and its effect on trace element speciation. *Limnology and Oceanography* 51: 1716–1728.

Photochemistry of Nitrogen

Photochemical reactions can release ammonium (NH_4^+) from dissolved organic nitrogen (DON) when the concentration of DON is high and that of NH_4^+ low (<6 μM). The proposed mechanism for photochemical release of NH_4^+ includes a photoreaction of organic amines with ketones and aldehydes to produce imines, which hydrolyze to NH_4^+. According to current knowledge NH_4^+ is the primary nitrogenous photoproduct accounting for perhaps *c.*90% of the total production of biologically labile N-photoproducts. The others are nitrite (NO_2^-), primary amines, and amino acids (**Figure 1**). Photochemistry can act with amino acids or peptides bound in humic substances. For example, when an amino acid (glycine) was mixed with humic substances its bioavailability decreased, but photoreactions converted the amino acid back into bioavailable form. The photochemical reactions mineralize more carbon than nitrogen, and therefore decrease the C/N ratio of DOM.

At the presence of high concentration of NH_4^+ (>8 μM) or NH_3, photochemical reactions can also incorporate these inorganic N-species into DON. In addition, photochemical reactions can convert aromatic amino acid (tryptophan) and protein (in the presence of CDOM) into less bioavailable forms.

Photochemistry of Phosphorus

Ferric (Fe^{3+}) iron binds phosphate ion tightly and can associate with organic matter to form Fe(III)–P_i–CDOM complexes. Photochemical reactions can release phosphate (PO_4^{3-}) from ferric iron–phosphorus–humic complexes (Fe(III)–P_i–CDOM). The phosphorus of these complexes is not readily available for organisms. When these complexes absorb solar radiation, ferric iron (Fe^{3+}) can be photoreduced into ferrous iron (Fe^{2+}) and a phosphate ion is released. Unlike many photochemical reactions of natural DOM, the photoreduction of ferric iron and subsequent release of phosphate can be a very rapid process (half-lives minutes–hours). The photochemical release of P_i from the Fe(III)–P_i–CDOM complexes may be an important pathway for converting nonbioavailable P into a bioavailable form.

Coupling of Photooxidation of Organic Matter to the Photoreduction of Metals

Metals can associate with organic matter and can become photoreduced via a ligand-to-metal charge-transfer pathway when exposed to natural solar radiation (**Figure 2**). When organic matter associates with iron, its absorption and the probability for photochemical transformation increases. Photochemical reactions can photoreduce ferric iron to more soluble ferrous iron and at the same time oxidize organic matter, e.g., via decarboxylation reactions, which can lead to the production of inorganic carbon and hydrogen peroxide (H_2O_2). Ferrous iron (II) is often quickly oxidized to ferric iron (III) in a dark process, which consumes dissolved oxygen or reactive oxygen species. If H_2O_2 is the oxidizer of Fe(II), the process produces hydroxyl radicals via Fenton reaction: Fe(II) + H_2O_2 → ·OH + Fe(III). When solar radiation produces both Fe(II) and H_2O_2, this reaction is called a photo-Fenton reaction.

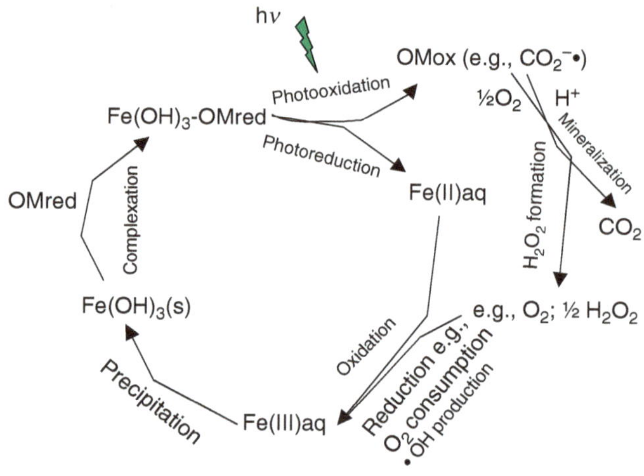

Figure 2 Schematic representation of the aquatic photoredox cycling of iron and its coupling to the oxidation of organic matter (OM) in illuminated surface waters. Fe(II)aq = dissolved ferrous iron; Fe(III)aq = dissolved ferric iron; $Fe(OH)_3(s)$ = ferric hydroxide; OM_{red} = reduced organic matter, which is a ligand for Fe(III); $Fe(OH)_3$–OM_{red} = a complex between reduced OM and Fe(III); OM_{ox} = oxidized organic matter; CO_2^- = carboxylic radical. The oxidation of Fe(II) with hydrogen peroxide is called the Fenton reaction (H_2O_2 + Fe(II) → Fe(III) + ·OH + OH^-) and it generates hydroxyl radicals. Adapted from Sulzberger B (1992) Heterogeneous photochemistry. In Stumm W (ed.) *Chemistry of the Solid–Water Interface*, pp 337–367. New York: John Wiley.

Heterogeneous Photochemistry

Photochemistry can modify particulate organic matter or organic matter adsorbed on the surfaces. The photochemical reactions associated with surfaces are called heterogeneous in contrast to the homogenous photochemistry of dissolved molecules.

Heterogeneous photochemical reactions take place on any surfaces exposed to photolytic solar radiation, e.g., on the surfaces of sediment or particles in suspension. The photochemical reactions on surfaces of natural mineral particles are largely similar to those in the dissolved phase. Photochemical reduction of iron(III) and dissolution of Fe takes place also on surfaces of colloidal or solid iron minerals. Concomitantly with the photoreduction of iron III, organic carbon adsorbed on particles is oxidized and O_2 consumed. Solar radiation can mineralize and decrease the molecular mass of organic matter adsorbed on the surfaces of mineral (e.g., clay) particles. This process decreases the adsorption of CDOM on surfaces and leads to the dissolution of DOC. Such DOC has larger bioavailability than the organic matter adsorbed on the surfaces of mineral particles.

Photochemical reactions also interfere with the decomposition of particulate dead organic matter (detritus) such as senescent phytoplankton or macrophytes. The (photosynthetic) pigments are primary absorbers of solar radiation and susceptible for photolysis in detrital plants with nonfunctional photoprotective apparatus. The photolysis of pigments in surface waters is an important mechanism for their decomposition, because pigments decompose ineffectively in dark anaerobic sediments.

The lignin of vascular plant detritus is highly susceptible for photolysis. The photochemical decomposition of lignin can break down the structural integrity of detrital plant material, enhance the leaching of particulate organic matter into DOM, and induce the loss of particulate organic matter. Photochemical reactions can also directly mineralize detrital plant material into CO_2. The exposure of detrital leaves to solar radiation can increase or decrease microbial activity on leaves possibly depending on the diagenetic state and the bioavailability of detrital leaves.

Photochemical Reactivity of Organic Matter

Photochemical reactivity of organic matter describes the susceptibility of organic matter to photochemical transformation and is quantitatively described by an apparent quantum yield (AQY or ϕ). The AQY for a photochemical reaction describes the magnitude of photochemical reaction (mole of photochemical product) in relation to the magnitude of photons absorbed by the sample (mole of absorbed photons).

The variability of AQYs is large among different photochemical reactions (**Figure 3**). For example, the AQY can be around 1 as is the case for the photochemical reduction of ferric iron of ferrioxalate, which can be coupled to the production of CO_2 (**Figure 3**). In this case, every photon absorbed by ferrioxalate reduces ferric iron to ferrous iron. When nitrate absorbs solar radiation, 1 mol of absorbed photons generates 0.017 mol of hydroxyl radicals and nitrogen dioxide.

$$NO_3^- + h\nu\ (+H^+) \rightarrow NO_2 + \cdot OH$$

The AQY for this reaction is 0.017. For known single chemical compounds (as for the nitrate and ferrioxalate above), the AQY is often assumed to be the same through that part of the spectrum, where solar radiation and the absorption of the compound interact. This is exemplified as a constant AQY for the photolysis of nitrate at wavelengths *c.*290–345 nm (**Figure 3**).

The AQYs for the photochemistry of CDOM are typically low but they vary also a lot and have a strong spectral dependence (**Figure 3**). The spectral dependence likely reflects the heterogeneity of chromophores and the photochemical reactivity of thousands of compounds included in CDOM. For example, when AQY was determined for the photochemical formation of CO in three fresh waters of North America, AQY ranged from 0.000 023 to 0.000 068 at 350 nm and showed an exponentially decreasing trend with increasing wavelength (**Figure 3**). Similar exponential spectral dependence has been found for the photochemical consumption of O_2, the photobleaching of CDOM, the photoproduction of CO_2, H_2O_2, ethene, and $\cdot OH$. Photochemical reactivity of DOM does not always follow exponential spectral dependence. Deviation of this behavior has been found in processes where iron has been involved for example for the (photo-) Fenton-type production of hydroxyl radicals. For the most photochemical reactions of CDOM, the magnitude and spectral dependence of AQY can be described by an exponential equation as illustrated by lines for the photochemical production of CO_2, BAPs, and NH_4^+ in **Figure 3**. One commonly used equation is

$$\phi_\lambda = a\ \exp(-b\lambda)$$

where ϕ_λ is the AQY at the wavelength λ, a is a pre-exponential factor, and b is a spectral slope coefficient (nm^{-1}). For example, the AQY spectrum for the photochemical mineralization of DOC ($\phi_{DOC,\lambda}$) had $a = 7.52$ (mol carbon per mol photons) and $b = 0.028$ (nm^{-1}) in the water of the humic Lake Valkea-Kotinen. From this equation one can calculate that $\phi_{DOC,300} = 1600$, $\phi_{DOC,350} = 400$, $\phi_{DOC,400} = 100$, and $\phi_{DOC,450} =$ 20 μmol carbon (mol photons at $\lambda)^{-1}$. In other words, 1 mol of photons absorbed by CDOM at 450 nm mineralizes 20 μmol DOC, while the mole of photons absorbed at 300 nm mineralizes 1600 μmol DOC.

By comparing AQYs among different photochemical reactions, one gets an idea about the relative importance of these photochemical reactions. For example, AQY for the formation of BAPs in Altamaha River ($\phi_{BAP,350} = 8 \times 10^{-4}$) is much larger than AQYs for the CO photoproduction in three North American fresh waters ($\phi_{CO,350} = 2.3\text{–}6.8 \times 10^{-5}$) or that for ethene photoproduction in the Parker River water (Florida; $\phi_{ethene,334} = 6.5 \times 10^{-8}$). Thus, photoproduced CO or ethene contributes only little to the

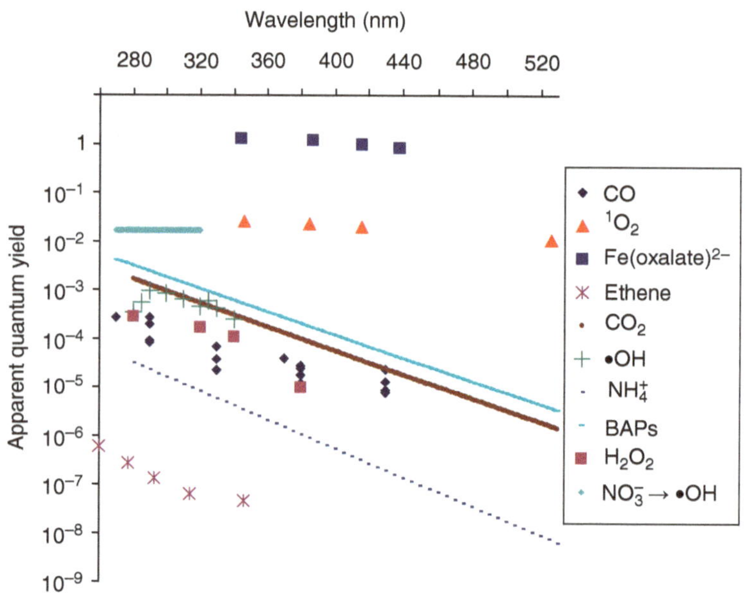

Figure 3 Apparent quantum yield (AQY) for the photochemical reactions of CDOM and those of nitrate and ferrioxalate. Photoproduction of carbon monoxide (CO) in Satilla and Suwannee River, Lake Kinoshe, Houghton wetland, Okefenokee Swamp (Valentine RL and Zepp RG (1993) Formation of carbon monoxide from the photodegradation of terrestrial dissolved organic carbon in natural waters. *Environmental Science and Technology* 27, 409–412. Gao H and Zepp RG (1998) Factors influencing photoreactions of dissolved organic matter in a coastal river of the southeastern United States. *Environmental Science and Technology* 32: 2940–2946). Photoproduction of singlet oxygen ($^1O_2$) in Lake Baldegg (Haag WR, Hoigné J, Gassman E, and Braun AM (1984) Singlet oxygen in surface waters – part II: Quantum yields of its production by some natural humic materials as a function of wavelength. *Chemosphere* 13: 641–650). Photoreduction of ferrioxalate ($Fe(oxalate)^{2-}$) (Demans JN, Bowman WD, Zalewski EF, and Velapoldi RA (1981) Determination of the quantum yield of the ferrioxalate actionometer with electrically calibrated radiometers. *Journal of Physical Chemistry* 85: 2766–2771. Faust BC and Zepp RG (1993) Photochemistry of aqueous iron(III) polycarboxylate complexes – roles in the chemistry of atmospheric and surface waters. *Environmental Science and Technology* 27: 2517–2522). Ethene photoproduction in Shark River (Riemer DD, Milne PJ, Zika RG, and Pos WH (2000) Photoproduction of nonmethane hydrocarbons (NMHCs) in seawater. *Marine Chemistry* 71: 177–198). CO_2 photoproduction in Lake Valkea-Kotinen (Vähätalo AV, Salkinoja-Salonen M, Taalas P, and Salonen K (2000) Spectrum of the quantum yield for photochemical mineralization of dissolved organic carbon in a humic lake. *Limnology and Oceanography* 45: 664–676). ·OH photoproduction in Suwannee River fulvic acid (Vaughan PP and Blough NV (1998) Photochemical formation of hydroxyl radical by constituents of natural waters. *Environmental Science and Technology* 32: 2947–2953). ·OH production from photolysis of nitrate (Blough NV and Zepp RG (1995) Reactive oxygen species in natural waters. In Foote CS, Valentine JS, Greenberg A, and Liebman JF (eds.), *Active Oxygen in Chemistry*, pp. 280–333. London: Blackie Academic & Professional). NH_4^+ photoproduction in the Baltic Sea (Vähätalo AV and Zepp RG (2005) Photochemical mineralization of dissolved organic nitrogen to ammonium in the Baltic Sea. *Environmental Science and Technology* 39: 6985–6992). Biologically available photoproducts (BAPs) in Altamaha River, USA. (Miller WL, Moran MA, Sheldon WM, Zepp RG, and Opsahl S (2002) Determination of apparent quantum yield spectra for the formation of biologically labile photoproducts. *Limnology and Oceanography* 47: 343–352). H_2O_2 photoproduction in Sharpes Bay (Scully NM, McQueen DJ, Lean DRS, and Cooper WJ (1996) Hydrogen peroxide formation: The interaction of ultraviolet radiation and dissolved organic carbon in lakes waters along a 43–75° N gradient. *Limnology and Oceanography* 41: 540–548).

photoproduction of biologically labile photoproducts. According to AQYs, the reactive oxygen species (·OH, H_2O_2, $^1O_2$) could potentially contribute in large extent to the photoproduction of CO_2. The recent studies, however, suggest that photoproduced ·OH contributes little (perhaps <10%) to the photoproduction of CO_2. The AQY for the production of singlet oxygen in Lake Baldegg water is much larger than ϕ_{CO_2}. This must mean that photoproduced singlet oxygen is a relatively inefficient decomposer of DOC. The AQY for the photoreduction of Fe(III)–polycarboxylate complexes is high, as exemplified by ferrioxalate in **Figure 3**. These reactions are often associated with the CO_2 production, and therefore the photochemistry of Fe(III)–polycarboxylate complexes can contribute to the photomineralization of DOC. This observation is in agreement with the findings that the photochemical production of CO_2 increases along with the concentration of iron.

Regulators of the Photochemical Reactivity of Organic Matter

The characteristics of DOM and environmental conditions affect the photochemical reactivity of DOM.

For example, the reported AQYs for the photochemical mineralization of DOC at 350 nm ($\phi_{DOC,350}$) have ranged from 230×10^{-6} to 2290×10^{-6} with median of 670×10^{-6} in the Satilla River, humic Lake Valkea-Kotinen, Lake Tuscaloosa reservoir, and inshore waters. The 10-fold difference in the $\phi_{DOC,350}$ indicates that the characteristics of DOM and environmental conditions affect the photochemical reactivity of DOM. The photoproduction of CO_2 is favored by low pH, as well as by high concentrations of iron and O_2.

The photochemical reactivity of DOM depends on the previous solar radiation exposure of DOM. DOM without previous solar radiation exposure can be considered most sensitive for the photochemical transformations. As solar radiation decomposes the most photoreactive parts of DOM (e.g., the aromatic structures) first, the photoreactivity of DOM decreases. The decrease can be rapid. For example, less than an hour-long irradiation of riverine DOM with 310-nm radiation decreased the ϕ_{310} for photochemical oxygen consumption by 50% in Shark River water (Florida). In contrast, the photoreactivity of DOM decreases much more gradually when DOM has been already exposed to solar radiation in past, e.g., in large lakes with long water retention times.

The photochemical rates and thus also the determined AQYs depend on temperature but in lesser degree than thermal chemical reactions. At the relevant temperature range in surface waters (0–25 °C), the rates are 1.5- and 3-times faster at 25 °C than at 0 °C with typical activation energies of 10 and 30 kJ mol^{-1} found for photochemical reactions, respectively.

Predicting the Rates of Photoreactions in the Environment

The photochemical transformation rate of DOM in an optically thin water layer at the depth z (pr_z, mol m^{-3} d^{-1}) can be expressed as

$$pr_z = \int_{\lambda_{min}}^{\lambda_{max}} Q_{s,z,\lambda}\, a_{CDOM,\lambda}\, \phi_\lambda \, d\lambda \quad [1]$$

where $Q_{s,z,\lambda}$ is the scalar photon flux density at λ and depth z (mol photons m^{-2} d^{-1} nm^{-1}), $a_{CDOM,\lambda}$ is the absorption coefficient of CDOM at λ (m^{-1} nm^{-1}), and ϕ_λ is the AQY for the photochemical reaction at wavelength λ (mol product [mol absorbed photons]$^{-1}$ nm^{-1}). Integration of eqn. [1] is over the wavelength range from λ_{min} to λ_{max}, the minimum and maximum wavelengths (nm), respectively, contributing to the photochemical reaction. Note that $Q_{s,z,\lambda}$ describes a photon flux density to a point from all directions and is different from vector photon flux densities, the most commonly measured form of solar irradiance. In most studies, $a_{CDOM,\lambda}$ has been measured from filtered water samples against H_2O blank, and in this case the measured $a_{CDOM,\lambda}$ includes also the absorption by inorganic constituents such as NO_3^- or iron associated with CDOM.

To visualize the use of eqn. [1], we calculate the photochemical mineralization of DOC in a humic Lake Valkea-Kotinen on a summer day (**Figure 4**). Panel (a) describes $Q_{s,z,\lambda}$ at three depths, (b) is $a_{CDOM,\lambda}$, and (c) describes the ϕ_λ for the photochemical mineralization of DOC. In this lake with high concentration of DOC (1 mmol DOC per liter), $Q_{s,z,\lambda}$ attenuates steeply into the water column and for example, solar radiation at 350 nm at the depth of 0.10 m is 8% of that found at the depth of 0.01 m (**Figure 4(a)**). When one multiplies $Q_{s,z,\lambda}$ with $a_{CDOM,\lambda}$, the result gives the rate of photon absorption by CDOM at depth z. When this photon absorption rate is multiplied with ϕ_λ, one will know how many of the absorbed photons resulted in the mineralization of DOC. The result in panel (d) shows that the photochemical mineralization of DOC is highest just below the surface but decreases with depth along the solar radiation. When the action spectra are integrated through the wavelengths, the photomineralization rates are 18, 6.1, and 2.1 mmol m^{-3} d^{-1} (or μmol l^{-1} d^{-1}) at the depths of 0.01, 0.05, and 0.10 m, respectively.

Over the entire water column, the rate for photochemical transformation of DOM (pr, mol m^{-2} d^{-1}) is

$$pr = \int_{\lambda_{min}}^{\lambda_{max}} Q_{a,\lambda}\, a_{CDOM,\lambda}/a_{tot,\lambda}\, \phi_\lambda \, d\lambda \quad [2]$$

where $Q_{a,\lambda}$ represents mole of photons absorbed by the water column (mol photons m^{-2} d^{-1} nm^{-1}) and the $a_{CDOM,\lambda}/a_{tot,\lambda}$ describes the contribution of CDOM to the total absorption coefficient, $a_{tot,\lambda}$. Note, that the downward photon flux density at the surface, the most commonly measured form of solar radiation can approximate $Q_{a,\lambda}$. For accurate estimates, one needs to account for the few percent losses due to the reflection of solar radiation from the air–water interface and the water leaving radiation from the water column. The use of shading factor, $a_{CDOM,\lambda}/a_{tot,\lambda}$, assumes that only the photons absorbed by CDOM will photochemically transform DOM. Thus, eqn. [2] assumes that the photons absorbed by H_2O and particulate matter do not contribute to the photochemical transformation of DOM.

To visualize eqn. [2], we continue to calculate the photochemical mineralization of DOC in Lake Valkea-Kotinen (**Figure 4(e)–4(h)**). Now, the $Q_{a,\lambda}$ is approximated by the downward vector photon flux density just below the surface, $Q_{v,0-,\lambda}$ (panel (e)); the $a_{CDOM,\lambda}/a_{tot,\lambda}$ estimates how much CDOM contributes to the total absorption of photons (panel (f)); and the same ϕ_λ estimates the fraction of the absorbed photons that contributes to the photomineralization

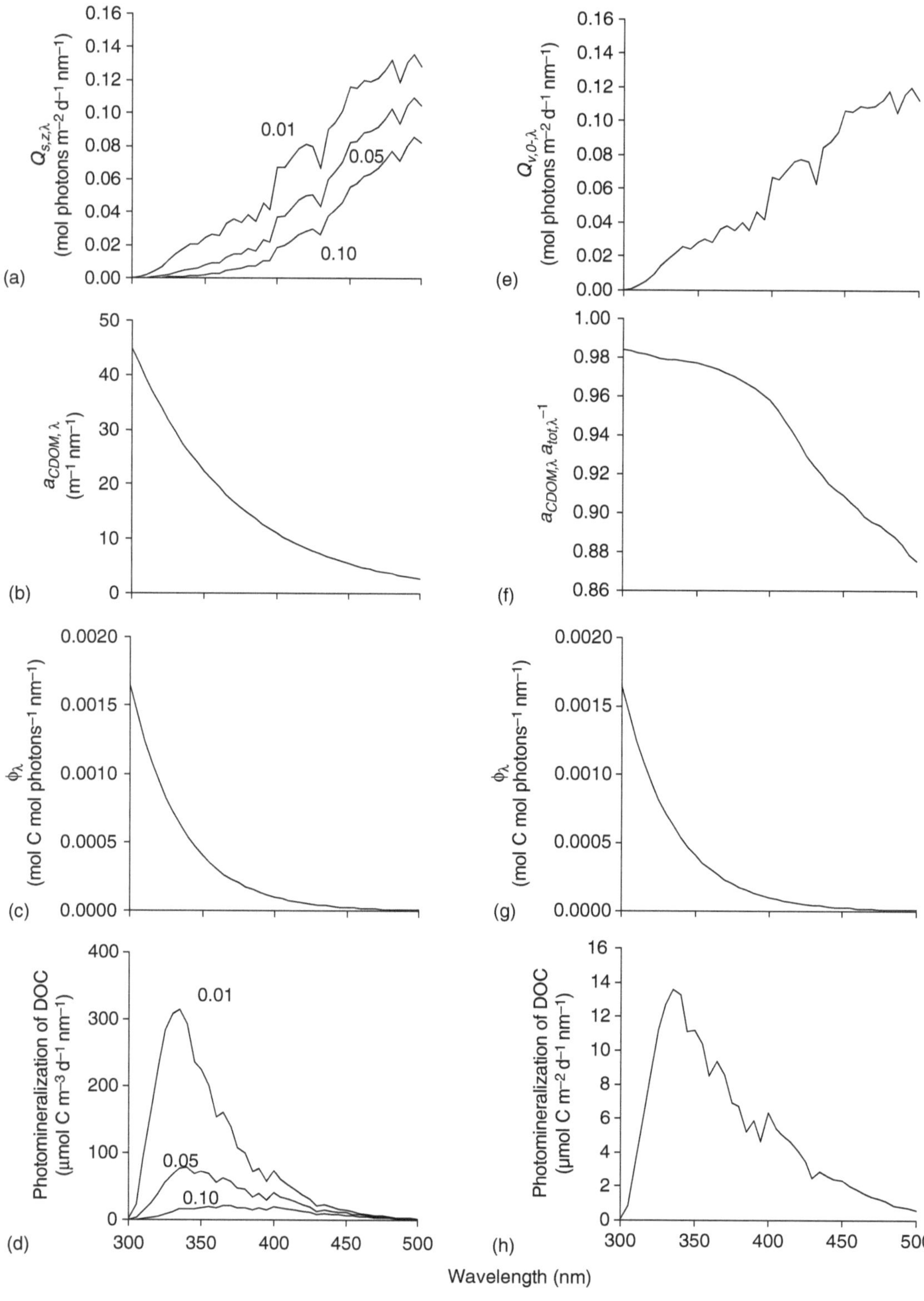

Figure 4 Calculating the photochemical production of CO_2 in Lake Valkea-Kotinen at three depths (panels (a)–(d)) or over the entire water column (panels (e)–(h)). CO_2 photoproduction at three depths: (a) Solar scalar photon flux density at three depths: 0.01, 0.05 and 0.10 m; (b) the absorption coefficient of CDOM; (c) the AQY spectrum for CO_2 photoproduction; (d) CO_2 photoproduction at three depths. CO_2 photoproduction over the entire water column: (e) Vector photon flux density just below the surface; (f) the proportion of radiation absorbed by CDOM; (g) the AQY spectrum for CO_2 photoproduction; (h) CO_2 photoproduction over the entire water column. Adapted from Vähätalo AV, Salkinoja-Salonen M, Taalas P, and Salonen K (2000) Spectrum of the quantum yield for photochemical mineralization of dissolved organic carbon in a humic lake. *Limnology and Oceanography* 45: 664–676.

of DOC (panel (g)). In Lake Valkea-Kotinen, CDOM absorbs >88% of solar radiation at 300–500 nm (**Figure 4(f)**). When integrated over wavelengths, the photomineralization rate over the entire water column is 1.1 mmol carbon per square meter per day.

Solar Radiation Spectrum Responsible for Photochemical Reactions

The solar UV and short wavelength visible part of the spectrum are mostly responsible for photochemical reactions in surface waters (**Figure 4(d)** and **4(h)**). Although the UV-B radiation (290–315 nm) is the most energetic part of the solar spectrum reaching surface waters, the irradiance in that range of spectrum is only 0.1% of the global radiation (290–3000 nm). Therefore, the less energetic but more abundant UV-A (315–400 nm, ~6% of global radiation) is often responsible for photochemical reactions. The solar radiation peaks at the visible range of the spectrum (400–700 nm covers ~50% of global radiation) but at that range of spectrum the energy of photons is lower and much of the radiation is absorbed by particles (e.g., plants) and H_2O. Very close to surface, the short wavelength solar UV radiation contributes (e.g., UV-B) most to photoreactions, but as these wavelengths attenuate steeply into water column, longer wavelength radiation (UV-A and visible radiation) becomes more important at deeper depths.

Because solar UV-A radiation is often mostly responsible for the photochemical transformation of DOM, the attenuation of solar UV-A radiation into the water column describes roughly also the attenuation of photochemical reactions. The optical characteristic of water column such as the concentration of CDOM, regulate the attenuation of solar UV radiation into water columns. The attenuation of UV-A radiation varies a lot among surface waters. For example, the vertical attenuation coefficient for 380 nm, $K_{d,380}$, ranged from 0.07 to 31.7 m^{-1} in 65 American lakes. In these lakes, the calculated depth where the irradiance at 380 nm and the photochemical transformation of DOM is 10% of the values at the surface ranges from 0.07 to 33 m. If the solar radiation reaches the solid surfaces such as plant detritus or sediments, much of solar UV radiation and photochemistry attenuates with K_ds up to 27 000 m^{-1} into a very narrow (<1 mm) surface stratum.

Rates of Photochemical Reactions in the Environment

The rates of photoreactions have been most often measured at the conditions representing those at the surface. For example, the photoproduction of CO_2 was 18 mmol m^{-3} d^{-1} at the depth of 1 cm in humic Lake Valkea-Kotinen and larger than the biological mineralization of DOC (bacterial respiration of 3 mmol m^{-3} d^{-1}). This example shows that in high-CDOM waters, the photochemical mineralization can exceed biological mineralization at the very surface. In high-CDOM waters, the photochemical reactions are attenuated very steeply. When the mineralization of DOC is calculated over the 1-m deep epilimnion of Lake Valkea-Kotinen, the photochemical mineralization of DOC (1 mmol m^{-2} d^{-1}) is 30% of the bacterial mineralization of DOC (3 mmol m^{-2} d^{-1}). Instead of looking at the rates at the narrow surface, it is more informative to examine the rates per area, as these rates give more information about the role of photoreactions on the biogeochemistry of the whole ecosystem.

The rates of photochemical reactions in the whole water column have been measured directly, or calculated according to eqn. [2], using mass balances, or based on the experimental determination of photochemical rates accounting for the attenuation of solar radiation into water column (**Table 1**). The summertime photomineralization rates have ranged from 1000 to 5000 μmol carbon per square meter per day in the 19 fresh waters reported in **Table 1**. In the perspective of the epilimnion or the whole lake, the photomineralizations have been 3–31% of the biological mineralization of DOC (i.e., bacterial respiration). This indicates that microbes dominate the mineralization of DOC in the water column, but also that the photochemical mineralization can contribute significantly to the total mineralization of DOC. The lowest value (3%) is in the 50-m-deep water column of Rio Negro River, where photochemical mineralizaton takes place in less than 1-m depth. Photochemical production of BAPs has ranged from 780 to 3800 μmol carbon per square meter per day in two lakes, where it had contributed 10–11% to the bacterial carbon demand in the epilimnion (**Table 1**). These values are similar to those in coastal seawater. Photochemical production of NH_4^+ from DON has ranged from 16 to 102 μmol nitrogen per square meter per day, and is much lower than the rates concerning DOC (**Table 1**).

Responses of Food Webs to the Photochemical Transformation of Organic Matter

Bacteria are the primary consumers of DOM and the photochemically produced BAPs affect first bacterial productivity, and subsequently the primary grazers of bacterioplankton such as flagellates and ciliates.

Table 1 Photochemical production of CO_2, BAPs, and NH_4^+ over the entire water column. The rates refer to those at summer expect for annual means in some cases

Rate (μmol m^{-2} d^{-1})	*Site (depth of integration)*	*Reference*
DOC → CO_2		
3700–4800(*) (>7–11% of BR)	In 1.5–7.5 m epilimnia of four lakes, Sweden	Granéli *et al.* (1996)
4200(*)(3% of BR)	In 50-m water column of Rio Negro-river, Brazil	Amon and Benner (1996)
2690(*)(11% of BR)	In 1-m epilimnion of Neusiedler See, Germany	Reitner *et al.* (1997)
1000(*)(30% of BR)	In 1-m epilimnion of Lake Valkea-Kotinen, Finland	Vähätalo *et al.* (2000)
1700–2250(*)	Lake Skervatjern, Norway	Salonen and Vähätalo (1994)
1000(c)	Lake Tuscaloosa, AL, USA	Vähätalo and Wetzel (2004)
3000–5000(c)	Seven lakes, Canada	Molot and Dillon (1997)
1100(c)	Parker River estuary	Pullin *et al.* (2004)
4200(c) (20% of BR)	Lake Öreträsket, whole lake, Sweden	Pers *et al.* (2001)
1280(c) (31% of BR)	Annual, Lake Plěsné, whole lake, Czech Republic	Kopáček *et al.* (2004)
DOC → BAPs		
3800(c) (10% of BCD)	In 3.5 m epilimnion of L. Skärhultsjön, Sweden	Bertilsson and Tranvik (1998)
780(c) (11% of BCD)	In 1-m epilimnion of L. Valkea-Kotinen	Vähätalo *et al.* (2003)
1600(c)	Parker River estuary	Pullin *et al.* (2004)
900–3000(c)	Coastal seawater 0–65°N	Miller *et al.* (2002)
570(c) (up to 20% of BCD)	Coastal mean annual	Moran and Zepp (1997)
DON → NH_4^+		
29–44(c)	A coastal lagoon	Buffam and McGlathery (2003)
71(c)	Humic Lake Valkea-Kotinen	Vähätalo *et al.* (2003)
16–102(c)	Offshore Baltic Sea	Vähätalo and Zepp (2005)
23–26(c)	Coastal Baltic Sea	Vähätalo and Järvinen (2007)

*= Measured; c = calculated.
BR = bacterial respiration = biological mineralization of DOC.
BCD = bacterial carbon demand in the epilimnion.
Sources
Granéli W, Lindell M, and Tranvik L (1996) Photo-oxidative production of dissolved inorganic carbon in lakes of different humic content. *Limnology and Oceanography* 41: 698–706.
Reitner B, Herndl GJ, and Herzig A (1997) Role of ultraviolet-B radiation on photochemical and microbial oxygen consumption in a humic-rich shallow lake. *Limnology and Oceanography* 42: 950–960.
Amon RMW and Benner R (1996) Photochemical and microbial consumption of dissolved organic carbon and dissolved oxygen in the Amazon River system. *Geochimica et Cosmochimica Acta* 60: 1783–1792.
Vähätalo AV, Salkinoja-Salonen M, Taalas P, and Salonen K (2000) Spectrum of the quantum yield for photochemical mineralization of dissolved organic carbon in a humic lake. *Limnology and Oceanography* 45: 664–676.
Salonen K and Vähätalo A (1994) Photochemical mineralisation of dissolved organic matter in Lake Skervatjern. *Environment International* 20: 307–312.
Vähätalo AV and Wetzel RG (2004) Photochemical and microbial decomposition of chromophoric dissolved organic matter during long (months–years) exposures. *Marine Chemistry* 89: 313–326.
Molot LA and Dillon PJ (1997) Photolytic regulation of dissolved organic carbon in northern lakes. *Global Biogeochemical Cycles* 11: 357–365.
Pullin MJ, Bertilsson S, Goldstone JV, and Voelker BM (2004) Effects of sunlight and hydroxyl radical on dissolved organic matter: Bacterial growth efficiency and photoproduction of carboxylic acids and other substrates. *Limnology and Oceanography* 49: 2011–2022.
Miller WL, Moran MA, Sheldon WM, Zepp RG, and Opsahl S (2002) Determination of apparent quantum yield spectra for the formation of biologically labile photoproducts. *Limnology and Oceanography* 47: 343–352.
Pers C, Rahm L, Jansson A, Bergström A.-K, and Janssen M (2001) Modelling dissolved carbon turnover in humic lake Örträsket, Sweden. *Environmental Modeling and Assessment* 6: 159–172.
Kopáček J, Brazáková M, Hejzlar J, Nedoma J, Porcal P, and Vrba J (2004) Nutrient cycling in a strongly acidified mesotrophic lake. *Limnology and Oceanography* 49: 1202–1213.
Bertilsson S and Tranvik LJ (1998) Photochemically produced carboxylic acids as substrates for freshwater bacterioplankton. *Limnology and Oceanography* 43: 885–895.
Vähätalo AV, Salonen K, Münster U, Järvinen M, and Wetzel RG (2003) Photochemical transformation of allochthonous organic matter provides bioavailable nutrients in a humic lake. *Archiv für Hydrobiologie* 156: 287–314.
Moran MA and Zepp RG (1997) Role of photoreactions in the formation of biologically labile compounds from dissolved organic matter. *Limnology and Oceanography* 42: 1307–1316.
Buffam I and McGlathery KJ (2003) Effect of ultraviolet light on dissolved nitrogen transformations in coastal lagoon water. *Limnology and Oceanography* 48: 723–734.
Vähätalo AV and Zepp RG (2005) Photochemical mineralization of dissolved organic nitrogen to ammonium in the Baltic Sea. *Environmental Science and Technology* 39: 6985–6992.
Vähätalo AV and Järvinen M (2007) Photochemically produced bioavailable nitrogen from biologically recalcitrant dissolved organic matter stimulates production of nitrogen-limited microbial food web in the Baltic Sea. *Limnology and Oceanography* 52: 132–143.

Two studies have shown that the stimulatory response from the photoproduced carbon substrates can be seen as an increased production of metazooplankton (*Daphnia*). These studies provide evidence that photochemically produced labile substrates can contribute to the whole heterotrophic food web of a lake. According to the current limited information, BAPs contribute 10–11% of bacterial carbon demand in two lakes (**Table 1**). One should notice that in both of these lakes, the photolytic stratum was ~10% of the depth of epilimnion over which the integration of bacterial carbon demand was made. Thus, one can expect even larger contributions of BAPs to bacterial carbon demand in lakes, where the depths of photolytic and mixing stratum are more equal.

Several recent studies have indicated that bacterial growth efficiency (BGE = produced biomass–C/consumed DOC) is lower in irradiated water samples than in dark control samples. Several factors may explain this finding. Although BAPs contain bioavailable carbon, their nutrient (N and P) content is low. The oxidation state of carbon is higher in the photoproducts than in bacterial biomass. The bacteria can respond negatively on the exposure to UV radiation and on the detrimental photoproduced H_2O_2. Additionally, solar radiation can decrease the exoenzymatic activity of bacteria and decrease the utilization of polymeric DOM.

The food web gaining from the BAPs can be considered to consist, e.g., from primary producers (I) → humification of organic matter → photochemical oxidation of organic matter → bacterioplankton (II) → flagellates (III) → ciliates (IV) → metazooplankton (V) → fish (VI). This kind of detritus-based food web can contain many trophic levels (e.g., six marked with I–VI in the above example, but only up to IV, if metazooplankton consumes bacterioplankton directly) and about 50% respiratory losses of carbon take place at each trophic transfer. As the photochemical oxidation additionally directly mineralizes a large portion of organic carbon to CO_2, the trophic efficiency of a food web based on photoproduced labile substrates is lower than that of a short herbivory-based food web (e.g., phytoplankton (I) → metazooplankton (II) → fish (III)). Thus, the gain from BAPs transfers relatively ineffectively to the higher trophic levels of freshwater ecosystems. However, because most (always >50%) of the primary production enters detritus food webs and CDOM is the major source of carbon in many freshwater ecosystems, the detritus-based food webs assisted by the photoproduction of BAPs can contribute significantly to the productivity of many aquatic ecosystems.

Photochemical transformation of organic matter can also decrease the bioavailability of organic matter. This phenomenon has been found out in particular with relatively fresh labile organic matter such as protein-, algal-, or macrophyte-derived organic matter or exudates. Solar radiation can also convert simple unsaturated fatty acids and NH_3 into humic substances. This photoreaction does reduce the bioavailability of both carbon (fatty acids) and nitrogen (NH_3) substrates. These studies suggest that in eutrophic lakes with high concentrations of labile organic matter, photoreactions can slow down the turnover time of organic matter by converting biologically labile organic matter into humic-like organic material. This photochemical formation of humic substances can be seen as a temporary break in the lake metabolism, which allows a part of the bioavailable organic matter to enter into a large pool of biologically recalcitrant organic matter. This kind of humic material can be transported downstream, where it is susceptible for slow biological consumption or photochemical transformation into more bioavailable forms. Thus, in the case of eutrophic lakes, photochemical transformation of bioavailable organic matter into less bioavailable humic type of material can be seen as a buffering mechanism, which balances the periods and sites of high production between the sites and the times of low productivity.

Photoproduced NH_4^+ can serve as a nutrient for both bacterioplankton and phytoplankton. Although the photochemistry of DOM produces more C-than N-products, the relatively small amount of the latter can have implications for the productivity of plankton under N-limiting conditions. Taking into account the low BGE with photoproduced carbon substrates, it has been suggested that the photoproduced NH_4^+ contributes more to bacterial production than photoproducted C-substrates do. Phytoplankton can directly assimilate the photoproduced NH_4^+ or respond indirectly to photoproduced NH_4^+ assimilated by bacterioplankton. Mixotrophic phytoplankton can graze on bacterioplankton and assimilate N bound in bacterial biomass. Heterotrophic grazers (flagellates and ciliates) retain a portion of the ingested particulate nitrogen in their biomass, while a portion becomes mineralized and can supply inorganic nitrogen for autotrophic primary producers. Thus the photochemical transformation of DOM can contribute to both heterotrophic and autotrophic production in surface waters.

Contribution of Photochemistry to the Biogeochemistry of Organic Matter

Photochemical decomposition of biologically recalcitrant but photochemically reactive organic matter has importance also at a landscape perspective. A part of the productivity in terrestrial ecosystems and

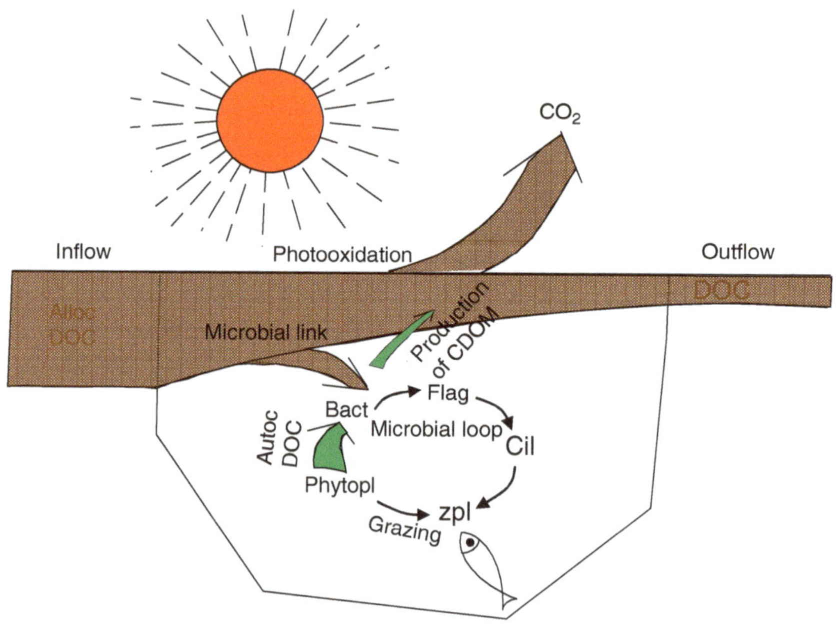

Figure 5 Schematic representation of the carbon cycling in a lake pelagial. The figure emphasizes the role of photooxidation as a major mineralizator of allochthonous biologically recalcitrant but photoreactive CDOM and the linkage of this matter into lacustrine food web (microbial link). The pelagial food web consists of bacteria (bact), flagellates (flag), ciliates (cil) and (macro)zooplankton (zpl). The productivity of food web bases both on allochthonous and autochthonous organic matter, the latter contributing little to the concentration of CDOM in the lake. Generated from the idea of Molot LA and Dillon PJ (1997) Photolytic regulation of dissolved organic carbon in northern lakes. *Global Biogeochemical Cycles* 11: 357–365.

wetlands is converted into CDOM, which is decomposed extremely slowly (turnover times of up to 100–1000 years) in dark waterlogged anaerobic soils. When such soils are hydrologically connected to surface waters, the imported allochthonous CDOM is exposed to solar radiation and becomes photochemically decomposed. The photochemical decomposition can remarkably accelerate the turnover of CDOM derived from terrestrial or wetland sources. For example, photochemical reactions mineralized 20% of introduced synthetic lignin mixed in humic lake water during a week-long exposure to surface solar radiation while no microbial mineralization of synthetic lignin was detected during the same time in darkness. In another example, the half-life of CDOM from Lake Tuscaloosa reservoir was 1 week under surface solar radiation but ~70 weeks in the presence of heterotrophic microbes in darkness. The photolytic half-lives of allochthonous CDOM in mixing stratum of lakes can be <1-year; and thus >100–1000 times faster than the turnover time of organic matter in soils of terrestrial and wetland ecosystems. Molot and Dillon (1997) calculated that the photooxidation of DOC to CO_2 was potentially fully responsible for the in-lake losses of inflowing allochthonous DOC in seven Canadian lakes (**Figure 5**). Their calculations suggest that photolytic decomposition is a primary mechanism for the removal of biologically recalcitrant but photochemically reactive CDOM in lakes. The photochemical mineralization of CDOM may thus convert allochthonous CDOM into CO_2 and contribute to the supersaturation of CO_2 and the transport of CO_2 from lake to atmosphere.

See also: Carbon, Unifying Currency; Gas Exchange at the Air-Water Interface; Interactions of Dissolved Organic Matter and Humic Substances; Iron and Manganese; Nitrogen; Organic Nitrogen; Phosphorus.

Further Reading

Blough NV and Zepp RG (1995) Reactive oxygen species in natural waters. In: Foote CS, Valentine JS, Greenberg A, and Liebman JF (eds.) *Active Oxygen in Chemistry*, pp. 280–333. London: Blackie Academic.

Bushaw KL, Zepp RG, and Tarr MA (1996) Photochemical release of biologically available nitrogen from aquatic dissolved organic matter. *Nature* 381: 404–407.

Caesar D, Granéli W, Kritzberg ES, and Anesio AM (2006) Stimulation of metazooplankton by photochemically modified dissolved organic matter. *Limnology and Oceanography* 51: 101–108.

Kieber RJ, Hydro LH, and Seaton PJ (1997) Photooxidation of triglycerides and fatty acids in seawater: Implication toward the formation of marine humic substances. *Limnology and Oceanography* 42: 1454–1462.

Mopper K and Kieber DJ (2002) Photochemistry and the cycling of carbon, sulfur, nitrogen and phosphorus. In: Hansell DA and Carlson CA (eds.) *Biogeochemistry of Marine Dissolved Organic Matter*, pp. 456–507. San Diego: Academic Press.

Moran MA and Zepp RG (1997) Role of photoreactions in the formation of biologically labile compounds from dissolved organic matter. *Limnology and Oceanography* 42: 1307–1316.

Tranvik LJ and Bertilsson S (2001) Contrasting effects of solar UV radiation on dissolved organic sources for bacterial growth. *Ecology Letters* 4: 458–463.

Vähätalo AV and Järvinen M (2007) Photochemically produced bioavailable nitrogen from biologically recalcitrant dissolved organic matter stimulates the production of nitrogen-limited microbial food web in the Baltic Sea. *Limnology and Oceanography* 52: 132–143.

Zafiriou OC, Joussot-Dubien J, Zepp RG, and Zika RG (1984) Photochemistry of natural waters. *Environmental Science and Technology* 18: A358–A371.

Zepp RG (2003) Solar ultraviolet radiation and aquatic carbon, nitrogen, sulfur and metals cycles. In: Helbling EW and Zagarese H (eds.) *UV Effects in Aquatic Organisms and Ecosystems*, pp. 137–183. Cambridge: Royal Society of Chemistry.

HYDROLOGY

Contents

Hydrological Cycle and Water Budgets
Atmospheric Water and Precipitation
Snow and Ice
Evapotranspiration
Vadose Water
Ground Water
Ground Water and Surface Water Interaction
Groundwater Chemistry
Fluvial Export
Fluvial Transport of Suspended Solids
Streams
Rivers
Springs
Wetland Hydrology

Hydrological Cycle and Water Budgets

T N Narasimhan, University of California at Berkeley, CA, USA

Introduction

The Earth's geology, its atmosphere, and the phenomenon of life have been profoundly influenced by water through geological time. One manifestation of this influence is the hydrological cycle, a continuous exchange of water among the lithosphere, the atmosphere, and the biosphere. The present-day hydrological cycle is characterized by a vigorous circulation of an almost insignificant fraction, about 0.01%, of the total water existing on the Earth. Almost all beings on land require fresh water for sustenance. Yet, remarkably, they have evolved and proliferated depending on the repeated reuse of such a small fraction of available water. The partitioning of water among the components of the hydrological cycle at a given location constitutes water balance. Water-balance evaluations are of philosophical interest in comprehending the geological and biological evolution of the Earth and of practical value in environmental and natural-resource management on various scales. The purpose here is to outline the essential elements of the hydrological cycle and water budgets relevant to inland waters and aquatic ecosystems.

Hydrological Cycle

The hydrological cycle is schematically shown in **Figure 1**. Atmospheric water vapor condenses and precipitates as rain or snow. A small portion of this is intercepted by vegetation canopies, with the rest reaching the ground. A portion of this water flows over land as surface water toward the ocean or inland depressions, to be intercepted along the way by ponds, lakes, and wetlands. Another portion infiltrates to recharge the soil zone between the land surface and the water table, and the groundwater reservoir below the latter. Pulled by gravity, ground water can move down to great depths. However, because of the presence of low permeability earth layers, the downward movement is resisted, and water is deflected up toward the land surface to be discharged in streams, lakes, ponds, and wetlands. Water escaping the influence of resistive layers and moving to greater depths encounters geothermal heat. Geothermal heating too has the effect of countering downward movement and impelling ground water toward the land surface. At the land surface, surface water and discharging ground water are subject to evaporation by solar radiation and to transpiration by plants as they consume water for photosynthesis. Collectively referred to as evapotranspiration, this transfer of water back to the atmosphere completes the hydrological cycle.

The components of the hydrological cycle, namely, atmosphere, surface water, and ground water (including soil water), are intimately interlinked over a variety of spatial scales (meters to thousands of kilometers)

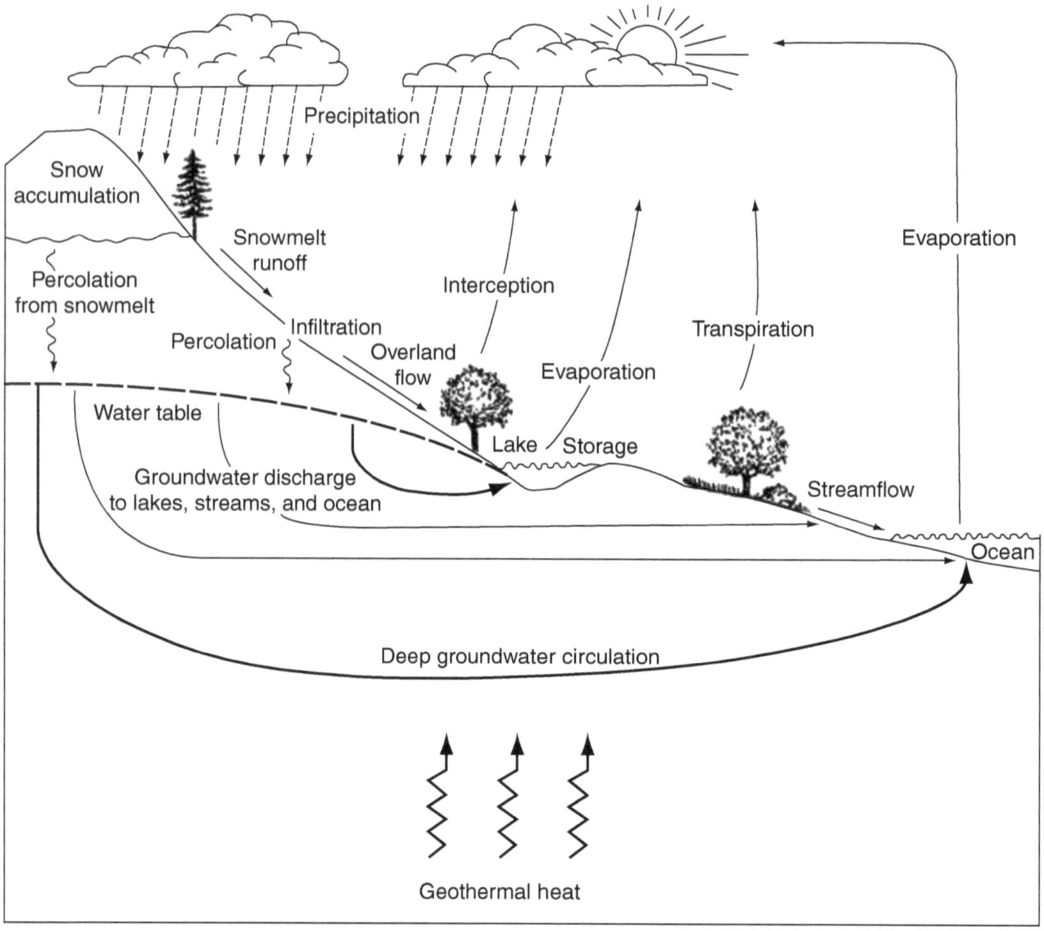

Figure 1 Schematic description of the hydrological cycle: adapted from T. Dunne and L. B. Leopold, 1978, *Water in Environmental Planning*, p. 5.

Table 1 Hydrological cycle: spatial and temporal scales

	Storage, % of Total[a,b]	*Spatial scale*[c]	*Residence Time*[d]
Atmosphere	0.001	Km to thousands of km	Days
Surface water[e]	0.01	Meters to hundreds of km	Weeks to years
Soil water	0.05	Meters to tens of meters	Weeks to years
Ground water	2.1	Tens of meters to hundreds of km	Days to millions of years
Oceans and seas	95.7	Km to thousands of km	Thousands of years
Ice caps and glaciers	2.1	Km to thousands of km	Tens of thousands of years

[a]Total volume of water on Earth, 1.43×10^9 km^3.
[b]From Unesco, 1971, *Scientific framework for World water balance*, p. 17.
[c]Distance over which cycle is completed.
[d]From Unesco, 1971, *Scientific framework for World water balance*, p. 17.
[e]Includes lakes and reservoirs, river and stream channels, swamps.

and temporal scales (days to millions of years). Information on the volume of water stored in each component of the hydrological cycle, and the relevant spatial and temporal scales are summarized in **Table 1**.

Water is a slightly compressible liquid, with high specific heat capacity and latent heats of melting and evaporation. It exists in solid, liquid, and gaseous phases within the range of temperatures over which life, as we know it, can sustain. Its bipolar nature enables it to form cage-like structures that can trap nonelectrolyte molecules as well as charged ions. For these reasons, water is an active chemical agent, efficient transporter of mechanical energy and heat, and a carrier of dissolved and suspended substances.

These attributes render water to be an extraordinary geological and biological agent that has endowed the Earth with features no other celestial object is known to possess.

The hydrological cycle is driven mostly by solar energy and to a minor extent by geothermal heat. The Earth's erosional and geochemical cycles exist due to water's ability to do mechanical work associated with erosion, chemically interact with rocks and minerals, and transport dissolved and suspended materials. Collectively, the hydrological, erosional, and geochemical cycles constitute the vital cycles that sustain life. The interrelationships among these vital cycles can be conveniently understood by examining the lithospheric components of the hydrological cycle.

Hydrological Cycle: Lithospheric Components

Surface water On the Earth's surface, water breaks down rocks physically and chemically through weathering, aided by solar energy and by actions of microbes, plants, and animals. The products of weathering are transported as sediments (bedloads and suspended loads) and dissolved chemicals. In addition, water also transports leaf litter and other decaying vegetation and animal matter. The sediments and organic matter together contribute to the cycling of life-sustaining nutrients. The habitats of flora and fauna along the course of a river depend, in very complex ways, on the texture of sediments as well as on their chemical makeup. A glimpse into the intricate influence of physical nature of sediments and the aquatic chemical environment on an organism's life cycle is provided by salmon, an anadromous fish. In the wild, salmon is hatched in gravelly stream beds that provide protection from predators and abundant supplies of oxygen to the eggs. Once hatched, the young fingerlings must have narrowly constrained aquatic chemical and thermal environment to survive as they migrate from a freshwater environment to a marine environment where they will spend their adult life.

Soil water The soil zone lies between land surface and the water table, where water and air coexist. Soil water, which is held in the pores by capillary forces, is not amenable for easy extraction by humans. However, plants have the ability to overcome capillary forces and extract water for their sustenance. Microbial populations constitute an integral part of the soil biological environment. With abundant availability of oxygen and carbon dioxide in the air, the soil is an active chemical reactor, with microbially mediated aqueous reactions.

In the soil zone, water movement is dominantly vertical, and a seasonally fluctuating horizontal plane separates vertically upward evaporative movement from downward directed gravity flow. Water moving down by gravity reaches the water table to recharge the groundwater reservoir. The journey of water from the time it enters the groundwater reservoir to the time it emerges back at the land surface may be referred to as regional groundwater motion. Regional groundwater motion constitutes a convenient framework for an integrated understanding of the formation of sedimentary rocks and minerals, and the areal distribution of soils and aquatic ecosystems on land.

Ground water Infiltrating water enters the groundwater reservoir at high elevations, and driven by gravity, moves vertically down in areas of groundwater recharge. Depending on topographic relief and the distribution of permeable and impermeable layers, the vertically downward movement is resisted sooner or later, and the movement becomes subhorizontal. With further movement, flow is deflected up toward the land surface in areas of groundwater discharge. Groundwater discharge typically occurs in perennial stream channels, wetlands, low-lying areas, and springs. In these discharge areas, surface water and ground water directly interact with each other, with important geological and biological consequences. For example, the spectacular tufa towers of Mono Lake in California represent precipitates of calcium carbonate resulting from a mixing of subaqueous thermal springs with the lake water. Hyporheic zones, which play an important role in stream ecology, are groundwater discharge areas where stream flow is augmented by groundwater discharge.

Regional groundwater flow provides a framework to interpret patterns of chemical processes in the subsurface. Water in recharge areas is rich in oxygen and carbon dioxide and has a significant ability to chemically break down minerals through corrosive oxidation reactions. However, available oxygen is consumed as water chemically interacts with the minerals along the flow path, and the oxidation potential of ground water progressively decreases along the flow path. In swamps and wetlands of discharge areas, water exists under strong reducing (anaerobic) conditions. Between these two extremes, ambient conditions of acidity (pH) and redox state (Eh) govern the chemical makeup of water as well as the types of minerals and microbial populations that are compatible with ambient water chemistry. In general, the cation content of ground water reflects the chemical make up of the rocks encountered along the flow path, and the anion content is indicative of the progress of chemical reactions.

The concept of hydrochemical facies denotes the diagnostic chemical aspect of aqueous solutions reflecting the progress of chemical processes within the framework of regional groundwater motion. Given the concept of regional groundwater motion, and that of hydrogeochemical facies, one can readily see how the spatial distribution of various types of soils, and the distribution of different types of ecosystems over a watershed, must represent the profound influence of the lithospheric segment on the hydrological cycle. Regional groundwater flow pattern in the Atlantic Coastal Plain as deciphered from hydrochemical information is shown in **Figure 2**.

Nutrient Cycling and Energy Balance

A discussion of the hydrological cycle is incomplete without examining its connections to nutrient cycling and solar energy balance.

A glimpse into connections to nutrient cycling can be gained by examining the role of water in the cycling of carbon, sulfur, and phosphorus. Almost all biological carbon originates in atmospheric carbon dioxide through photosynthesis by plants and phytoplankton. Water, essential for the photosynthesis process, is transferred as water by plants from the soil via leaves to the atmosphere, completing the hydrological cycle. In the lithosphere, water plays a dominant role in the decomposition and mineralization of organic carbon on diverse time scales, ultimately producing carbon dioxide or methane to be returned to the atmosphere. Sulfur is a multivalent, redox-controlled chemical species which plays an important role in metabolic reactions of plants. Under reducing conditions, sulfur is insoluble in water. Sulfate, its most oxidized form, is water soluble, and it is in this form that sulfur usually enters plant roots. Sulfide minerals constitute the principal source of sulfur in the lithosphere, and they are oxidized in the presence of bacteria to sulfate and become available for uptake by plants. In plants, sulfur is fixed in a reduced form. Thus, sulfur of dead organic matter is mobilized by oxidizing waters to sulfate to sustain the sulfur cycle. Phosphorus, which plays several important roles in the biological processes of plants and animals, is water soluble only under very narrow ranges of redox and pH. It does not readily form gaseous compounds. Therefore, phosphorus cycling is almost entirely restricted to the lithosphere. Phosphorus cycling effectively maintains biological habitats despite the severe aqueous constraints that limit its mobility.

The hydrological cycle is driven largely by solar energy. Just like water, solar energy is also subject to cyclic behavior. On the land surface, the energy received as incoming solar radiation (insolation) is balanced partly by outgoing longwave radiation, partly as sensible heat by convecting air columns and partly as latent heat transferred by water from

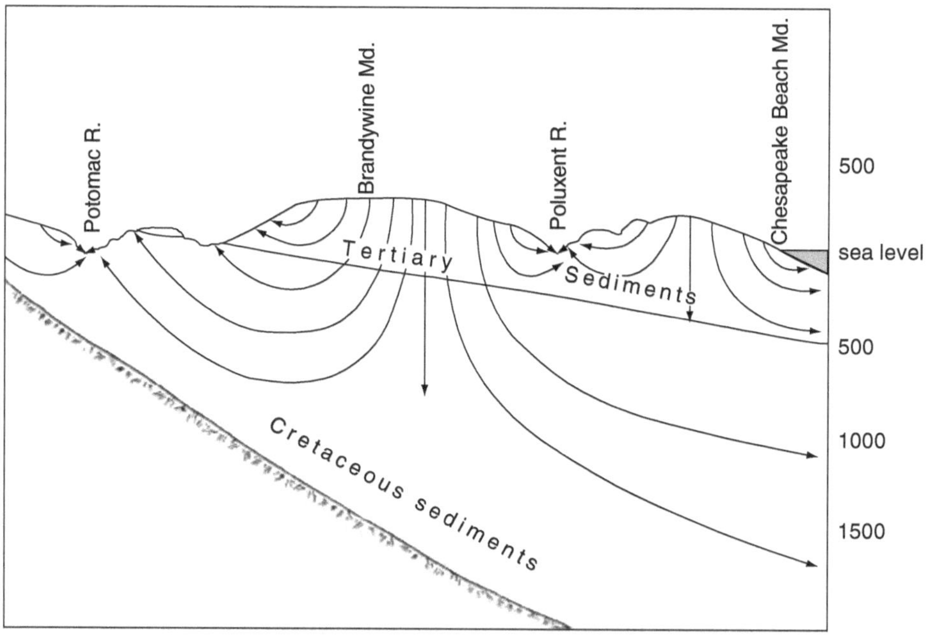

Figure 2 Groundwater flow patterns inferred from hydrochemical facies in the Atlantic coastal plain (W. Back, 1960, *Origin of Hydrochemical Facies of Groundwater in the Atlantic Coastal Plains*).

the land to the atmosphere. Of the total solar radiation received from the sun on land, the amount of energy returned by water to the atmosphere amounts to about 46%, a major fraction. Any significant perturbation of this contribution will have influence global climate.

Summary

The concept of hydrological cycle is elegantly simple. But, its importance in the functioning of the geological and biological Earth is profound, transcending water itself. It plays an overarching role in the cycling of solar energy, sediments, and chemical elements vital for the sustenance of life. Although it is clear that contemporary ecosystems reflect an evolutionary adaptation to the delicate linkages that exist among the various components of the hydrological cycle, it is also apparent that evolving life must have influenced the evolution of the hydrological cycle over geological time. Life, it appears, is simultaneously a product of the hydrological cycle and its cause.

Water Budgets

Framework

Despite advances in science and technology, climate remains outside human control and manipulation. Humans, just as plants and animals, have to pattern their existence submitting to the variability of climate. However, surface water, soil water, and ground water lie within reach of human control, to be managed for human benefit. In this context, the concept of water budgets becomes relevant.

Given a volume element of the Earth with well-defined boundaries, water budget consists in quantifying the relationships among inflow, outflow, and change in storage within the element. This simple concept is as valid over the Earth as a whole treated as a volume element, over a river basin, or over a small rural community. In a world of stressed water resources, water budget is assuming an ever increasing importance as a framework for wise and equitable water management.

Water is always in a state of motion, and its budget is governed by the simple notion that inflow must equal change in storage plus outflow. Symbolically, this may be stated as

$$P = E + R_{\mathrm{Su}} + R_{\mathrm{Gw}} + \Delta\mathrm{Su} + \Delta\mathrm{So} + \Delta\mathrm{Gw} + D_H \quad [1]$$

where P is precipitation, E is evapotranspiration, R is runoff, Su is surface water, Gw is ground water, So is soil water, D_H is diversion by humans, and Δ denotes change in storage. If the time interval of interest is smaller than a season, the terms involving change in storage cannot be neglected, the system being under transient conditions. If, however, the time interval of interest is a year or several years, seasonal increases and decreases in storage will effectively cancel out, and the water budget equation reduces to a steady-state balancing of inflow and outflow

$$P = E + R_{\mathrm{Su}} + R_{\mathrm{Gw}} + D_H \quad [2]$$

Implicit here is the assumption that precipitation constitutes the only inflow into the volume element, which is reasonable if one considers a watershed enclosed by a water divide, without any water import. Clearly, if the volume element of interest is defined by open boundaries, terms representing water import and export have to be added to the equations.

Assessment of Components

The simplicity of the above equations belies the difficulties inherent in assessing the different components involved. Perhaps the most widely measured quantities in water budgets are precipitation and surface runoff. Rainfall data from aerially distributed rain-gauging stations are integrated over space to arrive at the total volume of water falling over an area during a period of interest. Runoff estimated at a given location on a stream with flow meters or river-stage data supplemented by rating curves, represents outflow from the watershed upstream of that location.

Evapotranspiration Experience has shown that evapotranspiration constitutes a significant percentage of precipitation over the land surface. Yet, quantification of evapotranspiration is a difficult task. The gravimetric lysimeter provides a way of experimentally estimating evapotranspiration from a soil mass of the order of a few cubic meters in size. Although of much value as tools of research, lysimeters are helpful in estimating evapotranspiration only over small areas. For watersheds and river basins, it is customary to use a combination of empirical and theoretical methods. In one such approach, the concept of potential evaporation plays a central role. Potential evaporation is understood to be the height of column of water that would be evaporated from a pan at a given location, assuming unlimited supply of water, as from a deep lake. If precipitation at the location exceeds potential evaporation, the soil is assumed to hold a maximum amount of water in excess of gravity drainage. If precipitation is less than potential evaporation, then the actual evapotranspiration will be limited to what precipitation can supply. In this case, empirical curves are used to estimate soil moisture storage based on precipitation deficit and the maximum water-holding capacity of the soil. With the availability of instruments of increased sophistication and super computers, energy methods are increasingly sought after to estimate evapotranspiration from watershed scale to continental scale. In these methods, the goal is to carry out a solar energy budget and isolate the amount of energy that is transferred by water from the land surface to the atmosphere as latent heat. This estimate is then converted to evapotranspiration. To support this model, data are generated from detailed micrometeorological measurements such as short-wave and long-wave radiation, temperature, humidity, cloud cover, and wind velocity. Another method for estimating evapotranspiration is to carry out an atmospheric water balance in a vertical column overlying the area of interest. In this method, evapotranspiration is set equal to the sum of precipitation and change in water vapor content of the column, less the net flux of water laterally entering the column.

Soil-water storage In the field, water content of soils can be profiled as a function of depth with the help of neutron logs or by Time Domain Reflectometry. In principle, one can empirically estimate change in soil-water storage by carrying out repeat measurements with these instruments. However, these methods are of limited value when estimates are to be made over large areas.

The concept of field water capacity, used widely by soil scientists and agronomists, denotes the quantity of water remaining in a unit volume of an initially wet soil from which water has been allowed to drain by gravity over a day or two, or the rate of drainage has become negligible. The water that remains is held by the soil entirely by capillary forces. Field capacity depends on soil structure, texture, and organic content and is commonly measured to help in scheduling irrigation. Empirical curves presented by Thornthwaite and Mather (1957) provide correlations among field water capacity, water retained in soil, and the deficit of precipitation with reference to potential evaporation. These curves can be used to estimate change in soil-water storage.

Groundwater storage Changes in groundwater storage occur due to two distinct physical processes. At the base of the soil zone, as the water table fluctuates, change in storage occurs through processes of saturation or desaturation of the pores. In this case, change in groundwater storage per unit plan area is equal to the product of the magnitude of the water-level fluctuation and the specific yield of the formation, a parameter that is approximately equal to porosity. In the case of formations far below the water table, water is taken into storage through slight changes in the porosity, depending on the compressibility of the formations. In this case, change in groundwater storage can be estimated from the product of water-level fluctuation and the storage coefficient of the formations.

Groundwater runoff The movement of water in the subsurface is quantified with Darcy's Law, according to which the volume of water flowing through a given cross sectional area per unit time is equal to the product of the hydraulic conductivity of the formation, the gradient of hydraulic head, and the cross-sectional area. In the field, hydraulic gradients can be obtained from water table maps. These, in conjunction with the known hydraulic conductivity of the geological formations can be used to estimate groundwater runoff.

Table 2 Statewide water balance, California – m^3 (maf[a])

	Water year (Percent of normal precipitation)		
	1998 (171%)	*2000 (97%)*	*2001 (72%)*
Precipitation	4.07×10^{11}(329.6)	2.32×10^{11} (187.7)	1.72×10^{11}(139.2)
Imports: Oregon/Nevada/Mexico	9.00×10^{9} (7.3)	8.63×10^{9} (7.0)	7.77×10^{9} (6.3)
Total inflow	*4.16×10^{11} (336.9)*	*2.40×10^{11} (194.7)*	*1.79×10^{11} (145.5)*
Evapotranspiration[b]	2.58×10^{11} (208.8)	1.62×10^{11} (131)	1.53 (124.2)
Exports: Oregon/Nevada/Mexico	1.85×10^{9} (1.5)	1.11×10^{9} (0.9)	8.63×10^{8} (0.7)
Runoff	1.49×10^{11} (120.8)	8.46×10^{10}(68.6)	4.30×10^{10} (34.9)
Total outflow	*4.08×10^{11} (331.1)*	*2.48×10^{11} (200.8)*	*1.97×10^{11}(159.8)*
Change in surface water storage	8.88×10^{9} (7.2)	-1.60×10^{9} (−1.3)	-5.67×10^{9} (−4.6)
Change in groundwater storage	-1.72×10^{9} (−1.4)	-5.55×10^{9} (−4.5)	-1.20×10^{10} (−9.7)
Total change in storage	*7.15×10^{9} (5.8)*	*-7.15×10^{9} (−5.8)*	*-1.76×10^{10} (−14.3)*

[a]Million acre feet.

[b]Includes native plants and cultivated crops.

Two Examples

Global water balance Between 1965 and 1974, the International Hydrological Decade Program of UNESCO did much to focus attention on the imperative to judiciously manage the world's freshwater resources. An important contribution to the efforts of IHD by the Russian National Committee was the publication, *World Water Balance and Water Resources of the Earth* (Unesco, 1978), which provided detailed estimates of water balance for the different continents, and for the Earth as a whole. The general finding was that for the world as a whole, total annual precipitation is of the order of 113 cm, or about $5.76 \times 10^5\,km^3$. Globally, this precipitation is balanced by an equal magnitude of evapotranspiration. However, an imbalance exists between precipitation and evapotranspiration, if land and the oceans are considered separately. Over land, average annual precipitation is about 80 cm, or $1.19 \times 10^5\,km^3$. Of this, evapotranspiration constitutes 48.5 cm (60.6%) and runoff constitutes 31.5 cm (39.4%), indicating a deficit of precipitation in comparison to evapotranspiration. Over the oceans, the average annual precipitation is about 127 cm, or $4.57 \times 10^5\,km^3$, while evaporation is about 140 cm. The excess of evaporation over precipitation over the oceans is equal to the runoff from the land to the oceans.

California With a land area of 409 500 km^2, and spanning 10° of latitude and longitude, California exhibits remarkable diversity of physiography, climate, flora, and fauna. The Department of Water Resources of the State of California periodically prepares water balance summaries to aid state-wide water planning. The DWR's latest water balance estimates are instructive in that they provide comparison of water budget for an average year with those of a surplus year and a deficit year (California Department of Water Resources, 2005). Salient features are summarized in **Table 2.** It is interesting to note from the table that (1) ground water is being over pumped even during surplus years, (2) California experiences a deficit of about 3% even during an average year, and (3) evapotranspiration varies from 62% during a surplus year to as much as 85% during a drought year.

Epilogue

Modern science has shown that the observed behavior of the hydrological cycle can be understood and explained in terms of the laws of mechanics and thermodynamics. However, the ability of modern science to describe the hydrological cycle in precise detail and to predict the future behavior of components of the hydrological cycle with confidence is severely limited. The limitation arises from the many spatial and temporal scales in which the components interact, the complexity of processes, difficulties of access to observation, and sparsity of data, not to mention the role of living beings that defy quantification. Yet, we have to draw upon our best science so as to use the world's limited supplies of fresh water wisely and equitably. This goal will be best achieved if we recognize the limitations of science, moderate our social and economic aspirations, and use science to help us adapt to the constraints imposed by the hydrological cycle.

Throughout history, humans have been fascinated with water. Although modern science has been successful in elucidating the details of the functioning of the hydrological cycle, its essential features were astutely recognized and viewed with awe centuries (perhaps even millenniums) before Christ in China, India, Greece, and Egypt. It is therefore fitting to conclude this discussion of the hydrological cycle with a psalm from the Hindu scripture:

"The waters which are from heaven, and which flow after being dug, and even those that spring by themselves, the bright pure waters which lead to the sea, may those divine waters protect me here" (Rig-veda, VII 49.2).

See also: Atmospheric Water and Precipitation; Chemical Properties of Water; Evapotranspiration; Ground Water and Surface Water Interaction; Ground Water; Groundwater Chemistry; Phosphorus; Physical Properties of Water; Redox Potential; Streams; Vadose Water.

Further Reading

Encyclopedia Britannica (1977) Hydrological Cycle 9: 102–116.

Back W (1960) Origin of hydrochemical facies of groundwater in the Atlantic Coastal Plains, Report, *21st Session, Int. Geol. Congress*, Copenhagen, Pt. 1, pp. 87–95.

Freeze RA and Cherry JA (1979) *Groundwater.* Englewood Cliffs, New Jersey: Prentice Hall, 604 pp.

Narasimhan TN (2005) Pedology: A hydrogeological perspective. *Vadose Zone J.* 4: 891–898.

Thornthwaite CW and Mather JR (1957) Instructions and tables for potential evapotranspiration and water balance. *Publication in Climatology*, Vol. 10, No. 3. Centerton, New Jersey: Thornthwaite and Associates.

U.S. Geological Survey (2007) The Water Cycle, Complete Summary, http://ga.water.usgs.gov/edu/watercyclesummary.html.

Atmospheric Water and Precipitation

K Fienberg and E Foufoula-Georgiou, University of Minnesota, Minneapolis, MN, USA

Introduction

Atmospheric water and precipitation play a key role in the global water cycle, comprise a conduit between oceanic and inland waters, and provide the main forcing variable for surface hydrologic processes. The dynamics of water in the atmosphere can be categorized into three general processes within the hydrologic cycle: evaporation from the surface as water enters the atmosphere; transport of water by atmospheric processes; and precipitation as water returns to the surface. Of these, it is precipitation that is the direct driving force for inland water processes such as runoff, soil and groundwater storage, and then stream flow. Hence, from the view-point of surface hydrology, the final result of atmospheric water movements, and precipitation, is the quantity of greatest interest.

Precipitation forms the input of many hydrological models, and as such, measurements and model predictions of precipitation are vital in understanding inland waters. Precipitation has been measured for centuries, and has been the subject of scientific investigation at least since Pierre Perrault and Edme Marriotte related measured rainfall to the flow of the Seine in the seventeenth century. However, despite advances in both direct and remote measurement technologies, and the development of more powerful and sophisticated computer models, both measurement and modeling of precipitation remain a challenge. One of the major reasons for this is the high degree of variability in both time and space. Although water vapor has been estimated to have an average life cycle in the atmosphere on the order of 10 days, observations show variability and structure in precipitation patterns from time-scales as short as seconds to as long as multiyear cycles in some regions. As a general guideline, short-lived temporal patterns are generally associated with features of smaller spatial extend, while longer-term patterns are associated with processes that take place at larger spatial scales. It is clear that there are a large number of different processes causing this variability: from turbulent transport of raindrops at short time scales, to synoptic events with lifecycles measured in hours or days, to multiyear climatic patterns such as the El Niño-Southern Oscillation. Nevertheless, this range of different processes combines to create highly inhomogeneous precipitation fields with fluctuations at almost all wavelengths. This complex structure in space and time has long raised questions for the measurement of precipitation, i.e., if a rain gauge has an aperture measured in centimeters, how can hourly rainfall readings be extrapolated to regions surrounding the gauge? What about daily gauge totals? If one has satellite-based estimates of rainfall every few hours over a large area, what can one say about the precipitation intensities between observation times? How can observations with different instruments, which have different resolutions, be compared? Similarly, the variability over a wide range of scales also poses problems for the numerical modeling of rainfall, not only in the issue of comparison and initialization with measurements at different scales, but also on the fundamental questions of what processes, and at which scales, need to be included in the model, and how to accurately create closure schemes to account for those processes occurring at scales below the resolution of the model. As will be presented in the subsequent sections of this work, many of these questions have already been addressed using careful analysis of precipitation measurements and models that specify how the statistics of precipitation depend on scale. Other scale-related problems for measurements and models remain unresolved.

The question of the nature of rainfall variability at different scales is not only important for understanding precipitation itself, but also for modeling its effects on surface hydrology. Processes that are driven by precipitation, such as soil water-storage and runoff, are generally nonlinear and thought to depend on threshold levels of precipitation. Therefore, as studies have shown, modeling these processes requires knowledge of the space–time distribution of the precipitation field and not simply the mean values. Thus, the results of modeling these processes are dependent on both the resolution of the precipitation input and the scale (or resolution) of the model itself.

In light of the importance of the spatial and temporal patterns in precipitation, and their dependence on scale, this work surveys the scientific investigations of precipitation with a particular focus on spatial–temporal structure. First, in the following section, we briefly outline the various processes involved in precipitation formation to examine the origins of this structure. Subsequently, the techniques of precipitation measurement, each of which can be used to measure a different range of precipitation variability, are reviewed. Precipitation modeling is then described, with emphasis on scale-invariant statistical models that can describe the multiscale variability of precipitation in a relatively parsimonious manner.

Precipitation Formation

Water Vapor in the Atmosphere

Water enters the atmosphere via the process of evaporation from both the land-surface and the ocean. This includes the process of transpiration, which is the evaporation of water vapor into the atmosphere through the vascular system of plants. Evaporation from the land surface contributes, on a global average, approximately two-thirds of the total moisture available for precipitation over land. However, the evaporation rate at any particular time and place will depend on the amount of surface water present, the available energy to allow change of phase from liquid to vapor, and the atmospheric conditions to remove the evaporated molecules from the surface where they may condense again. Thus, the evaporation rate will change with location and time, creating differences in the level of water vapor in the atmosphere, and thus providing one cause of variability in precipitation. Over land, there are daily and seasonal cycles in evaporation due to the cycles of solar radiation input and temperature, and hence the available energy for evaporation. Variations in space are caused by changing surface water, land cover, and vegetation with location. Over the ocean, where surface temperatures are not as closely linked to the solar radiation cycles due to energy storage by the deep water, researchers have identified other cycles in sea-surface temperature that affect evaporation rates. One example is the multiyear El Niño phenomenon (part of the Southern Oscillation), in which changes in currents in the Pacific Ocean every 2–8 years move warm water from the west Pacific and the Indian Ocean to the east Pacific. Higher water temperature causes extra evaporation and drives rain processes, causing higher rainfall over the west coast of the Americas. Simultaneously, lower sea surface temperatures cause drier conditions in southeast Asia and northern Australia.

Once water vapor has entered the atmosphere via evaporation, it is transported by the atmosphere until it is released as precipitation. The concentration of water vapor decreases rapidly with height, with 50% of the total column water vapor being found in the first 1 or 2 km of the atmosphere. In the generally turbulent lower atmosphere, at least at scales above the homogeneity scale (on the order of millimeters), transport by molecular diffusion is inconsequential in comparison to advection by the wind velocity field. The turbulent velocity field $\boldsymbol{v}$ is described by the Navier–Stokes equations for incompressible fluids:

$$\frac{\partial \boldsymbol{v}}{\partial t} + (\boldsymbol{v}\cdot\nabla)\boldsymbol{v} = -\frac{1}{\rho}\nabla p + m\nabla^2\boldsymbol{v} + f$$

$$\frac{\partial \rho}{\partial t} + \nabla(\boldsymbol{v}\rho) = 0 \qquad [1]$$

where ρ is the fluid density, m is the viscosity, and f is the forcing term. This system has been studied for decades, with notable early contributions by Reynolds, Richardson, Kolmogorov, and Obhukhov, and for high Reynolds number, water vapor flux has been shown to be highly intermittent, nonstationary with stationary increments, have fat-tailed distributions, and show scale-invariant patterns. If the water vapor is assumed to be a passive scalar, the specific mass flux of water vapor is then given by $F = \rho_v v$, where ρ_v is the water vapor density. Thus, the water vapor flux should also be highly variable, nonstationary, and intermittent. Indeed, numerical studies for passive scalar advection have shown that they can be even more intermittent than the underlying velocity field.

In reality, of course, water vapor is not a passive scalar, since it affects the energy budget and thermodynamics of the atmosphere in a number of ways. For example, reduction (or increase) in water vapor through condensation (or evaporation) will release (or absorb) energy, thereby changing the temperature of the air particle, and making the pressure and forcing terms in eqn [1] functions of water vapor density. This process of energy release with condensation affects the stability of the atmosphere. The saturation vapor pressure of water is an increasing function of temperature (with the Goff–Gratch formula being the standard model to give exact values) and hence cooling can cause condensation if the vapor pressure is at saturation. This can affect atmospheric stability, since if the cooling associated with the rise of a parcel of air causes condensation and releases energy (to partially offset the cooling), the rate of change in temperature with height will be lower. Therefore the saturated or moist adiabatic lapse rate is lower than the dry adiabatic lapse rate of $9.8\,\mathrm{K\,km^{-1}}$. If the temperature decrease with elevation in the atmosphere is larger than the adiabatic lapse rate, the atmosphere is unstable, and rising air parcels will tend to continue rising. Rising air and condensation are important ingredients in cloud formation, and so the distribution of water vapor not only determines the amount and type of cloud formation through the supply of water, but also through its effects on atmospheric stability.

Cloud Formation

Since saturation vapor pressure is mainly a function of temperature, and water vapor will condense into water or ice particles if the vapor pressure is at or above saturation (supersaturation), there are essentially two ways to cause cloud droplet formation: introduce additional water vapor, or cool the air already containing some water vapor to reduce the saturation vapor pressure. Cooling is the most common cause of cloud formation, as the addition of water vapor only occurs through evaporation and usually happens close to the surface. This cooling can be achieved through radiative cooling, the air mass moving over a colder surface or the mixing of a warm moist air mass with a colder one, but in general these cooling mechanisms are inefficient and produce only light clouds or fogs. A more effective manner of cooling, and that which leads to the majority of precipitation, is through lifting, as the decrease in pressure with height leads to cooling through expansion, as can easily be shown through the adiabatic gas relations. This lift can be generated by air moving over mountains, by the air being heated from below, or by the collisions of air masses in a frontal system when warm moist air is forced upwards to move over colder denser air. The nature and scale of the uplift mechanism, along with the degree of water vapor present and the stability of the atmosphere, determine the type of cloud system formed. If the atmosphere is stable, or capped by an inversion, and the lifting mechanism is not overly strong but wide in extent, stratiform clouds may form: these clouds are vertically thin and horizontally wide. If the lifting mechanism is strong, such as from heating, and the atmosphere is unstable, which reinforces any lifting, convective cells may form that are of greater vertical extent, but may be smaller in the horizontal directions or at least have a greater degree of horizontal structure. On average, convective type clouds tend to produce more intense, but more localized, precipitation than stratiform clouds. In practice, cloud systems combine stratiform and convective features to different degrees and a thorough exploration of cloud types is beyond the scope of this article.

Once the cooling due to lifting has caused the air in some region to become supersaturated, water or ice particles will condense, depending on the temperature and the cloud condensation nuclei (CCN) available in the air. CCN are aerosol particles suspended in the air on which water can condense, and can consist of dust, smoke, salt, and a range of other substances such as pollutants produced by industry. The number and type of CCN affect the cloud droplet distribution and hence the chance of precipitation: a higher concentration of CCN can lead to the formation of many droplets. Because there is a finite amount of supersaturated water, all else being equal, a large number of droplets means that the average drop-size is smaller, and hence there is less chance of droplets growing large enough to fall out of suspension as precipitation. Conversely lower CCN concentrations produce fewer but larger droplets, all else being equal. This can be observed in the difference in cloud droplet concentration between marine cumulus clouds, which have fewer CCN (on the order of $10^2\,\mathrm{cm}^{-3}$), and continental cumulus clouds, which have a higher CCN concentrations (on the order of $10^3\,\mathrm{cm}^{-3}$). The presence of aerosols also affects the distribution of ice particles, because small droplets of pure water will not freeze until below $-39\,°\mathrm{C}$, but certain types of ice nuclei in the water can allow the freezing of ice particles with significantly higher temperatures: up to $-15\,°\mathrm{C}$ for some inorganic soil particles and up to $-4\,°\mathrm{C}$ for certain organic molecules from plants or plankton. Thus, the concentration and type of aerosol particles also influences the distribution and type of cloud particles and precipitation.

From Cloud Particles to Precipitation

Directly after forming, ice and liquid cloud particles have radii on the order of micrometers or less, making them easily small enough to remain suspended. Precipitation depends on whether the cloud particles can grow large enough to fall out of the updrafts and not to evaporate away during their fall. The exact size required for this will vary based on the meteorological conditions, but a radius on the order of 100 μm is usually a minimum. The growth of liquid-water cloud droplets is dominated by condensation when they are small, but growth above about 10 μm is driven by collision and coalescence of drops as they are advected by the moving air and later as they begin to fall under their increasing weight. Ice particles can grow larger than water droplets by deposition directly from the vapor phase, since the saturation vapor pressure over ice is lower than that over liquid water. Indeed, because of this, ice particles in mixed phase clouds below freezing temperatures can grow by robbing water from the supercooled liquid-water droplets by the Bergeron process. Ice particles can also grow by collision with supercooled water droplets, as the water droplets snap-freeze onto them – this is called riming, and is the process that produces graupel and in the extreme case, hail. Finally, ice particles can also grow to precipitable size through collision and aggregation with other ice particles.

Thus, the selective growth of some cloud particles to precipitation involves a number of competing processes, especially in commonly occurring mixed-phase clouds, which introduces another process of variability at small scales. Finally, just like the water vapor, the cloud particles and precipitation are being advected (to a greater or lesser extent depending on their mass) by the nonlinear, intermittent, and multiscale turbulent fluid flow in the atmosphere.

Precipitation Observations

Although it is possible to trace in a qualitative fashion the timeline of processes that lead to precipitation formation, from evaporation, advection of water vapor, to cloud formation and cloud droplet growth, the fact that these processes are complex, interconnected, nonlinear, and operating over a range of scales means that it is still an open problem of how to combine them into a model that reproduces the space–time distribution of precipitation. But measurements of precipitation allow direct observation of the final results of these processes: precipitation on the ground. The three main methods currently used for measuring precipitation are rain gauges, ground-based remote sensing using radar, and space-based remote sensing with a range of instruments.

Rain Gauges

The traditional method for measuring precipitation is through gauges that catch and measure rain or snow directly. Types of gauges range from a simple storage container that does not record the precipitation level itself, to weight measuring gauges, tipping bucket gauges that count number of bucket tips, and the modern optical precipitation gauge that records the entrance of each drop into the gauge as it disturbs a laser beam. For rainfall measurements, the major source of error is the distortion of the wind field by gauges mounted above the ground. The gauge forms an obstacle to the wind-flow, increasing the wind flow over the gauge and creating eddies around it. Since the precipitation is moved by the wind, this reduces the total entering the gauge by an amount depending on the wind speed, rain intensity, and gauge geometry, with losses on the order of 10% of the rainfall total being possible. While correction schemes for this have been developed by Sevruk and others, the World Meteorological Organization recommends using pit-gauges, i.e., gauges mounted in pits below the ground surface and hence sheltered from the wind, as the reference to calibrate other gauge types. Other errors for rain include smaller losses to splashing and evaporation.

Accurately measuring snow presents greater difficulties than measuring rainfall, even though standard gauges can be modified for snow by adding antifreeze to weighing gauges and heating tipping-bucket gauges to melt the snow. One reason for the difficulty is that the terminal fall velocity for snow flakes is much lower than that for water droplets, and they are much more easily transported by the wind, meaning that the error due to wind distortions around the gauge can be as high as 50% in worst-case conditions. As a result of these and other difficulties, it is recommended that a specially designed gauge for snow, such as the 'universal gauge' of Cox, be used, preferably in combination with wind-shielding.

Despite the various sources of error, gauges are the only direct measurements of precipitation, and remain the most accurate source of precipitation information at any point in space and time. In particular, gauges can give very accurate descriptions of the structure and distribution of precipitation in time. **Figure 1(a)** shows an example of high-resolution rain-gauge data from the Iowa Institute of Hydraulic Research, University of Iowa, which shows variability across a range of temporal scales. In particular, the power spectrum of the data shown in **Figure 1(b)** shows variability decreasing with decreasing scale (increasing frequency) in an approximately power law relationship (a straight line on the log–log graph) between scales of around 1 min and 1 h.

The limitation of gauge measurement is that gauges are usually sparsely spread over an area, and gauge apertures are on the order of centimeters, whereas hydrologic processes depend on the entire precipitation field over a catchment area that may have an area on the order of hundreds of kilometers. This is a long studied problem in hydrology, and a range of methods have been developed over the years to convert point gauge measurements to area values, beginning with Thiessen's polygon area-averaging method, and going through surface fitting methods using interpolation or smoothing, to objective analysis and kriging. All of these methods depend on assumptions about the smoothness or statistical distribution of the precipitation field, which has been shown in more recent times to be more intermittent than most of these methods postulate. For more information on the spatial variability of the precipitation field, measurements from instruments other than rain gauges are required.

Radar

Quantifying the spatial variability of precipitation fields requires estimating the precipitation rate over a large area and a relatively short sampling interval:

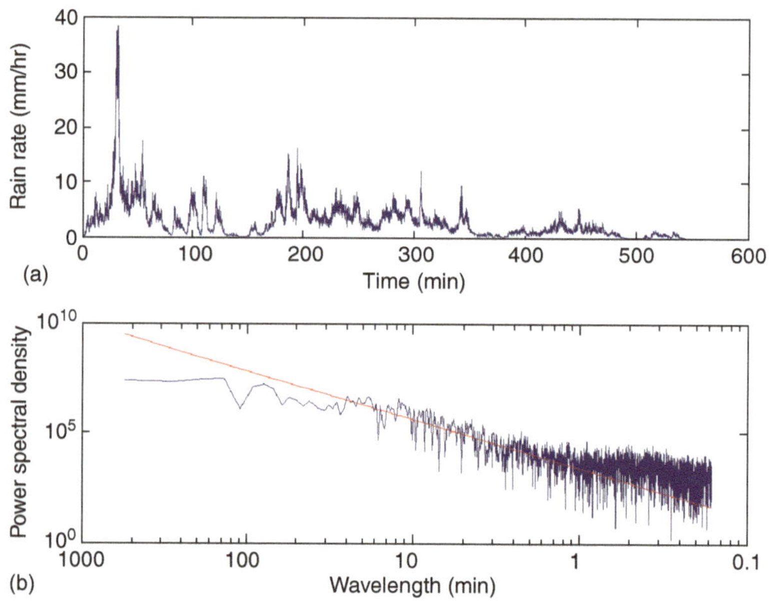

Figure 1 Rain gauge data from a high resolution optical gauge, with part (a) showing the data series in time, and part (b) showing the power spectrum of the data.

one of the most useful instruments to do this is land-based Doppler radars. Since rain and ice particles reflect electromagnetic radiation that has a wavelength on the order of centimeters (by the Rayleigh scattering process), radars emitting pulses at these wavelengths can estimate the amount of precipitation by measuring the reflected radiation. The fraction of radiated power that is reflected, or the reflectivity Z, has been found to relate to the precipitation rate R via the so-called Z–R relation, specifically:

$$Z = aR^{b} \qquad [2]$$

where a and b are constants that vary with the calibration of the radar, ground clutter, beam-broadening, and most importantly, the type of precipitation (rain or snow, convective or stratiform, continental or marine, etc.). The uncertainty due to precipitation-type can be reduced using dual-polarization radar, which adds an extra source of information because different precipitation types have different effects on the polarization of the reflected radiation. In this way, a Doppler radar can provide estimates of precipitation over an area of hundreds of kilometers, with a spatial resolution on the order of 1–5 km, every few minutes. An example of radar-retrieved precipitation rate is found in **Figure 2**. We can examine how the variability of rainfall depends on spatial scale by observing the power spectrum from radar data in **Figure 3**, which shows variability over a wide range of scales, and a power law over all observed scales.

A large fraction of the United States is covered by the NEXRAD system of 130 WSR-88D radar stations, which are S-band (or 10 cm wavelength) Doppler radars that have a range of 230 km, an average resolution of 2 km (actual resolution is a function of the distance from the radar), and record precipitation intensities every 6 min. Although initially they were set-up without polarization, the National Oceanic and Atmospheric Administration (NOAA) has a program underway to convert these stations to dual-polarization radars by 2010. Other developed countries have similar terrestrial radar networks, such as the Central European Weather Radar Network (CERAD), or are working on them. These radar networks provide high spatial and temporal resolution precipitation data that can be used as an input for hydrologic models, but there are still vast areas of the globe, especially in poorer or developing countries, that are not covered by terrestrial radar. There are also limits to the coverage of ground-based radar in some mountainous areas where the field of view is interrupted by the terrain. To investigate the distribution of precipitation in these areas, one must look to satellite observations.

Satellite Remote Sensing

There are a number of satellite-mounted instruments that are used to estimate precipitation levels, but they can be divided into two main groups. The first are those instruments that attempt to retrieve precipitation intensities through the active or passive measurement of microwave radiation. Remote sensing in the microwave frequencies is particularly applicable to

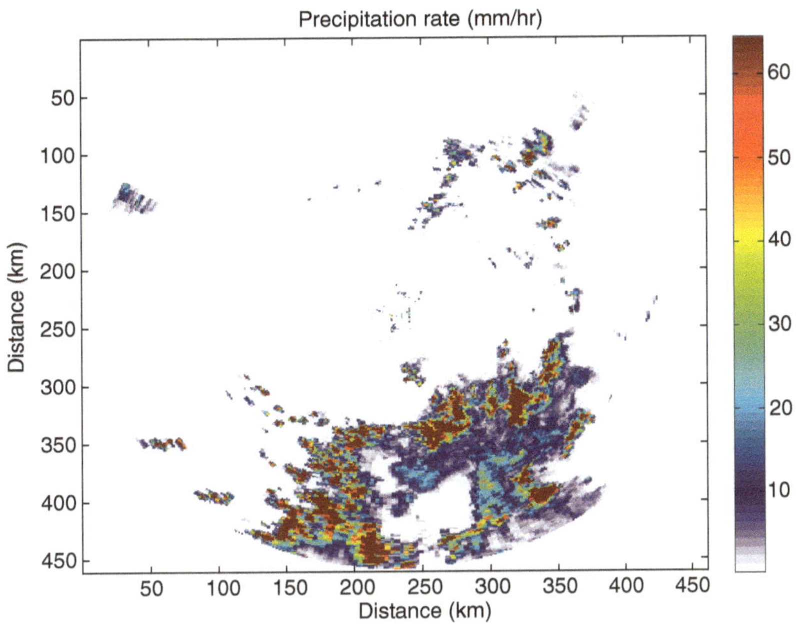

Figure 2 Precipitation rate from a single scan with a Doppler weather radar from the NEXRAD station KPBZ at Pittsburgh, PA, USA (coordinates 40.5317° N, 80.2181° W).

precipitation retrieval because microwave propagation is sensitive to the presence of large water and ice particles, but less affected by smaller cloud particles (unlikely to precipitate) and other atmospheric attenuation. Passive microwave sensors attempt to determine precipitation rates based on the increased natural emission of radiation, and hence higher observed brightness temperatures, due to the presence of precipitation. Alternatively, active sensors measure the scattering of an emitted beam. In this sense, they use the same general concept as earth-based radars, but with the added difficulties of the increased distance, the large column of atmosphere to penetrate to retrieve precipitation at the ground, and more background noise. Both active and passive microwave data must be corrected for these effects. Therefore, the uncertainties in the precipitation measurements are significantly higher than for earth-based radar. The advantage of the space-based measurements is their extremely large range of spatial coverage. Commonly used examples of microwave instruments used to measure precipitation from space are: the Special Sensor Microwave/Imager (SSM/I) that is part of NASA's Pathfinder Program; the Precipitation Radar (PR) and TRMM Microwave Imager (TMI) that are included in NASA's Tropical Rainfall Measuring Mission (TRMM); the Advanced Microwave Scanning Radiometer (AMSR-E) aboard the Aqua spacecraft; and the Cloud Profiling Radar (CPR) aboard the CloudSat satellite. An example of data collected by the TMI instrument on TRMM can be

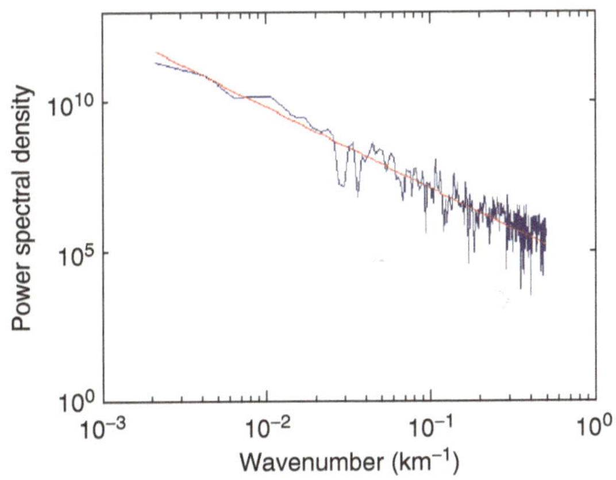

Figure 3 The power spectrum of precipitation field observed by a single scan with a ground-based radar from the NEXRAD station KPBZ at Pittsburgh, PA, USA (coordinates 40.5317° N, 80.2181° W).

seen in **Figure 4**, which shows the weekly accumulated precipitation over the entire globe at 0.25° resolution. Note the high degree of structure on very large scales that can be captured with satellite imaging, compared to the relatively local-scale detail observed by ground-based radar image (cf. **Figure 2**).

The second type of precipitation estimations made via satellite remote sensing uses passive imaging radiometers in the infrared, near-infrared, or visible parts of the spectrum. These wavelengths are more sensitive to cloud properties other than precipitation, such as cloud-top temperature, total cloud liquid water path, mean droplet radius, and cloud spatial

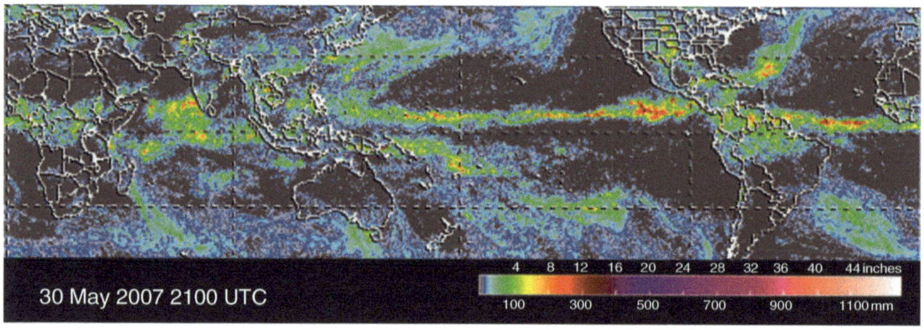

Figure 4 Total precipitation accumulated over a week from the microwave instruments on the TRMM satellite.

extent. The precipitation retrieval therefore relies on relationships between these other cloud properties and precipitation intensity, either through empirically derived look-up tables (or database inversion) or simplified precipitation models. Since they are less direct estimates, the precipitation readings from this method are considered more uncertain than the satellite microwave or radar measurements. However, these less direct estimates have the advantage of not requiring a specialized instrument, but instead using measurements from radiometers that are found on many operational satellites such as NOAA's Geostationary Operational Environmental Satellite (GOES) series or Japan's Geostationary Meteorological Satellites (GMS). This allows for a greater frequency of observations in time than is possible using the specialized microwave instruments currently mounted on only a relatively small number of satellites: specifically the visible and infrared satellite products are usually produced hourly, as opposed to a 3-h or longer gap between microwave precipitation images. NASA's TRMM mission now produces an hourly hybrid product that uses indirect infrared estimates that are calibrated every 3 h with the more accurate microwave data.

Satellite measurements of precipitation have the advantage of extremely wide spatial coverage over the globe, but are less accurate and have lower spatial and temporal resolutions than do ground-based weather radars. In particular, while spatial resolution seems to be increasing significantly with each generation of satellites, temporal resolution does not seem to be increasing at the same rate, as it depends not so much on the instrument but on the frequency of satellite overpass. This remains a challenge for the satellite remote sensing of precipitation.

Modeling Precipitation

Whether estimated by gauges, ground-based radars, or satellites, precipitation exhibits variability at all space–time scales: from seconds to years, and from meters to hundreds of kilometers. Understanding this variability is not only important for hydrologic applications, but even for the basic tasks of comparing, validating, and merging estimates of precipitation made at different scales by different sensors. Direct numerical models of clouds and rainfall are certainly possible, combining computational fluid mechanics of atmospheric flow with a model of the so-called microphysics of cloud and rain particles, and these cloud-resolving models (CRMs) form a strong area of ongoing research. They often involve nested model levels, with coarse resolution models to capture large-scale environmental forcings and provide boundary conditions for fine resolution models that include the cloud microphysics. This presents difficulties not only because of the high degree of computing power required, but because of challenges in developing accurate models or parameterizations of cloud microphysics. This is due to the fact that there are many complex processes involved in droplet formation, growth, and fall-out, and that direct observation of these in-cloud processes, for either validation or initialization purposes, is difficult. Despite these difficulties, research continues apace on improving CRMs and data-assimilation methods for quantitative precipitation forecasts. Also, at larger scales than it is currently feasible to run a fully detailed CRM, regional and global circulation models (GCMs) of the atmosphere include successively more coarse-grained parameterizations of cloud and precipitation processes. How to improve these parameterizations while maintaining reasonable model complexity and running-time is yet another challenge in precipitation modeling.

Scale-Invariant Statistical Models

For applications at scales below the ones well resolved by numerical models, and to even verify how well numerical models are in fact reproducing the space–time patterns of precipitation, statistical models of precipitation are required. Early recognition of the importance of small-scale variability of rainfall in space and time for hydrological applications led to stochastic point-process models or phenomenological spatiotemporal models of rainfall. However, these models tended to have a large number of parameters to fit, and to be scale-dependent in their application. From the late 1980s to early 1990s, the desire to unify the description of precipitation over scales, and reduce the number of model parameters, inspired the adoption of ideas and tools from turbulence modeling and fractal geometry. This led to the development of various scale-invariant models of precipitation. The basic meaning of (stochastic) scale invariance is that the variability of precipitation exhibits a statistically self-similar structure under changes of space and/or time scale. If the probability distribution function (PDF) of the fluctuations in rainfall ΔR_l at some scale l is $P_l(\Delta R_l)$, then the simplest form of scale invariance is if the PDF at a different scale l' is given by

$$P_{l'}(\Delta R_{l'}) = \lambda^{-H} P_l(\lambda^{-H} \Delta R_l) \qquad [3]$$

where l is the ratio of scales (l/l') and H is a constant. That is, the PDFs at the two scales are identical except for a rescaling factor that depends on the ratio of scales. It is then straightforward to show that the moments of the rainfall fluctuations are then a power-law function of scale, i.e.,

$$\langle |\Delta R_l|^q \rangle \propto l^{qH} \qquad [4]$$

for order q. This can be tested on observed data in a reasonably straightforward manner. This stochastic self-similar model can be referred to as a fractal, or monofractal, model, since in the limit as l tends to zero it would produce a field of singularities with a single fractal dimension H. In reality, the model does not apply all the way down to infinitely small scales, but over a finite range of scales. Within this scaling range, the model of how the PDF varies with scale can be used to compare data between scales, to perform down-scaling or up-scaling, to simulate precipitation fields, or in data-merging applications. The model can also be extended to multiscaling, in which the moments of the fluctuations are still power-law functions of scale as in eqn [4], but instead of the exponents being a linear function of q, they are given by

$$\langle |\Delta R_l|^q \rangle \propto l^{\tau(q)} \qquad [5]$$

where $\tau(q)$ is a continuous concave function of the moment-order q. In this case, the PDF does not maintain its shape between two different scales, but changes continuously in a way that still depends only on the ratio of scales. In general, the smaller the scale, the fatter the tail of the distribution, i.e., as scale $l \rightarrow 0$ the higher moments increase relatively faster than the lower-order moments. This multiscaling model is consistent with the statistics resulting from a multiplicative cascade processes, in which the values of the field at smaller scales are produced by multiplying the field value at a larger scale by a stochastic weight variable, as shown in **Figure 5**.

Multiplicative cascade models predict not only the one-point PDF of the rainfall fluctuations, but also the correlations between measurements made at different locations as a function of measurement scale, and even the correlation between measurements at different locations and different scales. This structure can be used for multiscale objective analysis and interpolation between sparsely recorded data, or for merging measurements at different scales and locations.

If there is a scale-invariant structure in both time and space, the time and space scaling can be combined into a single model, using a multiplicative factor to transform the time dimension to an equivalent spatial dimension. If a single velocity U is used at all scales to transform from time-scale t to space-scale l, i.e., $l = Ut$, then this is equivalent to Taylor's frozen turbulence hypothesis. However, it is also possible to use a different factor at each scale and maintain the scale-invariant structure in both space and time, as long as the velocity is also a power-law function of scale, which gives $(t/t') = (l/l')^z$ for constant exponent z. Then a single space–time model of scaling can be constructed with time considered to be an extra spatial dimension.

After almost two decades of examining measurements with different instruments, it has been shown, and is now generally accepted, that precipitation fields do indeed show scale-invariant behavior with multiscaling, at least over some range of spatial

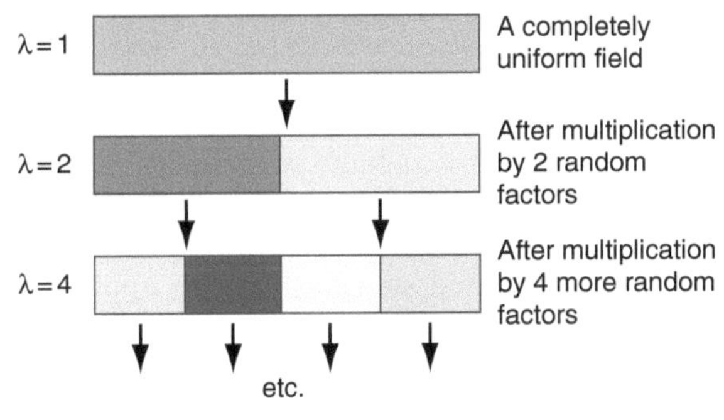

Figure 5 A schematic representation of the cascade process in one dimension, with shading used to represent the value of the scalar field. Each structure is broken up into two substructures, transferring some or all of its value to the substructures. This is modeled mathematically by multiplicative factors being applied at each level. The scale is labeled by the ratio λ = (maximum length)/(current grid size).

and temporal scales. What is not so clear is what this range of space–time scales is, or what the exact scaling exponents are, since the scaling range and scaling exponents vary between measurement instruments, and between different rainfall regimes and locations. This suggests that although the presence of (multi)scaling may be universal, the scaling range and parameters are dependent on the prevailing meteorological conditions, type of precipitation, and other local variables. Some research has been done along these lines: in 1996, Perica and Foufoula-Georgiou found that the scaling parameters for mid-latitude mesoscale convective systems could be predicted by the convective available potential energy (CAPE) before the storm; and more recently some work has been done to relate the scaling statistics of orographic rainfall to the surface elevation driving the precipitation system. However, it still remains a challenge to determine the dependence of the scaling parameters and the scaling-range for precipitation in different types of storms and different locations.

Conclusions

Precipitation is one of the most important links between weather and climate and the hydrological cycle, on time scales ranging from hours to decades or longer, and space scales from meters to hundreds or thousands of kilometers. Some of the most direct effects of climate change on humans are related to changes in the amount, or the distribution, of precipitation, including the distribution of extreme events. Many of the difficulties in modeling precipitation are the result of the space–time structure and intermittency of the precipitation field, which are due to the fact that it is influenced by many dynamic processes that act on a large range of scales. However, the improved spatial resolution provided by the modern technologies of radar and satellite measurements has allowed progress in the physical modeling of precipitation, as well as the development of scale-invariant statistical models that can encompass that variability with a relatively small number of parameters. Some challenges that lie ahead are to find the range of scales and conditions to which both the physical and statistical models apply accurately, and to relate the statistical scaling parameters to environmental conditions for various types of precipitation. For numerical cloud models, despite the computational resources now available, which allow very high resolutions, research shows that ignoring small scale processes and variability (which for practical purposes may not be of importance in themselves) leads to errors in larger-scales processes and quantities due to nonlinearities in the dynamics. Thus, cloud parameterizations and nested dynamically evolving grids that incorporate accurate subgrid statistics are active areas of research. Therefore, it seems that it would be of great benefit to incorporate research on statistical models with the numerical cloud resolving models, in an attempt to both improve the verisimilitude of the numerical models in reproducing precipitation fields with the correct statistics, and to study the dependence of (multi)scaling models on different environmental parameters.

Glossary

Adiabatic lapse rate – The rate of decrease in temperature with height for a parcel of air that does not exchange heat with the surrounding atmosphere i.e., that is assumed to be perfectly insulated.

Cloud condensation nuclei (CCN) – Aerosol particles suspended in the air on which water can condense.

Cloud resolving model (CRM) – Numerical model of the atmosphere that includes the modeling of cloud and precipitation processes.

Fractal – A geometrical object that is scale invariant, with structure at all scales and a simple one-parameter function to transform from one scale to another.

Multifractal – A generalization of the fractal to include more variable fields, and allow more complex scaling behavior so that the transform from one scale to another is a function of more than one parameter – in general a continuum of scaling exponents.

Passive scalar – A quantity or substance that is transported by moving fluid without having any effect on the dynamics of the fluid. Also known as a tracer.

Scale invariance – The property of being self-similar under changes in spatial or temporal scale, so that the probability distributions or other relevant properties of a field keep the same form (possibly with renormalization).

Stratiform – Forming a layer or arranged in layers. Used to refer to a broad class of clouds that form layers of large horizontal extent.

See also: Evapotranspiration; Hydrological Cycle and Water Budgets; Physical Properties of Water; Vadose Water.

Further Reading

Antolik MS (2000) An overview of the National Weather Service's Centralized Statistical Quantitative Precipitation Forecasts. *Journal of Hydrology* 239: 306–337.

Brutsaert W (2005) *Hydrology: An Introduction.* Cambridge: Cambridge University Press.

Doviak RJ and Zrnić DS (1993) *Doppler Radar and Weather Observations*, 3rd edn. San Diego, CA: Academic Press.

Droegemeier KK, Smith JD, Businger S, *et al.* (2000) Hydrological aspects of weather prediction and flood warnings: Report of the Ninth Prospectus Development Team of the U.S. Weather Research Program. *Bulletin of the American Meteorological Society* 81: 2665–2680.

Foufoula-Georgiou E (1997) On scaling theories of space–time rainfall: Some recent results and open problems. In: Gupta, *et al.* (eds.) *Stochastic Methods in Hydrology: Rain, Landforms and Floods*, pp. 25–72. Singapore: Word Scientific.

Foufoula-Georgiou E and Vuruputur V (2001) Patterns and organization in precipitation. In: Grayson R and Blöschl G (eds.) *Spatial Patterns in Catchment Hydrology – Observations and Modeling*, pp. 82–104. New York: Cambridge University Press.

Kidd C (2001) Satellite rainfall climatology: A review. *International Journal of Climatology* 21: 1041–1066.

Kummerow C, Olson WS, and Giglio L (1996) A simplified scheme for obtaining precipitation and vertical hydrometeor profiles from passive microwave sensors. *IEEE Transactions on Geoscience and Remote Sensing* 34: 1213–1232.

Randall DA, Khairoutdinov M, Arakawa A, and Grabowski W (2002) Breaking the cloud-parameterization deadlock. *Bulletin of the American Meteorological Society* 84: 1547–1564.

Rasmussen EM and Arkin PA (1993) A global view of large-scale rainfall variability. *Journal of Climate* 6: 1495–1521.

Veneziano D, Langousis A, and Furcolo D (2006) Multifractality and rainfall extremes: A review. *Water Resources Research* 42: W06D15,doi:10.1029/2005WR004716.

Wallace JM and Hobbs PV (1977) *Atmospheric science: an introductory survey.* New York: Academic Press.

Relevant Websites

http://www.weather.gov/ – National Weather Service.

http://www.noaa.gov/ – National Atmospheric and Oceanic Administration.

http://gpm.gsfc.nasa.gov/ – Global Precipitation Measurement.

http://trmm.gsfc.nasa.gov/ – Tropical Rainfall Measurement Mission.

http://ww2010.atmos.uiuc.edu – World Weather 2010 Project.

Snow and Ice

G Hornberger, Vanderbilt University, Nashville, TN, USA
T C Winter, US Geological Survey, Denver, CO, USA

The cryosphere is the portion of Earth's surface where water exists in solid form. In the winter, snow covers more than 60% of the Northern Hemisphere's land area whereas there is little snow cover in the Southern Hemisphere except for Antarctica (UNEP/GRID-Arendal, 2007). Snow and ice cover about 10% of the land area permanently. Snow and ice (primarily glaciers and ice sheets) store large amounts of fresh water; most of the Earth's fresh water resides in two major ice sheets, Greenland (area $1.75 \times 106\,km^2$) and Antarctica (area $12.1 \times 106\,km^2$).

Snow and ice affect essentially all aspects of the hydrological cycle. Frozen ground reduces infiltration of water into soils and can increase the runoff generated from melting snow. Permafrost severely reduces the amount of water that can be stored in soils. Seasonally and permanently frozen land surfaces interact significantly with the global weather and climate system, affecting surface albedo (i.e., the reflection of solar radiation) and latent energy exchanges (i.e., evaporation). Earth's glaciers and ice caps have been undergoing significant recession, with measurable impacts on sea level, water resources, and ecosystems. The state of the cryosphere, which has been cited as having 'a unique sensitivity to climate change at all spatial and temporal scales' (Slaymaker and Kelly, 2007), is viewed as an important indicator of climate change.

The melting of snow and ice has substantial effects on inland waters. Seasonal snow cover in many mountainous regions is of critical importance to water supply. The melting of snow and ice can affect hydrological processes and their linkages with ecosystems. Streams swollen by snow and ice melt in the spring can (1) scour streambeds and deposit new sand bars, resulting in new substrate for benthic and other organisms that rely on the changing geomorphic and fluvial conditions, and (2) overtop stream banks, depositing fresh sediments and providing water for riparian and floodplain wetlands. The melting of snow and glaciers in high mountains can sustain alpine lakes and wetlands as well as supply water to streams throughout summer. This relatively reliable source of water sustains both aquatic and terrestrial alpine ecosystems as well as riverine ecosystems associated with the ice-melt-sustained river flows.

Further Reading

Slaymaker O and Kelly R (2007) *The Cryosphere and Global Environmental Change.* Blackwell.

UNEP/GRID-Arendal. Cryosphere—winter seasons, Northern and Southern Hemispheres [Internet]. UNEP/GRID-Arendal Maps and Graphics Library; June 2007 [cited 2007 Nov 26]. Available at http://maps.grida.no/go/graphic/cryosphere-winter-seasons-northern-and-southern-hemispheres.

Relevant Websites

http://www.acia.uaf.edu/pages/scientific.html – Arctic Climate Impact Assessment (especially Chapter 6, Cryosphere and Hydrology).

http://www.nsidc.org – National Snow and Ice Data Center.

Evapotranspiration

G Katul and K Novick, Duke University, Durham, NC, USA

Introduction

Earth's water is highly dynamic and continuously in motion, and the terms 'water cycle' or 'hydrologic cycle' describe the continuous movement of water molecules on, above, and below the surface of the Earth. The water cycle concept may be traced back to the Greeks, evidenced for example in the Iliad (written around 800 BC), when Homer described the "oceans from whose deeps every river and sea, every spring and well flows..." suggesting interconnectedness of all of the Earth's water. Leonardo's Codex Leicester, written between 1506 and 1510, was a seminal document mostly focused on water, and also advanced the concept of a large-scale water cycle by offering keen observations on the dynamics and transport of water (and suspended particles) by streams and rivers originating in the mountains and continuing through the plains to the sea. More importantly, the Codex Leicester is one of the first 'Albums of Fluid Motion' or flow visualization studies, discussing many aspects of the hydrologic cycle and its connection to fossils, geology, and climate.

Because of its intrinsic role in the hydrologic cycle, the study of evapotranspiration (ET), the sum of evaporation (E) and plant transpiration (T), has a rich research history, and to discuss every nuance of the topic is beyond the scope of a single chapter. The focus here is on ET, the engine of the hydrologic cycle, crucial for determining usable water for humans and ecosystems. We explore how the projected climatic and land cover changes might alter ET over a hierarchy of scales ranging from global to continental to local. Throughout, E here refers to the movement of water to the atmosphere from sources such as the soil matrix, rainfall intercepted by plant canopies, and water bodies, while T refers to the loss of water in the form of vapor molecules passing through leaf stomata.

The first attempt to quantify the role of ET in the hydrologic budget is often attributed to John Dalton, who carried out the necessary calculations to construct hydrological balances of major rivers (including the Thames), and published them in 1802 in the manuscript titled "Experiments and observations to determine whether the quantity of rain and dew is equal to the quantity of water carried off by the rivers and raised by evaporation; with an inquiry into the origin of springs." Dalton is also known for his work on partial pressures, which lead to the first physically based quantitative model of evaporation (see **Box 1**).

In the next section, it is demonstrated that Dalton's seminal work, along with others in the nineteenth century, lead to developments of quantitative laws for E that find wide use today in constraining estimates of the acceleration of the hydrologic cycle because of projected increases in global air temperature. The term 'acceleration of the hydrologic cycle' refers to the fact that higher temperatures provide more kinetic energy to water molecules, leading to more evaporation and thus more precipitation.

Evaporation and the Projected Acceleration in the Global Hydrologic Cycle

It is now accepted that increases in greenhouse gas emissions lead to increases in air temperature. However, the effects on the hydrologic cycle are far more difficult to predict. It is appropriate to start with a first-order estimate of how much the global hydrologic cycle is expected to accelerate following an increase in global air temperature (δT_a) using only the nineteenth century equations presented by John Dalton, Rudolf Clausius, and Benoit Paul Emile Clapeyron (see **Box 1**). From **Box 1**, it can be shown that the combination of these nineteenth century laws lead to

$$\frac{\delta P}{P} = \frac{\delta E}{E} = 0.0675\delta T_a$$

where δP and δE are projected changes in global rainfall and global evaporation in response to δT_a, respectively. On the basis of the nineteenth century laws, a 1 °C warming (roughly commensurate with the warming trend experienced over the past century, which is estimated at 0.15 °C per decade) leads to a 6.8% increase in global rainfall (or E). Furthermore, a 4 °C projected warming trend, predicted to result from a doubling in atmospheric CO_2 concentration, can produce up to a 27% increase in rainfall (or E). These calculations may be compared with contemporary estimates obtained using the state-of-the-art high-resolution climate models that couple oceanic and atmospheric circulation (and stretch the best currently available supercomputing facilities) that predict

$$\frac{\delta P}{P} = 0.035(\delta T_a - 1.4)$$

This modern estimate suggests that a 4 °C projected warming will result in about 9% increase in rainfall.

Box 1 Acceleration of the Global Hydrologic Cycle with Increased Temperature

A Scaling Analysis Using Nineteenth Century Formulations

A first-order estimate of the acceleration in the hydrologic cycle in response to a global increase in air temperature (δT_a) is carried out using physical laws (in boxed quantities) derived by the nineteenth century scientists John Dalton, Rudolf Clausius, and Benoit Paul Emile Clapeyron. At the global scale, once sufficiently long periods of time have elapsed (e.g., decades or longer) it is safe to state that the global hydrologic balance can be reduced to $P \approx E$, where P is the global rainfall. Hence, any change in the global hydrological cycle due to an increase in air temperature must affect both rainfall and evaporation expressed as

$$\frac{\delta P}{P} = \frac{\delta E}{E}$$

where δP and δE are changes in global rainfall and global evaporation due to a δT_a. Using Dalton's law,

$$E \approx g_w D$$

where g_w is referred to as the conductance of the surface to water vapor, and D is the vapor pressure deficit defined as $e^*(T_a)(1 - \mathrm{RH})$, where e^* is the saturation vapor pressure at T_a and RH is the global air relative humidity. The Clausius–Clapeyron equation can now be used to relate e^* to T_a using,

$$e^*(T_a) = a \exp\left(\frac{bT_a}{T_a + c}\right)$$

where T_a is the temperature (°C), $a = 0.611$ kPa, $b = 17.5$ °C^{-1}, and $c = 249.93$ °C for typical atmospheric pressures. In existing climate simulations, the effects of increased greenhouse gases on δT_a do not lead to appreciable changes in RH, even across a wide range of climate scenarios. Hence, in a first-order analysis, assuming that RH maintains its present global value, we find that

$$\frac{\delta P}{P} = \frac{\delta E}{E} = \frac{\delta D}{D} = \frac{\delta e^*(T_a)}{e^*(T_a)} = \left(\frac{-bT_a}{(c + T_a)^2} + \frac{b}{c + T_a}\right)\delta T_a$$

An order of magnitude analysis demonstrates that $\frac{-bT_a}{(c+T_a)^2} \ll \left(\frac{b}{c+T_a}\right)$ resulting in

$$\frac{\delta P}{P} = \frac{b}{c + T_a}\delta T_a$$

Using the current global air temperature $T_a = 15$ °C, and substituting $b = 17.5$ °C^{-1}, and $c = 249.93$ °C results in

$$\frac{\delta P}{P} = \frac{\delta E}{E} = 0.0675\delta T_a$$

Note that this δP differs from the estimate in **Box 1** in slope (by about a factor of 2), and by the offset of 1.4 K that can be attributed to thermal inertia in the Earth system (i.e., some finite warming is necessary before the hydrologic cycle begins to be impacted). The differences in the slopes between these two estimates can be traced back to inefficiencies in the hydrologic cycle that are accounted for in climate models, as well as the many feedbacks not accounted for in the calculations shown in **Box 1**. Examples of such feedbacks include the formation of clouds in advance of rainfall, which block direct sunlight from arriving at the surface and reduce bulk conductance, thereby decreasing ET. Hence, according to **Box 1**, the projected acceleration in the global hydrologic cycle is primarily due to a global increase in ET, and nineteenth century laws provide a reasonable *upper limit* to constrain its value.

Evapotranspiration and the Continental-Scale Hydrologic Cycle

The problem of assessing how climatic changes propagate through various terms in the continental-scale hydrologic balance is complicated, than in the global case, by the addition of a new term – continental-scale runoff (R_o). Over long periods of time, the continental-scale hydrologic balance can be expressed as

$$p = ET + R_0$$

Nearly all studies investigating continental-scale trends in ET over the past 50–100 years suggest some change has occurred, but conflicting conclusions persist about the direction of this change. While continental-scale ET may be an order of magnitude smaller than oceanic E, replenishment of most water resources and ecosystem goods and services, as well as delivery of essential nutrients to marine estuaries, depends on the continental-scale R_0.

The arguments presented in **Box 1** suggest an increase in ET in a warmer climate, but a number of studies have documented an increase in continental-scale runoff in recent years. Hence, how ET may have changed over the past 50–100 years appears to be controversial, and reconstructing how continental-scale ET changed over the past 100 years is a logical test of current skills in predicting the future of the hydrologic cycle.

Three hypotheses have been promoted as plausible explanations for a decreasing ET over the past 50–100 years, each with certain limitations. The first is the so-called 'solar dimming' hypothesis. This hypothesis argues that a reduction in solar irradiance occurred because of an increase in cloud cover and aerosols concentration, the latter being consistent with measured increases in air pollution throughout the past 100 years (**Table 1**). Solar irradiance is a key forcing for the available energy that drives ET and influences bulk conductance through the effect of light on leaf photosynthesis (see **Box 2**). The decrease in pan evaporation rate measured over the last 50 years over much of the conterminous United States and Russia is used as indirect support for this hypothesis. A pan evaporameter is a simple device consisting of a cylindrical container about 1.2 m in diameter and 0.25 m deep filled with water, a water level measuring device, and a rain gage. Naturally, pan evaporation is influenced by complex micrometeorological

Table 1 Observed global changes in radiation

Dates of measurements	*Observed change in radiation*	*Source*
1958–1992	$-0.51\ W\ m^{-2}\ yr^{-1}$	1
1964–1980	$-0.41\ W\ m^{-2}\ yr^{-1}$ in densely populated areas, $-0.16\ W\ m^{-2}\ yr^{-1}$ in sparsely populated areas	2
1984–2001	$+0.24\ W\ m^{-2}\ yr^{-1}$	4, 5
1992–2002	$+0.66\ W\ m^{-2}\ yr^{-1}$	6

Sources

1. Stanhill G and Cohen S (2001) Global dimming: A review of the evidence for a widespread and significant reduction in global radiation with discussion of its probably causes and possible agricultural consequences. *Agricultural and Forest Meteorology* 107: 225–278.
2. Alpert P, Kishca P, Kauffman YJ, and Schwarzbard R (2005) Global dimming or local dimming?: Effect of urbanization on sunlight availability. *Geophysical Research Letters* 32: Art. No. L17802.
3. Roderick ML (2006) The ever-flickering light. *Trends in Ecology and Evolution* 21: 3–5.
4. Pinker RT, Zhang B, and Dutton EG (2005) Do satellites detect trends in surface solar radiation? *Science* 308: 850–854.
5. Wild M, Gilgen H, Roesch A, Ohmura A, Long CN, Dutton EG, Forgan B, Kallis A, Russak V, and Tsvetkov A (2005) From dimming to brightening: Decadal changes in solar radiation at Earth's surface. *Science* 308: 847–850.

Box 2 Changes in Continental-Scale ET

Extending Dalton's law to ET (see **Box 1**),

$$\mathrm{ET} \approx g_c D$$

where now g_c is the bulk conductance of the soil–plant system (i.e., it lumps conductances in the soil and plant). Hence,

$$\frac{\delta \mathrm{ET}}{\mathrm{ET}} = \frac{\delta D}{D} + \frac{\delta g_c}{g_c}$$

Any solar dimming, increased atmospheric CO_2 and its concomitant effect on bulk stomatal conductance, and the overall increase in global deforestation all are viewed as factors leading to a negative $\delta g_c/g_c$ that is greater than any expected increases in $\delta D/D$ with warming (see **Box 1**), resulting in a decline in $\delta \mathrm{ET}/\mathrm{ET}$ over continents. The basic challenge confronting the scientific community today in quantifying $\delta g_c/g_c$ over continents is that the relationships between g_c, light levels, atmospheric CO_2, soil moisture, and species composition are nonlinear and vary considerably across biomes, soil type, etc.

When quantifying $\delta g_c/g_c$, it is convenient to explore the individual conductances (i.e., stomatal and soil conductances) separately because they respond differently to environmental drivers, particularly elevated atmospheric CO_2.

processes like local wind flow and is often used as one indicator of 'potential' ET. Pan evaporation records are amongst the longest available hydrologic records, spanning some 100 years in several locations.

Some studies estimate that the measured reduction in pan evaporation is consistent with solar dimming rates of 2–4% per decade. The 2–4% per decade range was independently confirmed from observation for the period between 1960 to late 1980s, using the Baseline Surface Radiation Network (BSRN) of the World Climate Research Program (WCRP). However, solar dimming now appears to be giving way to the so-called solar 'brightening' at a rate of about 1.6% per decade (**Table 1**). This brightening is partly explained by the recovery from the large aerosol loadings associated with the 1991 Pinatubo eruption, and a decline in Eastern European aerosol emissions due to tighter air-quality regulations in those regions. Other authors question this 'continental' view of dimming and favor local-scale explanations. These studies reported that 'dimming' was four times more frequently observed near population centers (defined as centers with a population size exceeding 0.1 million) than in sparsely populated areas.

Irrespective of whether solar 'dimming,' 'brightening,' or even 'flickering' will be the scenario for the future, the contention that a reduction in pan evaporation can be correlated with actual reductions in ET is not universally accepted. The so-called complementary hypothesis argues that a reduction in pan evaporation actually corresponds to an *increase* in ET, particularly in water-limited ecosystems. This hypothesis is based on the prediction that higher ET increases humidity, cools the air, and reduces the vapor pressure deficit (D), thereby reducing potential (or pan) evaporation. Climatological studies across the conterminous United States suggest that D did not significantly increase over the past 50 years despite a decline in the pan evaporation record, thereby negating one of the assumptions of the complementary hypothesis. However, analysis of published precipitation and stream discharge data for several large basins across the conterminous United States show that ET rates, estimated as the difference between rainfall and runoff, have increased over the past 50 years. It is clear that further studies are necessary to resolve how the pan evaporation record needs to be interpreted and whether it can be used in a complementary formulation for actual ET.

The second hypothesis, promoted by sensitivity studies conducted using climate models, was aimed at exploring why continental-scale runoff increased in the past 50–100 years. This hypothesis argues that a reduction in stomatal conductance should occur following the 100 ppm increase in global atmospheric CO_2 concentration over the past 100 years (see **Box 2**). The response of plant stomata to elevated atmospheric CO_2 has been studied for over 30 years now and some experiments support a decrease of up to 50% with doubling of atmospheric CO_2. When such stomatal conductance reduction functions are directly incorporated into land-surface models embedded within the larger climate models, ET significantly declined and global runoff increased to levels consistent with runoff observations. These climate models are now routinely used as 'earth simulators' for addressing potential CO_2-induced interactions between terrestrial ecosystems and climate. However, these significant conductance reduction explanation due to elevated atmospheric CO_2 are not entirely supported by recent results from Free Air CO_2 Enrichment (FACE) experiments, which are designed to investigate how elevated atmospheric CO_2 affects both leaf and whole-ecosystem biosphere–atmosphere exchange rates. Several studies have explored how leaf stomatal characteristics are altered by elevated atmospheric CO_2 (**Table 2**). In particular, these studies examined the phenotypic response of stomatal index (SI), stomatal density (SD), and stomatal aperture (AP) to rising atmospheric CO_2 in 15 species after 4 years exposure to a field CO_2 gradient (200–550 ppm) or within three FACE sites. Along the CO_2 gradient experiments, SI and SD showed no evidence of a decline to increasing CO_2, while AP decreased slightly. It appears that without evolutionary changes, SI and SD may not respond to atmospheric CO_2 in the field and are unlikely to decrease in future climates characterized by high CO_2.

The third hypothesis argues that the decrease in continental-scale ET over the past 100 years is related to the large-scale land-use change, with deforestation being the 'dominant' trend. It is well known that clearing forests for development or agricultural purposes decreases ET and thus increases surface runoff. Direct experimental evidence of the impact of land cover conversion on ET was explored from long-term eddy-covariance measurements carried out at the Duke Forest, near Durham, North Carolina, at three stands experiencing similar climatic and edaphic conditions (**Figure 1**, **Table 2**). These measurements demonstrate that the difference between P (same for all sites) and ET is smallest for a pine plantation (PP), followed by the second-growth mixed hardwood forest (HW), followed by an abandoned agricultural field (OF) that is harvested at least once annually to prevent woody encroachment (**Figure 2**). Interestingly, the maximum difference in $P - \text{ET}$ for this experiment was 180 mm year^{-1} over this 5-year period here, which is comparable with the reported globally averaged decrease in streamflow following the *afforestation* of grasslands, shrublands, or croplands (227 mm

Table 2 Reported changes in E, T, and related variables and their attributed causes

Scale of study	*Region of study*	*Variable(s) of interest*	*Change in variable(s)*	*Proposed cause of change*	*Source*
Continental	United States	Pan evaporation	+90–150 mm yr^{-1} from 1960 to 1990	Solar dimming	1
Continental	United States	Pan evaporation/E	110 mm yr^{-1}/No trend from 1960 to 1990	Complementary hypothesis	2
Ecosystem	Eastern TN	$E/T/g_s/g_c$	No Change/−10%/−44%/−14%	Elevated CO_2	3
Ecosystem	Central NC	T/g_c	+4%/No change	Elevated CO_2	4
Global	Global	g_s	−27% to 40% in herbaceous plants	Elevated CO_2	5
Ecosystem	NC, TN, NV	SI, SD, AP	No change/no change/slight decrease	Elevated CO_2	6
Global	Global	R_o	+227 mm yr^{-1} in afforested watersheds	Land-use change	7
Ecosystem	Central NC	E/T	−28%/+58%	Conversion from a grass field to a PP	8

Sources

1. Roderick ML and Farquhar GD (2002) The cause of decreased pan evaporation over the past 50 years. *Science* 298: 1410–1411.
2. Golubev VS, Lawrinmore JH, Groisman PY, Speranskaya NA, Zharavin SA, Menne MJ, Peterson TC, and Malone RW (2001) Evaporation changes over the contiguous United States and the former USSR: A reassessment. *Geophysical Research Letters* 28: 2665–2668.
3. Wullschleger SD, Gunderson CA, Hanson PJ, Wilson KB, and Norby RJ (2002) Sensitivity of stomatal and canopy conductance to elevated CO_2 concentration – Interacting variables and perspectives of scale. *New Phytologist* 153: 485–496.
4. Schafer KVR, Oren R, Lai C-T, and Katul GG (2002) Hydrologic balance in an intact temperate forest ecosystem under ambient and elevated atmospheric CO_2 concentration. *Global Change Biology* 8: 895–911.
5. Field CB, Jackson RB, and Mooney HA (1995) Stomatal responses to increased CO_2—Implications from the plant to the global scale. *Plant Cell and Environment* 18: 1214–1225.
6. Reid CD, Maherali, H, Johnson HB, Smith SD, Wullschleger SD, and Jackson RB (2003) On the relationship between stomatal characters and atmospheric CO_2. *Geophysical Research Letters* 30: 1983–1986.
7. Jackson RB, Jobbagy EG, Avissar R, Roy SB, Barrett DJ, Cook CW, Farley KA, le Maitre DC, McCarl BA, and Murray BC (2005) Trading water for carbon with biological sequestration. *Science* 310: 1944–1947.
8. Stoy PC, Katul GG, Siqueira MBS, Juang J-Y, Novick KA, McCarthy HR, Oishi AC, Uebelherr JM, Kim, HS, and Oren, R (2006) Separating the effects of climate and vegetation on evapotranspiration along a successional chronosequence in the southeastern U.S. *Global Change Biology* 12: 2115–2135.

TN, NC, and NV refer to the states of Tennessee, North Carolina, and Nevada, respectively.

Figure 1 Experimental setup for the afforestation experiment at the Blackwood Division of the Duke Forest, near Durham, North Carolina showing the tower location at the grass site (OF), pine site (PP), and hardwood site (HW).

year^{-1} globally, or ~38% on average). Note that this reduction in ET due to land cover conversion from PP to OF is on the order of 20%, which is much larger than the 6.8% increase in $\delta D/D$ resulting from a 1 °C warming (see **Box 1**).

Evapotranspiration at Local Scales: Knowledge Gaps and Why the Problem of its Quantification Persists

Why does ET, studied for over thousands of years, still pose unique challenges to contemporary hydrologists? The answer may lie in the basic laws that describe water movement from the soil to the atmosphere. Movement of water in the soil–plant–atmosphere system begins with water migrating from wetter to drier soil pores adjacent to the root system moving along potential energy gradients. Once water reaches and enters the root system through a patchy and heterogeneous root-membrane, water flows through

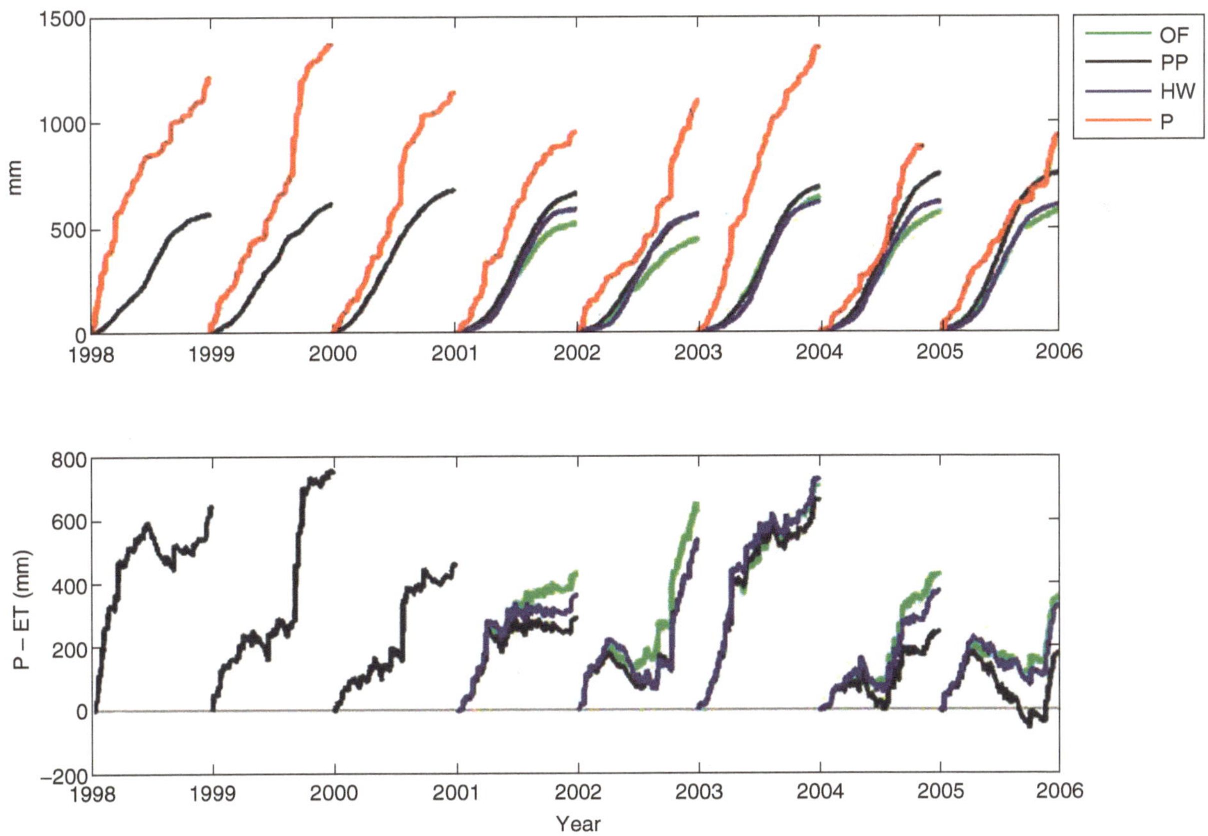

Figure 2 *Top*: Variations in eddy-covariance measured cumulative ET and *P* for the old-field (OF), pine plantation (PP), and hardwood forest (HW). *Bottom*: For reference, *P* – ET, a surrogate for water availability, is also shown.

a tortuous and complex network within the xylem. It experiences phase transition within the leaves, and exits to the atmosphere in the form of water vapor through patchy leaf stomata. The vapor molecules are then transported by turbulent eddies from within the canopy into the free atmosphere. The transporting energy and sizes of these eddies are partially determined by complex interactions among canopy attributes (e.g., leaf area and height), mesoscale forcing (e.g., geostrophic winds and weather patterns), and landscape heterogeneity. Resolving all spatial scales needed to describe the trajectory of water in the soil–plant–atmosphere system necessitates a three-dimensional simulation domain spanning 0.1 μm to tens of kilometers, equivalent to requiring $\sim(10^{10})^3$ computational nodes per time step. This time step must be sufficiently fine to resolve the fastest process, which is the action of viscous dissipation on turbulent fluctuations in the atmosphere ($\sim$0.001 s). This high 'dimensionality' in space and time is well beyond the capacity of any brute-force computation at present and in the foreseeable future. Furthermore, there are numerous insurmountable scale issues in attempting to relate water flow in the soil–plant system with its driving forces. For one, the constitutive laws now used to describe water movement in the soil, root, plant, and atmosphere systems do not share the same representative elementary volume (REV), defined here as the minimal spatial scale of representation for these laws. To elaborate further on these laws, we consider each of them separately for the three compartments of the soil–plant–atmosphere system:

1. *Soil*: Darcy's law, first proposed in 1856 and considered as one of the hallmarks of nineteenth century Earth sciences, describes the water flux, and when combined with the soil moisture conservation of mass equation (referred to as the continuity equation), leads to the so-called Richards equation. This equation describes water movement in unsaturated soils near the rooting-zone at an REV scale containing a sufficiently large number of pore spaces. Richards's equation is a nonlinear partial differential equation that provides a space–time description of water movement, but averages out variability of soil and root matrices at scales smaller than its REV. In soil physics, one of the major theoretical challenges to upscale this equation beyond the REV is how to include the effects of spatial heterogeneities in soil properties, macroporosity, and preferential flows of water and nutrient at discontinuities (e.g., large roots). Even the application of Darcy's law within the REV of a soil–system punctuated by complex rooting remains questionable. It is clear that the laws that describe water movement from the soil pores up to the rooting zone, proposed some 150 years ago, remain approximate.
2. *Plant*: Analogous problems arise at the plant level when describing water movement from the root to the leaf. Laminar flow equations (e.g., Hagen–Poiseulle's law for capillary tubes), based on the continuum assumption in fluid mechanics, are typically used to describe root and xylem water movement (or velocities). These assumptions are now being challenged by recent research in plant hydraulics. For example, prediction of the onset of embolism (cavitation) in the plant xylem requires microscale thermodynamic description of air and water micro-fluid dynamics not captured by macroscopic flow equations such as Poiseulle's law. The derivation of empirically measured embolism vulnerability curves for various plant organs from first principles has not yet been rigorously tackled and remains a topic of active research. Even the precise hydraulic pathways and connections between stomatal conductance and the plant–xylem system remain a subject of research awaiting novel experiments and theories.
3. *Atmosphere*: For the free atmosphere, being a single medium, the physical laws for mass, momentum, and energy exchanges are well described by the so-called Navier–Stokes equations, yet another hallmark of nineteenth century science. These are a set of nonlinear partial differential equations that are often described as the 'last frontier in classical mechanics.' They require detailed description of boundary conditions at the plant–atmosphere interface. Describing the boundary conditions for these equations remains complicated by stochasticity in the geometry and evolution at multiple scales. Randomness, beginning with patchiness at the stomatal level and progressing to patchiness in stomatal conductance, random leaf distribution, leaf area density, and onwards to the atmosphere must all be accounted for as dynamic boundary conditions. This boundary condition complexity does not diminish the complexity of solving these equations – even for simple static boundary conditions. The Navier–Stokes equations cannot be solved at all the necessary scales except in idealized cases.

Even when these constitutive equations (Richards equation, Poiseuille law, and Navier–Stokes equations) provide reasonable approximations at a particular scale, typically where microscopic heterogeneities can be averaged out within their respective REV, a major challenge remains in the derivation of effective parameters for simplified models at larger scales needed for addressing questions pertinent to the global and regional water cycle.

Conclusions

The importance of ET in sustaining the global and continental hydrologic cycle and the world's freshwater resources is rarely questioned in hydrology, meteorology, ecology, and soil science. Nonetheless, much uncertainty remains regarding the magnitude, and even the direction, of trends in continental-scale ET in the present day and over the course of the next century. Both state-of-the-art climate models and theoretical work indicate that global ET should increase in a warmer climate. Nonetheless, observational studies suggests that continental-scale ET may be increasing or decreasing as a result of a combination of forcings including warmer temperatures, decreased bulk conductance associated with rising CO_2 concentrations, and large-scale land-use change.

Attempts to resolve this uncertainty are challenged by the difficulty in integrating microscale processes, including water transport through soil pores and plant xylem, into a framework that can describe regional- and continental-scale patterns of ET. Constitutive laws such as Richards equation, Poiseulle's law, and the Navier–Stokes equations can describe water movement through soil, plants, and the atmosphere, respectively, under some circumstances. However, even these nineteenth century based hallmark equations, when applied to continental scales, challenge the computational limits. Given that the majority of fresh water available for use by both humans and ecosystems is governed by the difference between P and ET on regional to continental scales, it should be clear after this review that novel theoretical 'tactics' are needed to further the development of these constitutive laws and their upscaling for ET applications.

Acknowledgments

This study was supported by the US Department of Energy (DOE) through the Office of Biological and Environmental Research (BER) Terrestrial Carbon Processes (TCP) program (Grant nos. 10509–0152, DE-FG02–00ER53015, and DE-FG02–95ER62083), and by the National Science Foundation (NSF-EAR 0628342 and NSF-EAR 0635787).

Further Reading

Allen MR and Ingram WJ (2002) Constraints on future changes in climate and the hydrological cycle. *Nature* 419: 224–232.

Brutsaert W (1982) *Evaporation in the Atmosphere: Theory, History, and Applications 299*. Kluwer Academic Publishers.

Brutsaert W and Parlange MB (1998) Hydrologic cycle explains the evaporation paradox. *Nature* 396: 30.

Campbell GS and Norman JM (1998) *An Introduction to Environmental Biophysics*, pp. 286. Springer-Verlag.

Cox PM, Betts RA, Jones CD, Spall SA, and Totterdell IJ (2000) Acceleration of global warming due to carbon-cycle feedbacks in a coupled climate system. *Nature* 408: 184–187.

Easterling DR, Meehl GA, Parmesan C, Changnon SA, Karl TR, and Mearns LO (2000) Climate extremes: Observations, modeling, and impacts. *Science* 289: 2068–2074.

Gu L, Baldocchi DD, Wofsy SC, Munger JW, Michalsky JJ, Urbanski SP, and Boden TA (2003) Response of a deciduous forest to the Mount Pinatubo eruption: Enhanced photosynthesis. *Science* 299: 2035–2038.

Jury WA, Gardner WR, and Gardner WH (1991) *Soil Physics*, pp. 328. Wiley.

Katul GG, Porporato A, and Oren R (2007) Stochastic dynamics of plant–water interactions. *Annual Review of Ecology, Evolution, and Systematics* 38: 767–791.

Milly PCD, Dunne KA, and Veccia AV (2005) Global pattern of trends in streamflow and water availability in a changing climate. *Nature* 438: 347–350.

Parlange MB, Eichingwer WE, and Albertson JD (1995) Regional-scale evaporation and the atmospheric boundary layer. *Review of Geophyics* 33: 99–124.

Peterson TC, Golubev VS, and Grolsman PYa (1995) Evaporation losing its strength. *Nature* 377: 687–688.

Ramanathatn V, Crutzen PJ, Kiehl JT, and Rosenfeld D (2001) Aerosols, climate and the hydrological cycle. *Science* 294: 2119–2124.

Walter MT, Wilks DS, Parlange JY, and Schneider RL (2004) Increasing evapotranspiration from the Coterminous United States. *Journal of Hydrometeorology* 5: 405–408.

Vadose Water

J R Nimmo, U.S. Geological Survey, Menlo Park, CA, USA

Published by Elsevier Inc.

Introduction

The term vadose is derived from the Latin *vadosus*, meaning 'shallow.' In this sense, however, it refers to shallow depths beneath the land surface, not shallow portions of surface water bodies. The vadose zone is frequently called the unsaturated zone, and sometimes the zone of aeration, as its pore space usually contains air as well as water. The vadose zone extends from the land surface to the water table (the lowest water table if there is more than one).

Prediction of the transport rates of water and other substances within the vadose zone is critical to infiltration, runoff, erosion, plant growth, microbiota, contaminant transport, aquifer recharge, and discharge to surface water. Vadose-zone flow is fundamentally complicated by nonlinearity and hysteresis of unsaturated hydraulic properties, and extreme sensitivity to materials and hydraulic conditions. There is much variety in its natural constituents: soils, rocks, water, air, plants, animals, and microorganisms. Modern hydrology must consider interactions not only among these constituents themselves, but also with a wide variety of contaminants, including pesticides, fertilizers, irrigation wastewater, manure, sewage, toxic chemicals, radioactive substances, bacteria, mine wastes, and organic liquids.

The porous medium of the vadose zone is typically soil (**Figure 1**), but may also be porous rock, or any other material that occurs near the earth's surface. In general, the forces of molecular attraction are greater between solid and water than between solid and air. Consequently, water behaves as the wetting phase, and air as the nonwetting phase. Within the pores, water tends to cling to solid surfaces in films and in partially filled pores with curved air–water interfaces (**Figure 2**).

Fundamental Processes of Vadose Water

Unsaturated Hydrostatics

Water content The most basic measure of the water is volumetric water content, often symbolized θ, defined as the volume of water per bulk volume of the medium.

The most standard measurement of θ is the gravimetric method. The procedure is to dry a sample of the porous medium in an oven until the weight is constant, and then calculate how much water was in the soil from the difference between the dry weight and the initial wet weight. Other methods in widespread use have advantages, such as being less disruptive. Neutron scattering is commonly used for monitoring θ as a function of depth in the field. This method is based on the high effectiveness of water, among the various components of the wet soil, in slowing neutrons. Commercially available equipment has a neutron source and detector housed in a cylindrical probe that can be lowered to various depths in a lined hole. Another way to monitor θ in the lab or field is by measurement of the dielectric constant of the medium, usually by time-domain reflectometry (TDR). For most applications, TDR electrodes in the form of metal rods are inserted into the soil. Liquid water has a much higher dielectric constant than do other vadose zone constituents, so a measurement of this property can indicate the amount of water present within the volume sensed. A less common method is to measure electrical conductivity, which increases with θ. This principle can be applied tomographically for observing two- or three-dimensional details of changing water distributions in the field. The reflective or absorptive behavior of ground penetrating radar can also be used directly or tomographically to indicate water content distributions.

Water pressure and energy Matric pressure, usually symbolized ψ, may be thought of as the pressure of the water in a pore of the medium relative to the pressure of the air, in other words, the pressure difference across an air–water interface. The liquid–solid attraction that curves the air–water interfaces also causes capillary rise (**Figure 3**). Water that has risen to higher levels in the tube, because of the attraction of the tube walls for water, is at a lower pressure than the bulk water outside the tube. The narrower the tube, the stronger is the capillary effect. Similarly, in an unsaturated porous medium, water is generally at lower pressure than the air, so ψ is negative, the concave side of air–water interfaces is toward the air.

The most direct measurement of ψ is by a tensiometer. In firm contact with the porous medium, this device allows for equilibration of pressure between the water in unsaturated pores and the water in a

Figure 1 Particles and pores of a silt loam soil, scanning electron micrograph of a thin section. Reproduced from Lebron I, Schaap MG and Suarez DL (1999) Saturated hydraulic conductivity prediction from microscopic pore geometry measurements and neural network analysis. *Water Resources Research* 35: 3149–3158, with permission from the American Geophysical Union (http://www.agu.org/pubs/copyright.html).

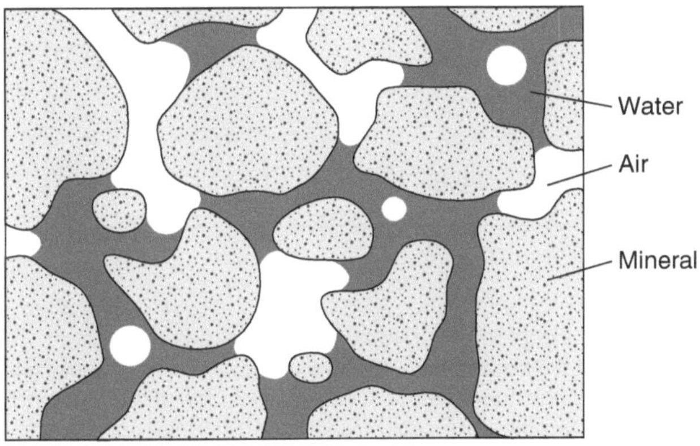

Figure 2 Pore space of a hypothetical unsaturated medium, illustrating the arrangement of solid, liquid, and gaseous phases. Few intergrain contacts appear in these figures because such contacts are essentially points in three-dimensional space, and mostly do not lie in the two-dimensional plane of the image.

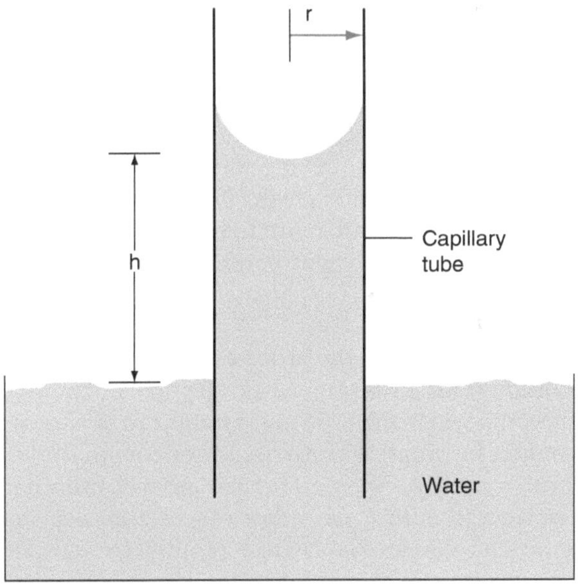

Figure 3 Capillary rise in a tube.

small chamber where a gauge or transducer reads the pressure (**Figure 4**). Its use is limited to relatively wet soils.

Other methods are available for relatively dry media and for easier application when less accuracy is acceptable. Some of these are based on the humidity of the air in soil pores. A low (strongly negative) ψ increases the pore water's effectiveness for absorbing water molecules out of the vapor in the soil air, resulting in a lower relative humidity. The effect is slight, however; a 0 to −15 atm range in ψ corresponds to a 100 to 99% range in relative humidity. Another class of methods uses an intermediary porous medium of known retention properties, typically gypsum blocks, nylon fabric, or filter paper. This medium is placed in contact with the medium to be measured so that ψ becomes equal in both. Then the water content of the intermediary medium is measured by other means (usually electrical conductivity, thermal diffusivity, or mass) and translated into a matric pressure using the known properties.

Water retention Analogously to capillary rise, the smaller pores of a medium hold water more strongly than the larger pores do. To extract water from a

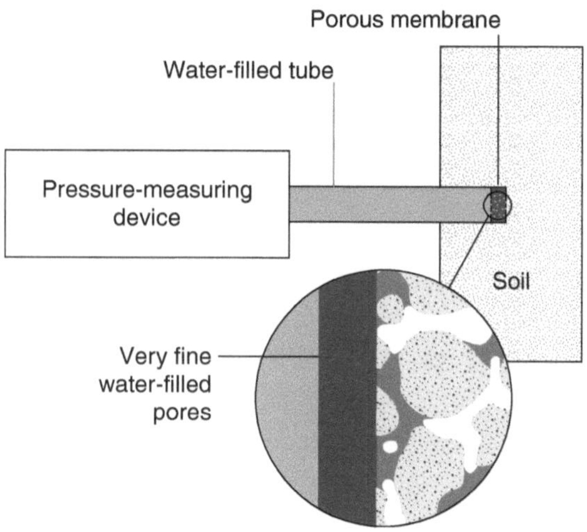

Figure 4 A tensiometer establishes a continuous water phase from the soil pores to a chamber for pressure is measured to indicate matric pressure.

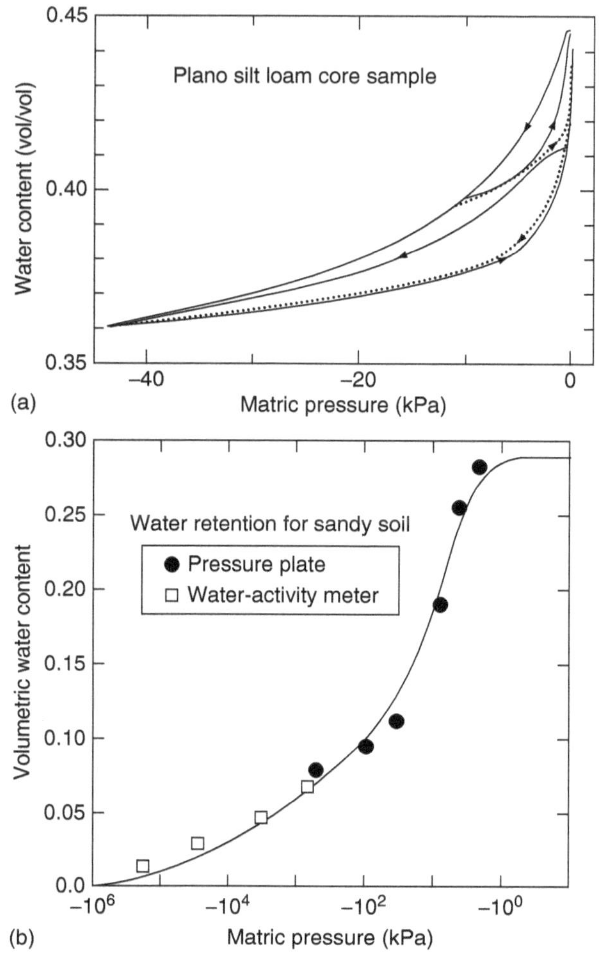

Figure 5 Water retention relations. (a) Hysteretic water retention of a silt loam soil, with arrows indicating the direction of change on each curve. Reproduced from Nimmo JR and Miller EE (1986) The temperature dependence of isothermal moisture vs. potential characteristic of soils. *Soil Science Society of America Journal* 50: 1105–1113, with permission from the Soil Science Society of America. (b) Drying retention curve for a sandy soil from the Amargosa Desert Research Site. The points are measurements by two different methods and the smooth curve is a fit of the model of Rossi and Nimmo (1994). Reproduced from Andraski BJ (1996) Properties and variability of soil and trench fill at an arid waste-burial site. *Soil Science Society of America Journal* 60: 54–66, with permission from the Soil Science Society of America.

small pore requires application of a more highly negative matric pressure. The volume of water held in the soil at different matric pressures therefore depends on the pore-size distribution, and is a characteristic of a particular porous material. This property, known as soil–water retention, is expressible as a set of θ vs. ψ curves for a given medium. When ψ is close to zero, air–water interfaces are broadly curved, nearly all pores are filled, and θ is high. If ψ is much less than zero, the interfaces are more tightly curved, they can no longer span the largest pores, and the pores have less water in them. Thus, greater θ goes with greater (less strongly negative) ψ.

Normally, during the process of wetting a porous medium, some air is trapped as the pores surrounding it become water-filled. In soil brought to $\psi = 0$, it is common for trapped air to occupy one-tenth or more of the total pore space. Trapped air will eventually dissolve and diffuse away, but vadose-zone moisture conditions commonly change rapidly enough that the pore space always contains some amount of air. Consequently, for most retention curves, when $\psi = 0$, θ has a value less than the total porosity.

In a granular medium, the particle-size distribution, or texture, relates in some way to the pore-size distribution. Larger particles may have larger pores between them. In addition to texture, the structure of the medium, especially as related to such features as aggregation, shrinkage cracks, and biologically generated holes, substantially influences the retention curve.

Examples **Figure 5(a)** shows a retention curve for a core sample of a silt loam soil from an apple orchard. Water retention is hysteretic; θ for a given ψ is different when measured for drying and wetting, and in general depends on the wetting/drying history of the medium. Thus, there is not a unique curve but a family of curves. The drying retention curve in **Figure 5(b)** is far from linear and covers five orders of magnitude in ψ. This enormous range requires multiple measurement techniques. In most cases investigators measure and plot only a single curve from the family of possible curves, usually a drying curve, and over only a portion of the range, usually at the wet end.

Considering the drying of soil from saturation, θ in **Figure 5(b)** stays high until a particular ψ value where it starts to decline. That ψ is called the air-entry value. By the capillary hypothesis, it is assumed to have a nonzero value because the largest fully wet pore of the medium will stay filled until the air–water pressure difference exceeds in magnitude the equivalent ψ value of capillary rise. In natural media, the air-entry value is usually poorly determined, as the decline in θ with ψ starts gradually, beginning at ψ nearly equal to zero. Artificial porous media, however, can be made in such a way that many pores are close to the size of the largest pore, so that air-entry is a sharp and sudden phenomenon.

Practical significance The water retention relation is important in quantifying soil moisture dynamics, as discussed later. Another area in which it is important is in soil–plant–water relations. Often termed matric potential because it is representative of the energy state of the soil water, ψ indicates the work that must be done by plant roots to extract water from the unsaturated soil. Plants wilt from inadequate soil moisture not in direct response to low θ, but rather because at low θ, ψ is low. In soil that is too dry, ψ is so highly negative that a plant is incapable of overcoming this energy barrier. Typically, the minimum ψ is about -15 atm for agricultural plants, though much lower in some plants, especially those native to arid regions.

Measurement or estimation of water retention Any system that makes independent simultaneous measurements of θ and ψ can indicate the water retention relation. In addition, there are methods specifically intended to measure this property. Many of these methods use a porous membrane, often ceramic, to permit equilibration of water pressure between the porous medium on one side of the membrane with bulk water on the other, as in tensiometers (**Figure 4**). The pressure of this bulk water (and hence the pore-water pressure) is controlled, as is the air pressure in the medium, in order to control ψ. The pressure, or less commonly the volume of water, is adjusted through a planned sequence, and paired values of ψ and θ (one of them controlled and the other measured) represent the retention curve.

Because various nonhydraulic properties of a medium, especially particle-size distribution, correlate in some way with water retention but are considerably easier to measure, property-transfer models have been developed for estimating water retention from other properties. One broad class of such models is based on theoretical relationships between pore sizes and particle sizes. Models of this type may work reasonably well for sandy media. Another class of such models uses statistically calibrated pedotransfer functions. The basis for this type of model is not a principle like the correlation of pore and particle size, but rather a database of measured water retention and other properties for a wide variety of media. Given a medium's particle-size distribution and other properties such as organic matter content, a pedotransfer function can estimate a retention curve with good statistical comparability to retention curves of other media in the database whose nonhydraulic properties may be similar. Whatever the choice of model, however, without any retention measurements for the medium in question, it is usually impossible to know whether the model result is a good representation of the retention curve.

Empirical formulas for water retention In general, a water retention curve can be represented by measured data, interpolated as needed. It is often convenient, however, to express the curve as a parametric empirical formula. Among the most widely used empirical formulas are that of Brooks and Corey

$$\theta = (\theta_{\max} - \theta_{\min})\left[\frac{\psi}{\psi_b}\right]^b + \theta_{\min} \qquad [1]$$

where ψ_b, b, and $\theta_{\min}$ are fitted empirical parameters and $\theta_{\max}$ is the maximum value of θ, and that of van Genuchten

$$\theta = (\theta_{\max} - \theta_{\min})\left[\frac{1}{1+\left(\frac{\psi}{\psi_c}\right)^{\nu}}\right]^{\mu} + \theta_{\min} \qquad [2]$$

where ψ_c, ν, μ, and $\theta_{\min}$ are fitted empirical parameters. Fundamentally, $\theta_{\min}$ should equal zero, but a finite value is often used to improve the fit in the higher-θ portion of the curve. Equations as simplified as these cannot represent the precise form of the θ–ψ relation of a natural medium, though they may serve for various practical purposes.

Diffuse Unsaturated Flow

Traditionally, unsaturated flow is considered as a continuum in which the average behavior of water in many pores within a compact region of space (a representative elementary volume (REV)) represents the characteristics of the medium point-by-point. In general, this leads to a conceptualization of moisture varying systematically throughout the medium (**Figure 6**).

In conventional unsaturated flow theory, two types of factors determine water flux: driving forces (chiefly gravity and matric pressure gradients) and properties of the medium. The matric forces sometimes greatly

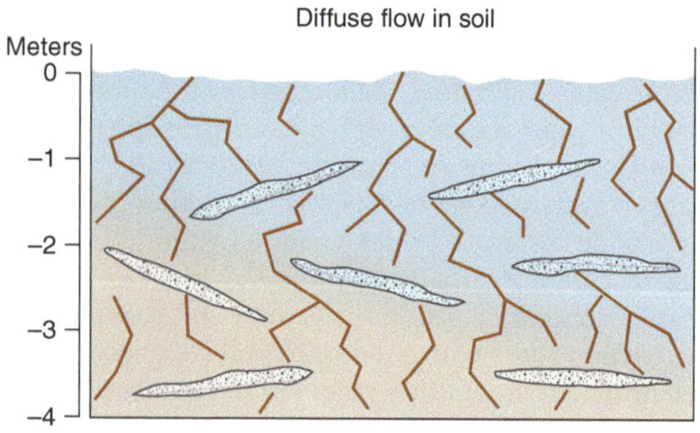

Figure 6 A diffuse distribution of water in soil. Blue shading indicates wetter soil over most of the upper half of this profile, especially to the right. Drier soil is lower and to the left. The transition between wet and dry soil is gradual and spread over a considerable volume.

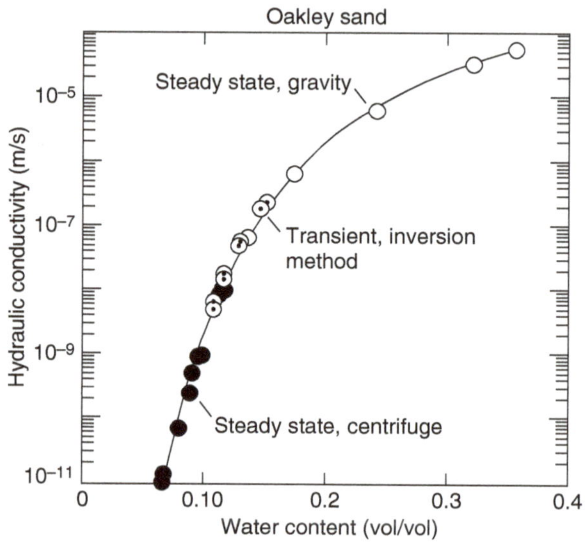

Figure 7 Hydraulic conductivity for a sandy soil (Oakley sand), measured by three methods.

exceed the gravitational force. Other forces may also drive flow under some conditions, as when temperature or osmotic gradients are significant.

Darcy's law for vadose water Unsaturated flow has its basic mathematical expression in Darcy's law, in a form such as

$$q = -\frac{K(\theta)}{\rho g}\left[\frac{\mathrm{d}\psi}{\mathrm{d}z} + \rho g\right] \quad [3]$$

where q is the flux density, K is the unsaturated hydraulic conductivity, ρ is the density of water, g is the acceleration of gravity, and z is upward distance. The conversion factor $1/\rho g$ is shown here explicitly so that this expression can be used directly with ψ in SI pressure units (kPa), and K in velocity units (m/s). In head units, ψ takes dimensions of length.

Unsteady diffuse flow In the general case of transient (nonsteady) unsaturated flow, the flow itself causes θ to change throughout the medium, which leads to continuously changing hydraulic conductivity and driving forces. These interacting processes can be accommodated mathematically by combining the equation of continuity

$$\frac{\partial\theta}{\partial t} = -\frac{\partial q}{\partial z} \quad [4]$$

with Darcy's law [3] to get Richards' equation, which for one-dimensional vertical flow within a medium in earth gravity can be written as

$$C\frac{\partial\psi}{\partial t} = \frac{1}{\rho g}\frac{\partial}{\partial z}\left[K\frac{\partial\psi}{\partial z}\right] + \frac{\partial K}{\partial z} \quad [5]$$

where C is the differential water capacity, a property of the medium defined as $\mathrm{d}\theta/\mathrm{d}\psi$. It is also possible to formulate this equation in terms of θ rather than ψ. In general the equation can be solved numerically.

Unsaturated hydraulic conductivity K of the medium depends on the whole set of filled pores, especially on their size, shape, and connectedness. In unsaturated media, as illustrated by the measurements in **Figure 7**, K depends very strongly on θ. Because the large pores empty first as θ decreases, the result is not only that fewer pores are filled to conduct water, but the remaining filled pores are smaller and therefore less conductive. With fewer pores filled, the paths of water flowing though the medium also become more tortuous. When the soil is quite dry, very few pores are filled, and the water moves mainly through poorly conducting films adhering to particle surfaces. The net effect of these factors is to reduce hydraulic conductivity by several orders of magnitude as the soil goes from saturation to typical field-dry conditions.

Measurement or estimation of unsaturated K The most accurate measurements of hydraulic conductivity are by steady-state methods. One technique is to establish constant (though not necessarily equal) pressures of water at two opposing faces of a porous medium, measure the flux density, and calculate K using Darcy's law. Another is to force water through the medium at a constant and known flux density, which lets ψ become uniform in part of the sample, then to compute K from the known flux density and force of gravity. With gravity as the main driving force, steady-state measurements are possible only for the high K values of fairly wet soil. Centrifugal force makes possible the accurate measurement of K at low θ.

Many techniques for measuring unsaturated hydraulic conductivity use unsteady flow. One of these is the instantaneous-profile method, useful in both laboratory and field. This method uses measurements of θ and ψ within a medium in which unsteady flow has been established, so that both the flux density and the ψ gradient can be computed at one or more instants of time. Another alternative for laboratory applications uses flow driven by evaporation. There are various indirect and inverse methods – a wide variety of situations where data are available describing water flow over time can provide information for an estimation of K. The tension infiltrometer method is in widespread use for field applications. This method uses the measured infiltration rate as a function of time for water applied at controlled ψ values to calculate the unsaturated hydraulic properties. It is often implemented as an inverse method.

Property-transfer models can be useful for estimating K. Usually these use water retention, not particle-size distribution, as the more easily measured type of data from which unsaturated K is calculated. If a transfer from particle size to K is needed, such a model may be combined with a water retention property-transfer model, though reliability is likely to be reduced because the particle-size distribution is less directly related to K.

Capillary theory provides an interpretation of the pores in the medium that relates to both retention and K. Models developed by Mualem and Burdine have become widely used for this purpose. A direct combination of an empirical formula for water retention, such as [1] or [2], into a capillary-theory formulation of unsaturated K can yield a convenient analytical formula for $K(\theta)$, and facilitate the combined treatment of water retention and unsaturated K.

Empirical formulas for unsaturated K As in the case of water retention, completely empirical formulas can represent unsaturated K. Gardner, for example, used

$$K(\theta) = A \exp(-\alpha\theta) \qquad [6]$$

where A and α are fitted empirical parameters. Such formulas have greater simplicity and sometimes lead to more realistic curve shapes than formulas developed for combined representation of K and water retention. The α parameter in [6] is used in developing and applying other models, such as analytical solutions of equations representing unsaturated flow.

Effects of dissimilar materials Layers that contrast in hydraulic properties impede vertical flow by various mechanisms. When water moves down from a coarse to a fine layer, as from coarse sand to silt, if both layers are near saturation, the fine layer has smaller hydraulic conductivity; therefore, flow slows when it reaches the fine layer. If, however, the coarse layer is nearly saturated but the fine layer is initially fairly dry, at first the flow may be temporarily accelerated while the flow is dominated by the sorptive nature of the fine medium, which tends to suck water out of the coarse material. When water moves down from a fine to a coarse layer it will also be impeded under many circumstances. Dry coarse material has an extremely small hydraulic conductivity; thus it tends not to admit water into the pores and exhibits a somewhat self-perpetuating resistance to flow. Water breaks into the coarse layer if the pressure at the layer contact builds to the point that the water-entry pressure (the minimum water pressure needed to fill an empty pore) of some of the large pores is exceeded. This can generate flow instabilities. Stable or not, water flow into the pores of the coarse medium increases that medium's hydraulic conductivity. With equal ψ values across the layer boundary, unsaturated K of the coarse layer is often less than that of the fine layer. In general, stable or diffuse flow through layers where fine overlies coarse is slower than it would be if both layers had the properties of the fine medium.

Preferential Flow

In recent decades it has become increasingly clear that much unsaturated-zone transport of importance, especially when water is abundant, occurs through a small fraction of the medium along preferential paths such as wormholes, fractures, fingers of enhanced wetness, and regions near contacts between dissimilar portions of the medium. This flow, for which accepted theory applies less well, occurs at rates typically some orders of magnitude faster than flow through the remainder of the medium. In many applications, its importance is redoubled because preferentially transported substances are exposed to only a small fraction of the soil or rock and only for limited time, reducing opportunity for adsorption or reactions.

Types of preferential flow Three basic modes of preferential flow (**Figure 8**) are (1) macropore flow, through pores distinguished from other pores by their larger size, greater continuity, or other attributes that can enhance flow; (2) funneled (or deflected or focused) flow, caused by flow-impeding features of the medium that concentrate flow in adjacent zones that are highly wetted and conductive; and (3) unstable flow, which concentrates flow in wet, conductive fingers.

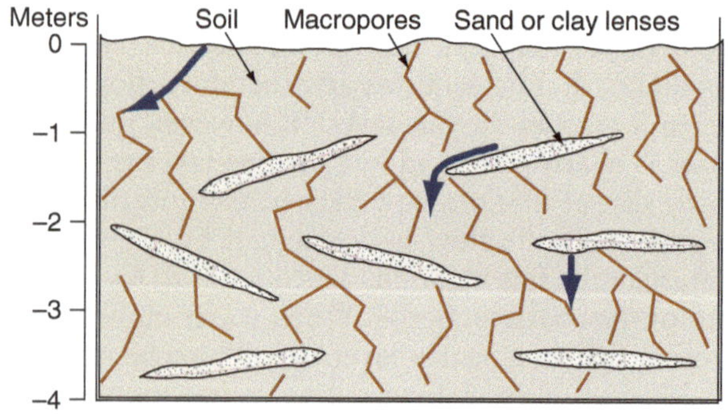

Figure 8 Three basic types of preferential flow. Arrows indicate narrow regions of faster flow than their surroundings. Macropore flow occurs through channels created by aggregation, biotic activity, or similar causes. Funneled flow occurs when flow is deflected by heterogeneities of the medium so as to create zones of higher water content and greater *K*. Unstable flow can be generated at layer boundaries such as the bottom of a sand lens at right, where flow into the lower layer moves in the form of highly wetted fingers separated by regions of relatively dry soil.

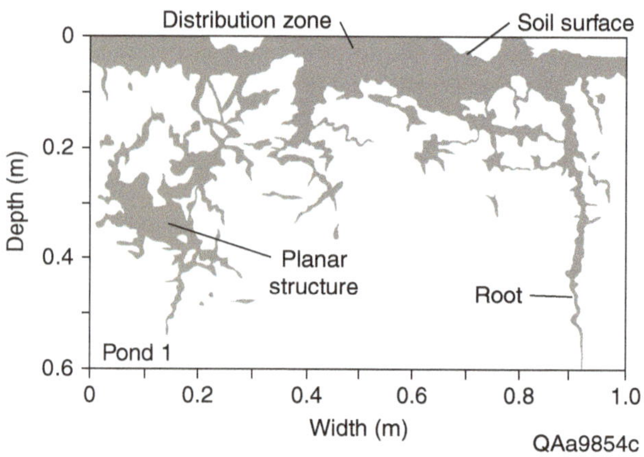

Figure 9 Macropore flow paths highlighted using a dye tracer. Reproduced from Scanlon BR and Goldsmith RS (1997) Field study of spatial variability in unsaturated flow beneath and adjacent to playas. *Water Resources Research* 33: 2239–2252, with permission from the American Geophysical Union (http://www.agu.org/pubs/copyright.html).

Common macropores include wormholes, root holes, and fractures (**Figure 9**). When macropores are filled with water, flow through them is fast. When they are empty, there may be essentially no flow through the macropores themselves though in some conditions film flow along macropore walls is significant. Macropores that are partly filled with water provide a variety of possibilities for the configuration and flow behavior of water.

Funneled flow commonly occurs with contrasting layers or lenses, where flow deflected in direction becomes spatially concentrated (**Figure 10**). The local increase in θ causes a corresponding increase in hydraulic conductivity and flux, and usually a change in the predominant direction of flow.

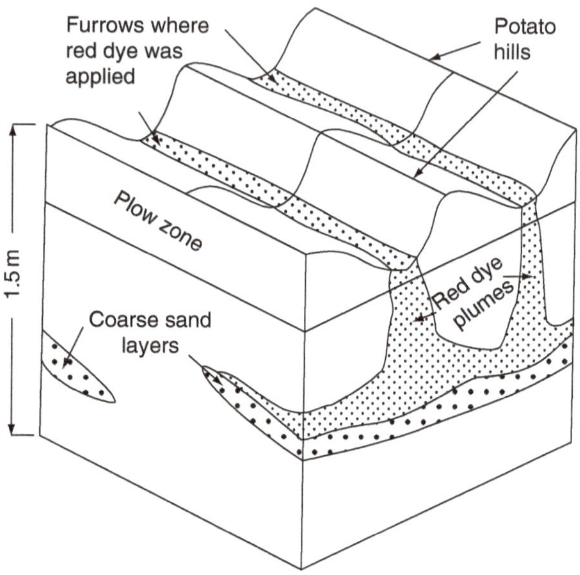

Figure 10 Funneled flow identified using a dye tracer. Reproduced from Kung KJS (1990) Preferential flow in a sandy vadose zone: 1. Field observation. *Geoderma* 46: 51–58, with permission from Elsevier.

Unstable variations in flow and water content, even within a uniform portion of the medium, can increase flow rates considerably. A typical case has a layer of fine material above the coarse material. Downward-percolating water builds up significantly at the interface, and breaks through into the coarse medium at a few points. The material near individual points of breakthrough becomes wetter and hence much more conductive. For some time thereafter, additional flow into the coarse material moves in the few fingers that are already wet (**Figure 11**). Between fingers, the medium can be relatively dry. In addition to textural contrasts, hydrophobicity (water repellency) and air trapping may cause flow instability.

Quantification of preferential flow One straightforward quantitative treatment is to represent preferential flowpaths with discrete conduits whose geometry, with appropriate laminar-flow expressions, predict the flow rate through the part of the medium they occupy. Usually this requires a statistical characterization of the set of conducting pathways, because the position, number, shape, orientation, and connectedness of the individual pathways are unknown.

Perhaps more widely used are various forms of equivalent-medium approach. The key assumption is that the effective hydraulic properties of a large volume of the medium that includes preferential pathways are equivalent to the average properties of a

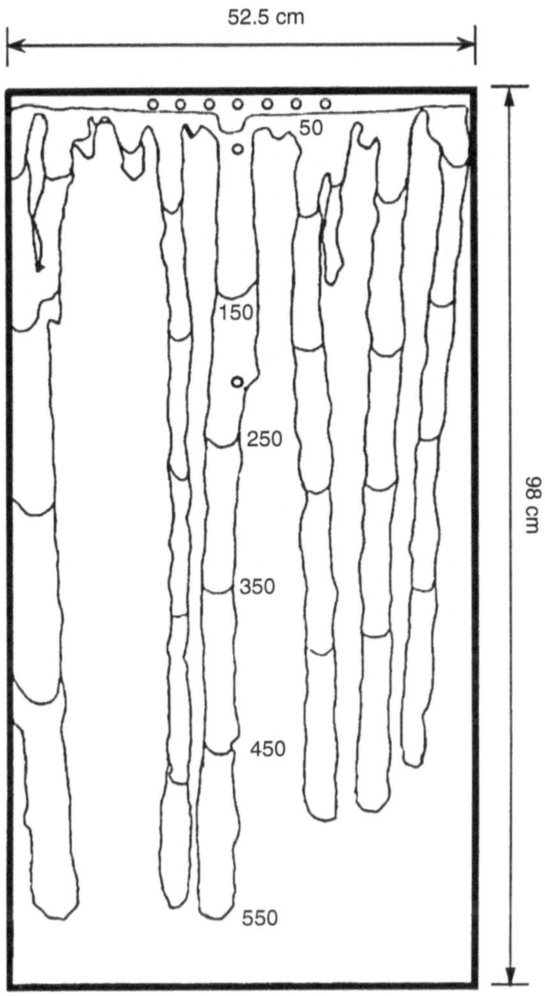

Figure 11 Fingers generated in unstable flow in a laboratory investigation, showing the extent of wetted soil as a sequence in time. Numerals on the diagram indicate the number of seconds since flow began. Reproduced from Selker J, Leclerq P, Parlange JY, and Steenhuis T (1992) Fingered flow in two dimensions 1. Measurement of matric potential. *Water Resources Research* 28: 2513–2521, with permission from the American Geophysical Union (http://www.agu.org/pubs/copyright.html).

hypothetical homogeneous granular porous medium. The effective hydraulic properties then can be applied directly in numerical simulators using Darcy's law and Richards' equation. A major advantage of such an approach is that the many existing theories, models, and techniques developed for diffuse flow in granular media can be applied to preferential flow. A major drawback is that preferential flow may deviate significantly from behavior describable using this type of medium, precluding reliable results. It is also a common practice to treat preferential flow differently from nonpreferential flow (often combined with a conventional Richard's equation for the matrix flow).

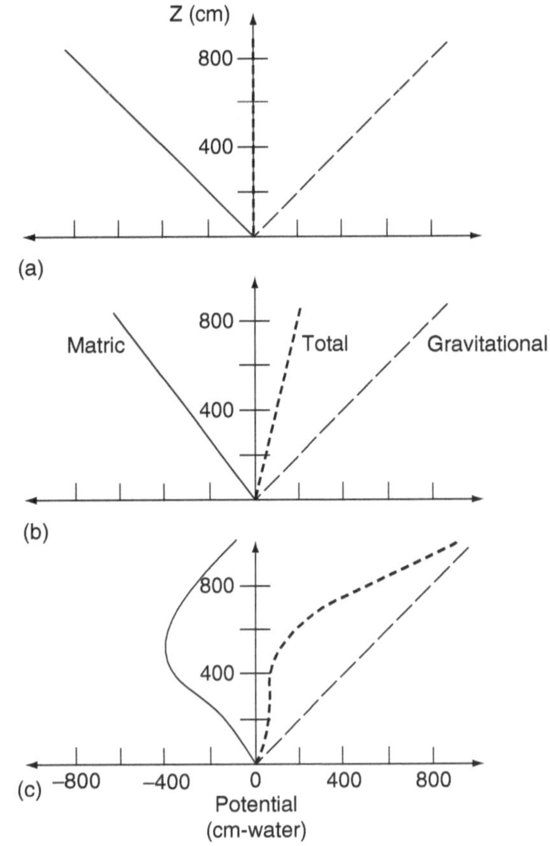

Figure 12 Profiles of matric, gravitational, and total potential for idealized situations of (a) static water, (b) steady downward flow, and (c) unsteady flow. The water table is at $z = 0$.

Vadose Water in the Hydrologic Cycle

Moisture State in the Vadose Zone

The water content and matric pressure within the vadose zone influence and in turn are influenced by conditions in the saturated zone and atmosphere. The distribution of vadose water at a given time also depends on the energy state (whose components include matric and gravitational potential), wetting/drying history, and dynamics of the water itself. If there is no flow, one can infer that the gradient of total potential is zero, so if the matric and gravitational components are the only significant ones, they add to a constant total potential. **Figure 12(a)** shows this type of hydrostatic profile for the case where a water table is present. Since the matric pressure in this case is linear with depth and θ is controlled by the water retention properties of the medium, for a uniform vadose zone, the θ profile (not shown here) mimics the shape of the water retention curve. Given time, approximate versions of such a hydrostatic profile may develop in portions of a profile where water movement is negligible. If water flows vertically downward at a steady rate in a homogeneous medium, the total gradient must be constant, but the matric pressure does not cancel out the gravitational potential, as illustrated in **Figure 12(b)**. In the general case of unsteady flow, the matric pressure profile cannot be determined so simply, and may take on an irregular form as in **Figure 12(c)**.

The uppermost part of the water distribution profile is sometimes described in relation to field capacity, defined as the water content of a soil profile when the rate of downward flow has become negligible 2 or 3 days after a major infiltration. This concept is used in agriculture to indicate the wettest soil conditions to be considered for plant growth, and sometimes is mentioned in hydrologic investigations related to soil moisture storage. The definition of field capacity requires some subjective judgment, for example, in deciding what flow is negligible. Field capacity is implicitly associated with the entire soil profile through the root zone, including preferential-flow characteristics and, especially, flow-retarding layers that enable layers above them to retain a high water content.

In a portion of the vadose zone immediately above the water table, it may happen that all pores are filled with water, held by capillary forces. The depth interval that is saturated but above the water table is called a capillary fringe. In a hydrostatic profile, this corresponds to a flat portion of the retention curve between saturation and an air-entry pressure. Some media do not have a significant capillary fringe because their retention characteristics have the air-entry pressure at essentially zero. Where the water table fluctuates, the hydrostatic equilibrium needed for a capillary fringe may take considerable time to establish. Soil–water hysteresis would make for a different capillary fringe with a falling water table than with a rising water table.

Moisture Dynamics in the Vadose Zone

Interactions at the land surface

Infiltration Infiltration is the downward movement of water through the land surface. If the soil is initially dry, ψ gradients may be the predominant downward driving force. When the soil is very wet to some depth, gravity may dominate instead. The usual case is that water infiltrates faster at the start and slows down as a zone of increased water content develops at the surface and expands. **Figure 13** shows actual infiltration rates varying over time in three soils. Mathematically, the decline of infiltration rate as the soil gets wetter is frequently represented by an inverse proportionality to the square root of time, as predicted by several models of infiltration. If water at the surface is abundantly available, but not under significant pressure, infiltration occurs at the infiltration capacity, a rate determined by the soil rather than the rate of application or other factors. If water arrives at the land surface faster than the infiltration capacity, excess water ponds or runs off. Like hydraulic conductivity, infiltration capacity is not single-valued for a given medium but varies with water content and other conditions. Conditions that complicate the ideal conception of infiltration include: variation of application rate with time, spatial variability of soil and surface properties, water repellency of the soil, air trapping, and variations of temperature.

Evapotranspiration The transport of water from soil through plants to the atmosphere (known as

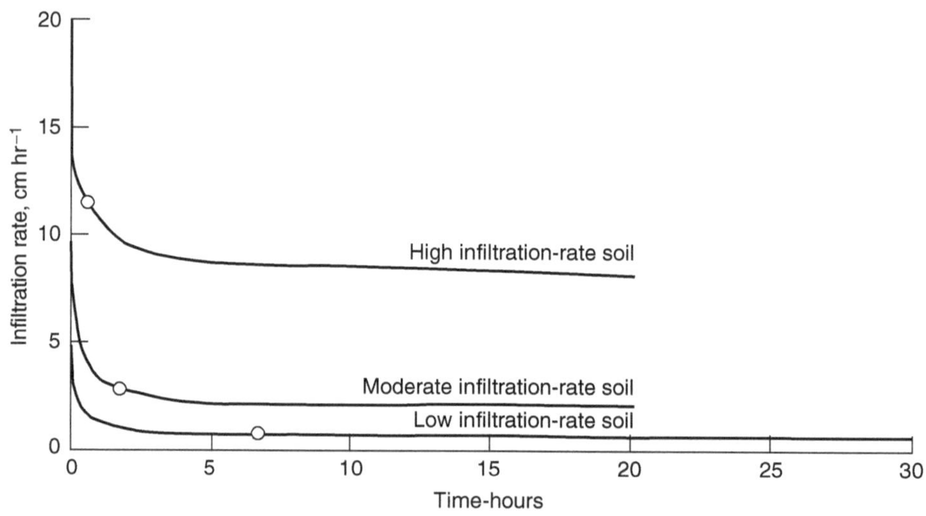

Figure 13 Measured infiltration rates over time for three different soils. Reproduced from Swarner LR (1959) *Irrigation on western farms*. Agriculture Information Bulletin 199, U.S. Department of Interior and U.S. Department of Agriculture: Washington, DC.

transpiration) and the direct transport from soil to atmosphere (known as evaporation) together constitute evapotranspiration. When the soil is wet enough, atmospheric conditions control the evaporation rate. When the soil is too dry to supply water at the maximum rate the atmosphere can absorb, the soil properties will control the evaporation rate. Thus there are at least two cases to consider: the atmosphere-dominated 'constant-rate' phase during which the transport mechanisms of the soil are ignored, and the soil-dominated 'declining-rate' phase, during which atmospheric effects are ignored. On vegetated land, transpiration typically far exceeds evaporation. Capillary forces can draw water up from the water table to depths from which it supplies the process of evapotranspiration, which can be a substantial loss mechanism from a water-table aquifer, especially where the vadose zone is thin.

Redistribution of infiltrated water After water has infiltrated, it redistributes, driven by gravity, matric pressure gradients, and possibly other forces. **Figure 14** illustrates θ distributions at various times during and after infiltration, in a mechanically

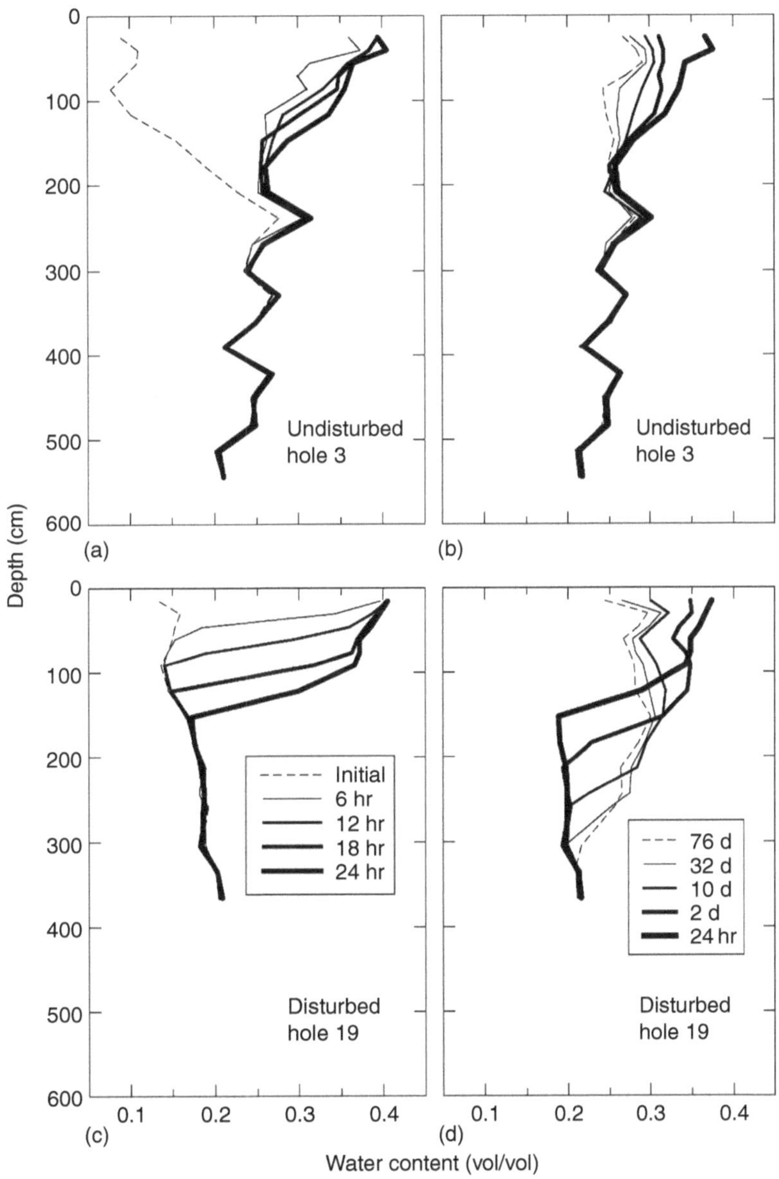

Figure 14 Measured water distributions during and after 24 h of flood infiltration in (a, b) an undisturbed soil on the Snake River Plain in Idaho and (c, d) nearby soil that was disturbed by temporary removal and replacement. Evaporation was inhibited by an impermeable cover at the land surface. Reproduced from Nimmo JR, Shakofsky SM, Kaminsky JF, and Lords GS (1999) Laboratory and field hydrologic characterization of the shallow subsurface at an Idaho National Engineering and Environmental Laboratory waste-disposal site. Water-Resources Investigations Report 99–4263, U.S. Geological Survey: Idaho Falls, Idaho.

disturbed soil and in a soil with intact natural structure. Redistribution continues until all forces balance out. Equivalently, the water may be considered to progress toward a state of minimal (and uniform) total energy of the earth–water–air system, i.e., equilibrium.

Normally hysteresis strongly influences redistribution because a wetting front progresses downward according to the wetting curves of water retention and conductivity, whereas θ in the upper portions of the wetted zone decreases according to the drying curves. Because a drying retention curve has greater θ for a given ψ, water contents remain higher in the upper portions than they would if there were no hysteresis. Thus one important consequence of hysteresis is to hold more water near the land surface where it is accessible to plants.

Usually, the above considerations need to be adjusted or reinterpreted with attention to preferential flow. Qualitatively, a major effect of preferential flow is to permit more rapid movement of water to significant depths. This would occur primarily under very wet conditions, and would be followed in the redistribution process by a slower flow of water into the regions between preferential flow channels.

A common phenomenon in layered media is perching, the accumulation of water in a region of the vadose zone to the point where it becomes saturated even though there is unsaturated material between that region and the water table. The high water content of a perched zone causes greater hydraulic conductivity and potentially faster transport through the three-dimensional system. The main effect is not a direct increase in vertical flow, though possibly in horizontal flow. New and different conditions may affect biological and chemical processes in a perched zone, e.g., reduced aeration.

A situation comparable to perching exists when a body of surface water has a vadose zone underneath it (**Figure 15**). This may be caused by a flow-restricting layer at or beneath a lakebed or streambed. The situation may alternatively be thought of as a perched water body directly under the lake or stream. The key condition is that the impeding layer must reduce the downward flow rate to less than the saturated hydraulic conductivity of the layer immediately below it. Another way this can happen is as a transient response to ephemeral surface water, below which an unsaturated state may persist for some time after standing water has come into the depression or channel.

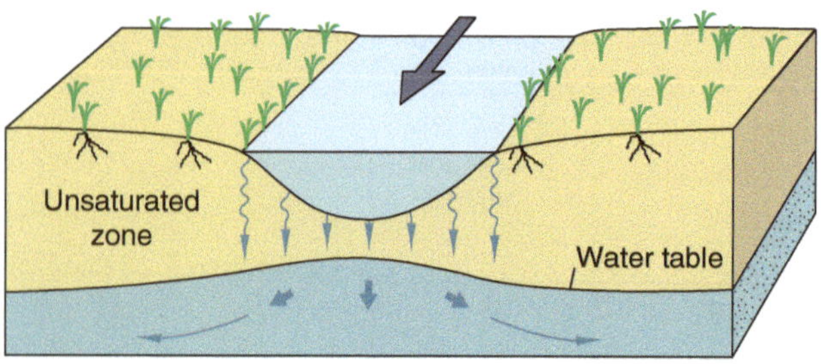

Figure 15 A stream, disconnected from the water table, so that interaction between surface water and the aquifer occurs through the unsaturated zone. Adapted from Winter TC, Harvey JW, Franke LO, and Alley WM (1998) Ground water and surface water – A single resource. Circular 1139, U.S. Geological Survey.

Aquifer Recharge

Aquifer recharge is water that moves from the land surface or unsaturated zone into the saturated zone. Quantitative estimation of recharge rate contributes to the understanding of large-scale hydrologic processes. It is important for evaluating the sustainability of groundwater supplies, though it does not equate with a sustainable rate of extraction. Where contamination of an aquifer is a concern, estimating the recharge rate is a first step toward predicting solute transport to the aquifer. Recharge may cause a short- or long-term rise of the water table. Artificial drainage, e.g., with horizontal porous pipes buried at a chosen depth, is sometimes used to maintain a minimal thickness of vadose zone for agricultural or other purposes.

Recharge rates vary considerably in time and space. Recharge often occurs episodically in response to storms and other short-term, high-intensity inputs. For a given amount of infiltration, temporal concentration enhances recharge because it entails shorter residence times for water in the portions of the soil from which evapotranspiration takes place. Similarly, a larger fraction will become recharge if it is concentrated in narrow channels such as fingers or macropores, not only because this tends to hasten its passage through the unsaturated zone, but also because the water then occupies less of the volume of soil from which evapotranspiration takes place.

Conclusion

The state and dynamics of vadose water are complicated by the interaction of multiple phases. At least three drastically different substances – water, air, and solid mineral – are critical to its nature and quantification. Unsaturated flow phenomena are extremely sensitive to the proportions of those phases, especially the fluid phases, as natural variations in the relative amounts of water and air can cause a property like hydraulic conductivity to vary over many orders of magnitude. When the flow of vadose water is diffuse in character, it can be treated quantitatively with Darcy's law adapted for unsaturated flow, and with Richards' equation. When it occurs within preferential pathways, there are various models, none yet generally accepted, to quantify the flow. The state and dynamics of vadose water control or contribute to a wide variety of processes within the hydrologic cycle, including infiltration, evapotranspiration, infiltration and runoff, and aquifer recharge.

See also: Atmospheric Water and Precipitation; Evapotranspiration; Ground Water; Hydrological Cycle and Water Budgets.

Further Reading

Bear J and Bachmat Y (1990) *Introduction to Modeling Phenomena of Transport in Porous Media.* Dordrecht, Netherlands: Kluwer.

Dane JH and Topp GC (2002) *Methods of Soil Analysis, Part 4 – Physical Methods.* Madison, WI: Soil Science Society of America.

Gardner WH (1986) Early soil physics into the mid-20th century. In: Stewart BA (ed.) *Advances in Soil Science,* vol. 4, pp. 1–101. New York: Springer-Verlag.

Germann PF and DiPietro L (1996) When is porous-media flow preferential? A hydromechanical perspective. *Geoderma* 74: 1–21.

Hillel D (1998) *Environmental Soil Physics.* San Diego: Academic Press.

Šimunek J, Jarvis NJ, van Genuchten MT, and Gärdenäs A (2003) Review and comparison of models for describing non-equilibrium and preferential flow and transport in the vadose zone. *Journal of Hydrology* 272: 14–35.

Jury WA and Horton R (2004) *Soil Physics.* New York: Wiley.

Nimmo JR (2005) Unsaturated zone flow processes. In: Anderson MG and Bear J (eds.) *Encyclopedia of Hydrological Sciences,* vol. 4, pp. 2299–2322. Chichester, UK: Wiley.

Nimmo JR (2007) Simple predictions of maximum transport rate in unsaturated soil and rock. *Water Resources Research* 43: W05426, doi:10.1029/2006WR005372.

Scanlon BR, Tyler SW, and Wierenga PJ (1997) Hydrologic issues in arid unsaturated systems and implications for contaminant transport. *Reviews of Geophysics* 35: 461–490.

Ground Water

W M Alley, U.S. Geological Survey, San Diego, CA, USA

Published by Elsevier Inc.

Introduction

Ground water occurs almost everywhere beneath the land surface and is an integral part of a complex hydrologic cycle that involves continuous movement of water on Earth. The widespread occurrence of potable ground water is a major reason for its use as a source of water supply worldwide. Furthermore, much of the world's food is produced by irrigated agriculture, which relies on ground water. Ground water plays a crucial role in sustaining streamflow during dry periods and is vital to many lakes and wetlands. In addition to human uses, many plants and aquatic animals depend upon the ground water that discharges to streams, lakes, and wetlands.

Ground water is a hidden resource. Information is gained at individual wells and spring locations, through indirect methods of measurement such as surface geophysics, and by measurements of flow and water level at hydrologically connected water bodies such as streams, lakes, and wetlands. These data are used to infer the occurrence, movement, and properties of ground water.

Aquifer Basics

Water-level measurement in a tightly cased well is a measurement of the hydraulic head (often simply referred to as head) in the aquifer at the depth of the screened or open interval of a well. Because the head represents the energy of water, ground water flows from locations of higher head to lower head. For simplicity, we use the terms water level and head interchangeably in this article.

Two general types of aquifers – unconfined and confined – are recognized (**Figure 1**). In unconfined aquifers, heads fluctuate freely in response to changes in recharge and discharge. Water levels measured in wells completed in the upper part of an unconfined aquifer help define the elevation of the water table, which is the top of the saturated zone. In confined aquifers, sometimes known as 'artesian' aquifers, water in the aquifer is confined under pressure by a geological body that is much less permeable than the aquifer itself. Many aquifers are intermediate between unconfined and confined conditions.

The removal of water by pumping sets up hydraulic gradients that induce flow from the aquifer to the well. The principal mechanism of aquifer drainage in response to pumping depends on whether an aquifer is unconfined or confined. Water drawn from storage in an unconfined aquifer comes from dewatering of the aquifer material by gravity drainage. Confined aquifers, which are filled by water under pressure, respond to pumping differently. The water for pumping is derived not from pore drainage but from aquifer compression and water expansion as the hydraulic pressure is reduced. Pumping from confined aquifers typically results in more rapid water-level declines covering much larger areas when compared with pumping the same quantity of water from unconfined aquifers. Drawdowns in water levels (heads) in the confined aquifer will induce leakage from adjacent confining units. Slow leakage over large areas can result in the confining unit supplying much, if not most, of the water derived from pumping. In some hydrogeologic terrains, the removal of water can cause compaction of fine-grained beds with resultant subsidence of the land surface. If water levels are reduced to the point where an aquifer changes from a confined to an unconfined condition (becomes dewatered), the source of water becomes gravity drainage as in an unconfined aquifer.

The equation that describes the movement of ground water within a porous medium is known as Darcy's law:

$$Q = -KAi,$$

where Q is the rate of flow or volume per unit of time $[L^3T^{-1}]$, K is the hydraulic conductivity $[LT^{-1}]$, A is the cross-sectional area $[L^2]$, and i is the hydraulic gradient (the change in head per unit of distance; dimensionless), which is a negative number in the direction of flow. The average linear velocity of the water can be calculated from Darcy's law by dividing by the cross-sectional area and effective porosity of the rock through which the water moves:

$$v = -Ki/n,$$

where v is the average linear velocity (referred to as velocity for the remainder of the text) of the ground water $[LT^{-1}]$ and n is the effective porosity (dimensionless).

The hydraulic conductivity, hydraulic gradient, and effective porosity are all important in determining the movement of ground water. The hydraulic conductivity, which represents the ability of the geologic framework to transmit water, is a property of the porous

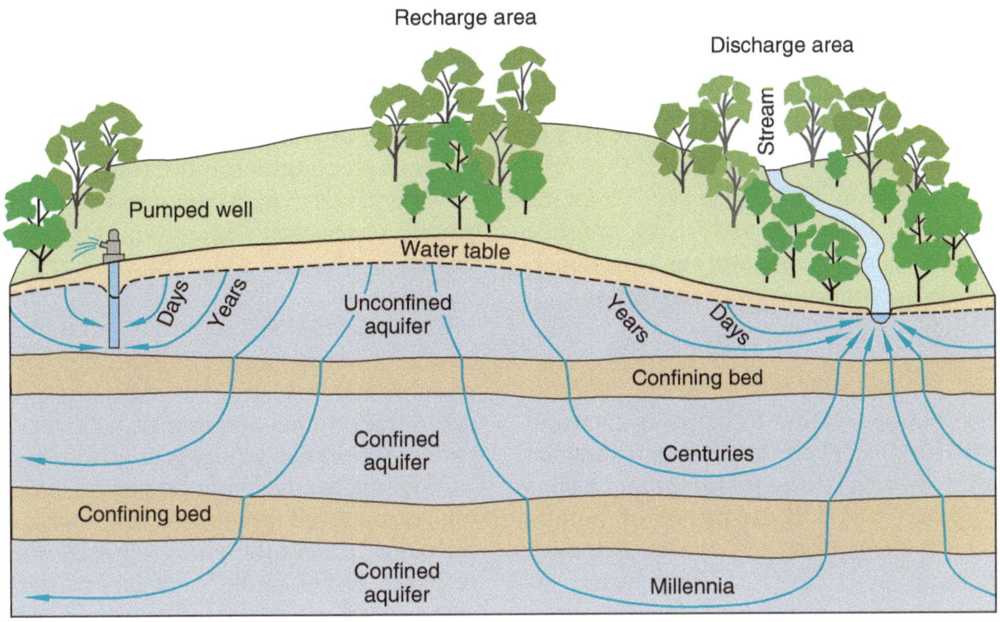

Figure 1 Ground water system showing generalized flow paths of groundwater movement and the relative age of the water since the time of recharge. Adapted from Heath RC (1983) Basic ground-water hydrology. U.S. Geological Survey Water-Supply Paper 2220.

medium and the fluid contained therein. The hydraulic conductivity of a groundwater system can vary over many orders of magnitude. The larger the hydraulic conductivity of a porous medium, the easier it is for water to flow through it. The porosity is the ratio of the volume of the voids divided by the total volume. The effective porosity is the volume of the voids that are interconnected and available for fluid transmission divided by the total volume. For ground water flow in cavernous karst terrain or in fractured-rock systems, the validity of Darcy's law, which was developed for porous media, may not be strictly applicable, and other methods for determining the velocity distribution may be required.

Groundwater Flow Systems

The three-dimensional body of Earth material saturated with moving ground water that extends from areas of recharge to areas of discharge (**Figure 1**) is referred to as a groundwater flow system (or more simply as a groundwater system). The areal extent of groundwater systems varies from a few square kilometers or less to tens of thousands of square kilometers. The lengths of groundwater flow paths range from a few meters to tens, and sometimes hundreds, of kilometers. A deep groundwater system with long flow paths between areas of recharge and discharge may be overlain by, and in hydraulic connection with, several shallow, more local, flow systems.

The age (time since recharge) of ground water varies in different parts of groundwater systems. The age of ground water increases along a particular flow path through the groundwater system from an area of recharge to an area of discharge. Travel times within groundwater systems can vary considerably (**Figure 1**). In shallow flow systems, ages of ground water at areas of discharge can vary from less than a day to a few hundred years. In deep, regional flow systems with long flow paths (tens of kilometers), ages of ground water may reach thousands or tens of thousands of years or more.

Fractured-rock systems in bedrock usually have smaller effective porosities than unconsolidated porous media such as sands and gravels, and flow velocities through fractured rock can be relatively fast. In more sluggish groundwater systems, long-term climate and geologic change need to be considered in understanding the movement of ground water over tens of thousands of years.

Tracer techniques have been applied widely to estimate the residence time of subsurface waters, as well as the amounts and timing of recharge and discharge. Most tracer techniques require knowledge (or assumption) of the time history of tracer input at the land surface or the water table. This temporal pattern is then correlated to a concentration–depth pattern in the subsurface at a point in time. Other approaches use information on decay products to determine age. Tracers can occur naturally (chloride, heat, the stable isotopes of hydrogen and oxygen in the water molecule), occur in the atmosphere as a result of human activities (tritium, CFCs), or be applied intentionally on the land surface (fertilizers, pesticides). Isotopes of elements, such as radon, dissolved from host rocks can also be used to estimate residence times and interactions with surface water. Over the past decade, research in age dating and tracking young ground water (<50 years old) using multiple tracers has resulted in major breakthroughs in understanding the dynamics of groundwater systems.

Recharge

Groundwater recharge includes recharge as a natural part of the hydrologic cycle and human-induced recharge, either directly through spreading basins or injection wells, or as a consequence of human activities such as irrigation and waste disposal. Artificial recharge with excess surface water or reclaimed wastewater is increasing in many areas, thus becoming a more important component of the hydrologic cycle.

Natural recharge to the water table can be diffuse or localized. Diffuse recharge is the widespread movement of water from land surface to the water table as a result of precipitation over large areas infiltrating and percolating through the unsaturated zone. Localized recharge refers to the movement of water from surface water bodies to the groundwater system and is less uniform in space than diffuse recharge. Most groundwater systems receive both diffuse and localized recharge. In general, the importance of diffuse recharge decreases as the aridity of a region increases.

Typically, most water from precipitation that infiltrates does not become recharge. Instead, it is stored in the soil zone and eventually returned to the atmosphere by evaporation and plant transpiration. The percentage of precipitation that becomes diffuse recharge is highly variable, being influenced by factors such as weather patterns, properties of surface soils, vegetation, local topography, depth to the water table, and the time and space scales over which calculations are made. Recharge to the water table can occur in response to individual precipitation events in regions having shallow water tables. In contrast, unsaturated zone water in some desert regions is estimated to have infiltrated the soil surface as long as tens of thousands of years ago.

Magnitudes of recharge fluxes are generally quite low and difficult to measure directly. Measurement of fluxes can be exacerbated by preferential flow (i.e., macropore flow) in the unsaturated zone, although preferential flow paths are of greatest concern as potential conduits for rapid contamination of aquifers. The above factors, in addition to temporal and spatial variability, greatly complicate the estimation of basin-wide recharge rates. Estimation methods include use of water budgets, tracers, geophysics, and simulation models. Because of the inherent uncertainties in any method, it is often advisable to apply multiple techniques for any study.

Water levels in many aquifers follow a natural cyclic pattern of seasonal fluctuation, typically rising during wetter, cooler months and declining during drier, warmer months. Superimposed on natural, climate-related fluctuations in groundwater levels are the effects of human activities that alter natural rates of groundwater recharge or discharge. For example, deforestation and draining of wetlands can expedite surface runoff and reduce groundwater recharge. Agricultural tillage, the impoundment of streams, and creation of artificial wetlands can increase groundwater recharge. Urbanization may either increase or decrease recharge. For example, built-up and paved areas promote runoff and inhibit infiltration. The enhanced runoff, however, may be channeled to a retention basin or infiltration gallery, resulting in relocation of recharge areas and the transition from slow, diffuse recharge to rapid, localized recharge. The effects of human-induced changes on groundwater recharge commonly are incremental, and the cumulative effects may not become evident for many years.

Groundwater Budgets

Water budgets are widely used to account for flow and storage changes in various hydrologic systems, including rivers, lakes, drainage basins, the land surface, and groundwater systems. In its most basic form, a water budget is expressed simply as

$$\text{Inflow} - \text{Outflow} = \text{Change in water storage.}$$

Each of the three main terms in the water-budget equation can include both natural (e.g., precipitation) and human-induced (e.g., imported water) components. Quantities in the budget apply to a predefined volume and can either be flow rates $[L^3T^{-1}]$ or volumes $[L^3]$ for a specified time period.

Understanding water budgets for groundwater systems is both critically important and challenging. Ground water divides (analogous to topographic divides on the land surface) do not necessarily coincide with surface water divides. Several aquifers may underlie the land surface at any given location with varying hydraulic properties and locations of recharge.

Under natural (predevelopment) conditions, a groundwater system is in long-term equilibrium. That is, averaged over some period of time (and in the absence of climate change), the amount of water entering or recharging the system is approximately equal to the amount of water leaving or discharging from the system.

Effects of Pumping on Groundwater Budgets

Withdrawal of ground water by pumping changes the natural (or predevelopment) flow system. Water that is withdrawn must be supplied by some combination of (1) more water entering the groundwater system (increased recharge), (2) less water leaving the system (decreased discharge), and (3) removal of water that was stored in the system. That is, the water pumped from the system must come from some change of flows and from removal of water stored in the system. An example of the change in groundwater budgets before and after development for the Gulf Coastal Plain aquifer system of the Southeastern United States is shown in **Figure 2**.

The amount of water coming from each of the potential sources to a discharging well changes through time until a new steady-state or equilibrium condition is established. **Figure 3** illustrates the sources of water in a simple idealized stream-aquifer system supplying one well. At the start of pumping, 100% of the water supplied to a well comes from groundwater storage. Over time, the dominant source of water to a well changes from groundwater storage to surface water. The source of water to a well can be either decreased discharge to the stream or increased flow (recharge) from the stream to the groundwater system. The streamflow reduction in either case is referred to as streamflow capture. The adjustments to pumping of an actual hydrologic system may take place over many years, decades or more, depending upon the physical characteristics of the aquifer, degree of connection between the stream and aquifer, and locations and pumping history of wells.

Similar effects occur with other surface water sources such as a lake, wetland, or estuary. The interactions of surface water bodies with groundwater systems are governed by the positions of the water bodies relative to the groundwater system, the characteristics of surface water beds and underlying geologic materials, and the climatic setting. These issues are discussed in more detail in the article on 'Ground water and Surface Water Interaction.'

Effects of Climate on Groundwater Budgets

Climate affects groundwater budgets in several ways. Most directly, climate influences the rates and distribution of recharge. Climate also affects human demands for ground water and affects plant transpiration from shallow ground water in response to changing inputs of solar energy and changing depths to the water table. Long-term droughts may be viewed as a natural stress on groundwater systems that in many ways have effects similar to groundwater withdrawals, namely, reductions in groundwater storage and accompanying reductions in groundwater discharge to streams and other surface water bodies.

Future climate variability and change may be important in determining ground water availability. For example, increasing temperatures in recent

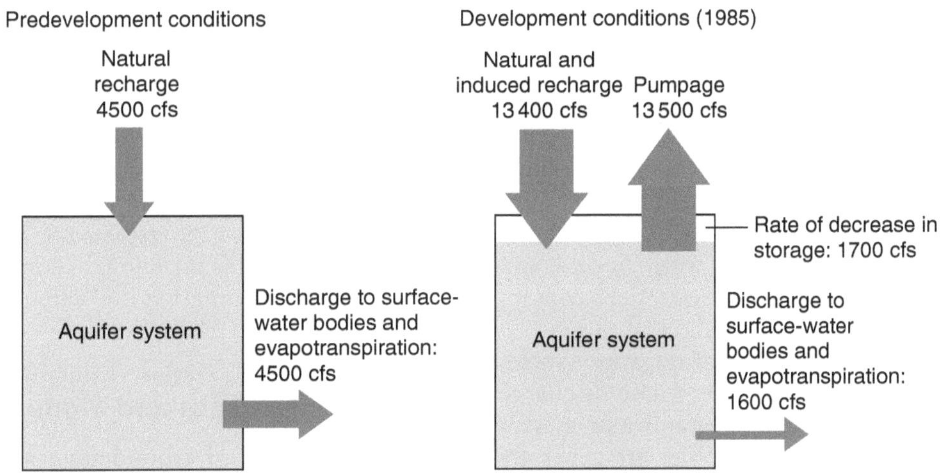

Figure 2 Groundwater budgets before and after development of the Gulf Coastal Plain aquifer system (all flows in cubic feet per second). The large withdrawals have been balanced by increases in recharge to the aquifer system and decreases in storage and discharge from the aquifer system. Adapted from Williamson AK and Grubb HF (2001) Ground-water flow in the Gulf Coast aquifer systems, South-Central United States. U.S. Geological Survey Professional Paper 1416-F.

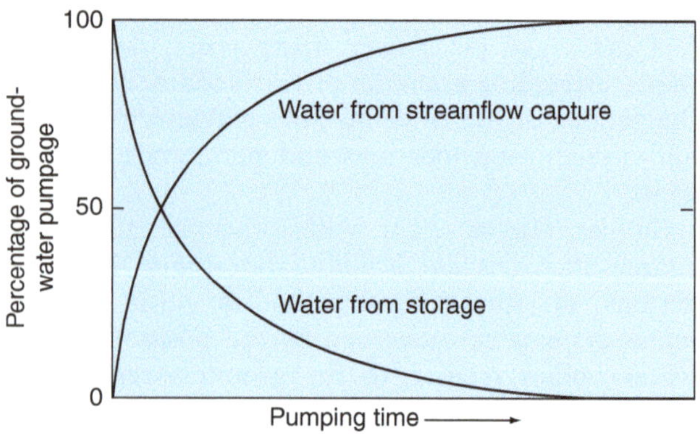

Figure 3 The principal source of water to a well can change with time from groundwater storage to capture of streamflow. Reproduced from Figure 14 in Alley WM, Reilly TE and Franke OL (1999) Sustainability of ground-water resources. U.S. Geological Survey Circular 1186.

decades in the mountains of the western United States have diminished the amount of precipitation falling as snow and caused earlier snowmelt runoff, which will likely change the quantity and distribution of recharge in the mountains as well as recharge downstream from mountain runoff. As climate changes, groundwater systems will respond accordingly, but the effects may take long periods of time to fully develop. Surficial aquifers that supply much of the water in streams, lakes, and wetlands are likely to be part of the groundwater system most sensitive to climate change.

Groundwater Quality

The sources of groundwater contamination are numerous and diverse. Contaminants can enter ground water through many routes, including downward percolation from land surface or shallow subsurface sources, direct entry of contaminants from the land surface through wells, crosscontamination between aquifers in wells open to more than one aquifer, flow of contaminated or saline water into freshwater aquifers as a result of pumping, interactions of ground water with surface water bodies, and interactions with geologic formations that contain natural contaminants, such as arsenic and radon.

Once contaminants have reached the water table, their movement to nearby surface water discharge areas or to deeper parts of the groundwater system is slow because groundwater velocities are generally low. Likewise, once parts of an aquifer are contaminated, the time required for a return to better water-quality conditions as a result of natural processes may be long after the original sources of contamination cease being active. Ground water quality remediation projects generally are very expensive and commonly only partly successful. In some settings, steep gradients caused by groundwater pumping can greatly increase the rate at which contaminants move to deeper ground water.

Contributing areas to wells often include surface water bodies, and increasing attention is being placed on surface water as a potential source of contamination to wells. Among the settings of greatest concern for contamination of ground water by streams are karst terrains where aquifers are hydraulically connected by sinkholes or other conduits that can channel river water directly into an aquifer with little or no filtration.

Because ground water commonly is a major component of streamflow, the quality of discharging ground water potentially can affect the quality of the receiving stream. Reductions in the quantity of ground water discharged to a stream as a result of pumping also may have consequences where this discharge significantly dilutes the concentration of contaminants introduced to streams from point sources and surface runoff. In many aquifers, large changes in chemical oxidation conditions, organic matter content, and microbial activity occur within a relatively thin (centimeter to meter scale) zone or interface between ground water and surface water. Thus, conditions near the interface between ground water and surface water can significantly affect the transport and fate of nutrients, metals, organic compounds, and other contaminants between the two resources.

The fresh groundwater resource is surrounded laterally and below by saline water. This is most evident along coastal areas where the fresh groundwater system comes into contact with the oceans, but it is also true in interior areas where deep saline water underlies the fresh water. Under natural conditions, the boundary between fresh water and saltwater maintains a stable equilibrium. The boundary typically is a gradation from fresh to saline water. When water is pumped from an aquifer that contains or is near saline ground water, the saltwater/fresh water boundary will move in response to this pumping. If the boundary moves far enough, some wells become saline, thus contaminating the water supply.

Groundwater Data and Models

An understanding of groundwater systems relies on data collection at various scales. Principal types of data commonly required are listed in **Table 1**. Some, such as precipitation data, are generally available and relatively easy to obtain at the time of a hydrologic

Table 1 Principal types of data and data compilations required for analysis of groundwater systems

Physical framework
Topographic maps
Geologic maps of surficial deposits and bedrock
Driller's logs
Results of surface and borehole geophysics
Hydrogeologic maps showing extent and boundaries of aquifers and confining units
Maps of tops and bottoms of aquifers and confining units
Saturated-thickness maps of unconfined aquifers
Average hydraulic conductivity maps for aquifers and confining units
Maps showing variations in storage coefficient for aquifers
Estimates of age of ground water at selected locations in aquifers
Hydrologic budgets and stresses
Precipitation data
Evaporation data
Streamflow data and measurements of gain and loss of streamflow between gaging stations
Maps of the stream drainage network showing extent of normally perennial flow, normally dry channels, and normally seasonal flow
Estimates of groundwater discharge to streams (baseflow)
Measurements of spring discharge
Measurements of surface water diversions and return flows
Quantities and locations of interbasin diversions
History and spatial distribution of pumping rates in aquifers
Well hydrographs and historical head (water-level) maps for aquifers
Location of recharge areas and estimates of recharge
Chemical framework
Geochemical characteristics of earth materials and naturally occurring ground water in aquifers and confining units
Spatial distribution of water quality in aquifers, both areally and with depth, and temporal changes in water quality
Sources and types of potential contaminants
Chemical characteristics of artificially introduced waters or waste liquids
Maps of land cover/land use at different scales, depending on study needs
Streamflow quality, particularly during periods of low flow

Source
Adapted from Table 2 in Alley WM, Reilly TE, and Franke OL (1999) Sustainability of ground-water resources. U.S. Geological Survey Circular 1186.

analysis. Other data and information, such as geologic and hydrogeologic maps, can require years to develop. Still other data, such as a history of water levels in different parts of groundwater systems, require foresight to obtain measurements over time, if they are to be available at all. Development of a comprehensive hydrogeologic database may include depths and thicknesses of hydrogeologic units from lithologic and geophysical well logs, water-level measurements to allow construction of predevelopment water-level maps for major aquifers as well as water-level maps at various times during development, ground water sampling to document changing water quality, and simultaneous measurements of streamflow and stream quality during low flows to indicate possible contributions of discharging ground water to surface water quality during the periods in which these contributions may have the greatest effects. Many types of data and data compilations listed in **Table 1** need to be viewed on maps. Thus, Geographic Information Systems (GIS) typically are an integral part of the database system to assist in organizing, storing, and displaying the substantial array of needed information.

Because ground water is hidden from view, scientists use idealized representations, or models, to gain insight into the behavior and interplay of various aspects of groundwater systems. Ground water data form the foundation of both conceptual and computer models of the groundwater system. A conceptual model is used to explain the extent of the groundwater system, the sources of water to the system, and the movement of water through the inferred hydrogeologic units. A computer model is a mathematical representation of the groundwater system based on the conceptual model and available data. The conceptual and computer models are subject to uncertainty and error and greatly simplify the real system. Nonetheless, these models are useful in that they synthesize information and assure consistency among aquifer properties, the water budget, and groundwater levels; can be used to estimate flows and other aquifer characteristics for which direct measurements are not available; can simulate the response of the aquifer system to different possible stresses; and can help identify areas where additional hydrologic information could improve understanding.

See also: Ground Water and Surface Water Interaction.

Further Reading

Alley WM, Healy RW, LaBaugh JW, and Reilly TE (2002) Flow and storage in ground-water systems. *Science* 296: 1985–1990.

Alley WM, Reilly TE, and Franke OL (1999) Sustainability of ground-water resources. U.S. Geological Survey Circular 1186. Also available at http://pubs.water.usgs.gov/circ1186.

Bear J, Cheng AH-D, Soreck S, Ouazar D, and Herrera I (eds.) (1999) *Seawater Intrusion in Coastal Aquifers – Concepts, Methods, and Practices.* Dordrecht, Netherlands: Kluwer Academic.

Cook PG and Herczeg AL (eds.) (1999) *Environmental Tracers in Subsurface Hydrology.* Boston, MA: Kluwer Academic.

Fetter CW (2001) *Applied Hydrogeology,* 4th edn. Upper Saddle River, NJ: Prentice-Hall.

Freeze RA and Cherry JA (1979) *Groundwater.* Englewood Cliffs, NJ: Prentice-Hall.

Heath RC (1983) Basic ground-water hydrology. U.S. Geological Survey Water-Supply Paper 2220.

Lohman SW (1972) Ground-water hydraulics. U.S. Geological Survey Professional Paper 708.

Todd DK and Mays LW (2005) *Ground-Water Hydrology*, 3rd edn. New York: John Wiley.

Winter TC, Harvey JW, Franke OL, and Alley WM (1998) Ground water and surface water – A single resource. U.S. Geological Survey Circular 1139. Also available at http://pubs.water.usgs.gov/circ1139.

Relevant Websites

http://water.usgs.gov/ogw/pubs.html – Selected U.S. Geological Survey ground-water publications.

http://www.groundwater.org/ – Groundwater Foundation.

http://www.agwt.org/ – American Ground Water Trust.

http://www.ngwa.org/ – National Ground Water Association.

Ground Water and Surface Water Interaction

H M Valett, Virginia Tech, Blacksburg, VA, USA
R W Sheibley, Washington Water Science Center, Tacoma, WA, USA

Introduction

Hydrologists have known for a long time that surface and ground waters are connected reservoirs involved in the hydrologic cycle. The hydrologic processes of infiltration, groundwater discharge, and recharge illustrate these linkages and emphasize the implicit connectivity between groundwater and surface water. Historically, the existence of streams has been argued to be a manifestation of ground water–surface water (GW–SW) interaction since they exist either as points of recharge or discharge. Hydrologic exchange occurs between aquifers and associated streams, rivers, floodplains, lakes, and wetlands.

This chapter reviews the basics of GW–SW exchange and how it influences the structure and function of aquatic ecosystems. Exchange is a component of ecosystem hydrology, defined as the study of how hydrologic fluxes influence the flow of energy and the exchange and transformation of materials in ecological systems. The chapter first addresses the multiple spatial and temporal scales that characterize exchange and provide a general framework for how GW–SW exchange has the potential to alter hydrologic, biogeochemical, and biological conditions within and among ecosystems. Next it addresses the basic methods and approaches employed to quantify GW–SW interaction. A topological approach then addresses GW–SW exchange and its ecological implications across streams, rivers, lakes, and wetlands. Finally, it briefly addresses human impacts on GW–SW interaction and needs for future research.

SW–GW Interaction: Scales and Effects

Aquifers are linked to above-ground bodies of water at multiple spatial and temporal scales that constrain and characterize the hydrologic character of GW–SW interaction. These concepts were clearly delineated in early work that recognized regional, intermediate, and local flow systems and characterized them by the temporal and spatial scales that link precipitation and recharge to groundwater discharge and channel runoff (**Table 1**). Many flow systems characterized by GW–SW interaction occur over smaller scales, are embedded within watershed-scale flow systems (**Figure 1**), and respond to topographic and geomorphic variation associated with bedforms and channel structure. This chapter focuses on these proximate flow systems ranging from 10s–100s of meters commensurate with most field studies of exchange that address how locations along flow paths organize ecological response.

Groundwater discharge can contribute all or most of the flow in streams and the resident volumes of lakes and wetlands. As such, GW–SW interaction is a critical hydrologic flux that creates and maintains aquatic habitats, including ponds and lakes, wetlands, headwater streams, and floodplain systems that include lakes, wetlands, and ponds. Such hydrologic exchange has dictated environmental conditions over evolutionary time frames, contributing to unique invertebrate groundwater fauna sensitive to the extent of GW–SW exchange. Thus, exchange acts as a main determinant of biotic habitat, a topic explored more fully later in this chapter.

At the ecosystem level, the direction of exchange is a critical aspect of GW–SW interaction because surface and groundwater environments often differ markedly in their physicochemical composition, including heat content, redox potential, chemical composition, and, potentially, pollutant load. Since surface water systems are characterized by turbulent flow, they are typically well oxygenated than subsurface environments where oxygen content is reduced by chemical and biological demand. Accordingly, GW–SW interfaces, or ecotones, are often characterized by steep gradients in concentrations of oxidized and reduced compounds, rendering them biogeochemical hotspots. Thus, the hydrology of proximate flow paths influences the magnitude and character of chemical loads and water residence times within ecosystems. Accordingly, GW–SW exchange has great potential to alter ecosystem processing and retention of solutes (**Figure 2**) while the spatial configuration of discharge and recharge influence responses by algae, aquatic plants, benthic macroinvertebrates, fishes, and water fowl that occupy surface environments in freshwater systems.

Measuring GW–SW Interaction: Methods and Approaches

There is a broad range of techniques available to address GW–SW exchange in freshwater ecosystems.

Table 1 Flow systems linking surface and ground water at multiple scales

Flow system	*Connection with surface recharge*	*Flow rates ($m\,day^{-1}$)*	*Oxygen status*	*Potential for transport from the surface*	*Example*
Regional	Virtually nonexistent	Almost stagnant	Anoxic/anaerobic	Virtually nonexistent	Deep basin/ petroleum reservoirs
Intermediate	Small	Slow 0.0001–0.001	Generally anoxic/ anaerobic	Moderate	Confined aquifer
Local	Extensive	Fast 0.001–0.01	Often oxic and aerobic	High	Water table aquifer
Proximate	Extensive	Extremely Fast 1–1000	Fully anoxic Fully oxic Mixed	Extremely High	Hyporheic zones, seepage lakes

Adapted from Chapelle FH (1993) *Ground-water microbiology and geochemistry.* New York: Wiley.

Here we provide an overview of the main approaches and emphasize that the determination of GW–SW exchange remains a major challenge owing to heterogeneities and the problems of integrating across scales of investigation.

Direct Measures of Exchange

Direct measures of water flow across the surface–subsurface boundary can be provided by seepage meters (**Figure 3(a)**). A seepage meter typically consists of a bottomless cylindrical chamber vented to a plastic bladder. Under conditions of groundwater discharge, flow from the substrate to the water column will be collected in the bladder at a measurable rate. Infiltration from the surface to subsurface may be measured as a loss of water purposefully placed into the bladder during chamber deployment. These types of seepage meters have been used extensively in lakes, estuaries, and reservoirs, and less frequently in streams.

Heat Tracer Methods

The differences in temperature between ground water and surface water can be used to assess GW–SW exchange. Groundwater temperatures are more stable throughout the year than surface water and mapping of temperature variation can reflect GW–SW exchange. In general, temperatures at the sediment–water interface in gaining reaches will be less variable than those in losing reaches. A time series of temperature in stream bed sediments can document the direction and extent of exchange. Downward flow results in the penetration of the zone of temperature variation while upward flow compresses the zone of variation. Thus, simple mapping of stream bed temperature can give semiquantitative assessment of the direction and magnitude of GW–SW exchange. More quantitative information can be obtained by applying heat transport equations that account for both conductive and convective transport. Using these models, seepage rates can be obtained from thermal profiles. Finally, heat balance equations can be used to address exchange at the reach scale. With this approach, stream temperature is the function of groundwater discharge rate, differences in stream and groundwater temperature, stream flow, and heat exchange from the stream surface. Thus, net heat (and water) flux from ground water is derived by difference.

Use of Darcy's Law

A number of methods rely on quantifying the components of Darcy's Law eqn. [1]

$$q = K\frac{v\mathrm{d}h}{\mathrm{d}l} \qquad [1]$$

where q = Darcy velocity ($\mathrm{L\,T^{-1}}$), K = hydraulic conductivity ($\mathrm{L\,T^{-1}}$), h is hydraulic head (L) and l is distance (L).

Hydraulic gradients can be measured using piezometers or mini-piezometers (**Figure 3(b)**). Piezometers are typically open-ended or screened pipes inserted into the sediments to measure hydraulic head at a particular depth (represented as the elevation of the water column within the pipe). The hydraulic head at depth can then be compared to the elevation of the free water surface to obtain information on the vertical component of GW–SW exchange. Differences in hydraulic head are normalized to piezometer depth to obtain the vertical hydraulic gradient (VHG). Positive values of VHG indicate groundwater discharge and negative VHGs are characterized by surface water flux into sediment (i.e., recharge).

Accurate measures of K are notoriously difficult to obtain owing to spatial heterogeneity and difficulties

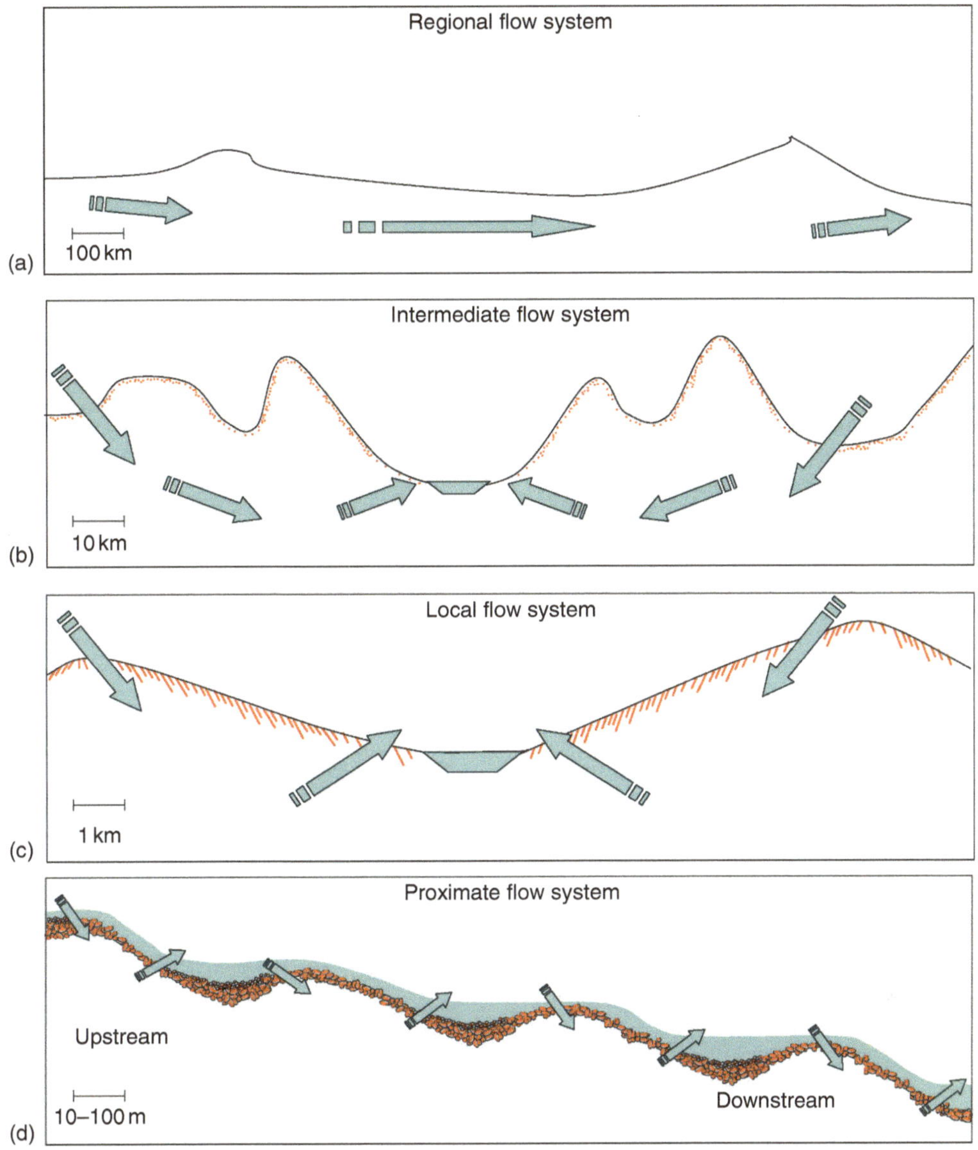

Figure 1 Flow systems linking surface and groundwater environments across multiple spatial scales. Regional flow systems are continental in scale. Intermediate flow systems include multiple topographic highs and lows while local flow systems are characterized by groundwater recharge at a topographic high and discharge at an adjacent low. Proximate flow paths occur within watershed boundaries in close proximity to surface water bodies where GW–SW exchange is promoted by interactions among groundwater potentials, surface water hydarulics, and channel or bed geomorphology. Modified from Toth J (1963) A theoretical analysis of ground water flow in small drainage basins. *Journal of Geophysical Research* 68(16): 4795–4812. © American Geophysical Union 1963. With permission from the American Geophysical Union.

associated with extrapolating point measures to larger spatial scales. Basic estimates of K can be derived from grain size analysis of stream bed sediments. Permeameters and piezometers can be used in conjunction with established equations to derive K. Other methods of assessing K involve pump tests and monitoring wells that provide information on a larger scale or sediment cores analyzed in the laboratory to provide much finer scale measures.

Mass–Balance Approaches

Mass–balance approaches rely on a budgetary approach to water flow along a defined reach or for a defined volume of water. Changes in the quantities of water are related to sources and sinks and the contribution of ground water can be identified and quantified. A combination of point measures of stream flow and dilution gauging (i.e., tracer dilution

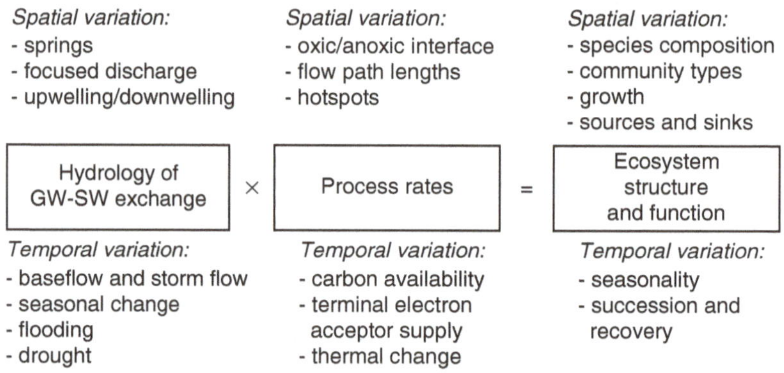

Figure 2 A general schematic model relating how hydrologic and both geochemical and biological processes dictate the influence of ground water–surface water exchange on various aspects of inland water environments. Determinants vary with space and time at multiple scales. Hydrologic influences include the magnitude, timing, and locations of interactions between ground water and surface water. Moreover, hydrologic variance dictates water residence times with implications for how process influence water availability and composition. Geochemical and biological processes alter water abundance and chemical quality and interact with hydrologic vectors to determine the net influence of exchange on physical, chemical, and biological aspects of inland aquatic ecosystems. Modified from Valett HM, Morrice JA, Dahm CN, and Campana, ME (1996) *Limnology and Oceanography* 41: 333–345. With permission from the American Society of Limnology and Oceanography.

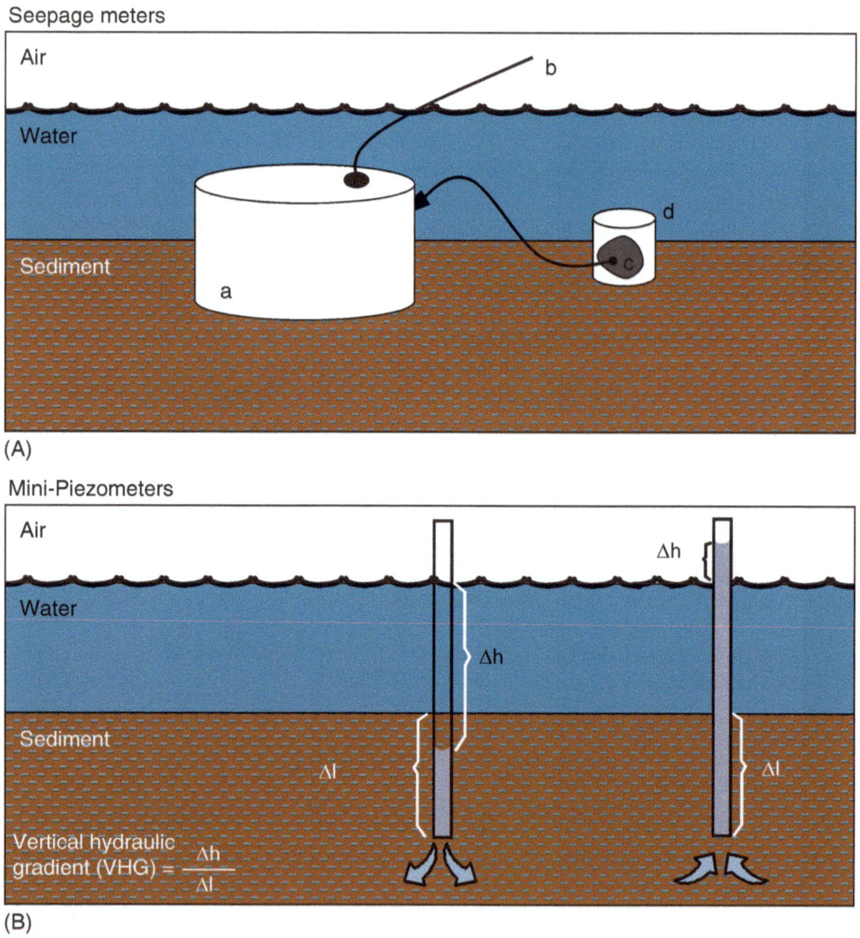

Figure 3 Schematic representations of the use of (panel A) seepage meters and (panel B) mini-piezometers to quantify ground water–surface water exchange in streams. Seepage meters include an open cylinder (a) inserted into stream sediments and vented to the atmosphere (b). Groundwater discharge is collected in a plastic bladder (c) placed in a stilling well (d). Mini-piezometers allow for comparison of hydraulic heads at depth with surface water potentials. The vertical hydraulic gradient (VHG) is then calculated as the head difference divided by the depth of the piezometer.

along a stream reach) can be used to quantify gross gains and losses along a given reach. Solute tracers can also be used to address bidirectional GW–SW exchange (addressed below). The same mass–balance approaches apply to lakes and wetlands but require accurate measures of loss from the water body to the atmosphere via evapotranspiration.

GW–SW Interactions across the Inland Hydrosphere

GW–SW interaction has the potential to influence the organisms and processes of both the surface and aquifer components they integrate. Here we emphasize the hydrology of GW–SW exchange and its ecological ramifications for aquatic environments typical of inland waters.

Streams

Along stream channels, surface water readily enters the streambed, courses through the near-stream subsurface, and returns to the channel. Purposefully borrowing terminology from oceanographers, ecologists refer to streambed areas of losing (i.e., influent) reaches as 'downwelling zones' and to gaining (i.e., effluent) reaches as 'upwelling zones' to emphasize the potential for nutrient delivery to alter benthic biological processes. Thus, the functional boundaries of streams have been extended to include the 'hyporheic zone' defined as the zone of saturated flow beneath and adjacent to a stream that receives recently advected water from the above-ground channel.

In running water systems, the hydrology and ecology of GW–SW exchange has been addressed by ecologist and hydrologist alike. Features that dictate exchange differ across scales (**Table 2**) such that unique characteristics are associated with GW–SW interactions at different points along fluvial networks. The majority of work has addressed headwater streams because they are easier to instrument and study than are larger rivers. Research in streams addressing how exchange is driven by changing stream bed slope has shown that downwelling is induced by areas of convex profile and upwelling by streambed concavities (**Figure 1(d)**). Other in-stream features induce exchange including plant hummocks, salmon redds or nests, woody debris, beaver dams, channel meanders, and other aspects of reach geomorphology (**Table 2**). Mini-piezometers in stream beds can document spatial variation in GW–SW exchange over short (i.e., 10s of meters) stream reaches (**Figure 4**). The direction and magnitude of exchange with the hyporheic zone also changes with time. Over diel time periods, GW–SW exchange may reflect the influences of evapotranspiration and display diel cycles of VHG. The character and location of proximate flow paths along streams can change with the hydrology of their larger catchments. In headwater systems, low flow conditions support both mildly gaining reaches or, with sufficient alluvial storage, losing reaches characterized by downwelling zones. Storms, floods, and snow melt may reverse negative VHGs and promote groundwater discharge along previously downwelling stream reaches (**Figure 4**). Similarly, prolonged drought can reverse the direction of exchange, turning upwelling zones to downwelling zones and eventually leading to spatial intermittency.

Historically, the hyporheic zone has been studied as habitat for unique invertebrate fauna. Recent classification systems rely on the evolutionary history of the biota and the nature of their affiliation with groundwater habitats. Subsurface environments influenced by recently advected surface water are considered hyporheic and macroinvertebrate assemblages in these environments (i.e., hyporheos or stygophiles) are dominated by insects, often similar to those of the benthos. In contrast, groundwater environments more isolated from proximate flow systems (i.e., phreatic water) support invertebrates with marine ancestry, predominantly crustaceans with obligate groundwater affinity (i.e., stygobites) frequently lacking eyes and pigment. Other invertebrates (i.e., amphibites), utilizing both groundwater and benthic environments, share taxonomic affiliation with surface assemblages and emerge from streams as adults, but also display evident evolutionary adaptation to groundwater residence.

Table 2 Multiple scales of GW–SW exchange in stream and river systems including spatial dimensions and examples of features that influence hydrologic interaction

Scale	*Spatial dimension*	*Exchange organizer*
Network	10s of kilometers	Alluvial valleys (beads on a string)
Sector	Kilometer	River bed slope (underflow or baseflow)
Reach	100s of meters	Constrained and unconstrained geomorphology
Channel unit	10s of meters	Riffles, pools, debris dams, gravel bars
Bed form	Meter	Redds, boulders, algal and macrophyte patches
Particle	Centimeter or less	Low conductivity sediments, particle size, permeability

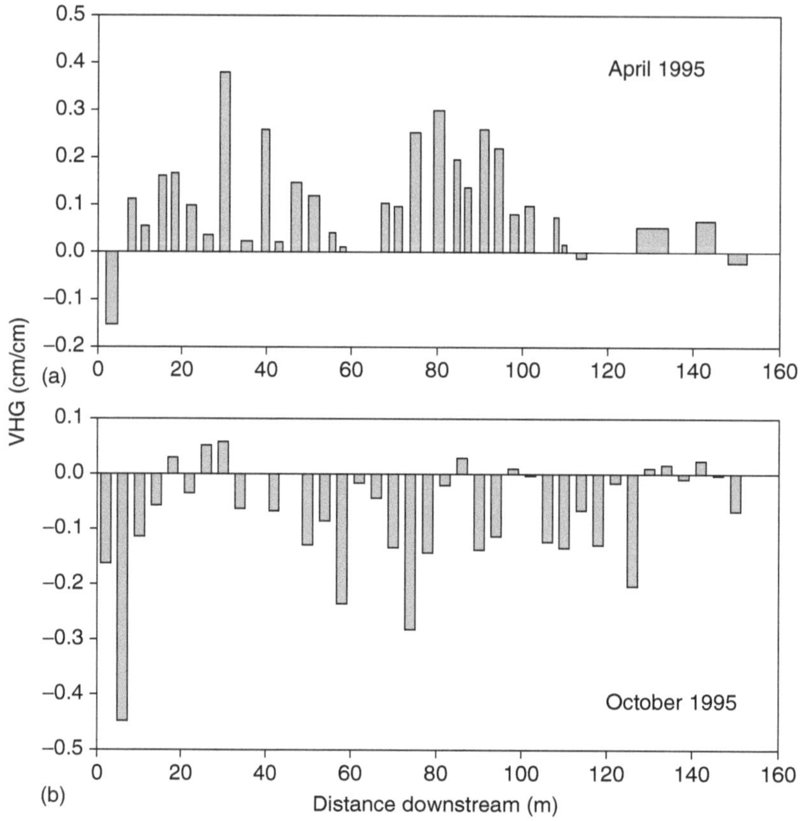

Figure 4 Vertical hydraulic gradient (VHG) along a 160 m reach of Rio Calaveras, a 2nd order stream in north central New Mexico during (a) snow melt high flow conditions in spring, and (b) baseflow in autumn.

At the ecosystem scale, hyporheic zones have been recognized as biogeochemically unique environments supporting unique biotic transformations resulting from the juxtaposition of electron donors and acceptors occurring at the GW–SW interface. In many systems, downwelling of water rich in organic carbon (C) supports metabolic activity in near-stream ground water while mineralization of organic material in the subsurface can increase inorganic nutrient pools. Accordingly, GW–SW exchange can influence biological composition and activity. For instance, in desert streams lacking abundant nitrogen, upwelling water rich in nitrogen promotes growth of algal forms preferring abundant nutrients (**Figure 5(a)**) while losing portions of the reach located downstream are dominated by nitrogen-fixing cyanobacteria (**Figure 5(b)**). Because surface environments are comparatively rich in organic matter content, downwelling zones may import carbon and represent hot spots of subsurface respiration and carbon mineralization.

The Hyporheic Zone and Transient Storage Modeling

The processes and effects of GW–SW exchange along streams have been addressed by a number of numerical modeling approaches and have been well studied using a transient storage zone modeling (TSM) approach. The TSM approach includes modified advective-dispersion equations (**Figure 6(a)**) that account for two-way hydrological exchange between the main channel and a nonadvecting storage zone (**Figure 6(b)**). Transient storage describes areas where the stream flow is slowed relative to the bulk advective velocity. Transient storage of stream flow may result from pools, backwaters, eddies, side channels and floodplains, as well as hyporheic exchange. Historically, storage has been interpreted to represent GW–SW exchange, but some work suggests that surface features (e.g., pools) may contribute to storage under appropriate conditions.

Since its inception, TSMs have been widely used for studying transient storage in streams, especially since the development of a number of user-friendly versions readily employed by hydrologists and ecologists alike. The general approach to estimating transient storage in streams involves the release of a conservative tracer (e.g., Cl, Br) and characterization of breakthrough curves (i.e., time course of solute concentrations through rising-limb, steady-state, and falling-limb conditions) at downstream transects.

A number of metrics have been derived from TSM parameters and they have been used to address how

Figure 5 Benthic algal responses to ground water–surface water exchange in Sycamore Creek, Arizona, USA. (a) Groundwater discharge in upwelling zones supplies nitrogen from nutrient-rich interstitial environments and supports luxurious blooms of the green alga *Cladophora*. (b) Just 150 m downstream, in a downwelling zone, surface nitrogen concentrations are lower and the benthic flora is dominated by nitrogen-fixing bluegreen bacteria of the genus *Nostoc*. Photographs by H.M. Valett.

$$\frac{\partial C}{\partial t} = -\frac{Q}{A}\frac{\partial C}{\partial x} + \frac{1}{A}\frac{\partial}{\partial x}\left(AD\frac{\partial C}{\partial x}\right) + \frac{q_L}{A}(C_L - C) + \alpha(C_S - C) \quad (1)$$

Equation 1: temporal change in main channel tracer concentrations where: Q = stream flow (L^3T^{-1}), A = main channel x.s area (L^2), C = tracer concentration in the main channel ($M\,L^{-3}$), D = dispersion coefficient (L^2T^{-1}), q_L = lateral inflow (L^2T^{-1}), C_L = tracer concentration in inflow (M L-3), α = exchange coefficient (T^{-1}), and C_S = tracer concentration in the storage zone (ML^{-3})

$$\frac{\partial C_S}{\partial t} = \alpha\frac{A}{A_S}(C - C_S) \quad (2)$$

Equation 2: temporal change in storage zone tracer concentrations where: A_s = area of the storage zone (L^2)

(a)

(b)

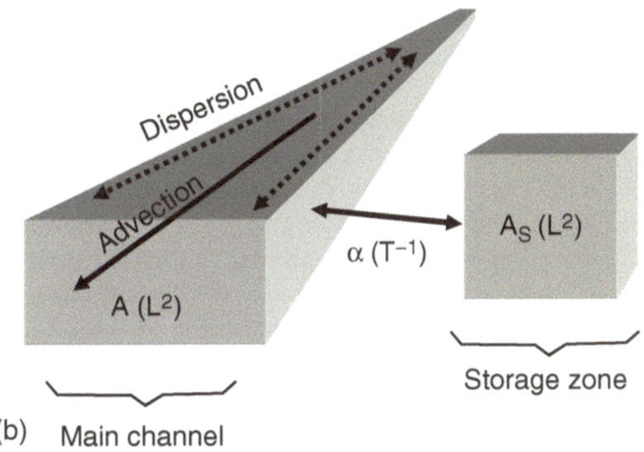

Figure 6 (a) Mathematical basis of the transient storage zone modeling (TSM) approaches to studies of ground water–surface water exchange in running water ecosystem. (b) A schematic representation of TSM modeling emphasizing that the model is physically-based with the processes of advection and dispersion occurring in the main channel compartment. Tracer is exchanged with a nonadvective storage zone via an exchange coefficient (α) which can remove tracer from the main channel or return it from storage depending on the relative concentrations of tracer in the storage zone (C_S) or channel (c) water.

stream hydrology and geomorphology influence transient storage and proximate flow paths along streams (**Table 3**). Some have argued that increased transient storage results in enhanced biological and geochemical processing of dissolved solutes with implications for chemical composition and resultant water quality (**Figure 2**).

Rivers and Floodplains

Streams differ from rivers in a number of ways other than size, but GW–SW remains a component of both low-order and higher-order lotic (i.e., flowing) systems. In rivers, the spatial and temporal scales of proximate flowpaths are increased compared to streams and they provide important links to adjacent floodplains, geomorphic entities absent from streams. Larger scale assessment of exchange in rivers indicates that rivers can be influenced by proximate, local, and intermediate scale flow systems. Two fundamental end-member types of flow paths exist: (a) underflow and (b) baseflow (**Figure 7**). Underflow occurs when groundwater flow paths parallel the river channel (**Figure 7(a)**), typically occurring in alluvial river valleys with a high channel slope. While baseflow is a term often used to describe flow conditions uninfluenced by recent precipitation (i.e., a temporal distinction) it also describes groundwater flow paths oriented roughly perpendicular to the longitudinal channel axis (**Figure 7(b)**). When river gradients are low and rivers are highly sinuous, groundwater flowpaths may intersect the stream channel, contributing to channel runoff, and loading river water with aquifer-derived materials. Under the same conditions, river water may leave the channel and generate proximate flow paths composed of a mixture of channel and subsurface water (i.e., hyporheic zones). In contrast to headwater reaches where the direction of GW–SW exchange may

Table 3 Metrics used to describe ground water–surface water exchange in streams and rivers

Parameter	*Equation*	*Description*
Normalized storage zone area ($L\,L^{-1}$)	A_s/A	The size of the storage zone area relative to the main channel area
Storage exchange flux ($L^2\,T^{-1}$)	$q_s = \alpha A$	The average water flux through the storage zone per unit length of stream
Hydrologic retention factor ($T\,L^{-1}$)	$R_h = \frac{T_{sto}}{L_s} = \frac{A_s}{Q}$	The storage zone residence time of water per unit of stream reach traveled
Turnover length (L)	$L_s = \frac{u}{\alpha}$	The average distance a molecule travels downstream prior to entering the storage zone; where $u = Q/A$
Hyporheic residence time (T)	$T_{sto} = \frac{A_s}{\alpha A}$	The average time a molecule remains in the storage zone before release into the main channel
F-med ($T\,T^{-1}$)	$F_{med} = (1 - e^{-L(\alpha/u)})\frac{A_s}{A+A_s}$	The fraction of median travel time of stream reach due to transient storage

Sources
1. Bencala KE and Walters RA (1983) Simulation of solute transport in a mountain pool-and-riffle stream: A transient storage model. *Water Resources Research* 19: 718–724.
2. Harvey JW, Wagner BJ, and Bencala KE (1996) Evaluating the reliability of the stream tracer approach to characterize stream–subsurface water exchange. *Water Resources Research* 32: 2441–2451.
3. Harvey JW and Wagner BJ (2000) Quantifying Hydrologic Interactions between Streams and Their Subsurface Hyporheic Zones. In Jones JB and Mulholland PJ (eds.) *Streams and Ground Waters*, pp 3–44. San Diego, CA: Academic Press.
4. Morrice JA, Valett HM, Dahm CN, and Campana ME (1997) Alluvial characteristics, ground water–surface water exchange and hydrological retention in headwater streams. *Hydrological Processes* 11: 253–267.
5. Mulholland PJ, Steinman AD, Marzolf ER, Hart DR, and DeAngelis DL (1994) Effect of periphyton biomass on hydraulic characteristics and nutrient cycling in streams. *Limnology and Oceanography* 42: 443–451.
6. Runkel RL (2002) A new metric for determining the importance of transient storage. *Journal of the North American Benthological Society* 21: 529–543.

be spatially variable under low flow conditions, rivers in mesic environments typically gain water during periods of low precipitation and runoff. Under high flow conditions, in-stream hydraulic head increases promoting the infiltration of river banks and recharge of the aquifer. During the dry season, release from bank storage can supply baseflow and buffer seasonal changes in river discharge.

In gravel-bed montane rivers, channel geomorphology promotes alluvial deposition in distinct segments that alternate with more constrained segments generating a succession of floodplains like 'beads on a string.' These floodplain areas frequently promote GW–SW exchange that intimately links the river and its alluvial aquifer with implications for aquatic biota and water quality. Researchers have documented the emergence of amphibitic stoneflies from wells and septic fields up to 1 km from the waters of Flathead River, Montana, USA. Within river channels, advective flows driven by bedforms like ripples or dunes generate GW–SW exchange between channel and pore water with flow paths of tens of meters even when rivers sediments are fine-grained sands or silts as is the case for the Mississippi and other large rivers.

As they do in stream systems, ecological conditions in river waters influence aquifer biogeochemistry. These linkages were illustrated by scientists working in Switzerland who showed that reduced phosphorus loading to Lake Greifen translated to less organic matter production in the Glatt River and a decrease in the extent of reducing conditions in the alluvial aquifer. As a result, aquifer concentrations of manganese and cadmium declined and water quality improved.

Unlike streams, many rivers have floodplains to which they are intimately linked. River flow regimes promote the establishment of floodplain lakes and ponds generated both by overland flow from rising water stages and by groundwater discharge as the water table rises above the land's surface in low elevations of the floodplain. These hydrologic links with ground water may seasonally connect lakes and ponds of the floodplain to main channel environments. At the same time, changing water table elevation resulting from connections between the aquifer and inundated floodplain can alter metabolism and nutrient cycling in soils and subsurface sediments, may accelerate or retard organic matter processing, and can influence productivity of riparian vegetation. Research on cottonwood growth rates has shown that riparian trees in gaining reaches influenced by baseflow GW–SW exchange support larger and longer-lived trees.

Hydrologic exchange between rivers and their aquifers also influence biological structure of the flora and fauna that occupy both habitats. Just as they do in smaller streams, gradient breaks that reflect the occurrence of riffles and pools generate flowpaths that penetrate into the subsurface or that deliver ground water to the stream channel. French scientists used distinctly different chemical signatures to distinguish surface (rich in sulfate) and groundwater (high nitrate

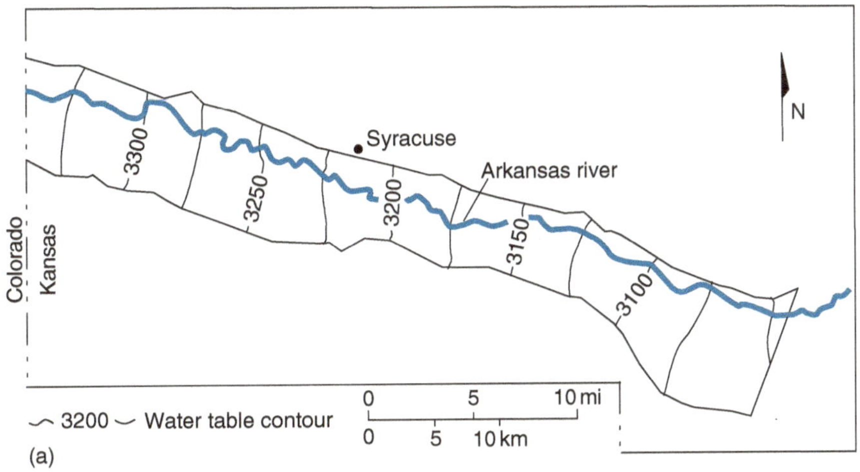

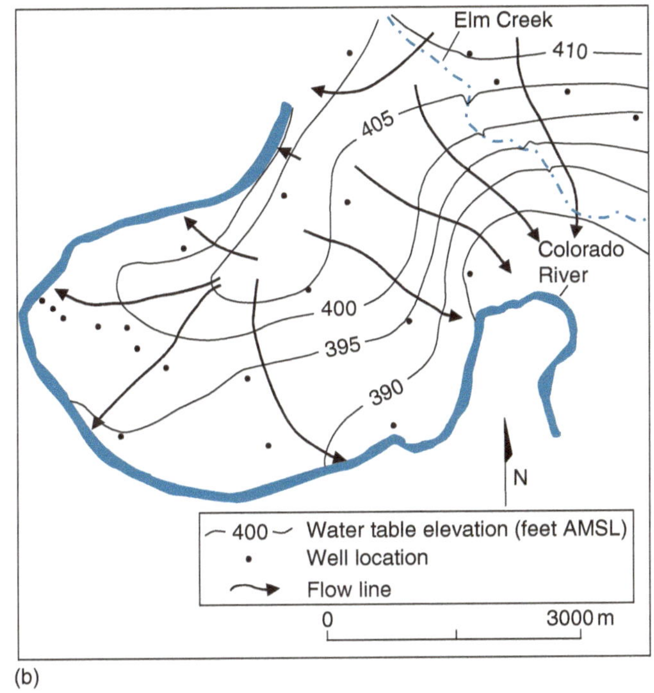

Figure 7 (a) Groundwater flow parallel to the Arkansas River channel illustrates 'underflow' conditions characteristic of higher valley gradients. (b) Flow paths oriented perpendicular to the Colorado River, Texas generate 'baseflow' exchange between the alluvial aquifer and river channel. Adapted from Larkin RG and Sharp Jr. RJM (1992) On the relationship between river-basin geomorphology, aquifer hydraulics, and ground-water flow direction in alluvial aquifers. *Bulletin of the Geological Society of America* 104: 1608–1620. With permission from Geological Society of America.

concentrations) sources to relate the composition of groundwater fauna to the direction of GW–SW exchange in the Rhone River. Over kilometers of distance, surface waters of the Rhone first infiltrated into the interstitial spaces of massive gravel bars while further downstream decrease in the river slope promoted groundwater discharge and shallow subsurface habitat was typical of deeper portions of the highly porous alluvial aquifer. This change in GW–SW exchange was reflected in the relative composition of groundwater fauna; downwelling zones at the head of the gravel bar were dominated by stygophiles including many forms of aquatic insects while downstream the assemblage included many forms of blind, unpigmented, stygobitic crustaceans.

Lakes and Wetlands

Hydrologic exchange between lake water and groundwater occurs in three basic ways that can be used to categorize GW–SW interaction in lake systems. Hydrologically mounded lakes receive virtually all of their water from precipitation and have seepage losses to ground water throughout their beds. Terminal groundwater-discharge lakes receive water from adjacent aquifers throughout their beds. Many lakes may gain and loose water at different locations and these throughflow lakes vary in the degree to which they receive or contribute water to or from the adjacent groundwater system. Because lakes exist in low points on the landscape, they may be influenced by flow systems ranging from local to regional while supporting proximate flow paths that influence their functioning as ecosystems.

In the process of GW–SW exchange, physical, chemical, and biological conditions are altered as water enters or leaves lakes. This has been well documented for throughflow lakes in glacial and dune terrain where they are underlain by highly permeable deposits and are characterized by groundwater inflow along one shore and outflow along the opposite shore. Biological processes within lakes may provide critical protection from processes like acid rain by generating acid neutralizing capacity that is exported through seepage and provided to downgradient water bodies. At the same time, groundwater inflow can provide physical and chemical conditions that promote the growth of littoral algae and plants that influence the distribution of invertebrates and water fowl within and among lakes. Under these conditions, throughflow patterns may be relatively stable, but the boundaries of inflow and outflow along the lake bed change with time as the water table rises or lowers reflecting the balance between GW–SW interaction, precipitation, and evaporation. Over varying time periods, precipitation and groundwater inflow may fail to match rates of evaporation, resulting in drastic declines in lake levels, concentration of dissolved solutes, and shifts in flow paths that alter GW–SW interaction. Accordingly, drought in the lake districts of Wisconsin and Minnesota, USA, resulted in very different responses by lakes that appear identical in most ways other than the character of GW–SW exchange.

Wetlands exist in climates and landscapes where groundwater discharge overwhelms rates of water loss via drainage or evapotranspiration. Their interaction with ground water is very similar to that of lakes since wetlands may be dominated by inflow, outflow, or some combination. Wetlands, however, do not always occur at low spots on the landscape and their location influences how they are influenced by GW–SW exchange and how they act as aquatic ecosystems. Typically, wetlands have been categorized by their hydroperiods (time of annual inundation and saturation). Wetlands such as fens may occur on slopes and act as groundwater donors to downgradient systems. Bogs typically occur where groundwater flow systems converge or divide and may have no appreciable drainage or throughput (i.e., act as groundwater receivers). Many wetlands have both groundwater inflow and outflow and, thus, act to some degree as groundwater conveyors. Wetlands are highly diverse hydrologically and biologically, occur in both simple and complex flow fields, may be associated with slopes, perched in bottom lands, or integrated via river floodplains. More in depth assessment of wetlands are available elsewhere in this encyclopedia.

Human Impacts and Future Research

Human influences are evident at all scales and disciplines of studies related to hydrology and ecology and this is equally true for the processes and effects of GW–SW exchange. We have a poor understanding of how GW–SW exchange responds to environmental change. Below, we address a few key areas of environmental change and how they influence GW–SW exchange and later address some areas of needed research.

Agricultural Development

The conversion of land from native vegetation to agricultural production includes three major processes that threaten aquatic systems: (1) erosion and sedimentation; (2) irrigation; and (3) application of agricultural chemicals. These processes may change ecosystem function by altering hydrologic or biogeochemical aspects of GW–SW exchange (**Figure 2**). Sedimentation to streams and other surface water bodies is a tremendous problem in agricultural areas in the United States and other countries. Erosion and sedimentation introduces fine-grained, often organic-rich, particles that may clog stream-sediment exchange, introduce sediment-bound contaminants, and alter biogeochemical function at the GW–SW interface. Irrigation often includes extensive canals and underground drains (i.e., tiles) that short circuit flow paths linking surface water to ground water. Irrigation increases recharge to local groundwater and elevates water tables that may result in enhanced groundwater inputs to surface bodies down gradient. Irrigation relying on groundwater sources, on the other hand, may result in groundwater depletion and induced recharge with the potential to reverse the direction of GW–SW exchange characteristic of natural flow. This often results in decreased natural discharge reflected as the drying of springs, marshes, and riparian areas. This can be particularly devastating in semi-arid lands where phreatophytic vegetation relies on aquifer sources of water. Agricultural chemicals include pesticides and herbicides that may enter ground water through mechanisms described above. In addition, ammonium, a common component of fertilizer is readily converted to nitrate, a mobile form of nitrogen that represents a wide-spread contaminant in agricultural areas. Point-source contaminants include feedlots and associated animal wastes. Agricultural contaminants can be introduced by municipal pumping of ground water where agricultural and urban developments co-occur.

Urbanization

Urbanization can markedly alter GW–SW exchange in streams through increases in impervious surface cover in the landscape. Impervious surfaces alter the link between precipitation and groundwater recharge resulting in more flashy storm flows, and reduced baseflows. The resulting alterations to GW–SW exchange can increase erosion, cause bank failures, and cause urban streams to become dry during low flow periods. Furthermore, urban runoff typically transports pulses of contaminants from sewage treatment plants, fluid storage tanks, landfills, industrial lagoons, and storm water drains. Inputs from urban sources can dominate surface water flows and degrade water quality downstream from developing population centers.

Hydrosphere Modifications

Human activities such as the construction of levees and impoundments have resulted in disconnection of rivers from their floodplains. Construction of reservoirs and diversion canals have redistributed waters in both small and large running water systems. Anthropogenic influences such as channelization of rivers, dam construction, and water diversions couple hydrologic and geomorphic change with strong implications for the magnitude, location, and direction of GW–SW exchange.

Future Research

There is a great potential for research on GW–SW interaction to help address the myriad of environmental issues that result from human activities including those listed earlier. At the same time, there are areas of fundamental research on the hydrology of GW–SW exchange that require investigation. In general, precise measures of the fluxes between aquifers and surface water bodies remain elusive, primarily reflecting heterogeneity and its influence on the ability to scale measures from small studies of sediment cores to assessment of the role of GW–SW exchange in river networks. The field requires continued rigorous efforts to quantify GW–SW fluxes, including the ability to incorporate temporal variation at seasonal and climatic time scales. Integration of hydrologic, ecological, and biogeochemical research has been a feature of GW–SW research and needs to be further emphasized across aquatic habitats.

Glossary

Amphibites – Invertebrates showing affinity to both surface and groundwater environments that may or may not display morphological adaptations to groundwater residence.

Hyporheos – Invertebrates occupying the hyporheic zone.

Mesic – Of, pertaining to, or adapted to an environment having a balanced supply of moisture.

Montane – Relating to mountainous habitats.

Phreatic – Groundwater environments distant from and slowly connected to surface water.

Phreatophytic – Long-rooted plants that absorb water from groundwater sources.

Stygobites – Invertebrates with obligatory groundwater association displaying evolutionary adaptation to subsurface residence.

Stygophiles – Invertebrates with affinity for ground water environments.

See also: Chemical Fluxes and Dynamics in River and Stream Ecosystems; Currents in Rivers; Ground Water; Hydrological Cycle and Water Budgets; Nitrogen; Phosphorus; Streams.

Further Reading

Chapelle F (2001) *Ground-Water Microbiology and Geochemistry*, 2nd edn. New York: Wiley.

Gibert J, Danielopol D, and Stanford J (1994) *Groundwater Ecology*. San Diego, CA: Academic Press.

Griebler C, Danielopol DL, Gibert J, Nachtnebel HP, and Notenboom J (eds.) (2001) *Groundwater Ecology: A Tool for Management of Water Resources*. Luxembourg: Office for Official Publications of the European Community.

Jones JB and Mulholland PJ (eds.) (2000) *Streams and Ground Waters*. San Diego, CA: Academic Press.

Kalbus E, Reinstorf F, and Schrimer M (2006) Measuring methods for ground water–surface water interactions: A review. *Hydrology and Earth Systems Sciences* 10: 873–887.

Larkin RG and Sharp JM Jr. (1992) On the relationship between river-basin geomorphology, aquifer hydraulics, and

ground-water flow direction in alluvial aquifers. *Bulletin of the Geological Society of America* 104: 1608–1620.
Sophocleous M (2002) Interactions between groundwater and surface water: The state of the science. *Hydrogeology Journal* 10: 52–67.
Valett HM, Morrice JA, Dahm CN, and Campana ME (1996) Parent lithology, groundwater–surface water exchange and nitrate retention in headwater streams. *Limnology and Oceanography* 41: 333–345.
von Gunten HR and Lienert Ch (1993) Decreased metal concentration in ground water caused by controls of phosphate emissions. *Nature* 364: 220–222.
Webster KE, Kratz TK, Bowser CJ, and Magnuson JJ (1996) The influence of landscape position on lake chemical responses to drought in northern Wisconsin. *Limnology and Oceanography* 41: 977–984.
Winter TC, Harvey JW, Franke OL, and Alley WM (1998) *Ground Water and Surface Water – A Single Resource.* Denver, CO: U.S. Geological Survey. USGS Circular 1139.

Relevant Websites

USGS Groundwater Information Page – http://water.usgs.gov/ogw/gwsw.html.
USGS OTIS Page – http://co.water.usgs.gov/otis/.
Surface-water Quality and Flow Monitoring Interest Group – http://smig.usgs.gov/SMIG/reading_refs.html.

Groundwater Chemistry

J W LaBaugh, U.S. Geological Survey, Reston, VA, USA

Published by Elsevier Inc.

Introduction

The study of inland waters, limnology, commonly has been thought of in terms of the study of surface waters. Characteristics of flow in subsurface, or ground water, as well as the chemical composition of ground water and the biogeochemical processes governing changes in the chemical composition of ground water have received relatively limited attention in limnological studies. In contrast, physical, chemical, and biological characteristics of ground water have been major foci of hydrogeology and geochemistry. The vast range of all aspects of groundwater chemistry subject to those foci is beyond the scope of this article. Instead, the emphasis herein will be on the various processes controlling groundwater chemical characteristics.

Chemical Composition Changes along Groundwater Flow Paths

Water that enters the atmosphere from evaporation and transpiration commonly resides there for a period of days before returning to land or ocean as atmospheric precipitation in the form of rain, sleet (frozen rain), or snow. In free-flowing rivers, that water commonly will remain in the rivers for a few weeks. The presence of reservoirs increases that residence time to more than a month. Water that enters lakes can reside there for days, months, years, or in the case of very large lakes, hundreds of years. The time water remains in the subsurface, in ground water, spans a few days to tens of thousands of years or more (**Figure 1**). The length of time water spends as ground water plays an important role in determining the chemical composition of ground water.

Chemicals found in rain water, or water from the melting of snow or ice, are present in small amounts. In some cases, these waters are acidic because of the presence of dissolved sulfur dioxide or nitrous oxide. As water infiltrates the ground, it moves through the soil, which can contain large concentrations of carbon dioxide produced by microbial decomposition of organic matter and by root respiration. Carbon dioxide dissolved in these waters produces carbonic acid. Minerals in the subsurface react with water as water enters the subsurface, through dissolution or interaction with the acids in solution. The outcome of this interaction is ground water that contains higher concentrations of dissolved chemicals than were present in the rain or water from melting snow or ice that originally entered the subsurface. At the same time, the hydrogen ion concentration, or pH (pH represents the negative logarithm of the hydrogen ion concentration), in ground water can change as a result of acid–base reactions.

The reactions that take place as ground water moves through the subsurface depend on the types of minerals present in the subsurface and the chemical characteristics of the water in contact with the minerals. In general, the longer that water is in contact with the minerals, the greater the concentration increase in the water of the chemicals that are being dissolved from the minerals. As a result, waters that have been in contact with the subsurface for longer periods of time generally will contain more dissolved chemicals than waters that have been in contact for shorter periods of time, depending on the reactivity of the minerals involved. More reactive minerals, such as carbonates, would likely contribute more dissolved chemicals in a short time than less reactive minerals, such as silicates, will over many years. Thus, surface waters receiving groundwater discharge from different length flow paths – waters of different 'ages' (**Figure 2**) – can receive waters containing different concentrations of chemicals derived from dissolution of minerals in the subsurface. A variety of indicators can be used to determine what those 'ages' are (**Table 1**). The groundwater age, in general, refers to the time since water entered ground water at the water table, such that, for a given parcel of ground water, the age will reflect the average age of all the molecules of water in that parcel.

Water moving through the subsurface also may encounter a variety of geologic materials (**Figure 1**) each containing distinct minerals. Ground waters flowing through different sequences of minerals will change chemical composition in response to the minerals encountered. A consequence of this is the fact that ground waters discharging to surface water from different flow paths may differ in the relative amounts of dissolved individual chemicals contained therein, as well as the absolute amount of all chemicals the ground waters contain.

At some point, concentrations of some chemicals within ground water may become high enough that the water is saturated with respect to specific minerals. Under such conditions, the minerals may precipitate out of solution, thereby changing the concentration of

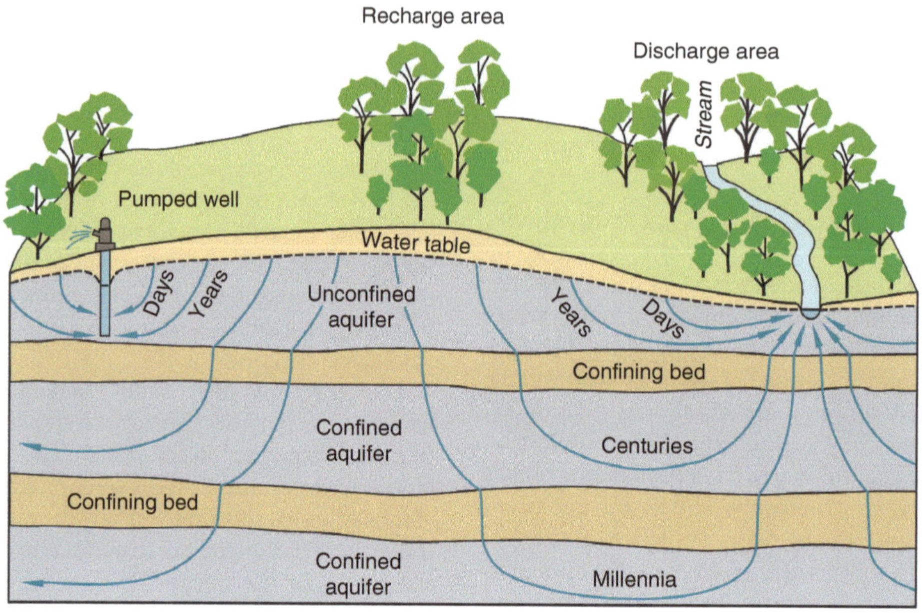

Figure 1 Groundwater flow paths vary greatly in length, depth and travel time from points of recharge to points of discharge in the groundwater system. From Heath RC (1983) *Basic Ground Water Hydrology*. U.S. Geological Survey Water Supply Paper 2220, http://pubs.er.usgs.gov/usgspubs/wsp/wsp2220, as modified by Winter TC, Harvey JW, Franke OL, and Alley WM (1998) Ground water and surface water a single resource. U.S. Geological Survey Circular 1139. Denver, CO: U.S. Geological Survey. http://pubs.usgs.gov/circ/circ1139/.

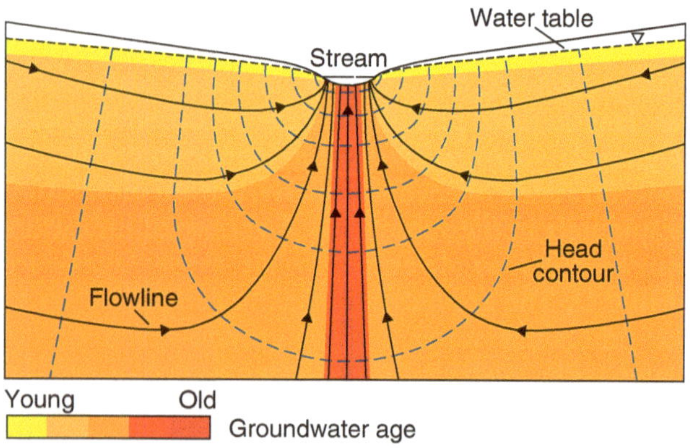

Figure 2 Conceptualized model of flow patterns and age of groundwater seepage to a stream channel. "In this model ground water that seeps into the stream near its banks originates near the stream, follows short flow paths, and requires short travel times. Ground water that seeps into the center of the stream channel originates from distant areas, follows relatively long flow paths, and requires longer travel times to move through the aquifer. As a result, the age of groundwater seepage across a section of the stream channel increases from its banks to its center. Ground water also becomes increasingly older with distance downstream." Modica E (1999) Source and Age of Ground-water Seepage to Streams. U.S. Geological Survey Fact Sheet FS-063–99.

the dissolved substances involved in the precipitation reaction. This is another way in which the relative concentration of chemicals in ground water can change over time. Some chemicals in ground water may become attached to the surface of the solid materials that make up the matrix of the subsurface, or particles suspended in the water, by the process of sorption. Examples of sorption are attachment of positively charged ions to negatively charged surfaces of clays or attachment of pesticides to solid surfaces. Concentrations within ground water may change through the process of ion exchange, whereby ions attached to solids are replaced by ions in ground water. A common example of ion exchange in ground water is the replacement of sodium or potassium ions attached to a solid by calcium and magnesium, resulting in an increase in the sodium and potassium concentrations in the ground water and a decrease in the calcium and magnesium concentrations. This process of ion exchange is the same as the process that occurs in water softening devices used in the home, whereby calcium and magnesium replace sodium in the water softener material. The process can be reversed in ground water, where saltwater enters what had been a fresher system and some of the sodium ions in the saltwater are exchanged for calcium ions attached to a solid. This process is similar to that used to renew the capability of a domestic water softener to remove calcium and magnesium from solution.

An outcome of the interaction between water and the geologic framework through which it flows in groundwater systems is that ground waters are quite varied in chemical composition. An example

Table 1 Indicators of groundwater ages

Type of indicator	*Indicator of age*	*Basis of age determination*
Radionuclide	Sulfur-35	87 day half-life
Radionuclide	Krypton-85	10.76 year half-life
Radionuclide	Tritium	12.32 year half-life Bomb pulse due to nuclear and thermonuclear weapons testing in the 1950s and 1960s
Noble gas	Helium-3	Tritium–helium-3 ratio is used to determine initial tritium content of water
Industrial chemical	Chlorofluorocarbons (CFCs)	Atmospheric lifetime of 50–100 years, finite period of use
Industrial chemical	Sulfur hexafluoride (SF6)	Very long atmospheric residence time, known period of anthropogenic production
Industrial chemical	Surfactants	Known period of production, presence indicates water is not older than dates of production
Industrial chemical	Herbicides	Known period of production, presence indicates water is not older than dates of production
Radionuclide	Silicon-32	140 ± 6 year half-life
Radionuclide	Argon-39	269 year half-life
Radionuclide	Carbon-14	5703 year half-life
Radionuclide	Krypton-81	2.29×10^5 year half-life
Radionuclide	Chlorine-36	$3.01 \times 10^5 \pm 4.0 \times 10^3$ years Bomb pulse due to thermonuclear weapons testing from 1954 to 1958
Radionuclide	Iodine-129	15.7 million year half-life

Information compiled from: Plummer LN, Michel, RL, Thurman EM, and Glynn PD (1993) Environmental tracers for age dating young ground water. In Alley WM (ed.) *Regional Ground-Water Quality*. pp. 255–294. New York, NY: Van Nostrand Reinhold; and Phillips FM and Castro MC (2004) Ground-water dating and residence-time measurements. In Drever JI (ed.) Volume 5: Surface and Ground Water, Weathering, and Soils. In Holland HD, and Turekian, KK (eds.) *Treatise on Geochemistry*, 10 vols. Oxford, UK: Elsevier Pergamon.

for such varied composition in a coastal zone is presented in **Figure 3**. In general, ground water that has been recently recharged from atmospheric deposition moving through the unsaturated zone will be relatively fresh, containing a few to less than $1000\,mg\,l^{-1}$ of total dissolved solids. Whereas ground water that has been in the flow system for some time, or has moved through geologic formations containing readily soluble minerals, may be considerably more saline, even to the point where salinity values are greater than $35\,000\,mg\,l^{-1}$ of total dissolved solids, which are also known as brines. Ground waters that are more saline also will have increased density, which in turn affects the movement of ground water within a flow system. Such density differences due to salinity differences are a factor in the flow of ground water in coastal areas, where less dense fresh water overlies more dense saltwater, with a transition zone of varying salinity between (**Figure 4**). Saline waters also occur in inland groundwater systems as a result of dissolution of minerals within the flow system or interaction with saline lakes.

Microbial activity in the subsurface plays a role in changing the chemical composition of ground waters. The process by which microorganisms decompose organic chemicals to obtain energy and carbon for growth is known as biodegradation. In addition, some microbes are able to use the energy released in the reactions that move electrons from electron donors (commonly organic carbon) to electron acceptors (e.g., molecular oxygen, nitrate, ferric iron, sulfate, carbon dioxide) to support life. Oxidation of sulfide minerals or ferrous iron also is a source of energy for microbial metabolism. The role of these microbes in affecting changes in the chemical composition of surface waters has been known for several decades, particularly in chemically stratified lakes – lakes containing waters of different densities resulting from differences in chemical concentrations within those waters. The role of bacteria in influencing the chemical composition of ground waters was recognized in the 1980s.

Stable isotope ratios are a tool that is useful in understanding processes affecting chemical characteristics of ground water, particularly for oxygen, hydrogen, carbon, nitrogen, and sulfur. The ratio (R) of the heavy to light isotope for a given element in a sample of water is compared with the ratio of the heavy to light isotope in a standard. The result of this comparison is reported as delta (δ) values in parts per thousand, where $\delta(‰) = ((R_{\text{sample}}/R_{\text{standard}}) - 1) \times 1000$. Isotope ratios change because of fractionation during phase changes, as water condenses, evaporates, and melts, or because of mixing of waters and/or solutes, or because of geochemical and biological reactions. The microbial transformation of chemicals in ground water can affect the isotopic ratios of the chemicals the microbes act upon. In general, the microbes use

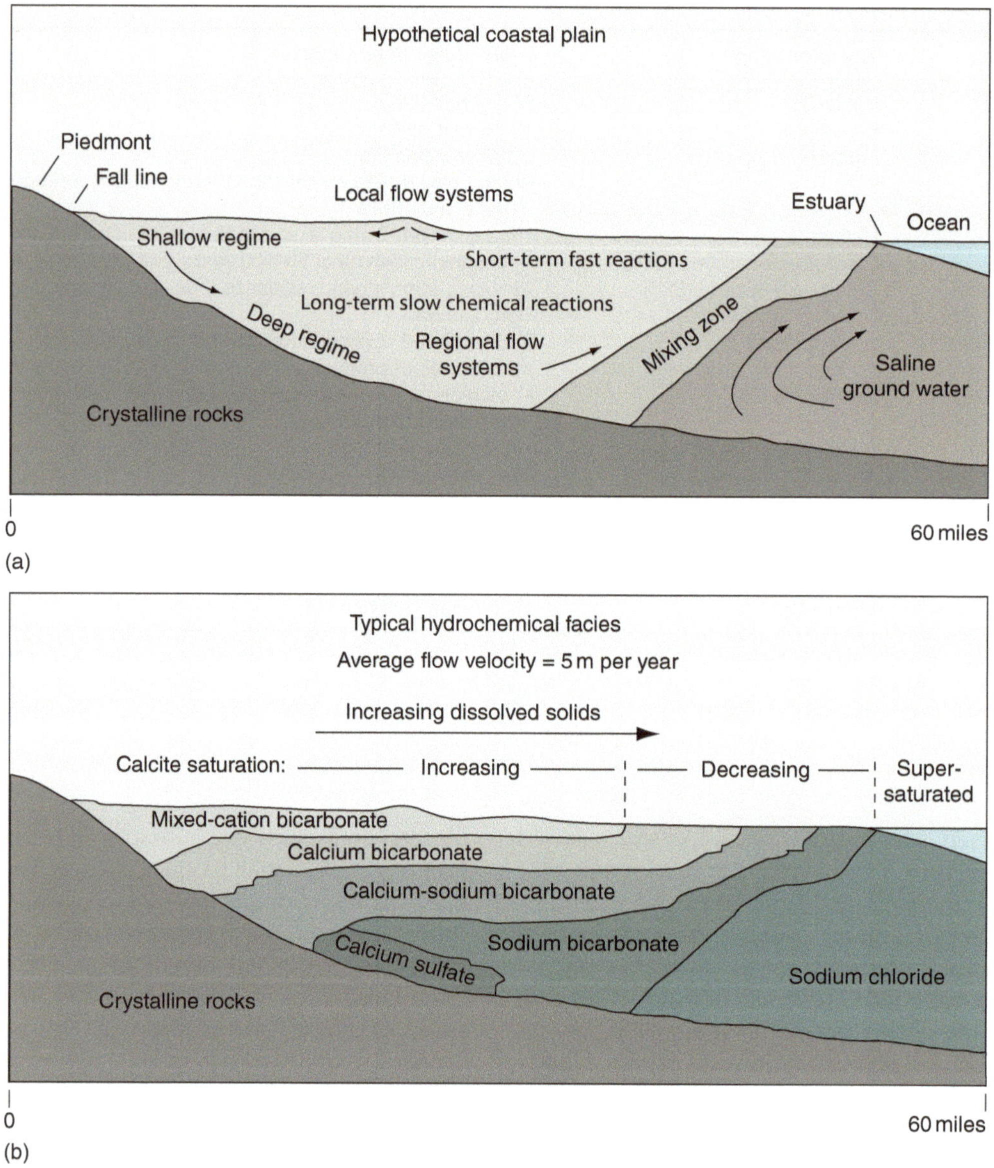

Figure 3 In a coastal plain, such as along the Atlantic Coast of the United States: (a) the interrelations of different rock types, shallow and deep groundwater flow systems (regimes) and mixing with saline water; (b) results in the evolution of a number of different groundwater chemical types. Adapted From Winter TC, Harvey JW, Franke OL, and Alley WM (1998) Ground water and surface water a single resource. U.S. Geological Survey Circular 1139. Denver, CO: U.S. Geological Survey. http://pubs.usgs.gov/circ/circ1139/. Modified from Back W, Baedecker MJ, and Wood WW (1993) Scales in chemical hydrogeology—A historical perspective. In Alley WM (ed.) *Regional Ground-water Quality*. New York, NY: Van Nostrand Reinhold.

more of the light isotope because it is energetically more efficient to do so. In the process, the isotopic ratio of the affected elements will change, providing evidence that a particular microbial process may be involved.

Gas concentrations provide the means to determine the temperature at which ground water was replenished or recharged by precipitation reaching the water table. As the noble gases are generally nonreactive, the temperature of the water at the time it enters the groundwater system at the water table determines the concentration of the dissolved gas in the water because of the solubility of the noble gas at that temperature. Concentrations of other dissolved gasses can be compared with the expected concentration based on the temperature at the time of recharge to determine whether more or less dissolved gas is present than expected. This can be of particular interest in the case of nitrogen, for example. The presence of more nitrogen gas dissolved in ground water than

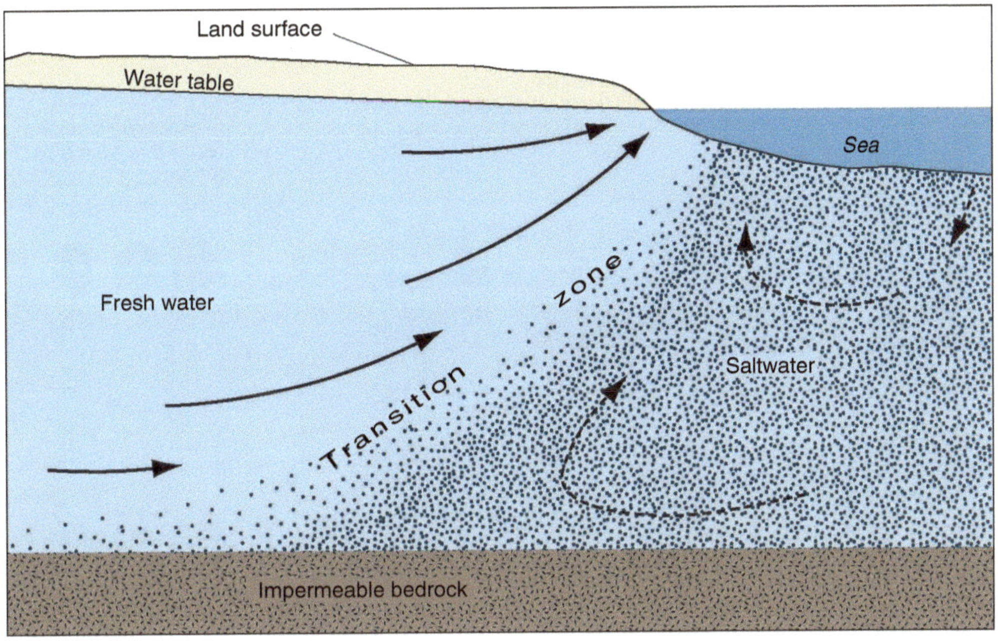

Figure 4 Groundwater flow patterns and the freshwater–saltwater transition zone in an idealized coastal aquifer. A circulation of saltwater from the sea to the transition zone and then back to the sea is induced by mixing of fresh water and saltwater in the transition zone. Adapted from Barlow PM (2003) Ground water in freshwater–saltwater environments of the Atlantic Coast. U.S. Geological Survey Circular 1262. http://pubs.usgs.gov/circ/2003/circ1262/. Modified from Cooper HH Jr. (1964) A hypothesis concerning the dynamic balance of fresh water and salt water in a coastal aquifer. U.S. Geological Survey Water-Supply Paper 1613-C.

expected, based on recharge temperature determined from the solubility of argon, may indicate that denitrification has occurred. Denitrification is the process by which nitrate is converted to nitrous oxide and nitrogen gas.

Understanding the processes resulting in observed concentrations of chemicals of interest in ground water is important when trying to relate land use to the quality of groundwater discharging to nearby surface waters. The importance of examining changes in concentration in ground water in relation to the age of the water, the solubility of dissolved gases, and the stable isotope ratios to understand processes controlling those concentrations is provided by the example of watersheds in the Atlantic Coastal Plain of Maryland (**Figure 5**). Streams receiving groundwater discharge from uplands that had received increased fertilizer applications over 60 years had lesser concentrations than that found in ground water beneath those uplands. Denitrification takes place at depth because of the presence of reducing sediments. Where more of these sediments are near the surface at Morgan Creek, more nitrate is removed. The presence of excess nitrogen gas in ground water moving to Morgan Creek and the isotopically heavier $\delta^{15}N$ in nitrate in the creek are indicators that more denitrification took place in ground water moving to Morgan Creek than took place in ground water moving to Chesterville Branch. This example reinforces the importance of the geological setting through which ground water flows in relation to chemical concentrations observed in the subsurface.

Chemical Composition Changes in Ground Water due to Human Activity

The example of fertilizer application to uplands in Maryland points out another way in which chemical composition of ground water is influenced – by human introduction of chemicals, such as pesticide residues, oils, phenols, solvents, and metals, into the environment (**Figure 6**). Human activities that introduce chemicals into the atmosphere also can affect chemical composition of ground water over broad areas through atmospheric deposition on land surfaces and infiltration to the water table. In the case of fertilizer application or application of other agricultural chemicals, the human introduction of chemicals occurs over a broad area as well, also known as nonpoint source contamination. Other nonpoint sources are animal feedlots, septic tank drain fields, runoff of water from highways, and mining operations. In the case of accidental spills, infiltration from

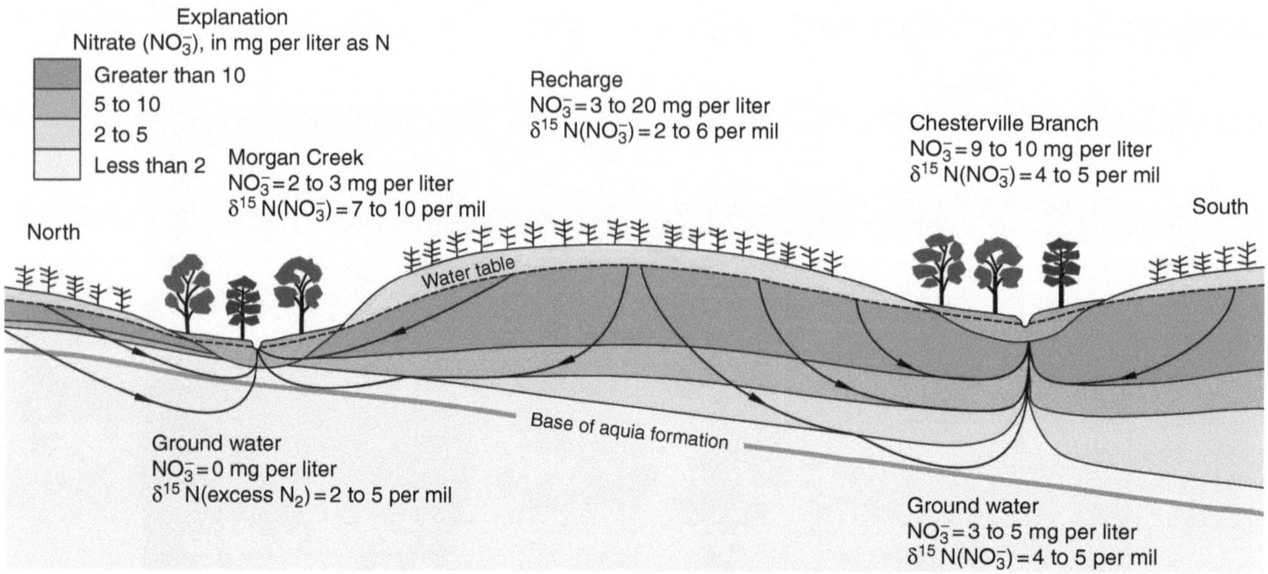

Figure 5 Denitrification had a greater effect on groundwater discharging to Morgan Creek than to Chesterville Branch in Maryland because a larger fraction of the local flow system discharging to Morgan Creek penetrated the reduced calcareous sediments near or below the bottom of the Aquia Formation than occurred in the flow system associated with the Chesterville Branch. Adapted from Winter TC, Harvey JW, Franke OL, and Alley WM (1998) Ground water and surface water a single resource. U.S. Geological Survey Circular 1139. Denver, CO: U.S. Geological Survey. http://pubs.usgs.gov/circ/circ1139/. Modified from Bolke JK and Denver JM (1995) Combined use of groundwater dating, chemical, and isotopic analyses to resolve the history and fate of nitrate contamination in two agricultural watersheds, Atlantic coastal plain, Maryland. *Water Resources Research* 31: 2319–2337.

landfills or disposal pits, underground storage tanks, injection wells (hazardous waste and brine disposal) or abandoned oil wells, the human introduction of chemicals into ground water occurs in localized areas, also referred to as point sources of contamination. Another human activity that can affect the chemical composition of ground water is through pumping, leakage from wells, or leakage between different geologic formations and aquifers due to leakage around the well casing, or leakage through well screens that are long enough to span those different formations and aquifers (**Figure 6**).

As human activities affect the chemical composition of ground water, the introduction of chemicals can act as a marker that indicates that ground water has been affected by human activity. Examples of such chemical markers are surfactants and optical brighteners used in detergents or even caffeine in wastewater (**Table 1**). Many of the pharmaceuticals used by humans can end up in ground water that has received wastewater, either directly from a septic system or indirectly from surface water containing those compounds, which then infiltrates ground water. Other similar markers are herbicides and pesticides, and their daughter products produced by decomposition of the parent chemical within ground water.

Wastewater infiltration to ground water may add organic carbon to ground water, thereby providing a source of electron donors for microbial processes, potentially changing the system from an oxidizing to reducing environment. In natural, uncontaminated environments, ground waters may contain sufficient organic carbon to affect oxidation–reduction reactions. Examples of such natural environments include ground waters that are recharged by surface waters, lakes, wetlands, and rivers, that may contain relatively large concentrations of dissolved organic carbon, or ground waters associated with peat or coal deposits. In any groundwater system, however, the introduction of additional organic carbon from human activity, such as wastewater, agricultural or industrial chemicals, petroleum, oil, or lubricants, can affect groundwater chemical characteristics. The solubility of the chemicals and their density determines how they will interact with ground water and move through the subsurface.

When viewed from the land surface, the movement of a chemical originating from a particular point in the subsurface is a plume moving from the point of larger concentration to lesser concentration in the groundwater flow field (**Figure 7**), along the direction of flow. Such plumes may intersect surface waters, thereby acting as a source of the chemical to

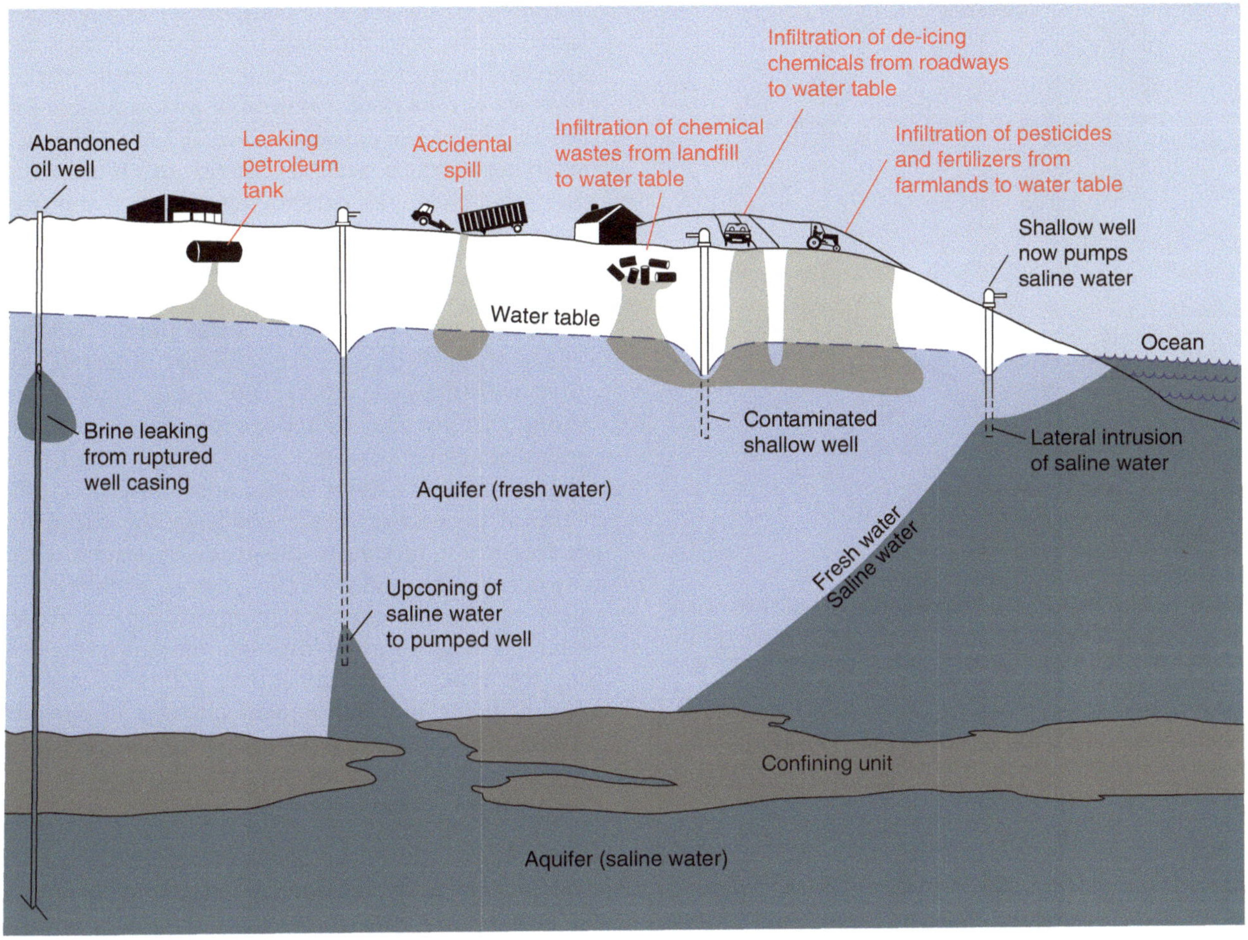

Figure 6 Mechanisms of groundwater contamination. Adapted from Johnston RH (1988) Factors affecting ground-water quality. U.S. Geological Survey, National Water Summary 1986—Hydrologic Events and Ground-Water Quality. U.S. Geological Survey Water-Supply Paper 2325, pp. 71–86. Washington, DC: United States Government Printing Office.

receiving surface waters. The movement of dissolved chemicals in ground water occurs by advection, diffusion, and dispersion. Chemicals transported with the ground water they are dissolved in move by the process of advection. Chemicals moving from areas of larger concentration to smaller concentration move by the process of diffusion. The mixing of waters of different concentrations due to differences in the velocity at which water travels at the pore level and through different strata is the process of dispersion.

The groundwater flow system is three-dimensional, thus the path of movement of a chemical originating from a point source should also be viewed at depth (**Figure 8**).

Chemicals dissolved in the ground water may be removed by groundwater flow once the source of the chemical is eliminated. Chemicals sorbed to solids in the subsurface, however, may remain, affecting chemical characteristics of ground water long after the source of the chemicals sorbed to solids are removed.

Some of the chemicals added to ground water by human activity are nonaqueous phase liquids (NAPLs), such as gasoline, whose concentrations exceed their water solubilities. Some NAPLs float on the surface of the water table, and slowly move in the direction of groundwater flow (**Figure 9**). Such compounds are less dense, or lighter than water (LNAPL). Some of the compounds associated with the NAPLs will volatilize and enter the unsaturated zone as a vapor. Some of the compounds in the NAPLs that are soluble or are colloidal will move within the groundwater flow system (**Figure 9**), providing a source of organic carbon to microbes. Some NAPLs are denser than water (DNAPL). These compounds will sink through the groundwater flow field until the compounds reach a confining layer or reach part of the groundwater flow field of like density.

As noted, the introduction of organic compounds affect the geochemical environment of ground water through the processes of biodegradation and

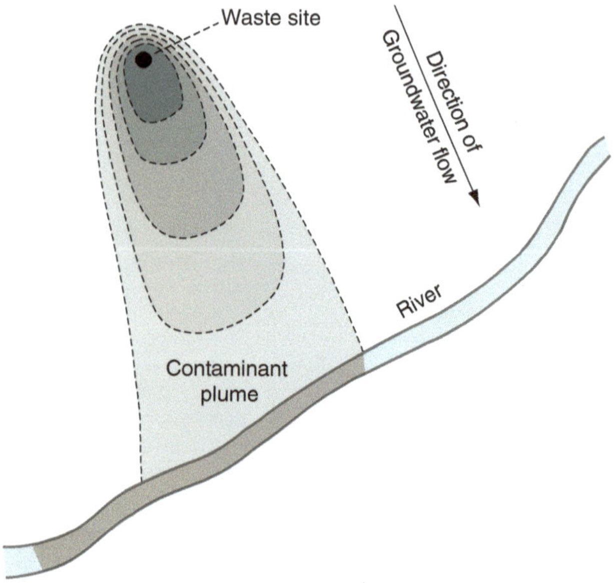

Figure 7 A highly simplified representation of a contaminant plume. The transport of contamination from a point source by ground water can cause contamination of surface water, as well as extensive contamination of ground water. Adapted from Winter TC, Harvey JW, Franke OL, and Alley WM (1998) Ground water and surface water a single resource. U.S. Geological Survey Circular 1139. Denver, CO: U.S. Geological Survey. http://pubs.usgs.gov/circ/circ1139/.

utilization of the organic compounds as a source of electrons, thereby affecting the oxidation or reduction condition of ground water. The result of these processes can be the formation of anaerobic zones in the vicinity of the carbon source (**Figure 10**) with a transition zone to aerobic conditions. Within each zone, the chemical compounds present and their concentrations will be affected by the oxidation or reduction potential of that parcel of ground water. When the electron activity (Eh), or oxidation–reduction potential, of ground water and the hydrogen ion activity (pH) change, the solubility of metals associated with the solid material of the subsurface may change, thereby altering the metals concentrations in ground water.

The chemical composition, solubility, and other chemical properties of the solids in the subsurface are important factors in the types of reactions that may occur in ground water, in general, and the consequent change in chemical composition of ground water, as water moves through the subsurface. So too, the type of reactions that will take place in ground water, and subsequent changes in chemical composition of ground water, when chemicals are introduced by human activity, are also a function of the chemical properties of the solids through which

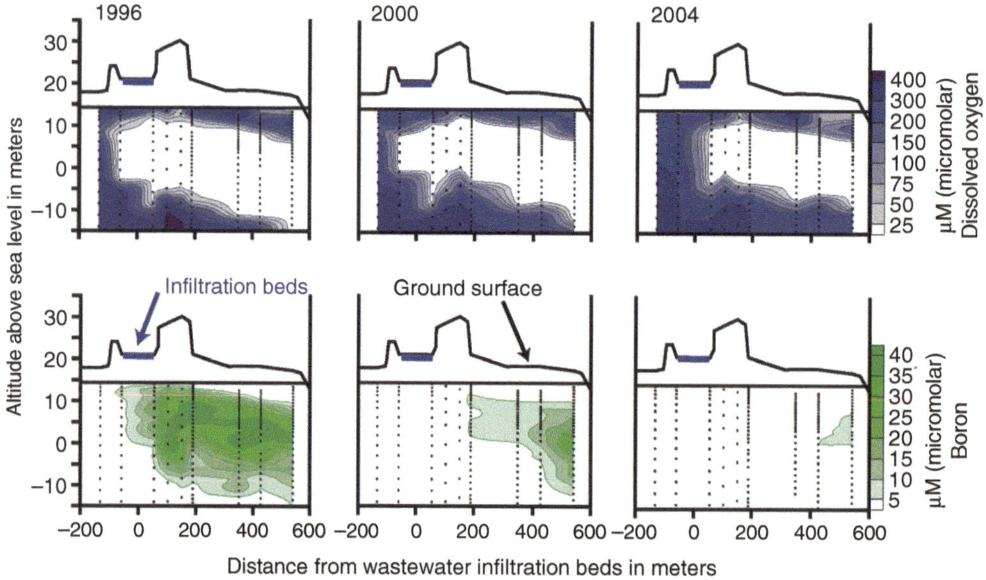

Figure 8 Vertical cross sections along the longitudinal axis of the wastewater plume on Cape Cod, MA. Each cross section gives a snapshot of the concentration of dissolved oxygen (upper three, blue) and boron (lower three, green) during 1996, 2000, and 2004. Concentrations are given in micromolar (μM, micromoles/liter). These cross sections show that mobile constituents, such as boron, have been flushed away from the infiltration beds by fresh ground water (flow is from left to right) upon cessation of inflilitration of wastewater, while dissolved oxygen concentrations have remained low. The oxygen in ground water flowing into the reservoir of adsorbed wastewater constituents from upgradient locations is quickly used up by microorganisms that are degrading the material leaching out of the sediments. Black dots show the positions of well screens and multilevel-sampler ports. The cross sections are a modified version of **Figure 3** from Repert DA, Barber LB, Hess KM, Keefe SH, Kent DB, LeBlanc DR, and Smith RL (2006) Long-term natural attenuation of carbon and nitrogen within a groundwater plume after removal of the treated wastewater source. *Environmental Science and Technology* 40:1154–1162. http://toxics.usgs.gov/highlights/gw_cessation.html.

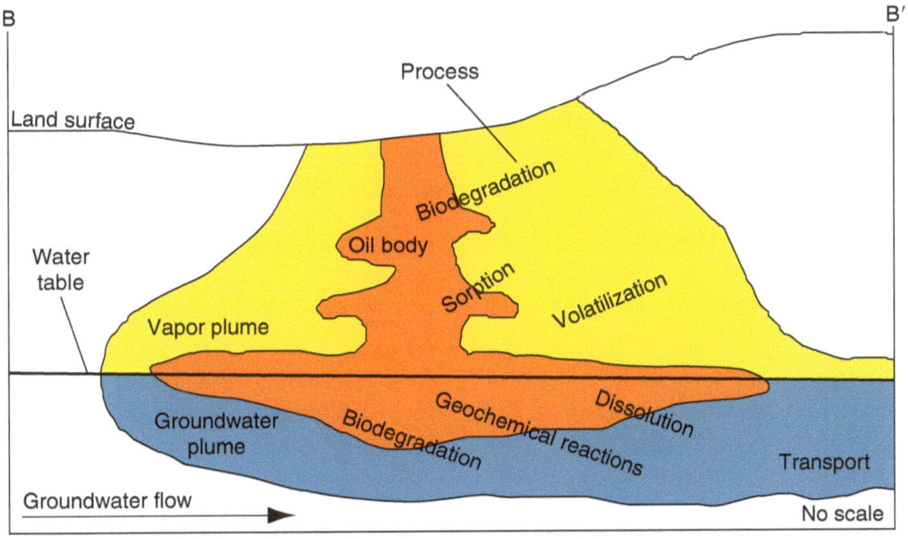

Figure 9 Processes critical to understanding the fate and transport of hydrocarbons in the subsurface at the Bemidji site. Adapted from Delin GN, Essaid HI, Cozzarelli IM, Lahvis MH, and Bekins BA (1998) Ground water contamination by crude oil near Bemidji, Minnesota. U.S. Geological Survey Fact Sheet FS-084–98, 4 p. http://mn.water.usgs.gov/bemidji/results/fact-sheet.pdf.

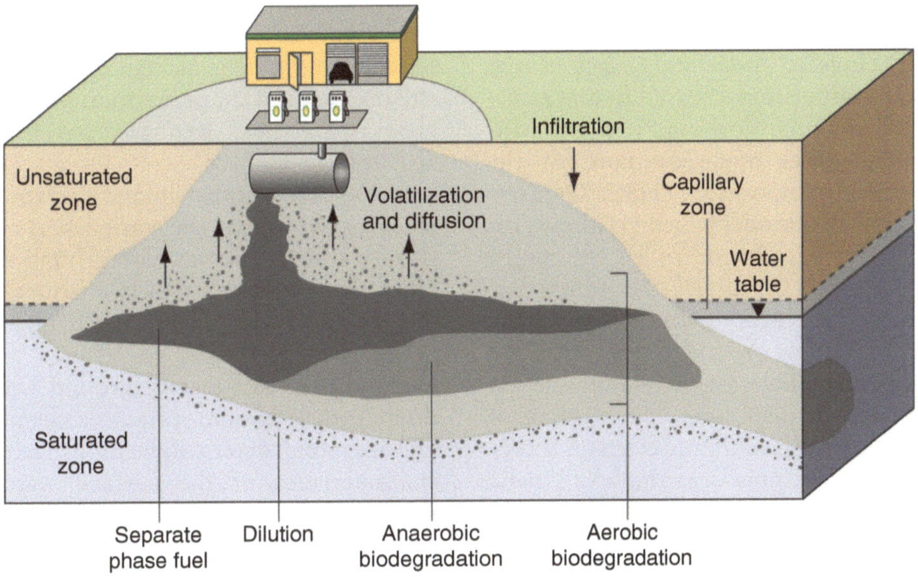

Figure 10 Conceptual illustration of the important natural attenuation processes that affect the fate of petroleum hydrocarbons in aquifers. Adapted from Bekins B, Rittmann BE, and MacDonald JA (2001) Natural attenuation strategy for groundwater cleanup focuses on demonstrating cause and effect. *Eos* 82: 57–58. Copyright 2001 American Geophysical Union. Reproduced/modified by permission of American Geophysical Union. http://toxics.usgs.gov/pubs/eos-v82-n5-2001-natural/e0s-v82-n5-2001-natural.pdf.

ground water moves. Acidic water containing large concentrations of metals is another example of human introduction of chemicals into the environment. When such water moves through solids containing minerals that will react under acidic conditions and large metals concentrations, the pH may be altered resulting in the precipitation or sorption of some metals, thereby decreasing concentrations of those metals in ground water. At the same time, metals with different solubilities or sorption characteristics may be released from the solid matrix as pH changes.

Geochemical Models and Mass Balance Examination of Chemical Composition Changes in Ground Water

Various low-temperature geochemical models, such as PHREEQC and NETPATH (http://wwwbrr.cr.usgs.gov/projects/GWC_coupled/), are tools used to help understand the complexities of the chemical composition of ground water. Such models allow the testing of hypotheses regarding the range of processes that have occurred that account for observed chemical composition of a parcel of water in the groundwater flow field. Determination of the saturation indices for various minerals based on thermodynamic data is a major component of these geochemical models. As such, the results provide information about which mineral phases are likely to dissolve or precipitate. Possible geochemical reactions that may occur along a flow path can be examined. Advective transport, fixed-pressure gas equilibria, surface complexation, and ion-exchange equilibria are used in PHREEQC. Use of geochemical models to determine speciation, reaction, and transport is known as forward modeling.

A second type of modeling that can be accomplished with the geochemical models is known as inverse modeling. Inverse modeling examines the observed chemical composition of two different aqueous solutions, and examines how reactions with different minerals and gases might account for the differences in chemical composition. PHREEQC (version II) is an example of a model which (1) allows for the evaluation of kinetically-controlled reactions, solid–solution equilibria; (2) allows the number of surface or exchange sites to vary with the dissolution or precipitation of minerals or kinetic reactants; and (3) includes isotope mole balances, fixed-volume gas phases, and diffusion or dispersion. The cycling of electrons between electron donors and electron acceptors by microbial metabolism is an example of kinetic reactions. Some observed changes in chemical composition of ground water not fully accounted for by equilibrium considerations, when examined in terms of mass balance, have revealed the importance of nonequilibrium reactions related to microbial activity. The oxidation–reduction state of ground water can be difficult to measure and determination of Eh alone may not reflect the full range of the microbial processes affecting chemical composition of ground water. Measurement of the changes in concentration of potential electron acceptors (oxygen, nitrate, ferric iron, and sulfate), the products of the microbial reactions (ferrous iron, sulfide, methane), and concentrations of dissolved hydrogen are good indicators of the oxidation–reduction state of ground water.

The Relation of Chemical Characteristics of Ground Water to Chemical Characteristics of Surface Water

The example of the interaction of ground water with the streams in Maryland (**Figure 5**) points out the importance of understanding groundwater flow paths and their relation to subsurface geological characteristics with regard to controls on chemical characteristics of ground water at different parts of the groundwater flow field and as that ground water flows into streams. As ground water flows into streams, it enters the flow field adjacent to the stream bottom, which may also include flow of stream water into and out of the stream bed. The area in which stream water flows into the subsurface and out again into the stream is known as the hyporheic zone. The area where ground water flows into the stream and interacts with the flow field of the hyporheic zone is a place where chemical transformations can be quite dynamic (**Figure 11**). The geomorphological setting also affects this interaction, such that the hyporheic flow field, and its relation to the interaction with ground water will vary according to the setting. Thus, when trying to relate chemical characteristics of ground water beneath uplands to the role those chemical characteristics play in affecting stream chemical characteristics, the hydrogeologic and geomorphological settings need to be taken into account.

In the case of other inland surface waters, such as lakes or wetlands, the varied ways in which these surface waters interact with ground water will determine the extent to which chemical characteristics of ground water will affect chemical characteristics of lakes or wetlands. In settings where a lake or wetland does not receive ground water, but instead discharges to ground water, the chemical characteristics of ground water will be influenced by the chemical characteristics of the surface water entering the subsurface.

Chemical characteristics of ground water can be quite varied, reflecting interaction with the matrix through which it flows and the time it takes for ground water to travel through that matrix. The chemical composition of water entering the groundwater flow system, either as water derived from rainfall or snow or ice melt, flow of water from surface water into the subsurface, or the introduction of chemicals through human activity, make different contributions to the chemical composition and reactions that take place in the receiving ground water. An understanding of the groundwater flow system, the geologic framework through which it flows, and the time needed for ground water to move along different flow paths is

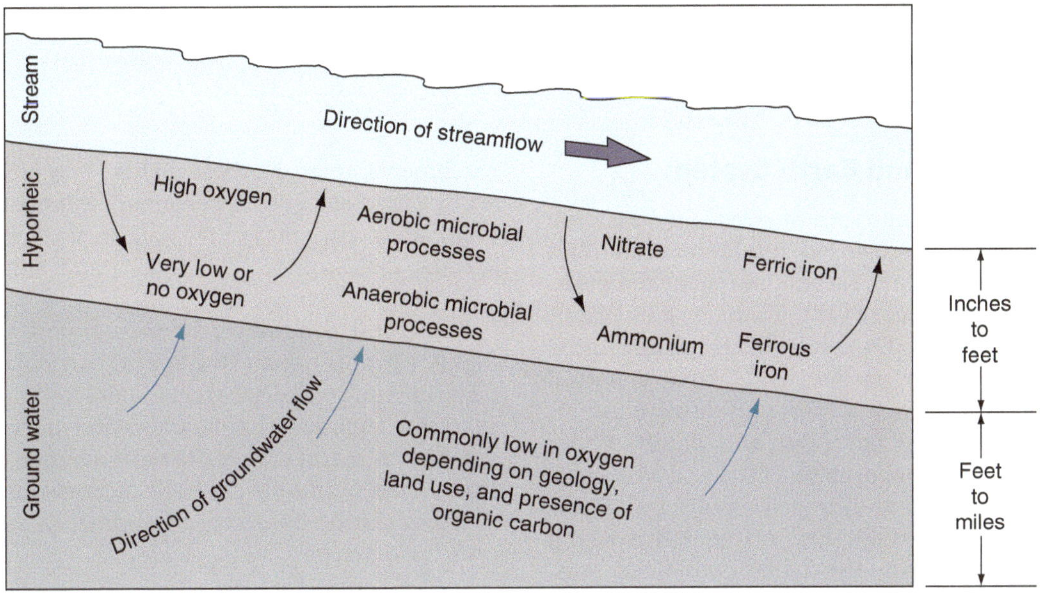

Figure 11 Microbial activity and chemical transformations commonly are enhanced in the hyporheic zone compared to those that take place in ground water and surface water. This diagram illustrates some of the processes and chemical transformations that may take place in the hyporheic zone. Actual chemical interactions depend on numerous factors including aquifer mineralogy, shape of the aquifer, types of organic matter in surface water and ground water, and nearby land use. Adapted from Winter TC, Harvey JW, Franke OL, and Alley WM (1998) Ground water and surface water a single resource. U.S. Geological Survey Circular 1139. Denver, CO: U.S. Geological Survey. http://pubs.usgs.gov/circ/circ1139/.

needed to understand all of the processes that contribute to the chemical characteristics of ground water observed at a point in the flow field.

See also: Ground Water.

Further Reading

Alley WM (ed.) (1993) *Regional Ground-Water Quality.* New York, NY: Van Nostrand Reinhold.
Chappelle FH (2000) *The Hidden Sea: Ground Water, Springs, and Wells,* 2nd edn. Westerville, OH: The National Ground Water Association.
Chapelle FH (2001) *Ground-Water Microbiology and Geochemistry.* New York: John Wiley.
Drever JI (ed.) (2004) Volume 5: Surface and Ground Water, Weathering, and Soils. In: Holland HD and Turekian KK (eds.) *Treatise on Geochemistry* 10 vols. Oxford, UK: Elsevier Pergamon. http://www.sciencedirect.com/science/referenceworks/0080437516.
Domenico PA and Schwartz FW (1998) *Physical and Chemical Hydrogeology.* New York: Wiley.
Fetter CW (1994) *Applied Hydrogeology.* Upper Saddle River, NJ: Prentice-Hall.
Jones JB and Mulholland PJ (eds.) (2000) *Streams and Ground Waters.* San Diego: Academic Press.
Miller JA (ed.) (2000) *Ground Water Atlas of the United States.* U.S. Geological Survey. http://capp.water.usgs.gov/gwa/gwa.html.
Mitsch WJ and Gosselink JG (2000) *Wetlands.* New York: Wiley.
Winter TC, Harvey JW, Franke OL, and Alley WM (1998) Ground water and surface water a single resource. U.S. Geological Survey Circular 1139. Denver, CO: U.S.Geological Survey. http://pubs.usgs.gov/circ/circ1139/.

Relevant Websites

http://water.usgs.gov/ogw/pubs.html – Selected U.S. Geological Survey ground-water publications.
http://www.groundwater.org – Groundwater Foundation.
http://www.agwt.org – American Ground Water Trust.
http://www.ngwa.org – National Ground Water Association.

Fluvial Export

M Meybeck, Université Pierre et Marie Curie, Paris, France

Riverine Fluxes and Earth System

River fluxes concern all types of river-borne material either particulate or dissolved: (i) total suspended solids (TSS), particulate organic carbon, nitrogen, and phosphorus (POC, PON, and POP), (ii) major ions (Ca^{2+}, Mg^{2+}, Na^+, K^+, Cl^-, SO_4^{2-}, and HCO_3^-, also reported as dissolved inorganic carbon, DIC), total dissolved solids (TDS = sum of major ions, plus dissolved silica, SiO_2 in milligrams per liter), nutrients (nitrate NO_3^-, ammonia NH_4^+, orthophosphate PO_4^{3-}, and SiO_2), dissolved organic carbon, nitrogen, and phosphorus (DOC, DON, and DOP). Bed material fluxes and fluxes of pollutants are also to be considered (e.g., heavy metals, PCBs, PAHs) but are not covered here for lack of reliable data at the global scale. Total phosphorus (TP) and total Kjeldahl nitrogen (TKN) are measured on unfiltered samples.

Fluxes are used in many fields of water and earth sciences:

- to establish chemical weathering rates (major ions and SiO_2),
- to assess the mechanical erosion and sediments transfers across the river networks to the coastal zone (TSS),
- to generate the nutrients and pollutants inputs to water bodies of any type and size (lakes, reservoirs, coastal zone, regional seas, global oceans),
- to control the pollution budgets within river basins,
- to establish the riverine component of the land to ocean biogeochemical fluxes for carbon, nitrogen, phosphorus, silica, and sulfur species (DIC, DOC, POC; NO_3^-, NH_4^+, PON, DON; PO_4^{3-}, DOP, POP, SO_4^{2-}).

On a given territory, fluxes *F*, i.e., mass per unit time, are obtained from a series of steps:

(i) at a given river station, instantaneous concentrations (C_i) and discharge (Q_i) are first multiplied to generate instantaneous fluxes F_i,
(ii) F_i are then integrated over the whole hydrological period to generate annual loads,
(iii) average loads (F_j) at stations are generated over multiyear periods, and
(iv) average loads (F_j) and yields (Y_j) at documented stations are extrapolated over greater basins or territories.

Riverine fluxes and water flows from catchments of different sizes are best compared when normalized to catchments area (A), these yields ($Y_j = F_j/A$) are generally expressed in tons per square kilometer per year (equivalent to $g\,m^{-2}\,year^{-1}$). For river flow, mean specific discharge (in $l\,s^{-1}\,km^{-2}$) and annual runoff (q in $mm\,y^{-1}$) are used.

In natural conditions, riverine fluxes are already highly variable in time and space, particularly in connection with river discharge. Cumulated human pressures on aquatic systems and water uses have now resulted in marked alterations of average concentrations, fluxes, and yields: both increase and decrease patterns can be observed depending on stations and on river material.

River Data Bases

The constitution of reliable data bases is the first step prior to any construction of river fluxes at any scale, from local to global. They should be representative and accurate for both average concentrations and water discharges. For a given station, each flux figure should be reported to a period of record and associated to a source reference, and it is highly recommended to describe precisely the flux method computation. For flux estimates on a given territory, e.g., a lake basin, it is also required to indicate the method used for the extrapolation of fluxes to undocumented river basins. The most used methods are (i) extrapolation of yields to unknown areas, (ii) extrapolation of concentrations to undocumented water flows, (iii) combinations of hydrological, biogeochemical, and anthropogenic sources models.

It is recommended that water discharge, river quality, and TSS concentrations are determined on similar periods and at the same stations to allow for comparison of fluxes (explained later). TSS fluxes need special attention, since they are still in many cases computed from water quality records, i.e., on surface samples. TSS may therefore differ from the suspended particulate water (SPM) determined by hydrological surveys and representative of the whole river section (weighted by water velocity from several vertical profiles sampled very frequently, ideally daily). Both TSS and SPM data are here merged as TSS.

World registers of river quality and/or fluxes are very limited. For *Q*, TSS, and TDS, the most complete dataset is the one of Milliman *et al.* (1995). The UNEP GEMS water program (http://www.gemswater.org) has held a global database on rivers and their tributaries since 1978 and a pdf version of Meybeck and Ragu (1996), who have listed the general river basin attributes, averages of major ions, carbon, and nutrients for the 500 major world rivers discharging to oceans. Few regional databases have also been set up as for the United States (http://waterdata.usgs.gov/nwis/qw), Europe by the European Environment Agency, or for the former Soviet Union.

Global river fluxes combine water chemistry databases and global water budgets. They have evolved rapidly with the development of global databases, GIS, and modeling (Vörösmarty and Meybeck, 2004). A complete set of global models for river carbon, nitrogen, and phosphorus fluxes, including those resulting from human impacts, was published in 2005. Currently, the most accepted figure of river inputs to oceans is 40 000 $km^3 year^{-1}$ for a potential drainage area to the oceans of 115.6 Mkm^2 termed exorheic drainage, in which large regions (39.5 Mkm^2) do not presently generate significant runoff. Such regions are termed arheic, defined here as a river basin in which average $q < 3$ mm $year^{-1}$ as commonly found in Sahara, Saudi Arabia, Namibia, Australia. The remaining nonglaciated area of the continents (17.4 Mkm^2) are drained towards internal regions, termed endorheic, as the Caspian Sea, Aral Sea, Lakes Chad (Central Africa), Eyre (Australia), Turkana (East Africa) and Titicaca (Bolivian Altiplano), the Great Basin (USA), and the Takla Makan and Gobi deserts in Central Asia. Apart from the Caspian Sea catchment most endorheic regions are also arheic excepted in their mountainous headwaters. The total area of arheic regions is 50.2 Mkm^2.

Riverine Fluxes at Stations

Daily to Seasonal Variations of Riverine Fluxes at Stations

Riverine fluxes vary at all time scales: flood event, seasonal biogeochemical cycles, annual hydrological year, interannual succession of dry and wet years, trends. The first control factor of these temporal variations is the river discharge, which commonly ranges over 1 order of magnitude or less in large rivers, to 2 and even 3 orders of magnitude in small catchments. For a given catchment size, seasonal variations are higher for the tropical monsoon, the Mediterranean, and the snowmelt regimes while interannual variations are higher for the monsoon and Mediterranean regimes.

Concentrations (C) vs. discharge (Q) relations, termed rating curves by engineers, depends greatly on the type of material:

1. Major ions tend to decrease with increasing discharge at most stations.
2. Nutrients have a station-specific pattern that depends on sources and sinks. Dissolved inorganic nitrogen and phosphorus species may seasonally decrease during spring and summer low flows because of algal bloom in eutrophic rivers. During low flow, nitrate can be decreased by an order of magnitude by riparian denitrification. Nutrients originating from point sources (e.g., NH_4^+, PO_4^{3-}, from municipal sewage or cattle farms) are diluted during high flows and present characteristic exponential decrease patterns with Q.
3. Suspended solids TSS increase faster than Q during floods at high water stages. A similar pattern is noted for concentrations (mg l^{-1}) of organic carbon (POC) and pollutants (PCBs, PAHs, heavy metals) attached to particles. However, their contents in river particulates (mg g^{-1}, μg g^{-1}, %) may remain stable or even decrease significantly during these events, as for POC.

Fluxes of TSS and their associated material (POC, PON, POP, particulate pollutants) are the most variable ones at a given station, since they increase faster than water discharge. Typically $C = aQ^b$, with $0 < b < 0.5$ for TP and TKN, $0.5 < b < 1.5$ for TSS. For TDS, the b exponent is generally negative, and can reach –0.6, also for NH_4^+ and PO_4^{3-} when they originate from point sources. Seasonal NO_3^- and PO_4^{3-} variations cannot be well quantified by b.

As an ideal dilution, which would result in a constant flux whatever the discharge may be ($b = -1$), is never observed, all river material fluxes increase with river flow.

At a given station, instantaneous fluxes of TSS extend over 3–6 orders of magnitude depending on b exponent and river regime. When normalized to the catchment area, daily TSS fluxes or yields vary from 0.1 to more than 100 000 kg km^{-2} day^{-1} and average long-term fluxes range from less than 10 to more than 10 000 kg km^{-2} day^{-1}. The range of variability of dissolved fluxes is 100–1000 less.

Flux Determination from Discrete Sampling

In medium and large river basins exceeding 1000 km^2, it is assumed that water quality data does not vary during 24 h, therefore instantaneous concentrations resulting from the analysis of one sample e.g., ions, nutrients or TSS, are attributed to average daily discharges. Ideally, the sample should be representative of the whole river section, vertically and laterally; this is rarely checked in most water quality surveys. As water quality surveys are performed at a weekly to monthly basis but not every day, the resulting set of discrete daily fluxes must be extrapolated to the whole hydrological year (preferred to a civil year), then yearly fluxes are ideally summed and averaged in order to account for dry and wet years variability over a suitable period of record (10 years for TDS, less for nutrients and more for TSS).

Flux uncertainties depend on

(i) representativity of the sample within the river sections as for particulate matter,
(ii) sampling frequencies,
(iii) relations between concentrations C_i and discharge Q,
(iv) other temporal variations such as seasonal uptake of nutrients, flood hysteresis of TSS,
(v) flux determination methods used.

Flux estimation procedures are numerous; the most comprehensive review has been made by Philipps *et al.* (1999). They should be adapted to the behavior of each type of river material, particularly to its relation with river flow and to sampling frequency. The most common ones are

(i) mean of instantaneous fluxes (F_i):$F = \sum(F_i/n)$ (constant flux hypothesis around the sample);
(ii) constant concentration hypothesis around the sample: each sample is associated to a water volume to which the sample concentration is attributed;
(iii) flow-weighted mean concentrations:$F = (\sum F_i/\sum Q_i) \times Q$ where Q_i and Q are the instantaneous and average discharges, respectively;
(iv) linear interpolation of concentrations or fitting of seasonal variations curves between discrete values;
(v) linear interpolation of fluxes;
(vi) rating curve using the C vs.Q relationship across the whole range of water discharge;
(vii) truncated rating curve using only the samples corresponding to the highest river flows (e.g., superior to median flow).

Rating curves are systematically used for suspended particulate fluxes (see reviews in Horowitz, 2003, and Williams, 1989). Methods (i)–(iv) are more used for dissolved salts and nutrients.

Flux Durations

Duration curves are established on daily fluxes, either measured (rarely performed) or reconstituted, e.g., on the basis of rating curves, which are ranked in decreasing order and then cumulated. The percent of cumulated fluxes versus cumulated elapsed time is termed the flux duration relation, which can be linearized in a double probability scale (**Figure 1**).

Flux duration curves can be very different according to river flow regime, to basin area and to the type of river-borne material. In the monsoon-fed Khong Mala River (Thailand), 70% of TSS fluxes are transported in 2% of time, i.e., a total of 7 days per year on average, during which 26% of river flow is carried. In the Rhone Lacustre (outlet of lake Geneva), 10.8% of TSS fluxes and 3% of river flow are transported during the same period. In all rivers, the TSS durations curve is above the river flow duration curve (**Figure 1**). For dissolved substances, the duration curves are generally below the river flow duration or close to this one. The percentage of river material that is carried in 2% of time (M_2) is a good indicator and can be used to determine the bias and uncertainty of river TSS, TDS, and nutrients fluxes when they are based on discrete sampling. It must be noted that for smaller catchments ($A < 1000\,km^2$), duration curves must be based on sub-daily datasets.

Other indicators of flux duration have been proposed as the percentage of time necessary to carry half of the river load (**Table 1**).

Interannual Variations of Fluxes

Large rivers with regular flow regimes (pluvial oceanic, snow melt, glacial melt) are characterized by limited year to year flux variations – e.g., less than 1 order of magnitude for TSS fluxes – while small catchments with irregular flow regimes (Mediterranean, hurricane/typhoon-influenced regimes) have the largest interannual variations, reaching 2 orders of magnitude. For instance in 'normal' years, the Oued Zeroud ($8950\,km^2$, Tunisia) does not reach the Mediterranean Sea and does not flow most of the year. In September and October 1969, an exceptional flood discharged $2.4\,km^3$ of water and 240 Mt of sediments to the sea, with maximum q reaching $1900\,l\,s^{-1}\,km^{-2}$ and TSS exceeding $10\,g\,l^{-1}$, making Oued Zeroud the no. 1 sediment discharge to the Mediterranean Sea that year, ahead of the Nile ($120\,Mt\,year^{-1}$, before Aswan Dam construction). Interannual variations for dissolved material are generally close to those of river discharge.

Geographic Variability of Riverine Fluxes

Particulate Fluxes and TSS Levels

Natural yields of suspended matter (Y_{TSS}) depend on multiple control factors: (i) basin size, (ii) river runoff, (iii) catchment relief, (iv) catchment lithology,

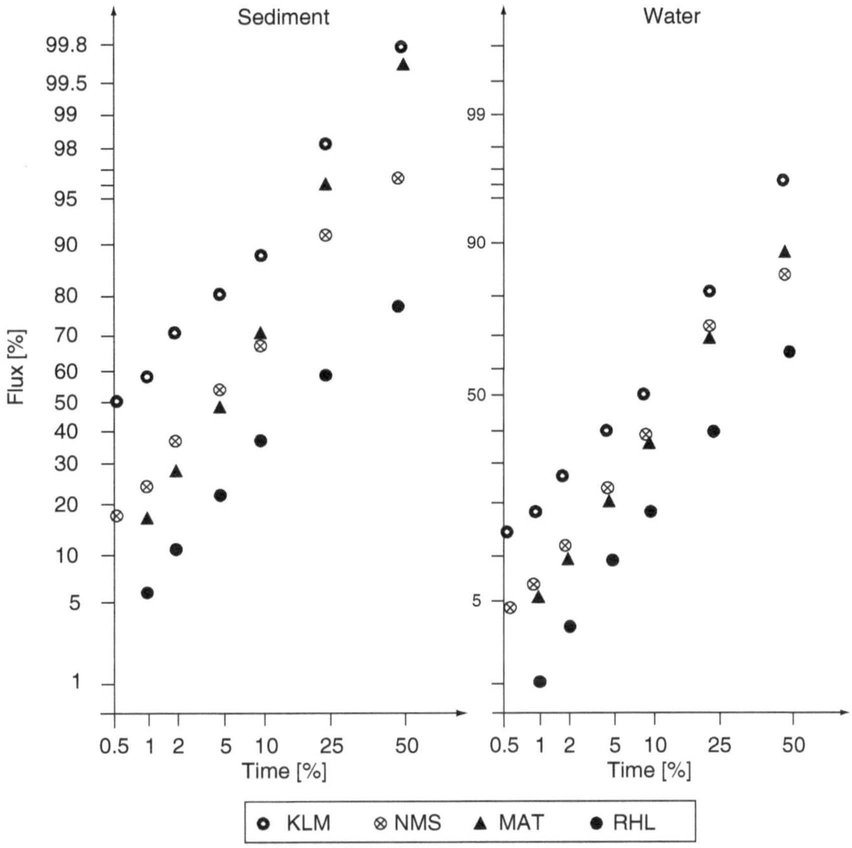

Figure 1 Flux duration curves for water and suspended particulate matter: percent of fluxes discharged in percent of time (Henry Law probability scale). Examples for the Khong Mala (KLM,Thailand) the Nam mae Pai (NMS, Thailand) the Matanuska (MAT, Alaska) and the Rhone Lacustre (RHL,Switzerland). Adapted from Meybeck M, Laroche L, Dürr HH, and Syvitski JPM (2003) Global variability of daily total suspended solids and their fluxes in rivers.*Global Planetary Change* 39: 65–93.

and (v) occurrence of lakes. Y_{TSS} increases with basin slope, and occurrence of most erodible rocks (loess > volcanic ash > loose sedimentary rocks > shales > sandstone > hard volcanic rocks > other crystalline rocks). Nonweathered rocks (e.g., those exposed after glacial erosion) are also less erodible than those exposed to weathering over long periods. Y_{TSS} also decreases with catchment size as the opportunity for redeposition in floodplain increases. Y_{TSS} is naturally very low at lake outlets, e.g., Saint Lawrence and Neva rivers (**Table 1**), because of particulate settling.

The sediment fluxes of rivers have greatly changed over time in response to human impact. On the one hand, sediment loads are increasing because of land clearance for agriculture and other land surface disturbance. Forest clear-cutting on steep slopes is responsible for the most abrupt changes, as observed in Papua New Guinea in the last 100 years. In other regions, the sediment increase started in earlier, some 2000 years ago (Danube) and 1000 years ago (Ebro River in Spain) (Walling, 2006). On the other hand, the trapping of sediments at the global scale by hundreds and thousands of small reservoirs (volume $V < 10^{-3}$ km^3) and a few dozen of very large ones (V > 10 km^3) is now widespread and has already had an impact at the global scale: more than 30% of river sediments generated by mechanical erosion is intercepted and stored by reservoirs.

The global range of median TSS concentrations in rivers (C_{50}), of discharge-weighted concentrations (C^*), and of their related daily sediment yields (Y_{50} and Y^*) are presented in **Table 1** together with the percentage of time necessary to carry half of the sediment flux ($T_{s50\%}$). The geographic variability is enormous, due to the hydrological regimes, relief, or basin size. It is also important to note the difference between median values (C_{50},Y_{50}) generally obtained from classical water quality surveys and discharge-weighted values (C^*,Y^*) based on daily surveys, particularly during floods.

The spatial distribution of interannual TSS concentrations and fluxes indicators is presented on **Table 1** for a set of rivers (exceeding 1000 km^2) that represent a wide variety of river regimes. All indicators vary over several orders of magnitudes, depending on local or regional conditions.

Table 1 Global range of concentrations and fluxes of suspended particulate matter (SPM) in rivers surveyed at the daily scale, medium and large basins

	Very low	*Low*	*Medium*	*High*	*Very high*	*Extremely high*
C_s^* (mg l^{-1})	Discharge-weighted total suspended solids					
	5–20	20–100	100–500	500–2000	2000–10 000	>10 000
		Bandama	Arve	Eel	Pecos	
	Annapolis	Chaudière	Huai Mae Ya	Matanuska	Piray	Colorado
	Rhone Lacustre	Loire	Red Deer	Mekong	Rio Grande	El Abid
	St. Lawrence	Sacramento	Stikine	Mississippi	Walla-Walla	Little Colorado
Y^* (kg km^{-2} day^{-1})	Average daily suspended solids yields					
	<10	10–50	50–200	200–1000	1000–5000	>5000
		Annapolis		Khlong Mala	Colorado	
	Gambia	Nzi	Chaudière	Mississippi	Eel	Alpine Rhine
	Rhone Lacustre	St. Lawrence	Middle Rhine	Peace	El Abid	Espejos
		Somme	Sacramento	Pecos	Matanuska	
					Salween	
C_s^*/C_{s50}	Weighted-suspended solids variability					
	1–2	2–5	5–10	10–20	20–50	>50
	Loire	Annapolis				
	Mississippi	Bani	Arve	Alpine Rhine	Peace	Eel
	Rio Grande	Fraser	Chaudière	Matanuska	Red Deer	Walla-Walla
	Seine	Lot	Salween	Stikine		
Y^*/Y_{50}	Suspended solids flux variability					
	<2	2–5	5–10	10–20	20–50	>50
	Mississippi	Bandama	Bani	Gambia	Alpine Rhine	Eel
	Rhone Lacustre	El Abid	Colorado	Khlong Sok	Peace	Matanuska
	Somme	Fyris	Fraser	Peace	Piray	Red Deer
	St. Lawrence	Rio Grande	Rhone Alpestre	South Saskat.	Stikine	Walla-Walla
$Ts_{50\%}$	Percentage of time needed to carry half of the sediment flux					
	>16.5%	16.5–8%	8–3.4%	3.4–1.4%	1.4–0.4%	<0.4%
	Middle Rhine	Bani	Adour	Chaudière	Eel	
	Mississippi	Fraser	Gambia	Red Deer	Grand Morin	Bermejo
	St. Lawrence	Middle Mekong	Sacramento	Stikine	Khlong Mala	Elvira
	Somme	Niger	Upper Mekong	Tuy	Nivelle	Walla-Walla

C_{50}: median concentrations(mg l^{-1}),C^*: discharge-weighted concentrations (mg l^{-1}),Y_{50}: median specific daily yields (kg km^{-2} day^{-1}),Y^*: long-term average specific yield (kg km^{-2} day^{-1}), $Ts_{50\%}$: percent of time during which half of the SPM flux is carried.

Source

Meybeck M, Laroche L, Dürr HH, and Syvitski JPM (2003) Global variability of daily total suspended solids and their fluxes in rivers. *Global Planetary Change* 39: 65–93.

Table 2 Ranked major riverine fluxes to oceans: water discharge (*Q*), annual runoff (*q*), sediment fluxes (F_{TSS}), dissolved fluxes (F_{TDS}) and river length (*L*)

River	River flow				Basin area		Sediment flux			Dissolved flux		Length	
	Q_{nat} (km³/y)	#	*q* (mm/y)	Q_p (km³/y)	*A* (Mkm²)	#	F_{TSS} (Mt/y)	#	F_{TSSP} (Mt/y)	F_{TDS} (Mt/y)	#	*L* (km)	#
Amazon	6590	1	1078		6.1	1	1200	1		287	1	6280	2
Congo (Zaïre)	1200	2	324		3.7	2	22.8	50		42.5	17	4370	8
Orinoco	1135	3	1032		1.1	15	150	13		28.2	21	2780	24
Chang Jiang (Yang Tse)	928	4	513		1.8	10	480	7	**	192.9	2	5520	4
Yenisey	620	5	239		2.59	7	5.9			65.8	7	3490	14
Parana	568	6	204		2.78	6	79	22	**	48.7	15	4700	5
Mississippi[a]	580	7	195	529	2.98	4	500	6	**305	146.1	3	5985	3
Lena	525	8	211		2.49	8	17.6			58.8	12	4400	6
Brahmaputra	510	9	879		0.58	32	540	4		51.4	14	3000	18
Ganges (Ganga)	493	10	470		1.05	17	520	5		91.5	5	2525	29
Irrawaddy (Ayeyarwady)	486	11	1180		0.41	39	260	8		97.7	4	2300	31
Mekong	467	12	587		0.79	24	150	12		46.1	16	4350	9
Ob	404	13	135		2.99	3	16.5		*	40.8	18	3650	12
Tocantins	372	14	491		0.76	25	75	23	*	15.2	39	2700	26
Zhu Jiang (Pearl)	363	15	831		0.44	37	69			58.4	13	2100	35
Amur	344	16	185		1.85	9	24.9	45		18.9	29	2820	20
Saint Lawrence	337	17	330		1.02	18	4.0			61.7	10	3060	17
Mackenzie	308	18	172		1.79	11	42	31		64.3	9	4240	10
Magdalena	237	19	1079		0.235	56	220	10		28	22	1530	
Columbia	236	20	353		0.67	28	15		**10	27.1	23	1995	
Salween (Thanlwin)	211	21	649		0.325	44	?			64.6	8	2820	21
Danube	207	22	250		0.817	23	68	26	*48	88.7	6	2800	22
Yukon	200	23	236		0.849	22	60	29		36.6	19	3186	15
Essequibo	178	24	1085		0.164	67	4.5			4.9		970	
Niger	154	25	128		1.2	13	40	34	**	9.1		4160	11
Ogooue	150	26	732		0.205	61	?			?		850	
Uruguay	145	27	604		0.24	53			11	10.9		1610	
Fly	141	28	2189		0.064	150	115	18		16.3	49	620	
Kolyma	132	29	200		0.66	29	16.1			9.7		2130	34
Pechora	131	30	404		0.324	46	6.5			9.8		1810	
Mamberano	130	31	1675		0.078	124	?			?		?	
Hong	123	32	1025		0.12	82	130	14	**	18	33	1200	
Sepik	120	33	1525		0.079	120	8.1?			13.7		700	
Fraser	114	34	510		0.22	58	20	50		10.4		1110	
Meghna	111	35	1388		0.08	115	?			?		950	
Northern Dvina	110	36	316		0.348	43	4.5		**	19.6	28	740	
Zambezi	106	37	80		1.33	12	20			8.5		2660	27
Godavari	105	38	335		0.31	47	170	11	**	20.3	27	1500	
Kapuas	101	39	1310		0.077	125	?		***	?		1010	
Sao Francisco	90	40	143		0.63	31			***6	[5.9]		2800	22
Indus	90	41	62	57	0.916	20	250	9	***59	[10.3]		3180	16

[a] Mississipi River proper and Red River are considered separately.

Q_{nat},Q_p: natural and present discharges after damming, irrigation, and/or diversion.

F_{TSS} and F_{TSSp}: natural and present sediment fluxes.

F_{TDS} and *L*: from Meybeck and Ragu (1996).

River names according to Time Atlas of the World.

*, **, *** Relative impacts of damming on river fluxes.

Source

Meybeck M and Ragu A (1996) River discharges to the oceans. An assessment of suspended solids, major ions and nutrients. Environment Information and assessment Rpt. UNEP, Nairobi, 250 pp .

Milliman JD, Rutkowski C, and Meybeck M (1995) River discharge to the sea. A global river index (GLORI), LOICZ reports and studies, Texel ND, pp. 125.

Major Ions and Total Dissolved Solids

The characteristics of major rivers draining to the oceans, and their fluxes of particulate (F_{TSS}) and dissolved matter (F_{TDS}) are listed in **Tables 2** and **3**.

In natural conditions, the factors controlling the major ions in rivers are basin lithology, river runoff, and geological history. Spatial ranges of average concentrations commonly range over 1–2 orders of magnitude and, for a given ion, yields are somewhat more variable. The comparison of river chemistry for a sample or at a station to a world average makes therefore little sense: local to regional references should be used.

The sum of cations in milliequivalent per liter (TZ^+) is commonly used by geochemists instead of total dissolved solids (TDS), used in water quality surveys, to compare samples and stations, and the proportions of dominant cations and anions (vs TZ^+ and TZ^-) are also considered. Granites, gneiss, and micaschists have waters with very low TZ^+ ($0.1 < TZ^+ < 1\,meq\,l^{-1}$). For sedimentary rocks TZ^+ is highly variable and ranges from $<0.5\,meq\,l^{-1}$ for quartz sandstone to $>5\,meq\,l^{-1}$ for gypsum and halite-bearing formations (up to $50\,meq\,l^{-1}$ in few rivers as the Urubamba, an upper Amazon tributary). Shales and marl provide river waters with $0.5 < TZ^+ < 3\,meq\,l^{-1}$, depending of their calcium carbonate contents. In chalk and limestone catchment rivers, TZ^+ are commonly between 3 and $5\,meq\,l^{-1}$. This TZ^+ distribution is masking the variations of dominant cations and anions: in dilute waters, proportions of Ca^{2+}, Mg^{2+}, Na^+, and K^+ are relatively close. When TZ^+ increases, Ca^{2+} generally dominates over Mg^{2+}, then Na^+ and K^+. In limestone-draining waters, Ca^{2+} exceeds 80% of TZ^+. Above $TZ^+ = 10\,meq\,l^{-1}$, calcite is precipitated, Ca^{2+} gradually decreases and Na^+ is dominating. Bicarbonates (HCO_3^-) are the most common anion in most waters excepted for the very dilute ones ($TZ^+ < 0.2\,meq\,l^{-1}$ in the Rio Negro, Amazonia) for which SO_4^{2-} and Cl^- may be dominant although in very low concentrations, and for the evaporites-draining waters ($TZ^+ > 5\,meq\,l^{-1}$) in which SO_4^{2-} and Cl^- are gradually dominant. Pyritic shales provide waters with high natural SO_4^{2-} levels. Examples of lithology influence on river ionic composition can be found in Hem (1989) and Meybeck (2003).

Silica originates from the weathering of silica minerals found in crystalline rocks particularly in volcanic rocks (olivine), and of amorphous silica found in some sedimentary rocks (chalk). High silica levels are thus found in volcanic catchments (**Table 4**) (Tone and Shinano rivers, Japan, see **Table 4**), in chalky catchments rich in amorphous silica (Thames), and downstream of hydrothermal fields (Waikato River, New Zealand). The lowest silica levels are noted in lake-dominated catchments (Churchill, Neva, St Lawrence rivers) because of the SiO_2 uptake by diatoms. Similar silica uptake is often noted downstream of reservoirs cascades in impounded catchments. There is a growing evidence of silica control by the terrestrial vegetation and soil phytoliths.

At global scale, annual average yields of major ions and silica increase with the river runoff; second-order variations result from the catchment lithology. In rivers draining arid and semiarid regions ($q < 30\,mm\,year^{-1}$), the evaporation leads to a progressive precipitation of salts and yields are very limited, as in Central Asia.

Major World Rivers and Their Ranked Fluxes

River flows and catchment areas are now well documented on most major rivers (**Table 2**). The Amazon and Congo rivers are respectively # 1 and 2 with regards to annual flow (Q in $km^3\,year^{-1}$; $1\,km^3\,year^{-1}$ is equivalent to an average discharge of $31.7\,m^3\,s^{-1}$), followed by the Orinoco, Chang Jiang (Yang Tse), Yenisey, and Parana. Figures listed in **Tables 2** and **3** include both rivers in pristine state and those already impacted by pollution, damming, irrigation, and diversion. Such dataset has some degree of uncertainty, also it is continuously revised:

(i) recent studies on the Amazon have lead to modifications for both Q and F_{TSS};
(ii) some natural fluxes estimates are relatively old, although they cannot be checked today due to their present land use change;
(iii) damming and irrigation are progressing in many regions of the world and therefore average Q and M_s are not stable; even in large river basins, erosion control measures can also be efficient and reduce sediment yields, as for the Yellow River (Huang He);
(iv) river TDS may be greatly affected by pollution;
(v) river drainage area in dry regions is difficult to be defined as large portions only contribute occasionally (e.g., Nile, Niger).

The rivers for which present sediment load is certainly much lower than natural figures are indicated by an asterisk. The concept of the 'world's greatest rivers' is not applicable; top-ranked rivers are very different for each river attribute: catchment area, length, flow, fluxes of dissolved salts, and particulate matter. Rivers drained to the oceans (exorheic, **Tables 2** and **3**) are only considered here.

Some rivers correspond to internal drainage (endorheic) as Volga (Russia), Chari (Central Africa),

Table 3 Miscellaneous river fluxes to oceans: water discharge (*Q*), annual runoff (*q*), suspended sediment fluxes (F_{TSS}), dissolved fluxes (F_{TDS}) and river length (*L*)

River	*River flow*				*Basin area*		*Sediment flux*			*Dissolved flux*		*River length*	
	Q_{nat} *(km³/y)*	#	q_{nat} *(mm/y)*	Q_p *(km³/y)*	*A (Mkm²)*	#	M_s *(Mt/y)*	#	M_{sp} *(Mt/y)*	M_d *(Mt/y)*	#	*L (km)*	#
Churchill (Hudson Bay)	40		87	25.8	0.298	48				3.45		1600	
Colorado (US/Mexico)	18.5		29	0.1	0.639	30	120	16	***0.1	[20.5]	[26]	2330	30
Copper	56.6		831		0.063		70	24				460	
Dnepr			106	53	0.5	34	2.3		***	14.6		2200	
Don			49	20.7	0.42	38	4.8		***	17.1	37	1870	
Elbe	23.7		162		0.146	70	0.84			(17.2)	(36)	1110	
Fuchun Jiang			686				50	30		0.18		494	
Hai Ho	9		34		0.264	51	660	3					
Huai	50		185		0.27	50	14					900	
Huang He (Yellow)			55	41	0.75	26	1100	2		18.9	30	4670	6
Indigirka	61		61		0.362	42	12.9			3.8		1726	
Jubba	17.2		23		0.75	27				5.7		1600	
Kaoping	8.5		2615		0.0033		36	33				170	
Khatanga	85.3		234		0.364	41	1.7			9.4		1636	
Krishna	30		116		0.259	52	64	27		9.5		1290	
Liao	16.2		74		0.219	60	41					1350	
Limpopo	26		59		0.44	36				6.2		1600	
Mahanadi			466	66	0.142	73	60	28		9.7		850	
Murray	23.6		7.9		1.06	16	30			(3)		3490	14
Narmada	40.7		394		0.99		125	15	30.7	11.1		1300	
Nelson	83		79		1.13	14				21	25	2600	28
Neva	84.4		285		0.282	49	0.82			5.4		74	
Nile	83.2		29	0.3	2.87	5	120	17	***2	[32]	[20]	6660	1
Olenek	35.8		163		0.219	59		35		4.06		2270	32
Orange			11	11.4	1	19	89	20	**17	[2.1]		1860	
Parnaiba	32.2		99	32	0.325	45						1450	
Po	46		699		0.017		15.2			17.3	35	680	
Purari	84	46	2750		0.0306		80	21		10.6		630	
Red River (Louisiana)[a]	55.5		23		0.24	54						2075	36
Rhine	69.4		310		0.224	57	3.4			(60.5)	(11)	1360	
Rhone	59.9		565		0.0196		31			18.2	32	810	
Rio Grande (Rio Bravo)	18		21	0.72	0.87	21			***20	14		2880	19
Senegal	24.4		55		0.44	35	21		***	[1.24]		1430	
Shatt el Arab			85	45.7	0.54	33	105	19	**	[18.3]	[34]	2760	25
Usumacinta	55.5		1164		0.048					13.3		430	
Volta			93	36.8	0.39	40	19		**	[2]		1600	
Weser	11.3		231		0.046		0.33			(26.1)	(24)	724	
Wisla (Vistula)	34.1		164		0.198	63	2.5			(18.9)	(31)	1090	
Yana	34.3		144		0.238	55	3.5			1.56		872	

Values in parentheses indicate rivers with marked increase of TDS fluxes resulting from human impacts.Values in square brackets indicate former TDS fluxes prior irrigation and diversion.

Source

Meybeck M and Ragu A (1996) River discharges to the oceans. An assessment of suspended solids, major ions and nutrients. Environnement Information and assessment Rpt. UNEP, Nairobi, 250 pp.

[a]Red River upstream of Atchafalaya connexion.

Table 4 Nutrients and organic carbon concentrations (mg l^{-1}) in world rivers; DOP, DON and DOC: dissolved organic phosphorous, nitrogen and carbon, respectively

River	*Period*	SiO_2	$N\text{–}NO_3^-$	$N\text{–}NH_4^+$	$P\text{–}PO_4^{3-}$	*DOP*	*DON*	*DOC*
Amazon	1990s	6.9	0.14	0.02	0.022	0.015	0.16	5.7
Amur		2.15	0.02		0.021			
Arno	1977–1983		1	1.19	0.5	0.08	1.2	
Chang Jiang	1980–1992	6.5	0.32	0.32	0.02			2.07
Churchill (Hudson)	1979–1992	1.4	0.01	0.1	0.006			
Citarum	1985–1994	30	0.43	0.53	0.032			
Columbia	1979–1993	9	0.2	0.01	0.014			
Congo	1980s	9.4	0.9	0.007	0.24		0.18	9.35
Danube	1979–1983	4.14	1.5		0.02			
Delaware	1979–1993	3.5	1	0.03	0.04			
Elbe	1979–1990		3.6	1.3	0.39			
Evros	<1984		1.4	0.33	0.285			
Huang He	1980–1992	7.7	2.2	0.01	0.02			
Khatanga	1980s	3.2	0.03	0.04	0.006	0.006	0.41	
Kiso	1979–1992	10	0.38	0.06	0.016			
Lena	1980–1990	4.2	0.03	0.08	0.204	0.022	0.46	6.6
Mackenzie		3.7	0.53		0.004		0.1	5.2
Mississippi	1980–1993	6.7	1.4	0.04	0.07			8
Murray	1979–1987	5	0.11	0.036	0.024			
Neva	1980–1991	0.1	0.23	0.03	0.03			
Ob	1980–1992	2.85	0.06		0.065			9.1
Olenek	1980s	2.7	0.03	0.05	0.003	0.006	0.41	
Orinoco	1980s	6.3	0.08	0.035	0.01	0.01	0.16	4.4
Parana	1979–1992	17.1	0.165	0.05	0.045			6.1
Po	<1985	4	1.43	0.27	0.075	0.001	0.32	2.4
Rhine	1975–1984	5.2	3.9	0.9	0.4			5.36
Rio Grande	1979–1993	14	0.15	0.03	0.03			
Sakarya	1980–1987	11.8	1.1	0.35	0.16			
Scheldt	1978–1991		4.6	7.7	0.81			7.9
Seine	1980–1986, 1990–1991	6.3	4.28	2.41	0.7		0.5	3.65
Shinano	1956/1957	15	0.27	0.07				
Shinano	1979–1992	17	0.78	0.08	0.047			
St. Lawrence	1979–1988	2.4	0.16	0.08			0.03	3.7
Susquehanna	1979–1993	3.8	1	0.03	0.01			
Thames	1980–1992	12.3	7	0.21	0.35			
Tone	1979–1992	21.5	1.56	0.2	0.033			
Trent	1980–1992	7.5	8.6	0.24	0.49			
Waikato	1979–1993	28.2	0.3	0.006	0.022			
Weser	1979–1990	4	5.08	0.125	0.37			
Yenisey	1980–1990	3	0.02		0.008			7.4
Zambezi	<1977	16.8	0.13	0.04	0.01			
Zhu Jiang	?	8.5	0.62	0.01	0.003			

Source
Kimstach V, Meybeck M, and Baroudy E (eds.) (1998) A Water Quality Assessment for the Soviet Union, pp. 611. E & FN Spon, London.
Meybeck M and Ragu A (1996) River discharges to the oceans. An assessment of suspended solids, major ions and nutrients. Environment Information and Assessment Rpt. UNEP, Nairobi, 250 pp.

Kerulen (Mongolia), Amu Darya, and Syr Darya (Central Asia). These rivers are among the top-ranked rivers for river length and drainage area. They have a low to very low runoff and are therefore most sensitive to climate change and to water uses, e.g., irrigation, diversion, waste inputs.

It must be noted that river runoff for the large exorheic catchments range from less than 30 mm $year^{-1}$ for the Nile River to more than 2000 mm $year^{-1}$ (Papuan rivers as Fly and Purari). Wet and humid tropics account for most of the greatest river flows (# 1 Amazon, # 2 Congo, # 3 Orinoco, # 9 Brahmaputra, # 10 Ganges, # 11 Irrawaddy or Ayerawadi, # 12 Mekong). For TDS fluxes, the Amazon is # 1, the Chang Jiang is # 2, the Mississippi is # 3, the Irrawaddy # 4, the Ganges # 5, the Danube # 6.

As expected, the river length (L, **Tables 2** and **3**) is not much related to riverine fluxes (Q,F_{TSS} and F_{TDS}) and has its own specific ranking with Nile as # 1, ahead of Amazon (# 2), Mississippi (# 3), and Chang Jiang (or Yang Tse, # 4). As observed by Milliman and Sivitski (1992), some very short rivers from South and East Asia, like Kaoping (Taiwan,L: 170 km), Fuchun Jiang (China, 494 km), Purari (New Guinea, 630 km), Kapuas (Indonesia, 1010 km), Sepik (New Guinea, 700 km), and Copper (Alaska, 460 km), should be taken into account in global budgets, particularly for suspended solids. According to a recent survey, the Amazon TSS flux is ranked # 1, Huang He # 2, Haï Ho # 3, Brahmaputra # 4, Ganges # 5, Mississippi # 6, Chang Jiang # 7, etc. in predamming conditions.

The sediments fluxes are only documented for river SPM (**Tables 2** and **3**) as bed loads are generally not measured or estimated. Most authors use a ratio of 1:10 between bed load and suspended load.

Human Impacts on Riverine Fluxes

Total Suspended Solids and Total Dissolved Solids

Present sediment flux figures are no longer natural, since damming has already affected some of the largest rivers, such as the Mississippi-Missouri, the Columbia, the Colorado, the Rio Grande/Bravo Rivers, the Sao Francisco, Danube, Volta, Nile, Zambezi, Orange, Indus, Huang He, Chang Jiang rivers, etc. (see reviews by Meade and Parker, 1985; Syvitski *et al.*, 2005; Walling, 2006).

River TDSs are also much affected by human impacts, which result in marked increase of some ions in the following order:

$$Cl^- = Na^+ > SO_4^{2-} > K^+ \gg Ca^{2+} = HCO_3^- = Mg^{2+}$$

In rivers affected by mining (rock salt, metal, coal mines), the salinity increase reaches an order of magnitude and more for Na^+, Cl^-, SO_4^{2-}, and K^+. Rivers exposed to direct salt mining impact rank in the first 40 rivers for TDS fluxes despite their medium size as the Rhine (rank # 11), Weser (# 24), Wisla (Vistula, # 31), and Elbe (# 36), (**Table 3**). The TDS fluxes for Colorado, Nile, and Rio Grande (Rio Bravo) listed in **Tables 2** and **3** are those estimated with natural runoff, i.e., prior irrigation or diversion.

In catchments where water is used for irrigation or is diverted, the river flows and all other riverine fluxes are very reduced (e.g., Indus, Huang He, Rio Grande) and even close to zero (Nile, Colorado, Amu Darya rivers). Once the # 1 river input to the Mediterranean Sea, the Nile is now not in the top 10 of inputs to this sea, since more than 95% of the Nile is now used for irrigation.

Nutrients in World Rivers: From Natural to Contemporary Loads

Nutrients are among the most sensitive river-borne material to human pressures. At the global scale nitrate, ammonia, and dissolved inorganic phosphorus concentrations in rivers are now commonly ranging over 2 orders of magnitude (**Table 4**), depending on natural control factors and on human pressures (fertilizers, urban and industrial wastes, and atmospheric deposition). It is very difficult to assess background nutrients levels in most of the rivers in the temperate regions, except in small-forested catchments where fertilizer use and wastewater inputs are limited. Some of the very large tropical rivers (e.g., Congo, Amazon, Orinoco) can still be considered as subpristine for nutrients levels. Global background levels usually combine selected datasets from small temperate catchments and large tropical rivers. Dissolved silica (SiO_2), dissolved organic nitrogen (DON), phosphorus (DOP), and carbon (DOC) are much less sensitive to pressures and their range is narrower.

The highest nitrate levels have been found in catchments with intensive use of fertilizers as in most medium-sized Western European rivers (Weser, Rhine, Thames, Trent, Seine). The N–NO_3^- level exceeds 5 mg l^{-1}, i.e., 20–50 times higher than in large rivers with very limited human pressures. In catchments with moderate urban population and limited fertilizer use, the dominant anthropogenic nitrogen sources (nitrate, ammonia) are often atmospheric (e.g., New England).

The highest ammonia levels are found in densely populated catchments ($d > 200$ p km^{-2}) where urban wastewaters are collected but not appropriately treated (e.g., Scheldt and Seine Rivers in the 1980s): these levels can reach 50–100 times the natural background levels. A similar pattern is noted for phosphorus, which originates from diffuse agricultural sources and from urban sources (e.g., phosphorus-containing detergents). Examples of maximum P levels are given in **Table 4** for some Europe an rivers (Arno, Rhine, Seine, Scheldt, Evros) and for the Sakarya River in Turkey.

Total phosphorus (TP) is generally measured on unfiltered samples as the sum of P–PO_4^{3-}, DOP and particulate phosphorus (PP), sum of organic (POP) and inorganic (PIP). POP originates from soil erosion and is generally linked to POC (POC/POP $= 22\,g\,g^{-1}$) while PIP originates from the erosion of rock minerals as apatite (PIP is seldom lower than 600 ppm in river suspended matter) and from the adsorption of phosphates ions onto fine minerals. Both POP and PIP contents increase downstream of urban sewage outfalls. PP is the dominant form of phosphorous during floods and in turbid rivers (e.g., PP in the Huang He exceeds $10\,mg\,l^{-1}$ because of the erosion of loess). TP budgets based on regular water quality surveys often underestimate the flashy PP inputs during floods. It is therefore recommended that analyses of particulate phosphorus are performed separately on river SPM and then combined to TSS fluxes to generate PP fluxes.

Nutrients Trends and Changes in Stoechiometric N:P:Si Ratios

Anthropogenic influences on nutrients sources and sinks can substantially increase the riverine fluxes of NO_3^-, NH_4^+, TKN, PO_4^{3-}, TP, or decrease those of dissolved silica. However, the DIC and DOC levels in most rivers are relatively stable on most long-term records (50 to 30 years). As a result the C:N:P:Si ratios in impacted rivers have greatly changed in some well-documented rivers over the last 50 years.

Freshwater algae require specific inorganic nutrient ratio (Redfield ratios) for proper nutrition: N:P = 16:1 (mol mol^{-1}) for all algae, and Si:N = 1:1 (mol mol^{-1}) for diatoms, the siliceous algae group. In natural conditions, the Si:N ratio exceeds largely 1:1 value, thus permitting diatom blooms, as dissolved SiO_2 is ranging between 50 and 400 $\mu mol\,l^{-1}$ and NO_3^- between 4 and 25 $\mu mol\,l^{-1}$. In impacted conditions, NO_3^- is increasing, sometimes by more than a factor of 10, and dissolved silica is retained by algal production within upstream reservoirs thus lowering the downstream SiO_2 content by a factor of 2 and more. In such impacted rivers, the Si:N ratio in riverine nutrient fluxes can be lower than 1.0 and silicon becomes the limiting factor of diatom growth. Such Si:N changes have been well demonstrated in the Mississippi River where, according to some authors, they favor alternate primary producers as cyanobacteria particularly in the coastal zone resulting in seasonal coastal hypoxia.

Trajectories of Riverine Fluxes Reflect Pressures Evolution and Water Quality Management

In many regions of the world, water quality records and TSS records have been available for more than 30 years (medium-term records), and a few long-term records (50–100 years) are available on some rivers (Mississippi, Volga, Rhine, Seine) or from sedimentary records in floodplain deposits (for persistent organic pollutants and for metals).

The typology of flux trajectories reflects the multiple human pressures and the resulting human responses when they are developed (**Figure 2**, trajectories A–K).

A. A *stable evolution* is noted for ions, DOC, and nutrients in most pristine rivers (i.e., without direct human pressures), although a slow modification of river DOC and DIC fluxes is expected from gradual climate change, because of permafrost melting and CO_2 increase. In catchments under pressure such trends are also noted for elements that are the least sensitive to human pressure such as Ca^{2+}, Mg^{2+}, and to a lesser extent, to bicarbonate and many

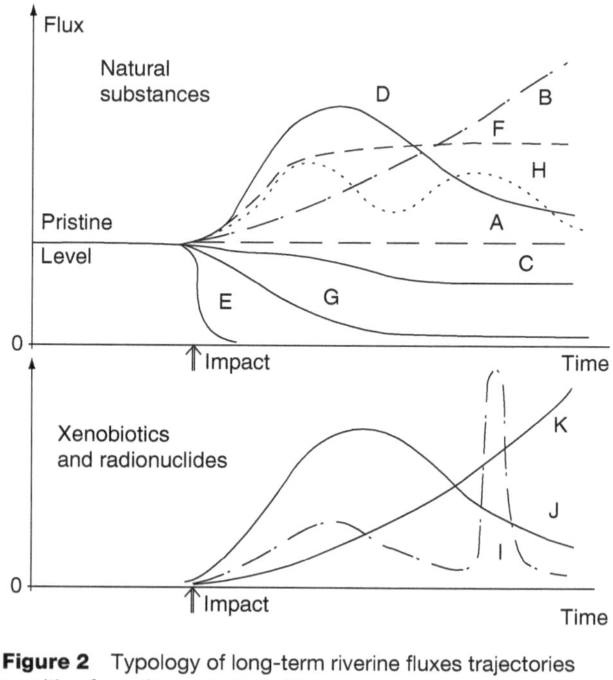

Figure 2 Typology of long-term riverine fluxes trajectories resulting from the evolution of human pressures and responses on river basins. Adapted from Meybeck M (2001) River basins under Anthropocene conditions. In von Bodungen B and Turner K (eds.) *Science and Integrated coastal Management*, Dahlem Conference Series, Wiley, 275–294.

particulate metals such as Al, Fe, Co, and V, which have very little impact from most human activities.

B. The *gradual increase* is associated to the progressive development of industries, cities (K^+, Na^+, Cl^-, SO_4^{2-}), and of industrial fertilizers in agriculture (K^+, SO_4^{2-}, NO_3^-, total P).

C. A *partial retention* of dissolved inorganic nutrients resulting from algal uptake is noted in many reservoirs (NO_3^-, PO_4^{3-}, SiO_2) (e.g., Mississippi, Danube, Sao Francisco Rivers).

D1. The *bell-shaped curve* characterizes a successful and gradual control of undesirable substances as for cadmium, PAHs, and for NO_3^-, NH_4^+, and PO_4^{3-} in some Western European and United States rivers.

D2. The *stepwise improvements* (not shown) are observed when major urban or industrial wastewater treatment are installed (few months responses). They are also linked to the reduction, or closure of economic activities following economic crises; e.g., in former socialist countries in Europe, the Elbe and Danube fluxes of nutrients and/or heavy metals, drastically dropped between 1989 and 1990.

D3. The *stepwise degradation* (not shown) is the symmetric evolution resulting from the installation of pollutants sources, or when wastewaters are collected to sewers without proper treatment (e.g., NH_4^+ increase from urban sewers development).

E. The complete *particulate matter* retention is caused by settling in reservoir, when residence time exceeds 3–6 months. It is also observed for particulate carbon, nutrients, and pollutants.

F. *Stabilized fluxes of contaminants and salts* (TDS, Cl^-) over decades may result from water quality regulation, particularly at international borders (e.g., Colorado River between the United States and Mexico, Rhine River between France and other member states of the Rhine Commission).

G. A *general decrease* of all fluxes is caused by water diversion and/or use for irrigation (e.g., Colorado, Rio Grande, Nile, Ebro, Amu Darya, Syr Darya, Indus, Huang He, Murray).

H. *Multiple cycles* are often noted in long-term trends as a result of the pressures/responses balance (e.g., Thames River organic matter between 1850 and 2000; metal fluxes in the Rhine and Meuse, as reconstructed from sediment archives).

I. *Fluxes of radioactive substances* are very seldom addressed or disclosed by water authorities. Since 1950, the development of nuclear industries, the nuclear weapon tests in the atmosphere, that peaked in 1962–1963 in the atmosphere, and their related leaks and accidents (as in 1986 at Chernobyl), have spread artificial radionucleides in all rivers, particularly in the Northern Hemisphere. The typical ^{137}Cs trajectory in West Europe present a double peak near 1962 and 1986 that is used to date alluvial sediments.

J. *Total ban (J)* is sometimes applied to some substances that do not occur naturally (xenobiotics) when they are found to be dangerous as for PCBs, used in industrial products and for DDT, banned in 1970, that gradually decreased in West Europe and North America rivers.

K. The *gradual contamination* of other types of xenobiotic substances as pesticides, drugs, some solvents is widely observed.

Further Reading

Colombani J and Olivry JC (1984) Phénomènes exceptionnels d'érosion et de transport solide en Afrique aride et semi-aride. *International Association of Hydrological Sciences Publications* 144: 295–300.

Gordeev VV (1983) *River Inputs to The Ocean and Features of Its Chemistry*, p. 160. Moscow: Nauka (in Russian).

Hem JD (1989) Study and interpretation of chemical characteristics of natural water. US Geological Survey Water Supply Paper 2254, USGS Reston, VA.

Horowitz A (2003) An evaluation of sediment rating curves for estimating suspended sediment concentrations for sub segment flux calculations. *Hydrological Processes* 17: 3387–3409.

Ittekkot V, Unger D, Humborg C, and TacAn N (eds.) (2006) *The Silicon Cycle*, SCOPE series 66, p. 275. *Washington: Island Press.*

Kimstach V, Meybeck M, and Baroudy E (eds.) (1998) *A Water Quality Assessment for the Soviet Union*, p. 611. London: E and FN Spon.

Ludwig W and Probst JL (1998) River sediment discharge to the oceans: present-day controls and global budgets. *American Journal of Science* 298: 265–295.

Meade RH and Parker RS (1985) Sediment in rivers of the United States. *National Water Summary, 1984. Water Supply Paper*, vol. 2275, pp. 40–60. Reston, VA: US Geological Survey.

Meybeck M (2001) River basins under Anthropocene conditions. In von Bodungen B and Turner K (eds.) *Science and Integrated coastal Management*, Dahlem Conference Series, pp. 275–294. Wiley.

Meybeck M (2003) Global analysis of river systems: from Earth system controls to Anthropocene syndromes. *Philosophical Transactions of Royal Society of London B* 358: 1935–1955.

Meybeck M (2003) Global occurrence of major elements in rivers. In Drever JI (ed.) *Surface and Groundwater, Weathering and Soils.* In: Holland HD and Turekian KK (eds.) *Treatise on Geochemistry*, vol. 5, pp. 207–223. Amsterdam: Elsevier.

Meybeck M and Ragu A (1996) River discharges to the oceans. An assessment of suspended solids, major ions and nutrients.

Environment Information and assessment Rpt, 250 pp. Nairobi: UNEP.

Meybeck M and Vörösmarty CJ (2005) Fluvial filtering of land to ocean fluxes: from natural Holocene variations to Anthropocene. *Comptes Rendus Géosciences* 337: 107–123.

Meybeck M, Laroche L, Dürr HH, and Syvitski JPM (2003) Global variability of daily total suspended solids and their fluxes in rivers. *Global Planetary Change* 39: 65–93.

Milliman JD and Syvitski JPM (1992) Geomorphic/tectonic control of sediment discharge to the ocean: the importance of small mountainous rivers. *Journal of Geology* 100: 525–544.

Milliman JD (2001) River inputs. In Steele JH, Turekian KK, and Thorpe SA (eds.) *Encyclopedia of Ocean Sciences,* vol. 4, pp. 2419–2427. Academic Press.

Milliman JD, Rutkowski C, and Meybeck M (1995) River discharge to the sea. A global river index (GLORI). LOICZ reports and studies, Texel, ND, pp. 125.

Moatar F and Meybeck M (2007) Riverine fluxes of pollutants, towards predictins of uncertainties by flux duration indicators. *Comptes Rendus Géosciences Hydrology-Hydrogeology* 339: 367–382.

Peters NE (ed.) (1996) Trends in water quality. *Hydrological Processes* 10: 127–356.

Philipps JM, Webb BW, Walling DE, and Leeks GJL (1999) Estimating the suspended loads of rivers in LOIS study area using infrequent samples. *Hydrological Processes* 13: 1035–1050.

Rabalais NN, Turner RE, Justic D, Dortch Q, Wiseman WJ, and SenGupta BV (1996) Nutrients changes in the Mississippi River and system responses on the adjacent continental shelf. *Estuaries* 19(2B): 386–407.

Rode M and Suhr U (2007) Uncertainties in selected river water quality data. *Hydrology and Earth System Science* 11: 863–874.

Salomons W, Kremer H, and Turner RK (2006) The catchment to coast continuum. In Crossland CJ, *et al.* (eds.) *Coastal Fluxes in Anthropocene*, pp. 145–200. Springer.

Seitzinger SP, Harrison JA, Dumont E, Beusen AH, and Bouwman AF (2005) Sources and delivery of carbon, nitrogen and phosphorus to the coastal zone: an overview of global nutrient export from watersheds (NEWS) models and their application. *Global Biogeochemical Cycles* 19. DOI: 10.1029/2005 GB 002606.

Syvitski JPM, Vörösmarty CJ, Kettner AJ, and Green P (2005) Impact of human on the flux of terrestrial sediment to the global coastal ocean. *Science* 308: 376–380.

Vörösmarty CJ and Meybeck M (2004) Responses of continental aquatic systems at the global scale: new paradigms, new methods. In Kabat P, *et al.* (eds.) *Vegetation, Water, Humans and the Climate, Global Change, IGBP series*, pp. 375–413. New York: Springer.

Walling DE (2006) Human impact on land-ocean sediment transfer by the world's rivers. *Geomorphology* 79: 192–216.

Williams GP (1989) Sediment concentrations versus suspended matter discharge during hydrologic events in rivers. *Journal of Hydrology* 111: 89–106.

Relevant Websites

http://waterdata.usgs.gov/nwis/qw – USA river water quality and sediment fluxes, US Geological Survey.

http://www.gemsstat.org/descstats.aspx – Gems Water program UNEP, global river water quality data.

Fluvial Transport of Suspended Solids

P Y Julien, Colorado State University, Fort Collins, CO, USA

Introduction

The fluvial transport of suspended solids is of great interest to living communities. It has been known for a very long time that the deposition of fine sediments on flood plains increases the fertility of farmlands.

In urban areas, high suspended sediment concentrations adversely impact the quality of drinking water and increase the operation cost of water treatment plants. During floods, the excessive suspended sediment concentrations can also cause major sedimentation problems resulting in aggradation, river navigation problems, and changes in river morphology. On the other hand, riverbed degradation from a lack of fine sediment in suspension can also undermine the stability of bridges and river protection structures.

Equilibrium Transport of Sediment Suspensions

There are two types of sediment sizes that contribute to the suspended sediment load of a river: (1) wash load and (2) bed material load. The difference between wash load and bed material load depends on whether the size fractions can be found in large quantities in the bed. The size fractions that are found in large quantities in the bed are referred to as bed material load. In practice, all size fractions that are finer than the d_{10} of the bed material will be considered wash load. The wash load does not depend on the sediment transporting capacity of the flow, but depends on the supply of sediment from upstream sources or from the river bank.

The quantity of wash load can only be determined from field measurements. Suspended sediment sampling is usually done with a point sediment sampler of the type P-61 or P-63. Point sediment samplers are designed to collect sediment through time at a given point along the stream vertical. The sampler weight is the primary difference between the P-61 (100 lb) and P-63 (200 lb) and heavy samplers must be used in deep and fast-flowing rivers. **Figure 1** shows a P-63 sampler on the Mississippi River.

The bed material load in suspension requires basic knowledge of the properties of the flow and of the transported sediment. In a very simplified form, two main properties describe suspended sediment transport in rivers: (1) the shear velocity of the flow and (2) the settling velocity of the bed material. The shear velocity $u_* = (ghS)^{0.5}$ is approximately equal to the square root of the product of gravitational acceleration g, flow depth h, and friction slope S.

The settling velocity is a property of the particle in its surrounding fluid. It can be directly calculated from the dimensionless particle diameter d_*, which is defined from the particle diameter d_s, the specific gravity G of sediment, the kinematic viscosity of the fluid v, and the gravitational acceleration g, as

$$d_* = d_s\left[\frac{(G-1)g}{v^2}\right]^{1/3} \qquad [1]$$

Simplified calculations are obtained with the grain diameter (m), the kinematic viscosity ($v = 1 \times 10^{-6}$ m^2 s^{-1}), g (9.81 m s^{-2}), and G (2.65). The settling velocity ω (m s^{-1}) of a sediment particle in clear water is then calculated from

$$\omega = \frac{8v}{d_s}\left\{\left[1+\frac{d_*^3}{72}\right]^{0.5} - 1\right\} \qquad [2]$$

The ratio of shear velocity u_* to settling velocity ω determines the primary mode of sediment transport. The bed material can be subdivided into three zones describing the dominant mode of transport: bed load, mixed load, and suspended load. In most rivers, bed load is dominant at values of u_*/ω less than about 0.4. Note that incipient motion corresponds to $u_*/\omega \cong 0.2$, which means that the bed material does not move when $u_* < 0.2\omega$. A transition zone called mixed load is found where $0.4 < u_*/\omega < 2.5$ in which both the bed load and the suspended load contribute to the total load. When $u_*/\omega > 2.5$, most of the sediment load is transported in suspension. Field measurements are necessary to determine the rate of sediment transport.

The concentration of suspended sediment in rivers varies with depth. The suspended sediment concentration C at an elevation z above the bed for the suspended load can be calculated from the Rouse equation as

$$C = C_a\left[\left(\frac{h-z}{z}\right)\left(\frac{a}{h-a}\right)\right]^{\omega/\kappa u_*} \qquad [3]$$

where C_a is the concentration at an elevation a above the bed, h is the flow depth, and κ is the von Karman constant ($\kappa \cong 0.4$). The exponent of this equation is called the Rouse number $Ro = \omega/\kappa u_*$, which varies with u_*/ω. The near-bed concentration can be obtained from point sediment concentration measurements near the bed, or in the lower part of the water column. The Rouse number can be experimentally obtained as the slope of the linear fit to the concentration profile obtained after a logarithmic transformation, as shown in **Figure 2**.

The sediment flux per unit area is the product of the flow velocity and sediment concentration. For instance, the volumetric flux of sediment is obtained from the product of the volumetric sediment concentration, the flow velocity, and the unit area. Because the flow velocity and sediment concentration vary with depth and width, it is necessary to integrate the velocity and concentration profiles along the vertical and across the entire width of a river to determine the sediment flux. This integral is very complex and is discussed in detail in Julien (1995). The depth integral of the sediment flux describes the unit sediment discharge or amount of sediment being transported per unit channel width q_{tx}. It represents a volume of sediment per unit width.

Figure 1 Suspended sediment sampling on the Mississippi River.

Nonequilibrium Transport of Sediment Suspensions

As rivers approach reservoirs, lakes, and estuaries, the reduced sediment transport capacity in the backwater areas causes deposition of the suspended sediment load. Owing to the continuity of sediment, the equation of conservation of mass determines the changes in vertical elevation from settling when there is a decrease in sediment transport in the downstream direction. This equation of conservation of mass shows that the settling sediment flux in the z direction causes a change in bed surface elevation z_b:

$$\frac{\partial z_b}{\partial t} = -\frac{T_E}{(1-p_0)}\frac{\partial q_{tx}}{\partial x} \qquad [4]$$

The porosity p_0 depends on the specific weight of sediment deposits and is approximately 0.43 for sand-bed rivers. The trap efficiency T_E describes the fraction of sediment that would deposit in a given river reach of length X. It is therefore a measure of how much sedimentation could take place in backwater flow conditions. Trap efficiency is a function of the reach length X, the river width W, the flow discharge Q, the mean flow velocity V, and the settling velocity ω as

$$T_E = 1 - \exp\left(-\frac{X\omega}{hV}\right) = 1 - \exp\left(\frac{-WX\omega}{Q}\right) \qquad [5]$$

This relationship for the trap efficiency of sediment can be useful. At a given flow discharge, the trap efficiency remains very small for very short reaches and for very fine sediment particles (low ω). Under

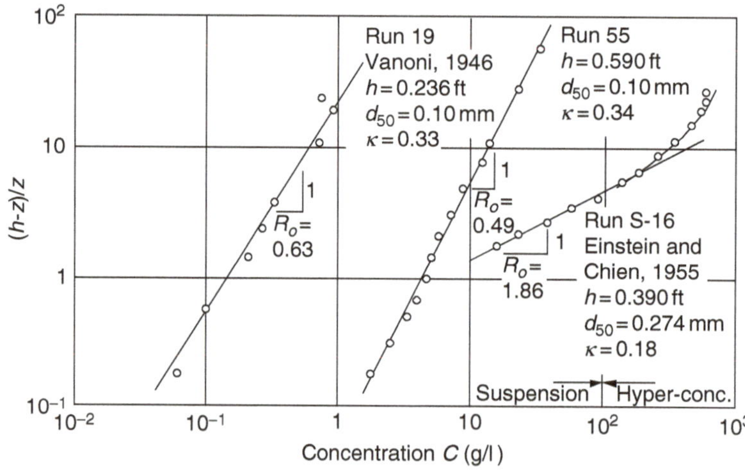

Figure 2 Examples of sediment concentration profiles (from Julien, 1995).

changes in sediment transport capacity in the downstream direction, the trap efficiency describes a greater potential for coarse sediment to deposit. It is also interesting to note that the trap efficiency at a given reach length and flow discharge increases with increasing sediment size ω and channel width W. It is therefore noticeable at a given discharge that river widening will induce settling of suspended sediment (Julien, 2002).

When calculating the trap efficiency of silt and clay particles in backwater areas like reservoirs and estuaries, careful consideration must also be given to density currents and possible flocculation. Flocculation of silts and clays is a complex subject, but in its essence, the settling velocity of floc of silts and clays is typically around 0.6 mm s^{-1}. The settling velocity of flocs increases with floc size but will rarely exceed 5 mm s^{-1}.

As an example to illustrate the concepts covered in this article, consider the Rhine River data from Julien (2002). The flow depth is approximately 10 m, the main navigable channel constrained between a series of dikes is about 260 m wide, the mean flow velocity is 1.68 m s^{-1}, and the friction slope 13.2 cm km^{-1}. The sediment concentration at mid depth is 38 mg l^{-1}, and the near-bed concentration is 400 mg l^{-1} at a distance of 0.5 m above the river bed. If the grain diameter of the sediment in suspension is 0.2 mm, the following parameters can be calculated from the methods covered in this article: (1) the shear velocity obtained from $u_* = (ghS)^{0.5}$ is approximately $u_* = 0.11$ m s^{-1}; (2) the dimensionless particle diameter is approximately 5 from eqn [1] and the settling velocity from eqn [2] is $\omega = 0.027$ m s^{-1}; (3) the ratio $u_*/\omega = 4.2$, and the Rouse number is 0.59 assuming $\kappa \cong 0.4$, thus most of the sediment is transported in suspension; (4) the suspended sediment concentration 1 m below the free surface is 19 mg l^{-1} obtained from Ca = 400 mg l^{-1} at $a = 0.5$ m, $h = 10$ m, and the concentration at $z = 9$ m from eqn [3]; and (5) the trap efficiency over a reach length of 500 m from eqn [5] is $T_E = 0.55$, which means that about half the suspended sediment load would deposit within half a kilometer if the transport capacity of this river would be suddenly reduced.

This example is quite instructive because most of the suspended sediment load would deposit very rapidly despite the fact that most of the sediment is fine and transported in suspension. The high T_E indicates that most of the sediment would also easily be trapped on the flood plain within short distance of the main channel during major floods. This river would be very dynamic and could change morphology if it were not constrained with a series of dikes. Finally, the suspended sediment concentration near the free surface is only 19 mg l^{-1} compared with 400 mg l^{-1} near the bed, and water intakes should definitely be located near the free surface.

Further Reading

Julien PY (1995) *Erosion and Sedimentation,* 280p. New York: Cambridge University Press.

Julien PY (2002) *River Mechanics,* 434p. New York: Cambridge University Press.

Streams

E Wohl, Department of Geosciences, Colorado State University, Ft. Collins, CO, USA

Introduction

Every point on the Earth's landmass lies within a drainage network formed of stream channels tributary to one another that eventually drain to an inland reservoir or to an ocean. The spatial arrangement of channels into a drainage network, the water and sediment moving from hillslopes and down streams, and the geometry of streams, all reflect climatic and geologic factors within the drainage basin.

Spatial Organization of Streams in Drainage Networks

A drainage network includes all the stream channels that drain toward a reference point. The network is bounded by a topographically defined drainage divide; precipitation falling on the far side of the divide flows down slope into an adjacent drainage network. A drainage network begins with first-order streams to which no other stream is tributary. In the most commonly used method of stream orders, a second-order stream begins at the junction of two first-order streams, a third-order stream begins at the junction of two second-order streams, and so on (**Figure 1**).

Patterns of drainage networks. The spatial distribution of streams within the network can be descriptively classified using terms including dendritic, rectangular, radial, and others. Dendritic drainages are the most widespread, taking their name from a resemblance to the outline of a tree (**Figure 2**). A dendritic drainage is commonly interpreted to reflect a relatively homogeneous substrate of moderate down slope gradients. A rectangular drainage, in contrast, has many right-angle tributary junctions that reflect a strong underlying control, such as joints in the bedrock, which influences the location of stream channels. A radial drainage network more likely reflects the underlying topographic control, such that individual streams radiate outward and down from a central high point such as a volcanic cinder cone. This descriptive classification for drainage networks is useful because it is readily apparent in aerial photographs, topographic maps, or digital elevation models of a landscape, and because the categories of the classification imply something about the geologic controls on the spatial arrangement of stream channels across a landscape.

Drainage density. Drainage networks can be quantitatively described using parameters such as drainage density, which is the ratio of total length of streams within a network to the surface area of the network (stream km/km^2 of drainage area). Drainage density reflects climatic controls, substrate on which the drainage network is formed, and age of the drainage network. The highest values of drainage density tend to occur in semiarid regions and in the seasonal tropics. In each of these regions, high-intensity rainfalls create sufficient erosive force to overcome the surface resistance of hillslopes and form stream channels. High values of drainage density can also be associated with very steep topography, with erodible substrates, and with patterns of land use such as deforestation that reduce hillslope resistance to surface erosion. Drainage networks initially form relatively rapidly on newly exposed landforms such as glacial or volcanic deposits. The rate of increase in drainage density then levels off with time as the network becomes fully integrated and the spacing of stream channels reflects the minimum surface area needed to produce sufficient runoff to support a channel.

Formation of stream channels. A stream channel can form as the result of either surface or subsurface processes, or some combination of the two. Heterogeneities in the surface and subsurface properties of hillslopes create zones of preferential flow during downslope movement of water. As water preferentially concentrates on the surface, the force exerted against the surface by the flowing water increases proportionally to the depth of the water. A self-enhancing feedback occurs such that an initial surface irregularity slightly concentrates surface flow on the hillslope, and the slightly deeper flow in this irregularity exerts more erosive force against the surface, thus deepening the irregularity, which then concentrates yet more flow as it widens and deepens. Eventually, the irregularity creates a spatially continuous downslope flow of water in the form of a rill. If one of a series of parallel rills enlarges faster than the neighboring rills, the master rill creates a secondary side slope between its channel and that of adjacent rills. This secondary slope facilitates shifting of the smaller rill so that it becomes tributary to the larger rill, and a drainage network begins to form.

An analogous process occurs in the subsurface, where differences in porosity and permeability create localized zones of greater flow that dissolve or

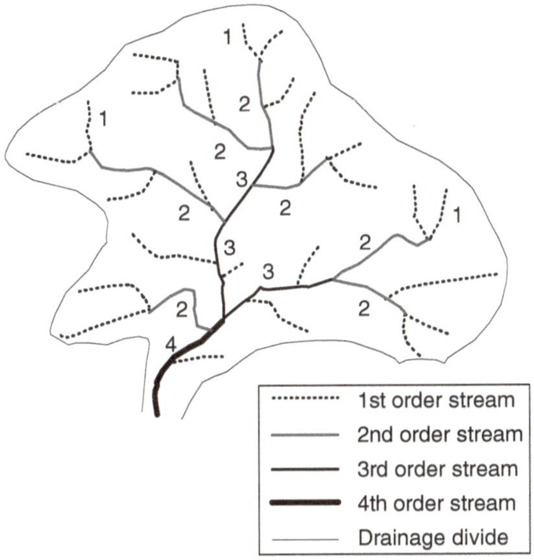

Figure 1 Schematic drawing of a drainage network, showing the ordering of streams, and delineation of the drainage divide.

Figure 2 Dendritic drainage network formed on a gently sloping surface with homogeneous underlying sediment, northwestern Australia. The trees in the photo are approximately 8–10 m tall.

physically erode material to create subsurface cavities. These cavities can form surface channels if the overlying material collapses into the cavity. The resulting sapping and piping networks have distinctive channels in which surface flow begins abruptly at an amphitheater-shaped depression in the ground surface.

Because the area of hillslope contributing flow to a stream channel increases downslope, thresholds for erosion and channel formation can be crossed at downstream portions of a slope first, and the stream channels then erode headward as the network of channels enlarges. If one set of channels erodes headward more rapidly than an adjacent network, the former channels can erode through the drainage divide and capture a portion of the adjacent network. This situation is occurring presently at the Casiquiare Canal, a naturally occurring channel along which a portion of the headwater drainage of the Amazon River of South America is capturing some of the headwater drainage of the adjacent Orinoco River.

Figure 3 Bedlands topography in Death Valley National Monument, California. Channels begin very close to the ridge crests, as can be seen most clearly along the dark brown ridge crest running across the lower third of the photograph.

The point along a hillslope at which stream channels begin to form depends on factors such as characteristics of precipitation, infiltration capacity of the surface, and erosional resistance of the surface (**Figure 3**). Regions with intense rainfall, low infiltration capacities, and highly erodible surfaces can have drainage networks that start very close to the crest of hillslopes, whereas other areas with less intense precipitation, higher infiltration, or greater surface resistance may have channel networks that begin much farther downslope.

Sources of Flow in Streams

The ultimate source of water flowing in any stream is snowmelt, rain-on-snow, or rainfall. Snowmelt generally produces regular seasonal patterns of stream flow during the onset of warmer temperatures when snow packs have accumulated during the winter melt. Snowmelt tends to be an important source of stream flow at higher latitudes and higher altitudes. Snow and rainfall can also enter streams after having been stored as ice in glaciers or icefields for periods of up to thousands of years. As with other forms of snowmelt, glacier melt is most pronounced during warmer seasons of the year, but can persist throughout the warm season (unlike snowmelt, which tends to be most pronounced during the early part of the warm season).

Rain-on-snow occurs when warmer temperatures cause rain to fall directly onto snow packs that have not yet completely melted. The warmer temperatures increase the melt rate of the snow which, when combined with rainfall, can create high rates of runoff and associated flooding. Rain-on-snow floods are particularly prevalent in low-to-moderate elevation catchments in coastal mountain ranges at middle latitudes.

The intensity, duration, and spatial extent of rainfall vary greatly among different types of climatic circulation patterns that generate rainfall. Convective storms create very intense rains that cover small areas ($1–10^2\ km^2$) for periods of up to a few hours. Frontal storms that last for days can extend across $10^4\ km^2$. The most extensive rains are associated with cyclonic storms such as hurricanes that last for days to weeks and monsoonal circulation patterns that last for months at a time. Both cyclonic and monsoonal storms can cover large areas of $10^5–10^7\ km^2$. Convective storms can generate enormously large stream flows within small drainage basins, but the effects of a small storm can be mitigated in large drainage basins where substantial portions of the basin remain unaffected by the storm. The more extensive frontal, cyclonic, and monsoonal storms can produce floods across much larger drainage basins.

The distribution of different types of precipitation reflects global-scale atmospheric circulation patterns, as well as regional topographic influences on the movement of air masses that bring moisture over a drainage basin. The regions with the greatest annual precipitation mostly lie within 30° north and south of the equator, where air masses moving across the warm surface of the tropical oceans pick up tremendous amounts of water vapor that is then transported inland to fall as precipitation. Smaller areas of very high precipitation can occur at higher latitudes where a mountain range forces moisture-bearing air masses to rise higher into the atmosphere, causing the water vapor within the air masses to condense and fall as precipitation, or where proximity to an ocean surface with relatively warm temperatures facilitates evaporation and inland transport of moisture from the ocean. Convective storms, which involve localized strong updrafts, are most common at latitudes 10° N–10° S. Frontal storms occur when the boundary between two air masses with different densities passes over a region and brings widespread precipitation. Monsoonal storms are associated with seasonal reversals of winds that draw moisture from adjacent oceans over land masses. Cyclonic storms, which have a strong rotational component, occur in two broad bands at approximately 10°–50° north and south of the equator.

Stream flow can also be dramatically affected by the failure of a natural or human-built reservoir. Lakes created when a landslide or debris flow blocks a stream commonly burst within a few days as the blockage is overtopped or weakened by seepage and piping. Water ponded upstream from glacial moraines or underneath glacial ice can also empty catastrophically when the moraine is overtopped or weakened within, or when the confined water builds sufficient pressure to lift the overlying glacial ice. Human-built fill and concrete dams can also fail by being overtopped or undermined. In each of these cases, sudden release of the ponded water initiates a catastrophic flood that continues until the reservoir is drained below the level of the remaining portions of the dam, or until the glacial ice once again shuts off the drainage path.

Patterns of stream flow reflect global and regional atmospheric circulation patterns and topography, as well as drainage area. Rivers in the equatorial and tropical latitudes commonly have the largest mean flow per unit drainage area because of the greater amounts of precipitation at these latitudes. Peak flow per unit drainage area tends to be greatest in relatively small rivers because the entire drainage area can be contributing runoff during intense precipitation events. Seasonal and interannual variability of flow tend to be largest in arid and semiarid regions, and in the seasonal tropics.

Movement of Water into Stream Channels

Precipitation falling across a landscape moves downward along various paths from hillslopes into stream channels. Precipitation can remain at the ground surface and move downslope relatively quickly as runoff or Hortonian overland flow. Precipitation can also infiltrate the ground surface and move downslope more slowly. Throughflow occurs when subsurface flow moves within the upper, unsaturated layers of sediment. Although the matrix as a whole remains unsaturated, concentrated zones of flow in pipes or macropores, or temporarily saturated zones, are particularly effective in moving water downslope into streams relatively rapidly. If the infiltrating water reaches the deeper, saturated layers of the subsurface, the water moves downslope with ground water. Hillslopes tend to be heterogeneous environments as a result of small-scale variations in surface topography and the porosity and permeability of subsurface materials. Throughflow moving downslope can concentrate in topographic irregularities and zones of limited porosity and permeability along the hillslope and return to the surface to move downslope as saturation overland flow. Overland flow and shallow, concentrated subsurface flow in pipes or macropores usually move downslope most rapidly, and these sources of runoff are together sometimes referred to by the descriptive term quickflow. Other forms of throughflow, as well as groundwater flow, move at slower rates (**Figure 4**).

The distribution of water among these four basic flow paths commonly varies across time and space. Rainfall that initially produces throughflow can subsequently create overland flow, for example, if infiltration capacity declines following prolonged rainfall or an increase in rainfall intensity. Convex portions of a hillslope can produce dominantly throughflow, whereas concave portions of the slope have saturation overland flow during the same rainstorm.

Spatial and temporal variability in the downslope movement of water reflects the characteristics of precipitation inputs and hillslope pathways. Precipitation intensity and duration exert particularly important influences on downslope movement of water. Higher intensities of precipitation are more likely to overwhelm infiltration capacity and produce overland flow, but prolonged precipitation at any intensity has the potential to exceed infiltration capacity.

Hillslope characteristics including vegetation cover, downslope gradient, and the porosity and permeability of materials at the surface and in the subsurface also strongly influence the downslope movement of water. Vegetation cover intercepts some precipitation, allowing snow or rain to evaporate or sublimate directly from the plant without reaching the ground, or reducing the force of impact when raindrops bounce from the plant onto the ground. Vegetation also sheds dead leaves and branches that can build up over time in a surface layer of duff with high infiltration capacity. Linear cavities left in the subsurface when plant roots die and decay can create macropores that facilitate rapid downslope movement of water in the subsurface. Steeply sloping surfaces can create large subsurface pressure differences that facilitate more rapid subsurface flow. Hillslope materials with high porosity (percent of void space) and permeability (interconnectedness of void spaces) also facilitate rapid infiltration and downslope movement of subsurface water. Porosity and permeability can result from spaces between individual grains in unconsolidated materials. Sand and gravel tend to have lower porosity but higher permeability than finer silt and clay-sized particles, with the result that sand and gravel commonly have higher infiltration and downslope transmission of water. Larger cavities in the form of pipes or macropores in sediments, or fractures in bedrock, also facilitate downslope flow. Pipes and macropores can result from biological processes including animal burrows or decayed plant roots. They can also form by erosion when subsurface flow concentrated above a less-permeable unit builds sufficient force to remove particles and create a continuous subsurface cavity (**Figure 5**).

In general, hillslopes with limited vegetation cover or surfaces disturbed by humans are likely to have

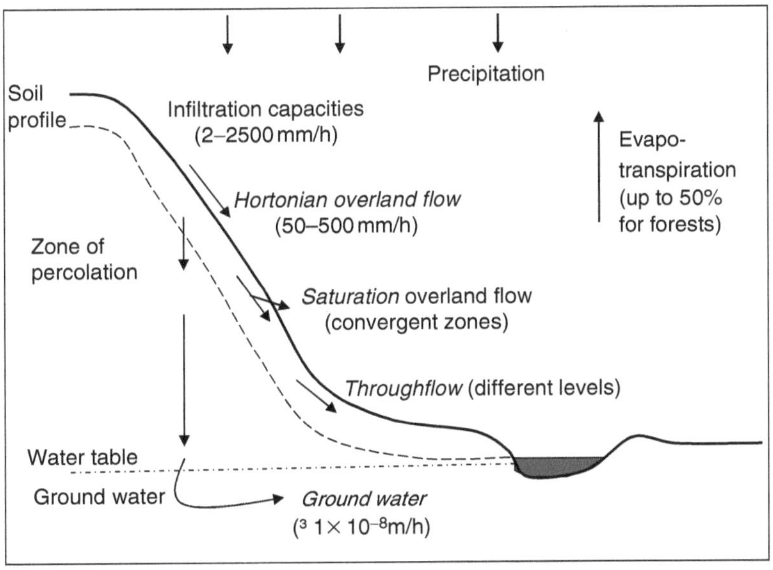

Figure 4 Schematic side view of hillslope illustrating four basic downslope pathways of water (italicized) and range of rates of movement.

Figure 5 Channel segments affected by subsurface piping along Cienega Creek in central Arizona. As subsurface pipes enlarge, the overlying sediment collapses into the cavity (upper photo) until eventually the collapse becomes longitudinally continuous, leaving a deeply cut channel with nearly vertical banks (lower photo).

more overland flow, whereas subsurface flow paths become more important with greater vegetation cover and deeper, more permeable soils. However, even a catchment with continuous, dense forest cover can have rapid downslope transmission of precipitation during conditions of high rainfall intensity or where thin soils and preferential subsurface flow paths such as pipes and macropores are present.

Movement of Sediment into Stream Channels

Sediment transported downstream can come from adjacent hillslopes, floodplains, and valley bottoms, and from erosion of the bed and banks within the stream. Hillslope sediment enters streams via gradual processes of slope erosion that occur through slope wash, soil creep, rill erosion, and other movements of individual sediment particles. Large volumes of sediment can also be introduced to streams during mass movements such as landslides, debris flows, and rock falls. Mass movements become progressively more important sources of sediment to streams where adjacent slopes are steeper and where episodic triggers such as intense rainfalls, seismic shaking, or wildfires periodically destabilize the hillslopes. Mass movements are particularly important in bringing sediment directly into headwater streams in mountainous terrains where narrow valley bottoms and spatially limited floodplains leave little storage space for sediment between the hillslopes and stream channels. Mountainous terrains around the world produce an estimated 96% of the sediment that eventually reaches the ocean basins, but occupy only 70% of the land area within river basins (**Figure 6**).

Floodplains adjacent to streams provide a very important source of sediment to streams, although the dynamics of sediment movement between streams and floodplains are spatially and temporally complex. Overbank flows that inundate floodplains can

Figure 6 A massive rockfall coming from the right enters a stream channel in the Nepalese Himalaya, causing the channel to become braided downstream.

deposit large volumes of sediment as particles carried in suspension settle from waters that move more slowly across floodplains. This sediment can remain in storage on the floodplain for periods ranging from hours to tens of thousands of years. The floodplain changes from a sink to a source of sediments when processes such as lateral stream migration cause the channel to move across the floodplain and reintroduce sediment from the floodplain into the stream. The rate and manner of floodplain deposition and erosion vary with stream type. Meandering channels tend to erode the outer portion of each meander bend, for example, creating more predictable directions and rates of floodplain erosion, whereas braided channels can shift abruptly back and forth across the valley bottom in a much less predictable fashion. Because many nutrients and contaminants travel adsorbed to silt and clay particles, the storage and remobilization of floodplain sediments can exert a strong influence on stream chemistry and ecological communities.

Erosion of the stream bed and banks provides a third primary source of sediment in stream channels. This form of erosion can be very temporary; most floods erode the channel boundaries while discharge is increasing, but then redeposit sediment during the falling limb of the flood when discharge is decreasing once more. Bed and bank erosion can also be more sustained when a stream is progressively incising downward in response to an increase in discharge, a decrease in sediment supply from other sources, or a drop in the base level (the lowest point to which the stream flows; the ocean is the ultimate base level). Most streams are continually adjusting to changes in water and sediment supply and base level. As a result, erosion of stream bed and banks is also continual in most streams, although this erosion may be balanced by deposition elsewhere along the stream, as when a migrating meander bend has erosion of the outer bend and simultaneous deposition of a point bar on the inner bend.

Characteristics of Flow in Streams

Hydrology of streams. One of the simplest ways to characterize flow in a stream channel is to quantify discharge through time. Discharge, usually expressed in cubic meters or cubic feet per second, is volume of flow per unit time. Discharge is calculated by measuring the velocity, or rate of flow (meters per second) within a cross-sectional area (square meters) calculated from mean width and depth of the flow. Continuous records of discharge come from stream-gaging stations where calibrated rating curves are used to convert measurements of stage, or flow depth, into discharge. These continuous records can then be used to construct a hydrograph, which is a plot of discharge versus time. A flood hydrograph represents a discrete event, whereas an annual hydrograph represents variations in discharge over the course of a year (**Figure 7**).

Hydrographs can be used to differentiate base flow, which is the relatively constant input of water to the stream from groundwater sources, from runoff that results from snowmelt and rainfall entering the stream via throughflow and overland flow. The shape of the hydrograph can be characterized by the relative steepness of the rising and falling limbs of higher flow, as well as the magnitude, duration, and frequency of occurrence of higher flow.

Hydrograph shape is influenced by the precipitation mechanism, the paths of downslope movement of water, and location within the drainage network. Higher intensity precipitation and greater overland flow produce more peaked hydrographs. Convective rainfall that results in Hortonian overland flow will produce a flash flood, for example, whereas snowmelt or prolonged gentle rain created by a low-pressure trough that results in throughflow will produce a lower magnitude, and a more sustained flood peak. Other factors being equal, smaller basins tend to have more peaked hydrographs because the close connections between hillslopes and streams, the narrow valley bottoms with limited floodplains, and the relatively short stream networks all facilitate rapid movement of water through the channel network. Larger basins that have broad, longitudinally continuous floodplains and longer travel times from headwaters to downstream measurement points produce floods that are less peaked and more sustained. Longer travel times occur both because water must

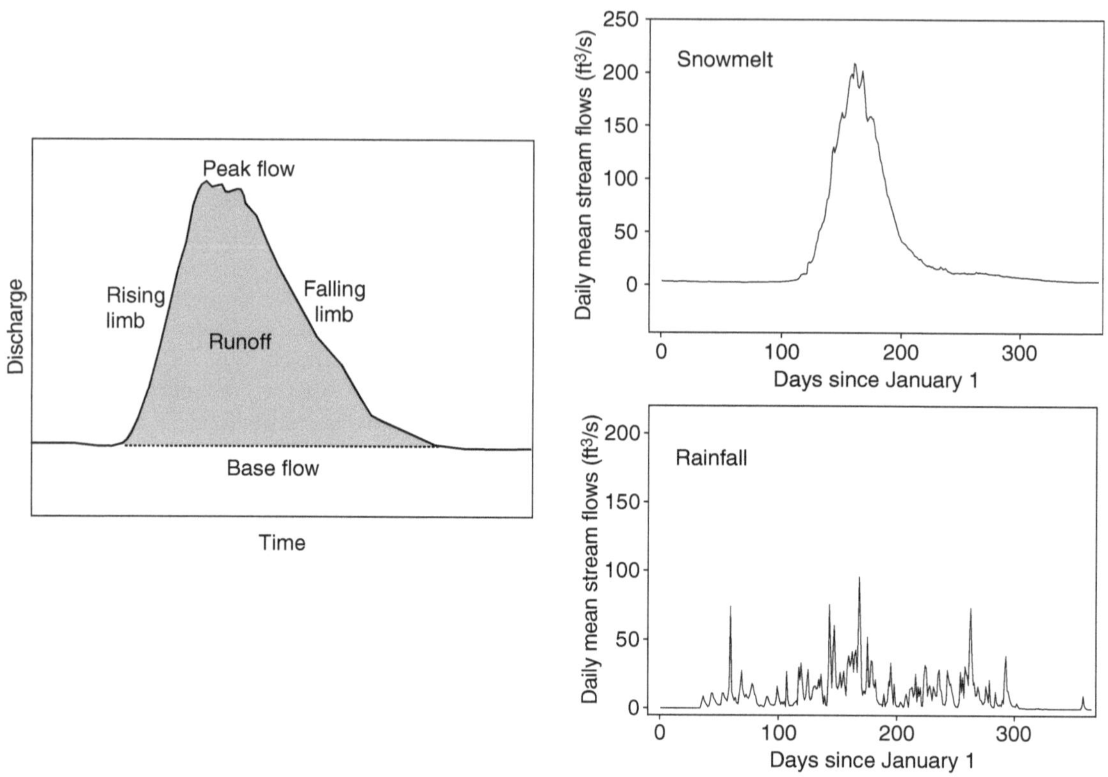

Figure 7 An idealized flood hydrograph (left), showing base flow, storm runoff (gray shading), and rising and falling limbs. Sample annual hydrographs (right) for a snowmelt-runoff stream (top) and a rainfall-runoff stream (bottom).

travel through a longer network of stream channels, and because more of the water travels slowly across floodplains rather than being concentrated within stream channels. The attenuation of flood discharges across floodplains is critical for depositing sediment and nutrients on floodplains, reducing flood hazards by limiting the magnitude of flood peaks, limiting channel erosion during floods by expending some of the flow energy, and nourishing floodplain wetlands and other ecosystems.

Flow duration curves, which plot magnitude of discharge versus the percent of time that discharge occurs, provide another means of characterizing the distribution of water in a stream through time. These curves graphically represent the variability of streamflow by the shape of the curve. Curves with low slope and high minimum values indicate a more ephemeral character and a quicker response to precipitation events. Flow duration curves are most frequently used for determining potential water supply for power generation, irrigation, or municipal use.

Flood-frequency curves indicate the average length of time, or recurrence interval, between floods of a similar magnitude. These curves are commonly used for predicting or mitigating flood hazards and for restoring streams in which the distribution of flow has been altered by dams, diversions, or other forms of flow regulation. Estimation of the recurrence interval of very large, infrequent floods, such as those that occur on average every hundred years, is particularly difficult because flood-frequency estimation is based on extrapolation from gage records that are commonly less than a century in duration. Supplementing gage records using information from historical, botanical, or geological sources can substantially improve the accuracy of estimated recurrence intervals for very large floods, or for streams with no gaging records.

Hydraulics of streams. Water flowing within a stream channel is converting potential energy to kinetic energy and heat. The amount of potential energy available for this conversion depends on the vertical drop as the water moves downstream, and on the mass of water moving downstream. Kinetic energy can be expended in overcoming external and internal resistance, and in transporting sediment. External resistance comes from roughness along the bed and banks of the stream. Individual grains that protrude into the flow create external roughness, as do bedforms such as ripples and dunes, coarse woody debris in the stream, irregularities in the channel banks, and downstream variations in channel shape such as meander bends or alternating pools and riffles. Internal resistance occurs when individual fluid elements do not follow all parallel flow paths and

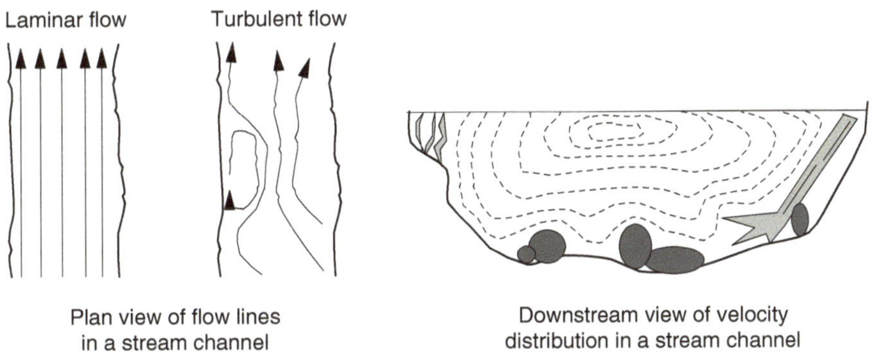

Figure 8 Simplified illustration of hydraulics in a natural channel. The plan view drawings illustrate laminar flow, in which all streamlines are parallel and water moves at equal rates, and turbulent flow, in which streamlines move at different rates, and flow has components of movement laterally across the channel and vertically within the channel, as well as downstream. The downstream view illustrates a natural channel with a slightly irregular cross-sectional form and sources of external resistance along the channel boundaries, including wood (right), cobbles and boulders (center), and submerged vegetation (left). The resulting isovels, or contours representing equal velocity distribution, are shown as dashed lines. The slowest velocities are along the sides and bottom of the channel.

move at the same velocity (laminar flow), but instead move at different rates with components of vertical and lateral movement as well as downstream movement (turbulent flow). Flow in all natural stream channels is turbulent to some extent because the water moving along the stream bed and banks encounters more external resistance and moves more slowly than water toward the top center of the stream (**Figure 8**).

Sediment transport in streams. Sediment can be transported in solution within streams. This dissolved or solute load constitutes a greater proportion of sediment transport during periods of base flow, when water that has moved slowly through the subsurface and had longer periods of time to react with the surrounding matrix, constitutes a greater proportion of stream discharge. Dissolved load is also relatively large for streams draining rocks such as limestone, which is susceptible to chemical weathering, and for streams in tropical regions that tend to have higher rates of chemical weathering for all types of rocks.

Sediment that is not dissolved in stream water can move in suspension within the water column or in contact with the stream bed. Washload is the finest portion of the suspended material, predominantly silts and clays that do not form a substantial portion of the sediment on the streambed. Washload requires so little energy to be transported that it tends to remain in suspension for hours or days even in areas of still water. Suspended load refers to the slightly coarser sands and pebbles that alternate between periods of moving in suspension and periods of moving along the bed and can settle from suspension relatively rapidly when velocity decreases. Bedload moves in nearly continuous contact with the streambed as larger particles roll, slide, and bounce downstream. Because the larger particles that constitute suspended and bedload require greater amounts of energy to move, much of this transport occurs during floods (**Figure 9**).

Figure 9 A flood from the Paria River (mouth at upper left) joins the Colorado River at Lees Ferry, Arizona. The Colorado River, at right, flows relatively clear, whereas the Paria carries high concentrations of suspended sediment.

Glossary

Bedforms – Regularly repetitive longitudinal alternations in streambed elevation, such as pools and riffles, steps and pools, or dunes.

Bedload – Sediment moving in nearly continuous contact with the streambed.

Dissolved load – Sediment transported in solution by stream flow.

Drainage density – A measure of the total length of stream channels per unit area of the drainage basin.

Drainage divide – A topographic high point or line that delineates the boundaries of a drainage network.

Drainage network – An integrated group of stream channels that drain toward a common point.

External resistance – Hydraulic resistance created by the channel boundaries.

Flood–frequency curve – A plot of flood magnitude versus recurrence interval.

Flow duration curve – A plot of discharge magnitude versus the percent of time that discharge occurs.

Glacier melt – Runoff created when glacial ice melts.

Groundwater flow – subsurface flow that occurs below the water table, or zone of saturation.

Hydraulics – The mechanical properties of liquids; for rivers, these properties are described by variables such as velocity.

Hydrograph – A plot of discharge versus time.

Internal resistance – Hydraulic resistance created by differences in the rate and direction of movement of individual fluid elements within a channel.

Laminar flow – Individual fluid elements follow parallel flow paths and move at the same velocity.

Overland flow (Hortonian, saturation) – Water moving across the ground surface; Hortonian overland flow has no infiltration, whereas saturation overland flow results from water that briefly infiltrates to shallow depths and then returns to the surface as the subsurface becomes saturated.

Piping – The processes whereby preferential flow in the unsaturated zone creates longitudinal cavities in the subsurface.

Rainfall – Liquid precipitation that results from different types of atmospheric circulation patterns.

Rain-on-snow – Rain falling directly on a snowpack, which increases the rate of snowmelt.

Reservoir failure – Collapse of a dam built by natural processes such as landslides, or by humans; the collapse results in rapid drainage of the water ponded behind the dam.

Rill – Channels that have no tributaries.

Sapping – The processes whereby preferential flow in the saturated zone creates longitudinal cavities in the subsurface.

Sediment transport – The movement of sediment in channels, includes dissolved, wash, suspended, and bedload.

Snowmelt – Runoff created when snowfall or snowpack melts.

Suspended load – Particulate material moving in suspension in stream flow and of a size that can settle relatively rapidly when velocity decreases.

Throughflow – Subsurface flow that occurs above the water table, or in the unsaturated zone.

Turbulent flow – Individual fluid elements move at different rates and exhibit lateral and vertical components of movement as well as moving downstream.

Washload – The smallest sizes of particulate material moving in suspension in stream flow; usually clay- and silt-sized particles that do not form a substantial portion of the sediment on the streambed.

See also: Currents in Rivers; Fluvial Export; Fluvial Transport of Suspended Solids; Small-Scale Turbulence and Mixing: Energy Fluxes in Stratified Lakes.

Further Reading

Milliman JD and Syvitski JPM (1992) Geomorphic/tectonic control of sediment discharge to the ocean: The importance of small mountainous rivers. *The Journal of Geology* 100: 525–544.

Poff NL, Allan JD, Bain MB, *et al.* (1997) The natural flow regime. *BioScience* 47: 769–784.

House PK, Webb RH, Baker VR, and Levish DR (2002) *Ancient Floods, Modern Hazards: Principles and Applications of Paleoflood Hydrology.* American Geophysical Union Press.

Hirschboeck KK (1988) Flood hydroclimatology. In: Baker VR, Kochel RC, and Patton PC (eds.) *Flood Geomorphology,* pp. 525–544. New York: John Wiley and Sons.

Knighton D (1998) *Fluvial Forms and Processes: A New Perspective.* Oxford University Press.

Leopold LB, Wolman MG, and Miller JP (1964) *Fluvial Processes in Geomorphology.* Freeman and Company.

Wohl EE (2000) *Mountain Rivers.* American Geophysical Union Press.

Relevant Websites

http://www.usgs.gov/hazards/floods/ – U.S. Geological Survey floods.

http://www.fema.gov/hazard/flood/index.shtm – Federal Emergency Management Agency floods.

http://www.noaa.gov/floods.html – National Oceanic and Atmospheric Administration floods.

http://www.dartmouth.edu/~floods/ – Dartmouth Flood Observatory.

http://waterdata.usgs.gov/nwis/sw – U.S. Geological Survey surface-water records.

http://nrrss.nbii.gov/ – National River Restoration Science Synthesis.

http://www.willametteexplorer.info/ – Willamette Basin (Oregon) explorer.

http://water.usgs.gov/osw/streamstats/index.html – U.S. Geological Survey StreamStats.

Rivers

P A Bukaveckas, Virginia Commonwealth University, Richmond, VA, USA

What is a River?

There is no strict definition to distinguish rivers from streams and therefore the designation 'river' encompasses flowing waters of widely varying size. Flowing waters may be ranked in size by various metrics that include discharge (glossary), catchment area, and length of channel. For example, the discharge of the Amazon River is six orders of magnitude greater than that of a small river. This range of variation is comparable with the range in volume observed among lakes worldwide. Rivers are sometimes defined as 'non-wadeable' flowing waters since this delineation has practical implications for the way sampling activities are carried out. Along the continuum from headwater streams to large rivers, there are gradients in channel slope, width, and depth. Idealized gradients in geomorphology provide a basis for understanding differences in the structure and functioning of streams vs. rivers. For example, the greater width of river channels reduces the importance of riparian inputs while greater depth lessens the influence of benthic processes. Rivers in their natural settings exhibit complex geomorphologies that give rise to a rich variation in channel form and function and provide diverse habitats for aquatic biota.

Hydrology and Geomorphology

Water Sources and Discharge

Water sources to rivers are principally surficial inputs via tributary streams (**Table 1**). Owing to their small surface area, direct atmospheric inputs are usually minor though ground water is important in some settings. For comparisons among river basins, discharge is converted to an areal water yield by dividing the volume of discharge by the area of the drainage basin. Water yields vary widely depending on the amount of precipitation relative to evapotranspiration (glossary). South American rivers such as the Amazon and Orinoco are notable for their high water yields, exceeding 1000 mm year^{-1} (**Table 2**). Arid and semiarid regions are characterized by low precipitation relative to evapotranspiration and water yields less than 100 mm year^{-1}. Arid regions occupy about one-third of the world's land area, including portions of several major river basins such as the Murray-Darling (Australia), Colorado (North America), Nile (Africa), and Ganges (Asia) Rivers.

Variation in river discharge arises from short-term, seasonal, and long-term variability in precipitation and evapotranspiration within the drainage basin. Over short time scales (days–weeks), discharge is affected by rain events associated with frontal passage. Though infrequent in occurrence, event-related discharge may account for a large proportion of the annual total. The frequency and magnitude of storm events is therefore an important factor influencing interannual variation in discharge. Event-driven and seasonal variations are superimposed upon long-term (decadal-scale) climatic cycles (e.g., El Niño Southern Oscillation), which may bring about extended periods of above- or below-average discharge. The combined effects of climatic variations occurring over multiple time scales results in a wide range of discharge conditions, which may exceed three orders of magnitude for a given site. Variation in discharge is typically larger that the variation in the concentration of dissolved and particulate substances such that the export of materials from the basin (flux rate) is principally determined by discharge.

Seasonal variation in rainfall and evapotranspiration give rise to predictable annual patterns in river discharge that are characteristic of climatic regions (**Figure 1**). In temperate-humid climates, rainfall may be distributed relatively uniformly throughout the year but seasonal changes in evapotranspiration give rise to variation in discharge. Warmer months are associated with high evapotranspiration, resulting in less runoff from the catchment and lower river discharge relative to colder months. Snowmelt may also contribute to a spring discharge pulse in climates that allow for winter accumulation of snow (including tropical rivers with mountainous catchments). The north-flowing rivers of Canada and Russia are representative of this hydrologic regime in exhibiting high year-round discharge but with a pronounced winter-spring peak. In tropical-humid climates, evapotranspiration is less variable throughout the year but rainfall is often strongly seasonal, particularly in regions affected by monsoons. Wet seasons are associated with elevated river stage and discharge and may be accompanied by extended periods of floodplain inundation. Most South American and African rivers are representative of the tropical unimodal hydrologic regime, which is characterized by an extended period of elevated discharge and floodplain inundation during the rainy season. Arid and

Table 1 Distinguishing characteristics of rivers, estuaries, and lakes

	Rivers	*Estuaries*	*Lakes*
Water movement	Unidirectional, horizontal	Bidirectional, horizontal	Vertical
Water forces	Gravitational	Tidal	Wind-induced
Water-level fluctuations	Large (seasonal)	Variable (daily, storm events)	Small (seasonal)
Water residence time	Days–weeks	Weeks–months	Months–years
Water sources	Runoff	Runoff, marine, precipitation	Runoff, ground water, precipitation
Stratification	Rare	Common (salinity)	Common (thermal)
Transparency	Low (nonalgal particulates)	Variable (particulates, dissolved color)	High (algae, dissolved color)

Table 2 Water and sediment delivery from large river basins of the world

River	*Drainage area (10^6 km^2)*	*Discharge (km^3 year^{-1})*	*Water yield (mm year^{-1})*	*Sediment load (10^6 t year^{-1})*	*Sediment yield (t km^{-2} year^{-1})*
Amazon	6.15	6300	1024	1200	195
Colorado	0.64	20	31	0.01	0.02
Columbia	0.67	251	375	10	15
Congo (Zaire)	3.72	1250	336	43	12
Danube	0.81	206	254	67	83
Ganges–Brahmaputra	1.48	971	656	1060	716
Huang He (Yellow)	0.75	49	65	1050	1400
Indus	0.97	238	245	59	61
Mackenzie	1.81	306	169	42	23
Mekong	0.79	470	595	160	202
Mississippi	3.27	580	177	210	64
Niger	1.21	192	159	40	33
Nile	3.03	30	10	0	0
Orinoco	0.99	1100	1111	150	152
St. Lawrence	1.03	447	434	4	4

Source: Milliman JD and Meade RH (1983) Worldwide delivery of river sediment to the oceans. *Journal of Geology* 91: 1–21.

semiarid regions occur in both temperate and tropical climates and occupy about one third of the world's land area. They are characterized by low precipitation relative to evapotranspiration and include portions of several major river basins, including the Murray-Darling (Australia), Missouri (North America), Nile (Africa), and Ganges (Asia) Rivers. River basins in arid regions exhibit sustained periods of low discharge interspersed with short periods of elevated discharge. For example, the Murray-Darling River is fed by infrequent summer monsoons which, coupled with high rates of evapotranspiration, result in an annual discharge equivalent to only 3% of annual rainfall.

Large river basins may span climatic and topographic regions and exhibit complex hydrologic regimes. For example, the Rhone is a snowmelt-dominated river in its upper, mountainous sections but is influenced by a Mediterranean climate in its lower course. The river exhibits a complicated flow regime with low discharge periods shifting from winter in the upper course to autumn in the lower course and floods occurring in all seasons. Despite the problems inherent in categorizing this continuum of variation, hydrologic regimes are useful for facilitating comparisons among river basins (e.g., in response to land-use and climate change effects).

Flooding

Rivers experience large and rapid fluctuations in surface water elevation (i.e., 'stage') in response to runoff. The rate and magnitude of rise in river stage is dependent in part on the morphometry of the channel (**Figure 2**). Low banks enable the river to escape the active channel and inundate lateral areas (floodplain). During flooding, the widening of the river lessens the stage response to runoff and reduces water velocity because the force of the water is distributed over a wider area. Flood-prone rivers are common in both temperate and tropical climates and exhibit considerable variation in the extent, timing, and frequency of flooding events. In some settings (e.g., Amazon River) the annual flood pulse is a defining feature of the riverscape, important not only to the life cycles of riverine biota but also in shaping floodplain communities. Floodplains are rare in naturally constricted

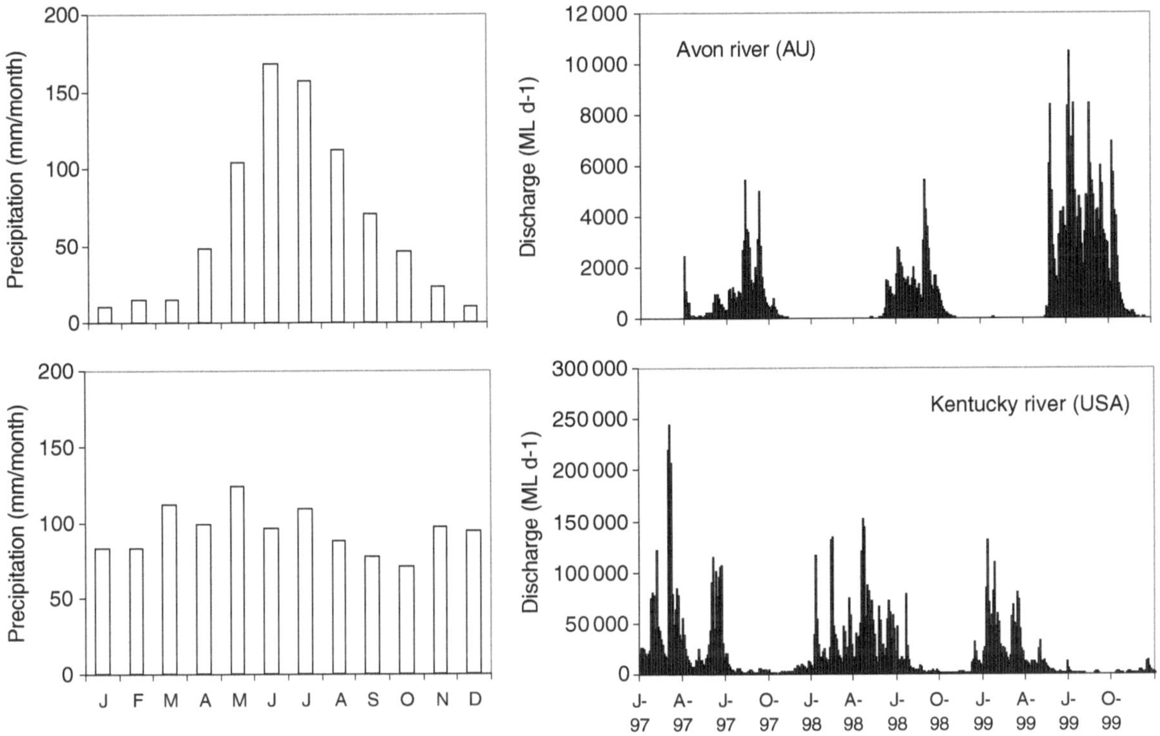

Figure 1 The hydrologic regimes of tropical and temperate rivers reflect differences in seasonal patterns of precipitation and evapotranspiration. The Avon River (Western Australia) experiences high evapotransipration throughout the year and variation in discharge is largely driven by seasonal patterns in rainfall. The Kentucky River (North America) receives similar rainfall throughout the year but variation in evapotranspiration results in similar seasonal patterns in river discharge (offset in northern vs. southern hemispheres).

rivers; or may be disconnected if lateral water regulation structures (i.e., levees) are present. In constricted and levied channels, the effects of runoff on river velocity and stage are accentuated because the ratio of water volume to bottom area increases with rising stage. Thus, the influence of frictional resistance in dissipating energy is lessened with rising stage. Hydrodynamics of river channels are often depicted using simulation models that describe water movements in one, two, or three dimensions (longitudinal, lateral, vertical). These models typically rely on input data describing channel geomorphometry (cross-sectional depictions of river bed and bank elevation) and calibrated using measured surface water elevation and discharge. The models predict surface water elevation under various discharge scenarios and are used to forecast the timing, severity and location of flood events.

Water Movement

Energy is required to move water and in the case of rivers, this energy is derived from gravitational forces acting along an elevation gradient. Rivers are similar to estuaries in that both are flow-dominated (advective) systems; in estuaries, however, the movement of water is bidirectional and driven by tidal forces (**Table 1**). Water movement in lakes is driven by comparatively weak forces associated with wind-induced vertical mixing. The slope of the channel and the frictional resistance imposed by its boundaries determine the velocity with which water is carried down the elevation gradient. The roughness of the channel reflects the composition of bed and bank materials and the presence of natural and artificial structures (e.g., woody debris, wing dams). Turbulence arises as force is dissipated by frictional resistance. This mixing energy maintains particulate matter in suspension and is sufficient to overcome differential heating of surface and bottom layers. Consequently, thermal stratification is rarely observed in rivers except in cases where impoundments are present.

The length of time that water resides within a given segment of the river determines in part the potential for physical, chemical, and biological processes to act upon the dissolved and particulate constituents in through-flowing water. Because of the unidirectional

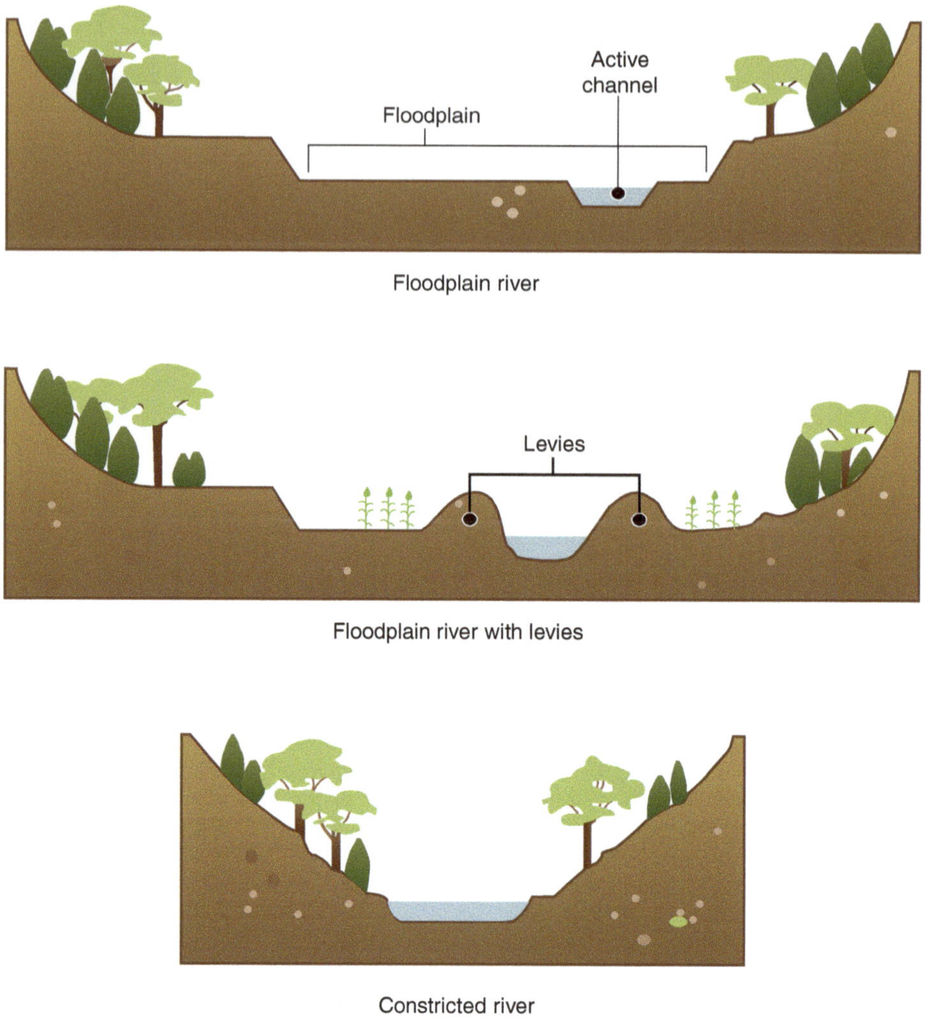

Figure 2 Cross-sectional morphologies of floodplain and constricted rivers. In floodplain rivers, rising river stage results in lateral inundation unless precluded by the presence of levees. Widening of the river during flood events increases frictional forces and reduces water velocity. In constricted rivers, lateral inundation is constrained by steep adjoining slopes resulting in rapidly increasing water velocity with rising river stage. (Illustration by Christopher O'Brion, VCU Design Services).

flow of water, transit time is a useful metric to characterize inter-river differences in the time required for water and materials to travel through a reach of specified length. Tracers such as dyes (rhodamine) or conservative solutes (chloride, bromide) are used to measure transit time by tracking the movement of labeled parcels of water. Tracer additions provide a reach-scale estimate that integrates longitudinal, lateral, and depth variations in water velocity. Application of this technique to larger rivers is problematic owing to the quantities of tracer required and the difficulty of achieving a laterally uniform addition. Transit time estimates may be obtained from hydrologic models using measured discharge and cross-sectional area to infer average (cross-sectional) velocity at multiple points along the channel. The coupling of transit time and nutrient uptake, termed nutrient spiraling (glossary), is a concept that has been widely used as a framework for understanding the interaction between hydrologic and biological processes in regulating nutrient retention. Transit time estimates are also used to design sampling programs in which a parcel of water is sampled repeatedly as it travels down the channel (termed LaGrangian sampling).

Geomorphology

At any point along a river course, channel morphology reflects the interplay between the force of water and the stability of bed and bank materials. Channel form is a quasi-equilibrium condition maintained by the dominant discharge and determined in part by the supply of sediment from upstream. Where rivers are not constrained by natural landforms or

Figure 3 Selective loss of fine materials may over time create channel reaches that are characterized by a predominance of large substrates such as the gravel bars illustrated here. Their presence in the river channel is important to the maintenance of biodiversity as some species colonize hard substrates or exploit interstitial spaces as a means of adapting to flowing environments. Gravel bars and other subsurface exchange zones are also important to ecosystem function such as nutrient retention. Photo of the Rio Apurimac in Peru by A. Aufdenkampe (see related paper by Aufdenkampe *et al.* (2007) *Organic Geochemistry* 38: 337).

water regulation structures, channels migrate laterally (meander) through erosion and redeposition of bank materials. Active channels are characterized by the ephemeral nature of their features (movement of bars and banks) and by their morphological complexity, which may include the presence of pools, riffles, side channels, and meanders (**Figure 3**). Constrained channels occur where natural landforms or water regulation structures limit lateral mobility. High discharge results in the erosion of bed materials leading to incised (entrenched) channels of low structural complexity and relatively uniform flow conditions. Channel forms and substrate conditions influence the structure and functioning of riverine food webs. For example, where flow conditions favor the deposition of fine materials, the accumulation of particulate organic matter enhances benthic microbial activity. Various schemes have been devised to categorize channel forms, though these efforts are often confounded by the continuous rather than discrete variation in channel features (e.g., width–depth ratio; size distribution of bed materials). Emerging technologies for sensing underwater environments hold much promise for linking biological and geophysical properties particularly in large rivers.

Water Regulation

Human activities have substantially affected the natural hydrologic cycles of rivers throughout the world. Land-use changes have indirect effects on river hydrology by altering the timing and quantity of runoff from the catchment. For example, urbanization creates impermeable surfaces that increase the volume and speed of storm runoff. Direct impacts include the abstraction (withdrawal) of river water for domestic supply and irrigation as well as the alteration of river channels by water regulation structures. Rivers have been altered through the construction of dams, levees and other channel modifications to accommodate local needs for flood protection, hydropower generation and navigation. Channelization (straightening) of river courses facilitates navigation but reduces channel and flow complexity thereby diminishing habitat diversity. Channelized rivers are subject to elevated flow velocities that cause erosion and necessitate bank stabilization. Levees preclude lateral exchange and thereby diminish the role of floodplains in material and energy cycles. In flood-prone rivers, biota are adapted to annual flood pulses that provide access to food and spawning areas within the floodplain. Among the most widespread of human impacts on rivers is the construction of dams, which currently number in excess of 45 000 worldwide. Together, their cumulative storage capacity is equivalent to 15% of global annual river runoff. Over half of the world's major rivers are affected by dams, most of which were constructed in the twentieth century. Dams induce pelagic conditions by increasing water storage and dissipating mixing energy. Pelagic conditions favor sediment deposition and biotic assemblages that differ from those occurring in flowing environments. The severity of water regulation effects varies according to the number and size of regulation structures along the river course. The cumulative effect of dams within a river basin can be gauged from their number and storage capacity expressed relative to river discharge (**Table 3**). Low dams (height < 10 m) are designed to maintain a minimum depth for navigation during low discharge and

Table 3 Water regulation effects vary according to the number and storage capacity of mainstem dams as illustrated by rivers of the central United States

River	*Mean annual discharge ($m^3\ s^{-1}$)*	*Storage capacity of mainstem impoundments (km^3)*	*Retention effect of impoundments (days)*
Kentucky	234	0.26	12.7
Green	314	1.11	41.0
Tennessee	1880	15.00	92.4
Cumberland	936	1.18	14.7
Wabash	800	0.19	2.8
Ohio	7811	8.92	13.2

The combined storage capacity of mainstem dams on the Tennessee River is equivalent to 26% of the river's annual discharge or approximately 92 days at average discharge. In contrast, the Wabash is a relatively free-flowing river with only a single mainstem impoundment that stores a volume less than 1% of its annual discharge.

thereby regulate stage but do not eliminate flowing conditions. High dams are designed for flood control and water storage. They inundate large areas and effectively create lake-like conditions, in some cases, resulting in thermal stratification of the water column.

Water Quality

Rivers integrate drainage waters from distant points in the landscape that may differ in topography, soils, vegetation, and land use. These differences give rise to widely varying water chemistry within river basins particularly where anthropogenic influences differ among sub-basins. Along the river course, water chemistry changes in response to inputs from these diverse sources and also reflects variable water residence times in channel, hyporheic, and lateral storage zones.

Particulate Matter

High concentrations of suspended particulate matter are a characteristic feature of rivers particularly during periods of elevated discharge. The upward component of water turbulence acts to maintain particulate matter in suspension, resulting in downstream transport. Particulate matter may originate within the channel through erosion of bed and bank materials, resuspension of sedimented materials, and biological production. Most particulate matter, however, is derived from sources outside the river channel that are transported via tributary streams. The rivers of Asia are particularly noted for their high sediment load. It is estimated that the Ganges, Brahmaputra, and Yellow Rivers contribute 20% of the total sediment load transported to the oceans (**Table 2**). High sediment production is attributed to natural factors affecting surface erosion (soil composition, steep slopes, and intensive rainfall) as well as anthropogenic effects associated with deforestation and urbanization. Riverine suspended matter is predominantly a fine-grained (<0.2 mm) mixture of mineral and organic particulates (e.g., clay and silt). Though recalcitrant, mineral particulates may undergo changes in their chemical composition through the selective sorption and desorption of dissolved substances. For example, proteins and other dissolved organic compounds adhere to the surfaces of mineral particulates, thereby altering both the bioavailability of these compounds and the chemical properties of particulate matter. Phosphate has a high sorption potential and is principally transported with the particulate fraction. The sorption capacity of particulate matter is determined by the number of available binding sites on the surfaces of the particles and their cumulative surface area (a function of particle density, shape, and size).

Particulate matter is the principal factor regulating water transparency in rivers, although light absorption by dissolved organic compounds may be important during periods of low discharge. When present in high concentrations, mineral particulates may have deleterious effects on filter-feeding organisms by interfering with feeding mechanisms or simply diluting the intake of the more nutritious organic fraction. This fraction includes phytoplankton and bacteria although these typically account for a small proportion of particulate organic matter. The bulk of the particulate organic matter is nonliving detrital material of terrestrial and aquatic origin. This material is of variable age and in varying stages of diagenesis, having been acted upon by both terrestrial and aquatic decomposers.

Dissolved Substances

River water contains dissolved inorganic and organic materials derived from mineral weathering and decomposition processes. Their concentration is largely determined by the types of soils and vegetation within the basin and the extent of interaction between runoff and soil. Low concentrations of dissolved substances occur where river basins are characterized by steep slopes and thin soils, particularly where soils are comprised of insoluble materials (e.g., sand, igneous rock). In these basins, river water is dilute (ion-poor) and similar in chemical composition to that of rain water. Gradual slopes and deeper soils allow for longer flowpaths and greater interaction between water and soil. In these settings, there is greater opportunity for biogeochemical processes to influence the chemistry of runoff especially where soils are dominated by easily-weathered materials (e.g., sedimentary rocks such as limestone). Temporal variation in dissolved ion concentrations is typically associated with rain and snowmelt events. High discharge is often characterized by lower concentrations of dissolved substances owing to rapid delivery of water via overland flow, shallow soil flowpaths, and short transit times in tributary streams. At the onset of rising discharge, rain or snow-melt waters may displace older ground water, resulting in an initial increase in ion concentrations. Thus, the relationship between discharge and concentration is often nonlinear and ion-specific.

Geologic differences among river basins will influence both the total amount of ions present and their relative proportions. Despite these differences, major ions are generally similar and include bicarbonate, sulfate, chloride, and the base cations (Ca, Mg, Na, K). Climatic factors also influence ionic strength and composition particularly in arid regions where evapoconcentration effects are large. A tea-colored appearance is an apparent feature of some ('blackwater') rivers owing to elevated concentrations of dissolved organic compounds. Their presence is associated with characteristic types of vegetation that leach humic and tannic acids and in some cases (e.g., in coastal areas) by the predominance of sandy soils, which have limited capacity to retain these compounds.

Nutrients

Rivers are not simply conduits for transporting watershed-derived materials but rather, riverine processes may exert considerable influence on water chemistry, particularly for those elements whose abundance is low relative to biological demand. Nitrogen is transported in rivers in dissolved inorganic form (NO_3, NH_4) and in dissolved and particulate organic forms. The latter include living cells and detrital matter as well as a diverse array of dissolved organic compounds that are released through exudation, excretion, and decomposition. Nitrate is a highly mobile ion owing to its low sorption potential. Therefore, it is readily transported through soils and is typically the dominant form of N in rivers, where agriculture and urbanization are prevalent. Elevated NH_4 concentrations may occur below wastewater discharge points. Dissolved organic N assumes greater importance in rivers with minimal human influence. Unlike N, phosphorus is principally transported in the particulate fraction. Concentrations of dissolved P (including PO_4 and other reactive forms) are low owing to biotic uptake and high sorption affinity for mineral particulates (e.g., clay). Sorption processes are reversible such that particle-bound P may desorb and enter the bioavailable pool. Anthropogenic impacts are associated with increases in the total amount of P and the proportion that is in the dissolved fraction. Within rivers, inorganic forms of nitrogen, phosphorus, and silica may be transformed to particulate organic forms (e.g., in algal and bacterial cells). Dissolved silica is converted to its biogenic form by diatoms, a common component of benthic and pelagic algal communities in rivers. Biogenic silica is relatively recalcitrant to remineralization (compared with N and P) such that autotrophic uptake results in progressive depletion of dissolved silica along the river course. Denitrification results in the loss of nitrogen to the atmosphere (as N_2) and is an important process determining N delivery from catchments.

Dissolved Gases

Dissolved gases, particularly oxygen and carbon dioxide, are of interest because their concentrations in river water are influenced by biological processes of photosynthesis and respiration. The solubility of dissolved gases is temperature dependent and therefore it is useful to express concentrations as a percent saturation; that is, relative to the expected concentration for a solution in atmospheric equilibrium. Departures from equilibrium concentrations occur when the rate at which gases are exchanged with the atmosphere is slow relative to rates at which gases are produced or consumed through biological activity. Atmospheric exchange is governed by the concentration gradient across the air–water interface, boundary layer thickness (a function of wind speed), the ratio of river surface area to volume, and factors related to agitation and turbulence of water (e.g., presence of waterfalls). Gas exchange occurs more rapidly in shallow and turbulent rivers relative to deeper, slow-moving rivers. In many rivers, dissolved oxygen is undersaturated while CO_2 is supersaturated (**Figure 4**). These departures from equilibrium reflect the heterotrophic nature of rivers in which community respiration exceeds autotrophic production. Respiration is supported in part by inputs of dissolved and particulate organic matter of terrestrial origin. Decomposition of terrestrial organic matter within the river results in a net production of CO_2 (i.e., in excess of photosynthetic C demand) and a net release of CO_2 from water to air. Diel variations in dissolved oxygen can be used to estimate production and respiration provided that re-aeration rates

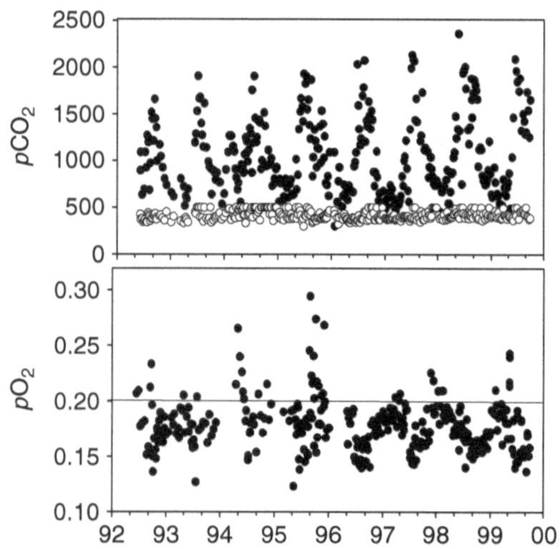

Figure 4 Partial pressure of dissolved carbon dioxide and oxygen in the Hudson River (NY, USA; atmospheric levels of CO_2 indicated by open circles). Persistent supersaturation of pCO_2 and undersaturation of pO_2 are indicative of net heterotrophic conditions whereby respiration exceeds net primary production (adapted from Cole and Caraco 2001, Marine and Freshwater Research).

can be reasonably estimated. Although undersaturation of dissolved O_2 is common, severe depletion (i.e., hypoxia – glossary) is rare in riverine environments because turbulent mixing promotes reaeration. Organic matter inputs from poorly-treated sewage effluent were once a wide-spread problem that resulted in chronic and severe oxygen depletion in rivers. Modern wastewater treatment plants are designed to minimize the biological and chemical oxygen demand of effluent.

Pollutants

Rivers integrate runoff over large areas of the landscape and therefore their pollutant loads reflect the cumulative effect of basin-wide releases. Macropollutants include a relatively short list of agents present in concentrations on the order of parts per million ($mg\ l^{-1}$) while micropollutants includes a much larger inventory of chemicals that occur at very low environmental concentrations (ppb or ppt; $\mu g\ l^{-1}$ or $ng\ l^{-1}$). The most common macropollutants are compounds of N and P, which originate in runoff from agricultural areas and from contamination by wastewater (including treated effluent and urban storm water overflow). Nitrogen and phosphorus often limit primary production in lakes and estuaries though their role in regulating the trophic state of rivers is less clear. Many rivers experience nutrient enrichment but biotic responses to elevated nutrient levels (i.e., eutrophication) may be muted by factors that constrain primary production (principally light and residence time). Other macropollutants include sulfate, chloride, and base cations; these are associated with atmospheric deposition, mining, wastewater, and de-icing. Their effects on river biota are less well studied compared with pollutants associated with eutrophication. Micropollutants are a diverse group of chemicals that have deleterious effects at low concentrations. They vary in their reactivity, mode of toxicity, and persistence in the environment and include inorganic pollutants such as metals as well as synthetic organic compounds (e.g., pharmaceuticals, detergents, pesticides). In rivers, the high throughput of water favors the rapid removal of pollutants in the dissolved form. Many pollutants, however, bind to particulates or enter the food chain, where they may persist over long periods of time in sediments and long-lived species such as fish. Regulatory policies aimed at mitigating pollution must take into account proximal effects on river biota as well as distant effects on receiving waters such as estuaries. In some cases (e.g., nutrients), the latter may exhibit greater sensitivity than rivers owing to their longer water residence time.

Biology of Rivers

Rivers owing to their diverse size, channel forms, and biogeographic settings differ greatly in their species assemblages. Constituent species include river specialists that rarely occur outside of flowing waters and habitat generalists that occur in both lentic and lotic waters. In coastal areas, marine species are seasonally important members of river food webs. Salmon and other anadromous fishes (glossary) serve as vectors for distributing marine-derived resources through drainage networks. Species inhabiting rivers face challenges imposed by the unidirectional flowing nature of their environment. Strategies include current avoidance in sheltered areas (along channel margins, behind debris dams or in interstitial spaces), and specialized adaptations such as attachment to hard substrates. Riverine species also share the benefits provided by water flow which supplies particulate matter to filter-feeding organisms, replenishes nutrients and oxygen at the cell boundary layer, and, during floods, allows periodic access to floodplain habitats.

Primary Producers

Attached algae (i.e., periphyton), phytoplankton, and macrophytes contribute to autotrophic production in rivers; their relative importance varies in accordance with river hydrogeomorphology. In shallow, fast-flowing rivers, benthic algae predominate particularly where rocks and woody debris provide stable substrates for colonization. Benthic algal abundance is determined by the availability of suitable substrates, light conditions (the extent of riparian shading), and flow regime (the frequency and severity of scour events). Nutrients and grazers may be important in some settings particularly where nutrient loading is associated with riparian disturbance and loss of canopy shading. In deep, slow-moving rivers, phytoplankton are often the dominant primary producers. Their abundance is principally determined by light availability. The average light intensity experienced by phytoplankton circulating within the river channel is determined by water transparency and the depth of the channel. Nutrients and grazing may be important particularly in regulated rivers and during low discharge conditions. Low flow velocities favor the accumulation of phytoplankton biomass owing to reduced washout (advective loss) and increased water transparency (due to sedimentation of nonalgal particulates). Phytoplankton communities are composed of taxa similar to those found in lentic environments but may also include detached benthic algae. Dominance by diatoms is often reported and may reflect their ability to tolerate the low light conditions in rivers (having a high light utilization efficiency) and the benefits of active mixing (to offset high sinking velocities). Channel morphometry is an important factor determining the species composition and areal coverage of submergent and emergent aquatic vegetation (**Figure 5**). Constricted and channelized rivers have steep shoreline areas, which provide little suitable habitat, whereas floodplain and low-gradient rivers allow for greater colonization in shallow-water areas. Substrate stability is likely a key factor determining the extent and persistence of macrophyte beds since perenniating structures (e.g., tubers, rhizomes) are vulnerable to displacement during periods of elevated discharge.

Figure 5 Aquatic macrophytes are common in rivers though usually they are restricted to channel margins and backwater areas, where flow conditions are reduced. Photo of Beaver River in the Adirondack Mountains of New York State (USA) by P. Bukaveckas.

Invertebrates

The diversity and productivity of invertebrates has received considerable attention in studies of riverine foods webs. For example, nearly one-third of Hynes' classic *Ecology of Running Waters* is devoted to benthic macroinvertebrates. Invertebrates are important to trophic energetics because they link primary sources of energy (autochthonous production and allochthonous inputs) to higher trophic levels such as fish. Pelagic invertebrates (zooplankton) are commonly found in regulated and deep rivers and at times in large numbers. Their abundance is determined by in situ production within the main channel and contributions from areas of reduced water velocity (**Figure 6**). Inputs from upstream reservoirs may also be important in some systems. River zooplankton assemblages are dominated by rotifers and small-bodied forms of cladocerans (e.g., *Bosmina*) and copepods, whereas larger zooplankters (*Daphnia*) are generally associated with lentic environments.

Benthic invertebrates are important components of river food webs and are widely used in habitat assessments owing to their sensitivity to water quality conditions. These include crustaceans such as amphipods and crayfish, mollusks (snails and bivalves), and a great variety of insects (dragonflies, damselflies, stoneflies, mayflies, midges, blackflies, caddisflies). The aquatic insects in particular draw attention to the productive nature of riverine environments through periodic emergence of adults in large numbers. Invertebrates may be grouped according to their feeding habits as predators, filtering and gathering collectors, deposit feeders, scrapers, and shredders. The River Continuum Concept predicts shifts in food resources and feeding habits along a gradient of stream order. In low- and middle-order streams, shredders and grazers rely on leaf litter inputs and benthic algal production whereas in rivers, collectors and filter-feeders utilize suspended particulate matter. Productivity is determined by water temperature, food quantity and quality and the presence of suitable habitat (e.g., hard substrates and snags). A variety of invertebrates, particularly oligochaetes, amphipods, chironomids and microcrustaceans, occur in large numbers in the subsurface zone (hyporheos; glossary) where they find refuge from predation and currents.

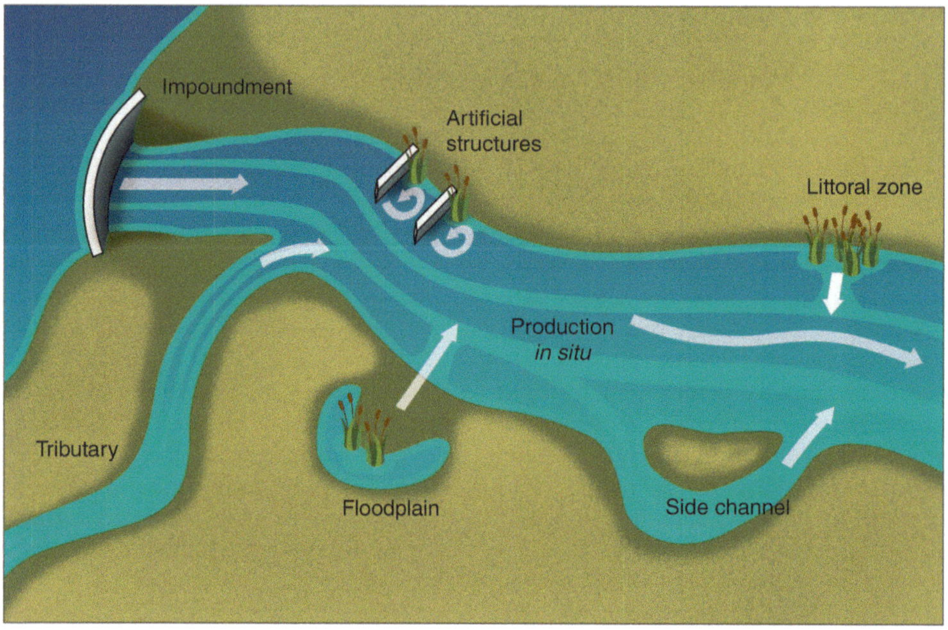

Figure 6 Sources of plankton to rivers include *in situ* production as well as inputs from tributaries, impoundments and near-shore areas of reduced water velocity (Illustration by John Havel and Christopher O'Brion).

Fishes

Fish are typically the top predators in river food webs and, like macroinvertebrates, are often used as 'bio-indicators' for habitat assessment. Many studies have focused on species that are important to commercial or recreational fisheries. However, quantitative estimates of abundance are difficult to obtain particularly in large and deep rivers. The lack of production and biomass estimates with which to compare against similar data for lower trophic levels greatly limits our understanding of food-web energetics. For example, the utility of using phosphorus or chlorophyll as a predictor of fish biomass, which is well-known for lakes, remains largely untested in rivers. In contrast, factors influencing the diversity and species composition of river fish communities are generally well studied. In both temperate and tropical rivers, the numbers of species increase with the size of the drainage basin. The dendritic form of river networks may foster high diversity (relative to contiguous water bodies of comparable area) by providing diverse habitat conditions and through isolation of populations in distant portions of the drainage basin. In many regions, rivers are ancient features of the landscape, thus providing opportunities for speciation among reproductively isolated populations. Anthropogenic influences generally act to make fish assemblages more similar within and among basins and lead to loss of biodiversity. In many rivers, the presence of water regulation structures has had a negative impact on species that prefer flowing conditions and in some cases, has restricted their ability to access former spawning areas. The introduction of nonnative fish species has also substantially altered fish communities in many rivers. As for other river biota, discharge is the key environmental factor structuring communities. In floodplain rivers, fish seek refuge from current velocities and utilize food resources in inundated areas. In levied and naturally constricted rivers, high discharge may cause high mortality, particularly of larval stages, due to elevated current velocities in the channel.

River Food Webs

Research on river food webs has focused on trophic energetics with the goal of understanding the sources of organic matter supporting secondary production. Several conceptual models have been advanced that relate the abundance of invertebrates and fishes to sources of organic matter from the catchment, the floodplain and the river itself. The most influential of these is the River Continuum Concept (RCC) published by Robin Vannote and his colleagues in 1980 and cited in over 1800 subsequent publications. The utility of the RCC model lies in its holistic view of drainage networks whereby changes in the physical template of the channel (morphometry and substrate composition) with increasing stream order is linked to corresponding changes in food resources and biotic communities. The model emphasizes the importance of terrestrial (allochthonous) inputs in supporting secondary production. Consumers in river environments are thought to benefit from allochthonous inputs to a greater extent than their lentic counterparts due to loading factors that reflect the large ratio of land to surface water area in river basins. Autochthonous inputs were thought to be of minor importance particularly in headwater reaches (where shading by the forest canopy limits primary production) and in large rivers (where turbidity and depth limit algal and aquatic plant growth). This viewpoint is supported by geochemical analyses of riverine particulate matter which show that it is predominantly of terrestrial origin. However, the utilization of allochthonous and autochthonous organic matter is determined not only by their relative availability but also by their suitability relative to consumer needs (e.g., edibility, digestibility, nutritional sufficiency). Allochthonous inputs are comprised of detrital materials low in nutritive value whereas organic matter of autochthonous origin is enriched in mineral nutrients (N, P) and important biochemicals (fatty acids, proteins, etc.). An alternative view of river food-web energetics (Riverine Productivity Model; RPM) is that higher trophic levels obtain a disproportionate fraction of energy (or key dietary factors) from autochthonous sources by selective feeding and preferential assimilation of the more nutritious algal component. Stable and radio isotopes of carbon are used to quantify inputs from various sources (e.g., aquatic vs. terrestrial) provided that the sources differ in their isotopic signatures. Stable isotope data have shown that various consumer groups in rivers rely on algal production despite the quantitative dominance of organic matter that is terrestrial in origin. While the RCC and RPM focus on transport and production within the main channel, the Flood-Pulse Concept (FPC) considers the contribution of floodplain areas in supporting riverine communities. The importance of floodplain resources depends on the duration, aerial extent and timing of floodplain inundation. In tropical regions, flooded areas may far exceed the size of the main channel, thereby allowing riverine consumers to utilize terrestrial resources over extensive areas (**Figure** 7). The growth of aquatic plants and algae in flooded areas may also augment terrestrial resources if the duration of flooding is sufficiently long and light-temperature conditions are favorable. Temperate rivers also experience periodic floods although these are typically of shorter duration and occur during periods when water temperature is low (e.g., in association with winter rains or spring snowmelt). The three models differ by their emphasis on longitudinal transport of terrestrial organic matter (RCC), autochthonous production within the channel (RPM) and floodplain resources (FPC). They share the common view that an appreciation of river hydrogeomorphology is central to understanding variations in the quantity and quality of food resources and, in turn, the energetic efficiency of river foods webs.

Figure 7 Inundation of the floodplain near a tributary of the Amazon River (Rio Unini). In many rivers, flood events follow a regular annual cycle to which riverine organisms and riparian communities are adapted. Flooding allows access by river organisms to terrestrial food resources in inundated areas. Photo by A. Aufdenkampe.

Global Biogeochemical Cycling

Rivers account for only a small proportion of land area worldwide but play an important role in regional and global biogeochemical cycles. Rivers are the principal means by which terrestrial-derived materials are transported to the ocean. Over 90% of the earth's landmass is drained by rivers; the 100 largest rivers drain 65% of global land area. Rivers are the most powerful erosive force on the planet, substantially modifying landscape features and transporting 20 gigatons of sediment to the coastal margin annually. The input of dissolved and particulate organic carbon from rivers is sufficient to account for the estimated replacement times of oceanic dissolved organic carbon (ca. 4000–6000 year). Much progress has been made in recent years to assess material export from rivers, but few studies have examined within-river processes and their significance in regional and global biogeochemical cycles. Work by Jeff Richey and his colleagues has shown that waters of the Amazon release 13 times more carbon through out-gassing (evasion) of respired CO_2 than is exported to the ocean. The respired carbon originates from terrestrial sources and suggests that the overall carbon budget of the rainforest is more closely balanced than would be inferred from terrestrial biomass accumulation and fluvial export losses alone. The cumulative effects of human activities within river basins have given rise to global-scale alterations in water and material fluxes. Anthropogenic inputs have enhanced the delivery of nitrogen and phosphorus by rivers to coastal environments and led to widespread problems with eutrophication. The combined storage capacity of the world's dams has increased water storage and sediment retention thereby partially offsetting erosion losses associated with watershed disturbance.

Glossary

Discharge – The volume of water moving past a given point in the river per unit time (typically, $l\ s^{-1}$)

Evapotranspiration – The movement of water from the Earth's land surface to the atmosphere via evaporation and plant transpiration.

Nutrient spiraling – The uptake and release of dissolved nutrients during downstream transport.

Anadromous – Fishes that live predominantly in marine waters but are seasonal residents of freshwater streams and rivers during spawning and larval development.

Hypoxia – A reduced concentration of dissolved oxygen in a waterbody.

Hyporheos – The zone beneath and lateral to the river bed where river- and groundwater mix.

See also: Chemical Fluxes and Dynamics in River and Stream Ecosystems; Currents in Rivers.

Further Reading

Finlay JC (2001) Stable-carbon-isotope ratios of river biota: Implications for energy flow in lotic foodwebs. *Ecology* 82: 1052–1064.

Hynes HBN (1970) *The Ecology of Running Waters*. Toronto: University of Toronto Press.

Junk WJ, Bayley PB, and Sparks RE (1989) The flood-pulse concept in river-floodplain systems. In: Dodge DP (ed.) *Proceedings of the International Large Rivers Symposium. Can. Spec. Publ. Fish Acquat. Sci.* 106: 110–127.

Kalff J (2002) Rivers and the export of materials from drainage basins and the atmosphere. In: *Limnology*, pp. 94–121. Upper Saddle River, NJ: Prentice-Hall.

Meybeck M (1982) Carbon, nitrogen and phosphorus transport by world rivers. *American Journal of Science* 282: 401–450.

Milliman JD and Meade RH (1983) Worldwide delivery of river sediment to the oceans. *Journal of Geology* 91: 1–21.

Nilsson C, Reidy CA, Dynesius M, and Revenga C (2005) Fragmentation and flow regulation of the world's large river systems. *Science* 308: 405–408.

Richey JE, Hedges JI, Devol AH, Quay PD, Victoria R, Martinelli L, and Forsberg BR (1990) Biogeochemistry of carbon in the Amazon River. *Limnology and Oceanography* 35: 352–371.

Richey JE, Melack JM, Aufdenkampe A, Ballester VM, and Hess LL (2002) Outgassing from Amazonian rivers and wetlands as a large tropical source of atmospheric CO_2. *Nature* 416: 617–620.

Syvitski JPM, Vorosmarty CJ, Kettner AJ, and Green P (2005) Impact of humans on the flux of terrestrial sediment to the global coastal ocean. *Science* 308: 376–380.

Thorp JH and Delong MD (1994) The riverine productivity model: an heuristic view of carbon sources and organic processing in large river ecosystems. *Oikos* 70: 305–308.

Vannote RL, Minshall GW, Cummins KW, Sedell JR, and Cushing CE (1980) The river continuum concept. *Canadian Journal of Fisheries and Aquatic Sciences* 37: 130–137.

Vorosmarty CJ, Fekete BM, Meybeck M, and Lammers RB (2000) Global system of rivers: Its role in organizing continental land mass and defining land-to-ocean linkages. *Global Biogeochemical Cycles* 14: 599–621.

Vorosmarty CJ, Sharma KP, Fekete BM, Copeland AH, Holden J, Marble J, and Lough JA (1997) The storage and aging of continental runoff in large reservoir systems of the world. *Ambio* 26: 210–219.

Wetzel RG (2001) Rivers and lakes – Their distribution, origins and forms. In: *Limnology: Lake and River Ecosystems*, pp. 15–42. San Diego, CA: Academic Press.

Springs

D S Glazier, Juniata College, Huntingdon, PA, USA

What are Springs?

Springs are places where underground water emerges onto the Earth's surface, often forming a stream, pond, or marsh. Small springs that arise by groundwater oozing or percolating from the soil are sometimes called seeps or seepage springs. Springs may also emerge beneath large bodies of water, including lakes (sublacustrine springs), rivers (subriverine springs), and oceans (submarine springs at geothermal vents and coastal karstic fracture points), but these relatively inaccessible springs are not discussed here. Also not discussed here are water wells, which are essentially artificial springs resulting from the tapping and (or) pumping of ground water by humans.

Since spring water is continuously replenished from thermally buffered underground sources, it often shows little variation in temperature on a daily, seasonal, and yearly basis. As result, living organisms are active year-round in many kinds of springs, even in those that occur in temperate regions where there are icy cold winters (**Figure 1**).

Sources describing details of many aspects of the nature and importance of spring systems are listed in an online Appendix.

Geological Origin of Springs

All spring water ultimately originates from precipitation (**Figure 2**). Rain and melted snow seep into the ground where they collect as ground water in porous rock material (aquifer) lying above a relatively impervious layer of rock (confining bed). The top of this layer of ground water is called the water table. By gravity, pressure, and (or) other forces, the ground water flows along paths of least resistance, often emerging as springs wherever the water table intersects the land surface. A spring may emanate from water saturated soil or porous rock, from a fracture, fissure or cave, or along a contact zone between two rock types of different permeability, often at the bottom of or along steep slopes of the land surface (including hills and mountains). In general, aquifers gain water (are recharged) by precipitation and may lose water (are discharged) by spring flow, though there is a lag time between these two processes because of underground water storage. For example, after a rain storm, a spring brook may show little immediate change in flow rate, whereas a nearby runoff stream will quickly (within a few hours or less) become engorged with rapidly flowing water (**Figure 3**). Springs associated with small, shallow aquifers or those having poor storage capacities (as in shale-siltstone areas) tend to show greater variability of flow than those associated with large, deep aquifers or those having high storage capacities and long water residence times (as in some karst terrains containing numerous underground pockets and passage ways in carbonate rocks such as limestone). Over relatively long periods of time (weeks, months or more), the amount of springwater flow (discharge) varies with aquifer (recharge) size and the local balance between precipitation and evapotranspiration (loss of water to the atmosphere by evaporation from vegetation, the landscape, and bodies of water). Some karst springs show daily or annual rhythms of outflow rate, the former being the result of the filling and draining of an internal siphon, and the latter the result of seasonal changes in regional water balance (precipitation minus evapotranspiration). Others show irregular changes in discharge rates. Because of variable storage times for ground water, spring water varies greatly in age from less than a day to over 10 000 (and possibly over a million) years old.

Kinds of Springs

Springs are very diverse and have been classified in numerous ways based on their geology, hydrology, water chemistry, water temperature, ecology, and human use (**Table 1**). The water chemistry of a spring is related to its geological setting. For example, in the Appalachian region of North America, limestone springs tend to have relatively alkaline hard water, rich in calcium and bicarbonate ions and low in hydrogen ions ($pH \geq 7$), whereas sandstone springs tend to have relatively acidic soft water, low in calcium and bicarbonate ions and high in hydrogen ions ($pH < 7$). The chemical composition of spring water may vary in many other ways depending on its geological origin. Principal cations (positively charged ions) may include sodium, potassium, calcium, magnesium, and heavy metals such as iron, zinc, and aluminum. Principal anions (negatively charged ions) may include chloride, carbonate, bicarbonate, sulfate, sulfide, nitrate, borate, phosphate, fluoride, iodide, and other oxides. Spring water often contains

Figure 1 An oasis of life in mid-winter: the outflow from Warm Spring, Huntingdon, Pennsylvania (USA) in January.

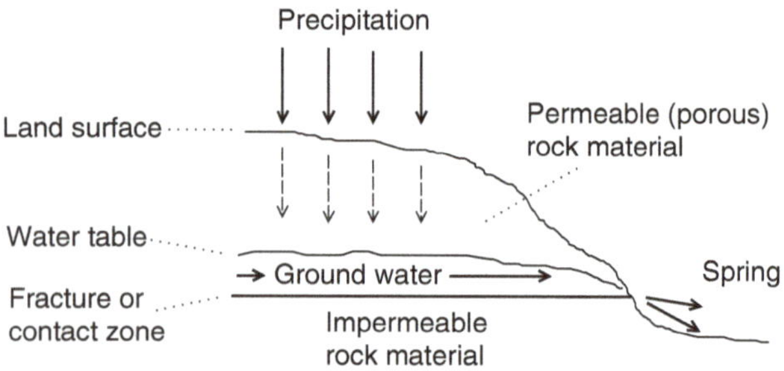

Figure 2 A common type of spring is depicted. A spring occurs where ground water emerges onto the land surface via a fracture, contact zone or other opening. Ground water occurs in permeable (porous) rock material above a layer of impermeable rock material, and is replenished (recharged) by rain or melted snow that seeps into the ground.

substantial amounts of dissolved silicon dioxide; and some springs even contain small concentrations of radioactive isotopes (e.g., of helium, potassium, radium, radon, or uranium). Dissolved gases found in spring water include nitrogen, oxygen, carbon dioxide, hydrogen sulfide, radon, helium, and other noble gases. Some springs are so carbonated (charged with dissolved CO_2) that they significantly elevate the CO_2 concentration of the surrounding air, which in turn significantly enhances the growth of nearby vegetation. The names of springs often reflect their chemical composition (e.g., mineral springs, soda springs, sulfur springs, etc.).

The hydrological properties of a spring also depend on the nature of its geological source. Diffuse springs that arise from numerous fissures, fractures, contact zones, or other pore spaces tend to have relatively stable water flow and chemistry, whereas conduit springs that arise from single or highly consolidated underground channels tend to be 'flashier', showing more variable flow rates and water chemistry. High-elevation 'overflow springs' also tend to have more variable, discontinuous flow than low-elevation 'underflow springs' that are constantly and continuously fed by the baseflow of an aquifer.

Conversely, the chemistry and (or) hydrology of a spring may affect its local geological setting. For example, travertine springs are characterized by extensive calcareous deposits resulting from the loss of CO_2 when highly carbonated ground water comes into contact with air at or near a spring source. Sediments or crusts called 'sinters' may also result from the precipitation of gypsum, sulfur, salts, silica, or heavy metals at other mineral-rich springs, as well. Microbes may be significantly involved in the formation of travertine and sinter deposits at thermal springs. In addition, sulfuric acid formed from hydrogen sulfide gas vented by hot springs may increase the erosion of nearby carbonate rocks.

Some springs emerge from single openings or vents, whereas others issue from multiple openings. Seepage springs, in particular, are characterized by oozing from numerous small openings in the soil.

Springs differ greatly in the magnitude of their discharge: some are a mere trickle, whereas others may supply enough water for a small town. The cause of a spring's flow usually involves gravity, but also may involve hydrostatic pressure (as in artesian springs), capillarity (as in seeps), or heated steam pressure (as in geysers). Geysers are especially spectacular because their explosive eruptions of hot water occur at intermittent intervals that vary from less than an hour to over a month.

The water temperature of many springs approximates the mean annual air temperature (MAAT) of a region. Traditionally these have been called 'cold springs', but recently it has been recommended that they be renamed 'ambient springs', whereas 'cold springs' should be considered to have temperatures below the MAAT. By contrast, 'thermal springs' (also called geothermal or superambient springs) have water temperatures significantly higher than that of the MAAT. Thermal springs have been classified in various ways, but it is commonly accepted that 'hot springs' have temperatures exceeding human body temperature ($\approx$ 37–38 °C or 98–100 °F). Hot springs discharge water that has been heated (up to boiling temperatures) by hot rocks deep in the interior of the earth (heated by radioactive decay and gravity-caused rock compression), or by magma (hot molten rock) near the earth's surface in volcanic regions. As a result, hot springs tend to be most common in mountainous regions (**Figure 4**). Like most springs, hot springs usually emanate as streams or pools, but some are geysers, fumaroles (holes in the ground giving off volcanic gases, some of which may be

Figure 3 Views of a brook emanating from Bouquet Spring and of a nearby runoff stream, Crooked Creek, one hour after a summer rainstorm (19 July 1996). Note that the flow and clarity of water in the spring brook has been little affected, whereas the water in the runoff stream has begun to flow rapidly and is heavily laden with eroded sediment. The streams are located on opposite sides of Route 26 running through McConnellstown, PA.

poisonous), or mudpots (bubbling pools of mud, also called mud volcanoes).

Biologists have further characterized springs according to the type of aquatic habitat that they create. Spring brooks and streams are called rheocrenes; spring pools and ponds are limnocrenes; spring marshes and bogs are helocrenes. Although spring seeps (sipeocrenes) intergrade with helocrenes, they tend to be smaller, more dominated by wet-loving terrestrial plants than by aquatic plants, may occur in places (e.g., rock faces and shaded forests) where typical marshy conditions are not possible, and often have distinctive species not found in other kinds of springs. Some springs have characteristics of all or some of the above types (e.g., rheohelocrenes and rheolimnocrenes). In addition, some artesian springs in arid areas are associated with mounds of peat and vegetation, and have been called 'mound springs' in Australia and 'artesian blister wetlands' in Africa.

Some computer databases have been constructed that classify regional springs according to several criteria, but a standard classification system that is applicable to springs worldwide has yet to be established and is much needed.

Biology of Springs

The study of life in springs has been called 'crenobiology' to emphasize the distinctive biota that lives in these habitats. Two major ecological zones are recognized: the 'eucrenon' at the spring source (springhead), and the 'hypocrenon' in the brook emanating immediately from a spring (springbrook). Various historical and ecological factors have affected the biotic composition of springs. A major ecological factor is water temperature: the biota of hot springs differs greatly from that of cooler, nonthermal springs.

Nonthermal Springs

Age of a nonthermal spring strongly influences its biotic composition. Intermittently flowing springs and geologically young springs in deglaciated regions tend to be numerically dominated by the larvae of various kinds of flying insects with high dispersal ability, whereas older, perennially flowing springs in unglaciated regions typically have several less vagile, abundant taxa of nonemergent (wholly aquatic) macroinvertebrates, such as amphipods, isopods, mollusks, and flatworms, as well. Crustaceans and mollusks are often especially common in alkaline, hard-water springs, where they can obtain enough calcium for shell or carapace formation, but are generally rare or absent in acidic, soft-water springs where insects again often dominate. Insects also tend to be the most common macroinvertebrates in high-elevation spring brooks (**Figure 5**).

Table 1 Some of the various ways that springs have been classified

Classification type	*Categories (examples)*	*Sources*
Geology		
Location of discharge	Subaerial (emerges onto land surface)	
	Subaqueous or submerged (emerges underneath a large body of water	
	Sublacustrine (emerges beneath a lake)	
	Subriverine (emerges beneath a river)	
	Submarine (emerges beneath an ocean)	
	Karstic openings on ocean floor (called vruljas)	
	Geothermal vents on ocean floor	1, 19, 33, 39
	Onshore springs (discharge above mean low-tide line; includes subaerial, sublacustrine, and subriverine springs)	
	Offshore springs (discharge below mean low-tide line; synonymous with submarine springs)	33
	Other (nonexclusive) types include base level springs that emerge at sea or lake level, or at valley floor; hanging springs that emerge above bodies of surface water, or above valley floor; buried springs that emerge below a valley floor; intertidal springs that emerge between high tide and low tide; and riverbed springs that emerge in riverbeds	1, 35
Bedrock	Limestone, dolomite, marls, sandstone, siltstone, shale, greywacke, gypsum, alluvium, coal, anhydrite, salt beds, conglomerates, and igneous (volcanic or plutonic) and metamorphic rocks	7, 20, 30, 39, 43
Geological structure	Alluviated springs	
	Artesian (Vauclusian) springs	
	Cave springs	
	Contact springs	
	Depression springs	
	Fissure or fracture springs	
	Glacial springs	
	Hidden springs	
	Volcanic springs	
	Springs in impervious rock (including tubular and fracture springs)	1, 5, 35
Topography	Depression, flat, flood plain, hillside, sinkhole, stream channel, terrace, undulating, valley, etc.	19
	Other (nonexclusive) types:	
	Artesian blister wetlands	2
	Moraine and glaciofluvial springs	15
	Mound springs	
	Travertine and sinter springs	39, 43
Geography		
Location	Longitude and latitude	
	Elevation	
	Hydrologic basin	19
Hydrology		
Magnitude of discharge	1st magnitude ($>2830\,l\,s^{-1}$)	
	2nd magnitude ($283–2830\,l\,s^{-1}$)	
	3rd magnitude ($28.3–283\,l\,s^{-1}$)	
	4th magnitude ($6.3–28.3\,l\,s^{-1}$)	
	5th magnitude ($0.63–6.3\,l\,s^{-1}$)	
	6th magnitude ($63–630\,ml\,s^{-1}$)	
	7th magnitude ($8–63\,ml\,s^{-1}$)	
	8th magnitude ($<8\,ml\,s^{-1}$)	23
	Low volume ($<0.01\,m^3\,s^{-1}$)	
	Medium volume ($0.01–0.5\,m^3\,s^{-1}$)	
	High volume ($>0.5\,m^3\,s^{-1}$)	9
Source(s) of discharge	Diffuse versus conduit (tubular)	34
	Single opening versus multiple openings	14, 19
	Seepage spring (discharge oozing from numerous small openings in porous rock material)	
	Vent spring (discharge flowing from an opening or openings larger than the mean pore space of surrounding rock material)	33

Continued

Table 1 Continued

Classification type	*Categories (examples)*	*Sources*
	Full-flow (fed by entire aquifer)	
	Underflow (fed constantly and continuously by baseflow of aquifer because of relatively low elevation of discharge point)	
	Overflow (fed variably and (or) discontinuously by overflow of aquifer during flood conditions because of relatively high elevation of discharge point)	
	Underflow–overflow (fed by baseflow and overflow because of intermediate elevation of discharge point)	12, 35, 45
Conduit type at spring	Open, horizontal conduit with airspace (gravity spring)	
	Water-filled, vertical conduit (pressure, Artesian or Vauclusian spring)	35
Size of aquifer and length of flow path	Local springs fed by small, shallow aquifers (short flow paths with fluctuating discharges)	
	Regional springs fed by large, extensive aquifers (long flow paths, with relatively constant discharges)	32
Location of aquifer	Above impermeable rock material (perched springs)	
	Below impermeable rock material (confined springs)	35
Source of recharge water	Originates from surface water that has flowed into underground reservoirs (allogenic aquifers) and resurfaced (resurgences)	
	Originates from atmospheric water that has seeped into underground reservoirs (autogenic aquifers) from which water travels to the surface (exsurgences)	14, 35
Direction of flow	Reversing (estavelles, often located near streams or rivers)	
	Nonreversing	14, 33, 35
Cause of discharge	'Free flow' by gravity (most springs)	
	Hydrostatic pressure (artesian springs)	
	Gaseous buoyancy and (or) heated steam pressure (e.g., geysers)	
	Capillarity (e.g., seeps)	19
Variation or persistence of flow	Perennial versus intermittent	19
	Variable ($v > 100\%$)	
	Subvariable ($v = 25$–100%)	
	Constant ($v < 25\%$)[a]	21, 26
	Impermanent (<5 years)	
	Apparently permanent (e.g., 100 years)	
	Permanent (>1000 years)	9
	Continuous	
	Steady (perennial: base flow)	
	Variable (perennial)	
	Periodic (rhythmic or episodic)	
	Discontinuous (intermittent)	
	Nonexistent (paleo- or abandoned spring)	1, 35
Water chemistry		
pH and hardness	Acidic, soft water	
	Alkaline, hard water	16
Total dissolved solids	Freshwater springs (<1000 mg l^{-1})	
	Mineral springs (1000–35 000 mg l^{-1})	
	High mineral (saline) springs (>35 000 mg l^{-1})	9, 39
Principal ions	Chloride spring water	
	Sulfate spring water	
	Sulfate–chloride spring water	
	Carbonate (tufaceous) spring water	
	Sulfate–carbonate spring water	
	Chloride–carbonate spring water	
	Chloride–sulfate–carbonate spring water	
	Siliceous spring water	
	Borate spring water	
	Nitrate spring water	
	Phosphate spring water	
	Chalybeate (iron hydroxide) spring water	
	Acid spring water	6, 35

Continued

Table 1 Continued

Classification type	*Categories (examples)*	*Sources*
Oxygen level	Aerobic (high oxygen)	
	Anaerobic (low oxygen)	25
Water temperature		
Temperature with respect to MAAT of a region, and human body temperature	Cold springs (water cooler than MAAT)	
	Ambient springs (water temperature approximates MAAT)	
	Superambient (thermal or geothermal) springs (water warmer than MAAT)[b]	
	Hot springs (water warmer than human body temperature ≈ 37–38 °C or 98–100 °F)	28, 29
	Nonthermal cold springs (water temperature < MAAT)	
	Warm springs (water temperature > MAAT, but <98 °F)	
	Hot springs (water temperature > 98 °F)	23
Absolute temperature range	Cold springs (<20 °C)	
	Warm springs (20–50 °C)	
	Hot springs (>50 °C)	38
	Cold springs (<10 °C)	
	Warm springs (10–40 °C)	
	Hot springs (>40 °C)	9
	Cold (<20 °C)	
	Hypothermal (20–30 °C)	
	Thermal (30–40 °C)	
	Hyperthermal (>40 °C)	27
	Cold or hypothermic (<18 °C)	
	Lukewarm (tepid) or hliarothermic (18–30 °C)	
	Warm or euthermic (30–50 °C)	
	Hot or akrothermic (50–70 °C)	
	Steaming hot or hyperthermic (>70 °C)	42
Variation and magnitude of temperature	Heterothermal springs with varying temperatures	
	Homothermal springs with same temperatures year-round	
	Cold springs (water cooler than or at annual mean temperature of place)	
	Tepid springs (water warmer than annual mean temperature, but cooler than mean maximum temperature of place)	
	Hot springs (water warmer than mean maximum temperature of place)	
	Relatively hot springs (springs without a special animal and plant community; temperature between mean maximum temperature of place and 40 °C)	
	Absolutely hot springs (springs with a special animal and plant community, or no life at all, excepting bacteria; temperature >40 °C).	37
Relation to fauna and flora	Hypothermophilous formations (<15 °C)	
	Mesothermophilous formations (15–30 °C)	
	Euthermophilous formations (30–80 °C)	42
Human sensation with therapeutic applications	Very cold (0–12 °C): pain sensation	
	Cold (12–18 °C): cold sensation, but tolerable	
	Fresh (18–27 °C): pleasant sensation, if short	
	Neutral (27–32 °C): pleasant sensation	
	Warm (32–36.5 °C): comfortable sensation	
	Hot (37–40 °C): erythema and warm sensation	
	Very hot (40–43 °C): Tolerable, if short	18
Ecology		
Habitat type	Rheocrene (brook or stream)	
	Limnocrene (pool or pond)	
	Helocrene (marsh or bog)	4, 36
	Other (nonexclusive) types:	
	Hygropetric rheocrene (thin layer of water flowing vertically over rocks)	3, 41
	Linear springs (diffuse seepage and increasingly continuous discharge along spring channel)	46
	Psammocrene (sandy-pebbly substratum)	8
	Rheohelocrene (seeping rheocrene with muddy substratum)	8, 11
	Rheolimnocrene (stream and pond)	24
	Rheopsammocrene (seeping rheocrene with sandy-gravel substratum)	11

Continued

Table 1 Continued

Classification type	*Categories (examples)*	*Sources*
	Sipeocrene (small seepage springs)	Present source
	Spring mires (spring-fed areas of soggy, muddy ground that may have peat and (or) mineral deposits)	44
	Types of helocrenes based on pH (acid to neutral), nutrient levels (eutrophic, mesotrophic or oligotrophic), and variability of discharge (constant or fluctuating)	40
	Types of rheocrenes based on geological setting (karst or alluvial) and presence or absence of travertine or lime sinters	46
Habitat type of thermal springs	Rheotherm (thermal rheocrene)	
	Limnotherm (thermal limnocrene with basin deeper than outlet)	
	Trybliotherm (thermal limnocrene with basin not deeper than outlet)	
	Choanotherm (thermal limnocrene with spring head at bottom of basin)	
	Helotherm (thermal helocrene)	37
Habitat type of spring seepages	Open seeps	
	Canopied seeps	
	Montane seeps	
	Pitcher plant seeps	14
Springside (riparian) habitat	Open land or meadow	
	Forest or woods	
	Bush or scrub land	
	Fen or swamp land	
	Human developed area	17
Proportion of species restricted to springs	Quantitative index of value (fidelity) of spring biota based on six categories of ecological distribution. Five value classes are recognized ('spring typical' to 'very spring foreign')[c]	10, 13
Conservation priority		
Relative value of resources at springs	Ranking criteria:	
	Presence of rare aquatic species	
	Rarity across landscape	
	Spring brook length	
	Scouring	
	Aquatic habitat persistence	
	Resource threats	
	Land ownership	
	Conflicting uses	31
Human use		
Drinking use	Water is potable (drinkable) or not	19
Based on yield	Household (0.044–0.310 $l\ s^{-1}$)	
	Household and farm (0.311–0.625 $l\ s^{-1}$)	
	Limited institutional or commercial (0.626–1.259 $l\ s^{-1}$)	
	Various uses, possibly limited public supply (1.260–2.145 $l\ s^{-1}$)	
	Small public supplies (2.146 to >4.416 $l\ s^{-1}$)	26
Based on purpose	Air-conditioning, bottling, brewing and distilling, commercial, domestic, industrial, institutional, irrigation, mining, religious, sightseeing, public supply, recreation, stock supply, fish culture, etc.	19, 35

[a]Variability index (v) = $[(m - l)/a]$ 100, where m = maximum flow, l = minimum flow, and a = mean flow.

[b]Some researchers propose that a spring should not be considered 'thermal' unless it exceeds the MAAT by a certain amount (e.g., by 5–10 °C: see source 39).

[c]Based on author's rough translation from the original German.

Sources

1. Alfaro C and Wallace M (1994) Origin and classification of springs and historical review with current applications. *Environmental Geology* 24: 112–124.
2. Ashley GM, Goman M, Hover VC, Owen RB, Renaut RW, and Muasya AM (2002) Artesian blister wetlands, a perennial water resource in the semi-arid Rift Valley of East Africa. *Wetlands* 22: 686–695.
3. Bonettini AM and Cantonati M (1996) Macroinvertebrate assemblages of springs of the River Sarca catchment (Adamello-Brenta Regional Park, Trentino, Italy). *Crunoecia* 5: 71–78.
4. Bornhauser K (1912) Die tierwelt der quellen in der umgebung Basels. *Internationalen Revue der Gesamten Hydrobiologie und Hydrographie Biologische Supplemente* 5: 1–90.
5. Bryan K (1919) Classification of springs. *Journal of Geology* 27: 552–561.
6. Clarke FW (1924) *The Data of Geochemistry*, 5th edn. United States Geological Survey Bulletin 770. Washington, DC: United States Government Printing Office.
7. Collier KJ and Smith BJ (2006) Distinctive invertebrate assemblages in rockface seepages enhance lotic biodiversity in northern New Zealand. *Biodiversity and Conservation* 15: 3591–3616.

8. Crema S, Ferrarese U, Golo D, Modena P, Sambugar B, and Gerecke R (1996) Richerche sulla fauna bentonica ed interstiziale di ambienti sorgentizi in area alpina e prealpina. *Report del Centro di Ecologia Alpina* 8: 1–104.
9. Danks HV and Williams DD (1991) Arthropods of springs, with particular reference to Canada: synthesis and needs for research. *Memoirs of the Entomological Society of Canada* 155: 203–217.
10. Fischer J (1996) Bewertungsverfahren zur quellfauna. *Crunoecia* 5: 227–240.
11. Gerecke R and Di Sabatino A (1996) Water mites (Acari, Hydrachnellae) and spring typology in Sicily. *Crunoecia* 5: 35–41.
12. Gillieson DS (1996) *Caves: Processes, Development and Management*. Oxford: Blackwell.
13. Hinterlang D (1996) Quellbewertung – verfahrensteil flora und vegetation, erste fortschreibung. *Crunoecia* 5: 241–253.
14. Hobbs HH (1992) Caves and springs. In: Hackney CT, Adams SM, and Martin WA (eds.) *Biodiversity of Southeastern United States: Aquatic Communities*, pp. 59–131. New York: Wiley.
15. Hoffsten P-O and Malmqvist B (2000) The macroinvertebrate fauna and hydrogeology of springs in central Sweden. *Hydrobiologia* 436: 91–104.
16. Hynes HBN (1970) *The Ecology of Running Waters*. Toronto: University of Toronto Press.
17. Krüger K (1996) Quellschutz im land Brandenburg. *Crunoecia* 5: 129–135.
18. Ledo E (1996) Mineral water and spas in Spain. *Clinics in Dermatology* 14: 641–646.
19. LaMoreaux PE and Tanner JT (2001) Springs and bottled waters of the world: Ancient history, source, occurrence, quality and use. Berlin: Springer-Verlag.
20. McColloch JS (1986) *Springs of West Virginia*. Charleston: West Virginia Geological and Economic Survey.
21. Meinzer OE (1923) Outline of ground-water hydrology, with definitions. *United States Geological Survey Water-Supply Paper* 494: 1–69.
22. Michaelis FB (1976) Physico-chemical features of Pupu Springs. *New Zealand Journal of Marine and Freshwater Research* 10: 613–628.
23. Meinzer OE (1927) Large springs in the United States. *United States Geological Survey Water-Supply Paper* 557: 1–94.
24. Nesterovich A (1996) Studies of the fauna of Belarusian springs. *Crunoecia* 5: 79–85.
25. Odum HT (1957) Primary production estimates in eleven Florida springs and a marine turtle grass community. *Limnology and Oceanography* 2: 85–97.
26. Otton EG and Hilleary JT (1985) *Maryland Springs – Their Physical, Thermal, and Chemical Characteristics*. Baltimore: Maryland Geological Survey Report of Investigations 42.
27. Parish LC and Lott TM (eds.) (1996) Balneology and the spa: the use of water in dermatology. *Clinics in Dermatology* 14: 547–692.
28. Pentecost A (2005) Hot springs, thermal springs and warm springs. What's the difference? *Geology Today* 21: 222–224.
29. Pentecost A, Jones B, and Renault RW (2003) What is a hot spring? *Canadian Journal of Earth Sciences* 40: 1443–1446.
30. Pérez ES (1996) Springs in Spain: classification according to their flows and lithologies and their hydraulic contributions. *Ground Water* 34: 1033–1041.
31. Sada DW and Pohlmann KF (2003) U.S. National Park Service Mojave Inventory and Monitoring Network Spring Survey Protocols: Level I (http://www.dmg.gov/documents/3-03-Draft-Level-I-Protocol.doc).
32. Sada DW, Williams JE, Silvey JC, Halford A, Ramakka J, Summers P, and Lewis L (2001) Riparian area management: a guide to managing, restoring, and conserving springs in the western United States. Technical Reference 1737–17. Denver, Colorado: United States Department of the Interior Bureau of Land Management.
33. Scott TM *et al.* (2004) Springs of Florida. *Florida Geological Survey Bulletin* 66: 1–658.
34. Shuster ET and White WB (1971) Seasonal fluctuations in the chemistry of limestone springs: A possible means for characterizing carbonate aquifers. *Journal of Hydrology* 14: 93–128.
35. Smart C and Worthington RH (2004) Springs. In: Gunn J (ed.) *Encyclopedia of Caves and Karst Science*, pp. 699–703. New York: Taylor and Francis.
36. Thienemann A (1922) Hydrobiologische untersuchungen an quellen. *Archiv für Hydrobiologie* 14: 151–190.
37. Tuxen SL (1944) The hot springs of Iceland: Their animal communities and their zoogeographical significance. In: Frioriksson A, *et al.* (eds.) *The Zoology of Iceland*, vol. I, part II, pp. 1–206. Copenhagen: Ejnar Munksgaard.
38. United States NOAA Geophysical Data Center.
39. van Everdingen RO (1991) Physical, chemical, and distributional aspects of Canadian springs. *Memoirs of the Entomological Society of Canada* 155: 7–28.
40. Verdonschot PFM (1996) Towards ecological spring management. *Crunoecia* 5: 183–194.
41. Verdonschot PFM and Schot JA (1986) Macrofaunal community types in helocrene springs. *Report of Research Institute of Nature Management* 1986/1987: 85–103.
42. Vouk V (1923) Die probleme der biologie der thermen. *Internationale Revue der Gesamten Hydrobiologie und Hydrographie* 11: 89–99.
43. White WB (1988) *Geomorphology and Hydrology of Karst Terrains*. New York: Oxford University Press.
44. Wolejko L (1996) Transformation of spring-mire vegetation in north-western Poland in relation to human impact. *Crunoecia* 5: 59–66.
45. Worthington SRH (1991) *Karst hydrogeology of the Canadian Rocky Mountains*. Hamilton, ON: McMaster University Dissertation.
46. Zollhöfer JM, Brunke M, and Gonser T (2000) The typology of springs in Switzerland by integrating habitat variables and fauna. *Archiv für Hydrobiologie Supplementband Monographic Studies* 121: 349–376.

The relative constancy of the water temperature and flow rate of many springs also appears to be importantly involved in determining their biotic composition. Nonemergent macroinvertebrates may have a competitive advantage over many insects in constant temperature springs of temperate regions, because there they can reproduce and maintain dense populations year-round, whereas insects must seasonally leave the water to breed as adults. Proper timing of emergence by many insects may also be disrupted by a lack of thermal cues in many springs. In addition, some nonemergent macroinvertebrates, such as amphipods, may prey on a variety of insects, thus keeping their populations relatively low. Accordingly, a noninsect species is the most abundant macroinvertebrate in 78% of the spring systems for which

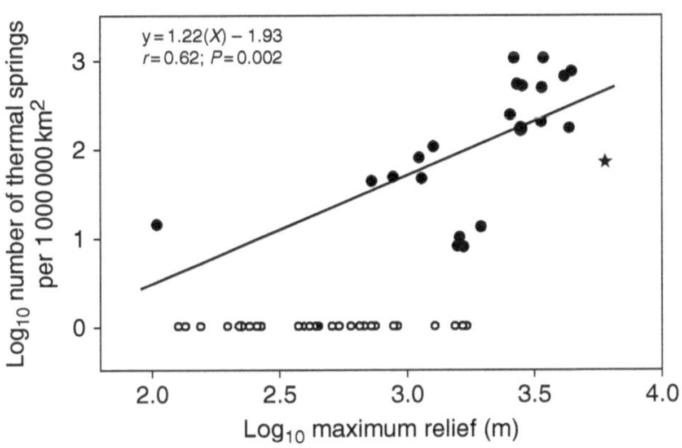

Figure 4 Log_{10} density of thermal springs in each of 50 states of the United States in relation to topographic relief (the log_{10} maximum range in elevation in a county). The positively significant regression line and equation (along with *r*, the correlation coefficient, and *P*, the probability that this relationship is due to chance) are depicted for 22 states (solid circles) that have recorded thermal springs. Alaska (star) was not included in this regression because it was considered an outlier. For comparison, the other 27 states with no thermal springs are plotted as 0 (arithmetic) values (open circles). Note that only 25% of the 36 states with a maximum relief less than 6000 m have thermal springs, whereas all of the 14 states with greater maximum relief have thermal springs. Data are from the United States National Geophysical Data Center (http://www.ngdc.noaa.giv/nndc/struts/describeField?t=100006&s=1&field=11029) and Maximum Relief Counties by State compiled by J. Brekhus (http://cohp.org/records/relief/Helman-counties.html).

population density data are available, and an amphipod or isopod crustacean is the most abundant in 62% (**Table 2**). In addition, spring-like faunal assemblages dominated by amphipods, isopods, and (or) mollusks occur in the flow- and temperature-constant tailwaters of many reservoir deep-release dams. Various factors hypothetically contributing to the dominance of noninsect macroinvertebrates in nonthermal, hard-water springs are depicted as a flow-diagram (**Figure 6**).

The abundance and (or) diversity of macroinvertebrates is often higher in perennially flowing springs than in intermittent springs. Large springs also tend to have more species than do smaller springs. Macroinvertebrate abundance, diversity, and microhabitat distribution may be influenced by substrate type and macrophyte density, as well. Macrophytes, such as watercress (*Nasturtium* spp.) which is characteristic of temperate hard-water springs worldwide, enhance macroinvertebrate abundance by supplying refuges from water currents and fish predation, and by trapping large amounts of detrital food particles. Amphipods, isopods, and other macroinvertebrates may also reach especially high densities in springs where there are no fish predators. In addition, many lotic species live in rheocrenes, whereas many lentic species inhabit limnocrenes.

Often the species diversity of macroinvertebrates increases from ground waters to surface spring habitats to downstream sites. This longitudinal pattern

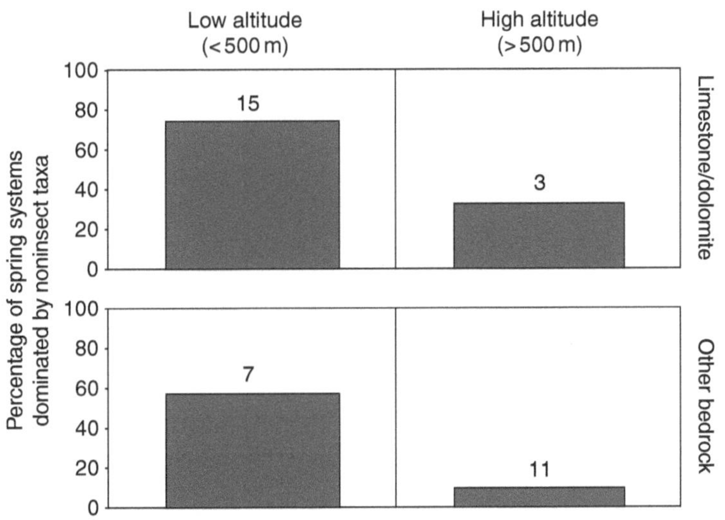

Figure 5 Percentage of spring systems (bars) in North America, Europe, and New Zealand numerically dominated by noninsect macroinvertebrates in relation to altitude and geology (number of springs above each bar). Note that noninsect dominance is most frequent in low altitude spring systems with calcium-rich bedrock, whereas insect dominance is most frequent in high altitude spring systems underlain by other kinds of bedrock. Data (with minor modification) from Barquín J and Death RG (2006) Spatial patterns of macroinvertebrate diversity in New Zealand springbrooks and rhithral streams. *Journal of the North American Benthological Society* 25: 768–786.

Table 2 Population density of the most abundant macroinvertebrate species in each of several nonthermal spring systems in Europe and North America

Country (state or province)	*Spring*	*Most abundant species*	*Mean population density (number m^{-2})*	*Source*
Canada				
(Ontario)	Valley	*Nemoura trispinosa* (P)	11 604	17
Denmark	Ravnkilde	*Nemourella picteti* (P)	16 561	6
	Rold Kilde	*Gammarus pulex* (A)	2320	5
Spain	Springs 1–6	*Echinogammarus* sp. (A)[a]	1028	1
Switzerland	Q1	*Gammarus fossarum* (A)	1850	15
	Q2	*G. fossarum* (A)[b]	600	15
	Q3	*G. fossarum* (A)	275	15
	Q4	*G. fossarum* (A)	1575	15
	Q5	*G. fossarum* (A)	1950	15
	Q6	*Pisidium* spp. (B)[b]	1300	15
	Q7	*G. fossarum* (A)	450	15
	Q8	*G. fossarum* (A)	5567	15
	Q9	*Pisidium* spp. (B)	3275	15
	Q10	*G. fossarum* (A)	3400	15
United Kingdom	Cowdale	*Nemoura erratica* (P)	385	10
	Kidtor	*Gammarus pulex* (A)	3583	10
	Ashwood Dale	*G. pulex* (A)[b]	400	10
	Woolow	*G. pulex* (A)[c]	945	10
	Topley Pike	*G. pulex* (A)	2273	10
	Wormhill, W	*Nemurella picteti* (P)	4188	10
	Wormhill, E	*N. picteti* (P)	2355	10
	Cheedale, W	*G. pulex* (A)	875	10
	Cheedale, E	*G. pulex* (A)[d]	123	10
	Cheedale Bridge	*G. pulex* (A)	1430	10
	Litton Mill	*G. pulex* (A)[b]	25	10
	White Cliff	*G. pulex* (A)	1818	10
	Lees Bottom 1	*G. pulex* (A)[e]	1135	10
	Lees Bottom 2	*G. pulex* (A)	950	10
	Lower Dimindale 1	*G. pulex* (A)	793	10
	Lower Dimindale 2	*G. pulex* (A)	1985	10
	Great Shacklow 1	*G. pulex* (A)[c]	303	10
	Great Shacklow 2	*G. pulex* (A)	2560	10
United States				
(California)	James[f]	*Gumaga nigricula* (T)	1694	9
(Idaho)	Paris	*Acrynopteryx signata* (P)	1520	14
(Iowa)	Cone	*Pentaneura* sp. (D)	504	13
(Kentucky)	Morgan's	*Gammarus minus* (A)	234	7
(Massachusetts)	Root[g]	*Limnodrilus hoffmeisteri* (O)	5465	12
(New Mexico)	Lander	*Gammarus desperatus (fasciatus)* (A)	10 416	8
(Ohio)	OZ	*Lirceus fontinalis* (I)	46	2
(Pennsylvania)	Ell	*Fontigens nickliniana* (G)	17 601	4
(Tennessee)	Root	*Stictochironomus sp.* (D)	8110	16
	Unnamed	*Caecidotea intermedius* (Asellus militaris) (I)[b]	178	11
(Utah)	Big	*Paludestrina* sp. (G)	1620	14
	Cascade	*Hyalella azteca* (A)	15 172	14
	China Row	*Baetis tricaudatus* (E)	344	14
	Clover Creek	*H. azteca* (A)	26 916	14
	Conrad	*Paludestrina* sp. (G)	13 286	14
	Ricks	*B. bicaudatus* (E)	586	14
	Thousand	*Paludestrina* sp. (G)	843	14
(Washington)	Tyee	*Planorbis* sp. (G)[b]	1424	3

The water in these systems is perennially flowing and has approximately neutral pH and is enriched with calcium and bicarbonate ions, except where noted. A = Amphipoda; B = Bivalvia; D = Diptera; E = Ephemeroptera; G = Gastropoda; I = Isopoda; O = Oligochaeta; P = Plecoptera; T = Trichoptera.

[a]Mean population density in six springs.

[b]Not counting the more abundant Chironomidae, which were not distinguished at the species level.

[c]Not counting the more abundant Oligochaeta, which were not distinguished at the species level.

[d]Not counting the more abundant Psychodidae, which were not distinguished at the species level.

[e]Not counting the more abundant Simuliidae, which were not distinguished at the species level.

[f]Water ceases flowing during droughts.
[g]Water mildly acidic (pH = 5.5–6.0).

Sources
1. Barquín J and Death RG (2004) Patterns of invertebrate diversity in streams and freshwater springs in northern Spain. *Archiv für Hydrobiologie* 161: 329–349.
2. Butler MJ and Hobbs HH (1982) Drift and upstream movement of invertebrates in a springbrook community ecosystem. *Hydrobiologia* 89: 153–159.
3. Davidson FA and Wilding JL (1943) A quantitative faunal investigation of a cold spring community. *American Midland Naturalist* 29: 200–209.
4. Gooch JL and Glazier DS (1991) Temporal and spatial patterns in mid-Appalachian springs. *Memoirs of the Entomological Society of Canada* 155: 29–49.
5. Iversen TM (1988) Secondary production and trophic relationships in a spring invertebrate community. *Limnology and Oceanography* 33: 582–592.
6. Lindegaard C, Thorup J, and Bahn M (1975) The invertebrate fauna of the moss carpet in the Danish spring Ravnkilde and its seasonal, vertical, and horizontal distribution. *Archiv für Hydrobiologie* 75: 109–139.
7. Minshall GW (1967) Role of allochthonous detritus in the trophic structure of a woodland springbrook community. *Ecology* 48: 139–149.
8. Noel MS (1954) Animal ecology of a New Mexico springbrook. *Hydrobiologia* 6: 120–135.
9. Resh VH (1983) Spatial differences in the distribution of benthic macroinvertebrates along a springbrook. *Aquatic Insects* 5: 193–200.
10. Smith H, Wood P, and Gunn J (2001) The macroinvertebrate communities of limestone springs in the Wye Valley, Derbyshire Peak District, UK. *Cave and Karst Science* 28: 67–78.
11. Stern MS and Stern DH (1969) A limnological study of a Tennessee cold springbrook. *American Midland Naturalist* 82: 62–82.
12. Teal JM (1957) Community metabolism in a temperate cold spring. *Ecological Monographs* 27: 283–302.
13. Tilly LJ (1968) The structure and dynamics of Cone Spring. *Ecological Monographs* 38: 169–197.
14. Van Gundy JJ (1973) *Factors Controlling the Diversity and Abundance of Macroinvertebrates in Non-thermal Springs*. Dissertation. Salt Lake City, University of Utah.
15. von Fumetti S, Nagel P, and Baltes B (2007) Where a springhead becomes a springbrook – A regional zonation of springs. *Fundamental and Applied Limnology* 169: 37–48.
16. Wilhm JL (1970) Some aspects of structure and function of benthic macroinvertebrate populations in a spring. *American Midland Naturalist* 84: 20–35.
17. Williams DD and Hogg ID (1988) The ecology and production of invertebrates in a Canadian coldwater spring-springbrook system. *Holarctic Ecology* 11: 41–54.

has been related to an increase in temperature fluctuations downstream, as well as to other factors, but is still not well understood. Nevertheless, macroinvertebrates can be very diverse in springs with heterogeneous substrates and microhabitats. In addition, in some springs (as in New Zealand) macroinvertebrate species diversity decreases with distance from the spring source. Furthermore, macroinvertebrate abundance (number of individuals) is often as high or higher in springs than at downstream sites or nearby runoff streams.

The abundance and diversity of other kinds of spring-dwelling organisms, such as algae, macrophytes, microinvertebrates, and vertebrates, may also be influenced by food availability, light conditions, substrate type, human influences, and (or) water temperature, chemistry, and flow rate. However, unlike the typical pattern seen for macroinvertebrates, diatoms appear to decrease in diversity from spring heads to downstream sites with faster flow rates.

The inhabitants of springs include species largely restricted to these habitats (crenobionts), species that commonly occur in spring habitats, but also occur elsewhere (crenophiles), species that are found in a variety of aquatic habitats and only occasionally occur in springs (crenoxenes), and subterranean species that have been washed into spring habitats from ground waters. Some crenobionts are so specialized for life in springs that this is reflected in their scientific name: e.g., the flatworm *Crenobia alpina* in Europe, midges of the genera *Krenopelopia*, *Krenopsectra*, and *Krenosmittia* in the Holarctic region, snails of the genus *Fontigens*, the fountain darter *Etheostoma fonticola*, and springfish of the genus *Crenichthys* in North America, the water mite *Torrenticola fontinale* in Costa Rica, and isopods of the genus *Crenoicus* in Australia, to name a few. The kingdom Crenarchaeota was even given its name because the first species discovered in this group of single-celled archaea were found in hot springs (see next section). The relative proportion of crenobionts in a spring depends on habitat type, persistence, isolation, and level of disturbance. Thermally constant springs in temperate regions have provided refuges for relict species (such as *C. alpina*) that have survived the great climatic shifts of the Pleistocene. In arid areas, relatively isolated, perennially flowing springs and their associated riparian habitats are also havens for rare or endemic animals and plants.

Thermal Springs

The hottest thermal springs with water temperatures exceeding 70 °C are exclusively populated by viruses, archaea, and bacteria (**Table 3**). Some archaea, which are low-energy microbes especially well adapted to extreme environmental conditions, can even live and grow in springs with boiling (temperatures >100 °C), highly acidic (pH < 2), or highly alkaline (pH > 9) water. Archaea are very diverse in hot springs and include types near the evolutionary root of all life. Viruses are also remarkably diverse featuring a variety

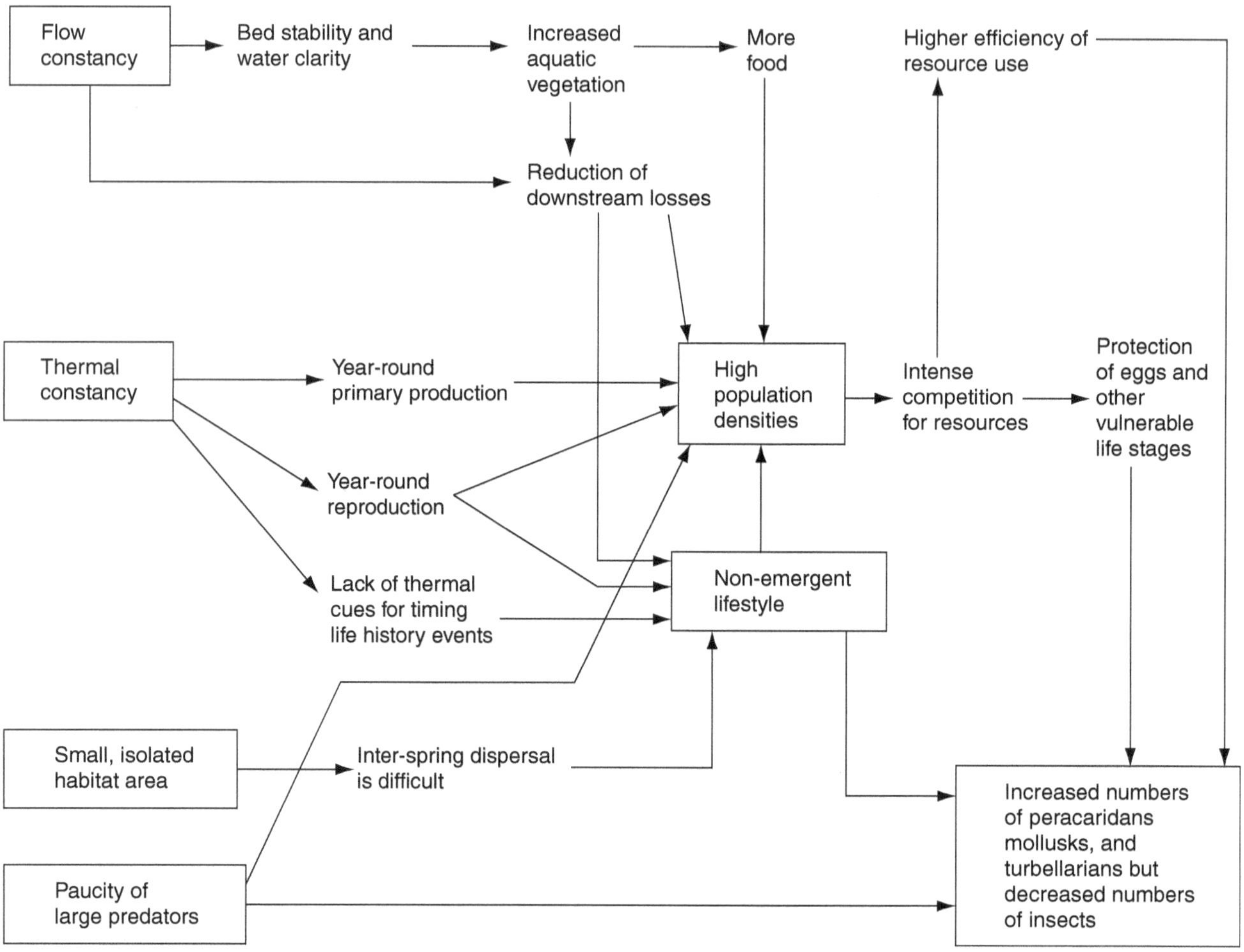

Figure 6 A hypothetical framework explaining the dominance of noninsect macroinvertebrates (peracaridan crustaceans including amphipods and isopods, mollusks including snails and clams, and turbellarian flatworms) in temperate nonthermal springs. The physical stability, small isolated area and paucity of large predators typical of these springs are considered to be the major environmental factors causing their distinctive macroinvertebrate fauna. Reproduced from Glazier DS (1991) The fauna of North American temperate cold springs: patterns and hypotheses. *Freshwater Biology* 26: 527–542, with permission from Blackwell.

of shapes including rods, spindles, filaments, droplets, spheres, polyhedrons, and forms resembling bottles and two-tailed lemons. They also have unique genes that are unknown elsewhere in the biosphere. Moreover, they provide insight into the common ancestry of viruses infecting species in all three of the domains of life (Archaea, Bacteria, and Eukarya).

Many kinds of bacteria with diverse biochemical mechanisms for harnessing energy abound in hot springs, as well. Some are autotrophic, producing their own food by using light energy (photosynthesis) or chemical energy from inorganic materials (chemosynthesis), whereas others are heterotrophic, using the fermentation products of the autotrophs. These microbial communities can be quite complex and often consist of extensive, visible mats dominated by filamentous cyanobacteria, other photosynthetic bacteria, and eukaryotic algae. However, photosynthetic organisms only occur at temperatures below 73 °C (**Table 3**). Relatively few multicellular plants and animals live in hot springs because they cannot tolerate temperatures above ~50–60 °C (**Table 3**). Animal, algal, and fungal diversities decrease as spring temperature increases (**Figures** 7 and 8). Longitudinal zonation of photosynthetic microbes is apparent in many thermal streams, with cyanobacteria dominating the spring head, and eukaryotic algae and diatoms being most abundant downstream where it is cooler.

Hot springs have attenuated food chains. Most aquatic animals, if present, are herbivores, grazing on the microbial mats. They are mostly insects, such as flies, bugs, beetles, and odonates, but a few species of nematode and annelid worms, rotifers, water mites, mollusks, crustaceans, fish, and amphibians may occur in thermal springs with temperatures between 30 and 55 °C. Air-breathing may facilitate survival of some insect larvae in hot springs with relatively low oxygen levels. The most thermophilic animal known is the

Table 3 Upper temperature limits of recorded occurrence of various groups of organisms in subaerial (terrestrial) springs

Group	*Upper temperature limit (°C)*
Prokaryotes	
Archaea	103
Bacteria	100
Cyanobacteria (blue-green algae)	73[a]
Photosynthetic bacteria	73
Eukaryotes	
Protists	
Algae	70
Protozoa	58[a]
Fungi	62
Plants	50
Animals	
Invertebrates	
Nematoda	61
Rotifera	46
Acari	51
Crustacea	55
Mollusca	46
Insecta	51
Vertebrates	
Fishes	40
Amphibia	41

Even higher limits have been found for prokaryotes at the hydrothermal vents of submarine springs.

[a]Some old reports of higher temperature limits have been discounted by Brock (1978).

Sources

Adams MWW and Kelly RM (1995) Enzymes from microorganisms in extreme environments. *Chemical and Engineering News* 73: 32–42.

Brock TD (1978) *Thermophilic Microorganisms and Life at High Temperatures*. New York: Springer-Verlag.

Brock TD (1985) Life at high temperatures. *Science* 230: 132–138.

Jana BB, Pal DN, and Sarkar HL (1982) Spatial distribution of the biotic community in the thermal gradients of the two hot springs. *Acta Hydrochimica et Hydrobiologica* 10: 101–108.

Kahan D (1969) The fauna of hot springs. *Verhandlungen der Internationalen Vereinigung für Theoretische und Angewandte Limnologie* 17: 811–816.

Kashefi K, Holmes DE, Reysenbach A-L, and Lovley DR (2002) Use of Fe(III) as an electron acceptor to recover previously uncultured hyperthermophiles: Isolation and characterization of *Geothermobacterium ferrireducens* gen. nov., sp. nov. *Applied and Environmental Microbiology* 68: 1735–1742.

Kristjansson JK (ed.) (1992) *Thermophilic Bacteria*. Boca Raton, Florida: CRC Press.

Obrdlik P (1988) The Longola Hot Springs of Zambia: The need for conservation. *Biological Conservation* 43: 81–86.

Poinar GO (2001) Nematoda and Nematomorpha. In: Thorp JH and Covich AP (eds.) *Ecology and Classification of North American Freshwater Invertebrates*, 2nd edn., pp. 255–295. San Diego: Academic Press.

Stetter KO (1998) Hyperthermophiles: Isolation, classification and properties. In Horikoshi K and Grant WD (eds.) *Extremophiles: Microbial Life in Extreme Environments*, pp. 1–24. New York: Wiley-Liss.

Winterbourn MJ (1968) The faunas of thermal waters in New Zealand. *Tuatara* 16: 111–122.

nematode *Aphelenchoides* sp., which can tolerate temperatures up to 61.3 °C. However, no vertebrate animal is known to live in spring water with a temperature exceeding 41 °C (**Table 3**).

Importance of Springs

Springs are important geologically and ecologically, and are of value to humans in several ways.

Geological Importance

The number of springs is not known for most geographical regions, and those few regional estimates that have been made (calculated here as density or number per area) vary by over three orders of magnitude (**Table 4**). However, if one extrapolates from these estimates, there are worldwide on land (not counting Antarctica) probably more than four springs per 10 km^2 (global total > 57 000 000) and more than seven thermal springs per 10 000 km^2 (global total > 100 000). These numbers are underestimates because they are based on surveys that usually ignore small springs and seeps, which greatly exceed the number of larger springs. For example in Spain, spring size, as estimated as flow volume ($l\ s^{-1}$), is inversely correlated with observed spring frequency (**Figure 9**). Because of their abundance, springs are obviously an important part of the hydrologic cycle, and they are also an important source of erosion and transport of dissolved minerals. In humid karstic areas, spring

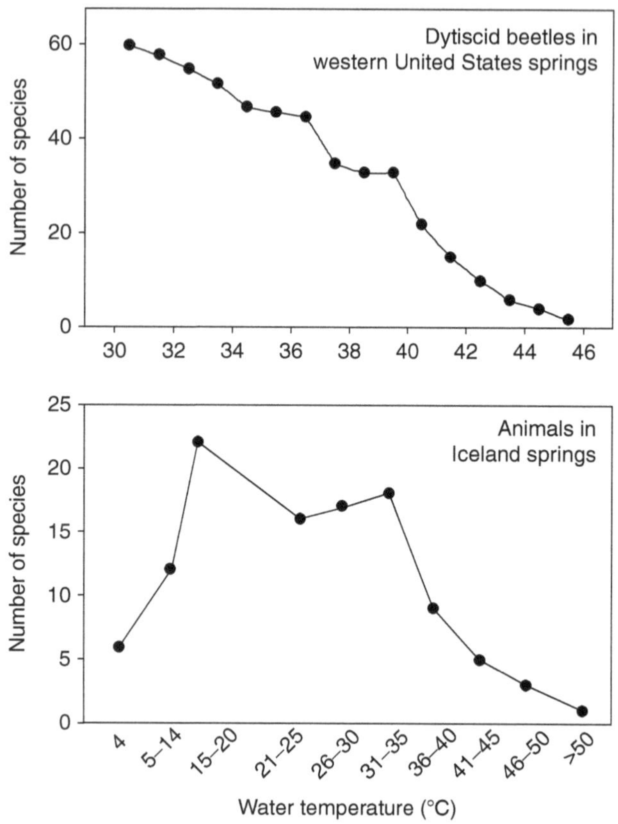

Figure 7 Number of species in relation to water temperature for dysticid beetles in western United States thermal springs [data from Brues CT (1932) Further studies on the fauna of North American hot springs. *Proceedings of the American Academy of Arts and Sciences* 67: 185–303] and for animals in Iceland springs [data from Tuxen SL (1944) The hot springs of Iceland; their animal communities and their zoogeographical significance. In: Frioriksson A, *et al.* (eds.) *The Zoology of Iceland, volume I, part II*, pp. 1–206. Copenhagen: Ejnar Munksgaard].

flow can account for up to one-half of the land erosion. In the United States, 30–40% of flowing surface waters has been estimated to be from ground water emerging from springs. Some springs also produce chemical and biological deposits that significantly alter the local landscape.

Ecological Importance

Springs provide many benefits to both aquatic and terrestrial life, including moisture, drinking water, food, minerals, shelter, thermal refuges, breeding sites, and travel corridors. Many kinds of plants and animals use thermally stable springs to endure hot summers or cold winters (e.g., **Figure 10**). Some springs have very high productivities, near the highest known in natural ecosystems. However, some acidic and sulfurous springs may have harmful effects on local ecological communities, including acidification and leaching of toxic metals that adversely affect downstream life, and production of sulfuric acid that can damage local vegetation. Highly carbonated springs emit CO_2 that may have both positive and negative effects on the local biota.

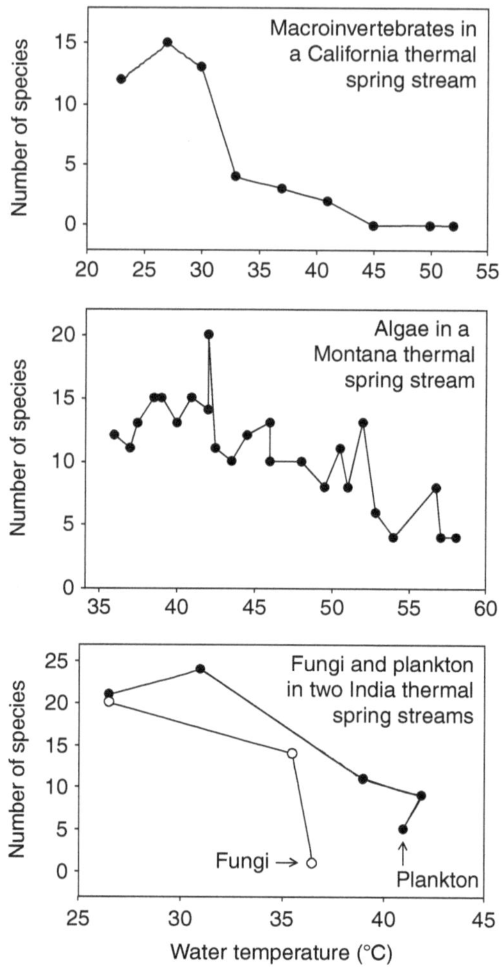

Figure 8 Number of species in relation to water temperature for macroinvertebrates in a California thermal spring stream [data from Lamberti GA and Resh VH (1985) Distribution of benthic algae and macroinvertebrates along a thermal stream gradient. *Hydrobiologia* 128: 13–21], for algae in a Montana thermal spring stream [Jackson Hot Springs; data from Kullberg RG (1971) Algal distribution in six thermal spring effluents. *Transactions of American Microscopical Society* 90: 412–434], and for fungi and plankton in two India thermal spring streams [data from Chandrashekar KR, Sridhar KR, and Kaveriappa KM (1991) Aquatic hyphomycetes of a sulphur spring. *Hydrobiologia* 218: 151–156; and Tanti KD and Saha SK (1993) Hydrobiological profiles along a thermal gradient of the hot springs of Rajgir (Bihar), India. *Journal of Freshwater Biology* 5: 107–117].

Scientific Importance

Springs and spring brooks are useful systems for studying a wide variety of scientific problems. Springs are places where interactions between land, water, air,

Table 4 Estimated number and density of springs in some different geographical regions of the world

Region	*Number of springs*	*Geographical area (km²)[a]*	*Density of springs (per 1000 km²)*	*Source*
Asia				
China	>2500[b]	9 578 678	>0.26[b]	12
India	>300[b]	3 287 590	>0.09[b]	11
Japan				
Vicinity of Mt. Fuji	~180			32
Vicinity of Aso volcano	>1500			32
Turkey				
Anatolia	>1800[b]	755 688	>2.4[b]	24
Yemen	>100[b]	528 038	>0.19[b]	16
Australia				
Great Artesian Basin	3000[c]		1.8[c]	20
	2000[d]	1 711 000	1.2[d]	20
Europe				
England				
White Peak area of Peak District	48	540	88.9	31
Finland (central)	25 000–30 000	19 700	1300–1500	22
Germany				
Berchtesgaden Alps	416	162	2568	6
Gütersloh District	203	220	923	8
North Rhine-Westphalia	~2000	34 069	~59	14
Pfälzerwald Mountains	141	1770	80	9
Ireland	29[b]	70 284	0.4[b]	1
Italy (central)	147[b]	~100 000	1.5[b]	15
Netherlands				
Northern part of province of Limburg	47			4
Portugal (northern)	>1500	650	>2300	18
Slovenia	28[b]	20 151	1.4[b]	22
Spain (62% of area)	17 305	342 428	50.5	19
Sweden				
Province of Göteborg & Bohus	~100	5141	~20	13
New Zealand				
South Island flood plains of three rivers	165	656.5	251.3	7
North America				
Mexico				
Yucatán	>3000	38 402	>78.1	33
United States	1661[b]	9 161 770	0.18[b]	27
Florida	>700	139 670	>5.0	5
Illinois	>200	144 120	>1.4	30
Kansas	'Thousands'			3
Missouri	>3000	178 414	>16.8	17
Nevada	~300	~4000	~75	21
Spring Mountains				
New Mexico				
Oscura and San Andres Mountains	276	~3500	~79	28
Texas	~2000[e]			2
Utah	~10 000	212 752	~50	29
Virginia	>1500	105 586	>14.2	10
	100[b]		0.9[b]	10
South America				
Argentina				
Jujuy	~40[b]	53 219	~0.75[b]	26
Chile	>240[b]	748 800	>0.32[b]	25

[a]Some areas are total areas; others are land area only.
[b]Thermal springs only
[c]Estimated number and density of springs before 1870.
[d]Estimated number and density of springs at present
[e]For only 183 of 254 counties in state

Sources

1. Beckett B, Tiorney A, and Emblow C (2003) Irish thermal springs (http://www.ecoserve.ie/projects/springs/index.html).
2. Brune G (2002) *Springs of Texas*, vol. I. College Station: Texas A & M University Press.
3. Buchanan R, Sawin R, and Lebsack W (2000) Water of the most excellent kind: historic springs in Kansas. *Kansas History* 23: 128–144.

4. Fellinger M and Verdonshot P (1996) In search of springs in the northern part of the province Limburg (the Netherlands). *Crunoecia* 5: 287–288.
5. Florida springs, Florida Department of Environmental Protection (http://www.dep.state.fl.us/springs/).
6. Gerecke R, Meisch C, Stoch F, Acri F, and Franz H (1998) Eucrenon–hypocrenon ecotone and spring typology in the Alps of Berchtesgaden (Upper Bavaria, Germany): A study of Microcrustacea (Crustacea: Copepoda, Ostracoda) and water mites (Acari: Halacaridae, Hydrachnellae). In: Botosaneanu L (ed.) *Studies in Crenobiology: The Biology of Springs and Springbrooks*, pp. 167–182. Leiden: Backhuys.
7. Gray DP (2005) *Braided River Springs: Distribution, Benthic Ecology and Role in the Landscape.* Canterbury, New Zealand: University of Canterbury Dissertation.
8. Gröver W (1996) Erfahrungen mit der umsetzung des quellschutzkonzeptes im Kreis Gütersloh. *Crunoecia* 5: 161–166.
9. Hahn HJ (2000) Studies on classifying of undisturbed springs in southwestern Germany by macrobenthic communities. *Limnologica* 30: 247–259.
10. Helfrich LA, Parkhurst J, and Neves R (2005) *Managing Spring Wetlands for Fish and Wildlife Habitat.* Virginia Cooperative Extension Publication 420–537. Blacksburg: Virginia Polytechnic Institute and State University.
11. Jana BB (1973) The thermal springs in Bakreswar, India: Physico-chemical conditions, flora and fauna. *Hydrobiologia* 41: 291–307.
12. Keshi A (1980) Thermal springs in China. *GeoJournal* 4.6: 507–513.
13. Lång L-O and Swedberg S (1995) Selection of springs for the groundwater monitoring program of the province of Göteborg and Bohus, SW Sweden. *Water, Air and Soil Pollution* 85: 1837–1842.
14. Laukötter G (1996) Vom AK – quellschutz zur quellschutzkampagne NRW. *Crunoecia* 5: 103–108.
15. Minissale A (2004) Origin, transport and discharge of CO_2 in central Italy. *Earth-Science Reviews* 66: 89–141.
16. Minissale A, *et al.* (2007) Thermal springs, fumaroles and gas vents of continental Yemen: their relation with active tectonics, regional hydrology and the country's geothermal potential. *Applied Geochemistry* 22: 799–820.
17. Missouri Department of Natural Resources (http://www.dnr.mo.gov/env/wrc/springsandcaves.htm).
18. Pacheco FAL and Alencoão AMP (2002) Occurrence of springs in massifs of crystalline rocks, northern Portugal. *Hydrogeology Journal* 10: 239–253.
19. Pérez ES (1996) Springs in Spain: Classification according to their flows and lithologies and their hydraulic contributions. *Ground Water* 34: 1033–1041.
20. Ponder WF (2002) Desert springs of the Australian Great Artesian Basin. Conference Proceedings. Spring-fed wetlands: Important scientific and cultural resources of the intermountain region (http://www.wetlands.dri.edu).
21. Sada DW, Fleishman E, and Murphy DD (2005) Associations among spring-dependent aquatic assemblages and environmental and land use gradients in a Mojave Desert mountain range. *Diversity and Distributions* 11: 91–99.
22. Šajna N, Haler M, Škornik S, and Kaligarič M (2007) Survival and expansion of *Pistia stratiotes* L. in a thermal stream in Slovenia. *Aquatic Botany* 87: 75–79.
23. Särkkä J, Levonen L, and Mäkelä J (1998) Harpacticoid and cyclopoid fauna of groundwater and springs in southern Finland. *Journal of Marine Systems* 15: 155–161.
24. Sayili M, Akca H, Duman T, and Esengun K (2007) Psoriasis treatment via doctor fishes as part of health tourism: A case study of Kangal Fish Spring, Turkey. *Tourism Management* 28: 625–629.
25. Spas and hot springs in Chile (http://www.visit-chile.org/activities/spas.phtml).
26. Thermal springs in Jujuy, Argentina (http://www.enjoy-argentina.org/adventure-travel-argentina-thermal-springs.php).
27. Thermal springs list for the United States, United States National Geophysical Data Center (http://www.ngdc.noaa.gov/nndc/struts/describeField?t=100006&s=1&filed=11029)
28. Thompson BC, Matusik-Rowan PL, and Boykin KG (2002) Prioritizing conservation potential of arid-land montane natural springs and associated riparian areas. *Journal of Arid Environments* 50: 527–547.
29. Utah Water Research Laboratory (http://www.engineering.usu.edu/uwrl/atlas/ch4/ch4springs.html).
30. Webb DW, Wetzel MJ, Reed PC, Philippe LR, and Harris MA (1995) Aquatic biodiversity in Illinois springs. *Journal of the Kansas Entomological Society* 68 (Supplement): 93–107.
31. Wood PJ, Gunn J, Smith H, and Abas-Kutty A (2005) Flow permanence and macroinvertebrate community diversity within groundwater dominated headwater streams and springs. *Hydrobiologia* 545: 55–64.
32. Yamamoto S. Springs of Japan. *Environmental Geology* 27: 118–119.
33. Yucatan today – the tourist guide (http://www.yucatantoday.com/destinations/eng-cenotes.htm).

and life are brought into sharp focus. Ground water interacts chemically and physically with subsurface rocks while traveling underground and then interacts with air and various organisms when it emerges above ground. As a result, several important geological phenomena can be investigated in spring systems, including subsurface hydrogeology, mineral deposition, geomicrobiology, and geothermal processes. In addition, the sediments deposited at springs enable past climatic and hydrogeological conditions of local regions to be estimated. And the chemical composition of spring water can provide clues to groundwater temperatures.

Spring systems are also wonderful natural laboratories for biologists and environmental scientists. The nearly constant water temperature, flow rate, and chemistry of many springs allow many biological processes to be studied under naturally controlled conditions that can also be relatively easily duplicated in the laboratory. Furthermore, springs differ significantly in many ways, even in local regions – e.g., in size, isolation, water temperature, flow rate, chemistry, substrate, and other ecological characteristics – thus providing excellent opportunities for comparative work. Since springs are often numerous in many places, they may supply natural replicates of various environmental conditions, thus also enabling useful field experiments to be carried out.

The stable conditions within springs, combined with significantly different conditions among springs,

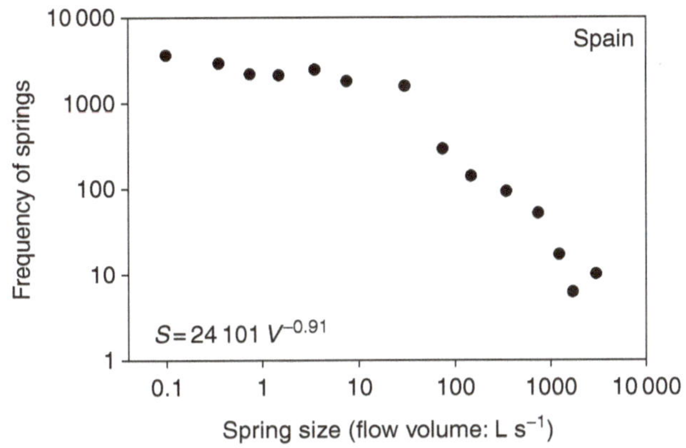

Figure 9 Inverse relationship between frequency of springs in Spain and their size (flow volume, based on class marks of intervals). The power function describing this relationship is shown. The data and equation are from Pérez ES (1996) Springs in Spain: classification according to their flows and lithologies and their hydraulic contributions. *Ground Water* 34: 1033–1041.

Figure 10 *Macaca fuscata* monkeys bathing in Jigokudani Hot Spring in Nagano Prefecture, Japan. Photo by Yosemite (14 February 2005) from http://commons.wikimedia.org/wiki/Image:Jigokudani_hotspring_in_Nagano_Japan_001.jpg, with permission from Free Software Foundation.

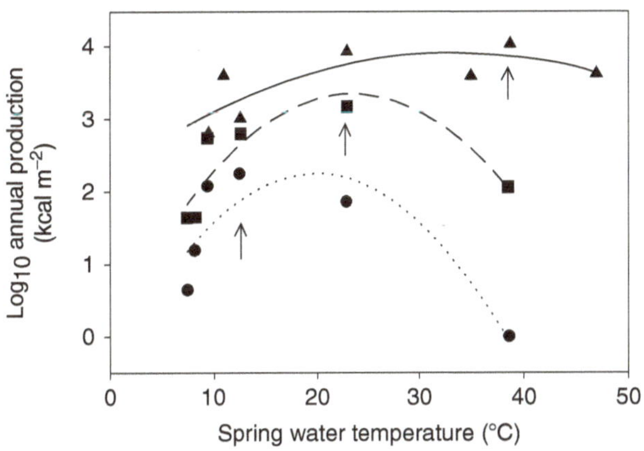

Figure 11 Log_{10} annual production of producers (autotrophic organisms, ▲), primary consumers (herbivores, ■) and secondary consumers (predators, •) in springs with different mean water temperatures. The consumers include macrofauna only. Polynomial regression lines (all with correlation coefficients >0.96) depict the apparent relationship between productivity and temperature at each trophic level (arrows indicate peak productivity values). The right most value for the secondary consumers is shown in arithmetic form (as a 0) to allow it to be plotted. Data are taken from Cushing CE and Wolf EG (1984) Primary production in Rattlesnake Springs a cold desert spring-stream. *Hydrobiologia* 114: 229–236; Iversen TM (1988) Secondary production and trophic relationships in a spring invertebrate community. *Limnology and Oceanography* 33: 582–592; Krno I, *et al.* (1998) The influence of organic inputs, acidification and fluctuating discharge on a spring ecosystem. In Bretschko G and Helešic J (eds.) *Advances in River Bottom Ecology*, pp. 99–106. Leiden: Backhuys; Naiman RJ (1976) Primary production, standing stock, and export of organic matter in a Mohave Desert thermal stream. *Limnology and Oceanography* 21: 60–73; Odum HT (1957) Trophic structure and productivity of Silver Springs, Florida. *Ecological Monographs* 27: 55–112; Stockner JG (1971) Ecological energetics and natural history of *Hedriodiscus truquii* (Diptera) in two thermal spring communities. *Journal of the Fisheries Research Board of Canada* 28: 73–94; Teal JM (1957) Community metabolism in a temperate cold spring. *Ecological Monographs* 27: 283–302; and Tilly LJ (1968) The structure and dynamics of Cone Spring. *Ecological Monographs* 38: 169–197.

are especially useful for incisive analyses of the in situ effects of various environmental factors on the temporal dynamics of biological systems at several hierarchical levels from single organisms to whole biotic communities. Unfortunately, however, these valuable research opportunities have not been sufficiently exploited. As one heuristic example, consider how a simple comparative analysis of springs with different mean water temperatures reveals novel patterns of how temperature affects rates of community-level production. At each trophic level of a spring community, production appears to increase and then decrease as temperature increases, with peak productivities occurring near 39 °C for producers, 23 °C for primary consumers, and 13 °C for secondary consumers (**Figure 11**). These previously unreported patterns require further study and verification before they can be fully understood, but presumably they are partly the result of increasing temperature increasing the rates of biosynthetic chemical reactions up to a point beyond which higher temperatures cause deleterious metabolic costs that reduce net production rates. However, unexpectedly the peak productivity of animal consumers does not coincide with the peak productivity of the producers from which they derive energy. Perhaps this is because many kinds of autotrophic microbes can grow exuberantly at temperatures exceeding 50 °C, whereas few heterotrophic animals can do so (**Table 3**; **Figures** 7 and 8). In addition, few animals can subsist on a diet of microbes alone, which would be all that is available at high temperatures. As a result, peak productivities occur at higher temperatures for producers than for consumers.

Other specific attributes of springs make them model systems for biologists, as well. Many springs are small, isolated and biotically simple, thus simplifying community and ecosystem analyses. As a result, classic studies on energy flow and trophic structure have been carried out in springs, including those used in the analysis described above. Along the length of spring-fed streams, gradients in water temperature, chemical composition, flow rate, or other environmental characteristics provide ideal conditions for precisely estimating the environmental tolerances of organisms. In addition, the clear water and physical stability of many springs enables in situ studies of animal behavior and microhabitat selection that are more difficult to do in many other murkier, more unstable aquatic habitats.

Island-like properties of springs suit them well for biogeographic and evolutionary studies. Many species have become highly genetically differentiated in these isolated habitats, and some have evolved major changes in lifestyle, including shifts from terrestrial to aquatic life as in the box turtle *Terrapene coahuila* of Cuatro Ciénagas (Coahuila, Mexico), and from benthic scavenging to pelagic filter-feeding as in the amphipod *Hyalella montezuma* of Montezuma Well (Arizona, USA). The historically spring-fed Lake Xochimilco near Mexico City, Mexico, is home to the axolotl, an entirely aquatic, gill-bearing salamander that no longer develops into a terrestrial adult. The axolotl has been a model system in developmental biology for over 100 years. In some places where springs occur in clusters, one can find significant adaptive radiations of microbes, water mites, snails, fishes, and other kinds of organisms. In Yellowstone National Park (Wyoming, USA), new major kinds of life have been discovered, including a new kingdom of life, the Korarchaeota in the domain Archaea, as well as other new major taxonomic groups of viruses and bacteria. Spring 'archipelagoes' also provide excellent systems for metapopulation studies, including the effects of immigration and extinction in local populations on the survival and evolution of species.

Some springs that occur in or constitute habitats that are extremely cold, hot, or chemically harsh offer unique opportunities to investigate environmental conditions resembling those of ancient earth or other planets such as Mars. For example, many hot springs contain prokaryotic microbial communities that are free of interactions with plants, animals, and other eukaryotic organisms, just as is thought to have occurred early in the history of life before eukaryotes had evolved. Stromatolites and other mineral-rich microbial deposits found in hot or carbonate-rich springs are reminiscent of those that were common in the Precambrian era. Hot springs may also provide insights into the origin of life because the earliest life forms evolved when the earth was warmer than it is now, and many hot-spring microbes occur near the bottom of life's evolutionary tree. In addition, the harsh environments of hot, acidic, sulfur-rich, and arctic-permafrost saline springs are supplying clues about the conditions under which extraterrestrial life may survive.

Valuable environmental science research is also possible at springs. Since spring water is emergent ground water, it provides a 'window' into the quality of underground water sources. Springs provide pure water that is used as a medium in laboratory ecotoxicology studies. Thermal and CO_2 springs are useful systems for investigating the effects of global warming and greenhouse gases on ecological processes. The chemical and biological composition of sediments deposited at springs can be used as indicators of the history of local environments, including climate, water quality, and land use. And conservation efforts in general have been energized by publicity relating to the plight of endangered spring-dwelling species, such as the pupfish *Cyprinodon diabolus*, which is only found in Devils Hole (Nevada, USA), the smallest known habitat of any vertebrate species. Threats to this habitat resulting from lowering of the groundwater table have alerted the public to the importance of protecting habitats, not just individual species.

Cultural and Societal Importance

For ages, springs have benefited humans, and their beauty and the purity of their life-giving waters have engendered mystical, reverent feelings in cultures all over the world. Springs have been regarded as sacred, magical places haunted by spirits, and thus have been the focus of special religious celebrations and ceremonies. In the Yucatan Peninsula of Mexico, the Mayans believed that sinkhole springs (cenotés) are bridges between the physical (surface) world and the invisible (spirit) world, with which they could connect by throwing gifts into them or by making ritual sacrifices. Similar beliefs in other regions of the world have led to the popular notions of 'wishing wells' and 'holy water', and in some places shrines have been constructed near or around springs.

Many people have believed that spring water has special healing powers and can prolong life, and hence the Spanish explorer Ponce de Leon's famous search for the 'fountain of youth' in Florida. Numerous medical studies have provided evidence that the mineral properties of some kinds of spring water have therapeutic value especially for skin, arthritic, respiratory, and gastrointestinal ailments. Bathing in warm, mineral spring waters at spas has been and continues to be very popular worldwide. The scientific field of balneology has been established to explore the health benefits of these activities.

Springs have been significant throughout human history as stopping points along trails and migration routes and for settlements. Need for water has resulted in the establishment of many villages and towns near springs. For example, Figeh spring supplies all of the water for Damascus (Syria), thought to be the oldest continually inhabited city in the world. Countless households, farms, and industries use spring water for drinking, cropland irrigation, livestock watering, brewing, bottling, processing of materials, and many other purposes. Hot springs are also a source of geothermal energy, which is exploited by many countries, especially in cold regions of the world such as Iceland. In Yugoslavia, flowing springs are used to generate electrical power. The economic importance of springs is enormous both regionally and globally. In 2002, the impact of four of the largest springs in Florida (USA) on local economies amounted to over $65 millions in spending, and over $1 billion in employment. The global economic importance of springs has not been estimated, but bottling of spring water alone generates revenue of billions of dollars per year.

The organisms found in springs may both benefit and harm humans. Springs have been used as hatcheries for edible sport fishes, such as trout. Hot springs, in particular, have supplied organisms that have important uses in medicine, industry, biotechnology, and bioremediation. For example, the doctor fish *Garra rufa*, which lives in hot springs of Kangel, Turkey, eats dead flaking skin and thus is being used to treat the skin disease psoriasis. In addition, the heat-tolerant bacterium *Thermus aquaticus* (Taq) isolated from a Yellowstone hot spring is the original source of the Taq polymerase used to amplify DNA in vitro in molecular biology laboratories around the world. This polymerase chain reaction (PCR) now supports a biotechnology industry worth hundreds of million dollars per year. However, on the downside, some thermal springs may harbor pathogens, such as those causing pneumonia, meningitis, and legionellosis.

Many springs are ideal for swimming, fishing, boating, and other forms of recreation. Some attract thousands of tourists every year, such as Silver Springs (Florida) famous for the teeming aquatic life in its crystal clear waters that can be easily viewed through glass-bottom boats. Silver Springs has even been the location for several movies and television shows.

Conservation of Springs

Springs contribute greatly to the aquatic biodiversity of an area, especially in arid regions. Many endemic and endangered species are found in springs. For example, a large proportion of the estimated ~450 species of native freshwater snail species in North America are spring-dwellers, including over 120 species of springsnails in the hydrobiid genus *Pyrgulopsis* alone. A recent survey has shown that nearly 80% (158 of 199) of the aquatic animal species endemic to the Great Basin Region of the United States primarily inhabit springs, and many of these species are declining due to human disturbance.

Throughout the world, springs are not being properly managed and conserved, and are disappearing rapidly, especially in arid regions (**Figure 12**). Major threats to springs and their inhabitants include human draining and diversion of groundwater sources, pollution, habitat alteration, and invasive species. Public awareness of the loss of these valuable resources is much needed, to avoid another potential tragedy of 'silent springs.'

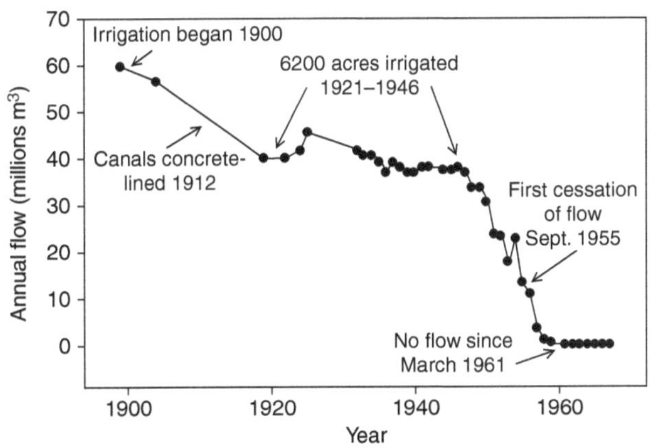

Figure 12 The demise of Comanche Springs at Fort Stockton, Pecos County, Texas, USA. Although this spring had supplied abundant water to the Comanches and other American Indians for ages, excessive pumping of water from its aquifer caused it to stop flowing permanently in the 1960s. Data from Brune G (2002) *Springs of Texas, volume I*. College Station: Texas A&M University Press.

Glossary

Age of spring water – The time it takes water to move from recharge to discharge areas.

Aquifer – Underground porous rock and cavities where water is stored. Aquifers receive water from precipitation and surface waters, and supply water to springs.

Archaea and bacteria – These mostly single-celled organisms are the two major types of prokaryote organisms (characterized by simple cells lacking nuclei and other organelles). They constitute two of the three major branches (domains) of life; the third branch is the eukarya, which includes the protists, fungi, plants and animals, all of which are made of more complex eukaryotic cells that have nuclei and other organelles. The archaea and bacteria are found almost everywhere on earth, though the archaea are especially common in extreme environments, including very hot and chemically caustic habitats.

Balneology – Study or practice of using mineral (spring) water to treat and cure diseases and other ailments. Typically involves bathing.

Base flow – Water flowing through a spring that comes from the storage reservoir of an aquifer during periods of low precipitation (recharge). Base flow tends to be more constant than 'overflow' that results from flooding events during periods of high precipitation (recharge).

Biota – Living organisms of all kinds in a specific area.

Contact zones – Flat surfaces separating different rock layers (strata). Also called 'bedding planes'.

Cyanobacteria – Photosynthetic bacteria sometimes called 'blue-green algae'.

Discharge – The volume of water flowing into and through a spring (or other flowing waters).

Dolomite – A soluble, sedimentary rock chiefly composed of the mineral $Ca,Mg(CO_3)_2$. Often found in karstic landscapes.

Fauna – Animal life in an area.

Fissures and fractures – Cracks, breaks or discontinuities in rocks. Include 'faults' along which rock has moved, and 'joints' along which no movement of rock has occurred.

Flora – Plant life in an area.

Hydrologic cycle – The circulation of the Earth's water molecules between the atmosphere, soil, living organisms, bodies of surface water, and reservoirs of ground water (also called the water cycle). Precipitation moves water from the atmosphere to the Earth's surface (including soil and surface waters); infiltration and gravity-driven recharge moves water from the Earth's surface to ground waters; absorption and other uptake processes move water from the atmosphere, soil and surface and ground waters to living organisms, discharge moves water from ground waters to soil and surface waters, evaporation moves water from soil, surface waters and other objects on the land surface to the atmosphere; and evaporation, transpiration and excretion move water from living organisms to the atmosphere, soil, and surface and ground waters.

Hydrology – The study of water resources and the hydrologic cycle, including the properties, movement, distribution and effects of water on geological features throughout the Earth.

Interstitial – Spaces between sediment particles.

Karst terrains – Characterized by the strong effects of the aqueous dissolving of bedrock minerals, resulting in landscape depressions and complex surface and underground drainage patterns (including caves, sinkholes, springs, and sinking streams). Usually composed largely of carbonate rocks, chiefly calcareous such as limestone and dolomite, but may also arise in gypsum, salt, silica and other rocks.

Lentic – Associated with lakes and ponds.

Limestone – A soluble, sedimentary rock chiefly composed of the mineral calcite ($CaCO_3$). Often found in karstic landscapes.

Lotic – Associated with flowing waters, such as streams and rivers.

Macroinvertebrates – Animals without backbones that usually grow longer than 0.5 mm in body length and can readily be seen with the naked eye.

Macrophytes – Aquatic plants that are partially or completed submerged in water.

Metapopulation – Regional cluster of patchy populations of the same species that engage in interchange of individuals by migration.

Microinvertebrates – Animals without backbones that usually do not grow longer than 0.5 mm in body length and are generally best seen with a microscope.

Recharge – The gain of water by an aquifer from precipitation and surface waters.

Runoff – Water appearing in streams directly from precipitation on land surfaces. Runoff streams tend to have more variable flow rates than springs fed by groundwater reservoirs.

Sinkhole – A natural depression in the landscape often containing exposed ground water at its bottom. Caused by the subsidence or collapse of surface soil, sediment or rock as a result of the dissolving of underlying rock strata (e.g., limestone and dolomite) by ground water.

Stromatolites – Multi-layered sedimentary structures that grow as domes, columns, cones or branching structures as a result of the activity of microorganisms (especially cyanobacteria) and (or) various abiotic processes. Stromatolites were common in the Precambrian era and their fossils represent some of the earliest life forms in the geological record. Stromatolites are presently rare, growing in harsh environments where animal grazing is excluded (e.g., hot springs and highly saline lakes and marine lagoons).

Trophic structure – Feeding relationships among species in an ecological community. Includes the network of flow of energy and materials from autotrophic organisms, which make their own food from inorganic materials (= producer trophic level), to heterotrophic herbivores, predators, parasites and detritivores, which use other organisms or organic materials as food (= consumer trophic levels).

Vertebrates – Animals with backbones, such as fish, amphibians, reptiles, birds and mammals.

See also: Acidification; Alkalinity; Aquatic Ecosystems and Human Health; Atmospheric Water and Precipitation; Bioassessment of Aquatic Ecosystems; Biological-Physical Interactions; Carbon, Unifying Currency; Chemical Fluxes and Dynamics in River and Stream Ecosystems; Chloride; Currents in Rivers; Dissolved CO_2; Distribution and Abundance of Aquatic Plants – Human Impacts; Flow Modification by Submerged Vegetation; Fluvial Export; Fluvial Transport of Suspended Solids; Gas Exchange at the Air-Water Interface; Ground Water; Ground Water and Surface Water Interaction; Groundwater Chemistry; Hydrological Cycle and Water Budgets; Light, Photolytic Reactivity and Chemical Products; Major Cations (Ca, Mg, Na, K, Al); Mercury Pollution in Remote Fresh Waters; Methane; Natural Organic Matter; Pollution of Aquatic Ecosystems I; Pollution of Aquatic Ecosystems II: Hydrocarbons, Synthetic Organics, Radionuclides, Heavy Metals, Acids, and Thermal Pollution; Rivers; Salinity; Silica; Streams.

Further Reading

Botosaneanu L (ed.) (1998) *Studies in Crenobiology: The Biology of Springs and Springbrooks*. Leiden: Backhuys.

Cantonati M, Gerecke R, and Bertuzzi E (2006) Springs of the Alps – sensitive ecosystems to environmental change: From biodiversity assessments to long-term studies. *Hydrobiologia* 562: 59–96.

Chapelle FH (1997) *The Hidden Sea: Groundwater, Springs, and Wells*. Tucson, AZ: Geoscience Press.

Ferrington LC (ed.) (1995) Biodiversity of aquatic insects and other invertebrates in springs, *Journal of the Kansas Entomological Society* 68 (Supplement): 1–165.

Glazier DS (1991) The fauna of North American temperate cold springs: Patterns and hypotheses. *Freshwater Biology* 26: 527–542.

Hinterlang D (ed.) (1996) 1st European Symposium of Spring Ecology and Conservation, *Crunoecia* 5: 1–304.

Hynes HBN (1970) *The Ecology of Running Waters*. Toronto: University of Toronto Press.

LaMoreaux PE and Tanner JT (eds.) (2001) *Springs and Bottled Waters of the World: Ancient History, Source, Occurrence, Quality and Use*. Berlin: Springer-Verlag.

Marsh PC (ed.) (1984) Proceedings of a special symposium: Biota of Cuatro Ciénegas Coahuila, Mexico, *Journal of Arizona-Nevada Academy of Science* 19: 1–89.

Meyer JL, Strayer DL, Wallace JB, Eggert SL, Helfman GS, and Leonard NE (2007) The contribution of headwater streams to biodiversity in river networks. *Journal of the American Water Resources Association* 43: 86–103.

Renaut RW, Jones B, *et al.* (2003) Special issue: Sedimentology of hot springs. *Canadian Journal of Earth Sciences* 40: 1439–1738.

Reysenbach AL, Voytek M, and Mancinelli R (eds.) (2001) *Thermophiles: Biodiversity, Ecology and Evolution*. New York: Kluwer.

Sada DW, Williams JE, Silvey JC, Halford A, Ramakka J, Summers P, and Lewis L (2001) Riparian area management: A guide to managing, restoring, and conserving springs in the western United States, Technical Reference, pp. 1737–17. Denver, CO: United States Department of the Interior Bureau of Land Management.

Smart C and Worthington RH (2004) Springs. In: Gunn J (ed.) *Encyclopedia of Caves and Karst Science*, pp. 699–703. New York: Taylor and Francis.

Waring GA, Blankenship RR, and Bentall R (1965) Thermal springs of the United States and other countries of the world – A summary. Geological Survey Professional Paper 492. Washington, DC: United States Government Printing Office.

Williams DD and Danks HV (eds.) (1991) Arthropods of springs, with particular reference to Canada, *Memoirs of the Entomological Society of Canada* 155: 1–217.

Zeidler W and Ponder WF (eds.) (1989) *Natural History of Dalhousie Springs*. Adelaide: South Australian Museum.

Relevant Websites

http://ga.water.usgs.gov/edu/watercyclesprings.html – Springs and the water cycle.

http://faculty.juniata.edu/glazier/researchlink/rl2kpage1.html – Freshwater springs and their uses in research and teaching.

http://desertfishes.org/australia/habitats/springs/springen.shtml – Australia's desert springs, with special emphasis on fishes.

http://biology.usgs.gov/s+t/noframe/f126.htm – Biota of Illinois caves and springs.

http://www.srwmd.state.fl.us/water+data/springs/what+is+a+spring.htm – Springs of the Suwannee River Water Management District in Florida, with much basic information.

http://www.answers.com/topic/hot-spring – Information about hot springs.

http://www.bact.wisc.edu/bact303/b20 – Life at high temperatures, especially in the hot springs of Yellowstone National Park.

http://faculty-staff.ou.edu/K/Lee.R.Krumholz-1/nsfzodletone-page03.html – Microbial observatory at sulfide- and methane-rich Zodletone Spring, Oklahoma.

http://www.floridasprings.org/ – Florida's springs: protecting nature's gems.

http://www.ext.vt.edu/pubs/fisheries/420-537/420-537.html – Managing spring wetlands for fish and wildlife habitat.

http://biology.usgs.gov/s+t?SNT/noframe/sw156.htm – Rare aquatic snails in springs of the southwestern United States.

Wetland Hydrology

R W Tiner, University of Massachusetts, Amherst, MA, USA

Introduction

Wetland hydrology – frequently occurring prolonged inundation and/or soil saturation (waterlogging) – is the driving function that creates and maintains wetlands and provides wetlands with unique qualities and significant ecological functions that are highly valued by society. Wetlands may be generally defined as shallow water areas or lands that are periodically flooded or saturated long enough to support hydrophytic vegetation and/or other forms of aquatic life. Conceptually, wetlands lie between dry land and deep water and as a result have often been referred to as ecotones (a transitional habitat; part land, part water). Depending on the wetland type, the wetland may be subjected to flooding or soil saturation or a combination of both. From a hydrologic standpoint, wetlands encompass a wide range of wetness conditions from permanent to periodic inundation or waterlogging. Differences in climate, geologic setting, and other factors have created a diversity of wetland types globally with varied hydrologies that affect plant and soil development, their use by wildlife, their functions and values. This article is a general introduction to wetland hydrology for a nontechnical audience; for more advanced coverage, consult 'Further Reading.' Since the focus of this encyclopedia is on inland waters, the discussion of wetland hydrology emphasizes that of inland wetlands (nontidal wetlands and tidal freshwater wetlands) and does not address marine and estuarine wetlands.

Water Sources

Nontidal wetlands receive water from meteoric sources (precipitation, snow, sleet, hail, fog, and mist) or telluric sources (ground water), while tidal wetlands receive a significant inflow of water from tides in addition to the other sources. Meteoric sources affect all lands, but for wetlands to form, water must persist either on the surface or in the soil for sufficient time to promote the colonization, growth and survival of hydrophytic vegetation and the development of hydric soils and substrates, and to create environmental conditions that support other aquatic life.

Factors Contributing to Wetland Hydrology

Topography (landform), landscape position (proximity to a water source), soil properties, geology, and climatic conditions are important factors in wetland formation. Depressions and broad flats with poor drainage are places where water can accumulate in sufficient quantities to create wetlands. Mountainous areas tend to have less wetland than coastal and glaciolacustrine plains largely due to drainage properties (e.g., rainwater drains readily from slopes and collects on flats). Wetlands in mountainous areas likely receive considerable groundwater inflow (groundwater wetlands), while wetlands on broad flats in areas of high rainfall may be supported by rainwater (surface water wetlands; **Figure 1**) or ground water. Inland wetlands occur (1) along the shores of lakes and ponds where high water levels and the presence of a permanent waterbody lead to permanent inundation of shallow water zones and periodic inundation of low-lying neighboring areas, (2) on floodplains where they are subject to seasonal inundation, (3) in depressions that receive runoff from adjacent areas and groundwater discharge, (4) on broad flats of coastal plains or glaciolacustrine plains where drainage is poor, (5) at toes of slopes where subsurface water reaches the surface, (6) on slopes associated with springs and seeps where ground water discharges to the surface, (7) in paludified landscapes where low evapotranspiration and an excess of water allow peat mosses to grow over once dry land covering the landscape with peat (e.g., blanket bogs), (8) in permafrost areas where frozen soils serve as an impermeable layer that perches water at or near the surface, and (9) in areas below glaciers and snowfields where the seasonal flow of meltwater creates wet conditions. Soils with low hydraulic conductivity such as clayey soils and soils with restrictive layers (e.g., hardpans) near the surface favor wetland development over sandy soils that tend to have good internal drainage due to large effective porosity (high hydraulic conductivity). Sandy soils become wet if external drainage is poor or if periodically flooded for long duration or saturated by water from external sources (e.g., regional water tables). Geological features, such as contacts between different

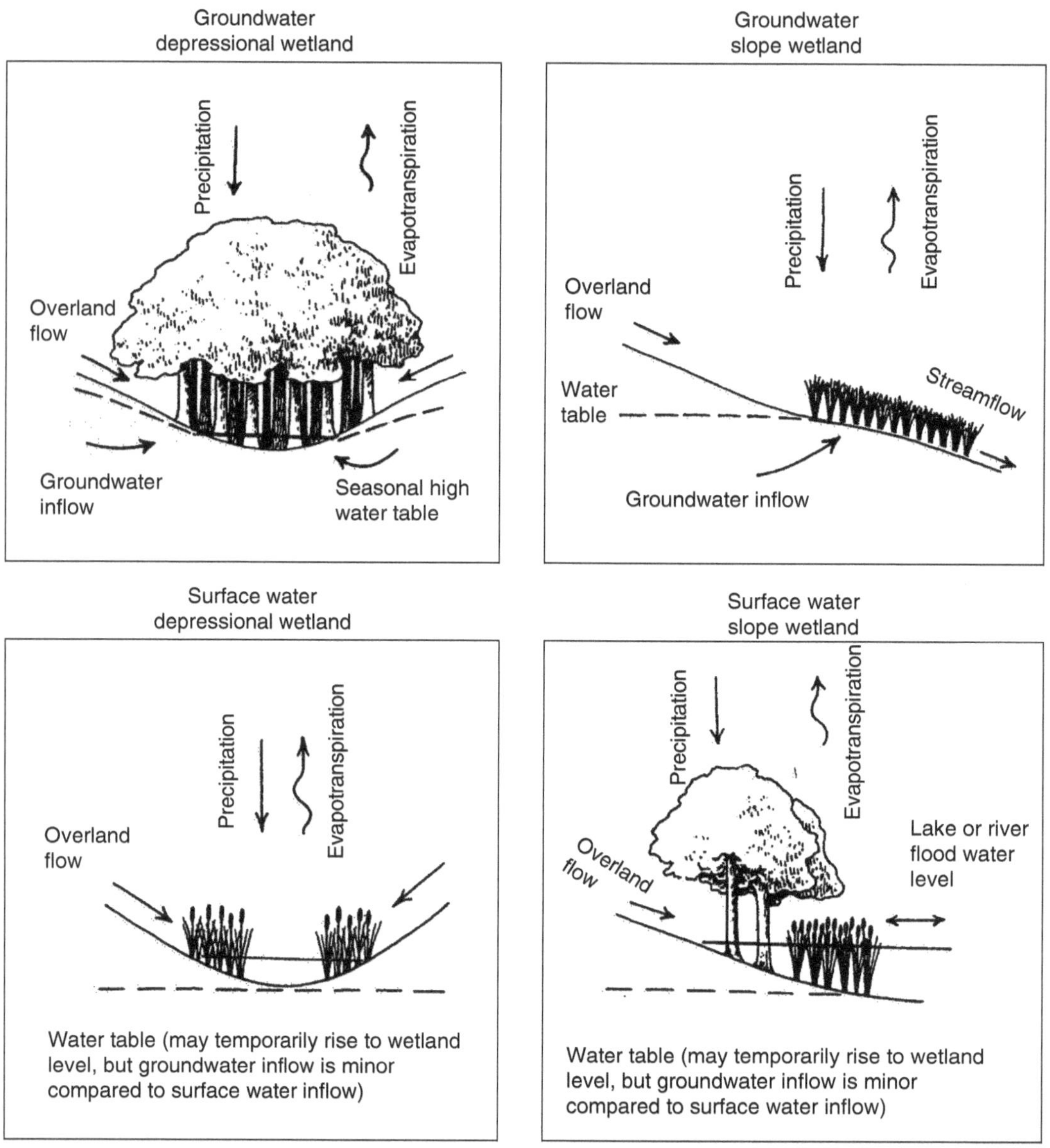

Figure 1 Four types of wetlands defined by wetland hydrology and topography: groundwater depressional wetland, groundwater slope wetland, surface water depressional wetland, and surface water slope wetland. Following this approach, wetlands on broad flats may be considered surface water flat wetlands; they are not inundated (except in microdepressions) but have seasonal high water tables. Source: Tiner RW (1988) Field guide to nontidal wetland identification. Newton Corner, Massachusetts, USA: U.S. Fish and Wildlife Service and Maryland Dept. of Natural Resources. Reprinted by Institute for Wetlands and Environmental Research, Inc., Leverett, MA; redrawn from Novitski RP (1982) Hydrology of Wisconsin wetlands, Information Circular 40. Reston, VA: U.S. Geological Survey.

rock types that outcrop along hillsides, create conditions favoring the formation of seeps and associated wetlands where ground water or near-surface interflow intersects the land surface. Desert spring-fed wetlands are the result of groundwater discharge from regional aquifers. Dissolution of limestone formations allow for the establishment of wetlands in karst terrain. Morainal deposits in glaciated regions typically create deranged (nonintegrated) drainage patterns that produce isolated depressions on low-permeability substrates where wetlands develop, whereas kettle basins in outwash deposits will also produce wetlands where such basins are in contact with aquifers. Precipitation patterns significantly influence wetland formation. Wetlands are naturally more abundant in regions with warm humid climates than in hot arid climates for obvious regions. The latter regions may support 'ephemeral wetlands' during extremely wet years that create conditions in landscape positions that typically support wetlands in humid regions.

How Wet is a Wetland?

One would think it would be relatively easy to define a wetland from a hydrologic standpoint; after all if an area is wetland, it must be wetter than dry land. Yet to fully describe and understand hydrology requires long-term measurements and for wetlands this involves monitoring water tables (soil saturation) as well as surface water. Given the diversity of wetland types, that wetlands are transitional habitats between dry land and open water, and the need to conduct studies over long time periods, it is little wonder that such information is lacking for most wetlands and that it is not a simple matter to define the minimum wetness of wetland.

Both surface and groundwater sources need to be considered in determining minimum wetness for defining wetland hydrology. Scientists in the United States probably have spent more time contemplating this topic because wetlands on both public and private lands are regulated by the federal government and by many state and local governments. Such regulation requires identifying specific limits of wetlands on the ground. To be a wetland for jurisdictional purposes, an area typically must have a positive indicator of hydrophytic vegetation, hydric soils, and wetland hydrology. The former two features are manifestations of wetland hydrology and serve as valid indicators of wetlands in the absence of drainage. Yet, many situations are encountered where some drainage has been performed, thereby raising the question – is the area still wet enough to be identified as wetland? In such areas, the plants and soils may better reflect past hydrologic conditions and may no longer be valid indicators of current site wetness. So to answer this question, wetlands hydrology must be defined and a lower threshold of wetness established. (Note: The upper threshold of wetness clearly is permanent inundation or saturation to the surface.)

Wetland Hydrology Defined

A group of distinguished American wetland scientists studied the topic of wetland delineation for 2 years and came up with the following definition:

> An area has wetland hydrology when it is saturated within one foot of the soil surface for two weeks or more during the growing season in most years (about every other year on average).

Depth of Saturation

Roots supply plants with nutrients and water needed for growth and reproduction. Heavy rains that saturate the soil for as little as a few hours can cause root hairs to die of oxygen starvation as evidenced by wilting of vegetable crops (e.g., squash and cabbages) after rainfall. Prolonged saturation and accompanying anaerobic conditions adversely affect root function, causing changes in root morphology (aerenchyma development), root rot and dieback, and the formation of shallow roots (near the surface). Since they live in areas of frequent soil saturation, wetland plants have most of their roots located in the upper, partly aerated zone of the soil. This zone is typically within 30 cm (1 ft) of the soil surface.

Duration of Wetness

Although flooding for as little as a day can create anaerobic conditions under special circumstances in some soils, most plants need to be wet longer to adversely affect their growth and survival. Some wetland plants begin to develop morphological properties (e.g., aerenchyma tissue, adventitious roots, and hypertrophied lenticels) within 1–2 weeks of flooding or waterlogging. Soil scientists have identified long and very long durations as important periods of flooding for soils. Long duration is defined as inundation from 7 to 30 days, while very long duration is longer than 30 days. Flooding (from overbank flows or runoff) or ponding (standing surface water in a closed depression) for long duration or more during the growing season in most years can be used to identify a hydric (wetland) soil regardless of its morphological properties.

Frequency of Prolonged Wetness

In order to be a wetland under most definitions, an area has to be frequently wet, which has been defined as every other year on average. This definition works for wetlands in humid regions where average conditions may have some significance. Yet for arid and semiarid regions, this definition is problematic, for such regions may experience long-term droughts that have a profound effect on the 'average.' In these regions, a series of wet years can create conditions long enough for wetlands to become established. While there are typical wetlands supported by springs and river overflows in these areas, some 'wetlands' are ephemeral types showing up only during periods of extreme wetness. The term 'episodic' has been applied to describe such wetness and these infrequently wet areas are viewed as wetlands in some countries (e.g., Australia). Soils in arid regions may develop hydromorphic properties when saturated for less than 5 weeks at a frequency of once every 3 years.

Growing Season

Although the wetland hydrology definition is explicit in stating the frequency and duration of wetness and the depth of saturation, the seasonality of wetness – 'growing season' – can be defined in many ways which could lead to different interpretations of wetland hydrology. For example, the growing season has been traditionally used in agriculture to assist in determining planting times for crops like corn, wheat, and rice. As such, the growing season could be defined as the frost-free period where there would be virtually no risk of crop failure due to frost or freeze. This concept is not useful for natural plant communities that are adapted to local environmental conditions. For example, by the time the frost-free period arrives, many native plants have already flowered and leafed-out. So, from an ecological standpoint, the growing season has commenced well before there is no risk of frost. The growing season is actually a concept that is best applied to a particular plant. What is the growing season for red maple (*Acer rubrum*) or corn (*Zea mays*)? Such a definition would be based on when they initiate new growth after a period of dormancy. Yet, this concept is also not useful for defining wetland hydrology. Given that the focus of wetland hydrology is often to define conditions that affect plant establishment, growth, and survival, and sometimes the formation of hydric soils, the time of year should, at least, be related to plant activity and possibly to soil formation processes. From the botanical standpoint, one could say the start of the growing season for wetlands should be predicated on the vegetation growing in wetlands. Observations of plant growth in wetlands would therefore serve as valid indicators that the growing season has begun. The earliest of the plants that leaf out or flower in spring (e.g., willows – *Salix* spp., alders – *Alnus* spp., marsh marigold – *Caltha palustris*, leatherleaf – *Chamaedaphne calyculata*, and skunk cabbage – *Symplocarpus foetidus*) would be the best indicators for determining this. A more generalized approach would be to consider the growing season for an area to have commenced when any plants (wetland or upland) show signs of growth (e.g., budbreak, leaf emergence, or blooming). The end of the growing season should be defined by the end of plant growth in wetlands or in the local area. Recognizing that the fall is an important time for root growth, the growing season extends beyond the time when leaves fall off deciduous-leaved plants. In all likelihood, it continues until the ground freezes. Minimum temperatures for root growth may be from slightly above 0–7 °C (32–44.6 °F), with optimum temperatures ranging from 10 to 25 °C (50–77 °F). Soil scientists have used the term 'biologic zero' to define conditions that relate to microbial activity in the soil: 5 °C (41 °F) degrees measured at a depth of 50 cm (20 in.) in the soil. This concept may have some utility in temperate and tropical regions, but it is not valid for subarctic and arctic regions where permafrost or frozen soil occurs at shallow depths and therefore no growing season would exist.

Despite the significance of flooding and soil saturation on growth of nonevergreen plants, there are other factors to consider when defining wetland hydrology (i.e., should the focus be on growing season or year-round conditions?). For example, what are the needs of wetland-dependent animals and are significant wetland functions being performed outside the 'growing season'. Consider the following:

1. Evergreen plants and persistent graminoids (grass and grasslike plants) continue to grow during the 'dormant period' for nonevergreens and saturation during this time should have some effect on these and competing species.
2. Water conditions during the dormant period have a profound influence on hydrologic conditions during the early part of the growing season and may prevent winter dessication of some plants.
3. Hydric soil properties have developed under reducing conditions that extend beyond the growing season.
4. Critical activities of some animals require dormant season flooding or soil saturation (e.g., woodland vernal pool breeders and pond animals).
5. Aquatic animals like fish need water year round and are active year round.
6. Wetland functions such as nutrient transformation and cycling, shoreline stabilization, surface water detention, and sediment retention are independent of the growing season.
7. Wetness limitations during the dormant period also affect the potential uses of the land.

Defining wetland hydrology based on year-round conditions appears to be justified from ecological and functional perspectives. This approach also would better reflect how wet some of the drier-end wetlands really are. Many of these drier-end wetlands (e.g., wet flatwoods of the southeastern United States) are wet for significant periods (months) during the year, but are saturated near the surface for relatively short periods (weeks) during the growing season.

Does Prolonged Saturation Guarantee Anaerobic and Reducing Conditions?

Although the definition of hydric soil emphasizes anaerobic reducing conditions and most wetlands are exposed to periodic anaerobiosis, there may be some situations where wetlands exist in aerobic environments. Also, soils saturated for long periods are not always reduced; there must be a source of organic matter, sufficiently high temperatures to support microbial activity responsible for reduction, and a population of reducing microbes present. (Note: The latter is typically present if the first two conditions are satisfied.) Possible 'aerobic' wetlands occur in seepage areas where oxygenated water is continuously flowing downhill, especially in colder climates and high-elevation sites, and along coldwater mountain streams on cobble-gravel or sandy substrates. Saturated soils may not be reduced under the following circumstances: (1) in cold climates with average temperatures of less than 1 °C (33.8 °F), (2) in very saline waterlogged desert soils where salinity restricts growth of reducing microbes, (3) in areas with little or no organic matter and moderate to high levels of calcium carbonate (e.g., irrigated rice basins in northwest India lack a low chroma matrix soil), and (4) in areas subject to groundwater discharge where dissolved oxygen is present in water (e.g., in areas of moderate relief and soils on the edges of valleys).

Wetland Water Regimes

The hydrology of wetlands can be described in numerous ways. The duration of flooding and soil saturation can be defined by various water regimes (**Table 1**). The flow of water can be classified as inflow (water coming into a wetland with no outlet; a sink), outflow (water flows out of wetland; source), throughflow (water comes into and exits wetland), and bidirectional flow (water levels rise and fall in wetlands due to tides, lake or pond levels). If the wetland is surrounded by dry land and there is no known flow into or out of the wetland (besides runoff and near-surface flow from adjacent upland), the wetland is hydrologically isolated from a surface-water perspective, but it may not be hydrologically isolated from a groundwater perspective. The topographic position of 'isolated wetlands' may determine whether they are sources, throughflows, or sinks based on their location in groundwater flow systems. In semiarid regions like the North American Prairie Pothole Region, topographic position and local geology influence water salinity and vegetation patterns, with freshwater wetlands at the highest elevations or levels of the groundwater flow system (sources), the most saline wetlands at the lowest levels (sinks or sumps), and flow-through wetlands with intermediate salinities in between (**Figure 2**). Because salt tolerance is an adaptation possessed by certain plants, vegetation can be used to infer the hydrologic function in these regions.

Hydrographs for Different Wetland Types

Due to variations in climate, soils, vegetation, geologic setting, and other factors, the presence of water and its location in wetlands varies around the globe. Even within local areas, the hydrology of wetlands differs

Table 1 Inland wetland hydrology descriptors based on the U.S. Fish and Wildlife Service's wetland classification system

Water regime modifiers	*General definition*
Permanently flooded*	Inundated continuously, year-round in all years
Intermittently exposed	Inundated year-round in most years, but exposed during extreme droughts
Semipermanently flooded*	Inundated throughout the growing season in most years
Seasonally flooded*	Inundated for extended periods during the growing season, but usually not flooded later in the growing season; water table may be near the surface for much of the time when not flooded (seasonally flooded/saturated) or may be well below the surface
Temporarily flooded*	Inundated for brief periods during the growing season (usually a couple of weeks or less early in the growing season), with the water table typically well below the surface for extended periods thereafter
Intermittently flooded	Inundated for variable periods with no detectable seasonality; area is usually exposed
Saturated	Water table is at or near the surface for most of the growing season and surface water usually absent; when soil is saturated only seasonally, usually early in the growing season, the hydrology is referred to as seasonally saturated
Artificially flooded	The frequency and duration of inundation is controlled by humans; in the strictest sense, the control is purposeful by means of pumps or siphons, but more generally, wetlands flooded by any artificial means qualify, including irrigation

An asterisk (*) denotes water regimes that can also be modified to describe freshwater tidal wetland hydrology by adding '-tidal' to the term (i.e., permanently flooded-tidal, semipermanently flooded-tidal, seasonally flooded-tidal, and temporarily flooded-tidal).

Adapted from Cowardin LM, Carter V, Golet FC, and LaRoe ET (1979) *Classification of Wetlands and Deepwater Habitats of the United States*. Washington, DC: U.S. Fish and Wildlife Service, FWS/OBS-79/31.

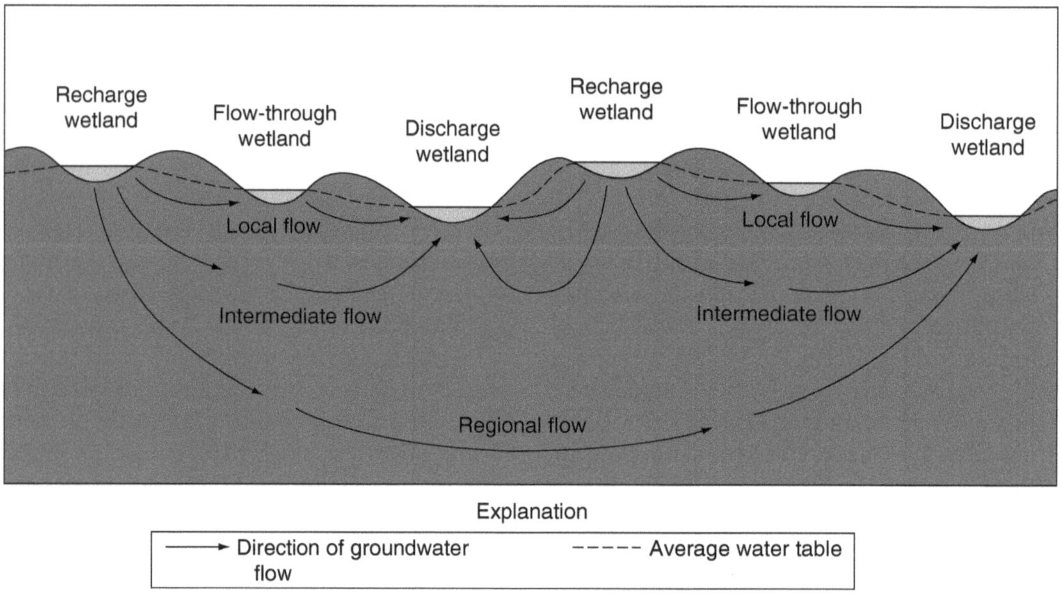

Figure 2 Groundwater flow patterns in North American prairie pothole wetlands. Characteristics of the three types shown differ: recharge wetlands (that recharge ground water) are more fresh and have standing water for only a few months, whereas discharge wetlands (that receive ground water) are the most saline and are permanently flooded and flow-through wetlands are intermediate in salinity and duration of surface water. Source: Berkas WR (1996) North Dakota wetland resources. In: Fretwell JD, Williams JS, and Redman PJ (eds.) *National Water Summary on Wetland Resources*, Water-Supply paper 2425, pp. 303–307. Reston, VA: U.S. Geological Survey.

by wetland type. Some wetlands are permanently flooded or nearly so, others are never flooded but permanently saturated, and the rest are either periodically inundated or seasonally saturated. The fluctuations of the water table and water levels in wetlands may be depicted graphically by hydrographs (**Figure 3**).

Changing Water Levels

Site wetness varies seasonally, annually, and long-term. Seasonal changes are reflected in the hydrographs that clearly show the wet season and dry season for certain wetland types. Some types are permanently flooded or saturated near the surface with the latter showing slight changes in water table during the dry season (e.g., summer in the northeastern United States). Other types show marked fluctuations in the water table during the year. Wetlands also experience changes from year to year. During wet years, water tables are higher than normal and water may persist on the surface for longer periods, while in dry years they are lower (**Figure 4**).

The long-term hydrologic cycle encompasses years of normal precipitation, above normal precipitation and below normal precipitation. If wetland hydrology were monitored for decades, the effect of these precipitation patterns on wetland water levels and water tables could be readily seen. Unfortunately such data are lacking for most wetland types. Water levels in wetlands along the shores of North America's Great Lakes fluctuate with changes in lake levels which vary from about 1–2 m (3.5–6.5 ft) from extremely wet years to extremely dry years. These changes have a significant impact on plant communities with aquatic beds predominating in high water years and wet meadows in low water years (**Figure 5**). Many woody plants that colonize these wetlands during dry years are killed by high water during wet years. These wetlands are among the most dynamic of inland wetlands in North America from a plant composition standpoint. Wetlands in arid and semi-arid regions (e.g., prairie potholes and playas) also experience somewhat similar vegetation changes due to variability in regional precipitation patterns.

Water Budget

The water budget of an area, wetland or nonwetland, is an accounting of water inflows (gains or inputs) and outflows (losses or outputs) to determine the change in storage (**Figure 6**). Inputs include water sources (precipitation, surface water, and ground water),

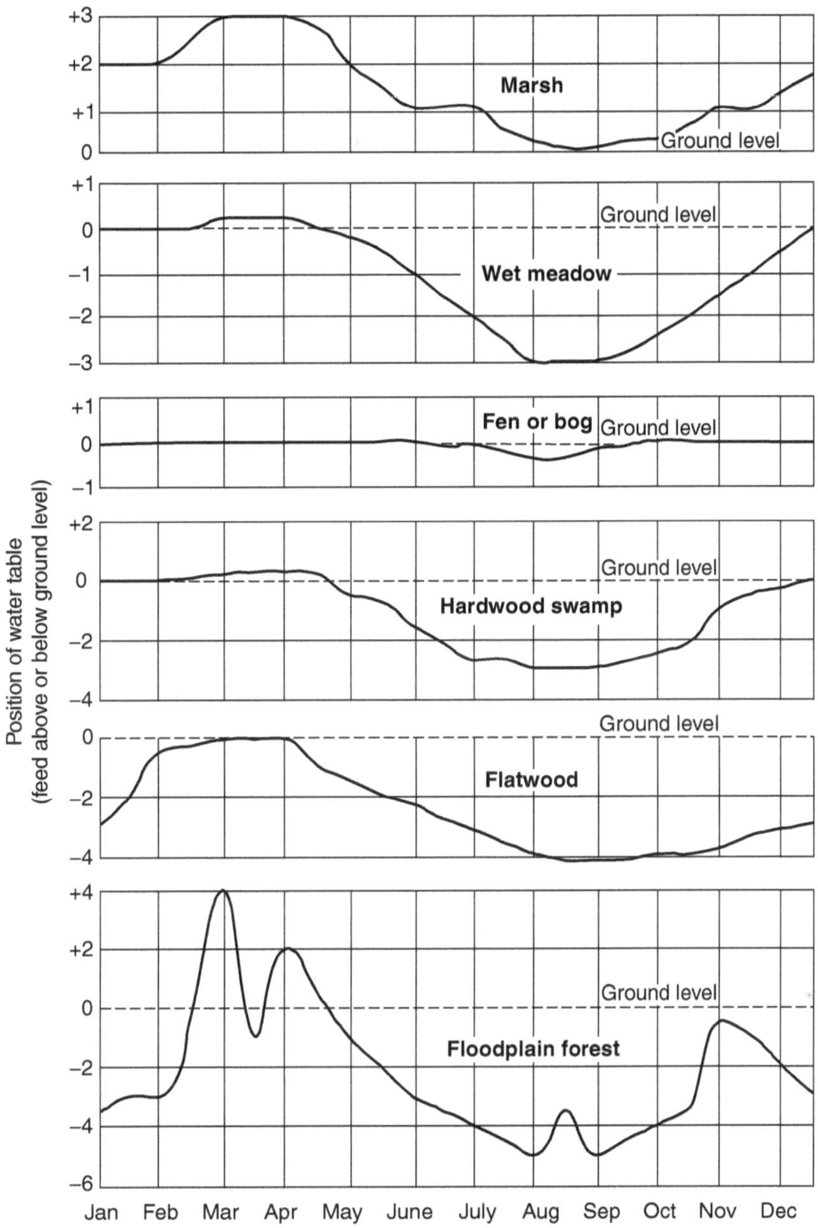

Figure 3 Hydrographs of some common wetland types in the northeastern United States. Note that there is considerable variability within types that are not reflected in these hydrographs and that these hydrographs are intended to represent general tendencies in water levels for illustration purposes. Source: Tiner RW (2005) *In Search of Swampland: A Wetland Sourcebook and Field Guide*. New Brunswick, NJ: Rutgers University Press.

while outputs or water losses are attributed to evaporation, transpiration by plants, surface and subsurface water runoff, and groundwater recharge. Because wetlands have an excess of water, at least seasonally, the inputs are greater than the outputs at such time. The water budget equation is used by hydrologists and other scientists to evaluate the net change of the volume of water in a defined area over time. Change in volume is the sum of the inputs minus the sum of the outputs as expressed by the following equation:

$$\Delta V = [P + Si + Gi + Ti] - [ET + So + Go + To]$$

Inputs

The four sources of water are precipitation (P), surface water inflow (Si), groundwater inflow (Gi), and

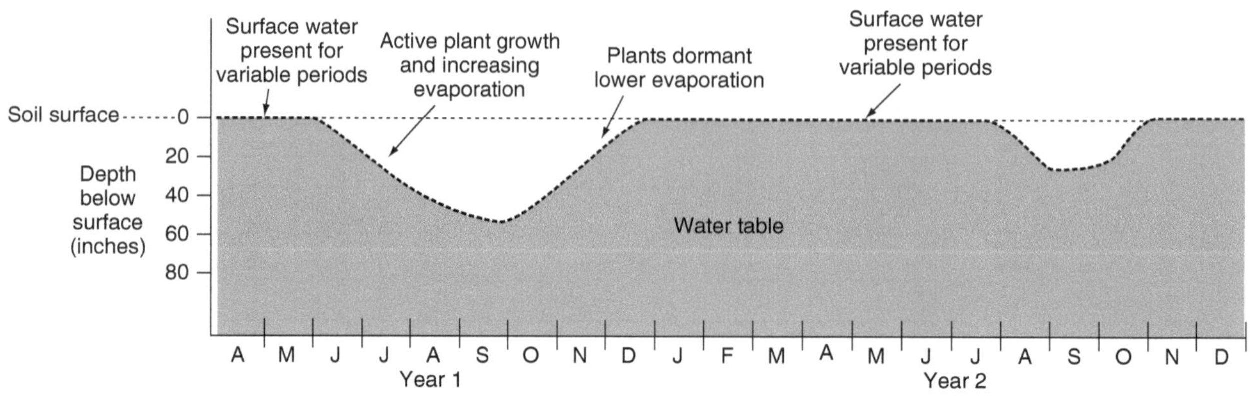

Figure 4 Seasonal and annual differences in water levels in a forested wetland in the northeastern United States. Note that Year 1 is a year of normal precipitation, while Year 2 represents a wet year, with corresponding changes in water table levels. Sources: Tiner RW, and Burke DG (1995) Wetlands of Maryland. National wetlands inventory cooperative publication. Hadley, MA: U.S. Fish and Wildlife Service, Northeast Region and Maryland Dept. of Natural Resources; based on data from Lyford WH (1964) Water table fluctuations in periodically wet soils of central New England, Harvard forest paper No. 8. Petersham, MA: Harvard University Forest.

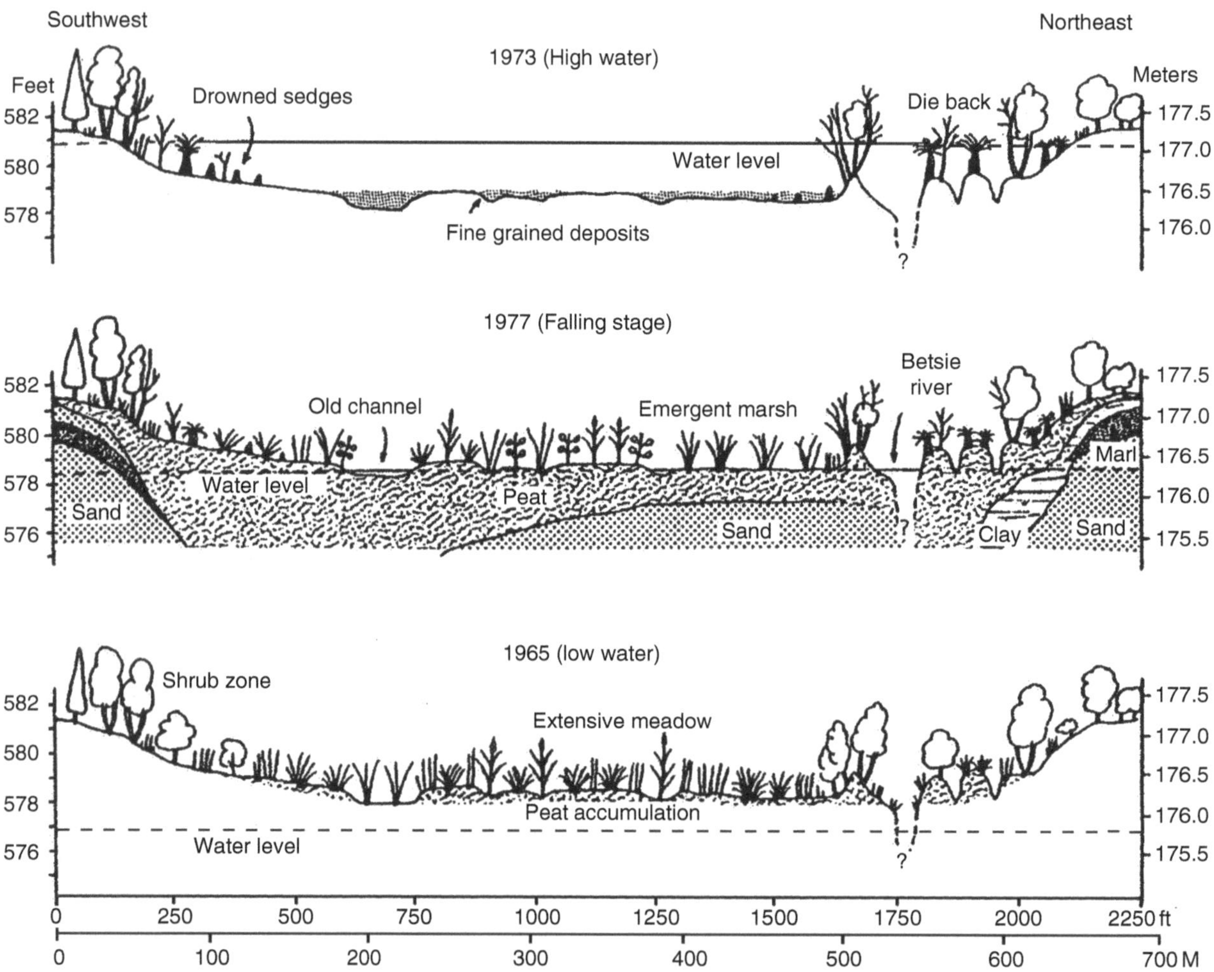

Figure 5 Vegetation dynamics in a Great Lakes coastal wetland in response to changing lake levels. Source: Herdendorf CE, Hartley SM, and Barnes MD (eds.) (1981) *Fish and wildlife resources of the Great Lakes coastal wetlands within the United States, vol. 1: overview*. Washington, DC: U.S. Fish and Wildlife Service, Biological Services Program, FWS/OBS-81/02-v1.

Figure 6 Components of the wetland water budget: P, precipitation; SWI, surface water inflow; GWI, groundwater inflow; ET, evapotranspiration; SWO, surface water outflow; GWO, groundwater outflow; and ΔS, change in storage. Sources: Carter V (1996) Wetland hydrology, water quality, and associated functions. In Fretwell JD, Williams JS, and Redman PJ (eds.) *National Water Summary on Wetland Resources*, Water-Supply Paper 2425, pp. 35–48. Reston, VA: U.S. Geological Survey.

incoming or flood tides (Ti). The contributions of these sources vary daily, seasonally, and yearly. Inflows are typically natural, but can be human induced by releases of water from dams and similar inputs.

Outputs

Water is lost through evapotranspiration (ET), surface water outflow or runoff (So), groundwater outflow or recharge (Go), and outgoing or ebb tides (To). Evapotranspiration is a combination of evaporation and plant transpiration. Evaporation increases with rising air temperature and exposure of the land or water surface to the sun, while plant transpiration is the natural uptake of water from the soil by plants and eventual loss to the atmosphere in the form of water vapor. Groundwater outflow could recharge ground water supplies or result from water withdrawals by humans.

Annual Water Budgets

Water budgets may be calculated for a wetland or a watershed and for different time periods. As mentioned, wetlands form in areas inundated or waterlogged for long periods of time; the wet season could be from winter to early spring (as it is in the eastern United States), summer to fall (as in Florida and the southwest United States), or in winter (in Mediterranean climates like that of California). During the wet season, the inputs would exceed the outputs, so that water is stored in the wetland. When evaluated over a longer period, however, outputs may exceed inputs, especially for wetlands that are nearly permanently wet. Some examples of annual water budgets are shown in **Table 2**.

Wetland Hydrology Indicators

In the absence of site-specific hydrological data (e.g., hydrographs or recorded data from water-level gages or observation wells), various features can be used to verify recent flooding or waterlogging. These features may be direct observations of water on the surface or near the soil surface, indirect evidence such as features left by recent flood or saturation events (**Table 3**), or inferred from soil properties (e.g., hydric soil indicators, presence of a hard pan, dense clay layer, or permafrost layer near the surface) or vegetation (e.g., presence of obligate hydrophytes or plants with certain morphological adaptations to excessive wetness like hypertrophied lenticels and water roots).

Table 2 Examples of annual water budgets for selected wetland in North America

Wetland (location)	*Inputs*			*Outputs*			*Net storage*
	P	*Si*	*Gi*	*ET*	*So*	*Go*	
Bog (Massachusetts)	100	–	–	70	8	8	14
Cypress River Swamp (Illinois)	23	70	7	22	71	6	1
Okefenokee Swamp (Georgia)	76	23	–	55	43	2	–1
Prairie Pothole Marshes (North Dakota)	48	52	–	78	–	22	0
Hidden Valley Marsh (Ontario)	11	53	36	12	35	50	3
Arctic Fen (Northwest Territories)	22	52	26	27	40	<1	>32
Experimental Marsh (Ohio)	3	97	–	2	72	26	0

A positive storage value indicates a net gain in water.

Sources

Carter V (1996) Wetland hydrology, water quality, and associated functions. In Fretwell JD, Williams JS, and Redman PJ (eds.) *National Water Summary on Wetland Resources, Water-Supply Paper 2425*, pp. 35–48. Reston, VA: U.S. Geological Survey.

Winter TC (1989) Hydrologic studies of wetlands in the northern prairie. In Van der Valk A (ed.) *Northern Prairie Wetlands*, pp. 16–54. Ames, IA: Iowa State University Press.

Zhang L and Mitsch WJ (2002) Water budgets of the two Olentangy River experimental wetlands in 2001. In Mitsch WJ and Zhang L (eds.) *Annual Report – Olentangy River Wetland Park*, pp. 23–28. Columbus, OH: Ohio State University.

Note that numbers reflect percent of inputs or outputs.

Table 3 Potential wetland hydrology indicators for North American wetlands

Direct Evidence
Visual observation of surface water
Visual observation of a water table within 30 cm (12 in.) of the surface of nonsandy soils
Visual observation of a water table within 15 cm (6 in.) of the surface of sandy soils
Visual observation of soil glistening, or shaking or squeezing pore water out of the soil within 30 cm in nonsandy soil or within 15 cm in a sandy soil
Positive reaction to a ferrous iron test with 30 cm of the surface in a nonsandy soil or within 15 cm of a sandy soil
Observed soil color change when exposed to air due to oxidation of ferrous iron
Sulfidic odor (rotten egg smell) from soil sample within 30 cm of the surface
Indirect Evidence of Flooding or Ponding
Water marks (e.g., blackish stains or silt lines)
Drift or wrack lines (piles of water-carried debris)
Sediment deposits
Algal crusts on or near the ground
Drainage patterns (e.g., braided streams, network of minor streams, scoured areas, scouring around roots, and living plants bent over or lying in the direction of water flow)
Water-stained leaves
Mud cracks or surface polygons
Live or dead remains of aquatic invertebrates
Presence of crayfish burrows
Presence of periphyton (aufwuchs) growing on plants
Aerial photo showing flooding or ponding
Indirect Evidence of Recent Soil Saturation
Oxidized rhizospheres within 30 cm of soil surface (e.g., iron oxide plaques on living roots or redox concentrations in soil surrounding roots)
Presence of muck or mucky mineral soil on surface
Presence of deep soil cracks in clayey soils (e.g., Vertic soils)
Presence of salt deposits on soil surface
Presence of redox features in soil horizon despite bioturbation (i.e., mixing of soil by animals like earthworms)
Presence of deep impressions in soil left by heavy objects (e.g., vehicles or livestock)
Observed water table between 30 and 60 cm (12–24 in.) deep during dry season or dry year

Note that all these indicators are not of equal stature in verifying that the site still has wetland hydrology.

Source

Noble CV, Martel DJ, and Wakely JS (2005) A national survey of potential wetland hydrology regional indicators. Vicksburg, MS: U.S. Army Corps of Engineers, Waterways Experiment Station, ERDC TN-WRAP-05-1.

Glossary

Adventitious roots – Roots formed above ground; in wetland and aquatic plants they are induced by prolonged inundation.

Aerenchyma – Air-filled tissue in plants (typically in wetland and aquatic plants) that facilitates air movement from aboveground plant parts to roots.

Aufwuchs – Plants and animals attached to plants, rocks, pilings, or other erect materials in water.

Biologic zero – A term used by soil scientists to refer to the temperature threshold that generally causes soil microbes to become relatively inactive, so that reducing conditions do not readily develop in saturated soils; defined by the soil temperature of 5 °C or 41 °F measured at a depth of 50 cm (20 in.) below the soil surface.

Flatwood – Low flat, forested landscape typical of coastal or glaciolacustrine plains; soils vary from poorly drained to well drained with slight changes in topography.

Flooding – In general terms, a condition where an area is inundated (covered by water); soil scientists tend to restrict the term to inundation resulting from overbank flooding of a river or stream.

Floodplain – Nearly level alluvial land subject to periodic inundation (overflow from river or stream).

Glaciolacustrine plain – Low flat landscape associated with a former glacial lake, the exposed bed of a former glacial lake.

Graminoids – Grasses (members of the Family Poaceae – true grasses) and grasslike plants (typically members of the families – Cyperaceae: sedges, bulrushes, spikerushes, beakrushes, and cotton-grasses and Juncaceae: rushes, plus other herbaceous plants with long narrow grasslike leaves).

Hydraulic conductivity – Measure of the ability of water to flow through the soil; low conductivity resists flow, while high conductivity favors flow.

Hydric soils – Soils formed under frequent and prolonged reducing conditions due to excessive wetness; soils typical of wetlands.

Hydrology – The scientific study of water properties, distribution, and circulation; also the dynamics of water presence and movement in a particular habitat (e.g., wetland hydrology, lake hydrology, or forest hydrology) or the study of these patterns.

Hydrophytic vegetation – Plants adapted for life in permanently to periodically flooded or waterlogged substrates; plants growing in water and wetlands.

Hypertrophied lenticels – Expanded, enlarged corky pores on woody plants typically the result of prolonged inundation.

Isolated wetlands – Better referred to as 'geographically isolated wetlands,' wetlands that are surrounded by upland (nonhydric soils), with no surface water connection to other wetlands or waters; these wetlands may be connected hydrologically to ground water.

Karst terrain – A landscape formed in a limestone region where the topography is shaped by dissolution of limestone (or dolomite, gypsum, or salt), characterized by caves, springs, seeps, sinkholes, and disappearing streams.

Lenticels – Corky roundish pores or lines on bark of woody plants that facilitate gas exchange between inner plant tissue and the atmosphere.

Meteoric water – Water precipitating from the atmosphere as rain, snow, sleet, hail, fog, or mist.

Morainal deposit – Unsorted rocky and soil material (till) carried by and deposited by glaciers, typically marking the extent of glacier advance (terminal moraine) or the sides of the glacier (lateral moraine).

Paludified landscape – Peat-dominated landscape in regions of high rainfall and low evapotranspiration (often cool, wet climates) formed by the process of paludification where peat mosses (Sphagnum spp.) grow over once dry land converting it to bog (peatland; blanket bogs).

Ponding – A term mostly used by soil scientists to describe inundation resulting from surface water runoff into a closed depression or water accumulating in a depression from precipitation or high ground water.

Porosity – The state of having pores or space filled with gases or liquids; also a measure of the volume of pores in a material relative to the total volume of the material; sandy soils have higher porosity (a higher ratio of pore space to a given volume) than clayey or other fine-grained soils.

Sink (or sump) – From a hydrological standpoint, an area lacking an outflow where water accumulates or is absorbed, including terminal basins associated with watersheds in some arid regions (e.g., the Great Basin of the southwestern United States).

Telluric water – Water from the earth, ground water.

Throughflow – A condition where water both enters and exits an area; water moves through the wetland, for example.

Vernal pool – A type of open-water wetland where water is present seasonally (spring in temperate regions); vernal pools may be imbedded in woodlands as in the eastern United States or in grasslands as in the western United States (e.g., California, Oregon, and Washington); in Mediterranean climates, these wetlands are typically inundated in winter.

Waterlogging – A condition where the substrate is saturated at or near the surface for extended periods; in wetlands, saturation is usually long enough to create anaerobic (low oxygen) and reducing soil conditions that affect plant growth.

Wetland – In general terms, a shallow-water ecosystem or at least, periodically wet ecosystem subject to frequent inundation or prolonged soil saturation (waterlogging) that is often characterized by hydrophytic vegetation, other aquatic organisms, and hydric soils/substrates; a variety of specific definitions have been created for a host of purposes (legal and scientific) including land use regulation, habitat protection, and natural resource inventories.

Wetland hydrology – The recurrent, sustained saturation of substrates at or near the surface by either surface or ground water sufficient to create conditions that support aquatic life including the growth of hydrophytic vegetation, and the formation of hydric soils or substrates; the dynamics of water presence and movement in wetlands.

See also: Atmospheric Water and Precipitation; Evapotranspiration; Ground Water; Hydrological Cycle and Water Budgets.

Further Reading

Carter V (1996) Wetland hydrology, water quality, and associated functions. In: Fretwell JD, Williams JS, and Redman PJ (eds.) *National Water Summary on Wetland Resources*, Water-supply paper 2425, pp. 35–48. Reston, Virginia, USA: U.S. Geological Survey. http://water.usgs.gov/nwsum/WSP2425/hydrology.html.

Fretwell JD, Williams JS, and Redman PJ (eds.) (1996) *National Water Summary on Wetland Resources*, water-supply paper 2425. Reston, VA: U.S. Geological Survey.

Gilman K (1994) *Hydrology and Wetland Conservation*. Chichester, England: John Wiley & Sons.

Ingram HAP (1983) Hydrology. In: Gore AJP (ed.) *Mires: Swamp, Bog, Fen, and Moor. Ecosystems of the World 4A, General Studies*, ch. 3, pp. 67–158. Amsterdam, The Netherlands: Elsevier Science.

Jackson CR (2007) Wetland hydrology. In: Batzer DP and Sharitz RR (eds.) *Ecology of Freshwater and Estuarine Wetlands*, ch. 3, pp. 43–81. Berkeley, CA: University of California Press.

Mitsch WJ and Gosselink JG (2000) *Wetlands*. New York, NY: John Wiley & Sons.

National Research Council, Committee on Characterization of Wetlands. (1995) *Wetlands: Characteristics and Boundaries*. Washington, DC: National Academy Press.

Price JS, Branfireun BA, Waddington JM, and Devito KJ (2005) Advances in Canadian wetland hydrology, 1999–2003. *Hydrological Processes* 19: 201–214.

Richardson JL, Arndt JL, and Montgomery JA (2001) Hydrology of wetland and related soils. In: Richardson JL and Vepraskas MJ (eds.) *Wetland Soils: Genesis, Hydrology, Landscapes, and Classification*, ch. 3, pp. 35–84. Boca Raton, FL: Lewis Publishers.

Stone AW and Lindley Stone AJ (1994) *Wetlands and Groundwater in the United States*. Concord, NH: The American Ground Water Trust.

Tiner RW (1999) *Wetland Indicators: A Guide to Wetland Identification, Delineation, Classification, and Mapping*. Boca Raton, FL: Lewis Publishers, CRC Press.

Tiner RW (2005) *In Search of Swampland: A Wetland Sourcebook and Field Guide*. New Brunswick, NJ: Rutgers University Press.

Williams TM (1998) Hydrology. In: Messina MG and Conner WH (eds.) *Southern Forested Wetlands: Ecology and Management*, pp. 103–122. Boca Raton, FL: Lewis Publishers.

Winter TC (1989) Hydrologic studies of wetlands in the northern prairie. In: Van der Valk A (ed.) *Northern Prairie Wetlands*, pp. 16–54. Ames, IA: Iowa State University Press.

Winter TC and Woo MK (1990) Hydrology of lakes and wetlands. In: Wolman MG and Riggs HC (eds.) *Surface Water Hydrology of North America*, vol. 1, pp. 159–187. Boulder, CO: Geological Society of America.

Relevant Websites

http://www.wcc.nrcs.usda.gov/wetdrain.
http://www.srs.fs.usda.gov/pubs/2083.
http://www.nap.usace.army.mil/cenap-op/regulatory/water_monitor_technote.pdf.
http://el.erdc.usace.army.mil/elpubs/pdf/tnwrap06-2.pdf.
http://el.erdc.usace.army.mil/elpubs/pdf/tnwrap05-1.pdf.
http://www.gret-perg.ulaval.ca/Price_et_al_HP18_2005.pdf.
http://www.lk.iwmi.org/ehdb/wetland/displayallreferences.asp.
http://www.info.usda.gov/CED/ftp/CED/EFH-Ch19.pdf.

HYDRODYNAMICS AND MIXING IN LAKES, RESERVOIRS, WETLANDS AND RIVERS

Contents

Biological-Physical Interactions
Density Stratification and Stability
The Surface Mixed Layer in Lakes and Reservoirs
Small-Scale Turbulence and Mixing: Energy Fluxes in Stratified Lakes
The Benthic Boundary Layer (in Rivers, Lakes, and Reservoirs)
Currents in Rivers
Currents in Stratified Water Bodies 1: Density-Driven Flows
Currents in Stratified Water Bodies 2: Internal Waves
Currents in Stratified Water Bodies 3: Effects of Rotation
Currents in the Upper Mixed Layer and in Unstratified Water Bodies
Flow in Wetlands and Macrophyte Beds
Flow Modification by Submerged Vegetation
Hydrodynamical Modeling

Biological-Physical Interactions

C S Reynolds, Centre of Ecology and Hydrology and Freshwater Biological Association, Cumbria, UK

Introduction

As formal branches of science, limnology and ecology are each around a century in age. Both disciplines feature prominently in the evolving understanding of inland waters, where they are invoked to explain observable phenomena and their role in shaping the abundance, structure, and variability of the biotic communities. It is also true that, from the pioneer studies onwards, much of the scientific investigation has been focused on the mutual relationships among the biota and their chemical environments, most especially with regard to the important nutrient elements. Physical factors were not overlooked altogether: it is well known how the specific heat of water, its curiously variable coefficient of thermal expansion, and its transparency each influence the stability and duration of seasonal thermal stratification or to the underwater distribution of macrophytic plants and photosynthetic algae. The benefit to aquatic organisms of the mechanical support contingent upon the high density of water is also generally understood. On the other hand, the mathematics of fluid motion in lakes proved to be less amenable to solution, so its impact on the evolutionary ecology of the pelagic biota – those of the open water, mostly living independently of shores or the bottom oozes – for a long time remained wholly intuitive. Although broad patterns of wind- and gravity-generated currents could be described and modeled, the smaller scales of water movement and the quantitative description of turbulence were, until the last twenty years or so, relatively intransigent to solutions relevant to the function, selection, and evolutionary ecology of aquatic biota, or even to such matters as the transport and dispersal of organisms, particles, and solutes. This chapter reviews briefly the physical properties of fluid motion at the macro and microscale in inland water bodies, seeking to establish the spatial and temporal scales that impinge directly and indirectly on the organisms that live there, as well as on the functional adaptations that enable them to do so.

Water Movements – The Concept of Scale

From the major oceanic circulations down to the (Brownian) behavior of the finest colloidal particles and the diffusion of solutes, water characteristically comprises molecules in motion. Curiously, the propensity of water molecules to polymerize into larger 'aquo complexes,' which is responsible for the relatively high density and viscosity of liquid water, makes them resistant to movement, so there is a constant battle between, on the one hand, the energy sources driving motion (the Earth's rotation, gravitational flow, convectional displacement due to thermal expansion and contraction and, especially, the work of wind forcing applied at the water surface of lakes; and, on the other, the resistance of internal viscous forces. Thus, the energy of external forcing is dissipated through a cascade of *turbulent eddies* of diminishing size and velocity, to the point that it is overwhelmed at the molecular level.

Turbulence in the upper water column affects pelagic organisms through several mechanisms operating at a range of temporal and spatial scales. Mixing influences the distribution of regeneration and recycling of dissolved nutrients, the dispersion of zooplankton, the location of rewarding feeding grounds for fish, and the resuspension of detrital particles, including biotic propagules. In relation to the depth of light penetration, the penetration of mixing may constrain the exposure of entrained photosynthetic algae and bacteria to light and to regulate their primary production. At the microscale, turbulence is relevant to individual organisms, conditioning their suspension in the mixed layer, the interaction with their own intrinsic motility and the fluxes of dissolved nutrients and gases to their cells. It is to the latter influences that this article is particularly addressed.

Small-Scale Turbulence

The measurement of turbulence or its convenient components, such as the shear or friction velocity (symbolised as u_*), is not the concern of the present chapter. Turbulence in the upper water column is induced by wind, heat loss, and wave breaking. When wind is the predominant cause of turbulence, the turbulent velocity can be approximated from the shear stress at the air–water interface; if convective heat loss is the dominant driver, then a similar turbulent velocity scale, w_*, comes from the velocity of the resultant convectional plumes; if several processes are operating simultaneously, all are included in the calculation, and the resultant turbulent velocity scale is sometimes called the turbulent intensity u. Deeper in the water column, turbulence is often caused by breaking internal waves. Technically, the turbulent intensity is the root mean square velocity of the velocity fluctuations in a turbulent flow field. This measurement has only recently been applied in limnological studies. More commonly, turbulence is obtained from microstructure profiling as the rate of dissipation of the turbulent kinetic energy ε with the assumption that turbulence production and dissipation are in balance. When turbulence is induced by wind in the surface layer, the turbulent velocity scale is roughly proportional to the square root of the quotient of the applied force per unit area (τ, in kg $m^{-1}\,s^{-2}$) and the density of the water (ρ_w, in kg m^{-3}). Then

$$u_* = \left(\tau \rho_w^{-1}\right)^{1/2} \qquad [1]$$

The units are in $m\,s^{-1}$. The rate of dissipation of turbulent kinetic through the spectrum of eddy sizes (ε) is correlated to the dimensions of the largest eddies in the turbulence field (l_e) and their velocities (u) through eqn. [2]:

$$\epsilon = (u)^3 l_e^{-1} \qquad [2]$$

The units are thus $m^2\,s^{-3}$. The energy is lost, as heat, through progressively smaller eddies. If the dissipation rate is measured, and the root mean square sizes of turbulent eddies with microstructure profiling, then the turbulent velocity scales are obtained. Eventually, the spectrum collapses at the point where the driving energy is finally overcome by viscosity: the size of the smallest eddy (l_m) is predicted by:

$$l_m = \left[(\eta/\rho_w)^3 \epsilon\right]^{1/4} \qquad [3]$$

where η is the absolute viscosity of the water (units: kg $m^{-1}\,s^{-1}$). These various equations have been used to calculate that the turbulence generated in the open waters of the unstratified Bodensee (Lake of Constance) by winds of 5–20 $m\,s^{-1}$ drive a spectrum of eddies penetrating to depths of between 45–180 m, dissipating at rates (ε) of between 1.4×10^{-8} to $2.2 \times 10^{-7}\,m^2\,s^{-3}$ and culminating in eddy sizes (l_m) of between 2.9 and 1.5 mm. In stratified lakes, where the density gradient acts as a barrier to penetration by weak eddies, and in shallow lakes, where the water column is unable to accommodate the unrestricted propagation of turbulence, the same driving energy must be dissipated within a smaller spatial extent and, hence, at a faster rate and to a smaller spatial limit. The energy of a 20 $m\,s^{-1}$ wind applied to Lough Neagh (mean depth < 9 m), is calculated to be dissipated at $\sim 4.3 \times 10^{-6}\,m^2\,s^{-3}$, i.e., at nearly twenty times the rate in Bodensee, under the same wind forcing, and culminating in an eddy size of ~0.7 mm. In the most aggressively mixed estuaries and fluvial rapids, ε may approach $5.5 \times 10^{-4}\,m^2\,s^{-3}$, with eddies as small as 0.2 mm across. The highest values observed in lakes are near the air–water interface and are of order $10^{-5}\,m^2\,s^{-3}$ but typical high values are of order $10^{-6}\,m^2\,s^{-3}$.

The capacity of turbulent motion to entrain particles, including living organisms, depends broadly upon the magnitude of the relation of the turbulent velocity scale to the intrinsic rates of gravitational settling of the particles in water (w_s, in $m\,s^{-1}$). Whereas a stone always drops almost unimpeded through water, particles of clay (<5 μm in diameter but of similar density) persist in suspension for long periods and remain dispersed by the aggressive mixing evident in tidal estuaries. The sinking velocities of planktic diatoms such as *Asterionella* ($w_s \leq 10\,\mu m\,s^{-1}$), the cells of which form stellate colonies measuring 150–200 μm across, are so trivial compared to the shear velocity of turbulence generated at the lake surface by a wind of (say) $8\,m\,s^{-1}$ ($\sim 1 \times 10^{-2}\,m\,s^{-1}$) that their entrainment is near complete ($u/w_s \sim 10^3$); the algae go more or less where the flow takes them. In metalimnia, or in the boundary layer adjacent to the lake bottom, wherein turbulent velocities are abruptly depressed to below $10^{-3}\,m\,s^{-1}$, the same diatoms are readily disentrained, sinking almost unimpeded. For smaller and less dense algae and bacteria, the threshold of disentrainment may be an order of magnitude or two smaller. Equally, the rather larger cells of some dinoflagellates, such as *Ceratium*, and the colonies of the Cyanobacterium, *Microcystis*, that aspire to rising velocities of $\leq 300\,\mu m\,s^{-1}$, have a correspondingly enhanced capacity to escape from weakly turbulent flows. The generally recognised criterion for the entrainment of particles is that $\psi = w_s/(15\ u) \leq 1$; based on the proportionality of the velocities of intrinsic settling and shear, however, entrainment can never be wholly complete. There is a persistent tendency for particles to be disentrained gradually into adjacent structures.

Physical Influences on the Size and Structure of Organisms

The scaling of turbulence, observations as to orders of magnitude of turbulence, and understandings of boundary layer fluid dynamics provide insight into a number of biological problems. For instance, in the context of evolutionary adaptation to physical characteristics of the aquatic medium, it is plain that the foliage of submerged macrophytic plants and algae require tensile strength rather than rigidity; live biomass is supported by the density of the water but turbulent stresses are accommodated by their typically, flexible, flat, much-dissected or filamentous leaves. Self-evidently, macroinvertebrates and fish must be adequately robust to maintain their integrity in the face of turbulence, some of which is, of course, generated by their own movements. In contrast, the often delicate and unstrengthened substance of planktic protistans, algae and cyanobacteria, most of which are smaller or much smaller than 1 mm across, is normally protected within the viscous range of the eddy spectrum. While they have no power to prevent their entrainment by turbulence, they rarely experience its physical stresses. When larger algae are exposed to intense turbulence, however, their sizes relative to the smallest eddies expose them to shear; mostly they are insufficiently robust to resist fragmentation and disruption.

Generally, the adaptive problem for small organisms is rather to cope with viscosity – especially in the context of motility and the foraging requirement of phagotrophic protists to encounter food. The analogy has been made of a person collecting bananas suspended in a swimming pool filled with a fluid of the consistency of molasses or setting concrete. Reciprocating swimming strokes are of less use than are boring or grapple lines, favouring slender flagella over flat paddles. In contrast, mesozooplanktic crustaceans are of a size (in the range 0.2 to 2 mm) that coincides with the transitional scale, in which swimming and the generation of shear currents are feasible; these currents may entrain small particles towards the feeding apparatus where it may be collected as food. In the filter-feeding cladocerans (including *Daphnia*), the flattened, paddle-like abdominal limbs (phyllopods) beat rhythmically within a feeding chamber formed by the abdomen and the carapace: an inhalant current is thus drawn into a chamber, where the particles are strained out by bristle-like setae on the phyllopods, and carried in viscous flow to the mouth. To pursue the analogy, another person immersed in the same fluid might gain more satisfaction through collecting kidney beans from a vessel closer to the size of a bath tub. It is only by being big enough and strong enough to escape the drag of viscosity and to be able to generate turbulent currents that the rate of particle encounter can be significantly increased.

Open turbulence is also constrained in the vicinity of solid surfaces, for instance, at the bottom and sides bounding the flow field, as well as the bodies of aquatic animals. Boundary layers are regions of reduced flow adjacent to the solid surfaces, characterized by a gradient of velocities perpendicular to the surface. Depending upon current velocity in the main direction of flow, boundary gradients may be compressed to fractions of a millimeter or extend over several meters. Their presence influences the structural and behavioral adaptations of organisms in a number of ways. A well-known example is the dorsoventral flattening of Ecdyonurid mayfly larvae that can apply themselves to the surfaces of stones in fast flowing streams. By presenting a protrusion of barely 2 mm above the stone surface, animals are able to exploit the boundary layer as a physical refuge from the flow, whilst they graze on the diatoms and other algae attached to the surface. These algae are of a small size (generally <0.3 mm) so they also experience minimal shear stress; if they are sufficiently dense, their presence augments the boundary effect by extending it further into the water. It should be noted that boundary layers are not necessarily free of turbulence: even weak eddies potentially increase the delivery of food particles to the vicinity of benthic filter feeders. In the case of stream-dwelling larvae of the black fly, *Simulium*, the firm attachment and attitude of the abdomen ensure the optimum presentation of the filtering mouth parts to the flow.

Subject to other conditions, broader boundary layers, variously adjacent to stony or silty river beds or within beds of submerged macrophytic plants, offer refuges of weakened turbulence exploitable by algae, fungi and macroinvertebrates that together comprise the *Aufwuchs* communities. There is no exact equivalent word in English – but the substratum-specific subdivisions of epilithic, epipelic, and epiphytic associations together convey the analogous concept; the term *periphyton* is also close but it should refer strictly to the microhabitats and species associated with aquatic macrophytes. Periphyton and the plants in shallow water with which it is associated together constitute some of the most productive habitats in the biosphere, in some instances turning over more than 1.5 kg C m^{-2} annually.

Turbulent Extent and the Pelagic Habitat

Just as open-water turbulence dominates the environment of the planktic organisms in the near-surface layer, so the vertical extent of the entraining mixed layer has far-reaching consequences for the photosynthetic activity of phytoplankton. The vertical attenuation of light penetrating the surface, even in clear water, is such that significant net photosynthetic carbon fixation is severely restricted at depths greater than 60 m. At low angles of incidence and under cloud cover, this photic depth is further reduced. However, coloration of the water (due, for instance, to humic substances in solution) and the suspension of clay or other fine particles, as well as the presence of significant concentrations of phytoplankton, may restrict the photic depth to the order of a few meters or even centimeters only. In these instances, turbulent entrainment may well carry photosynthetic organisms to depths where net photosynthesis cannot be sustained. Under steady wind forcing, the probability is that the same chlorophyll-containing organisms will soon be carried back to the depth range wherein production may be resumed. Calculation of the proportionality of time during which net production is actually possible is complex, not least because, within limits, the photosynthetic apparatus itself is able to compensate to the average light exposure ('light adaptation'). However, the product of the day length and the ratio of the volumes of the photic and the wind-mixed layers (in large lakes, the ratio of their depths, h_p/h_m is convenient) is adequately illustrative. Ignoring, for the present, the variables determining the evaluation of h_p and focusing on the variability in wind forcing on the water columns of deep, non-stratified lakes, there is an approximate proportionality among wind velocity, u_* and h_m, winds of 5–20 m s^{-1} being theoretically capable of driving turbulence to depths of about 44–177 m. For a mixed layer depth of 44 m and a dissipation rate of 10^{-7} m^2 s^{-3}, as would be found on a windy, cloudy day, it would require, probabilistically, an average of 42 minutes for an alga entrained in open turbulence to be transported through the entire mixed layer. In mixed columns truncated by pycnoclines or the bottom of the lake, the mixing time is proportional to the (constrained) depth. In mixed layers of the order of one meter in depth, mixing times are in the order of one minute.

The interaction between h_m and h_p can be used to approximate the instantaneous phytoplankton-carrying capacity of the mixed layer to support a productive phytoplankton. Whereas an otherwise clear mixed layer of 10 m in depth might support a chlorophyll concentration $\leq$150 mg m^{-3}, deepening it to 40 m reduces the maximum supportable concentration to <20 mg m^{-3}; mixing to 80 m takes the capacity to under 1 mg m^{-3}. It follows that, in deep, temperate lakes like Bodensee, phytoplankton growth and net population increase in the winter months is weak, pending longer days, more surface heating and the onset of thermal stratification. Equally, as heat is lost in the autumnal months, so the stability of stratification and the resistance to wind mixing weaken, the mixed layer increases in depth and the downwelling light energy is increasingly diluted.

Mixed Layer Depth and the Maintenance of Non-motile Plankton

Weak winds and a shrinking mixed layer depth are not, however, an unmitigated benefit to phytoplankton. For non-motile plankton, as well as fine particles, the thickness of the entraining mixed layer has an important relationship to the intrinsic rates of their settlement (w_s) and, thus, to the probability of their entry into the lower, less turbulent basal layer. Depletion from suspension is a first-order process, analogous to dilution; the rate of the decline in the suspended population is described by an exponent ($-r_S$), equivalent to $-(w_s/h_m)$. Plainly, the greater is the particle-specific w_s, the greater is the dependence upon a deep mixed layer for its maintenance in suspension and the greater is its sensitivity to density stratification and the consequent contraction of h_m. It may seem paradoxical that planktic diatoms such as

Asterionella (w_s typically in the range 0.3–1.0 m day^{-1}) require a shrinking mixed layer to promote their growth but then have to grow fast enough to balance sinking loss rates from a shrinking mixed layer! A pycnocline reaching to within 3 m of the surface will impose a sinking loss rate of $r_S \geq -0.1$ day^{-1}, which must be countered by a comparable rate of cell recruitment (r') if a net collapse of the standing population is to be avoided. This is well within the capacity of the maximum performance of *Asterionella* but low temperature, shortage of nutrients and other losses (flow displacement, grazing by zooplankton) often make this hard to attain. Nevertheless, the onset of density stratification, especially in smaller lakes, is generally the principal correlative of the culmination of diatom abundance (as, for instance, in the so-called 'spring bloom') and settlement from suspension contributes prominently to their subsequent demise.

Not all plankton is heavier than water; diminution of the surface mixed layer promotes different strategies somewhat dependent upon the trophic status of the lake. In oligotrophic lakes, small, near-neutrally buoyant algae may proferate. In mesotrophic and eutrophic lakes, buoyant plankton may predominate. The most extreme case is also well known to freshwater ecologists: under calm conditions, physical entrainment may be insufficient to prevent suitably buoyant Cyanobacteria from accumulating at the water surface, where, if present in the water in significant amounts, they may constitute a striking surface scum. The exaggerated view of their abundance and the rapidity with which such 'water blooms' form have given them a notoriety among water users; that many of these are now known also to be toxic to livestock and to human consumers has increased the impetus to devise measures to better manage the frequency of their occurrence.

Physical Influences on Nutrient Fluxes

A major problem for planktic photoautotroph is to accumulate from the suspending water a sufficient stock of the essential nutrients to be able not just to maintain itself but to accumulate the material requirements of the next generation (effectively harvesting its own mass again of carbon, nitrogen, phosphorus, etc. within each generation). There are at least twenty elements required in the production of living cells, several of which (C, N, P) are typically regarded as being in short supply relative to the needs and, hence, likely to limit the capacity to support biomass. For autotrophic phytoplankton, these elements are obtained essentially from relatively simple compounds and ions, dissolved in the water. In most lakes, the main proximal sources of these elements (even the gaseous ones, like carbon dioxide) are dissolved in rainfall and, especially, inflowing streams draining the catchment. There is thus a tendency for supplies to fluctuate seasonally, and often quite uncoupled from the demands of production. Moreover, while the growth of phytoplankton removes nutrients from the water in the surface circulation, they are unlikely to be released again only after cells have died, including through consumption by animals and through progressive settlement of intact cells beyond the photic zone and the mixed layer. Other things being equal, the net flux of fecal pellets, detritus and disentrained phytoplankton is to greater depth. At the basin scale, there is an inevitable tendency for the resources of the lake to gravitate towards the bottom and for the surface waters to be depleted. Density stratification of the water masses only amplifies this segregation.

As a physical counter to vertical segregation, the mechanisms for transporting nutrient-rich water from depth may assume considerable biological importance. Increased wind action may expand the circulation of the surface mixed layer, depressing the pycnocline downwind, raising the shear stress of the return current and increasing the intensity of internal wave generation. Erosion and entrainment of deep, relatively nutrient-rich water into a deepened surface mixed layer can have a significant effect in refreshing nutrient availability and stimulating phytoplankton growth. Such macroscale events may pass, with the system returning to something approaching the earlier physical state. Alternatively, they may persist and, aided by surface cooling and seasonally more frequent wind episodes, lead eventually to the breakdown of stratification and substantial vertical mixing throughout most of the water column.

Nutrient Fluxes at the Microscale

Biological-physical interactions influence the nutrient relationships at the microscale of individual phytoplankton cells. The resources present within healthy living cells are typically about one million times more concentrated than they are in the medium. This formidable gradient cannot be overcome by passive movement of molecules into the cell – they would move rapidly in the opposite direction – an elaborate apparatus and system of reactivity captures, retains and transports specific molecules from the exterior of the cell to the intracellular sites of deployment and assimilation. The operation involves kinases and the consumption of ATP; the cell expends significant energy in order to assimilate its essential resources.

Suspended in the relative vastness of the water mass, however, the phytoplankton cell can have no such influence on the external supply of nutrients or the frequency of the molecular encounters that it requires to satisfy its uptake demands. Within its immediate viscous environment, the cell is dependent upon diffusion of target molecules through the medium, in place of those taken up across the cell membrane. Delivery is governed by Fick's laws: the number of moles (n) of a solute that will diffuse across an area (a) per unit time, t, is a function of the gradient of solute concentration, C (i.e., dC/dx), and the coefficient of molecular diffusion of the substance (m). Then, introducing appropriate units, uptake per unit time is:

$$n = a\ m(dC/dx)\ \text{mol m}^{-2}\ \text{s}^{-1} \qquad [4]$$

Taking a small, spherical alga like *Chlorella* (diameter, $d \sim 4 \times 10^{-6}$ m; volume $\sim 33 \times 10^{-18}$ m^3; surface area $\sim 50 \times 10^{-12}$ m^2) and given that (i), for a small-sized molecule such as that of carbon dioxide, $m \sim 10^{-9}$ m^2 s^{-1}, (ii), the thickness of the water layer adjacent from which CO_2 molecules can be sequestered is equal to the cell radius, and (iii) that the carbon dioxide concentration in the water beyond is at air equilibrium (11 μmol l^{-1}, or 11×10^{-3} mol m^{-3}), eqn. [4] may be solved to show diffusion should sustain a rate of acquisition by the cell equivalent to 275×10^{-18} mol s^{-1}. Given a cell carbon content of 0.63×10^{-12} mol carbon and that the doubling requirement is the uptake of a further 0.63×10^{-12} mol carbon, it can be calculated that eqn. [4] reveals a theoretical capability of fulfilling the demand within ~2300 s, or just over 38 min. This is rather less than the generation time (≥9 h at 20 °C), so there is a considerable scope for CO_2 depletion before it impinges on the growth rate of the cell.

For comparison, a turbulent velocity of 10^{-2} m s^{-1} would deliver CO_2 molecules at a the same concentration of 11×10^{-3} mol m^{-3} through an area equal to the area projected by the same *Chlorella* cell (~12.6×10^{-12} m^2) at a rate of 1386×10^{-18} mol s^{-1}. The capacity of open turbulence to fulfill a minimum frequency of encounter with nutrient molecules becomes especially important at very low resource concentrations. Taking its requirement for phosphorus instead of carbon (say, one hundredth the number of carbon atoms), the *Chlorella* mother cell has to accumulate some 0.006×10^{-12} mol if the cell is to sustain adequately the next doubling of biomass. Like most freshwater phytoplankters, *Chlorella* has a high capacity for the uptake of phosphate (maximum $\sim 13.5 \times 10^{-18}$ mol cell^{-1} s^{-1}), with a sufficient affinity for phosphate molecules for the uptake to be half-saturated by an external concentration of soluble, reactive phosphorus of about 0.7×10^{-3} mol P m^{-3} (about 20 μg P l^{-1}). Given these conditions, the requirement could be met in less than 15 min). On the other hand, the external concentration required to sustain the maximum rate of growth (at 20 °C, 1.84 day^{-1}) may be sequestered from an external concentration of just 6.3×10^{-6} mol P m^{-3}, about 0.2 μg P l^{-1}, before the rate of growth may be supposed to become phosphorus-limited.

Ultimately, the rate of renewal of the local water and the delivery of fresh nutrients to the vicinity of the algal cell affects the capacity of molecular diffusivity to meet growth demands. Being generally smaller than the turbulent eddies, most phytoplankton cells are eventually reliant upon movement of themselves and their viscous packets of water relative to the flow field to realize the environmental capacity to support their growth. The effect of advective transport of a given flow field, relative to diffusion, on a phytoplankton cell can be calculated from the ratio of the respective dimensionless Péclet numbers (Pe) of particles phytoplankton cells, of diameter d, either sinking passively in turbulence-free water or transported advectively through the water.

$$Pe = w_s\ d\ D^{-1} \qquad [5]$$

where D is the appropriate diffusion coefficient in water. In the absence of turbulence, D is substituted for m. The calculation with respect to advective transport is difficult but solutions have been presented by Riebesell and Wolf-Gladrow (see Further Reading). Their calculations showed Pe values for smaller phytoplankton generally fall within the range 0.1–10, the larger values applying to larger algae moving more rapidly through the water. The Sherwood Number (Sh) expresses the ratio of the fluxes of nutrients arriving at the cell surface in the presence of motion with those of diffusion. For the small cells embedded deep within the eddy spectrum ($Pe < 1$), Sh is always close to 1, signifying that any effect of motion is marginal. For larger organisms ($1 < Pe < 100$), the non-linear dependence of the nutrient supply on advection is approximated by

$$Sh = 1/2 + 1/2(1 + 2Pe)^{1/3} \qquad [6]$$

The effect of turbulence on the nutrient supply to cells of varying diameters (d) is also scaled to the Péclet numbers, calculated from the turbulent shear rate dissipation rate, as

$$Pe = (d/2)^2[\epsilon/(\eta/\rho_w)]^{1/2} m^{-1} \qquad [7]$$

Then, in the range $0.01 \le Pe \le 100$, the Sherwood scale conforms to:

$$Sh = 1.014 + 0.15Pe^{1/2} \qquad [8]$$

These derivations are interesting in the context of selection of evolutionary adaptations. It is at once apparent that the absolutely greater nutrient requirements of larger cells are more difficult to glean from low concentrations of specific nutrients in the medium, without the intervention of significant levels of turbulence. The further deduction is made that this dependence leaves larger phytoplankters less likely than smaller ones to be able to fulfill all their resource requirements in unit time, except under strongly turbulent conditions. With a less prevalent constraint on the development of picoplanktic and nanoplanktic than on the growth of large algae, their frequently-observed dominance over microplankton in chronically nutrient-deficient oligotrophic systems finds a compelling physical explanation. There are other alternative, probably additive, reasons for the dominance of picophytoplankton, invoking the inability of most filter-feeding mesozooplankters to harvest them adequately from low concentrations. Interestingly, it generally requires the intervention of a macroscale physical event, such as storm deepening, to raise nutrient levels for a period long enough for larger algae to flourish. This principle, that deep turbulent embedding and passive delivery of nutrients provides a superior strategy for the survival of photoautotrophs in an extensive, chronically nutrient-deficient medium, contrasts with what happens in the face of nutrient depletion in the epilimnion of a seasonally-stratified eutrophic lake. Here, the selective advantage passes to large-celled or large coenobial, motile, phytoplankton that are the most readily disentrained from weak turbulence and are capable of undertaking vertical migrations to scavenge the residual nutrient resources of the water column.

This regulation by nutrient fluxes of the size selection of planktic organisms is an especially satisfying instance of demonstrable physical–biological interaction in open waters. It is clear that the interactions between fundamental organismic processes and the physical motion of the fluid in which they may be suspended do not stop with transport and entrainment. Physical processes play a large part in sustaining nutrient fluxes in chemically rarefied pelagic environments and in underpinning their relative exploitability as habitats of autotrophic microorganisms.

See also: Currents in Stratified Water Bodies 1: Density-Driven Flows; Currents in Stratified Water Bodies 2: Internal Waves; Currents in Stratified Water Bodies 3: Effects of Rotation; Currents in the Upper Mixed Layer and in Unstratified Water Bodies; Fluvial Transport of Suspended Solids; Physical Properties of Water; Small-Scale Turbulence and Mixing: Energy Fluxes in Stratified Lakes; The Surface Mixed Layer in Lakes and Reservoirs.

Further Reading

Huisman J, van Oostveen P, and Weissing FJ (1999) Critical depth and critical turbulence: two different mechanisms for the development of phytoplankton blooms. *Limnology and Oceanography* 44: 1781–1787.

Hudson J, Schindler DW, and Taylor W (2000) Phosphate concentrations in lakes. *Nature* 406: 504–506.

Imberger J (1985) Thermal characteristics of standing waters: An illustration of dynamic processes. *Hydrobiologia* 125: 7–29.

Karp-Boss L, Boss E, and Jumars PA (1996) Nutrient fluxes to planktonic osmotrophs in the presence of fluid motion. *Oceanography and Marine Biology* 34: 71–107.

MacIntyre S (1993) Vertical mixing in a shallow eutrophic lake – Possible consequences for the light climate of phytoplankton. *Limnology and Oceanography* 38: 798–817.

Mann KH and Lazier JRN (1991) *Dynamics of Marine Ecosystems.* Oxford, UK: Blackwell.

O'Brien KR, Ivey GN, Hamilton DP, *et al.* (2003) Simple mixing criteria for the growth of negatively buoyant phytoplankton. *Limnology and Oceanography* 48: 1326–1337.

Reynolds CS (1998) Plants in motion: Physical–biological interaction in the plankton. *Coastal and Estuarine Studies* 54: 535–560.

Reynolds CS (2006) *Ecology of Phytoplankton.* Cambridge, UK: Cambridge University Press.

Riebesell U and Wolf-Gladrow DA (2002) Supply and uptake of inorganic nutrients. In: Wiliams J.leB, Thomas DN, and Reynolds CS (eds.) *Phytoplankton Productivity,* pp. 109–140. Oxford, UK: Blackwell Science.

Rothschild BJ and Osborn TR (1988) Small-scale turbulence and plankton contact rates. *Journal of Plankton Research* 10: 465–474.

Vogel S (1994) *Life in Mixing Fluids,* 2nd edn. Princeton, NJ: Princeton University Press.

Density Stratification and Stability

B Boehrer and M Schultze, UFZ – Helmholtz Centre for Environmental Research, Magdeburg, Germany

Introduction

In most lakes, water properties change from the surface to greater depth, i.e., these lakes show a vertical stratification of their water masses at least for some extended time periods. Heat exchange with the atmosphere and the forming of gradients of dissolved substances controls internal waves and the vertical exchange of water within the lakes. This has decisive impact on the evolution of water quality and, as a consequence, on the community of organisms living in the lake. This article deals with processes contributing to the stratification of lakes and the forming of layers. The most common numerical approaches for the quantitative evaluation of stratification-relevant physical quantities, e.g., electrical conductivity, are included. The final section lists quantities for stability of density stratification and what conclusion can be drawn from them.

Circulation Patterns

Surface temperatures of lakes show a pronounced temperature cycle over the year (**Figure 1**), in most latitudes. This is a consequence of heat exchange with the atmosphere and the seasonal variation of meteorological parameters, such as incoming solar radiation. The temperatures in the deep water follow the surface temperatures only for the time when the lake is homothermal, as in our example Lake Goitsche, Germany (**Figure 1**), during winter, from November until April.

Throughout summer, temperatures vary from the surface to the lake bed, and the lake remains stratified. Warmer and less dense water floats on top of colder, denser water (**Figure 1**). Thus, Lake Goitsche is called stably stratified, as overturning water parcels would require energy. On the contrary, during winter, no density differences obstruct the vertical transport. These seasons are commonly referred to as the stagnation period and the circulation period. Lakes that experience a complete overturn during the year are called holomictic.

During the circulation period, dissolved substances, such as oxygen or nutrients, get distributed over the entire water body (**Figure 2**). Hence, the circulation pattern is a decisive factor for the evolution of water quality and the biocenosis of the lake. In conclusion, the commonly used classification of lakes is according to their circulation patterns.

- *Holomictic lakes* overturn and homogenize at least once a year.
- In *meromictic lakes*, the deep recirculation does not reach the deepest point of the lake. As a consequence, a chemically different layer of bottom water is formed, the monimolimnion (see below), and remains there for at least 1 year.
- *Amictic lakes* do not experience a deep recirculation. Usually permanently ice-covered lakes are included in this class. Lakes, however, can circulate underneath an ice sheet by external forcing, such as solar radiation that penetrates to the lake bed and geothermal heat flux, or salinity gradients created when ice is forming on a salt lake.
- *Lakes with episodic partial deep water renewal* do not experience a complete overturn. The deep water however is partially replaced in episodic events.

Holomictic lakes are subdivided into classes indicating the frequency of complete overturn.

- *Polymictic* lakes are not deep enough to support a continuous stratification period throughout summer. The entire lake is mixed by sporadic strong wind events over the year or even on a daily basis in response to strong diurnal temperature variation.
- *Dimictic* lakes are handled as the prototype of lakes in moderate to cold climates. A closer look at the lakes, however, reveals that in most cases an ice cover or a great maximum depth is required to guarantee a stratification period during the cold season (see **Figure 2**). Between ice cover and summer stratification, the lake can be circulated completely in the vertical, the easiest when surface temperatures traverse the temperature of maximum density at 4 °C.
- *Monomictic* lakes possess one circulation period in addition to the stratification period. Many lakes in the temperate climate zone belong into this class, if they do not develop an ice cover during winter. Sometimes such lakes are also referred to as warm monomictic to distinguish them from cold monomictic lakes, which show an ice cover for most of the year and circulate during the short period without ice.
- *Oligomictic* lakes circulate less frequently than once a year, normally at irregular intervals, triggered by extreme weather conditions such as unusually cold winters for the respective location.

As a consequence of the natural variability of the weather conditions between years, the circulation patterns of the lakes also vary. A usually monomictic lake, for example, can show a dimictic circulation pattern when it freezes in an unusually cold winter. As another example, late during the twentieth century, Mono Lake turned meromictic for intermittent periods of 5 or 7 years, respectively, because of inflowing fresh water, but in other years showed a holomictic circulation.

Density Differences and Formation of Layers

Temperature Stratification

Although the surface water is exposed to solar radiation and thermal contact with the atmosphere, the deeper layers are shielded from major sources of heat. Diffusive heat transport on a molecular level is very slow and requires a month for the transport of heat over a vertical distance of 1 m. A much more efficient heat transport can be accomplished by turbulent transport. The energy for the turbulence is mainly supplied by wind stress at the lake surface and transferred via instabilities through friction at the side walls and internal current shear.

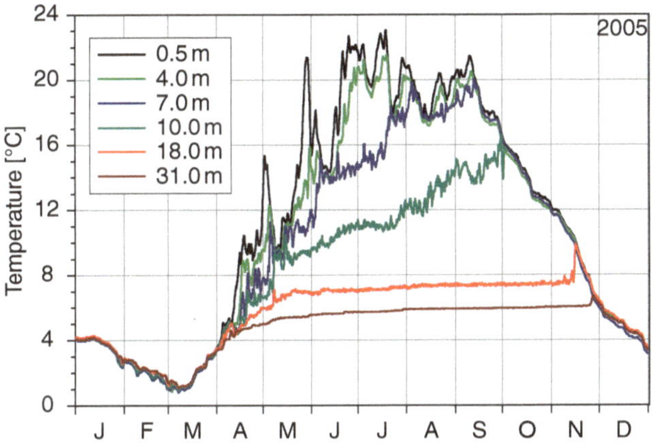

Figure 1 Temperatures (24 h mean) on several depths in Lake Goitsche near Bitterfeld, Germany during the year 2005. Reproduced from Boehrer B and Schultze M (2008) Stratification of lakes. *Reviews in Geophysics*, 46, RG2005, doi:10.1029/2006RG000210, with permission from American Geophysical Union.

Heating a lake over 4 °C at the surface results in a stable stratification. As a consequence, transport of heat to greater depths requires energy. The limited budget of kinetic energy available for mixing limits the depth to which a certain amount of heat can be forwarded over the stratification period. In sufficiently deep lakes, the thermal stratification holds until cooler autumn and winter temperatures permit a deeper circulation. The warm surface water layer is called *epilimnion*, while the colder water layer beneath, which has not been mixed into the epilimnion is called *hypolimnion*. A sharp temperature gradient (*thermocline*) separates both layers (**Figure 3**).

Epilimnion and atmosphere are in thermal contact and exchange volatile substances with each other. In addition, the epilimnion is recirculated by wind events or periods of lower temperatures during the stratification period. During those periods, dissolved substances are distributed within the epilimnion. On the contrary, the hypolimnion is insulated from exchange with the atmosphere during the stratification period. Transport of dissolved matter across the vertical density gradient of the thermocline usually is small.

In general, wind determines the thickness of the epilimnion, with few exceptions, e.g., where light penetrates beyond the mixing depth because of wind, or where the stratification is determined by

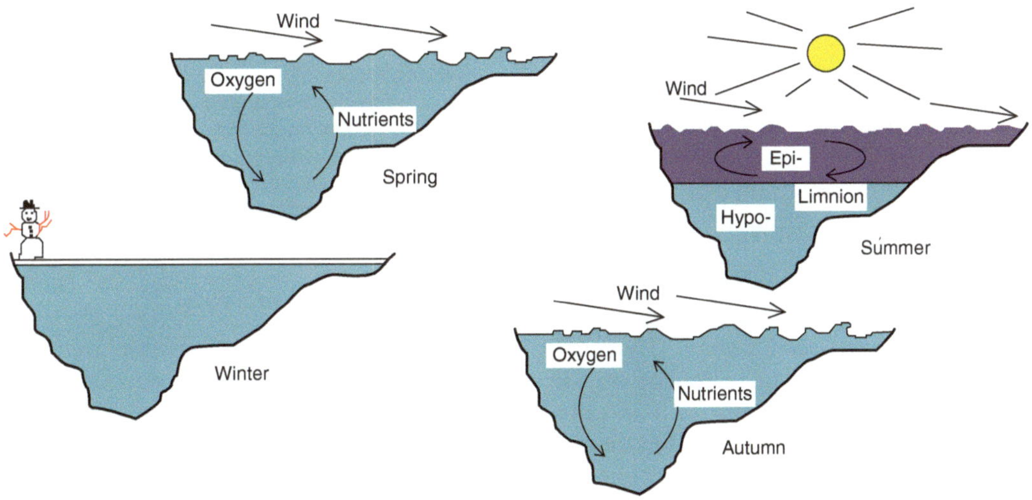

Figure 2 Annual cycle of a dimictic lake with ice cover during winter.

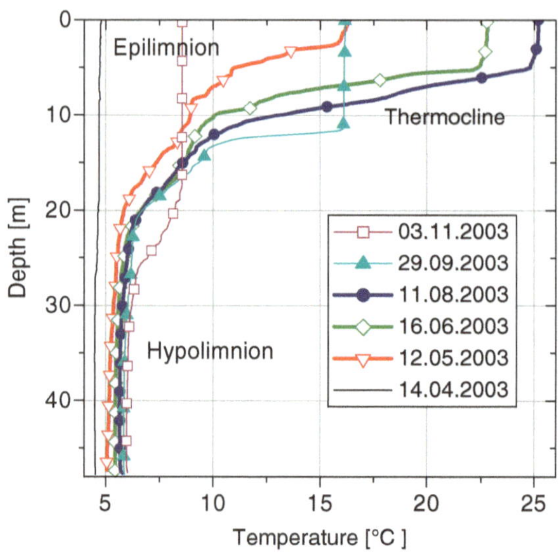

Figure 3 Temperature profiles of Lake Goitsche/Germany in station XN5 in 2003. Symbols are added for every sixteenth data point to distinguish between acquisition dates.

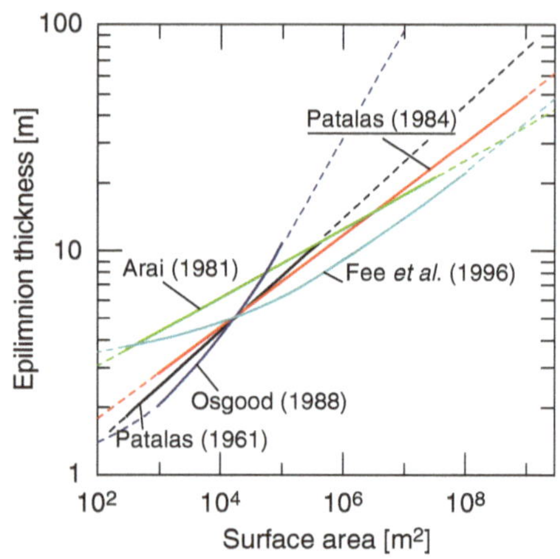

Figure 4 Graphical representation of several approximations of epilimnion thickness z_{epi} versus surface area of the respective lakes. Adapted from Jöhnk KD (2000) 1D hydrodynamische Modelle in der Limnophysik – Turbulenz, Meromixis, Sauerstoff. Habilitationsschrift, Technical University of Darmstadt, Germany.

inflow and water withdrawal (reservoirs). As inferred from **Figure 3**, the thickness of the epilimnion is not constant over the stratification period. In spring, a thin layer is formed, which gradually thickens over the summer because of the cumulative input of wind energy and diurnal heating and cooling. It takes until autumn, when colder temperatures at the lake surface can erode the stratification. During this later period of thermal stratification, substances dissolved in hypolimnetic waters, such as nutrients, become available in the epilimnion again. Eventually epilimnion and hypolimnion are homogenized.

The epilimnion thickness is a crucial factor for living organisms. Hence limnologists have tried to correlate epilimnion thickness with lake morphometry to achieve an a priori estimate (**Figure 4**). The most central regression is $h_{epi} = 4.6 \times 10^{-4} A^{0.205}$, which includes the higher energy input from winds over lakes with larger surface area A.

Because of its high gradients, the thermocline forms a special habitat. Organisms controlling their density can position themselves in the strong density gradient. Also, inanimate particles can accumulate on their level of neutral buoyancy and motile organisms dwell in the thermocline to profit from both layers, epilimnion and hypolimnion. As a result, a layer of distinctive properties can form, called *metalimnion*. Especially in nutrient-rich lakes, the decomposition of organic material can deplete oxygen resulting in a so-called metalimnetic oxygen minimum (**Figure 5**). On the contrary, if light can penetrate to the thermocline and photosynthesis can overcome the oxygen consumption locally, a metalimnetic oxygen maximum occurs.

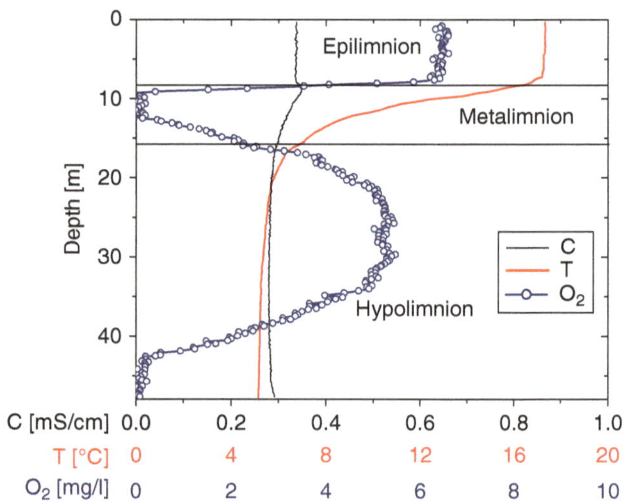

Figure 5 Profiles of temperature (T), (in situ) conductivity (C), and concentration of dissolved oxygen (O_2) from 6 September 2000 in Arendsee/Germany. The boundaries between layers were drawn along the gradients in the oxygen profiles. Oxygen concentration numerically corrected for response time of 7.5 s of the sensor. Adapted from Boehrer and Schultze, 2005, *Handbuch Angewandte Limnologie*, Landsberg: ecomed.

Thermobaric Stratification

Cold water is more compressible than warmer water, in the range of temperatures encountered in lakes. As a consequence, the temperature of maximum density T_{md} decreases as pressure, i.e., depth, increases (by about 0.2 K over 100 m water depth). Hence in cold enough regions, very deep freshwater lakes can show temperature profiles during summer stratification

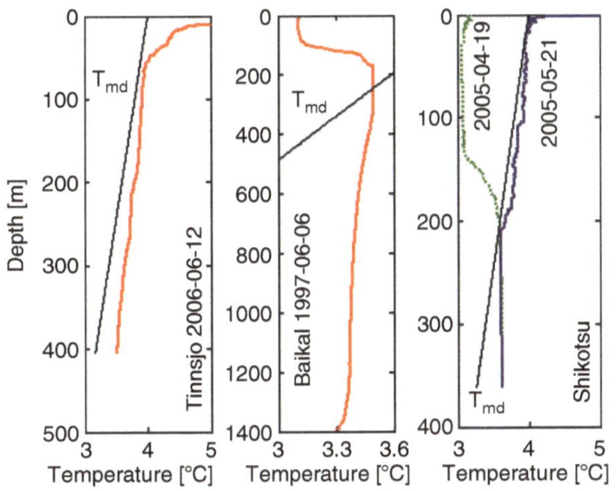

Figure 6 Temperature profiles of thermobarically stratified lakes: Left panel: Tinnsjø, Norway, during (early) summer stratification; Central panel: Lake Baikal, Siberia, Russia, during (late) winter stratification with the vertical transition through T_{md}; Right panel: Lake Shikotsu, Hokkaido, Japan, with the nearly isothermal deep water body below the T_{md} transition. Reproduced from Boehrer B and Schultze M (2008) Stratification of lakes. *Reviews in Geophysics*, 46, RG2005, doi:10.1029/2006RG000210, with permission from American Geophysical Union.

that extend below 4 °C, though limited to the cold side by the T_{md} profile (see **Figure 6**, left panel). During winter, surface temperature can be lower than T_{md}, while at greater depth the temperatures may be above T_{md} (**Figure 6**, central panel). Stable density stratification is achieved in these profiles, if above the T_{md} intersection, colder temperatures overlie warmer temperatures, while below this it is the opposite. At the intersection itself, the vertical temperature gradient must disappear. The water body below the T_{md} profile is not directly affected by the annual temperature cycle (**Figure 6**, right panel).

Nevertheless, observations in such lakes (e.g., Lake Baikal, Russia; Crater Lake, USA; Lake Shikotsu, Japan) show deep waters well supplied with oxygen. This can in part be attributed to the fact that temperatures are low, and productivity and depletion of oxygen happen at a slow rate. However, it is also an indication for a considerable amount of mixing between mixolimnion and the deep water. As a consequence, chemical gradients do not appear in these lakes, and scientists have refrained from calling these lakes meromictic, although a complete overturn does not occur.

Salinity Stratification

A considerable portion of lakes is salty. As there is no compelling boundary, lakes are called salt lakes, if the salt content lies above 3 g in a kilogram of lake water; i.e., $3\ \mathrm{g\ kg^{-1}} = 3‰$. From this concentration, humans can clearly taste the salt, and ecological consequences become obvious. The salt content in lakes can be as high as $300\ \mathrm{g\ kg^{-1}}$. However, salinities normally lie below $0.5\ \mathrm{g\ kg^{-1}}$. Even smaller salinity gradients can determine the circulation of lakes.

Many large salt lakes, e.g., Caspian Sea, Issyk-Kul, Aral Sea, Lake Van, Great Salt Lake, and the Dead Sea, are located in endorheic basins, i.e., areas on the Earth without hydraulic connection to the world ocean at the surface. Salt lakes also occur outside these areas, as solar ponds or basins filled with sea water that lost the connection to the sea. In addition, some inland lakes are fed by saline ground water.

Salt modifies the properties of lake water. As a quantitative expression, the familiar magnitude of salinity has been transferred from oceanography to limnetic water. One kilogram of ocean water contains about 35 g of salt. Brackish water, i.e., water mixed from sea water and fresh water, shows a similar mix of dissolved substances, while the composition of salts in lakes can greatly deviate from ocean conditions. Consequently, salinity is better replaced by total dissolved substances TDS in the limnetic environment. For quantitative investigations, usually the physical quantity of electrical conductance is suited much better (explained later).

In some lakes, dissolved substances raise the density of the deep waters enough that part of the water column is not recirculated at any time during the annual cycle. The remaining bottom layer, the monimolimnion, can show very different chemical conditions (**Figure 7**). Such lakes are termed meromictic. Well-known examples are the meromictic lakes of Carinthia, Austria, and Lake Tanganyika, or Lake Malawi. Some small and deep maar lakes as well as natural lakes in southern Norway and Finland are permanently stratified by small concentration differences between mixolimnion and monimolimnion. In addition, deep pit lakes tend to be meromictic.

The monimolimnion is excluded from the gas exchange with the atmosphere over long time periods. Diffusive and turbulent exchange across the chemocline usually is small. As a consequence, anoxia establishes in most cases after sufficient time. Under these chemical conditions, nitrates and sulphates serve as agents for the microbial oxidation of organic material and substances can be produced that would chemically not be stable in the mixolimnion. Permanently exposed to the hydrostatic pressure, gases (CO_2, H_2S, and others) can accumulate in monimolimnion in concentrations far beyond the concentrations encountered in mixolimnia (e.g., Lake Monoun in Cameroun, Africa, **Figure 8**).

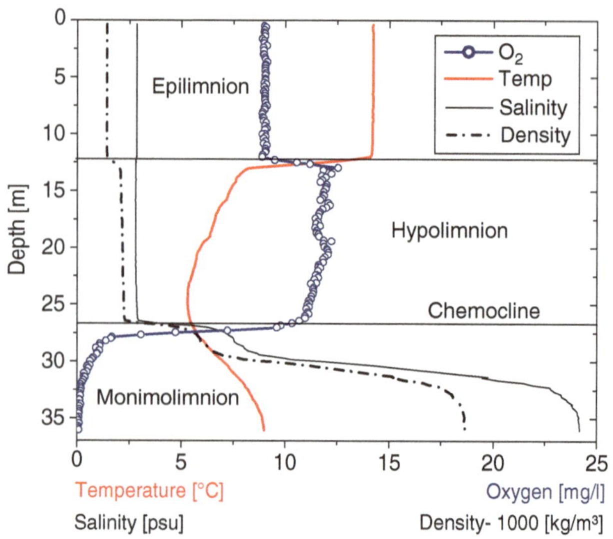

Figure 7 Profiles of temperature, salinity, dissolved oxygen and density from Rassnitzer See in former mining area Merseburg-Ost on 7th October 2003. Oxygen concentrations are numerically corrected for a sensor response time of 7.5s. Adapted from Boehrer and Schultze, 2005, *Handbuch Angewandte Limnologie*, Landsberg: ecomed.

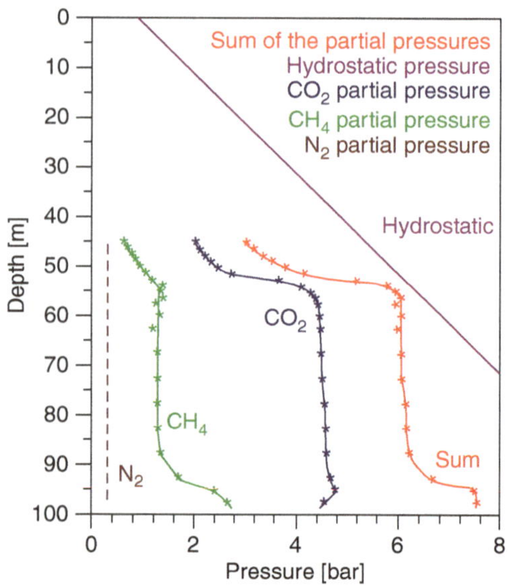

Figure 8 Profiles of partial pressures of dissolved gases in deep water of Lake Monoun, Cameroun, in direct comparison with hydrostatic pressure (solid line). Reproduced from Halbwachs *et al.* (2004) *EOS* 85(30): 281–288, with permission of American Geophysical Union.

In many meromictic lakes, the recirculation of the mixolimnion erodes the monimolimnion leaving a sharp gradient at the end of the circulation period. The transition of all water properties happens within few decimeters from mixolimnetic to monimolimnetic values (see **Figure 7**). This sharp gradient is called *halocline*, *chemocline*, or *pycnocline*, depending on whether the salinity, chemical, or density gradient is referred to. From observations, cases of intensive colonization with only few different species are known, where some plankton species obviously take advantage of such gradients (e.g., Lago Cadagno in the Swiss Alps, Lake Bolvod, Gek Gel, and Maral Gel).

Processes Forming Gradients of Dissolved Substances

In general, diffusive and turbulent diffusive processes distribute dissolved substances more equally throughout a lake over time. On the contrary, inflows of different concentration of dissolved substances or internal processes can produce gradients within a lake. Fresh waters can flow onto a salt lake or enter the epilimnion during the stratification period and hence induce a difference between epilimnion and hypolimnion. These gradients contribute to the density gradients and if strong enough they may even control the circulation pattern of the lakes. In extreme cases, saline waters have been captured in deep layers of lakes for several thousands of years (e.g., Rorhopfjord, Norway; Salsvatn, Norway; Powell Lake, Canada). Such stable stratifications have been induced on purpose in mine lakes to confine heavy metals in the salty monimolimnion for further treatment (Island Copper Mine Lake, Vancouver Island, BC, Canada).

Also, the opposite case of lakes being exposed to high evaporation can encounter salinity gradients in the water column with the saltier layer above. In the case of Lake Svinsjøen (Norway), salty water from deicing roads has removed the previously present permanent stratification. In another case, when ice is forming on salt lakes the residual water can be highly loaded with salts, while after ice melt, relatively fresh water floats on more saline water (lakes in Antarctica). Also, groundwater inflows can form such gradients of dissolved substances (e.g., Lago Cardagno in Swiss Alps, Rassnitzer See; see **Figure 7**). Especially, lakes in volcanic areas are known for the continuous recharge of dissolved substances (e.g., Lake Nyos and Lake Kivu in East Africa, Lake Monoun see **Figure 8**). In particular, the latter case is interesting, as the dissolved gases contribute decisively to the stable density stratification, but they also supply the buoyancy for catastrophic limnic eruptions, in which poisonous gases escape abruptly from a lake with disastrous consequences for living organisms in the area.

Chemical reactions and biological activity in preferred layers can locally change the composition of dissolved substances and impact on the density structure. Although in most lakes at most depths physical transport mechanisms prevail, in some cases the density stratification is controlled by chemical and biological transformations of dissolved material and even control the circulation pattern of lakes. For example, photosynthetically active plankton uses incoming solar radiation for the production of organic material. In addition, allochthonous material is carried into the lake by surface inflows and wind. A portion of the organic material settles on the lake bed. Its decomposition is facilitated by the presence of oxygen or other oxidizing agents and the end products CO_2 and HCO_3^- dissolve in the deep layers of the lake where they contribute to the density. Also iron (and manganese) cycling, calcite precipitation, and sodium sulfate precipitation have been documented to control density in lakes.

In some cases, gradients are strong enough to prevent overturns. These meromictic lakes are classified according to the dominant process sustaining the density difference between mixolimnion and monimolimnion in ectogenically (surface inflow; e.g., Rorhopfjord and Salsvatn in Norway; Powell Lake and Island Copper Mine Lake in Canada, Lower Mystic Lake in the United States), crenogenically (groundwater inflow; e.g., Lago Cardagno, Rassnitzer See, Lake Nyos), or biogenically (or endogenically) meromictic lakes (decomposition of organic material; e.g., Woerther See and Laengssee in Austria, iron cycle: Lakes in Norway, mine lakes, manganese: Lake Nordbytjernet, Norway; calcite: Lake La Cruz, Central Spain; sodium sulfate: Canadian prairie lakes). Also, basin depth and basin shape play an important role in the erosion of a monimolimnion and the formation of deep water renewal. Often monimolimnia are found in well-defined depressions in a lake bed, which are only marginally impacted by basin scale currents in the water body above.

Episodic Partial Deep Water Renewal

A number of lakes do not experience a complete overturn. Episodic events replace parts of the deep water. For example, by cooling, water parcels of high density are formed within the mixolimnion and manage to proceed through the surrounding waters down into the deep water. This process is similar to the deep ocean circulation. Thus, it is not surprising that some of the largest lakes undergo this process. Issyk-Kul (central Asia) recharges its deep water by surface cooling and channelling the cold waters through submerged valleys to the abyss, while in Lake Baikal, wind forcing can push water below the compensation depth so that (temperature-dependent) compression under high pressure raises the density enough compared with surrounding water that its buoyancy becomes negative; the water parcel continues to proceed deeper. Issyk-Kul and Lake Baikal do not show significant chemical gradients and hence are not termed meromictic. On the contrary, Lake Malawi (East Africa) shows an anoxic monimolimnion, and hence is called meromictic, although the deep water formation is similar to that of Issyk-Kul.

In parallel to heat, concentration gradients of dissolved substances also can force deep water renewal in perennially stratified water bodies. Salinity is increased when water is exposed to high evaporation in a side basin (Dead Sea before 1979), or salt accumulates underneath a forming ice cover (Deep Lake, Antarctica). As soon as water parcels become dense enough, they can proceed along slopes into deep parts of the lake.

Quantifying Stability

Temperature

Temperatures recorded in lakes are so-called in situ temperatures. Without any further annotation, temperature data will be understood as such. Nearly all calculations refer to this value, as it is the physically, chemically, and ecologically relevant magnitude. However, if detailed considerations on stability and vertical temperature gradients are envisaged, the reference to potential temperature may be useful. This latter quantity includes the effect of energy required for the expansion, when a water parcel is transferred to atmospheric pressure:

$$\left(\frac{dT}{dz}\right)_{ad} = \frac{g\alpha(T + 273.15)}{c_p} \quad [1]$$

where α is the thermal expansion coefficient for a water parcel of temperature T along the path from depth z to the surface. In lakes where the deep water is close to temperatures of maximum density T_{md}, the thermal expansion coefficient is very small, $\alpha \approx 0$, and, as a consequence, the difference between in situ temperature and potential temperatures is small $\Theta \approx T$. In lakes with warmer deep waters, α can be considered constant. **Figure 9** shows a monotonous potential temperature profile in Lake Malawi, Africa, which indicates stable stratification by temperature only.

Salinity, Electrical Conductivity and Electrical Conductance

Many substances in lake water are dissolved as ions. Hence electrical conductivity has been used to

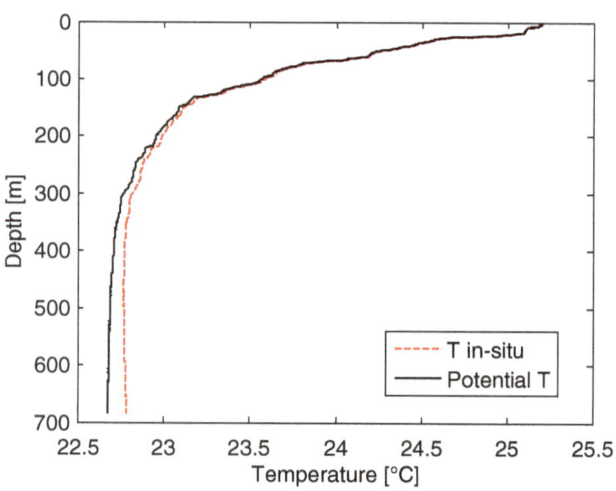

Figure 9 Profiles of (in situ) temperature T and potential temperature near the deepest location of Lake Malawi on 13 September 1997. Reproduced from Boehrer B and Schultze M (2008) Stratification of lakes. *Reviews in Geophysics*, 46, RG2005, doi:10.1029/2006RG000210, with permission from American Geophysical Union.

quantify dissolved substances. For compensation purposes, the temperature dependence of electrical conductivity of a water sample is recorded, while scanning the relevant temperature interval. In most cases, a linear regression $C(T) = aT + b$ is satisfactory to define the conductance, i.e., the electrical conductivity $\kappa_{ref} = C(T_{ref}) = aT_{ref} + b$ at a certain reference temperature T_{ref}. Most commonly, 25 °C is used for the reference:

$$\kappa_{25} = \frac{C(T)}{\alpha_{25}(T - 25°C) + 1} \text{ where } \alpha_{ref} = (T_{ref} + b/a)^{-1} \quad [2]$$

In most surface waters, a value close to $\alpha_{25} = 0.02\ K^{-1}$ is appropriate. Electrical conductance is used for a bulk measurement of concentrations of ionically dissolved substances, quantifying transports from changes in the conductance profile, and to base density regression curves on.

Oceanography uses electrical conductivity and temperature to calculate salinity in practical salinity units (psu), which gives a good indication for dissolved salt in grams per kilogram for ocean water and brackish water (water mixed from ocean water and fresh water). In limnetic systems, however, the composition of dissolved substances differs from that of the ocean. Even within some lakes, there are pronounced vertical gradients. As a consequence, salinity can only be used with reservation in limnic systems.

Density

As direct density measurements in the field are not accurate enough, indirect methods based on easy to measure temperature and conductivity are implemented. As in most practical applications the difference between in situ and potential temperature is small; we use T for temperature. For lakes of low salinities (<0.6 psu), density can be approximated:

$$\rho = \rho(S, T) = \sum_{i=0}^{6} a_i T^i + S \cdot \sum_{i=0}^{2} b_i T^i \quad [3]$$

using

$$a_i = [999.8395; 6.7914 \times 10^{-2}; -9.0894 \times 10^{-3}; 1.0171 \times 10^{-4}; -1.2846 \times 10^{-6}; 1.1592 \times 10^{-8}; -5.0125 \times 10^{-11}]$$

$$b_i = [0.8181; -3.85 \times 10^{-3}; 4.96 \times 10^{-5}]$$

In lakes of a composition of dissolved substances similar to the ocean, the so-called UNESCO formula may be applied (e.g., Rassnitzer See in **Figure 6**), which is applicable for salinities above 2 psu. In cases where salinity cannot be used, calculation of density may directly be based on measurements of temperature and conductivity. For Lake Constance – Obersee, the following formula was proposed, where 20 °C was used as reference temperature for conductance κ_{20}:

$$\rho = \rho_T + \Gamma = 999.8429 + 10^{-3} \times (0.059385\,T^3 - 8.56272\,T^2 + 65.4891\,T) + \Gamma \quad [4]$$

adding the conductivity contribution in separate

$$\Gamma = \gamma\kappa_{20} \text{ and } \gamma = 0.67 \times 10^{-3}\ \mathrm{kg\,m^{-3}\,mS^{-1}\,cm} \quad [5]$$

Alternatively, if the dissolved substances are known, e.g., from chemical analysis, density can be calculated by adding the separate contributions:

$$\rho = \rho_T\left(1 + \sum_n \beta_n C_n\right) \quad [6]$$

where C_n is the concentration of the substance n (g kg^{-1}). A short table of coefficients

$$\beta_n = \frac{1}{\rho}\left(\frac{\partial \rho}{\partial C_n}\right)_{\Theta,p,C_m} \quad m \neq n$$

is given in **Table 1**. In most limnological applications, density is used for stability considerations. Hence, density refers to potential density, i.e., the density of a certain water parcel under normal atmospheric conditions (1013 hPa). As a consequence of the (small) adiabatic compressibility of water, in situ density increases with pressure, i.e., water depth, by about 5×10^{-10} Pa. This means that at 200 m depth, (potential) density and in situ density differ by about 10^{-3}.

Stability

Stability of a water column derives from the density increase in the vertical. Hence, it is a measure for the

Table 1 Contribution of dissolved or suspended substances to the density of water

Substance	β_n[(kg/kg)]
$Ca(HCO_3)_2$	0.807
$Mg(HCO_3)_2$	0.861
$Na(HCO_3)$	0.727
$K(HCO_3)$	0.669
$Fe(HCO_3)_2$	0.838
$NH_4(HCO_3)$	0.462
CO_2	0.273
CH_4	−1.250
Air	−0.090

Modified from Imboden, DM and Wüest A (1995). *Physics and Chemistry of Lakes*, pp. 83–138. Berlin: Springer-Verlag.

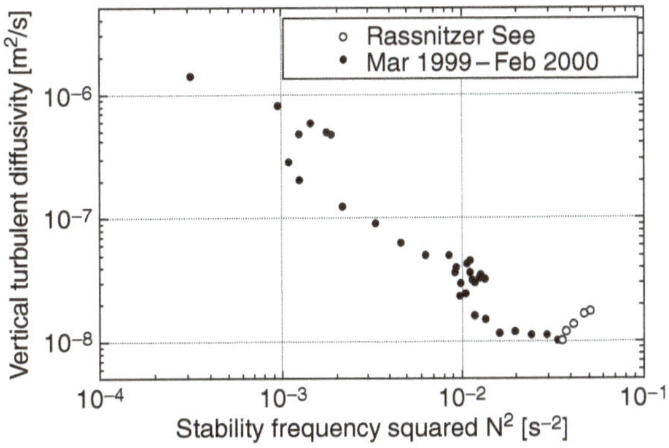

Figure 10 Turbulent diffusive transport of an artificial tracer (SF_6) in the strongly stratified monimolimnion of Rassnitzer See, versus density gradient, $N^2 = -g/\rho\ d\rho/dz$. Adapted from von Rohden and Ilmberger (2001) *Aquatic Sciences* 63: 417–431.

potential energy required for vertical excursion and for overturning of water parcels. Stability considerations can be made for an interface in the water column or for the entire stratified water body as a whole. If an energy source is known, the ratio of required and supplied energy yields a nondimensional number.

Differential Quantities The stability of a density stratification is quantified by

$$N^2 = -\frac{g}{\rho}\frac{d\rho}{dz} \qquad [7]$$

where g is the acceleration due to gravity, and z is the vertical coordinate. The magnitude N is also called stability frequency or Brunt–Väisälä frequency (s^{-1}), which indicates the maximum frequency (ω) for internal waves that can propagate in the respective stratification. N^2 indicates how much energy is required to exchange water parcels in the vertical.

As a consequence, chemical gradients can only persist for longer time periods where density gradients limit the vertical transport of dissolved substances (see **Figure 10**). Lake basin Niemegk of Lake Goitsche (Germany) has been neutralized by introducing buffering river water to the epilimnion. During summer 2000, the vertical transport through the temperature stratification was limited and a chemical gradient in pH could be sustained (**Figure 11**). However, in winter the temperature stratification vanished, vertical transport was enhanced, and consequently, chemical gradients were removed. Gradients in the pH close to the lake bed were stabilized by increased density because of higher concentration of dissolved substances.

In a stratified water column, a current shear can supply kinetic energy for producing vertical

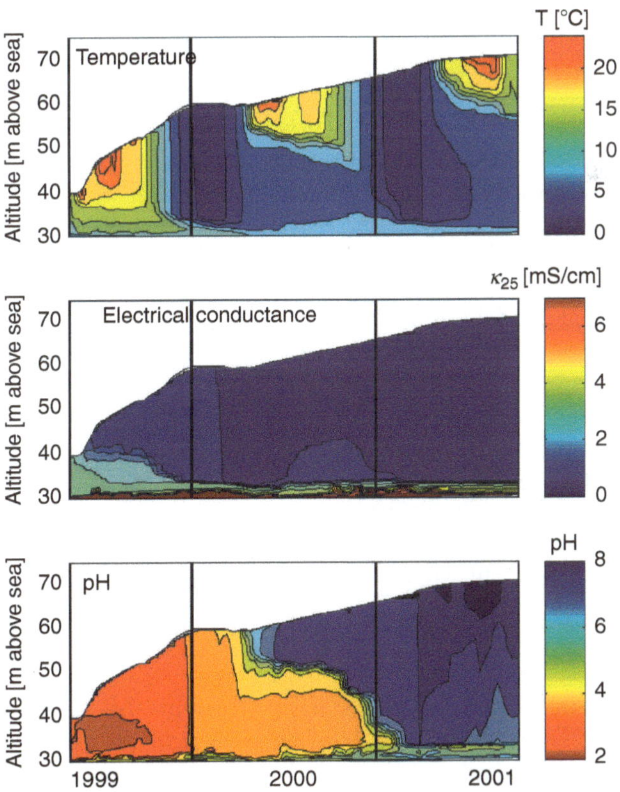

Figure 11 Contour plot of temperature, electrical conductance and pH value versus time and depth in mining Lake Goitsche (station XN3 in Lake basin Niemegk); period of neutralization by flooding with river water; the rising water level is represented by the increasing colored area.

excursions and overturns. A comparison between density gradient and current shear yields the nondimensional gradient Richardson number:

$$Ri = \frac{N^2}{(du/dz)^2} \qquad [8]$$

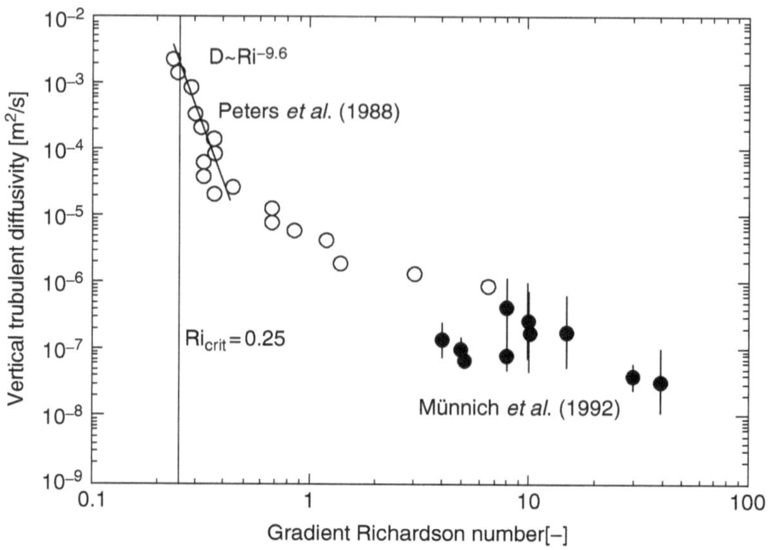

Figure 12 Relation between diapycnal diffusivities and gradient Richardson number. Adapted from Imboden DM and Wüest A (1995) *Physics and Chemistry of Lakes*, pp. 83–138. Berlin: Springer-Verlag.

where $u = u(z)$ represents the horizontal current velocity profile. The critical value of $Ri = 1/4$, when the shear flow supplies enough energy to sustain overturning water parcels, is found by considering the energy balance in the centre of mass frame. As a consequence, diapycnal transports rapidly increase, if Richardson numbers get close to 0.25 or even fall below this critical value (see **Figure 12**). Although in zones of high shear, e.g., in the bottom boundary layer, supercritical Richardson numbers can be found, they appear only sporadically in the pelagic region of lakes, at least if measured on a vertical scale of meters. The measurements in **Figure 12** suggest a correlation between vertical transport coefficients and gradient Richardson number of

$$D = 3 \times 10^{-9} Ri^{-9.6} + 7 \times 10^{-6} Ri^{-1.3} + 1.4 \times 10^{-7} (\mathrm{m^2 s^{-1}}) \quad [9]$$

if the value of molecular diffusivity of heat is included. On the basis of this, vertical diffusivities can be calculated from gradient Richardson number measurements, which are displayed in **Figure 13**, where gradient Richardson number was measured in pelagic waters over a depth resolution of 3–5 m.

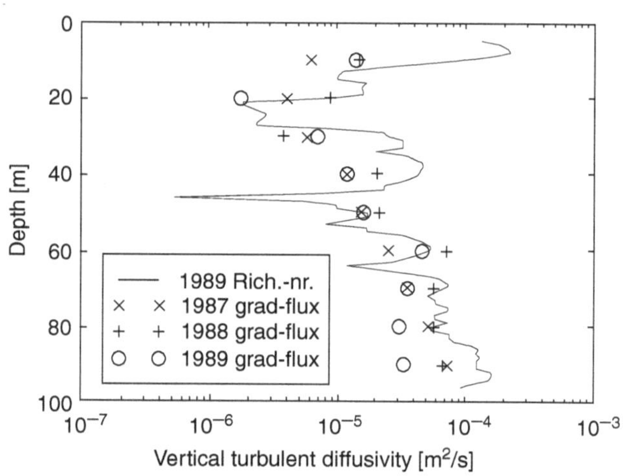

Figure 13 Profiles of vertical transport coefficients in Lake Constance during the stratification period, based on gradient Richardson number measurements in the area of high shear at the Sill of Mainau (solid line), in comparison with results of the gradient flux method for the entire lake based on the evolution of temperature profiles in Überlinger See in years 1987, 1988 or 1989. Adapted from Boehrer *et al.* (2000) *Journal of Geophysical Research* 105(C12): 28,823.

Bulk Quantities For the bulk stability of a stratified water body, various quantities have been proposed, based on potential energy integrated from lake bottom z_b to the surface z_s (e.g., Birge work). We list the definition of Schmidt stability S_t as the most often used reference for the work required for mixing a stratified lake:

$$S_t = \int_{z_b}^{z_s} (z - z_V)(\rho(z) - \bar{\rho}) A(z) \mathrm{d}z \quad [10]$$

where $z = z_V = \frac{1}{V}\int_{z_b}^{z_s} zA(z)\mathrm{d}z$ is the vertical position of the centre of lake volume V, and $\bar{\rho}$ is the density of the hypothetically homogenized lake.

In a two-layer system, like thermally stratified lakes, it is reasonable to compare the potential energy needed for vertical excursion with the wind stress applied to the surface as done by the Wedderburn number W:

$$W = \frac{g' h_{\mathrm{epi}}^2}{u_*^2 L} \quad [11]$$

where $g' = g\frac{\Delta\rho}{\rho}$, with $\Delta\rho$ representing the density difference between epilimnion and hypolimnion, $u_*^2 = \tau/\rho$ is the friction velocity resulting from the surface stress τ implied by the wind, and L stands for the length of the fetch.

Although the common use of the Wedderburn number is connected to its simplicity, the more sophisticated Lake number $L_N = M_{bc}/\left(z_V \int_A \tau \mathrm{d}A\right)$ compares the wind stress applied to the lake surface with the angular momentum M_{bc} needed for tilting the thermocline, and hence represents the integral counterpart of the Wedderburn number.

Small values of both, W and L_{N}, indicate that wind stress can overcome restoring gravity forces because of density stratification. Under such conditions, upwelling of hypolimnion water is possible and intense mixing of hypolimnion water into the epilimnion can be expected. A typical consequence of ecological importance facilitated by this process is the recharging of nutrients in the epilimnion from the hypolimnion.

Nomenclature

a, a_i, b, b_i	coefficients
A	area, especially surface area of a lake (m^2)
c_p	specific heat (J K^{-1} kg^{-1})
C	electrical conductivity (mS cm^{-1})
C_n	concentration of substance (g kg^{-1})
D	vertical turbulent diffusivity (m^2 s^{-1})
h_{epi}	thickness of epilimnion (m)
g	acceleration due to gravity (m^2/s)
g'	reduced acceleration due to gravity (m^2/s)
L	length of lake or wind fetch (m)
L_N	lake number
M_{bc}	angular momentum (N m)
N	stability frequency (s^{-1})
Ri	gradient Richardson number
$Ri_{\mathrm{crit}} = 0.25$	critical gradient Richardson number
$[\,]_{\mathrm{ref}}, [\,]_{25}$	at reference temperature, mostly 25°C
S	salinity for fresh water or ocean conditions (psu)
S_{t}	Schmidt stability (kg m)
T	(in situ) temperature (°C)
T_{md}	temperature of maximum density (°C)
T_{ref}	reference temperature (°C)
u	horizontal current velocity (m s^{-1})
u_*	friction velocity (m s^{-1})
V	lake volume (m^3)
W	Wedderburn number
z	vertical coordinate (m)
z_b, z_s, z_V	vertical coordinate of lake bed, surface, centre of volume (m)
$\alpha = \left(\frac{\partial\rho}{\partial T}\right)_{\mathrm{S,p}}$	thermal expansion (K^{-1})
$\alpha_{\mathrm{ref}}, \alpha_{25}$	coefficient at reference temperature, at 25 °C (K^{-1})
β_n	coefficient for specific density contribution of salts
γ	conductivity specific (potential) density contribution (kg m^{-3} mS^{-1} cm)
Γ	(potential) density contribution by dissolved substances (kg m^{-3})
$\kappa_{ref}, \kappa_{25}$	electrical conductance at reference temperature, at 25 °C (mS cm^{-1})
ρ	(potential) density (kg m^{-3})
$\rho_{\mathrm{in\ situ}}$	(potential) density (kg m^{-3})
ρ_{T}	(potential) density of pure water (kg m^{-3})
$\bar{\rho}$	reference density (kg m^{-3})
Θ	potential temperature (K)
τ	surface stress (Pa)
ω	wave frequency (rad/s)

See also: The Benthic Boundary Layer (in Rivers, Lakes, and Reservoirs); Chemical Properties of Water; Currents in Stratified Water Bodies 1: Density-Driven Flows; Currents in Stratified Water Bodies 2: Internal Waves; Currents in Stratified Water Bodies 3: Effects of Rotation; Currents in the Upper Mixed Layer and in Unstratified Water Bodies; Effects of Climate Change on Lakes; Physical Properties of Water; Salinity; Small-Scale Turbulence and Mixing: Energy Fluxes in Stratified Lakes; The Surface Mixed Layer in Lakes and Reservoirs.

Further Reading

Boehrer B and Schultze M (2008) Stratification of lakes. *Reviews in Geophysics*. 46, RG2005, doi:10.1029/2006RG000210.

Boehrer B Ilmberger J, and Münnich K (2000) Vertical structure of currents in western Lake Constance. *Journal of Geophysical Research* 105(C12): 28,823–28,835.

Chen CTA and Millero FJ (1986) Precise thermodynamic properties for natural waters covering only the limnological range. *Limnology and Oceanography* 31: 657–662.

Gat JR (1995) Stable isotopes of fresh and saline lakes. In: Lerman A, Imboden D, and Gat J (eds.) *Physics and Chemistry of Lakes*, pp. 139–165. Berlin: Springer-Verlag.

Halbwachs M, Sabroux J-C, Grangeon J, Kayser G, Tochon-Danguy J-C, Felix A, Beard J-C, Vilevielle A, Vitter C, Richon P, Wüest A, and Hell J (2004) Degassing the 'Killer Lakes' Nyos and Monoun, Cameroon. *EOS* 85(30): 281–284.

Hongve D (2002) *Endogenic Meromixis: Studies of Nordbytjernet and Other Meromictic Lakes in the Upper Romerike Area*. PhD thesis, Norwegian Institute of Public Health, Oslo Norway.

Hutchinson GE (1957) *A Treatise on Limnology* vol. 1. New York: Wiley.

Imberger J and Patterson JC (1990) Physical limnology. *Advances in Applied Mechanics* 27: 303–475.

Imboden DM and Wüest A (1995) Mixing mechanisms in lakes. In: Lerman A, Imboden D, and Gat J (eds.) *Physics and Chemistry Lakes*, pp. 83–138. Berlin, Germany: Springer-Verlag.

ISO standard 7888 (1985) Water quality: Determination of electrical conductivity. International Organization for Standardization. www.iso.org.

Jöhnk KD (2000) 1D hydrodynamische Modelle in der Limnophysik – Turbulenz, Meromixis, Sauerstoff. Habilitationsschrift, Darmstadt, Germany, Technical University of Darmstadt.

Kalff J (2002) *Limnology*. Upper Saddle River, NJ: Prentice Hall.

Kjensmo J (1994) Internal energy, the work of the wind, and the thermal stability in Lake Tyrifjord, southeastern Norway. *Hydrobiologia* 286: 53–59.

Sorenson JA and Glass GE (1987) Ion and temperature dependence of electrical conductance for natural waters. *Analytical Chemistry* 59(13): 1594–1597.

Stevens CL and Lawrence GA (1997) Estimation of wind forced internal seiche amplitudes in lakes and reservoirs, with data from British Columbia, Canada. *Aquatic Science* 59: 115–134.

von Rohden C and Ilmberger J (2001) Tracer experiment with sulfur hexafluoride to quantify the vertical transport in a meromictic pit lake. *Aquatic Sciences* 63: 417–431.

Relevant Websites

http://www.ilec.or.jp/ – International Lake Environment Committe; data on various lakes on Earth.

http://www.ioc.unesco.org/ – Intergovernmental Oceanographic Commission of UNESCO; provides on-line calculator for salinity following the so-called UNESCO formula.

http://www.cwr.uwa.edu.au/ – Centre for Water Research (CWR) at The University of Western Australia. Research Institution focussing of physical limnology.

http://www.eawag.ch/ – Swiss Federal Institute of Aquatic Science and Technology. Institution dealing with water related issues.

The Surface Mixed Layer in Lakes and Reservoirs

S G Monismith, Stanford University, Stanford, CA, USA
S MacIntyre, University of California, Santa Barbara, CA, USA

Background

The physical environment of the surface waters of lakes and reservoirs reflects the dynamic interplay of wind stresses, radiative heating and cooling, and heat and moisture exchanges with the atmosphere. Often this results in the development of a nearly isothermal (or isohaline in the case of salt lakes) layer, known as the surface mixed layer (SML), extending from the water surface to a depth determined by the strength and history of heating, cooling, and winds (**Figure 1**). In classical limnology, the surface mixed layer is also referred to as the epilimnion. Below the SML there is usually a layer (or layers) across which the temperature varies rapidly, the thermocline, and below that a layer that is less strongly stratified, the hypolimnion, which may include another well-mixed layer on the bottom.

The fundamental physics of the surface mixed layer is a competition between the elimination of stratification by turbulent mixing produced either by the wind acting on the lake surface or by cooling of the surface, and the creation of stratification by (mostly) heating of the water column by the absorption of incident shortwave radiation. The key feature is that density stratification brings into play buoyancy forces that tend to suppress or kill turbulent mixing such that a frequently energized surface mixed layer can become physically isolated from the more quiescent and strongly stratified water column below. As discussed in detail elsewhere in this encyclopedia (refer 'see also' section), a measure of the *local* importance of stratification is the Richardson number, defined as

$$Ri = -\frac{g\partial\rho/\partial z}{\rho(\partial U/\partial z)^2} \qquad [1]$$

where z is height, U is horizontal velocity, and ρ is lake water density. When $Ri > 0.25$ turbulent mixing essentially shuts down; thus, it is not the strength of the temperature stratification *per se* that matters, but rather its strength relative to the shear that acts to maintain turbulence. In addition, heating in the day creates buoyancy forces that resist shear-induced mixing, but cooling due to heat loss at night, heat loss from evaporation, or heat loss during the passage of cold fronts or during fall cooling induces turbulence and, by reducing stratification, either facilitates mixing by shear, and in some cases, may be the dominant process causing mixing.

Biological rate processes, species composition, water quality, and the fate of pollutants are all tied to mixing dynamics in the upper water column. Turbulence in the SML circulates phytoplankton cells, determines their light exposure, and thus affects the magnitude of primary productivity. If mixing is limited, cells may be exposed to sufficient irradiance to require cellular repair. Pollutants can sorb to dissolved organic carbon (DOC). Their persistence in the water column versus return to the atmosphere depends upon light history. The frequency and intensity of turbulence in the upper water column determines whether marine and lake snow form thus affecting microbial activity and fluxes of particulates out of the SML, moderates phytoplankton community structure as some species are favored in environments with minimal variations in irradiance, regulates feeding success of zooplankton and larvae, and may physically influence size of colonial organisms or damage heterocystic dinoflagellates. Finally, the presence or absence of turbulence determines whether physical conditions are optimal for buoyant (and undesirable) cyanobacteria like *Microcystis aerginosa.* The rate of mixed layer deepening influences supply of nutrients from the hypolimnion to phytoplankton in the upper, well lit portions of the water column, and persistence of anoxia within the hypolimnion with concomitant affects on water quality.

Mixed Layer Motions and Turbulence

Basic Ideas of Turbulence (refer 'see also' section)

Conceptually, motions in the SML can be decomposed into three components – an organized, mean flow component, $\vec{U}$, a turbulent component, $\vec{u}'$, and a wave component, $\vec{u}_w$, i.e., the instantaneous velocity (or any other variable) can be written as

$$\vec{u}(\vec{x},t) = \vec{U}(\vec{x}) + \vec{u}'(\vec{x},t) + \vec{u}_w(\vec{x},t) \qquad [2]$$

Note that $\vec{U}$ can vary slowly in time (e.g., over hours or days) as the overall winds etc. change such that it is averaged over a short time, e.g., 10 min. This decomposition is based on what is referred to as Reynolds averaging, which is simply averaging in time, although it better thought of as low-pass filtering.

The time-averaged effect of the turbulence is the net transfer of momentum by correlations of fluctuating

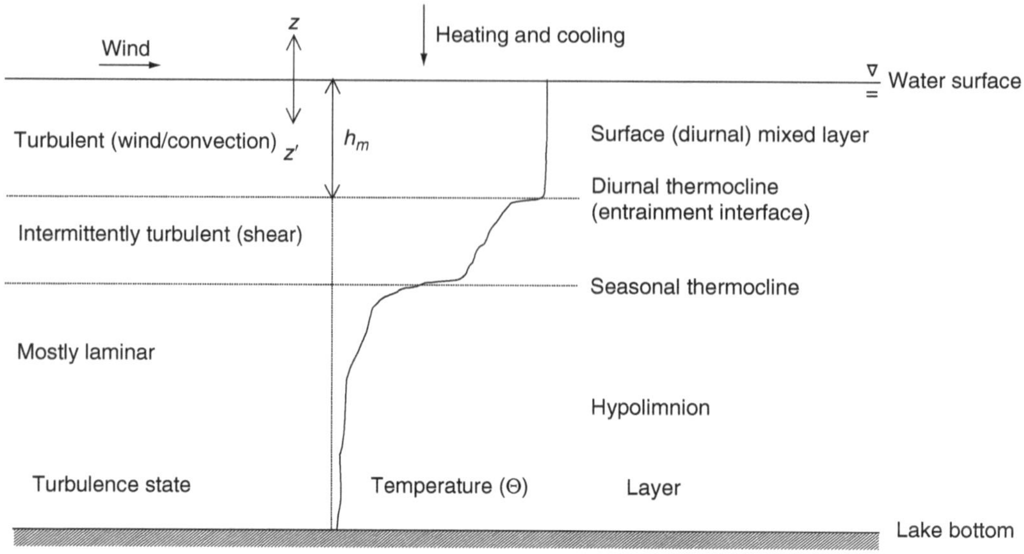

Figure 1 Sketch of lake thermal structure showing the surface mixed layer as well as layers below.

velocities known as Reynolds stresses such as $\overline{u'w'}$, where the overbar indicates the averaging of turbulent fluctuations, and w represents the vertical velocity. Reynolds stresses are often parameterized by the eddy viscosity formulation

$$-\overline{u'w'} = \nu_t \frac{\partial U}{\partial z} \quad [3]$$

In the presence of stratification ν_t is reduced, and is sometimes written as a function of Ri with $\nu_t \to 0$ for $Ri > 0.25$.

Much of turbulence theory is premised on an energetic balance between the creation of turbulence, or turbulent kinetic energy $\text{TKE} = \frac{1}{2}\left(\overline{u'}^2 + \overline{v'}^2 + \overline{w'}^2\right)$, by shear production

$$P = -\overline{u'w'}\frac{\partial U}{\partial z} = \nu_t \left(\frac{\partial U}{\partial z}\right)^2 \quad [4]$$

and its destruction by dissipation, ε (units = W kg^{-1}). Density variations come into play through the buoyancy flux

$$J_b = -\frac{g\overline{w'\rho'}}{\rho} \cong g\alpha\overline{w'\theta'} \quad [5]$$

where α is the thermal expansivity. J_b describes the rate of work associated with vertical turbulent motions of fluid parcels with densities different from their surroundings. For example when $w' > 0$ and $\rho' > 0$, turbulence lifts relatively heavy fluid particles up; in this case, J_b is negative, representing a loss of energy by the turbulence. This is the case when turbulence is mixing a stably stratified fluid. The opposite case is found when the surface of the lake is cooled, leading to convection. Here $w' < 0$ and $\rho' > 0$, meaning that relatively heavy fluid particles are falling through the fluid, generating turbulence; in this case, B is positive, representing a gain of energy by the turbulence.

Finally when turbulence is inhomogeneous, i.e., varies spatially, as is the case in the SML, there can be redistribution of turbulent kinetic energy by the turbulence. This is accomplished by pressure-velocity correlations (like in a engine cylinder!) and by triple correlations of fluctuating velocities. While the details of redistribution are more obscure, what is observed is that the transfer term, T, acts to move TKE from high turbulence regions to low turbulence regions.

Thus, the final accounting for TKE for a one-dimensional water column as we will describe below reads:

$$\frac{\partial(\text{TKE})}{\partial t} = P + J_b - \varepsilon + \frac{\partial T}{\partial z} \quad [6]$$

This relation is fundamental to lake turbulence dynamics; in what follows, we will emphasize its use in understanding the dynamics of and modeling the SML.

Winds on Lakes

When wind blows over the lake surface it applies a force, i.e., imparts momentum, to the fluid, driving mean currents, generating surface waves, and producing turbulence, either through instability of the shear flow created by the wind or through wave breaking (**Figure 2**). In terms of U_{10}, this force, represented as a stress (force/area) is given as

$$\tau_0 = \rho_a C_D U_{10}^2 = \rho u_*^2 \quad [7]$$

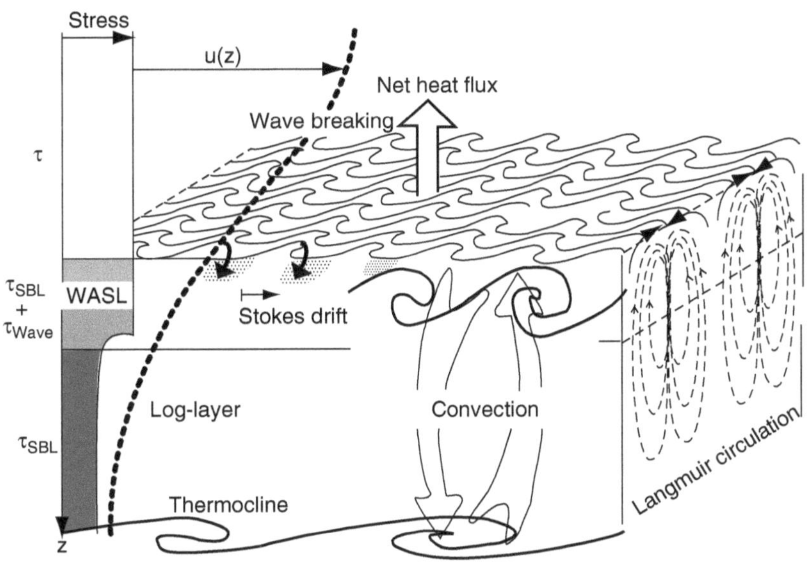

Figure 2 Sketch of SML including waves, Langmuir circulations etc. (Figure 2 in Wuest and Lorke used with permission.) WASL refers to the wave affected surface layer, i.e., the region in which enhanced turbulence due to breaking is important.

where ρ_a is the air density, $C_D = O(10^{-3})$ is the drag coefficient, U_{10} is the wind speed measured at 10 m above the lake surface, ρ is the water density, and u_* is the shear velocity in the water. Besides atmospheric stability, for lakes, C_D can depend on fetch (the distance over which wind-generated waves propagate), especially including topographic sheltering by surrounding hills or even trees in the case of small lakes, and on wind speed.

The transfer of momentum from the wind to the water at the air-water interface depends on how the airflow over wind-generated surface waves is modified by the presence of those waves, a modification that depends both on the height of the waves and their wavelength. These two features of surface waves increase with increasing fetch. Consequently, C_D depends on fetch and wind speed. The drag relation, eqn. [7], represents the total momentum input from the wind. However, only part of it goes directly into the underlying flow; the rest goes to making the waves grow. This 'wave stress' can be 50% of the total stress. Ultimately, the waves reach a state wherein the rate of input from the wind is matched by the loss of momentum to the mean flow by wave breaking, a state in which all of the wind stress is applied to the mean flow. Practically, given the fact that the winds often vary considerably over the surface of a lake, and that often even in the best of circumstances, only one wind station might be available on the lake, it often suffices to assume that $C_D \approx 1.3 \times 10^{-3}$.

The shear velocity in the water, u_*, plays a central role in representing flows in the SML. In many cases the mean velocity has been found to vary logarithmically with distance, z', measured down from the water surface, i.e., the difference between the velocity at the water surface, U_s, and at any depth satisfies the well established 'law of the wall' that describes flows in turbulent boundary layers:

$$\vec{U}(z') = U_s - \frac{u_*}{\kappa}\ln\left(\frac{z'}{z_0}\right) \qquad [8]$$

where $\kappa \simeq 0.41$ is the von Karman constant, and z_0 is the apparent roughness of the near surface flow. Most studies have found that z_0 is proportional to the surface wave height albeit much smaller. Given the complex physics operant at the air-water interface, it is hardly surprising that the reasons for this behavior remain unclear. Although difficult to measure, laboratory and field observations suggest $U_s \approx 25u_*$. The law of the wall description is useful in that it has well-known properties, e.g., the eddy viscosity and thus eddy diffusivity (the mixing coefficient for scalars like dissolved oxygen) vary parabolically within the SML and are proportional to u_*:

$$\nu_t \cong \kappa u_* z'\left(1 - \frac{z'}{h_m}\right) \qquad [9]$$

The dependence of ν_t on u_* comes about because the turbulent velocity scale $q \sim u_*$, whereas the turbulence length scale, which varies with position within the SML, is proportional to h_m but depends on position within the mixed layer. From the standpoint of TKE dynamics, near the water surface, P and ε are in balance, and since $-\overline{u'w'} = u_*^2$ is nearly constant:

$$\varepsilon \approx P = \frac{u_*^3}{\kappa z'} \qquad [10]$$

However, observations of the SML show that it differs from simpler boundary layers in two important ways. First, the presence of waves brings into play two phenomenon not found in canonical law of the wall flows: wave-breaking, an extra source of TKE near the surface; and Langmuir cells or circulations (LCs), large-scale, streamwise vortices thought to be produced by wave-current interactions. Secondly free-convection due to surface cooling can also introduce TKE into the SML. Furthermore, continued heating of the upper water column can modify the energy balance above such that some of the energy introduced into the SML is lost in working against buoyancy forces. Modifications to the predicted law of the wall scaling are described in eqns. [15] and [16].

When surface waves break they transfer momentum and energy into the underlying flow. In terms of turbulence dynamics, this can be viewed as providing, an extra source of TKE near the surface that diffuses down into the underlying flow such that the TKE balance changes to one in which $\varepsilon \approx \partial T/\partial z$. Empirically, it is found that the flux of TKE from the water surface, i.e., the breaking region, is 60–100 u_*^3, which gives rise to values of ε much larger than those predicted by eqn. [10]. The thickness of this layer of enhanced ε is $O(u_*^2/g)$; inside this layer ε decays rapidly, i.e., $\varepsilon \propto z'^{-3}$. Nonetheless, below this region, ε tends to wall values. Because of this behavior, it appears that wave-breaking is a critical aspect of air-water gas exchange and near-surface turbulence but may not be important to processes at work deeper in the SML such as those responsible for deepening of the SML.

Besides injecting turbulence into the SML by breaking, surface waves also modify horizontal transport because of their associated Stokes drift velocity, $\vec{U}_{\text{Stokes}}$, the difference between mean Lagrangian velocity of water particles (i.e., the average velocity of particular particles) and the mean Eulerian velocity (i.e., the average measured at fixed points). As discovered by Stokes in 1848, $\vec{U}_{\text{Stokes}}$ comes about because even though the wave motion at any point may have zero mean (as is this case with surface waves), as fluid particles move under the waves, they sample the velocities of points in the fluid that are slightly different from the velocity where they started, resulting in drift in the direction the waves are moving. Written for waves with wavelength that is short relative to the SML depth, i.e., $\lambda = \frac{2\pi}{\|\vec{k}\|} < h_{\text{m}}$, where λ is the surface wave wavelength and $\vec{k}$ is the wavenumber vector, the Stokes drift velocity is

$$\vec{U}_{\text{Stokes}} = (ak)^2 \exp(2kz)\vec{C}_{\text{p}} = U^{(0)}_{\text{Stokes}} \exp(2kz)\frac{\vec{C}_{\text{p}}}{\|\vec{C}_{\text{p}}\|} \qquad [11]$$

for wave height a and phase velocity

$$\vec{C}_{\text{p}} = \frac{2\pi g}{T_{\text{w}}}\frac{\vec{k}}{\|\vec{k}\|} \qquad [12]$$

Here T_{w} is the wave period.

In addition to its importance to the horizontal transport of floating materials, waves, through the Stokes drift also can drive LCs, aka 'windrows,' arrays of streamwise vortices roughly aligned with the wind that collect floating materials (foam, garbage, jellyfish…) at their convergences on the surface (**Figure 3**). As first revealed in studies by the chemist Irving Langmuir made in Lake George, New York, LCs can have a profound effect on the overall flow in the SML; in particular they can rapidly transport neutrally buoyant materials, e.g., phytoplankton cells, to the base of the SML.

Current theory views LCs as being driven by the averaged interaction of surface waves with the mean flow. This interaction leads to an extra net force on the mean flow referred to as the Craik-Leibovich force (after the theory's developers) that is the cross product of the Stokes drift velocity and the mean vorticity of the flow. Analysis of the stability of flows driven by wind stresses in the presence of Stokes drift, shows that when $\frac{\partial \vec{U}_{\text{Stokes}}}{\partial z} \cdot \frac{\partial \vec{U}}{\partial z} > 0$, LCs can develop. This condition is generally satisfied in lakes since both the waves and wind and wind-driven flow tend to be co-aligned. While LCs can exhibit complex pairing and branching behavior (**Figure 3**), and some substantial unsteadiness, generally the end state of the instability is a set of rolls that have a dominant wavelength across the wind that is twice the SML depth, and

Figure 3 Langmuir circulations visualized by flotsam lines on the Great Salt Lake, Utah (USA). Photo taken from a small airplane by Dr. Steve Robinson. The main row spacing is comparable to the depth. Used by permission.

produce a velocity in the plane perpendicular to the wind $u_{LC} \sim O(U_s^{(0)} u_*)$ that can be several cm/s, resulting in relatively rapid stirring of the upper mixed layer.

Recent numerical studies of turbulent LCs make use of the fact that the Craik-Leibovich force also describes the interactions of waves with low-frequency turbulence, suggesting that LCs can be considered to be an organized form of turbulence. Questions about the role of LCs in SML dynamics remain: For example, given that LCs can transport significant momentum downwards, it is conceivable (although not yet shown observationally) that they may dominate the downwards transport of horizontal momentum, thus altering the fundamental physical basis of the law of the wall velocity profile discussed above. Moreover, it is clear that LCs in shallow flows over a solid, no-slip bottom seem to be more stable and more clearly defined than those in the SML (see **Figure 3**), which, in effect, overlies a slippery bottom.

Besides waves, the SML also differs from the generic wall layer in that there are significant heat and thus buoyancy fluxes through the air–sea interface. When cooled from the surface, the SML will experience free convection; the surface of the lake will be 0.01 °C–0.1 °C or so cooler than the water immediately beneath leading to gravitational instabilities with parcels of fluid intermittently breaking away from the free surface and dropping through the SML to the base of the SML. By continuity, these plumes must be matched by upward rising fluid elements at the same time. For a surface buoyancy flux

$$J_{b0} = \frac{\alpha g \widetilde{H}}{\rho c_p} \quad [13]$$

where $\widetilde{H}$ is the surface heat flux (positive outwards for cooling) and c_p is the heat capacity (4.2×10^3 J °C^{-1} kg^{-1}), the characteristic velocity of these plumes (assuming that $J_{b0} > 0$) is

$$u_f = (J_{b0} h_m)^{1/3} \quad [14]$$

Thus, in the presence of cooling, wind stress and surface waves, there are potentially 3 velocity scales for turbulent motions in the SML, u_*, u_{LC} , and u_f.

With both buoyancy and shear (but omitting wave effects), the lacustrine SML bears a strong similarity to the atmospheric boundary layer, where both buoyancy and shear can be important. In this case, a second length scale in addition to the mixed layer depth comes into play: the Monin Obukhov length,

$$L_{MO} = -\frac{u_*^3}{\kappa J_{b0}} \quad [15]$$

a length scale that characterizes the importance of the buoyancy flux relative to the wind stress (or equivalently in the absence of Langmuir circulations) the near surface shear. When $L_{MO} > 0$, the equilibrium SML depth produced by mixing in the presence of a stabilizing buoyancy flux will be L_{MO}. When $L_{MO} < 0$, L_{MO} is the depth at which buoyancy effects equal shear effects. Indeed, careful measurements of ε made in lake SMLs match the depth variation observed in atmospheric flows, i.e.,

$$\frac{\varepsilon \kappa z'}{u_*^3} = 1.14\left(1 + 0.46(z'/|L_{MO}|)^{2/3}\right)^{3/2} \quad [16]$$

Note that for small values of $z/|L_{MO}|$, eqn. [16] reverts to the law of the wall whereas the dissipation associated with convective plumes tends to be nearly independent of depth and approximately equal to J_{b0}. Although there are no suitable measurements that show this behavior, one can imagine that cooling with wave breaking should lead to a three layer structure – a near surface wave layer with very high values of ε, a law of the wall layer and a deeper convective layer, the latter two described by eqn. [16]. Of course, convection also modifies the shear flow, with the enhanced vertical mixing reducing the vertical shear, while for small stabilizing buoyancy fluxes the vertical shear is enhanced, behaviour that has been found for the atmospheric case to be represented by a modified law of the wall.

Surface Energy Exchanges

Heat fluxes through the air-water interface are due to net shortwave radiation (Q_{sw}), net longwave radiation (Q_{lw}), latent heat transfer (H_l) and sensible heat transfer (H_s). The total surface heat exchange can be written as

$$\widetilde{H} = Q_{sw} + Q_{lw} + H_l + H_s \quad [17]$$

where the usual sign convention is that $\widetilde{H}$ is negative for heating of the water column and positive for cooling.

The downward shortwave radiation Q_{sw} is usually measured directly, and the upward component due to reflection is either measured directly or accounted for by a calculation based on the variation of albedo with sun angle. Unlike the other components of the surface heat flux, Q_{sw} penetrates into the water column according to Beer's law:

$$Q_{sw}(z) = Q_{sw}(0)\exp(\beta z) \quad [18]$$

where β^{-1} is the effective extinction depth for light. Often eqn. [18] is modified to be the sum of terms reflecting different extinction lengths for different portions of the irradiance spectrum. The diffuse attenuation coefficient measured in the visible part of the spectrum, β_d, depends on what is in the water column, e.g., fine sediment particles, dissolved organic matter, or phytoplankton; consequently it varies daily, weekly, seasonally, and inter-annually in response to changing sediment inputs and chemistry and biology of the lake. Q_{sw} acts to stratify a water column except when water temperatures are below the temperature of maximum density.

The net long-wave radiation Q_{lw} is the difference between incoming infrared radiation from the sky and outgoing infrared radiation from the water surface. These values can be measured directly but in the absence of instrumentation, the net heat loss from the lake can be calculated from observations of air temperature, lake surface temperature and cloud cover using any one of a number of formulations. One popular model for Q_{lw} is:

$$Q_{lw} = 5.23 \times 10^{-8} T_s^4 - 5.18 \times 10^{-13} T_a^6 (1 + 0.2C^2) \quad [19]$$

where T_s is the water surface temperature (°K), T_a is the atmospheric temperature (°K)(where the height of measurement depends on the thickness of the atmospheric boundary layer; typically 10 m for large lakes and 2–3 m for small), and C is the fraction of sky covered with clouds. The dependence on clouds is the result of blackbody radiation by water moisture in the atmosphere. The coefficients include emissivity of air, which is proportional to T_a^2, and water as well as Stefan Boltzman constant.

The latent heat flux, H_l is the surface heat loss due to evaporation. H_l is parameterized in terms of air density, ρ_a, latent heat of evaporation, L_w, the wind speed U_{10} and relative humidity, r, both measured at 10 m, and the saturation humidity at the water surface temperature, $q_s(T_s)$

$$H_l = \rho_a L_w C_l U_{10}(q_s - q(r, T_a)) \quad [20]$$

The exchange coefficient $C_l \approx 0.0015$ is an empirical function of fetch and atmospheric stability. In a like fashion, the sensible heat flux is given as

$$H_s = \rho c_p C_s U_{10}(T_s - T_a) \quad [21]$$

where $C_s \approx C_l$ is also an empirical constant.

Given the surface heat fluxes, the first law of thermodynamics describes the subsequent evolution of the temperature field. Written for a one-dimensional water column, this reads:

$$\frac{\partial \Theta}{\partial t} = -\frac{1}{\rho c_p}\frac{\partial Q_{sw}}{\partial z} - \frac{\partial\left(\overline{w'\theta'}\right)}{\partial z} \quad [22]$$

where the surface value of the turbulent heat flux is equal to the total surface heat flux minus the short wave radiation, i.e.:

$$-\overline{w'\theta'}|^0 = H_{LW} + H_L + H_S \quad [23]$$

It is the vertical variation in shortwave radiation that creates stratification, hence the extinction scale β^{-1} plays a fundamental role in determining SML depth.

An example of time series of surface meteorology, the resulting energy fluxes at the air-water interface, and the sum of the surface energy fluxes and the effective heat flux are presented in **Figure 4** for a small arctic lake during summer. The resulting evolution of the thermal structure is illustrated in **Figure 5**. The forcing of the surface layer during periods with warm and cold fronts differs considerably with consequences for thermal structure. During cold fronts, clouds are prevalent and reduce the short wave inputs, air temperatures drop below surface water temperatures thus setting the stage for increases in sensible heat losses, and relative humidity increases. Winds came up during the second half of the cold front illustrated here, day of year (doy) 210–215 such that sensible and latent heat fluxes both increased with total heat loss of 400 W m^{-2}. Due to the high cloud cover, the sum of the terms in eqn. [24] is positive for only brief periods during the day and overall the lake loses heat. In contrast, when warm air masses are present, cloud cover is lower, air temperatures diurnally increase and decrease to temperatures just above and just below surface water temperatures, and winds often have a diel periodicity. In consequence, the terms in eqn. [24] vary from negative in the day (the lake is gaining heat) to positive at night (the lake loses heat). The consequences of these variations in surface forcing are readily observed in the lake thermal structure. During periods with warm air masses, the upper layer heats on a daily basis and a diurnal thermocline forms. In contrast, during the initial part of a cold front when winds were light, day 210–212, the surface layer slowly loses heat. Temperatures decrease much more rapidly as the heat losses increase and the mixed layer deepens from 3 to 6 m.

On a global scale, latent heat fluxes tend to dominate cooling from the tropics to the poles although net longwave radiation can be significant particularly on cloud free nights. Sensible heat fluxes can either heat or cool but are generally smaller than the other terms. Experience has shown that given a weather station located on the lake that measures windspeed, relative humidity, air temperature, and incident solar radiation, the above relations are likely accurate to 10–20 W m^{-2}. The intensity of turbulence and the size of turbulent eddies over a diurnal period with light wind forcing are illustrated in **Figure 6**. Heat loss at night, accompanied by light breezes, generates eddies which fill the SML and turbulence is measured throughout. During the day, due to positive effective heat flux, the turbulence is damped at

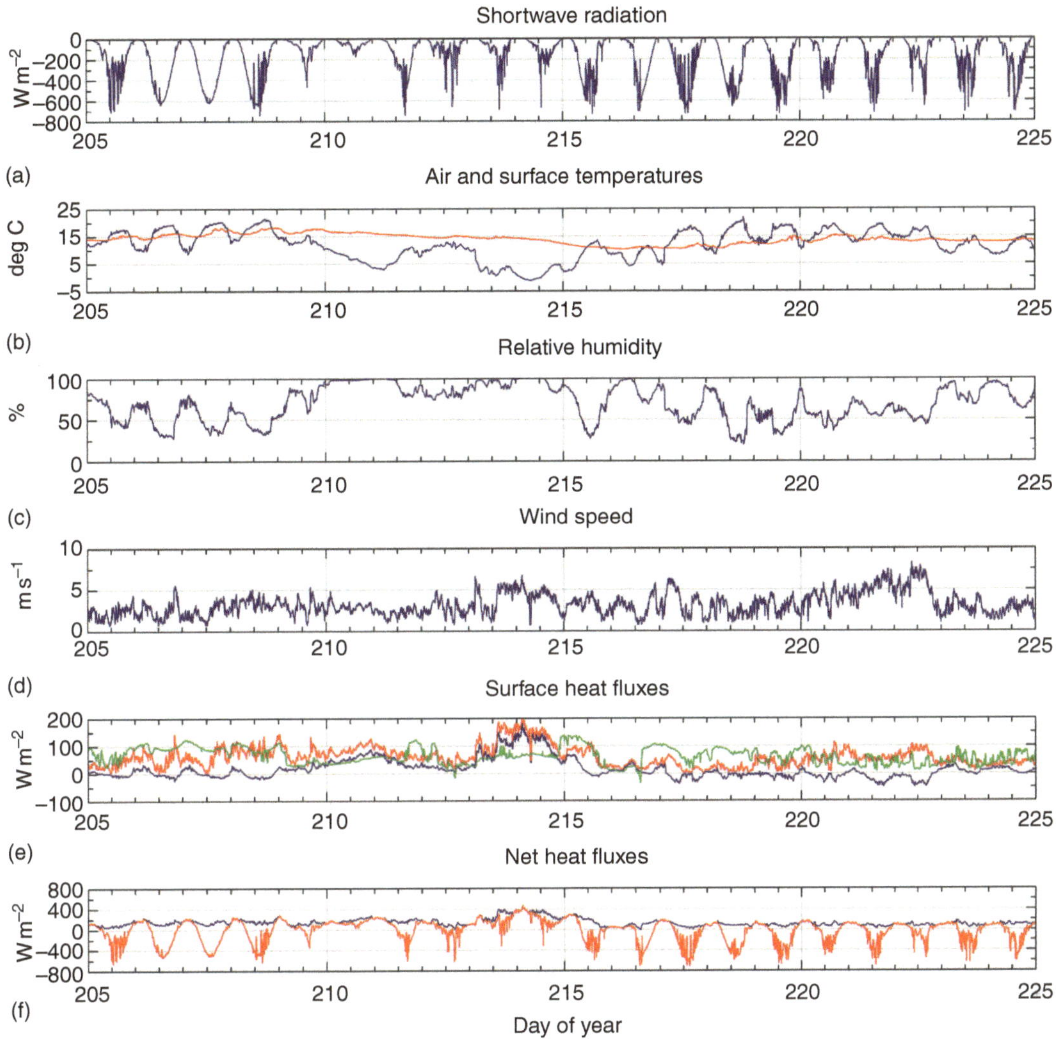

Figure 4 Meteorological forcing of Lake E5, an arctic kettle lake, from mid-June through mid-August 2004: (a) net shortwave radiation; (b) Air (blue) and surface water (red) temperatures; (c) relative humidity; (d) wind speed; (e) surface heat fluxes comprised of sensible heat flux (blue), latent heat flux (red) and net longwave heat flux (green); and (f) sum of the surface heat fluxes (blue) and total heat flux (green) computed as the sum of surface heat fluxes and net short wave absorbed in the surface mixing layer. (MacIntyre S, unpublished data).

depths below a shallow surface layer. Light afternoon winds only tend to cause mixing in a small portion of the SML. Nocturnal cooling again energizes the full SML. During a cloudy, windy period, the SML would remain turbulent throughout the day.

Mixed Layer Deepening and Entrainment

Mixing and Turbulence Energetics

It is the combination of wind stress, convection and surface heating that leads to the formation of the SML and partially sets its depth h_m. Qualitatively this can be viewed as follows: energy for mixing (TKE) is introduced at the surface and is partially (or mostly) dissipated within the SML; what remains at the mixed layer base is available to lift colder, heavier fluid up into the mixed layer, thus increasing the potential energy of the water column. The sketch of the SML shown in **Figure 7** shows an additional aspect of the problem: besides a near-discontinuity in temperature, there is also a strong velocity shear across the mixed layer base. This shear, which arises from the direct effect of the wind and from seiching in the lake (refer 'see also' section), has two consequences: (1) it results in the additional production of TKE at the mixed layer base, and (2) it leads to Kelvin Helmholtz billowing. The consequences for turbulence at the base of the mixed layer can be seen in late afternoon in **Figure 6**. Most current models of

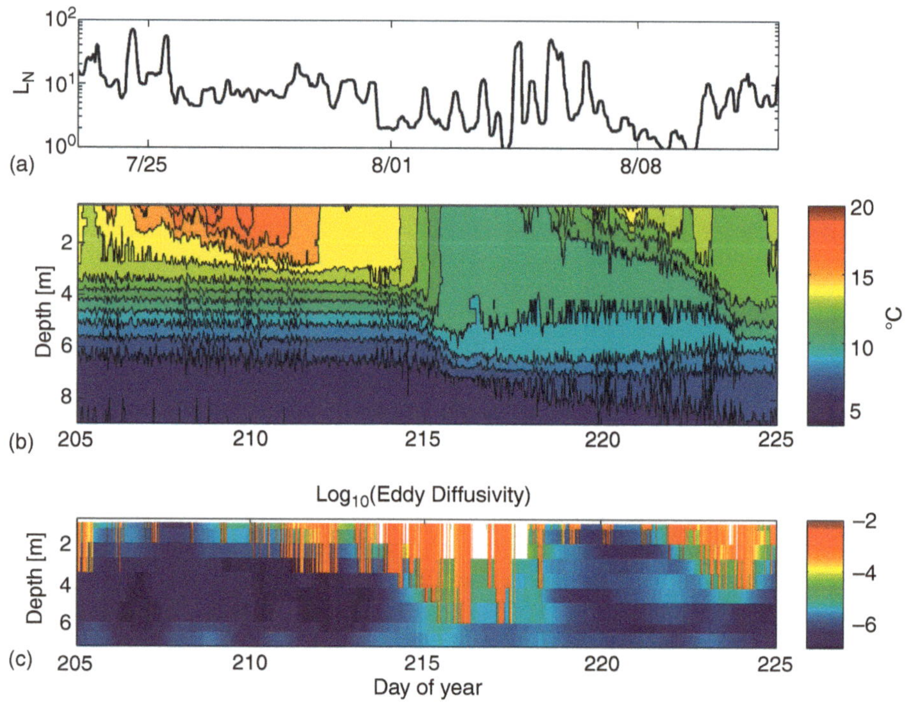

Figure 5 Time series of (a) Lake Number, (b) thermal structure as 0.5 °C isotherms, and (c) coefficient of eddy diffusivity at Lake E5, Alaska, from mid-June through mid-August 2004. The coefficient of eddy diffusivity K_z ($m^2 s^{-1}$) is computed for the SML as the product of $c_1 q^* \, l$ where q^* is obtained from the turbulent velocity scales for heat loss and wind shear in eqn. [25], l is the size of the energy containing eddies here defined as the depth of the actively mixing layer and c_1 is estimated as 0.1. Below the SML, K_z is estimated as from a heat budget method. Depths where assumptions for calculations were not met are white. (MacIntyre S, unpublished data)

SML dynamics incorporate these ideas through integration of the TKE balance over the entire SML depth in order to arrive at a prediction of the rate of deepening of the mixed layer for given forcing. This is also known as the entrainment velocity. While the details of the integration are complicated, the result can be represented as a simple energy balance:

$$\frac{g' h_m}{2}\frac{dh_m}{dt} = \frac{C_F}{2}E_s^{3/2} + \frac{C_s \Delta U^2}{2}\frac{dh_m}{dt} - \frac{E_s}{2}\frac{dh_m}{dt} \qquad [24]$$

where $g' = g\Delta\rho/\rho_0$ is the effective gravity associated with the density jump $\Delta\rho$ across the mixed layer base.

The LHS of eqn. [24] is the rate of change of potential energy, i.e., the rate at which work must be done to lift heavy fluid from the base of the mixed layer up to the centroid of the SML. The terms on the RHS represent respectively the flux of TKE from the surface that is available for mixing, the generation of TKE in the interface by shear, and the rate at which energy must be added to make entrained fluid turbulent. Aside from the constant, the formulation of mixed layer shear production is exact in that when a slower fluid from below the mixed layer is entrained there is a loss of mean kinetic energy. Turbulence in the SML is parameterized by the combination of turbulence associated with the wind and cooling:

$$E_s = \frac{1}{(C_E + C_F)^{2/3}}\left(u_f^3 + C_N^3 u_*^3\right)^{2/3} \qquad [25]$$

The constants in eqns. [24] and [25] reflect efficiencies of mixing, e.g., only a small fraction of the TKE introduced to the SML ends up increasing the potential energy of the fluid. As determined from laboratory, numerical and field experiments, they are given in **Table 1**.

While eqn. [24] is relatively complete, it includes simpler, limiting cases such as the slow deepening of the mixed layer due to wind stresses when ΔU and u_f are negligible:

$$\frac{dh_m}{dt} = \frac{C_F E_s^{3/2}}{g' h_m} = \frac{C_F C_N^3}{C_E + C_F}\frac{u_*^3}{g' h_m} = 0.42\frac{u_*}{Ri} \qquad [26]$$

The inclusion of interfacial shear production, which has much higher efficiency because it takes place in the entrainment interface, gives an entrainment law, a relation between mixed layer parameters and deepening rate that depends on shear as is often observed in laboratory experiments. An interesting limit is one in

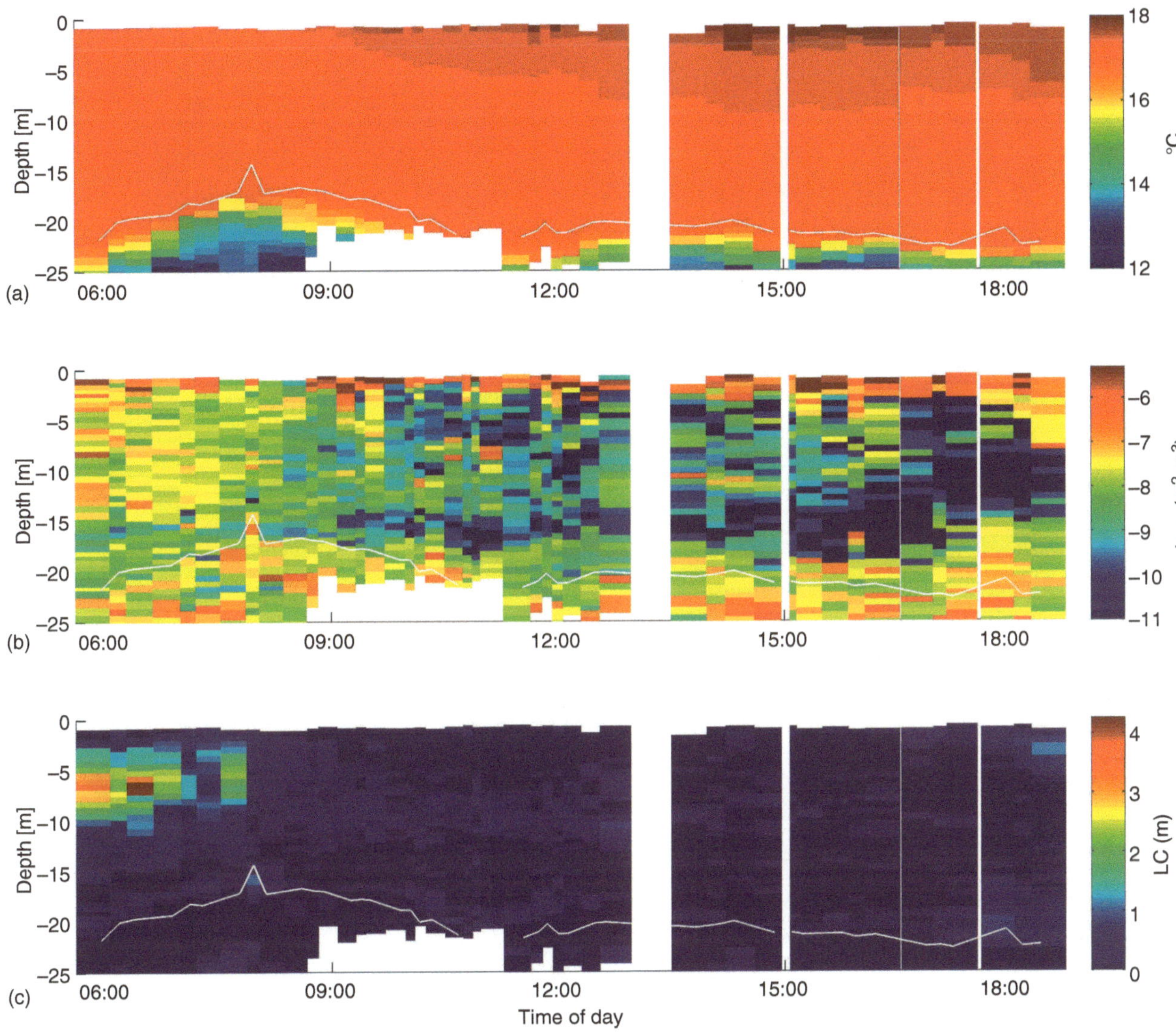

Figure 6 Time series of (a) temperature, (b) rate of dissipation of turbulent kinetic energy (ε), and (c) size of turbulent eddies in Lake Tahoe, CA in Sept. 2001. Data were obtained with a SCAMP microstructure profiler (www.pme.com). Winds were light (<6 m s^{-1}) throughout the period.

which shear dominates; in this case, the mixed layer depth is set by a stability condition similar to (1):

$$h_{\mathrm{m}} = C_{\mathrm{s}} \frac{\Delta U^2}{g'} \qquad [27]$$

In contrast to the 'stirring' limit, eqn. [27] implies a strong coupling between mixed layer motions and deepening, e.g., showing that the entrainment law depends on the rate at which the shear changes in time.

In lakes, shear and wind stirring are not independent. While the details depend on lake geometry and stratification, and the importance of rotation, the simple model of a two-layered lake with constant depth and a suddenly imposed wind provides a good basis for understanding when and how shear can be important. In this case, the velocity shear builds linearly in time until the effects of the edges of the lake are felt, cutting off the shear. Analysis of this case shows that the importance of shear production to entrainment can be assessed using a parameter known as the Wedderburn number:

$$W = \frac{g' h_{\mathrm{m}}^2}{u_*^2 L} \qquad [28]$$

where L is the length of the lake. When $W < 1$, shear production is important and deepening, whereas as W is increased above 10, shear production becomes less and less important and deepening is slower (**Figure 8**).

Mixed layer shear also brings into play the shearing instability known as the Kelvin Helmholtz instability, with its distinctive spiral eddying motion, Kelvin Helmholtz (KH) billows (**Figure 9**). KH billows

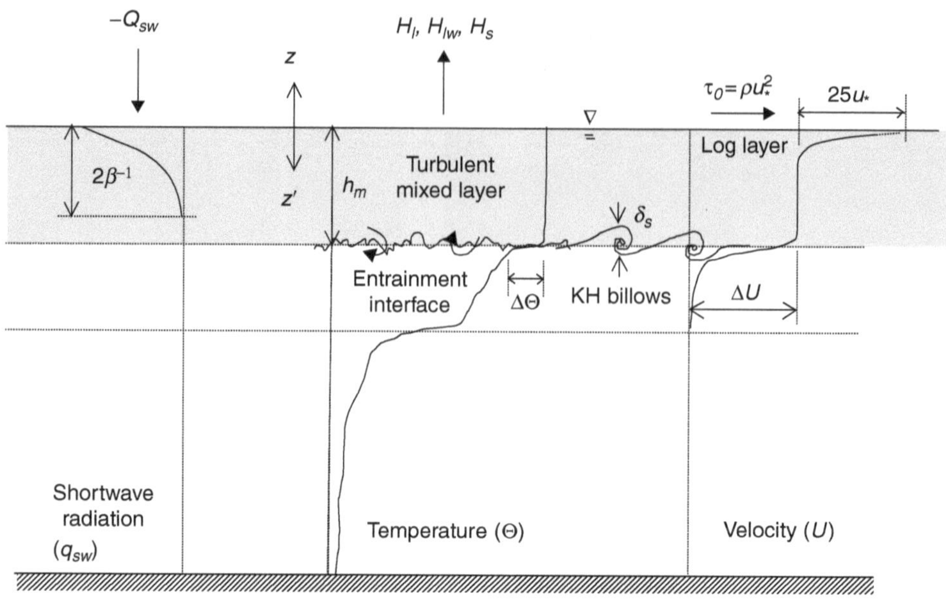

Figure 7 Sketch of SML showing turbulence dynamics for mixed layer deepening – Adapted from Spigel RH, Imberger J, and Rayner KN (1986) Modeling the diurnal mixed layer *Limnology and Oceanography* 31: 533–556.

Table 1 Constants in the mixed layer model (after Imberger, 1985)

Constant	C_E	C_F	C_N	C_S
Estimated value	1.33	0.25	1.33	0.24

develop when $Ri < 0.25$, and grow until limited by stratification they collapse, locally mixing fluid such that Ri afterwards is >0.25. Empirically, this can be represented by the simple prescription: billowing of a sheared interface will lead to a stable interface thickness

$$\delta_s = 0.3\frac{\Delta U^2}{g'} \qquad [29]$$

Arguably one of the most accurate rules applying to stratified turbulence, when applied to the two layer lake with $W \sim 1$, eqn. [29] predicts that besides rapid deepening of the mixed layer, mixed layer shear will lead to billowing that can give interface thicknesses that occupy a large fraction of the water column.

Upwelling

Given that the steady slope of the interface under the action of the wind is $O(u_*^2/g'h_m)$ (refer 'see also' section), W also measures the upwind displacement of the interface relative to the at-rest mixed layer thickness. Thus, when $W \sim 1$, the interface should surface at the upwind end of the lake, i.e., there will be upwelling (**Figure 10**). This is important because upwelling fundamentally changes how mixing takes place. In the absence of upwelling, entrainment at the mixed layer base does not affect the fluid below. In contrast, during upwelling, fluid from below the mixed layer is mixed with water in SML, creating intermediate density fluid. While upwelling is underway, the steady wind-driven flow disperses the upwelled fluid downwind, in a process, once described as 'edge leakage,' that is formally analogous to shear-flow dispersion. At this time, horizontal density gradients develop in the SML. When the wind ceases, gravity induces this layered fluid to flow as intrusions into the thermocline resulting in a thickening rather than a sharpening of the thermocline. This behavior would not have been predicted solely on the basis of a one-dimensional model of the lake.

Finally, while the two-layer model requires relatively strong winds to create upwelling, in reality since the stratification in lakes is continuous not layered, upwelling from within the thermocline, rather than from the hypolimnion is relatively common. A second more general number, one that is based on the actual stratification and lake hypsography, rather than an assumed layered distribution and constant depth, has been termed the Lake number (usually written as L_N),:

$$L_N = \frac{gS_t\left(1 - \frac{z_T}{z_m}\right)}{\rho u_*^2 A_0^{3/2}\left(1 - \frac{z_g}{z_m}\right)} \qquad [30]$$

where z_T is the thermocline depth, z_M is the maximum depth, z_g is the depth of the center of volume, A_0 is the surface area, and

$$S_t = \int_0^{zm} (z - z_g)A(z)\rho(z)dz \qquad [31]$$

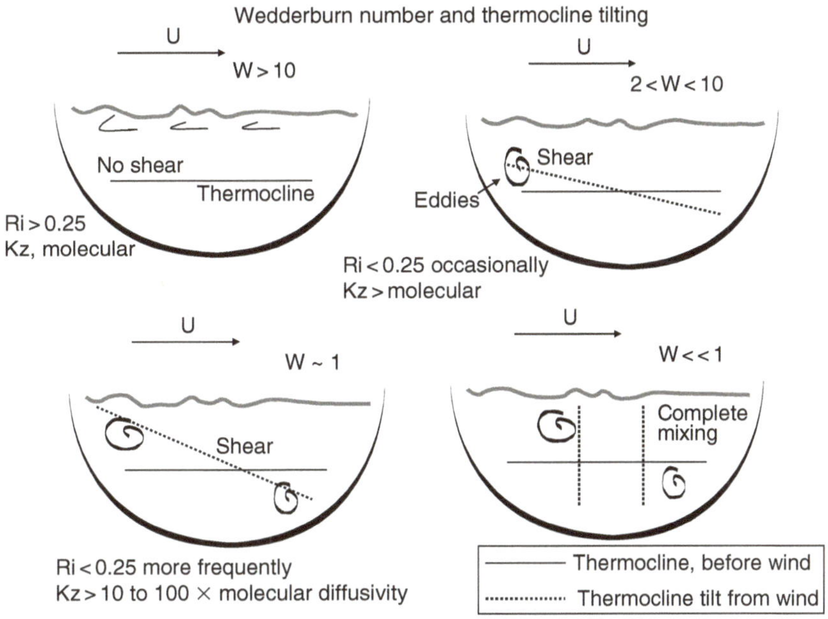

Figure 8 Tilting of the thermocline as a function of Wedderburn number or related Lake number and anticipated changes in *Ri* and measured changes in metalimnetic K_z (Conceptualization developed from Spigel and Imberger (1980) and Monismith (1986) and changes in K_z based on observations from numerous field experiments). Adapted and used with permission from MacIntyre S (2008) Describing fluxes in lakes using temperature arrays and surface meteorology. *Verh. Internat. Verein. Limnol.* 30(3).

Figure 9 Kelvin Helmholtz billows created in a two-layer laboratory shear flow. The upper layer is moving to the left whereas the lower (denser) layer is moving to the right. Time increases going from the first panel to the fourth and the sequence shows the build up, roll over, and finally the turbulent mixing of the billows. Photos by Prof. W. Debler; used by permission.

is the Schmidt stability of the lake. Note that in general, $z_T \neq h_m$, since the main thermocline may be deeper than the diurnal thermocline where active entrainment takes place. By analogy to *W*, when $L_N < 1$, upwelling comes from the hypolimnion, while even for L_N up to ~10, some form of upwelling is important, i.e., mixing by upwelling is important to overall stratification and SML dynamics. A time series of L_N illustrates the dynamic nature of this variable over the stratified period. When it drops to values near 1, the increased shear in the metalimnion is indicated by increases in coefficient of eddy diffusivity, a term indicative of the turbulence in lakes and oceans, in the top of the metalimnion as well as deeper in the water column (**Figure 5**). The accompanying deepening of the SML and diurnal thermocline, particularly that during the heating period from days 218–223, illustrates the enhanced shear when L_N is order 1.

Synthesis

Overall, this process-based view of SML dynamics can be used to explain both diurnal, seasonal, and latitudinal variations in SML depth and stratification in lakes (refer 'see also' section). In small sheltered temperate lakes the conceptual model is as follows: During the spring warming, the lake stratifies, with possible episodes of complete mixing and low L_N. As the stratification strengthens through summer, L_N tends to be larger (>10) and mixing is driven more by stirring and convection at night. In cases where a shallow diurnal SML develops over a deeper seasonal thermocline, deepening of the diurnal SML can be

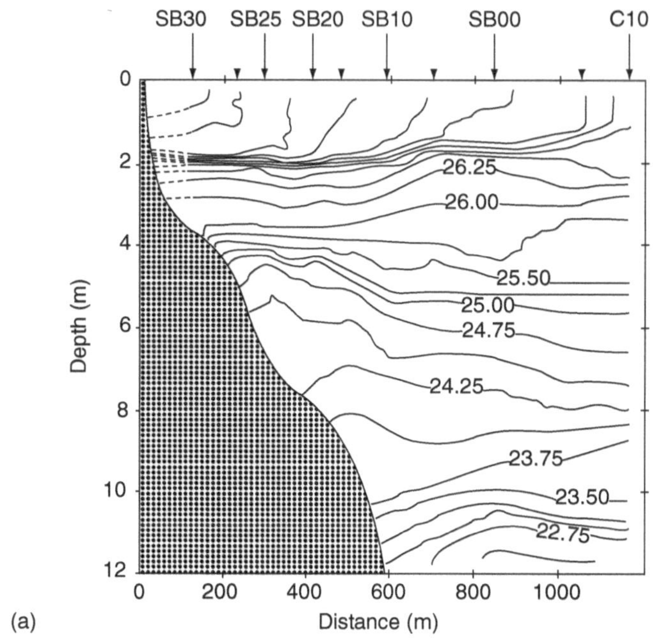

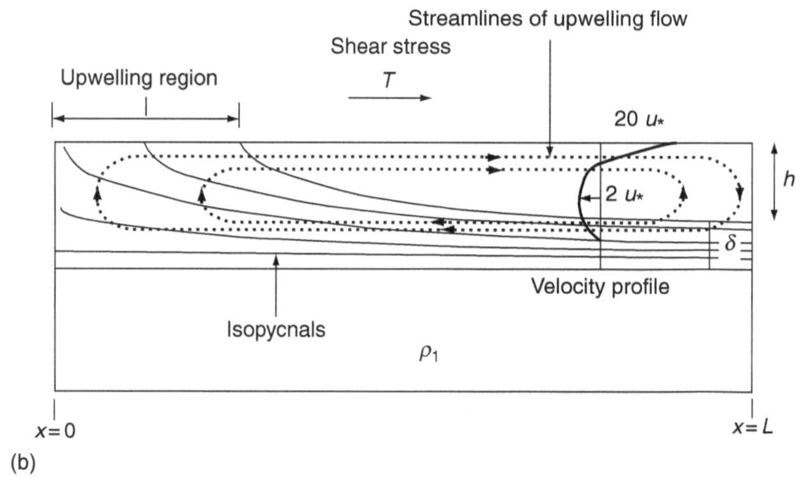

Figure 10 (a) Temperature structure during upwelling in Wellington Reservoir, Feb 1985 – figure taken from Monismith *et al.* (1990) – used with permission by ASLO. In this case the wind is blowing from the right to the left of the picture. The small triangles mark the locations of CTD casts from which the contoured temperature field was produced. (b) Conceptual model of circulation and mixing due to upwelling – figure taken from Imberger J and Monismith SG (1986) Appendix to Monismith (1986). Again, isotherms are shown, but in this case the wind stress is in the opposite direction from what is shown in a.

driven by shear and heat loss particularly from evaporation, at least until the SML base reaches the main thermocline. For larger lakes and small unsheltered lakes, the summer stratified period includes multiple episodes with low L_N. These commonly occur during the passage of cold fronts, so mixed layer deepening can be rapid due to both shear and heat loss (**Figures 5** and **6**). In fall, cooling will deepen the mixed layer, but events with $L_N < 1$ will lead to more rapid homogenization.

An important aspect of both stirring and shearing is that the overall buoyancy of the SML ($g'h_m$) never decreases and generally increases as the mixed layer gets deeper. Thus, as the mixed layer deepens due to 'mechanical' forcing, the interface itself becomes more stable. Since billowing spreads the interface, it can work hand in hand with stirring to produce a more rapid deepening of the SML than might be achieved by stirring alone. On the other hand, cooling that drives convection, also weakens the buoyancy in the mixed layer and thermocline; as cooling-driven deepening proceeds, the interface becomes less stable, thus accelerating mixed layer deepening. This process occurs in the night during summer and year round in the tropics. For temperate lakes, it occurs to an even larger extent during the passage of cold fronts during the stratified period, and during fall cooling.

The connection of mixed layer shear and mixing offers a dynamical explanation of why larger lakes tend to have deeper mixed layers than smaller ones if exposed to the same wind stress: they are more likely to have small values of L_N and thus to experience greater rates of mixed layer deepening in response to wind forcing. However, addressing this problem correctly is complicated and the mixed layer depth of large lakes is likely to be deeper than predicted based on dynamical arguments alone. Small lakes are sheltered by trees and thus have much reduced fetch. In addition, wind speeds in the center portions of large lakes are much larger than over adjacent land and evaporation rates will also be higher in large lakes. Thus, while the SML will be deeper in larger lakes in the same region, it may be even deeper than predicted based on calculations of Lake number (**Figure 11**). It is important to note that this classification only applies to lakes that are smaller in horizontal dimension that the first-mode baroclinic Rossby radius, something that is typically the order of 10 km. For larger lakes, the picture is more complicated since the nearshore and offshore regions behave somewhat differently (refer 'see also' section). In the middle of large lakes, mixed layer shear is limited by Coriolis forces rather than by boundary effects such that the response of the lake to starting or stopping the wind is to generate inertial oscillations rather than seiches. However, other mixed layer processes, e.g., LCs, wave breaking, etc. appear to be identical.

Understanding the dynamical implications of W and L_N is also important for differentiating mixing dynamics in lakes of the same size across latitudinal gradients (refer 'see also' section). For instance, small temperate lakes such as Lawrence Lake, MI (45° 42′ N, 84° 52′ W), have W and $L_N > 100$ throughout the summer stratified period. Thus, all the mixing in the

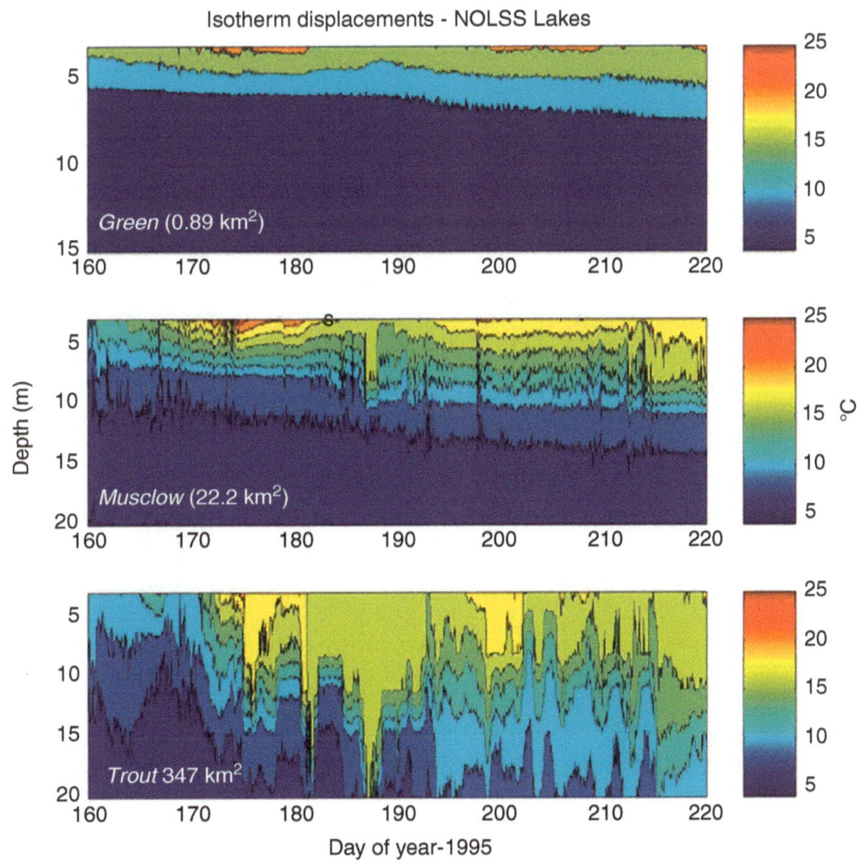

Figure 11 Thermal structure in a set of nearby lakes located in Ontario but having different surface areas. The small lakes are stratified to the surface, the moderate sized lake has a shallow mixed layer except when wind forcing increases for brief periods, and the depth of the SML in large lake is highly variable owing to wind forcing, higher evaporation rates, and the oscillations of internal waves. The onset of stratification was also delayed in the larger lake. Adapted from Kratz TK MacIntyre S, and Webster KE (2005) Causes and consequences of spatial heterogeneity in lakes. In: Lovett GM, Jones CG, Turner MG, and Weathers KC (eds.). *Ecosystem Function in Heterogeneous Landscapes*. pp. 329–347. Springer, NY, – used by permission.

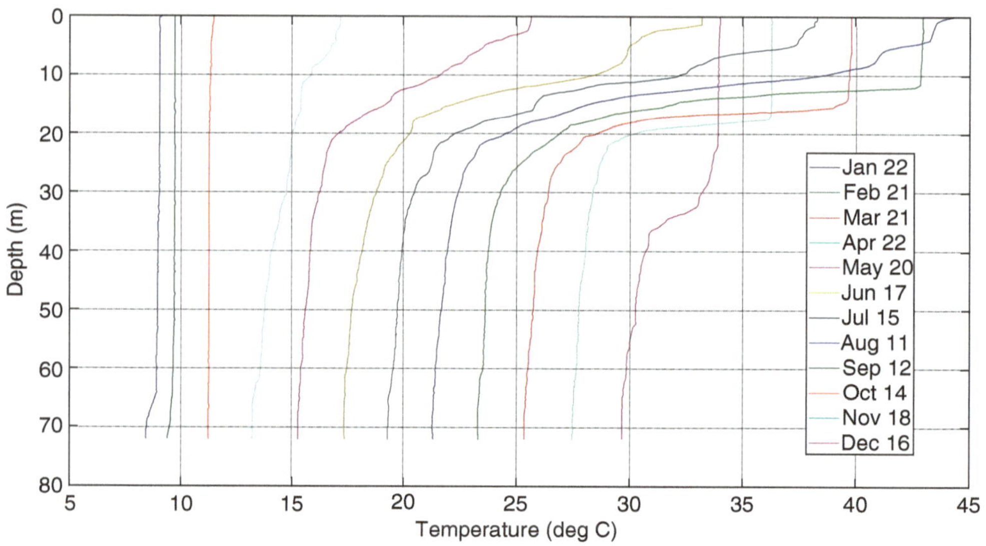

Figure 12 Annual variation of temperatures in Lake Biwa (S. Endoh, unpublished data). Taken from http://www.edu.shiga-u.ac.jp/~endoh/doc/endoh.htm and used by permission. Each profile is progressively offset by 2 °C.

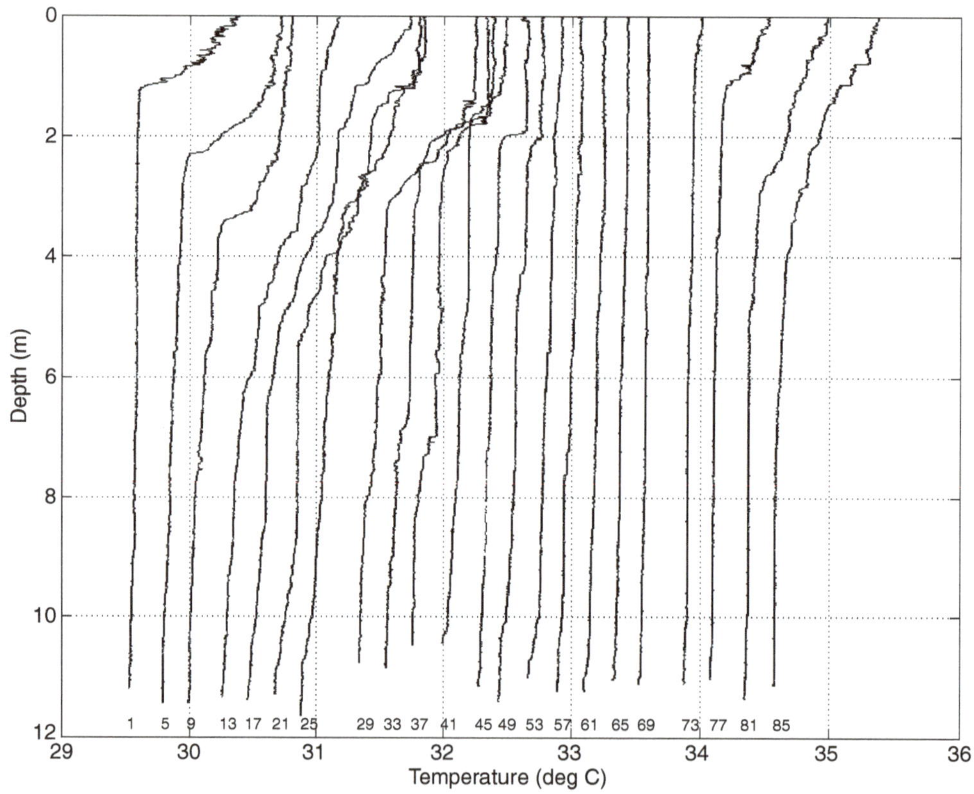

Figure 13 Daily variation in a tropical lake: Kranji Reservoir Singapore March 23–24, 2006 (Hench J, Monismith S, Xing Z, and Lo E, unpublished data). The numbers refer to which cast a given profile corresponds. The profiles shown are about an hour apart, with each profile shifted horizontally 0.2° h^{-1} relative to the first profile, which was taken at ca. 1300 hours. The variation seen above for one day is comparable to what is seen over the entire year for Kranji Reservoir. The 'squiggle' seen in many profiles are symptomatic of active turbulent mixing. Note that a uniform, clearly defined SML is only evident in *some* of the profiles.

mixed layer is caused by stirring and heat loss and deepening is dominated by heat loss. In contrast, similar sized lakes in the arctic frequently experience events with $L_N \sim 1$ (**Figure 5**). In such lakes, mixed layer deepening is driven by the combination of heat loss and shear and deeper mixed layers may result. Furthermore, evaporation is larger in warmer water bodies exposed to the same wind forcing. Hence, we predict that mixed layer depth will be deeper in the tropics than the temperate zone for lakes of similar sizes and light penetration, although this effect may be offset by the fact that the thermal expansivity is considerably larger for tropical lake temperatures which can often be in the range 25 °C – 30 °C than it is for colder, higher latitude lakes.

The case of $L_N \leq 1$ is particularly relevant to shallow, polymictic lakes where strong stratification does not develop. Besides shallow lakes, lakes in the tropics less than 100 m deep tend to have weak stratification in the morning and low L_N because annual variations in heating and cooling are often minimal compared with the diurnal variation in forcing. That is, the cold water found in the hypolimnion of a temperate lake is not present in a tropical lake, and the annual variations in stratification seen in the temperate zone (**Figure 12**) may occur over the course of a day in the tropics (**Figure 13**). Thus, the dynamical balances of heating, cooling, wind mixing, and shear vary with lake size, latitude and by season thus creating large but predictable differences in turbulence and the habitat of organisms.

Summary

The SML of lakes and reservoirs is a dynamic region in which wind, wave, and cooling produced turbulence competes with the stabilizing effects of shortwave radiation to determine mixing characteristics and rates of mixed layer deepening. It is generally much more energetic than most of the rest of the lake. Flows in the SML seem to follow scaling laws that parallel those found for turbulent wall flows, particularly those that apply to the atmospheric boundary layer, although there is the added complication that Langmuir cells, large-scale organized structures not generally found in atmospheric or engineering flows, may be important. In addition, mixed layer dynamics must take into account cooling at the air–water interface. Finally, while the traditional model of the SML and of lake stratification is a one-dimensional view, i.e., temperatures etc. only vary vertically, it is clear that during strong wind events, horizontal variability can become significant, especially in the case of upwelling.

See also: The Benthic Boundary Layer (in Rivers, Lakes, and Reservoirs); Currents in Stratified Water Bodies 1: Density-Driven Flows; Currents in Stratified Water Bodies 2: Internal Waves; Currents in Stratified Water Bodies 3: Effects of Rotation; Currents in the Upper Mixed Layer and in Unstratified Water Bodies; Density Stratification and Stability; Small-Scale Turbulence and Mixing: Energy Fluxes in Stratified Lakes.

Further Reading

Anis A and Singhal G (2006) Mixing in the surface boundary layer of a tropical freshwater reservoir. *Journal of Marine Systems* 63: 225–243.

Carmack EC and Farmer DM (1982) Cooling processes in deep temperate lakes: A review with examples from two lakes in British Columbia. *Journal of Marine Systems* 40: 85–111.

Dillon TM and Powell TM (1979) Observations of a surface mixed layer. *Deep Sea Research* 26A: 915–932.

Fischer HB, List EJ, Koh RCY, Imberger J, and Brooks NH (1979) *Mixing in Inland and Coastal Waters,* 483 pp. San Diego: Academic Press.

Gal G, Imberger J, Zohary T, Antenucci J, Anis A, and Rosenburg T (2003) Simulating the thermal dynamics of Lake Kinneret. *Ecological Modelling* 162: 69–86.

Imberger J (1985) The diurnal mixed layer. *Limnology and Oceanography* 30: 737–771.

Imberger J and Patterson J (1990) Physical limnology. *Advances in Applied Mechanics* 27: 303–475.

Kundu PK and Cohen IM (2004) *Fluid Mechanics,* 3rd edn., 759 pp. San Diego: Academic Press.

MacIntyre S (1993) Vertical mixing in a shallow eutrophic lake: Possible consequences for the light climate of phyoplankton. *Limnology and Oceanography* 38(4): 798–817.

Monismith SG (1986) An experimental study of the upwelling response of stratified reservoirs to shear stress. *Journal of Fluid Mechanics* 171: 407–439.

Monismith SG and Magnaudet JM (1998) On wavy mean flows, strain, turbulence and Langmuir cells. In: Imberger J (ed.) Physical processes in Lakes and Oceans. *Coastal and Estuarine Studies* 54: pp. 101–110. Washington DC: American Geophysical Union.

Spigel RH and Imberger J (1980) The classification of mixed layer dynamics in lakes of small to medium size. *Journal of Physical Oceanography* 10: 1104–1121.

Stefan H and Ford DE (1975) Temperature dynamics of dimictic lakes. *Journal of Hydraulic Division, American Society of Civil Engineering* 101(HY2): 334–354.

Thorpe SA (1977) Turbulence and Mixing in a Scottish Loch. *Philosophical Transactions of Royal Society of London Series A* 286: 125–181.

Wuest A and Lorke A (2003) Small-scale hydrodynamics in lakes. *Annual Review of Fluid Mechanics* 35: 373–412.

Small-Scale Turbulence and Mixing: Energy Fluxes in Stratified Lakes

A Wüest, Eawag, Surface Waters – Research and Management, Kastanienbaum, Switzerland
A Lorke, University of Koblenz-Landau, Landau/Pfaly, Germany

Introduction

Density Stratification and Mixing – the Basin Scale

Nearly all lakes, reservoirs, and ponds that are deeper than a few meters, experience cycles of density stratification and destratification. Most important for this variation is the temperature-dependence of water density. During spring/summer – or the wet season in the tropics – the water is heated from above and a surface layer (SL: typically a few m thick) with warmer and hence lighter water develops on top of the cooler and heavier water below (**Figure 1**). In addition, although more important in saline lakes than fresh water ones, biological and hydrological processes may strengthen the density stratification by generating a vertical gradient in the concentration of dissolved substances (salinity). The resulting stratification is usually depicted by a strong density gradient (also called pycnocline), separating the SL from the deeper reaches of the water column (indicated as metalimnion and hypolimnion in **Figure 1**). Mixing of heavier water from greater depth with lighter water from the SL implies that water parcels of different densities are exchanged in the vertical direction (**Figure 2**). It is evident that mechanical energy is needed to move these water parcels against the prevailing density gradient, which forces lighter water up and heavier water down. The amount of energy needed to overcome vertical density stratifications is therefore determined by the potential energy ΔE_{pot} (**Figure 1**) stored in the stratification. ΔE_{pot} is calculated from the vertical separation of the centre of volume of the water body and its center of mass. Density stratification results in a lowering of the centre of mass by the vertical distance Δh_M (**Figure 1**) and the energy needed to overcome the stratification and to mix the entire water column is $\Delta E_{pot} = H\rho g \Delta h_M$ ($J\,m^{-2}$), where H is the average depth of the water body, g is the gravitational acceleration, and ρ is the density. Density stratification thus imposes stability on the water column and reduces – or even suppresses – vertical mixing.

Besides convective mixing in the SL – caused by seasonal or nocturnal surface cooling – in most lakes and reservoirs, the major source of energy for vertical mixing is the wind, whereas river inflows usually play a minor role (**Figure 1**). As water is 800 times denser than air and as momentum is conserved across the air–water interface, SLs receive only about 3.5% of the wind energy from the atmosphere above. Surface waves transport a portion of this energy to the shore where it is dissipated; the remaining energy causes large-scale currents, with surface water flows of 1.5–3% of the wind velocity. Moreover, surface currents cause a stratified water body to pivot with warm water piling up at the downwind end (causing downwelling) and deep-water accumulating at the upwind end (causing upwelling). After the wind ceases, the water displacement relaxes and various internal waves develop – including basin-scale seiches – inducing motion even in the deepest layers.

These deep-water currents are usually one order of magnitude less energetic than those in the SL. Typical deep flows of a few centimeters per second (or $\sim 1\,J\,m^{-3}$) with energy dissipation of less than $1\,mW\,m^{-2}$ are able to reduce the potential energy of the stratification by only ~ 0.01–$0.05\,mW\,m^{-2}$. Compared to the potential energy stored in the stratification (order of $1000\,J\,m^{-2}$; **Figure 1**) it would take much longer than one season to entirely mix a moderately deep lake. This implies that wind energy input (**Figure 1**) forms the vertical hypolimnion structure at times of weak stratification (beginning of the season), whereas the wind is not able to substantially change the vertical structure once the strong stratification is established. Therefore, in most regions on Earth, only very shallow waters (less than a few meters deep (such as Lake Balaton, Hungary) are found to be entirely nonstratified, even during the summer season. The majority of lakes and reservoirs deeper than a few meters are thus only 'partially' mixed to a limited depth, which is basically defining the SL. For those lakes that show a pronounced SL, its maintenance is mostly supported by night-time cooling. In this article, we focus on the 'limited' mixing below the SL, which occurs in the metalimnion and hypolimnion (**Figure 1**).

Density Stratification and Mixing – the Small Scale

The same concepts of stability and mixing – described in the preceding section for the entire water body – also apply locally within the water column for small-scale vertical mixing of stratified layers. Local stability of the density stratification is quantified by the

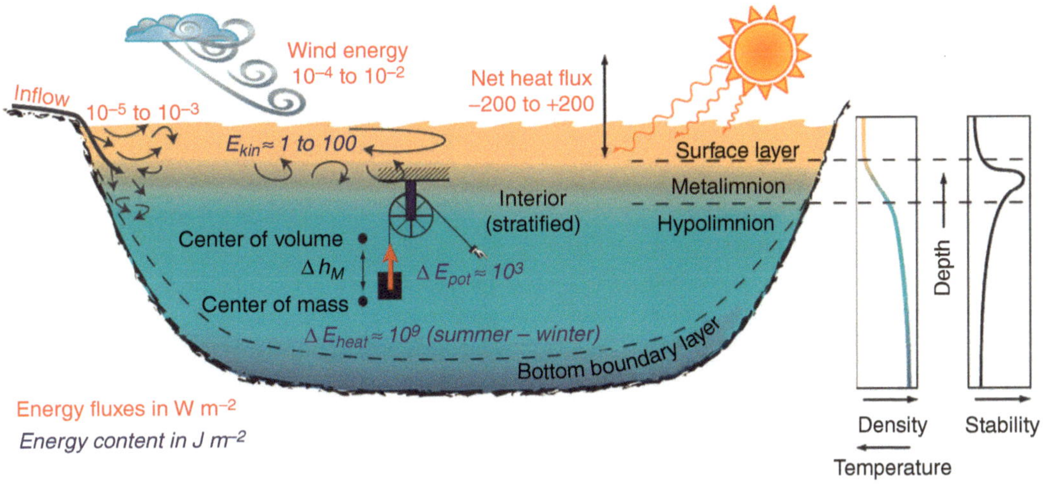

Figure 1 Energy fluxes (heat, wind, and river inflow; in red) into the water (W m^{-2}) and energy content (heat, kinetic energy, potential energy; in blue) stored in the lake water body (J m^{-2}). Note that the energy fluxes and contents related to heat are many orders of magnitude larger than those of kinetic and potential energy. The effect of mixing by the river is only local and less effective than wind. The stratified part of the lake (below surface layer) has historically been divided into the metalimnion (see large stability, right) and the deep hypolimnion (weak stratification) below. The lower water column can also be differentiated into an interior region (away from the boundaries) which is quiescent except during storms and a bottom boundary layer where turbulence is enhanced. Adopted from Imboden DM and Wüest A (1995) Mixing mechanisms in lakes. In: Lerman A, Imboden D, and Gat JR (eds.) *Physics and Chemistry of Lakes*, vol. 2, pp. 83–138. Berlin: Springer-Verlag.

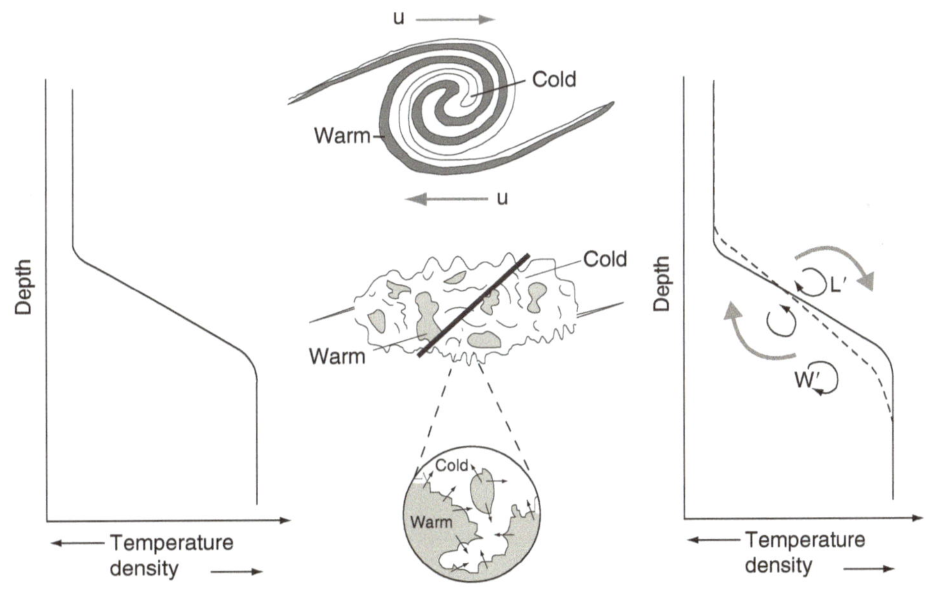

Figure 2 The effect of turbulent mixing in a stable stratification: if the vertical gradient of horizontal currents (current shear $\partial u/\partial z$) is stronger then the stability of the water column (eqn. [1]), Kelvin Helmholtz instabilities can develop (top of middle panel) bringing warmer (lighter) and cooler (heavier) water in close proximity (bottom of middle panel). Finally, heat (or any other water constituent) is mixed by molecular diffusion across the manifold small-scale interfaces, which are generated by turbulence. The turbulent exchange of small water parcels leads to a fluctuating vertical heat flux (see example in **Figure 3**) which averages to a net downward heat flux. As a result, the original temperature profile (left) is modified (right): the gradient is weakened and expanded vertically with heat transported from top to bottom, and density vice versa, across the interface. Figure after the idea of Winters KB, Lombard PN, Riley JJ, and D'asaro EA (1995) Available potential-energy and mixing in density-stratified fluids. *Journal of Fluid Mechanics* 289: 115–128. Experiments were first performed by Thorpe SA (1973) Experiments on instability and turbulence in a stratified shear-flow. *Journal of Fluid Mechanics* 61: 731–751; and the phenomenon of sheared stratification in lakes was reported by Mortimer CH (1952) Water movements in lakes during summer stratification; evidence from the distribution of temperature in Windermere. *Philosophical Transactions of the Royal Society of London B: Biological Sciences* 236(635): 355–398; and by Thorpe SA (1977) Turbulence and Mixing in a Scottish loch. *Philosophical Transactions of the Royal Society of London A: Mathematical Physics and Engineering Sciences* 286(1334): 125–181.

Brunt-Väisälä frequency (also buoyancy frequency) N (s^{-1}), defined by:

$$N^2 = -\frac{g}{\rho}\frac{\partial \rho}{\partial z} \qquad [1]$$

z is the depth (positive upward). As a result of wind-forced motions, a vertical gradient of the horizontal current u (shear $\partial u/\partial z$) is superimposed on the vertical density gradient $\partial \rho/\partial z$. Depending on the relative strength of N compared to the current shear $\partial u/\partial z$ such a stratified shear flow may eventually become unstable and develop into turbulence (**Figure 2**).

Although the large-scale (advective) motions are mainly horizontal, the turbulent eddies are associated with random velocity fluctuations in all three dimensions (u', v', w'). Turbulent kinetic energy (TKE) ($J\,kg^{-1}$) is defined as the energy per unit mass of water which is contained in these velocity fluctuations:

$$\text{TKE} = \frac{1}{2}(\overline{u'^2} + \overline{v'^2} + \overline{w'^2}) \qquad [2]$$

In stratified turbulence, the vertical velocity fluctuations w' are of particular importance as they transport water parcels and their contents in the vertical direction (**Figure 3**). The product of the vertical velocity fluctuations w' and the associated density fluctuations ρ' describes an instantaneous vertical flux of density ($w'\rho'$ ($kg\,m^{-2}\,s^{-1}$)). Resulting from many irregular and uncorrelated fluctuations (**Figure 3**) the averaged flux $\overline{w'\rho'}$ leads to a net upward mass flux, which is usually expressed as a buoyancy flux J_b:

$$J_b = -\frac{g}{\rho}\overline{w'\rho'} \qquad [3]$$

Therefore, we can interpret vertical mixing as an upward flux of mass, which causes a change of the potential energy of the stratification (**Figures 1** and **2**), expressed as a buoyancy flux (eqn. [3]). The required energy originates from the TKE, which is itself extracted from the mean (horizontal) flow. Approximately 90% of the TKE, however, does not contribute to the buoyancy flux (and hence to vertical mixing) but is instead dissipated into heat by viscous friction, without any further effect. By defining local rates of production P ($W\,kg^{-1}$) and viscous dissipation ε($W\,kg^{-1}$) of TKE, the simplest form of TKE balance can be formulated as:

$$\frac{\partial}{\partial t}\text{TKE} = P - \varepsilon - J_b \qquad [4]$$

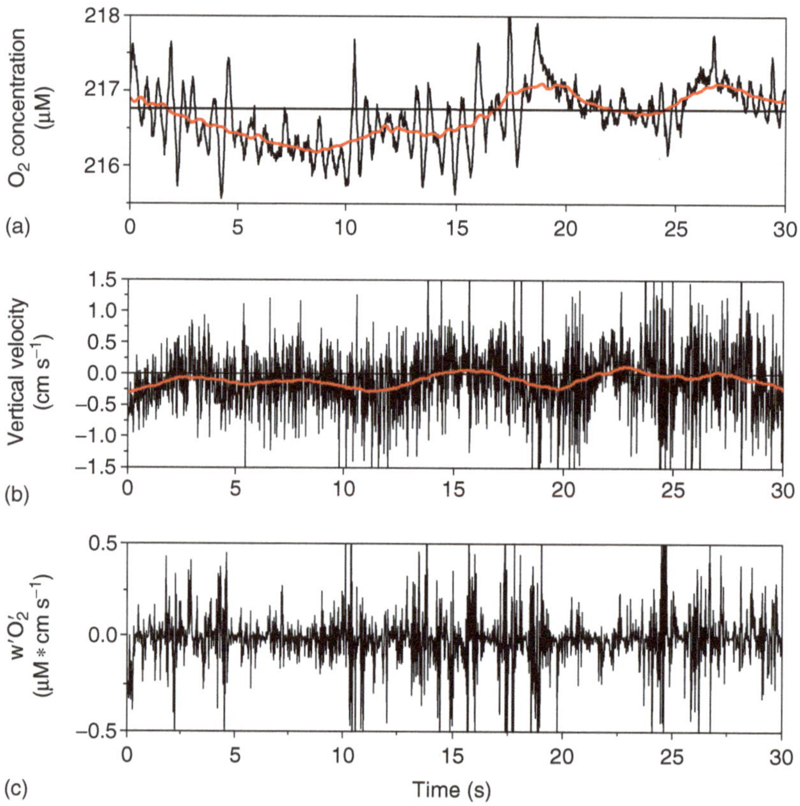

Figure 3 Time series of O_2 concentration (thin line, a) and vertical velocity w' (thin line, b; positive = upward), as measured 10 cm above the sediment in reservoir Wohlensee (Switzerland) at a frequency of 64 Hz. Red lines indicate the temporally varying averages, determined as running mean, whereas the black horizontal line marks the averages. Panel (c) shows the instantaneous eddy flux – covariance of w' and O_2': The average downward O_2 flux over the 30 s (~1900 data pairs) is $-6.4\,mmol\,m^{-2}\,day^{-1}$. Data source: Claudia Lorrai, Eawag.

As mentioned above, the dissipation rate is usually much larger than the buoyancy flux, and hence the mixing efficiency γ_{mix}, which is defined as the ratio

$$\gamma_{mix} = \frac{\mathcal{J}_b}{\varepsilon} = \frac{-g\rho^{-1}\overline{w'\rho'}}{\varepsilon} \qquad [5]$$

is much smaller than 1. A number of studies in stratified lakes and reservoirs have revealed typical mixing efficiencies in the range of 10–15%.

Density Stratification and Mixing – the Turbulent Transport

The local flux of a water constituent is given by the product of the velocity times the concentration. In stratified waters, the time-averaged vertical velocity is often close to zero (negligible) and thus, the vertical fluxes stem only from the fluctuations of velocity and concentration, such as explained above for the vertical mass flux $\overline{w'\rho'}$ caused by the turbulence. This concept holds for any other water constituent, such as for oxygen, as exemplified in **Figure 3**, where the in situ measured w', O_2' and the product $w'O_2'$ is shown for a 30-s-long record. Although the momentary fluxes up and down are almost of equal variations and amounts, the averaging $\overline{w'O_2'}$ reveals slightly larger fluxes downwards to the sediment, where the oxygen is consumed.

Until recently, direct measurements of turbulent fluxes had not been possible and therefore turbulent fluxes in stratified waters are commonly expressed using the eddy diffusivity concept. Applied to the mass flux $\overline{w'\rho'}$ it implies assuming that (i) a well-defined local density gradient $\partial\rho/\partial z$ exists (due to the stratification) and (ii) the flux – in analogy to molecular diffusion – can be expressed by the eddy (or turbulent) diffusivity K_z ($m^2 s^{-1}$) multiplied by this local gradient:

$$\overline{w'\rho'} = -K_z\frac{\partial\rho}{\partial z} \qquad [6]$$

In this formulation, K_z describes the vertical transport of density caused by turbulent velocity fluctuations w' over a typical eddy distance L' given by the level of turbulence and the strength of the stratification. Therefore, in contrast to the molecular diffusion process, eddy diffusivity is neither a function of medium (water) nor of the water constituents (particulate or dissolved), but rather a property of the turbulent flow within the stratified water itself. In particular, K_z reflects the extent of the velocity fluctuations w' and the eddy sizes L': K_z can be interpreted as the statistical average $\overline{w'L'}$ of a large number of eddies, which exchange small water parcels as a result of the turbulent flow (**Figures 2** and **3**).

In addition to density, all other water properties – such as temperature or substances – are transported and mixed in the same way via the turbulent exchange of small eddies or parcels of water (**Figure 3**). The eddy diffusivity concept can be applied to any dissolved or particulate substance and the associated vertical fluxes F can be readily estimated in analogy to eqn. [6] by

$$F = -K_z\partial C/\partial z \qquad [7]$$

where C is the appropriate concentration.

Assuming steady-state conditions, i.e., by neglecting the left-hand side of eqn. [4], and combining eqns. [1], [5], and [6] yields:

$$K_z = \gamma_{mix}\frac{\varepsilon}{N^2} \qquad [8]$$

This equation provides an expression to estimate K_z from field measurements of ε and N^2 and, moreover, it demonstrates the direct proportionality of K_z on the level of turbulence (ε) and the inverse proportionality on the strength of stratification (N^2). In the last decades, two fundamentally different approaches have been used for the estimation of K_z: (i) the microstructure method and (ii) the tracer method. Method (i), is based on eqn. [8] where the dissipation of TKE or of temperature variations are measured by usually free-falling profilers which measure either temperature or velocity over small spatial scales. For example, spectral analysis of the temperature gradient signal provides estimates of ε and the local buoyancy frequency is obtained from density computed from the temperature and salinity profiles. For the application of tracer method (ii), one has to measure the three-dimensional spreading of a tracer (artificial or natural) and then infer the diffusivities (K_z) from the observations. Heat is also used as a tracer and K_z is obtained by computing a time series of the heat budget below the respective depth of a lake. Typical values in stratified natural waters are listed in **Table 1** and **Figures 4** and **5**.

Turbulence is caused by current shear, breaking surface waves, and instabilities in the internal wave field. Currents induce shear near boundaries regardless of whether the flow is stratified. Thus, the concept of eddy diffusivity is also applied to surface mixing layers and to nonstratified systems such as rivers.

Turbulence and Mixing in Stratified Lakes and Reservoirs

Turbulence Production in the Surface and Bottom Boundaries

There are fundamentally two mechanisms generating turbulence in the SL: (i) the action of wind causing

Table 1 Typical values of dissipation, stability and vertical diffusivity in stratified waters

	Dissipation[a] ε *(W kg^{-1})*	*Stability N^2 (s^{-2})*	*Diffusivity* K_z (m^2 s^{-1})*
Ocean thermocline	10^{-10}–10^{-8}	$\sim 10^{-4}$	$(0.3\text{–}3) \times 10^{-5}$
Surface layer	10^{-6}–10^{-9}	0–$\sim 10^{-5}$	10^{-5}–10^{-2}
Lake interior only (without BBL)	10^{-12}–10^{-10}	10^{-8}–10^{-3}	10^{-7}–10^{-5}
Metalimnion (basin scale)	10^{-10}–10^{-8}	$\sim 10^{-3}$	$(0.5\text{–}50) \times 10^{-7}$
Near-shore metalimnion	10^{-10}–10^{-6}	$\sim 10^{-3}$	$(0.3\text{–}3) \times 10^{-4}$
Deep hypolimnion (basin scale)	10^{-12}–10^{-10}	10^{-8}–10^{-6}	$(0.03\text{–}3) \times 10^{-4}$

[a]During storm events values are larger by orders of magnitudes for short.

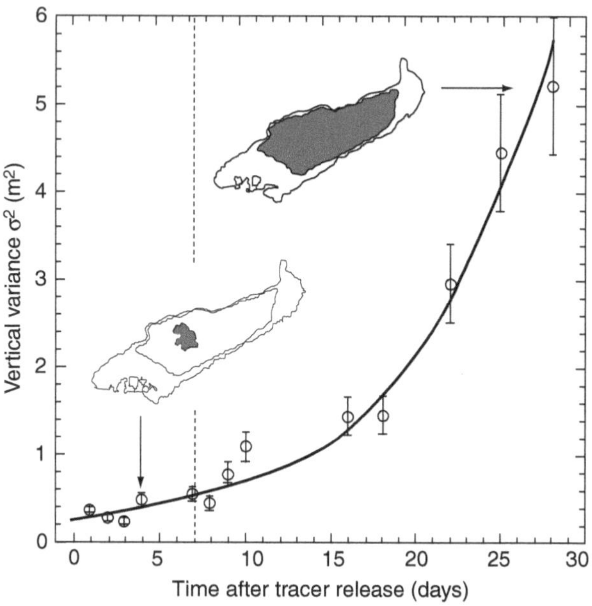

Figure 4 Vertical spreading of the tracer Uranine after injection at 25 m depth in Lake Alpnach (Switzerland). The vertical line demarcates the initial period of 7 days, during which Uranine resided in the interior of the stratified deep water. The two insets show the lake area at the surface and at the depth of the Uranine injection, as well as the horizontal distribution of the Uranine cloud (shaded in gray) after 4 and 28 days. The slow growth of the spreading in the first 7 days illustrates the quietness in the interior. The fast growth of the vertical spreading after day 7 is due to the increasing contribution of BBL mixing after the tracer has reached the sediment at 25 m depth. Reproduced from Goudsmit GH, Peeters F, Gloor M, and Wüest A (1997) Boundary versus internal diapycnal mixing in stratified natural waters. *Journal of Geophysical Research* 102: 27903–27914, with permission from American Geophysical Union.

wave breaking and shear in the top few meters of the water column and (ii) surface cooling causing the sinking of heavier water parcels. Temperature-driven mixing (case (ii)) leads to homogenization of the SL and therefore to nonstratified conditions – at least for a few hours or days before heat fluxes from/to the atmosphere restratify the SL. This process is discussed in detail elsewhere in this encyclopedia. Only in shallow ponds or basins with relatively high through-flow will turbulence have other case-specific sources.

For wind-driven mixing (case (i)), the crucial parameter governing the dynamics of turbulence in the SL is the surface shear stress τ (N m^{-2}), the force per unit area exerted on the water by the wind. This stress is equal to the downward eddy-transport of horizontal momentum from the atmosphere. Part of τ is consumed in the acceleration and maintenance of waves (τ_{Wave}), whereas the remaining momentum flux τ_{SL} generates currents and turbulence in the SL. By assuming a constant stress across the air–water interface, the two momentum fluxes on the water side equal the total wind stress ($\tau = \tau_{\text{SL}} + \tau_{\text{Wave}}$).

Immediately below the waves, the momentum flux, τ_{SL}, drives the vertical profiles of horizontal velocity $u(z)$ in the SL. If the wind remains relatively constant for hours, quasi-steady-state conditions may develop in the SL: $u(z)$ then depicts the Law-of-the-Wall $\partial u/\partial z = u_*(\kappa z)^{-1} = (\tau_{\text{SL}}/\rho)^{1/2}(\kappa z)^{-1}$, where $u_* = (\tau_{\text{SL}}/\rho)^{1/2}$ is the frictional velocity and κ (=0.41) is the von Karman constant. Because the buoyancy flux in the SL (defined in eqn. [3]) is not a large contribution in eqn. [4], we can assume a balance between the production of TKE and the rate of viscous dissipation (ε) of TKE. This local balance between production and dissipation of turbulence determines the turbulence intensity as a function of depth throughout the SL. Under those assumptions, the dissipation

$$\varepsilon = (\tau_{\text{SL}}/\rho)\ \partial u/\partial z = u_*{}^3(\kappa z)^{-1} \qquad [9]$$

is only a function of the wind-induced stress τ_{SL} (here expressed as u_*) and of depth z. Several experiments have demonstrated that dissipation is indeed inversely proportional to depth (eqn. [9]), if averaged for long enough. However, one has to be critical about the validity of eqn. [9] for two reasons: First, at the very top of the water column, breaking waves, in addition to shear stress, produce a significant part of the turbulence in the SL. This additional TKE generation at the surface can be interpreted as an injection of TKE from above. Therefore, in the uppermost layer, the turbulence exceeds the level described by eqn. [9], depending on the intensity of the wave breaking. Second,

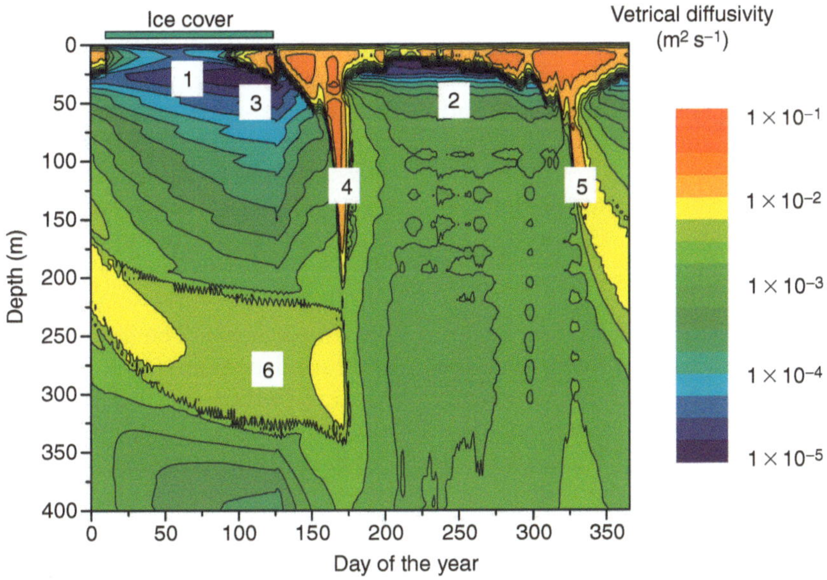

Figure 5 Vertical diffusivities in Lake Baikal simulated with a k-epsilon model. The numbers (1–6) on the contour plot indicate the main features of the seasonal stratification and changes in diffusivity: the formation of thermal stratification with weak mixing (1) during winter under the ice and (2) during summer; (3) the formation of a convectively mixed layer in spring under the ice; the deep convective mixing in (4) June and (5) November; and (6) the formation of a mixed layer near the temperature of maximum density. Here the emphasis is on the temporal and vertical structure of the turbulent diffusivity and not on the absolute accuracy, which may be difficult to achieve with turbulence modeling better than a factor of 2–3. Reproduced from Schmid M *et al.* (2007) Sources and sinks of methane in Lake Baikal: A synthesis of measurements and modeling. *Limnology and Oceanography 52:* 1824–1837, with permission from American Society of Liminology and Oceanography.

eqn. [9] relies on quasi-steady-state conditions which may hold applicable for limited episodes only.

Despite these restrictions, eqn. [9] gives a good estimate of the diffusivity in the SL, if it is weakly stratified. Equations [8] and [9] reveal that the rate of mixing increases substantially within the SL as the surface is approached. The corresponding stability N^2 decreases at the surface and maintains rapid mixing. Therefore, gradients of temperature, nutrients, and particulates are usually smallest at the surface and increase with depth. During sunny days, diurnal thermoclines form with mixing reduced below them. On cloudy, windy days, the SL may mix fully and may even deepen depending upon the surface forcing. Factors that affect the depth of mixing are discussed elsewhere in this encyclopedia. It is typically a few m during the warm season and a few tens of meters during the cold season. Below, a strong density gradient (pycnocline) can develop leading to the separation between the SL and the metalimnion/hypolimnion. In the stratified interior (away from the BBL; see below), the effect of wind is shielded and the mixing regime is completely different.

As discussed in greater detail (see **The Benthic Boundary Layer (in Rivers, Lakes, and Reservoirs)**), turbulence generation and mixing along the bottom boundaries of water bodies can be described in analogy to the SL. Under steady-state conditions, the resulting bottom boundary layer (BBL) follows a similar vertical structure of (i) current shear (see above), (ii) rate of TKE dissipation (eqn. [9]) and (iii) rate of vertical mixing. Although the original indirect driving force for turbulence in the BBL is also the wind, it is not the direct turbulent momentum flux from the atmosphere to the water which is the cause. Rather, the mechanism is indirectly induced by wind which causes large-scale currents and basin-wide internal waves (such as seiches) which act as intermediate energy reservoirs that generate TKE by bottom friction. Along sloping boundaries in particular, the breaking of propagating internal waves and convective processes – a secondary effect of bottom friction – can produce additional TKE, leading to dissipation and mixing in excess of that predicted by eqn. [9]. As with the SL, the BBL is also usually partly (and weakly) stratified. Again, mixing (eqn. [8]) increases substantially when approaching the sediment and often a completely homogenized layer a few m thick develops at the bottom.

Internal Waves and Turbulence in the Stratified Interior

In the lake interior, away from surface and bottom boundaries (**Figure 1**), the water body is stratified and quiescent, and it does not feel the direct effects of the turbulence sources at the surface and above the sediment. This stratified interior consists of an upper region, the metalimnion where gradients in temperature and density are strongest, and a lower region, the hypolimnion, which is only weakly stratified and most water properties are homogeneous. Internal waves are prevalent.

The rate of mixing in the interior water body is low because (i) currents and shear are weak and the resulting turbulence production is reduced and (ii) stratification suppresses the turbulent mixing. The mechanical energy originates mainly from basin-scale internal currents and waves (see above), whereas the waves of smaller scale and higher frequencies – potentially generated at a few specific locations – are not contributing much to the energy budget of the deep-water. At the transition between small- and large-scale waves are the near-inertial currents, which can carry – especially in large lakes – a significant portion of the mechanical energy typically in the order of $\sim 1\,\mathrm{J\,m^{-3}}$. Given that observed energy residence time-scales are days (small lakes) to weeks (deepest lakes), the dissipation of the internal energy is $\sim 10^{-12}$–$\sim 10^{-10}\,\mathrm{W\,kg^{-1}}$ (**Table 1**). Considering typical values for stratification N^2 (eqn. [1]) of 10^{-8}–$10^{-3}\,\mathrm{s^{-2}}$ and $\gamma_{mix} \approx 0.1$ (eqn. [5]), interior diffusivities of 10^{-7}–$10^{-5}\,\mathrm{m^2\,s^{-1}}$ can be expected (**Table 1**; **Figure 4**). The stratified interior – away from the SL and the BBL – is by far the most quite zone in lakes.

Important for the generation of small-scale mixing are local instabilities related to internal (baroclinic) motions, such as illustrated in **Figure 2**. Instabilities occur mostly where the usually weak background shear is enhanced by nonlinear steepening of internal waves or by superposition of the shear with small-scale propagating internal waves.

Direct observations of turbulence and mixing, using microstructure and tracer techniques, confirm that turbulence is indeed very weak in the stratified interior. Typically, only a few percent of the water column is found to be actively mixing. The occurrence of such turbulent patches is highly intermittent in space and time. During periods when the fluid is nonturbulent, we can expect laminar conditions and thus the dominance of molecular transport. The observable average diffusivity can be considered the superposition of a few turbulent events separated by molecular diffusion for most of the time. The resulting transport in the stratified interior will therefore be close to molecular. Tracer experiments and microstructure profiling conducted in small and medium-sized lakes confirm these quiet conditions in the interior and enhanced turbulence in the bottom boundary. In **Figure 4** the vertical spreading of a tracer, injected into the hypolimnion, is shown for the interior (first few days) and for a basin-wide volume including the BBL (after a few days). From **Figure 4** it is evident that turbulent diffusivity in the interior is at least one order of magnitude lower than in the basin-wide deep-water volume, including the bottom boundary. In addition to these spatial differences, one has to be aware of the temporal variability. During storms, turbulence can be several orders of magnitude larger for short episodes. The transition from quiescent to actively mixing occurs rapidly once winds increase above a certain threshold relative to the stratification. The internal wave field is energized and turbulence can develop. But the greatest increases occur in the benthic BBL. It is during such storms that most of the vertical flux takes place.

The turbulent patches, where vertical fluxes are generated (as exemplified in **Figure 3**) vary in size depending in part upon the turbulence intensity ε and the stratification N^2. Several length scales have been developed to characterize the sizes of turbulent eddies. One is the Ozmidov scale

$$L_O = (\varepsilon/N^3)^{1/2} \qquad [10]$$

and the other is the Thorpe scale, L_T, which is based on direct observations of the size of unstable regions. The ratio of the two numbers varies depending upon the strength of stratification and is useful for predicting the efficiency of mixing, γ_{mix} in eqn. [5] Typical values for L_O and L_T range from a few centimeters to a meter but for weak stratification eddies are larger and on scales of tens of meters to 100 m as found in weakly stratified Lake Baikal.

The spatial and temporal dynamics of mixing challenges not only the experimental estimation, but also the numerical simulation of its net effect, in terms of a turbulent diffusivity K_z. Local measurements of K_z following eqn. [8] often neither resolve its spatial nor its temporal dynamics and the coarse grid sizes used in numerical simulations do not capture the small scales relevant for mixing processes in the interior.

Turbulent Energy Flux through the Water Column – Synthesis

From the discussion above, we can draw the following overall scheme of the energy flux through the stratified waters of a lake. The origin of the energy for turbulent mixing is usually wind, which is imposing momentum onto the surface of the water. Approximately 3% of the wind energy from the atmosphere reaches the epilimnion in the form of horizontal currents and about 10% thereof is finally transferred to the stratified water body underneath. The major part of the energy is dissipated by bottom interaction, and the minor part is dissipated in the interior by shear instabilities and breaking of internal waves. Of this dissipated energy, only about 10% produce buoyancy flux (mixing efficiency γ_{mix}, eqn. [5]) increasing the potential energy of the stratification. Compared to the wind energy flux in the atmosphere, only a small fraction of ~0.0003 actually causes the mixing against the stratification in the deep water, whereas the large fraction of ~0.9997 is dissipated somewhere along the flux path. Although this partitioning depends on many factors, the overall scheme likely holds within a factor of 2–3 based on comparisons between different lakes.

The small amount of energy available for mixing – compared to the potential energy stored in the stratification – explains why lakes deeper than a few meters remain permanently stratified during the warm season. Consistent with this conclusion, the enhanced turbulence in the surface and bottom boundary layers cannot erode the stable – and partly very strong – stratification in the interior. Turbulent patches are intermittent and the eddies within them are small compared to the depth. As an example, the timescales to transport heat, solutes or particulates over a distance of 10 m would be $(10\,\text{m})^2\,K_z^{-1}$; i.e., several years for a $K_z = 10^{-6}\,\text{m}^2\,\text{s}^{-1}$ in the metalimnion (**Table 1**). Therefore, two-dimensional processes, such as upwelling become important for vertical exchanges as well.

See also: The Benthic Boundary Layer (in Rivers, Lakes, and Reservoirs); Currents in Rivers; Currents in Stratified Water Bodies 1: Density-Driven Flows; Currents in Stratified Water Bodies 2: Internal Waves; Currents in Stratified Water Bodies 3: Effects of Rotation; Currents in the Upper Mixed Layer and in Unstratified Water Bodies; Density Stratification and Stability; Hydrodynamical Modeling; The Surface Mixed Layer in Lakes and Reservoirs.

Further Reading

Goudsmit GH, Peeters F, Gloor M, and Wüest A (1997) Boundary versus internal diapycnal mixing in stratified natural waters. *Journal of Geophysical Research* 102(C13): 27903–27914.

Mortimer CH (2005) *Lake Michigan in Motion: Responses of an Inland Sea to Weather, Earth-Spin, and Human Activities.* University of Wisconsin Press. ISBN 978-0299178345.

Imberger J (1998) *Physical Processes in Lakes and Oceans. Coastal Estuarine Studies,* vol. 54. Washington, DC: American Geophysical Union.

Imberger J and Ivey GN (1991) On the nature of turbulence in a stratified fluid. Part II: Application to lakes. *Journal of Physical Oceanography* 21: 659–680.

Imberger J and Patterson JC (1990) Physical limnology. *Advances in Applied Mechanics* 27: 303–475.

Ivey GN, Winters KB, and Koseff JR (2008) Density stratification, turbulence, but how much mixing? *Annual Review of Fluid Mechanics* 40: 169–184.

Kantha LH and Clayson CA (2000) *Small Scale Processes in Geophysical Fluid Flows. International Geophysical Series,* vol. 67. London: Academic Press. ISBN-10: 0-12-434070-9.

Imboden DM and Wüest A (1995) Mixing mechanisms in lakes. In: Lerman A, Imboden D, and Gat JR (eds.) *Physics and Chemistry of Lakes*, pp. 83–138. Berlin: Springer-Verlag.

MacIntyre S, Flynn KM, Jellison R, and Romero JR (1999) Boundary mixing and nutrient flux in Mono Lake, CA. *Limnology and Oceanography* 44: 512–529.

Schmid M, *et al.* (2007) Sources and sinks of methane in Lake Baikal: A synthesis of measurements and modeling. *Limnology and Oceanography* 52: 1824–1837.

Thorpe SA (2007) *An Introduction to Ocean Turbulence.* Cambridge, UK: Cambridge University Press. ISBN: 978-0-521-85948-6.

Winters KB, Lombard PN, Riley JJ, and D'asaro EA (1995) Available potential-energy and mixing in density-stratified fluids. *Journal of Fluid Mechanics* 289: 115–128.

Wüest A and Lorke A (2003) Small-scale hydrodynamics in lakes. *Annual Review of Fluid Mechanics* 35: 373–412.

Wüest A and Lorke A (2005) Validation of microstructure-based diffusivity estimates using tracers in lakes and oceans. In: Baumert HA, Simpson J, and Sündermann J (eds.) *Marine Turbulence – Theories, Observations and Models.* Cambridge, UK: Cambridge University Press. ISBN: 0521837898.

The Benthic Boundary Layer (in Rivers, Lakes, and Reservoirs)

A Lorke, University of Koblenz-Landau, Landau/Pfaly, Germany
S MacIntyre, University of California, Santa Barbara, CA, USA

Introduction

Definition and Relevance of the Benthic Boundary Layer

The benthic boundary layer (BBL) of lakes, reservoirs, and rivers constitutes that part of the water column that is directly influenced by the presence of the sediment–water interface. Similar to the surface mixed layer, it represents a hot spot not only of dissipation of kinetic energy, but also of biological activity and of geochemical transformation processes. These different processes are strongly coupled and interact with each other: While the hydrodynamic conditions are modified by biological activity, which changes the structure of the sediment surface, the release of dissolved solids from the sediment can modify the density stratification in the BBL. Moreover, the actual sediment surface cannot always be regarded as rigid since the BBL flow does not only modify the structure of the sediment surface, but it can also bring sediment particles into suspension, whereas at other sites or at other times, the particles resettle.

The BBL definition provided here and the more detailed discussions later explicitly refer to *direct* influences of the sediment surface. From a more general point of view, the BBL is of great importance for the entire water body, almost independent of the dimensions of the basin. Strong turbulence and mixing along the boundaries are known to be important for vertical mixing, and transport on a basin scale and biogeochemical processes at the sediment surface or within the sediment effect the distribution of relevant water constituents on scales much larger than the actual dimensions of the BBL. These larger-scale effects, however, require additional transport processes for energy and matter into or out of the BBL and are considered elsewhere.

A major characteristic of the BBL is the magnitude and the temporal dynamics of the physical forcing, i.e., the current velocity at the top of the BBL. Although in most rivers this forcing can be regarded as a steady-state unidirectional flow, its nature in deep and stratified lakes and reservoirs is more complex. In these, usually stratified, water bodies the major energy is provided by surface waves in the shallow littoral zone, by high-frequency internal waves at the depth of the thermocline and by basin-scale internal waves (seiches or Kelvin and Poincare waves) in the hypolimnion. Hence, the magnitude and temporal dynamics of the different forcing mechanisms range from current velocities of some 10 cm s^{-1} and time scales of seconds for surface waves, to typical current velocities of a few centimeters per second and time scales of several hours to days for basin-scale internal waves (**Figure 1**).

Structure of the BBL

The BBL can be structured vertically according to the physical processes governing the vertical transport of momentum and solutes (**Figure 2**). In an outer layer (turbulent BBL) up to several meters above the sediment surface, this transport is governed by turbulent eddies and the associated mixing rates are high. While approaching the sediment surface down to scales where viscous forces suppress overturning turbulent motions, the vertical transport of momentum is governed by molecular viscosity and a viscous sublayer with a typical height of O(1 cm) develops. The exchange of heat and dissolved solids and gases is eventually controlled by a diffusive sublayer with a height of O (1 mm) directly at the sediment–water interface.

The Transport of Momentum

The Turbulent BBL

The equation for total average shear stress τ in a turbulent boundary layer is

$$\tau = \mu \frac{du}{dz} + \rho \overline{u'w'} \qquad [1]$$

where μ is dynamic viscosity, du/dz is the vertical velocity gradient, u' and w' are the fluctuating horizontal and vertical velocities, and the overbar denotes temporal averaging. While the first term on the right describes viscous shear, the second term is related to momentum transport by turbulent velocity fluctuations. In most aquatic systems, the Reynolds number associated with near-bottom flows is sufficiently high to sustain a turbulent boundary layer. Under such conditions, the first term on the right-hand side of eqn [1] may be negligible and turbulent shear stress is likely to dominate the shear stress computation.

On the basis of dimensional arguments, it can be assumed that the shear stress, τ, on the sediment

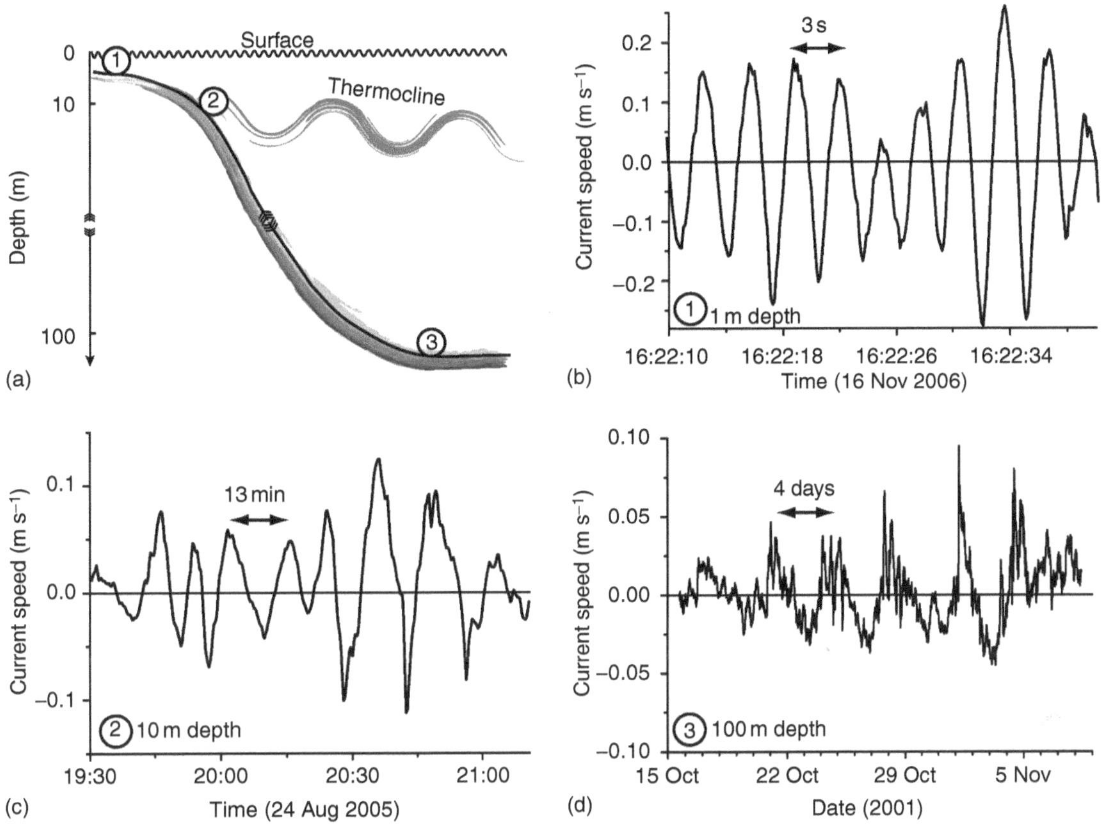

Figure 1 Typical near-bottom current velocities measured at various locations (depths) in a large Lake (Lake Constance) emphasizing the different periods and magnitudes of BBL forcing. (a) A schematic cross-section of the lake with the three sampling sites indicated by numbers. Near-bottom current velocities induced by surface waves at 1-m depth are shown in panel (b). Typical periods of surface waves in lakes are in the order of seconds. (c) Near-bottom current velocities measured at the depth of the seasonal thermocline (10-m depth). The observed current velocities are driven by propagating internal waves with periods between 8 and 20 min. Near-bottom currents at 100-m depth (d) are mainly driven by basin-scale internal waves. The major period of about four days is associated with a Kelvin wave. Note that several other basin-scale modes of oscillation (e.g., 12 h) are superimposed on this four-day period.

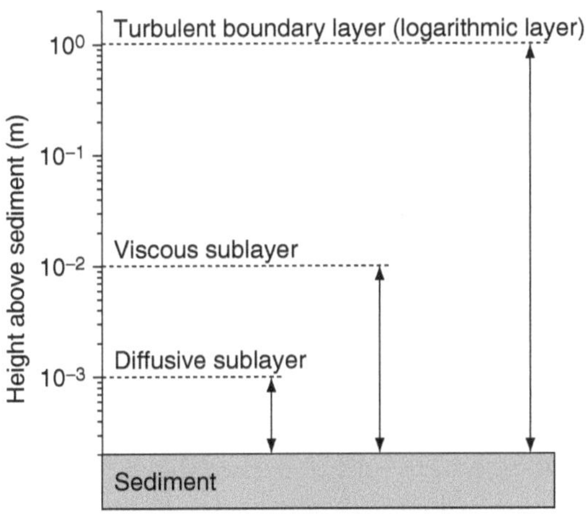

Figure 2 Idealized structure of the BBL on a flat sediment surface. Note that the heights provided by the logarithmically scaled axis represent order of magnitude estimates for typical conditions found in inland water bodies.

surface is related to the current speed at a certain height above the sediment:

$$\tau = \rho C_D U_{1\text{m}}^2 \qquad [2]$$

where ρ is the density, C_D an empirical constant ($\approx 1.5 \times 10^{-3}$), the so-called drag coefficient, and $U_{1\text{ m}}$ refers to the current speed measured at a height of 1 m above the sediment surface. Note that C_D depends on the reference height where the current speed was measured and a standard height of 1 m is assumed from now on. The bed shear stress τ is assumed to be constant throughout the boundary layer (constant stress layer) and it is used to define a turbulent velocity scale u_*, the so-called friction velocity:

$$u_* = \sqrt{\frac{\tau}{\rho}} \qquad [3]$$

Dimensional analysis can then be used to deduce the velocity distribution $u(z)$ near the sediment surface, where z is the distance from the sediment surface. For a layer far enough from the boundary so that the direct effect of molecular viscosity on the flow can be neglected (the outer layer), the analysis results in

$$u(z) = \frac{u_*}{\kappa} \ln \frac{z}{z_0} \qquad [4]$$

where $\kappa \approx 0.41$ is von Karman's constant and z_0 is the roughness length, which is related to the drag coefficient in eqn [2] by

$$C_{1m} = \left(\frac{\kappa}{\ln(1\,\text{m}) - \ln(z_0)} \right)^2 \qquad [5]$$

and will be discussed later.

Equation [4] is called 'law of the wall,' and by assuming a local steady-state equilibrium between production and dissipation of turbulent kinetic energy (TKE), it can be used to estimate the vertical distribution of the turbulence dissipation rate ε

$$\varepsilon = \frac{u_*^3}{\kappa z} \qquad [6]$$

Thus, in analogy to the wind-forced surface layer, the level of turbulence increases with decreasing distance from the boundary. This increasing turbulence leads, again in analogy to the surface mixed layer, to the development of a well-mixed boundary layer of up to several meters in height.

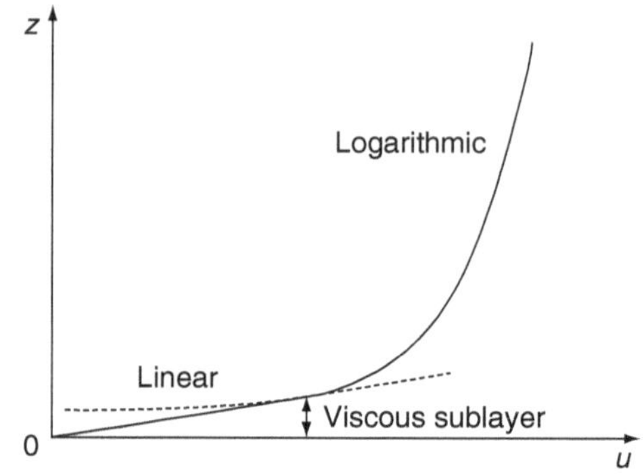

Figure 3 Velocity distribution above a smooth and rigid bottom (solid line). Within the viscous sublayer the velocity u increases linearly with distance from the sediment surface z. Above the viscous sublayer the velocity distribution follows the law of the wall (eqn [4]) and increases logarithmically with distance from the surface. Extrapolated continuations of the linear and logarithmic velocity distributions are shown as dashed lines.

The Viscous Sublayer

It should be noted that eqn [4] is strictly valid for turbulent flows only for which the vertical transport is governed by cascading turbulent eddies. The maximum (vertical) size of the turbulent eddies is determined by the distance from the sediment surface, and by approaching the sediment surface down to scales where overturning turbulent motions are suppressed by the effect of molecular viscosity, the momentum transport becomes governed by viscous forces (first term on the right-hand side of eqn [1]). Within this layer, which is called the viscous sublayer, current shear becomes constant and the resulting linear velocity profile can be described by

$$u(z) = \frac{u_*^2}{\nu} z \qquad [7]$$

On a smooth sediment surface, the viscous sublayer extends to a height δ_v of about $10\nu/u_*$, which is comparable to the Kolmogorov microscale describing the size of the smallest turbulent eddies (typically O (1 cm), cf. **Figure 2**).

Since viscosity is reduced to its molecular value, current shear within the viscous sublayer is greater than that in the turbulent layer above (cf. **Figure 3**), a fact which has major consequences for organisms living within the viscous sublayer on the sediment surface because they have to withstand these strong shearing and overturning forces. It is further interesting to note that an appreciable amount of energy entering the BBL is dissipated within this layer (about 40%).

Effects of Bottom Roughness

The roughness length z_0 in eqn [3] determines the effective height above the bottom z at which the current velocity approaches zero. It is determined by the topographic structure of the sediment surface and hence by the typical height, width, and spacing of individual roughness elements on a stationary bed. When the scale of these roughness elements z_S is on the order of the height of the viscous sublayer δ_v or less, z_0 is solely determined by δ_v and $z_0 \approx 0.1\nu/u_*$. This flow regime is called *smooth*. When the size of the roughness elements exceeds δ_v, the flow regime is called *rough* and the corresponding roughness length is given by $z_0 \approx z_S/30$. Note that the drag coefficient $C_D \approx 1.5 \times 10^{-3}$ provided earlier (eqn [2]) corresponds to a roughness length $z_0 \approx 2.5 \times 10^{-5}$ m (eqn [5]) and hence is valid for smooth flows unless u_* exceeds 0.4 m s^{-1} or $U_{1\,m}$ exceeds 10 m s^{-1}.

In addition to the shear stress derived from viscous forces as described above (the so-called *skin friction*), larger-scale roughness elements can cause a *form drag*, which results from pressure gradients between the upstream and downstream side of particular roughness elements. Although skin friction is important for the lower part of the turbulent BBL and for the viscous sublayer, form drag resulting from, e.g., ripples, sand waves, or submerged vegetation is more important for the upper part of the turbulent BBL and for the total drag on flow. When form drag is significant, the turbulent BBL may consist of more than one logarithmic layer, described by different roughness lengths z_0, respectively.

Oscillatory Boundary Layers

The turbulent BBL equations derived here are based on steady-state conditions, i.e., on a local balance between production and dissipation of TKE, which is in equilibrium with the applied forcing. As described later, however, many forcing mechanisms for near-bottom flows are related to surface or internal waves and are hence associated with oscillatory flows. Well above the viscous sublayer, such oscillatory BBL show, similar to the effect of form drag, deviations of the of the velocity distribution from its steady-state logarithmic pattern. One major characteristic of oscillatory BBL is a pronounced maximum of the current speed at some decimeters or meters above the bed. The analytical solution to this problem (Rayleigh flow or Stokes' second problem) is an exponentially damped vertical oscillation of the current profile with a vertical wave number of $\sqrt{\omega/2\nu}$, where ω is the forcing frequency and ν the turbulent viscosity, which, however, is a function of time and distance from the sediment surface. Depending on the overall energetics of the BBL flow, characteristic current speed maxima at 2–3 m above the bed could be observed in lakes where the internal wave forcing had a period as long as 24 h. Another major consequence of oscillatory BBL is that the maximum in turbulent intensity near the bed does not coincide with the maximum of the current speed, at the top of the BBL.

Stratified BBL

Effects of Density Stratification

Density stratification affects turbulent mixing in the outer layer by causing buoyancy forces that damp or even suppress overturning turbulent eddies. The vertical distribution of velocity and turbulence described earlier for unstratified BBL may hence change significantly under stratified conditions. In addition, the vertical structure of density stratification along with the presence of sloping boundaries can introduce additional mixing phenomena in BBL of enclosed basins.

Similar to the surface mixed layer, increased production of TKE along the boundaries of a water body often leads to the generation and maintenance of a well-mixed BBL of height h_{mix}. On a flat bottom (away from the slopes) and where a logarithmic boundary layer occurs, h_{mix} can be estimated by applying scaling laws from the wind-mixed surface layer

$$h_{mix} = 2^{3/4}\frac{u_*}{\sqrt{Nf}} \quad [8]$$

where u_* is the friction velocity in the BBL (eqn [2]), N is the Brunt–Väisälä frequency, and f is the Coriolis parameter. In small- to medium-sized water bodies, where the effect of Earth's rotation is unimportant, f has to be replaced by the respective forcing frequency, e.g., the frequency of internal seiching.

In productive water bodies, in particular, the sediment can be a significant source of remineralized nutrients as a result of microbial degradation of organic matter at the sediment surface or within the sediment. The diffusion of solutes across the sediment–water interface (see Solute Transport and Sediment–Water Exchange section) has the potential to set up density stratification within the BBL, which could suppress turbulent mixing. Hence, whether or not a turbulent and mixed BBL can be developed and maintained depends not only on the amount of available TKE, which is typically extracted from basin-scale motions, but also on the buoyancy flux from the sediment that has to be overcome by turbulent mixing. Geothermal heating, in contrast, can result in unstable stratification and convective mixing in the BBL. A mean geothermal heat flux of 46 mW m^{-2} results in a mean vertical temperature gradient of about -8×10^{-2} K m^{-1}, which can be observed when the BBL is chemically stratified and turbulent mixing is suppressed.

2-Dimensional Mixing Processes in Enclosed Basins

The occurrence of mixed BBL (in terms of density) is a straight consequence of the application of a zero-flux boundary condition at the sediment surface, i.e., no exchange of heat and dissolved solids across the sediment–water interface. This boundary condition forces the isopycnals (or isotherms if density changes are mainly caused by temperature) to intersect the boundary at a right angle, leading to a mixed density or temperature profile in the vicinity of the boundary. Enhanced mixing along the boundaries is thus not a necessary requirement for the development of such mixed layers. Along the sloping boundaries of enclosed basins, these mixed BBL cause horizontal density gradients and hence drive horizontal currents – a process which is believed to have important consequences for basin-wide diapycnal transport.

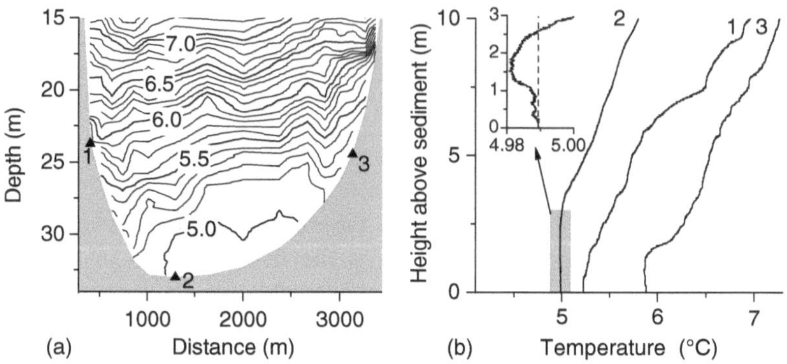

Figure 4 (a) Isotherm depths along the main axis of Lake Alpnach (Switzerland) calculated from repeated CTD profiling. The increment between neighboring isotherms is 0.1 °C and the numbers refer to the respective temperature (in °C) of the isotherms plotted as thick lines. Note that the figure is not to scale and only the lower portion of the water column is shown. The numbered symbols show the position of temperature profiles shown in (b). (b) Near-bottom temperature profiles at selected positions along the transect shown in (a). Numbers at the top of the profiles refer to the positions indicated by symbols in (a). The inset emphasizes the inverse temperature stratification observed in the BBL of profile 2. Panel (a) is adopted from Lorke A, Wüest A, and Peeters F (2005) Shear-induced convective mixing in bottom boundary layers on slopes. *Limnology and Oceanography* 50: 1612–1619, with permission from American Society of Limnology and Oceanography.

Measurements, however, have revealed that h_{mix} is not constant along the sloping boundaries, as demonstrated in **Figure 4**. There the upper limit of the mixed BBL can be defined by the depth of the 5 °C isotherm (chemical stratification can be neglected in this particular lake), leading to $h_{mix} \simeq 2.3$ m in the central part of the lake, whereas $h_{mix} \leq 1$ m further up on the slopes (**Figure 4(b)**). The pronounced tilt of the isotherms in **Figure 4(a)** further emphasizes the importance of basin-scale internal waves for local estimates of h_{mix} on the slopes because the associated currents push the well-mixed BBL from the central part of the lake up and down the respective slopes during the course of the seiching.

The flow within the BBL remains parallel to the sediment surface on the sloping boundaries because the vertical velocity component must vanish at a rigid surface. Hence the flow is no longer in parallel to the isopycnals (or isotherms in **Figure 4(a)**) and, in combination with the fact that the flow velocity is increasing with increasing distance from the sediment surface (eqn [4]), convective instabilities can occur on the slopes when heavier water is moved on top of lighter water. Unstable stratification, as shown in the inset of **Figure 4(b)**, can only be generated when the current is directed up-slope, a down-slope current leads to a stabilization of the BBL by the same principle. Thus, in a periodic, internal wave-driven flow, stratified and convectively mixing BBL occur alternately on the two opposing slopes of the water body. This shear-induced convection provides an additional source for mixing in BBL on slopes. Its general importance for BBL turbulence, however, is not yet fully understood.

Turbulence Induced by Internal Wave Interactions with Bottom Boundaries

Turbulence production in stratified regions of lakes is linked with instabilities in the internal wave field. A considerable portion of the turbulence occurs within or near the BBL and can be due to internal wave breaking at critical frequencies or internal wave steepening. Recent studies have shown that the form and degree of nonlinearity of internal waves in the thermocline can be predicted from the Wedderburn and Lake numbers, two dimensionless indices which indicate the balance between buoyancy forces and shear forces and which further take into account basin morphometry. As illustrated in **Figure 5**, the hypolimnion is also an internal wave field with similar wave forms to those observed in the thermocline. When Lake numbers, L_N, drop below 3, turbulence associated with the internal wave field increases (**Figure 5**). Thus, not only is turbulence induced in the thermocline when nonlinear waves form, but also in the hypolimnion with the greatest increases in the BBL. Values of the coefficient of eddy diffusivity increase 1–3 orders of magnitude above molecular. In small lakes, flow speeds in the BBL increase from a few millimeters per second to ~2 cm s^{-1} with the decreases in Lake number.

Solute Transport and Sediment–Water Exchange

The Diffusive Sublayer

In the immediate vicinity of the sediment surface, the vertical transports of momentum and solutes

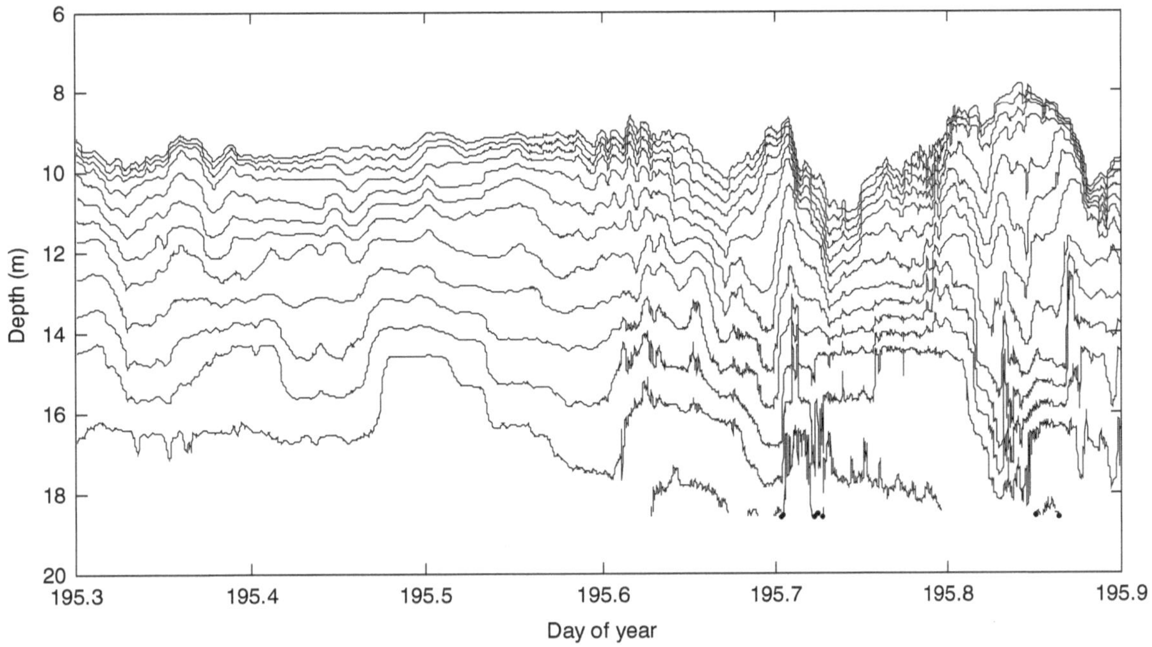

Figure 5 Isotherms illustrating the internal wave field in the hypolimnion of Toolik Lake, Alaska, prior to and after wind forcing increased to 9 m s^{-1} (day 195.6) and the Lake number decreased to 1.5. Isotherms are at 0.1 °C intervals with uppermost isotherm 6 °C. Thermistors were 80 cm apart between 10.2 and 12.6 m depth and 2 m apart deeper in the water column. Deepest thermistor was within 50 cm of the lake bottom. Increased temperature fluctuations after L_N decreases below 1.5 are indicative of turbulence either beginning or increasing in the lower water column (unpublished data, S. MacIntyre).

(dissolved gases, solids, and heat) are reduced to their respective molecular levels, as described above. In analogy to the viscous sublayer, where momentum transport is governed by molecular viscosity, a diffusive sublayer exists, where the vertical transport of heat and solutes is governed by molecular diffusion. Since the molecular diffusivity of solutes D is about 3 orders of magnitude smaller than the molecular viscosity ν (the Schmidt number Sc, defined as $Sc = \nu/D$, is about 1000), the height of the diffusive sublayer δ_D is with $\delta_D = O(1\ \text{mm})$ considerably smaller than the height of the viscous sublayer δ_ν (**Figure 2**). A typical profile of dissolved oxygen concentrations measured through the sediment–water interface is shown in **Figure 6**. Although turbulent transport is already suppressed within the viscous sublayer, straining of concentration gradients by viscous shear results in an efficient vertical transport of solutes and to typically well-mixed solute distributions within most of the viscous sublayer. Concentration gradients are hence compressed to the diffusive sublayer overlaying the sediment surface and the constancy of the molecular diffusivity results in a linear concentration gradient $C(z)$ (**Figure 6**). At the top of the diffusive boundary layer, the concentration gradient decreases gradually to zero and the solute concentration reaches its constant bulk value C_∞.

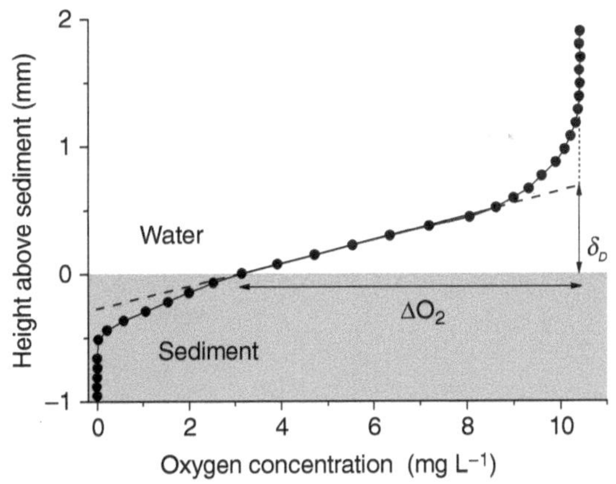

Figure 6 Oxygen concentration profile across the sediment–water interface measured in Lake Alpnach (Switzerland). The diffusive sublayer is characterized by the linear concentration gradient above the sediment surface where transport is governed by molecular diffusion. The effective height of the diffusive sublayer δ_D is defined by the intersection of the extrapolated linear concentration gradient (dashed line) with the constant oxygen concentration above the diffusive sublayer.

Within the sediment, molecular diffusivity is reduced by the porosity (reduction of surface area) and by turtosity (increase of diffusion path length), and the concentration profile is additionally determined by chemical and microbial production and loss processes. In the case of oxygen (**Figure 6**), a reaction–diffusion model with rather simple zero-order kinetics of oxygen consumption resulting in a parabolic oxygen profile provides surprisingly good agreement with measured oxygen distributions.

The fluxes F of solutes through the diffusive boundary layer can be derived from Fick's first law:

$$F = \frac{D}{\delta_D}(C_\infty - C_0) \quad [9]$$

where C_0 is the solute concentration at the sediment surface. Hence, for a given concentration gradient $(C_\infty - C_0)$ and by ignoring the temperature dependence of the molecular diffusivity, the magnitude of the flux is determined by the thickness of the diffusive sublayer δ_D. It has been demonstrated in numerous laboratory and field measurements that δ_D depends strongly on the flow regime in the turbulent BBL above. With increasing levels of turbulence (e.g., with increasing u_*) the diffusive boundary layer becomes more compressed, and according to eqn [9], the fluxes increase. The thickness of the diffusive sublayer can be related empirically to u_* or to the thickness if the viscous sublayer, which in turn is related to u_*, as described earlier:

$$\delta_D = \delta_\nu Sc^a \quad [10]$$

where the Schmidt number Sc accounts for the different kind of 'solutes' (e.g., heat or dissolved oxygen) and observed Schmidt number exponents a range between 0.33 and 0.5. As u_* is not always an appropriate parameter for describing BBL turbulence (e.g., in oscillatory BBL or in the presence of form drag or density stratification), δ_D can also be described in terms of the Batchelor length scale, which describes the size of the smallest fluctuations of a scalar tracer in turbulent flows as a function of the turbulence dissipation rate.

It is most interesting to note that the sediment–water exchange in productive water bodies is often flux-limited by the 'bottleneck' of the diffusive sublayer and that it is actually the wind acting at the water surface that provides energy for turbulence within the BBL and hence effects the magnitude of the sediment–water fluxes by controlling the thickness of the diffusive sublayer.

Effects of Small-Scale Sediment Topography

Increased roughness of the sediment surface (e.g., due to biological activity) affects the sediment–water exchange by increasing the mass and momentum transfer as well as by increasing the surface area of the sediment–water interface. Detailed observations have demonstrated that the diffusive sublayer tends to smooth out topographic structures that are smaller than the average height of the sublayer, but smoothly follows larger roughness elements (**Figure 7**). The detailed structure of the oxygen distribution within the diffusive sublayer is then not only determined by diffusion (normal to the local sediment surface) but also by advection with the flow (in parallel to the local sediment surface) and it can be expected that the degree of smoothing increases with decreasing flow velocities. Detailed comparisons of measured 3-dimensional fluxes over rough topography with

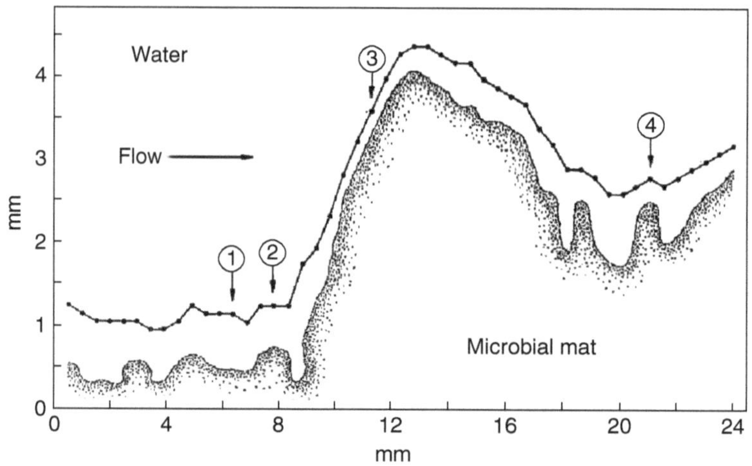

Figure 7 Horizontal transect along the direction of flow showing how the upper limit of the diffusive sublayer (solid line with data points) follows the surface topography of a microbial mat. The diffusive sublayer limit was defined by the isopleth of 90% air saturation of oxygen. Notice different vertical and horizontal scales. Flow velocity at a height of 1 cm was 4 cm s^{-1}. Numbers indicate specific measuring positions discussed in the original publication. Reproduced from Jørgensen BB and Des Marais DJ (1990) The diffusive boundary layer of sediments: Oxygen microgradients over a microbial mat. *Limnology and Oceanography* 35:1343–1355, with permission from American Society of Limnology and Oceanography.

the fluxes calculated from the respective average diffusive sublayer heights and concentration gradients are enhanced by factors up to 49%. It must be noted, however, that fluxes estimated from concentration profiles measured at one particular location on a rough sediment surface can severely overestimate or underestimate the average flux, as demonstrated in **Figure 7**.

Nondiffusive Fluxes

Besides the diffusive fluxes, there exist several additional pathways for the exchange of solutes across the sediment–water interface. Convectively driven transport of pore water through the interface can occur in shallow waters where shortwave solar radiation penetrates the water column to the sediment surface and heats the sediment. Similarly, changes in temperature of the water overlaying the sediment surface, e.g., due to internal waves in stratified water bodies, have been observed to drive convective transport across the sediment–water interface. The existence of larger roughness elements, such as ripples, on permeable sediments can further result in advective pore water exchange driven by pressure differences. Higher dynamic pressure at the upstream side of such topographic structures give rise to the transport of water into the sediment, whereas the lower pressure at the downstream side sucks pore water out of the sediment.

Bioturbation and *bioirrigation* are processes by which benthic fauna or flora enhances the sediment–water exchange. Whereas bioturbation mainly refers to the displacement and mixing of sediment particles by, e.g., worms, bivalves, or fish, bioirrigation refers to the flushing and active ventilation of burrows with water from above the sediment surface. These processes are particularly important in oligotrophic water bodies where the sediment surface remains oxic and provides suitable conditions for a diverse benthic fauna. In more eutrophic systems the emanation of gas bubbles (mainly methane or carbon dioxide) which are formed by biogenic production and a resulting supersaturation of pore water with these gases may have similar effects.

In Situ Flux Measurements

The sediment oxygen demand or the release of nutrients from the sediment can be of major importance for the overall productivity and for the geochemical composition of a particular water body and quantification of sediment–water fluxes is often essential for understanding biogeochemical cycles within the water column. As these fluxes depend strongly on the hydrodynamic conditions in the BBL and as these conditions have a strong spatial and temporal dynamics, in situ measurements are often desirable. The measurement of concentration profiles through the sediment–water interface capable of resolving the diffusive sublayer are one way for estimating the fluxes. From a measured profile of dissolved oxygen, as shown in **Figure 6**, the sediment–water flux can be readily estimated by applying eqn [9]. Such measurements are carried out using microelectrodes, which are available for a large number of solutes, mounted on a benthic lander system. However, there are two major problems associated with this method: First, although these microelectrodes have tiny tip diameters (down to 10 μm or less for oxygen sensors), they were demonstrated to disturb the concentration distribution within the diffusive boundary layer while profiling. The second and more severe problem results from the complexity of the spatial distribution of the fluxes resulting not only from the small-scale sediment topography (cf. **Figure 7**) but also from the strongly localized effects of advective pore water exchange and bioturbation. To overcome these problems, the flux can be measured within the turbulent BBL at some distance from the actual sediment surface. By neglecting any sources or sinks within the water column between the sampling volume and the sediment surface, this flux represents an areal average of the sediment–water flux including all nondiffusive flux contributions. The turbulent flux F_{turb} is determined by the cross-correlation of turbulent vertical velocity (w') and turbulent concentration (C') fluctuations:

$$F_{turb} = \overline{w'C'} \qquad [11]$$

where the overbar denotes temporal averaging.

Particle Dynamics

BBLs are often characterized by enhanced concentrations of suspended particles as compared to the water column above. Such nepheloid layers are generated by resuspension of particles from the sediment surface and subsequent upward transport. The potential of a turbulent flow to entrain sediment particles of size D is often described in a Shields diagram where empirical thresholds of sediment motion are provided as a function of a nondimensional shear stress $\theta = \rho u_*^2/((\rho_p - \rho)gD)$ and a particle Reynolds number $Re_* = u_* D/\nu$, in which ρ_p is the particle density and g the gravitational constant. The θ can be interpreted as the ratio between the lift force provided by the turbulent shear stress defined in eqn [3] and the gravitational force acting on the particle. While in suspension, the fate of the particle is determined by the balance between upward transport by turbulent diffusion and Stokes settling.

The quantitative characterization of resuspension and particle transport, however, is often complicated by cohesive properties of the particles. Cohesive particles require greater shear stresses to become resuspended; moreover, they tend to form aggregates when in suspension, which alters their settling velocities.

The resuspension–settling cycles increase the contact area between particle surfaces and water and thus enhance the fluxes from and to the particles. In addition, suspended particles contribute to water density and locally enhanced resuspension, generated, e.g., by high near-bottom current velocities in the littoral zone or at the depth of the thermocline (cf. **Figure 1**), may lead to the formation of turbidity currents.

Glossary

Dissipation rate of TKE – Rate of dissipation of TKE per unit volume of water and per unit time. This energy is dissipated into heat by internal friction among fluid elements described by viscosity.

Nepheloid layer – Particle-rich layer above the sediment. This layer is sustained by a balance between gravitational settling and turbulent vertical diffusion counteracting it.

Pore water – Water that fills the interstitial space between sediment grains.

Reynolds number – The dimensionless Reynolds number *Re* is the ratio of inertial to viscous forces acting on a fluid element, obstacle in the flow, or submerged organism and describes the transition from laminar to turbulent flow regimes.

Shear stress – Force per unit area acting in parallel (tangential) to the surface of a fluid element or interface (e.g., bed shear stress).

Turbulent eddies – Turbulence is composed of eddies: patches of zigzagging, often swirling fluid, moving randomly around and about the overall direction of motion. Technically, the chaotic state of fluid motion arises when the speed of the fluid exceeds a specific threshold, below which viscous forces damp out the chaotic behaviour (see also *Reynolds number*).

Turbulent kinetic energy – Kinetic energy per unit volume of water, which is contained in the random fluctuations of turbulent motions. Turbulent velocity fluctuations u' can be separated from the mean current velocity $\bar{u}$ by Reynolds decomposition of the actual velocity u ($u = \bar{u} + u'$). Turbulent kinetic energy (TKE) is then defined by, $TKE = 1/2\rho u'^2$, where ρ denotes density of water.

See also: Biological-Physical Interactions; Currents in Rivers; Currents in Stratified Water Bodies 1: Density-Driven Flows; Currents in Stratified Water Bodies 2: Internal Waves; Currents in Stratified Water Bodies 3: Effects of Rotation; Currents in the Upper Mixed Layer and in Unstratified Water Bodies; Density Stratification and Stability; Flow Modification by Submerged Vegetation; Small-Scale Turbulence and Mixing: Energy Fluxes in Stratified Lakes; The Surface Mixed Layer in Lakes and Reservoirs.

Further Reading

Ackerman JD, Loewen MR, and Hamblin PF (2001) Benthic–Pelagic coupling over a zebra mussel reef in western Lake Erie. *Limnology and Oceanography* 46: 892–904.

Berg P, *et al.* (2003) Oxygen uptake by aquatic sediments measured with a novel non-invasive eddy-correlation technique. *Marine Ecology Progress Series* 261: 75–83.

Boudreau BP and Jørgensen BB (2001) *The Benthic Boundary Layer.* New York: Oxford University Press.

Caldwell DR and Chriss TM (1979) The viscous sublayer at the sea floor. *Science* 205: 1131–1132.

Gloor M, Wüest A, and Münnich M (1994) Benthic boundary mixing and resuspension induced by internal seiches. *Hydrobiology* 284: 59–68.

Gundersen JK and Jørgensen BB (1990) Microstructure of diffusive boundary layers and the oxygen uptake of the sea floor. *Nature* 345: 604–607.

Lorke A, Müller B, Maerki M, and Wüest A (2003) Breathing sediments: The control of diffusive transport across the sediment-water interface by periodic boundary-layer turbulence. *Limnology and Oceanography* 48: 2077–2085.

Lorke A, Umlauf L, Jonas T, and Wüest A (2002) Dynamics of turbulence in low-speed oscillating bottom-boundary layers of stratified basins. *Environmental Fluid Mechanics* 2: 291–313.

Lorke A, Wüest A, and Peeters F (2005) Shear-induced convective mixing in bottom boundary layers on slopes. *Limnology and Oceanography* 50: 1612–1619.

Mellor GL (2002) Oscillatory bottom boundary layers. *Journal of Physical Oceanography* 32: 3075–3088.

Miller MC, Mccave IN, and Komar PD (1977) Threshold of sediment motion under unidirectional currents. *Sedimentology* 24: 507–527.

Wüest A and Gloor M (1998) Bottom boundary mixing: The role of near-sediment density stratification. In: Imberger J (ed.) *Physical Processes in Lakes and Oceans. Coastal and Estuarine Studies*, pp. 485–502. American Geophysical Union.

Currents in Rivers

A N Sukhodolov and H-P Kozerski, Institute of Freshwater Ecology and Inland Fisheries, Berlin, Germany
B L Rhoads, University of Illinois at Urbana-Champaign, Urbana, IL, USA

Introduction

A characteristic feature of fluvial systems is the distinctive directed motion of water masses, or current, caused by gravitational forces. Currents in fluvial systems also differ from other geophysical flows (atmospheric, oceanic, and limnetic) primarily by a presence of irregular flow boundaries, or river channels, developed in bedrock or alluvium. In alluvial channels, cohesive or noncohesive materials are subjected to erosion, transport, and deposition, shaping channels by currents and causing alterations in the structure of the currents. For example, nonuniformities in flow structure produce local erosion at banks that with time evolve into the large-scale channel deformations – meander bends or loops. Flow curvature in bends produces centrifugal forces and counteracting pressure-gradient forces, thereby generating helical motion and secondary currents, which enhance channel deformation. Feedbacks among currents, transport of alluvium, and channel deformation result in self-regulating adjustments that dynamically sustain the form of natural fluvial systems.

Apart from their significance for processes of erosion and deposition, currents in rivers represent an abiotic component of fluvial ecosystems. Flow rates and patterns of currents determine transport and mixing of oxygen, nutrients, and pollutants. Moreover, distinctive flow patterns create specific habitats for various forms of aquatic life. On the other hand, biota can substantially influence currents, shaping the structure of habitat to favor conditions for dominant species. A good example is the interaction between biota and flow in vegetated river reaches.

Although the qualitative and quantitative assessment of river currents has attracted the attention of scientists for centuries, detailed understanding of these currents has proven elusive due in part to the lack of a general theory of turbulent flows. Therefore, available methods of characterizing river flow quantitatively are based either on case-specific computational models or purely empirical techniques. This article provides an abbreviated overview of the essential physical processes associated with river currents. The simple case of currents in wide and straight channels is considered first because it provides a theoretical framework and represents the basic (primary) class of currents. Then the effects of nonuniformity in morphology or composition of the riverbed that result in the development of secondary currents are considered along with the effects of channel curvature responsible for the secondary currents of centrifugal origin. Further, complications in the pattern of flow currents are considered using the example of flows through river confluences – essential components of river networks. The influence of human actions on river currents in the form of the complex structure of flow around groynes – transverse dikes that deflect the flow from erodible banks and promote navigability of river reaches – are discussed, followed by analysis of navigation-induced currents generated by commercial vessels. Conceptual and theoretical principles are illustrated with the examples of original field studies completed by the authors on rivers in Germany and the United States.

Controlling Factors and Classifications of Currents

Currents in rivers originate at a defined source (channel head or the junction of two streams) and evolve under the mutual influence of gravitational (G) and frictional (F) forces. At the most basic level, currents in rivers can be classified according to whether or not the bulk rate of flow remains constant over time (*steady flow*) or it changes over time (*unsteady flow*), and whether or not the bulk rate of flow remains constant over space (*uniform flow*) or it changes over space (*varied flow*). In the case of steady, uniform flow, gravitational and frictional forces are equal ($G = F$). However, in unsteady flow, the forces are unbalanced ($G \neq F$) over time, whereas in varied flow they are unbalanced over space. If $G > F$ the flow accelerates, whereas if $G < F$ the flow decelerates.

Motion of water in fluvial systems is a continuous physical process of energy transformation. Potential energy of liquid $\rho g h$ (where ρ is density of water, g is gravity acceleration, and h is flow depth) is transformed into kinetic energy ρU^2 (U is bulk flow velocity). The ratio between these two forces $Fr = U/\sqrt{gh}$, the Froude number, provides the basis for further classification. The river current can be in a *subcritical* ($Fr < 1$), critical ($Fr = 1$), or supercritical ($Fr > 1$) state. Subcritical flows are typical for lowland rivers and are characterized by smooth, undisturbed water surfaces. In the critical regime, the surface of the stream develops standing waves, and in supercritical conditions the surface of the water may become distorted into breaking waves. The critical and supercritical regimes are characteristic of mountain torrents and flow around or over engineering structures (dykes, dams), but can also develop in lowland rivers during large floods. Transitions between subcritical and supercritical regimes produce hydraulic drops, or abrupt decreases in flow depth, whereas transitions between supercritical and subcritical regimes result in hydraulic jumps, or abrupt increases in flow depth.

Rivers originate in uplands and flow downhill into lakes, seas, or oceans. Therefore, rivers flow within channels with longitudinal gradients, or slopes (S). The shear stress associated with the gravitational force per unit area oriented along the inclined plane of the channel bed is ρghS. A simple expression for the mean velocity of the flow current can be derived from assumptions of uniform flow as: $c_f = ghS/U^2$, where c_f is a friction factor. This equation can be rearranged as $U = C\sqrt{hS}$, which is known as the Chezy formula and C is the Chezy coefficient. A related empirical formula, $U = h^{2/3}S^{1/2}/n$, is known as the Manning–Strickler formula and n is the Manning coefficient ($C = h^{1/6}/n$). It can be seen that the empirical channel resistance coefficients are related to the friction factor as $c_f = g/C^2$ and $c_f = n^2 g/h^{1/3}$. Values of friction coefficients have been determined empirically and are summarized in standard manuals for open-channel flow computations. Values of the Chezy coefficient vary in rivers from 30 to 70, and the Manning coefficient ranges from 0.020 (lowland rivers) to 0.2 (flow over floodplains with terrestrial vegetation).

Although the theory of uniform flow is capable of describing bulk characteristics of currents in rivers, flow in rivers exhibits significant spatial variability because of zero local flow velocity near riverbeds and banks. This variability is a distinctive feature, providing diversity of habitat for aquatic life and is therefore a key factor determining patchiness in the community structure of aquatic organisms and plants communities. The following sections illustrate spatial patterns of currents in rivers and the possibilities of quantifying the processes producing these patterns.

Currents in Fluvial Channels

Because the permeability of riverbeds and banks is relatively low, velocity at these boundaries can be assumed to be zero. Therefore, velocities in a river cross-section reduce to zero values at the bottom and sides, and are maximal at the surface in the center of the channel (**Figure 1**). For steady uniform flow not close to the river banks, the gravitational shear-stress component ρghS is balanced by boundary friction causing shear stress within the water column that can be expressed as $\tau(z) = -\rho\overline{u'w'}$, where τ is shear stress, and u', w' are turbulent fluctuations of velocity in the streamwise and vertical directions, and z is the distance from the riverbed. It can be shown that in uniform flow shear stresses are linearly distributed over the flow depth $\tau = \tau_0(1 - z/h)$ with a maximum bed shear stress $\tau_0 = \rho ghS$, at the riverbed. A characteristic velocity scale, shear velocity, can be expressed respectively as $U_* = \sqrt{ghS} = \sqrt{\tau_0/\rho}$.

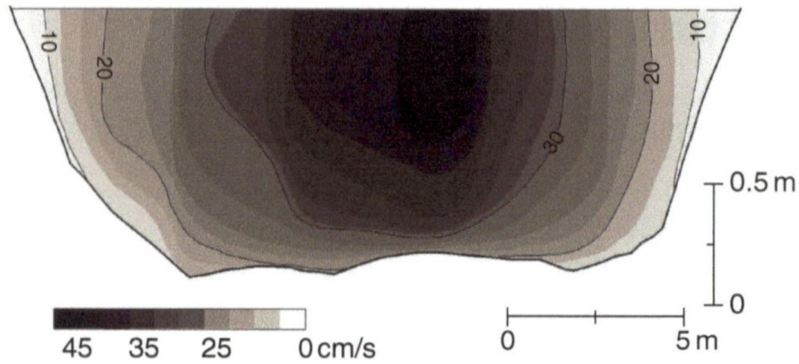

Figure 1 Distribution of time–mean streamwise velocity in a river cross-section (the Spree River, Germany).

Relating turbulent velocity fluctuations $-\overline{u'w'}$ to time–mean velocity $U(z)$ at certain distance z from the bed provides a simple model of turbulence and allows shear stresses to be expressed as $\tau = \rho\nu_t\, dU/dz$, where ν_t is turbulent viscosity. This relationship can be integrated to obtain the vertical velocity distribution, but first requires an estimate of ν_t. The assumption of a parabolic distribution of turbulent viscosity over depth, $\nu_t = \kappa U_* z\sqrt{1 - z/h}$, when substituted into the expression for shear stress, yields a logarithmic distribution for mean velocity over river depth

$$U(z) = \frac{U_*}{\kappa}\ln\frac{z}{k_s} + B_0 \qquad [1]$$

where $\kappa = 0.4$ is an universal constant, k_s is a characteristic height of roughness elements, and $B_0 = 8.5$ is a constant of integration. Alternatively, eqn [1] can be expressed as

$$\frac{U(z)}{U_*} = \frac{1}{\kappa}\ln\left(\frac{z}{z_0}\right) \qquad [2]$$

where $z_0 = \exp(\ln k_s + \kappa B_0)$ is hydrodynamic roughness parameter. Integrating [2] over river depth provides a logarithmic function expressing the influence of riverbed resistance on the depth-averaged velocity (U_a):

$$\frac{C}{\sqrt{g}} = \frac{U_a}{U_*} = \frac{1}{\kappa}\left(\ln\frac{h}{z_0} - 1\right) \qquad [3]$$

To illustrate the performance of logarithmic law [2], experimental data from some rivers are presented in **Figure 2** in nondimensional coordinates $z/h = \exp[U\kappa/U_* - \ln(h/z_0)]$.

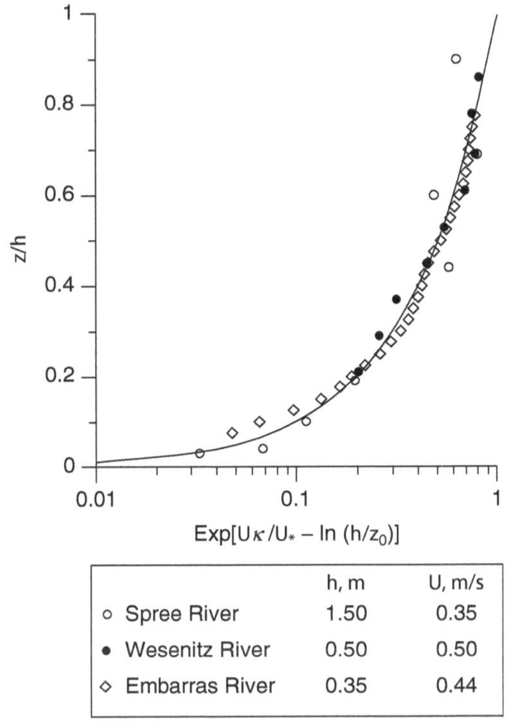

Figure 2 Comparison of measured time–mean streamwise velocities over the river depth (symbols) with predicted logarithmic law (line).

Secondary Currents

So far only the streamwise velocity, or the velocity component parallel to centerline of the channel, has been considered for straight river reaches with uniform cross-sectional geometry and riverbed material. However, natural stream channels usually meander and exhibit complex morphology and distributions of riverbed material. These variations produce components to currents that have significant magnitudes perpendicular to the streamwise component. These components, referred to as secondary currents, result in substantial three dimensionality of the overall pattern of currents in streams.

Depending on their genesis, secondary currents are classified into two categories: secondary currents of first and second kinds. Secondary currents of the first kind are produced by large-scale nonuniformities of channel pattern – for example, river bends. Centrifugal forces that develop in a curved channel produce superelevation of the water surface along the outer bank channel, which generates a counteracting pressure-gradient force. Local imbalances between these forces over the flow depth produce outward motion at the surface, downward motion along the outer bank, and upward, inward motion along the bed (**Figure 3**).

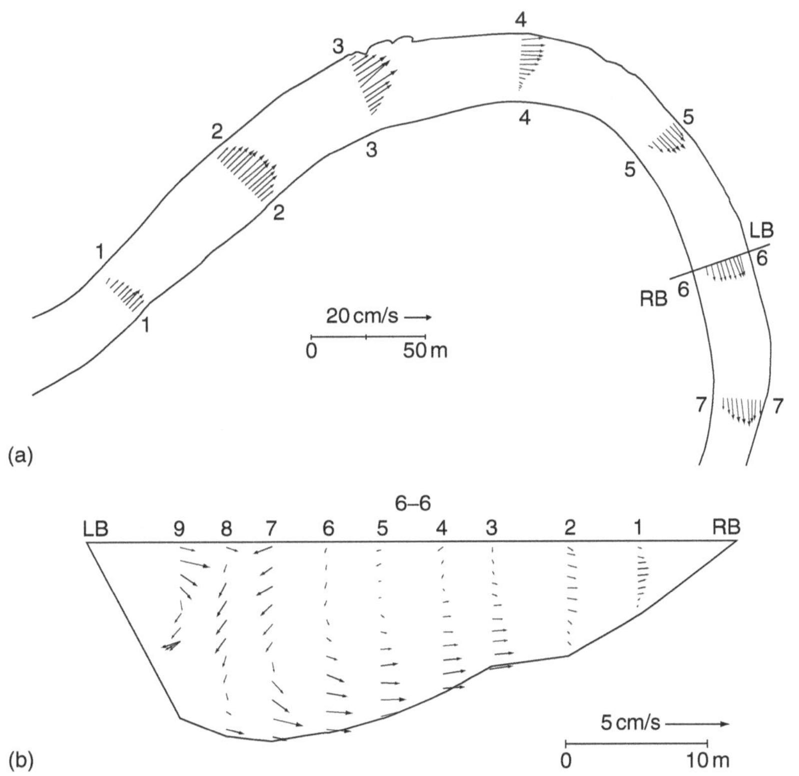

Figure 3 Distribution of time–mean velocity vectors near the free surface of the flow (a), and (b) secondary currents in the river cross-section, depth is enlarged five times (the Spree River, Germany).

The resulting pattern of helical motion redistributes momentum shifting the zone of maximum streamwise velocity towards the concave bank near the bend apex (**Figure 3**). In curvilinear systems of coordinates r (radial), θ (tangential), and z (vertical) dynamical equation of flow are presented in the following form:

$$u_r\frac{\partial u_r}{\partial r}+\frac{u_\theta}{r}\frac{\partial u_r}{\partial \theta}+w\frac{\partial u_r}{\partial z}-\frac{u_\theta^2}{r}=-gS_r+\frac{\partial}{\partial z}\left[\nu_t\frac{\partial u_r}{\partial z}\right] \quad [4]$$

$$u_r\frac{\partial u_\theta}{\partial r}+\frac{u_\theta}{r}\frac{\partial u_\theta}{\partial \theta}+w\frac{\partial u_\theta}{\partial z}+\frac{u_r u_\theta}{r}=gS_\theta+\frac{\partial}{\partial z}\left(\nu_t\frac{\partial u_\theta}{\partial z}\right) \quad [5]$$

where u_r, u_θ, and w are radial, tangential, and vertical mean velocities, and S_r, S_θ are radial and tangential slopes. Systems [4] and [5] can be solved analytically for the radial component, which represents the secondary current, if the distributions of tangential velocities and of turbulent viscosity are presented in an analytical form. For natural streams with large radii of channel curvature the distribution of mean velocities usually differs little from the logarithmic law and the parabolic distribution of turbulent viscosity applies $\nu_t=\kappa U_* z\sqrt{1-z/h}$, then

$$u_r=\frac{1}{\kappa^2}U\frac{h}{r}\left[F_1(\eta)-\frac{\sqrt{g}}{\kappa C}F_2(\eta)\right],$$

$$F_1(\eta)=\int_0^1\frac{2\ln\eta}{\eta-1}d\eta, \text{ and } F_2(\eta)=\int_0^1\frac{\ln^2\eta}{\eta-1}d\eta \quad [6]$$

where $\eta=z/h$. Comparison of values predicted with eqn [6] and measured in a typical lowland river meander bend show good agreement (**Figure 4**). The magnitude of secondary currents of the first kind can be up to 20–30% of the magnitude of streamwise velocity component. These currents are crucial for shaping riverbed relief in channels with loose alluvium.

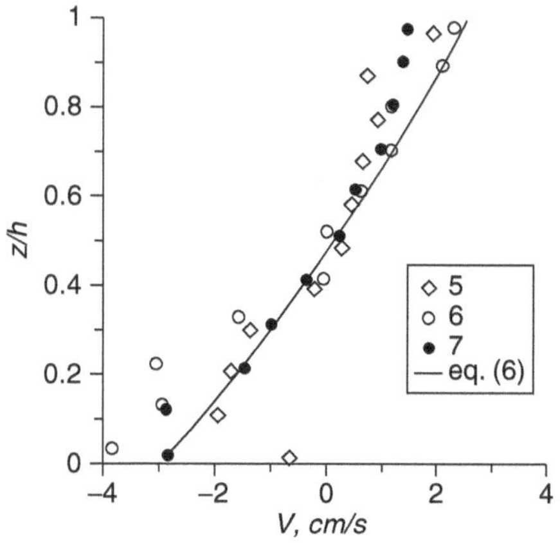

Figure 4 Comparison of measured (symbols mark verticals, **Figure 3(b)**), and predicted (line) time–mean radial velocities (the Spree River, Germany).

Genesis of secondary currents of the second kind is attributed to the nonuniformity in distributions of roughness or morphology of the riverbed in straight river reaches. These secondary currents have much smaller magnitudes, about 5–10% of the primary current, and thus are similar to the magnitudes of turbulent fluctuations. Despite their small magnitudes, these currents are responsible for lateral redistributions of fine sediments on the channel bed, forming longitudinal ridges and thus shaping habitats of benthic invertebrates. Some researchers explicitly associate these currents with turbulent structures and use turbulence anisotropy terms as the driving force in models describing the formation of secondary currents in straight river reaches. However, quantitative methods describing such currents are still unavailable because of a lack of knowledge about river turbulence, and particularly about coherent structures.

Structure of Currents at River Confluences

River confluences, or the locations where two rivers join one another, are integral and ubiquitous features of river networks. Currents within confluences are marked by highly complex three-dimensional patterns of flow that include a zone of stagnant, recirculating flow at the upstream corner of the junction of the two rivers, a region of strong flow convergence within the center of the confluence, a shear layer between the merging flows, and in some cases a zone of separated flow near one or both of the banks (**Figure 5**). It is generally acknowledged that flow structure within confluences is influenced by the junction planform, junction angle, momentum flux ratio, and degree of concordance of the channel beds at the entrance to the confluence. A characteristic pattern of currents within a confluence is the presence of two discrete zones of maximum velocity associated with flow from the two upstream rivers (**Figure 6(a)** and **6(b)**). Between these

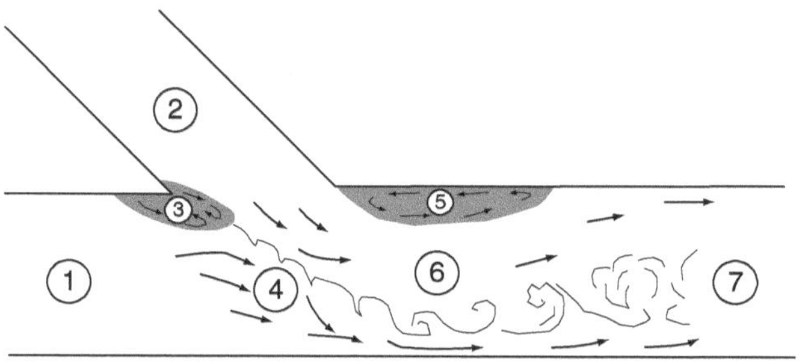

Figure 5 Schematic representation of a confluence.

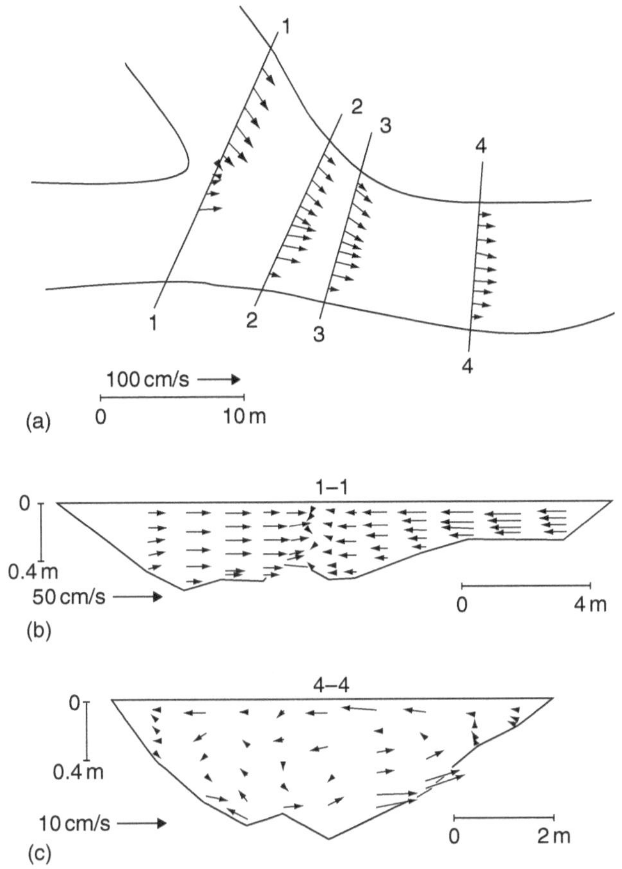

Figure 6 Distribution of time–mean velocity vectors near the free surface of the flow (a), and (b, c) secondary currents in the river cross-sections (confluence Kaskaskia-Copper Slough, USA).

zones is either a stagnation zone (a region of recirculating, separated flow characterized by negative downstream velocities) or farther downstream, a shear layer (a zone of intense turbulence along the interface between the converging flows).

The most prominent characteristic of the cross-stream velocity fields is the opposing orientation of transverse and vertical velocity vectors on each side the confluence (**Figure 6(b)**). The magnitudes of the cross-stream vectors reflect the momentum ratio of the two confluent streams with the largest vectors located on the side of the confluence corresponding to the dominant tributary. As the flow moves downstream, the two distinct zones of maximum downstream velocity gradually converge. Gradual convective acceleration of flow occurs within the low-velocity region between the two maxima until a uniform downstream velocity field with a single zone of maximum velocity develops downstream of the confluence. A pattern of helical flow, similar to secondary currents of the first kind developing in river bends, can also be present at confluences where curvature of flow from a lateral tributary into the downstream channel generates an effect similar to that which occurs in bends (**Figure 6(c)**).

Although river confluences have been actively studied during the past two decades, the complexity of currents at these locations hinders accurate theoretical descriptions. Numerical simulation models currently represent the most sophisticated tools for trying to characterize the complexity of river currents at confluences. However, ongoing studies of shallow mixing layers and free recirculating flows suggest that generalized theoretical models may emerge in the next decade.

Currents at Engineering Structures

Lateral nonuniformity of river currents is associated with natural riverbank protrusions and various engineering structures, among which groynes (spur dykes) are the features most widely used to support navigation and protect banks against erosion. Groynes are usually placed in sequences so that the area between successive groynes is referred to as a groyne field. Flow separates at the tip of the protruding groyne, or groyne head, and forms a rotating current in groyne fields depicted by large-scale vortexes with a vertical axis of rotation called gyres. Since water is forced to recirculate within the groyne field in spiraling trajectories, the local residence time for suspended particulate matter can increase substantially and may be sufficient to maintain local phytoplankton reproduction. Therefore, understanding of these complex currents can have important ecological implications.

Specific patterns of flow in a groyne field are controlled by the geometry of the field: the aspect ratio between the groyne length (L_g) and the streamwise length of the groyne field (L_f). Observations indicate that a two-gyre circulation pattern develops when the aspect ratio is less than a critical value ($L_g/L_f < 0.5$), and a one-gyre circulation forms in groyne fields aspect ratios greater than the critical value ($L_g/L_f > 0.5$) (**Figure 7**).

The distribution of mean velocities within the flow in groyne fields can also be reasonably described by a shallow mixing layer model and hyperbolic tangential equation. A canonical free mixing layer (family of free-turbulent flows) evolves in coflowing liquids of different densities or flows of different mean velocities, and can be described by a simple model

$$\frac{\partial \delta}{\partial x} = \alpha\lambda \quad [7]$$

$$\lambda = \frac{\Delta U}{U_1 + U_2} = \frac{\Delta U}{2U_c}, \quad \Delta U = U_1 - U_2 \quad [8]$$

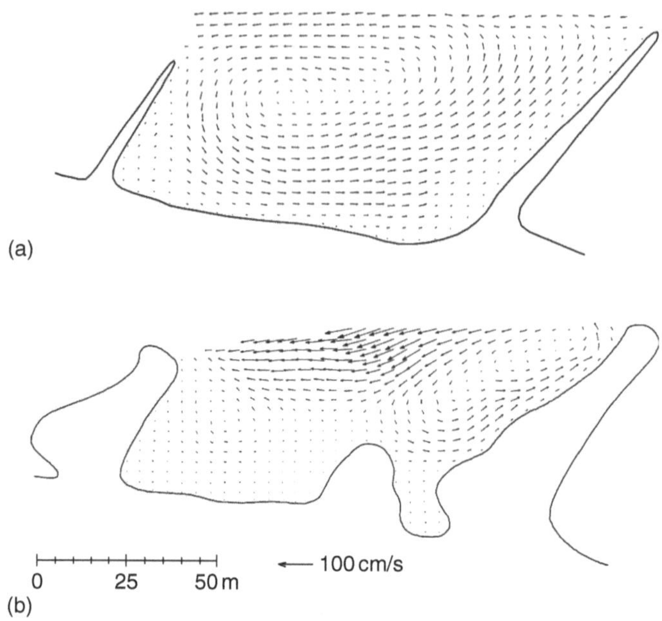

Figure 7 Distribution of depth-averaged time–mean velocity vectors (interpolated from measurements) in the groyne field with aspect ratio (a) 0.6, and (b) 0.4 (the Elbe River, Germany).

where α is the spreading rate (a constant in canonical mixing layer, equals 0.18), δ is the width of mixing layer, x is the downstream coordinate, λ is the velocity ratio, U_1 is the velocity in free part (above) of flow, and U_2 is inside the stand, and U_c is the velocity in the centre of the mixing layer. Mean velocity profiles in mixing layers have been shown to comply with a hyperbolic tangential distribution

$$\frac{U}{U_s} = 1 + \tanh\left(\frac{2y}{L}\right) \qquad [9]$$

where $U_s = \Delta U/2$, y is the distance across the layer, and $L = \gamma\delta$ is a characteristic length scale normally proportional to the width of the mixing layer. An example of the distribution of depth-averaged velocity across the groyne field and its interface with the mean flow is shown in **Figure 8**.

The specific pattern of recirculation has important implications for distribution of deposited fine sediments within the groyne field. The low-velocity area in the centre of gyres promotes accumulation of relatively fine sediments. The thickness of the layer of deposited fines decreases toward the gyre margins, where flow velocities increase.

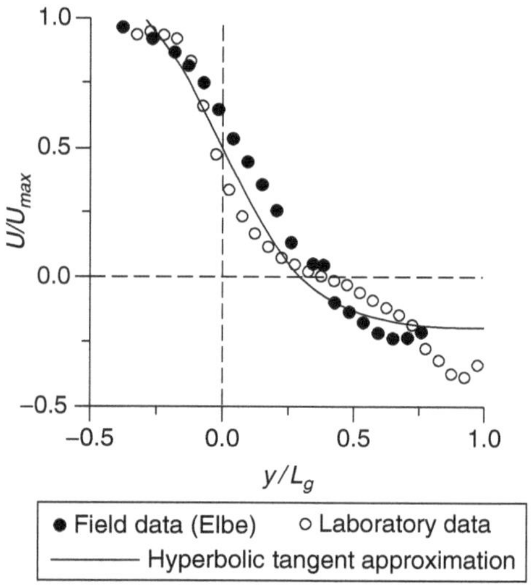

Figure 8 Measured distributions of depth-averaged time–mean velocities (section through the gyre center) in the mixing layer between main flow and groyne field, and their approximation with hyperbolic tangent (eqn [10]).

Navigation-Induced Currents

Rivers have been always heavily exploited as the inland waterways for commercial and recreational navigation. A vessel moving along a river channel expends energy to overcome resistance of water. The energy is transformed into complex pattern of currents and waves in the lee of the vessel. In width-restricted channels, when commercial tugs towing loaded barges cruise with speed close to the navigational limits, the navigation-induced currents maintain extremely large velocities (**Figure 9**). An analytical framework for assessment of such currents was deduced from the balance of kinetic energy and represented by a relationship

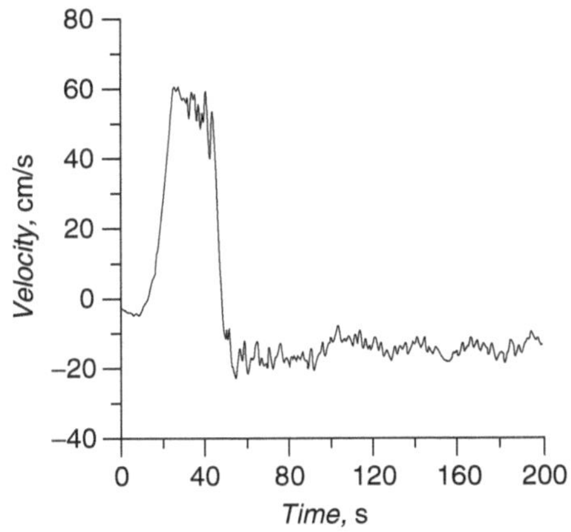

Figure 9 Navigation-induced current from a towing barge measured 15 cm above the riverbed at 2 m distance from the water edge (Oder-Havel Canal, Germany).

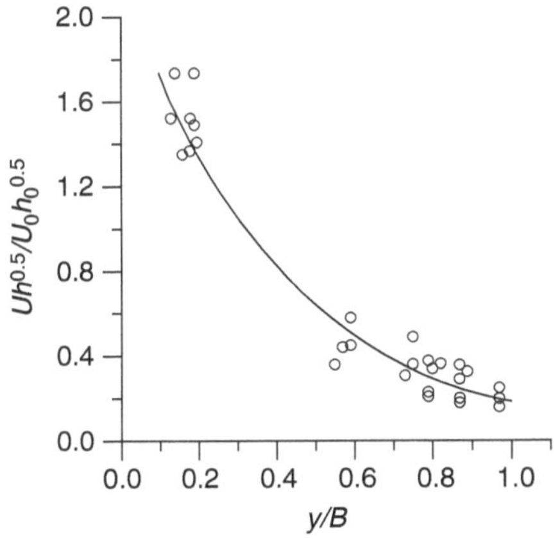

Figure 10 Measured (circles) and predicted (line) values of return currents.

$$\frac{U}{U_0} = \alpha_0 \sqrt{\frac{h_0}{h}} \exp\left[-\beta \frac{y}{B}\right] \qquad [10]$$

where α_0 and β are parameters depending on the characteristics of kinetic energy transfer (dispersion coefficient, wave celerity, and channel width B), U is maximal value of depth-averaged velocity in navigation-induced current, U_0 is the speed of the vessel, h is the flow depth, h_0 is the draught of the vessel, and y is the transverse distance from the vessel towards waterway bank. Performance of the relation [10] is illustrated in **Figure 10**, where the results of field measurements are compared with values predicted by the model [10].

Conclusions

The major factors controlling currents in rivers are the macro- and microscale geometry of the river channel, the joining of streams induced by the structure of fluvial networks, properties of alluvium composing the riverbed material, biological features, and anthropogenic influences. Although quantitative descriptions of currents in rivers are based on well-known theoretical principles, a universal theory of river currents has yet to be developed mainly because of the lack of a universal theory for turbulence aspects of river flows. The problem of quantitative description is presently solved with application of case-specific models and relies substantially on the application of empirical knowledge.

Mutual interactions among flow, the river channel, and biota at different spatial and temporal scales ranging from a sediment grain to the scale of a river reach, and from milliseconds to hundreds of years complicate unambiguous studies of currents. Therefore, available data on important parameters in models of river currents include significant scatter that leads to uncertainties in prediction of magnitudes and patterns of flows. Some of this uncertainty reflects the fact that most theoretical, laboratory, and field studies investigate or assume uniform steady flow, while in nature such flows are always an idealization.

Nomenclature

B	width of a channel (m)
B_0	integration constant
C	Chezy coefficient ($m^{1/2}$ s^{-1})
c_f	friction factor
F	friction force (kg m s^{-2})
Fr	Froude number
F_1, F_2	integral functions
G	gravity force (kg m s^{-2})
g	gravity acceleration (m s^{-2})
h	mean depth (averaged over cross-section) (m)
h_0	draught of a vessel (m)
k_s	height of roughness elements (m)
L	length scale (m)
L_g	length of groyne (m)
L_f	length of groyne field, m
n	Manning–Strickler coefficient ($m^{-1/3}$ s)
r	radial coordinate (m)
S	longitudinal gradient of the water surface
S_r	radial slope
S_θ	tangential slope
U	time–mean velocity (m s^{-1})
U_c	time–mean velocity in the centre of the mixing layer (m s^{-1})
U_0	cruising velocity of a vessel (m s^{-1})
u'	streamwise velocity fluctuations (m s^{-1})
u_r	radial mean velocity (m s^{-1})

u_θ	tangential mean velocity ($m\ s^{-1}$)
U_*	shear velocity ($m\ s^{-1}$)
U_s	mean velocity half of velocity difference in mixing layer ($m\ s^{-1}$)
U_1, U_2	velocity in fast and slow flows ($m\ s^{-1}$)
w'	vertical velocity fluctuations ($m\ s^{-1}$)
y	transverse distance (m)
z	vertical coordinate (m)
z_0	hydrodynamic roughness parameter (m)
ΔU	velocity difference ($m\ s^{-1}$)
α	spreading rate
α_0	parameter
β	parameter
γ	dimensionless coefficient
δ	width of the mixing layer (m)
η	dimensionless distance
θ	tangential coordinate (grad.)
κ	von Karman parameter
λ	velocity ratio
ν_t	turbulent viscosity ($m^2\ s^{-1}$)
ρ	density of water ($kg\ m^{-3}$)
τ	shear stress ($kg\ m^{-1}\ s^{-2}$)
τ_0	bottom shear stress ($kg\ m^{-1}\ s^{-2}$)

See also: Flow in Wetlands and Macrophyte Beds; Flow Modification by Submerged Vegetation.

Further Reading

Cunge JA, Holly FM Jr, and Verwey A (1980) *Practical Aspects of Computational River Hydraulics*. London: Pitman.

Best JL and Reid I (1984) Separation zone at open-channel junctions. *Journal of Hydraulic Engineering* 110: 1588–1594.

Ghisalberti M and Nepf H (2002) Mixing layers and coherent structures in vegetated aquatic flows. *Journal of Geophysical Research* 107(C2). doi 10.1029/2001JC00871.

Gordon ND, McMahon TA, and Finlayson BL (1992) *Stream Hydrology: An Introduction for Ecologists*. Chichester: Wiley.

Leopold LB (1994) *A View of the River*. Cambridge: Harvard University Press.

Nezu I and Nakagawa H (1993) *Turbulence in Open-Channel Flow*. Rotterdam: Balkema.

Rhoads BL and Sukhodolov AN (2001) Field investigation of three-dimensional flow structure at stream confluences: 1. Thermal mixing and time-averaged velocities. *Water Resources Research* 37(9): 2393–2410.

Rozovskii IL (1961) *Flow of Water in Bends of Open Channels*. [trans.]. Jerusalem: Israel Program for Scientific Translations.

Schlichting H and Gersten K (2000) *Boundary Layer Theory*. Berlin: Springer.

Sukhodolov AN, Uijttewaal WSJ, and Engelhardt C (2002) On the correspondence between morphological and hydrodynamical patterns of groyne fields. *Earth Surface and Landforms* 27: 289–305.

Sukhodolov AN and Sukhodolova TA (2006) Evolution of mixing layers in turbulent flow over submersible vegetation: field experiments and measurement study. In: Ferreira RML, Alves ECTL, Leal JGAB, and Cardoso AH (eds.) *RiverFlow2006*, pp. 525–534. London: Balkema.

Wolter C, Arlinghaus R, Sukhodolov AN, and Engelhardt C (2004) A model of navigation-induced currents in inland waterways and implications for juvenile fish displacement. *Environmental Management* 35: 656–668.

Yalin MS (1992) *River Mechanics*. Oxford: Pergamon.

Relevant Websites

http://cwaces.geog.uiuc.edu – Center for Water as a Complex Environmental System (CWACES), Urbana-Champaign, Illinois, USA.

http://www.wldelft.nl – WL|Delft Hydraulics, the Netherlands.

http://www.nced.umn.edu – National Center for Earth-surface Dynamics, Minneapolis, USA.

http://www.niwascience.co.nz – National Institute of Water and Atmospheric Research, New Zealand.

http://www.ifh.uni-karlsruhe.de – Institute for Hydromechanics in Karlsruhe, Germany.

http://www.igb-berlin.de – Institute of Freshwater Ecology and Inland Fisheries, Germany.

Currents in Stratified Water Bodies 1: Density-Driven Flows

F Peeters, Universität Konstanz, Mainaustrasse, Konstanz, Germany
R Kipfer, Swiss Federal Institute of Environmental Science and Technology (Eawag), Swiss Federal Institute of Technology (ETH), Ueberlandstr, Duebendorf, Switzerland

Introduction

Vertical transport of dissolved substances and heat in lakes mainly results from two different mechanisms: (a) mixing by turbulence that is usually described as a diffusive transport and (b) density-driven exchange that can be considered as an advective transport. A typical example of the latter is convection owing to surface cooling in fall, which often leads to isothermal conditions in shallow lakes of the temperate zone. Because entrainment of ambient water limits the depth of convective plumes, density-driven transport to large depth in deep lakes usually occurs along the lake boundaries and is often the result of specific and localized processes, which are discussed later.

The important role of density-driven transport for vertical exchange in lakes becomes evident if one considers that temperature stratification typical for most lakes is characterized by a decrease in water temperature with increasing water depth. Turbulent diffusion causes heat to flow from high to low temperatures and hence typically leads to a gradual continuous warming of cold deep-water regions. Thus, on a long-term average, advective processes transporting cold surface water downwards must be sufficient to compensate for the heat flux due to turbulent diffusion. The low temperatures in the deep water are usually either the remnant of isothermal conditions generated by buoyancy-driven overturn during the cold season or originate from cold density currents propagating to largest depth. Because vertical transport due to density currents plays an important role in overall deep-water renewal and heat exchange, density driven exchange processes are central to the understanding of oxygenation and nutrient transport especially in deep lakes.

In the world's largest and deepest water bodies several processes have been identified that lead to advective deep-water renewal by density currents: river inflow, e.g., in Lake Constance, Lake Geneva, and Lake Baikal; inter-basin exchange, e.g., in Lake Lucerne, Lake Baikal, or even in the Caspian Sea; differential cooling, e.g., in Lake Geneva, Lake Constance, Lake Issyk-Kul, and Lake Malawi; thermal-bar mixing, e.g., in the Lake Ontario, Lake Ladoga, Lake Michigan, and Lake Baikal; and transport due to thermobaric instabilities, e.g., in Lake Baikal and possibly in Crater Lake. All these processes have been shown to significantly contribute to deep-water renewal in lakes, although advective transport to the lake bottom was not conclusively demonstrated in all cases. More details on the different processes are given below.

In the following, we first describe the principal characteristics of density currents and the associated signals of intrusions in vertical profiles of water constituents and temperature. Then, we present several mechanisms that lead to the generation of density plumes in deep freshwater lakes and discuss which of these processes can also be responsible for deep-water renewal in tropical and saline lakes. Finally, we discuss the potential impact of changes in the catchments of lakes and in the meteorological conditions on deep-water renewal by density currents.

Characteristics of Density Currents

Density currents are driven by differences in water density, which can result from gradients in water temperature, salinity, dissolved uncharged substances, or suspended particles and are also affected by pressure. If a water mass with higher density is situated above a water mass with lower density, the stratification is unstable and buoyancy causes the upper water mass to sink. In the sinking process ambient water is mixed into the sinking water mass and thus alters its density (**Figure 1**), thereby reducing the density difference between the density plume and the ambient water.

Furthermore, the properties of the ambient water change along the path of the sinking water mass. Hence the buoyancy of the sinking plume changes continuously as it sinks into deeper depth. Eventually, a depth is reached where the density of the density plume and the density of the surrounding water become equal. At this depth the sinking process ceases and the plume water spreads out laterally into the ambient water forming an intrusion (**Figure 1**). In many cases the sinking plume meets the lake boundary and then continues to sink along the lake boundary (**Figure 1**). In this case, entrainment of ambient water is limited to the upper side of the density plume and less ambient water is entrained per unit sinking depth. Hence the characteristic properties of the water within the density plume (e.g., temperature, salinity, suspended particles, dissolved oxygen) change more

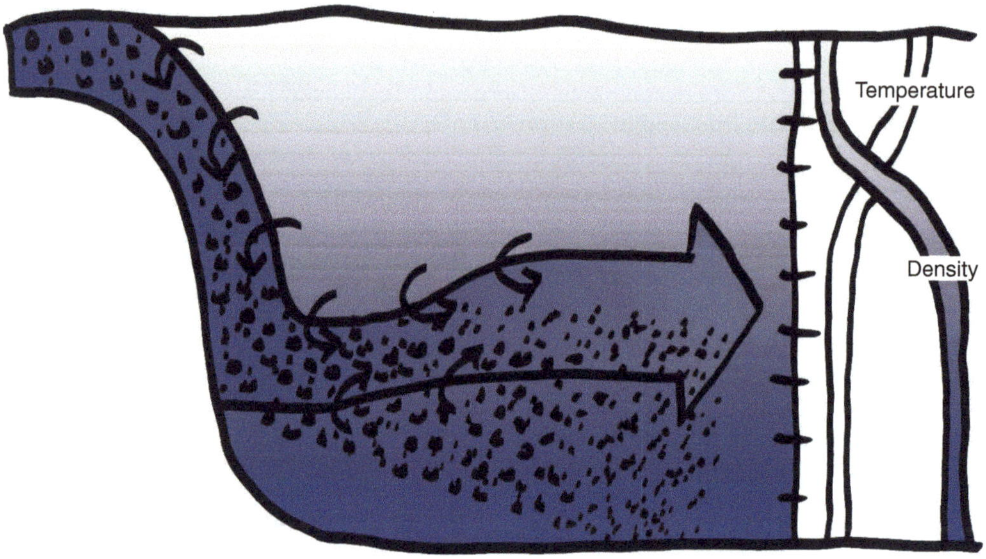

Figure 1 Schematic illustration of density currents. The shading on the left-hand side of the figure indicates an increase in density with increasing depth.

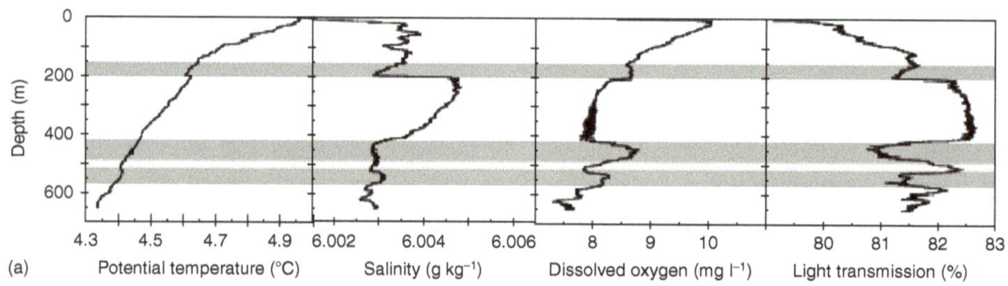

Figure 2 Intrusions as indicators of density currents. Vertical profiles of temperature, salinity, dissolved oxygen, and light transmission measured in Lake Issyk-Kul. The distinct features in these profiles suggest intrusions from density currents. Grey bars mark depth regions with high concentrations of dissolved oxygen and low light transmission, suggesting water originating from shallower depth regions. Redrawn from **Figure 2** in Peeters FD, Finger M, Hofer M, Brennwald DM, and Livingstone R Kipfer (2003). Deep-water renewal in Lake Issyk-Kul driven by differential cooling. *Limnology & Oceanography* 48(4): 1419–1431.

slowly and the density plumes typically can propagate to larger depths than would be possible in the open water.

The occurrence of density plumes propagating from shallow to deep water are indicated by intrusions that can be identified in CTD-profiles (conductivity as measure of salinity, temperature, and depth) and in profiles of dissolved substances and suspended particles. **Figure 2** presents an example from Lake Issyk-Kul (Kyrghystan), where intrusions are characterized by a higher dissolved oxygen concentration and a lower light transmission than is observed in the ambient water.

High oxygen levels in the intrusions indicate oxygen-rich surface water that must have been transported recently because oxygen levels have not yet been significantly reduced by degradation processes. The low light transmission in the intrusions indicates water with a high load of suspended particles suggesting either, that the water that generated the density current was enriched in suspended particles and thus may have originated from river inflow, or, that the density plume responsible for the intrusions has propagated along the lake boundary and caused resuspension of sediments during the sinking process. Temperature is usually not a good indicator of intrusions because it is a key parameter determining plume density. Thus, at the depth of the intrusion, plume water and surrounding water often have about the same temperature. However, in cases where the density plume propagates down to the largest depths, as it is sometimes the case e.g., in Lake Baikal, temperature anomalies at the lake bottom can be used to identify density plumes.

Density Plumes Generated by External Inputs

River Inflows

The density differences required to drive density plumes originate from processes that generate horizontal or vertical gradients in water properties. An obvious example is river inflow (**Figure 3**).

River water usually contains an increased load of suspended particles and has a different temperature and salinity than the lake water. Hence, river inflow is commonly associated with density plumes propagating from the river mouth to larger depths. The kinetic energy associated with the inflow of the river water is usually rapidly dissipated and the horizontal density gradients resulting from the different densities of river water and lake water are the main cause of river induced transport of water masses in lakes. During summer, the density plumes induced by river inflow typically intrude at some depth within the thermocline of freshwater lakes. Because of the large temperature gradients in the thermocline, water densities change significantly within a rather narrow depth range in the lake. Hence, the probability that the density of the plume water agrees with the density in the water column of the lake is especially large within the thermocline. This fact explains that river water typically intrudes in this depth range. The depth reached by the density plumes varies during the course of the year since water properties and hence the density of lake and river water changes seasonally (**Figure 3**). Density currents containing a high load of

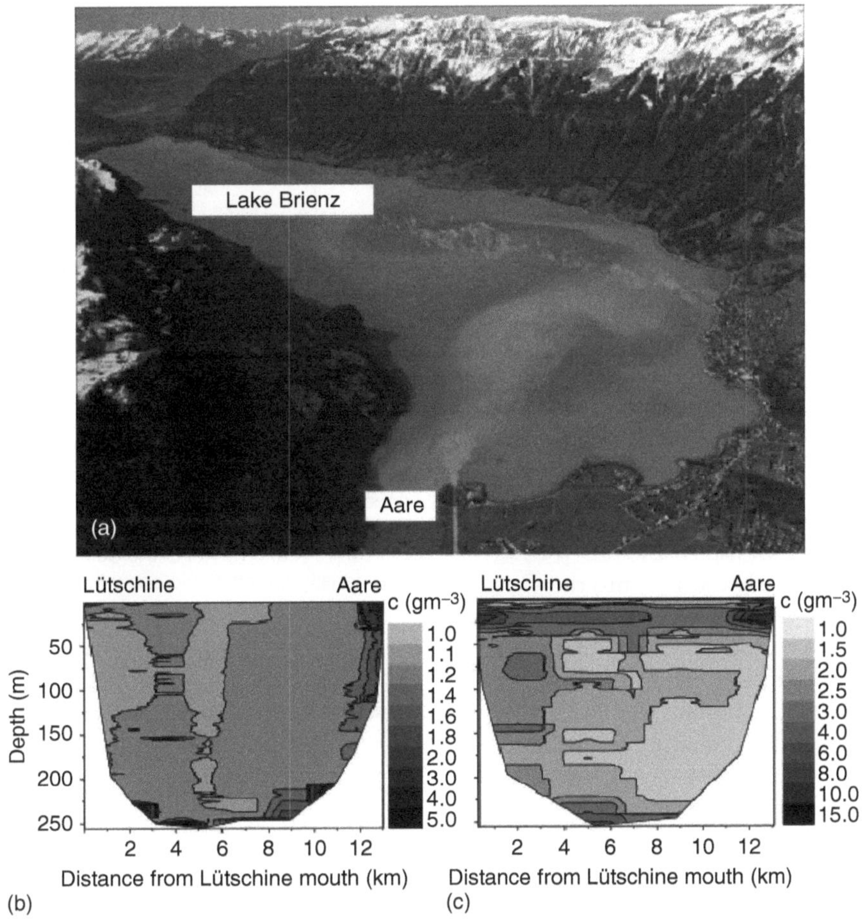

Figure 3 Density currents generated by river inflow. (a) Water of river Aare indicated by high turbidity intruding near the surface of Lake Brienz. The sharp boundaries of this surface plume indicate plunging of river water to larger depth. (b, c) Suspended particle distribution inferred from light transmission measurements in a longitudinal cross-section of Lake Brienz measured in February (b) and October (c). The particle distributions suggest that, in February water introduced by the river Aare (inflow on the right-hand side) sinks as density plume along the lake bottom towards largest depth (b). In October river Aare and river Lütchine both intrude at intermediate depth (c). (**Figure 3(a)** was provided by Ueli Ochsenbein; **Figure 3(b)** and **(c)** are redrawn from **Figure 7(a)** and **7(d)** in Finger D, Schmid M, and Wüest A (2006). Effects of upstream hydropower operation on riverine particle transport and turbidity in downstream lakes. *Water Resources Research* 42, W08429, doi:10.1029/2005WR004751. Reproduced/modified by permission of American Geophysical Union.

suspended particles are often called turbidity currents. Sedimentation of particles out of intrusions resulting from turbidity currents can reduce the density of the intruding water sufficiently that the depth of the intrusions becomes shallower over time. Density currents induced by river inflows usually propagate along the sloping bottom boundary before the water intrudes laterally. If the concentrations of solutes and particles are very different between the river and the lake water, the plumes can propagate down to the deepest parts of the lake. In cases where the sinking water is confined to underwater channels, e.g., density plumes propagating down the Kukui Canyon of the Selenga delta in Lake Baikal, entrainment of ambient water is reduced, and the density plume can propagate over a depth range of more than 1000 m down to 1640 m (see also **Figure 4(a)**).

Interbasin Exchange

River inflows not only result in localized density plumes propagating from the river mouth to larger depths but also can generate subtle large scale gradients in water properties that affect the horizontal density distribution on a basin scale. These small density gradients on a large spatial scale may contribute to the generation of density currents far from the river inflow that occur especially in the vicinity of sills separating sub-basins of the lake. This mechanism is exemplified in **Figure 4** for two lakes of different size, Lake Baikal and Lake Lucerne, that both are structured into several sub-basins separated by sills.

In both lakes large scale horizontal salinity gradients are generated by river inflows introducing water with different ion concentrations into the different sub-basins. In the case of Lake Baikal, the River Selenga introduces more saline water into the Central Basin than the Upper Angara River introduces into the Northern Basin. In case of Lake Lucerne, the River Sarner Aa introduces more saline water into the sub-basin Lake Alpnach than the River Reuss introduces into the sub-basin Lake Uri. Because of the salinity gradients, the density of the water in the different sub-basins differs if temperature is the same. In Lake Lucerne the densest water can be found in sub-basin Lake Alpnach when winter cooling reduces surface water temperature to ~4 °C. Horizontal transport of the dense water from Lake Alpnach to the sub-basin Lake Vitznau across the sill separating the two sub-basins induces a density current renewing the deep-water of sub-basin Lake Vitznau (**Figure 4(b)**). The density plume causes upwelling of cold dense water within Lake Vitznau. Horizontal transport of water across the sills between sub-basin results into a cascading of density driven transport within all

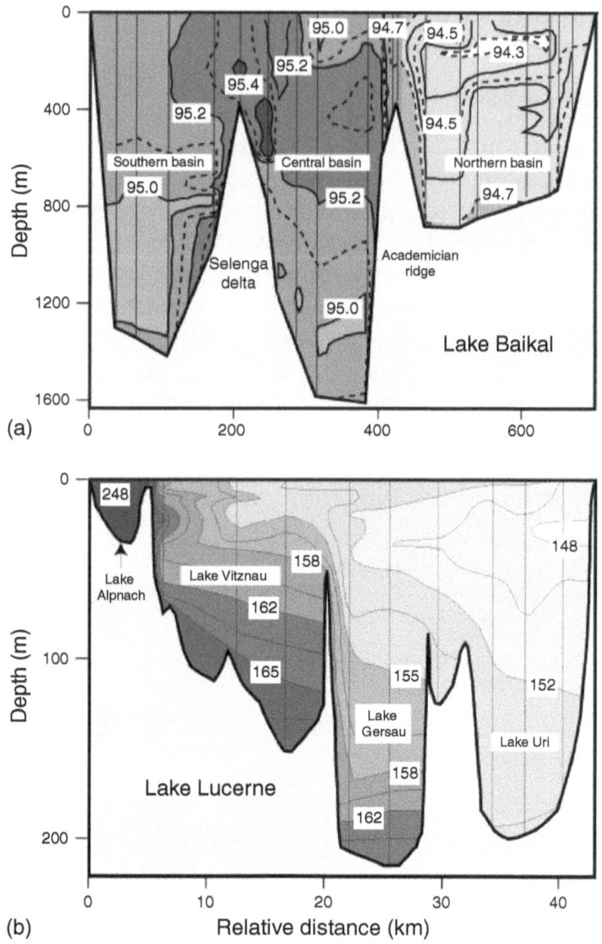

Figure 4 Density currents generated at sills between sub-basins. Vertical transects of salinity in Lake Baikal (a) and in Lake Lucerne (b). In both lakes the horizontal gradients in salinity are generated by river inflows introducing water with different ion concentration. The salinity distributions suggest that density currents are not only generated directly by river inflow as is the case in Lake Baikal at the Selenga delta (see panel a) but that density currents also occur in both lakes at the sills between sub-basins most likely driven by horizontal transport across the sill. Contours depict salinity in mg kg^{-1}. **Figure 4(a)**: redrawn from Kipfer R and Peeters F (2000). Speculation on consequences of changes in the deep water renewal in Lake Baikal, in K Minoura (ed.) *Lake Baikal a mirror in time and space for understanding global change processes*, pp. 273–280. Amsterdam, Netherlands: Elsevier. **Figure 4(b)**: drawn using data from Aeschbach-Hertig W, Kipfer R, Hofer M, Imboden DM, and Baur H (1996) Density-driven exchange between the basins of Lake Lucerne (Switzerland) traced with the ^3H-^3He method. *Limnology and Oceanography* 41: 707–721.

sub-basins (**Figure 4(b)**). A similar process is operating in the different basins of Lake Baikal (**Figure 4(a)**). Prerequisite of these density currents generated at sills between sub-basins is (1) the structuring of the lake basin into sub-basin that prevents homogenization of water properties by horizontal mixing and (2) a heterogeneous input of water properties, as e.g., the salinity by the river inflows in **Figure 4**.

Subsurface Inflows

Besides the input from rivers, external water sources derived from groundwater inflows and from hydrothermal vents can generate density currents depending on the depths at which the inflows are located. Ground water and hydrothermal water is usually highly enriched in ions and thus can cause salinity driven density plumes. Groundwater inflows into lakes are common in artificial lakes such as mining lakes or gravel-pit lakes or occur in karstic environments. Density currents due to hydrothermal vents have been reported for instance in Lake Baikal, where hydrothermal water is introduced in Frohliha Bay at a depth of 200–400 m and propagates as a bottom following density current down to 1400 m depth. In this specific case the salinity of the hydrothermal water is sufficiently large to compensate the decrease in density due to the increased water temperatures in the hydrothermal water.

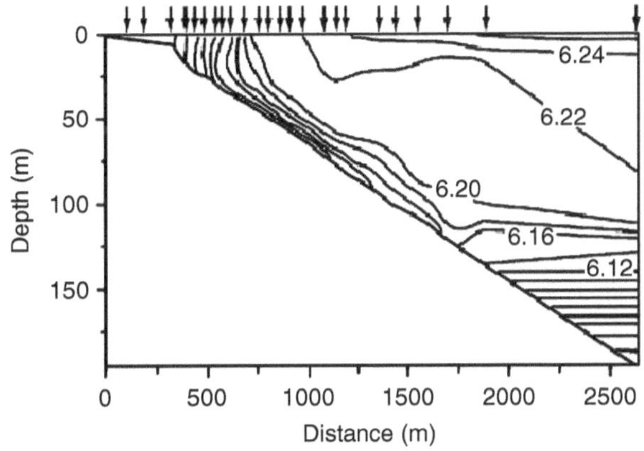

Figure 5 Density current generated by differential cooling. Contour lines represent isotherms in °C indicating a density plume generated by differential cooling in Lake Geneva. Note that the temperatures are well above 4 °C. The isotherms are constructed from CTD-data collected at the locations indicated by the arrows. Redrawn from **Figure 2(b)** in Fer I and Lemmin U. Winter cascading of cold water in Lake Geneva. *Journal of Geophysical Research* 107, NO. C6, 10.1029/2001JC000828, 2002. Reproduced/modified by permission of American Geophysical Union.

Density Plumes Generated by Internal Processes

Differential Cooling

A key parameter affecting water density is temperature. However, for density currents to be induced on the basis of the temperature of water, horizontal gradients in temperature are required. Horizontal temperature gradients are generated by external surface and subsurface inflows (see earlier text) but also by internal processes. In most lakes the heat flux at the lake surface (expressed per unit area) can be considered as horizontally homogeneous because meteorological parameters and radiation do not vary significantly at the length scale of the lake basin. Nevertheless, differential cooling can generate significant temperature differences within lakes and thus generate density currents (e.g., Wellington Reservoir, Lake Geneva, Lake Constance, Lake Banyoles). Heat loss at the lake surface causes vertical convection and thus mixing of surface water with water from layers below. This process continuously mixes the cooled surface water with water from deeper layers containing heat stored during the warm season. In shallow-water regions, the reservoir of warmer deep water is exhausted earlier than in regions with large water depth. Hence, in shallow-water regions heat loss at the lake surface leads to a faster cooling of the water column than in deep-water regions. Because the cold water in the shallow-water regions has a larger density than the warmer water in the pelagic, deep-water regions, the cold water propagates downwards as a density current. Such density plumes often occur only sporadically. They are typically generated during night-time cooling in fall as has been demonstrated in e.g., Lake Constance and Lake Geneva (**Figure 5**) or during events that also induce cooling such as cold fronts or monsoons.

Differential cooling can result in density driven currents from any shallow region in a lake basin. Therefore it may affect a large volume of water and thus may significantly contribute to overall vertical transport. The process is particularly effective if large shelf regions are located around a deep basin. The density currents induced by differential cooling propagate along the lake bottom and can reach large depths especially if channels exist along which the density current can propagate without significant entrainment of ambient water, as is the case e.g., in Lake Issyk-Kul. Note, that in freshwater lakes differential cooling can only generate density plumes if water temperatures are above the temperature of maximum density (T_{md}) which is about 4 °C at the lake surface. Cooling below T_{md} implies a decrease in water density and thus prohibits density plume development by differential cooling. Hence, in freshwater lakes the temperature of density plumes associated with differential cooling is at least 4 °C or higher.

Thermal Bar

Temperature-driven density currents can also result from horizontal mixing of two adjacent surface water masses, one having a temperatures above and the other below T_{md}. Because of the non-linear temperature dependence of the equation of state, mixing of water masses with different temperatures always results in an increase in the mean density of the water. This process is called cabbeling. If the temperature of the mixed water is closer to the T_{md} than the water below, it sinks as a density plume. A so-called thermal bar develops which is characterized by a vertically isotherm water column with a temperature close to T_{md} separating an open water region with temperatures below T_{md} from a warmer shore region with temperatures above T_{md} (**Figure 6**).

The downward flow of dense surface water at the thermal bar can cause significant renewal and oxygenation of deep water. As time progresses the thermal bar moves further away from the shore to the open water. The dynamics of this process depends on the morphometry of the lake and on the atmospheric forcing. Mixing associated with the thermal bar has been

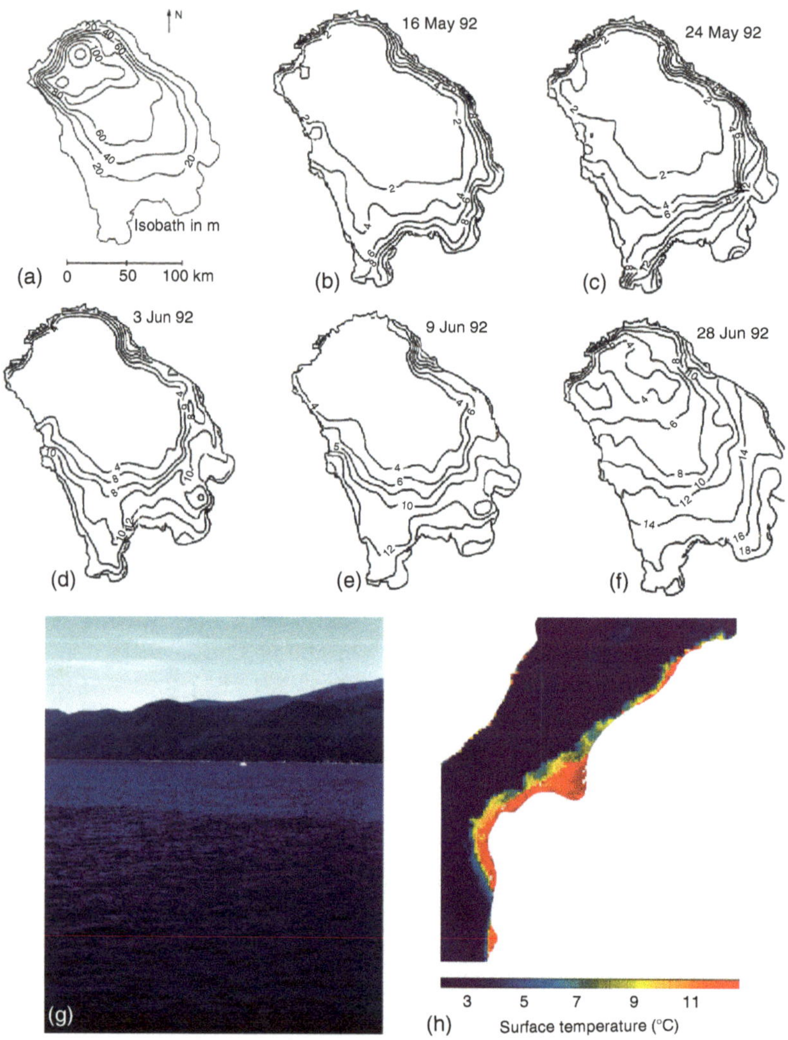

Figure 6 The thermal bar. The development of a thermal-bar in Lake Ladoga indicated by surface temperatures measured with satellites (a–f) and the observation of a thermal bar near Selenga delta in Lake Baikal (g and h). Panel (a) provides the morphometry of Lake Ladoga with depth contours given in m. Panels (b–f) depict surface temperatures with isotherms in °C. Panels (b–f) show how the thermal bar, which is located at the 4 °C isotherm, moves towards deeper water as the season progresses. The position of the thermal bar changes more rapidly in the gently sloping shallower south-eastern part of the lake than in the steep and deep northern part. In (g), the sharp boundary between near shore water and open water indicates the position of the themal bar located near the Selenga delta in Lake Baikal. The color differences result from differences in the load of suspended particles. The water trapped near shore by the thermal bar has an increased load of suspended particles owing to the nearby inflow of the Selenga River. Panel (h) shows an image of the surface temperatures near Selenga delta derived from satellite data. Density currents are generated at the sharp transition between warm shore water and cold open water characterized by a temperature close to 4 °C. White areas are land. **Figure 6(a–f)** are redrawn from **Figures 1** and **3** in Malm J, Mironov D, Terzhevikl A, and Jiinsson L (1994) Investigation of the spring thermal regime in Lake Ladoga using field and satellite data. *Limnology and Oceanography* 39(6): 1333–1348. Copyright 2000 by the American Society of Limnology and Oceanography, Inc.

reported for Lake Ontario, Lake Ladoga, Lake Michigan, and Lake Baikal. Because the thermal bar requires open-water temperatures below T_{md}, it occurrs mainly in lakes which become ice covered in winter.

The horizontal temperature gradients required for thermal bar development are generated by differential heating during spring warming, a process similar to differential cooling. Initially in spring, when surface-water temperatures are below T_{md}, an increase of water temperature during spring warming leads to an increase in the density of the surface waters and thus to convection. Because convection mixes the warmer surface water with colder water from below and the reservoir of the cold water from below is smaller in shallow near-shore than in deeper open-water regions the temperature in the shallow-water regions increases faster than in the deep off-shore regions. When the temperature in the shallow regions exceeds 4 °C the water column is stratified and further influx of heat is not connected to convection, but leads to an even faster increase in the water temperature. Thus, a horizontal temperature gradient typical for a thermal bar situation develops with temperatures below T_{md} at the surface of the open-water region and temperatures above T_{md} at the surface in the shallow near-shore regions. A reverse thermal bar situation with temperatures below T_{md} in the shallow-shore region and temperatures above T_{md} in the open-water region could develop in fall as a consequence of differential cooling if cooling in the shore region progresses to temperatures below T_{md}. However, in the case of the reverse thermal bar exchange processes will be dominated by horizontal density gradients below the surface (see the section on 'Differential cooling').

Thermal Baricity

Another process that can generate density plumes as a consequence of the nonlinearity of the equation of state of fresh water is the thermobaric effect. The thermobaric effect results from the fact that T_{md} decreases with increasing pressure. The generation of density currents by the thermobaric effect requires a very specific temperature stratification that occurs only in few stably stratified deep freshwater-lakes, e. g., Lake Baikal or Crater Lake. To generate density currents by the thermobaric effect, water temperatures must be below 4 °C throughout the water column. The temperature in the surface layer must increase with increasing depth, whereas the temperature in the deep-water must decrease with increasing depth. Then, the temperature profile has a maximum at intermediate depth, the so-called mesothermal maximum (**Figure 7**).

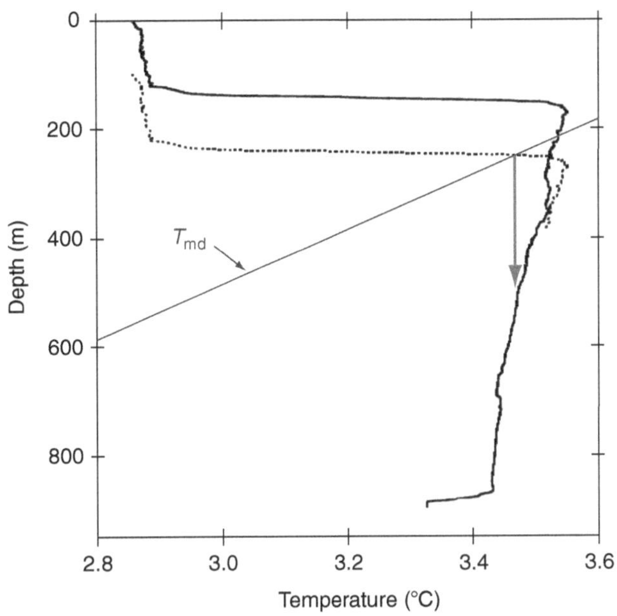

Figure 7 Schematic on the generation of density currents by the thermobaric effect. The temperature profile presented has been measured in the northern basin of Lake Baikal. The temperature of maximum density as function of depth shown for comparison is labelled with T_{md}. Vertical displacement of the temperature profile (indicated by the dashed line) leads to a density driven vertical transport that is self supporting over the depth range indicated by the arrow.

A water column with such a temperature profile is stably stratified because of the effect of pressure on fresh water density. If the water column is displaced downwards or pressure is increased, the temperature at the mesothermal maximum is higher than the local T_{md} and the water column becomes unstable. Cold water from above the mesothermal maximum can sink downwards as a density plume (**Figure 7**). Similarly, a water mass from the cold upper layer can be pushed downwards below the mesothermal maximum to a depth where its temperature is closer to T_{md} than the temperature of the ambient water. Then, it will continue to sink driven by its buoyancy until the surrounding water has the same temperature as the sinking water mass. Besides the specific temperature profile in the water column the exchange due to the thermobaric effect also requires a mechanism by which the water pressure is altered substantially and/or the water is locally pushed downwards across the depth of the mesothermal maximum. Hence density plume generation by the thermobaric effect is not very common. Several investigations have claimed that the thermobaric effect may be important for deep-water oxygenation in Lake Baikal. However, the mechanism that could cause the required downward displacement in the open water column remained unclear. Recently, wind-driven Ekman transport near the coast of the

Southern Basin of Lake Baikal has been suggested to cause thermobaric instabilities.

Turbidity Currents Generated by Waves

Internal processes not only can generate temperature gradients but also can cause gradients in suspended particles that are sufficient to drive density currents. Shear stress at the lake bottom associated with surface waves and high-frequency internal waves can induce sediment resuspension and thus increase the load of suspended particles in the water column. If the density increase induced by the change in particle load is sufficiently large to compensate vertical density stratification, turbidity currents are generated. Because the turbidity gradient is generated by the interaction of waves with the sediment, the resulting turbidity currents usually originate either near the shore at the lake surface (surface waves) or at the depth of the thermocline (high-frequency internal waves). The turbidity currents usually propagate along the lake bottom to larger depth until they intrude into the open water.

Horizontal Density Currents Generated under Ice Cover

Temperature-driven density currents can also occur under ice cover due to differential heating by solar radiation. This process is invoked if the optical properties of the ice and snow cover vary horizontally either due to a variation in the ice structure (e.g., white ice and black ice) or in the thickness of snow cover. Then, penetration of solar radiation through the ice and snow cover varies horizontally and thus induces differential heating in the surface water below the ice. This process causes convection below the ice and results in horizontal density gradients that can drive horizontal density currents. Such under-ice currents are believed to be essential for the development of algal blooms in early spring in Lake Baikal.

Density Currents in Tropical and Saline Lakes

In tropical lakes and also in saline lakes several of the processes mentioned above do not occur. Because the development of a therma bar requires a lateral transition of water temperature from above to below T_{md} at the lake surface, a thermal bar never occurs in tropical lakes where water temperatures are above 4 °C all year round and thus always exceed T_{md}. Because T_{md} decreases with increasing salinity reaching freezing temperature at about $25\,g\,kg^{-1}$, thermal bar development and the associated density currents also do not play an important role for the vertical exchange in saline lakes. The same arguments exclude the thermobaric effect as a significant cause of density currents in tropical and saline lakes.

In tropical lakes temperature gradients play the dominant role in the generation of density currents because at high water temperatures, slight temperature gradients imply large differences in density. Therefore, salinity gradients play a smaller role for density plume generation in tropical lakes than for lakes of temperate regions during the cold season. Convection due to night-time cooling and density currents due to differential cooling and river inflows can be expected to be the most important processes for advective deep-water renewal in tropical lakes.

In saline lakes, however, river inflows usually cannot drive density currents because rivers typically introduce fresh water that in most cases has a much lower density than the saline lake water, even if the inflow has a low temperature. For example water with a salinity of $4\,g\,kg^{-1}$ and a temperature of 24 °C has a greater density than fresh water with 4 °C. Hence, river inflows can only directly drive density currents in saline lakes, if the riverine water carries a substantial load of suspended particles. Nevertheless, in saline lakes river inflows can generate large scale differences in salinity and thus may indirectly lead to density currents at sills induced by inter basin exchange as is probably the case in the Caspian Sea. Another process that is likely to cause density currents in saline lakes is differential cooling. In contrast to freshwater lakes this process can trigger density currents with temperatures well below 4 °C and even down to freezing temperature, because T_{md} can be substantially reduced or even does not exist depending on the salinity of the water.

Impact of Changes in the Environmental Conditions on Density Currents and Deep-Water Renewal

Changes in climatic conditions and human activities in catchments may affect density currents and thus vertical mixing in lakes. An increase in precipitation and in the percentage of the land made impervious by human development typically leads to a higher discharge of rivers. Enhanced river discharge is usually associated with increased erosion and a higher load of suspended particles in the river water such that density currents will propagate to larger depth. Enhanced deep-water renewal is thus anticipated in freshwater lakes. In saline lakes, however, the increase in fresh water input associated with increased precipitation reduces the density of the surface waters and thus can significantly limit the generation of density currents. Consequently, deep-water renewal may be reduced or suppressed. Deep-water exchange decreased drastically as a consequence of increased riverine discharge in the Caspian Sea and in Mono Lake, CA.

Water storage dams in the catchments of lakes have the opposite effect on density currents than soil sealing. Retention by water storage dams reduces peak discharge and the load of suspended particles in downstream rivers. This results in a decrease in the intensity of density currents and in the depth of intrusions in lakes located downstream of dams.

Climate warming can lead to a reduction of deep-water renewal in lakes because additional input of heat at the lake surface may result in an increase in density stratification of the water column and in an extension of the stratified period. Persistence of the increased stratification over many years likely depends on the lake's latitude and depth and whether warming is intensified in winter or summer or, for tropical lakes, during the monsoon period or during less windy periods.

In a warmer climate, however, mixing in freshwater lakes due to density currents associated with the thermal bar and/or the thermobaric effect may cease, if climate warming leads to an increase in surface water temperature to values above 4 °C all year round, i.e., to values above T_{md}. Because density is a nonlinear function of temperature, warming of surface water may also shift the relative importance of turbidity and salinity gradients towards temperature gradients as agent to drive density plumes.

The potential consequences of environmental change on density currents are exemplified for Lake Baikal, the deepest lake on earth. Because of the peculiar temperature profile with the mesothermal temperature maximum (see **Figure 7**), deep-water renewal in Lake Baikal is predominantly driven by salinity differences between river and lake water and between the basins of the lake. The salinity differences result in density plumes associated with riverine inflows and inter-basin exchange. Hence, changes in the catchments leading to an increase in the concentration of dissolved ions and suspended particles in river inflow will intensify deep-water mixing by density plumes. Climate warming on the other hand will not severely affect density plumes and thus deep-water renewal in Lake Baikal, as long as the lake has an annual ice cover. Higher air temperatures most likely result in a shift of ice break-up to earlier times in the year, but will not have an affect on the thermal conditions immediately after ice break-up. Hence, the conditions required to generate density plumes will not change, but may occur earlier in the season.

In summary, density currents significantly contribute to deep renewal, especially in deep and very deep lakes. The density currents can result from a variety of processes. Which of these processes are relevant in a specific lake depends on the temperature regime of the lake, its salinity and also its morphometry. The environmental conditions, e.g., precipitation in the catchments and heat flux at the lake surface affect the occurrence of density currents and the depth reached by the density plumes. Hence, changes in the environmental conditions will have consequences for deep-water renewal and oxygenation of deep lakes not only because of a change in turbulence levels, but also by their effect on the intensity of density currents.

See also: The Benthic Boundary Layer (in Rivers, Lakes, and Reservoirs); Density Stratification and Stability.

Further Reading

Aeschbach-Hertig W, Kipfer R, Hofer M, Imboden DM, and Baur H (1996) Density-driven exchange between the basins of Lake Lucerne (Switzerland) traced with the ^{3}H-^{3}He method. *Limnology and Oceanography* 41: 707–721.

Fer I and Lemmin U (2002) Winter cascading of cold water in Lake Geneva. *Journal of Geophysical Research* 107: NO. C6, 10.1029/2001JC000828.

Finger D, Schmid M, and Wüest A (2006) Effects of upstream hydropower operation on riverine particle transport and turbidity in downstream lakes. *Water Resources Research* 42: W08429, doi:10.1029/2005WR004751.

Fischer HB, List EJ, Koh RCY, Imberger J, and Brooks NH (1979) *Mixing in Inland and Coastal Waters*. San Diego, CA: Academic Press.

Hamblin PF and Carmack EC (1978) River induced currents in a fjord lake. *Journal of Geophysical Research* 83: 885–899.

Hohmann R, Kipfer R, Peeters F, Piepke G, Imboden DM, and Shimaraev MN (1997) Processes of deep water renewal in Lake Baikal. *Limnology and Oceanography* 42: 841–855.

Malm J, Mironov D, Terzhevikl A, and Jiinsson L (1994) Investigation of the spring thermal regime in Lake Ladoga using field and satellite data. *Limnology and Oceanography* 39: 1333–1348.

Monismith SG, Imberger J, and Morison ML (1990) Convective motions in the sidearm of a small reservoir. *Limnology and Oceanography* 35: 1676–1702.

Peeters F, Finger D, Hofer M, Brennwald M, Livingstone DM, and Kipfer R (2003) Deep-water renewal in Lake Issyk-Kul driven by differential cooling. *Limnology and Oceanography* 48: 1419–1431.

Weiss RF, Carmack EC, and Koropalov VM (1991) Deep-water renewal and biological production in Lake Baikal. *Nature* 349: 665–669.

Zilitinkevich SS, Kreiman KD, and Terzevik AY (1992) The thermal bar. *Journal of Fluid Mechanics* 236: 27–42.

Currents in Stratified Water Bodies 2: Internal Waves

L Boegman, Queen's University, Kingston, ON, Canada

Introduction

In the surface layer of lakes, exchange with the atmosphere and energetic mixing from wind and convection provides oxygen and light, thus enabling growth of plankton and other aquatic life. The nutrients required to sustain surface layer ecology are primarily found in the benthos, where sediment resuspension, nutrient release from the sediments, and oxygen consumption occur. In the interior of lakes, seasonal stratification of the water column suppresses vertical mixing, effectively isolating the surface layer from the sediments. However, the stratification simultaneously provides an ideal environment for internal waves, whose oscillatory currents energize a quasi-steady turbulent benthic boundary layer (TBBL) that drives vertical biogeochemical flux.

The wave motions in lakes are initiated by the surface wind stress. Waves will occur on both the free surface and internal stratifying layers (e.g., the thermocline) and these are referred to as surface or barotropic and internal or baroclinic motions, respectively. The waves are categorized according to their length scale. Basin-scale waves have wavelengths that are of the same order as the lake diameter and are manifest as standing wave modes – or seiches. Sub-basin-scale waves have wavelengths of 10–1000 m. These waves are progressive in nature and will break where they shoal on sloping topography at the depth of the thermocline.

Characteristic Geometry and Water-Column Stratification

The internal waves described in this chapter are in lakes that are not affected by the Coriolis force due to the Earth's rotation: for example, small lakes in the arctic, mid-sized lake (i.e., diameter $>\sim 5$ km) in the mid latitudes, and large lakes near the equator. These lakes have a Burger number >1. As we shall see later, the wave modes that are supported in such lakes depend upon the nature of the water column stratification.

We limit the analysis to several characteristic types of stratification that are commonly observed to occur. The simplest case is that of a homogenous lake of length L and depth H, as may be typical for a shallow system or one that has recently experienced a turnover event (**Figure 1(a)**). During the summer months, solar heating causes a lake to become stratified with a layered structure consisting of an epilimnion, metalimnion, and hypolimnion (see **The Surface Mixed Layer in Lakes and Reservoirs**). If the vertical density gradient is abrupt through the metalimnion, the lake may be approximated as a simple two-layer system of thickness h_1 and density ρ_1 over thickness h_2 and density ρ_2, where $H = h_1 + h_2$ is the total depth (**Figure 1(b)**). In lakes where a strong diurnal thermocline is present or the metalimnion is thick, the vertical density structure may be approximated with three contiguous fluid layers of density ρ_1, ρ_2, and ρ_3 with thicknesses $H = h_1 + h_2 + h_3$ (**Figure 1(c,d)**). The layered model for the stratification is inappropriate for shallow lakes ($H < \sim 15$ m), where the entire water column may be composed of weakly stratified water (e.g., western Lake Erie). In these lakes a transient diurnal thermocline may still occur. Shallow weakly stratified lakes are best characterized as having a continuous stratification (**Figure 1(e)**). Very deep lakes (with a thick laminar region between the metalimnion and TBBL) and those with a significant chemical (saline) component will also have a continuous stratification beneath the metalimnion (**Figure 1(f)**). In general, the strength of the stratification is measured according to the Brunt-Väisälä or buoyancy frequency $N = \sqrt{-(g/\rho_o)d\rho/dz}$, where z is the vertical coordinate direction, g is the gravitational constant, and $\rho_o = 1000\,\mathrm{kg\,m^{-3}}$ is the characteristic water density; in the thermocline of lakes the maximum $N \sim 10^{-2}$ Hz.

Surface Momentum Transfer and Wind Set-Up

Wind Set-Up of the Free Surface

The action of the wind across the lake surface results in frictional momentum transfer from the wind to the water (see **The Surface Mixed Layer in Lakes and Reservoirs**). This transfer occurs in the form of a stress (N m^{-2}) applied at the free surface. The stress may be parameterized as

$$\tau = C_D \rho_a U_{10}^2$$

where C_D is the drag coefficient, $\rho_a = 1.2\,\mathrm{kg\,m^{-3}}$ is the air density, and U_{10} the wind speed measured at 10 m above the water surface. Typically $C_D = 1.3 \times 10^{-3}$, but this value may vary by ±40% depending upon the wind speed, water depth, and relative temperature

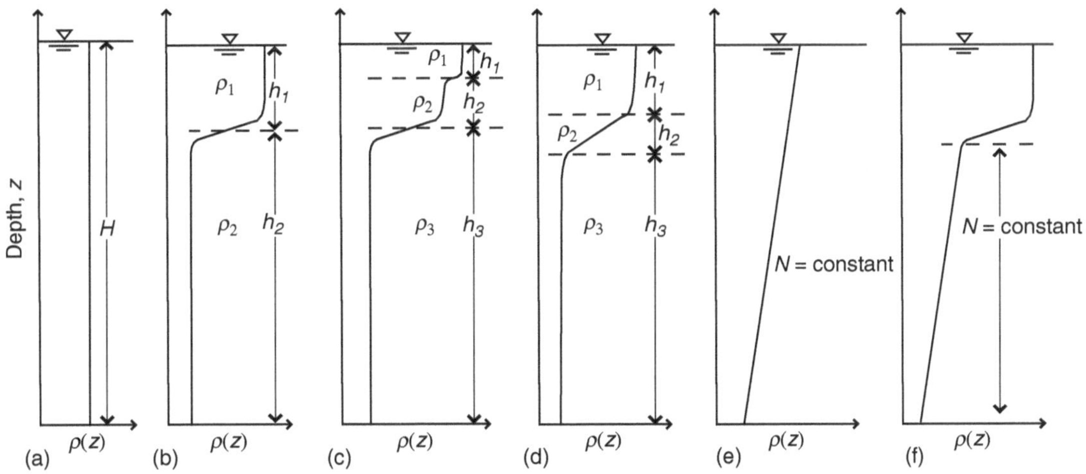

Figure 1 Characteristic continuous water-column stratifications as found in lakes and typical layer approximations. (a) Homogeneous water-column of constant density. (b) Two-layer approximation of the continuous stratification, where the layer separation occurs at the thermocline. (c) Three-layer approximation of the continuous stratification, where the layer separation occurs at the diurnal and seasonal thermoclines. (d) Three-layer approximation of the continuous stratification, where the layer separation occurs at the upper and lower surfaces of the metalimnion. (e) Continuous stratification throughout the water column with constant buoyancy frequency. (f) Continuous stratification where the hypolimnion is characterized by a constant buoyancy frequency.

difference between water surface and adjacent air column.

The momentum transfer associated with steady winds will push the surface water to the leeward shore, causing a displacement of the free surface due to the presence of the solid boundary (**Figure 2a**); for long and shallow lakes this may be as large as several meters (e.g., ~2 m in Lake Erie) (see **Currents in the Upper Mixed Layer and in Unstratified Water Bodies**).

This displacement is called wind set-up. If the wind stress is applied for sufficient time (one quarter of the fundamental seiche period as defined below), a steady-state tilt of the free surface will occur where there is a balance between the applied wind force ($\tau \times$ surface area) and the hydrostatic pressure force due to the desire of the free surface to return to gravitational equilibrium. Balancing these forces at steady state, given the equation for the slope of the free surface

$$\frac{\partial \eta_s}{\partial x} = \frac{u_*^2}{gH}$$

where $u_* = \sqrt{\tau/\rho_o}$ is the surface wind shear velocity, $\eta_s(x,t)$ is the interfacial (surface) displacement from the equilibrium position, and x is the longitudinal coordinate. The equation for the free-surface slope may be integrated to give the maximum interfacial displacement, as measured along the vertical boundary

$$\eta_s(t = 0, x = 0, L) = \pm \frac{u_*^2}{gH}\frac{L}{2}$$

Wind Set-Up of the Internal Stratification

In a manner analogous to the set-up at the free surface, wind induced displacements can also occur along the thermocline. Consider a simple two-layered lake. Water piled-up at the leeward shore by windward drift simultaneously pushes down the thermocline while pushing up the free surface (**Figure 2(b)**, **Figure 3**). The free surface remains nearly horizontal owing to a return flow that develops in the hypolomnion, leading to vertical velocity shear through the metalimnion. A corresponding upwelling occurs at the windward shore (**Figure 3**). The steady-state slope of the free surface is given by a balance between the baroclinic gravitational pressure force from the tilted thermocline and the force due to the wind-stress acting through the epilimnion

$$\frac{\partial \eta_i}{\partial x} = \frac{u_*^2}{g' h_1}$$

where $g' = g(\rho_2 - \rho_1)/\rho_2$ is the reduced gravity across the interface (thermocline) (see **Currents in Stratified Water Bodies 1: Density-Driven Flows**). The equation for the thermocline slope may be obtained through integration over the basin length

$$\eta_i(t = 0, x = 0, L) = \pm \frac{u_*^2}{g' h_1}\frac{L}{2}$$

The effect of buoyancy can be seen by decreasing the density difference between the two layers resulting in a decrease in g' and corresponding increase in η_i.

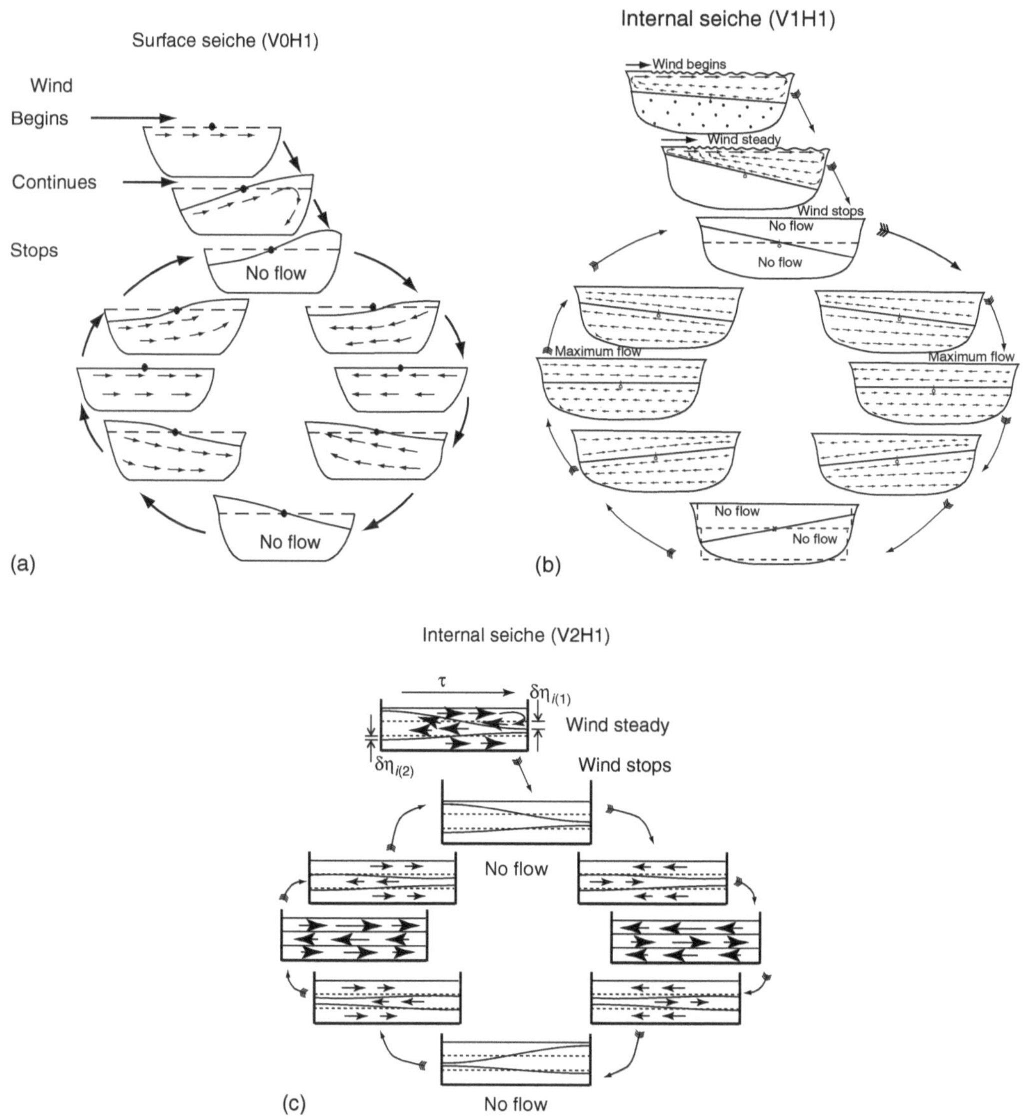

Figure 2 Movement caused by steady moderate wind stress on a hypothetical layered lake and subsequent internal seiche motion neglecting damping. (a) Horizontal mode one surface seiche in a homogeneous one-layered system, (b) horizontal mode one vertical mode on internal seiche in a two-layered system, both adapted from Mortimer CH (1952) Water movements in lakes during summer stratification: Evidence from the distribution of temperature in Windermere. *Proceedings of the Royal Society of London Series B*. 236: 355–404 and (c) horizontal mode one vertical mode two internal seiche in a three-layered system. Arrows denote distribution and magnitude of water particle velocities. At $t = 0$, $(1/2)T_1$, T_1, $(3/2)T_1$, etc. the wave energy is purely in the potential form, isotherms are at their maximum tilt and there is no seiche induced flow, while at $t = (1/4)T_1$, $(3/4)T_1$, $(5/4)T_1$, etc. the energy is purely kinetic, giving rise to strong horizontal currents within the lake-basin and horizontal isotherms.

A simple comparison $\eta_s/\eta_i \sim (\Delta\rho/\rho_o)(h_1/H)$ shows that for weakly stratified deep systems, internal displacements (~1–100 m) may be more than an order of magnitude greater than their surface counterparts (~0.01–1 m); for example in Lake Baikal $\eta_s/\eta_i \sim 0.11\,\mathrm{m}/75\,\mathrm{m} \sim 10^{-3}$.

The available potential energy (APE) embodied in the tilted interface is readily calculated for a two-layer system by integrating the interfacial displacement over the basin length

$$\mathrm{APE} = g\frac{\rho_2 - \rho_1}{2}\int_0^L \eta_i^2(x,t)\,dx$$

which may be integrated for the initial condition of a uniformly tilted thermocline

$$\mathrm{APE} = \frac{1}{6}gL(\rho_2 - \rho_1)\eta_i^2$$

After the thermocline tilt has reached steady state, work done by continued winds is either dissipated

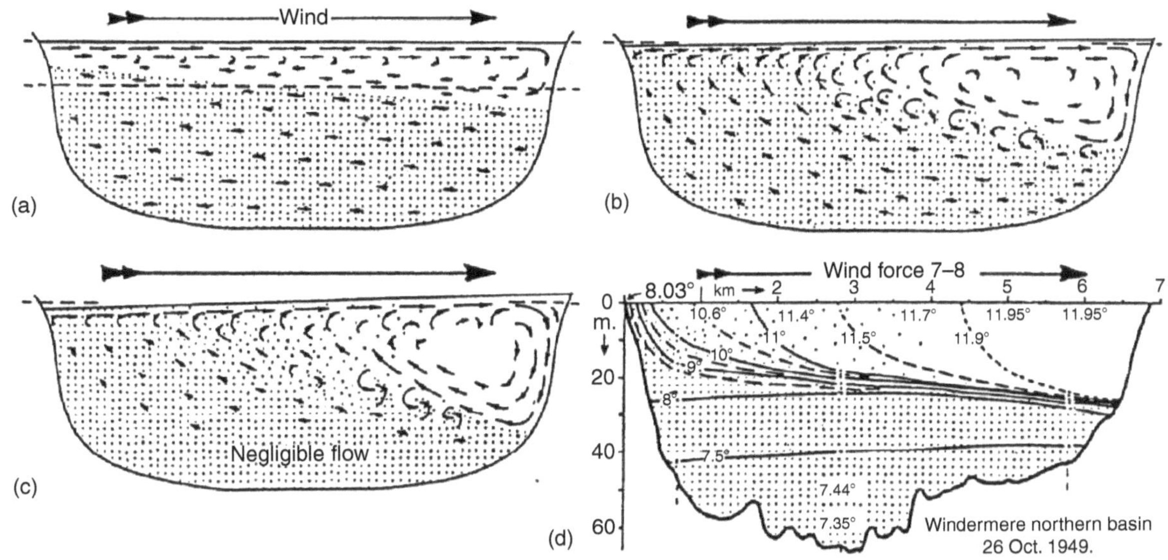

Figure 3 Schematic showing the response of a stratified lake to a surface wind stress. (a–c) show the stages of development of a steady state thermocline tilt. The hypolomnion is shaded and the arrows show the relative speed and direction of the flow. (d) Isotherm distribution and temperatures in Lake Windermere, northern basin, after a steady wind for 12 h. Reprinted from Mortimer CH (1954) Models of the flow-pattern in lakes. *Weather* 9: 177–184.

internally as heat or acts to mix the water column by further deepening the surface layer. Surface layer deepening has been characterized into four distinct regimes based on the strength of the stratification and winds (**Figure 4** and see **The Surface Mixed Layer in Lakes and Reservoirs**). For strong stratification and weak winds (Regime A) the thermocline set-up is small, internal seiches persist for long times, mixing is weak and the thermocline remains sharp. If the stratification is weaker or the winds stronger (Regimes B and C), seiche amplitudes increase and become a predominant feature, shear instabilities (e.g., Kelvin–Helmholtz billows) form leading to entrainment of the metalimnion into the epilimnion, enhanced mixing and causing rapid damping of the internal seiches. For weak stratification and strong winds (Regime D), shear instabilities are strong; the thermocline becomes diffuse with a steep slope and rapidly deepens toward the lake bed. This creates a sharp downwind interface and a broad upwelling at the upwind shore. The upwelled fluid creates a longitudinal temperature gradient, which subsequently mixes the lake horizontally. The colder upwelled water is nutrient rich and as it mixes rapid fluctuations in temperature and biogeochemistry result. In deeper lakes, stratification can be strong during summer and upwelling of metalimnetic (partial upwelling) or hypolimnetic (full upwelling) water is unlikely. In these lakes upwelling is favored just after spring turnover or prior to fall turnover when the thermal stratification is weak or near the surface.

Wedderburn and Lake Numbers

The degree of tilt of the base of the surface layer resulting from an applied wind stress may be quantified using the dimensionless Wedderburn number *W*, which as the ratio of the wind disturbance force to the gravitational baroclinic restoring force is given by

$$W \sim \frac{g' h_1^2}{L u_*^2}$$

Here, g' is evaluated across the base of the surface layer. The cases of $W \gg 1$, $W \sim 1$ and $W \ll 1$ correspond to Regimes A, B/C, and D, respectively.

For idealized laboratory studies and back-of-the-envelope calculations, substitution of η_i into the equation for W leads to a leads to two-layer form

$$W \sim \frac{h_1}{\delta\eta_i}$$

where $\delta\eta_i$ is the steady wind induced vertical displacement of the seasonal/diurnal thermocline measured at the boundary. Due to the order of magnitude scaling, the factor of 2 has been dropped as is commonly found in the scientific literature. Moreover, the somewhat counterintuitive nature of $W \to \infty$ as $\delta\eta_i \to 0$, leads to frequent use of the inverse form of the Wedderburn number ($W^{-1} \sim \delta\eta_i / h_1$).

For lakes which are not well approximated using a two-layer stratification, *W* has been generalized into the Lake Number, L_N (see **Density Stratification and Stability**). This accounts for the depth

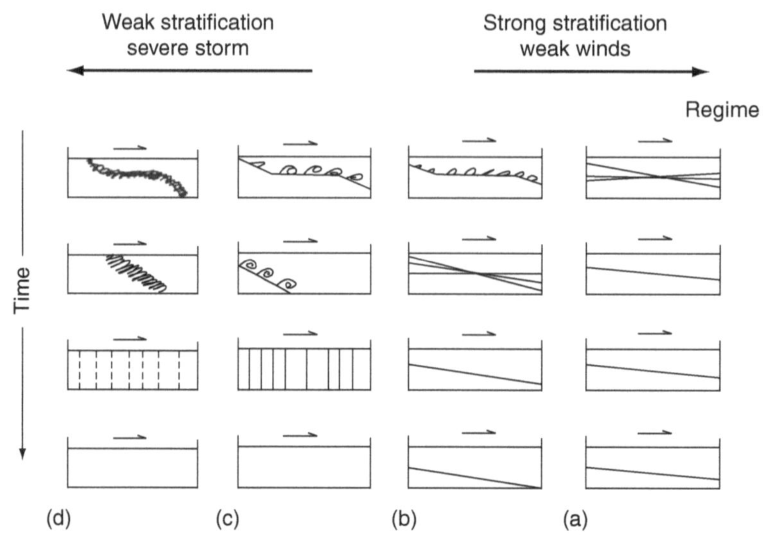

Figure 4 Schematic showing the mixed layer deepening response of a lake to wind stress. (a) Regime A: internal waves; (b) Regime B: internal waves and slight billowing; (c) Regime C: strong billowing and partial upwelling; (d) Regime D: intense billowing and full upwelling. Adapted From Fischer HB, List EJ, Koh RCY, Imberger J, and Brooks NH (1979). *Mixing in Inland and Coastal Waters*. San Diego, CA: Academic Press. After Spigel RH, and Imberger J (1980) The classification of mixed layer dynamics in lakes of small to medium size. *Journal of Physical Oceanography* 10: 1104–1121.

dependence of stratification and horizontal area. For a constant wind stress over a lake with an arbitrary basin shape and stratification

$$L_N = \frac{S_t(H - h_2)}{u_*^2 A_o^{3/2}(H - h_v)}$$

Here A_o is the surface area of the lake and h_v is the height from the lake-bed to the centre of volume of the lake. The stability of the lake

$$S_t = \frac{g}{\rho_o}\int_0^H (z - h_v)A(z)\rho(z)dz$$

incorporates the variable stratification $\rho(z)$ and irregular bathymetric area $A(z)$. For large Lake numbers, the stratification will be severe and dominate the forces introduced by the wind stress. The isotherms will be horizontal, with little or no seiching and associated turbulent mixing in the benthic boundary layer and interior. Changes in S_t with latitude and season cause L_N to vary spatially and temporally around the globe. Under comparable wind conditions L_N is maximal in the mid-latitudes during summer.

Basin-Scale Standing Wave Motions (Seiches)

Interfacial Waves in a Layered Stratification

Horizontal modes When steady winds cease and the surface stress condition is relaxed, the gravitational restoring force associated with the tilted interface (water surface or thermocline) becomes unbalanced. The available potential energy embodied in the tilt is released under the action of gravity and converted to kinetic energy as the interface oscillates in the form of a sinusoidal standing waves or seiche. Antinodes are found at the basin end walls and nodal points in the basin interior (**Figures 2** and **5**). Seiches are commonly called linear waves because the evolving wave-field is well described in space and time by the linear wave equation

$$\frac{\partial^2 \eta}{\partial t^2} = c_o^2 \frac{\partial^2 \eta}{\partial x^2}$$

where $\eta(x,t)$ is the interfacial displacement and c_o the linear shallow water phase speed (speed at which the crests/troughs propagate). This equation is equally applicable to interfacial waves travelling on the free-surface or thermocline by applying the appropriate form of $c_o = \sqrt{gH}$ or $c_o = \sqrt{g'h_1h_2/H}$, for the cases of surface and internal seiches, respectively. Due to the reduced effect of gravity across the thermocline relative to the free surface ($g' \ll g$), surface waves travel at ~50 times the speed of internal waves.

The familiar standing wave patterns associated with seiches forms as symmetric progressive waves of equal amplitude and wavelength, but opposite sign, propagate from the upwelled and downwelled fluid volumes at the opposite ends of the basin (**Figure 2**). These waves are most commonly represented with cosine functions (**Figure 5**), which have central node(s) and antinodes at the basin walls. Summing cosine equations for waves propagating in opposite directions,

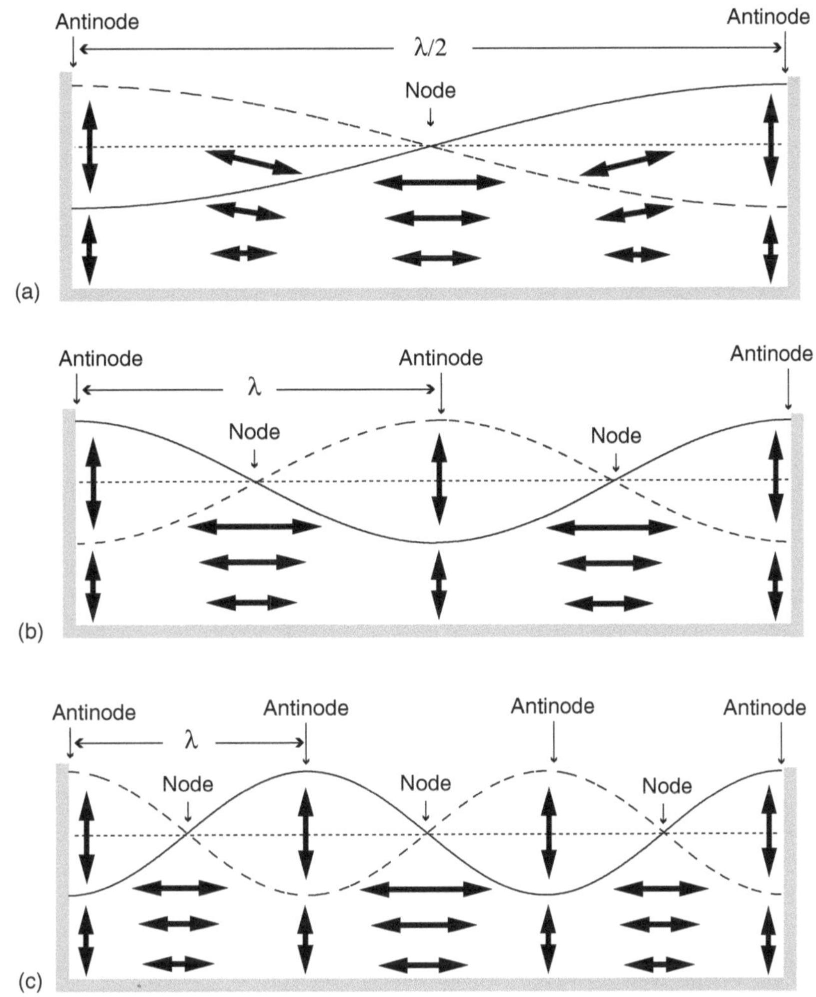

Figure 5 Schematic diagram showing the first three horizontal interfacial seiche modes: horizontal mode one ($n = 1$), mode two ($n = 2$), and mode three ($n = 3$). Arrows denote direction of water particle velocities. Solid and dashed lines denote the interfacial displacement at one-half period intervals. Upper layer velocities for the baroclinic case are not shown and can be inferred from symmetry.

gives the equation for the horizontal mode one (H1) standing wave pattern (**Figure 5(a)**)

$$\eta(x,t) = a\cos(kx + \omega t) + a\cos(kx - \omega t) = 2a\cos kx \cos \omega t$$

The component wave amplitude $a = \delta\eta_i/2$ or $\delta\eta_s/2$ depending on the interface under consideration, the angular frequency $\omega = c_0 k$, where $k = 2\pi/\lambda$ is the wavenumber and λ the wavelength and $T = 2\pi/\omega$ the wave period. For an enclosed basin, there is one half wavelength of an H1 seiche in a lake, giving $\lambda = 2L$ and a period of

$$T_n = \frac{2L}{nc_0}$$

where $n = 1$ for a H1 seiche is the number of nodal points or half wavelengths in the horizontal direction.

The layer-averaged horizontal velocities associated with the H1 seiche are maximum the centre of the basin and are given by

$$U_1 = g'\frac{h_2}{H}\frac{\delta\eta_i}{L/2}t$$

$$U_2 = g'\frac{h_1}{H}\frac{\delta\eta_i}{L/2}t$$

These velocities are zero at the vertical boundaries, where the motion is purely vertical (**Figure 5**). Similarly for the surface seiche, the mid-lake depth-averaged velocity is

$$U = g\frac{\delta\eta_s}{L/2}t$$

The oscillatory seiche currents are low-period and quasi-steady. Observations from a variety of lakes

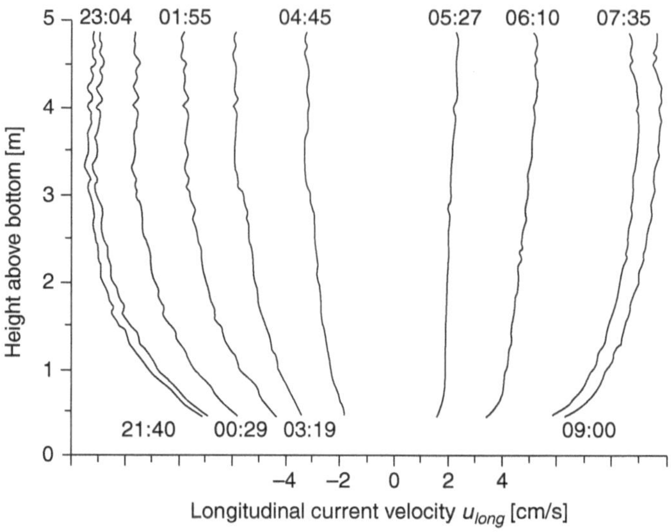

Figure 6 Near-bed velocity profiles in a small lake showing the oscillatory nature and no-slip boundary associated with seiche currents. Observations are over one-half of the seiche period taken at times as indicated. The profiles are offset and are all plotted with the given velocity scale. From Lorke A, Umlauf L, Jonas T, and Wüest A (2002) Dynamics of turbulence in low-speed oscillating bottom boundary layers of stratified basins. *Environmental Fluid Mechanics* 2: 291–313.

and reservoirs show the surface and internal seiche currents to have a typical range of 0.02–0.20 m s^{-1}, with a maximum ~0.2 m s^{-1} during storms and to approach zero at the no-slip sediment boundary where the flow is impeded by friction (**Figure 6**).

Higher horizontal mode seiches ($n > 1$) are also observed in lakes. These are described by a more general solution of the linear wave equation, where the initial condition is that of a uniformly tilted interface, made up of the superposition of all higher horizontal modes. The general solution is

$$\eta(x,t) = \sum_{n=1}^{\infty}\left(\frac{-8\delta\eta}{n^2\pi^2}\right)\cos\left(\frac{n\pi}{L}x\right)\cos\left(\frac{c_o n\pi}{L}t\right) \quad n = 1, 3, 5, \ldots$$

$n = 1$ for the horizontal mode-one (H1) seiche, $n = 2$ for the horizontal mode-two (H2) seiche, $n = 3$ for the horizontal mode-three (H3) seiche (**Figure 5**). An infinite number of modes are possible, each with decreasing amplitude and energy as the modal number increases. The fundamental solution is composed only of odd modes (i.e., $\eta(x,t) = 0$ when n is even) as is intuitively expected because only odd odes have a nodal point at the mid-basin location where there is zero displacement associated with an initial uniform initial tilt (**Figures 2** and **3**). By calculating the APE associated with each mode, it can be shown that that more than 98% of the wave energy is contained in the H1 mode, but the energy distribution between modes may be significantly affected by resonant forcing and basin shape.

Examples of surface and internal seiche periods for various horizontal modes are given in **Table 1**. Energy will pass between potential and kinetic forms as the wave periodically oscillates with time (**Figure 2**). At $t = 0$, $(1/2)T_1$, T_1, $(3/2)T_1$, etc. the wave energy is purely in the potential form, while at $t = (1/4)T_1$, $(3/4)T_1$, $(5/4)T_1$, etc. the energy is purely kinetic, giving rise to horizontal currents within the lake-basin (**Figure 6**). For a non dissipative system, the modal energy distributions represent the sum of kinetic and potential energy and are independent of time. Dissipative processes will lead to a decrease in wave amplitude, but not period, with time (**Table 1**); unless there is sufficient mixing across the thermocline to cause a change in the stratification and hence c_o. For surface seiches the decay in amplitude with each successive period can range from 3% (Lake Geneva) to 32% (Lake Erie).

Vertical modes When the vertical density structure may be approximated with three or more fluid layers (**Figure 1c,d**), in addition to vertical mode-one, horizontal mode-one (V1H1) seiches (**Figure 2b**), higher vertical mode seiches are supported; for example V2H1, etc. (**Figure 2c**). For a three-layer system c_o becomes

$$c_o = \frac{1}{2H}\left(\gamma - \sqrt{\gamma^2 - 4\alpha H}\right)$$

where

$$\gamma = (1 - \rho_1/\rho_2)h_1h_2 + (1 - \rho_1/\rho_3)h_1h_3 + (1 - \rho_2/\rho_3)h_2h_3$$

and

$$\alpha = h_1h_2h_3(1 - \rho_1/\rho_2)(1 - \rho_2/\rho_3).$$

Table 1 Observed surface and internal seiche periods from a variety of lakes and the associated amplitude decay

Lake and location	T_1 *(h)*	T_2 *(h)*	T_3 *(h)*	T_4 *(h)*	T_5 *(h)*	*Fractional decrease in amplitude with each successive period*
Surface seiches						
Tanganyika (Africa)[1]	0.075	0.038	0.028			
Loch Earn (Scotland)[1]	0.24	0.14	0.10	0.07	0.06	
Yamanaka (Japan)[1,2]	0.26	0.18	0.09			0.099
Garda (Italy)[1,2]	0.72	0.48	0.37	0.25	0.20	0.045
Geneva (Switzerland–France)[1,2]	1.2	0.59				0.030
Vättern (Sweden)[1,2]	3.0	1.6	1.3	0.97	0.80	0.113
Baikal (Russia)[1]	4.6					
Michigan (Canada–USA)[1]	9.1	5.2	3.7	3.1	2.5	
Erie (Canada–USA)[1,2]	14.3	9.0	5.9	4.2		0.322
Internal seiches						
Baldegg (Switzerland)[3]	9.3	4.6	3.1	2.1		
Lugano (Switzerland–Italy)[4]	24	12	8.0	6.2		
Windermere (England)[3]	24	13	9			
Zurich (Switzerland)[5]	45	24	17			
Loch Ness (Scotland)[3]	57	27	18	14	11	
Geneva (Switzerland–France)[3]	74	46	30	22	18	

[1]Wilson W (1972) Seiches. In Chow VT (ed.) *Advances in Hydroscience* 8: 1–94.
[2]Wüest AJ and Farmer DM (2003) Seiches. In *McGraw-Hill Encyclopedia of Science and Technology*, 9th Edition. New York: McGraw-Hill.
[3]Lemmin U and Mortimer CH (1986) Tests of an extension to internal seiches of Defant's procedure for determination of surface seiche characteristics in real lakes. *Limnology and Oceanography* 31: 1207–1231.
[4]Hutter K, Salvadè G, and Schwab DJ (1983) On internal wave dynamics in the northern basin of the Lake of Lugano. *Geophysical and Astrophysical Fluid Dynamics* 27: 299–336.
[5]Horn W, Mortimer CH and Schwab DJ (1986) Wind-induced internal seiches in Lake Zurich observed and modeled. *Limnology and Oceanography* 31: 1232–1254.

Substitution into the equation for the wave period $T_n = 2L/nc_0$ gives the period of a vertical mode-two wave, where the horizontal modal structure is defined by n.

Vertical mode-two seiches can be generated when there is an asymmetry in the tilting of upper and lower interfaces of the stratifying layers (the diurnal and seasonal thermoclines or the upper and lower boundaries of the metalimnion). Laboratory and limited field data shows that such asymmetries are introduced from the compression and expansion of the metalimnion that occurs along the downwind and upwind shores, respectively, under upwelling conditions (e.g., **Figure 3(d)**). The strength of the vertical mode two response has been hypothesized to depend upon the relative values of the Wedderburn and Lake Numbers. A V2 response occurs for small W and large L_N (strong tilting of the upper interface, large shear across the base of the surface layer and a relatively undisturbed lower interface); whereas a V1 response occurs for small W and large L_N (comparable tilts of both interfaces and a strong velocity in the hypolimnion).

Higher vertical mode basin-scale internal waves have been observed in several lakes, generally after a sudden wind pulse has excited an initial V1H1 response, which then evolves into a V2 seiche (e.g., Wood Lake, Upper Mystic Lake, Lake Constance). Resonance with the wind forcing (e.g., Lake Alpnach), sloping basin topography and unequal density differences between stratifying layers can cause this preferential excitation of higher vertical modes.

Internal Modes in a Continuous Stratification

The two-layer assumption for the stratification in lakes is inappropriate for the many shallow lakes ($H < \sim 15$ m; e.g., Frains Lake) and in the hypolomnion of lakes that are very deep (e.g., Lake Baikal) or where the stratification has a significant chemical (saline) component (e.g., Mono Lake). These systems are better modeled using a continuous stratification where the tilting of the isopycnals due to wind set-up is captured by L_N. Upon relaxation of a wind stress, a continuous stratification will support a spectrum of vertical basin-scale modes. The frequency associated with each mode is given by

$$\omega = \frac{N}{\left(1 + \frac{n^2}{m^2}\frac{L^2}{H^2}\right)^{1/2}}$$

which is dependent upon the basin geometry and may be used to calculate the wave period $T = 2\pi/\omega$. The structure of each vertical mode m is described by the wavefunction ψ_m, which is obtained from the linear long-wave equation with constant N

$$\psi_m(z) = \sin\left(\frac{m\pi}{H}z\right) \text{ and } c_o = \pm\frac{NH}{m\pi} m = 1, 2, 3, \ldots$$

and is a measure of the wave-induced vertical displacement of the internal strata at a particular depth. For example, the V1 wave has a maximum internal displacement at mid-depth, whereas the V2 wave has a positive displacement at $1/4H$ and a negative displacement at $3/4H$ (**Figure 7a**); showing the characteristic opening of the strata.

The wavefunction contains no knowledge of the actual wave amplitude and consequently is often normalized $-1 \le \psi_m \le 1$. The utility of ψ_m is that it may be numerically calculated from the Taylor–Goldstein equation for an arbitrary $N(z)$ profile, thus determining the wave modes that are supported by a particular water column stratification. The basin-scale vertical modal structure, either a single mode or combination of modes, may be calculated when the wave amplitude is known $\eta(z, t) \approx \psi_m(z)a(t)$. The vertical modal structure can then be projected in space and time throughout the basin by assuming a wave profile $\eta(x, z, t) \approx \psi_m(z)\eta(x, t)$; this is typically a cosine for seiches or sech^2 function for solitary waves (described later).

The horizontal velocity profile induced by the wave motion scales with the gradient of the wavefunction $c_o d\psi_m/dz$ (**Figure 7(b)**). Analytical expressions for the velocity field can be found in stratified flow texts. The continuous velocity profiles in **Figure 7** are consistent with the depth averaged currents presented in **Figure 2**. In both models, although technically incorrect (**Figure 6**), a free-slip bottom boundary has been assumed. Accordingly the velocities from these models are not representative of flow near the bottom boundary.

Degeneration of Basin-Scale Internal Waves in Lakes

Understanding the factors leading to wave degeneration has been a major goal of limnologists. This is because internal waves ultimately lose their energy (degenerate) to dissipation (viscous frictional heating of the fluid at mm scales) and diapycnal mixing (mixing of fluid perpendicular to isopycnals or surfaces of constant density) in regions where the flow is turbulent. In turn, mixing drives biogeochemical fluxes. Turbulence is produced directly from the seiche induced currents through fluid straining in the lake interior and TBBL, and from processes that are uncoupled from seiche generation, such as surface wave breaking and inflows, which also act to disrupt the seiche motion. As a general rule, for deep lakes, the period of internal

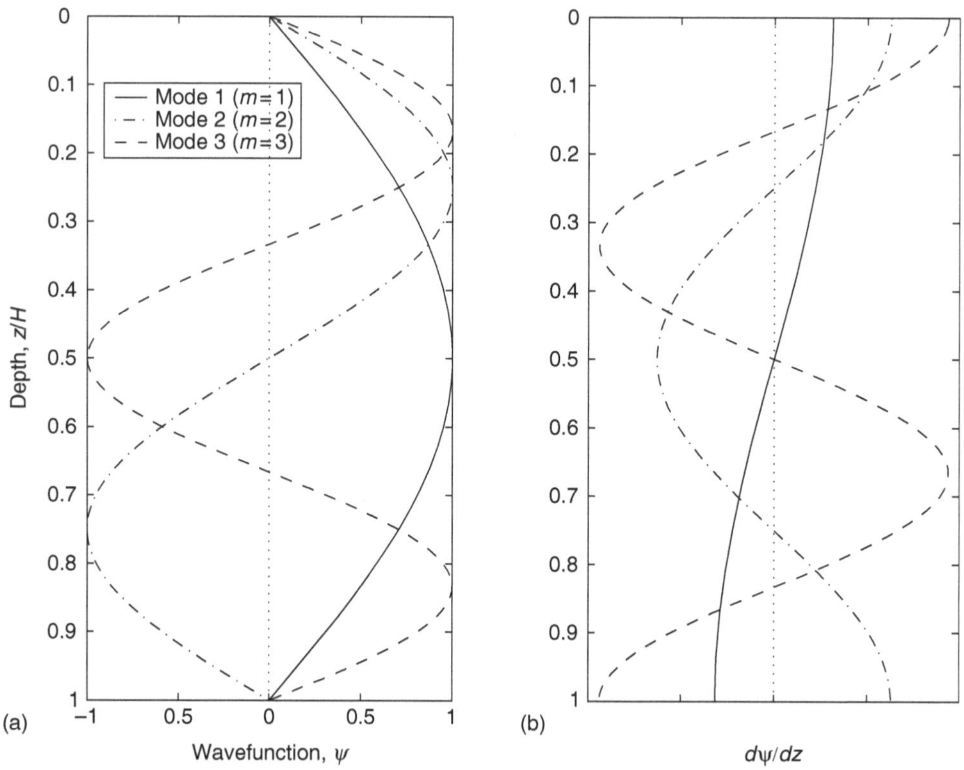

Figure 7 (a) Wavefunction profiles for vertical modes one ($m = 1$), two ($m = 2$), and three ($m = 3$) supported by a constant N stratification. (b) Characteristic velocity profiles $c_o d\psi/dz$ for wavefunctions shown in panel (a).

seiche decay is about 1 day per 40 m of water-column depth. For very deep lakes, e.g., Lake Baikal where $H = 1637$ m, this equates to a decay period of more than one month.

Field observations show the degeneration of basin scale internal waves to occur primarily as a result of turbulence production in the TBBL rather than the interior; observations of dissipation and mixing are more than ten times greater in the TBBL than the interior (see **The Benthic Boundary Layer (in Rivers, Lakes, and Reservoirs)**). The degeneration can occur through four possible mechanisms: (1) viscous damping of seiche currents in the TBBL, (2) the formation of shear instabilities in the interior, (3) the production of nonlinear internal waves that will break on sloping topography; and (4) the formation of internal hydraulic jumps. By calculating the timescales over which each mechanism will occur, regimes have been delineated in which a particular mechanism will dominate (**Figure 8**). The regimes are defined according to the inverse Wedderburn number $W^{-1} = \delta\eta_i/h_1$ and the depth of the seasonal thermocline (h_1/H). Although strictly applicable to long narrow lakes that match the rectangular system used in the analysis, the regime diagram has been shown to suitably predict field observations from a variety of lakes (**Table 2**). The degeneration regimes are described below.

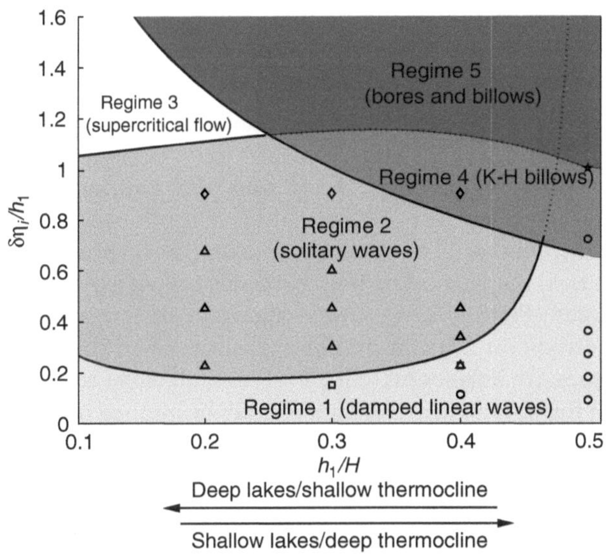

Figure 8 Analytical regime diagram showing the degeneration mechanisms of seiches in long rectangular lakes. The regimes are characterized in terms of the normalized initial forcing scale $W^{-1} = \delta\eta_i/h_1$ and the depth of the seasonal thermocline (h_1/H). Laboratory observations are also plotted (*, Kelvin–Helmholtz (K-H) billows and bore; ◇, broken undular bore; Δ, solitary waves; □, steepening; ○, damped linear waves). From Horn DA, Imberger J, and Ivey GN (2001) The degeneration of large-scale interfacial gravity waves in lakes. *Journal of Fluid Mechanics* 434: 181–207.

Regime 1: Damped Linear Waves

Under relatively calm conditions ($W^{-1} < \sim 0.2$) weak seiches develop, which are damped by viscosity in the TBBL. Seiche amplitudes and currents are not sufficient for solitary wave production, shear instability and/or supercritical flow. This regime corresponds to regime A in **Figure 4** (see **The Surface Mixed Layer in Lakes and Reservoirs**). The time scale associated with viscous damping of a basin-scale seiche T_D is estimated from the ratio of the seiche energy to the rate of energy dissipation in the benthic boundary layer ϵ_{TBBL}:

$$T_D \sim \frac{\text{Seiche energy}}{\epsilon_{TBBL}} \times \frac{\text{Lake volume}}{\text{TBBL volume}} \sim 1 \text{ to } 10 \text{ d(for moderately sized lakes)}$$

The energy dissipation in the lake interior $\epsilon_{Interior}$ is neglected because observational studies show that $\epsilon_{TBBL} > 10\epsilon_{Interior}$ (see **The Surface Mixed Layer in Lakes and Reservoirs**). More complex models for viscous seiche decay may be found in references below.

Regime 4: Kelvin–Helmholtz Billows

Kelvin–Helmholtz instabilities can form due to a number of mechanisms. Under strong forcing conditions and as the thermocline approaches mid-depth (> ~0.8 and > ~0.3), the seiche-induced currents will be sufficiently strong to overcome the stabilizing effects of stratification. Shear instabilities will occur before solitary waves can be produced and/or the seiche is damped by viscosity. Shear from other processes, such as the surface wind stress, can augment that from seiche induced currents and induce instabilities under weaker forcing conditions.

Shear instabilities manifest themselves as high-frequency internal waves that are common features through the metalimnion of lakes and oceans. The waves are the early stages of growth of the instabilities and have sinusoidal profiles, frequencies $\sim 10^{-2}$ Hz, wavelengths ~10–50 m and amplitudes ~1–2 m. Depending on the particulars of the stratification and velocity shear profiles, the shear instabilities may be classified according to their profile (e.g., Kelvin– Helmholtz billows, Holmboe waves or combinations thereof). Instabilities grow exponentially from small random perturbations in the flow, leading to rapid degeneration as patches of localized turbulent mixing or billowing (see **Small-Scale Turbulence and Mixing: Energy Fluxes in Stratified Lakes**). This process is shown schematically in **Figure 9**.

In a continuously stratified flow, the stability behavior is governed by the Taylor–Goldstein

Table 2 Comparison between predicted regime and observations in several lakes

Lake	*Dates*	*Observations*	*W* (W^{-1})	*Regime*	*Source*
Loch Ness	Oct. 2–3, 1971	'Pronounced front or surge'	3 (0.3)	2	1
Lake of Zurich	Sept. 11–14, 1978	'Steep fronted solitary wave'	3 (0.3)	2	2
Windermere	Aug. 14–20, 1951	'Damped harmonic oscillations'	5 (0.2)	1–2	3
	Sept. 13–17, 1951	'Oscillatory waves' with some steepening	3 (0.3)	1–2	
Babine Lake	July 5–10, 1973	'Surges'	2 (0.5)	2	4
	Aug. 12–15, 1973	'Steep shock front'	3 (0.3)	2	
	Oct. 2–7, 1973	'Surges' and 'solitary waves'	2 (0.5)	2	
Seneca Lake	Oct. 14–21, 1968	'Surges' consisting of trains of 'solitons'	2 (0.5)	2	5
Kootenay Lake	July 13–Aug. 17, 1976	'Surges' consisting of waves resembling 'solitons'	2 (0.5)	2	6
Balderggersee	Nov. 1–15, 1978	'Asymetrical waves'	9 (0.1)	2	7
	Nov. 16–22, 1978	'Steepened wave front. . .described as an internal surge'	2 (0.5)	2	
Lake Biwa	Sept. 4–13, 1993	'Undular bores and solitary waves'	1 (1)	2	8

Most observations fall in regime 2 ($W^{-1} > \sim 0.3$) and the internal seiche will degenerate into nonlinear internal waves. For Windermere, the baroclinic tilting is weaker ($W^{-1} < \sim 0.3$) and the predominant response is a damped seiche. After Horn DA, Imberger J, and Ivey GN (2001) The degeneration of large-scale interfacial gravity waves in lakes. *Journal of Fluid Mechanics* 434: 181–207.

Sources

1. Thorpe SA, Hall A, and Crofts I (1972) The internal surge in Loch Ness. *Nature* 237: 96–98.
2. Mortimer CH and Horn W (1982) Internal wave dynamics and their implications for plankton biology in the Lake of Zurich. *Vierteljahresschr. Naturforsch. Ges. Zurich* 127(4): 299–318.
3. Heaps NS and Ramsbottom AE (1966) Wind effects on the water in a narrow two-layered lake. *Philosophical Transactions of the Royal Society of London A* 259: 391–430.
4. Farmer DM (1978) Observations of long nonlinear internal waves in a lake. *Journal of Physical Oceanography* 8: 63–73.
5. Hunkins K and Fliegel M (1973) Internal undular surges in Seneca Lake: A natural occurrence of solitons. *Journal of Geophysical Research* 78: 539–548.
6. Wiegand RC and Carmack E (1986) The climatology of internal waves in a deep temperate lake. *Journal of Geophysical Research* 91: 3951–3958.
7. Lemmin U (1987) The structure and dynamics of internal waves in Baldeggersee. *Limnology and Oceanography* 32: 43–61.
8. Saggio A and Imberger J (1998) Internal wave weather in a stratified lake. *Limnology and Oceanography* 43: 1780–1795.

equation, from which it can be shown that a gradient Richardson number

$$Ri_g = \frac{N(z)^2}{(du/dz)^2} < \frac{1}{4}$$

is a necessary but not sufficient condition for instability (Miles-Howard criterion). Billowing occurs along thin layers in the flow (~10 cm thick) where Ri_g is low and there are interfaces with sharp density gradients. Through billowing, the interfaces become more diffuse and are replaced by shear layers of thickness $\delta \sim 0.3(\Delta U)^2/g'$, where ΔU is the velocity jump over the interface. As a result of billowing, the flow becomes stable unless ΔU increases (e.g., due to increasing wind stress) or δ decreases (e.g., due to mixed layer deepening). A spectrum of growing instabilities are theoretically possible, but the most unstable mode wave will have a wavelength $\lambda = 2\pi/k \sim 7\delta$.

Application of the Taylor–Goldstein equation to field observations shows that the frequency of the most unstable mode is just below the maximum buoyancy frequency through the metalimnion. This is because a fluid parcel displaced vertically from its equilibrium density position, as occurs during the growth of an instability, will be subjected to buoyancy forces arising from the sudden density anomaly with respect to its surroundings. The fluid parcel will oscillate as a wave at frequency N until the motion is frictionally damped by viscosity or evolves into a billow and collapses into turbulence. Fluid parcels will not naturally oscillate at frequencies greater than N, and so waves will not propagate at these frequencies; N is thus the limiting high-frequency cut-off for internal wave motions.

Internal seiches generate substantial vertical shear due to the baroclinic flow reversals across layer interfaces (**Figures 2–5**). The magnitude of this shear is periodic and has a maximum value when the interfaces are horizontal (e.g., $T_1/4$) and all wave energy is in the kinetic form. A bulk Ri may be applied over the interfaces separating discrete layers

$$Ri = \frac{g'\delta}{(\Delta U)^2} < \frac{1}{4}$$

to predict the formation of instabilities. If the background flow is time-variable, either the condition $Ri = 1/4$ must be maintained for longer than the growth and billowing period of the instability $T_b \sim 20(\Delta U)/g'$ or Ri must be $<<1/4$. Values of T_b in lakes are of the order of minutes or less.

Shear instabilities can occur at the nodal locations in lakes, where the vertical shear is greatest, at the

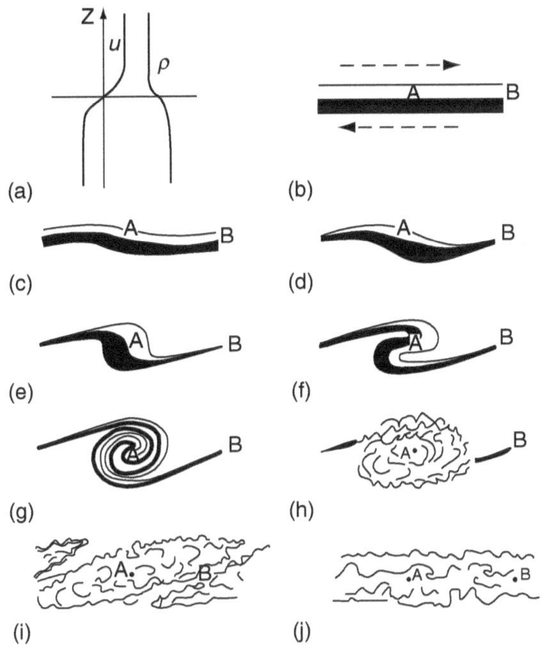

Figure 9 Schematic showing the growth and turbulent degeneration of a Kelvin–Helmholtz shear instability leading to diapycnal mixing of the stratified fluid. The condition is shown as velocity (u) and density (ρ) profiles in (a) and as isopycnal surfaces in (b), where the arrows denote the flow direction. In (b–j), A and B are fixed points in the flow and the lines represent surfaces of constant density (isopycnals). Adapted from Mortimer CH (1974). Lake hydrodynamics. *Mitteilungen Int Ver Limnol* 20: 124–197, after Thorpe SA (1987) Transitional phenomena and the development of turbulence in stratified fluids: A review. *Journal of Geophysical Research* 92: 5231–5248.

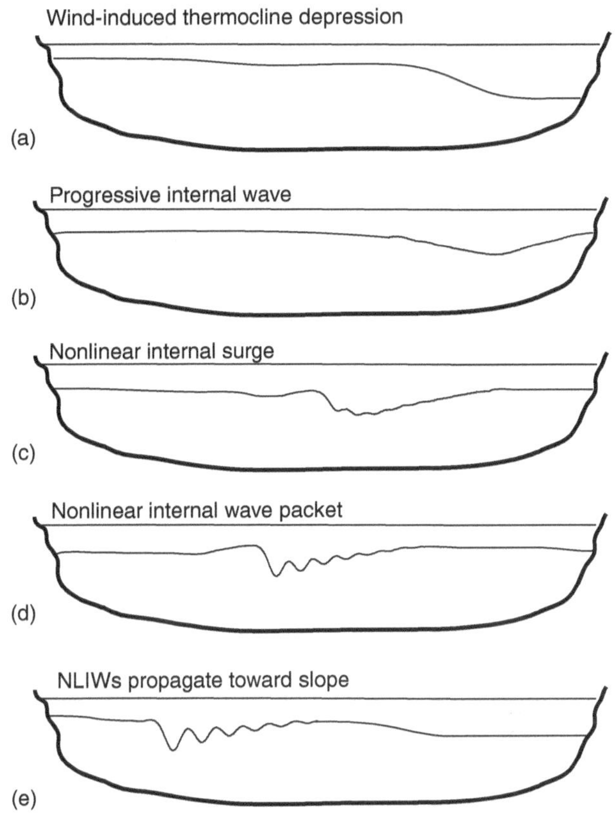

Figure 10 Schematic showing the evolution of a NLIW packet. (a) Initial wind induced thermocline depression, (b–c) formation of a progressive surge through nonlinear steepening, (d–e) evolution of a dispersive NLIW packet at $t = T_s$. For the case shown, the wind had been blowing for less than $T_1/4$ and a steady state tilt of the entire thermocline was not achieved (i.e., upwelling did not occur). This is a common occurrence in long (~100 km) narrow lakes (e.g., Seneca Lake, Babine Lake, etc.).

base of the surface layer under strong wind conditions, on the upper and lower surfaces of thermocline jets that result from vertical mode-two compression of the metalimnion, near river influents and reservoir withdrawal layers, and in regions where there is flow over rough topography.

Regime 3: Supercritical Flow

Internal hydraulic jumps occur in stratified flows at the transition from supercritical ($Fr_1^2 + Fr_2^2 > 1$) to subcritical ($Fr_1^2 + Fr_2^2 < 1$) flow conditions, where the upper and lower layer Froude numbers are defined as $Fr_1^2 = U_1^2/g'h_1$ and $Fr_2^2 = U_2^2/g'h_2$. Although they are more commonly observed in the ocean as a result of tidal flow over topographic features (e.g., Knight Inlet sill); internal hydraulic jumps can occur in lakes.

Progressive jumps form when supercritical flow (resulting from a gravity current, inflow, wind event or thermocline jet) propagates into an undisturbed region. Flow over a topographic feature can lead to a stationary jump in the lee of the obstacle. Localized energy dissipation and mixing occur near jumps and waves radiate from the critical point where $Fr = 1$. The impacts and distribution of hydraulic phenomena in lakes are not well understood.

Regime 2: Solitary Waves

In small- to medium-sized lakes subjected to moderate forcing ($0.3 < W^{-1} < 1.0$), nonlinearities become significant and the linear wave equation no longer completely describes wave evolution. In addition to the linear standing wave (composed of symmetric cosine components combined in a standing wave pattern), asymmetrical nonlinear wave components are generated from the wind induced thermocline tilt. The asymmetric components combine into a progressive internal wave pattern. The downwelled fluid becomes a dispersive packet of sub-basin scale internal waves of depression (**Figure 10**) called nonlinear internal waves (NLIWs), while upwelled fluid evolves into a progressive nonlinear basin-scale wave, referred to as a rarefaction or internal surge.

The weakly nonlinear Korteweg-de-Vries (KdV) equation mathematically describes the generation and unidirectional progression of NLIW from the wind-induced thermocline setup

$$\frac{\partial \eta}{\partial t} + c_o \frac{\partial \eta}{\partial x} + \alpha \eta \frac{\partial \eta}{\partial x} + \beta \eta \frac{\partial^3 \eta}{\partial x^3} = 0$$

where the nonlinear coefficient $\alpha = (3c_o/2)(h_1 - h_2)/h_1 h_2$ and the dispersive coefficient $\beta = c_o h_1 h_2/6$. Initially, the internal surge propagates under a balance between the unsteady $(\partial\eta/\partial t)$ and nonlinear $\alpha\eta(\partial\eta/\partial x)$ terms. As nonlinearities become more apparent as the waveform steepens and the wavefront approaches vertical (**Figure 10c**). This occurs at the steepening time scale

$$T_s = \frac{L}{\alpha \delta \eta}$$

Wave steepening causes the dispersive term $\beta\eta(\partial^3\eta/\partial x^3)$ to become significant, eventually balancing nonlinear steepening at $t = T_s$, leading to the production of high-frequency NLIWs (**Figure 10d**).

In many lakes, NLIWs have a wave profile that matches a particular solution to the KdV equation

$$\eta(x,t) = a\,\mathrm{sech}^2\left(\frac{x - ct}{\lambda}\right)$$

These are called solitary waves. The maximum amplitude of the solitary wave a, the solitary wave speed c, and characteristic horizontal length-scale λ are given by

$$c = c_o + \frac{1}{3}\alpha a \text{ and } \lambda^2 = 12\frac{\beta}{a\alpha}$$

The KdV equation reveals some interesting characteristics of solitary waves. In a two-layer system they will always protrude into the thicker layer and so are generally observed as waves of depression upon the thermocline. If the interface occurs at mid-depth, $\alpha \to 0$; thus preventing nonlinear steepening and subsequent solitary wave generation. Moreover, the dependence of α on $h_1 - h_2$ demonstrates that the degree of nonlinearity depends not only on the magnitude of the interfacial displacement, but also on the relative heights of the stratifying layers.

The dispersive nature of the wave packet is evident from the relationship between wave amplitude a and wave speed c; a spectrum of waves in a particular packet will be rank ordered according to amplitude (**Figure 10e**) and will disperse with time as they propagate. An estimate of the number of solitary waves and their amplitudes, while beyond the scope of this article, may be obtained from the Schrödinger wave equation.

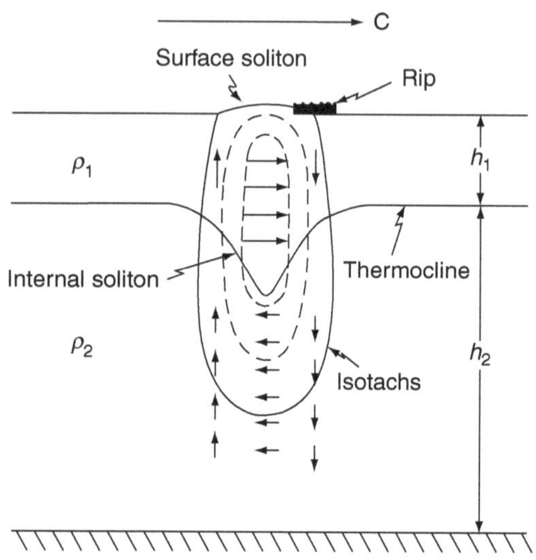

Figure 11 Schematic showing the passage of an internal solitary wave in a two-layer stratified fluid. Dashed lines are contours of water particle speed (isotachs) and arrows indicate the magnitude and direction of the flow. A small surface solitary wave of amplitude $\sim(\rho_2 - \rho_1)a$ accompanies the solitary wave and causes the rip or surface slick. From Osborne AR and Burch TL (1980) Internal solitons in the Andaman Sea. *Science* 208 (4443): 451–460.

Laboratory experiments show that NLIWs can contain as much as 25% of the energy (APE) introduced to the internal wave field by the winds. In large lakes they may have amplitudes and wavelengths as large as ~20 m and ~50–1000 m, respectively, and travel at $c \sim 0.5$–$0.75\ \mathrm{ms}^{-1}$. The velocity field associated with large amplitude NLIWs, commonly found in the ocean, will form a slick (rip) on the water surface (**Figure 11**) allowing them to be located and tracked using shore-based, aerial or satellite imagery.

Shoaling of Nonlinear Internal Waves

Progressive nonlinear internal waves travelling along the thermocline will break when they shoal upon the sloping lake bottom (**Figure 12**). Wave breaking irreversibly converts wave energy to mixing and dissipation, thus contributing to localized intense turbulence in the TBBL. The internal breaking process is similar to surface wave breaking at a beach and may be interpreted in the same manner using an internal form of the Iribarren number form of the Iribarren number ξ, which is the ratio of the boundary slope(s) to the offshore wave slope (a/λ)

$$\xi = \frac{s}{(a/\lambda)^{1/2}}$$

The relationship between ξ and the breaking mechanism is an ongoing area of active research. Laboratory experiments show that for small ξ (**Figure 13(a)**),

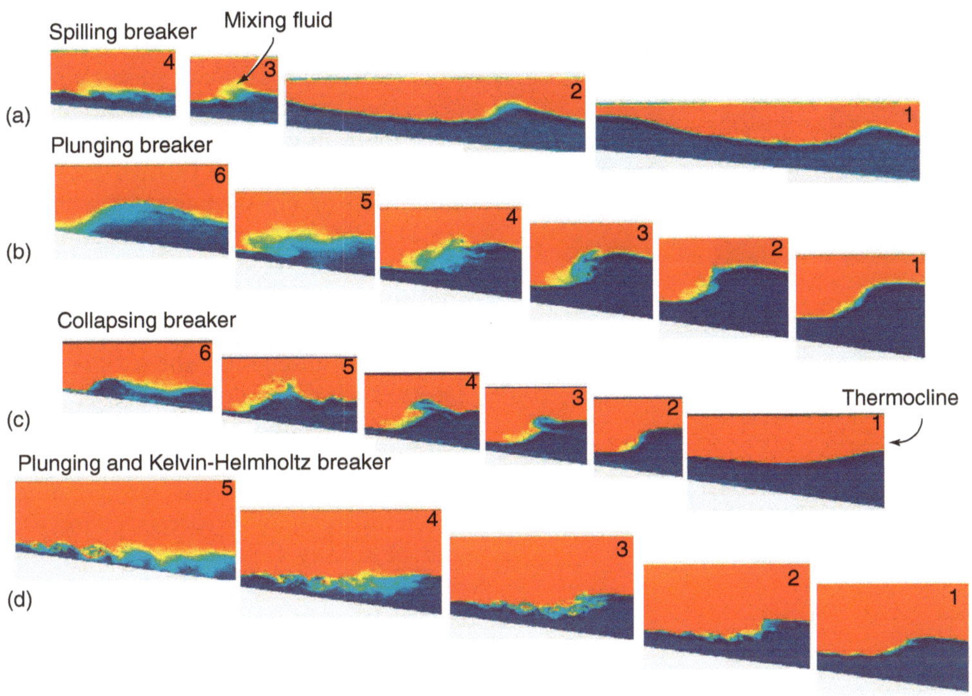

Figure 12 False color images showing types of internal wave breakers: (a) spilling breakers, (b) plunging and (c) collapsing breakers, (d) mixed plunging and Kelvin–Helmholtz breaking. Note the steepening of the rear face of the incident wave of depression, the transformation to a wave of elevation and subsequent breaking. Small numbered figures denote different stages of the breaking process. Adapted from Boegman L, Ivey GN, and Imberger J (2005) The degeneration of internal waves in lakes with sloping topography. *Limnology and Oceanography* 50: 1620–1637.

spilling breakers occur when small-scale shear instabilities form on the wave crests prior to breaking, causing mixing to be suppressed by viscosity. The mixing efficiency $R_f = b/(b+\epsilon)$, where b is the energy lost irreversibly to mixing and ε the energy dissipated by viscosity as heat, is less than 15%. As ξ increases, plunging (**Figure 12(b)**) then collapsing (**Figure 12(c)**) breakers occur. In the plunging region, wave inertia dominates and the most energetic overturns approach the Ozmidov scale (the largest scale where inertia can overcome buoyancy); the potential energy available for mixing is maximized and $R_f > 15\%$. For collapsing breakers, the wave breaking processes is not sufficiently energetic to overcome the stratification, mixing is suppressed by buoyancy and $<R_f$ 15%. For $\xi \rightarrow 1$ surging breakers occur, $R_f \rightarrow 0$ and the wave energy is reflected from the slope. Mixed-mode convective and shear-driven breaking is also possible when the wave shoals through a strong background flow field (**Figure 12(d)**).

Progressive Internal Wave Rays in a Continuous Stratification

We have conveniently described standing waves in a continuous stratification in terms of wave modes. A continuous stratification will also support progressive sub-basin scale internal waves; however, it is more insightful to describe these waves in terms of rays (both methods of analysis can be shown to be equivalent). Unlike NLIWs, which require a thermocline waveguide, progressive waves in a continuous stratification are described by linear equations and occur in regions of the water column where $N > 0$ and slowly varying (e.g., **Figure 1(e,f)**). A disturbance in the flow (e.g., flow over rough topography) with a particular excitation frequency will generate a range of wavelengths that will radiate from the source at the same frequency. The wave rays will propagate through the fluid at a fixed angle to the horizontal β given by the dispersion relation

$$\omega = N \sin \beta = \frac{kN}{K}$$

where ω is the wave frequency and the wavenumber vector $K = \sqrt{k^2 + m^2}$ has horizontal k and vertical m components (**Figure 13(b)**). The angle at which the rays propagate is chosen such that the vertical component of their frequency matches N, leading to a four-ray St. Andrew's cross pattern (**Figure 13(a)**). From the dispersion relation, the wave frequency is independent of the magnitude of the wavelength and only depends

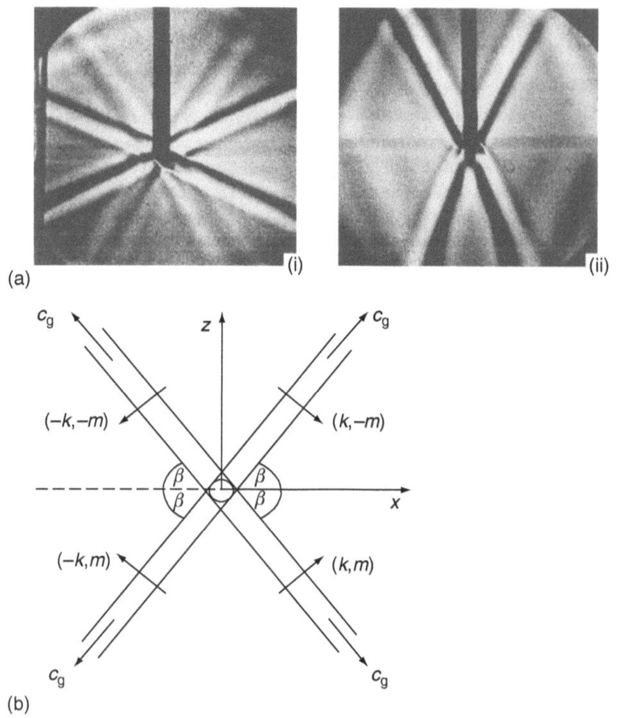

Figure 13 Propagation of internal wave rays in a stratified fluid with constant *N*. (a) Laboratory images showing internal wave rays propagation from an oscillating cylinder. The light and dark bands are lines of constant phase (wave crests and troughs). In (i) and (ii) $\omega = 0.4N$ and $0.9N$ giving $\beta = 25°$ and $64°$, respectively. From Mowbray DE and Rarity BSH (1967) A theoretical and experimental investigation of the phase configuration of internal waves of small amplitude in a density stratified liquid. *Journal of Fluid Mechanics* 28: 1–16. (b) Schematic showing the direction of the group velocity c_g, wavenumber vectors (showing the direction of phase propagation), and angle of inclination of the rays relative to the horizontal β, for the experiments in (a). Adapted from Thorpe SA (2005) *The Turbulent Ocean*. Cambridge, UK: Cambridge University Press.

upon β. This property is quite different than for interfacial waves, where the wave frequency and period depend only on the magnitude of the wavelength. Equations for the velocity and density perturbations induced by the wave passage are beyond the scope of this review and can be found in physical oceanography texts.

Another property of the dispersion relation is that the wave frequency must lie in the range $0 < \omega < N$, mathematically showing that N is indeed the cut-off frequency for internal waves. Excitation at frequencies $\omega > N$ generates motions that are exponentially damped. Wave energy will propagate from an excitation region at the group velocity c_g of the wave envelope, which is perpendicular to the phase velocity c; the wave rays carry the energy at right angles to the motion of the crests and troughs! These waves are difficult to visualize, internal waves generated by a local source do not have the concentric circle pattern of crests and troughs familiar to those who observe a stone thrown into a pond, but are composed of crests that stretch outward as spokes (rays) carrying energy radially from the source. The wave crests and troughs slide perpendicularly across the rays, seeming to appear from and disappear to nowhere.

From the discussion above, it is not surprising that internal wave rays do interesting things as they reflect from sloping topography and propagate into regions of variable $N(z)$. The intrinsic frequency ω is always conserved causing rays propagating into depths of diminishing $N(z)$ to refract towards the vertical and be totally reflected at the turning depth where $\omega = N(z)$. If generated in the seasonal thermocline these waves can be trapped between the upper and lower surfaces of the metalimnion where $\omega < N(z)$. Waves propagating into depths of increasing $N(z)$, such as toward the seasonal thermocline, will refract toward the horizontal.

Upon reflection from the lake surface or a sloping bottom with angle α, the intrinsic wave frequency is conserved and, by the dispersion relation, the reflected ray must propagate at the same angle β as the incident ray (**Figure 14**). The wavelength and group velocity will change as is evident in the change in concentration of wave rays upon refection. If $\alpha = \beta$, the rays are reflected parallel to the slope and have zero wavelength and group velocity. A turbulent bore will from and propagate along the slope and wave energy is rapidly converted into local dissipation and mixing. In this case, both the slope angle and wave frequency are considered critical.

Subcritical waves $\beta < \alpha$ will be reflected back in the direction from which they came and the may escape to deeper water (**Figure 14(c,d)**). However, supercritical waves $\beta > \alpha$ will continue in the same direction (**Figure 14(a)**) and if propagating towards shallower water in the littoral zone may thus be trapped, repeatedly reflecting off the surface, lakebed and turning depths (**Figure 15**). Eventually the rays will break when on a critical slope $\alpha = \beta$, where they have a critical frequency $\omega = N \sin \alpha$.

Progressive internal waves are produced by small localized disturbances such as flow over rough topography, patches of shear and turbulence and wave-wave/wave-flow interactions, when the excitation frequency $\omega < N$. Progressive waves are found ubiquitously in lakes and those in the 10^{-5} to 10^{-3} Hz bandwidth generally have critical frequencies relative

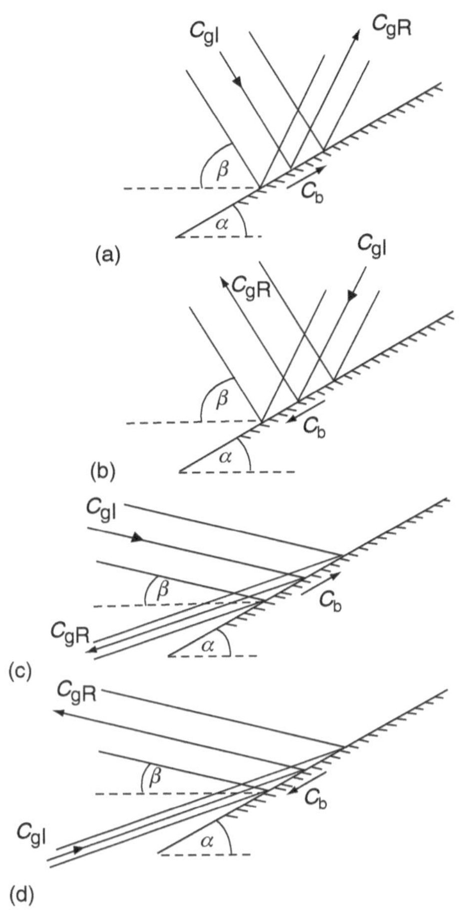

Figure 14 Schematic showing the reflection of internal wave rays from a uniform slope. The lines indicate crests and troughs (lines of constant phase) and c_{gI} and c_{gR} the direction of the incident and reflected group velocity, respectively; c_b is the direction of phase propagation on the slope. (a–b) show the supercritical case where $\alpha < \beta$ and rays continue upslope or downslope. (c–d) show the subctitical case $\alpha > \beta$ where rays are reflected out to deeper water. The distance between rays is proportional to the wavelength and can increase or decrease upon reflection. As $\beta \rightarrow \alpha$, the reflected rays become parallel to the slope and their wavelength goes to zero. Adapted from Thorpe SA and Umlauf L (2002) Internal gravity wave frequencies and wavenumbers from single point measurements over a slope. *Journal of Marine Research* 60: 690–723.

to the sloping boundaries found where the metalimnion intersects the lake bed.

Resonant and Forced Internal Waves

The periodicity inherent in weather patterns creates over-lake wind fields that occur at regular frequencies. For example, the winds over Lake Erie have periodicities of 10 d and 24 h, associated with frontal weather systems and diurnal land/sea breeze phenomena, respectively. When the forcing frequency matches one of the natural frequencies of the basin-scale wave modes, resonant amplification will occur. Seasonal heating and deepening of the thermocline will adjust the natural frequencies relative to the forcing frequencies, thus tuning the system into and out of resonance.

Laboratory experiments show that the relative frequencies of the wind forcing f_w and H1 internal seiche f_{H1} can be used to model internal wave response to periodic forcing conditions (**Figure 16**). When the forcing frequency of the wind stress is less than the natural frequency of the H1 internal seiche ($f_w < \sim 2/3f_{H1}$), the phase of the basin-scale oscillations are reset with each wind event and a forced internal seiche is generated at the same frequency as the wind forcing. A resonant H1 seiche will occur when the frequency of the wind forcing is near the frequency of the H1 seiche ($2/3f_{H1} < f_w < 2f_{H1}$) and when the forcing frequency is greater than the natural frequency of the H1 seiche ($f_w > \sim 2f_{H1}$), higher-mode horizontal seiches are generated. Resonance appears to be particularly effective in amplifying the response of the second vertical mode (e.g., Lake Alpnach) and the even horizontal modes that are not naturally energized by a wind-induced interfacial setup.

Resonance between the wind and the H1 internal mode leads to increased seiche amplitudes under relatively weak wind forcing conditions. For example, when $W^{-1} > 0.03$, nonlinearities become significant (compared to $W^{-1} > 0.3$ under non-resonant conditions), favoring the formation of steepened internal wave fronts (e.g., Loch Ness) and progressive NLIWs. If the wind induced tilting increases, $W^{-1} > 0.2$ will cause Kelvin–Helmholtz instabilities to form within the progressive NLIWs, leading to significant diapycnal mixing within the basin interior. The energy content of the H1 seiche, NLIWs and shear instability modes thus appear to be capped and continued energy input via resonant amplification is transferred between these discrete modes, ultimately being lost to dissipation and mixing at turbulent scales.

Analysis of Timeseries Data

The various wave, instability and turbulence processes described in this chapter are shown schematically in **Figure 17**. These processes occur beneath the lake surface and so practicing limnologists do not have the luxury of being able to directly observe them in the field. Some insight regarding their spatial structure is gained from idealized laboratory and computational models (see **Hydrodynamical Modeling**) but for the most part, limnologists must resort

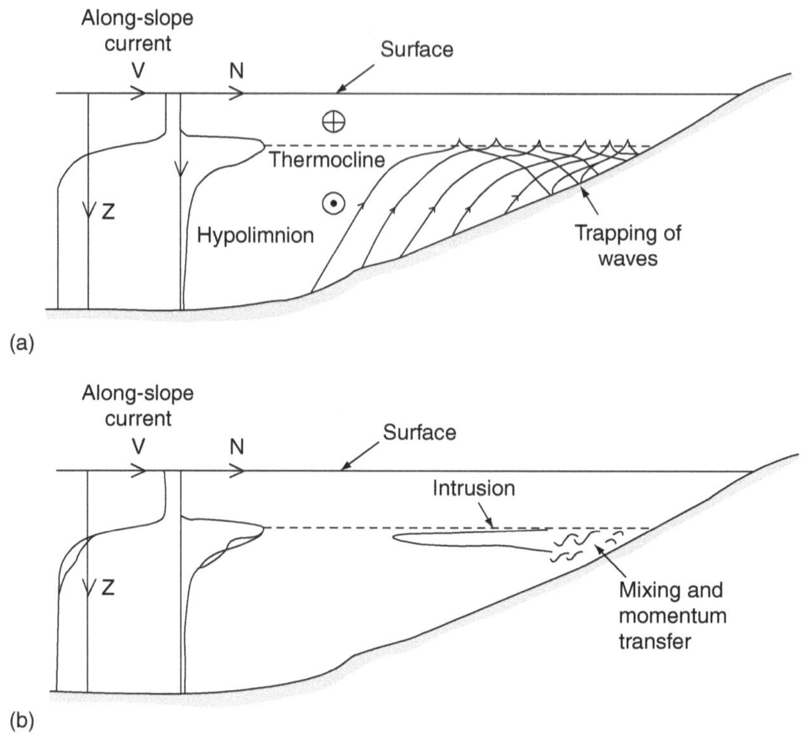

Figure 15 Schematic showing wave rays radiating from along-slope flow over bottom topography and propagating toward the littoral zone. (a) The waves are trapped between the mixed surface layer and the lake bed and will eventually break where the bed slope α is critical and the wave frequency $\omega = N\sin\alpha$. (b) The mixed fluid created during wave breaking collapses and intrudes into the lake interior, carrying sediment and nutrients from the littoral zone. The local strength of the stratification N' is reduced and the along-slope flow is modified. Adapted From Thorpe SA (1998). Some dynamical effects of internal waves and the sloping sides of lakes. Coastal and estuarine studies: Physical processes in lakes and oceans. *American Geophysical Union* 54: 441–460.

to deciphering timeseries data from thermistor chains, which are the most useful tools in their arsenal. **Figure 18** shows many of the processes from **Figure 17** as they would appear on thermistor data, which has been contoured to show isotherm displacement timeseries with depth.

During the strong wind event, shear instabilities form at the base of the surface layer (**Figure 18(d)**). Shortly thereafter, a thermocline jet occurs within a compressed region of the metalimnion and causes a rapid expansion of the strata and localized mixing (**Figure 18(e)**). The basin-scale internal wave energized by the wind event has steepened into an internal surge supporting large amplitude NLIWs of depression and series of step like features resembling internal hydraulic jumps (**Figure 18(c)**). The surge wave interacts with the lake bed leading to significant mixing over the bottom 20 m of the water column.

The motions described above all result from a single intense wind event. Considering the periodic nature of surface winds, it is not surprising that periodic wave motions occur in lakes over a range of length scales and frequencies. Spectral frequency analysis is used to conveniently analyze timeseries data and determine the relative amount of energy found at each frequency.

Observations from many lakes suggest the existence of a universal frequency spectrum model for lakes. The main features of the internal wave spectrum are shown for several lakes in **Figure 19**. These lakes range in diameter from 20 km (Lake Biwa) to 10 km (Lake Kinneret) to 1 km (Lake Pusiano). Unlike the spectral energy cascade occurring in turbulent flows, internal waves are generated at discrete frequencies throughout the spectrum. Motions are bounded at the low frequency end of the spectrum by H1 seiches that contain the most energy with frequencies between zero and 10^{-4} Hz. At the high-frequency end of the spectrum, motions are bounded by the high-frequency cut-off N. Shear instabilities cause a sub-N spectral peak with frequency $\sim 10^{-2}$ Hz and five orders of magnitude less energy than the basin-scale seiches. The middle portion of the spectrum contains freely propagating linear and nonlinear internal waves. The NLIWs are generated under moderate forcing conditions with sech^2 or solitary wave profiles and frequencies $\sim 10^{-3}$ Hz. These waves are short lived because they break upon shoaling

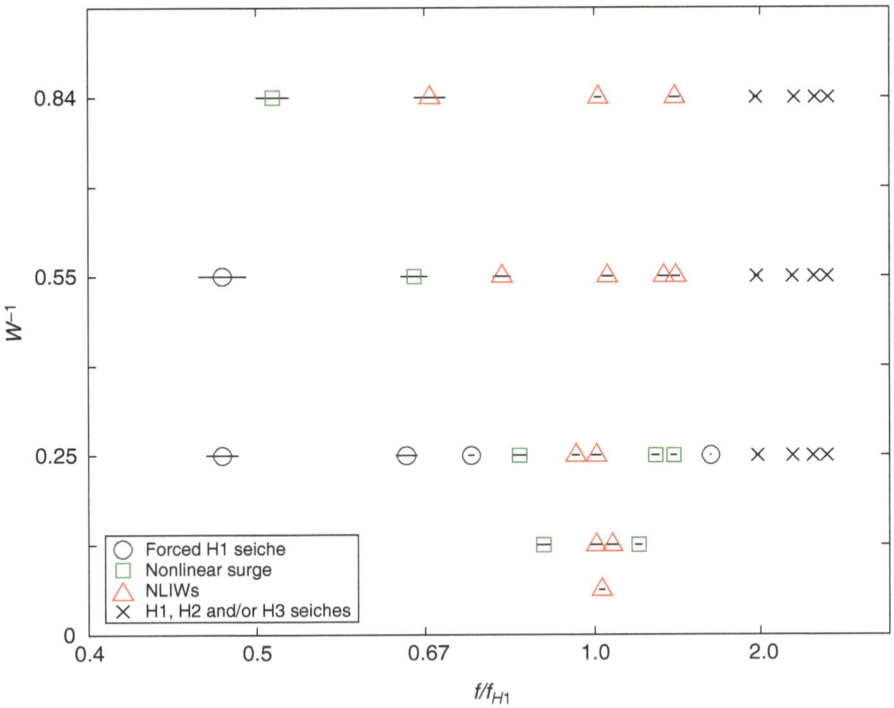

Figure 16 Regime diagram showing the dominant internal wave response under periodic forcing conditions in a long rectangular laboratory tank. Error bars denote the variation in forcing during an experiment (mean ± standard deviation). After Boegman L and Ivey GN (2007) Experiments on internal wave resonance in periodically forced lakes. In *Proceedings of the 5th International Symposium on Environmental Hydraulics*, 4–7 Dec. 2007, Tempe, Arizona.

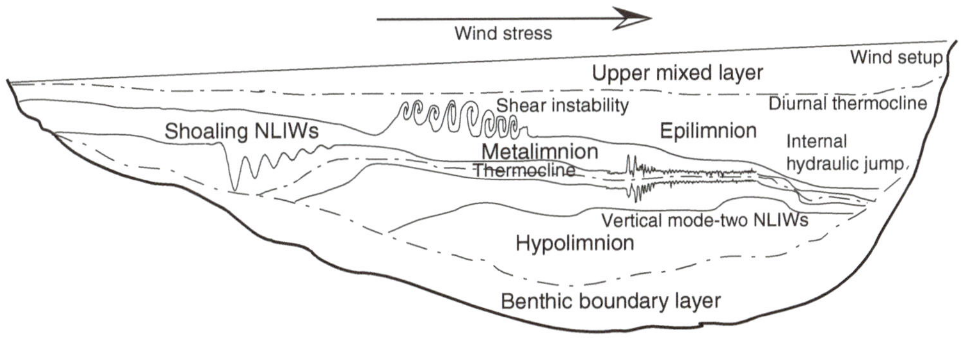

Figure 17 Pictorial representation of the various regions of a lake and some of the wave-like physical processes that occur. See also **Figure 6.13** in Fischer HB, List, EJ, Koh RCY, Imberger J, and Brooks NH (1979) *Mixing in Inland and Coastal Waters*. San Diego, CA: Academic Press; **Figure 15** in Imberger J (1985) Thermal characteristics of standing waters: An illustration of dynamic processes. *Hydrobiologia* 125: 7–29; and **Figure 7** in Imboden DM and Wüest A (1995). Mixing mechanisms in lakes. In Lerman A, Imboden DM, and Gat J (eds.) *Physics and Chemistry of Lakes*. pp. 83–138. Berlin: Springer.

topography at the depth of the thermocline. The portion of the spectrum between the basin-scale seiches and NLIWs (i.e., $\sim 10^{-4}$ Hz) appears to consist of freely propagating gravity waves that have linear or sinusoidal profiles; similar to the broadband background internal wave field associated with the Garrett-Munk spectrum in the ocean. These waves are generated by disturbances within the flow field (radiation from flow over rough topography, wave–wave interactions, mixing regions, intrusions, nonlinear surges and internal hydraulic jumps) where gravity acts as restoring force on fluid parcels displaced from their equilibrium position. The waves in this frequency bandwidth are interacting with one another making it difficult to identify their source. The bandwidth limits on the spectrum depend only upon the stratification and basin size and are independent of the strength of the wind forcing. Stronger winds lead to sharper energy peaks with higher energy content (larger amplitude seiches and more shear instabilities).

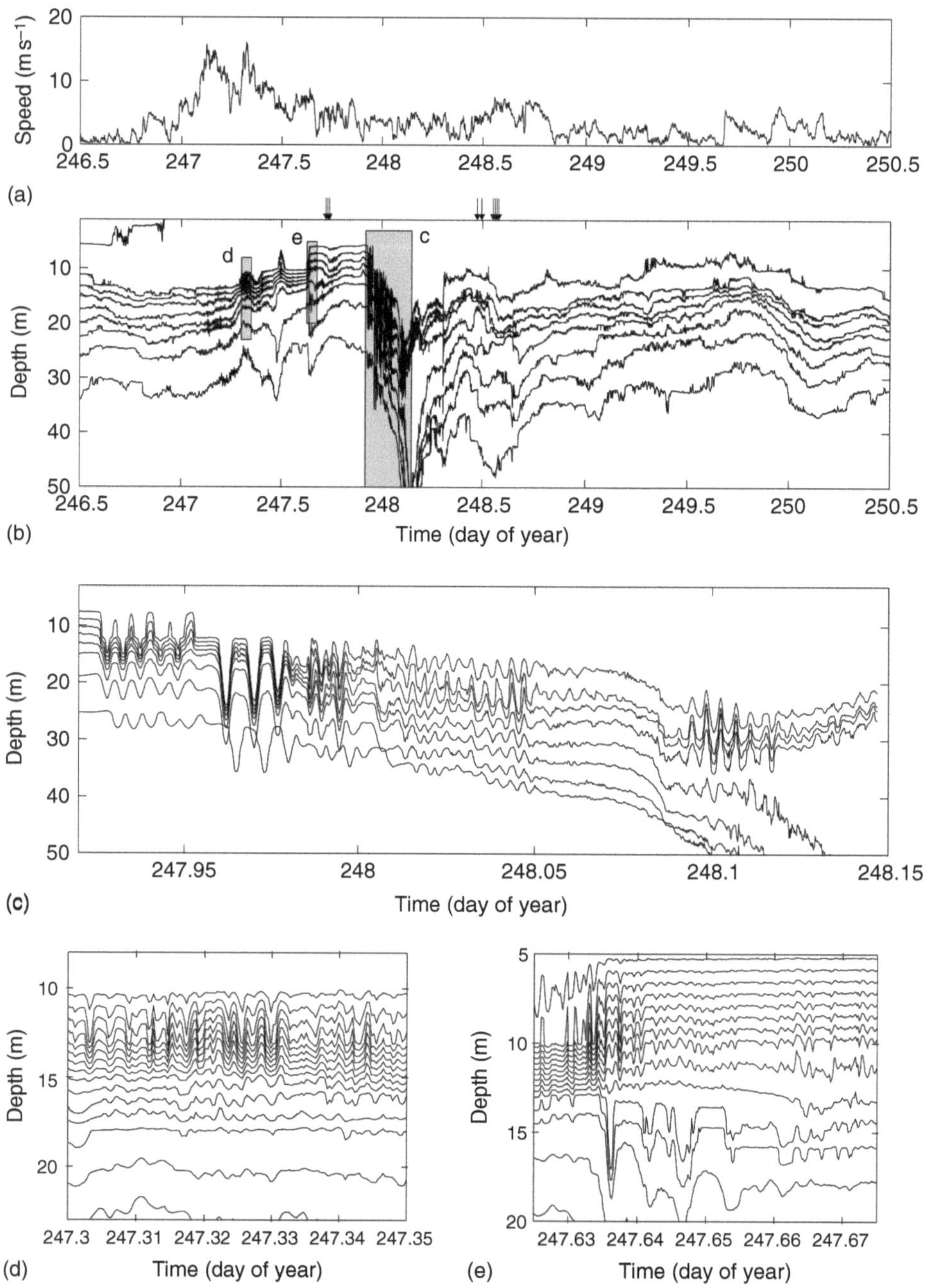

Figure 18 Observations from Lake Biwa (Japan) in 1993 showing steepened nonlinear basin-scale internal wave front and associated nonlinear internal waves. (a) Wind speed collected at 10-min intervals showing the passage of a storm event; (b) nonlinear response of basin-scale internal wave field showing steepened wave front (2 °C isotherms); (c) magnified view of shaded region in panel b showing details of NLIWs (2 °C isotherms). (d) Magnified view of shaded region d in panel b showing shear instabilities resulting from enhanced shear at the base of the surface layer during the strong wind event (1 °C isotherms); (e) Magnified view of shaded region e in panel b showing a V2 expansion of the metalimnion resulting from a thermocline jet that forms after a period of metalimnion compression (1 °C isotherms). The bottom isotherm in panels b and c is 10 °C. Adapted from Boegman L, Imberger J, Ivey GN, and Antenucci JP (2003). High-frequency internal waves in large stratified lakes. *Limnology and Oceanography* 48: 895–919.

Summary

The dynamical interplay between stratification, waves, and wind can is best summed up in using the conceptual model proposed by J. Imberger in 1990 and supported by the more recent measurements by A. Wüest and others. A lake behaves like an engine that is powered by the wind and does work against the potential energy gradient embodied in the

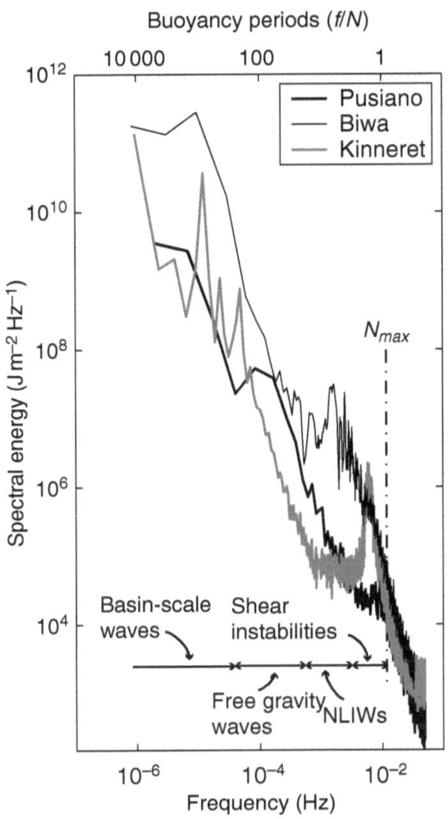

Figure 19 Spectra of the vertically integrated potential energy signals from Lakes Pusiano, Kinneret and Biwa showing basin-scale seiches (0–10^{-4} Hz), freely propagating nonlinear wave groups with sinusoidal profiles ($\sim10^{-4}$ Hz) and solitary wave profiles ($\sim10^{-3}$ Hz), and shear instabilities ($\sim10^{-2}$ Hz). N_{max} denotes the maximum buoyancy frequency. Adapted from Boegman L, Ivey GN, and Imberger J (2005). The degeneration of internal waves in lakes with sloping topography. *Limnology and Oceanography* 50: 1620–1637.

stratification. Approximately 2% of the wind energy flux enters the lake; of this ~80% is dissipated in the surface layer and ~20% is transferred to the basin-scale internal wave field. The basin-scale seiches energized by the wind are frictionally damped as they swash along the lake bed and by degeneration into progressive high-frequency internal waves, shear instabilities, and eventually turbulence. Approximately 90% of the seiche energy is lost energizing turbulent dissipation and mixing in the TBBL; $<1/4$ of which first passes through the nonlinear internal wave field prior to wave breaking and $>3/4$ of which is lost to frictional swashing in the TBBL. The remaining ~10% of the seiche energy results in intermittent shear instability in the basin interior. The overall mixing efficiency of the turbulence is ~15% leading to an upwards buoyancy flux that works to weaken the stratification and raise the centre of gravity of the lake. The lake engine is extremely inefficient with only ~0.06% of the wind work acting to irreversibly mix the stratification; the bulk of the wind work is lost to frictional viscous dissipation, which due to the large heat capacity of water has no significant effect on the lake temperature. From this model, it is clear that while internal waves control biogeochemical mixing and transport within a stratified waterbody, they are not able to significantly influence the stratification upon which they propagate. Consequently, their existence depends entirely upon the seasonal stratification cycle set up by the surface thermodynamics.

See also: The Benthic Boundary Layer (in Rivers, Lakes, and Reservoirs); Currents in Stratified Water Bodies 1: Density-Driven Flows; Currents in Stratified Water Bodies 3: Effects of Rotation; Currents in the Upper Mixed Layer and in Unstratified Water Bodies; Density Stratification and Stability; Hydrodynamical Modeling; Salinity; Small-Scale Turbulence and Mixing: Energy Fluxes in Stratified Lakes; The Surface Mixed Layer in Lakes and Reservoirs.

Further Reading

Boegman L, Imberger J, Ivey GN, and Antenucci JP (2003) High-frequency internal waves in large stratified lakes. *Limnology and Oceanography* 48: 895–919.

Boegman L, Ivey GN, and Imberger J (2005) The degeneration of internal waves in lakes with sloping topography. *Limnology and Oceanography* 50: 1620–1637.

Fischer HB, List EJ, Koh RCY, Imberger J, and Brooks NH (1979) *Mixing in Inland and Coastal Waters*. San Diego, CA: Academic Press.

Helfrich KR and Melville WK (2006) Long nonlinear internal waves. *Annual Review of Fluid Mechanics* 38: 395–425.

Horn DA, Imberger J, and Ivey GN (2001) The degeneration of large-scale interfacial gravity waves in lakes. *Journal of Fluid Mechanics* 434: 181–207.

Imberger J (1998) Flux paths in a stratified lake: A review. Coastal and estuarine studies: Physical processes in lakes and oceans. *American Geophysical Union* 54: 1–18.

Imberger J and Patterson JC (1990) Physical limnology. *Advances in Applied Mechanics* 27: 303–475.

Imboden DM and Wüest A (1995) Mixing mechanisms in lakes. In: Lerman A, Imboden DM, and Gat J (eds.) *Physics and Chemistry of Lakes*, pp. 83–138. Berlin: Springer.

Lazerte BD (1980) The dominating higher order vertical modes of the internal seiche in a small lake. *Limnology and Oceanography* 25: 846–854.

MacIntyre S, Flynn KM, Jellison R, and Romero JR (1999) Boundary mixing and nutrient fluxes in Mono Lake, California. *Limnology and Oceanography* 44: 128–156.

Mortimer CH (1974) Lake hydrodynamics. *Mitteilungen Internationale Vereinigung Limnologie* 20: 124–197.

Mortimer CH (2004) *Lake Michigan in Motion*. 310 pp. Madison: The University of Wisconsin Press.

Münnich M, Wüest A, and Imboden DM (1992) Observations of the second vertical mode of the internal seiche in an alpine lake. *Limnology and Oceanography* 37: 1705–1719.

Thorpe SA (1998) Some dynamical effects of internal waves and the sloping sides of lakes. Coastal and estuarine studies: Physical processes in lakes and oceans. *American Geophysical Union* 54: 441–460.

Thorpe SA (2005) *The Turbulent Ocean.* Cambridge, UK: Cambridge University Press.

Turner JS (1973) *Buoyancy Effects in Fluids.* Cambridge, UK: Cambridge University Press.

Wüest AJ and Farmer DM (2003) Seiches. *McGraw-Hill Encyclopedia of Science and Technology*, 9th ed. New York: McGraw-Hill.

Wüest A and Lorke A (2003) Small-scale hydrodynamics in lakes. *Annual Review of Fluid Mechanics* 35: 373–412.

Wüest A, Piepke G, and Van Senden DC (2000) Turbulent kinetic energy balance as a tool for estimating vertical diffusivity in wind-forced stratified waters. *Limnology and Oceanography* 45: 1388–1400.

Currents in Stratified Water Bodies 3: Effects of Rotation

J P Antenucci, University of Western Australia, Nedlands, WA, Australia

Introduction

In this article, we outline the role of the Earth's rotation in modifying currents in inland waters. The first investigation into these dynamics was conducted by Lord Kelvin in the nineteenth century, and the analytical model he developed is still useful in describing some of these dynamics today. More recently, theoretical developments, laboratory experimentation, and field measurements have allowed for the development of a relatively complete picture of the role of the earth's rotation in inland waters.

It is important to note that the effect of the earth's rotation on currents in stratified lakes is predominantly through periodic, oscillatory motions such as gravity waves and vorticity waves, which will be defined later. It is not possible to discuss one without the other, and so in this article we discuss both waves and the currents they induce simultaneously, remembering that waves can be identified through fluctuations in potential energy (typically measured as thermocline or isotherm oscillations) or kinetic energy (typically measured as currents). It is therefore essential that the reader has a good understanding of the material presented in the preceding article. We follow on from this material by investigating how the wind induces motion in large lakes, where the rotation of the earth cannot be ignored.

We begin by defining several parameters that will assist in describing the impacts of the earth's rotation. The most important parameter is the Coriolis parameter (or the inertial frequency)

$$f = \frac{4\pi}{T}\sin\theta \qquad [1]$$

where T is the period of rotation of the earth (1 day or 86 400 s), θ is the latitude, and the units of f are radians per second. This parameter is zero at the equator (meaning that the effects of the earth's rotation on internal waves and currents can be ignored at the equator), and reaches a maximum value at the poles. The inertial period is defined as

$$T_I = \frac{2\pi}{f} \qquad [2]$$

which is infinite at the equator, and has a minimum value of 12 h at the poles. We also define the Rossby radius of deformation

$$R = \frac{c}{f} \qquad [3]$$

where c is the celerity (speed) of long gravity waves in the water body of interest, where we define long waves as those whose wavelength is far greater than the water depth. For surface ('barotropic') waves $c = \sqrt{gH}$, where g is the gravitational constant ($9.8\ \mathrm{m\,s^{-2}}$) and H is the water depth, as described in the preceding article.

Note that it is possible to represent a stratified system as an equivalent depth of homogenous fluid so that the internal ('baroclinic') dynamics can be represented by the same equations. For example, for a two-layer stratification, we can define the equivalent depth as

$$H_e = \frac{\rho_2 - \rho_1}{\rho_2}\frac{h_1 h_2}{h_1 + h_2} \qquad [4]$$

where ρ is the density and h the depth and the subscripts refer to the upper and lower layer. This allows for simple definition of the barotropic phase speed as $c = \sqrt{gH}$, and the baroclinic phase speed as $c_i = \sqrt{gH_e}$, and also allows us to define an internal Rossby radius of deformation that applies to baroclinic processes (those due to stratification) as

$$R_i = \frac{c_i}{f} \qquad [5]$$

The equivalent depth can also be defined for a continuous stratification, using c_m in eqn. [17] discussed later in this chapter.

We also define the Burger number

$$S = \frac{R}{L} \qquad [6]$$

where L is a length characterizing the basin length and/or width and R can represent either the Rossby radius or the internal Rossby radius. This dimensionless number is used to help classify both vorticity and gravity waves, and has been called by various names in the literature, such as the stratification parameter or nondimensional channel width. It is simply the ratio of the length scales at which rotation effects become important to the length scale of the lake in question, so for $S \to 0$ rotation is very important for the dynamics, and for $S \to \infty$ rotation can be ignored as the lake is physically small. Note that there is no abrupt transition, where the effects of rotation are suddenly felt at $S=1$, but a gradual transition – this will be discussed later in this article.

Governing Equations

The dynamics described herein are based solely on the linear inviscid equations of motion for a homogenous fluid. The x-momentum, y-momentum, and conservation of mass equations are

$$\frac{\partial u}{\partial t} - fv = -g\frac{\partial \eta}{\partial x} \quad [7]$$

$$\frac{\partial v}{\partial t} + fu = -g\frac{\partial \eta}{\partial y} \quad [8]$$

$$\frac{\partial \eta}{\partial t} + \frac{\partial}{\partial x}(Hu) + \frac{\partial}{\partial y}(Hv) = 0 \quad [9]$$

where η is the height of the water surface above equilibrium, H is the water depth, and f is the Coriolis parameter. The momentum equations listed here are nothing more than the application of the Newton's famous equation $F=ma$, where the terms on the left-hand side represent acceleration terms (unsteady and Coriolis, respectively), and the term on the right-hand side represents the restoring force due to gravity. The same equations can be applied for barotropic and baroclinic motions, where for the baroclinic case, we replace the actual water depth H by the equivalent depth H_e described earlier.

In describing the effects of the earth's rotation on currents in inland waters, we consider two classes of motion based on the above equations. As we are interested in rotational effects, we first assume $f \neq 0$, such that we are sufficiently far away from the equator. For the first class of motions, which we will term 'gravity waves,' we also assume that the body of water under consideration is sufficiently small such that f can be considered constant (i.e., the lake is at a constant latitude) and that the bottom is flat, and therefore the restoring force is due to gravity only. For the second class of motions, which we will term 'vorticity waves,' we assume f is constant as for gravity waves, but we allow for variable water depth. This variable water depth allows for waves that arise due to the conservation of angular momentum. Dynamically, the second class of motions have similar characteristics to planetary Rossby waves in the ocean and atmosphere (i.e., where the latitude is not considered constant). In most cases, gravity waves dominate the dynamics of lakes and hence are explained later in detail. Only a brief summary description is given of the dynamics of vorticity waves, and for additional information the reader is referred to the references in Further Reading.

Gravity Waves

Of the two classes of periodic motions outlined earlier, gravity waves are the most well-studied and best understood in inland waters. We will consider only linear waves, that is, motions where the amplitude of the oscillations of the thermocline is small compared with the depth of the surface and bottom layer. This is not a major restriction on the analysis, as the inclusion of nonlinear effects has been shown in most cases to require only a minor correction to the linear approximation. In this article, we focus on cases where the Burger number is ~1, such that rotational effects can be expected. For surface (barotropic) waves, this would typically require lakes of more than 300 km width, of which there are very few. For baroclinic motions, where the phase speed c is far less than for barotropic motions, there are many lakes in which the Burger number is ~1. For typical values of the baroclinic internal wave phase speed (0.05–0.4 m s^{-1}), the internal Rossby radius is ~1–5 km (**Figure 1**), indicating that internal gravity waves in lakes of this scale (or larger) should experience the rotational effects of the earth. Note also that unlike the barotropic phase speed, which depends on water depth alone, the baroclinic phase speed varies as a function of stratification and so changes through the year. Rotation may therefore play a more important role in the internal wave dynamics during the strongly stratified period when the internal Rossby radius (and therefore the Burger number) are minimal than at other times of the year.

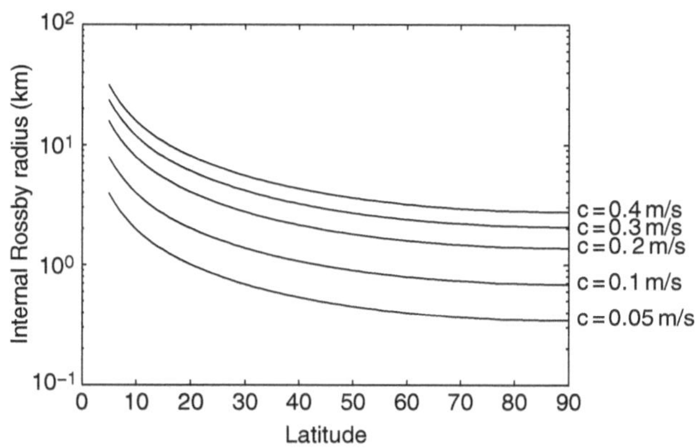

Figure 1 Internal Rossby radius as a function of latitude for several internal wave speeds. For horizontal length scales similar to or greater than the internal Rossby radius, rotational effects should be observed.

To understand the form of these motions in inland waters, it is instructive to build up our knowledge from simpler systems. We begin in a rotating system without boundaries, such as in the middle of the ocean far from the coast, where the classic gravity wave solutions are called plane progressive Poincaré waves. The amplitude (η) and velocity structure (u,v) associated with these waves can be described by

$$\eta = \eta_0 \cos(kx - \omega t) \qquad [10]$$

$$u = (\omega\eta_0/kH)\cos(kx - \omega t) \qquad [11]$$

$$v = (f\eta_0/kH)\sin(kx - \omega t) \qquad [12]$$

where u is the velocity in the direction of propagation of the wave, v is the velocity in the transverse direction, η_0 is the maximum amplitude, k is the wave number ($=2\pi/\lambda$, where λ is the wavelength), ω is the wave frequency ($=2\pi/T$, where T is the wave period), H is the water depth (or equivalent water depth H_e for an internal wave), and f is the inertial frequency. The fluid particle trajectories (in plan view, as the vertical motion is small due to the linear wave assumption) are ellipses with major axes in the direction of propagation, with the ratio of the ellipse axes equal to ω/f and the direction of rotation anticyclonic (i.e., opposite to the direction of rotation of the earth). For short waves, which have high frequency, ω/f is large and so the trajectory ellipses are long and thin. Long waves are the opposite, with ω/f approaching one as the wave frequency is low and therefore the particle tracks are circular and trace out the well-known 'inertial circles' in the ocean (**Figure 2**). The radius of these circular tracks is U/f, which can be reformulated using eqns. [11] and [12] as η_0/kH.

An important aspect of the dynamics of internal waves influenced by the earth's rotation is that energy is generally not equally split between kinetic and potential forms. For the plane progressive Poincaré wave (**Figure 2**), the mean kinetic energy per unit area is

$$\mathrm{KE} = \frac{1}{4}\left(\frac{\omega^2 + f^2}{\omega^2 - f^2}\right)\rho g \eta_0^2 \qquad [13]$$

where ρ is the water density, and the potential energy per unit area is

$$\mathrm{PE} = \frac{1}{4}\rho g \eta_0^2 \qquad [14]$$

such that the ratio of potential to kinetic energy is

$$\frac{\mathrm{PE}}{\mathrm{KE}} = \frac{\omega^2 - f^2}{\omega^2 + f^2} \qquad [15]$$

This indicates that waves with a frequency much greater than the inertial frequency will have a potential

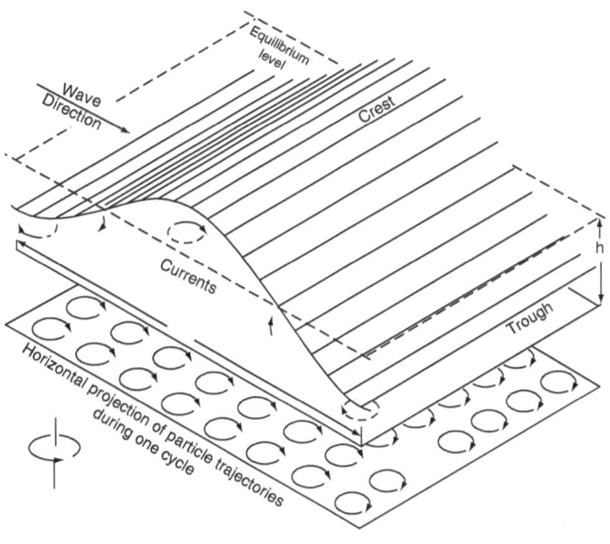

Figure 2 A long plane progressive Poincaré wave, in an infinite ocean, where $\omega \sim f$. Note the rotation of the current vectors is opposite to the direction of the earth's rotation. Reproduced from Mortimer CH (1974) Lake hydrodynamics. *Mitt. Int. Ver. Theor. Angew. Limnol.* 20: 124–197, with permission from E. Schweizerbart (http://www.schweizerbart.de/).

to kinetic ratio approaching one (as in the nonrotating case where $f \to 0$), and that waves close to the inertial frequency will have close to zero potential energy signal (such as the wave shown in **Figure 2**). This has implications for measurement of these waves, as they will only generally be observed by current measurements (a measure of kinetic energy) and not by fluctuations in stratification (a measure of potential energy variation).

The introduction of a boundary allows for the existence of Kelvin waves. The classical Kelvin wave solution is one in which the velocity perpendicular to the shore is considered to be zero (**Figure 3**). These waves propagate parallel to the boundary with the maximum amplitude at the shore, where the waves crests to the right (in the Northern Hemisphere) when looking along the direction of propagation. The amplitude decreases exponentially offshore at a rate equal to the Rossby radius of deformation R,

$$\eta = \eta_0 e^{-y/R}\cos(kx - \omega t) \qquad [16]$$

where x is both the alongshore direction and the direction of propagation, and y is the offshore direction (**Figure 3**). Note that the phase speed of the wave is $c = \sqrt{gH}$, the same as for a wave in a nonrotating system. Current vectors, by definition, are rectilinear and oscillate in the alongshore direction only. As with waves in a nonrotating frame, the ratio of potential to kinetic energy is unity. For internal Kelvin waves, the dynamics are the same, except that the baroclinic phase speed applies and the wave amplitude decreases

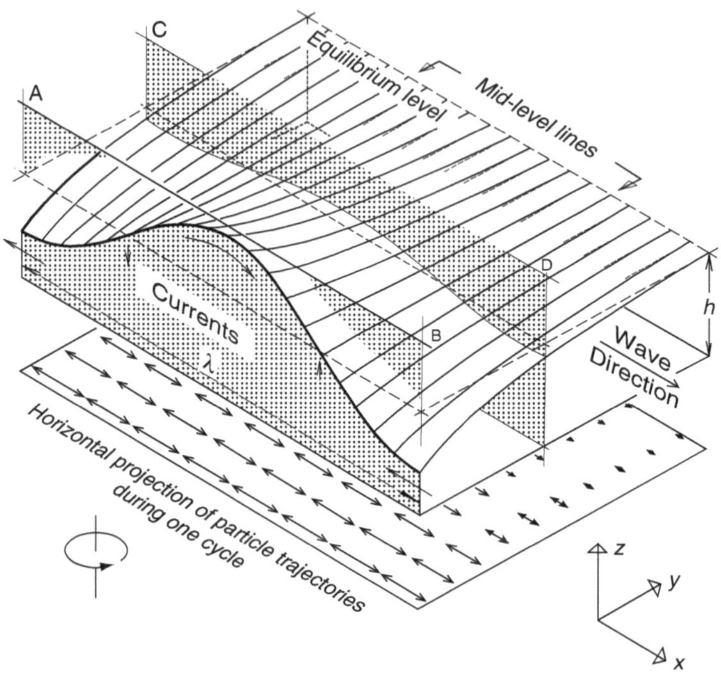

Figure 3 A long Kelvin wave progressing in the x-positive direction, with the shore located at $y=0$. Channel walls can be placed vertically at any point of constant y, for example indicated by the planes AB and CD. Reproduced from Mortimer CH (1974) Lake hydrodynamics. *Mitt. Int. Ver. Theor. Angew. Limnol.* 20: 124–197, with permission from • • • (http://www.schweizerbart.de/).

exponentially offshore with the internal Rossby radius of deformation R_i.

The introduction of a second boundary significantly complicates the waves supported in a rotating system. A channel (defined as two parallel walls with open ends) is able to support progressive Poincaré waves, made up of an obliquely incident plane progressive Poincaré wave with its reflection, and standing Poincaré waves, consisting of two progressive Poincaré waves traveling in opposite directions. These waves consist of cells similar to those presented in **Figures 2** and **3**, however the velocity at the border of each cell approaches zero. Closing a basin, and therefore creating a 'lake,' significantly complicates the wave field. For a rectangular basin, due to the complexity of the corners, an incident plus a reflected Kelvin wave is required along with an infinite number of Poincaré waves of the same frequency to satisfy the boundary conditions. Far simpler solutions can be found by assuming lakes to be represented by circular or elliptical basins of uniform depth, which we will use in the following discussion. As outlined earlier, a key nondimensional parameter controlling this response is the Burger number S. Based on this parameter, it is possible to determine the wave frequency for both circular and elliptic basins, the ratio of potential to kinetic energy in the wave response, and the response of a basin to external forcing. We will also rely on the simplified Kelvin and plane progressive Poincaré waves described earlier to assist in interpreting the results.

To understand the spatial structure of the waves (and the currents they induce) in a rotating system, it is helpful to consider the two end points: strong rotation ($S \to 0$) and no rotation ($S \to \infty$). For inland waters, these might also be considered the 'large lake' and 'small lake' case. For $S \to 0$, in the lake interior, we might expect plane progressive Poincaré waves to be present as outlined earlier (**Figure 2**), where the frequency approaches the inertial frequency, the current vectors rotate anticyclonically and the majority of energy is in the kinetic form. At the lake boundary, we might expect the classical Kelvin wave solution, where the offshore decay in amplitude is exponential at a rate R, the velocity at the boundary is parallel to the shore, the ratio of potential to kinetic energy is unity and the frequency approaches zero (**Figure 3**). Data collected from the North American Great Lakes support this conceptual model, with motion in the interior dominated by near-inertial frequencies and motion at the boundary appearing in the form of 'coastal jets,' which are the manifestation of the Kelvin wave solution. As the lake gets smaller (i.e., for $S \to \infty$), we should expect waves that are similar to the nonrotating case, where the ratio of potential to kinetic energy is unity, and the offshore decay of amplitude is no longer exponential, and the current vectors become rectilinear (i.e., the ellipses become long and thin).

How the characteristics of these waves vary as a function of Burger number and aspect ratio of the lake is graphically presented in **Figure 4**. The nondimensional frequency ω/f and the ratio of potential to kinetic energy are presented for analytical solutions to the circular and elliptical basin case for cyclonic (Kelvin-type) and anticyclonic (Poincaré-type) waves. Note that the potential to kinetic energy ratios are integrated over the entire lake, and do not represent the character at a particular point in space. We first consider waves in a circular basin (where the aspect ratio by definition is 1:1). In the strong rotation case ($S\rightarrow0$), the cyclonic wave frequency goes to zero and the energy ratio approaches unity. This is the Kelvin wave limit for a semi-infinite boundary outlined earlier (**Figure 3**). For the anticyclonic wave, the frequency approaches the inertial frequency and the energy is predominantly kinetic – the plane progressive Poincaré wave solution outlined earlier (**Figure 2**). As S increases, the wave frequency for both types of waves increases and slowly converges as the lake gets smaller. The energy ratio for the anticyclonic wave also increases and asymptotically approaches unity, the nonrotation limit ($S\rightarrow\infty$). For the cyclonic wave, the energy ratio increases to a maximum of ~1.5, before asymptotically approaching unity as $S\rightarrow\infty$. Importantly as $S\rightarrow\infty$ these two solutions have the same characteristics (frequency, energy ratio, cross-basin structure), except that they rotate in opposite directions. They will thus manifest themselves at high S as a standing wave.

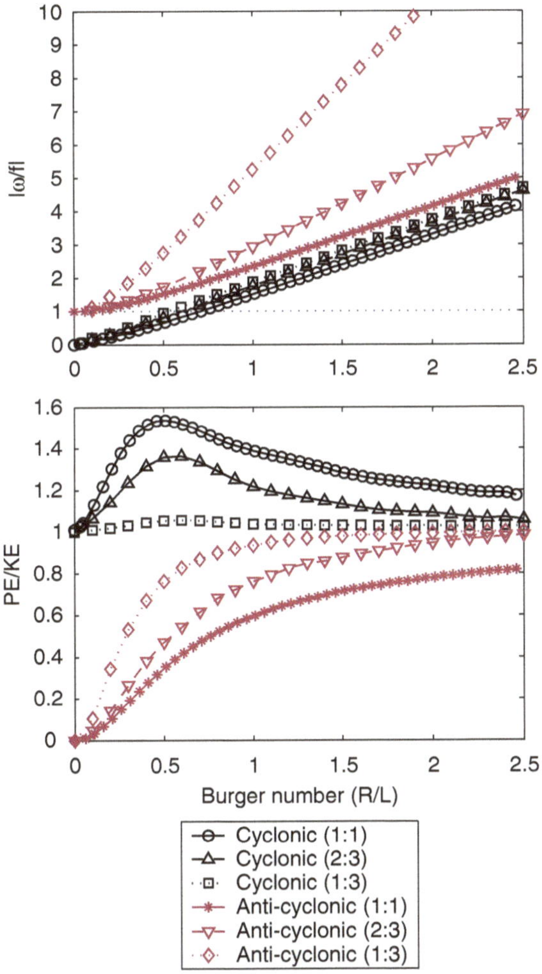

Figure 4 Nondimensional frequency (upper panel) and ratio of potential to kinetic energy (lower panel) as a function of wave type and aspect ratio, where the numbers in parentheses refer to the aspect ratio. The absolute value of the nondimensional frequency is presented as the results are independent of hemisphere. Reproduced from Antenucci JP and Imberger J (2001) Energetics of long internal gravity waves in large lakes. *Limnology and Oceanography* 46: 1760–1773, with permission from the American Society for Limnology and Oceanography.

The distribution of potential (i.e., thermocline oscillations) and kinetic energy (i.e., currents) in the basin also changes as the importance of rotation changes (**Figure 5**). For the strong rotation case ($S\rightarrow0$), the cross-shore potential energy structure of the cyclonic waves has the exponential decay associated with Kelvin waves propagating along a shoreline, where we can rewrite eqn. [14] $\eta = \eta_0 e^{-y/SL}$ so that for small S the exponential decay is rapid relative to the lake width (**Figure 5(a)**). The kinetic energy is also predominantly located close to the shore (**Figure 5(c)**), hence the term 'coastal jet' being applied to these motions in the North American Great Lakes. As the importance of rotation decreases (S increases), there is a stronger signal of the cyclonic waves present in the interior. It is important to note that for this case, the currents are not parallel to the boundary everywhere in the lake – next to the shoreline they remain parallel as in **Figure 3**; however, towards the interior, the current ellipses become more circular and actually rotate in a cyclonic direction. For the anticyclonic (Poincaré-type) waves, the structure changes very little as the importance of rotation changes (**Figure 5(b)** and **5(d)**). Note that as $S\rightarrow\infty$, the distribution of the potential and kinetic energy in the cyclonic wave approaches that of the anticyclonic wave.

We now consider the impact of changing the aspect ratio by moving towards elliptical basins from a circular basin shape. Note that the Rossby radius is defined in the elliptical basin based on the length of the major axis, not the minor axis, in **Figure 4**. The effect of decreasing the aspect ratio is that the system approaches the nonrotating case for lower values of S. Unlike the case of the circular basin, the frequencies of

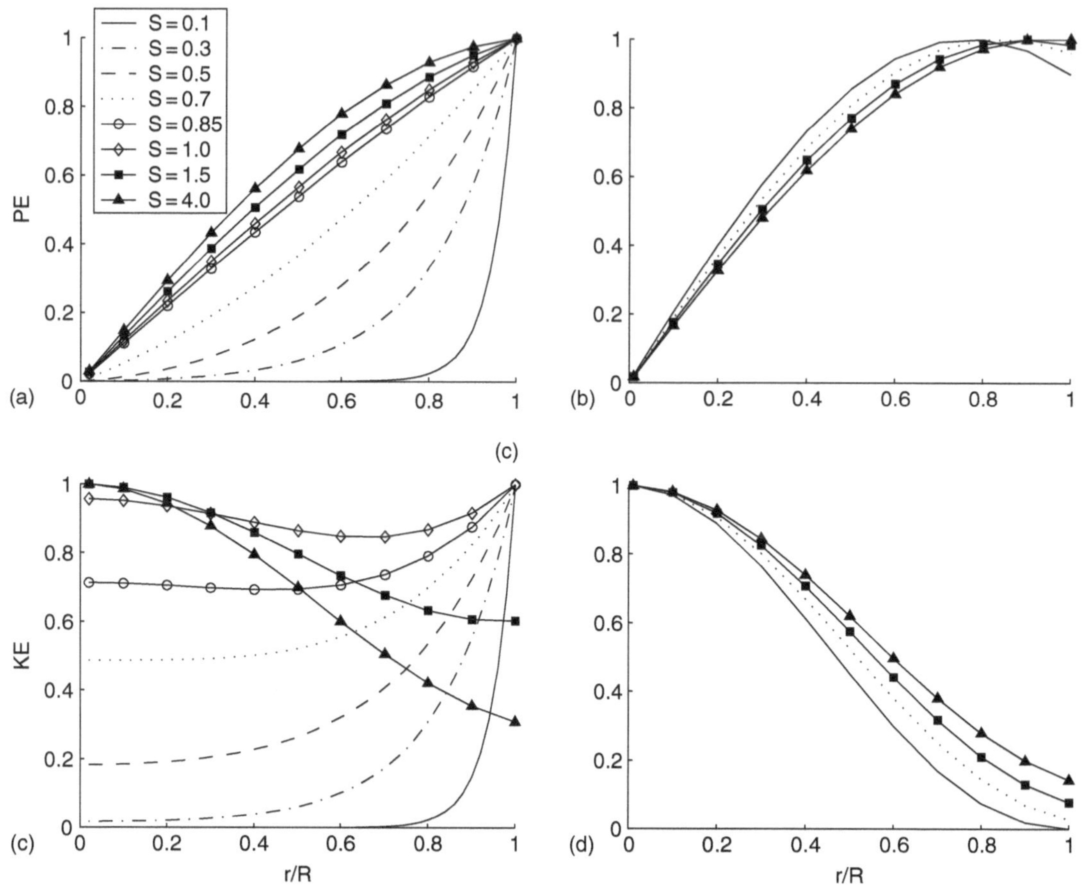

Figure 5 Radial structure of cyclonic and anticyclonic wave energy distribution as a function of Burger number for a circular lake for the lowest frequency motion (fundamental mode). The cyclonic wave structure is shown in panels (a) and (c). The anticyclonic wave structure is shown in panels (b) and (d). The radial structure for each Burger number has been normalized by its maximum value. Note that not all Burger numbers are shown in panels (b) and (d). Note the exponential decay in (a) is the same as that represented in **Figure 3** and eqn. [16]. Reproduced from Antenucci JP and Imberger J (2001) Energetics of long internal gravity waves in large lakes. *Limnology and Oceanography* 46: 1760–1773, with permission from the American Society for Limnology and Oceanography.

each cyclonic and anticyclonic wave pair diverge rather than converge. In the limit of $S \rightarrow \infty$, the cyclonic waves transform into longitudinal seiches, whereas the anticyclonic waves become the transverse seiche solution. It is for this reason that wind-forcing in the transverse direction has been observed to more easily generate anticyclonic, Poincaré-type, waves.

Current Structure and Measurement

As we have made the linear wave assumption, the vertical velocities induced by these motions are small; however, the horizontal current structure can show significant complexity both in the horizontal and vertical dimension. The complexity in the horizontal direction is due primarily to the presence of boundaries (and hence the horizontal structure of the waves), whereas the vertical complexity is due to both the stratification and the vertical mode.

It is important to tie the vertical position of the measurement location with the likely motion that dominates the flow. The simplest method to determine the likely points of maximum displacement and maximum current is to solve the long linear internal wave problem in a rotating system

$$\frac{d^2 w_m(z)}{dz^2} + \frac{N^2(z)}{c_m} w_m(z) = 0 \qquad [17]$$

where w_m is the vertical velocity eigenfunction for waves of vertical mode m, z the vertical dimension, $N^2(z)$ is the vertical profile of the square of the buoyancy frequency,

$$N(z)^2 = -\frac{g}{\rho_0}\frac{\partial \rho(z)}{\partial z} \qquad [18]$$

where g is the gravitational acceleration, ρ_0 is a reference density (typically the maximum density), $\rho(z)$ is the vertical profile of density, and c_m is the internal wave phase speed for the particular vertical mode in question. This is an eigenvalue problem, such that an infinite number of solutions exist for an infinite number of vertical modes m. For the case of a constant N^2, the equation has the sinusoidal solution

$$w_m(z) = \sin\left(\frac{m\pi}{H}z\right) \qquad [19]$$

$$u_m(z) = u_m^0 \cos\left(\frac{m\pi}{H}z\right) \qquad [20]$$

$$c_m = \frac{H}{m\pi}N \qquad [21]$$

where u_m is the horizontal velocity induced by the wave and u_m^0 is a constant. It is quite clear from the above that the position of maximum vertical displacement is offset from the position of maximum horizontal current, which the selection of measurement points needs to take into account. Note that the equivalent depth for each vertical mode in a continuously stratified system can be calculated as $H_{em} = c_m/\sqrt{g}$.

For nonconstant $N^2(z)$, eqn. [17] is relatively easily solved numerically as it takes the form of a Sturm–Liouville equation, so from a depth profile of temperature a vertical profile of N^2 can be computed and fed into the eigenvalue solver, which will return both the eigenvalues c_m and the eigenvectors w_m. These eigenvectors will indicate which region of the water column will experience the maximum isotherm oscillations, and the derivate of this eigenvector with respect to z will give the position of the maximum horizontal velocity fluctuations. It is at these locations that thermistors and current meters should be concentrated, respectively. For most vertical modes, concentrating instruments in and around the thermocline is sufficient, except for capturing the velocity signal of the first vertical mode in which current measurements are best made either near the surface or bottom.

A key method to link currents with the wave motion described earlier is the use of rotary spectra of currents. By analyzing the different direction of propagation of the currents at different depths and points in space, it is possible to understand not only the predominant frequency of oscillation but also the predominant direction of rotation. **Figure 6** shows data from Lake Kinneret in which the basin-scale wave field is dominated by a cyclonic vertical mode one Kelvin wave of period ~24 h, an anticyclonic vertical mode one Poincaré wave of period ~12 h, and an anticyclonic vertical mode two Poincaré wave of period ~22 h. Analysis of current data collected at this location indicates that the Kelvin wave effect on currents in the thermocline (where the 24 °C isotherm is located) is weak at this station as the current in the 20–24 h bandwidth is dominated by anticyclonic rotation.

Vorticity Waves

From the above equations, allowing either f to vary as a function of y (the β-plane) or allowing H to vary as a function of x and y allows a similar class of waves to exist. In the ocean, where f does vary, a class of waves called Rossby waves (or planetary Rossby waves) exist because of the conservation of angular momentum.

In inland waters such as lakes, these effects can be ignored as they are generally smaller than 500 km and f can be assumed to be constant. However, variations in water depth H result in a similar type of wave being possible, again due to the conservation of angular momentum. The structure of these motions is typically more complex than planetary Rossby waves as variations in water depth can occur in all directions, whereas variations in f are limited to the north–south

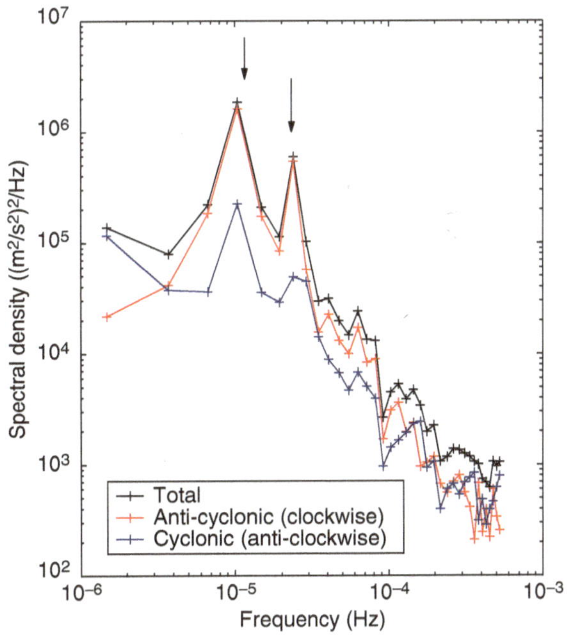

Figure 6 Spectra of currents along the 24C isotherm in Lake Kinneret during summer 1998 at station T3 on the western margin, showing the total spectrum, the component due to anticyclonic motion ('Poincaré waves'), and the component due to cyclonic motion ('Kelvin waves'). The arrows denote periods of 24 and 12 h from left to right. Adapted from Antenucci JP, Imberger J, and Saggio A (2000) Seasonal evolution of the basin-scale internal wave field in a large stratified lake. *Limnology and Oceanography* 45: 1621–1638.

direction. In this section, we use the term 'vorticity waves' to describe these motions, though they are also called 'topographic waves,' 'vortical modes,' 'second class waves,' or 'quasi-geostrophic waves.' These waves have been observed in large lakes such as Lake Ontario, Lake Michigan, Lake Zurich, and Lake Lugano.

The frequency of these motions is always subinertial (i.e., less than the inertial frequency at that latitude), and the frequency depends primarily on the topography of the basin as it is the topography (through the variation in $H(x,y)$) that causes changes in angular momentum. Importantly, the frequency of these motions is not a function of stratification, and so does not vary on a seasonal basis. This simplifies the measurement of these waves as they are existing at the same frequency year-round.

These waves propagate their phase cyclonically (anticlockwise in the Northern Hemisphere); however, the currents measured rotate both cyclonically and anticyclonically dependent on the horizontal structure of the wave. The currents induced by these waves in the bottom layer consist of a barotropic component only, whereas in the surface layer both a barotropic and a baroclinic current component exists, provided the upper layer is relatively thin than the lower layer. The exact structure of these waves is difficult to determine as there are multiple solutions that have similar frequencies, though it is recognized that the fundamental modes are the most likely to be generated.

Practical Guide to Measurement of these Waves

On the basis of the aforementioned disucssion, a step-by-step guide is provided on the method that should be applied when investigating these waves-

Gravity Waves:

1. Compute inertial frequency for the latitude of the lake in question using eqn. [1].
2. Make a simple two-layer approximation to the stratification, and compute the equivalent water depth H_e using eqn. [4].
3. Compute the internal wave speed using $c = \sqrt{gH_e}$. Typically this value will be between 0.1 and 0.3 m s^{-1}.
4. Compute the internal Rossby radius using eqn. [5].
5. Based on the dimension of the lake, compute the Burger number using eqn. [6]. If the lake is approximately circular, use the radius for the length-scale L. If the lake is approximately elliptical, use the major axes half-length for the length-scale L. If the Burger number is greater than 2, rotational effects will be minimal. If the Burger number is less than 1, rotational effects will be very important.
6. From **Figure 4(a)**, read off the nondimensional frequency ω/f for both the fundamental (lowest horizontal mode) cyclonic and anticyclonic wave for the aspect ratio of your lake.
7. Compute the angular frequency ω from the nondimensional frequency ω/f and the inertial frequency f for each of these two fundamental modes.
8. Compute the period T of these two waves from $T = 2\pi/\omega$.
9. Install measuring equipment (thermistor chain and/or current meters) for a sufficient period to measure more than 10 cycles of each wave. For example, if $T = 2$ days, at least 20 days of measurement will be required to achieve significant confidence in the data analysis. The location of these instruments should be carefully selected and is nontrivial. Typically the best location is halfway between the lake center and the lake boundary. Multiple sampling points are generally necessary – if two stations are deployed they should *not* be placed 180° apart as the direction of propagation can not be determined. It is best to orient stations such that they are 45–135° offset.
10. Compute spectra of temperature signals, isotherm depths, or integrated potential energy to determine the dominant frequencies in the field.
11. Compute rotary spectra of currents to determine the dominant rotation direction.
12. Compute phase and coherence between stations to assist in determining rotation direction and spatial structure. This can be done graphically by simply overlaying signals from the two stations or by using spectral analysis techniques.

Vorticity Waves:

1. Compute inertial frequency for the latitude of the lake in question using eqn. [1].
2. Deploy current meters for many periods longer than the inertial period, typically several months of record will be required. Two stations are required at the minimum, as with gravity waves they should *not* be placed 180° apart. Current meters should be placed in both the upper and lower layer, away from the thermocline.
3. Compute spectra of current signals to determine dominant frequencies, whether these frequencies change as a function of stratification, and whether the upper and lower layer current structure differs.

Armed with this information, it should be possible to determine which waves are dominating the temperature and current signals. Going into greater detail would typically require the reader to conduct further reading.

Glossary

Anticyclonic – Rotating in the opposite direction as the earth's rotation (clockwise in the Northern Hemisphere, anticlockwise in the Southern Hemisphere).

Baroclinic – A flow in which lines of constant pressure are not parallel with lines of constant density, in the context of lake dynamics this applies to all motion that is dependent on stratification.

Barotropic – A flow in which lines of constant pressure are parallel with lines of constant density, in the context of lake dynamics this assumption can be made when considering surface motion only.

Cyclonic – Rotating in the same direction as the Earth's rotation (anticlockwise in the Northern Hemisphere, clockwise in the Southern Hemisphere).

Equivalent depth – Depth of homogeneous fluid where the long-wave barotropic phase speed $c = \sqrt{gH_e}$ is equal to the long-wave baroclinic internal wave phase speed. As internal waves travel slower than surface waves, the equivalent depth is always smaller than the actual depth.

Gravity wave – Oscillatory motion where the restoring force is due to gravity.

Rectilinear motion – Movement in a straight line.

Vertical mode – Number of maxima in the vertical velocity structure for a particular wave.

Vorticity – The curl of the velocity field, or circulation ('spin') per unit area about a local vertical axis.

Vorticity wave – Oscillatory motion that results from the conservation of angular momentum in flow over varying topography.

Further Reading

Antenucci JP and Imberger J (2001) Energetics of long internal gravity waves in large lakes. *Limnology and Oceanography* 46: 1760–1773.

Antenucci JP and Imberger J (2003) The seasonal evolution of wind/internal wave resonance in Lake Kinneret. *Limnology and Oceanography* 48: 2055–2061.

Antenucci JP, Imberger J, and Saggio A (2000) Seasonal evolution of the basin-scale internal wave field in a large stratified lake. *Limnology and Oceanography* 45: 1621–1638.

Csanady GT (1967) Large-scale motion in the Great Lakes. *Journal of Geophysical Research* 72: 4151–4162.

Csanady GT (1972) Response of large stratified lakes to wind. *Journal of Physical Oceanography* 2: 3–13.

Gill AE (1982) *Atmosphere-Ocean Dynamics*. San Diego: Academic Press.

Huang JCK and Saylor JH (1982) Vorticity waves in a shallow basin. *Dynamics of Atmospheres and Oceans* 6: 177–186.

Hutter K (ed.) (1984) *Hydrodynamics of Lakes*, CISM Courses and Lectures, vol. 286. Wien: Springer-Verlag.

Mortimer CH (1974) Lake hydrodynamics. *Mitteilungen Internationale Vereinigung für Theoretische und Angewardte Limnologie* 20: 124–197.

Mysak LA, Salvade G, Hutter K, and Scheiwiller T (1985) Topographic waves in a stratified elliptical basin, with application to the Lake of Lugano. *Philosophical Transactions of the Royal Society of London A* 316: 1–55.

Simons TJ (1980) Circulation models of lakes and inland seas. *Canadian Bulletin of Fisheries and Aquatic Sciences* 203: 146.

Stocker R and Imberger J (2003) Energy partitioning and horizontal dispersion in a stratified rotating lake. *Journal of Physical Oceanography* 33: 512–529.

Stocker T and Hutter K (1987) *Topographic Waves in Channels and Lakes on the f-Plane*. Lecture Notes on Coastal and Estuarine Studies, vol. 21, 176 pp. Springer-Verlag.

Thomson SW (Lord Kelvin) (1879) On gravitational oscillations of rotating water. *Proceedings of the Royal Society of Edinburgh* 10: 92–100.

Currents in the Upper Mixed Layer and in Unstratified Water Bodies

F J Rueda and J Vidal, Universidad de Granada, Granada, Spain

Introduction

Water parcels in lakes and reservoirs change their position at a wide range of spatial and temporal scales, describing trajectories that are the result of random and periodic displacements superimposed on more orderly or coherent patterns. It is the coherent and orderly pattern of motions with long-(seasonal) time scales and large-(basin) spatial scales, what we refer to as circulation (or currents). Random motions, in turn, acting at subseasonal and sub-basin scales are responsible for mixing and diffusion processes. Our focus, in this chapter, is on the large-scale circulation in homogeneous or unstratified water bodies (here on HB) and in the upper mixed layer (here on SML) of stratified lakes. The behavior of HB and SML are considered together, since both are homogeneous layers directly forced at the free surface by wind. The exact circulation that develops in response to wind forcing depend on the specific spatial and temporal patterns characterizing the wind field over the lake, together with morphometric characteristics of the layers themselves. In HB, currents interact with the bottom and variations in topography help create gyres and other secondary flows. Under stratified conditions, in turn, currents in the SML are influenced by the time-varying topology of the thermocline (the lower limit of the SML), which is controlled by internal or density driven motions in complex interaction with the wind field and the Earth's rotation (Coriolis forces). Hence, in generating the currents in the SML, both internal wave motions and wind forcing are intricately linked. In this chapter we will focus on the response of homogenous layers to wind forcing, leaving aside any discussion on the time evolution of their lower boundaries, i.e., whether they remain constant (as is the case of the bottom of the lake in HB) or they change with time, as a consequence of internal wave motions, as is the case of the SML. The temporal evolution of the thermocline, or in general, that of isopycnal surfaces is discussed in depth in other chapters.

The goal of this chapter is to introduce concepts and tools that are needed to understand the mechanisms by which circulation in these homogenous layers is generated in response to external forcing, mainly wind. Wind forcing over lakes is episodic in nature, characterized by a sequence of events of varying intensity and duration interspersed with periods of calm. To describe the response of a lake to such forcing, we will assume that, to first order, lakes behave as linear systems. Under that assumption, the state of motion in a given lake and at any given instant can be described as the result of superimposing the responses to all individual wind events that have acted over its free surface in the past (i.e., a convolution exercise). Owing to frictional losses, water bodies have 'limited' memory, and only those wind events in the 'closest' past (within a frictional adjustment time scale) will effectively determine the circulation patterns exhibited at any given point in time. Here, our focus will be on describing the currents that, according to the linear theory, will develop in an initially quiescent lake in response to a suddenly imposed wind.

Our rationale for using the linearized equations of motion as the starting point for the description of circulation (and not the full non-linear Navier-Stokes equations, governing the motion of fluids in nature) is that linear theory can accurately predict the spatial scales of the large-scale motions, along with their build up or decay time. Furthermore, considerable insight can be gained into the mechanisms involved in the generation of currents by using a simplified set of equations. However, the reader should be aware that the description of the dynamics of circulation provided by the linear theory is, at most, approximate and 'other' features appear as a consequence of the nonlinearity of the fluid motion. The nonlinearities make the problem of studying circulation patterns intractable with analytical tools. The analysis of the nonlinear dynamics of motion needs to be approached with sophisticated numerical models that simulate the hydrodynamic behavior of lakes. Much of the research in the last few years in the study of lake circulation has been in this direction. Considerable advances, also, have been done in the identification of circulation patterns in lakes, with the help of new observational technologies not previously existing, as high-resolution remote sensing (**Figure 1**), autonomous satellite-tracked drogues, acoustic Doppler velocimetry, and others. We will review some of the studies conducted in several lakes located throughout the world in which the large-scale circulation has been described and studied.

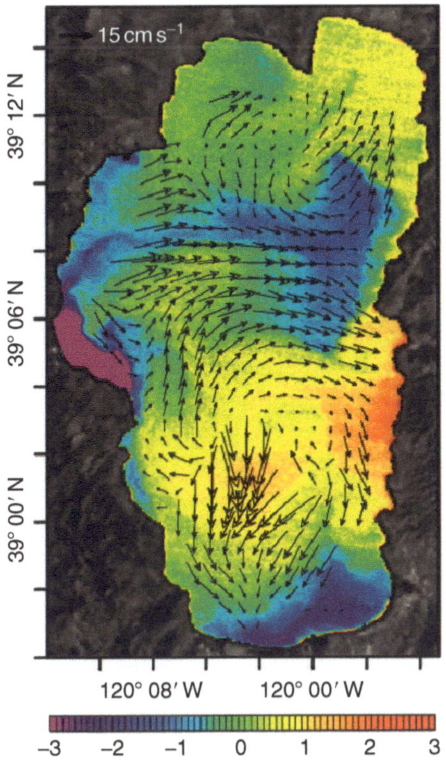

Figure 1 ETM+ Band 6 (high gain) temperature anomaly, June 3, 2001 18:28 UTC, in Lake Tahoe. The ETM+ image was interpolated to a 90 m grid using bilinear interpolation. The satellite image-derived surface current vector field is overlaid on the image. Adapted from Steissberg TE, Hook SJ, and Schladow SG (2005). Measuring surface currents in lakes with high spatial resolution thermal infrared imagery. *Geophysical Research Letters* 32: L11402, doi:10.1029/2005GL022912, 2005. Copyright (2005) American Geophysical Union. Reproduced with permission from the American Geophysical Union.

Fundamental Concepts

Shallow Water Equations

The starting point in our analysis of currents in the SML or HB is the linear set of equations governing the motion for a shallow layer of homogeneous fluid in direct contact with the atmosphere, i.e.,

$$\frac{\partial u}{\partial t} = fv - g\frac{\partial \eta}{\partial x} + \frac{1}{\rho}\frac{\partial \tau_{xz}}{\partial z} \; x\text{—momentum} \quad [1]$$

$$\frac{\partial v}{\partial t} = -fu - g\frac{\partial \eta}{\partial y} + \frac{1}{\rho}\frac{\partial \tau_{yz}}{\partial z} \; y\text{—momentum} \quad [2]$$

$$\frac{\partial(\eta - \eta')}{\partial t} + \frac{\partial}{\partial x}(Hu) + \frac{\partial}{\partial y}(Hv) = 0 \text{ continuity} \quad [3]$$

Here, t is time, u and v are the velocity components along the x- and y- Cartesian directions, g is the acceleration of gravity, f is the Coriolis parameter (or inertial frequency), $H(x, y)$ is the depth of the layer in equlibrium, η and η' denote the vertical displacement of the free surface and the bottom boundary from the equilibrium level, and $h = H + \eta - \eta'$. The symbols τ_{yz} and τ_{xz} represent the x- and y-components of the shear stress applied to a horizontal surface at a vertical location z. The z- coordinate is here considered positive upwards. These symbols will be used consistently throughout this chapter. Equations [1] and [2] are expressions of the second Newton's Law (i.e., force equal mass times acceleration). The first term in their right-hand sides represents the Coriolis force, through which the Earth's rotation influences the dynamics of geophysical flows; the second term accounts for the effects of pressure gradients on fluid motion; and, finally, the third term represents the transfer of horizontal momentum in the vertical z direction by turbulent diffusion. It is presumed, in the derivation of eqns [1]–[3], that the vertical dimension of the layers is much smaller than their horizontal length dimension L, i.e., $H << L$. Consequently the ratio of vertical to horizontal velocities is small (the motions are quasi-horizontal) and the distribution of pressure is hydrostatic in the layer. Furthermore, it will be presumed that $\eta << H$ and $\eta' << H$. Only the components τ_{xz} and τ_{yz} of the stress tensor are considered here, since they are probably the most important in determining the dynamics of the homogenous layers. It will be presumed that they can be modelled using a gradient transport relationship, i.e.,

$$\tau_{xz} = \rho K\frac{\partial u}{\partial z} \quad \text{and} \quad \tau_{yz} = \rho K\frac{\partial v}{\partial z} \quad [4]$$

where ρ is water density and K is an eddy momentum diffusivity of (or kinematic viscosity), which in the most general case is dependent on the vertical z coordinate. Boundary conditions at the free surface and the bottom of the homogenous layers are

$$\tau_{xz}|_{z=0} = \tau_x^w \quad [5]$$

$$\tau_{yz}|_{z=0} = \tau_y^w \quad [6]$$

$$\tau_{xz}|_{z=-h} = \tau_x^b \quad [7]$$

$$\tau_{yz}|_{z=-h} = \tau_y^b \quad [8]$$

where (τ_x^b, τ_y^b) are the x- and y- components of the stresses applied to the bottom surface of the layer $\boldsymbol{\tau}^b$ and (τ_x^w, τ_y^w) represent the x- and y- components of the wind stress $\boldsymbol{\tau}^w$.

Lateral and Vertical Circulations

Consistent with the linear approach and for the clarity of presentation, the circulation is considered as the result of superimposing a depth-integrated (or 'lateral') circulation and a vertically-varying horizontal motion (vertical or overturning circulation). The vertical distribution of horizontal velocities can be estimated by solving the momentum equations (eqns [1] and [2]), once a suitable approximation to a distribution of $K(z)$ has been adopted and using the gradients of the free-surface elevation at any (x, y) location as external inputs. This problem is referred to as the 'local problem'. The remaining problem (referred to as the 'global' problem) is to calculate the basin-wide distribution of pressure or lake level (see wind setup). This, in turn, involves solving the set of equations that result from integrating in depth eqns [1]–[3]. The depth-integrated form of eqs. [1]–[3] are also referred to as the 'transport' equations, and they are posed in terms of the horizontal transports $U = \int_{-H}^{0} u\, dz$ and $V = \int_{-H}^{0} v\, dz$.

Rotational Effects in Shallow Layers

Rotational effects need to be taken into account when analysing the circulation of lakes with horizontal dimensions larger than the Rossby radius of deformation R_0 and when considering processes with time scales which are on the order of f^{-1}or larger. The Rossby radius of deformation is estimated as the ratio between c, a characteristic speed at which information is propagated, and f, i.e., $R_0 = c/f$. For example the speed at which long- surface waves presure perturbations in the free surface of shallow layers travel can be estimated as

$$c = \sqrt{gH} \qquad [9]$$

For typical values of $f = 10^{-4}\ \mathrm{s}^{-1}$ and $H = 10$ m, R_0 for long surface waves is approximately 100 km, larger than the dimensions of most lakes around the world. Hence, the rotational effects can be safely ignored in most lakes when analysing free surface dynamics. Internal waves (which modify η' in eqn [3]) however, have periods which are on the order of the inertial period ($2\pi/f$) and their Rossby radius (referred to as the internal Rossby radius) is of order 1000 m. Moreover, surface currents are typically on the order of 10^{-1} ms^{-1}, and R_0 is also $\approx$ 1000 m). Hence, it is likely that circulation is affected by rotational effects in most lakes. Only in very small or narrow lakes, with widths $b << R_0$ can the influence of the Earth rotation be safely neglected and the motion of water can be described using the same governing eqns. [1]–[3] but setting $f = 0$.

Wind Setup

The response of a homogenous layer, initially at rest (i.e., with its free surface being horizontal), to a suddenly imposed wind of constant speed and direction consists of two parts: (1) a steady response, manifested by the piling up of water against a leeward coast (referred to as wind setup) and, as consequence, by tilted isopycnals and free surfaces, and (2) a transient or oscillatory response (seiching) of those same surfaces, as a consequence of the interplay between inertial and gravity forces. The period of the oscillatory responses will vary from lake to lake according to their geometry, and for stratified lakes, depending upon the vertical density distribution. The seiching motion of isopycnals (internal waves) in stratified waters and will not be discussed here. As for the surface seiches, their frequency σ can be, in most cases estimated using Merian's formula

$$\sigma_n = \frac{n\pi c}{L} \qquad [10]$$

where c is the speed of propagation of long surface waves in a basin of depth equal to the averaged depth of the basin H, L is the length of the basin along the direction of the wind and n is the mode of the oscillation. In lakes of complex geometry, such as those with multiple basins, the frequency of the surface seiches needs to be determined through more complicated procedures, but its magnitude is, in any case, on the same order as given by eq. 10. For example, in a lake with $H \approx 10$ m and $L \approx 10$ km, the frequency σ_1 of the gravest first mode is $O(10^{-3})$. Being their frequency much larger than f, surface seiches are seldom affected by rotational effects. They are subject, though, to large dissipative losses. Velocity observations collected in straights (for example connecting large lakes and embayments or several basins in multibasin lakes) are often contaminated and even dominated by large-amplitude oscillations affecting the whole water column, which are associated to the seiching motion of the free surface. The wind setup is characterized by a constant-slope free surface whose magnitude follows from a simple stress pressure-gradient balance (the two terms in the right hand side in eqns [1] and [2]). For a lake of constant depth H acted upon a uniform wind stress in the y-direction, the free surface displacement can be estimated as

$$\eta = \frac{u_*^2}{gH} y \qquad [11]$$

where $y = 0$ at the center of the lake and u_* is the shear velocity at the surface of the lake. Equation [11] is obtained by presuming that the lake is sufficiently narrow (i.e., $b/R_0 << 1$, where b is the width of the basin and the Rossby radius is evaluated for long surface waves) so that both components of the transport (U, V) vanish everywhere in the basin and Coriolis effects can be effectively neglected. Coriolis forces will only affect the wind setup solution in the limit of $b/R_0 >> 1$.

Lateral Circulation

Generation of Circulation in Homogenous Layers

To characterize the spatial patterns of the horizontal currents use is customarily made of a variable called vorticity ζ defined as the curl (a mathematical operation, denoted by the symbol ∇x, involving spatial derivatives) of the horizontal velocity vector **u**, i.e.,

$$\zeta = \nabla \times \mathbf{u} = \frac{\partial v}{\partial x} - \frac{\partial u}{\partial y} \quad [12]$$

A flow field where a fluid particle describes counterclockwise (or cyclonic) loops will have a positive vorticity; a flow field with negative vorticity, in turn, will make a fluid particle to rotate clockwise (or anticyclonically). The magnitude of ζ is equal to twice the angular velocity of a fluid parcel: the larger ζ becomes the higher will be the speed at which the particle circulates. Here, the equation governing the evolution of the vorticity ζ of the depth-averaged flow is horizontal velocity analyzed to understand the mechanisms generating circulation in HBs and SMLs. Ignoring advection and diffusion of vorticity, the governing equation for ζ in a homogenous layer of thickness $h(x, y, t)$ can be written as

$$\frac{\partial \zeta}{\partial t} = \underbrace{\left(\frac{f+\zeta}{h}\right)\left(\frac{\partial h}{\partial t} + u\frac{\partial h}{\partial x} + v\frac{\partial h}{\partial y}\right)}_{(1)} + \underbrace{\nabla \times \left(\frac{\boldsymbol{\tau}^w}{Ph}\right)}_{(2)} - \underbrace{\nabla \times \left(\frac{\boldsymbol{\tau}^b}{Ph}\right)}_{(3)} \quad [13]$$

The three terms on the right hand side of eq. 13 represent the sources and sinks of vorticity in a closed basin. The generation of cyclonic or anticyclonic vorticity depends on the balance of those terms, which represent:

1. generation of vorticity by temporal changes in the layer thickness, or by flow running into regions with larger or smaller thickness (e.g., a sloping bottom), which results in the stretching and/or squashing of the fluid columns (**Figure 2.1**);
2. flux of vorticity at the top boundary caused by spatial variations of the wind stress $\boldsymbol{\tau}^{\mathrm{w}}$ and/or the bottom bathymetry (**Figure 2.2**);
3. flux of vorticity at the bottom boundary caused by the curl of the ratio of the bottom stress ($\boldsymbol{\tau}^{\mathrm{b}}$) to the depth of the basin.

Term 3 is the only sink of vorticity. All other terms are sources. Term (1) is associated with spatial and/or temporal changes (usually of oscillatory nature) in the thickness h of the homogenous layer. In a stratified water body, it is intricately linked to the existence of internal wave motions. Under the influence of oscillatory currents induced by internal waves, water parcels will describe open trajectories, with a small net displacement after each oscillation. The net basin-wide circulation created by Term (1), hence, is characterized by mean velocities (when averaged over several wave periods) that are much less than the instantaneous values. Such circulation is referred to as *residual circulation*, and it is cyclonic in nature. Residual circulations are defined in contrast with *direct circulations*, which are driven directly by spatially variable winds (Term 2). Water parcels in this type of circulations will move continuously in large gyral patterns, and the mean velocities of water parcels are similar in magnitude and direction to the instantaneous velocities. In contrast with the residual circulations, direct circulation can be either cyclonic or anticyclonic depending on the spatial variations of the wind over the lake surface (see Clear Lake in **Table 2**).

Topographic Gyres in Homogenous Basins

The evolution of ζ in a HB acted upon by a suddenly imposed wind can be to first order, described as a result of the spatial variations of wind and the lake bathymetry (i.e. the second term in the right hand side of eqn [13]. This description is only valid for a period of time after the onset of the wind stress which must be short compared to f^{-1}, so that Coriolis forces (see Fundamental Concepts) and bottom friction can be safely neglected. The extent to which circulation is controlled by topographic effects or the spatial variability of wind can be assessed using simple scaling arguments, by breaking the second source term in eqn [13] as follows

$$\nabla \times \left(\frac{\tau^w}{h}\right) = \frac{1}{h}\nabla \times \tau^w + \tau^w \times \nabla\left(\frac{1}{h}\right) \approx O\left(\frac{\Delta\tau^w}{h}\right) + O\left(\frac{\tau^w}{\Delta h}\right) \quad [14]$$

Here, the symbol Δ applied to a variable refers to changes in that variable's magnitude across the basin. The first term represents the changes in magnitude and/or direction of the wind stress over the lake. The second is the result of wind acting in a direction perpendicular to depth variations, and it is referred to as the topological moment. The spatial variability of wind stress will dominate over the topological moment when

$$\left(\frac{\Delta\tau^w}{\tau^w}\right) > \left(\frac{\Delta H}{H}\right) \quad [15]$$

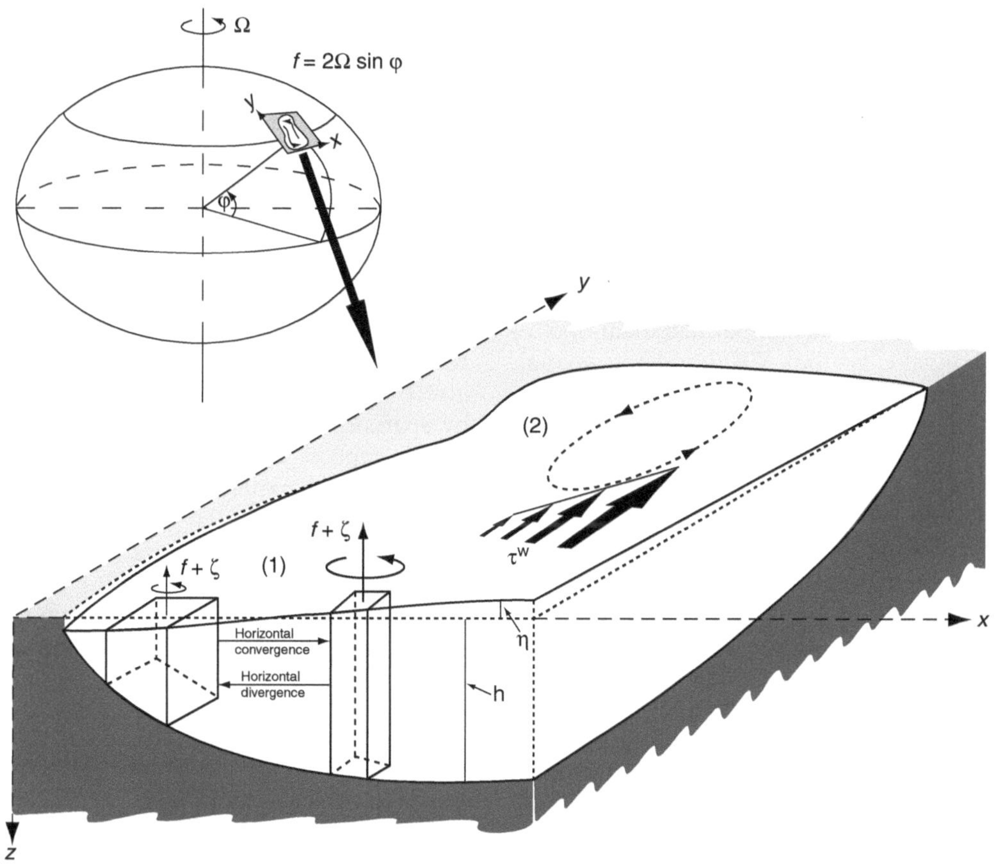

Figure 2 Three dimensional section of a lake. (1) Conservation of volume and circulation of a fluid parcel undergoing squeezing or stretching. Vorticity decreases as a fluid parcel travels to shallower regions, and vice versa. (2) Vorticity induced by the curl of the wind stress τ^W.

In flat basins ($\Delta H \approx 0$) the main source of circulation is the curl of the wind stres. In lakes of more realistic varying bathymetry, though, both wind stress and depth variations control the lake circulation. For the case of nearly uniform winds acting along the main axis of a narrow and elongated lake with sloping sides, the only source of circulation is the topological moment and the resulting circulation is characterized by in a double gyre pattern: to the right of the axis in the direction of the wind the curl of τ^w/h is positive and cyclonic vorticity is produced, whereas to the left anti-cyclonic vorticity is produced (**Figure 3**). Bottom friction (not included, so far in our analysis of circulation) tends to cancel the vorticity input by the wind, but only after the topographic gyres have become established.

Topographic Waves in Homogenous Bodies

As a consequence of the Coriolis forces acting on an initial double-gyre pattern, this will tend to rotate anticlockwise (see **Figure 4**). This mode of motion is referred to as *topographic or vorticity wave*. It is free, since it does not depend on the existence of external forcing (i.e., it only requires that the initial double gyre has been previously established), and has a characteristic frequency σ which scales as

$$\sigma \approx \lambda flk \qquad [16]$$

Here λ is a constant of order unity, l is a length scale characterizing the size of the sloping sides in a lake, and k is the wave-number of the wave, which for the gravest first mode, is given by $k = 2\pi/P$ (P being the perimeter). For typical values $f = 10^{-4}$ s^{-1}, $l = 10^4$ m and $k = 10^{-5}$ m^{-1}, σ is of the order 10^{-5} s^{-1}, hence, $\sigma << f$.

Even though the possiblity of vorticity waves occurring in lakes has been recognized in classical hydrodynamics, evidence of their existence in real lakes, revealed by the cyclonic rotation of the velocity vector at a frequency given by eqn [16], is sparse and weak (see **Figure 4**). This is, in part, due to the fact that winds are almost never absent and will, most likely, interact with existing topographic waves. Moderate and strong winds with enough impulse can change existing circulation patterns, and establish a new basin-scale topographic gyre. Furthermore,

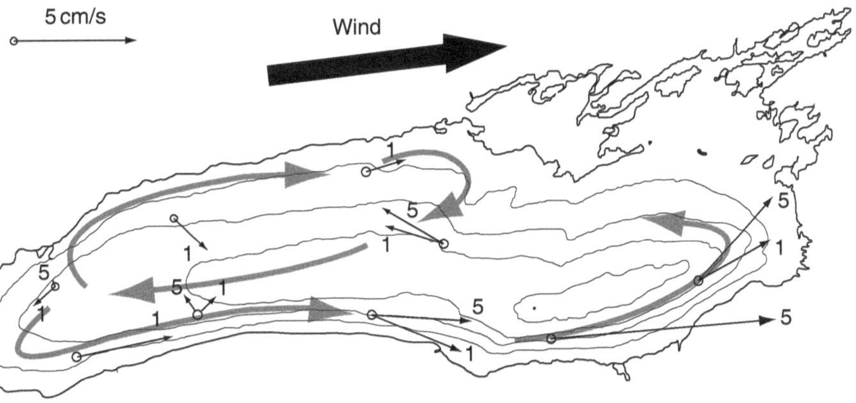

Figure 3 Large-scale winter circulation in Lake Ontario, reconstructed from observations collected ad different stations (shown as black thin arrows). The numbers in the figure indicate the depth at which measurements were taken: 1 = 15 m and 5 = 75 m. The size of the arrows shows the magnitude of the currents. The gray thick and curved arrows represent the circulation interpreted from the data. Black thick arrow indicates the predominant wind direction. From Beletsky D, Saylor JH, and Schwab DJ (1999) Mean Circulation in the Great Lakes. *Journal of Great Lakes Research* 25: 78–93. Reproduced with permission from the International Association for Great Lakes Research.

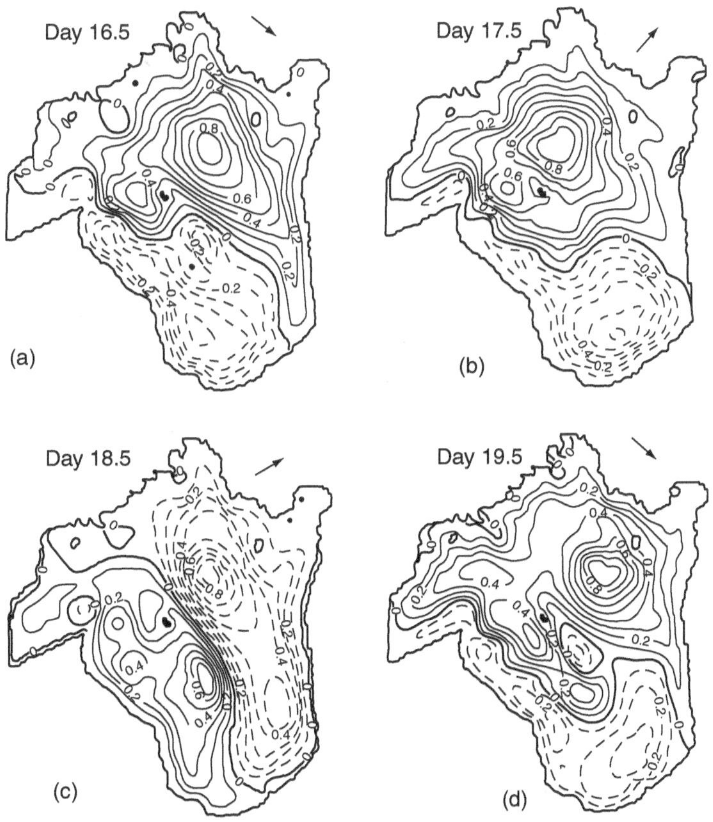

Figure 4 Normalized stream function calculated with a 2D-model of lake circulation in the Gulf of Riga, at (a) 16.5, (b) 17.5, (c) 18.5 and (d) 19.5 days after the start of simulations. Currents will be parallel to lines of equal stream function. Continuous lines show the cyclonic gyre while dashed lines show the anticyclonic gyre of the double-gyre. Bold arrows show the direction of wind. Adapted from Raudsepp U, Beletsky D, and Schwab DJ (2003) Basin-scale topographic waves in the Gulf of Riga. *Journal of Physical Oceanography* 33: 1129–1140. Copyright 2003 American Meteorological Society. Reproduced with permission from the American Meteorological Society.

time-varying wind directions can also affect circulation patterns: cyclonically rotating winds have been shown in numerical experiments to reinforce the basin-scale topographic waves, while anticyclonically rotating winds tend to destroy the wave. In shallow areas, where the direct wind forcing and bottom friction dominate the vorticity balance, the topographic waves are not likely observed. Hence, only in the deepest points, evidence of topographic waves can be found.

Spatial Variability of Wind Forcing

The specific wind-driven circulation patterns that develop in the SML or HB are tightly linked to the spatial and temporal variations of the wind stress over the lake (see eqn [13]). While the time variability of wind stress at a single point in space can be characterized with high-resolution wind sensors (e.g., sonic anemometers), characterizing its spatial variability, though, has proved to be a difficult task. Considerable effort in the recent years has been devoted to characterize the spatial variability of the wind stress field over lakes. This is done either by applying dynamic models of atmospheric circulation or by measuring wind speed and direction in arrays of wind sensors located on and around the lake (see, for example, **Figure 5** in Lake Kinneret). Bulk aerodynamic formulations are typically used to derive wind stress values from the wind speed. Technologies developed to characterize wind stress fields directly (scatterometry) have, so far, only applied to oceanic scales, given the very low resolution of existing sensors.

Studies conducted to characterize wind fields over lakes ranging in size from small to large demonstrate that a considerable degree of spatial variability exists both on synoptic and local scales. *Synoptic* scale variability of the wind field will only affect lakes of large dimensions (e.g., Great Lakes). On a *local* scale, factors such as spatial variations in the land surface thermal and/or moisture properties, surface roughness, or the topography can modify and even generate flows in the atmospheric boundary-layer. All of them are, most probably, at play over or in the immediate vicinity of all lakes. The most significant effect of the topography is the aerodynamic modification of ambient synoptic winds. The topographic features, existing around lakes will, among some other effects, cause the ambient wind to change direction (deflection effect) and will create areas of momentum

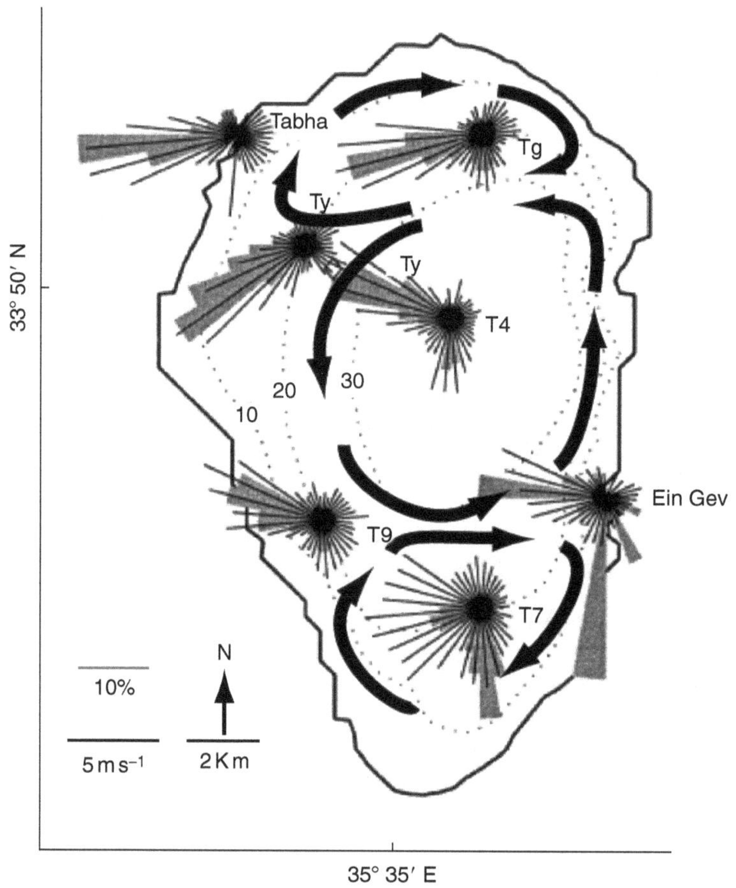

Figure 5 Lake Kinneret, Israel with 10-, 20-, and 30-m depth contours; wind measurements over the lake and on the shore line. Average wind speed and frequency of occurrence in 108 direction bins, from which the wind is coming (meteorological convention), during days 170–183 are plotted as black lines and grey bars, respectively. Adapted from Laval B, Imberger J, and Hodges BR (2003) Modelling circulation in lakes: Spatial and temporal variations. *Limnology and Oceanography* 48(3): 983–994. Copyright 2003 by the American Society of Limnology and Oceanography, Inc. Reproduced with permission from the American Society of Limnology and Oceanography.

deficit or wakes (sheltering effect). These effects are particularly relevant for lake applications, as they induce spatially variable wind fields over the lower levels in the regional topography where lakes are usually located.

Other Sources of Circulation

Shoreline irregularities (headlands, peninsulas, islands, bays, etc) or inflow/outflow features (i.e., open boundaries) can also act as sources of vorticity in the SML and HB (see **Figure 6**). The extent to which shoreline irregularities or inflow/outflow features can create vortex structures will depend on geometrical and hydraulic characteristics of the flow such as

- the length scale L of the features; in the case of the peninsulas this length scale, compared with the width of the lake determines the extent to which the flow is blocked and the peninsula effectively seperates the basin in sub-basins;
- their sharpness (i.e., whether the changes in shoreline direction are abrupt or smooth), which can by characterized be a radius of curvature R_w;
- the inflow velocity or, more generally, the spatial gradients of horizontal velocity induced by shoreline features:
- the rate at which momentum is transferred horizontally; and
- the Coriolis force, which represents the rate at which flow tend to veer in response to the Earth's rotation, which is characterized by the inertial frequency f.

For smooth features and small velocities gradients, currents will be diverted but they will tend to follow the shoreline, and no gyral patterns will develop.

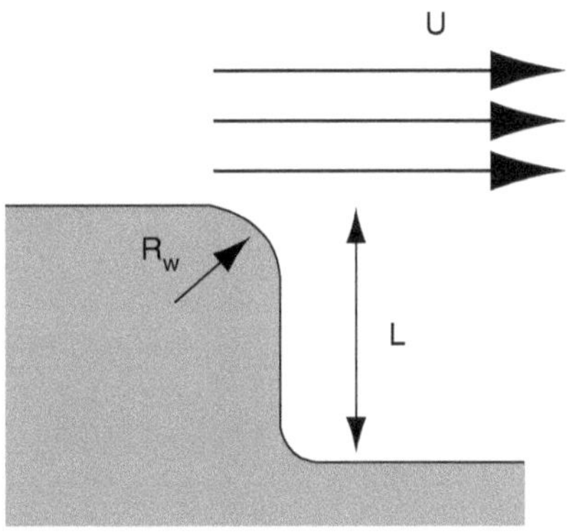

Figure 6 Schematic representation of a shoreline feature where velocity gradients form. The magnitude of the undisturbed velocity is U; L represents a characteristic length scale for the feature and Rw characterizes its sharpness. This scheme could represent a portion of a river entering a lake, part of a peninsula, or a section of a bay.

For sharp features and large velocity gradients, gyral patterns will, most likely, develop. Dimensional analysis and laboratory experiments suggest that gyral patterns will develop near inflows or shoreline irregularities if the sharpness of the exit corner is larger than the inertial radius, i.e.,

$$\frac{u}{R_w f} >> 1 \qquad [17]$$

The dividing line between gyre formation and shoreline attachment is not well established and may depend on details such as the actual current profile across in the main body or in the inflowing water. In any case, the behavior of currents near shoreline irregularities is controlled by the nonlinear nature of water motions.

The horizontal extent of the gyres generated near shoreline irregularities will most likely scale with the size of the irregularities, and in most cases, these gyres will have a only be local (not basin) scale (see **Figure 7**). In the case of peninsulas or islands that are large compared to the total width of the lake, the large-scale circulation can be modified to the extent that the basins on each side of the obstruction respond to wind as if they, effectively, were independent, with gyral structures similar to those described above for individual lakes. Vortices associated to river jets will only drive basin-scale gyral patterns for very large inflows. For the Upper Lake Constance, for example, it has been estimated that the observed basin-scale cyclonic gyres (12 000 m wide) can only be driven by extreme inflow events of about 1500 $m^3\ s^{-1}$ from the Alpine Rhine.

Vertical Circulation

The wind stress applied to the lake surface is directly transmitted through internal friction to a surface shear layer which, in general, is much thinner than the water column. In stratified water bodies, the thickness of that shear layer is determined by stratification: the vertical flux of momentum beyond the bottom of the SML, by viscous or turbulent processes, is considerably reduced by the existence of large density gradients. In homogeneous water bodies, those shear layers are referred to as Ekman layers and their thickness D is calculated as $(2K/f)^{1/2}$. For turbulent flows, D is empirically found to scale as

$$D \propto u_* f^{-1} \qquad [18]$$

where u^* is the friction velocity at the free surface (defined as the square root of the result of dividing wind stress by water density), and the proportionality factor ranges from 0.1 to 0.4. For typical values of

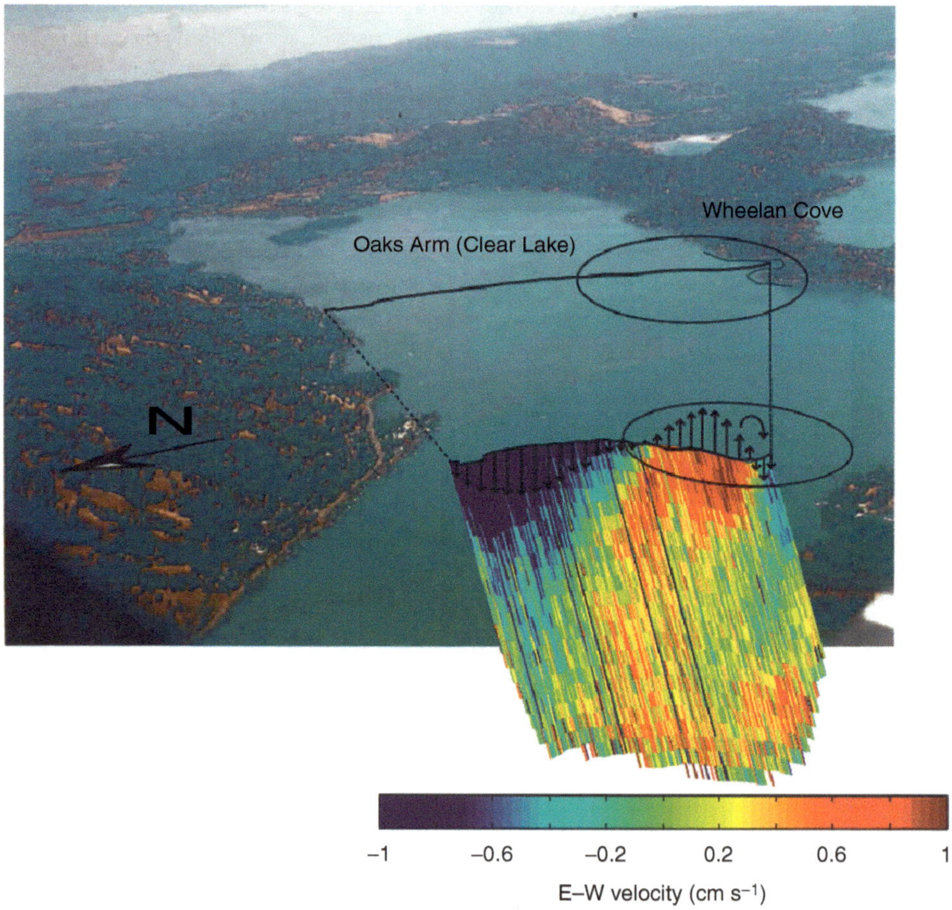

Figure 7 ADCP velocity measurements taken in the Oaks Arm of Clear Lake, California in the course of a field experiment (Data taken and provided by S.G. Monismith, Stanford University). The ADCP was mounted upside-down on a boat and operated on bottom-tracking mode. The depth of the water column was 10 m. The transect starts on the N-end of the basin and ends in Wheelan Cove at its southern shore. The East–West (EW) velocity component of the velocity field is shown as a cross-section in the lower part of the figure. Color marks the magnitude of the EW velocity: blue color represents velocities towards the West; red color represents velocities directed towards the East. Black arrows represent superficial velocity. Note that at the southern end, within Wheelan cove the measurements reveal the existence of a vortex structure, which forms as a result of the shoreline irregularity. The photograph was provided by S.G. Schladow, University of California – Davis.

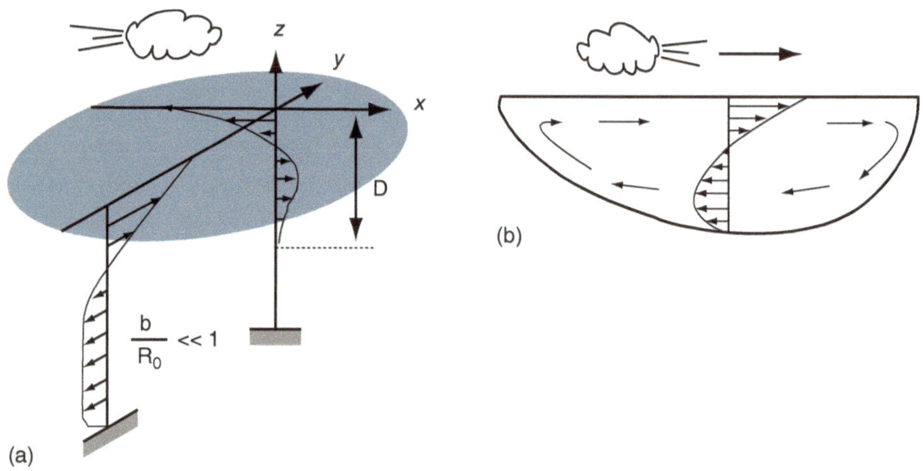

Figure 8 (a) Interior velocities caused by wind acting in the negative *x*-direction in a deep basin of limited horizontal extent. Symbols defined in the text. (b) Vertical circulation in a small narrow lake.

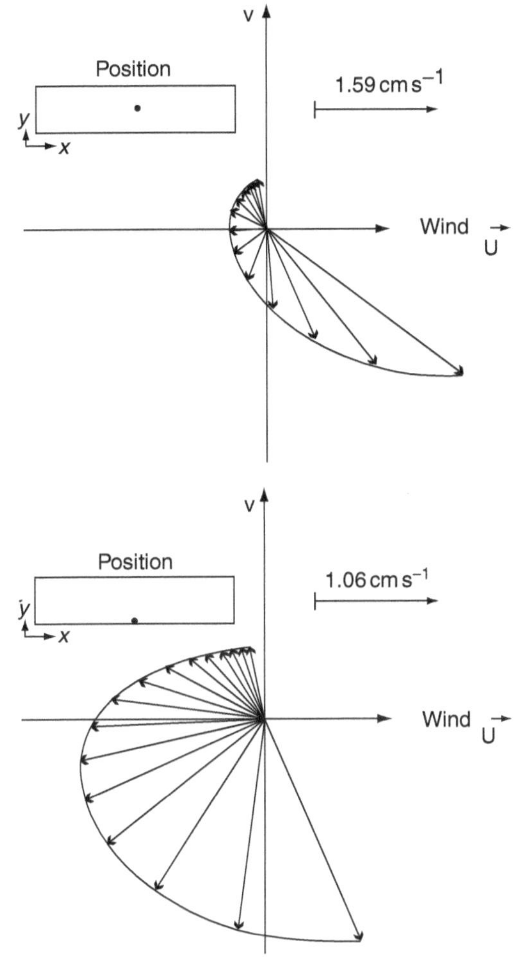

Figure 9 Simulated steady state velocity profiles at two points in a 100 m deep rectangular flat bottom basin, with sides of length 65 km (in the x-direction) and 17 km (in the y-direction). The arrows correspond to velocity vectors and are shown at 5-m depth intervals from the free surface to the bottom. These are results of simulations conducted with three-dimensional hydrodynamic model, where $K(z)$ is set constant and equal to 0.02 $m^2\ s^{-1}$. The top frame corresponds to a point located near the center of the basin while the lower frame is for a point located near the southern end. Adapted from Hutter K, Bauer G, Wang Y, and Guting P (1998) Forced motion response in enclosed lakes. In Imberger J (ed.) *Physical Processes in Lakes and Oceans*. American Geophysical Union, Washington D.C. pp 137–166. Copyright (1998) American Geophysical Union. Reproduced with permission of American Geophysical Union.

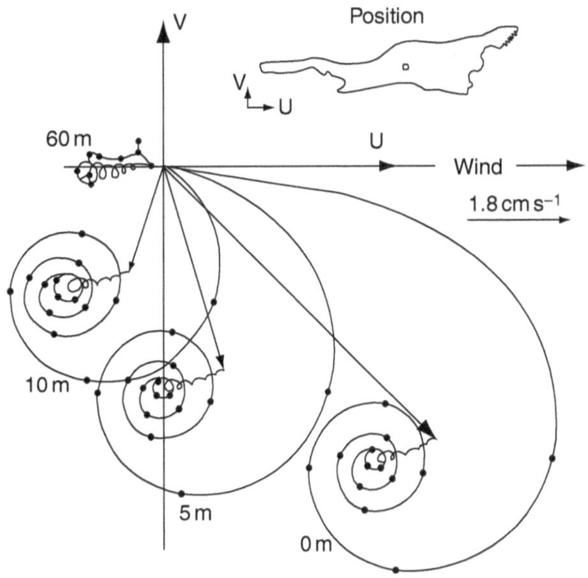

Figure 10 Hodographs, i.e., time series of the horizontal velocities in the middle of the Obersee in Lake Constance, at 0, 5, and 10 m depths. The motion is set up from rest. The small circles mark time intervals of approximately four hours. Adapted from Wang Y, Hutter K, and Bäuerle E (2001) Barotropic response in a lake to wind-forcing. *Annales Geophysicae* 19: 367–388. Reproduced with permission from European Geophysical Society.

$u_* = 10^{-2}\ \mathrm{ms}^{-1}$ in mid-latitude lakes ($f = 10^{-4}\ \mathrm{s}^{-1}$) the Ekman boundary layer depth is of the order 10 m. Within the Ekman layer, the stresses reduce to zero, and both velocity magnitude and direction undergo significant changes.

The simplest possible model of velocity change with depth for lakes, is that of the steady-state local solution of eqns [1]–[3] (with K constant), for a flat bottom deep basin of limited horizontal dimensions acted upon by a constant and uniform wind stress in the y-direction, is (see **Figure 8(a)**)

$$u = -\frac{u_*}{fh} + \frac{u_*}{fD} e^{z/D}\left[\cos\frac{z}{D} - \sin\frac{z}{D}\right] \qquad [19]$$

$$v = \frac{u_*}{fD} e^{z/D}\left[\cos\frac{z}{D} + \sin\frac{z}{D}\right] \qquad [20]$$

For z >> D, i.e., outside the Ekman layer, the velocity is in geostrophic balance with the pressure gradient that develops in response to wind forcing (or wind setup, eqn [11]). Closer to the surface, the velocity is a sum of the geostrophic velocity and the Ekman layer velocity (the exponentially decaying terms in eqns [19] and [20]). In the derivation of eqns [19] and [20], bottom stresses have been ignored. For deep basins, this is a reasonable assumption since the velocity near the bottom from eqn [19] and [20] is, in this case, negligible u_*/fh.

In basins of intermediate depth, the velocity profiles will have both a surface (driven by wind stress) and bottom (driven by bottom stress or currents) Ekman layers. In both layers, the horizontal velocity vectors would tend to rotate clockwise (in the northern hemisphere) as one moves out of the boundary. The rate at which velocity vectors rotate with vertical

Table 1 Circulation patterns of HB in natural systems (lakes and reservoirs): observations and models

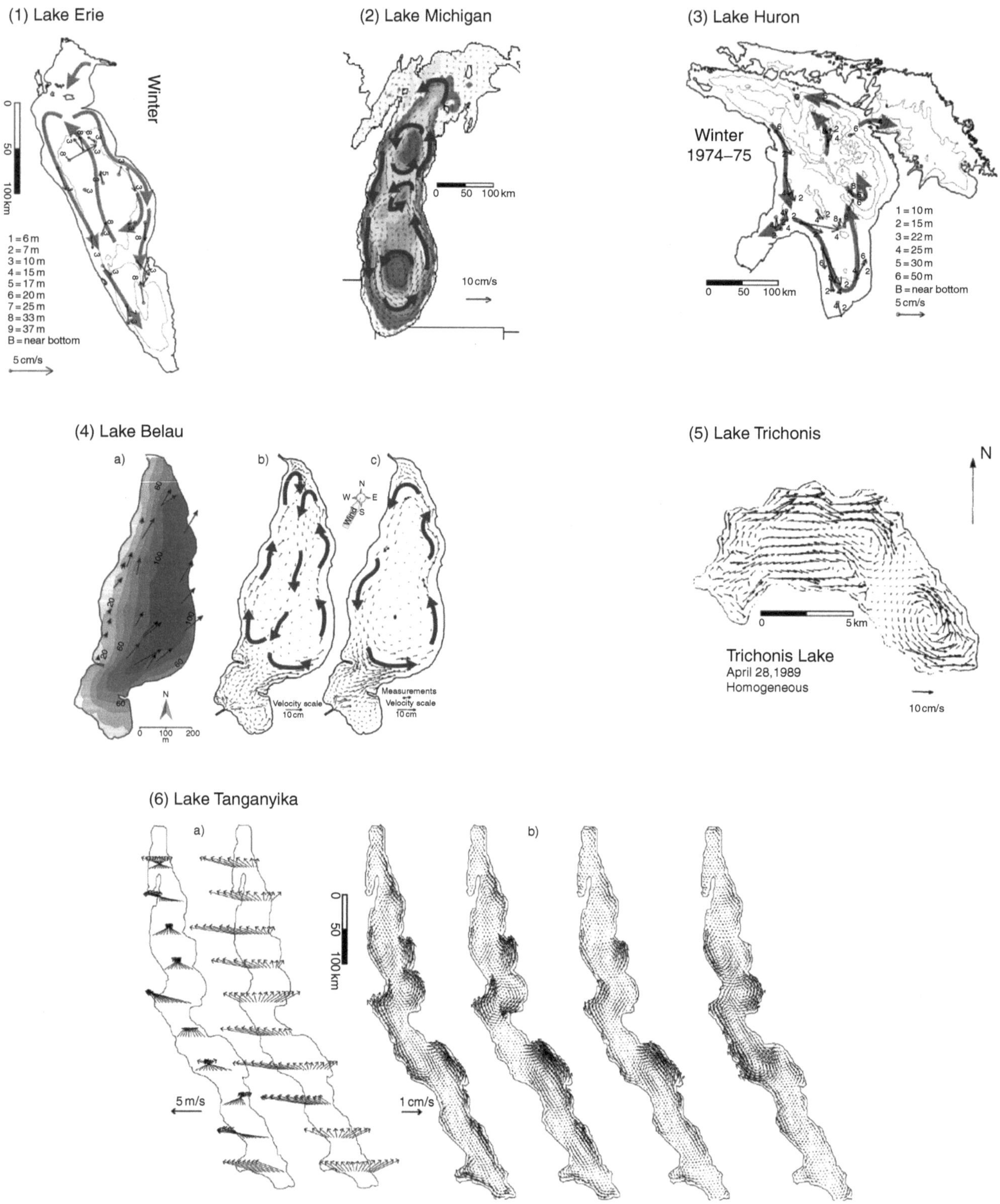

distance away from the boundaries (and, in general, the velocity profiles) is very sensitive to the vertical distribution of the eddy viscosity *K*. In deriving eqns [19] and [20], *K* has been presumed constant, for simplicity. However, this assumption is not strictly correct since *K* is known to vary rapidly in the vertical direction within 'wall layers' existing near the top and bottom boundaries and where the velocity varies logarithmically. Only outside these wall layers, but within an homogenous turbulent shear flow region, such as the top and bottom mixed shear layers, the eddy viscosity is approximately constant and proportional to the velocity and the virtual extent of the flow. In any case, the velocity profiles are very sensitive to the magnitude of *K* and the location (near or far from lateral boundaries) where they are monitored (**Figure 9**). In the limit of very shallow layers ($U^*/fh >> 1$) or, equivalently, for very large Ekman numbers ($K f^{-1} h^{-2} >> 1$) it can be assumed that the effects of Coriolis forces in the vertical distribution of velocities are negligible. Large Ekman numbers occur also in narrow or small lakes. In those cases, the balance between pressure and frictional forces (the last two terms in eqns. [1] and [2]) results in a velocity profile in the direction of the wind (no currents perpendicular to the wind) of the form shown in **Figure 8(b)**. For constant *K* the following analytical expression can be derived that represents the vertical distribution of flow velocities

$$v(z) = \frac{u_*^2}{4Kh} z(3z - 2h) \qquad [21]$$

Note that at the bottom (here $z = 0$) the velocity is zero. Observed velocity profiles in laboratory experiments are, however, closer to a double-logarithmic curve, which corresponds to a parabolic distribution of the eddy viscosity *K*.

In response to a suddenly imposed wind stress on its free surface, the vertical velocity distribution in a homogenous layers of thickness $h >> D$, consists of a steady-state Ekman spiral and inertial oscillations (of frequency *f*) that penetrate to gradually increasing depths. At time *t* inertial oscillations exists up to a depth of order $(2Kt)^{1/2}$, which is the depth to which momentum diffusion penetrates in a nonrotating system. However, the magnitude of the inertial oscillations will decrease with time as

$$\frac{u_*^2}{f\sqrt{2Kt}} \qquad [22]$$

Thus, after a long period, the inertial oscillations are imperceptibles, even if we do not consider frictional losses at the bottom (see **Figure 10**).

Circulation Patterns in Natural Systems (Lakes and Reservoirs)

Large-scale circulation patterns for a suite of lakes of different sizes and geometries are illustrated in **Tables 1** (for HB) and **2** (for SML). The patterns were obtained from observations using drogues, velocimetry or other measurements, numerical model simulations, or both.

(1) Double-gyre wind-driven circulation in Lake Erie. Figure shows the main winter circulation from observed patterns.
- Source: Beletsky D, Saylor JH, and Schwab DJ (1999) Mean circulation in the Great Lakes. *Journal of Great Lakes Research* 25(1), 78–93. Reproduced with permission from the International Association for Great Lakes Research.

(2) Cyclonic Gyres in Lake Michigan. Figure shows the winter averaged currents from numerical model simulations. Colored shading represents stream function values; yellow is positive (generally anticyclonic vorticity) and the green-blue-purple areas are negative stream function (generally cyclonic vorticity).
- Source: Schwab DJ, and Beletsky D, Relative effects of wind stress curl, topography, and stratification on large-scale circulation in Lake Michigan. *Journal of Geophysical research* 108(C2): 3044, doi:10.1029/2001JC001066, 2003. Copyright (2003) American Geophysical Union. Reproduced with permission from the American Geophysical Union.

(3) Circulation in Lake Huron. Figure shows the main winter circulation from observed patterns.
- Source: Beletsky D, Saylor JH, and Schwab DJ (1999) Mean circulation in the Great Lakes. *Journal of Great Lakes Research* 25(1): 78–93. Reproduced with permission from the International Association for Great Lakes Research.

(4) Cyclonic circulation induced by wind sheltering in Lake Belau. Figure (a) shows the sheltering effect of the wind in the lake. Arrows indicate the measured wind and direction. (b) Double-gyre wind-driven circulation obtained by numerical model using an uniform spatially wind field. (c) Cyclonic circulation obtained with the model and observations using the measured, spatially variable, wind of (a).
- Source: Reprinted from Podsetchine V and Schernewski G (1999) The influence of spatial wind inhomogeneity on flow patterns in a small lake. *Water Research* 33(15): 3348–2256. Copyright (1999). Reproduced with permission from Elsevier.

(5) Gyres in the wind-driven circulation in Lake Trichonis. Figure shows the simulated water circulation.
- Source: Reproduced from Zacharias I and Ferentinos G (1997) A numerical model for the winter circulation in Lake Trichonis, Greece. *Environmental Modelling & software* 12(4): 311–321. Copyright (1997). Reproduced with permission from Elsevier.

(6) Gyres induced by diurnally varying winds in Lake Tanganyika. Figure (a) shows the simulated near-surface winds over the lake at midnight (left) and midday (right). Figure (b) shows the depth-averaged flow field, from left to right, at 00:00, 06:00, 12:00, and 18:00, for an average July day.
- Source: Podsetchine V, Huttula T, and Savijärvi H (1999) A three dimensional-circulation model of Lake Tanganyica. *Hydrobiologia* 407: 25–35. Copyright (1999). Reproduced with kind permission of Springer Science and Business Media.

Table 2 Circulation patterns of the SML in natural systems (lakes and reservoirs): observations and models

(1) Lake Constance

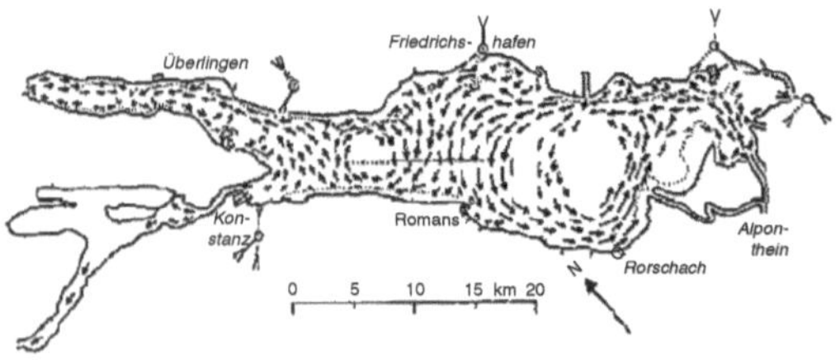

(2) Lake Kinneret

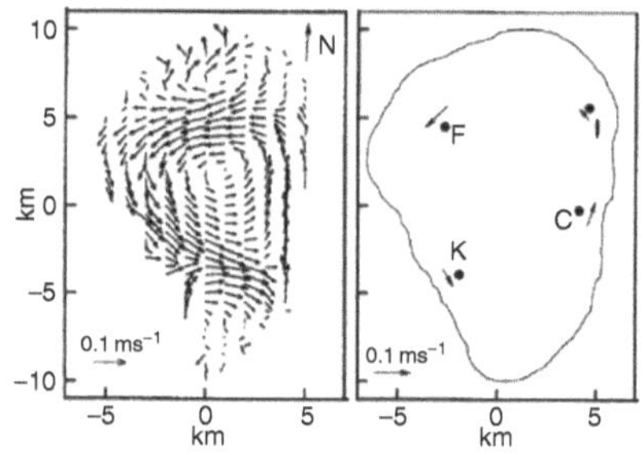

(3) Clear Lake

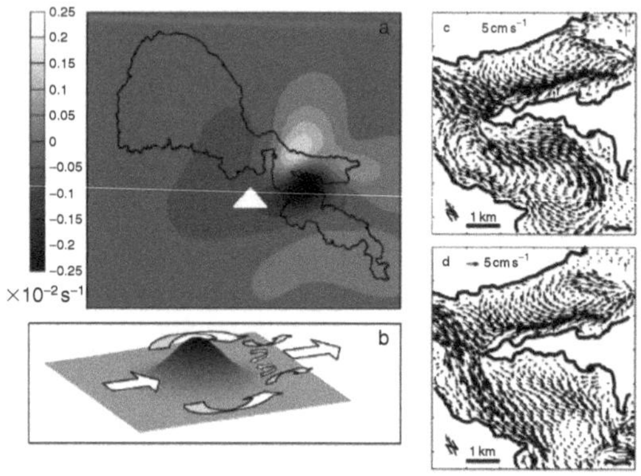

(4) Lake Geneva

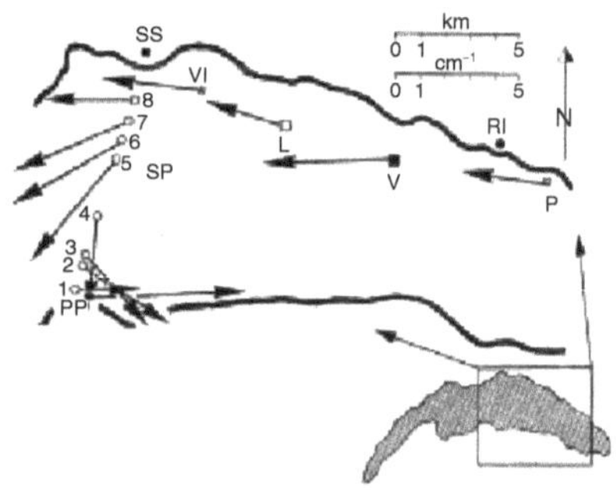

(5) Lake Biwa

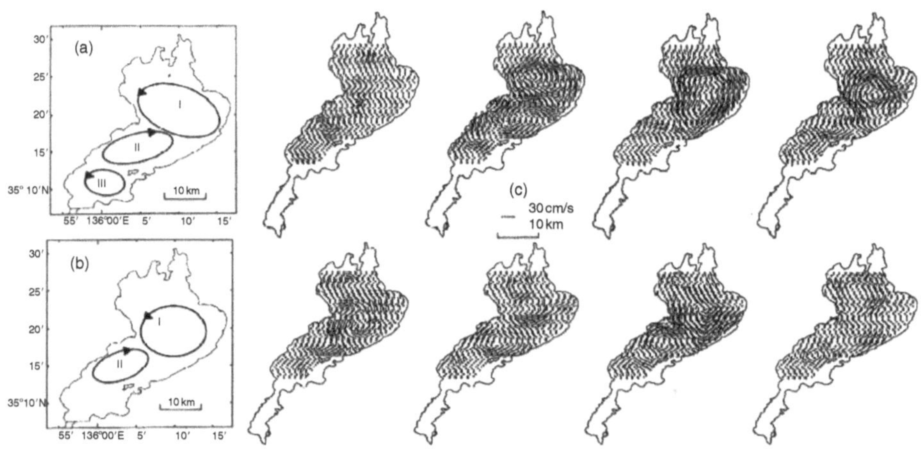

(1) Gyres induced by wind and inflow-driven currents in Upper Lake Constance. Figure shows a qualitative map of typical surface currents. The inflow may induce synoptic eddies; however only extreme discharges could induce the largest eddies of about 12 Km diameter.

- Source: Hollan E. Large inflow-driven vortices in Lake Constance. In Imberger J (ed.) *Physical Processes in Lakes and Oceans*, pp 123–136. AGU, 1998. Copyright (1998) American Geophysical Union. Reproduced with permission of the American Geophysical Union.

(2) Gyres in Lake Kinneret. Figure shows daily mean currents simulated with a numerical model (left panel) and measured (right panel).

- Source: Pan H, Avissar R, and Haidvogel DB (2002) Summer circulation and temperature structure of Lake Kinneret. *Journal of Physical Oceanography*, 32: 295–313. Reproduced with permission from the American Meteorological Society.

(3) Circulation induced by windfield vorticity in Clear Lake. Figure shows windfield vorticity generated by Mount Konocti (white triangle) in (a) and (b); and, model simulation of velocity field 2.5 m below free surface, with (c) spatially variable wind and (d) with a uniform wind field.

- Source: Rueda FJ, Schladow SG, Monismith SG, and Stacey MT (2005) On the effects of topography on wind and generation of currents in a large multibasin lake. *Hydrobiologia* 532: 139–151. Copyright (2005). Reproduced with kind permission from Springer Science and Business Media.

Conclusions and Future Research Needs

Currents in homogenous bodies and the upper mixed layers of stratified lakes are mainly the result of wind forcing applied at the free surface. The specific basin-scale patterns of motion that develop in response to wind depend on the specific spatial and temporal patterns characterizing the wind field (the frequency, intensity, duration, and location at the time of the events), interacting with the topography, internal waves, and Coriolis effects. By using a simplified and linear set of governing equations and boundary conditions, we have analyzed the mechanisms by which some specific patterns of motion, observed in the field, are generated. However, the governing equations of motion are not linear and the description presented here is, at most, approximate. In shallow lakes, stratification is weak and one cannot easily determine where the bottom surface of the upper mixed layer is. Even if one could decide where that surface is, upwelling events frequently occur in these lakes, when bottom water surfaces, and the upper mixed layer disappears from the upwind end of the lake. These are cases when the nonlinear effects become important. In those cases, an appropriate description of currents involves using the fully nonlinear governing equations, which can only be solved with numerical methods. Even if the response of lakes were linear (or one had a 'perfect' numerical model), one major difficulty that needs to be faced in interpreting and analyzing quantitatively circulation and currents in HB and SML is to characterize the wind stress field (i.e., define realistic boundary conditions). Partly, because of this difficulty, there continues to be a considerable amount of uncertainty on what is the relative importance of wind stress curl, topography, and internal waves in generating the large-scale circulation in lakes. Another difficulty that we also face in analyzing the velocity field in lakes is the fact that any description (or observation) of such fields is, at most, incomplete. Many of the techniques used are based on the study of discrete points (Eulerian current meters, Lagrangian drifters, etc.), at single locations or at a series of locations along a transect.

One of the most promising techniques in describing and analyzing the currents in lakes consists in integrating the numerical solution of nonlinear equations and field observations (data-assimilation) in creating a coherent description of circulation. These techniques are now used in the ocean but very few studies have applied this technique to lakes.

See also: The Benthic Boundary Layer (in Rivers, Lakes, and Reservoirs); Currents in Stratified Water Bodies 1: Density-Driven Flows; Currents in Stratified Water Bodies 2: Internal Waves; Currents in Stratified Water Bodies 3: Effects of Rotation; Hydrodynamical Modeling; Small-Scale Turbulence and Mixing: Energy Fluxes in Stratified Lakes; The Surface Mixed Layer in Lakes and Reservoirs.

Further Reading

Beletsky D, Saylor JH, and Schwab DJ (1999) Mean circulation in the Great Lakes. *Journal of Great Lakes Research* 25(1): 78–93.

Bormans M and Garret C (1989) A simple criterion for gyre formation by the surface outflow from a straight, with application to the Alboran Sea. *Journal of Geophysical Research* 94(C9): 12637–12644.

Csanady GT (1982) *Circulation in the Coastal Ocean*. Dordrecht, Holland: D. Reidel Publishing Company.

Chubarenko BV, Wang Y, Chubarenko IP, and Hutter K (2000) Barotropic wind-driven circulation patterns in a closed rectangular basin of variable depth influenced by a peninsula or an island. *Annales Geophysicae* 18: 706–727.

Hunter JR and Hearn CJ (1987) Lateral and vertical variations in the wind-driven circulation in long, shallow lakes. *Journal of Geophysical Research* 92(C12): 13106–13114.

Laval B, Imberger J, and Hodges BR (2003) Modelling circulation in lakes: Spatial and temporal variations. *Limnology and Oceanography* 48(3): 983–994.

Lemmin U and D'Adamo N (1996) Summertime winds and direct cyclonic circulation: observations from Lake Geneva. *Annales Geophysicae* 14: 1207–1220.

Mathieu PP, Deleersnijder E, Cushman-Roisin B, Beckers JM, and Bolding K (2002) The role of topography in small well-mixed bays, with application to the lagoon of Mururoa. *Continental Shelf Research* 22: 1379–1395.

(4) Cyclonic circulation in Lake Geneva. Figure shows the observed mean current vectors recorded at around 10 m depth in the central basin of Lake Geneva.

- Source: Lemmin U and D'Adamo N (1996) Summertime winds and direct cyclonic circulation: Observations from Lake Geneva. Ann. Geophysicae, 14: 1207–1220. Reproduced with permission from European Geophysical Society

(5) Spatio-Temporal evolution of Gyres in Lake Biwa. Figures (a) and (b) shows the described gyres I, II, and III at different stages; while figure (c) from left to right and up to down shows the monthly evolution (April–Dec 1994) of the epilimniun velocities measured with ADCP transects. Gyre I was described to move south as temperature stratification became stronger and north again as the stratification weakened. Gyres II and III were not stable, varying position and size accordingly to the wind field.

- Source: Kumagai M, Asada Y, and Nakano S (1998) Gyres Measured by ADCP in Lake Biwa. In Imberger J (ed.) *Physical processes in lakes and oceans*, pp. 123–136. Washington, DC: AGU. Copyright (1998) American Geophysical Union. Reproduced with permission from the American Geophysical Union.

Raudsepp U, Beletsky D, and Schwab DJ (2003) Basin-scale topographic waves in the Gulf of Riga. *Journal of Physical Oceanography* 33: 1129–1140.

Rubbert S and Köngeter J (2005) Measurements and three-dimensional simulations of flow in a shallow reservoir subject to small-scale wind field inhomogeneities induced by sheltering. *Aquatic Sciences* 67: 104–121.

Rueda F, Schladow GS, Monismith SG, and Stacey MT (2005) On the effects of topography on wind and the generation of currents in a large multi-basin lake. *Hydrobiologia* 532: 139–151.

Schwab DJ and Beletsky D (2003) Relative effects of wind stress curl, topography, and stratification on large-scale circulation in Lake Michigan. *Journal of Geophysical Research* 108(C2): doi:10.1029/2001JC001066.

Strub PD and Powell TM (1986) Wind-driven surface transport in stratified closed basins: direct versus residual circulations. *Journal of Geophysical Research* 91(C7): 8497–8508.

Wang Y, Hutter K, and Bäuerle E (2001) Barotropic response in a lake to wind-forcing. *Annales Geophysicae* 19: 367–388.

Flow in Wetlands and Macrophyte Beds

A M Folkard, Lancaster University, Lancaster, UK
M J Coates, Deakin University, Warrnambool, Vic., Australia

The presence of vegetation in aquatic environments has a wide range of important implications from physical, biological, and chemical perspectives. This chapter follows on from **Flow Modification by Submerged Vegetation** by focussing primarily on the first of these, by discussing the hydrodynamics of aquatic vegetation, but also highlights ecological and biogeochemical phenomena, which are hydrodynamically influenced in this context. The physical implications of the presence of aquatic vegetation can be summarized into a few general areas that are common to most or all settings:

- changes to flow and wave fields;
- changes to turbulence characteristics;
- changes to the dispersion of heat and dissolved matter concentrations;
- changes to sediment fluxes and budgets.

From chemical and biological perspectives, the impacts of hydrodynamic interactions with aquatic vegetation can also be summarized into a few key areas:

- supply, accumulation, and removal of nutrients and waste material;
- changes to light availability and rate of photosynthesis;
- plants' physiological and community-level responses to mechanical stresses;
- plant patchiness, species diversity, and mesohabitat creation.

The nature of plant–hydrodynamics interactions depends fundamentally on the functional form of the plants, which can be characterized into four main categories: emergent vegetation (which has part of its biomass below the water surface and part above), submerged vegetation (which lies entirely beneath the water surface), rooted plants with surface floating leaves (the height of which is thus equal to the water depth), and free-floating vegetation (which has a floating element and a system of roots that trail in the water column but do not reach the substrate) (**Table 1**). These are sketched in **Figure 1** (see also Figure 2 in **Flow Modification by Submerged Vegetation**).

Aquatic vegetation occurs in several different contexts, all of which tend to be relatively shallow, physically stable, and of moderate hydrodynamic energy. This chapter is structured around these contexts and covers lakes (and by implication reservoirs, ponds, and other lentic waterbodies), nontidal wetlands (including swamps, marshes, inland deltas, and constructed wetlands for wastewater treatment), rivers (and by implication streams, dykes, ditches, and other natural, altered, or constructed conveyance channels), saltmarshes, and mangroves. Note that, in the following, the term 'aquatic vegetation' is used for convenience to cover all forms of plant life found in natural waterbodies, and 'shoot' is used rather than 'stem' or 'plant' to describe an individual member of a vegetation canopy.

Lakes

In lakes, aquatic vegetation usually occurs only in the littoral zone fringing the lakeshore, where light can penetrate to the bed and enable rooted plants to photosynthesize and grow (**Figure 2**). Species of vegetation found in lake littoral zones include relatively rigid, emergent reeds and rushes such as *Phragmites* spp. and *Juncus* spp., floating-leaved plants such as water lilies (*Nymphaeaceae*), and submerged waterweeds such as *Elodea spp.* Their distribution is often spatially and temporally structured. For example, floating-leaved and submerged vegetation are usually found lakeward of emergent plant stands. Less obviously, submerged, rooted macrophytes occur most commonly in sites with intermediate levels of exposure to waves. At high exposure sites, the physical stresses applied by the waves limit the macrophytes' extent, whereas at low exposure sites, they have to compete with macroalgae. At some sites, macroalgae are transitory pioneers, establishing initially, but being removed by high wind stress events because of their lack of roots, allowing rooted macrophytes to take over.

Unlike other aquatic vegetation contexts, there is generally an absence of permanent, significant unidirectional currents in lake vegetation. Because the littoral zone tends to lie entirely within the well-mixed surface layer of lake water, there is also usually an absence of direct interactions between the vegetation and internal waves. Thus, two other hydrodynamic processes tend to dominate, namely (i) surface wind stress, which generates surface waves and turbulent mixing that diffuses downwards and creates a surface mixed layer; and (ii) convection currents caused by heterogeneous heating and cooling.

Table 1 List of selected common types of aquatic vegetation and a summary of their hydrodynamic characteristics

Common name	*Example species*	*Locational preference*	*Functional form*	*Hydrodynamic characteristics*
American (or Canadian or common) waterweed	*Elodea canadensis* *Elodea nuttallii*	Lakes, rivers, ponds, and ditches. *E. nuttallii* also in slightly brackish water. Prefers silty sediments and nutrient-rich water	Submerged and rooted, with relatively broad leaves in spirals around stems. Often reaches water surface in dense mats	Thickness of leaves and density of canopies will heavily resist flow and ttenuate turbulence. Fairly flexible, so will pronate in significant flows
Bur-reed	*Sparganium* spp.	Most commonly found in marshland, although can occur in other aquatic environments such as slow-flowing parts of rivers and lake littoral zones	Semirigid, which can be found in submerged or emergent state. Typically rather sparsely distributed	In flowing water, their sparse distribution implies that flow resistance will be dominated by processes at the individual plant scale, where wakes will reduce turbulent eddy scales to the order of the shoot separation
Common Reed	*Phragmites australis*	Moist sites throughout temperate and tropical regions of the world. Especially common in lake littoral zones and fringing river channels	Tall (up to 6 m), rigid plants forming extensive stands (reed beds). Leaves are 20–50 cm long and 2–3 cm broad	Often the dominant emergent vegetation, extensive reedbeds will provide very significant flow resistance, even changing the direction of riverflow, for example. In lake littoral zones, will provide extensive shading and almost complete attenuation of wind-induced waves and mixing
Cordgrass	*Spartina* spp.	Saltmarshes, where they are often the dominant plants in pioneer zones because of their strongly halophytic (salt tolerant) nature	Relatively rigid-leaved plants, often forming dense, monospecific communities. Initial colonization is usually in patches, which then merge as the saltmarsh accretes	Strongly attenuate waves and tidal currents, causing strong sedimentation at their seaward edge
Duckweed	*Lemna gibba* *Lemna minor* *Spirodela polyrhiza*	Lakes, ponds, other lentic water. Prefer moderately rich to eutrophic water	Floating leaves, roots trailing in water	Shade water from solar heating; attenuate wind-stress-induced waves and turbulent mixing
Frogbit	*Hydrocharis morsus-ranae*	Lakes, ponds, other lentic water. Prefer moderately rich to eutrophic water	Floating leaves, roots trailing in water (may anchor to bed in shallow water where roots reach bottom)	Shade water from solar heating; attenuate wind-stress-induced waves and turbulent mixing
Glasswort	*Salicornia* spp.	Saltmarshes, where they are highly salt tolerant and can survive complete inundation for parts of the tidal cycle	Relatively low-lying, flexible plant (cf. *Spartina* spp.)	Usually overshadowed by cordgrass in terms of hydrodynamic influence, but where it is dominant, creates extensive meadows, which significantly raise turbulent shear stress maximum above bed and thus protect bed sediment from erosion

Continued

Table 1 Continued

Common name	*Example species*	*Locational preference*	*Functional form*	*Hydrodynamic characteristics*
Hornwort/ coontail	*Ceratophyllum demersum*, *Ceratophyllum submersum*	Ponds, marshes, and quiet streams in tropical and in temperate regions	Submerged, usually not rooted and floating on the surface. Feathery leaves arranged in whorls on the stem, resembling a racoon's tail	Shade water from solar heating; attenuate wind-stress-induced waves and turbulent mixing. Abundant, narrow leaves will strongly alter turbulence
Lake quillwort	*Isoetes lacustris*	Stony or sandy bottoms of clear, usually slightly acidic ponds, typically in mountain tarns, growing at 5–300 cm depth of water	Long, narrow leaves 8–20 cm long and 0.5–2 mm broad. Very small stem, roots with leaf-like appendages (rhizomorphs). Upper leaves sprout in clumps	Where emergent, will strongly attenuate waves and turbulent mixing. To a lesser extent, it will shade the water surface from solar radiation. Where submerged, will protect bed sediment effectively and act as a bed roughness element in the same way as shoreweed (see earlier)
Pondweed	*Potamogeton* spp.	At least 80–90 species found throughout the world in lentic or lotic waters	Rhizomatous (attached to the bed). Either fully submerged or with floating leaves. Leaves very narrow in some species, broader and more leathery in others	Floating-leaved species will shade water from solar heating; attenuate wind-stress-induced waves and turbulent mixing. Rhizomes and near-bed stems will protect bed sediment from erosion. Submerged leaves will alter turbulence
Red mangrove	*Rhizophora mangle*	Shallow marine water in bays and estuaries	Trees, supported by curving roots. May form dense thickets	Large size and density of stands implies that this species is particularly effective at protecting coastline from storm waves and tidal currents
Rushes	*Juncus* spp.	All wet environments, but rarely in the tropics. Typically found in moist areas along streams and rivers	Grows in large clumps to ~1.5 m. Rooted, emergent plant, often found at water's edge and only surrounded by water during flooding	Rigid emergent plant that will provide significant flow resistance and thus significantly increase sediment retention in its immediate vicinity and provide a significant wake in flowing water
Sedges	*Carex* spp.	Most commonly found in marshes, where flow is stagnant or very slow and ground water is important for water fluxes	Rooted, rigid plants, often growing in dense clumps	Where present in wetlands or marshes, their dominance will determine patterns of throughflow and spatial variability of residence times
Shoreweed	*Littorella uniflora*	Lake littoral zones and slow-moving areas of rivers	Rooted tufts of short, rigid leaves a few centimeters long. Often forms dense swards in lake littoral zones	Dense swards will protect bed sediment from resuspension. Given its short height in relation to typical water depths, in unidirectional flow, flow resistance due to this species can be accurately represented using the Manning equation

Continued

Table 1 Continued

Common name	*Example species*	*Locational preference*	*Functional form*	*Hydrodynamic characteristics*
Stoneworts	*Chara* spp.	A green alga that grows in fresh water, most commonly in stagnant or slow-moving water such as lakes or ponds	Submergent, floating alga that can become encrusted with calcium carbonate	Where encrustation is heavy, they will strongly shade water from solar radiation, and heavily damp wind-induced waves and turbulent mixing. Where they form dense near-bed meadows, they will very effectively protect bed sediment from resuspension
Water crowfoot	*Ranunculus* spp.	Still or running water, most often in relatively fast-flowing, unpolluted, well-oxygenated streams, on gravel, base-rich substrates	Highly flexible, submerged species, which can grow up to the water surface. Thread-like leaves underwater, broader floating leaves at surface. Can form dense mats on the water, which can clog waterways	Where isolated plants occur, narrow leaves and high shoot flexibility imply low flow resistance. Where dense mats are formed, creates semi-impermeable blockages within river channels
Water lily	*Nymphaea* spp.	Lake littoral zones and other slow-flowing or stagnant water environments	Rooted plants with characteristic large floating leaves	Very effective at shading water surface from solar radiation and reducing wind-induced waves and turbulent mixing
Water milfoil	*Myriophyllum* spp.	Prefers lakes, ponds, and slow-moving streams, but can also grow in fast-moving water. Tolerates wide range conditions, even with salinity up to 10‰. Able to overwinter in frozen lakes and to grow in tropical bays. Prefers cool, slightly alkaline water with high calcium content	Rooted, with stout stems almost entirely underwater. Whorled, featherlike leaves	Emergent sections will shade water from solar heating; attenuate wind-stress-induced waves and turbulent mixing. Flexible stems will pronate in significant flow, reducing their flow resistance. Stem is quite brittle, so will break at high flow speeds
Water starwort	*Callitriche* spp.	Most common in slow-flowing waters, muddy and silty sediments. Will grow in still and fast-flowing conditions up to 1 m deep. Intolerant of inorganic pollution	Rooted plants that reach the water surface. Slender stems and long, thin submerged leaves, which expand when they reach the surface. In slow-flowing conditions they form dense monospecific stands	Thickness of leaves and density of canopies will heavily resist flow and attenuate turbulence. Fairly flexible, so will pronate in significant flows

Note: list is strictly nonexhaustive, but illustrative. Information taken in part from the following: Rodwell JS (ed.) (1998) *British Plant Communities: Aquatic Communities, Swamps and Tall-herb Fens*, vol. 4 (British Plant Communities). Cambridge, UK: Cambridge University Press. http://www.ceh.ac.uk (UK Natural Environment Research Council Centre for Ecology and Hydrology, Centre for Aquatic Plant Management); http://www.ecy.wa.gov/ (Washington State Department of Ecology); http://plants.ifas.ufl.edu/ (University of Florida Institute of Food and Agricultural Sciences, Center for Aquatic and Invasive Plants); http://www.habitas.org.uk/ (HabitasOnline: The web site of the Ulster Museum Sciences Division); http://www.plant-identification.co.uk/skye/index.htm (Skye Flora); http://academic.emporia.edu/aberjame/wetland/wet_veg/veg_index.htm (Emporia State University, Kansas, USA: Wetland Vegetation pages of JS Aber and TA Eddy).

In lakes where wind stress is the dominant forcing factor, the primary roles of aquatic vegetation are to reduce the fetch for wind-driven flows and wave fields and attenuate wind-induced mixing, thus strengthening temperature stratification and reducing the surface mixed layer depth compared with the open water. In dynamical terms, the mixing caused by turbulent kinetic energy (TKE) imparted to the

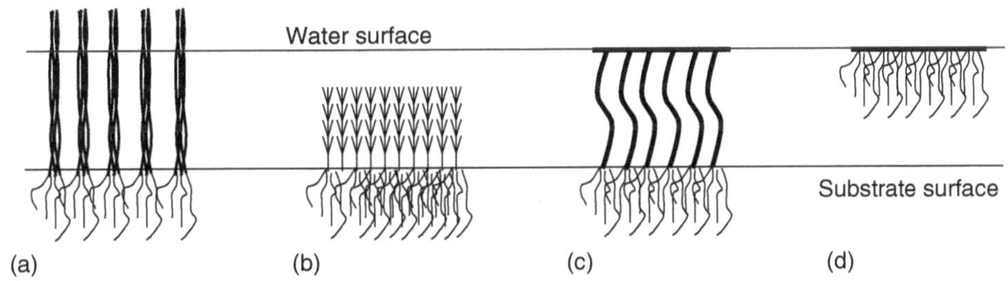

Figure 1 Sketches of the configurations of (a) emergent vegetation; (b) submerged vegetation; (c) rooted plants with surface floating leaves; (d) free-floating vegetation in relation to the water surface.

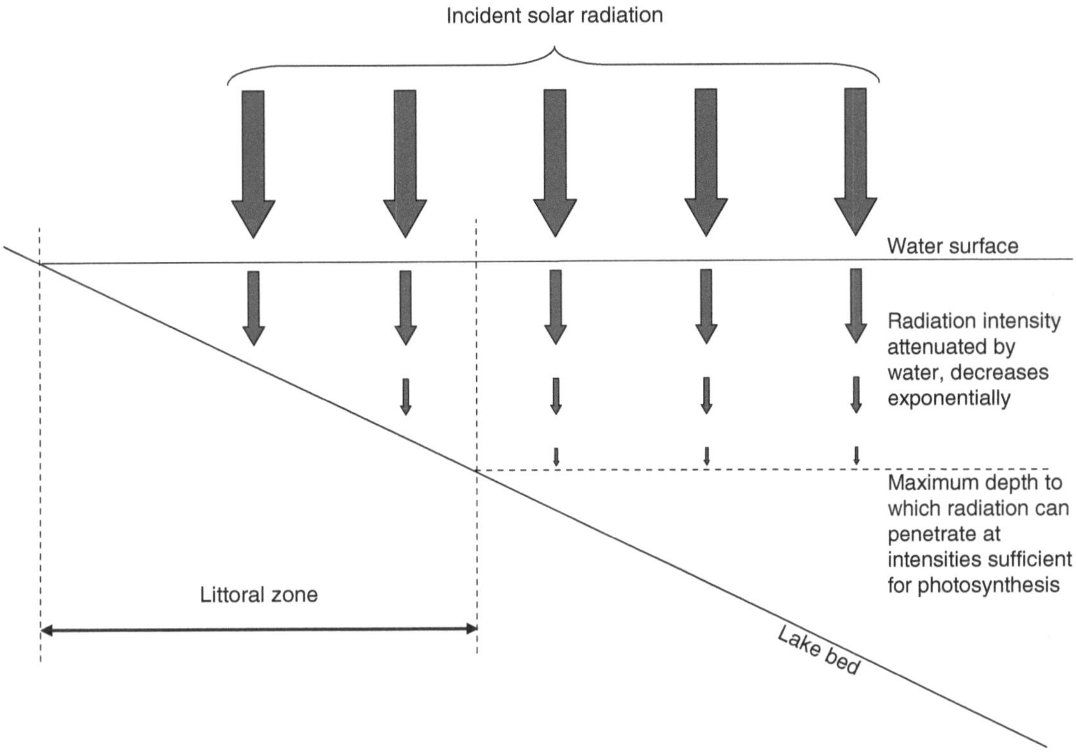

Figure 2 Definition sketch of the littoral zone within lakes.

surface water by the wind stress is counteracted in open water by stratification caused by solar heat flux-induced buoyancy production. In vegetation, the mixing is also attenuated by TKE dissipation caused by the drag of the plant canopy.

Vegetation canopies attenuate wind-induced waves and mixing, thus reduce sediment resuspension at the sediment–water interface. Water quality may be improved by the reduction in particulate load. The implications for nutrient recycling are less clear and likely depend upon the amount of plant material shed and whether anoxia develops owing to the remineralization and reduced exchange of water. The factors that lead to reduced sediment resuspension vary with vegetation type. Emergent and floating-leaved plants reduce sediment resuspension by attenuating wind stress, whereas dense, near-bed meadows of submerged species (e.g., *Chara* spp.) do so by lifting the turbulent shear stress maximum above the bed. Although plant density and individual species' morphology will have an effect, in general, floating-leaved plants appear to be less effective at reducing resuspension rates than are emergent or submerged macrophytes.

Differential heating of the littoral and pelagic zones in lakes has the potential to drive horizontal convection currents between these two regions even in the absence of vegetation. During the day, the warmer, littoral zone water flows lakeward near the surface, and the cooler counter-current of pelagic zone water

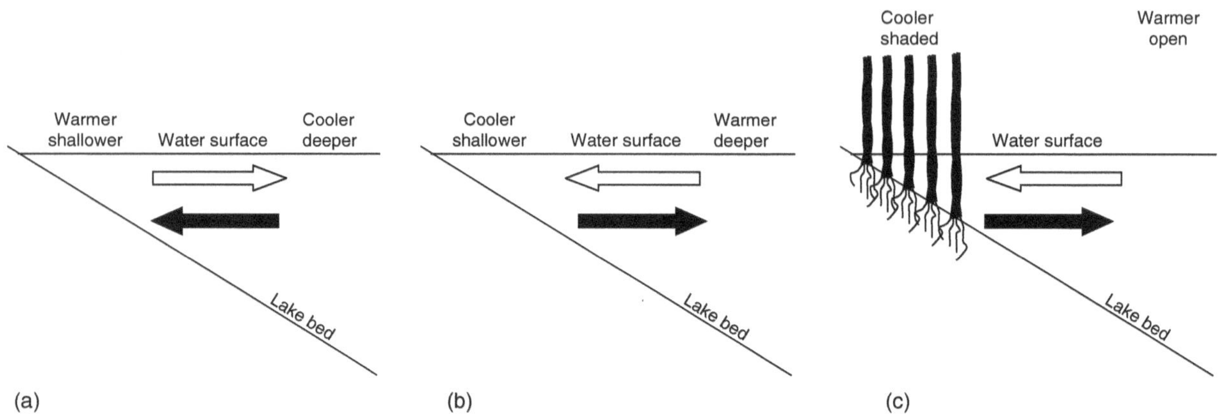

Figure 3 Sketches of the density-driven exchange flows (a) in open, shallow water during the day; (b) in open, shallow water during the night; (c) between vegetated and open water during the day.

intrudes into the littoral zone at depth (**Figure 3(a)**). During the night, the directions are reversed because of more rapid cooling in the littoral zone (**Figure 3(b)**). Where vegetation is present, its shading reduces thermal stratification and creates a region of cooler water during the day, which sets up a similar convective exchange, but with the water from within the vegetation intruding out into the lake at depth and the return current flowing near the surface (**Figure 3(c)**). Whether they affect significantly nocturnal convection is not yet clear. Nevertheless, daytime convection plays an important role in flushing the often nutrient-rich waters within the vegetation out to the open water. Clearly, it will be most important on sunny days with low winds, but typically, this mechanism results in current speeds of the order of $10^{-2}\,\mathrm{ms}^{-1}$ and flushes canopies in time scales of a few hours.

Nontidal Wetlands

There is an important distinction between tidal wetlands (saltmarshes and mangroves) – whose hydrodynamics are dominated by waves and tidal currents and which are covered below – and nontidal wetlands. The latter are discussed in this section and include a wide range of environments, from extensive freshwater wetlands such as the Florida Everglades and the Okavango Delta, to constructed wetlands, which are increasingly popular as wastewater treatment plants. In these nontidal wetlands, surface water flow is typically very slow and dominated by interactions with vegetation. Indeed, within these wetlands, vegetation has many functions, including retention of particulates, reduction in turbulence, stabilization of sediments, and provision of habitats for epiphytes. A characteristic of the hydraulics of these wetlands is the high spatial variability in residence time (the period spent by water within the system), the hydraulic patterning being a combination of dead zones and preferential flow paths, which are closely related to the spatial distribution of vegetation (**Figure 4**). Given that two major concerns in managing nontidal wetlands are maintenance of water quality and habitat diversity, and that the main function of constructed wetlands is wastewater contaminant removal, control of this residence time via vegetation management is of paramount importance here.

Another key issue is the prediction of dispersion rates for scalar parameters (such as contaminant concentration) within vegetation. Dispersion can be separated into longitudinal and lateral components. Longitudinal dispersion is controlled by canopy morphology. It is caused by velocity gradients owing to individual shoot wakes in the near field, and to spatial variations in canopy frontal surface area (which often varies vertically and horizontally) in the far field (further details of longitudinal dispersion in vegetation may be found in **Flow Modification by Submerged Vegetation.** Lateral dispersion is caused by two mechanisms: turbulent diffusion due to shoot wakes, and mechanical diffusion caused by water particles having to take laterally extended paths through the matrix of shoots within the canopy (**Figure 5**). Turbulent diffusivity can be predicted via a modified random walk model, in which the lateral diffusion outside of shoot wakes is considered negligible compared with that within the wakes. Mechanical diffusivity can be predicted by assuming that it is analogous to Fickian diffusion in a porous medium. Together, these give the following expression for the (nondimensionalized) total lateral diffusivity

$$\frac{K_{\mathrm{tot}}}{Ud} = \alpha\left[\overline{C_{\mathrm{D}}}ad\right]^{1/3} + \left[\frac{\beta^2}{2}\right]ad \qquad [1]$$

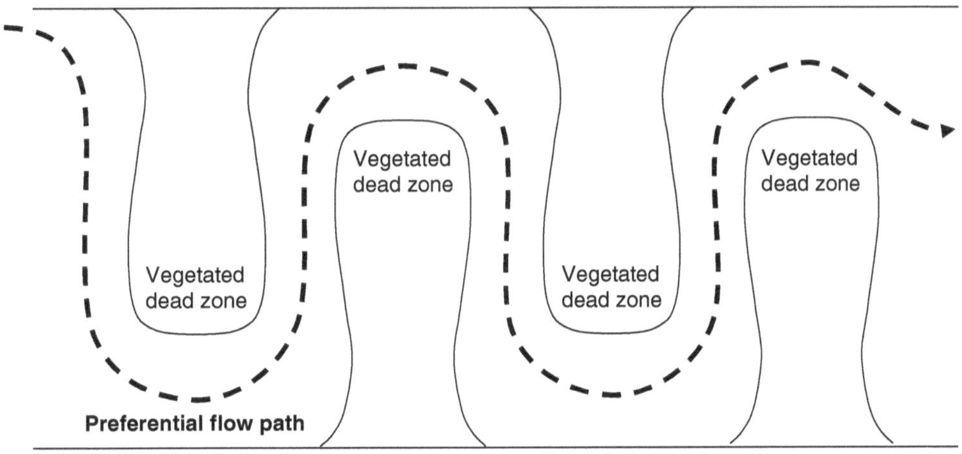

Figure 4 Planview sketch of typical patterning of dead zones and preferential flow paths in wetlands constructed for wastewater treatment.

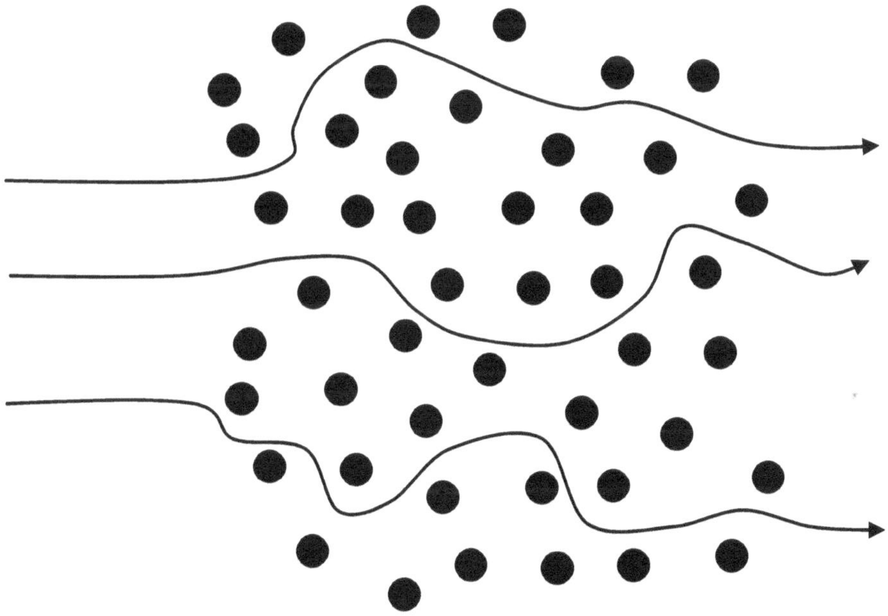

Figure 5 Planview sketch of the mechanical lateral diffusion process, in which water particles take laterally extended paths through the matrix of shoots within a canopy (adapted from Nepf, 1999).

Here, K_{tot} ($m^2\ s^{-1}$) is the total lateral diffusivity, U is the mean flow speed, d is the vegetation shoot diameter, α and β are empirically derived coefficients, $\overline{C_D}$ is the bulk drag coefficient of the vegetation, and a is the canopy frontal surface area per unit volume. The terms on the right-hand side of eqn. [1] represent the turbulent and mechanical diffusivity, respectively.

Note that dispersion rate is a function of Reynolds number: in low Reynolds number conditions typical of nontidal wetlands, shoot wakes will be non- or weakly-turbulent, so the turbulent diffusivity will be small, and the velocity gradients that determine longitudinal dispersion will be similarly weak. In higher Reynolds number environments, such as channel flow (see later text), turbulence, shear, and thus dispersion will be stronger and will lead to the longitudinal dispersion processes explained in **Flow Modification by Submerged Vegetation.**

Rivers

In rivers and other conveyance channels, longitudinal flow dominates. As noted in **Flow Modification by Submerged Vegetation**, the challenge is to determine the variables that cause flow resistance when vegetation is present so that discharge rate, mean speed, and depth can be predicted. This understanding is required to maintain water supplies and habitat for a diversity of organisms and to predict flooding.

Generally, vegetation reduces mean flow energy, but bryophytes and similar organisms can reduce flow resistance because they make the surface of sediment grains smoother. Flow resistance characteristics may be derived by taking a force balance approach. Here, a steady state is assumed and the force applied to the vegetation by the flow is equated to the resistive force owing to the vegetation, the bed resistive force being assumed to be negligible. Most commonly, however, the approach taken is to derive relationships between measurable parameters of the flow and vegetation and a roughness coefficient such as those used in the well-known Darcy–Weisbach or Chezy equations or, most commonly, the Manning equation:

$$U = \frac{R^{2/3} S^{1/2}}{n} \qquad [2]$$

where R is the hydraulic radius ($= A/P$, where A is the flow cross-sectional area and P is the wetted perimeter), and S is the channel slope and n is Manning's coefficient – which can be thought of as a measure of the channel roughness. Relatively recently, more general forms of resistance equation have been proposed of the form

$$q = KD^{\beta} S^{\alpha} \qquad [3]$$

where q is the discharge per unit width, K is a roughness coefficient and D is flow depth (which for typical channels, where channel width is much greater than channel depth, is approximately equal to the hydraulic radius R found in the Manning equation). Nevertheless, the problem remains to parameterize the roughness coefficient.

Strictly speaking, Manning's equation is only applicable in channel flow with a classical logarithmic boundary layer velocity profile (see Figure 2c in **Flow Modification by Submerged Vegetation**). Thus where the vegetation height is much less than the flow depth, as in grassed channels for example, its use is appropriate, although n will become a decreasing function of flow speed as the vegetation bends in strong flows. However, in many vegetated flow conditions, the vegetation height equals or exceeds the water depth (Figure 2(a) and 2(b) in **Flow Modification by Submerged Vegetation**). This is common in lowland, alkaline rivers, which can become heavily choked with vegetation in summer months. In these environments, use of Manning's and similar equations is not justified. Nevertheless, their use has persisted as a convenient method for quantifying flow resistance: values of roughness coefficients are often quoted for heavily vegetated channels (though their accuracy is disputed), and there have been many parameterizations developed for these conditions. An example is the widely-used Petryk–Bosmajian equation

$$n = \sqrt{\frac{C_D \sum A_i}{2gAL}} R^{4/3} \qquad [4]$$

where C_D is the drag coefficient due to an isolated vegetation shoot, A_i is the projected area of the ith shoot, A is the cross-sectional area of the whole flow, and L is the length of the vegetated channel reach being considered (note that this form of the equation assumes that the contribution to n due to bed shear stress is negligible). The problem with these relationships is that they tend to lack portability to a wide range of plant morphologies or contexts or require logistically difficult measurements for each application. As such, there is wide recognition that robust representation of vegetative flow resistance in heavily-vegetated environments requires determination of relationships involving a wide variety of parameters encompassing shoot geometric properties (e.g., shoot density or average distance between shoots; shoot height, shoot diameter, frontal area, submerged biomass, solid volume fraction) shoot mechanical properties (e.g., modulus of elasticity, moment of inertia, surface roughness, drag coefficient), and flow properties (e.g., flow velocity, depth, hydraulic radius, energy gradient, shear velocity, and fluid viscosity). For these relationships to be accurate, it is essential to understand the physics behind flow resistance induced by vegetation.

When flow enters a vegetation canopy, bed shear stress rapidly becomes unimportant, as the resistance of the vegetation becomes the dominant source of drag. As noted in **Flow Modification by Submerged Vegetation**, there are two scales of flow resistance owing to vegetation canopies: at the shoot scale, mean flow energy is extracted into turbulent wakes, and at the patch scale, mean flow energy is extracted within shear layers between vegetated and nonvegetated flow regions. For both to have equal influence, the discharge through and around the vegetation must be equal. In dense structures, such as those formed by starwort (*Callitriche* spp.), discharge around the vegetation is much greater than that through the vegetation, so patch scale phenomena dominate. In sparse vegetation such as bur-reed (*Sparganium* spp.) for example, flow through the vegetation dominates, so the shoot scale resistance is more important. For the latter case, shoot structural details are important in determining resistance. For example, leaves on willow trees increase their drag by two to three times over bare willow trees. Spatial arrangement also significantly affects flow resistance, in that linearly arranged shoots allow roughly straight preferential flow paths to develop, while staggered or randomly arranged shoots increase mechanical lateral dispersion of flow energy (**Figure 6**).

The resistive action of vegetation is amended by its flexibility, which decreases its drag as flow speed increases (see also **Flow Modification by Submerged Vegetation**, Figure 6). Thus, rather than the standard quadratic relation between velocity and drag force, this relationship is typically linear in flexible vegetation. However, this tendency may be at least partially counteracted by the increase in turbulent energy losses caused by the waving of fronds or leaves in the flow. Shoots can also vary their stiffness, altering their flow resistance. They tend to be stiffer during the day, when excess oxygen production due to photosynthesis increases their buoyancy, and are more rigid in midsummer. However, these are small effects which are only important for very high density vegetation. At very high velocity, the flow will break the shoots, the speed at which this occurs (the 'lodging velocity') being dependent on the geometric and mechanical nature of the shoots. From the perspective of the vegetation, there is a trade-off to be made between rigidity, which will enable them to reduce hydrodynamic energy and thus enable beneficial sediment deposition, and flexibility, which will enable them to survive high energy flow regimes.

Often in channel flow environments, vegetation growth is patchy. In rivers, vegetation patches can act as pseudo-riffles or can create pseudo-braided channels. The spatial distribution and shape of patches affects their flow resistance, as well as their longitudinal dispersion as explained in **Flow Modification by Submerged Vegetation**. Species diversity also complicates the hydraulics as each species will have its own mechanical and geometric properties. Where vegetation occurs in densely populated patches within a channel, it can be treated as an impermeable blockage, and flow resistance calculated on this basis. The blockage factor may be calculated as the fraction of the flow that is occupied by vegetation either in cross-sections, in the plan view of the river channel, or in a volumetric sense. The spatial heterogeneity of vegetation also has important ecological implications. Fluvial vegetation thrives and has the highest diversity of species, where a variety of habitats are present, for example where there is a range of sediment from fines to clean gravel, and a range of flow conditions, a condition which is enhanced by patchiness in the vegetation distribution.

Often, vegetation is arranged within river channels as longitudinal strips and in these cases, the shear that is developed at lateral boundaries between the slow flow within the vegetation and the faster, non-vegetated flow becomes a crucial element of flow resistance within the channel. At these boundaries, vortices are generated owing to inflectional instability, analogous to eddies formed at the top of canopies owing to Kelvin Helmholtz instability (see **Flow Modification by Submerged Vegetation**). The flow field at these boundaries is complex and not yet fully understood, but bed shear stress and sediment deposition both appear to peak near them.

Within the vegetation canopy, the turbulent field becomes strongly altered, canopies of different morphologies having widely different effects on

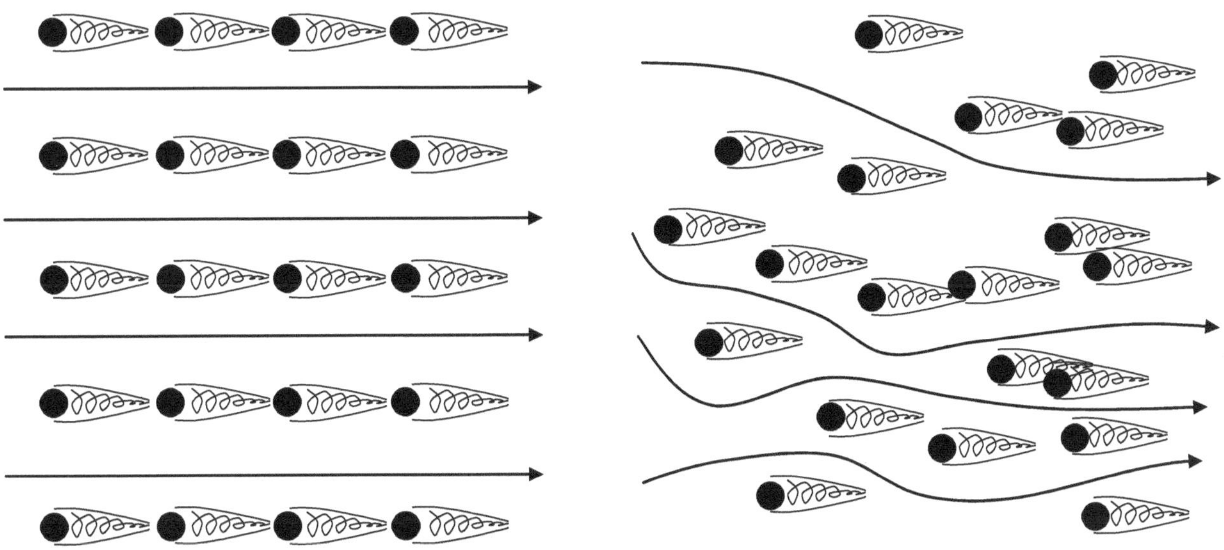

Figure 6 Planview sketch illustrating how spatial arrangement of shoots affects flow resistance.

turbulence. As in the case of drag, when the flow enters a canopy, turbulence production rapidly becomes dominated by shoot wakes rather than bed shear stress. In general, meadows tend to damp the large eddies that are found in unvegetated flow and the eddy scale becomes of the order of the shoot separation. Thus diffusivity is reduced within the vegetation, even though turbulent intensity may be increased. Where vegetation density is sparse, turbulence intensity is generally increased, but where density becomes high, turbulence intensity is reduced because of the reduction in the mean flow speed. Furthermore, the vertical component of turbulence tends to be more attenuated than the horizontal components. The result of this increasingly 2D turbulence is that suspended sediment concentrations are decreased within the canopy. An important aspect of the turbulence field within vegetation canopies is the generation of coherent eddy structures, both at the shoot scale in individual wakes and, as noted earlier, at the patch scale at canopy boundaries.

A key role of vegetation in flowing water is its capture of suspended sediment particles: retention by vegetation of up to 80% of transported sediment has been reported. There are three mechanisms whereby this happens: by reduction of discharge and thus flow competence due to vegetative drag, direct capture of particles on the plants, and by the plants acting as a substrate for filter feeders which convert the suspended matter into faecal pellets, which are then deposited. Vegetation also reduces bedload transport due to absorption of momentum by vegetative drag. From an ecological perspective, this can be viewed as the vegetation acting as an 'ecological engineer', enhancing sediment deposition to its own advantage.

Saltmarshes and Mangroves

As noted above, coastal wetlands – saltmarshes in temperate zones and mangroves in tropical zones – differ radically in their hydraulics from inland wetlands in that they are dominated by tidal currents and wave action. Although numerical models of flow in mangroves have been developed successfully, saltmarshes have been much more thoroughly studied. For this reason, the latter and in particular their pioneer zones – their most hydrodynamically-active regions – are focused on here. Pioneer zone plant species such as cordgrass (*Spartina* spp.) are typically rigid plants, which form dense patches and meadows, and can therefore withstand the relatively high mechanical stress levels encountered in these environments.

As in fluvial environments, flow entering a saltmarsh vegetation canopy adjusts rapidly. Clearly, the extent and nature of these changes will vary between different canopy morphologies. Typically, the vegetation takes over from the bed shear stress as the primary source of drag after the first few centimetres, and a significant proportion of the incident mean velocity and turbulent kinetic energy are lost in first few metres of canopy. Saltmarsh vegetation canopies also have significant effects on surface wave fields: wave energy dissipation rates over vegetation canopies are typically two or three times higher than over bare mudflats because of the increase in surface friction. Within canopies, these dissipation rates vary according to water depth and relative incident wave height (the ratio of wave height to water depth).

A key aspect of vegetated flow in saltmarshes is the extent to which the presence of vegetation alters sediment deposition patterns, since it is via this process that saltmarshes initiate and evolve. Sediment deposition is closely linked to the velocity profile within the vegetation canopy, which in turn is strongly influenced by the canopy morphology, such that flow speed is dependent on vegetation density at each height. Sedimentation rates and suspended sediment concentrations decrease logarithmically across marshes, being highest at the vegetation boundary, because of the rapid reduction in flow speed and turbulent energy with the marsh canopy. As a result, sediment deposition in the interior of marsh canopies is typically less than over open mudflats during fair weather conditions. This implies that vertical growth of saltmarshes is due more to sediment protection during storms than enhanced deposition during normal conditions.

Saltmarsh vegetation is typically heterogeneous in density and thus sedimentation patterns are closely linked to vegetation distribution. During stages of the tide when flow depth is less than the canopy height, flow is faster over unvegetated areas of marshes, with vegetated areas being flooded from them by flow perpendicular to the vegetation boundary (see **Figure 7**). As the flow depth becomes greater than the vegetation height, there is a change to a large scale sheet flow and sedimentation patterns become more homogeneous. Thus vegetation cover distribution is the key to the long-term geomorphic development of marshes and it is this that leads to the structure of tidal channels and vegetated platforms in between the channels, which is characteristic of mature marshes.

Summary

Although the basic physics of hydrodynamic processes in vegetation canopies in inland waters is understood, it is difficult to develop generalities. Species diversity of aquatic vegetation is high and leads to

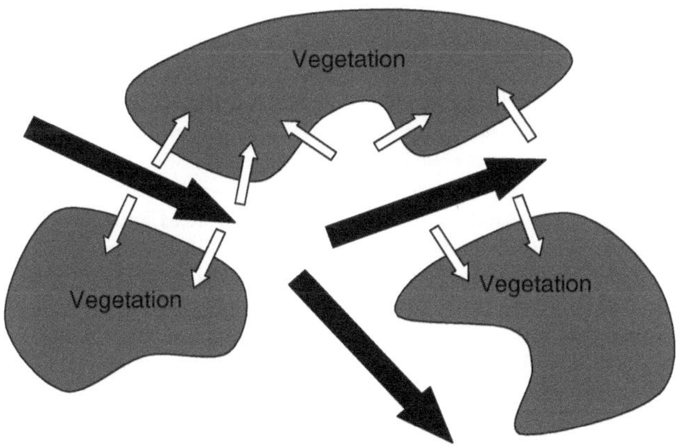

Figure 7 Planview sketch illustrating the process of saltmarsh flooding during stages of the tide when flow depth is less than the canopy height. Dark arrows indicate the main flow path through the unvegetated areas; white arrows indicate the subsequent vegetation-flooding flow paths perpendicular to the vegetation patch edges.

a wide range of complex geometric and mechanical plant properties. Deriving robust mathematical models of the hydrodynamics of aquatic vegetation applicable to more than a very limited subset of species is problematic. Nevertheless, there are several overarching generalities:

- where unidirectional flow is dominant, flow resistance from the vegetation often exceeds bed shear stress within a few centimeters of the canopy and thus becomes the primary source of drag. Flow resistance occurs at the shoot scale, where mean flow energy is transferred into shoot wakes, and at the patch scale, where it is transferred into coherent eddying structures within the shear layer between the canopy and the unvegetated flow (e.g., **Flow Modification by Submerged Vegetation**, Figure 2 (b) and 7);
- the shoots' flexibility decreases their drag as flow speed increases (Figure 6 in **Flow Modification by Submerged Vegetation**, and linearly-arranged shoots provide less resistance than staggered or randomly arranged ones;
- where wind stress is the dominant forcing factor, emergent and floating aquatic vegetation reduces fetch and attenuates wind-induced waves and mixing, thus strengthening temperature stratification;
- where incident waves are dominant, emergent and floating vegetation canopies enhance wave energy dissipation rates typically two- or three-fold;
- where solar radiation-induced thermal forcing is dominant, emergent and floating aquatic vegetation shades the water column, reducing thermal stratification during the day and setting up convection currents;
- vegetation canopies alter turbulence fields that enter them. Eddy scales become of the order of shoot separation, so that diffusivity is reduced. In sparse vegetation, turbulence intensity is generally increased, but in dense canopies it is reduced because the mean flow slows. Turbulence also becomes more 2D (horizontal), contributing to the reduction in sediment suspension. Coherent eddy structures are formed, both in individual shoot wakes and at canopy boundaries;
- dense vegetation reduces suspended sediment concentrations, because sediment is preferentially deposited at the upstream end of the canopy through three mechanisms: reduction of discharge, direct capture of particles, and conversion to depositional faecal matter by epiphytes. These processes occur throughout less dense canopies. Vegetation also reduces bedload transport because it reduces bed shear stress;
- vegetation disperses scalar parameters. Longitudinal dispersion is controlled by velocity gradients created by the canopy morphology (**Flow Modification by Submerged Vegetation**, Figure 10). Lateral dispersion is caused by turbulent diffusion in shoot wakes and mechanical diffusion is caused when water is forced to take laterally extended paths through the matrix of shoots (**Figure 5**);
- vegetation canopies are typically patchy and often made up of a variety of species of different geometric and mechanical properties. These impact hydrodynamic and sedimentation patterns, particularly where plant height is greater than water depth, leading to spatial patterning of preferential flow paths and dead zones and a matching variability in residence time (**Figure 4**).

Nomenclature

Convection – Motion of water (in this case) caused by thermally induced pressure gradients.

Diffusion – Transfer of suspended or dissolved matter down its concentration gradient, independent of motion of the supporting medium (which, in this case, is water).

Drag coefficient – Parameter that describes the tendency of an obstacle in a flow to apply a decelerating force to that flow.

Epiphytes – Organisms that live on, or near plants and depend on the plants for their survival.

Hydraulic radius – A measure of the proportion of the water at a cross-section within a channel that is in contact with the bed and/or banks and thus subject to the flow resistance caused by their roughness. Calculated as the ratio of the cross-sectional area to the wetted perimeter (the length of the bed and/or banks within the cross-section that are in contact with the water).

Lentic – Non-flowing, stagnant (cf. lotic).

Littoral zone – region of a lake or other waterbody, where light reaches the bed, enabling rooted plants to photosynthesize and grow (cf. pelagic zone).

Lotic – Flowing (cf. lentic).

Macroalga – large algal organism that anchors itself through surface attachment to the bed, rather than subsurface roots.

Macrophyte – Member of the macroscopic plant life (of a water body).

Pelagic zone – Region where light does not reach bed beneath water column, so that photosynthesizing organisms need to float within the water column to exist (cf. littoral zone).

Reynolds number – Nondimensional parameter that quantifies the ratio of the inertial force (i.e., the force causing the water to move) to the viscous force (i.e., the forces within the water itself resisting motion, due to its 'stickiness'). Equal to the product of the speed of the flow and the relevant length scale (which may be the flow depth, or the plant stem diameter, or vegetation patch size), divided by the dynamic viscosity of water ($\approx 10^{-6} m^2 s^{-1}$ at standard temperature and pressure). Used to determine whether a flow will be turbulent or laminar (nonturbulent).

See also: The Benthic Boundary Layer (in Rivers, Lakes, and Reservoirs); Biological-Physical Interactions; Currents in Rivers; Currents in Stratified Water Bodies 1: Density-Driven Flows; Currents in the Upper Mixed Layer and in Unstratified Water Bodies; Flow Modification by Submerged Vegetation; Small-Scale Turbulence and Mixing: Energy Fluxes in Stratified Lakes.

Further Reading

Green JC (2005) Modelling channel resistance in vegetated streams: review and development of new theory. *Hydrological Processes* 19(6): 1245–1259.

Herb WR and Stefan HG (2005) Dynamics of vertical mixing in a shallow lake with submersed macrophytes. *Water Resources Research* 41: W02023, doi:10.1029/2003WR002613.

James CS and Makoa MJ (2006) Conveyance estimation for channels with emergent vegetation boundaries. *Proceedings of the Institution of Civil Engineers Water Management* 159: 235–243.

Leonard LA and Croft AL (2006) The effect of standing biomass on flow velocity and turbulence in *Spartina alterniflora* canopies, *Estuarine, Coastal and Shelf Science* 69: 325–336.

Moller I, Spencer T, French JR, Leggett DJ, and Dixon M (1999) Wave transformation over salt marshes: A field and numerical modelling study from north Norfolk, England. *Estuarine, Coastal and Shelf Science* 49: 411–426: Article No. ecss.1999.0509.

Nepf HM (1999) Drag, turbulence, and diffusion in flow through emergent vegetation. *Water Resources Research* 35: 479–489.

Nepf HM (2004) Vegetated flow dynamics. In: Fagherazzi S, Marani M, and Blum LK (eds.) *The Ecogeomorphology of Tidal Marshes, Coastal and Estuarine Studies 59*. American Geophysical Union.

Oldham CE and Sturman JJ (2001) The effect of emergent vegetation on convective flushing in shallow wetlands: Scaling and experiments. *Limnology and Oceanography* 46(6): 1486–1493.

Pitlo RH and Dawson FH (1990) Flow resistance of aquatic weeds. In: Pieterse AH and Murphy KJ (eds.) *Aquatic Weeds: The Ecology and Management of Nuisance Aquatic Vegetation*, pp. 74–84. Oxford, UK: Oxford University Press.

Temmerman S, Bouma TJ, Govers G, Wang ZB, De Vries MB, and Herman PMJ (2005) Impact of vegetation on flow routing and sedimentation patterns: Three-dimensional modeling for a tidal marsh. *Journal of Geophysical Research (Earth Surface)* 110 (F4): Art. No. F04019.

Tsihrintzis VA and Madiedo EE (2000) Hydraulic resistance determination in marsh wetlands. *Water Resources Management* 14: 285–309.

Vermaat JE, Santamaria L, and Roos PJ (2000) Water flow across and sediment trapping in submerged macrophyte beds of contrasting growth form. *Archiv für Hydrobiologie* 148: 549–562.

Vogel S (1996) *Life in Moving Fluids*. 2nd edn. Princeton: Princeton University Press.

Wharton G, Cotton JA, Wotton RS, *et al.* (2006) Macrophytes and suspension-feeding invertebrates modify flows and fine sediments in the Frome and Piddle catchments, Dorset (UK). *Journal of Hydrology* 330: 171–184.

Wotton RS (1994) *The Biology of Particles in Aquatic Systems*. 2nd edn. London: CRC Press.

Relevant Websites

http://sofia.usgs.gov – provides information on vegetation hydraulics research projects in the Florida Everglades, USA.

http://www.uidaho.edu/ecohydraulics – the Center for Ecohydraulics Research at the University of Idaho, USA.

http://ccru.geog.cam.ac.uk/ – the Cambridge Coastal Research Unit at the University of Cambridge, UK.

http://web.mit.edu/nepf/www/ – Environmental Fluid Mechanics Research Laboratory at Massachusetts Institute of Technology, USA.

Flow Modification by Submerged Vegetation

H M Nepf, Massachusetts Institute of Technology, Cambridge, MA, USA

Introduction

The main impact of submerged vegetation is an increase in flow resistance and subsequent reduction in conveyance capacity. Indeed, vegetation has often been removed from channels to accelerate the passage of peak flow during floods. However, aquatic canopies have a positive influence on water quality and clarity, suggesting that significant harvesting can be ecologically damaging. This article describes the mean and turbulent flow fields in the presence of submerged canopies, considering the roles of canopy density and stem flexibility. This is followed by a discussion of water retention for canopies of different morphology.

Distribution of Flow in and above the Vegetation

We define the principle direction of the current, U, along the x-coordinate, with y- and z-coordinates denoting the lateral and vertical directions, respectively. The bed is located at $z=0$. Because water is forced to move around stems and branches, the flow within the canopy is both three-dimensional and highly heterogeneous at this scale. This scale is defined by the mean stem size (diameter), which we denote as d, and the mean stem (branch) spacing, ΔS. To understand processes occurring at the stem scale, e.g., fluxes across the leaf boundary layer, the intricate stem-scale velocity distribution must be resolved. However, to understand processes that occur at scales larger than individual stems, e.g., advection by mean flow, this resolution is not needed, and in fact is quite cumbersome. We thus simplify the description of flow by averaging over both time (to eliminate temporal variation associated with turbulence) and then over space (to eliminate spatial variation associated with stem-scale canopy structure). Within this article, the velocity (U, V, W) and pressure (P) fields denote values that have been averaged in this fashion, with the spatial average taken over a horizontal plane of scale ∂x and $\partial y \gg \Delta S$ (shaded plane in **Figure 1**). The plane has a vertical dimension, ∂z, which is negligibly small. Within this averaging scheme, we can define a mean leaf-area density, a,

$$a = d/\Delta S^2 \qquad [1]$$

to represent the plant area projected into the plane perpendicular to the flow per unit volume. If the canopy morphology varies in space, a will be a function of x, y, or z. From [1], the volume fraction occupied by the canopy elements is $(\pi/4)ad$. The canopy porosity is $n=1-(\pi/4)ad$. The range of canopy porosity observed in the field is quite large. Using typical stem densities and stem diameters for *Spartina alterniflora*, $ad=O(0.001\text{–}0.01)$, and $n\approx 1$. In a typical freshwater wetland, $ad\approx 0.1$ and $n\approx 0.9$. In mangroves, the solid volume fraction can be as high as 45%, i.e., $n=0.55$.

The vertical distribution of mean flow in and above a submerged canopy has a range of behavior depending on the canopy's relative depth of submergence, defined as the ratio of canopy height, h, to flow depth, H (**Figure 2**). First consider the emergent canopy (**Figure 2(a)**). The force balance for steady conditions can be written

$$\rho g\left[\frac{\partial H}{\partial x}+\frac{\partial z_b}{\partial x}\right]n = \frac{1}{2}\rho C_D a U^2 + \frac{\partial(\tau n)}{\partial z} \qquad [2]$$

Here, ρ is water density, g is gravitational potential, and z_b is the position of the bed relative to an absolute datum. The first term on the right-hand side is the hydraulic resistance associated with the canopy. The drag coefficient, C_D, is a function of the stem morphology as well as stem Reynolds number, defined as Ud/ν, with ν the kinematic viscosity. The second term on the right-hand side is the gradient of stress, τ arising from intermolecular forces (producing viscous stress) or turbulent motions (producing turbulent stress). Except close to a boundary, the viscous stresses are negligible relative to the turbulent (or Reynolds') stresses. Furthermore, within an emergent canopy the turbulent stresses tend to be small, because the length scale of turbulent eddies within the canopy is small compared with the water depth (see discussion later). Typically, the vegetation provides the dominant hydraulic resistance for $ad \geq 0.01$, and other stresses can be ignored. Then, for an emergent canopy, equation [2] reduces to

$$U(z) = \sqrt{\frac{2gSn}{C_D a(z)}} \qquad [3]$$

For simplicity, the sum of the pressure gradient and bed slope is abbreviated to $S=\partial(H+z_b)/\partial x$. Because the flow forcing, represented by gS, is not a function of the vertical coordinate, z, eqn [3] implies that the velocity profile $U(z)$ will vary inversely with the canopy density, i.e., U is highest where $C_D a$ is lowest.

For canopies with distinct basal stem regions, this leads to a local maximum in velocity close to the bed, below the level at which branching begins. For example, a near-bed maximum is often observed in *S. alterniflora*. In contrast, the more vine-like *Atriplex portuloides* has leaves that are evenly distributed over depth, and the resulting velocity profile is uniform over depth.

Next consider a submerged canopy, $H/h > 1$ (**Figure 2(b)** and **2(c)**). When a canopy is submerged, a strong shear (velocity gradient) is generated by the sharp discontinuity in resistance at the top of the canopy. This produces large turbulent stresses at the top of the canopy. Therefore, this term cannot be neglected in eqn [2], and the velocity structure will not be a simple inverse function of canopy density, as in eqn [3]. Rather, the turbulent stress is the dominant driver of flow in the upper region of the canopy. The Raupach–Thom model for the stress-driven, in-canopy flow was derived for terrestrial canopies but

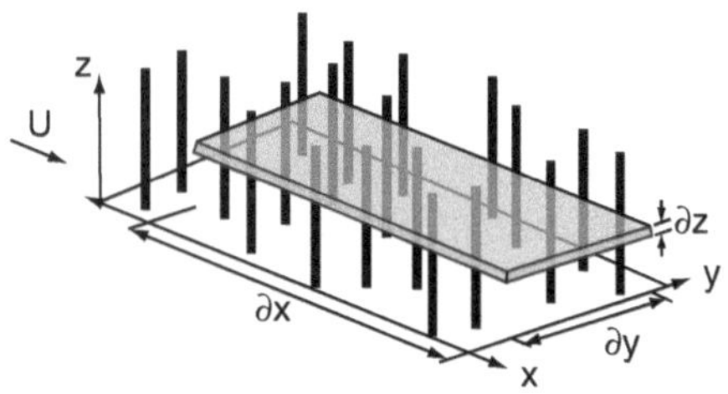

Figure 1 Conceptual canopy. The main velocity component, *U*, is in the positive *x* direction. Stem-scale heterogeneity is eliminated by considering a spatial average over the horizontal scales ∂x and ∂y, both larger than the stem spacing (ΔS). The vertical scale of the averaging volume, ∂z, is kept small to resolve vertical variation in canopy morphology.

can be applied to aquatic canopies. With U_h the velocity at the top of the canopy (at $z = h$), the profile within the canopy is

$$U(z) = U_3 + (U_h - U_3)\exp(\beta(z/h - 1)) \quad [4]$$

U_3 is the velocity in the lower canopy, given by eqn [3] (see **Figure 3**). Recall that $z = 0$ at the bed and is positive upward, so that this expression gives a maximum velocity at the top of the canopy (equal to U_h), and an exponential decay of velocity downwards into the canopy (**Figure 3**). The attenuation of the velocity within the canopy is characterized by the coefficient $\beta \sim C_D ah$, which reflects the momentum absorption of the canopy. In many cases, C_D is of the order of 1, so that

$$\beta \approx ah \quad [5]$$

For submergence depth $H/h > 1.5$, the velocity profile above the canopy is logarithmic, described by the zero-plan displacement z_m, the friction velocity, u_*, and the roughness height, z_0.

$$U(z) = \frac{u_*}{\kappa}\ln((z - z_m)/z_0) \quad [6]$$

The empirically derived von Karman constant κ has the value 0.4. The zero-plane displacement, z_m, represents the vertical position within the canopy past which momentum from the overflow cannot penetrate. Thus, $H - z_m$ is the effective depth of the overflow, and the friction velocity for the overflow region is defined as

$$u_* = \sqrt{g\frac{\partial H}{\partial x}(H - z_m)} \quad [7]$$

That is, ρu_*^2 is the equivalent bed stress provided by the canopy to the overflow and appearing to act

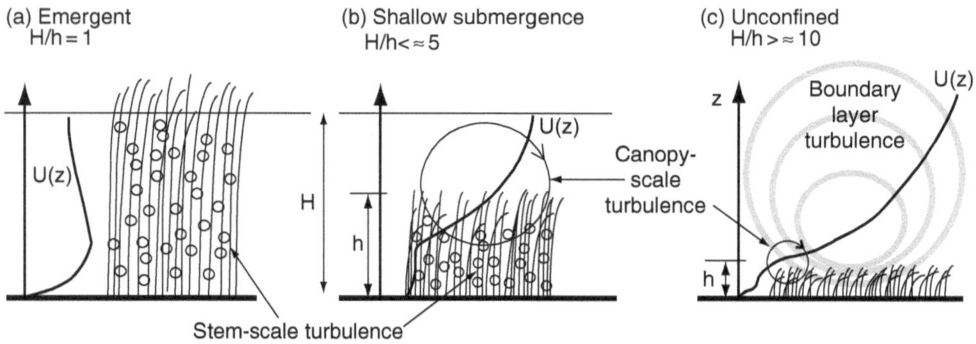

Figure 2 Vertical profiles of velocity, *U*(*z*), and turbulence scales for three depths of submergence. *H* is water depth and *h* is canopy height. (a) For an emergent canopy, $H/h = 1$, flow is driven predominantly by potential gradients from bed-slope and water-surface slope. Turbulence is limited to the stem scale. (b) With shallow submergence, $H/h < 5$, flow is driven by both potential gradients and turbulent stress. A region of strong shear at the top of the canopy generates canopy-scale turbulence. (c) An unconfined canopy, $H/h > 10$, behaves like a terrestrial canopy. Flow within the canopy is predominantly driven by the penetration of turbulent stress from above. Three scales of turbulence are present, at the stem, canopy, and boundary-layer scales.

at $z = z_m$. Both z_0 and z_m depend on the plant height, pronation, stiffness, and stem density.

The velocity profile over the full flow depth is achieved by matching the models for above-canopy and in-canopy flow at the top of the canopy, i.e. matching U_h (**Figure 3**). The transition region at the top of the canopy, i.e., between the velocity models, eqns [4] and [6], is often called the roughness sublayer (**Figure 3**).

Turbulence Structure with Submerged Canopies

The vegetative drag damps out channel-scale turbulence within the canopy. At the same time, however, the canopy may generate two new scales of turbulence. First, the wakes and shear layers that form behind and between individual stems and branches within the canopy generate stem-scale turbulence (**Figure 2**). Second, if the canopy is sufficiently dense

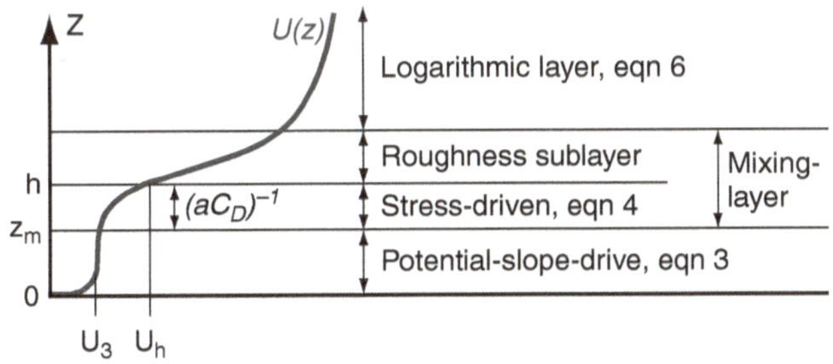

Figure 3 Velocity profile in and above a submerged canopy. Flow in the upper portion of the canopy flow is predominantly stress-driven. Turbulent stress penetrates downward into the canopy over attenuation scale $(aC_D)^{-1}$. Below this flow is driven by potential gradients associate with water-surface and/or bed-slope. At the top of the canopy the discontinuity in drag generates a region of strong shear. Above this the profile transitions to a logarithmic boundary-layer profile.

(see Sparse and Dense Canopies section), the discontinuity in drag at the top of the canopy produces a velocity profile that is strongly sheared and resembles a mixing layer (or a shear layer), including an inflection point located near the canopy interface. Large coherent vortices form via the Kelvin–Helmholtz (K–H) instability associated with the inflection point (**Figure 4**). These vortices generally grow to a scale comparable with the canopy height, and are called the canopy-scale turbulence (**Figure 2**). They dominate the exchange of scalar and momentum between the canopy and overflow. There is one important difference between free- and vegetated mixing layers. In a free shear-layer K–H vortices grow continually downstream. In a vegetated shear layer, however, the vortices reach a fixed scale and a fixed penetration into the canopy at a short distance from the canopy's leading edge. The penetration of the vortices into the canopy decreases as the stem area density increases. When the vortices penetrate only a fraction of the canopy height, the canopy is divided into two distinct regions. The upper canopy is flushed rapidly by the canopy-scale turbulence (**Figure 2(b)**). The lower canopy, however, experiences slower flushing, because only stem-scale turbulence exists in this region (**Figure 2(b)**). As a result of the difference in turbulence scale, the vertical diffusivity in the lower canopy can be an order of magnitude smaller (e.g., of order $0.1\,cm^2\,s^{-1}$) than the vertical diffusivity in the upper canopy (e.g., of order $1\,cm^2\,s^{-1}$).

The character of the mixing-layer and canopy-scale turbulence also depends on the canopy's depth of submergence (H/h). For unconfined canopies ($H/h >$ 10), the mixing-layer exists within a boundary layer of much greater scale (**Figure 2(c)**). Interference from much larger and stronger boundary-layer eddies cause the canopy-scale turbulence to be intermittent.

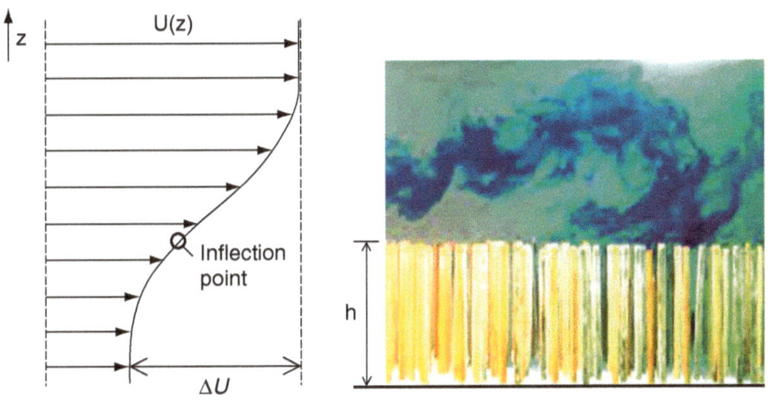

Figure 4 The drag discontinuity at the top of the canopy creates a region of shear with an inflection point, which in turn generates large coherent vortices via Kelvin–Helmholtz (K–H) instability. The velocity of vortex rotation scales with the velocity difference between the canopy and overflow, ΔU. Photo by Marco Ghisalberti.

In contrast, for shallow submergence (**Figure 2(b)**, $H/h < 5$), the overflow depth is insufficient to support large-scale boundary-layer turbulence. Under these conditions, the K–H vortices are very persistent, and the canopy-scale turbulence is the dominant turbulent scale in the overflow.

Sparse and Dense Canopies

The mean and turbulent flow discussed in this article describes the conditions in and above a dense canopy. A dense canopy is defined as having sufficient momentum absorption (drag) to generate an inflection point. In the presence of a sparse canopy, the velocity follows a rough boundary-layer profile over the full flow depth, with $z_m = 0$ in eqn [6]. Because sparse canopies do not generate an inflection point, they also do not generate the distinct canopy-scale turbulence described earlier. The transition between sparse and dense canopies occurs around $C_D ah = 0.1$. A conceptual picture of this transition is shown in **Figure 5**.

Flexible Canopies

Many submerged canopies are both buoyant and flexible, which results in a wide range of plant-flow interaction (**Figure 6**). When the flow depth is shallow relative to the blade length, and the current is weak, the canopy fills the water column and behaves as an emergent canopy (**Figure 6(a)**). As flow speed increases (**Figure 5(b)**), hydraulic drag pushes the blades over, clearing the upper water column and producing a two-layer velocity structure. High-speed flow occurs over the canopy and low-speed flow occurs within the canopy. This bifurcation of flow occurs whenever the water depth extends above the canopy height, even if the canopy is not severely pronated. When the canopy is in a steady, prone position (**Figure 5(b)**), the region within the canopy is sheltered from the overflow. In this position, both light penetration and vertical exchange of scalars, such as larvae, may be limited. If, however, the shear at the top of the canopy becomes sufficiently strong (**Figure 5(c)**), canopy-scale vortices will form via K–H instability, as discussed earlier. The vortices travel along the canopy at speed U_v that is greater than the mean velocity at the top of the canopy, U_h. The passage of the vortices causes a local depression in the canopy. The depression travels along the canopy interface, in synch with the traveling vortices. The progressive wave traveling along the canopy is called a *monami*. In between the depression, the canopy is lifted, relative to the steady prone condition.

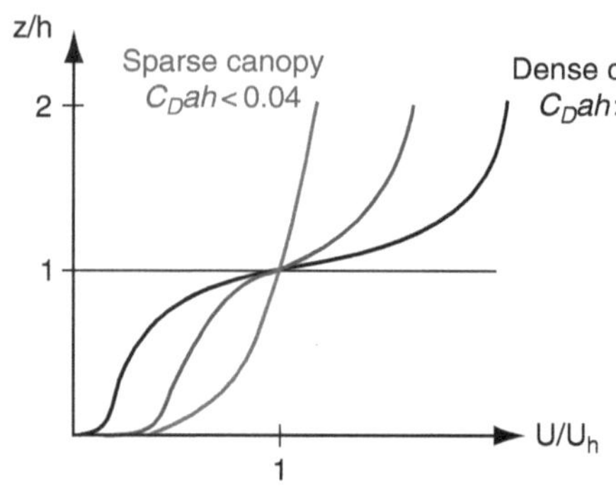

Figure 5 The shape of the velocity profile depends on the magnitude of the total canopy drag, characterized by $C_D ah$. When the canopy drag is high ($C_D ah > 0.1$), the profile resembles a mixing layer, with an inflection point near the top of the canopy. Under these conditions the velocity profile is modeled with a composite of three profiles, as shown in **Figure 3**. When the canopy drag is low ($C_D ah < 0.04$), the velocity profile is logarithmic, as in eqn [6] with $z_m = 0$.

Canopy Water Renewal

The hydrodynamic structure described here can be used to construct a physical model for water renewal

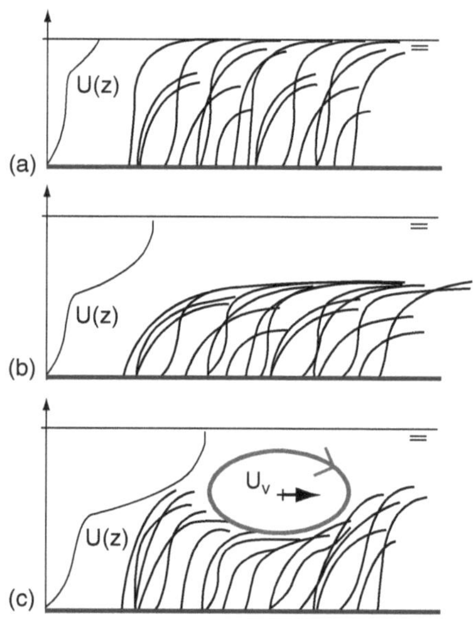

Figure 6 There is a wide range of plant-flow interaction for flexible canopies. Here, a flexible canopy responds to an increase in flow depth and flow magnitude. (a) For weak flow the grass, which is buoyant, is essentially emergent. (b) As flow depth and magnitude increase, flow accelerates over the canopy. With steady pronation, the blades shelter the interior of the canopy. (c) As the flow depth and magnitude increase further, the vortices generated by the inflection point at the top of the canopy become sufficiently strong to vigorously move the canopy blades. The passage of the street of vortices, each moving at speed U_v, generates a progressive waving of the canopy that is called "monami".

within a submerged canopy. Recall that the presence of the canopy generates two distinct scales of turbulence, the stem scale, which exists throughout the canopy, and the canopy scale, which is generated at the top of the canopy. The penetration of the canopy-scale vortices into the canopy is limited by the canopy drag, which is proportional to the stem area density. Specifically, the length scale of penetration measured from the top of the canopy is approximately

$$\delta_e \approx \frac{0.2}{C_D a} \qquad [8]$$

This scale divides the canopy into two regions (**Figure 7**). The upper canopy ($h - \delta_e < z < h$) flushes rapidly through the action of the canopy-scale vortices. The lower canopy ($z < h - \delta_e$), however, flushes more slowly, because only stem-scale turbulence exists in this region.

Exchange between the upper canopy and the overflow is controlled by the canopy-scale vortices, which have a velocity scale ΔU, which is the velocity difference between the canopy and the overflow. The canopy–overflow exchange is described by an exchange velocity k_e,

$$k_e = \Delta U/40 \qquad [9]$$

Exchange between the lower canopy and upper canopy is controlled by the stem-scale turbulence

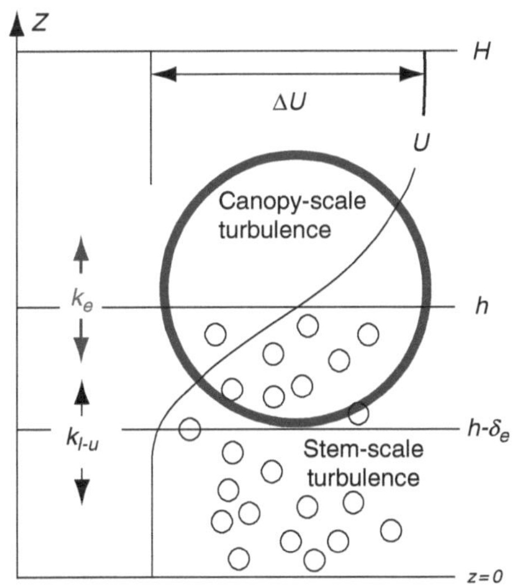

Figure 7 Length-scale δ_e divides the canopy into two regions. In the upper region ($h - \delta_e < z < h$) canopy-scale turbulence (grey) facilitates exchange with the overflow, described by exchange velocity k_e (eqn [9]). If the canopy-scale turbulence does not penetrate to the bed, the transport in the lower canopy ($z < h - \delta_e$) is controlled by stem-scale turbulence (black), which sets the rate of exchange between the lower and upper canopy, described by the exchange velocity $k_{l\text{–}u}$ (eqn [11]).

(**Figure 7**), which produces a vertical diffusivity with the form, $D_z \sim Ud$. Adopting the scale-constant observed in field canopies.

$$D_z = 0.17Ud \qquad [10]$$

U is the local, in-canopy velocity, which, in the lower canopy, can be predicted from eqn [3]. Since the slower vertical diffusion within the lower layer controls the exchange between the lower and upper canopy, the exchange velocity between these two zones can be approximated by

$$k_{l\text{–}u} = \frac{D_z}{h - \delta_e} \qquad [11]$$

The vertical exchanges described above renew the water in the canopy over a time-scale $T_{\text{vert.exch}}$, which depends upon the exchange between the upper and lower canopy (first term in eqn [12]) as well as the exchange between the upper canopy and the overflow (second term in eqn [12]).

$$T_{\text{vert.exch.}} = \frac{(h - \delta_e)}{k_{l-u}} + \frac{\delta_e}{k_e} \qquad [12]$$

Using the equations given above, the residence time of water within a submerged canopy can be predicted for a range of field conditions: $h = 25$ and 100 cm, $H/h = 5$, $S = 10^{-3}$ and 10^{-5}, and $C_D ah = 0.1$–10 (**Figure 8**). Note that the model does not extend to $C_D ah < 0.1$, because below this value canopy-scale vortices are not produced and vertical exchange between the canopy and overflow will not follow eqn [9]. The field conditions shown in **Figure 8** have a transition at $C_D ah \approx 0.3$. This marks the value of bulk canopy drag ($C_D ah$) at which the control of the vertical exchange switches between the upper and the lower canopy. For low canopy drag, $C_D ah < 0.3$, the K–H vortices penetrate through all or nearly all the canopy, with $\delta_e/h \geq 0.75$. Under these conditions, the first term in eqn [12] is negligible, and the vertical exchange time scale is set by the action of the canopy-scale vortices ($T_{\text{vert.exch}} = \delta_e/k_e$). For $C_D ah \gg 0.3$, the first term in eqn [12] is dominant, and the slow transport in the lower canopy controls the time scale for vertical exchange.

The range of field conditions considered here produce canopy renewal times of minutes to an hour, with the shortest, ~1 min, occurring for the short canopy, $h = 25$ cm, under the stronger forcing, $S = 10^{-3}$. The longest time scale, of order 10 h, occurs for the tall canopy, $h = 100$ cm, under the weaker forcing, $S = 10^{-5}$ (**Figure 8**). Surprisingly, the depth of submergence, H/h, has little influence on the exchange time scale for $H/h > 2$. These time scales are comparable with the retention time of sand bed forms, (10^2 s) and gravel bars (10^3 s), but are

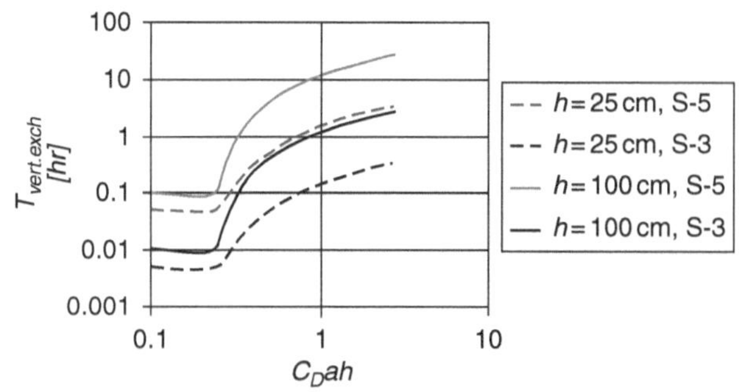

Figure 8 Time scale for vertical flushing of canopy for canopy height $h = 25$ cm (dashed lines) and $h = 100$ cm (solid lines). Two values of potential gradient are considered: $S = 10^{-3}$ (black lines) and $S = 10^{-5}$ (grey lines).

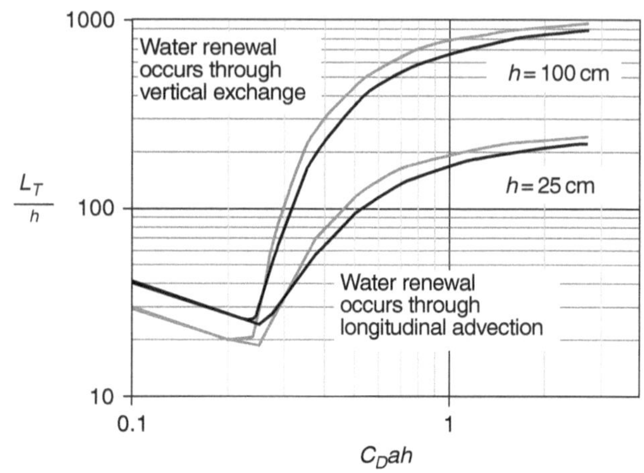

Figure 9 Length of canopy at which the control of water renewal changes from longitudinal advection to vertical exchange (L_T) for canopy height $h = 25$ and 100 cm. Two values of submergence depth are considered: $H/h = 2$ (black) and $H/h = 5$ (grey). The potential slope (S) has no influence on L_T, such that the curves $S = 10^{-3}$ and 10^{-5} overlap.

less than the retention time associated with bank or point bars (10^5 s). These comparisons suggest that submerged plants contribute to hydrologic storage at levels comparable with small hyporheic features (10–100 cm).

The above discussion focused on water renewal achieved by vertical exchange. However, water within a canopy may also be renewed by longitudinal advection, i.e. water entering at the leading edge of a canopy and flowing through the canopy to the trailing edge. For a canopy of length L, this process occurs over time-scale $T_{adv} = L/U_c$, with U_c the mean in-canopy velocity. It is useful to distinguish the conditions for which vertical exchange or longitudinal advection control the canopy water renewal. For this purpose, we define a transition length scale, $L_T = U_c T_{vert.exch}$. If the canopy length $L < L_T$, then $T_{adv} < T_{vert.exch}$ and water renewal (retention) is controlled by longitudinal advection. If $L > L_T$ water renewal is controlled by vertical exchange.

Similar to $T_{vert.exch}$, L_T has a minimum close to $C_D ah \approx 0.3$ (**Figure 9**). The greatest influence on L_T comes from the canopy height, because $T_{vert.exch}$ increases as h increases for all conditions. The depth of submergence ($H/h = 2$ or 5, grey and black, respectively, in **Figure 9**) has only a small influence on L_T. The potential gradient, S, has negligible impact, because the canopy velocity $U_c \sim S^{1/2}$, but $T_{vert.exch} \sim S^{-1/2}$ so that the dependencies cancel in $L_T = U_c T_{vert.exch}$. The curves for $S = 10^{-3}$ and 10^{-5} entirely overlap. The range of conditions considered in **Figure 9** suggest a minimum value of $L_T/h \approx 20$, indicating that a canopy whose length is less than 20 times its height, $L < \approx 20h$, will nearly always be flushed by longitudinal advection. The transition to renewal dominated by vertical exchange is dependent on the canopy drag ($C_D ah$), as this controls the penetration of the canopy-scale vortices into the canopy. At low values of $C_D ah$, vertical exchange dominates when $L/h > \approx 50$. But, for high values of $C_D ah$, the transition does not occur until $L/h = 200$–1000. Real canopies range from $L/h = 4$ to over 100, indicating that canopies flushed either by longitudinal advection or vertical exchange occur in the field.

Summary

Submerged vegetation modifies the flow profile by creating a region of diminished flow within the canopy. The velocity profile is divided into three zones. First, above the canopy, the flow is logarithmic, with roughness length (z_0) and displacement height (z_m) dictated by the canopy morphology. Second, in the upper canopy, flow is driven by turbulent stress from above, and the velocity decays exponentially into the canopy. The abrupt change in fluid drag at the top of the canopy creates a region of strong shear, which in turn generates coherent turbulent structures that scale with the canopy height. The canopy-scale vortices rotate with a velocity proportional to the velocity difference between the canopy and overflow, ΔU. These vortices control the exchange of momentum and mass between the canopy and the overflowing water, producing an exchange velocity $k_e \sim \Delta U$. The penetration of vortices into the canopy is dependent on the planting density and blade stiffness, which are reflected in the canopy drag parameter $C_D a$, specifically the penetration scale $\delta_e \sim 1/(C_D a)$. This penetration scale dictates the vertical extent of the upper canopy region. Finally, in the lower canopy, i.e., below the penetration of the shear-scale turbulence, flow is driven by potential gradients. In this region, the velocity profile shape is inversely proportional to the local canopy density. In addition, vertical transport in this region is dominated by stem-scale turbulence, with vertical diffusivity $D_z \sim Ud$, the local velocity, and stem scale, respectively. The time scale for water renewal within a submerged canopy depends on the canopy density and height. It ranges from minutes to an hour. Canopies are flushed by vertical exchange when their length, $L \gg 20h$. Shorter canopies are flushed by longitudinal advection.

Acknowledgments

The author is grateful to the work of graduate students Brian White, Marco Ghisalberti, and Enda Murphy. Some of this material is based on work supported by NSF Grants EAR 9629259, EAR 0125056, EAR-0309188. Any opinions expressed in this material are those of the author and do not necessarily reflect the views of the NSF.

Further Reading

Ghisalberti M and Nepf H (2002) Mixing layers and coherent structures in vegetated aquatic flow. *Journal of Geophysical Research* 107(C2): 1–11.

Harvey J, Conklin M, and Koelsch R (2003) Predicting changes in hydrologic retention in an evolving semi-arid alluvial stream. *Advances in Water Research* 26: 939–950.

Lopez F and Garcia M (2001) Mean flow and turbulence structure of open-channel flow through non-emergent vegetation. *Journal of Hydraulic Engineering ASCE* 127(5): 392–402.

Nepf H (2004) Vegetated flow dynamics. In: Fagherazzi S, Marani M, and Blum LK (eds.) *The Ecogeomorphology of Tidal Marshes, Coastal and Estuarine Studies,* vol. 59. A GU.

Nepf H, Ghisalberti M, White B, and Murphy E (2007) Retention time and dispersion associated with submerged aquatic canopies. *Water Research* in press.

Poggi D, Porporato A, Ridolfi L, Albertson J, and Katul G (2004) The effect of vegetation density on canopy sub-layer turbulence. *Boundary Layer Meteorology* 111: 565–587.

Raupach M, Finnigan J, and Brunet Y (1996) Coherent eddies and turbulence in vegetation canopies: The mixing-layer analogy. *Boundary Layer Meteorology* 60: 375–395.

Raupach M and Thom A (1981) Turbulence in and above plant canopies. *Annual Reviews in Fluid Mechanics* 13: 97–129.

Sukhodolov A and Sukhodolova T (2006) Evolution of the mixing layers in turbulent flow over submersed vegetation. In: River Flow 2006, International Conference on Fluvial Hydr., A4001, Lisbon, Portugal, Sept. 6–8.

Hydrodynamical Modeling

B R Hodges, University of Texas at Austin, Austin, TX, USA

Introduction

We model lakes to visualize and quantify fluid flow, mass transport, and thermal structure. Understanding the evolving physical state (e.g., surface elevation, density, temperature, velocity, turbidity) is necessary for modeling fluxes of nutrients, pollutants, or biota in time-varying fields of one, two, or three space dimensions (1D, 2D, or 3D). Hydrodynamic modeling provides insight into spatial–temporal changes in physical processes seen in field data. For example, **Figure 1** shows temperature profiles simultaneously recorded at different stations around Lake Kinneret (Israel). Extracts from model results (**Figure 2**) provides a context for interpreting these data as a coherent tilting of the thermocline. A time series of the thermocline can be animated, showing the principal motion is a counter-clockwise rotation of the thermocline. The complexities of the thermocline motion can be further dissected by using spectral signal processing techniques to separate wave components, illustrating a basin-scale Kelvin wave, a first-mode Poincaré wave, and a second-mode Poincaré wave.

It is said we build by 'measuring with micrometer, marking with chalk, then cutting with an axe'. However, hydrodynamic modeling inverts this process: we take an axe to the real world for our governing equations; we chalk a model grid on our lake, then numerically solve to micrometer precision. Thus, the governing equation approximations, grid selection, and numerical method all affect how a model reflects the physical world. Selecting an appropriate model requires understanding how model construction may affect the model solution.

As a broad definition, hydrodynamic modeling is the art and science of applying conservation equations for momentum, continuity, and transport (**Figure 3**) to represent evolving velocity, density, and scalar fields. The modeling science is founded upon incompressible fluid Newtonian continuum mechanics, which can be reduced to (1) any change in momentum must be the result of applied forces, and (2) the net flux into or out of a control volume must balance the change in the control volume. The modeling art is in selecting approximations, dimensionality, and methods that fit the natural system and provide adequate answers to the question asked.

Tables 1–4 list some of the 1D, 2D, and 3D lake modeling work from the mid-1990s to the present in the refereed literature. Unfortunately, much of the details of model development have been relegated to technical reports that are often either unavailable or difficult to obtain. Similarly, modeling applications conducted by or for government agencies often does not reach refereed publication. However, technical reports are generally detailed and valuable resources for applying and analyzing results; the reader is encouraged to seek out these publications before applying any model.

Dimensionality and Capabilities

Lake and reservoir models range from simple representation of thermocline evolution to multidimensional modeling of transport and water quality. Averaging (or integrating) the governing equations across one or more spatial dimensions provides representation of larger areas with less computational power. Such reduced-dimension models require less boundary condition data but more parameterization/calibration data. The simplest lake models average over horizontal planes (i.e., x and y directions) to obtain a 1D-model of the vertical (z-axis) lake stratification. Narrow reservoirs are modeled in 2D by laterally-averaging across the reservoir, thereby representing both vertical stratification and horizontal gradients from the headwaters to the dam. In shallow lakes, 2D-models in the x-y plane are used to examine depth-averaged circulation (without stratification). These reduced-dimension approaches cannot directly simulate variability in an averaged direction. However, such variability may be parameterized where processes are sufficiently well understood; in contrast, where processes are not understood or cannot be parameterized, the missing variability affects calibration and model results. Modelers must be careful not to simply parameterize or calibrate a poorly understood process by arbitrarily altering model coefficients.

Increasing model dimensionality and complexity reduces the time-scale over which a lake can be modeled. Typically, 1D-models can be applied for decades; 2D-models over multiple years to decades; 3D-models over days/weeks/months (but have been applied for longer in a few cases). This inverse relationship between dimensionality and time is not simply a function of computational power, but is inherent in the effects of stratification, mixing parameterizations, and model data requirements. For 1D-models,

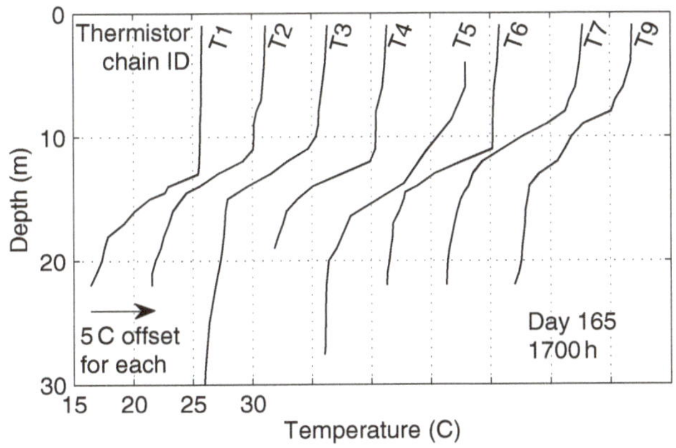

Figure 1 Temperature profiles collected in Lake Kinneret in 1997 (data of J. Imberger, Centre for Water Research, University of Western Australia).

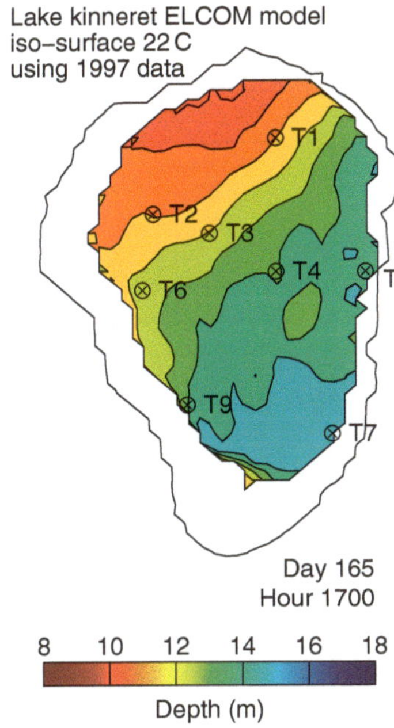

Figure 2 Modeled depth of temperature isosurface in the thermocline. Results from 3D-model at same time as field data collected in **Figure 1**.

vertical mixing is readily parameterized or calibrated to maintain sharp thermal stratification. However in 2D- and 3D-models, vertical mixing is caused by both the turbulence model and numerical diffusion of mass (a model transport error). Numerical diffusion always leads to weakening the thermocline, which affects modeled circulation and mixing that further weakens the thermal structure in a positive feedback cycle. Thus, longer-term 2D- and 3D-models require careful setup and analysis or results may be dominated by model error that accumulates as excessive mixing across the thermocline, resulting in poor prediction of residence time, mixing paths, and lake turnover.

For 1D-models (**Table 1**), we distinguish between turbulent mixing models derived from energy principals (e.g., DYRESM) or averaging transport equations (e.g., k-epsilon models) and pure calibration models that simply fit coefficients to nonphysical model equations (e.g., neural networks). Between these extremes are vertical advection/diffusion models (e.g., MINLAKE), which represent hydrodynamics by an advection/diffusion equation that requires site-based calibration. To the extent that more mechanistic models have all physical processes represented and correctly parameterized without site-specific calibration, they can be reasonably used for long-term predictions and readily transferred from lake to lake. Models relying solely on parameter fitting are questionable outside the calibration range, but are often easier to apply when sufficient calibration data is available. Although mechanistic models are arguably preferred, we rarely have sufficient data for a completely calibration-free mechanistic model; thus, in practical application such models are generally calibrated to some extent.

Laterally-averaged 2D-models (**Table 2**) are the workhorse of reservoir hydrodynamic/water quality modeling. For a narrow reservoir, the lateral-averaging paradigm is successful in capturing the bulk thermal/hydrodynamic processes, which are dominated by the large pelagic volume. However, where water quality processes are dependent on concentrations in shallow littoral regions, such models must be used carefully and with some skepticism. For example, a littoral algae bloom may depend on high nutrient concentrations that result from reduced circulation in the shallows. To correctly represent the bloom, a 2D water quality model must distort the biophysical relationships between growth rate and concentration. Furthermore, any scalar (e.g., toxic spill) represented simply by a concentration will automatically be diffused all the way across the reservoir, whether or not there is sufficient physical transport. Thus, a laterally averaged model will represent a toxic spill that mixes as a function of the reservoir width rather than physical processes.

Although 3D-models (**Table 3**) provide good representations of lake physics, they are notoriously complicated to set up and run. Although their operation is becoming easier with more established models, the progression to 'black box' modeling (i.e., where the user is not intimately familiar with the model itself) remains problematic. The interaction of the lake physics with the numerical method, governing equation approximations, time step, grid size, and initial/boundary condition data provides a wide scope for model inaccuracies. The effectiveness of a 3D-model

Momentum:

$$\underbrace{\frac{\partial u_i}{\partial t}}_{\text{Unsteady velocity}} + \underbrace{\sum_{j=1}^{3} u_j \frac{\partial u_i}{\partial x_j}}_{\text{Nonlinearity}} = \underbrace{-g_i}_{\text{Gravity force}} \underbrace{-\frac{1}{\rho_0}\frac{\partial P_{nh}}{\partial x_i}}_{\text{Non-hydrostatic pressure gradient}} \underbrace{-g\frac{\partial \eta}{\partial x_i}}_{\text{Free-surface or 'barotropic' pressure gradient}} \underbrace{-\frac{g}{\rho_0}\frac{\partial}{\partial x_i}\int_{z}^{\eta} \Delta\rho \, dx_3}_{\text{Stratification or 'baroclinic' pressure gradient}} + \underbrace{\sum_{j=1}^{3} \nu \frac{\partial^2 u_i}{\partial x_j^2}}_{\text{Viscous force}} \quad : i=1, 2, 3$$

Continuity:

$$\frac{\partial u_1}{\partial x_1} + \frac{\partial u_2}{\partial x_2} + \frac{\partial u_3}{\partial x_3} = 0$$

Free surface evolution

$$\frac{\partial \eta}{\partial t} + \frac{\partial}{\partial x_1}\int_{B}^{\eta} u_1 dx_3 + \frac{\partial}{\partial x_2}\int_{B}^{\eta} u_2 dx_3 = 0$$

x_i	Cartesian space ($i=1, 2$ are horizontal; $i=3$ is vertical)
u_i	Velocity
η	Free surface elevation
B	Bottom elevation
g	Gravitational acceleration
ν	Kinematic viscosity
P_{nh}	Non-hydrostatic pressure
ρ_0	Reference density
$\Delta\rho$	Difference between local density and reference density

Figure 3 General 3D incompressible flow equations (with the Boussinesq approximation) that are the basis for most hydrodynamic models. Hydrodynamics in lake modeling also requires transport equations for temperature, salinity (if important), and an equation of state for density.

presently depends on the user's understanding of the model capabilities and limitations. It is doubtful that we will see scientifically dependable black box models for at least another decade or so. Development of such models depends on development of expert systems that can replace a modeler's insight and experience in diagnosing different error forms.

Boundary and Initial Conditions

A hydrodynamic model is a numerical solution of an initial-and-boundary-value problem of partial differential or integral–differential equations. The model solution is never better than the initial and boundary conditions used for model forcing. Lake boundary conditions include spatially-varying wind field, thermal and mass exchange with the atmosphere, river inflows/outflow, groundwater exchanges, local catchment runoff, and precipitation. Boundary conditions may be poorly known, so understanding the model sensitivity to possible perturbations boundary conditions is necessary for setting the bounds of model believability.

Initial conditions are a snapshot of the system at time $t=0$ (the model start time). Problems arise from our inability to obtain the full data set necessary to initialize a model (a problem that increases with model dimensionality). These problems can be somewhat reduced by providing sufficient model 'spin-up' time so that the boundary forcing dilutes the initial condition error. However, spin-up is only successful when (1) the initial conditions are a reasonable approximation of $t=0$ and (2) the boundary forcing dominates the initial conditions by the end of spin-up. For example, in 3D-models the velocity initial condition is usually zero and the spin-up time is approximated by the 'spin-down' time from typical circulation velocities (i.e., the time over which inertia can be expected to keep water moving). However,

Table 1 Examples of 1D hydrodynamic lake models

Model	*Name/source*	*Notes*	*Details*	*Applications*
AQUASIM		tu	15, 31	15, 31, 32
DLM	Dynamic Lake Model	tu	23	7, 23, 24, 27, 33
DYRESM	Dynamic Reservoir Simulation Model	tu	14, 21	1, 5, 6, 14, 16, 17, 18, 22, 26, 35, 36
MINLAKE	Minnesota Lake Model	adv/dif	9, 20, 25	8, 9, 10, 11, 12, 13, 19, 34
Other		adv/diff	4, 30, 37	2, 3, 4, 30, 37
Other		tu	28	28, 29

tu: turbulent transport model; adv/diff: calibrated advection/diffusion transport model.

Sources

1. Balistrieri LS, Tempel RN, Stillings LL, and Shevenell LA (2006) Modeling spatial and temporal variations in temperature and salinity during stratification and overturn in Dexter Pit Lake, Tuscarora, Nevada, USA. *Applied Geochemistry* 21(7): 1184–1203.
2. Bell VA, George DG, Moore RJ, and Parker J (2006) Using a 1-D mixing model to simulate the vertical flux of heat and oxygen in a lake subject to episodic mixing. *Ecological Modelling* 190(1–2): 41–54.
3. Bonnet MP and Poulin M (2004) DyLEM-1D: A 1D physical and biochemical model for planktonic succession, nutrients and dissolved oxygen cycling application to a hyper-eutrophic reservoir. *Ecological Modelling* 180(2–3): 317–344.
4. Bonnet MP, Poulin M, and Devaux J (2000) Numerical modeling of thermal stratification in a lake reservoir. Methodology and case study. *Aquatic Sciences* 62(2): 105–124.
5. Bruce LC, Hamilton D, Imberger J, Gal G, Gophen M, Zohary T, and Hambright KD (2006) A numerical simulation of the role of zooplankton in C, N and P Cycling in Lake Kinneret, Israel. *Ecological Modelling* 193(3–4): 412–436.
6. Campos H, Hamilton DP, Villalobos L, Imberger J, and Javam A (2001) A modelling assessment of potential for eutrophication of Lake Rinihue, Chile. *Archiv Fur Hydrobiologie* 151(1): 101–125.
7. Coats R, Perez-Losada J, Schladow G, Richards R, and Goldman C (2006) The warming of Lake Tahoe. *Climatic Change* 76(1–2): 121–148.
8. Fang X and Stefan H G (1996) Long-term lake water temperature and ice cover simulations/measurements. *Cold Regions Science and Technology* 24(3): 289–304.
9. Fang X and Stefan HG (1997) Development and validation of the water quality model MINLAKE96 with winter data, Project Report No. 390, 33 pp. Minneapolis, MN: St. Anthony Falls Laboratory, University of Minnesota.
10. Fang X and Stefan HG (1998) Temperature variability in lake sediments. *Water Resources Research* 34(4): 717–729.
11. Fang X and Stefan HG (1999) Projections of climate change effects on water temperature characteristics of small lakes in the contiguous US. *Climatic Change* 42(2): 377–412.
12. Fang X, Stefan HG, and Alam SR (1999) Simulation and validation of fish thermal DO habitat in north-central US lakes under different climate scenarios. *Ecological Modelling* 118(2–3): 167–191.
13. Fang X, Stefan HG, Eaton JG, McCormick JH, and Alam SR (2004) Simulation of thermal/dissolved oxygen habitat for fishes in lakes under different climate scenarios – Part 1. Cool-water fish in the contiguous US. *Ecological Modelling* 172(1)
14. Gal GJ, Imberger T Zohary, Antenucci J, Anis A, and Rosenberg T (2003) Simulating the Thermal Dynamics of Lake Kinneret. *Ecological Modelling* 162 (1–2): 69–86.
15. Goudsmit GH, Burchard H, Peeters F, and Wuest A (2002) Application of k-epsilon turbulence models to enclosed basins: The role of internal seiches. *Journal of Geophysical Research-Oceans* 107(C12): 13.
16. Hamblin PF, Stevens CL, and Lawrence GA (1999) Simulation of vertical transport in mining pit lake. *Journal of Hydraulic Engineering-ASCE* 125(10): 1029–1038.
17. Hamilton DP, Hocking GC, and Patterson JC (1997) Criteria for selection of spatial dimension in the application of one- and two-dimensional water quality models. *Mathematics and Computers in Simulation* 43(3–6): 387–393.
18. Han BP, Armengol J, Garcia JC, Comerma M, Roura M, Dolz J, and Straskraba M (2000) The thermal structure of Sau Reservoir (NE: Spain): A simulation approach. *Ecological Modelling* 125(2–3): 109–122.
19. Herb WR and Stefan HG (2005) Dynamics of vertical mixing in a shallow lake with submersed macrophytes. *Water Resources Research* 41(2): 14.
20. Hondzo M and Stefan HG (1993) Lake water temperature simulation model. *Journal of Hydraulic Engineering-ASCE* 119(11): 1251–1273.
21. Imberger J and Patterson JC (1981) A dynamic reservoir simulation model: DYRESM 5. In Fischer HB (ed.) *Transport Models for Inland and Coastal Waters*, pp. 310–361. New York: Academic Press.
22. Kusakabe M, Tanyileke GZ, McCord SA, and Schladow SG (2000) Recent pH and CO_2 profiles at Lakes Nyos and Monoun, Cameroon: Implications for the degassing strategy and its numerical simulation. *Journal of Volcanology and Geothermal Research* 97(1–4): 241–260.
23. McCord SA and SG Schladow (1998) Numerical Simulations of Degassing Scenarios for CO_2-Rich Lake Nyos, Cameroon. *Journal of Geophysical Research-Solid Earth* 103(B6): 12355–12364.
24. McCord SA, Schladow SG, and Miller TG (2000) Modeling Artificial Aeration Kinetics in Ice-Covered Lakes. *Journal of Environmental Engineering-ASCE* 126(1): 21–31.
25. Riley MJ and Stefan HG (1988) MINLAKE – A dynamic lake water-quality simulation-model. *Ecological Modelling* 43(3–4): 155–182.
26. Romero JR, Antenucci JP, and Imberger J (2004) One- and three-dimensional biogeochemical simulations of two differing reservoirs. *Ecological Modelling* 174(1–2): 143–160.
27. Rueda FJ, Fleenor WE, and de Vicente I (2007) Pathways of river nutrients towards the euphotic zone in a deep-reservoir of small size: Uncertainty analysis. *Ecological Modelling* 202(3–4): 345–361.
28. Sahlberg J (2003) Physical modelling of the Akkajaure reservoir. *Hydrology and Earth System Sciences* 7(3): 268–282.
29. Sahlberg J and L Rahm (2005) Light limitation of primary production in high latitude reservoirs. *Hydrology and Earth System Sciences* 9(6): 707–720.
30. Salencon MJ (1997) Study of the thermal dynamics of two dammed lakes (Pareloup and Rochebut, France), using the EOLE model. *Ecological Modelling* 104(1): 15–38.

31. Schmid M, Lorke A, Wuest A, Halbwachs M, and Tanyileke G (2003) Development and sensitivity analysis of a model for assessing stratification and safety of Lake Nyos during artificial degassing. *Ocean Dynamics* 53(3): 288–301.
32. Schmid M, Halbwachs M, and Wuest A (2006) Simulation of CO_2 concentrations, temperature, and stratification in Lake Nyos for different degassing scenarios. *Geochemistry Geophysics Geosystems* 7: 14.
33. Sherman B, Todd CR, Koehn JD, and Ryan T (2007) Modelling the impact and potential mitigation of cold water pollution on Murray cod populations downstream of Hume Dam, Australia. *River Research and Applications* 23(4): 377–389.
34. Stefan HG, Fang X, and Hondzo M (1998) Simulated climate change effects on year-round water temperatures in temperate zone lakes. *Climatic Change* 40(3–4): 547–576.
35. Straskraba M and Hocking G (2002) The effect of theoretical retention time on the hydrodynamics of deep river valley reservoirs. *International Review of Hydrobiology* 87(1): 61–83.
36. Wallace BB and Hamilton DP (2000) Simulation of water-bloom formation in the cyanobacterium *Microcystis aeruginosa*. *Journal of Plankton Research* 22(6): 1127–1138.
37. Wiese BU, Palancar MC, Aragon JM, Sanchez F, and Gil R (2006) Modeling the Entrepenas Reservoir. *Water Environment Research* 78(8): 781–791.

a lake model with temperature initial condition that does not reflect the initial real-world stratification cannot recover through spin-up.

Calibration

Ideally, hydrodynamic models should not require calibration; i.e., with sufficient data for boundary conditions, initial conditions and turbulence coefficients, a model should adequately represent the physics of the real system. Unfortunately, our data and parameterizations are often inadequate. Calibration may be either through adjusting turbulence coefficients (e.g., the various '*c*' values in a k-epsilon model) or by adjusting boundary conditions. Modelers often jump straight into adjusting a turbulence model rather than examining the sensitivity of the model to inaccuracies in the boundary conditions. For example, wind-driven lakes may have unknown spatial gradients of the wind, or the wind sensor may be biased (e.g., in the wind shadow of a building). If the applied wind data under-predicts the actual wind forcing, then calibrating the turbulence model could lead to the 'right' answer for the wrong reason! There should be evidence that the calibrated process is the data mismatch problem (not just the solution). Furthermore, naïve calibration of turbulence coefficients can lead to unphysical values (e.g., an efficiency greater than unity should be a warning sign that something has been missed).

Hydrostatic Approximation

Horizontal length scales are larger than vertical scales in lakes and reservoirs, so the hydrostatic approximation is generally employed. This approximation neglects vertical acceleration ($\partial u_3/\partial t$) and nonhydrostatic pressure gradients ($\partial P_{\mathrm{nh}}/\partial x_i$). Note that vertical transport may be reasonably represented in a hydrostatic model, even while vertical acceleration is neglected. Vertical transport has multiple causes: continuity applied to divergence/convergence in the horizontal plane, turbulent mixing, and vertical inertial effects; only the latter is hydrostatically neglected.

Although large-scale free surface motions and the resulting circulations are well-represented by a hydrostatic model, internal seiches are more problematic. Tilting of a pycnocline (e.g., a thermocline) may be relatively steep and ensuing basin-scale waves may evolve in a nonhydrostatic manner. However in a hydrostatic model, the numerical dispersion errors may mimic nonhydrostatic behavior. Thus, physically correct wave dispersion may be serendipitously achieved when numerical dispersion is similar to physical dispersion. Such results must be used with caution as they are highly grid-dependent and the serendipitous confluence of errors may disappear when the model grid is refined. When model results show greater disagreement with the physical world as the model grid is made finer, this type of error may be a suspect.

Applying nonhydrostatic models for large-scale natural systems requires significantly more computational time, model complexity, and modeling expertise than for similar hydrostatic models. An extremely fine model grid and time step is necessary resolve vertical accelerations and nonhydrostatic pressure gradients. Nonhydrostatic lake and ocean models are actively under development, but their widespread application is not imminent.

Model Grid

Overview

Hydrodynamic modeling requires discretizing physical space on a model grid. The size and characteristics of the grid determine the scales of what a model can and cannot represent. In the horizontal plane, there

Table 2 Examples of 2D hydrodynamic lake models

Model	*Name/Source*	*Notes*	*Details*	*Applications*
CE-QUAL-W2	U.S. Army Corps of Engineers	fd, la, Ca		1, 2, 3, 7, 8, 9, 10, 17, 19, 20, 21, 24
RMA2	Research Management Associates	fe, da, cu		14, 23
HYDROSIM	Hydrodynamic Simulation Model	fe, da, cu	11	18
others		la	5, 26	5, 12, 13, 26
others		da	4, 16, 22	4, 6, 16, 22

Numerical Method: fd = finite difference; fe = finite element
Horizontal Grid: Ca = Cartesian grid; cu = curvilinear grid
2D form: da = depth-averaged; la = laterally-averaged
Sources

1. Adams WR, Thackston EL, and Speece RE (1997) Modeling CSO impacts from Nashville using EPA's demonstration approach. *Journal of Environmental Engineering-ASCE* 123(2): 126–133.
2. Bartholow J, Hanna RB, Saito L, Lieberman D, and Horn M (2001) Simulated limnological effects of the Shasta Lake temperature control device. *Environmental Management* 27(4): 609–626.
3. Boegman L, Loewen MR, Hamblin PF, and Culver DA (2001) Application of a two-dimensional hydrodynamic reservoir model to Lake Erie. *Canadian Journal of Fisheries and Aquatic Sciences* 58(5): 858–869.
4. Borthwick AGL, Leon SC, and Jozsa J (2001) Adaptive quadtree model of shallow-flow hydrodynamics. *Journal of Hydraulic Research* 39(4): 413–424.
5. Botte V and Kay A (2000) A numerical study of plankton population dynamics in a deep lake during the passage of the Spring thermal bar. *Journal of Marine Systems* 26(3–4): 367–386.
6. Boudreau P, Leclerc M, and Fortin GR (1994) Modelisation Hydrodynamique du lac Saint-Pierre, fleuve Saint-Laurent: l'influence de la vegetation aquatique. *Canadian Journal of Civil Engineering* 21(3): 471–489.
7. Gelda RK and Effler SW (2007) Modeling turbidity in a water supply reservoir: Advancements and issues. *Journal of Environmental Engineering-ASCE* 133(2): 139–148.
8. Gelda RK and Effler SW (2007) Testing and application of a two-dimensional hydrothermal model for a water supply reservoir: implications of sedimentation. *Journal of Environmental Engineering and Science* 6(1): 73–84.
9. Gu RR and Chung SW (2003) A two-dimensional model for simulating the transport and fate of toxic chemicals in a stratified reservoir. *Journal of Environmental Quality* 32(2): 620–632.
10. Gunduz O, Soyupak S, and Yurteri C (1998) Development of water quality management strategies for the proposed Isikli reservoir. *Water Science and Technology* 37(2): 369–376.
11. Heniche M, Secretan Y, Boudreau P, and Leclerc M (2000) A two-dimensional finite element drying-wetting shallow water model for rivers and estuaries. *Advances in Water Resources* 23(4): 359–372.
12. Holland PR, Kay A, and Botte V (2001) A numerical study of the dynamics of the riverine thermal bar in a deep lake. *Environmental Fluid Mechanics* 1: 311–332.
13. Holland PR, Kay A, and Botte V (2003) Numerical modelling of the thermal bar and its ecological consequences in a river-dominated lake. *Journal of Marine Systems* 43(1–2): 61–81.
14. Jennings AA (2003) Modeling sedimentation and scour in small urban lakes. *Environmental Modelling & Software* 18(3): 281–291.
15. Kim Y and Kim B (2006) Application of a 2-dimensional water quality model (CE-QUAL-W2) to the turbidity interflow in a deep reservoir (Lake Soyang, Korea). *Lake and Reservoir Management* 22(3): 213–222.
16. Kramer T and Jozsa J (2007) Solution-adaptivity in modelling complex shallow flows. *Computers & Fluids* 36(3): 562–577.
17. Kuo JT, Lung WS, Yang CP, Liu WC, Yang MD, and Tang TS (2006) Eutrophication modelling of reservoirs in Taiwan. *Environmental Modelling & Software* 21(6): 829–844.
18. Martin C, Frenette JJ, and Morin J (2005) Changes in the spectral and chemical properties of a water mass passing through extensive macrophyte beds in a large fluvial lake (Lake Saint-Pierre, Quebec, Canada). *Aquatic Sciences* 67(2): 196–209.
19. Martin JL (1988) Application of two-dimensional water-quality model. *Journal of Environmental Engineering-ASCE* 114(2): 317–336.
20. Nestler JM, Goodwin RA, Cole TM, Degan D, and Dennerline D (2002) Simulating movement patterns of blueback herring in a stratified southern impoundment. *Transactions of the American Fisheries Society* 131(1): 55–69.
21. Saito L, Johnson BM, Bartholow J, and Hanna RB (2001) Assessing ecosystem effects of reservoir operations using food web-energy transfer and water quality models. *Ecosystems* 4(2): 105–125.
22. Sanmiguel-Rojas E, Ortega-Casanova J, del Pino C, and Fernandez-Feria R (2005) A Cartesian grid finite-difference method for 2D incompressible viscous flows in irregular geometries. *Journal of Computational Physics* 204(1): 302–318.
23. Shrestha PL (1996) An integrated model suite for sediment and pollutant transport in shallow lakes. *Advances in Engineering Software* 27(3): 201–212.
24. Sullivan AB, Jager HI, and Myers R (2003) Modeling white sturgeon movement in a reservoir: The effect of water quality and sturgeon density. *Ecological Modelling* 167(1–2): 97–114.
25. Wu RS, Liu WC, and Hsieh WH (2004) Eutrophication modeling in Shihmen Reservoir, Taiwan. *Journal of Environmental Science and Health Part A-Toxic/Hazardous Substances & Environmental Engineering* 39(6): 1455–1477.
26. Young DL, Lin QH, and Murugesan K (2005) Two-dimensional simulation of a thermally stratified reservoir with high sediment-laden inflow. *Journal of Hydraulic Research* 43(4): 351–365.

Table 3 Examples of 3D hydrodynamic lake models

Model	*Name*	*Notes*	*Details*	*Applications*
CH3D	Curvlinear Hydrodynamics in 3-Dimensions	cfd, cu, zl/sg, ms	10, 34	16
EFDC	Environmental Fluid Dynamics Code	cfd, cu, zl, ms		9, 12, 13, 14, 15
ELCOM	Estuary and Lake Computer Model	cfd, Ca, zl, si	11	2, 7, 8, 11, 17, 18, 19, 20, 21, 23, 25, 26, 27, 28
GLLVHT	Generalized Longitudinal Lateral Vertical Hydrodynamic and Transport Model	fd, cu, we, si		24, 36
POM; ECOM	Princeton Ocean Model; Estuary and Coastal Ocean Model	cfd, cu, sg/zl, ms/si	1	1, 3, 4, 5, 33, 35, 37
RMA10	Research Management Associates 10	fe, un, zl		6, 29
SI3D	Semi-Implicit 3D	cfd, cu, zl, si	31	30, 31, 32

Numerical Method: cfd – conservative finite difference; fd – finite difference; fe – finite element.
Horizontal Grid: Ca – Cartesian grid; cu – curvilinear grid; un – unstructured grid.
Time-stepping: ms – mode-splitting; si – semi-implicit.
Vertical Grid: zl – z-level vertical grid; sg – sigma grid.
Sources

1. Ahsan A and Blumberg AF (1999) Three-dimensional hydrothermal model of Onondaga Lake, New York. *Journal of Hydraulic Engineering-ASCE* 125(9): 912–923.
2. Appt J, Imberger J, and Kobus H (2004) Basin-scale motion in stratified upper Lake Constance. *Limnology and Oceanography* 49(4): 919–933.
3. Beletsky D (2001) Modeling wind-driven circulation in Lake Ladoga. *Boreal Environment Research* 6(4): 307–316.
4. Blumberg AF, Khan LA, *et al.* (1999) Three dimensional hydrodynamic model of New York harbor region. *Journal of Hydraulic Engineering-ASCE* 125(8): 799–816.
5. Chen CS, Zhu JR, Kang KY, Liu HD, Ralph E, Green SA, and Budd JW (2002) Cross-frontal transport along the Keweenaw coast in Lake Superior: A Lagrangian model study. *Dynamics of Atmospheres and Oceans* 36(1–3): 83–102.
6. Cook CB, Orlob GT, and Huston DW (2002) Simulation of wind-driven circulation in the Salton Sea: Implications for indigenous ecosystems. *Hydrobiologia* 473(1–3): 59–75.
7. Dallimore CJ, Hodges BR, and Imberger J (2003) Coupling an underflow model to a three-dimensional hydrodynamic model. *Journal of Hydraulic Engineering-ASCE* 129(10): 748–757.
8. Dallimore CJ, Imberger J, and Hodges BR (2004) Modeling a plunging underflow. *Journal of Hydraulic Engineering-ASCE* 130(11): 1068–1076.
9. Elci S, Work PA, and Hayter EJ (2007) Influence of stratification and shoreline erosion on reservoir sedimentation patterns. *Journal of Hydraulic Engineering-ASCE* 133(3): 255–266.
10. Gessler D, Hall B, Spasojevic M, Holly F, Pourtaheri H, and Raphelt N (1999) Application of 3D mobile bed, hydrodynamic model. *Journal of Hydraulic Engineering-ASCE* 125(7): 737–749.
11. Hodges BR, Imberger J, Saggio A, and Winters KB (2000) Modeling basin-scale internal waves in a stratified lake. *Limnology and Oceanography* 45(7): 1603–1620.
12. Jin KR and Ji ZG (2005) Application and validation of three-dimensional model in a shallow lake. *Journal of Waterway Port Coastal and Ocean Engineering-ASCE* 131(5): 213–225.
13. Jin KR, Hamrick JH, and Tisdale T (2000) Application of three-dimensional hydrodynamic model for Lake Okeechobee. *Journal of Hydraulic Engineering-ASCE* 126(10): 758–771.
14. Jin KR, Ji ZG, and Hamrick JH (2002) Modeling winter circulation in Lake Okeechobee, Florida. *Journal of Waterway Port Coastal and Ocean Engineering-ASCE* 128(3): 114–125.
15. Jin KR, Ji ZG, and James RT (2007) Three-dimensional water quality and SAV modeling of a large shallow lake. *Journal of Great Lakes Research* 33(1): 28–45.
16. Kim SC, Cerco CF, and Johnson BH (2006) Three-dimensional management model for Lake Washington, part I: Introduction and hydrodynamic modeling. *Lake and Reservoir Management* 22(2): 103–114.
17. Laval BE, Imberger J, and Findikakis AN (2005) Dynamics of a large tropical lake: Lake Maracaibo. *Aquatic Sciences* 67(3): 337–349.
18. Laval B, Hodges BR, and Imberger J (2003) Reducing numerical diffusion effects with pycnocline filter. *Journal of Hydraulic Engineering-ASCE* 129(3): 215–224.
19. Laval B, Imberger J, and Findikakis AN (2003) Mass transport between a semienclosed basin and the ocean: Maracaibo System. *Journal of Geophysical Research-Oceans* 108(C7).
20. Laval B, Imberger J, Hodges BR, and Stocker R (2003) Modeling circulation in lakes: Spatial and temporal variations. *Limnology and Oceanography* 48(3): 983–994.
21. Leon LF, Lam DC L, Schertzer WM, Swayne DA, and Imberger J (2007) Towards coupling a 3d hydrodynamic lake model with the Canadian Regional climate model: Simulation on Great Slave Lake. *Environmental Modelling & Software* 22(6): 787–796.
22. Leon LK, Imberger J, Smith REH, Hecky RE, Lam DCL, and Schertzer WM (2005) Modeling as a tool for nutrient management in Lake Erie: A hydrodynamics study. *Journal of Great Lakes Research* 31: 309–318.
23. Marti CL and Imberger J (2006) Dynamics of the benthic boundary layer in a strongly forced stratified lake. *Hydrobiologia* 568: 217–233.
24. Na EH and Park SS (2006) A hydrodynamic and water quality modeling study of spatial and temporal patterns of phytoplankton growth in a stratified lake with buoyant incoming flow. *Ecological Modelling* 199(3): 298–314.
25. Okely P and Imberger J (2007) Horizontal transport induced by upwelling in a canyon-shaped reservoir. *Hydrobiologia* 586: 343–355.

26. Romero JR and Imberger J (2003) Effect of a flood underflow on reservoir water quality: Data and three-dimensional modeling. *Archiv Fur Hydrobiologie* 157(1): 1–25.
27. Romero JR, Antenucci JP, and Imberger J (2004) One- and three-dimensional biogeochemical simulations of two differing reservoirs. *Ecological Modelling* 174(1–2): 143–160.
28. Romero JR, Antenucci JP, and Imberger J (2004) One- and three-dimensional biogeochemical simulations of two differing reservoirs. *Ecological Modelling* 174(1–2): 143–160.
29. Rueda FJ and Schladow SG (2002) Quantitative comparison of models for barotropic response of homogeneous basins. *Journal of Hydraulic Engineering-ASCE* 128(2): 201–213.
30. Rueda FJ and Schladow SG (2003) Dynamics of large polymictic lake. Ii numerical simulations. *Journal of Hydraulic Engineering-ASCE* 129(2): 92–101.
31. Rueda FJ, Schladow SG, and Palmarsson SO (2003) Basin-scale internal wave dynamics during a winter cooling period in a large lake. *Journal of Geophysical Research-Oceans* 108(C3), 3097, doi:10.1029/2001JC000942.
32. Rueda FJ, Schladow SG, Monismith SG, and Stacey MT (2005) On the effects of topography on wind and the generation of currents in a large multi-basin lake. *Hydrobiologia* 532: 139–151.
33. Schwab DJ and Beletsky D (2003) Relative effects of wind stress curl, topography, and stratification on large-scale circulation in Lake Michigan. *Journal of Geophysical Research – Oceans* 108(C2).
34. Sheng YP (1990) Evolution of a three-dimensional curvilinear-grid hydrodyamic model for estuaries, lakes and coastal waters: CH3D. In *Estuarine and Coastal Modeling: Proceedings of the Conference*, Newport, Rhode Island, November 15–17, 1989. ASCE.
35. Song Y, Semazzi FHM, Xie L, and Ogallo LJ (2004) A coupled regional climate model for the Lake Victoria basin of East Africa. *International Journal of Climatology* 24(1): 57–75.
36. Wu J, Buchak EM, Edinger JE, and Kolluru VS (2001) Simulation of cooling-water discharges from power plants. *Journal of Environmental Management* 61(1): 77–92.
37. Zhu JR, Chen CS, Ralph E, Green SA, Budd JW, and Zhang FY (2001) Prognostic modeling studies of the Keweenaw current in Lake Superior. Part II: Simulation. *Journal of Physical Oceanography* 31(2): 396–410.

are three grid systems: Cartesian, curvilinear, and unstructured. For models including a vertical dimension, the vertical grid may be terrain-following (sigma coordinate), Cartesian (z-level) or isopycnal (Lagrangian). Unstructured grids can also be used in the vertical plane, but have not been widely adopted.

Grid Size and Convergence

The local grid size controls the 'resolution' of local processes; e.g., a single grid cell has only a single velocity on a simple Cartesian finite-difference grid. Thus, the grid mesh is a top-level control on the resolvable physics and transport. For example, if only two grid cells are used across a narrow channel the transport may be theoretically either unidirectional or bidirectional; however, two grid cells cannot represent a deep center channel flow with return flows along both shallow banks. A useful exercise is to consider how many grid cells are necessary to represent 1.5 periods of a sine wave: although three cells is clearly the minimum, the resulting discrete pattern will not be particularly sinusoidal. Arguably, 10–15 grid cells should be the minimum resolution for most important flow features. An effective model grid resolves the key physical features at practical computational cost. Grid design should be an iterative process wherein model results at different grid scales are compared to gain insight into model performance. A model grid is 'converged' when further grid refinement does not significantly change model results. Unfortunately, obtaining a converged grid is not always practical; indeed, most large-scale models suffer from insufficient grid resolution. Such models may still have validity, but grid-scale effects may dominate physical processes.

Horizontal Grid Systems

Cartesian grids are obtained with a square or rectangular mesh (**Figure 4(a)**). The mesh structure allows simple model coding since a grid cell's neighbors are easily determined. For multidimensional models, simple Cartesian grids cannot be applied with fine resolution in some regions and coarse resolution in others. These deficiencies can be addressed with 'plaid' structured meshes (i.e., nonuniform Cartesian grid spacing), domain decomposition or nested grid (e.g., quadtree) techniques. To use an efficient rectangular mesh on a sinuous reservoir, the topography may be straightened along the channel centerline before applying the Cartesian mesh.

Curvilinear grids in the horizontal plane are structured meshes (similar to a Cartesian grid) that smoothly distort the quadrilateral elements throughout horizontal space (**Figure 4(b)**). The distortion between physical (x,y) space and curvilinear (ξ,η) space requires transformation of the governing equations. Curvilinear meshes allow fine grid resolution in one area and coarse resolution in another, as long as the mesh changes smoothly between regions. The smoothness and orthogonality of the mesh (as seen in physical space) will affect the model solution. Reasonable rule-of-thumb criteria are (1) adjacent grid

Table 4 Model applications

Lake	*1D*	*2D*	*3D*
Akkajaure Reservoir (Sweden)	28		
Lake Alpnach (Switzerland)	15		
Lake Baldegg (Switzerland)	15		
Lake Balaton (Hungary)		4	
Lake Baikal (Russia)		5, 12	
Bassenthwaite Lake (UK)	2		
Lake Belau (Germany)		22	
Lake Beznar (Spain)	27		
Brenda Pit Lake (Canada)	16		
Brownlee Reservoir (USA)		24	
Lake Burragorang (Australia)	26		26, 27, 28
Cheatham Lake (USA)		1	
Clear Lake (USA)		23	30, 32
Lake Constance (Germany/Switzerland)			2
Cummings Lake (Canada)	24		
Dexter Pit Lake (USA)	1		
East Dollar Lake (Canada)	24		
Entrepenas Reservoir (Spain)	37		
Lake Erie (Canada/USA)		3	22
Flint Creek Lake (USA)			36
Great Slave Lake (Canada)			21
Hartwell Lake (USA)			9
Hume Reservoir (Australia)	33		
Isikli Reservoir (Turkey)		10	
Kamploops Lake (Canada)		13	
Lake Kinneret (Israel)	5, 14		11, 18, 20, 23
Lake Ladoga (Russia)			3
Lake Maracaibo (Venezuela)			17, 19
Lake Michigan (Canada/USA)			33
Lake Monoun (Cameroon)	22		
Mundaring Weir (Australia)			25
Lake Neusiedl		16	
Lake Nyos (Cameroon)	22, 23, 31, 32		
Lake Ogawara (Japan)			7
Lake Okeechobee (USA)			12, 13, 14, 15
Onondaga Lake (USA)			1
Orlik Reservoir (Czech Republic)	35		
Otter Lake (USA)	19		
Lake Paldang (South Korea)			24
Pareloup Reservoir (France)	30		
Pavin Crater Lake (Canada)	16		
Prospect Reservoir (Australia)	26		
Lake Rinihue (Chile)	6		
Rochebut Reservoir (France)	30		
Lake Saint Pierre (Canada)		6, 18	
Salton Sea (USA)			6
Sau Reservoir (Spain)	18		
Schoharie Reservoir (USA)		7, 8	
Lake Shasta (USA)		2, 9, 21	
Lower Shaker Lake (USA)		14	
Shihmen Reservoir (Taiwan)		25	
Slapy Reservoir (Czech Republic)	35		
Lake Soyang (Korea)		15	
J. Strom Thurmond Lake (USA)		20	
Lake Superior (Canada/USA)			5, 37
Lake Tahoe (USA)	7		31
Te Chi Reservoir (Taiwan)		17, 26	
Tseng-Wen Reservoir (Taiwan)		17	
Lake Victoria (Kenya/Tanzania/Uganda)			35
Villerest Reservoir (France)	3, 4		
Lake Washington (USA)			16
Wellington Reservoir (Australia)			8
Lake Yangebup (Australia)	36		

Numbers correspond to notes from **Tables 1, 2,** and **3** for 1D, 2D and 3D models, respectively.

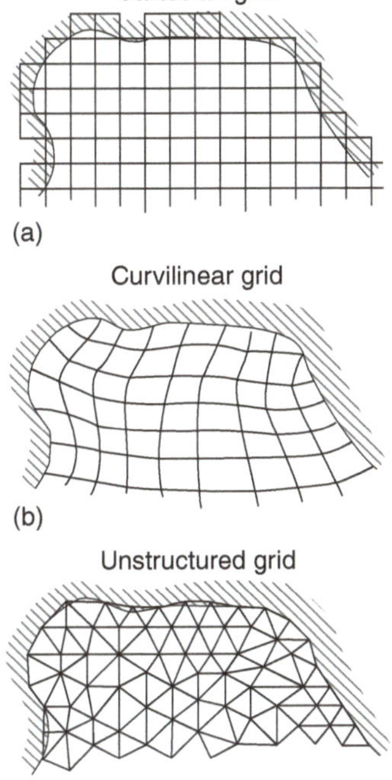

Figure 4 Plan view illustrating different horizontal grid systems.

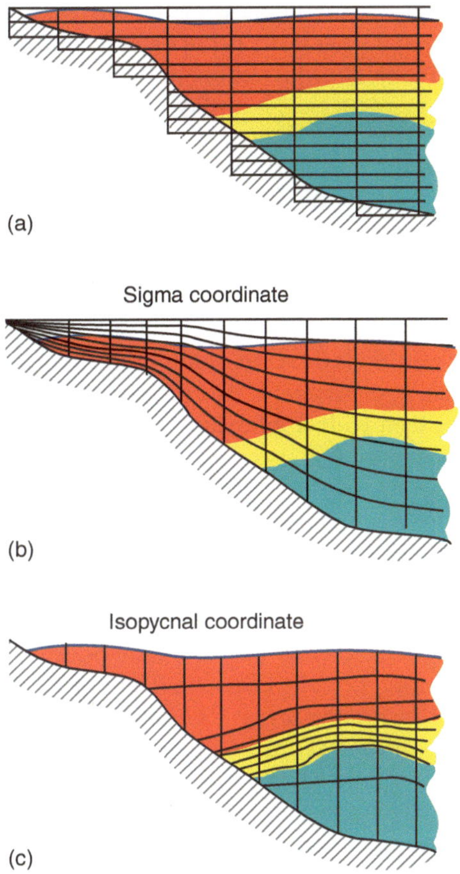

Figure 5 Elevation view illustrating different vertical grid systems relative to a stratified lake with warm (red) surface water, thermocline (yellow) and cooler (blue) hypolimnetic water.

cells should have increased/decreased volume by no more than 10% and (2) off-orthogonal transformation metrics should be an order of magnitude smaller than orthogonal metrics. Smooth curvilinear meshes can be manually designed with simple drafting tools, but stand-alone mesh creation software is generally used. Some models require orthogonality or near-orthogonality for the mesh, which severely constrains mesh creation.

Unstructured grids in the horizontal plane are composed of n-sided polygons (**Figure 4(c)**); triangular and quadrilateral elements are typically preferred or required. An unstructured mesh is the easiest for fitting complicated topography and arguably has the greatest flexibility for providing fine resolution in some areas with coarse resolution in others. However, model solutions are still affected by local gradients of grid cell volume and grid orthogonality. Creating a good unstructured grid is an art, requiring separate grid creation software and a lengthy trial-and-error process. It is often necessary to carefully examine model performance on several different unstructured grids to gain an understanding of how different grid choices affect the solution. Finite difference and finite volume models for unstructured meshes are relatively recent developments, but have not yet seen extensive use in lakes or reservoirs.

Vertical Grid Systems

Z-level grids are the simplest vertical system, using layers with whose thickness is uniform across the horizontal plane (**Figures 5(a)**). Layer thicknesses may vary in the vertical, but should do so smoothly (i.e., no more than about 10% expansion of thickness in adjacent layers). Z-level grids are generally preferred for 2D and 3D lake models due to their simplicity. A disadvantage is that steep bottom slopes are represented as discrete stair steps, which distorts along-slope flow. Coupling a 2D- or 3D-model with a benthic boundary layer model can overcome the stair-step problem, albeit by increasing model complexity and introduction of empiricism and ad hoc coupling mechanisms.

Sigma-coordinate (terrain following) vertical grid systems are commonly used in oceanic-scale modeling (e.g., the Great Lakes), but have significant drawbacks for inland waters. Sigma-coordinate systems divide each water column into a fixed number of layers, resulting in thick layers in deep water and thin layers in shallow water (**Figures 5(b)**). For sloping boundaries, the sigma-coordinate system must be truncated or a singularity occurs where the depth goes to zero. Sigma-coordinates are preferred for modeling along-slope processes, but may distort internal wave dynamics along the slope.

Isopycnal coordinate systems require a moving grid that tracks the Lagrangian movement of predefined isopycnals (**Figures 5(c)**). This approach is common for 1D lake models as a means of easily tracking stratification creation and destruction. Multidimensional isopycnal models have been developed for ocean simulations to limit numerical diffusion that otherwise weakens stratification; these models have not seen wide application in lakes or reservoirs.

Time Step

Unsteady models take an initial density/velocity field and advance these forward in time (subject to the boundary conditions of the system). A model that is stable at a large time step is often prized as being more computationally efficient. The model time step is generally limited by a Courant–Lewy–Friedrichs (CFL) condition, defined as $u\Delta t\Delta x^{-1} < C_a$, where u is a velocity (fluid or wave), Δt is the time step, Δx is the grid spacing (in the same direction as u), and C_a is a constant that depends on the numerical method (typically $C_a \leq 1$). Some models also have a viscous limitation controlled by the turbulent vertical eddy viscosity (ν_z) such that $\nu_z\ \Delta t\ \Delta z^{-2} < C_\nu$. It is possible to design stable numerical methods for $C_a > 1$ or $C_\nu > 1$; however stability at large time step does not imply accuracy. For example, a reservoir that is 10 m deep × 10 km length will have a surface seiche period of ~30 min; the physics of this seiche cannot be modeled with a 20 min time step, even if the model is stable. Thus, the model time step should be chosen *both* for model stability and to accurately resolve the time-scale of processes. In particular, a large model time step will mask the cumulative effect of nonlinearities from short-time-scale processes.

Numerical Methods

There are three basic methods for discretizing the governing equations on a model grid; in order of increasing complexity these are: (1) finite difference, (2) finite volume, and (3) finite element. *Finite difference methods* represent spatial derivatives by discrete gradients computed from neighboring grid cells. *Finite volume methods* pose an integral form of the governing equations for conservative cell-face fluxes. Both finite difference and volume methods provide a set of discrete algebraic equations representing the continuous governing equations. For a model with a sufficiently refined grid and time step, the solution of the discrete equations is an approximate solution of the continuous equations. In contrast, *finite element methods* directly approximate the *solution* of the governing equations rather than the governing equations themselves. Finite element methods are often characterized as being appropriate for unstructured grids, whereas finite difference methods are often characterized as appropriate for structured grids; this outdated canard needs to be put to rest. Finite difference and finite volume methods have both been successfully applied on unstructured grids, and finite element methods can also be successfully applied to structured grids. The choice of grid and numerical method are entirely independent in model development. However, most models are designed for only one type of grid.

The finite element method is mathematically appealing but requires considerable computational effort, especially for density-stratified flows. Because temperature gradients are directly coupled to momentum through density and hydrostatic pressure gradients, a pure finite element discretization requires simultaneous solution of momentum, temperature transport, and an equation of state. As a further complication, global and local conservation is not always achieved in finite element methods; i.e., local scalar transport fluxes into and out of an element may only approximately balance the scalar accumulation in the element, and the integrated global scalar content may not be conserved. These effects can create problems for water quality models that are directly coupled to finite element hydrodynamic models as source/sink water quality terms may be dominated by numerical nonconservation. Note that consistent finite element methods may be implemented for global scalar conservation, but many existing codes have not been tested or proven consistent.

Finite difference and finite volume methods are conceptually quite different, but may be very similar in the model code. Finite differences are often described as point-based discretizations, whereas finite-volume methods are described as cell-based. However, most multidimensional hydrodynamic models apply a hybrid approach: momentum is discretized with finite differences, but continuity is discretized on a staggered grid, which is discretely equivalent to a volume integral (i.e., a finite volume approach). Thus, these hybrid or 'conservative finite difference' methods ensure exact volume conservation for fluxes into and out of a grid cell. This exact local and global scalar transport conservation, along with their simplicity, has made these methods the most popular 3D-modeling approaches.

Order of Accuracy

Multidimensional models are often judged by the 'order of accuracy' of their time and space discretizations for the governing equations. This order reflects how the error changes with a smaller time step or smaller grid spacing. For example with 2nd-order spatial discretization, model error reduces by two orders of magnitude for each magnitude reduction in grid size. Higher-order methods are generally preferred, although they are more computationally expensive than low-order methods for the same number of grid cells. There is a trade-off when computational power is limited: a higher-order method may only be possible with a larger time step and/or grid cell size than a lower-order model. It is generally thought that for converged grids the absolute error of a higher-order method on a coarse grid will be less than the absolute error of a lower-order method on a fine grid. However, this idea presupposes that both the model grids provide converged solutions. When the grid cannot be fully converged due to computational constraints (often the case for practical problems), the comparative efficiency of high-order or low-order methods must be determined by experimentation.

As a general rule, 1st-order spatial discretizations (e.g., simple upwind) are too numerically diffusive for good modeling. Spatial discretizations that are 2nd-order (e.g., central difference) often have stability issues, so 3rd-order (e.g., QUICK) is generally preferred. The best 3rd-order spatial methods include some form of flux limiting (e.g., TVD or ULTIMATE) to reduce unphysical oscillations at sharp fronts. Fourth-order and higher spatial discretization methods can be found in the numerical modeling literature, but have not been applied in any common lake models.

For time discretizations, 2nd-order methods (e.g., Crank–Nicolson) are preferred, but many models are only 1st-order due to the complexity of higher-order methods. In general, if one process is modeled with a 1st-order time-advance, then the entire scheme is 1st-order. As a note of caution, some semi-implicit 2nd-order methods may be only 1st-order accurate (albeit stable) for $CFL > 1$.

Model Errors

We separate the idea of 'model error' from 'data error'; the latter is associated with incorrect or unknown boundary/initial conditions, while the former is inherent in the model itself. Model errors are not randomly distributed. Instead, models provide an exact solution of an approximation of the governing equations, so the errors are determined by the discrete approximations. Three different types of fundamental errors will occur in any sufficiently complicated transport field: numerical diffusion of mass, numerical dissipation of energy and numerical dispersion of waves.

Numerical diffusion of mass occurs when advection of a sharp density gradient causes the gradient to weaken (as if mass diffusivity were greater). In a stable model, this error is has a net bias towards weakening sharp gradients and can be a significant problem for representing the evolution of stratification when an active internal wave field is modeled.

Numerical dissipation of energy occurs when momentum is numerically diffused (as if viscosity were greater). This effect is generally referred to as 'numerical viscosity'. It typically occurs near sharp velocity gradients and tends to weaken the gradients. A stable model requires nonnegative numerical dissipation, as negative (or anti-) dissipation leads to positive feedback and the exponential growth of kinetic energy (i.e., the model 'blows up').

Numerical dispersion of waves occurs when a model propagates a wave component (free surface or internal) at the wrong speed. This effect can have interesting consequences for hydrostatic models (as discussed in Hydrostatic Approximation above).

In general, higher-order models have smaller errors, but may lead to antidiffusion (i.e., artificial resharpening of a gradient) or antidissipation (i.e., artificial increase in energy) that can destabilize a model. For any model to be reliable, the numerical diffusion of mass should be an order of magnitude smaller than turbulent mixing, and numerical dissipation of energy should be an order of magnitude less than turbulent dissipation.

Modeling Turbulence and Mixing

The governing equations for lake and reservoir hydrodynamic modeling are generally the Reynolds-Averaged Navier Stokes (RANS) equations, although some Large-Eddy Simulation (LES) methods may be suitable for future applications. With either method, processes smaller than the grid and time scales are empirically-modeled rather than directly simulated. Local values for eddy viscosity and eddy diffusivity are generally used to represent the nonlinear turbulent advection of momentum (viscosity) and scalars (diffusivity). As turbulence varies in both time and space, constant and uniform values of eddy viscosity are rarely appropriate. In particular, the ability of stratification to suppress vertical turbulence and mixing leads to nonuniform profiles with near-zero values at strong stratifications. A wide variety of RANS turbulence models are in use, the most popular being $k-\varepsilon$, $k-l$, and mixed-layer approaches, which must be modified to account for stratification. Performance of turbulence models may be highly dependent on the model grid resolution, so grid selection must be combined with selection of the appropriate turbulence model and settings. A key difficulty is that discretization on a coarse model grid (often required due to computational constraints) leads to high levels of numerical dissipation and diffusion. Indeed, it is not unusual to find that the model error dominates the turbulence model, particularly in the horizontal flow field. The relative scales of numerical dissipation and diffusion may also have an impact. If numerical dissipation is dominant, then internal waves may be damped before they cause significant numerical diffusion of mass. Thus, a 2D- or 3D-model that artificially damps internal waves may provide a 'better' long-term representation of the thermocline, but at the cost of poorly representing the 2D or 3D transport processes!

Similarities and Differences between Lake and River Modeling

Although the focus of this article is on lake models, many of the underlying discussions of model types and errors are equally applicable to river modeling. Such models can also be 1D, 2D, or 3D, may be hydrostatic or nonhydrostatic, and have difficulties with turbulence modeling and grid resolution (especially at finer scales). River models are perhaps easier to validate because there is a single major flux direction (downstream) that quantitatively dominates the hydrodynamics; this directionality is in dramatic contrast to the unsteady oscillatory forcings in a lake that make collecting sufficient validation data a complex and time-consuming task. On the other hand, the higher flow rates typical of rivers lead to bed motion and sediment transport that may strongly affect the flow patterns. At high flows, rivers may be geomorphically active and the use of simple fixed-bed models (appropriate for lakes over shorter time scales) may be entirely unsuitable. Thus, knowledge gained in lake modeling cannot always be transferred directly to rivers or vice versa – each discipline has its own key challenges. For lakes, modeling evolving temperature stratification is the critical requirement; for a river model, the correct representation of the riverbed geometry and its geomorphologic evolution is critical.

Summary and Future Directions

Selecting whether to use a 1D-, 2D-, or 3D-model depends on the water body, available computational power, available field data and the type of answers desired. Applying 1D-models is always fastest and simplest, whereas 3D-models are computationally intense and require the greatest user skill and effort. 2D- and 3D-models need extensive field data to drive spatially-varying model boundary conditions and provide validation. In contrast, 1D-models need less extensive boundary condition data, but may require field studies to parameterize variability in the averaged directions. Whether a 3D-model is 'better' than a 1D-model will depend on the physics of interest. For example, if the physics of internal waves in a lake are unknown, a 3D-model may be needed to understand their effects. However, if the basic internal wave physics are already understood, then a 1D lake model (appropriately parameterized) may be adequate. The ideal conjunction of 1D and 3D lake models has yet to be attempted: the strength of 3D-models lies in quantifying the short-time, space-varying lake response to an event. Theoretically, such a model could be used to develop better parameterizations of 1D-models, increasing our understanding of how short-term events modify longer-term system behavior.

In considering hydrodynamic models coupled to water quality models, the ability to adequately capture bulk transport of hydrodynamic fields (e.g., velocity, temperature) should not be taken as proof of the ability to capture greater complexities in scalar biogeochemical distributions. Modeling the temperature is relatively easy because the problem is bounded and provides negative feedback. That is, lake temperatures are typically between 4 and 35 °C with the warm side facing up, and any attempt to turn the warm side downwards leads to horizontal density gradients and pressure forces that oppose overturning. Similarly, warming of the lake surface leads to increased heat loss to the atmosphere, which tends to moderate and limit errors. Velocity is also subject to large-scale forcing (wind) and is a bounded problem as unphysically large velocities will cause a model to blow up. Furthermore, dissipation is a limiting mechanism that works everywhere and at all times to bring the velocity towards zero. Thus, both velocity and temperature have preferred 'rest' states and model error cannot accumulate indefinitely without the results becoming obviously wrong. In contrast, scalar dispersion is driven by local turbulence and advection, without any global bounds to limit model error accumulation. Thus, even while the large-scale velocity and temperature fields look reasonable, a model may produce localized features that lead to unrealistic

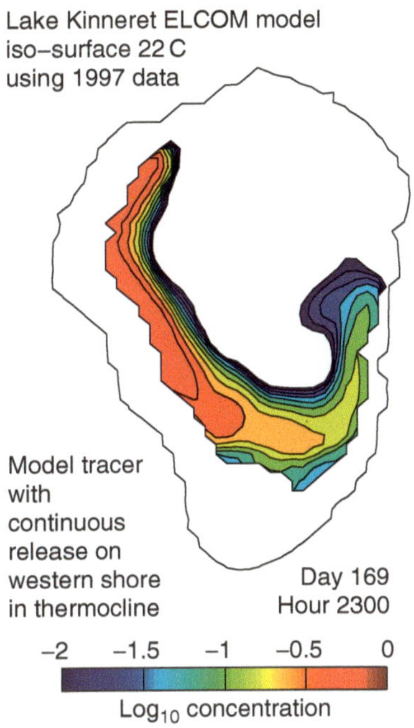

Figure 6 Modeled passive tracer concentrations in the thermocline of Lake Kinneret for a tracer concentration of 1.0 continuously released from the western boundary. This tracer motion is principally due to a basin-scale Kelvin wave.

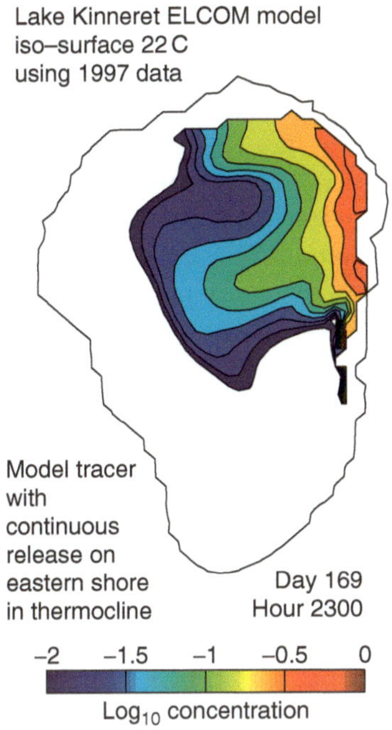

Figure 7 Modeled passive tracer concentrations in the thermocline of Lake Kinneret for a tracer concentration of 1.0 continuously released from the eastern boundary. This tracer motion is due to the combination of a Kelvin wave and a 2nd-mode Poincaré wave.

transport of scalars. Even simple passive tracer transport leads to complicated model-predicted gradient features as illustrated in **Figures 6** and 7 and associated animations. Although such tracer fields illustrate model-predicted transport, there are relatively few field studies or methods for effective validation. These problems become even more pronounced for water quality models as biogeochemical scalar concentrations (such as phytoplankton biomass) are locally forced by nutrient concentrations, do not have a preferred 'rest' state, and have source/sink behaviors that may be affected by model transport errors. As such, 2D and 3D hydrodynamic/water modeling without validating field data should be considered cartoons that may be informative, but are also speculative and may be simply wrong!

As computers grow more powerful, there is a tendency to throw more grid cells at a system to improve model results. However, as the model grid is made finer, there is some point where neglect of the nonhydrostatic pressure is inconsistent with the grid scale – i.e., the model provides a better solution to the wrong equations. As a reasonable rule of thumb, if the horizontal grid scale is substantially smaller than the local depth of water, then the hydrostatic approximation may be inappropriate. Where internal wave evolution is important, nonhydrostatic pressure gradients should be included in future models. Although nonhydrostatic models presently exist, they have not yet been practically demonstrated for large-scale lake modeling.

Model calibration should be used carefully and in conjunction with sensitivity analyses. Indeed, the difference between an uncalibrated hydrodynamic model and field data may provide greater insight into the physical processes than a calibrated model. A careful modeler will estimate the uncertainty in various boundary conditions and conduct model sensitivity tests to understand how the uncertainty may affect results. Unnecessary calibration can be avoided by gaining a better understanding of the model error characteristics. Before applying any 3D-model to a lake or reservoir, the model should be tested on 2D rectangular domains at similar scales; e.g., simple models of internal waves, river inflows, and wind-driven mixing can provide relatively rapid insight into the relationship between model error, grid scale, time step and physics.

The horizontal grid for lake models may be Cartesian, curvilinear, or unstructured; these methods have different strengths, weaknesses and complexities, such that the practical choice depends on the system, model availability and the modeler's bias. Where fine grid resolution is needed over a part of a domain (e.g., littoral zones), future developments in automated quadtree meshing of Cartesian grids may be easier to use than either curvilinear or unstructured grids.

Both *z*-level and sigma-coordinate vertical grids have significant drawbacks that remain unaddressed in the literature. Boundary layer sub-models have attempted to patch these problems, but are relatively complicated to develop and apply. Isopycnal methods may provide some future improvement, but it is not clear that they will be a panacea. Although a few isopycnal simulations have been made in lakes, we presently lack a thorough analysis of how isopycnal models represent internal wave dynamics at lake scales and along sloping boundaries.

There have been significant advances in ocean and estuarine modeling that have not yet appeared in lake models, but one must be careful about generalizing their applicability. Lake modeling faces two key problems: (1) long residence times allows model error to accumulate, unlike error that washes out with the tide in an estuary, and (2) the forcing is inherently unsteady in direction/amplitude, and may have sharp spatial gradients. Thus, methods suitable for a strong tidal exchange or a unidirectional ocean current with a smoothly varying wind field may not be effective for weak, unsteady forcing of a lake in the wind-shadow of a mountain. Indeed, despite our advances there remains significant work ahead before the art of hydrodynamic modeling is replaced by simple engineering.

Further Reading

Chen XJ and Sheng YP (2005) Three-Dimensional Modeling of Sediment and Phosphorus Dynamics in Lake Okeechobee, Florida: Spring 1989 Simulation. *Journal of Environmental Engineering-ASCE* 131(3): 359–374.

Edinger JE (2001) *Waterbody Hydrodynamic and Water Quality Modeling*, 215 pp. Reston, VA: ASCE Press.

Ferziger JH and Peric M (2002) *Computational Methods for Fluid Dynamics*, 423 pp. Berlin: Springer.

Hodges BR, Laval B, and Wadzuk BM (2006) Numerical error assessment and a temporal horizon for internal waves in a hydrostatic model. *Ocean Modelling* 13(1): 44–64.

Hodges BR, Imberger J, Saggio A, and Winters KB (2000) Modeling basin-scale internal waves in a stratified lake. *Limnology and Oceanography* 45(7): 1603–1620.

Kowalik Z and Murty TS (1993) *Numerical Modeling of Ocean Dynamics*, 481 pp. Singapore: World Scientific.

Leon LF, Lam DCL, Schertzer WM, Swayne DA, and Imberger J (2007) Towards coupling a 3d hydrodynamic lake model with the Canadian regional climate model: Simulation on Great Slave Lake. *Environmental Modelling & Software* 22(6): 787–796.

Rueda FJ, Schladow SG, and Palmarsson SO (2003) Basin-scale internal wave dynamics during a winter cooling period in a large lake. *Journal of Geophysical Research-Oceans* 108(C3): 3097, doi:10.1029/2001JC000942.

Schwab DJ and Beletsky D (2003) Relative effects of wind stress curl, topography, and stratification on large-scale circulation in Lake Michigan. *Journal of Geophysical Research-Oceans* 108(C2): 3044, doi:10.1029/2001JC001066.

INORGANIC CHEMICALS: CYCLES AND ECOSYSTEM DYNAMICS

Contents

Chemical Fluxes and Dynamics in River and Stream Ecosystems
Dissolved CO_2
Alkalinity
Major Cations (Ca, Mg, Na, K, Al)
Chloride
Iron and Manganese
Micronutrient Elements (Co, Mo, Mn, Zn, Cu)
Biogeochemistry of Trace Metals and Mettaloids
Nitrogen Fixation
Nitrogen
Phosphorus
Silica
Salinity

Chemical Fluxes and Dynamics in River and Stream Ecosystems

W M Lewis, University of Colorado, Boulder, CO, USA

Introduction

Rivers and streams have in common a unidirectional flow of water passing over a solid substrate. Thus, biogeochemical events that occur at a given location inevitably influence biogeochemistry downstream but have very little influence on biogeochemistry upstream. This is a profound contrast with lake ecosystems, wherein biogeochemical processes at a given location may influence later biogeochemical processes at any other location in the lake.

The unidirectional nature of flow in streams and rivers causes longitudinal chemical and biological gradients in these ecosystems. Examples include progressive biological conversion of coarse particulate matter to fine particulate matter in the downstream direction or accumulation of dissolved solids and nutrients in the downstream direction.

The continual passage of water through a reach of stream or river may suggest that biogeochemical processes occur mainly within a given parcel of water as it travels downstream. This would be incorrect, however, in that a high proportion of biogeochemical activity in streams and rivers is associated with organisms that maintain position within a stream channel or occupy habitats beyond the boundaries of the channel where the water is not moving rapidly. Thus, biogeochemical influences on the moving water of streams and rivers mostly originate from stationary habitats over which the water passes.

From the viewpoint of biogeochemistry, a stream or river consists of four major biogeochemical zones (**Figure 1**). Flowing waters of the channel are the most obvious zone, but only a small portion of total stream metabolism occurs within this zone.

Autotrophs (algae) typically do not prosper when suspended in flowing waters because they are removed so rapidly that they cannot accumulate biomass. Under some circumstances, however, very low velocity allows suspended algae (phytoplankton) to become abundant in streams or rivers. Bacterial decomposers are better suited than algae for growth in flowing water in that they grow with sufficient speed that they can accumulate biomass by degrading some portion of dissolved and particulate organic matter in transit. Even so, the main microbial action in streams occurs on solid surfaces and not in the flowing water of the channel.

A second zone is the channel surface (rock, sand, or clay), along with any stationary organic debris that it may contain (e.g., tree limbs, leaves). The solid surfaces are coated with bacteria, fungi, protozoans, and attached algae. These small organisms, along with the associated nonliving organic matter and mineral particles deposited onto the surfaces, can be called a

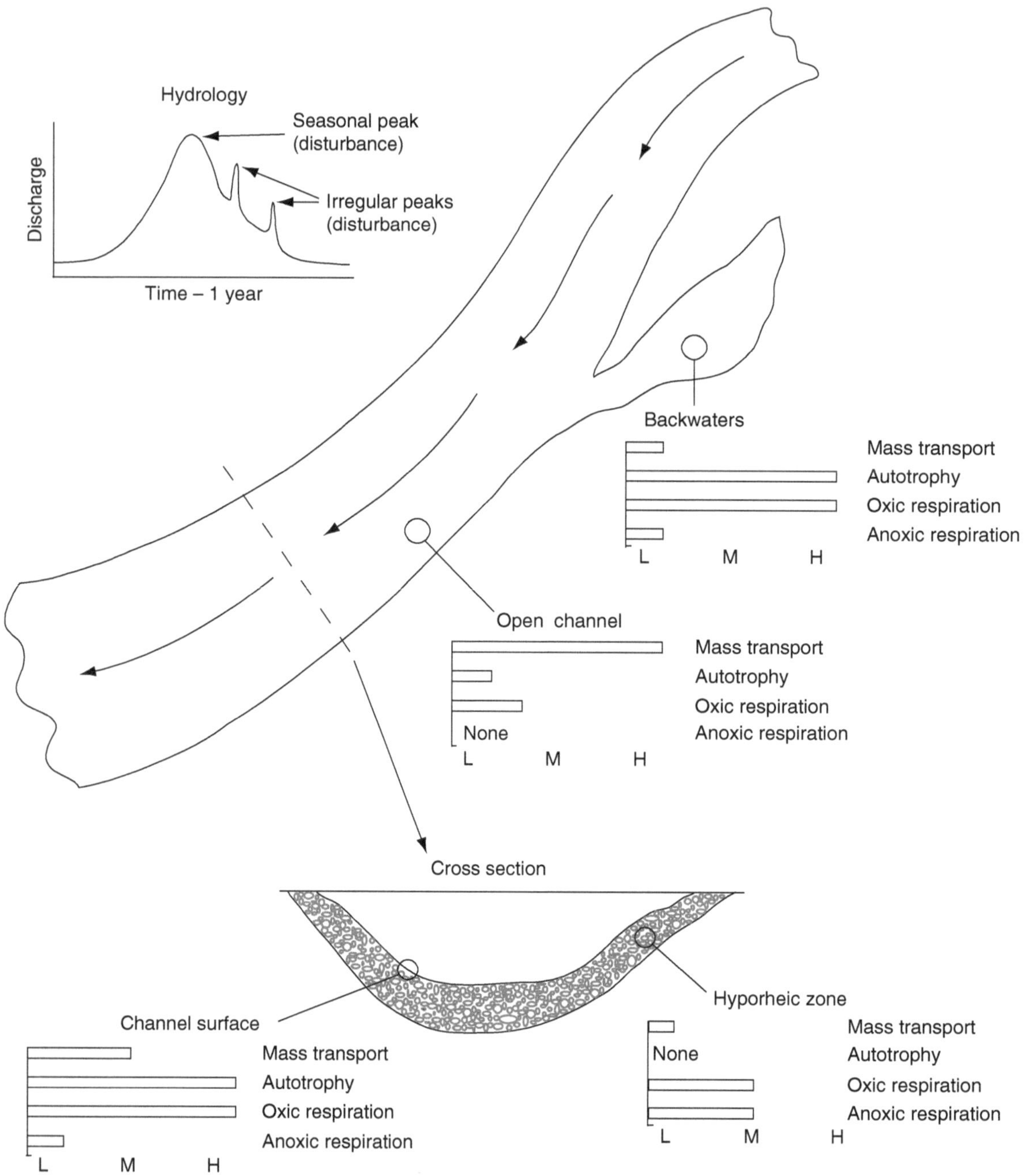

Figure 1 The four biogeochemical zones of a stream or river and a graphic indication for each of the magnitude of processes that influence biogeochemistry (L = low, M = moderate, H = high). Seasonal and irregular changes in flow that influence biogeochemistry are represented in the upper left.

'biofilm.' The biofilm has both an autotrophic component (algae) and heterotrophic component (bacteria, protozoans), and supports a wide diversity of invertebrates that live within or near it. Biofilms of the channel typically are oxic because they are in direct contact with moving water, which efficiently distributes dissolved oxygen entering from the atmosphere, and because they often produce substantial oxygen through photosynthesis. Therefore, biofilms on the channel surface usually promote oxygen-based biogeochemistry and readily support photoautotrophs and heterotrophic eukaryotes (protozoans, invertebrates), which require oxygen. Interior regions of thick biofilms may be anoxic in some cases, however.

Beneath the channel is the hyporheic zone (**Figure 1**), which consists of all substrate below the channel that has a continuous exchange of water with the overlying channel. The hyporheic zone may be very well defined where the channel sediments are coarse, or may be nonexistent where the channel is lined with bedrock or fine clay. The hyporheic zone can be partly or completely hypoxic or anoxic, even when the channel water is saturated with dissolved oxygen. Because the particles that make up the hyporheic zone impede flow, movement of water through the hyporheic zone is much slower than movement of water in the channel. Furthermore, there is no photosynthesis in the hyporheic zone because it is dark. Therefore, water within the hyporheic zone may lose some or all of its oxygen, and oxygen concentrations often vary spatially within the zone.

The surfaces of particles within the hyporheic zone develop a biofilm, but its composition differs from that of the channel surface biofilm. No photoautotrophs occupy hyporheic biofilms. If oxygen is present, hyporheic biofilms will contain bacteria, protozoans, and fungi; invertebrates will live within or around the biofilm. Where concentrations of oxygen are low or nil, hyporheic biofilms are without eukaryotes; they are made up of the many kinds of bacteria that can metabolize actively in the absence of oxygen. Biogeochemically speaking, production of energy in the absence of oxygen occurs by oxidation involving electron acceptors other than oxygen, through bacterial metabolism.

Yet another biogeochemical zone of rivers and streams consists of backwaters, which include overflow zones that may be wet only seasonally or may be always wet but change volume markedly according to season. In addition, backwaters typically include irregularities along the sides of the channel that are in contact with waters of the channel but move very slowly or are completely stagnant (**Figure 1**). Because the movement of water within backwaters is either slow or negligible most of the time, accumulation of biomass is possible. Therefore, backwaters may have phytoplankton blooms or dense macrophyte populations, develop large populations of small crustaceans and rotifers (zooplankton), and serve as a nursery for species of fish or invertebrates that must be away from strong currents in order to reproduce.

In addition to the linearity of flow and the presence of surprisingly diverse biogeochemical conditions in streams and rivers, direct influences of the watershed and riparian corridor on biogeochemical processes in streams and rivers must be taken into account. Presence of riparian plants, as well as their type and seasonality, affects processes in streams through organic-matter subsidy (leaves, branches), addition of solid surfaces from wood or leaves falling into the channel, and shading of the water surface. In addition, the entire watershed transmits water containing organic matter and inorganic solids to stream channels through the soil with shallow ground water and occasionally, during very wet weather, with surface runoff. Thus, the chemical feedstocks of a stream in its upper reaches consist almost entirely of organic matter from terrestrial vascular plants and dissolved or suspended inorganic solids from soils. As water ages longitudinally, the biogeochemical influences of the stream itself may grow substantially through biologically driven chemical transformations, but the chemical imprint of the watershed is always present.

Changes through time in the amount of flow in the channel (hydrographic change) also derive from the watershed and have strong effects on the biogeochemistry of streams and rivers. In almost all streams and rivers, there is a seasonal pattern of discharge that reflects the seasonality of precipitation or snowmelt in the watershed (**Figure 1**). The seasonal pattern can vary substantially from year to year, and the hydrograph also shows irregularities that reflect nonseasonal precipitation events (**Figure 1**). High flow, whether seasonal or irregular, moves the bed sediments, thus disturbing biofilms to depths of a few centimeters to as much as 10 or 15 cm, depending on the amount of flow. Surface biofilms that are buried or scoured subsequently are replaced by new biofilms when the flow has declined. Periods of high flow also remove biomass from backwaters and are responsible for most of the annual mass transport of both dissolved and suspended materials through the channel. Only the deeper parts of the hyporheic zone are minimally affected by high flows.

Humans influence the biogeochemistry of streams directly through water pollution and indirectly through manipulation of watersheds or alteration of atmospheric chemistry. Byproducts of human activity may include increased concentrations of nutrients, introduction of toxins, increase or decrease in the organic-matter subsidy reaching streams from watersheds, increase in sediment load, and alteration of the light climate through changes in the riparian zone. Also included under this general heading are physical changes in channels and construction of artificial impoundments, which break the longitudinal continuity of streams and rivers.

Transport

Mass transport in flowing waters can be given as mass per unit time (e.g., $kg\,d^{-1}$) passing a given cross-section of the stream channel. Mass transport is the product of concentration (e.g., $mg\,L^{-1}$) and volume of flow at any given time (e.g., $m^3\,s^{-1}$). Average mass transport for a day, month, or year can be calculated as the time-weighted average of mass transport measured at specific intervals.

Mass transport is often related to watershed area through calculation of specific yield, which is mass transport at a given point on the stream divided by the area of the watershed above that point. Units are mass per unit area per unit time (e.g., $mg\,m^{-2}\,year^{-1}$).

Flowing waters of very different size can be compared on a basis that is not directly influenced by size on the basis of specific yield or discharge-weighted mean concentrations. For example, the Amazon River has a huge mass transport of both dissolved and suspended substances, but its discharge-weighted mean concentrations are low, as are its specific yields.

Mass transport, discharge-weighted mean concentrations, and specific yields can be calculated for any substance that is carried by water. Transport involves both particulate matter (solids) and dissolved substances (dissolved solids and dissolved gases). For solids, two forms of transport are recognized: suspended solids (washload) and bedload (bed movement). Transport of solids is erratic because suspended solids may settle to the bed during periods of low discharge and then be mobilized when flows are high. At times of the highest flows, solids that have settled in areas of low velocity are lifted to become part of the suspended solids in transit, and larger particles that are part of the channel surface tumble along the bed, comprising the bedload. Thus, transport of particulate solids is difficult to quantify accurately.

Suspended solids include mineral particles (clay, silt, and sometimes sand) as well as particulate organic matter, most of which is nonliving (detritus). In addition, all of the elemental components of the suspended particles (silicon, iron, phosphorus, and other elements) as well as materials that are adsorbed onto small mineral particles (phosphorus, organic matter) would comprise part of the total transport in any estimate of transport for specific elements. The same would be true of bedload. For example, a comprehensive study of organic carbon transport would take into account particles of detritus, living organisms, organic coatings on suspended silts and clays, and organic matter attached to rocks in the bedload. Under specific circumstances, however, the transport associated with some of these components would be very small and could be ignored.

Transport of dissolved solids is more easily quantified than transport of particulate substances, and often is the dominant mode of transport for a given substance. Major dissolved solids in water often are quantified, as are less abundant dissolved substances of special interest in relation to biotic metabolism (**Table 1**). Some elements of very low concentration are also quantified in connection with studies of toxicity or nutrition. Dissolved gases also are quantified, but transport calculations must take into account atmospheric exchange, which can be a major influence on the inventory of any gas in transit.

Elements and compounds in solution may be changed in transit by abiotic processes (e.g., flocculation, chelation), but biotic processes are often dominant. For example, phosphorus and nitrogen could be taken up as nutrients by attached algae, dissolved organic carbon could be assimilated and converted to CO_2 by bacterial respiration, and dissolved oxygen could be produced by autotrophs and consumed by all organisms.

It is common to refer to substances that have a low rate of transformation relative to transport rate as being 'conservative,' whereas constituents that have a high transformation rate are designed as 'labile.' In reality, no element is perfectly conservative, but some substances, such as calcium, magnesium, or chloride are transformed very slowly, and therefore are controlled mainly by transport processes. Other elements, such as carbon, phosphorus, or nitrogen, are in high demand biotically and may show high rates of transformation under biotic influence in relation to their transport rates (**Table 1**). For example, calcium (Ca^{2+}) concentrations in a stream could remain relatively constant on a given day, whereas concentrations of dissolved oxygen or specific categories of carbon (e.g., organic carbon) might change substantially through time or spatially along the river in response to biological activity. Thus, concentrations of substances in rivers and streams are affected by transformation as well as transport.

Streams and rivers show drastic changes in concentration as discharge varies seasonally or in response to storms (**Table 2**). Concentrations of dissolved substances may either increase or decrease as discharge increases. Transport of suspended solids and rate of bedload movement typically increase exponentially with increasing flow because they are mobilized and maintained in motion by the energy of flow. In contrast, inorganic dissolved substances often decrease in concentration as flow increases because rain or snow contains lower concentrations of most constituents than does water in prolonged contact with soil. Thus, the presence of rain or snowmelt in superficial soil layers may temporarily dilute the concentration of these constituents. There are some exceptions, however. Dissolved organic matter often shows an increase in concentration on the rising limb of the hydrograph (when discharge of a stream or river rises in response to precipitation or snowmelt). This increase probably reflects mobilization of accumulated stocks of dissolved organic matter that are readily flushed from the superficial soil layers.

Table 1 Biogeochemically important elements in flowing waters

Element	*Main dissolved forms*	*Lability*	*Comments*
Most abundant (>1 mg L^{-1})			
Ca	Ca^{2+}	Low	Major salinity component
Mg	Mg^{2+}	Low	Major salinity component
Na	Na^{+}	Low	Major salinity component
K	K^{+}	Low	Major salinity component
Cl	Cl^{-}	Low	Major salinity component
C	H_2CO_3, HCO_3^-, DOC[a]	High	Supports photosynthesis
S	SO_4^{2-}, S^{2-}	Moderate	Supports biotic redox conversions
O	O_2[b]	High	Dominant electron acceptor (respiration)
Intermediate abundance (0.1–1 mg L^{-1})			
N	NO_3^-, NH_4^+, DON[a]	High	Commonly limits autotrophs
Si	H_4SiO_4	Moderate	Diatom nutrient
Low abundance, nutrients (0.001–0.1 mg L^{-1})			
P	$H_2PO_4^-$, DOP[a]	High	Commonly limits autotrophs
Cu	Cu^{2+}	Low	Toxic when high
Zn	Zn^{2+}	Low	Toxic when high
Fe	Fe^{3+}, Fe^{2+}	Moderate	Supports biotic redox conversions
Mn	Mn^{4+}, Mn^{2+}	Moderate	Supports biotic redox conversions
Mo			
Low abundance, toxins (0.001–0.1 mg L^{-1})			
As	As^{3+}	Low	
Cd	Cd^{2+}	Low	
Cr	Cr^{3+}, Cr^{6+}	Low	
Pb	Pb^{2+}	Low	
Hg	Hg^{2+}	Low	Biomagnified in food chains
Ni	Ni^{2+}	Low	
Se	Se^{4+}	Low	
Ag	Ag^{+}	Low	

Heavy metals and metalloids may be present mostly as oxides or hydroxides rather than being truly dissolved.
[a]DOC, DON, DOP = dissolved organic carbon, nitrogen, phosphorus.
[b]Oxygen is also present as O^{2-} in combination with other elements, such as carbon (e.g., CO_2).

Oxic Processes

Elemental components of protoplasm are removed from or added to water through the complementary processes of photosynthesis, respiration, growth, and mortality. Components of protoplasm include all elements that are required to synthesize biomass (growth), many of which are listed in **Table 1**.

Oxic photosynthesis is conducted by a subset of organisms, the photoautotrophs, which take up protoplasmic constituents from water. For example, an algal cell attached to a rock in a stream will remove from the passing water or biofilm carbon dioxide, phosphorus, nitrogen, and other substances that are needed to synthesize protein, nucleic acids, lipids, and carbohydrates, which are the main constituents of protoplasm. Thus, photoautotrophs convert dissolved gases and dissolved solids into biomass; conditions that favor autotrophs (large amounts of light reaching solid surfaces in the channel) magnify the uptake of elements that are required for synthesis of autotrophic biomass.

In oxygenated environments, such as the channel surface, the channel water, or backwaters of a stream or river, oxic respiration dominates; it is characteristic of all groups of organisms, including autotrophs, invertebrate consumers, and bacteria. Oxic respiration always involves conversion of organic matter to CO_2 and water, and is accompanied by synthesis and maintenance reactions within organisms, during which metabolic byproducts may be produced and released into the environment (excretion). Support for respiratory processes may also involve processing of food leading to elimination (fecal output), which releases both dissolved and particulate organic substances to the water, as does mortality of organisms through mechanisms not involving consumption.

The bacterial decomposers, which are abundant on solid surfaces in streams, degrade organic matter in an oxic environment not only as a means of acquiring protoplasmic constituents to increase their own mass, but also as a means of supporting respiration. Because respiration requires only organic carbon (CH_2O) and not the other components of protoplasm (P, N, etc.), respiration by bacteria often is accompanied by large releases of protoplasmic components such as phosphorus and nitrogen. For this reason,

Table 2 Percent change in concentration corresponding to a 100% (doubling) increase in discharge (flow) for the Orinoco River, Venezuela (data from W.M. Lewis and O.F. Saunders, 1989, *Biogeochemistry* 7: 203–240.)

Constituent	*Change %*
	Discharge
Water	+100
	Major inorganic dissolved solids
Ca^{2+}	−16
Mg^{2+}	−13
Na^{+}	−16
K^{+}	−7
HCO_3^-	−13
SO_4^{2-}	−19
Cl^-	−19
H_4SiO_4	−3
	Nutrient inorganic dissolved solids
NO_3^-	−27
NH_4^+	11
$H_2PO_4^-$	19
	Dissolved organic substances and fractions
DOC[a]	15
TDN[a]	4
TDP[a]	15
	Suspended solids and solid fractions
TSS[a]	74
PC[a]	41
PN[a]	4
PP[a]	62

Note that most inorganic dissolved solids show a dilution effect for increased discharge, whereas organic matter and organic fractions as well as particulate matter and particulate fractions show a purging effect (increase) in response to higher discharge.

[a]D = dissolved, T = total, P = particulate (except that $H_2PO_4^-$, P = elemental phosphorus), O = organic, SS = suspended solids.

the decomposition process can be referred to as mineralization; it is a source of nutrient elements for autotrophs, which take them up from water.

In summary, simultaneous occurrence of photosynthesis and oxic respiration on a daily basis in the oxidized portions of streams and rivers is accompanied by both uptake and release of protoplasmic constituents such as CO_2, N, and P. These processes may not be balanced at any given moment in time. If not, a progressive accumulation or reduction in the concentration of a biomass constituent may be observed through time or along the length of a stream.

Anoxic Processes

Anoxic metabolism is possible for many types of microbes (mostly bacteria) but not for other organisms. Anoxic metabolism can occur within thick biofilms, even when oxygen may be present on the surface of the biofilm, or within regions of the hyporheic zone where bacteria have depleted oxygen, the preferred electron acceptor for respiration.

Table 3 Chemical forms produced and consumed by oxidation and reduction (redox) reactions

Primary forms	
Oxidized	Reduced
O_2	O^{2-} (CO_2)
NO_3^-	N_2
Mn^{4+}	Mn^{2+}
Fe^{3+}	Fe^{+2}
CH_2O	CH_3O
SO_4^{2-}	S^{2-}
CH_2O	CH_4

The pairs of constituents are shown in approximate order of transformation assuming progressively more reducing conditions (increasing electron availability).

Anoxic metabolism involves both respiration and growth. The effect of anoxic growth (which involves only microbes) is similar to the effect of growth under oxic conditions: organisms take up and retain elements that are needed for the synthesis of biomass. Because oxygen is absent, however, an oxidizing agent (electron acceptor) other than oxygen is needed to support respiration. Any substance that can readily accept electrons is a potential oxidizing agent. **Table 3** lists the substances that commonly accept electrons under anoxic conditions with the assistance of microbial metabolism. Thus, for example, nitrate has substantial oxidizing potential when oxygen is absent. Microbes use this oxidizing potential by converting nitrate (which has a nitrogen atom with a +5 oxidation state) to dinitrogen (N_2, oxidation state zero).

Electron acceptors are used by microbes in anoxic environments in a progressive sequence, beginning with the most effective oxidizing agents and moving to weaker oxidizing agents as the stronger oxidizing agents are exhausted. The sequential nature by which microbes use electronic acceptors in an anoxic environment leads to a progression of biogeochemical changes. First, oxygen is depleted. As oxygen is nearing depletion, oxidized manganese (Mn^{4+}) and nitrate (NO_3^-) will be converted to reduced manganese (Mn^{2+}) and N_2 gas. If respiration continues (which requires the presence of labile organic matter), oxidized iron (Fe^{3+}) may be converted to reduced iron (Fe^{2+}). In fact, prolonged anoxia accompanied by a strong supply of organic matter may bring the anoxic environment through the entire sequence of biogeochemical changes indicated in **Table 3**. In this case, methane is eventually produced from organic matter, which is used by microbes in highly reduced environments as an electron acceptor.

It is common in streams and rivers for biogeochemical changes in anoxic sediments not to pass through the entire sequence of potential changes indicated in **Table 3**. Several factors may interrupt progression through the full sequence. For example, anoxic conditions in a hyporheic zone may be associated specifically with low stream flow. At high flows, the anoxic environment may be flushed with oxic water, thus reinitiating the sequence of changes from the beginning. In addition, an anoxic environment may contain only a small amount of labile organic matter for microbes to use. In this case, the progression of biogeochemical changes ceases when the labile organic matter is used up by the microbes. Also, some of the potential electron acceptors shown in **Table 3** may be present in very small amounts, which would prevent any significant biogeochemical changes involving these electron acceptors. For example, sulfate is present only in small quantities in many inland waters. Production of sulfide from sulfate under anoxic conditions involving little sulfate would be trivial.

Under anoxic conditions, conversions summarized in **Table 3** set the stage for a reversal of biogeochemical changes when oxygen reaches an anoxic environment. For example, if reduced iron is flushed from an anoxic hyporheic environment into a river channel, microbes will reoxidize the iron, extracting energy in the process. In general, the presence of reduced chemical forms in oxygenated zones is temporary; reduced forms are converted to oxidized forms either spontaneously or by microbial processes.

Because of the diversity and importance of redox reactions that are dependent on the oxidation state of a given biogeochemical zone, water samples from different zones may show very different chemical composition. A sample from the channel or a backwater, for example, might be rich in oxygen and oxidized chemical constituents, whereas a sample from an anoxic portion of a hyporheic zone in the same stream might be rich in some or all of the reduced chemical substances shown in **Table 3**.

The importance of anoxic metabolism is related not only to the production of reduced chemical substances in the absence of oxygen, but also to the continued breakdown of organic matter by microbes even when oxygen is absent. Although it may seem that an anoxic environment is biologically hostile, microbial decomposition of organic matter continues under anoxic conditions because of the ability of many types of bacteria to use electron acceptors other than oxygen to support respiration.

Overview

In a lake ecosystem, where the longitudinal component of flow is typically smaller than the circulation of water, the concept of element cycling is highly applicable. A nutrient such as phosphorus, for example, may pass from the water column into an autotroph such as an algal cell, be consumed by an herbivore, and then pass to a carnivore or be released from an herbivore or carnivore back to the water column as fecal output or excretion. In addition, phosphorus that becomes attached to detritus (e.g., fecal pellets) might pass to the bottom of a lake and subsequently be released back to the water column. Thus, cycling is an accurate means of describing the movement of a given substance within a lake.

In a stream, the cycling concept is more difficult to apply because of the dominance of longitudinal movement. Although longitudinal movement is constant, an element such as phosphorus may be taken up as biomass, thus converting it from the dissolved to the particulate (solid) form. In the dissolved form, phosphorus moves longitudinally, but in the particulate form it typically may not move if it becomes attached to a solid surface as part of an organism or perhaps even to a grain of sediment that is immobile at low flow. Subsequently, phosphorus comes into motion again as it is released in soluble form or when a high flow moves it in particle form. Because longitudinal movement is added to cycling between mobile and nonmobile forms, the concept of 'spiraling' has been applied to elements in streams. The spiraling concept can be quantified in terms of the average distance (spiraling length) that a given element travels when it moves through one full cycle of the high mobility/low mobility sequences. Elements that are in high biological demand have tight spirals, whereas those that are in lower biological demand have loose spirals.

The distinct biogeochemical zones of flowing waters, in combination with concepts related transport under the control of hydrology and watershed effects, including pollution, produce a mechanistic understanding of the chemistry of flowing waters and the underlying biogeochemical processes that affect chemistry as water travels through a drainage network. The concept of spiraling unifies the longitudinal movement under control of flow with the processes that control the forms and transformations of elements in a stream or river.

Glossary

Anoxic – Devoid of free oxygen.

Backwaters – Waters adjoining a stream or river channel that are sometimes inundated by the channel at high flow but typically have very low rates of water exchange or are completely separated from this channel.

Biofilm – The combination of bacteria, fungi, and algae, along with associated nonliving organic and inorganic particulate matter, that coats solid surfaces within a stream or river. Invertebrates live within or around biofilms.

Conservative substance – Any element or substance whose rates of biogeochemical transformation are low relative to rates of transport.

Hyporheic zone – Sediments containing water that is constantly exchanged with water in the channel of a stream or river.

Labile constituent – Any element or substance that has high rates of biogeochemical transformation in relation to its rates of transport.

Nutrient spiraling – Movement of any element or substance within a stream or river under the simultaneous influence of longitudinal flow and repeated temporary immobilization through incorporation into a particle, living organism, or nonflowing zone.

Oxic – Containing free oxygen.

See also: Fluvial Transport of Suspended Solids; Redox Potential; Streams.

Further Reading

Drever JI (1997) The Geochemistry of Natural Waters. *Surface and Groundwater Environments,* 3rd ed. Upper Saddle River, NJ: Prentice Hall.

Hauer RR and Lamberti GA (2006) Methods in Stream Ecology. 2nd ed. London: Elsevier.

Sterner RW and Elser JJ (2002) Ecological Stoichiometry. *The Biology of Elements from Molecules to the Biosphere.* Princeton and Oxford: Princeton University Press.

Stumm W and Morgan JJ (1996) Aquatic Chemistry. *Chemical Equilibria and Rates in Natural Waters,* 3rd ed. New York, NY: John Wiley.

Dissolved CO_2

J J Cole, Cary Institute of Ecosystem Studies, Millbrook, NY, USA
Y T Prairie, Universite de Quebec à Montreal, Montreal, QC, Canada

The Dissolve Inorganic Carbon System

All water contains some amount of inorganic C in solution. The total dissolved inorganic C content is referred to as DIC (total CO_2 by some authors). DIC, however, is made up of several chemical species: the bicarbonate ion (HCO_3^-), the carbonate ion (CO_3^{2-}), and aqueous carbon dioxide, which consists of two pools (dissolved free CO_2 and its hydrated form H_2CO_3). At equilibrium with each other, H_2CO_3 is about 1/1000th of the concentration of free CO_2. Neither CO_2 nor H_2CO_3 is a charged species; they interchange readily and behave as one pool in chemical reactions. Thus, all of the constants and equations treat $CO_2 + H_2CO_3$ as a single pool and refer to it as either $H_2CO_3^*$ or simply as dissolved or aqueous CO_2.

In the ocean, the concentration of DIC in surface waters is in a narrow range (near 2400 μM) and varies with salinity. In inland waters, DIC ranges from <20 μM in acidic soft waters to more than 5000 μM in alkaline hard waters and even higher in endorheic, saline lakes. In most freshwaters DIC is between 100 and 1000 μM in the surface and typically the concentration increases with increasing depth in stratified systems.

In most natural waters HCO_3^- makes up the major part of DIC with aqueous CO_2 and CO_3^{2-} being relatively less abundant constituents. In 'soft waters' (fresh waters with low amounts of alkalinity and low pH), CO_2 can be co-equal to or larger than HCO_3^-, and CO_3^{2-} vanishingly small in comparison. The proportions of the constituents of DIC relative to each other determine the pH of water. Similarly, if one knows the pH, the relative magnitudes of the constituents of DIC are known. For example, at pH 5.3, only CO_2 is present in significant quantities; at pH 6.3, CO_2 and HCO_3^- are in equal quantities and CO_3^{2-} is effectively absent; at pH 7.3 there is ten times as much HCO_3^- as CO_2 and CO_3^{2-} is still effectively absent; at pH 8.3 there is one hundred times as much HCO_3^- as CO_2 and CO_3^{2-} present but is still very low proportionally; at pH 10.3, HCO_3^- and CO_3^{2-} are in equal proportions and the relative proportion of CO_2 is now extremely small. A good diagram of the relationship between pH and the constituents of DIC in shown in elsewhere in this encyclopedia.

Dissolved CO_2 – A Master Variable

Dissolved CO_2 is a master variable in aqueous chemistry. It is the most dynamic constituent of DIC and is usually the major acid in inland waters that have not been impacted by acid rain. In watersheds dissolved CO_2, originating from the respiration of plants and microbes in the soil, is the major chemical responsible for the dissolution of rocks, particularly carbonate rocks and aluminosilicate minerals. These reactions convert dissolved CO_2 into carbonate and bicarbonate ions, the main source of alkalinity in surface waters. Thus, dissolved CO_2, in concert with the parent rock material affects the entire chemical balance in surface waters that receive this runoff, and is an important regulator of pH.

Within the aquatic system the amount of dissolved CO_2 is the net result of several dynamic processes. CO_2 is a gas. As such, CO_2 in surface waters exchanges with CO_2 in the atmosphere. If lakes were lifeless vessels, with no reactive chemicals, and only open to the atmosphere (call this the 'Teflon lake'), the concentration of CO_2 would be dictated entirely by that of the atmosphere and the physics of gas exchange. Henry's Law states that partial pressure of a gas in water in equilibrium with the air will be equal to the partial pressure of the gas in air. In such situations, all lakes would have the same of quantity of dissolved CO_2, except for differences in solubility at different temperatures and altitudes (which affects pressure). Note that some authors prefer the term 'fugacity' to partial pressure for gasses dissolved in water. Fugacity can be defined as a thermodynamic or potential pressure. For a perfectly ideal gas, fugacity and pressure are equal. Most authors working on CO_2 use partial pressure (e.g., pCO_2) and we will continue with that convention.

At equilibrium with the overlying atmosphere, then

$$pCO_{2(\text{water})} = pCO_{2(\text{atmosphere})} \quad [1]$$

In reality, the concentration of CO_2 in the waters of lakes, reservoirs, rivers, and streams is quite dynamic both in time and among ecosystems and often the pCO_2 of surfaces waters is far from that of the overlying atmosphere. Clearly then 'Teflon' lake or river model is not appropriate. The vast majority of lakes, rivers, streams, and reservoirs have a higher or much higher concentration of surface-water

CO_2 than can be explained by equilibrium with the overlying atmosphere. Such systems are said to be 'supersaturated' in CO_2 with respect to the atmosphere. Where surface waters have less CO_2 than at equilibrium with the atmosphere, they are termed 'undersaturated' (**Figure 1**).

Atmospheric Exchange of CO_2

Because dissolved CO_2 is a gas, it can exchange with gas in the atmosphere. When surface waters are supersaturated in CO_2, they are a net source of CO_2 to the atmosphere; the opposite is true for undersaturated waters. The magnitude of this flux can be described in terms of Fick's first law of diffusion which states that a diffusive flux is proportional to the concentration (in this case the partial pressure) gradient existing at the interface. It can be described by a simple equation:

$$\text{Flux} = k \times K_h \left(p\text{CO}_{2(\text{water})} - p\text{CO}_{2(\text{atmosphere})} \right) \quad [2]$$

where K_h is the temperature-dependent Henry's constant. $K_h \times p\text{CO}_{2(\text{water})}$ gives the concentration of CO_2 in water. Similarly, $K_h \times p\text{CO}_{2(\text{atmosphere})}$ gives the concentration the water would have were it in equilibrium with the atmosphere. Thus eqn. [2] can be rewritten as

$$\text{Flux} = k \left(\text{CO}_{2(\text{water})} - \text{CO}_{2(\text{atmosphere})} \right) \quad [3]$$

Where CO_2 is now expressed in concentration units (e.g., $\mu\text{mol l}^{-1}$) rather than as a partial pressure of a gas and $\text{CO}_{2(\text{atmosphere})}$ represents the dissolved CO_2 concentration the water would have were it in equilibrium with the atmosphere.

k is the gas exchange velocity, often termed piston velocity. The piston velocity is a way to represent the physical rate of gas exchange with the atmosphere. k can be thought of as the height of water that exchanges gas per unit time with the atmosphere, as if a piston pushed the excess CO_2 (over saturation) through that height of water. k is the result of turbulent energy exchange with the atmosphere and increases with factors such as wind and current speed. While not straightforward, k can be measured directly (see Further Reading). From these measurements, we know that k is greatly affected by factors affecting the turbulent regime of surface waters, such as wind and current speeds. While several other factors are known to influence gas exchange velocities, the most widely used models predict k from wind speed. If we express k in m d^{-1} and realize

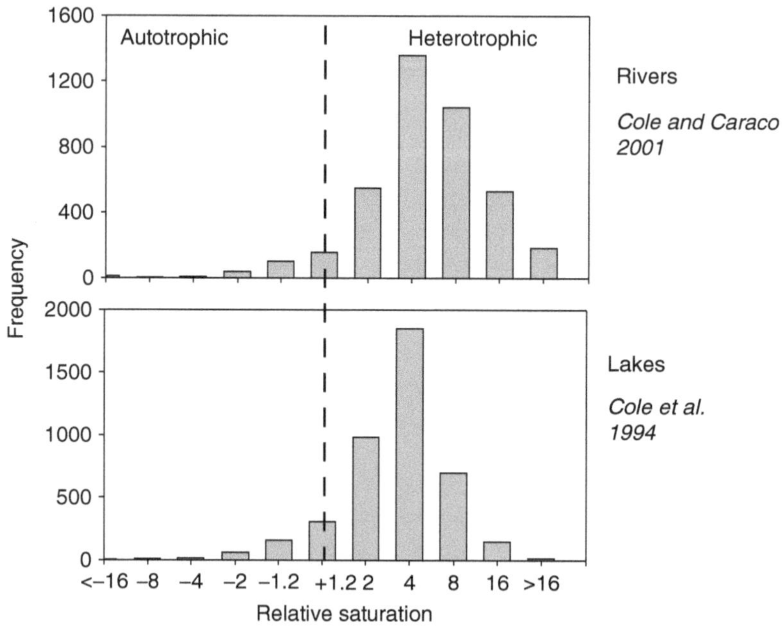

Figure 1 Dissolved CO_2 in the surface waters of lakes (upper) and rivers (lower) with a global distribution. Relative saturation is the ratio of actual CO_2 to the CO_2 the water would have were it in equilibrium with the atmosphere. Lakes with surface waters near atmospheric equilibrium would have relative saturation values of 1.0 on this scale. Positive values denote supersaturation; negative values denote undersaturation. A value of 4 means the water body has 4 times as much dissolved CO_2 as could be explained by atmospheric equilibrium; −4 means the water body has 4 times less (or one-fourth as much dissolved CO_2 as could be explained by atmospheric equilibrium. Lake data is for about 5000 lakes from Cole *et al.* (1994); River data is for 80 of the world's major rivers from Cole and Caraco (2001). See also Duarte and Prairie (2005).

that CO_2 in μmol l^{-1} is the same as mmol m^{-3}, Flux, in eqn. [3], has the useful units of mmole CO_2 per square meter per day. When $CO_{2(water)} > CO_{2(atmosphere)}$ the flux is from the water to the air and the water body is a source of CO_2 to the atmosphere. The opposite is true when $CO_{2(water)} < CO_{2(atmosphere)}$.

Figure 1 shows that most water bodies are sources of CO_2 to the atmosphere. This means that the processes generating CO_2 exceed those consuming it. What are these processes?

Regulation of Dissolved CO_2 in Aquatic Ecosystems

The balance of photosynthesis and respiration explains the excess CO_2 in many lakes. Gross primary production (GPP) is the total amount of photosynthesis occurring in an ecosystem independent of its fate. Total respiration (R) is the biological oxidation of organic matter to CO_2 by all pathways and includes the respiration of plants (primary producers), consumers, and microbes. In most lakes and deep, slow rivers, R exceeds GPP and this imbalance explains most of the excess CO_2 in these systems. R exceeds GPP in these systems because of the input and subsequent metabolism of material from the terrestrial watershed. In oxic surface waters, the diel and annual pattern of CO_2 concentrations is the mirror image of that of dissolved oxygen when the biological metabolic balance is the major cause of the CO_2 supersaturation. Abiotic oxidation of terrestrially derived organic matter, principally dissolved organic matter, by UV mediated reactions is another way that excess CO_2 can be formed.

In small streams, and lakes with short hydrologic residence times, the input of CO_2 in ground water can be the main reason why the water body has excess CO_2. This CO_2 in ground water is the result of respiration in soil that dissolved in water rather than being degassed out of the soil into the air.

There are other chemical and physical reactions that can lead to CO_2 concentrations that are out of equilibrium with the atmosphere. These reactions are certainly important in some systems, but are less important overall than the mechanisms above. Changes in the acid or base status can affect CO_2. In most natural waters, the pH is strongly controlled by the ratios of CO_2: HCO_3:CO_3 but other factors, such as the input of other acids or bases also affect the ratios of the C species. Acids can enter externally in rain water, or can be formed by the oxidation of other chemicals (e.g., sulfides) within the system. If we add acid to water with DIC in it, we push the equilibrium leading to an increase in the amount of CO_2 relative to HCO_3 and CO_3. If we add a base (e.g., some hydroxide or calcium oxide lye), we push the equilibrium to the right and lower the amount of CO_2. If a system receives an input of solid carbonates (e.g., limestone) CO_2 will be consumed in the dissolution of carbonate, forming more HCO_3. Somewhat counter intuitively, the precipitation of dissolved carbonate ion into solid phase carbonate (as in the formation of coral or the shells of mollusks) causes the release of CO_2. Carbonate precipitation also occurs abiotically in both the oceans and in marl lakes (lakes with a lot of dissolved carbonate) and can be significant in the C balances of these systems.

Regulation of pH by Dissolved CO_2

In most inland waters, the carbonate system regulates pH. In the exceptional situation described above where lake water with elevated alkalinities (HCO_3 and CO_3) receives a source of acidity (such as acid rain), one can construe pH as controlling the pCO_2. However, in general, it is more appropriate to view pCO_2 as controlling the pH. Below pH of about 5, effectively all DIC is in the form of CO_2 and changing the amount of CO_2 has only small effects on pH. From pH values of about 5.5 and above, however, changes in CO_2 cause major changes in pH. Let us first consider the case when the pH is less than about 7.5. This is simpler because the amount of CO_3^{2-} is small enough to ignore.

$$K_{a1} = [H^+][HCO_3^-]/[CO_2]$$

$$K_{a2} = [H^+][CO_3^{2-}]/[HCO_3^-]$$

K_{a1} is the temperature-dependent 'first-dissociation constant' which expresses the ratios between the species on the right side of the equation. At 20 °C in dilute waters, $K_{a1} = 10^{-6.38}$. K_{a2}, the 'second dissociation constant' expresses the ratios between HCO_3^- and CO_3^{2-}. At 20 °C in dilute waters $K_{a2} = 10^{-10.377}$. Unless the pH is high (≫7.5), we can ignore K_{a2}.

We can rearrange the equation to show how CO_2 affects pH.

$$[H^+] = K_{a1}[CO_2]/[HCO_3]$$

If CO_2 increases, $[H^+]$ increases and the pH goes down. In a productive system pH varies over the course of the day, increasing during the daylight hours (when photosynthesis dominates over respiration) and decreasing at night when respiration continues but photosynthesis stops. To calculate the quantitative effect of CO_2 on pH it is important to remember that CO_2 does not affect alkalinity (also called acid neutralizing capacity or ANC).

$$ANC = [HCO_3^-] + 2[CO_3] + [OH^-] - [H^+]$$

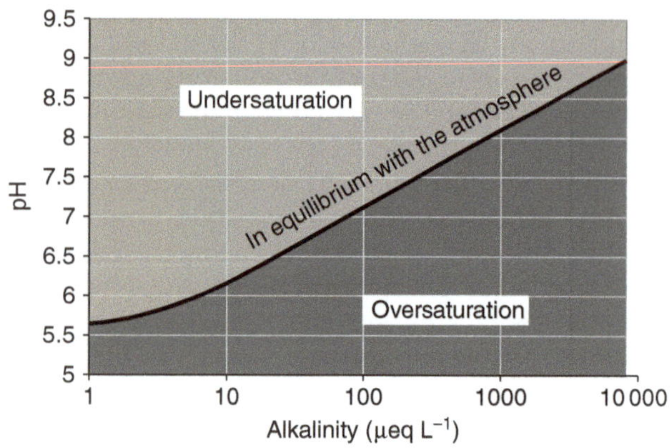

Figure 2 A theoretical diagram relating pH, alkalinity and dissolved CO_2 in inland waters. For any fixed value of CO_2 and ANC, there is a unique value for pH. The solid line shows the relationship between alkalinity (ANC) on the *X*-axis and pH on the *Y*-axis for surface waters in CO_2 equilibrium with the atmosphere at 20 °C. The diagram can be used to quickly asses if a water body is supersatured or undersaturated in CO_2. Most lakes and rivers will fall above the solid line on this diagram (see **Figure 1**). ANC and pH (for a given value of CO_2) are related as:
$ANC = (K_h \times pCO2)/\alpha_0 \times (\alpha_1 + 2\alpha_2) + K_w/H^+ - H^+$
where $\alpha_0 = 1/(1 + K_{a1}/H^+ + K_{a1}K_{a2}/(H^+)^2)$,
$\alpha_1 = 1/(1 + H^+/K_{a1} + K_{a2}/H^+)$,
$\alpha_2 = 1/(1 + H^+/K_{a2} + (H^+)^2/(K_{a1}K_{a2}))$.
All other terms are defined above.

ANC can be calculated from pH and DIC or measured directly by titration. At pH < 7.5, ANC can be simplified to

$$ANC = [HCO_3^-] - [H^+]$$

Across all pH values ANC and DIC can be related as follows:

$$ANC = [DIC] \times (K_{a1}[H^+] + 2K_{a1} \times K_{a2})/([H^+]^2 + K_{a1} \times [H^+] + K_{a1} \times K_{a2}) + K_w/[H^+] - [H^+]$$

where K_w is the temperature-dependent dissociation constant for water ($10^{-14.16}$ at 20 °C). With the equations above and a little algebra one can derive **Figure 2** which shows the relationship between pH and ANC when the surface waters are (solid line) and are not (denoted areas) in equilibrium with the atmosphere for CO_2.

Measurement and Calculation of Dissolved CO_2

Measurement

Under most conditions the most accurate determination of CO_2 is by direct measurement. Several approaches in both the lab and field exist. The basic idea is to directly measure the partial pressure of CO_2 (pCO_2) using either a gas-chromatograph or an infrared gas analyzer (IRGA). In theory, pCO_2 would be the partial pressure of CO_2 in an infinitely small gas head space, equilibrated with an infinitely large volume of water. Field measurements, using an in situ equilibrator directly in the water body accomplish this precisely (**Figure 3**). Alternatively, one can take discrete water samples in gas-tight bottles and equilibrate a large volume of water (1 or 2 l) with a small volume of headspace gas (50 ml). In this case, one needs to take care that the temperature of the water is held at ambient (or some known temperature) and one has to make corrections because the volume of water is not infinitely large nor the headspace infinitely small. For a given amount of CO_2 in solution, the pCO_2 will change with temperature. CO_2 is more soluble at lower temperatures and exerts a smaller partial pressure for the same concentration at low temperatures. Thus, K_h is larger at lower temperatures. Suppose you had a gas tight bottle of lake water at 10 °C that had 15 μM dissolved CO_2. At the ambient temperature, $K_h = 0.0537\ \mu mol\ l^{-1}\ \mu atm^{-1}$. So pCO_2 would be undersaturated with respect to the atmosphere at 279 μatm. Suppose we let our sealed water bottle get warm to 25 °C where K_h is now $= 0.0338\ \mu mol\ l^{-1}\ \mu atm^{-1}$. Now the same 15 μM dissolved CO_2 exerts a partial pressure of 442 μatm, and is supersaturated with respect to the atmosphere. So, since one measures pCO_2 and calculates dissolved CO_2 from it, the temperature is critical.

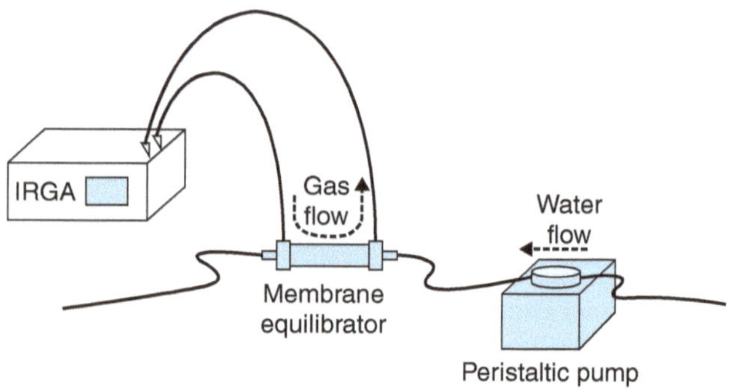

Figure 3 Diagram of direct, in situ measurement of pCO_2. The components include an IRGA which measures CO_2 gas concentrations; a membrane equilibrator which is deployed in the water; a peristaltic pump that pumps water through the equilibrator. The IRGA circulates gas (labeled gas flow) in a closed loop through the membrane equilibrator. This gas comes rapidly to equilibrium with the free CO_2 dissolved in the water. Since the volume of water is essentially infinitely large (e.g., the lake) compared to the volume of gas in circulation, one is directly measuring the partial pressure of CO_2 in the water. Both gas standards and samples of ambient atmosphere need to be measured frequently as well.

Calculation

Dissolved CO_2 can be calculated from temperature and either total DIC and pH, ANC and pH, or ANC and DIC. There are some advantages to the calculation over direct measurement. First, one can calculate dissolved CO_2 from existing data that reports the more commonly measured variables needed to calculate it. Second, it is a good check on direct measurements. Further, in many cases a continuous high-frequency record of pH and temperature (as from a sonde) can be used to compute CO_2, if one has starting and ending DIC or ANC and the ANC has not changed significantly, and if the DIC or ANC can be assumed to change monotonically and smoothly over the time interval of interest. Equations representing the various relationships can be found in several textbooks but a simple solution is:

$$[CO_2] = [DIC] \times [H^+]^2/(K_{a1} \times K_{a2} + K_{a1} \times [H^+] + [H^+]^2)$$

The temperature is needed to calculate the correct values of the dissociation constants (K_{a1} and K_{a2}). If DIC is entered in micromoles, CO_2 will be in micromoles. There can be serious problems with calculated values of CO_2. The calculation of CO_2 is very sensitive to having accurate pH measurements, which are difficult to perform in dilute waters, such as most lakes and rivers. A protocol for high quality pH measurements is given in the reading by Stauffer (below). Further, there are complications, particularly at pH values above 8 and especially in waters with high ionic strength because of the formation of solid phase carbonates (usually calcium carbonate) and because of 'ion pairing' which affects the chemical activities and the apparent values of the key constants in the above equation. There are approximations that deal with these effects as a function of salinity, but the underlying assumption is that the water behaves as sea water, more or less diluted with deionized water. Other reactions, such as the hydration of CO_2 directly with hydroxide ions and water molecules occur at high pH. This effect, known as chemical enhancement, increases the rate of exchange of CO_2 with the atmosphere beyond what would be expected from turbulent diffusion.

Glossary

Henry's law – Henry's law states that at a given temperature, the amount of a given gas dissolved in a given type and volume of liquid is directly proportional to the partial pressure of that gas in equilibrium with that liquid.

Partial pressure – The partial pressure of a gas dissolved in a liquid is the partial pressure of that gas which would be generated in a gas phase in equilibrium with the liquid at the same temperature.

Gas piston velocity – *k*, a measure of the physical rate of gas exchange. *k* can be thought of as the height of water that equilibirates gas with the atmosphere per unit time.

Alkalinity – Also called acid neutralizing capacity (ANC).

Hydrologic residence time – Also called water renewal time. This is the amount of time it takes, in theory, for the inputs of water to replace the volume of water in a given water body.

Ion pairing – Some ions have strong interactions with each other such that the activities of these ion are significantly different than 1.0. Both HCO_3^- and CO_3^{2-} form ion pairs with calcium, magnesium and other cations. These ion pairs are the main reason that the dissociation constants (K_{a1} and K_{a2}) are strikingly different in sea water and fresh water, see Butler (1992).

See also: Alkalinity.

Further Reading

Butler JN (1992) *Carbon Dioxide Equilibria and Their Applications.* Chelsea, Michigan: Lewis Publishers.

Cole JJ and Caraco NF (2001) Carbon in catchments. *Marine Freshwater Research* 52: 101–110.

Cole JJ, Caraco NF, *et al.* (1994) Carbon dioxide supersaturation in the surface waters of lakes. *Science* 265: 1568–1570.

Duarte CM and Prairie YT (2005) Prevalence of heterotrophy and atmospheric CO_2 emissions from aquatic ecosystems. *Ecosystems* 8(7): 862–870.

Matthews CJD, St Louis VL, *et al.* (2003) Comparison of three techniques used to measure diffusive gas exchange from sheltered aquatic surfaces. *Environmental Science & Technology* 37(4): 772–780.

Stumm W and Morgan JJ (1996) *Aquatic Chemistry: Chemical Equilibria and Rates in Natural Waters.* New York: Wiley.

Alkalinity

M D Mattson, Massachusetts Department of Environmental Protection, Worcester, MA, USA

Introduction

Alkalinity is a measure of the amount of bases in a solution, which can neutralize hydrogen ions H^+ from strong acids. A closely related and more encompassing term is Acid Neutralizing Capacity (ANC), which may include any type of buffering, including buffering action of particulates suspended within unfiltered samples. In most fresh waters, the major buffers in the carbonate system, bicarbonate $[HCO_3^-]$, and carbonate ions $[CO_3^{2-}]$, with bicarbonate ion being the dominant buffering agent that moderates large variations in the pH of the system.

Chemistry of Buffers

Chemical Background

Although pH is considered as reference to the intensity of acids, analogous to the concentration of protons, alkalinity can be considered the capacity of the system to buffer additions of strong acids. In common usage, the terms alkaline ($pH > 7$) and acidic ($pH < 7$) refer to the intensity of protons in relation to neutral ($pH = 7$) solutions. Acidity is more rigorously defined as a capacity term as the amount of acids in a solution that can be titrated with strong base, and the term buffering capacity may be applied to either acid or base buffering capacity.

In a simple freshwater system, alkalinity [Alk] can be expressed as a concentration in terms of acid equivalents per liter and defined as the amount of bases in excess of acid protons:

$$[\mathrm{Alk}] = [HCO_3^-] + 2[CO_3^{2-}] + [OH^-] - [H^+] \quad [1]$$

In more complex systems other buffers may contribute. In marine systems borate $[B(OH)_4^-]$ may contribute about 4% to the total alkalinity. In some systems, additional weak buffers may include phosphate, silicate, weak organic acid anions, or even more rarely uncharged bases such as unionized ammonia.

In effect, alkalinity is the imbalance, or difference between strong acids and bases and can also be written as

$$[\mathrm{Alk}] = 2[Ca^{2+}] + [Na^+] + [K^+] + 2 * [Mg^{2+}] \ldots - [Cl^-] - 2[SO_4^{2-}] - [NO_3^-] \quad [2]$$

Again with all concentrations expressed in molar units multiplied by their respective valences. In practice, these equations of alkalinity are essentially identical because of charge balance:

$$[H^+] + [\text{base cations}] = [OH^-] + [\text{weak acid anions}] + [\text{strong acid anions}] \quad [3]$$

Because it is a balance, the difference between the bases and the acids, it is not a simple mass quantity, and in fact the alkalinity can become a negative quantity when acids exceed bases. Alkalinity is a conservative property of water with respect to changes in carbon dioxide, and thus alkalinity can be used as a tracer of water masses or as a tracer of various acid/base reactions as well as many oxidation reduction reactions (redox reactions).

Measurement

In practice, alkalinity is measured by acid titration of a water sample to an endpoint, typically near pH 4.5 for the carbonic acid equivalence point of typical fresh waters. The actual endpoint depends on the amount of buffers present, which in fresh waters is largely a function of the amount of total dissolved inorganic carbon. A log C vs. pH diagram detailing changes in the ionic composition of alkalinity as a function of pH for an example system with a total inorganic carbon concentration of 0.001 $\mathrm{mol\,l^{-1}}$ and end point pH of 4.65 is shown in **Figure 1**. Note that alkalinity is dominated by and essentially identical to the bicarbonate ion in the figure, except at very high pH where carbonate and hydroxide dominate, and at low pH where alkalinity drops to zero and hydrogen ion begins to dominate. At pH values below the end point negative alkalinities are shown as mineral acidity.

Historically, titrations were carried out in two parts with different colorimetric indicator solutions. First, phenolphthalein indicator with color change of pink to clear at pH 8.3, was used on high pH samples to indicate the portion of alkalinity contributed by the carbonate ion (**Figure 1**). Second, the titration continued with methyl orange indicator that changes color from clear to shades of pink at a pH of about 4.25 to indicate total alkalinity, although this may result in an over-titration error of about 55 $\mu\mathrm{eq\,l^{-1}}$ in many low alkalinity systems. Because limestone is the most common freshwater buffering agent, the resulting acid neutralizing equivalent of the titration is reported in terms of calcium carbonate as milligram per liter of $CaCO_3$. Currently samples are most often

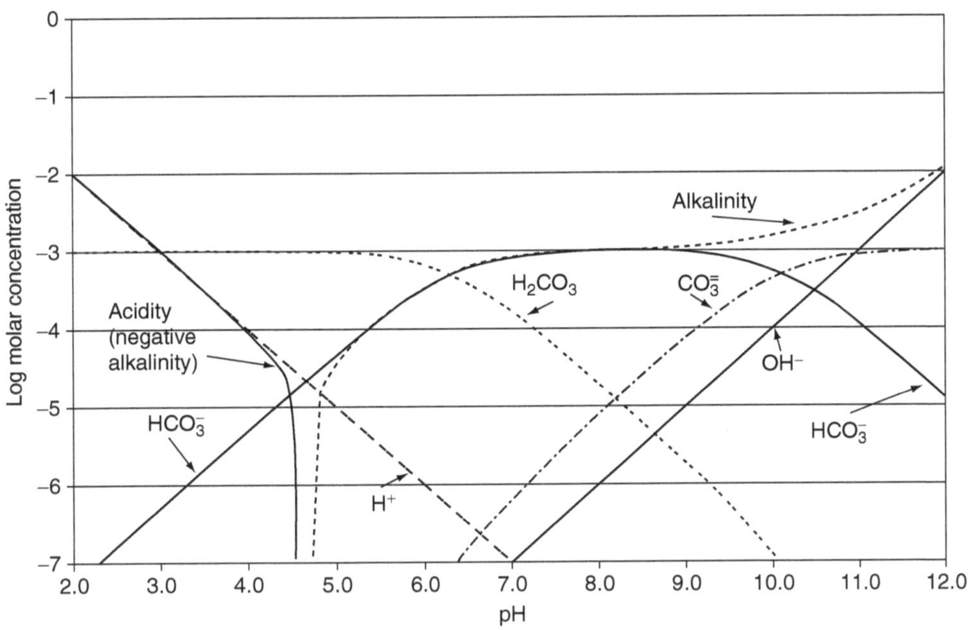

Figure 1 Alkalinity diagram for carbonate system. Total carbonate = 0.001 mol l^{-1}.

titrated on unfiltered samples and the result is expressed as ANC in milliequivalent per liter (meq l^{-1}) or microequivalent per liter (μeq l^{-1}), rather than total alkalinity, because ANC includes the neutralizing effects of suspended particulate matter as well as dissolved species. Neglecting the suspended solids, the conversion between the units is 1 mg l^{-1} $CaCO_3$ = 0.020 meq l^{-1} or 20 μeq l^{-1}.

Rather than titrating to a fixed endpoint, the most accurate procedure is to use the Gran titration, which also measures other weak acids with dissociation constants greater than 4.5 and typically also includes particulates on unfiltered samples. In this method, the titration is carried out by adding acid in a series of additions and recording pH until all of the buffers are exhausted (i.e., below pH 4) and subsequently added protons exhibit a linear accumulation in titrations that continue in steps down to a pH of 3.5 or less. The endpoint is easily identified by graphically back extrapolating the linear portion of a Gran titration curve to the endpoint.

Effect of Carbon Dioxide

The addition or subtraction of carbon dioxide has no direct effect on alkalinity. Although CO_2 gas does react with water to form bicarbonate, and further reacts to form carbonate, equal amounts of protons are released in the reaction and thus alkalinity is not affected directly.

$$[CO_2] + [H_2O] \leftrightarrow [H^+] + [HCO_3^-] \leftrightarrow 2[H^+] + [CO_3^{2-}] \quad [4]$$

The increase in carbon dioxide gas, either through respiration or increased burning of fossil fuels, can have a large indirect effect on alkalinity, because weathering is enhanced at the lower pH and calcium carbonate minerals present may dissolve and release alkali as follows:

$$2[H^+] + [CO_3^{2-}] + CaCO_3 \leftrightarrow Ca^{2+} + 2[HCO_3^-] \quad [5]$$

From eqn. [1], one might assume the addition of CO_2 would tend to increase the carbonate ion concentration in the system and result in a greater tendency to precipitate calcium carbonate in the presence of calcium ions, thus removing alkalinity. However, the addition of protons from the carbonic acid in the equation above lowers the pH sufficiently, such that a shift occurs in the balance of the carbonate system toward bicarbonate and carbonic acid, resulting in less carbonate ions and an increased tendency to dissolve calcium carbonate. If calcium carbonate is available and does dissolve, it increases alkalinity until an equilibrium is re-established. Thus, the concern for global increases in CO_2 in lowering of pH and dissolution of calcium carbonate deposits, particularly in the marine environment.

The reverse of the reaction may also occur in freshwaters of high calcium content as a result of photosynthesis. In this case, high photosynthesis results in CO_2 uptake, which raises the pH and subsequently results in a lowering of alkalinity as precipitation of calcium carbonate forms white deposits called marl in lakes. A similar process occurs when carbon dioxide rich ground water degasses in the lake in equilibrium with the atmosphere and results in calcium carbonate marl deposits.

Origin of Alkalinity

Alkalinity in freshwater systems is derived from several sources: weathering of rocks and soil, exchange reactions in soils, biological uptake and reduction of strong acid anions, evaporation and precipitation of minerals, and atmospheric deposition of dust. Precipitation has little if any alkalinity, and in most systems, weathering is typically the dominant source of alkalinity for inland waters. In relatively unpolluted regions, precipitation is only slightly acidic because the carbon dioxide in the atmosphere in equilibrium with pure water results in a weak carbonic acid solution with zero alkalinity but a pH of about 5.6. In industrialized areas, the sulfates and nitrates from emissions may result in precipitation with a pH of about 4.3 and an alkalinity of about negative 50 $\mu eq\,l^{-1}$. In some arid regions, dry deposition of alkaline dusts may result in atmospheric alkalinity inputs to the inland waters.

Weathering

Most of the large variations in the alkalinity of inland waters can be attributed to the amount of weathering of bedrock material and soils derived from the bedrock. Rock types can be divided into three broad categories: slow weathering noncarbonate rock (igneous granite, gneiss) with resulting thin soils, low cation exchange capacity, and low pH; moderate weathering of rocks with low carbonate content, but deeper soils; and rapid weathering carbonate bedrock (e.g., sedimentary limestone), deep soils with high cation exchange and high soil pH. In areas where bedrock is deeply buried, soils play the major role in alkalinity of surface waters. The weathering of rock provides carbonates and to a lesser extent silicates as weak acid anions along with base cations such as calcium and magnesium from the rock. Even in areas dominated by silicate rocks, however, the more rapidly weathered carbonates, even if present in small amounts, tend to be the predominant source of alkalinity. Carbon dioxide in the air or water tends to increase weathering reactions because the carbonic acid formed lowers the pH. The additional weathering, particularly of carbonates, results in higher alkalinity export to surface waters.

In rare cases, the weathering and oxidation of reduced iron and sulfur containing minerals such as pyrite found in some bedrocks and coal deposits may result in reductions in alkalinity as ferric hydroxides and sulfuric acid are produced. The oxidation of sulfides and accumulation of sulfuric acids in volcanic lakes may lead to extremely acid conditions with negative alkalinities.

Cation Exchange

Cation exchange can increase alkalinity whenever hydrogen ions in solution exchange on surfaces for base cations. The effect is generally reversible, and thus the process may not contribute to long-term increases in alkalinity once the cation exchange sites are depleted. In fact, cation exchange can act in reverse if base cations are added from sea spray or road salt to a soil solution, causing temporary acidification and loss of alkalinity. Nevertheless, soils with large cation exchange capacities can act as a large buffering reserve against relatively short-term acidification events.

Assimilatory Uptake

The uptake of inorganic carbon and other cations and anions during plant growth are generally balanced with little net effect on alkalinity with the general exception of nitrogen uptake. The uptake of nitrate during plant growth tends to raise alkalinity of the solution. Nitrate is assimilated by plants and reduced to amine groups R-NH_2 with the net result being the removal of nitric acid. A general equation for the reaction is

$$106CO_2 + 16H^+ + 16NO_3^- + H_3PO_4 + 122H_2O \rightarrow (CH_2O)_{106}(NH_3)_{16}H_3PO_4 + 138O_2 \quad [6]$$

Forests, for example, may produce net gains in alkalinity in stream water by the uptake of nitrates found in acid precipitation. The opposite may occur in some systems, where ammonium [NH_4^+] is the dominant form of nitrogen uptake by plants or algae and results in loss of alkalinity as a hydroxide [OH^-] is also taken up to balance the charges.

Dissimilatory Redox Reactions

A similar reaction may occur with dissimilatory uptake of acid anions. In this case anaerobic bacteria do not use the ions to build cellular material, but instead use the nitrate and sulfate ions as electron acceptors during anaerobic decomposition of organic matter. A simplified equation for sulfate reduction is

$$SO_4^{2-} + 2(CH_2O) \rightarrow H_2S + 2HCO_3^- \quad [7]$$

This process strips the oxygen from the acid anions to produce CO_2 and in effect removes the strong acids from solution. The reduced end products from the processes of denitrification and sulfate reduction are nitrogen gas and hydrogen sulfide, respectively. As long as the reduced end product is not reoxidized, the alkalinity gain is permanent. Nitrogen gas is stable and will not readily oxidize, and similarly, hydrogen sulfide gas may be lost to the atmosphere as a gas release or permanently stored in the sediments (e.g., as iron sulfide), thus preventing reoxidation.

Similarly, the oxidized forms of metals such as iron and manganese can be used by bacteria as electron acceptors during anaerobic decomposition of organic matter. Again, this process produces alkalinity as long as the reduced metal end products are removed from the system and do not become reoxidized.

In general, any chemical reduction of noncarbon compounds, such as minerals, metals, as well as reduction of sulfates and nitrates will increase the alkalinity of the system. Oxidations such as nitrification are a similar bacterial processes that operate in reverse. In the presence of oxygen, the bacterial oxidation of ammonium to nitrate provides energy for bacterial growth, which results in a loss in alkalinity.

The amounts of alkali produced in any process can be estimated based on the coefficients of balanced chemical redox reactions. In summary, the in situ generation of alkali has the potential to neutralize acid precipitation inputs in some lakes that have long retention times. In quickly flushing lakes, however, the alkali generated within the lake is generally flushed out, and thus, the alkalinity of the lake is largely determined by processes within the lake's terrestrial watershed.

Evaporation and Precipitation

The amount of alkali produced by chemical precipitation reactions depends on the cations and anions involved. As noted earlier, precipitation of calcium carbonate as calcite removes two equivalents of alkalinity from solution because the base cation calcium is removed by a weak acid anion carbonate. However, the precipitation of gypsum, calcium sulfate $CaSO_4$, has no effect because equal equivalents of base cation and strong acid anion are removed together.

Just as dilute low alkalinity waters can mix with and reduce alkalinities of surface waters, such as occurs during large rainfall or snowmelt events, evaporation and transpiration can concentrate waters, resulting in seasonal and daily fluctuations in alkalinity. Evaporation has both a direct and indirect effect on alkalinity. Although evaporation per se does not alter the mass of dissolved ions, it does concentrate them, and thus if alkalinity is present (in either a positive or negative amount), it will be magnified by the concentration effect. The evaporation process also has indirect effects on the alkalinity of some systems, such as salt lakes. In lakes where evaporation greatly exceeds freshwater inputs, the solubility product of many minerals may be exceeded and precipitates will form. Typically the first minerals to precipitate are calcium and magnesium carbonates. In areas dominated by silicate bedrock weathering, other monovalent cations such as the alkali metals, sodium and potassium may be abundant enough to create alkaline soda lakes as the polyvalent cations quickly precipitate out of solution.

Biology of Alkaline and Acidic Lakes

The pH of a system is often referred to as a master variable in aquatic systems, but it is the alkalinity, particularly bicarbonate ion, which buffers the effects of this master variable. The pH of a solution influences many of the most important chemical processes including acid/base reactions, precipitation and dissolution, oxidation and reduction, and adsorption/desorption exchange reactions. These reactions result in a cascade of effects on plants and animals, including chemical speciation, weathering, nutrient and toxicant uptake, and ionic regulation. These processes have impacts at the subcellular level up to the ecosystem level in determining the distribution of plants and animals as a function of alkalinity from calcareous fens with rare endemic plants to acidic lakes where mussels and fish are excluded. Alkalinity is what defines the acidic/basic character of ecosystems and provides the buffers to control the potential for wide daily swings in pH.

Natural Buffering

The alkalinity of most lakes and rivers ranges between about 100 and 5000 $\mu eq\,l^{-1}$ with a pH between 6 and 9. For comparison, seawater is typically pH 8.2 with an alkalinity of about 2300 $\mu eq\,l^{-1}$. As noted earlier, most of the variation in alkalinity can be accounted for by variations in bedrock. As shown in **Figure 2**, the low alkalinity areas are in the Adirondacks of New York, the Northeast, and in some of the mountainous regions of the west with granitic bedrock containing little carbonates. Such areas extend to cover large areas of Canada as part of the bedrock of the Canadian Shield. Higher alkalinities are found in the Midwest and West as well as parts of Africa, which generally contain more carbonates and often have deeper soils and higher rates of evapotranspiration.

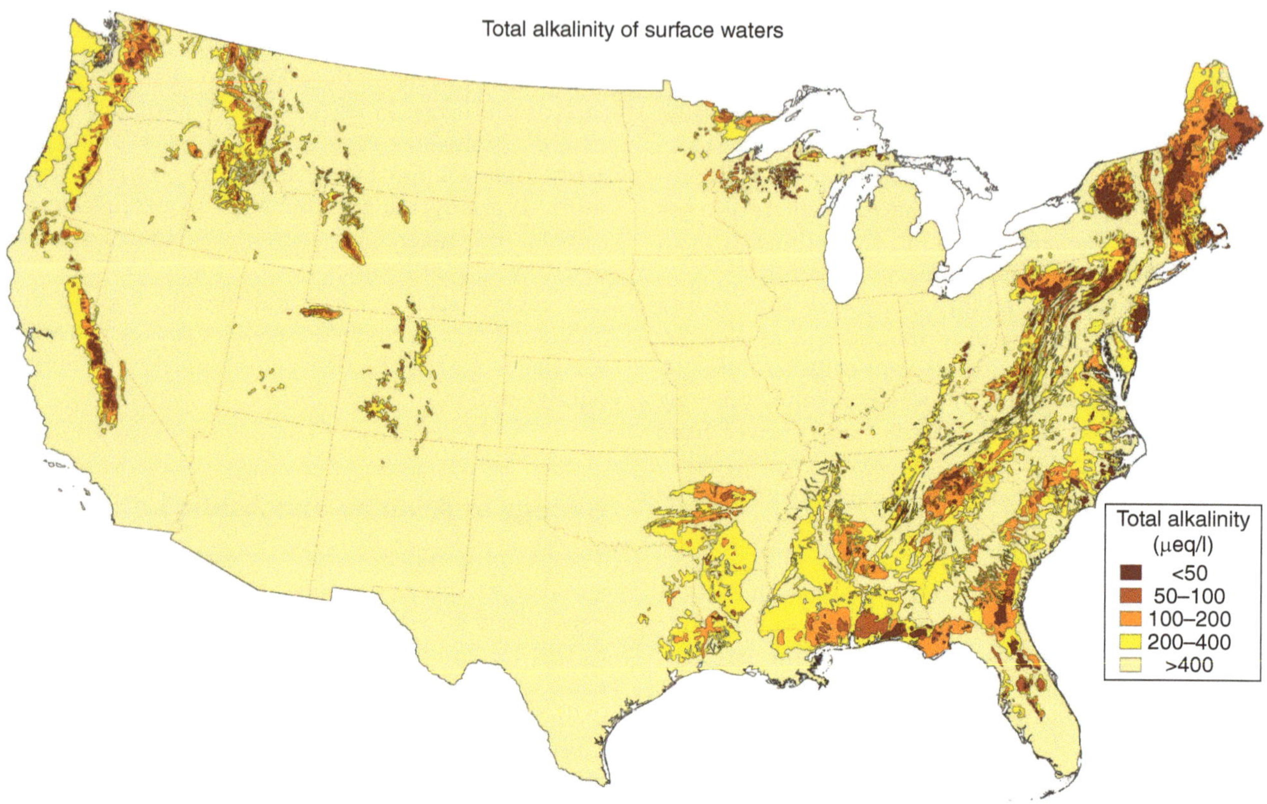

Figure 2 Alkalinity of surface waters of the United States. Reprinted from Omernik JM, Griffith, GE, Irish JT, and Johnson CB (1988) Total alkalinity of surface waters: a national map. Corvallis Environmental Research Laboratory, U.S. Environmental Protection Agency, Corvallis, OR.

Circumneutral waters support the diversity of species of algae, plants, and higher organisms commonly seen in inland waters. Because higher alkalinities are associated with increased weathering and generally higher ionic concentrations overall, it is not surprising that these waters are also higher in nutrients and tend to have higher productivities. Species diversity often increases up to a pH of about 8 and some animals with calcareous shells, such as freshwater mussels and snails, are restricted to higher alkalinity waters, which also have higher levels of calcium. Similar increases in productivity and shifts in species composition are seen in freshwater wetlands. Acidic 'poor' fens, for example, typically have less diversity of vascular plants than do alkaline 'rich' fens.

Low Alkalinity Acidic Waters

Aquatic systems having less than $100\,\mu eq\,l^{-1}$ alkalinity are defined as low alkalinity or sensitive systems and such systems are also typically low in nutrients as well. These systems generally have a pH of less than 6. When acids exceed the buffers and alkalinity drops to zero or into the negative range, the systems are called acidified, and are characterized by a pH of less than 5.6. Typically there are two classes of low alkalinity or acidified systems, those with clear water and mineral acids such as sulfuric and nitric acids from acid precipitation, and those with dark, tea-colored water acidified by natural organic acids such as fulvic, tannic, and/or humic acids, which are often found in poorly drained soils and wetlands and bogs in particular. Sphagnum mosses in bogs produce weak acids and also tend to reduce alkalinity by uptake of base cations. These bogs and dark water streams and lakes are often nitrogen poor and provide growth conditions unfavorable to many species of plants and animal but result in a variety of rare acid-tolerant species, such as sundews and pitcher plants and bladderworts. These carnivorous plants are able to tolerate the low alkalinity, acidic conditions, and obtain nitrogen by trapping and digesting insects or small invertebrates.

The clear water acid lakes may be associated with high concentrations of mineral acids, such as within volcanic craters or within sulfide-bearing mineral formations, or due to anthropogenic acidic precipitation of sulfuric and nitric acids. Because fish are typically absent from waters with a pH below 4.5 this may allow insects such as water boatmen (Corixidae) and whirligig beetles (Gyrinidae) to thrive without significant predation.

Highly Alkaline Lakes

Moderately alkaline surface waters may be found in areas of carbonate bedrock, such as limestone or 'chalk' streams and lakes. Although these are often nutrient-rich systems, the solubility of limestone alone does not lead to excessively high pHs. The most highly alkaline waters are typically formed in arid regions, where evaporation in 'pans' results in very high alkalinities found in salt lakes and 'soda' lakes such as found in dry areas of Africa and the American Southwest. In some cases, these lakes are extremely large such as the Caspian Sea, Aral Sea, Lake Balkhash, and Great Salt Lake. The salt lakes also tend to exclude fish and as a result are home to rare species such as the brine shrimp.

See also: Acidification.

Further Reading

Wetzel RG and Likens GE (2000) *Limnological Analyses*, 3rd edn. New York: Springer.

Stumm W and Morgan JJ (1996) *Aquatic Chemistry*, 3rd edn. New York: John Wiley and Sons.

Relevant Websites

http://www.standardmethods.org/.

http://en.wikipedia.org/wiki/Alkalinity.

http://or.water.usgs.gov/alk/.

http://water.usgs.gov/owq/FieldManual/Chapter6/section6.6/.

Major Cations (Ca, Mg, Na, K, Al)

C S Cronan, University of Maine, Orono, ME, USA

Fresh waters contain the chemical signatures of diverse element cycling processes that transform dilute precipitation inputs to a watershed and eventually generate surface waters with a wide range of distinctive ionic compositions. Depending on the integrated outcome of these different biogeochemical processes, a given lake or stream may exhibit a chemical profile that is enriched or depleted in specific solutes, thereby influencing ecological health, biotic diversity, and aquatic productivity. Although many different parameters may be considered when an aquatic ecosystem is assessed for its water quality and ecological integrity, one important chemical indicator or measurement is an analysis of the major cations Ca, Mg, Na, K, and Al.

Cation Functions and Stoichiometry

Each major cation has important biological functions, biogeochemical relationships, and ecological roles that influence the ability of an aquatic ecosystem to sustain life forms and active food webs. Calcium (Ca) is essential for osmotic balance, cell signaling, and control of muscle contractions in many aquatic organisms. It contributes to structural support in animals as an integral part of vertebrate bones and in plants as a component of cell walls. As a contributor to alkalinity, calcium helps to govern acid–base buffering in surface waters and participates in the formation of calcium carbonate shells for aquatic invertebrates.

Magnesium (Mg) is a divalent cation that is required in chlorophyll molecules for plant photosynthesis, and serves as an enzyme co-factor in several reactions controlling cellular metabolism and synthesis. It can also function as an electrolyte and contributes to osmotic balance. Finally, Mg^{2+} is another major ion that contributes to buffering and alkalinity in aquatic ecosystems.

Sodium (Na) and potassium (K) are both monovalent cations that fill pivotal roles as electrolytes in cells. These ions function in the maintenance of membrane pumps and are especially important in the propagation of action potentials in nerve cells. Potassium also serves as a key regulator of guard cell activity in vascular plants. In many surface waters, sodium is a main contributor to buffering and alkalinity.

Aluminum (Al) is a trivalent cation that is nonessential for virtually all living organisms, but serves as an important geochemical component in many ecosystems. Most soils and sediments are rich in aluminum oxides, aluminosilicate clays, and Al-silicate minerals. Clays and Al hydroxides in watershed soils and aquatic ecosystems influence acid–base buffering, phosphorus availability, and solubility of ionic Al, thus affecting water quality conditions for aquatic organisms. From a biological perspective, inorganic ionic Al is a potential stressor for aquatic organisms, because it can act as a toxin for sensitive biota and can interfere with normal gas exchange processes in fish gill membranes.

Cation concentrations and stoichiometric ratios in aquatic organisms are variable, with Ca representing roughly 0.5–1.5% of dry mass, Mg and Na amounting to 0.1–0.2% of dry mass, K ranging from 0.4–1.0% of dry mass, and Al occurring at ppm levels. For aquatic plants, molar ratios of Ca:Mg:Na:K in live biomass are on the order of 3:2:1:5, compared with ratios on the order of 10:1:2:2 in aquatic vertebrates.

Sources of Major Cations

Concentrations of major cations in surface waters are determined by the interactions of various source and sink processes. Atmospheric inputs of wet and dry deposition are a potentially important source of soluble and particulate forms of cations for watersheds and their associated surface waters. As illustrated with data from the U.S. National Atmospheric Deposition Network, concentrations and fluxes of cations in precipitation vary as a function of geographic location and proximity to emission sources for cations (**Table 1**). Comparing the four regions represented in **Table 1**, cation concentrations vary by as much as eight to ten-fold (e.g., Ca and Na) and annual deposition fluxes differ by up to a factor of twenty (e.g., Na). Atmospheric inputs of elements such as Na may be elevated at locations near sources of marine aerosols, whereas atmospheric deposition may be enriched with other cations as a result of dust and industrial emissions generated in the airshed upwind of an aquatic receptor.

One of the major sources for cations in surface waters is mineral weathering in the soils, surficial

Table 1 Comparison of weighted mean concentrations and annual fluxes of cations in precipitation collected during 2004 in Florida, Illinois, Colorado, and Washington at sites monitored by the U.S. National Atmospheric Deposition Program

	Mean concentration (mg l^{-1})				*Deposition flux (kg ha^{-1} $year^{-1}$)*			
	Ca	Mg	Na	K	Ca	Mg	Na	K
Florida	0.07	0.04	0.34	0.02	1.0	0.5	4.9	0.3
Illinois	0.26	0.05	0.09	0.03	1.9	0.4	0.7	0.2
Colorado	0.18	0.02	0.03	0.04	1.0	0.1	0.2	0.2
Washington	0.03	0.01	0.1	0.01	0.6	0.2	1.9	0.2

Table 2 Estimates of weathering contributions to watershed annual export fluxes of major cations in different regions (based on stream outputs minus precipitation inputs)

	Annual weathering flux to surface waters (kg ha^{-1} $year^{-1}$)			
	Ca	Mg	Na	K
Coweeta, NC	0.7	1.8	4.3	2.0
Walker Branch, TN	134	75	0.6	3.7
Hubbard Brook, NH	11.7	2.7	5.9	1.5
Birkenes, Norway	11.4	2.7	11.2	−0.7

Data based on Likens GE and Bormann FH (1995) *Biogeochemistry of a Forested Ecosystem*. New York: Springer.

deposits, and bedrock of drainage basins. In some watersheds, weathering of calcareous limestone and dolomite generates substantial releases of soluble Ca, Mg, and associated alkalinity for drainage waters feeding lakes and streams. In other locations, dissolution of aluminosilicate minerals can release variable mixtures of Ca, Mg, Na, K, and Al to surface waters. For example, weathering breakdown of feldspars such as albite ($NaAlSi_3O_8$), anorthite ($CaAl_2Si_2O_8$), and orthoclase ($KAlSi_3O_8$), along with minerals such as tremolite ($Ca_2Mg_5Si_8O_{22}(OH)_2$) and biotite ($KMg_3AlSi_3O_{10}(OH)_2$) produces a wide range of different stoichiometric ratios of soluble Ca, Mg, Na, K, and Al. **Table 2** presents a comparison of annual watershed export fluxes of cations derived from mineral weathering.

Exchangeable ions adsorbed to clays and organic colloids are a third important source of soluble Ca, Mg, Na, K, and Al for surface waters. In watersheds with ample alkalinity generation, clays and humus will tend to be dominated by exchangeable Ca and Mg, whereas soils and sediments that have experienced acidification from anthropogenic or natural processes tend to exhibit elevated concentrations of exchangeable Al ions. These pools of exchangeable cations react continuously with soluble ions in drainage waters and surface waters, releasing variable mixtures of the major cations.

A final major source of soluble cations to surface waters is microbial mineralization of detritus. In senescent leaf litter, the elemental stoichiometric ratios of Ca:Mg:Na:K:Al in mass units are generally in the range of 30:5:1:10:1. Thus, decomposition of detritus is a potentially large source for releases of Ca, Mg, and K, but is a much smaller source for Na and Al.

Biological and Geochemical Sinks for Major Cations

Inputs of soluble cations in watershed runoff, groundwater, and atmospheric deposition cycle through freshwater ecosystems at rates that are governed by interactions between hydrologic processes and biogeochemical sinks. During the growing season, biological demand for Ca, Mg, K, and Na results in strong assimilation of these nutrients into the biomass and productivity of aquatic organisms. Once absorbed, the nutrients may be used and recycled within the water column or may be transferred to the sediments as detrital particles. When the growing season is over, the strength of this biological sink tends to decline.

Two of the major cations, Ca and Al, are subject to the influence of geochemical precipitation reactions that generate solid-phase mineral sinks for these elements in aquatic ecosystems. Under conditions of elevated pH and alkalinity, Ca may precipitate from the water column as $CaCO_3$ when the ion activity product of $[Ca^{2+}] \times [CO_3^{2-}]$ exceeds the solubility product of this compound under current pH and alkalinity conditions. This process may transfer Ca to the sediments for shorter or longer term storage and retention within the system. In other types of watersheds containing sources of mobile ionic Al, inputs of dissolved inorganic Al may precipitate from the water column as $Al(OH)_3$ when the ion activity product of $[Al^{3+}] \times [OH^-]^3$ exceeds the solubility product of aluminum trihydroxide in equilibrium with current lake chemistry. When $Al(OH)_3$ forms in the water column, it may scavenge soluble phosphorus and thus exert a strong influence on the availability of this critical nutrient. In **Figure 1**, solubility lines for $CaCO_3$ and $Al(OH)_3$ are shown as a function of solution pH; it is apparent that

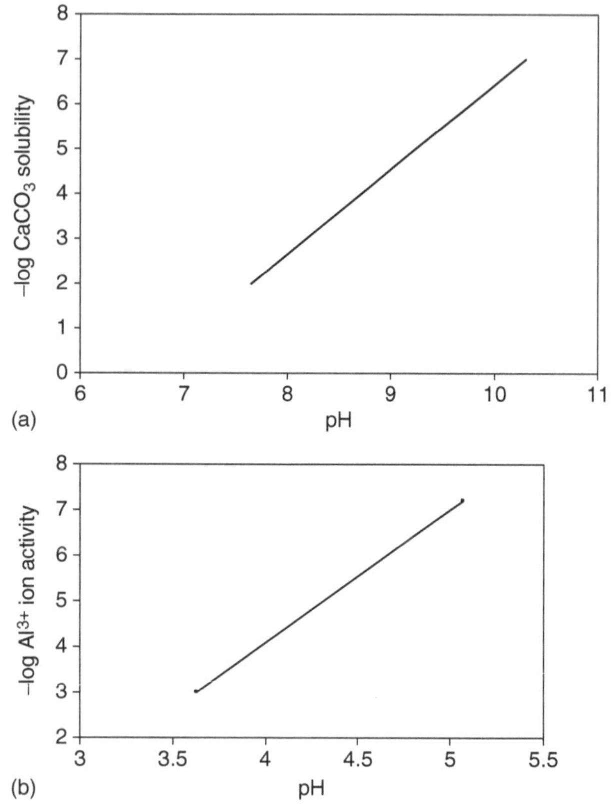

Figure 1 Molar solubility of (a) calcium carbonate $CaCO_3$ and (b) aluminum trihydroxide. $Al(OH)_3$ expressed as a function of solution pH. The solubility of each mineral is inversely related to pH.

precipitation of each mineral is more likely as the pH increases in an aquatic system.

Analytical Distinctions among Cations

The cations in a lake, stream, or river occur in a variety of chemical forms or species that may be more or less available to biota or geochemically reactive. Divalent Ca and Mg may be distributed into insoluble colloidal forms of suspended particulates, ionic Ca^{2+} or Mg^{2+}, and soluble complexes formed through the binding of either ionic Ca or Mg with various organic or inorganic ligands. In comparison, monovalent Na and K largely exist as soluble ionic Na^+ and K^+ and as insoluble colloidal particles in the water column. The chemical speciation of soluble Al in surface waters is perhaps the most complex, and detailed analytical methods are required to distinguish the different forms of this element. Ionic inorganic Al may exist as Al^{3+}, $AlOH^{2+}$, or $Al(OH)_2^+$ – three charged species that are toxic to some organisms. Soluble Al may also occur as complexes in which the metal is bound to a humic ligand or an inorganic ligand such as fluoride or sulfate. Finally, Al may exist as acid-soluble colloidal particles in the water column.

Cation Chemistry in Fresh Waters

How do the chemical signatures of freshwater lakes and rivers vary across the landscape? If one surveys a wide cross section of freshwater lakes, streams, and rivers, it is apparent that normal ranges of concentrations for Ca are 10–700 μmol l^{-1} (where 100 μmol l^{-1} = 4.0 mg l^{-1}), for Mg are 5–350 μmol l^{-1} (where 100 μmol l^{-1} = 2.4 mg l^{-1}), for Na are 20–280 μmol l^{-1} (where 100 μmol l^{-1} = 2.3 mg l^{-1}), for K are 1–85 μmol l^{-1} (where 100 μmol l^{-1} = 3.9 mg l^{-1}), and for Al are <0.5–25 μmol l^{-1} (where 100 μmol l^{-1} = 2.7 mg l^{-1}). By comparison, Livingstone reported in 1963 that mean concentrations of base cations in fresh waters of the world have the following values: Ca = 375 μmol l^{-1}, Mg = 170 μmol l^{-1}, Na = 275 μmol l^{-1}, and K = 60 μmol l^{-1}.

Patterns of base cation chemistry in fresh waters can be illustrated with examples from the U.S. EPA National Stream Survey for eastern North America and a comparative data set for lakes in North America and Europe (**Figure 2**). As shown in **Figure 2a**, streams in the eastern U.S. range from dilute fresh waters such as those draining the southern Blue Ridge Mountains with SBC (sum of base cation charge equivalents) <400 μmol_c l^{-1} to more concentrated streams such as those in the southern Appalachian Mountains where SBC >1300 $\mu mol_c\,l^{-1}$. By comparison, the Norwegian Gaula River is one of the more dilute rivers in the world, with a mean SBC <120 $\mu mol_c\,l^{-1}$. In **Figure 2b**, a cross-section of lake chemistries is presented to illustrate how this class of freshwater ecosystems exhibits a comparable range of concentrations and element ratios of base cations. These examples vary from small ponds with SBC <100 $\mu mol_c\,l^{-1}$ to large lakes with SBC concentrations approaching 1000–2000 $\mu mol_c\,l^{-1}$. Both sets of stream and lake data are plotted in **Figure 3** with bar graphs that illustrate the percentages of molar charge contributed by each of the four major base cations.

Despite the obvious variations in cation chemistry illustrated with the examples depicted in **Figures 2** and **3**, there are some general patterns that can be noted. For the majority of fresh waters, Ca is the most abundant cation on a molar charge basis and K is consistently the least abundant base cation, whereas Mg and Na are typically the second and third most abundant cations. A major exception to that pattern occurs in surface waters that are located near marine sources of Na such as the Florida streams in **Figure 2a** and the coastal ANP lakes and Lake Gardsjon

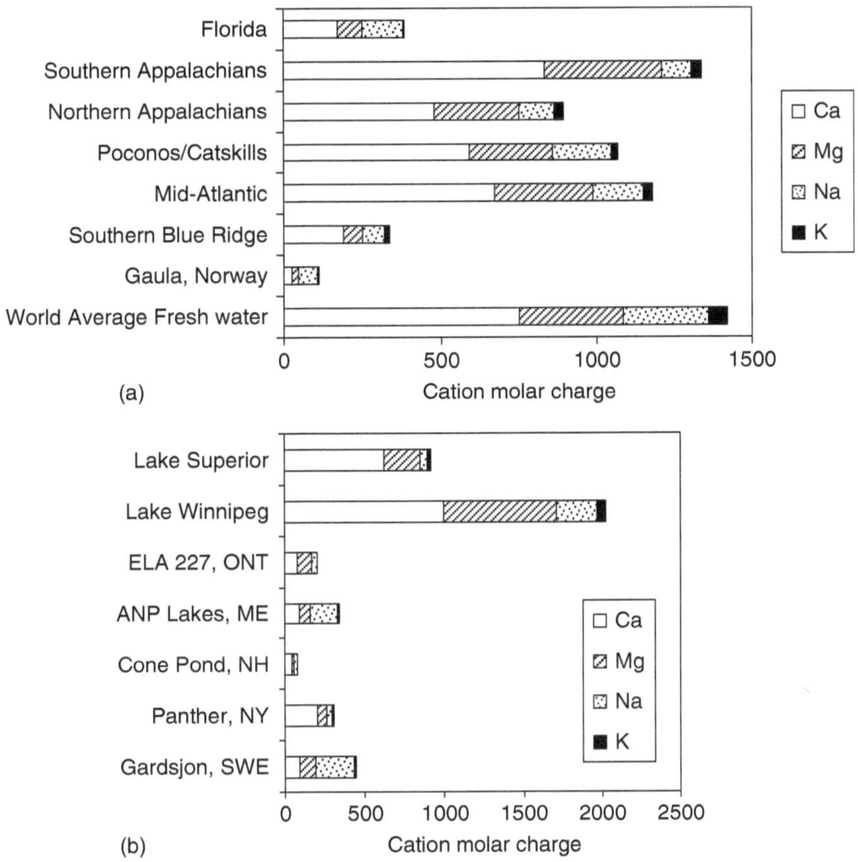

Figure 2 Examples of base cation concentrations in $\mu mol_c\ l^{-1}$ for a range of (a) streams and rivers and (b) lakes in North America and Europe. Data sources: U.S. EPA (1988) *Chemical characteristics of streams in the Mid-Atlantic and Southeastern United States* – Volume II. EPA/600/3–88/021b; Henriksen A (1982) *Preacidification pH values in Norwegian rivers and lakes*. Norwegian Institute for Water Research Report 3/1982; Armstrong FAJ and Schindler DW (1971) Preliminary chemical characterization of waters in the Experimental Lakes Area, northwestern Ontario. *Journal of the Fisheries Research Board Canada* 28: 171–187; Livingstone DA (1963) *Chemical Composition of Rivers and Lakes. U.S. Geological Survey Professional Paper 440-G*, Washington, DC; Kahl JS, Andersen JL, and Norton SA (1985) Water resource baseline data and assessment of impacts from acidic precipitation, Acadia National Park, ME, *National Park Service Tech. Rpt. 16*; Buso DC, Martin CW, and Hornbeck JW (1984) Potential for acidification of six remote ponds in the White Mountains of New Hampshire. *Water Resources Research Center Research Rpt. 43*, Durham, NH; Hultberg H (1985) Budgets of base cations, chloride, nitrogen, and sulphur in the acid Lake Gardsjon catchment, SW Sweden. *Ecological Bulletins* 37: 133–157; Driscoll CT, Newton RM, Gubala CP, Baker JP, and Christensen SW (1991) Adirondack Mountains, pp. 133–202. In: Charles, DF (ed.) *Acidic Deposition and Aquatic Ecosystems-Regional Case Studies*. New York: Springer.

depicted in **Figure 2b**. There, marine-derived Na serves as one of the dominant cations. As shown in **Figure 4**, the relationship between Ca and Mg concentrations is remarkably consistent across a wide range of fresh waters, with roughly a 2:1 or 3:1 molar ratio between the divalent cations.

Aluminum is a cation that occurs at very low concentrations in most surface waters with pH values of 5.0 or above. However, in acidified lakes and streams with pH < 5.0, Al concentrations may exceed 10–20 $\mu mol_c\ l^{-1}$, thereby resulting in a proportionately large molar charge contribution by this element. In **Figure 5**, examples of four streams with pH values in the range of 4.5–4.9 are presented showing the comparative molar charge contributions from Ca and Al. As indicated, streams at the Birkenes watershed (Norway) and Gardsjon watershed (Sweden) contain elevated Al concentrations, so that Ca and Al concentrations are nearly equivalent on a molar charge basis. Even more striking are the conditions in streams affected by acid mine drainage (e.g., West Virginia), where soluble Al concentrations reach levels of 100 $\mu mol\ l^{-1}$ and the ionic charge contributed by Al is as high as 250–300 $\mu mol_c\ l^{-1}$. In contrast to these examples, most surface waters with pH values >5.0 typically contain low Al concentrations that are less than or comparable to background K concentrations.

Efforts have been made to summarize the imprint of different geologic conditions on the base cation

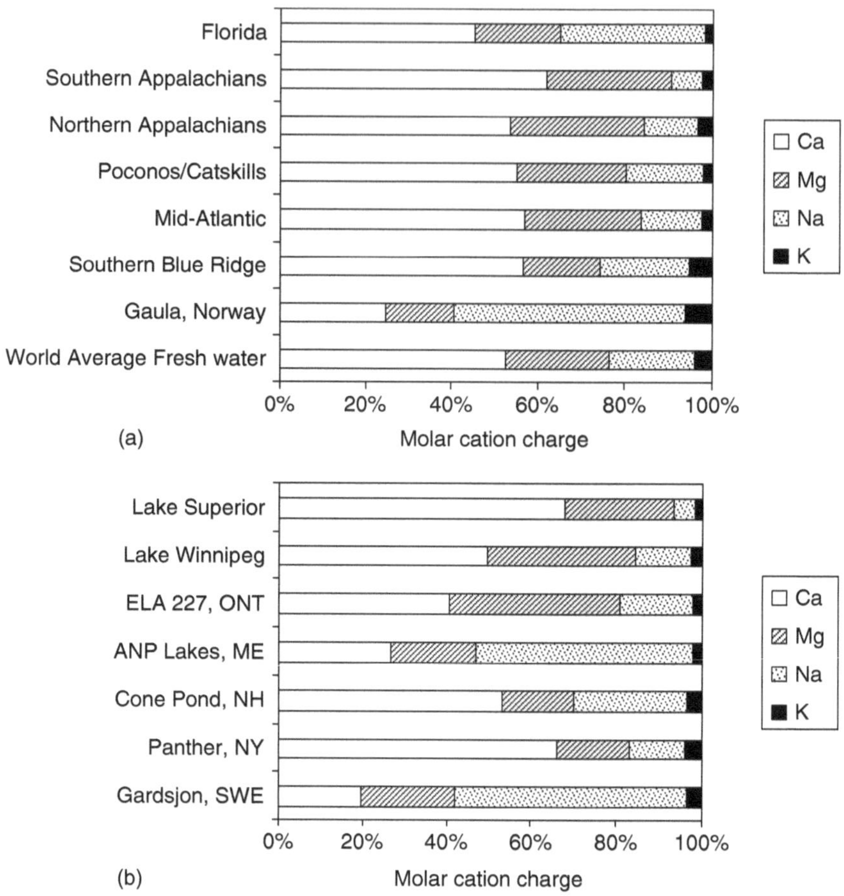

Figure 3 Relative proportion of ionic charge equivalents contributed by each of the four base cations in the (a) streams and (b) lakes illustrated in the previous figure.

chemistry of surface water runoff. As illustrated in **Table 3**, waters draining contrasting igneous, metamorphic, and sedimentary geologic source areas may exhibit pronounced chemical differences, especially with respect to Ca and Mg concentrations. For example, drainage waters dominated by carbonate materials in the watershed may contain 60-fold more Ca and 20-fold more Mg than waters derived from granitic source areas (**Table 3**).

As might be expected from the principles of electrical neutrality and charge balance, the base cations which dominate most surface waters exhibit a strong relationship with major anion concentrations. In general, surface waters can be separated into three groups based on anion chemistry and the relative dominance of strong acid anions, organic anions, and bicarbonate alkalinity in overall solution chemistry. As such, one can distinguish specific groups of surface waters where concentrations of base cations are positively correlated with increases in concentrations of those dominant anionic constituents. In acidic surface waters, concentrations of soluble Al are often strongly correlated with dissolved organic carbon (DOC) or strong acid anions such as sulfate or nitrate. In well-buffered waters, concentrations of base cations are often tightly correlated with bicarbonate alkalinity.

One of the interesting spatial patterns observed in river drainage systems is the tendency for downstream longitudinal changes in cation chemistry. For example, a longitudinal survey of Falls Brook in New Hampshire demonstrated a progressive increase in base cations and decrease in monomeric Al with increasing distance downslope from the headwaters located at an elevation of 812 m (**Figure 6**). This pattern is a reflection of changing geochemical and hydrologic factors that differentially influence the generation of acid neutralizing capacity, inputs of soluble cations, and solubility of inorganic Al. In a separate example, investigators Robert Reynolds and Noye Johnson sampled a stream draining an alpine glacier in the Cascade Mountains, WA, and found that pH increased from 5.2 to 6.9 and divalent cation concentrations increased ten-fold from 12 to 105 $\mu mol_c\,l^{-1}$ over a descent of 1400 m from the glacier to the valley below.

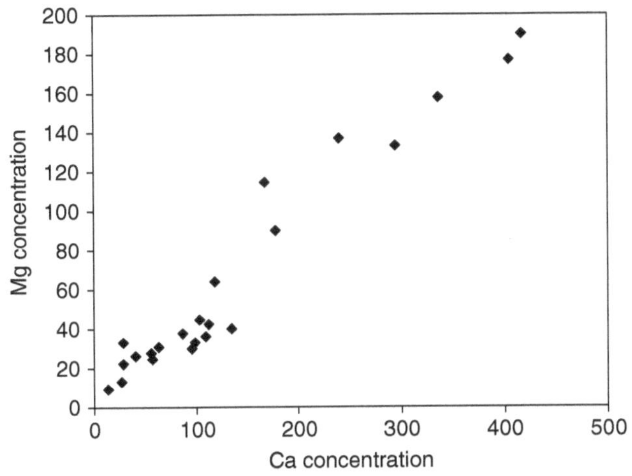

Figure 4 Relationship between Ca and Mg concentrations (μmol l^{-1}) for a range of streams in the U.S., Canada, and Europe based on 17 regional data sets. Data sources: U.S. EPA. (1988) *Chemical Characteristics of Streams in the Mid-Atlantic and Southeastern United States – Volume II*. EPA/600/3–88/021b; Henriksen A (1982) Preacidification pH values in Norwegian rivers and lakes. *Norwegian Institute for Water Research Report 3/1982*; Kahl JS, Andersen JL, and Norton SA (1985) Water resource baseline data and assessment of impacts from acidic precipitation, Acadia National Park, ME, *National Park Service Tech. Rpt. 16*; Stoddard JL and Murdoch PS (1991) Catskill Mountains, pp. 237–271. In: Charles DF (ed.) *Acidic Deposition and Aquatic Ecosystems-Regional Case Studies*. New York: Springer; Cosby BJ, Ryan PF, Webb JR, Hornberger GM, and Galloway JN (1991) Mountains of Western Virginia, pp. 297–318. In: Charles DF (ed.) *Acidic Deposition and Aquatic Ecosystems-Regional Case Studies*. New York: Springer; Haapala H, Sepponen P, and Meskus E (1975) Effect of spring floods on water acidity in the Kiiminkijoki area, Finland. *Oikos* 26: 26–31.

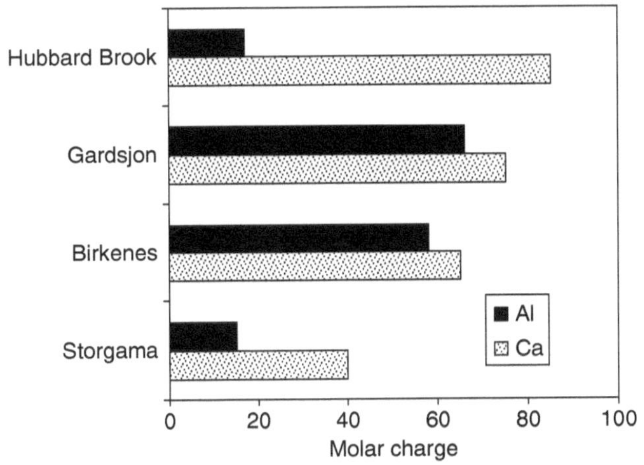

Figure 5 Comparison of molar charge contributions by Ca and Al ions (μmol$_c$ l^{-1}) in four acidic streams, including Hubbard Brook, NH (pH 4.9), Gardsjon, SWE (pH 4.2), Birkenes, NOR (pH 4.5), and Storgama, NOR (pH 4.5). Based on data from Hultberg H (1985) Budgets of base cations, chloride, nitrogen, and sulphur in the acid Lake Gardsjon catchment, SW Sweden. *Ecological Bulletins* 37: 133–157.

Table 3 Comparison of mean cation concentrations in drainage waters originating in different geologic source areas

Rock Type	*Ca*	*Mg*	*Na*	*K*
Granite	20	16	87	8
Gneiss	30	29	80	10
Sandstone	45	33	51	21
Volcanic	78	82	105	14
Shale	202	119	105	20
Carbonates	1283	321	34	13

Concentrations have been corrected to remove sea salt inputs and are expressed in μmol l^{-1}. Data are based on Meybeck *et al.* (1989) as cited in Kalff J (2002) *Limnology*. NJ: Prentice-Hall.

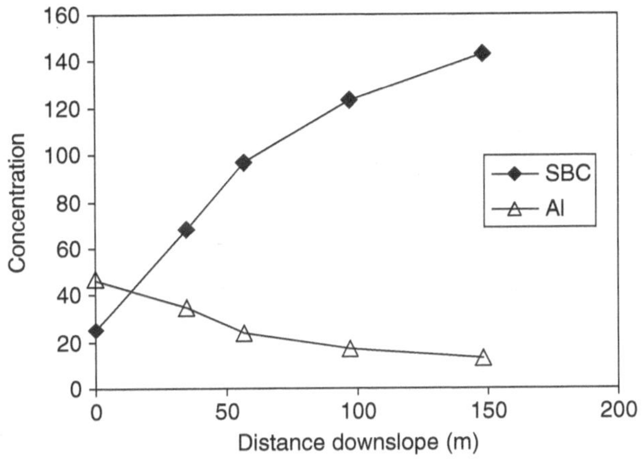

Figure 6 Changes in concentrations of base cations (SBC) and monomeric Al over a longitudinal sample gradient beginning at the headwater of Falls Brook, NH at 812 m elevation. Units are μmol$_c$ $l^{-1}$and data are based on Johnson NM, Driscoll CT, Eaton JS, Likens GE, and McDowell WH (1981) Acid rain, dissolved aluminum, and chemical weathering at the Hubbard Brook Experimental Forest, NH. *Geochimica et Cosmochimca Acta* 45: 1421–1437.

Environmental and Anthropogenic Influences on Cation Chemistry

Cation chemistry in surface waters may respond to various environmental factors and influences in the watershed or airshed surrounding a lake or river. As an example, Na concentrations may become elevated in surface waters affected by episodic or chronic atmospheric deposition of marine aerosols. The Florida streams and ANP lakes depicted in **Figures 2 and 3** reflect the influence of this type of process. In contrast, there are other instances of surface waters in cool temperate regions where Na concentrations become enriched as a result of anthropogenic road salting activities. A study of Adirondack lakes sampled in the U.S. EPA eastern lake survey indicated that lakes with >20 μmol$_c$ l^{-1} Cl^- showed evidence of disturbance from road salting and contained up to

3–4 times more Na than unimpacted lakes in the survey.

Seasonal hydrologic changes can alter drainage water flowpaths in ways that shift the proportions of interflow and ground water contributing to surface waters. As a result, the chemistry of lakes and runoff can experience dynamic shifts in cation chemistry. One of the general patterns observed in headwater streams is the tendency for higher storm flow conditions to generate runoff that is more acidic and enriched in soluble Al, whereas low flow conditions correspond with runoff that is more enriched with base cations and acid neutralizing capacity derived from ground water. In our studies of the Aroostook River, ME, results indicated that during peak flow conditions, concentrations of Ca (175 $\mu mol\,l^{-1}$) and Mg (55 $\mu mol\,l^{-1}$) were roughly one-third of the levels observed during baseflow, when Ca and Mg concentrations were 490 and 150 $\mu mol\,l^{-1}$, respectively. As such, these observations suggest that river runoff is dominated by ground water enriched in Ca and Mg during baseflow conditions, whereas peak storm flow runoff is generated with a large contribution from shallow interflow drainage water that contains lower concentrations of base cations than ground water.

The chemistry of surface runoff and lake water may be affected by watershed disturbance and management practices. Liming soils and surface waters to control acidity can elevate Ca and Mg concentrations and associated acid neutralizing capacity. In some cases, fires in the surrounding watershed can promote the transport of alkaline ash enriched with base cations into surface waters. Land conversions from forest cover to agricultural cropping may be accompanied by increased transport of particulate and soluble Ca, Mg, Na, and K to surface waters. Likewise, forest harvesting activities may disturb watershed nutrient cycles in ways that accelerate the transport of soluble and particulate phases of cations to linked aquatic ecosystems.

Nomenclature

ANC or alkalinity – The acid neutralizing capacity of a solution or the sum of proton acceptors.

Base cations – The major cations Ca, Mg, Na, and K that are capable of forming bases such as NaOH or KOH.

Molar charge units – $\mu mol_c\,l^{-1}$ refers to the micromoles of ionic charge per liter of solution; a mole of a monovalent ion contains a mole of ionic charge, whereas a mole of a divalent ion contains two moles of ionic charge.

Further Reading

Charles DF (1991) *Acidic Deposition and Aquatic Ecosystems – Regional Case Studies*. New York: Springer.

Kalff J (2002) *Limnology*. NJ: Prentice-Hall.

Magnuson JJ, Kratz TK, and Benson BJ (2006) *Long-Term Dynamics of Lakes in the Landscape*. New York: Oxford University Press.

Stumm W and Morgan JJ (1970) *Aquatic Chemistry*. New York: Wiley-Interscience.

Relevant Websites

http://water.usgs.gov/data
http://waterdata.usgs.gov/nwis

Chloride

S S Kaushal, University of Maryland Center for Environmental Science, Solomons, MD, USA

Introduction: Chloride and Salinization

Salinization of inland waters has been recognized as an environmental problem in arid and semiarid environments in the United States and throughout the world and its temporal and spatial extent because of land use and land cover change have also now been documented in humid regions. Salinization describes an increase in the sum concentration of all the ionic constituents dissolved in inland waters, and salinities are typically described in units of milligrams per liter or millequivalents per liter. Chloride, an important anion in many salts, can comprise ~5–17% of total solids in river waters globally with highest relative contributions to total solids in river waters in Australia. Chloride typically comprises ~1.9% the mass of sea water, where it is most abundant in the Earth's surface waters. It is the dominant natural form of the element chlorine with atomic number 17, which is formed when the element chlorine picks up an electron to form an anion, and is necessary to most forms of life.

The salinity of inland waters has a world average concentration of ~120 mg l^{-1}, according to synthetic work by Wetzel, and this is thought to be slightly elevated above a natural range because of human activities. Salinity in inland waters is dominated by the four major anions: bicarbonate (HCO_3^-), carbonate (CO_3^-), sulfate (SO_4^{2-}), and chloride (Cl^-). In inland waters that are experiencing minimal anthropogenic disturbance, concentrations of chloride are generally no more than a few milligrams per liter, with some local or regional variations of higher salinity. The low salinity and chloride concentrations of many fresh waters have influenced the distribution of organisms, and a long history and evolutionary adaptation to the osmotic conditions of low salinity.

High chloride concentrations as a result of salinization can negatively influence freshwater life at the cellular level either from lack of water or an excess of ions (or both) that can result in a range of toxic effects, however. Cells of plants and animals can suffer from water deficit because of difficulty in extracting water from the surrounding environment. Microbial communities are capable of adjusting their osmotic potential through the production of solutes within cells, but also may show salt stress and community shifts in response to changes in increased chloride concentrations and salinity.

In soils, chloride concentrations above 30 mg l^{-1} have been shown to damage land plants. Acute increases in salinity up to 1 g l^{-1} have been shown to have lethal and sublethal effects on some aquatic plants and invertebrates in fresh water. Other work has also reported that acute toxicity levels for stream macroinvertebrates ranges from 1 to 13 g l^{-1}, which are similar to concentrations reported recently in some urban streams of the United States by Kaushal *et al.* and other areas of the world reported by Williams. Chronic concentrations of chloride as low as 250 mg l^{-1} have been recommended for the protection of some sensitive taxa of freshwater life and human drinking water by federal agencies in the United States and Canada, although there are no formal regulations for chloride concentrations in inland waters.

Natural and Anthropogenic Sources of Chloride to Inland Waters

Under conditions of minimal anthropogenic disturbance, chloride is typically not the dominant anion in streams and lakes. Streams and lakes near marine environments can receive substantial inputs of chloride from atmospheric deposition from marine sources. Similarly, a major source of salinity and chloride to many dilute inland waters of arid regions can be the ionic content of atmospheric precipitation and particulate deposition. Continental rain contains typically more sulfate as an anion than chloride, but chloride ion concentrations tend to increase with proximity to the sea. Other natural sources of chloride to fresh waters can include weathering of rocks and ion exchange in soils. Chloride-containing minerals are usually only found in greater abundance in dry climates or deep underground because chloride is soluble in water. Some of the common chloride-containing minerals include halite (sodium chloride), sylvite (potassium chloride), and carnallite (potassium magnesium chloride hexahydrate). The effects of these environmental factors can produce some temporal and spatial variability in chloride concentrations in streams, rivers, and lakes because of changes in hydrology and geologic setting. It has been demonstrated that substantial amounts of chloride can also be contained in vegetation and in roots in watersheds, and that it can be re-released from decaying plant biomass into forest soils and eventually streams. In many soft waters, sodium (Na) and chloride (Cl) typically occur in approximately equivalent concentrations. The equivalency may be disrupted in streams and rivers experiencing differences in anthropogenic sources though (**Figure 1**).

In arid environments, salts and chloride in freshwater ecosystems are primarily concentrated by evaporation of irrigation water used in agricultural practices, which ultimately leach into soils and ground water. In areas with limited rainfall, an accumulation of sodium and chloride may also occur. When salinity reaches threshold concentrations of 1 g l^{-1}, freshwater resources of inland waters become largely useless for agriculture. Water with high solute concentrations can travel to streams and rivers contributing to further salinization of freshwater resources and high chloride concentrations. In 2001, Williams estimated that the annual economic costs of salinization in Australia attributed to loss of agricultural production were $50 million, and another US $90 million loss arising from degradation of infrastructure and loss of freshwater resources. Williams predicted that salinized landscapes will increase approximately sixfold within the first few decades of the twenty-first century in Australia, although predicted changes in drought intensity and frequency could enhance increased chloride and salt concentrations.

It was previously thought that anthropogenic salinization of fresh water was not significant in humid and temperate climatic regions, where a majority of the global population lives and where land use change is increasingly prevalent. It is now known that increasing land use/land cover change can increase chloride concentrations and contribute to salinization of fresh water in humid environments. Jobbagy and Jackson reported increasing salinization and chloride concentrations because of evapotranspiration by plants and alterations in groundwater levels as a result of increasing tree plantations in Argentina. There have also been recent documented increases in chloride concentrations in fresh waters of the Amazon River basin because of rapid population growth, urban expansion, and increases in wastewater and septic discharges.

Road salt represents a considerable source of chloride to many inland waters in colder regions (**Figure 2**), although road salt can lose its melting ability at temperatures below −15 to −20 °C. Approximately 51% of the world output of salt production is now used by cold countries to deice roads in winter both in grit bins and spread by winter service vehicles. Benbow and Merritt estimate that some 10–15 million tons of road salt (predominantly halite or 'rock salt') are used each year in the United States. Road salt melts ice because it forms a eutectic mixture, or mixture at such proportions that the melting point is as low as possible, and that all the constituents crystallize simultaneously at this temperature from liquid solution.

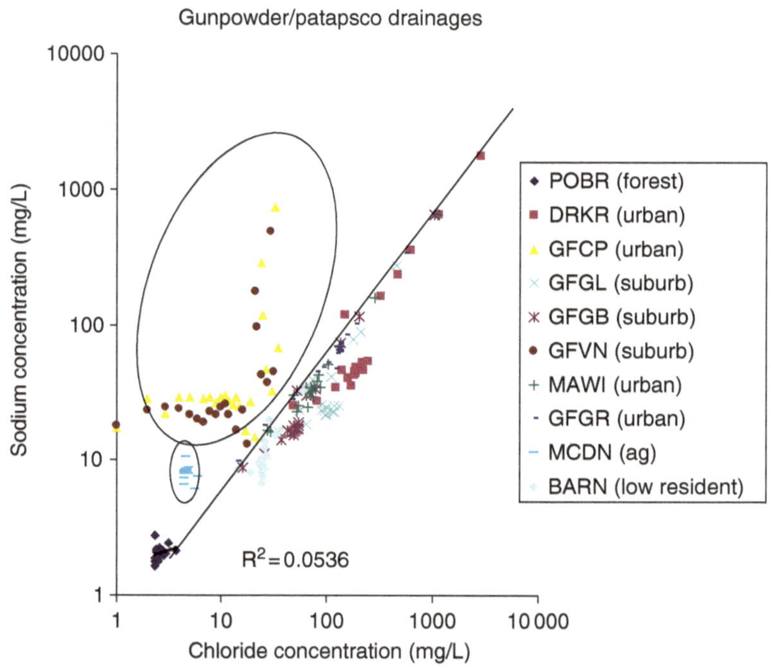

Figure 1 Relationships between sodium and chloride in streams of the Gunpowder and Patapsco River watersheds in Baltimore, MD, USA, draining differing land uses. Most ratios of sodium and chloride in steams are equivalent to halite or NaCl, but data from the circled streams suggests differences in sources of salt to other streams due to other human sources of salt in the watersheds. Data collection supported by Maryland Water Resources Research Center.

Previous work has shown that there have been sharp increases in concentrations of sodium and chloride in aquatic systems of the rural northeastern and Midwestern United States over decades because of road salt use. Concentrations of chloride have been rapidly increasing in tributaries of the Hudson River since the 1950s because of road salt, and concentrations of chloride have increased by over an order of magnitude in a mountain lake in the Hubbard Brook Experimental Forest of New Hampshire after the opening of highways in the 1970s. Bubeck *et al.* showed that chloride concentrations increased at least fivefold in Irondequoit Bay over a two-decade period. The chloride concentration in Lake Erie increased threefold over 50 years, with ~70% of the chloride inputs estimated from industrial sources, runoff from road salt, and municipal wastewaters. Jackson and Jobbagy estimated that the amount of Na and Cl in road salt surpassed the importance of Na and Cl deposition for the continental United States some time in the early 1960s. Long-term concentrations of chloride in rural areas in the northeast are presently lower than in the suburban and urban areas and other more developed areas of the eastern United States; suburban and urban areas with greater road density and impervious surface coverage may be showing advanced stages of salinization of inland waters. Surprisingly, long-term concentrations of chloride are also increasing in areas of the United States where urbanization and deicer use has roughly remained the same possibly because of groundwater storage over decades.

Figure 2 Road salt crystals near a storm drain entering a major drinking water supply in Baltimore, MD, USA. (Photo courtesy of Ken Belt.)

Kaushal *et al.* demonstrated an empirical relationship between increasing impervious surface coverage in the mid-Atlantic United States and increasing mean annual chloride concentration found in stream water (**Figure 3**). Not only small streams have been affected by increased chloride concentrations because of land use change, larger rivers that serve as important habitat for aquatic life and drinking water supplies for humans in metropolitan areas of the eastern United States have also shown marked increases in chloride concentrations since the 1950s. Recent long-term records from drinking water intakes compiled by the U.S. Geological Survey and U.S. Environmental Protection Agency have demonstrated chloride concentrations have increased markedly from the early twentieth century, suggesting that major supplies of

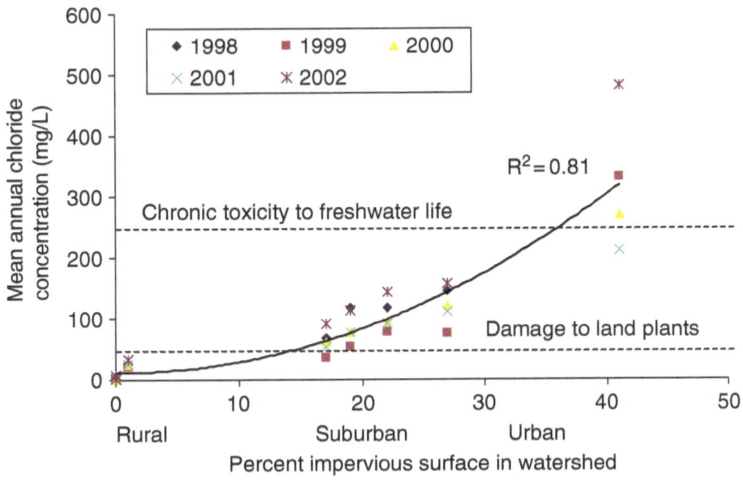

Figure 3 Relationship between impervious surface and mean annual concentration of chloride in streams of the Baltimore Long Term Ecological Research site during a 5-year period (R^2 = 0.81). Sites are located along a gradient of urbanization. Dashed lines indicate thresholds for damage to some land plants and for chronic toxicity to sensitive freshwater life (Kaushal *et al.*, 2005).

drinking water in the eastern United States are contaminated from road salt. Kaushal *et al.* similarly reported that chloride concentrations in major drinking water supplies of Baltimore, MD, USA, have increased fourfold over the last several decades (**Figure 4**). Long-term increases in chloride may be indicative of regional increases in the importance of roadway chemicals on aquatic ecosystems and salinization of drinking water supplies.

Ecological Implications of Increased Chloride Concentrations

Increased salinization and chloride concentrations can induce a variety of ecological effects within both aquatic and terrestrial ecosystems. It can lead to the acidification of streams, mobilize toxic metals from soils through ion exchange, affect mortality and reproduction of aquatic plants and animals, alter community composition of plants in riparian areas and wetlands, facilitate the invasion of saltwater species into previously freshwater ecosystems, and interfere with the natural mixing of lakes. At relatively low concentrations, salt has also been shown to alter the structure of microbial communities. It can also inhibit the process of denitrification, a microbial process critical for removing nitrate and maintaining water quality, in inland waters that are receiving inputs of salt for the first time. Previous work has also shown that increased chloride can inhibit nitrification and respiration of organic matter. Undesirable effects of increasing salinity to particular taxa may directly and indirectly influence broader ecosystem processes in inland waters related to primary productivity, decomposition, nutrient spiraling and cycling, nutrient recycling in lakes because of changes in mixing regime, and the trophic complexity of food webs. The effects of salinization on these ecosystem level properties in inland water are surprisingly less known in both arid and humid environments, but are under current investigation.

In addition to the consistently elevated concentrations of chloride, extreme fluctuations in concentrations of chloride over intraannual periods due to drought or deicer inputs may also provide a harsh environment for many freshwater organisms and disrupt ecosystem functions. Drought may concentrate salt concentrations in ground water and surface waters. The runoff of salts and other deicing compounds containing chloride applied to roads can also undergo large seasonal fluctuations and large variations preceding and following the snow events (**Figure 4**). Fluctuations in chloride concentrations may have compounded effects on biota in addition to elevated concentrations because of difficulties in regulation of osmotic balance. Previous work has documented changes in fish assemblages in prairie streams, where variations in salinity are greatest. Previous work has also shown that monovalent Cl^- is more toxic to aquatic animals than to other forms, and minimum tolerance by organisms decreases as fluctuations in its concentration increase. Salt toxicity is directly related to water temperature, and its effects may be pronounced in suburban and urban streams

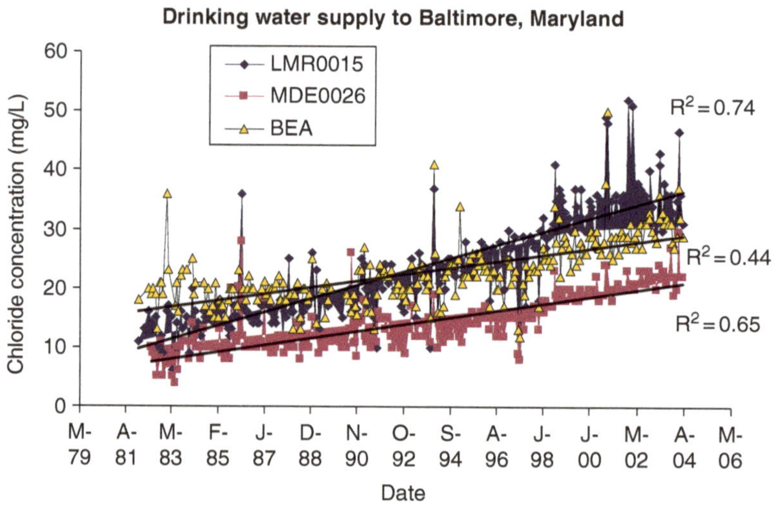

Figure 4 Examples of significant, long-term increases in baseline concentration of chloride in streams in Baltimore, MD, USA. The R^2 values are given for linear regressions. Streams are located in rural areas, but contain roads within their watersheds. LMR0015 (Little Morgan Run), MDE0026 (Middle Run), and BEA (Beaver Run) are sampling stations for tributaries to Liberty Reservoir, a major drinking water supply for Baltimore, MD.

receiving warm runoff from pavements or in arid environments experiencing a greater intensity and frequency of warmer temperatures.

Retention of Chloride within Watersheds and Inland Waters

As previously noted, concentrations of chloride have increased in inland waters of many arid environments because of agricultural practices, and inputs of sodium and chloride have also steadily increased in many watersheds of the United States as a result of increasing land use change and road deicer use. Other work has shown that a relatively large fraction of chloride entering watersheds is surprisingly retained by soils, soil water, and ground water. It has been estimated that over 50% of anthropogenic salt inputs from deicer use to a watershed can be retained in shallow ground water within watersheds, and others have predicted the flushing of chloride to freshwater ecosystems can take up to half a century. In addition, average annual ionic budgets estimated by Likens in a lake in New Hampshire receiving road salt inputs showed considerable higher retention of chloride relative to other ions such as Ca, Mg, Na, K, and H^+ during a 6-year monitoring period. Further, work in the same research area in the Hubbard Brook Experimental Forest of New Hampshire hypothesized that substantial amounts of chloride can be retained in organic plant biomass and re-released following clear-cutting of trees.

The abiotic and biotic retention and storage of chloride may be of additional importance for consideration in hydrologic and ecosystem studies, where chloride is typically used as a conservative tracer to delineate hydrologic flow paths and biogeochemical processing in watersheds and inland waters. A conservative tracer is a solute that is not readily used by the biota or otherwise transformed from state to state by physical and chemical processes and pass unaltered through inland water ecosystems. Given the growing body of work demonstrating chloride retention in watersheds (e.g., increased chloride concentrations during summer months in watersheds receiving road salt applications in winter months, increased chloride in groundwater wells, potential organic transformations of chlorine in watersheds of New Hampshire and Sweden, etc.), it may be necessary to further understand the abiotic and biotic mechanisms contributing to hindering the downstream passage of chloride due to potential uptake or temporary storage.

Although factors and mechanisms influencing watershed storage of chloride are less known compared with other anions of concern in pollutant studies (relative importance of storage in deep ground water, shallow ground water, soils and sediments, or organic transformations and vegetative uptake), accumulation of salt and chloride in close proximity to hydrological flow paths to streams, rivers, and lakes may contribute to increasing long-term salinity in streams, rivers, and lakes. There also maybe storage and flushing of chloride during variability in hydrologic conditions as reported for other mobile anions with implications for interannual variations in both mass transport and concentrations.

Conclusions

Salinization of inland waters represents a major environmental problem. Chloride is the dominant natural form of the element chlorine, and is a constituent of many salts in the environment. Long-term concentrations of chloride have increased substantially in many regions of the world as a result of land use and land cover change. Large amounts of chloride entering watersheds and inland waters can be retained within biota and abiotically in sediments and ground water. In particular, road salt applications and accumulation in humid environments may now be contributing to increased salinization of important drinking water supplies for humans and habitat for aquatic organisms. Further research on factors related to the retention of chloride in inland waters may be useful in elucidating mechanisms explaining long-term increases in baseline concentrations of chloride in surface waters, and it may be useful in ecosystem restoration and reclamation of inland waters from anthropogenic salinization.

Knowledge Gaps

An elucidation of the relationship between increasing land use/land cover change and dynamics of chloride in inland waters are necessary in forecasting the time periods of salinization of many freshwater resources and/or potential trajectories of recovery from salinization. In addition, interactive effects between land use/land cover change and climate variability (particularly increased frequency and severity of drought) will become increasingly important in influencing regional patterns of salinity and chloride in inland waters of both arid and humid environments. In addition, loss of vegetation can lead to the process of desertification in dry lakebeds and may influence the quantity and chemical composition of regional groundwater inputs to inland waters.

More work is needed to investigate factors influencing retention of chloride, using input–ouput budgets and mass balances of chloride in watersheds of differing land use, topography, and watershed size. Information on increasing chloride inputs from natural and anthropogenic sources (e.g., atmospheric deposition, numbers of septic systems, water softeners, fertilizer application rates, and road salt use) can be compared with hydrologic outputs in streams and rivers. Net retention can be determined from the difference between inputs and outputs. Work using mass balance has suggested that, surprisingly, chloride can be retained biologically by plants and soils in the Hubbard Brook Experimental Forest, and other research on chloride dynamics in Swedish watersheds have also suggested the importance of organic transformations of chlorine and chloride.

Because chloride is typically used as a hydrologic tracer of water and human wastewater inputs, determining sources, ages, and hydrologic flow paths of chloride is an active area of research and uses a combination of stoichiometric (elemental ratios) and isotopic tracer approaches. It can also be used to identify different anthropogenic sources of chloride to aquifers and inland waters. Current work uses ratios of Na^+, Cl^-, Br^-, and I^- to infer sources of salt to streams and lakes. It has been proposed that the ratio of sodium to chloride can be used as a tracer of surface vs. subsurface flowpaths to inland waters because sodium and chloride travel differentially (based on charge) through soils; chloride is typically considered more mobile than sodium. In addition, isotopic ratios of chloride, which can differ based on atmospheric vs. deep geologic sources, have been used to study the dynamics of chloride in inland waters. The primary isotopes of interest are the stable isotopes, ^{35}Cl (75.77% abundant) and ^{37}Cl (24.23% abundant), and the trace amounts of radioisotope ^{36}Cl found in the environment. In addition, to natural formation of ^{36}Cl in the atmosphere by spallation of ^{36}Ar by interactions with cosmic ray protons and neutron capture by ^{35}Cl or muon capture by ^{40}Ca, ^{37}Cl was formed in large quantities by irradiation of sea water during atmospheric detonations of nuclear weapons between 1952 and 1958. These isotopic tracers of chloride can provide further information on the relative importance of deep ground water vs. surface water flow paths and also be useful in identifying anthropogenic sources of chloride in streams and lakes.

Finally, there is a growing need for understanding in-stream and riparian chloride dynamics. Chloride is often used as a conservative tracer in studies of nutrient cycling in streams, and relationships between chloride dynamics and stream channel geomorphology may be important in predicting solute storage and residence time in streams and rivers. This can be accomplished through use of chloride injection experiments in streams or it can be investigated using transects of wells placed along hydrologic flowpaths from ground water to streams and lakes. This is particularly important given that stream and river restoration is an important industry and strategies to increase hydrologic exchange between stream channels and rivers may also lead to the accumulation of road salt in banks.

Although further work is needed to understand the dynamics of chloride, there is no doubt that it is increasing in many inland waters due to anthropogenic activities. Future work should elucidate the fluxes, flow paths, and stores of salt in the environment as a result of anthropogenic activities and the biological effects of increased chloride concentrations and their variability in inland waters.

Further Reading

Benbow M and Merritt RW (2005) Road salt – WQ-831. In: Lehr JH and Keeley J (eds.) *Water Encyclopedia*. New York: Wiley.

Biggs TW, Dunne T, and Martinelli LA (2004) Natural controls and human impacts on stream nutrient concentrations in a deforested region of the Brazilian Amazon basin. *Biogeochemistry* 68: 227–257.

Environment, Canada (2001) Assessment, Report – Road, Salts.

Godwin KS, Hafner SD, and Buff MF (2003) Long-term trends in sodium and chloride in the Mohawk River, New York: The effect of fifty years of road-salt application. *Environmental Pollution* 124: 273–281.

Groffman PM, Gold AJ, and Howard G (1995) Hydrologic tracer effects on soil microbial activities. *Soil Science Society of America Journal* 59: 478–481.

Hale R and Groffman PM (2006) Chloride and nitrogen dynamics in forested and urban stream debris dams. *Journal of Environmental Quality* 35: 2425–2432.

Jackson RB and Jobbágy EG (2005) From icy roads to salty streams. *Proceedings of the National Academy of Sciences USA* 102: 14487–14488.

James KR, Cant B, and Ryan T (2003) Responses of freshwater biota to rising salinity levels and implications for saline water management: a review. *Australian Journal of Botany* 51: 703–713.

Jobbágy EG and Jackson RB (2007) Groundwater and soil chemical changes under phreatophytic tree plantations. *Journal of Geophysical Research* 112: GO2013, doi: 10. 1029/2006JG000246.

Kaushal SS, Groffman PM, Likens GE, *et al.* (2005) Increased salinization of fresh water in the northeastern U.S. *Proceedings of the National Academy of Sciences* 102: 13517–13520.

Likens GE, Eaton JS, Johnson NM, and Pierce RS (1985) Flux and balance of water and chemicals. In: Likens GE (ed.) *An Ecosystem Approach to Aquatic Ecology. Mirror Lake and Its Environment*, pp. 135–155. New York: Springer-Verlag.

Ohren A (2001) The bioenergetic basis for the decrease in metabolic diversity at increasing salt concentrations: Implications for the functioning of salt lake ecosystems. *Hydrobiologia* 466: 61–72.
Rosenberry DO, Bukaveckas PA, Buso DC, Likens GE, Shapiro AM, and Winter TC (1999) Movement of road salt to a small New Hampshire lake. *Water Air and Soil Pollution* 109: 179–206.
Wetzel RG (2001) Salinity of Inland Waters. In: *Limnology: Lake and River Ecosystems*, pp. 169–186. San Diego, CA: Academic Press.
Williams WD (2001) Anthropogenic salinisation of inland waters. *Hydrobiologia* 466: 329–337.

Relevant Websites

http://www.ec.gc.ca/nopp/roadsalt/en/index.cfm – Environment Canada Road Salts.
http://www.freshwater.org/ – The Freshwater Society.
http://www.sciencebulletins.com/bio/s/salty_roads.20060116/index.php – American Museum of Natural History.
http://www.csiro.au/science/Salinity.html – Salinity and Acidity, Commonwealth Scientific and Industrial Research Organization (CSIRO).

Iron and Manganese

A E Giblin, Marine Biological Laboratory, Woods Hole, MA, USA

Introduction

Iron (Fe) and manganese (Mn) are present in relatively high concentrations in freshwater ecosystems and in many respects can be viewed as 'master metals' owing to their ability to strongly influence the cycles of other biologically important elements through both microbially mediated and abiotic chemical reactions (**Table 1**). Iron and manganese play major roles in determining the forms and cycling of phosphorous (P), sulfur (S), and trace metals in lakes, rivers, streams, and bogs (**Figure 1**). In some systems they also serve as important alternative oxidants for the degradation of carbon. Both Fe and Mn are also essential micronutrients although they rarely appear to limit primary production in freshwater ecosystems.

Iron and manganese are both transition elements and belong in the subgroup VIIA of the periodic table. While there are important differences between iron and manganese, they share several chemical properties and, in many respects, exhibit similar behavior in natural waters. However, in most aquatic systems iron is present in higher concentrations than manganese, and iron also interacts more strongly with the P and S cycles.

Iron and Manganese Chemistry

Redox Reactions

Iron and Manganese can exist in both oxidized (Fe(III), Mn(III) and Mn(IV)) and reduced (Fe(II), Mn(II)) oxidation states in natural waters (**Table 2**). Therefore the forms and cycling of these elements are strongly determined by the oxidation/reduction (redox) conditions in the environment.

Redox transformations of manganese and iron are mediated by a large number of chemical and microbial reactions with other biologically important elements, including oxygen (O), nitrogen (N), sulfur (S), and carbon (C).

The most reduced forms of these metals, ferrous Fe^{2+}, and manganous Mn^{2+} are quite soluble under anoxic conditions. Iron and manganese can both be reduced by reactions with organic matter, hydrogen sulfide (and other reduced sulfur compounds), and H_2. Although there are a few organic carbon compounds that may react spontaneously with Fe and Mn, the reduction of Fe and Mn by organic matter is usually catalyzed by bacteria that derive energy from the degradation of organic matter using metals the electron acceptor. In contrast, hydrogen sulfide abiotically reduces Fe and Mn. Nitrite and Fe^{2+} can also serve as additional reductants for manganese but are of minor importance in most systems. Light can also photochemically reduce Fe^{3+} to Fe^{2+} and this reaction may play an important role in making iron available to phytoplankton.

By far the most important oxidant for Fe^{2+} and Mn^{2+} in the aquatic environment is oxygen. The chemical oxidation of Fe^{2+} in the presence of oxygen proceeds quite rapidly, within minutes to hours, at circum-neutral pH. It is only at low pH values (<5) that microbially catalyzed Fe^{2+} oxidation becomes important. In contrast, the chemical oxidation of Mn^{2+} by oxygen is orders of magnitude slower than that of Fe^{2+}. This allows for a much great mobility for manganese than for iron in aquatic systems. However, Mn oxidation can be greatly accelerated by microbes under favorable conditions, such as where stable redox gradients exist in the water column or in the sediment. Microbial oxidation rates for Mn^{2+} in natural waters can approach those of iron, and oxidation rates of a day or less are commonly reported.

Nitrate and nitrite can serve as oxidants for both Fe^{2+} and Mn^{2+} and manganese oxides will react with Fe^{2+}. These reactions are far less important than those with oxygen but do occur in some freshwater ecosystems.

Major Chemical Forms in Natural Waters

Reduced, dissolved forms of Mn and Fe are often present in high concentrations in sediments and in anoxic waters. Reduced iron and manganese are also present in a variety of clay minerals. Fe and Mn in clays are released slowly through weathering but only undergo significant biological or chemical transformations over geologic time scales.

Minerals that form in situ by precipitation are referred to as authogenic minerals. Reduced metals may form a variety of authogenic carbonate, sulfide, or phosphate minerals. Manganese sulfides and phosphates are quite soluble and only rarely found in lakes and rivers. In contrast, iron sulfides are far less soluble and have been reported from a wide variety of lakes and rivers worldwide. Ferrous phosphate (vivianite), ferrous carbonates (siderite), and manganese

Table 1 Examples of some environmentally relevant reactions involving iron and manganese

Oxidation
$Fe^{2+} + 1/4O_2 + H^+ \rightarrow Fe^{3+} + 1/2H_2O$
$Fe^{3+} + 3H_2O \rightarrow Fe(OH)_3 + 3H^+$
$Mn^{2+} + 1/2O_2 + H_2O \rightarrow MnO_2 + 2H^+$
Reduction
$FeOOH + 1/2\ H_2S = 4H^+ \rightarrow S^o + 2Fe^{2+} + 4H_2O$
$CH_3COO^- + 8FeOOH + 3H_2O \rightarrow 2HCO^- + 8Fe^{2+} + 15OH^-$
$CH_3COO^- + 4\ MnO_2 + 3H_2O \rightarrow 2HCO^- + 4Mn^{2+} + 7\ OH^-$
$MnO_2 + 2H^+ + H_2 \rightarrow Mn^{2+} + H_2O$
Precipitation
$Fe^{2+} + HS^- \rightarrow FeS + H^+$
$Mn^{2+} + HCO_3^{-} + \rightarrow MnCO_3 + H^+$
Pyrite oxidation
$FeS_2 + 7/2O_2 + H_2O \rightarrow Fe^{2+} + 2SO_4^{-2} + 2H^+$
$FeS_2 + 14Fe^{3+} + 8H_2O \rightarrow 15Fe^{2+} + 2SO_4^{-2} + 16H^+$
Photoreduction
$FeOH{+}2 + hv \rightarrow Fe^{2+} + {}^*OH$ (hydroxyl radical)
$FeOOH + L$ (organic ligand) $\rightarrow Fe(III)L + hv \rightarrow Fe^{2+} + L$

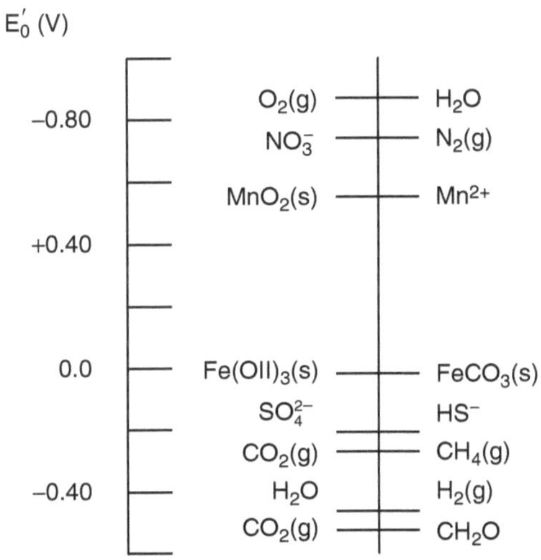

Figure 1 Electrode potentials for important redox reactions involved in the iron and manganese cycling. Values assume a pH is 7.0 and the temperature is assumed to be 25 °C. (redrawn from Canfield *et al.*, 2005).

carbonates (rhodocrosite) have all be found in fresh waters but are often not found as pure minerals.

Oxidized ferric Fe^{3+}, and manganic Mn^{4+}, are not found in high concentrations in dissolved forms in circum-neutral waters containing oxygen because they precipitate to form a variety of relatively insoluble oxides, hydroxides, and oxyhydroxides. Hence, oxidized particulate forms of Fe and Mn dominate in oxic water columns and surface sediments. These vary in chemical composition, grain size, and the degree of crystalinity, which in turn, affects their reactivity. However, very acidic ($pH < 5$) waters may contain very high concentrations of oxidized iron. The intermediate oxidation state of manganese, Mn^{3+}, is relatively unstable

Table 2 Some important forms of iron and manganese found in lakes and streams dissolved primarily Fe^{2+} and Mn^{2+}; Fe^{3+} and Mn^{4+} at low pH

Dissolved	Primarily Fe^{2+} and Mn^{2+}; Fe^{3+} and Mn^{4+} at low pH
Colloidal oxides	Fe^{3+} colloidal
	Ferric (Fe(III)) oxides and oxyhydroxides such as FeOOH, goethite; $Fe(OH)_3$, ferricrite; Fe_2O_3 hematite
	Mixed Fe^{3+}/Fe^{2+}, Fe_3O_4, magnetite
	Manganese (Mn(IV)) oxides MnO_2, pyrolusite, vernadite
	Mn^{4+} oxides containing Na and other cations; todorokite, birnessite,
	Mn^{3+} oxides, MnOOH, manganite
Silicates	Fe and Mn in clays, often not in pure mineral forms and including Fe^{2+} in olivines, garnets, smectites, and hornblend, Fe(II) and Fe(III) in micas, (Mn(II),Fe(II),Ca,Mg)SiO_3 in rhodonitie
Sulfides	Ferrous (Fe(II)) sulfides including FeS (amph, mackinawite), Fe_3S_4 (gregite), and FeS_2 pyrite
	Manganeous (Mn(IV)) sulfides including MnS, alabandite, and MnS_2 haurite
Carbonates	Fe(II)CO_3, siderite
	Mn(IV)CO_3, rhodocrosite
Phosphates	$(Fe(II))_3(PO_4)_2$, vivianite
	Ferric phosphates
	$(Mn(II))_3(PO_4)_2$

in natural waters and is usually only found in low concentrations at oxic/anoxic interfaces or in some minerals.

Chemical Interactions through Adsorption onto Fe and Mn Oxides

Freshly precipitated Fe and Mn oxides are excellent scavengers of a variety of trace metals. Fe oxides also strongly bind phosphate so while Fe and P can react to form a variety of minerals in fresh waters, the abiotic sorption and release of phosphate by iron oxides is usually a much more important factor controlling phosphorous availability than mineral formation and dissolution. The amount of phosphate adsorbed per mole of iron oxide is highly dependent upon the type of oxide, the pH, organic matter content, and the presence of other ions.

Fe and Mn in Rivers and Streams

The World's Major Rivers

Particulates usually make up more than 90–99% of the total Fe found in large rivers while more than 80% of the Mn is found in particulate form. In the world's major rivers, iron concentration of particulates averages about 4.5% of dry weight and concentrations average about 300 μg l^{-1}. Mn in rivers is somewhat enriched compared with the 50:1 Fe/Mn molar ratio in the crust (**Figure 2**) but Mn concentrations are much lower than Fe, averaging about 10 μg l^{-1}.

As a global average, particulate Fe in rivers is almost equally divided between iron that is chemically reducible (primarily amorphous iron oxides and some crystalline iron oxides) and much less easily extracted iron held in silicates and associated with some crystalline oxides such as magnetite. However, there is large variation between rivers, and the reducible oxide component may make up as much as 70% or as little as 20% of the total iron. On a regional basis, the percentage of reducible oxides generally increases with indicators of weathering intensity, such as runoff. The percentage of Mn in the oxide form is higher than that for Fe, and was found to average about 75% in rivers of the United States.

Smaller Streams and Rivers

There is a much wider variety in the forms of Fe and Mn in smaller rivers and streams. Ground water passing through stream side channels and under the stream bed may encounter zones where oxygen is very low or absent. In these systems dissolved Fe and Mn concentrations can reach mM concentrations so dissolved inputs to streams become increasingly important. In streams high in dissolved organic matter, a significant amount of Fe may be complexed by organic ligands. However, over time the majority of

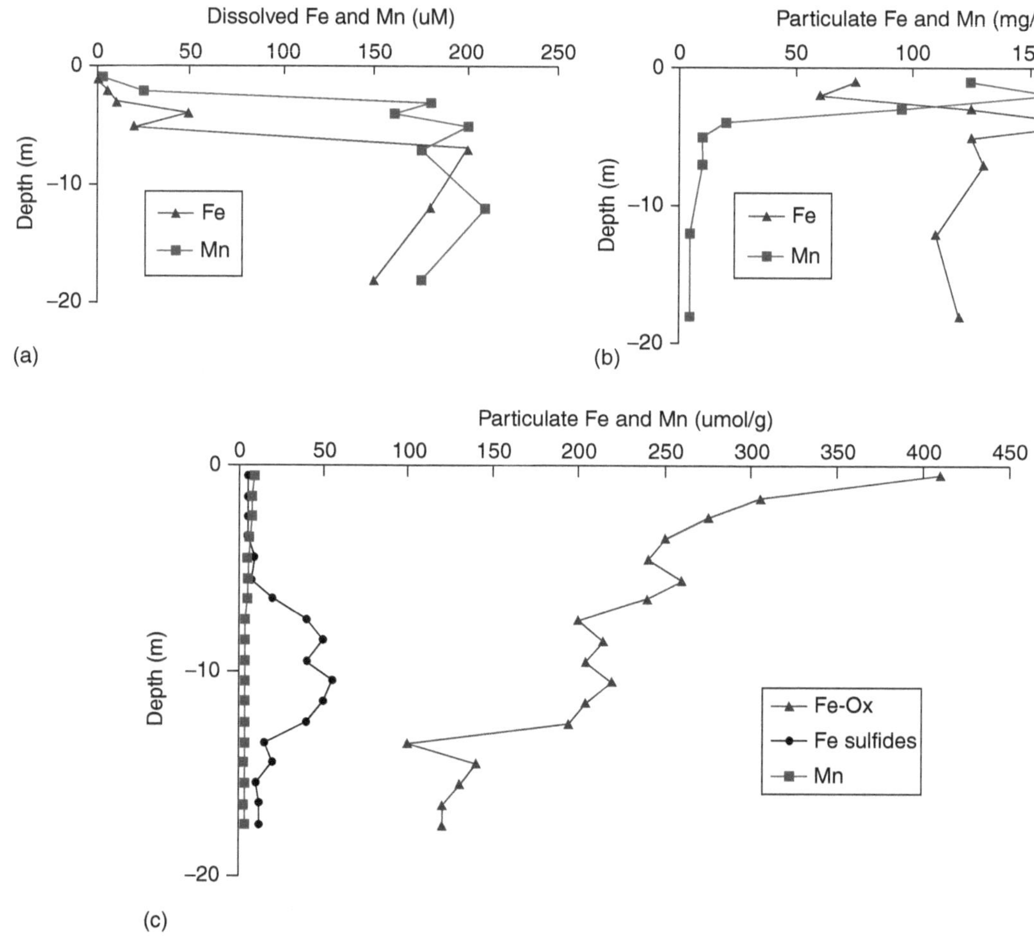

Figure 2 (a) Dissolved Fe and Mn in sediments from Toolik Lake, an oligotrophic lake in northern Alaska; Core was from 20 m depth. (b) Particulate Fe and Mn in the same sediments as 2(a). Note that in both cases Mn concentrations are found higher in the sediment column than Fe. This lake is characterized by well defined metal crusts with Mn oxides found above iron oxide crusts. Reduced sulfur concentrations in the sediments were less than 0.1 mg g^{-1} and not shown (redrawn from Cornwell and Kipphut (1992) *Hydrobiologia* 240: 45–59). (c). Particulate Fe and Mn from a core taken at 10 m in Mirror Lake New Hampshire. This mesotrophic lake has ferrous sulfide concentrations that exceed 50 μmol within the shallow subsurface sediments. Mn concentrations are low but slightly enriched near the surface when compared with 20 cm (data from Giblin *et al.* (1990) *Limnology and Oceanography* 35: 852–869).

dissolved, Fe^{2+} quickly precipitates upon reaching oxic waters except under quite acidic conditions. Hence, the majority of the Fe carried in even small streams will be particulate although the composition differs from that of the larger rivers with a higher proportion found as easily reducible Fe and a lower proportion of the Fe associated with Si minerals.

For manganese the proportion of found in dissolved forms ranges from 0% to 100% and overall dissolved forms are more important in smaller rivers and steams than in the world's major rivers. The importance of dissolved Mn also increases with acidity and the majority of Mn is often found in dissolved forms even in mildly acidic streams. For example, a study in the Adirondack Lakes region of the United States found that the majority of Mn entering lakes from mildly acidic streams was dissolved and less than 10% was in particulate forms.

Fe and Mn in Lakes

Inputs

Atmospheric inputs of Fe and Mn to lakes are negligible and watersheds provide the vast majority of Fe and Mn through streams and rivers. Iron is largely supplied in particulate forms and consist of the same mixture of clay bound and particulate metal oxides previously described for rivers. Metals in clay lattices, and oxides associated with clay particles, sink rather rapidly, while small particles ($<0.5\ \mu m$) may have a much longer residence time in the water column. Manganese associated with clays also sinks rapidly but, as discussed earlier, a significant amount of the manganese entering lakes may be dissolved or present in colloidal forms.

Losses

Fe and Mn undergo extensive cycling within lake sediments and within the water column. In spite of this extensive recycling, many lakes are extremely retentive of iron and often the majority of the Fe coming into a lake is lost to the sediments. Retention is highest in oligotrophic lakes with long residence times. In contrast, retention of Mn by lakes is not nearly as high as for iron. Mn retention is higher in oligotrophic lakes and increases with lake depth and residence time. Shallow lakes and mildly acidic lakes may show little or no net Mn retention. As a result of the differences in retention, the Fe/Mn ratio in lake sediments tends to be considerably higher than for the Fe/Mn ration of the inputs and this ratio has been used to assess historical changes in water depth and sediment redox conditions.

Cycling within Sediments

Lakes

The processes affecting the sedimentary release of Fe and Mn from sediments were first described by Mortimer in the 1940s for seasonally anoxic lakes. Fe and Mn cycling within a lake is highly influenced by lake mixing (**Figure 3**), lake trophic status, and lake water chemistry.

When the waters overlying sediments are well oxygenated, iron and manganese reduction takes place in sediments where anoxic conditions are usually reached within just a few millimeters to centimeters of the sediment water interface. As long as the overlying water remains oxic, the fate of the Fe and Mn will largely depend upon the depth of the oxic layer in the sediment, and the presence of other ions. In soft water lakes with low sulfate concentrations, most of the Fe reduced in the deeper sediments will diffuse upward, and nearly all of it is reoxidized and precipitated near the sediment–water interface. Manganese reduces at a high redox potential than iron, and oxidizes more slowly. Depending upon the depth of the oxic layer and the degree of sediment mixing, Mn oxides will precipitate closer to the sediment water interface than iron oxides, or dissolved Mn may escape into the overlying water. In cases where Mn precipitates within the sediments, Mn and Fe oxides may be very enriched in surface sediments and there may be a distinct layering of the oxides. In rare cases, the Mn and Fe oxides may form distinct structures such as nodules or plates (**Figure 4**). In more acidic lakes, or those with more reducing sediments, substantial amounts of Mn^{2+} may escape from the sediments.

A different pattern will form if the overlying water is oxic, but there is active production of sulfides within the sediments. In these cases some amount of the Fe^{2+} produced will be precipitated in the anoxic portions of the sediments as iron sulfides. Iron sulfides can make up a substantial percentage of the iron in some sediments and reach concentrations of greater than 75 μmol per gram dry weight. This formation of iron sulfides limits the amount of iron that can diffuse to near surface sediments and form oxides. This trapping of iron at depth, and the reduction of the oxide layer can reduce the efficiency of P trapping within sediments and increase internal P loading. However, some of the iron sulfide formed within sediments may be subsequently reoxidized if particles are carried to the oxic layer, or if the sediment oxic layer deepens. MnS is quite soluble so it is rarely found in freshwater sediments.

In lakes which stratify seasonally, the overlying oxygen concentration in the hypolimnion begins to decrease over time. Fe and Mn oxides in the surface

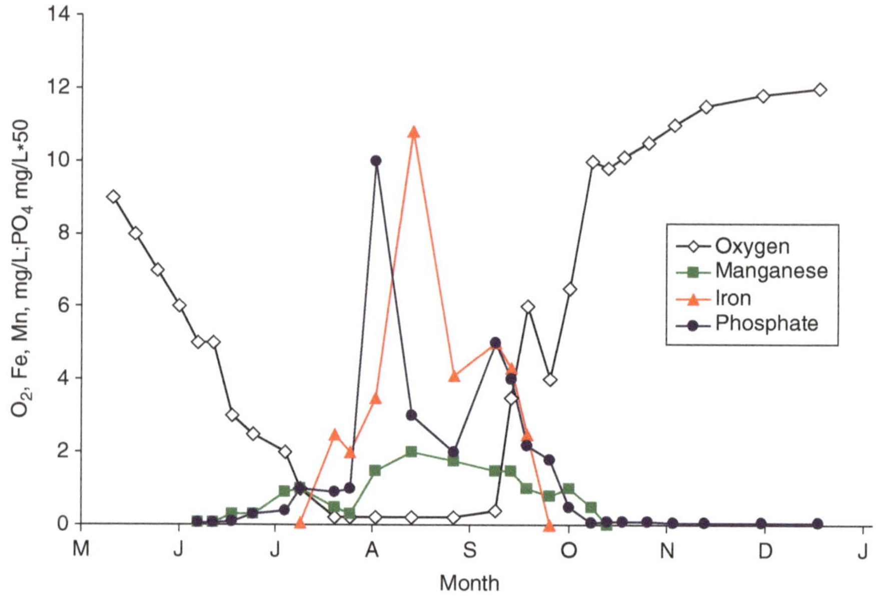

Figure 3 Seasonal change in the water at 30 cm above the sediments in Esthwaite Water, England. Station was in 14 m of water. Sampling began in May and ended in December. The lake partially turned over in mid-September and completely mixed in mid-October. Note that manganese was released from the sediments before iron and that manganese release continued longer into the fall than iron release. The majority of the phosphate release occurs during the anoxic period and closely follows iron. Redrawn from Mortimer CH (1971) *Limnology and Oceanography* 16: 387.

Figure 4 Examples of ferromanganese concretions found in lakes; (top) lattice type nodule (Photo by Celeste A. Asikainen), and (bottom) plate type nodule (Photo by Sean Werle).

sediments are reduced and dissolved, and Fe and Mn are released to the water column. Mn reduction precedes Fe reduction so Mn concentrations usually increase in bottom waters before Fe, and often before oxygen is completely depleted. Following anoxia iron begins to also increase in the bottom water. The reduction of the available metal oxides is seldom complete and may take place over a substantial time period. Iron and manganese oxides can persist in anoxic sediments by either being chemically protected or because of a lack of suitable reducing agents. Iron in particular often shows a lag between the time anoxia is reached in the bottom water and when Fe release begins. When systems are sufficiently reduced and when sulfate is available, sulfate reduction takes place and sulfide concentrations begin to increase in the water column. In these systems dissolved iron in bottom waters may stay constant or decrease as solid phase iron sulfide minerals form and settle onto the sediment surface.

As iron oxides are reduced, adsorbed phosphate is released. This phosphate may not escape to the overlying water if there is sufficient adsorption capacity in the remaining oxides. The ratio of Fe oxides to P at which P is released varies significantly between lakes. Laboratory studies suggest that a minimum inorganic Fe to P atomic ratio of 7 is needed for P retention to occur. This is relatively close to values found in field studies of Danish lakes, where a Fe/P ratio above 8 was found to limit P sediment release. Factors such as pH and conductivity strongly regulate this ratio, and other studies have found that P release may begin at Fe/P ratios as high as 10 or 15. It also needs to be stressed that in hard water lakes with calcareous minerals in the sediments, Fe often plays a minor role in controlling P release owing to the importance of the formation of calcium phosphate carbonates in binding P.

Many of the Fe and Mn phosphate and carbonate minerals that may be formed in lakes exhibit very slow precipitation kinetics. Hence, when water chemistry data is examined, it is often found that sediment porewaters are apparently oversaturated with respect to a number of Fe and Mn phosphate and carbonate minerals, yet the minerals are not found in pure forms in the sediments. Carbonate and phosphate minerals may be absent or Fe and Mn may be found in mixed carbonate precipitates that may contain calcium and other cations. However, there are freshwater sediments where vivianite and siderite are found in concentrations exceeding 1% dry wet. Vivianite, is normally associated with fairly eutrophic lakes and bogs, but it has been also reported to be present in oligotrophic lake sediments where there are very high iron concentrations.

It is important to recognize that the observed profiles of Fe and Mn in sediments may have been highly influenced by dissolution, migration, and precipitation. Therefore, sediment Fe and Mn profiles in many lakes do not faithfully record the history of Fe and Mn deposition to the sediments. Iron and Mn migration may also significantly affect the profiles of other trace metals and phosphate, especially when sedimentation rates are low. For lakes with low sedimentation rates, migration of Fe and Mn must also be taken into account when using ^{210}Pb dating techniques. However, in spite of their mobility, changes in the forms of Fe and Mn within lakes can sometimes be useful in understanding past lake history. For example, the burial of pyrite (FeS_2) and other FeS minerals in lake sediments can be meaningful indicators of changes in both overlying water sulfate concentrations and trophic status.

Streams and Rivers

In the depositional areas of streams where there is little scouring, cycling within stream and river sediments can be similar to that described for lakes. An additional important sedimentary zone of Fe and Mn cycling is the hyporheic zone beneath the stream bed. In cases where oxic overlying water mixes with anoxic ground water there may be substantial precipitation of Fe and Mn within the hyporheic zone. These zones may not be permanent sinks for metals, as they may become depleted in oxygen during times of low flow and later release Fe and Mn to the overlying water. In other cases, the hyporheic zone is metabolically quite active and overlying water becomes reduced and serves as a source of metals, which may precipitate in areas where oxic overlying or ground water mixes within the zone.

Cycling within the Water Column

Lakes

In stratified lakes, sharp redox boundaries move from within the sediment up into the water column as the hypolimnion is depleted in oxygen. Many of the same metal cycling processes that occurred in the sediments also occur in the water column. The major difference is that both dissolved and particulate substances in the water column are subjected to transport through water movement and turbulent mixing; newly precipitated metals in particles are subjected to sinking. As a result Fe and Mn cycling at redox boundaries in the water column may be much more dynamic than in the sediment and opportunities for metal losses are greater.

The water column cycling of iron in lakes where an oxic–anoxic boundary is present has been classically described as a 'ferrous wheel' (**Figure 5**). Iron is reduced in anoxic sediments and in the water columns and is transported by eddy diffusion to the oxic/anoxic boundary. As Fe^{2+} crosses into water containing oxygen it is oxidized and precipitates as amorphous Fe(III) oxyhydroxides. Newly precipitated iron oxides within the water column scavenge phosphorous and in stratified systems may nearly completely limit the transfer of P from the hypolimnion to the epilimnion. Some proportion of these particles are very small or colloidal and lost from the lake through the outflow but the remainder aggregate and form particles that sink back into the anoxic water column and sediments. Here Fe^{3+} is reduced again to form Fe^{2+} and complete the cycle.

Not all lakes with an anoxic hypolimnion show this cycle. In lakes that undergo significant sulfate reduction, sulfide may build up in bottom waters and begin to trap Fe^{2+} in sediments and deeper water. In some anoxic lakes, sulfide concentrations build up so fast that iron never builds up in the bottom waters. In others, the supply of oxidants such as nitrate and manganese, may be so much larger than the supply of reductants that Fe never appears in the water column.

Manganese undergoes a cycle similar to Fe in the water column, but the efficiency of the cycle varies greatly between lakes. In some lakes Mn^{2+} oxidation rates at the oxic/anoxic boundary are a day or less

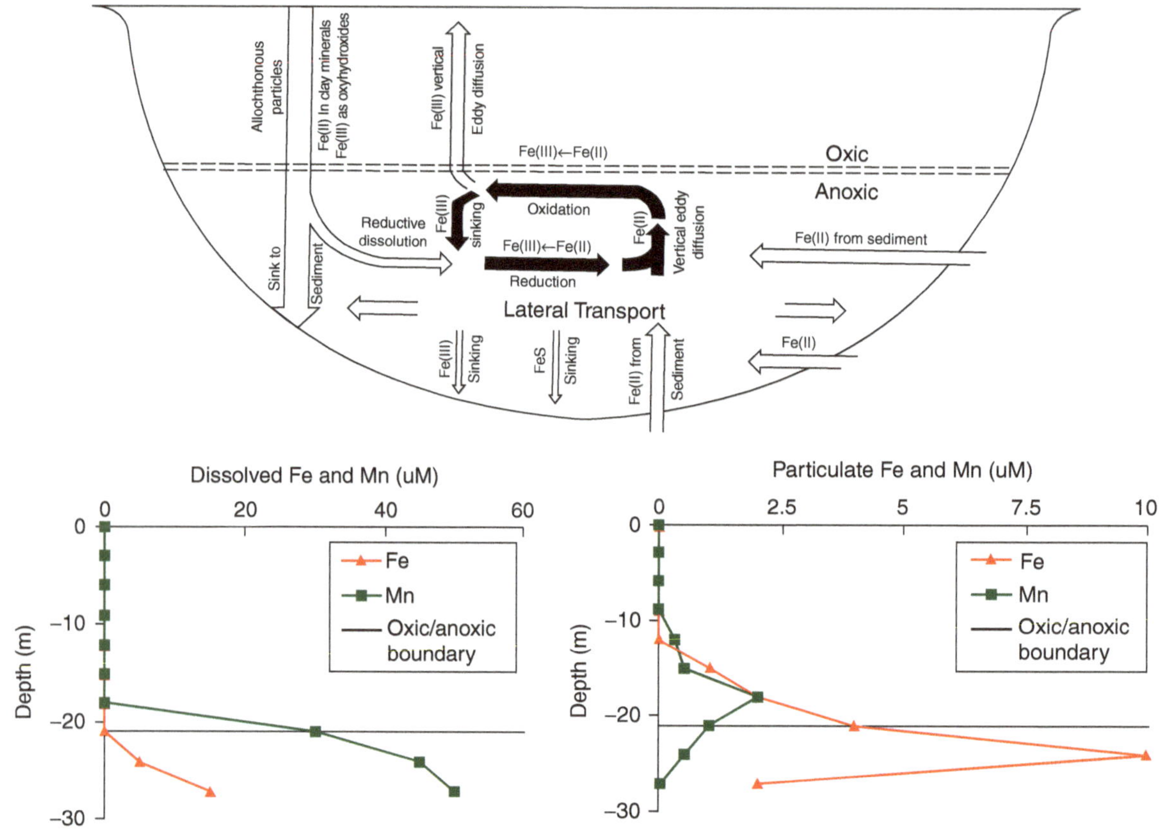

Figure 5 A drawing of a 'ferrous' wheel in a stratified water column (*top*) from Davidson (1993). Similar processes occur for Mn although Mn oxidizes higher in the water column and dissolved Mn can persist in oxic waters. (*bottom*) An example of how Mn and Fe differ in where they are found in the water column can be seen in the dissolved (*right*) and particulate (*left*) Fe and Mn in the water column of Lake Sammsmish. The oxic/anoxic boundary indicates the depth at which oxygen went to zero. Note that dissolved Mn persists higher in the oxidized water column than Fe and that the peak of particulate Fe is below that of particulate Mn. Redrawn from Balistriere *et al.* (1992) *Limnology and Oceanography* 37: 510–528.

and catalyzed by microbes. In these lakes Mn precipitation effectively traps Mn within the lake and Mn undergoes cyclic reduction in the bottom water and oxidation near the redox boundary as seen with iron. However, because the oxidation of Mn in the water column is sensitive to inhibition by low pH, Mn oxidation rates in even mildly acidic waters is quite slow and Mn oxide precipitation may be quite limited. Mn water column concentrations are rarely influenced by precipitation as sulfides but in some lakes Mn carbonate precipitation within the water column has been observed.

Streams and Rivers

Reactions between Fe, light, and organic matter are more important in streams and rivers than in lakes. These reactions can occur with suspended colloidal Fe hydrous oxides in the water column or with Fe coatings on the streambed. Photochemical Fe reduction can occur through the direct reaction with light or may be catalyzed by organic matter and does not appear to be mediated through microbial reactions. Photochemical iron reduction has been well studied in more acidic, iron rich waters where there are significant diurnal changes in Fe^{2+} concentrations in the water. In circum-neutral waters the reactions also occur but the Fe^{2+} produced is quickly reoxidized and does not accumulate in the water column but the reaction is believed to be important for making Fe available to phytoplankton.

Shallow streams can also exhibit significant diurnal fluctuations in Mn. During the day as photosynthesis proceeds stream water pH rises, which speeds the oxidation of Mn and dissolved concentrations decrease. During the night the pH falls and Mn concentrations increase.

Fe and Mn Oxides and Trace Metals

Fe and Mn oxides are excellent scavengers of other trace metals but the degree to which they can control the cycling of other elements in the water column varies widely between systems. The cycling of a variety of heavy metals is tightly coupled to Fe and Mn dynamics in metal-rich acidic streams but in circum-neutral waters they do not always control trace metal cycling. This is because in many natural waters organic compounds or other ligands also strongly complex trace metals and the adsorption of metals onto Fe and Mn oxides is strongly pH dependent.

In areas with very slow Fe and Mn accumulation rates, such as where Mn and Fe nodules or crusts are forming, significant trace metal enrichment occurs. Ferromanganese nodules and crusts tend to be highly enriched in barium, copper, chromium, zinc, and nickel than the sediment.

Special Environments

Waters Receiving Acid Mine Drainage or Acid Rock Drainage

Pyrite (FeS_2) and other iron sulfides are often present in significant quantities in many sedimentary deposits, including those associated with coal and commercially important metal ores. When sedimentary deposits containing large quantities of sulfide minerals are exposed to air and water, they rapidly oxidize to produce iron and other metals, sulfate, and acid. This results in streams and lakes significantly impacted with highly acidic and iron-rich waters (**Figure 6**). The greatest impacts come from stream and lakes near mining activities (acid mine drainage)

Figure 6 (a, b) The Rio Tinto (Red River) in southwest Spain. The area has been mined for iron, manganese, and copper for over 5000 years. The waters are highly acidic and stained deep red from iron (photos by Erik Zettler). (c) Iron deposits along the Animas River near Silverton, Colorado (Mike Gooseff ASLO Image Library (http://www.aslo.org/education/library.html)). (d) Iron deposits in Waterton Creek, a pristine stream (Michael Melford/National Geographic Stock).

but some streams, lakes, and wetlands are naturally acidic owing to local geology (acid rock drainage).

Pyrite oxidation begins with the oxidation of sulfur by oxygen yielding sulfate, ferrous iron, and acidity. The rate-limiting step is the oxidation of Fe^{2+} to Fe^{3+}, which occurs only very slowly at low pH chemically, but is catalyzed by bacteria that speed up the reaction rate by a factor of 100 000. Once underway, the process of pyrite oxidation may continue until the source is exhausted. Acid mine drainage is a serious environmental issue in some areas. Inland waters affected by acid mine drainage may have pH values well below 3 or 4, Fe concentrations are often above 1 mM. Depending upon the nature of the mineral deposit the waters may also contain very high concentrations of Mn, arsenic, cadmium, zinc, copper, and aluminum.

Hydrothermally Influenced Water Bodies

Lakes and streams influenced by hydrothermal sources often have exceedingly varied chemistries ranging from exceedingly acid to alkaline (**Table 3**). Metal concentrations tend to be highest in the most acidic solution and iron and manganese in the very acidic ($pH < 3$) hydrothermal fluids can exceed 10 mM.

Table 3 Composition of a variety of hydrothermal waters

Source		*pH*	*Mn mM*	*Fe mM*
2	Alluvium Creek, trib, USA	1.7	1.070	5.200
2	Alluvium Creek site 1, USA	2.1	2.200	8.510
2	Calcite Spring area, USA	2.1	0.582	46.200
2	Alluvium Creek site 2, USA	2.7	3.000	11.100
1	Taiwan	3.2	NR	24.429
1	California, USA	4.7	25.455	40.893
1	Japan	4.9	NR	9.071
2	Canary Spring, USA	6.5	0.014	<0.002
1	Ngawha, NZ	7.4	0.000	0.002
1	Wairakei, NZ	8.3	0.000	0.001
2	Azure Spring, USA	8.8	<0.001	0.037
1	Iceland	9.6	0.000	0.002

Overall dissolved Fe and Mn concentrations are highest when pH values are the lowest but hydrothermal solutions exhibit a wide range of concentrations and Fe/Mn ratios. NR: not reported.

Source

1. Sabadell JE and Axtmann RC (1975) *Environmental Health Perspective* 12: 1–7.
2. McCleskey RB, Ball JW, Nordstrom DK, Holloway JM, and Taylor HE (2005) *Water Chemistry Data for Selected Hot springs, Geysers and Streams in Yellowstone National Park, Wyoming 2001–2002*. Reston, VA: US Geological Survey. USGS open file report 2004–1316.

Often hydrothermal solutions enter streams and rivers through active geysers, hot springs, and fumaroles. These may undergo periodic releases resulting in inputs of metals and large deposition of Fe and Mn minerals downstream (**Figure 7**). In a few cases, seeps of hydrothermal sources enter lakes and lead to the formation of Fe and Mn deposits similar to the ferromanganese nodules. The growing interest in extracting and using hydrothermal solutions for heating could increase the number of water bodies influenced by these fluids and make them a more important source of iron and manganese to surface waters.

Figure 7 (a) A ferromanganese concretion collected from Yellowstone Lake, USA. Note the laminations on the surface, which are alternating layers of Mn and Fe oxides (from Aguilar C *et al.* (2002) In: *Proceedings of the 6th Biennial Conference on the Greater Yellowstone Ecosystem*, Yellowstone Center for Resources and the George Wright Society, pp. 1–18). (b) Black waters draining from Boiling Lake in the Valley of Desolation, Dominica. The black waters are caused by the presence of iron monosulphide (FeS) and are common in many geothermal areas. (Photo courtesy of The University of the West Indies Seismic Research Center.)

Glossary

Authogenic minerals – Minerals that form by in situ precipitation.

Epilimnion – The less dense water that forms the surface layer in a stratified lakes.

Hyporheic Zone – The area below the stream bed where stream water mixes with ground water.

Hypolimnion – The denser water found in the bottom zone of a stratified lake.

Redox – Stands for *reduction/oxidation* reactions and these are chemical reactions involving the transfer of electrons.

Weathering – The physical, chemical, and biological processes that break rock down into smaller sized particles and alter its chemical constituents.

See also: Biogeochemistry of Trace Metals and Metta-loids; Micronutrient Elements (Co, Mo, Mn, Zn, Cu); Phosphorus.

Further Reading

Canfield DE, Kristensen E, and Thamdrup B (2005) *The iron and manganese cycles*. In: Aquatic Geomicrobiolgy. Oxford, UK: Elsevier-Academic Press.

Cornwell JC and Kipphut GW (1992) Biogeochemistry of manganese- and iron-rich sediments in Toolik Lake, Alaska. *Hydrobiologia* 240: 45–59.

Davidson W (1993) Iron and manganese in lakes. *Earth Science Reviews* 34: 119–163.

DiToro DM (2001) *Sediment Flux Modeling*. New York: Wiley.

Giblin AE, Likens GE, White D, and Howarth RW (1990) Sulfur storage and alkalinity generation in New England lake sediments. *Limnology and Oceanography* 35: 852–869.

Mmenegger L, Schonenberger R, Sigg L, and Sulzberger B (2001) Light induced redox cycling of iron in circumneutral lakes. *Limnology and Oceanography* 46: 49–61.

Hongve D (1997) Cycling of iron, manganese and phosphate in a meromictic lake. *Limnology and Oceanography* 42: 635–647.

McKnight DM, Kimball BA, and Bencala KE (1988) Iron photoreduction and oxidation in an acidic stream. *Science* 240: 637–640.

Mortimer C (1941) The exchange of dissolved substances between mud and water in lakes. *Journal of Ecology* 29: 280–329.

Poulton SW and Raiswell R (2002) The low temperature geochemical cycle of iron: From continental fluxes to marine sediment deposition. *American Journal of Science* 302: 774–805.

Stumm W and Morgan JJ (1996) *Aquatic Chemistry: Chemical Equilibria and Rates in Natural Waters*. 3rd edn. New York: Wiley.

Tessier A, Rapin F, and Carignan R (1985) Trace metals in oxic sediments: Adsorption onto iron oxyhydroxides. *Geochimica Cosmochima Acta* 49: 183–194.

Younger PL, Banward SA, and Hedin RS (2002) *Mine Water: Hydrology, Polution, Remediation*. London: Kluwer, p. 442.

Wetzel RG (2001) *Limnology: Lake and River Ecosystems*. 3rd edn. New York: Academic Press.

White JR and Driscoll CT (1987) Manganese cycling in an acidic Adirondack lake. *Biogeochemistry* 3: 87–103.

Micronutrient Elements (Co, Mo, Mn, Zn, Cu)

C R Goldman, University of California, Davis, CA, USA

Introduction

Interest in trace elements began many years ago when the importance of micronutrients to human health and agriculture was first recognized. Micronutrient deficiencies were discovered to cause various illnesses, defects, and failure to thrive in both humans and livestock. Perhaps the most influential limnologist of the last century, G. Evelyn Hutchinson was probably the first to suggest that the trace elements might also prove important to natural aquatic ecosystems as described in his famous *Treatise on Limnology* (1957, Volume I). Still earlier, algal culturists had determined that supplementation with soil extract was beneficial to laboratory cultures. This inspired others to devise artificial media containing combinations of trace elements essential for healthy algal growth. The field expanded as researchers delved into the functional aspects of inorganic micronutrients in green plants and, eventually, in whole-lake systems.

With the author's development of a carbon-14 bioassay technique for natural phytoplankton populations in 1957, it became possible to rapidly determine the major and minor nutrients most limiting to algal growth in lakes. In the author's studies of three Alaskan Peninsular lakes, the bioassay approach focused on nitrogen and phosphorus limitation, but also utilized a trace element mixture. In the early 1960s, bioassays with trace element additions to the natural phytoplankton of Castle Lake, California, and a series of New Zealand lakes yielded similar results. The extremely rapid and sensitive carbon-14 bioassay technique shows that trace element mixtures were extremely stimulating to carbon uptake by some phytoplankton populations. Once a response from the mixture is noted, the elements can be added individually to determine which is limiting carbon fixation by that phytoplankton.

The following discussion describes some of the eight essential trace elements, their role in phytoplankton growth and health, and the dynamic equilibria that determines the form and availability of each micronutrient in aquatic ecosystems. **Table 1** contains average concentrations of the soluble form of the trace elements manganese, copper, zinc, cobalt, and molybdenum in a variety of natural surface waters. The ranges of concentration for these trace elements can be extensive, varying both seasonally and in response to natural sources as well as pollution. See **Biogeochemistry of Trace Metals and Mettaloids** for more discussion of the toxic effects of trace metals.

Cobalt

The trace element cobalt is a catalyst in nitrogen fixation and is essential to vitamin B-12 synthesis in various forms of algae. Studies have shown that the trace element requirements of algae may be far less than the concentration found in the surrounding medium. In fact, high levels of cobalt have been found to have a toxic effect on certain types of algae, including *Chlamydomonas reinhardtii*. Sources of cobalt vary and include coal combustion and waste products from nuclear power plants, where cobalt is in the form of the isotope ^{60}Co. Cobalt is also a natural component of soils, with concentrations up to 0.26 ng m^{-3} in the Sahara.

Cobalt has chemical properties similar to manganese in terms of its solubility in water with varying reduction/oxidation (E_h) values. In Lake Sammanmish, Washington, and Esthwaite Water in the UK, soluble Co^{2+} levels are highest in the anoxic conditions that occur in the hypolimnion during August. Cobalt concentrations are greatest on the lake bottom when a reduced hypolimnetic condition occurs in summer. During the fall overturn cobalt levels then decrease. Dissolved Co^{2+} can often form cobalt sulfide (CoS), which is insoluble and cause a decrease in soluble Co^{2+} in the hypolimnion from August to September, even though fall turnover has not yet occurred. Cobalt sulfide was found to occur well below the oxic/anoxic interface in Lake Sammanmish. Behaving in similar fashion to manganese, cobalt is often insoluble in oxic conditions and instead, occurs in particulate form. For example, in Waquoit Bay, cobalt was found to coexist with manganese in manganese nodules.

Cobalt can also combine with organic ligands. In some cases 60% of the cobalt forms organometallic complexes. Co(III) has the ability to form more stable organic complexes than does Co(II), which usually occurs in oxic environments. Such organic complexes often compete with the adsorption onto particulate surfaces.

While pH and salinity do not affect the chemistry or solubility of cobalt, they do have an impact on the ability of algae to absorb cobalt. One example of an algal species that utilizes cobalt is *Chlorella salina*.

Table 1 Average concentration of key trace elements (μg l^{-1}) in natural waters (surface). (Modified from Wetzel, 2001)

	Mn	*Cu*	*Zn*	*Co*	*Mo*
World surface lakes	35	10	10	0.9	0.8
Alpine lakes, California	0.3	1.2	1.5	<0.3	0.4
Northern German lakes	28.6	2.9	6.6	0.05	0.39
Northeastern Indiana lakes	21.3	<2	12.9	<2	30.0
Linsley Pond, Connecticut	140	53	–	0.05	0.19
South American lakes	15.6	1.7	8.7	0.11	0.6
Lake Constance, Southern Germany	11.7	1.7	6.2	–	–
Three Lakes, Victoria, Australia	–	2.4	5.8	–	–
Lago Maggiore, Italy	5.5	8.3	39	–	–
Tjeukemeer, The Netherlands	29	1.3	3.3	0.3	–

Positively charged ions often compete with soluble cobalt (Co^{2+}) for binding sites on Chlorella cells. High concentrations of H^+ ions at low pH values in combination with the Na^+ ions in highly saline waters cause a reduction in cobalt sorption. Nevertheless, for some algal species such as Chlamydomonas, pH has little effect on the sorption of cobalt.

Molybdenum

Molybdenum is used by cyanobacteria for nitrogen fixation (see **Nitrogen Fixation**) and it is required by algae to reduce nitrates to a usable form. This was described in some detail by Goldman in 1960 in relation to the molybdenum deficiency in Castle Lake, California. In Castle Lake, a major stimulation of the food chain, including increased trout production, resulted from an addition of only about 3 kg of sodium molybdenum to this small, 20 ha, subalpine lake. Midwestern lakes in the United States are seldom nitrogen-deficient while unpolluted western lakes are more likely to be nitrogen-limited or co-limited by nitrogen and phosphorus. In recent years atmospheric deposition of nitrogen has greatly increased the nitrogen to phosphorus ratio in western lakes like Tahoe. Lakes in the midwestern region of the United States tend to have high molybdenum concentrations (**Table 1**) and are therefore less likely to have their photosynthetic activity stimulated by molybdenum additions. In freshwater environments where molybdenum is available for biological uptake, cyanobacteria are able to fix nitrogen, which in turn tends to make more productive freshwater ecosystems phosphorus-limited. Lakes such as Castle Lake are not necessarily nitrogen-limited, but available nitrogen is mainly in the form of nitrate. In such situations, molybdenum is instrumental in the nitrification process, where nitrate is reduced and thus converted to ammonium, making nitrogen readily available for phytoplankton uptake. For agricultural applications, ammonia is the preferred fertilizer since it does not require reduction. To utilize nitrate, a plant requires the enzyme nitrate reductase and the cofactor molybdenum, which activates the enzyme within its cells. Nitrogen limitation can occur simply because molybdenum-deficient phytoplankton are unable to reduce nitrate to ammonium, which is essential for their growth.

The Spanish karst lakes near El Tejo, Cuenca, also show evidence of nitrogen limitation and have low molybdenum levels similar to those found in Castle Lake. On the other hand, rivers and lakes in the Truckee River region of the Eastern Sierra, which are associated with ancient granitic rocks and Cenozoic volcanic material, tend to have high levels of molybdenum. Other sources of molybdenum include the release and mineralization of sedimenting organic detritus.

Molybdate uptake by cyanobacteria can be inhibited by tungstate and sulfate owing to the similar stereochemistries of these anions. Sulfate levels are quite high in seawater and in some salt lakes, and the inhibitory effect of sulfate on molybdate uptake can greatly slow rates of growth by the cyanobacteria. This slow growth rate makes the cyanobacteria vulnerable to grazing by zooplankton. As a result, nitrogen-fixing cyanobacteria are absent from the plankton of most estuaries and of high-sulfate saline lakes (see **Nitrogen Fixation**). This is true even when these ecosystems are strongly nitrogen limited, and one might expect high rates of nitrogen fixation to occur if it were not for the inhibitory effect of sulfate (see **Nitrogen**).

The solubility and availability of molybdenum is also dependent on the redox potential. Under oxygenated conditions, molybdenum is found in the form of the molybdate anion (MoO_4^{-2}). Molybdate is quite soluble, but as noted earlier, its biological availability can be limited by high concentrations of sulfate found in estuaries and some saline lakes. Under anoxic conditions the molybdate anion is reduced to cationic forms of molybdenum. If these stay in solution, their biological availability can be quite high, which helps explain the high rates of nitrogen fixation often associated with reducing sediments (see **Nitrogen Fixation**). The chemistry of reduced forms of molybdenum in nature is complicated, though. These forms can coprecipitate with iron monosulfides and become unavailable, but reduced forms of molybdenum can also be held in solution with soluble polysulfide complexes.

Manganese

Manganese activates enzymes that contribute to chlorophyll formation, making it essential for photosystem II, the oxygen-evolving process in photosynthesis. Trace additions of manganese have been shown to increase photosynthesis in some algal populations. Manganese is present in lake systems worldwide (**Table 1**) and has sources ranging from industrial pollutants to the deposition of soil and dust particles. Manganese toxicity may result from mining wastes. In the Mediterranean region, Saharan dust episodes can deposit up to 13 ng m^{-2} on land and water surfaces. The Great Lakes and others of the Northeastern United States (**Table 1**) are in close proximity to coal-burning power plants, including one in Buffalo, New York. The coal used in these power plants contains up to 74.5 mg kg^{-1} Mn, one of the highest concentrations reported globally. Some of this manganese is released into the atmosphere in fly ash and is deposited on the land and lakes downwind of the plant.

As shown in **Figure 1**, the chemistry of manganese, like iron, depends on both pH and the redox potential (E_h) in water systems. In lakes, manganese behaves in a fashion similar to iron but is slightly more soluble at reduced oxygen levels. At high oxidation potentials and pH values, manganese forms precipitates including MnO_2. These compounds are insoluble and are responsible for the formation of manganese nodules such as those found in the oceans and in the Great Lakes especially around Lake Michigan. In Lake Washington, Metallogenium and similar varieties of bacteria obtain energy from oxidizing manganese to form MnO_2. Manganese oxide often flocculates as particles and forms crusts on the bacterial cells themselves. In the Waquit Bay off the coast of Massachusetts, these bacteria often form manganese-rich deposits in the sediment leading to the formation of nodules similar to those in Lake Michigan.

In nutrient-rich eutrophic lakes, the hypolimnion and the associated sediments are often anoxic. Metallogenium bacteria, often present at the oxic/anoxic interface, oxidize the Mn^{2+} which reduces again as it settles back into the anoxic hypolimnion. Similar processes occur in Lac Léman (Lake Geneva) where anoxic sediments serve as a ready source of Mn^{2+} for Metallogenium bacteria. In certain lake systems, manganese-reducing bacteria present in the hypolimnion can accelerate the formation of soluble Mn^{2+}. However, at high pH values Mn^{2+} can be insoluble in the form of $MnCO_3$, as evident in **Figure 1**.

The uptake of manganese by algae for the photosynthetic electron transport process is unaffected by pH values above 5. However, absorption of manganese by algae decreases as pH values decline below 5. This is attributed to the fact that, at higher pH values, manganese tends to form complexes with organic ligands which dissociate to form Mn^{2+} at low pH values.

Zinc

Zinc is a cofactor for the enzyme carbonic anhydrase which is essential to carbon uptake during photosynthesis. While important for phytoplankton in

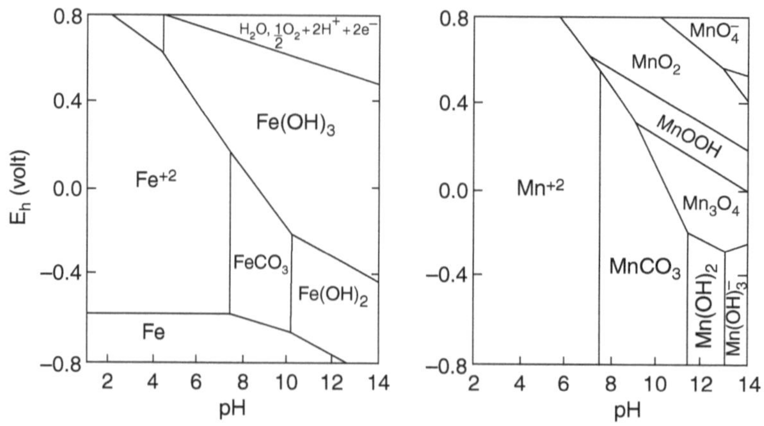

Figure 1 Pourbaix diagrams illustrating the distribution of possible equilibrium phases of Fe and Mn based on pH and redox potential (E_h) in an aqueous system. Modified from Stumm and Morgan, 1996.

terms of photosynthesis, zinc concentrations above $0.05\,mg\,l^{-1}$ are often considered toxic and impair photosynthesis. Nevertheless, many forms of algae including green algae (Microspora, Ulothrix, and Hormidium) and red algae (Lemanea and Batrachospermum) appear to be resistant to zinc toxicity.

Zinc sources are largely anthropogenic resulting from industrial activities and automobile exhaust. Coal combustion also releases zinc, since coal may contain concentrations up to $272\,mg\,kg^{-1}$ Zn. Additional sources of zinc pollution include mining wastes and zinc-smelting processes.

Soluble zinc concentrations in freshwater systems are to some extent influenced by redox potential. At low E_h values iron sulfides are abundant and instrumental in scavenging zinc out of solution. As a result, zinc concentrations are low when the hypolimnion becomes anoxic during the summer. Once the lake becomes oxic again, mineralization of organic detritus becomes important in restoring zinc concentrations in the hypolimnion. Nevertheless, Metallogenium often precipitate zinc in the form of ZnO under oxic conditions which in turn can reduce the amount of free zinc as observed in Lac Léman. At high pH values zinc also occurs in the form of $Zn(OH)_2$. Zinc often combines with other compounds to form chelates that are usually stimulatory to algal growth since the trace metal is maintained in its soluble form.

Copper

Copper is a component of enzymes involved in nitrogen transformation. It also plays a role in respiration and photosynthetic reactions occurring in plankton, aquatic plants, and animals. Copper has long been utilized as an algaecide since, at high levels, it interferes with cell division and photosynthesis. Additionally, copper and zinc can act synergistically to inhibit photosynthesis in plant and phytoplankton species.

Smelting operations are the primary anthropogenic source of copper pollution in the form of contaminant-laden air emissions, wastewater, and slag. Concern has been expressed that the great Siberian Lake Baikal could fall prey to this potent algaecide if copper smelting is conducted upwind of the lake. Bioassays that the author ran at Baikal in 1993 indicated the serious toxic potential of copper pollution on the endemic phytoplankton population. Atmospheric deposition of copper from mining has been suspected as the source of copper to Lake Superior. However, other lakes in the vicinity that are more distant from copper industrial sites have a similar copper content. This suggests other possible sources of copper in the Great Lakes region. The presence of copper can sometimes be attributed to combustion of coal which may contain up to $50\,mg\,kg^{-1}$ Cu.

Like zinc, the form of copper depends on its oxidation state. When copper is present at low redox potentials it is scavenged out of the water just as molybdenum is precipitated by iron sulfides. Unlike zinc, copper is not taken up by Metallogenium and maintains a constant depth profile under oxic conditions. At high pH values under oxic conditions, copper forms $Cu(OH)_2$ compounds and insoluble CuO. At lower pH values copper can form complexes with organic compounds. These complexes contribute to loss of copper from solution by attaching to particulate matter such as organic detritus and humus.

Conclusion

Trace elements have been shown to be important and essential in the metabolism of both plants and animals. Owing to the increasing pollution of the world's surface waters and to widespread transport of aerial contaminants, micronutrient limitations can diminish in certain areas while concentrations may eventually reach toxic levels in others. Sources of copper from smelting operations or coal-fired electrical generation are examples of trace element pollution. Additions of individual trace elements to cultures of the Castle Lake phytoplankton population showed that algal productivity was limited by the trace element molybdenum. This suggests that lakes of undesirable low fertility or those involved in aquaculture may be responsive to trace element additions. Very small additions of a limiting trace element could increase productivity at minimal cost. When eutrophication is more of a problem than low fertility in lakes worldwide, the desirability of trace element additions to natural lakes and reservoirs would more likely be the exception rather than the rule. Further studies and improved analysis of trace element concentrations in natural water will doubtlessly lead to a better understanding of the current and future role of trace elements in the world's lakes.

Glossary

Chelate – In chemistry, a ligand forms a chelate complex in which the metal ion is bound to two or more atoms of the chelating agent.

Eutrophication – The process by which a lake becomes increasingly productive as the result of additional nutrient input from the watershed, airshed, or agricultural or domestic pollution. The result is typically oxygen reduction in the deeper water of the lake and the development of cyanobacterial blooms.

Detritus – Dead organic matter such as dead leaves, algae, or animal remains.

Ligand – In chemistry, a ligand is an atom, ion, or molecule that bonds to a central metal.

Metallogeniume – A type of bacteria or a structure of microbial size found in the depths of some lakes and is often associated with the formation of magnesium nodules.

Nitrate reductase – A metalloenzyme, containing molybdenum, that reduces nitrate to nitrite.

See also: Biogeochemistry of Trace Metals and Mettaloids; Iron and Manganese; Nitrogen; Nitrogen Fixation.

Further Reading

Axler RP, Gersberg RM, and Goldman CR (1980) Stimulation of nitrate uptake and photosynthesis by molybdenum in Castle Lake, California. *Canadian Journal of Fisheries and Aquatic Sciences* 37: 707–712.

Balistrieri LS, Murray JW, and Paul B (1992) The biogeochemical cycling of trace metals in the water column of Lake Sammamish, Washington: Response to seasonally anoxic conditions. *Limnology and Oceanography* 37(3): 529–548.

Camacho A, Wurtsbaugh WA, Miracle MR, Armengol X, and Vicente E (2003) Nitrogen limitation of phytoplankton in a Spanish karst lake with a deep chlorophyll maximum: a nutrient enrichment bioassay approach. *Journal of Plankton Research* 25(4): 397–404.

Goldman CR (1960) Primary productivity and limiting factors in three lakes of the Alaskan Peninsula. *Ecological Monographs* 30(2): 207–230.

Goldman CR (1960) Molybdenum as a factor limiting primary productivity in Castle Lake, California. *Science* 132(3433): 1016–1017.

Goldman CR (1965) Micronutrient limiting factors and their detection in natural phytoplankton populations. In: Goldman CR (ed.) *Primary Productivity in Aquatic Environments, Memorie dell'Istituto Italiano di Idrobiologia,* 18(suppl.): 121–135.

Goldman CR (ed.) (1966) Primary productivity in aquatic environments. In: Proceedings of an International Biological Programme Symposium, Pallanza, Italy, Berkeley: University of California Press, 464 pp.

Horne AJ and Goldman CR (1994) *Limnology,* 2nd edn, 576 pp. New York: McGraw-Hill.

Hutchinson GE (1957) *A Treatise on Limnology, Vol. 1: Geography, Physics and Chemistry,* 1015 pp. New York: Wiley.

Itoh N, Tani Y, Nagatani T, and Soma M (2003) Phototrophic activity and redox condition in Lake Hamana, Japan, indicated by sedimentary photosynthetic pigments and molybdenum over the last ~250 years. *Journal of Paleolimnology* 29(4): 403–422.

Jaquet J-M, Davaud E, Rapin F, and Vernet J-P (1982) The manganese cycle in Lac Léman, Switzerland: the role of Metallogenium. *Hydrobiologia* 91(1): 323–340.

Marino R, Howarth RW, Chan F, Cole JJ, and Likens GE (2003) Sulfate inhibition of molybdenum-dependent nitrogen fixation by planktonic cyanobacteria under seawater conditions: a non-reversible effect. *Hydrobiologia* 500: 277–293.

Marquart A, Flockerzi V, Achterberg EP, van den Berg CMG, Boussemart M, and Davison W (1997) Speciation and cycling of trace metals in Esthwaite Water: A productive English lake with seasonal deep-water anoxia. *Geochimica et Cosmochimica Acta* 61(24): 5233–5253.

Stumm W and Morgan JJ (1996) *Aquatic Chemistry: Chemical Equilibria and Rates in Natural Waters,* 3rd edn, 1040 pp. New York: Wiley.

Wetzel RG (2001) *Limnology: Lake and River Ecosystems,* 3rd edn, 1006 pp. San Diego, CA: Academic Press.

Biogeochemistry of Trace Metals and Mettaloids

C Gilmour and G Riedel, Smithsonian Environmental Research Center, Edgewater, MD, USA

Introduction

Trace elements are simply elements that are present in low concentrations in the environment. Trace element concentrations in inland surface waters can range from a few parts per million (iron concentrations in some highly enriched systems), down to parts per trillion or less (e.g., Cd and Hg). Accurate measurement of these low concentrations has been an ongoing challenge in trace metal biogeochemistry. Many trace elements are essential micronutrients (i.e., Fe, Mn, Cu, Ni, Mo, Se, and Zn), acting as catalysts in enzyme systems. Others, like Hg and Pb, serve no known physiological function in organisms. However, all of these elements exert toxic effects on organisms as concentrations increase. Trace elements may impact ecosystem productivity and community structure either as limiting nutrients or as toxicants.

This article focuses on the biogeochemistry of trace metals and metalloids that are toxic at relatively low concentrations and are common contaminants of inland aquatic ecosystems. Trace elements are ubiquitously, if unevenly, distributed in the Earth's crust. The more abundant, like Fe, may occur at percentage concentrations, while a large number, including Zn and Cu generally occur at part per million (ppm) concentrations, and a few, like Hg and Ag, occur at part per billion (ppb) or even lower levels.

The chemical properties of each trace element are unique, although patterns of similarities exist across and down the periodic table, causing each to partition differently in the environment, and to interact differently with organisms. Furthermore, organisms substantially influence the redox status and complexation of trace metals, and therefore their fate and transport in the environment. The level of complexity is such that we cannot comprehensively cover the subject of trace element biogeochemistry in inland waters here, but we hope to highlight the importance and the wide variety of these interactions.

Essential and Nonessential Trace Elements

Trace elements can be roughly divided into two groups based on their biological requirements. Essential elements are necessary for the metabolism of organisms.

Examples include Fe, Cu, Mn, and Zn, which virtually all organisms use in the active sites of a significant number of important enzymes and other active proteins. Some elements, for example Se, Co, and Cr have a few specific uses in most if not all organisms; others like Cr, Ni, Mo, and W, seem to have specific essential roles for a limited class of organisms. For example, Co is essentially universally used in vitamin B12 and similar compounds while Ni functions in nitrogen metabolism in some microorganisms and plants. Nonessential elements include common contaminants such as Ag, Pb, and Hg. These elements are often toxic at lower concentrations than the essential elements. The physiological role of a number of trace elements (e.g., Bo, Sb, and V) remains open to conjecture.

Low concentrations of essential trace elements may limit algal growth (primary production) in some inland waters; however, trace metal limitation is generally less severe in fresh waters than in the more dilute oceans. The need for organisms to take up essential trace elements and to detoxify others leads to significant changes in the speciation and complexation of trace metals in natural waters.

Sources of Trace Elements

Geological Sources

Erosion and dissolution of rock and soils release trace elements into aquatic environments. Natural ores release metals when exposed to water and oxygen. In most cases, the natural concentrations of trace metals in natural waters are low and well tolerated, or even limiting for biological growth, but in some instances, geologic background levels can be toxic. In the American west, for example, rocks and soils of the Colorado Plateau leach high concentrations of Se, which threaten native fish.

Trace metal concentrations are generally elevated in geothermal areas. Microorganisms in these areas develop resistance to toxic trace metals and some are uniquely adapted to utilize trace metals in chemolithotrophic energy-generating processes. The passage of ground water through aquifers and soils prior to surfacing can also bring trace elements into surface waters. For example, in many regions of the world, aquifers used as drinking water contain arsenic at levels that can adversely affect human health.

Human activity can exacerbate the mobility and toxicity of trace metals with natural geologic sources. For example, acid deposition increases aluminum leaching from soils, leading to fish toxicity in streams. Evapo-concentration of Se in agricultural drainage waters in California has caused significant toxicity to breeding birds. In Bangladesh, wide scale fertilization of soil has led to anaerobic conditions, resulting in As solubilization into ground water used for drinking.

Anthropogenic Sources

Human activities have mobilized trace elements from geologic reservoirs into the surface environment for thousands of years. Some of the first polluted sites known are prehistorical Cu mining and refining sites, which date back at least 7000 years. Trace metal pollution sources may be localized or diffuse, and may release metals to aquatic ecosystems, to land, or to the atmosphere. Metals emitted to the atmosphere are redeposited to watersheds. Efforts to control, assess, and monitor metal contaminants have been on the rise in the developed world for at least two decades. Sediment and ice cores, which provide an historical record of metal contamination, suggest that the highest levels of metal contamination in Europe and North America occurred in the mid-to-late twentieth century, and that metal releases have generally declined since that time. Progress has been slower for the developing world, where metal pollution continues to increase in many countries.

Point source pollution Releases from isolated sources are known as 'point source pollution.' Point source releases from mining, smelting, refining, and other industry have contaminated harbors and industrial areas worldwide, resulting in some of the most acute incidents of trace metal toxicity in natural waters. Contamination from mining includes contaminated spoil heaps, abandoned mines, and contaminated floodplain sediments. Most urban harbors contain substantially elevated levels of metals in sediments, making relocation of dredge spoils problematic.

Nonpoint source pollution Metal pollution from diffuse sources to natural waters is more difficult to control than metals from point-sources. The widespread use of metals in construction materials, batteries, vehicles, personal care products, clothing, and many other materials leads to moderate contamination of watersheds in populated and industrialized areas. Wastewater treatment effluent, combined sewer overflows, and urban runoff all contain elevated concentrations of metals that are very difficult to control. Control of metals in an upstream manufacturing process can reduce metals loads effluent.

Common diffuse sources of metal contamination include lead-based paints; lead pipes; CCA (Cu, Cr and As)-treated lumber; batteries (Zn, Pb, and formerly Hg); Cd in phosphate fertilizers; Hg from switches, thermometers, and dentistry; Cu and tributyl tin from antifouling paints; and Cu in biocides. Rubber tires and brake pads contain substantial quantities of zinc, which wears off over time as particulates. Galvanized (Zn) or Cd-coated steel slowly weathers releasing Zn or Cd. Copper roofs and guttering also oxidize and slowly erode. The use of Ag in dentistry and black and white photography has lead to widespread Ag contamination of municipal waste water with Ag (not to mention cleaning Grandma's silver).

Sources to the atmosphere Metals may be released to the atmosphere as gases, aerosols, or particulates, and subsequently deposited to natural waters and watersheds. Ice cores provide a record of trace metals emissions and deposition through time. They suggest at least one order of magnitude increases in emissions of Cd, Cu, Zn, Bi, Cu, and Hg in industrial regions relative to preindustrial times. Sources of metals to the atmosphere include ore smelting, industrial metal processing, combustion source such as coal-burning power plants, vehicles, refuse incineration, and cement manufacture; and road and soil dust. Mercury and Pb are two metals for which significant atmospheric emissions have led to global contamination of soils and sediments. Tetraethyl lead added to gasoline was a widespread airborne pollutant, until it was phased out in many countries in the 1970s and 1980s. A large fraction of Hg is emitted to the atmosphere during combustion. Many nations are now implementing control technologies to reduce Hg emissions from combustion processes.

Biogeochemistry of Trace Metals in Fresh Waters

The basic chemical properties of the elements control their interactions with the biological, geological, and chemical environment and determine the structure of their biogeochemical cycles. Trace elements are present in the environment in a plethora of forms, including different oxidation states, and a wide variety of organic and inorganic complexes in the dissolved, colloidal, and solid phases. These forms are partitioned among phases in natural waters, including solid, aqueous, and colloidal phases.

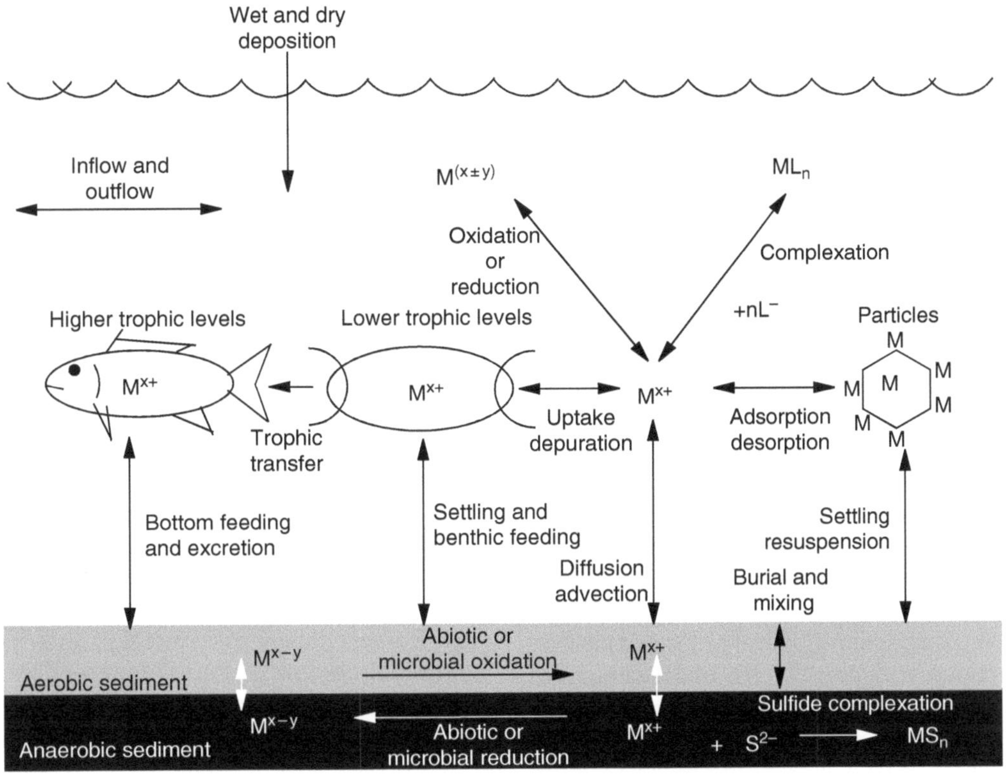

Figure 1 Some of the major processes affecting the behavior of trace element cycling between water, sediments, and organisms.

Figure 1 shows the major types of reactions in the biogeochemical cycle of a hypothetical metal. All metals do not go through all of the reactions in this theoretical cycle. Microorganisms may participate in all of these reactions. The major reactions include:

- aqueous complexation – which influences bioavailability, toxicity, transport, and partitioning behavior
- adsorption/desorption – which determines the loss of metals to sediments, and affects bioavailability and toxicity
- reduction and oxidation – which affects solubility, partitioning, and toxicity, and for Hg, remission to the atmosphere
- biotic uptake
- methylation and demethylation – for a subset of metals and metalloids, either via addition of a methyl group directly to a metal cation or to a metalloid oxyanion.

The following sections discuss each of these aspects of trace metal biogeochemistry.

Aqueous Complexation

Metals bind reversibly to a wide variety of compounds present in natural waters, called ligands. The equilibrium reaction between a generic free metal cation (M) and a simple anionic ligand (L) can be written as

$$M^{(x+)} + nL^{(-)} \leftrightarrow ML_n^{(+x-n)}$$

where the number of ligand molecules in the complex, and x is the charges of metal and cation.

The strength of the interaction can be defined via the equilibrium association constant

$$K_a = [ML_n]/[M][L]^n$$

where K_a is the equilibrium association constant, [M] is the free molar concentration of the trace element, [L] is the free molar concentration of the ligand, and $[ML_n]$ is the molar concentration of the trace metal ligand complex. It is important to note that metals and ligands are often not at equilibrium in natural waters.

Metals and metalloids form complexes with inorganic and organic ligands. Equilibrium constants for most common metals cations with common inorganic anions (e.g., Cl^-, SO_4^{2-}, Br^{2-}, CO_3^{2-}) are well-defined, as are the equilibrium coefficients for anionic metalloids with the abundant dissolved inorganic cations (e.g., K^+, Na^+, Ca^{2+} and Mg^{2+}). If the concentrations of metals and ligands, including pH, are known, the equilibrium concentrations of their complexes can be calculated by solving all of the equilibrium equations simultaneously. A number of commonly-used equilibrium speciation software programs (e.g., MINEQL,

MINETAQ, and PHREQC) provide numerical solutions, and help to automate the calculation.

However, the dominant metal complexes in natural waters are often complexes with natural organic matter (NOM). Complexation of metals and metalloids with organic ligands significantly complicates the calculation of metal speciation. Metal complexation with organic ligands ranges from ionic interactions with small dissolved ligands to complex interactions with colloidal organic matter. The interaction between metals and organic matter is a particularly active research area. Filter passing NOM, often called dissolved organic matter (DOM), is composed of relatively high-molecular weight organic compounds, such as humic and fulvic acids, the exact composition of which is difficult to determine. DOM interacts with a variety of dissolved metals and metal complexes, and with major cations (like Ca^{2+}) that compete for binding sites with metals, to form complex colloids.

Metals interact in a number of ways with DOM, including ligand interactions, hydrophobic and electrostatic interactions with metal ions, and interactions with neutrally-charged metal complexes. Weaker ligands, such as those dominant in humic and fulvic acid, are largely carboxylate sites. NOM also contains strong metal ligands at lower concentrations. These strong ligands are poorly defined, but likely include thiol, amine, phenol, pyrrole, or pyridine moieties, all of which are used by cells internally to complex metals. Complexing agents may have single or multiple active sites on a single molecule. For example, ethylenediaminetetraacetic acid (EDTA), a common synthetic complexing agent, contains two amine complexing sites and four carboxylate sites in a single molecule. Complexing agents with more than one active site are often called 'chelators' from the Greek word 'chela' for claw. In natural waters, with relatively low metal concentrations, nearly all of the filterable metal pool may be bound up in these ligands, making metals largely unreactive and unavailable to cells. Exceptions are molecules specifically excreted by organisms to capture and aid in the transport of essential trace elements into cells (e.g., siderophores, compounds excreted by fungi and bacteria to promote dissolution and uptake of iron). Weaker binding sites on DOM are filled after the strong sites are saturated. Electrostatic interactions are more difficult to model; however, numerical models (e.g., the Windermere humic aqueous model WHAM) have been specifically designed to model these interactions.

Phase Partitioning

Complexation or other interactions with the solid phase is often called partitioning. All trace metals are particle reactive to some extent, meaning they readily bind to organic colloids and suspended particles in natural waters, and are removed via particle settling. This means that most metals are highly concentrated in sediments and wetland soils relative to surface waters. Particle reactivity is highly variable among metals, and depends on nature and strength of the interaction. This tendency of trace elements to adsorb to particles is commonly measured as a 'distribution coefficient' or k_d

$$k_d = C_s/C_w$$

where C_s is the concentration of the element in the solid phase and C_w is the concentration of the element dissolved in the water, in the same units. Partition coefficients range from on the order of 10^5 to 10^6 for Hg and Pb, to as little as 10^2 or even less for chromate, selenate, and selenite. Partition coefficients also depend strongly on the water quality, particularly pH, and the concentrations of other competing ions. Theoretically, the partition coefficient for a metal in an unsaturated system should be concentration independent. In practice, this is not always the case.

Oxidation, Reduction, and Speciation

Trace metals may undergo reduction and oxidation via photochemical, microbial, and geochemical processes. In the oxidizing environment of most surface waters, a relatively oxidized valence state is often the most thermodynamically stable; however, biological processes and slow oxidation kinetics sometimes result in significant concentrations of reduced forms. In sediments and anoxic waters, microbial respiration can produce highly reducing conditions through oxygen depletion and by the production of reduced end-products like sulfide, hydrogen, and methane. In reducing, sulfidic environments, many divalent metals (i.e., Fe, Zn, Cd, and Cu) can form relatively insoluble complexes with sulfides and precipitate, while other metals are themselves reduced. For some metals and metalloids, reduction increases solubility while the opposite is true for others. Dredging, eutrophication, and other anthropogenic influences can change the redox status of sediments, soils, and surface waters, and thus alter the mobility of trace metals.

Iron and chromium are examples of metals that change oxidation states between oxic surface waters and anoxic sediments. Under oxic conditions, iron is generally found as ferric iron, Fe(III), which forms insoluble hydroxyl complexes (essentially rust). However, dissolved and colloidal organic matter may hold Fe(III) in the aqueous phase. Under reducing conditions in anoxic water or sediments, Fe(III) is reduced to ferrous iron Fe(II), which is much more soluble and mobile.

Chromium, a common pollutant metal, is another example of a metal that readily changes valence. In natural waters, Cr is present in two oxidation states, Cr(VI), generally as soluble CrO_4^{2-} (chromate) or $Cr_2O_7^{2-}$ (dichromate) depending on pH, or Cr(III), which forms insoluble hydroxides. In oxygenated water, Cr(VI) should be favored, while in reducing environments, Cr(III) should be favored. However, the kinetics of the reactions between these two forms is very slow, and oxygenated surface waters can contain both. The two valences of Cr also have strikingly different biological effects. Cr(III) has relatively low toxicity, probably because Cr^{3+} is relatively poor at penetrating biological tissue. Cr(VI) species are highly toxic, because the anionic forms are readily taken up by cells, possibly by sulfate transporters.

Microbial Metal Reduction and Oxidation

Most microbes (and indeed, all nonphotosynthetic organisms) obtain their energy by facilitating reactions that are thermodynamically favored, but kinetically slow, and capturing a part of that energy. Indeed, respiration is merely the facilitation of the natural (but very slow) oxidation of organic matter by atmospheric oxygen. Some bacteria can use trace metals and metalloids as electron acceptors in place of oxygen in respiration, coupling metal reduction to organic matter oxidation for energy generation. Iron and manganese reduction are the best studied of these processes, but uranium, chromium, technetium, selenium, arsenic, gold, and others can also be used. Microbial metal reduction may serve as a remediation route for metals that are more mobile in their oxidized forms, like Cr and U. Metals may be reduced via electron-shuttling compounds excreted outside of cells, or by uptake and reduction within cells. Most of the bacteria that can use metals as electron acceptors in metabolism are found in among the iron-reducing bacteria and the sulfate- or sulfur-reducing bacteria. Nonmetabolic metal reduction can also occur, among a wider range of microorganisms. These reduction processes may have evolved as metal or metalloid resistance mechanisms.

Chemolithotrophic ('rock eating') bacteria can use the zero-valent (metallic) or reduced forms of trace metals as electron donors in energy generation, often coupled to the reduction of inorganic anions like nitrate, selenate, or arsenate. These types of organisms are common in acid-mine drainage and in metal corrosion environments. Acid-mine drainage occurs when mine wastes with high concentrations of sulfide are exposed to oxygenated water. Sulfide is oxidized to sulfate with microbial help, making sulfuric acid, and metals are solubilized in the resulting acid solution. Iron is probably the most important metal in this behavior, because it is the most common, but other elements can be used as well, including Mn, As, Se. Similar, if less marked, bacterial action occurs when reduced sediments or soils are exposed to oxygen.

Biological Uptake, Accumulation, and Transformation of Metals

Uptake of trace elements by algae and bacteria is generally the first step in metal accumulation into higher trophic levels. Microorganisms use active cation transport systems to take up most essential trace elements. These usually operate by binding the metal at the cell surface with strong ligands on the transport protein and then spending energy in some form (ATP, NADP, proton gradient, etc.) to transport the element into the cell. Although dedicated transport systems have evolved for some trace elements, others may be transported via major cation transport systems.

Metal uptake by algae and bacteria also plays a critical role in the cycling and transport of trace elements in the inland waters. Fixed metals may be transferred through food webs, released back into the aqueous phase in a variety of forms as cells decay, or sequestered in sediments following deposition. A number of microbial uptake pathways have evolved for acquisition of essential metals. Pathways have also evolved for protection against toxic metals, including a variety of redox, speciation, and phase transformations. These include enzyme systems that produce less reactive or more volatile forms of toxic elements (i.e., Se, Hg, As), systems that sequester metals (metallothioneines, phytochelatins, and mineral granule precipitates are examples), and excretion systems. Only a small fraction of the dissolved pool of any trace metal is generally available for uptake. Early models for metal toxicity and uptake assumed the entire dissolved metal pool was available.

Over time, models for bioavailability have evolved to reflect the idea that only certain metal complexes are available for uptake. The free metal ion activity model is one paradigm for metal uptake by bacteria and algae. It postulates that metal uptake is a function of free (uncomplexed) metal activity in bulk solution. To apply the free ion activity model, equilibrium geochemical models are used to calculate the concentration of free metals cations in the milieu of ligands and competing cations in natural waters.

The Biotic Ligand Model (BLM) is a commonly-used mathematical framework for determining the bioavailability and toxicity of metals in natural waters. It is an extension of the free ion activity model, based on the idea that metal transport proteins on the surfaces of cells act as competing ligands for free metal cations, so that metal uptake can be modeled using geochemical equilibrium models. The BLM was derived from models for metal uptake by fish gills, and is often used to generate site-specific ambient water quality criteria for metals.

However, some metal uptake mechanisms are not captured by the free ion activity and biotic ligand models. These include uptake mechanisms that use an extracellular complexing agent to capture metals for transport (e.g., siderophores or biofilms). Cells may also take up metals via diffusion of neutrally-charged complexes (Hg, Ni, and Ag are examples) or through the uptake of metals bound to amino acids (e.g., Se). Metalloids are dominantly found as oxyanions in surface waters, and are taken up in that form. They can be taken up via transport systems for analogs like phosphate (arsenate) or sulfate (selenate), or through metalloid-specific transport systems. Therefore, metalloid bioavailability is determined by the concentration of metalloid oxyanions, competing oxyanions, and the nutrient status of organisms.

Metal Resistance Mechanisms

A wide variety of metal-resistant mechanisms have evolved in microorganisms. In bacteria, the most common metal resistance mechanism is removal from cells through energy-dependant efflux mechanisms. Metals present as divalent cations (Ag, Cd, Co, Cu, Ni, and Zn) are often detoxified in this way. Less commonly, enzymatic reduction, oxidation, chelation (often with metallothioneins), or sequestration are used for metal detoxification. Metalloid oxyanions (i.e., arsenate, selenite) can be detoxified by methylation to produce less toxic and/or more volatile forms. Long-term exposure of microbial populations to high-metal environments selects for resistant strains. Metal resistance genes may be specific to one metal or a group of similar metals, and are often carried on mobile genetic elements with antibiotic resistance genes. Human use of metals may enhance the number and type of metal- and antibiotic-resistant microorganisms.

Example: Arsenic and Selenium

Arsenic and selenium speciation, transport, and toxicity in natural waters are strongly affected by microoganisms. Arsenic is present in the surface waters in at least two valences, As(V) (primarily as arsenate) and As(III) (as arsenite), and a variety of organic arsenic compounds, nominally As(−III). Oxic waters favor As(V), and sufficiently reducing conditions favor As(III) (**Figure 2**). However, in oxic fresh and marine waters, high concentrations of arsenite and monomethyl and dimethyl arsine are often observed in conjunction with phytoplankton blooms. Phytoplanktons take up arsenate (AsO_4^{3-}) as a phosphate (PO_4^{3-}) analog, especially under conditions where phosphate limits phytoplankton growth. Many algae and bacteria have developed chemical pathways to reduce arsenate to arsenite and/or volatile methylated arsenic compounds, and excrete these compounds. In bacteria, the ars operon (one of the most widely distributed operon systems) confers arsenic resistance. It commonly includes a regulator gene and an efflux pump. Other arsenic oxidation and reduction systems also exist in bacteria, both for detoxification and arsenate respiration.

Selenium is taken up by plants, including algae, as a required nutrient. Selenium is a partial sulfur mimic, and forms seleno-amino acids in cells. Seleno-cysteine is an essential part of the active sites of a few important enzymes. Cells release a poorly defined group of organic selenium compounds, which can constitute a significant fraction of the total Se in surface waters. Selenium can be accumulated through food webs in amino acid complexes. Algae and bacteria can produce volatile methylated Se compounds, dimethylselenide and dimethyldiselenide (similar to As compounds), which can be lost from surface waters. Some anaerobic bacteria can also reduce selenate and selenite to elemental selenium, coupling Se reduction to energy generation, in process analogous to microbial sulfate reduction. This can be a major Se detoxification mechanism in sediments and wetland soils. Bacterial and algal Se reduction can also occur through processes that do not generate energy, and may confer resistance.

Example: Mercury

Mercury is an example of a metal whose toxicity and bioaccumulation in natural waters is largely controlled by microbial transformation to methylmercury. The biogeochemistry of Hg is elaborate and contributes greatly to its role as a contaminant (**Figure 3**). It is the only metal found as a liquid at room temperature, has a high vapor pressure and significant cycling in the atmosphere. It is readily reduced and oxidized, so that microbial, chemical, and photochemical redox processes play important roles in its cycling.

A small percentage of the inorganic Hg in the environment is converted to the highly toxic and highly bioaccumulated monomethyl mercury (MeHg). Worldwide, MeHg contamination of fisheries has resulted in more consumption advisories than any other contaminant. For humans, concern has centered

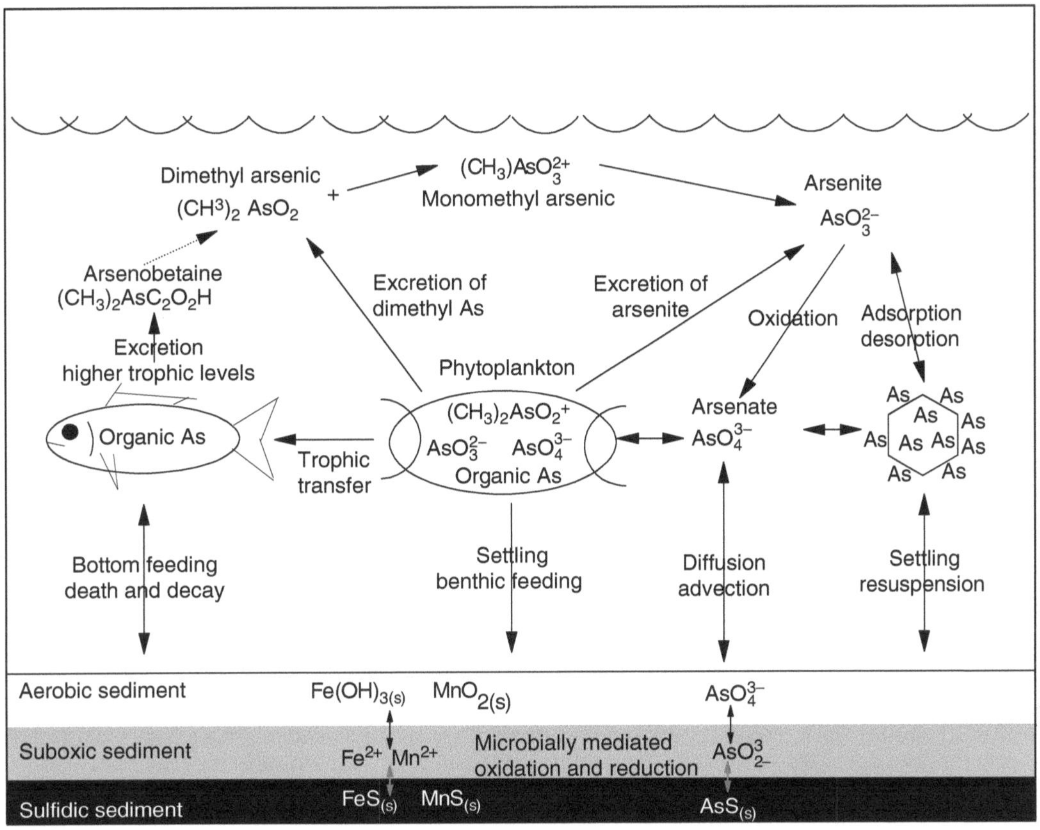

Figure 2 A diagram showing many of the specific forms and major processes known to affect the cycling and speciation of arsenic in aquatic ecosystems.

on developmental effects, particularly reductions in IQ, which have constituted the basis for risk assessments and public health policies. However, there are health hazards associated with exposure to MeHg throughout the lifespan, including potential cardiovascular and endocrine impacts. In wildlife, MeHg has significant impacts on the reproductive success of a number of species at concentrations commonly found in atmospherically-contaminated ecosystems.

MeHg Production

The sensitivity of an aquatic ecosystem to mercury is determined by the ability of that ecosystem to transform inorganic Hg load into MeHg in biota. For Hg to impact an aquatic ecosystem, mercury must be transported from sites of deposition to sites of methylation, it must undergo methylation, and the MeHg produced must be accumulated in food webs. Transport of Hg through watersheds is complex, and retention of Hg is watersheds may be very long (decades to centuries). Key controlling factors include watershed size, with decreasing Hg yields as watershed size increases, and land development and disturbance, which enhance Hg transport downstream.

Methylmercury production is a microbial process, occurring mainly under the anoxic conditions found in aquatic sediments, wetlands, and other hydric soils. It appears to be a inadvertent side reaction of normal metabolism. Methylation is not a detoxification mechanism nor does it confer Hg resistance. Although a strong link to microbial sulfur cycling is clear, the pathways and distribution of Hg-methylating organisms have not been well characterized. Dissimilatory sulfate-reducing bacteria (DSRB) are important mediators of MeHg production in aquatic ecosystems. Sulfate stimulation of methylation has been demonstrated in studies that range from pure bacterial culture to field amendments to lakes and wetlands. Control of sulfur sources to freshwater ecosystems – including control of sulfur emissions and control of sulfur in agricultural runoff – can be an effective strategy to reduce MeHg risk. A few other bacterial groups that produce MeHg have been identified, particularly iron-reducing bacteria, and the role of these organisms in methylation in natural systems is being investigated.

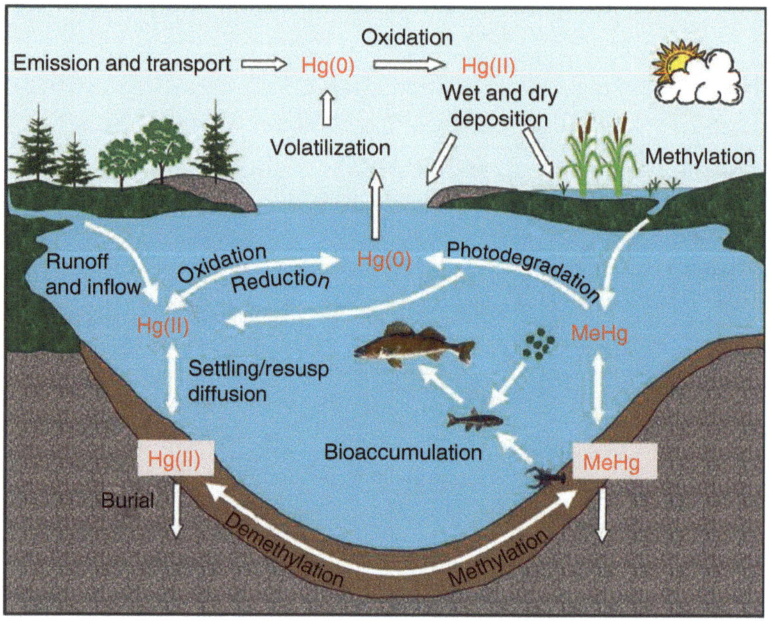

Figure 3 Mercury cycling in a lake and its watershed. Mercury emissions are transported long distances, primarily as gaseous elemental mercury [Hg(0)], oxidized in the atmosphere to reactive gaseous mercury [Hg(II)], and deposited in precipitation and by surface contact (dry deposition). Anaerobic bacteria convert a small portion of the incoming Hg(II) to methylmercury (MeHg), which is then bioconcentrated in the aquatic food chain. Various biotic and abiotic reactions interconvert the different forms of Hg, affecting uptake, burial, and evasion back to the atmosphere. From Engstrom D (2007) Fish respond when mercury rises. PNAS 104: 16394–16395. www.pnas.org/cgi/doi/10.1073/pnas.0708273104.

MeHg Demethylation and Hg Reduction

Demethylation of MeHg is largely a microbial process in sediments and soils, although photo-demethylation is important in lighted surface waters. Mechanisms to detoxify Hg and MeHg via reduction and demethylation are common among bacterial communities, and the genetics and distributions of these systems are fairly well understood. In bacteria, demethylation and reduction often proceed via the highly specific and evolved *mer* operon system, although other, less well-characterized pathways exist.

Temporal and Spatial Controls on Net Hg Methylation

Wetlands can be particularly active zones of MeHg production, although highly dependant on flow paths that govern the location and extent of zones of microbial activity. Reservoir formation, soil drying, and rewetting cycles (including seasonal cycles in wetlands) can increase net MeHg production because of increased decomposition rates, sulfur oxidation during drying, and development of anoxia after rewetting. In freshwater aquatic ecosystems, warm, shallow sediments and the anoxic bottom waters of stratified lakes and estuaries are important zones of net methylation. Estuarine and coastal sediments are also significant net producers of MeHg.

MeHg Accumulation in Food Webs

Methylmercury concentrations increase by roughly a factor of 10 at each step through a food chain, and MeHg fish tissue concentrations increase substantially with age. Of the 'persistent bioaccumulative pollutants,' only PCBs exhibit comparable bioconcentration factors. Although field and experimental evidence clearly indicate that Hg loading rates affect fish Hg concentrations, MeHg concentrations in fish can vary by an order of magnitude or more in regions with similar atmospheric Hg deposition rates as a result of the factors that control production and accumulation of MeHg. Comparisons of MeHg bioaccumulation across water bodies must include consideration of species, food web structure, and fish size.

Solving the Mercury Problem

Fortunately, Hg availability for methylation and bioaccumulation decreases as it ages in place in sediments and soils. This means that recently deposited mercury contributes disproportionately to current MeHg levels in fish, and that Hg emissions control should help reduce MeHg levels in fish in the near-term, despite 150 years' of accumulated mercury in our watersheds.

Recently, the United States and other developed countries have begun to implement direct emissions controls on coal-fired utilities. Most new Hg emissions worldwide come from combustion processes: coal burning, trash burning, and cement production. Controls on major point sources of Hg have been in place for some time in developed nations. Efforts to reduce Hg emissions in North America and Europe may be overshadowed by the growing energy demands of the developing nations. Controls on diffuse sources of Hg – such as batteries, switches, and thermometers – have also been implemented. Fluorescent light bulbs contain small amounts of Hg that are difficult to contain because of lack of recycling programs. Their increased use must be balanced against Hg emission reductions that result from more efficient lighting. Metallic mercury use in gold mining in developing nations remains a direct human health risk.

The complexity of the atmospheric and biogeochemical cycles of this element, and the widespread impact of Hg on people and ecosystems, necessitates the development of carefully designed Hg monitoring programs to assess the effectiveness of our regulatory actions.

Further Reading

Andren AW and Bober TW (eds.) (2002) *Silver in the Environment: Transport, Fate, And Effects*, pp. 192. Pensacola, FL, USA: Society of Environmental Toxicology and Chemistry (SETAC).

Campbell PCG (1995) Interactions between trace metals and aquatic organisms: A critique of the free-ion activity model. In: Tessier A and Turner DŽ (eds.) *Metal Speciation and Bioavailability in Aquatic Systems. IUPAC Series on Analytical and Physical Chemistry of Environmental Systems*, pp. 45–102. Chichester: Wiley.

Fasfous II, Yapici T, Murimboh J, *et al.* (2004) Kinetics of trace metal competition in the freshwater environment: Some fundamental characteristics. *Environmental Science and Technology* 38: 4979–4986.

Lindberg S, Bullock R, Ebinghaus R, *et al.* (2007) A synthesis of progress and uncertainties in attributing the sources of mercury in deposition. *Ambio* 36: 19–33.

Lloyd JR (2003) Microbial reduction of metals and radionuclides. *FEMS Microbiology Reviews* 27: 411–425.

Mason RP, Abbott ML, Bodaly RA Jr, *et al.* (2005) Monitoring the response to changing mercury. *Environmental Science and Technology* 39: 14A–22A.

Meyer JS, Clearwater SJ, Doser TA, Rogaczewski MJ, and Hansen JA (eds.) (2007) *Effects of Water Chemistry on Bioavailability and Toxicity of Waterborne Cadmium, Copper, Nickel, Lead, and Zinc to Freshwater Organisms*, pp. 352. Pensacola, FL, USA: Society of Environmental Toxicology and Chemistry (SETAC).

Monastersky R (1996) Ancient metal mines sullied global skies. *Science News* 149: 230.

Munthe J, Bodaly R, Branfireun B, *et al.* (2007) Recovery of mercury-contaminated fisheries. *Ambio* 36: 33–44.

Nelson WO and Campbell PGC (1991) The effects of acidification on the geochemistry of Al, Cd, Pb and Hg in fresh-water environments – A literature-review. *Environmental Pollution* 71: 91–130.

Oremland RS, Stolz JF, and Hollibaugh JT (2004) The microbial arsenic cycle in Mono Lake, California. *Fems Microbiology Ecology* 48: 15–27.

Riedel GF, Williams SA, Riedel GS, Gilmour CC, and Sanders JG (2000) Temporal and spatial patterns of trace elements in the Patuxent River: A whole watershed approach. *Estuaries* 23: 521–535.

Santschi PH, Lenhart JJ, and Honeyman BD (1997) Heterogeneous processes affecting trace contamination distribution in estuaries: the role of natural organic matter. *Marine Chemistry* 58: 99–125.

Scheuhammer AM, Meyer MW, Sandheinrich MB, and Murray MW (2007) Effects of environmental methylmercury on the health of wild birds, mammals, and fish. *Ambio* 36: 12–19.

Stolz JE, Basu P, Santini JM, and Oremland RS (2006) Arsenic and selenium in microbial metabolism. *Annual Review of Microbiology* 60: 107–130.

Stumm W and Morgan JJ (1996) *Aquatic Chemistry*, 3rd. edn. New York: John Wiley and Sons.

Relevant Websites

http://www.webelements.com/ – WebElements.

http://www.atsdr.cdc.gov/toxfaq.html – ASTDR ToxFAQs.

http://www.dartmouth.edu/~toxmetal/TX.shtml – Dartmouth Toxic Metals Program.

http://nadp.sws.uiuc.edu/mdn/ – Mercury Deposition Network.

http://www.chem.unep.ch/chemicals/default.htm – UNEP Chemicals Programme.

http://www.epa.gov/mercury/ – EPA Mercury homepage.

http://www.epa.gov/waterscience/criteria/ – EPA Water Quality Criteria.

Nitrogen Fixation

R W Marino and R Howarth, Cornell University, Ithaca, NY, USA

Introduction to Biological Nitrogen Fixation in Aquatic Systems

Nitrogen fixation is the bacterial process whereby molecular N_2 gas is converted to reactive, biologically available forms of nitrogen. The vast majority of nitrogen on Earth is present as molecular N_2, and is not available to living organisms other than those that can fix nitrogen for growth. Before the advent of the agricultural and industrial revolutions, biological nitrogen fixation was the only significant process creating reactive nitrogen to support fixation of carbon into enery-rich organic matter (primary production) in Earth's ecosystems. Since nitrogen is an essential element for life as well as an element frequently limiting primary productivity in ecosystems, at the global scale the process of nitrogen fixation is essential for life.

There is a wide variety of prokaryotic organisms capable of nitrogen fixation, including bacteria that utilize organic carbon (heterotrophs), and photosynthetic bacteria that fix inorganic carbon into biomass (autotrophs). The latter group includes the cyanobacteria, so named because of the blue colored photosynthetic pigment they contain, other photoautotrophs (such as the purple sulfur bacteria that use hydrogen sulfide rather than water as the reducing agent in photosynthesis), and chemo-autotrophic bacteria that use inorganic chemical energy rather than light energy. The latter two types of nitrogen-fixing bacteria only live in somewhat hostile environments that are highly specialized for their needs, whereas cyanobacteria can thrive in a wide range of aquatic habitats and have mechanisms which help them grow and persist during periods of adverse environmental conditions. Nitrogen-fixer species can either live freely in the environment, or symbiotically in association with other organisms. The cyanobacteria have a unique ecological role in aquatic ecosystems, as they are the only group of organisms on Earth that can fix both inorganic carbon and nitrogen in an oxic (oxygen-containing) environment.

In inland waters heterotrophic bacteria and cyanobacteria (primarily from the order Nostocales) are responsible for most of the nitrogen fixation that occurs. The rates of nitrogen fixation and the resulting nitrogen input to an ecosystem from this process vary greatly among lakes and other inland water bodies. In general, when biological nitrogen fixation is an important contributor to the nitrogen economy of an aquatic ecosystem, high rates of nitrogen fixation by cyanobacteria living in the plankton of the water column are the primary reason. Cyanobacteria and other autotrophic bacteria living in the benthos (sediments) or in assemblages attached to surface substrate of shallow water bodies, marshes and streams can also fix nitrogen at high rates and provide a locally important source of nitrogen to the communities they support. Benthic nitrogen fixation by photoautotrophs in very low nutrient lakes, where light can peneterate to much of the bottom area, can constitute a major source of nitrogen to these systems. Most of this chapter is therefore devoted to a discussion of cyanobacterial nitrogen fixation in inland water environments. Nonetheless, nitrogen fixation by heterotrophic bacteria is fairly ubiquitous in sediments and can be significant in the nitrogen cycle of some aquatic ecosystems. A consideration of the controls on this process help us to understand the energy-related costs of nitrogen fixation.

Before considering the rates of biological nitrogen fixation in inland waters and general factors which influence this process, it is useful to understand a bit about how nitrogen fixation is estimated in natural systems. The most commonly used method remains the acetylene reduction assay, introduced in the late 1960s. This relatively simple and highly sensitive technique indirectly assays nitrogen fixation by measuring the conversion of acetylene gas, a structural analogue for N_2, to ethylene gas. Both the N_2 and acetylene reactions use the nitrogenase enzyme present in all organisms capable of biological nitrogen fixation. It is necessary to assume a fixed relationship between acetylene and N_2 reduction by the enzyme. Variation in this ratio due to physiological and environmental factors is the primary source of uncertainty in estimating field nitrogen fixation rates using the acetylene reduction method, although sampling and incubation techniques can significantly affect measured rates. Nitrogen fixation can be also assayed directly by techniques which measure the incorporation of a stable isotopic form of N_2 into biomass using a high-precision mass spectrometer.

Rates of biological nitrogen fixation vary tremendously across the wide range of aquatic ecosystems where measurements have been made. In addition to the variation and uncertainty inherent in the methodologies used, a number of environmental factors acting at levels from the cellular up to the ecosystem and the region act to determine the rates and relative importance of nitrogen fixation in different inland water environments. Nonetheless, patterns and generalities have emerged.

Nitrogen Fixation by Heterotrophic Bacteria in Sediments

Heterotrophic bacteria fix nitrogen in virtually all sediments. The rates reported for the sediments in lakes and estuaries tend to be low to moderate, from 0.002 to 0.3 g N m^{-2} $year^{-1}$. Such rates are low compared with those mediated by cyanobacteria in many lakes or rates by heterotrophic bacteria found in many terrestrial ecosystems. Nitrogen fixation rates are generally higher when the organic matter content of the sediment is greater. In marine and freshwater ecosystems where vascular plants are present (including marshes, swamps, bogs, and submerged macrophyte and seagrass beds) nitrogen fixation rates in the sediments are more variable, and can be much higher than unvegetated sediments in similar environments. For Temperate Zone systems measured rates generally range from 0.5 to 2 g N m^{-2} $year^{-1}$, values which are more typical of many terrestrial systems. Nitrogen fixation rates in the sediments of seagrass beds in subtropical and tropical coastal water bodies are often higher than in temperate systems, and range from approximately 1–20 g N m^{-2} $year^{-1}$. The higher rates of fixation in wetlands and seagrass beds as compared with unvegetated lake or coastal marine sediments are probably a result of the greater availability of labile organic matter associated with the plants as a substrate to fuel the process. Although rates of nitrogen fixation are fairly similar in saltmarshes and freshwater wetlands, the resulting nitrogen is a much more significant input to freshwater systems because the latter are much more closed systems. For example, nitrogen fixation was estimated to contribute 1–2% of the total nitrogen input to a salt marsh in New England (USA), as compared with 60% of the nitrogen input to a freshwater bog in the same region.

In sediments, the microbial process that counters nitrogen fixation as a nitrogen input is denitrification, where fixed inorganic nitrogen is converted back to a gas and lost from the system. Rates of denitrification also tend to increase with greater organic matter in sediments, and so the processes of nitrogen fixation and denitrification tend to increase in parallel. Denitrification rates are generally found to be similar to or greater than heterotrophic nitrogen fixation rates in aquatic ecosystems, where both processes are measured. As a result, sediments tend to either show no net nitrogen flux or serve as a net nitrogen sink to inland aquatic ecosystems, despite heterotrophic nitrogen fixation. In some low nutrient lakes and coastal marine systems, there can be additional nitrogen fixation associated with sediments owing to the activity of autotrophic nitrogen-fixing bacteria on the surface sediments (see 'Nitrogen Fixation by Benthic and Epiphytic Cyanobacteria' later in this chapter).

The enzyme that catalyzes nitrogen fixation (nitrogenase) is readily poisoned by oxygen. Oxygen seldom penetrates into sediments below the first millimeter or two, and so does not inhibit nitrogen fixation by heterotrophic bacteria in most sediments. On the other hand, when oxygen is present in an environment, it may severely constrain nitrogen fixation by these bacteria. For example, essentially no nitrogen fixation by heterotrophic bacteria occurs in the oxygenated water columns of lakes, probably because of the deleterious effect of oxygen on nitrogenase activity. Some nitrogen fixation is associated with heterotrophic bacteria on leaf litter and other organic matter in streams, but rates are very low, again perhaps due to the high level of oxygen in flowing water.

Nitrogen fixation is often considered energy-intensive relative to many other biological processes, as it requires breaking a N–N triple bond using a specific enzyme complex. As such, the process of synthesizing nitrogenase and reducing N_2 to create a source of fixed nitrogen for growth is often viewed as energetically 'expensive' to the organism when compared with the uptake of fixed inorganic nitrogen present in the environment or the synthesis of other cellular compounds. However, this appears not to be the case for N_2 fixation by heterotrophic bacteria in anoxic (no oxygen present) sediments. Moderate to high rates are observed to occur in the presence of elevated concentrations of ammonium (NH_4^+) typically present in porewaters as a result of decomposition and remineralization. Since bacteria can readily assimilate the ammonium and thereby have a plentiful supply of nitrogen, the fact that they also fix N_2 indicates that it 'costs' them little to do so. However, ammonium at high levels has been shown to inhibit the rate of nitrogen fixation in some sediments, perhaps keeping these rates lower than otherwise might occur. As discussed later, nitrogen fixation by cyanobacteria and particularly those in the plankton, may be a much more energy expensive process to the organisms, in part because of the need to build and maintain structures that protect the nitrogenase enzyme from deactivation by oxygen.

Nitrogen Fixation by Planktonic Cyanobacteria in Freshwater Lakes

Nitrogen fixation by planktonic cyanobacteria varies greatly among lakes, and rates are related to the overall nutrient (trophic) status. Rates of N_2 fixation range from 0 to 0.003 g N m^{-2} $year^{-1}$ in oligotrophic (very low nutrient) lakes, from 0.01 to 0.1 g N m^{-2} $year^{-1}$ in mesotrophic lakes, and from 0.2 to 10 g N m^{-2} $year^{-1}$ in eutrophic (highly enriched) lakes. Planktonic nitrogen fixation is generally not an important term in the nitrogen budgets of oligotrophic and mesotrophic lakes, and contributes less than 1% of the nitrogen inputs to most of these ecosystems. On the other hand, nitrogen fixation can be very important in the nitrogen economy of eutrophic lakes, with studies showing from 5% to 80% of the total nitrogen load to the ecosystem from this source.

The most important species of planktonic cyanobacteria that fix nitrogen in lakes belong to the genera *Anabaena*, *Aphanizomenon*, and *Nostoc*, which can rapidly reproduce and form large blooms (high biomass) under optimal conditions. These species form long chains or filaments of cells, which can differentiate to allow photosynthesis to occur in most cells (vegetative) and nitrogen fixation to occur at the same time, in specialized, nonphotosynthetic cells called heterocysts (**Figure 1**). Heterocysts form in response to low intracellular concentrations of nitrogen, as a result of low availability of nitrogen in the environment. The cells have a thick wall that protects nitrogenase from oxygen and so allows nitrogen fixation to occur in fully oxic environments. The energy necessary for nitrogen fixation comes from carbon fixed in the photosynthetic cells, which is transferred to the heterocysts. To support the energy required for nitrogen fixation, many photosynthetic cells are needed per one heterocyst cell.

The heterocysts develop in regularly spaced intervals of vegetative cells. Both cyanobacterial growth and nitrogen fixation rates are generally higher when there are more than 10–20 vegetative cells per heterocyst cell. As a result, filamentous heterocystic cyanobacteria can be very sensitive to grazing: zooplankton can quickly shorten chains (and thereby reduce the number of photosynthetic cells available to support nitrogen fixation in a heterocyst) both by breaking chains into pieces and by eating along the end of a chain. Such grazing by zooplankton can constrain the development of a cyanobacterial bloom in nitrogen-depleted waters, by limiting nitrogen fixation and so nitrogen availability for further growth. Significant reduction in cyanobacterial abundance by some types of zooplankton and filter-feeding bivalves has been observed in some inland waters, although many nitrogen-fixing cyanobacteria species often contain toxic chemical compounds as a protection against grazing. These defense compounds may be less of a factor for grazers in the early stages of a cyanobacteria bloom, when filaments are developing and available nutrients and energy are needed for producing more vegetative cells rather than such accessory protective compounds. Also some of the pigments in cyanobacteria cells at this stage of growth contain nitrogen-rich amino acids, making the cells a potentially high quality food source. In addition to the ecological factors such as grazing that influence the extent of planktonic cyanobacterial growth and nitrogen fixation, there are other chemical and physical factors recognized to be important (**Figure 2**). These factors include light and the related parameters

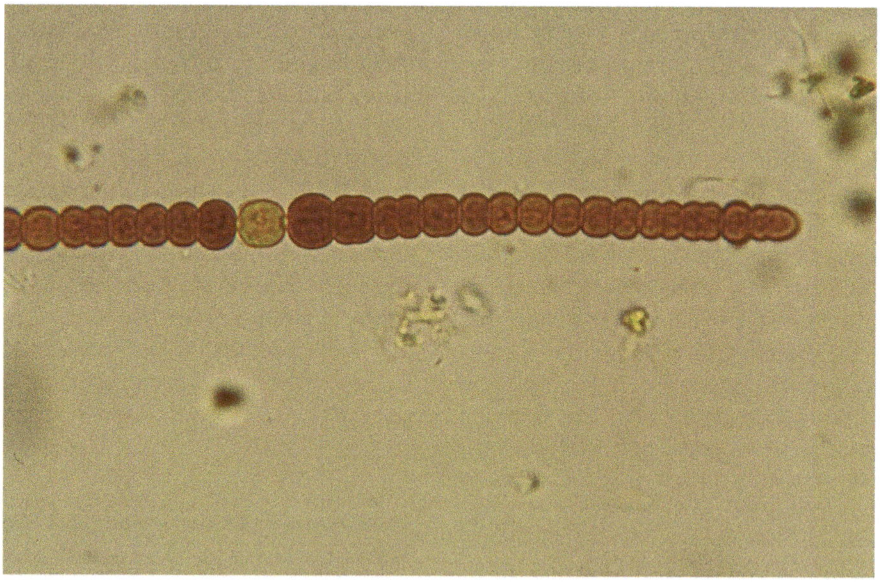

Figure 1 Photograph of an *Anabaena* filament with heterocyst, stained for contrast. Heterocyst appears yellow, vegetative cells are orange-brown. Courtesy of Colleen Kearns, Cornell University.

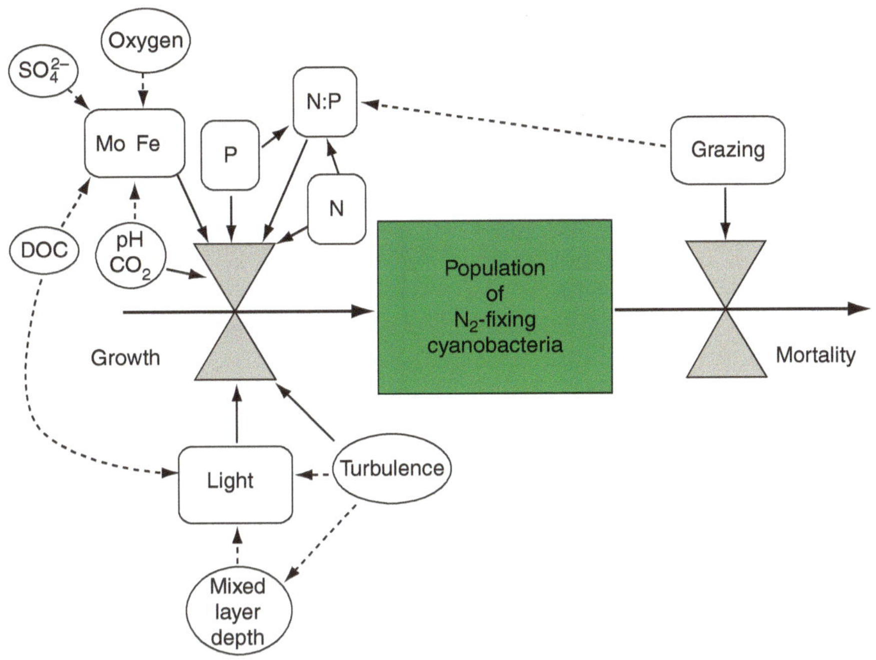

Figure 2 Schematic representation of the factors that control growth and N_2 fixation by planktonic cyanobacteria in lakes and estuaries. Thickness of arrows represents the relative strength of influence; dashed lines indicate indirect effects by acting to alter the availability of the direct factor.

of turbulence and mixing depth, and the availability of essential nutrients nitrogen and phosphorus and trace metals (iron and molybdenum). Other characteristics of the system, such as pH or dissolved organic carbon concentration, can interact with one or more of the primary factors to influence cyanobacterial dynamics. The relative importance of any one factor in controlling nitrogen fixation in a water body varies depending on the size, depth and external inputs to the system.

In contrast to nitrogen fixation by heterotrophic bacteria in anoxic sediments, nitrogen fixation by planktonic cyanobacteria appears to be an energetically more expensive process for the cells. Most of the energy cost of fixation is not in the catalysis of the reaction per se but rather in synthesizing and then protecting nitrogenase from oxygen, and perhaps the cost of assimilating trace metals that are required for nitrogenase: iron and molybdenum. The biological availabilities of both these metals are likely to be much greater in anoxic sediments than in an oxic water column. For iron, the greater availability under anoxic conditions simple reflects the higher solubility of reduced iron compared to oxidized iron. For molybdenum, the chemically stable form found under oxic conditions in the pH range of most natural waters is molybdate. This anion can be difficult for bacteria to assimilate because of its close similarity in effective size and stereochemistry to the sulfate anion. Sulfate is far more abundant in inland waters than is molybdate, ranging from four to six orders of magnitude higher in concentration. As such, the cellular uptake system needs to be more selective (higher affinity) for molybdate, which likely requires more energy for synthesis than a lower affinity transport system. In anoxic environments, molybdenum is present in different chemical forms, which may be more biologically available.

It is only advantageous to planktonic cyanobacteria to invest energy in heterocyst and nitrogenase synthesis to obtain nitrogen via the nitrogen fixation process when other sources of available nitrogen are low. Concentrations of a even few micromolar nitrate or ammonium can inhibit the synthesis of nitrogenase in an oxic environment and stop further nitrogen fixation (again in contrast to the relative lack of such inhibition in anoxic sediments). Further, even when concentrations of inorganic nitrogen are low, nitrogen fixation by planktonic cyanobacteria is not likely to occur if the N:P ratio of inorganic nitrogen and phosphorus in the environment is well above that needed for synthesis of phytoplankton biomass (the so-called Redfield ratio of 16:1, moles N to mole P, after Alfred Redfield, the scientist who first described this relatively constant relationship. Under these conditions, the production of the cyanobacteria is more limited by phosphorus than by nitrogen. Nitrogen-fixing cyanobacteria generally have a high overall requirement for phosphorus, and the biomass of planktonic cyanobacteria in freshwater lakes is often positively correlated with high concentrations of phosphorus in the water. The highest rates of nitrogen fixation are found when available nitrogen concentrations are low, available phosphorus concentrations are high, and the molar N:P ratio is below 16:1.

In a series of classic experiments in the 1970s at the Experimental Lakes Area in Canada, David Schindler and colleagues demonstrated that planktonic nitrogen fixation can help maintain phosphorus limitation of net primary productivity in eutrophic lakes. For several years, they fertilized a lake with nitrogen and phosphorus, with an N:P ratio of total inputs of 32:1 (by moles). Under these conditions, there is no advantage for cyanobacteria that can fix nitrogen, as phytoplankton growth is phosphorus-limited and sufficient dissolved inorganic nitrogen is present to meet the needs of the phytoplankton. No cyanobacteria were observed in the plankton. Then the experimental protocol was changed. The same amount of fertilizer was added to the lake, but the nitrogen input was reduced relative to phosphorus, so that the N:P ratio of total inputs was 14:1. This was expected to induce a slight limitation of nitrogen to phytoplankton growth, but this did not occur. Rather, large populations of planktonic cyanobacteria grew in the lake and fixed a sufficient quantity of nitrogen such that net primary productivity in the lake remained limited by phosphorus.

Nitrogen Fixation by Planktonic Cyanobacteria in Saline Waters

Many of the world's lakes are composed of saline waters. Nitrogen-fixing planktonic cyanobacteria are common in many of these lakes, and sometimes fix nitrogen at very high rates. Surprisingly, high rates have been observed even in oligotrophic salt lakes. For example, cyanobacteria in Pyramid Lake, Nevada, USA, fix nitrogen at a rate of $2\ g\ N\ m^{-2}\ year^{-1}$, a rate more characteristic of eutrophic than oligotrophic freshwater lakes. In some years, nitrogen fixation contributes roughly 80% of the total nitrogen input to Pyramid Lake, again a very high amount. However, nitrogen-fixing cyanobacteria are not present in the plankton of many saline lakes even when conditions seem optimal – primary production is limited by nitrogen, concentrations of nitrate and ammonium are low, concentrations of dissolved inorganic phosphorus are high, and the N:P ratio is well below the Redfield ratio of 16:1.

The controls on nitrogen fixation in saline lakes have received relatively little study, but the chemistry of the lakes is one major factor explaining variation in rates of nitrogen fixation. The specific dissolved substances that dominate the salinity in salt lakes varies greatly. In some saline lakes, sulfate is the dominant anion, while in others, chloride, carbonate, or other anions dominate. Growth and nitrogen fixation by planktonic cyanobacteria can be adversely affected by the high concentrations of sulfate found in some lakes because of an inhibition of molybdenum uptake, as discussed earlier. Thus, nitrogen-fixing cyanobacteria are most likely to be abundant in the plankton of salt lakes where the sulfate concentration is low. Salinity itself does not appear to be a constraint on growth and nitrogen fixation, and nitrogen-fixing cyanobacteria are abundant in some high salinity lakes characterized by anions other than sulfate.

Nitrogen-fixing cyanobacteria are present in the plankton of some estuarine ecosystems as well, although they are surprisingly rare, and generally are found only when the salinity is less than 10 parts per thousand (the salinity of open ocean seawater is approximately 35 parts per thousand). Much of the Baltic Sea and its estuaries has a salinity below 10 parts per thousand, and nitrogen-fixing cyanobacteria are often present in high abundances. A few other estuaries have reported cyanobacteria that persist at higher salinities, but these are ecosystems that tend to experience wide changes in salinity over the season; the cyanobacteria blooms typically originate during times when there is high freshwater inflow and the salinity is at or near zero parts per thousand, and then persist as the salinity increases. For the vast majority of estuaries that have been studied, including all those in North America and in Europe outside of the Baltic Sea, nitrogen-fixing cyanobacteria have not been reported at salinities much above 6 or 8 parts per thousand. This is a marked contrast with freshwater lakes, as many of these higher salinity estuaries typically have conditions that strongly favor nitrogen limitation of phytoplankton.

What processes act to exclude nitrogen-fixing cyanobacteria from the plankton of most estuaries? Part of the answer again lies with sulfate, the second most abundant anion in seawater. Sulfate is a conservative constituent of seawater and estuarine waters, and so the sulfate concentration always varies in direct proportion to the overall salinity. The concentrations of sulfate found in estuaries where the salinity is greater than 8–10 parts per thousand can significantly inhibit molybdate assimilation by heterocystic cyanobacteria. This is not an absolute constraint on nitrogen fixation, but the sulfate does make molybdate uptake energetically expensive by requiring more specificity (higher affinity) in the transport system, and therefore slows the growth rate of the cyanobacteria. Experimentally measured maximum growth rates for planktonic cyanobacteria isolated from estuarine systems tend to be an average of four times lower than those for the same genera (*Anabaena, Aphanizomenon*) from freshwater systems.

As discussed earlier, filamentous cyanobacteria growth can also be susceptible to grazing by zooplankton, because of the need to have many photosynthetic cells per heterocyst in a filament. A slower growth rate increases this vulnerability to grazing and is a negative feedback to the overall rates of production of cyanobacterial biomass and nitrogen fixation. Experiments have demonstrated that when grazing by zooplankton is suppressed in high salinity estuarine water, heterocystic cyanobacteria can grow in the plankton and fix nitrogen, even when they are not observed under natural (and nitrogen limited) conditions. The interaction of sulfate-induced slow growth rate and grazing by zooplankton is an important factor acting to exclude blooms of nitrogen-fixing cyanobacteria from the plankton of saline estuaries.

Nitrogen Fixation by Benthic and Epiphytic Cyanobacteria

When sufficient light reaches the bottom of an aquatic ecosystem to support photosynthesis, benthic nitrogen-fixing cyanobacteria frequently occur. This is true in both fresh water and saline inland waters, and in streams, particularly where current velocities are low and so hydrodynamic disturbance to the mats is minimal. There is relatively little information on nitrogen fixation by benthic cyanobacteria in streams, however, the conditions most favorable to high rates appear to be similar to those in other fresh waters: low inorganic nitrogen availability and high phosphorus, high light, and warmer water temperatures. In lakes, rates of nitrogen fixation by benthic cyanobacteria tend to be fairly low and range from 0 to 0.3 g N m^{-2} $year^{-1}$. Rates are frequently higher in oligotrophic lakes than in mesotrophic and eutrophic lakes, since the light intensities at the sediment surface tend to be higher owing to lower phytoplankton productivity and lower concentrations of dissolved organic substances, which can attenuate light. Benthic nitrogen fixation by cyanobacteria generally constitutes a small proportion of the nitrogen budget of mesotrophic and eutrophic lake ecosystems, but fixation may be important in oligotrophic lakes. For example, benthic nitrogen fixation by cyanobacteria contributes an estimated 6% of the total nitrogen inputs to Mirror Lake (New Hampshire, USA) and 32% of the total nitrogen inputs to Lake Tahoe (California and Nevada, USA). Some studies have found that grazing on benthic cyanobacteria can reduce nitrogen fixation, although often the effect is small. Other studies have shown a stimulative effect from grazing, sometimes because the light availability to the cyanobacteria increased as overlying sediment and competing phototrophic organisms are removed, or because the supply of phosphorus increased.

Rates of nitrogen fixation in marine cyanobacterial mats can be very high, ranging from 1 to 75 g N m^{-2} $year^{-1}$. The nitrogen fixation rates measured at the high end of this range occur in tropical marine environments and are some of the highest rates reported for any type of ecosystem in the world. High rates of nitrogen fixation (10–20 g N m^{-2} $year^{-1}$) have also been reported associated with cyanobacterial mats in temperate coastal environments, in salt marshes in both England and the northeastern United States. However, these well developed cyanobacterial mats are usually fairly limited in area, and so nitrogen fixation is generally not an important term in the overall nitrogen budgets of the larger ecosystems in which the mats exist. The factors that lead to the sometimes extremely high rates of nitrogen fixation in coastal marine cyanobacterial mats remain poorly known, but reducing conditions within the mats may result in chemical species of iron and molybdenum that are more readily available to the cyanobacteria, as well as limit oxygen inactivation of nitrogenase. Thus, as with heterotrophic fixation in anoxic sediments, the energetic cost of the fixation to these cyanobacteria may be low. Such physiologically favorable conditions, combined with high light in the mat environments, likely favor high rates of nitrogen fixation.

Nitrogen fixation by cyanobacteria also occurs in seagrass beds, an important coastal marine habitat. These cyanobacteria grow as epiphytes on seagrass leaves and on surface sediments in between the grasses. The seagrass leaves provide a high surface area substrate in the photic zone for the autotrophic N fixers, and overall fixation rates tend to be fairly high even in dense beds where there is significant shading. It is not clear whether nitrogen fixation by autotrophic nitrogen fixers is more significant per unit area than that of heterotrophic bacteria in the sediments of these systems. The relative contribution of each to the overall nitrogen economy of the beds likely depends on a number of factors including seagrass growth patterns and overall biomass, nitrogen and other nutrient availability, and other bacterial processes in the sediments which influence nitrogen fixation, such as sulfate reduction.

Conclusions

The process of nitrogen fixation is carried out by only a relatively few species of bacteria, capable of living in a very wide diversity of aquatic conditions. In most inland waters where biological nitrogen fixation is an important component of the nitrogen economy of the ecosystem, the fixation occurs in either the sediments or the water column, and is fueled either directly (cyanobacteria) or indirectly (heterotrophic species) by light energy. Heterotrophic bacteria fix nitrogen in virtually all lake and estuarine sediments, at low to moderate rates relative to cyanobacteria. Sediment nitrogen fixation is generally higher when the organic matter content of the sediment is greater, and is often balanced or exceeded by nitrogen losses through denitrification. Nitrogen fixation by benthic (sediment and mat) and epiphytic autotrophs can be high under optimal conditions of light and nutrient availability and is important to the local environment of the organisms. However, the area of this type of habitat is limited and so such fixation usually constitutes only a small portion of the total nitrogen input to the ecosystem. An exception is oligotrophic lakes, where nitrogen fixation by benthic cyanobacteria can be a significant overall nitrogen input.

Nitrogen fixation by planktonic cyanobacteria varies greatly among lakes and is generally not a large term in the nitrogen budgets of oligotrophic and mesotrophic lakes, but can be very important in the nitrogen economy of eutrophic lakes. Heterocystic species of cyanobacteria are the major nitrogen-fixers in both freshwater and saline lakes and estuarine environments. There is a marked contrast in the significance of heterocystic, N-fixing cyanobacteria in the plankton of freshwater versus estuarine systems, which are similarly nitrogen-limited, with large blooms often observed to occur and alleviate nitrogen deficits in the former, but rarely if ever in the latter. This is in part owing to an inhibitory effect of sulfate, present in much higher concentration in estuaries and some saline lakes than in fresh waters, on molybdate uptake. Molybdenum is necessary for synthesis of the most abundant and efficient form of nitrogenase and thus for growth on N_2; when growth is slowed by lower molybdenum availability, the abundance of filamentous cyanobacteria is more susceptible to control by grazing. Several additional chemical, physical, and ecological factors and interactions also influence the extent of planktonic cyanobacterial growth and fixation of 'new' nitrogen in inland waters. As population and land use pressures around inland waters increase and nutrient enrichment of these systems accelerates, the relative controls on this important biological process may shift. Current management strategies for controlling eutrophication and problematic algal blooms may need to be altered as a result.

Glossary

Anoxic – Oxygen-free, lacking oxygen.

Autotroph – An organism that fixes carbon dioxide and converts it into organic matter; photoautotrophs use light as the energy source to fix the carbon dioxide; photo-autotrophs include plants, algae, cyanobacteria, and other photosynthetic bacteria that use reduced substances such as hydrogen sulfide rather than water as an electron donor; chemo-autotrophs do not use light but rather use the energy released from oxidizing reduced chemical substances to fix carbon dioxide.

Benthic – In or on bottom sediments.

Cyanobacteria – Photosynthetic bacteria, often called 'blue green algae;' these organisms can use fixed nitrogen in the environment and many, but not all, species are also capable of fixing atmospheric N_2.

Denitrification – A bacterial respiration process that takes place in the absence of oxygen; nitrate is converted to N_2 gas through a series of steps involving nitrogen oxides.

Eutrophication – The process of increased growth and production of algae in an aquatic ecosystem in response to nutrient inputs.

Nitrogenase – The enzyme that catalyzes the fixation (reduction) of N_2 to NH_3.

Heterocyst – A specialized, nonphotosynthetic cell that is the site of nitrogen fixation in many species of cyanobacteria.

Heterotroph – An organism that gains its energy for life's processes by consuming organic carbon compounds produced by autotrophic organisms. All animals, all fungi, and many bacteria are heterotrophs; animals and many fungi and bacteria consume the organic matter by respiring it to carbon dioxide, using oxygen as the electron acceptor; some bacteria use other electron acceptors instead, such as nitrate or sulfate; and some bacteria and fungi gain energy through fermenting organic matter, where no inorganic electron acceptor is used and instead electrons are transferred from one type of organic molecules to another.

Nitrogen fixation – The process of converting molecular N_2 to reactive, biologically available forms.

Nutrient limitation – The condition where the rate of growth and production of biomass by photoautotrophs is less than maximum owing to a lack of one or more nutrients, usually phosphorus or nitrogen.

Oxic – Having oxygen present.

Planktonic – Living in the water column of an aquatic ecosystem and largely dependent upon motion of the water for movement; bacteria and archaea are prokaryotes.

Prokaryotic – Describing an organism that lacks a cell nucleus or other organelles bound by membranes; bacteria and archea are prokaryotes.

Salinity – The content of dissolved salt and other minerals in water, usually used in the context of ocean waters or salt lakes; salinity is generally reported in units of parts per thousand, or grams of dissolved salts per kilogram (or liter) of water.

Stereochemistry – The aspect of chemistry relating to the spatial arrangement of atoms within molecules. Stereochemistry is important in the reactivity of molecules.

See also: Iron and Manganese; Micronutrient Elements (Co, Mo, Mn, Zn, Cu); Nitrogen; Salinity.

Further Reading

Carr NG and Whitton BA (eds.) (1982) *The Biology of Cyanobacteria*. Berkeley: University of California Press.

Chan FC, Pace M, Howarth RW, and Marino R (2004) Bloom formation in heterocystic nitrogen-fixing cyanobacteria: The dependence of colony size and zooplankton grazing. *Limnology & Oceanography* 49: 2171–2178.

Howarth RW, Marino R, and Cole JJ (1988) Nitrogen fixation in freshwater, estuarine, and marine ecosystems. 2. Biogeochemical controls. *Limnology & Oceanography* 33: 688–701.

Howarth RW, Marino R, Lane J, and Cole JJ (1988) Nitrogen fixation in freshwater, estuarine, and marine ecosystems. l. Rates and importance. *Limnology & Oceanography* 33: 669–687.

Marino R, Chan F, Howarth RW, Pace M, and Likens GE (2006) Experimental tests of ecological constraints on planktonic N fixation in saline estuaries: 1. Nutrient and trophic controls. *Marine Ecology Progress Series* 309: 25–39.

Marino R, Chan F, Howarth RW, Pace M, and Likens GE (2002) Ecological and biogeochemical interactions constrain planktonic nitrogen fixation in estuaries. *Ecosystems* 5: 719–725.

Paerl HW (1996) A comparison of cyanobacterial bloom dynamics in freshwater, estuarine, and marine environments. *Phycologia* 35: 25–35.

Postgate J (1998) *Nitrogen Fixation*, 3rd edn. Cambridge, UK: Cambridge University Press.

Sprent JI and Sprent P (1990) *Nitrogen Fixing Organisms: Pure and Applied Aspects*. London: Chapman and Hall.

Schindler DW (1977) The evolution of phosphorus limitation in lakes. *Science* 195: 260–262.

Vitousek PM, Cassman K, Cleveland C, *et al.* (2002) Towards an ecological understanding of biological N fixation. *Biogeochemistry* 57–581–45.

Nitrogen

R Howarth, Cornell University, Ithaca, NY, USA

Introduction

Nitrogen is an essential element for life and is the fourth most abundant element in the living biomass (by moles) after hydrogen, carbon, and oxygen. Nitrogen is in all amino acids and nucleotides, and therefore in all proteins and nucleic acids. Nitrogen is also a major component of the chitin that makes up the cell walls of fungi and the exoskeletons of aquatic insects and crustaceans. Together with phosphorus, nitrogen limits rates of primary production in most of the ecosystems on Earth, including inland waters. However, unlike phosphorus, nitrogen has a very active oxidation–reduction cycle, and nitrogen in nature exists in valence states from −3 to +5. Curiously, and unlike the biogeochemical cycle of any other major element, most of the nitrogen on Earth is present with an intermediate valence state of 0 (as molecular N_2). More than any other major element cycle, human activity has accelerated the nitrogen cycle leading to many critical changes in the aquatic ecosystems.

Forms and Transformations of Nitrogen

Nitrogen occurs as both organic and inorganic nitrogen in aquatic ecosystems. Frequently, the organic forms dominate, including both particulate organic nitrogen (PON) and dissolved organic nitrogen (DON). The PON not only includes the nitrogen in living organisms, but also large amounts of nitrogen in detritus or dead organic matter. The DON consists of a wide range of organic substances, including simple substances such as free amino acids. Much of the DON, however, consists of higher molecular weight compounds. Most of the DON in natural waters has never been chemically characterized because of the analytical challenge of measuring thousands of unknown substances, each present at relatively low concentrations.

The inorganic nitrogen in aquatic ecosystems includes dissolved N_2 gas, oxidized ions such as nitrate (NO_3^-) and nitrite (NO_2^-), the reduced ammonium ion (NH_4^+), and the reduced ammonia gas (NH_3). Nitrate is the most oxidized form of nitrogen (valence state of +5), while ammonia and ammonium are most reduced (valence state of −3). Ammonium is a weak acid that is in equilibrium in a solution with ammonia gas, which is a base. The equilibrium constant for this relationship is $10^{-9.3}$. Thus, ammonium is more dominant whenever the pH is less than 9.3, as is generally true in aquatic ecosystems. At pH = 8.3, the ammonium concentration is 10-fold greater than the ammonia concentration. At pH = 7.3, the ammonium concentration is 100-fold greater than the ammonia concentration. Since ammonia is a gas, it can be volatilized to the atmosphere. The rate of loss is a function of the ammonia concentration, and so this process is much greater at higher pHs, where higher concentrations of ammonia are favored.

Most of the nitrogen on the Earth is present as N_2 gas. This becomes biologically available only through bacterial nitrogen fixation, fixation by lightning or volcanic activity, or fixation by human activity. Before the industrial revolution, bacterial nitrogen fixation was by far the major mechanism for the creation of reactive, biologically available forms of nitrogen on the planet. Increasingly, human activity is fixing nitrogen, and this dominates the nitrogen cycle in many regions, as discussed below.

The primary forms of reactive nitrogen assimilated by algae, rooted plants, fungi, and bacteria are nitrate, nitrite, ammonium, and ammonia. Once taken up, nitrate and nitrite are reduced to ammonium in a process called assimilatory nitrate or nitrite reduction. Ammonium – whether taken up directly or formed by assimilatory reduction in the organism – is used by plants, algae, and microorganisms to produce organic nitrogen compounds. The organic nitrogen of plants, algae, and microorganisms can flow through a food web to animals, and detrital PON and DON is decomposed by bacteria and fungi. The organic nitrogen eaten by animals or decomposed by microorganisms is excreted as ammonium or sometimes as urea, a low-molecular weight compound that is quickly hydrolyzed to ammonium in water. These processes of releasing nitrogen back to the environment are called nitrogen mineralization (**Figure 1**).

The forms of inorganic, reactive nitrogen are converted from one to another in aquatic ecosystems through a variety of bacterially mediated processes. Ammonium is oxidized to nitrate in a process called nitrification, an energy-yielding process. The nitrifying bacteria that catalyze the reaction gain energy and use this energy to fix carbon dioxide into new bacterial biomass, a process called chemosynthesis. The energy yield of the reaction is low compared with many chemosynthetic processes based on oxidizing sulfur or iron compounds, and so the growth of nitrifying bacteria is slow. Nitrification rates are in part a function of the population size of nitrifying bacteria; where mortality on the bacteria is high due to grazing, the slow growth result keeps population sizes low, and rates of nitrification can be very low, allowing ammonium to accumulate.

Nitrate is reduced to nitrite and nitrite is reduced to N_2 in a process called denitrification or dissimilatory nitrate reduction. The bacteria that catalyze these reactions are heterotrophic bacteria that gain their energy from the degradation or organic matter; they use the nitrate or nitrite as an electron acceptor, very much as plants and animals and many other microorganisms use oxygen as an electron acceptor for respiration. Since the energy yield of respiring organic matter using nitrate as the electron acceptor is somewhat less than when using oxygen as the electron acceptor, denitrification tends to occur only when oxygen is absent or present at very low levels. This condition is frequently the case in aquatic sediments, and often in the bottom waters of stratified lakes and estuaries. Denitrification is the major sink for reactive nitrogen in natural ecosystems. At the global scale, denitrification serves to balance nitrogen fixation and maintains N_2 gas as the major form of nitrogen on the planet (**Table 1**).

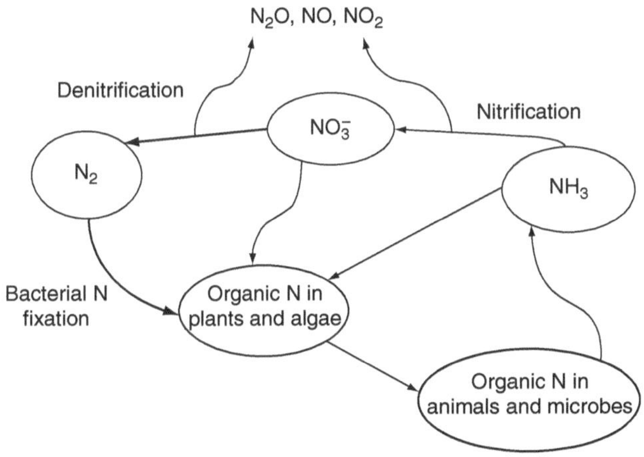

Figure 1 Simplified diagram of the nitrogen cycle in aquatic ecosystems. Reprinted from Howarth (2002).

Other nitrogen cycle processes have been discovered in the past few decades. One of these is denitrification based on chemosynthetic oxidation of sulfide or reduced iron rather than respiration of organic matter. Another is the anaerobic oxidation of ammonium to N_2 (ANAMOX), which is also a chemosynthetic process. And a third is the dissimilatory reduction of nitrate to ammonium (DNRA). The DNRA process is sometimes a form of respiration, with organic matter being consumed using nitrate as the electron acceptor; however, the nitrogen is reduced to ammonium rather than to N_2. Dissimilatory reduction of nitrate to ammonium can also occur through fermentation of organic matter or through chemosynthesis associated with oxidation of sulfides. Unlike denitrification, DNRA conserves the nitrogen in the ecosystem in a reactive, biologically available form. The relative importance of these newly discovered processes in natural ecosystems remains quite uncertain. The amount of organic matter, the carbon to nitrogen ratio of the organic matter, and the amounts of sulfides and iron in sediments may all play a role in determining which of these bacterial nitrogen-cycle processes is dominant (**Figure 2**).

Table 1 Summary characteristics of some key nitrogen-cycle processes

Nitrification:	$NH_4^+ + O_2 \rightarrow NO_3^- + H_2O + 2H^+$ Energy-releasing reaction; the rather small quantity of energy released is used to support the growth of chemo-autotorophic nitrifying bacteria that catalyze the reaction. Two protons worth of acidity are produced for every ammonium ion that is oxidized, so this process tends to make the environment more acidic.
Denitrification:	Organic matter $+ NO_3^- + H^+ \rightarrow 5/4\ CO_2 + 1/2\ N_2 + 5/4\ H_2O$ Energy releasing reaction, with the energy coming from the respiration of organic matter; the bacteria that carry out this reaction use nitrate rather than oxygen as an electron acceptor, and gain approximately 90% of the energy yield compared to oxygen-based respiration. One proton of acidity is consumed for every nitrate ion consumed, so this process tends to raise the pH of the environment.
Nitrogen fixation (by photosynthetic cyanobacteria):	$1/2\ N_2 + 3/2\ H_2O + H^+ \rightarrow NH_4^+ + 3/4\ O_2$ Nitrogen fixation requires an input of energy; in this case, the energy is coming from photosynthesis, which cleaves oxygen to gain electrons and produces oxygen. Note that the cost of the reaction is only part of the cost of nitrogen fixation. The total cost varies with the environment, and can include substantial costs for building and protecting enzymes and assimilating necessary trace metals. One proton of acidity is consumed for every atom of nitrogen fixed, so this process tends to raise the pH of the environment.

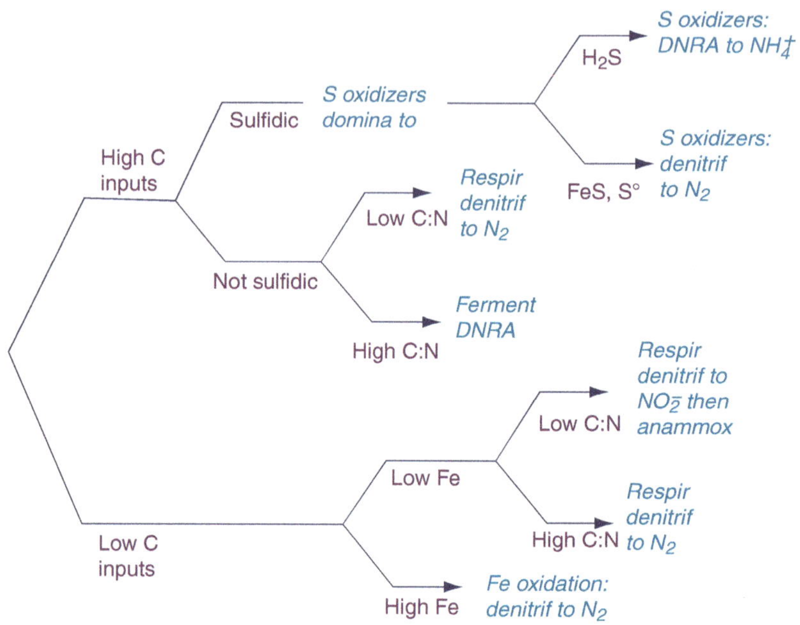

Figure 2 Hypothesized controls on denitrification, DNRA, ANAMOX, and similar processes in aquatic sediments based on the level of organic matter inputs to the sediments, the amount of sulfides present, the amount of iron present, and the ratio of carbon to nitrogen in the organic matter in the sediments. Reprinted from Burgin and Hamilton (2007).

Nitrogen Cycling at the Ecosystem Scale

The nitrogen budget of most inland water ecosystems is dominated by nitrogen inputs from upstream systems, including inputs from streams and rivers and in ground waters. As discussed elsewhere within this encyclopedia, bacterial nitrogen fixation can be an important source, but in most aquatic ecosystems this is far less than the inputs from upstream sources. Direct deposition from the atmosphere can also be an important input of nitrogen, that is nitrogen inputs in rain and snowfall as well as uptake of nitrogen gases by the water body from the atmosphere and deposition of nitrogen particles onto the water surface. However, only in aquatic ecosystems with very small watersheds does the direct deposition of nitrogen from the atmosphere rival the input of nitrogen in stream and river flow and in ground waters.

In most aquatic ecosystems, the rate of nitrogen uptake by plants, algae, and microorganisms is far greater than the rate of inputs from external sources, and the nitrogen needs of primary production and uptake by bacteria and fungi is supported by rapid recycling of nitrogen within the ecosystem. In oligotrophic lakes, the rate of uptake of nitrate and ammonium by phytoplankton is generally 50- to 100-fold more than the rate of input of nitrogen from external sources. In highly productive eutrophic lakes, the rate of nitrogen uptake by phytoplankton is often 10- to 30-fold greater than the rate of external inputs. In some very productive coastal salt marshes with huge inputs of nitrogen delivered by twice-daily tides, nitrogen uptake by the primary producing organisms is roughly equal to the external nitrogen inputs, but this is an extreme case of very high inputs.

In streams and rivers, the uptake of nitrogen and other nutrients can be considered in the context of spiraling distances, or the average distance over which an atom flows before being assimilated by plants, algae, bacteria, or fungi. In streams that are not heavily impacted by humans, this distance may be just meters to perhaps a hundred meters. After assimilation, the nitrogen atom is eventually remineralized and moves downstream again; again, on average, it moves the same spiraling distance. In larger rivers, the distances can be much greater, up to tens or hundreds of kilometers or even greater. And in landscapes with heavy human influences, the distances can be much greater; for instance, the spiraling distance for nitrogen in a ditch draining agricultural lands may be kilometers or more.

Denitrification can be a major sink of nitrogen in aquatic ecosystems, including wetlands, lakes, reservoirs, estuaries, streams, and rivers. In lakes, reservoirs and estuaries, rates of denitrification can be very high when the water column is anoxic, as frequently occurring in the bottom waters of stratified systems. Rates can also be high in the sediments of virtually all inland water ecosystems due to low oxygen levels. Commonly in many sediments and in wetland soils, nitrification and denitrification co-occur in close proximity in microzones where oxygen is present (for nitrification) and absent (for denitrification). This coupled nitrification–denitrification can account for a majority of the denitrification in aquatic ecosystems. The coupling of these processes is favored by the bio-irrigating activities of animals and by the rhizosphere of vascular plants.

In aquatic ecosystems with an oxygenated water column, the percentage of the nitrogen inputs that are denitrified is well predicted as a function of the water residence time and the depth of the water. This reflects the interaction time of the nitrogen with the sediments, the site of denitrification. The percentage loss of nitrogen is greater in ecosystems that are shallower and that have longer water residence times. At the landscape scale, the water residence time is frequently more important. Thus, the loss of nitrogen through denitrification tends to be greater in lakes and reservoirs than in streams and rivers (**Figure 3**).

Nutrient Limitation of Net Primary Production

Relative to the need for other essential elements by rooted plants and algae, nitrogen is frequently in short supply. As a result, nitrogen limits rates of net primary production in many types of both aquatic and terrestrial ecosystems. Either nitrogen or phosphorus limits primary production in almost all ecosystems on Earth, except for some 10% of the area of the world's oceans where iron may be limiting. For inland waters, both nitrogen and phosphorus are often important limiting factors, and some interesting patterns of nutrient limitation occur across different types of inland waters. For instance, in the Temperate Zone in ecosystems of moderate to high productivity, phosphorus is more generally limiting in freshwater

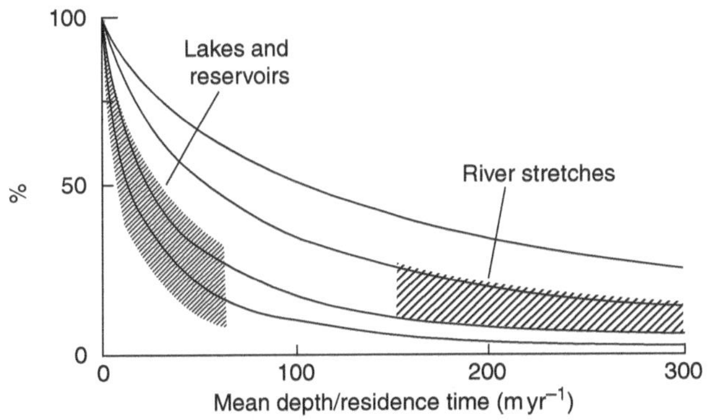

Figure 3 The percentage of nitrogen inputs to aquatic ecosystems that is lost through denitrification as a function of depth and residence time of the system. Modified from Howarth *et al.* (1996).

lakes and nitrogen in estuaries. This has important ramifications for the management of eutrophication (excess production) in these ecosystems. As is discussed in the section below, human activity has greatly accelerated the nitrogen cycle, leading to massive increases in nitrogen inputs to estuaries in many parts of the world. Nitrogen is now the largest source of pollution in estuaries, and in the United States, two-thirds of estuaries are moderately to severely degraded from nitrogen pollution and the eutrophication it causes.

Whether nitrogen or phosphorus is more limiting is a function of the relative availabilities of these two elements in the ecosystem. Although there is some plasticity in the elemental composition of phytoplankton, the variation is surprisingly small and the ratio of nitrogen to phosphorus in phytoplankton biomass is usually in the range of 16:1 by moles (the 'Redfield ratio', named after Alfred Redfield who described the relationship between water chemistry and phytoplankton composition in the world's oceans in the 1930s). In ecosystems where the ratio of available nitrogen to phosphorus in the water is well above 16:1, phosphorus is in relatively short supply and will be more limiting to net primary production. In ecosystems where the nitrogen:phosphorus ratio of available nutrients is well below 16:1, nitrogen is in relatively short supply (compared with the needs of the phytoplankton) and will be more limiting. The ratio of dissolved inorganic nitrogen (the sum of nitrate, nitrite, ammonium, and ammonia) to dissolved inorganic phosphorus is a reasonable indicator of the ratio of dissolved nitrogen and phosphorus in many cases and can be used to infer whether nitrogen or phosphorus is more limiting, if the ratio is sufficiently above or below 16:1. However, dissolved inorganic nutrients often recycle quite quickly in the water column of aquatic ecosystems, and so concentrations are not always a reliable indicator of supply. Further, most analytical methods for measuring dissolved inorganic phosphorus overestimate this quantity by some unknown amount due to inclusion of some organic phosphorus and positive interference from arsenic.

The ratio of available nitrogen to phosphorus within an ecosystem is a function of both the ratio in the external inputs of these elements and the processing of nitrogen and phosphorus within the ecosystem. Often, the nitrogen to phosphorus ratio in nutrient inputs from watersheds are well above 16:1, which tends to drive the aquatic ecosystems towards phosphorus limitation. In contrast to lakes, estuaries receive nutrients both from upstream watersheds and from ocean waters. The ocean waters are frequently phosphorus rich and somewhat depleted in nitrogen (relative to the Redfield ratio). This is one of the reasons that estuaries are more prone to nitrogen limitation.

A variety of processes within an ecosystem can change the availabilities of nitrogen and phosphorus. Of particular importance for phosphorus is the adsorption to sediment particles. This lowers concentrations of dissolved inorganic phosphorus, and therefore makes phosphorus limitation more likely. The presence of competing ions in seawater makes adsorption less in estuaries than in fresh waters, and inorganic phosphorus that is adsorbed to suspended sediment particles in rivers is largely desorbed and becomes soluble in estuaries as the salinity increases. This tends to make phosphorus limitation less likely (and nitrogen limitation more likely) in estuaries than in freshwater ecosystems (within the constraints set by the inputs of nitrogen and phosphorus, and the influences of other biogeochemical processes).

Both nitrogen fixation and denitrification can have large influences on the availability of nitrogen in aquatic ecosystems. Nitrogen fixation is discussed in detail elsewhere within this encyclopedia. Briefly, nitrogen fixation by planktonic cyanobacteria is an important process in many lakes that helps to alleviate nitrogen shortages and contributes to phosphorus limitation in fresh waters. In stark contrast, nitrogen fixation by planktonic cyanobacteria seldom, if ever, occurs in even strongly nitrogen-limited estuaries at salinities greater than 10 parts per thousand. This is probably a result of the high sulfate concentrations in seawater interfering with the uptake of molybdenum by planktonic cyanobacteria in estuaries; molybdenum is required for nitrogen fixation. It is interesting to note that sulfate levels are sometimes high and sometimes not in inland saline lakes. As a result, nitrogen fixation is a major process in some saline lakes but not in others. In saline lakes where nitrogen fixation occurs, the process helps to alleviate nitrogen shortages and maintain phosphorus limitation, while high sulfate lakes without significant nitrogen fixation are often nitrogen limited.

We have an imperfect understanding of whether nitrogen or phosphorus is more limiting in many types of aquatic ecosystems, including oligotrophic (low-productivity) lakes and many tropical reservoirs, particularly for those systems where the ratio of dissolved inorganic nitrogen to phosphorus is too close to the Redfield ratio of 16:1 to allow for unambiguous interpretation. Often, researchers will use short-term bioassays to evaluate nutrient limitation. In this approach, nitrogen, phosphorus, or some other element is added to water samples in bottles and any change in phytoplankton biomass or productivity over a few days to a week is noted. While this can provide very useful information, the approach has many difficulties. One of the greatest difficulties is that the time frame may not allow for some important biogeochemical processes, such as nitrogen fixation, to occur. In experiments with nutrient additions to whole lakes over growing seasons and years, researchers have demonstrated that temporary shortages of nitrogen that result from fertilizing with phosphorus and nitrogen at ratios below 16:1 can be alleviated by planktonic nitrogen fixation, and that the lakes remain phosphorus limited over these longer time frames. But such a response would not be observed in a short-term bioassay, which might instead suggest nitrogen limitation. The nitrogen limitation would in fact be only a short, transient feature.

The most robust information on nutrient limitation in aquatic ecosystems has come from longer-term, whole-ecosystem experiments and experiments with large tanks or bags designed to mimic important components of the ecosystem over time periods of a growing season or longer (mesocosm experiments). Such experiments have clearly shown a prevalence for phosphorus limitation in moderately productive lakes and for nitrogen limitation in estuaries in the Temperate Zone. The lack of such experiments in tropical systems and in low-productivity temperate systems to date constrains our understanding of nutrient limitation.

Human Acceleration of the Nitrogen Cycle

Human activity has altered the nitrogen cycle more than that of any other major element on Earth. Prior to the industrial and agricultural revolutions, the vast majority of reactive nitrogen on the planet was created by bacterial nitrogen fixation, and this rate of creation was balanced over geological time by denitrification. Now the creation of nitrogen by humans to produce synthetic nitrogen fertilizer through the Haber–Bosch process rivals the rate of natural fixation on the continents. The inadvertent creation of reactive nitrogen during fossil fuel combustion also adds to the global nitrogen cycle.

The influences of human activity on the nitrogen cycle vary across the regions of the globe. Because of the high chemical reactivity and high biological demand for nitrogen, most forms of reactive nitrogen (i.e., nitrogen other than N_2 gas) do not cycle globally, but rather at the scale of large regions. The natural rates of nitrogen fixation in terrestrial ecosystems are much higher in the tropics, whereas most of the fertilizer use and fossil fuel combustion to date has occurred in temperate regions. This is changing somewhat as fertilizer use and fossil fuel combustion increase in South Asia and tropical South America. Nonetheless, to date the largest changes in nitrogen cycling have occurred in the developed countries in the Temperate Zone. The flow of nitrogen in large rivers in these areas has increased 15-fold or more due to human activity, with much of this increase occurring in the past three to four decades (**Figure 4**).

Globally, the use of synthetic nitrogen fertilizer has driven most of the increase in nitrogen cycling. However, in some regions such as the northeastern United States, the combustion of fossil fuel has been a larger contributor. This fossil fuel combustion leads to pollution of the atmosphere with oxidized nitrogen compounds. Their deposition onto the landscape contributes to the flux of nitrogen to aquatic ecosystems. Much of the nitrogen is deposited onto forests, which retain a great deal of it, and some falls on urban and suburban lands, where retention is much less.

The largest consequence on aquatic ecosystems of the human acceleration of the nitrogen cycle has been the eutrophication of estuaries and other coastal waters. The resulting eutrophication has led to oxygen-deprived waters (anoxic and hypoxic zones, sometimes called 'dead zones') in many areas, as well as habitat degradation, loss of biotic diversity, and increased incidences and extent of harmful algal blooms. In the Temperate Zone, nitrogen has probably not led to eutrophication in many freshwater ecosystems simply because phosphorus limitation is so much more prevalent. However, increased nitrogen inputs to some tropical lakes such as Lake Victoria in Africa may well be contributing to eutrophication.

A larger impact of accelerated nitrogen cycling on freshwater ecosystems in the Temperate Zone comes from acid rain. Both sulfuric and nitric acids contribute to acid rain. Much of the focus historically has

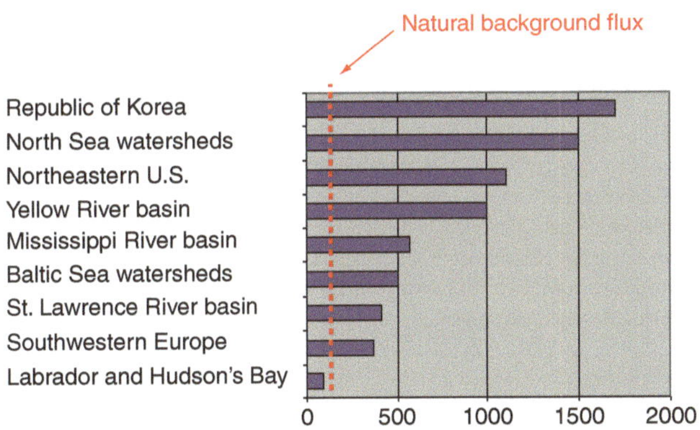

Figure 4 The flux of nitrogen in rivers expressed per area of watershed for contrasting areas in the Temperate Zone. Units are kg N km^{-2} yr^{-1}. The natural background flux in Temperate Zone regions without human disturbances is estimated to be roughly 100 kg N km^{-2} yr^{-1}. Based on information in Howarth *et al.* (2005).

been on sulfuric acid, and that has led to some measurable progress in reducing sulfur emissions to the atmosphere. Reductions in NOx, the precursor for nitric acid, have been far less, and nitric acid is therefore growing as an overall source of acid rain. This nitric acid – which comes from fossil fuel combustion and is the same nitrogen deposition that contributes to nitrogen flows to estuaries – is the major source of acid rain now in some regions.

The deposition of ammonia gas and ammonium particles can also contribute to the acidification of inland waters. Most of the ammonia in the atmosphere comes from volatilization from agricultural sources, and particularly from large animal feedlot operations. Ammonia is a base, and so acts to raise the pH of precipitation. In this technical sense, ammonia counteracts acid rain. However, the ammonia and ammonium deposited onto terrestrial ecosystems or directly onto aquatic ecosystems can be nitrified to nitrate, as discussed above. This chemical transformation produces large quantities of acid – 1 mole of proton acidity generated for every mole of nitrate created. Consequently, the deposition of ammonia and ammonium also can contribute substantially to the acidification of inland waters.

Nitrate in drinking water for humans is also of concern, and a standard of 10 mg nitrate-N per liter is set in most of the world's nations, based on advice from the World Health Organization. Such a standard has existed for many decades in the United States, and was originally set over concern about 'blue-baby syndrome,' a disease in infants where nitrate binds to hemoglobin in the blood, interfering with oxygen transport, and turning the blood blue. Studies over the past few decades have suggested that nitrate in drinking water poses other health risks, including risk of a variety of cancers and a variety of developmental issues, and perhaps at levels below the drinking water standard. This topic has seen much debate in the past decade, with some arguing that nitrate is a natural occurrence in saliva that the human body may actually manipulate to reduce bacterial infection. However, evidence shows a strong carcinogenic risk from nitrate in drinking water, probably due to the formation of nitrosomines, which are highly carcinogenic compounds. The best current evidence suggests a strong interaction between nitrate exposure and other dietary considerations: nitrate in vegetables such as spinach seems to pose little risk, while nitrate in drinking water combined with a high red-meat and low vegetable and fruit diet poses a major risk. Nitrate in drinking water at levels well below the current drinking water standard can pose a risk, in association with poor diet. Unfortunately, many of the drinking water sources in the world – including the developed world – exceed the current drinking water standard, and concentrations of nitrate in drinking water are rising.

Nitrogen poses health risks beyond those of direct toxicity or carcinogenicity from nitrate, to both humans and natural animal populations. The toxicity of ammonia to aquatic animals is well known and is a major problem in waters heavily polluted from sewage or animal wastes. A more subtle aspect is the indirect effect of nitrogen on animal vectors that carry disease. Recent work has shown that parasitic diseases of amphibians can increase markedly as eutrophication proceeds: nutrient pollution (probably phosphorus) in fresh waters in Temperate-Zone areas can lead to increased parasite releases from snails, which are an intermediary host, and cause massive disease in frogs and salamanders. The disease-causing organism is closely related to schistosomiasis, a major disease for humans in many tropical areas. An increase in snail populations as a result of nitrogen-based eutrophication in tropical reservoirs and lakes could lead to a similarly large increase in schisotomiasis. The Temperate-Zone studies on amphibian disease show an increased risk of disease resulting not only from an increase in snail populations as eutrophication progresses, but also a huge increase in the release of parasites from each snail. In another study, malaria has increased in some coastal areas of Brazil due to nitrogen-induced increases in eutrophication in mangrove swamps leading to increased abundances of mosquitoes that carry the malaria parasite.

One aspect of the human acceleration of the nitrogen cycle demands particular attention: the gas nitrous oxide (N_2O) presents severe challenges to the global environment. Laughing gas, as it is sometimes called, is an effective pain-killing drug sometimes used by dentists. It is also a long-lived gas in the Earth's atmosphere, with a residence time estimated at 120 years. Concentrations are rising rapidly. The gas catalyzes the destruction of ozone in the stratosphere, and increasingly nitrous oxide is responsible for ozone holes that allow excessive UV-B radiation to penetrate to the Earth's surface. Nitrous oxide is also a major absorber of infra-red energy, and therefore is a potent greenhouse gas that contributes to global warming. The Intergovernmental Panel on Climate Change considers nitrous oxide to be nearly 300-fold more powerful than carbon dioxide in terms of its long-term effect on global warming. Nitrous oxide is produced as a leakage-produce of both nitrification and denitrification. The rapid increase in the concentration of this gas in the atmosphere is clearly a side effect of the global acceleration of the nitrogen cycle by human activity.

Glossary

Anoxic – Oxygen-free, lacking oxygen.

Denitrification – The process of converting nitrate to molecular N_2; also called dissimilatory nitrate reduction.

Eutrophic – Characterized by high levels of primary production and high algal biomass.

Hypoxic – Having oxygen present at only low concentrations (generally less than 2 mg/l).

Nitrification – The oxidation of ammonium to nitrite and nitrate by bacteria.

Nitrogen fixation – The conversion of molecular N_2 into reactive nitrogen compounds such as ammonium.

Nutrient limitation – The condition of rate of growth of production by photo-autotrophs being less than maximum due to a constraint by one or more nutrients, usually phosphorus or nitrogen.

Oligotrophic – Characterized by low levels of primary production and low algal biomass.

Oxic – Having oxygen present.

Primary production – The rate of accumulation of new biomass by plants, algae, and photosynthetic bacteria through photosynthesis.

Rhizosphere – The roots, rhizomes, and associated micro-organism of a vascular plant.

See also: Acidification; Alkalinity; Micronutrient Elements (Co, Mo, Mn, Zn, Cu); Nitrogen Fixation; Phosphorus; Salinity.

Further Reading

Burgin AJ and Hamilton SK (2007) Have we overemphasized the role of denitrification in aquatic ecosystems? A review of nitrate removal pathways. *Frontiers in Ecology & Environment* 5: 89–96.

Carpenter SR, Caraco NF, Correll DL, Howarth RW, Sharpley AN, and Smith VH (1998) Nonpoint pollution of surface waters with phosphorus and nitrogen. *Issues in Ecology* 3: 1–12.

Galloway JN, Aber JD, Erisman JW, Seitzinger SP, Howarth RH, Cowling EB, and Cosby BJ (2003) The nitrogen cascade. *BioScience* 53: 341–356.

Howarth RW (2002) The nitrogen cycle. In: Mooney HA and Canadell JG (eds.) *Encyclopedia of Global Environmental Change. The Earth System: Biological and Ecological Dimensions of Global Environmental Change,* Vol. 2, pp. 429–435. Chichester: Wiley.

Howarth RW and Marino R (2006) Nitrogen as the limiting nutrient for eutrophication in coastal marine ecosystems: Evolving views over 3 decades. *Limnology and Oceanography* 51: 364–376.

Howarth RW, Billen G, Swaney D, Townsend A, Jarworski N, Lajtha K, Downing JA, Elmgren R, Caraco N, Jordan T, Berendse F, Freney J, Kueyarov V, Murdoch P, and Zhao-liang Zhu (1996) Riverine inputs of nitrogen to the North Atlantic Ocean: Fluxes and human influences. *Biogeochemistry* 35: 75–139.

Howarth RW, Ramakrishna K, Choi E, Elmgren R, Martinelli L, Mendoza A, Moomaw W, Palm C, Boy R, Scholes M, and Zhao-Liang Zhu (2005) Chapter 9: Nutrient Management, Responses Assessment. In: *Ecosystems and Human Well-being, Policy Responses, the Millennium Ecosystem Assessment*, Vol. 3, pp. 295–311. Washington, DC: Island Press.

Seitzinger SP, Styles RV, Boyer EW, Alexander R, Billen G, Howarth R, Mayer B, and van Breemen N (2002) Nitrogen retention in rivers: model development and application to watersheds in the northeastern US. *Biogeochemistry* 57&58: 199–237.

Townsend AR, Howarth R, Bazzaz FA, Booth MS, Cleveland CC, Collinge SK, Dobson AP, Epstein PR, A Holland E, Keeney DR, Mallin MA, Rogers CA, Wayne P, and Wolfe AH (2003) Human health effects of a changing global nitrogen cycle. *Frontiers in Ecology & Environment* 1: 240–246.

Vitousek PM, Aber J, Bayley SE, Howarth RW, Likens GE, Matson PA, Schindler DW, Schlesinger WH, and Tilman GD (1997) Human alteration of the global nitrogen cycle: Causes and consequences. *Ecological Applications* 7: 737–750.

Phosphorus

N Caraco, Cary Institute of Ecosystem Studies, Millbrook, NY, USA

Introduction

Inland waters include lakes, reservoirs, rivers, streams, and estuaries, and these systems combined provide a large number of ecosystem services to humans throughout the globe. Included in these services are drinking water supply, fisheries production, and recreation. Phosphorus (P) plays a crucial role in controlling the value of all these ecosystem services. The basis for this importance is the key role that P plays in controlling production of inland waters. Inland waters with low P concentrations generally have not only very low primary production, but also low secondary production of invertebrates and fish. Conversely extremely high P concentrations often lead to eutrophic (high productivity) conditions that are associated with low water clarity, decline of rooted plants, anoxic bottom waters, fish kills, and algal blooms. These consequences of eutrophication have triggered much of the research on P in inland waters over the past 50 years. This work like much of the research on inland waters extends past the boundaries of the aquatic system and includes controls of inputs from the watershed via surface water and groundwater inputs as well as inputs from the atmosphere. Within the aquatic system work includes both the rates and controls of P cycling in well-lit surface waters and the movement of P between these waters and underlying bottom waters and sediments (**Figure 1**). The P cycling in inland waters and the supply from the watershed is controlled by the interaction of hydrology, chemistry, and biology. This article is primarily a review of the state of knowledge of P supply and cycling in inland waters but starts with some background on the history of P and its general properties.

History of P Research and Basic Chemical Properties of P

Phosphorus literally means 'bearer of light' owing to the properties of elemental P (P^o). P^o, and in particular white phosphorus, glows in the presence of oxygen and can easily ignite. P^o was isolated from urine in the 16th century by the German alchemist, Brandt. It quickly gained favor for its entertainment value and for a perceived potential gold-producing ability in alchemy. In the following centuries P^o was used to make matches and both defensive and offensive weapons. The interest in and utility of P^o lead in part to studies of P abundance in tissues of animals and plants and to methods of P extraction form these tissues, particularly the $CaPO_4$ of bone. Soon the extraction of P moved from these tissues to rock P (primarily apatite) and today these rocks still supply the major source of industrial P for human use. Today very little of the P extracted from rocks is converted to the elemental form; rather it is primarily used in the oxidized form (phosphates) in fertilizer, feed additives, and detergent additives. These uses of P by humans have profoundly influenced inputs to inland waters (later).

P potentially has several oxidation states ranging from -3 to $+5$ (P^o is of course 0). However, unlike N, C and S, whose cycles in inland waters could not be described in the absence of changes in oxidation state, P exists almost exclusively in its most oxidized form, PO_4^{3-}. This is true if P is in its dissolved or particulate or organic, or inorganic forms. That is, whether it is found in mineable apatite deposits, soils, sediments, bones, or in living organisms as part of ATP, DNA and phospholipids or detrital organic matter, P is in the $+5$ (oxide) form. A related difference between P and many other biologically important elements is that a gaseous phase is almost absent. The only exception to this statement is the potential formation and presence of phosphine and phosphides in reduced sediments of wetlands and ponds. For most systems, however, the contribution of phosphine and phosphides to the P cycle appears to be small.

Despite the fact that P does not itself have an active REDOX cycle and has only a minor gaseous form, the cycling of P in both terrestrial and aquatic systems can be complex owing to the large number of biotic and chemical pathways controlling its movement (e.g., transport vs burial) or the form of P found in an ecosystem. The chemical control is due to the fact that P is a major component of a large number of minerals, including Ca-phosphates, Fe-phosphates, and Al-phosphates. Additionally, P readily binds to siliceous clays, humic material, iron, and aluminum oxides. This chemical binding is sufficient under many conditions to inhibit strongly phosphorus release from soils and sediments. The chemical reactivity of phosphorus to particles partly explains the relatively low concentration of P in most ground water and surface water. Also contributing to these low concentrations is biotic uptake. All organisms require P for cell division, energy transformations, and cell maintenance as phosphorus is a critical component of nucleic acids, ADP, and phospholipids. The P requirement for

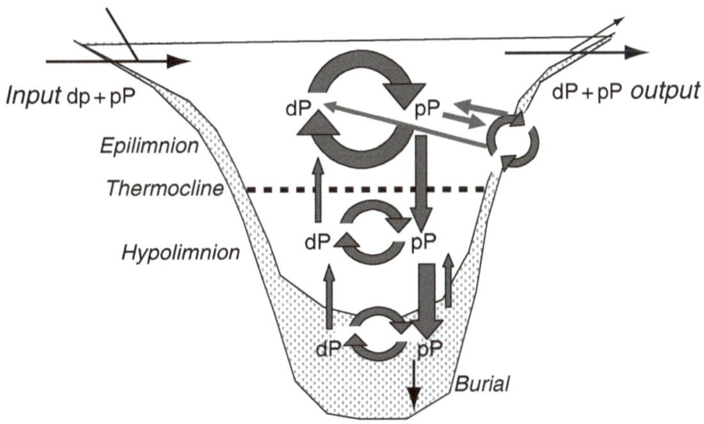

Figure 1 Phosphorus exchange and cycling within inland waters. The 'black box' budget of P includes only exchange (black arrows), including watershed and atmospheric inputs and outputs as well as permanent burial in sediments. Within inland waters, P is cycled between dissolved forms (dP) and particulate forms (pP) and moves within the water column and the underlying sediments. These exchanges are influenced by biotic, chemical, and physical processes.

photosynthetic organisms, bacteria, and fungi is met completely or in part by the uptake of dissolved P.

P Limitation of Production in Inland Waters

Primary production in aquatic systems occurs both within the water column and on the sediments. All primary producers take up inorganic P from the environment to produce the organic compounds containing P needed for growth and reproduction. Small planktonic algae (phytoplankton) generally dominate production within the water column of inland waters and these and it is well established that for a large number of systems P is important in limiting the primary production of these organisms. This limitation has been demonstrated by adding P alone or in conjunction with nitrogen result and demonstrating significant and sometimes very large increases in production and biomass of phytoplankton. Conversely declines in P input can result in marked declines in phytoplankton production and biomass. This control of P input on phytoplankton production has been particularly well demonstrated in low turbidity temperate lakes. For some other inland waters the strong link between P status of the aquatic system and phytoplankton production is less clear and may occur only seasonally, in some parts of the system, or may not occur at all. For example, in several estuaries P concentrations are high relative to nitrogen concentrations (e.g., N:P ratios are substantially lower than 16:1, **Figure 2**) and a substantial stimulation of primary production may occur by nitrogen additions alone.

In addition to the primary production that occurs in the water column of aquatic system is the production

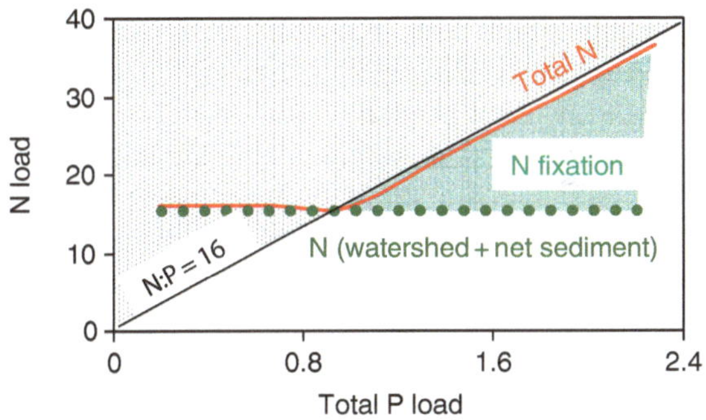

Figure 2 Example and P and N loading to inland waters (in atoms). In this example the P load from sediment and water varies along the *x*-axis from 0.2 to 2.2 while inputs of N from the same sources remains constant at 16. When P inputs exceed 1 the ratio of N:P in the water body would fall below 16:1 and the system would become N limited if it were not for the additional input from N fixation (stippled blue area). With this input the total N load is sufficient so that N:P does not fall below 16:1. This scenario can occur if high P conditions select for high nitrogen fixation by cyanobacteria.

occurring on the sediments. This benthic primary production includes both algal production and true aquatic plants. Benthic primary production generally dominates total system primary production in wetlands, shallow lakes, and estuaries, and streams. Although benthic primary producers clearly require P to grow, the actual impact of P additions on these producers is not always positive and can in some cases be quite negative. For well-lit streams in particular, several studies have shown that benthic algal biomass and production increase with increased P inputs. P additions have been shown to increase production of freshwater wetland plants. In shallow lakes and some estuaries on the other hand declines in benthic production may occur with increased P inputs.

This decline may seem surprising but mechanistically makes sense in systems where phytoplankton production is stimulated to such a large degree that light levels reaching the benthos become too low to sustain primary production. Interestingly, once a system has lost benthic plants owing to nutrient enrichment it can be difficult to restore the system's benthic plants, and it may remain a phytoplankton-dominated system even after nutrients decline.

Beyond the relationship to primary production is the role that phosphorus plays in selection for specific types of primary producers. As discussed earlier, broadly the system can switch from domination of benthic producers to planktonic producers. Additionally, however, benthic or planktonic producer that dominates can be altered. Of particular interest is the switch to dominance by cyanobacteria (blue-green algae) that seems to occur within the phytoplankton community. This switch has received a great deal of attention as cyanobacteria have several representatives that produce toxins that can kill or make ill animals, including humans and livestock who ingest the waters. An added significance is that several cyanobacteria have the capacity for nitrogen fixation, and fixation by these organisms can be a significant nitrogen input to many inland waters but not in most estuaries or coastal waters. The potential for high P inputs to result in increased nitrogen fixation and N inputs to aquatic systems has lead to the argument that P can or even should ultimately limit production even when nitrogen loads from the watershed and sediments are low relative to those of P (**Figure 2**).

In addition to limiting production of primary producers, P can also limit production of consumers (secondary production). These heterotrophic organisms that include bacteria, fungi, and grazers can use organic P created by primary producers to meet their P requirements. Additionally, however, both bacteria and fungi can subsidize this organic P with inorganic P taken up from the environment, and bacteria have been shown to be effective competitors for inorganic P with phytoplankton. Despite the two possible modes of P uptake, however, heterotrophic bacteria have been shown to be P-limited in many aquatic systems. This limitation can translate into lower bacterial production as well as lower decomposition rates of organic matter in low P systems. The limitation of decomposition can be especially severe for terrestrial plant material entering aquatic systems (e.g., leaves, twigs) as this material has very low P content compared with the demands of bacteria or fungi.

There is some evidence that grazers (e.g., crustacean zooplankton, benthic invertebrates, etc.) in aquatic systems can also be P limited. Lower growth rates and lower reproductive rates can result when grazed organic matter has very low P content as compared with the P content of consumers. P content of Phytoplankton varies substantially with P limitation status, which means that low P content in inland waters may lead to lower grazer abundance both due to lower overall primary production and to a poorer food quality of this production.

Global Regional and Landscape Transfers of P

At the global scale P transfers are long distance (100s–1000s of km) and cycles are of long duration (millenia). This global scale transfer follows the creation of new inputs of P into the environment from weathering of rocks to the burial in the ocean rock formation from pressure and heat and eventual uplift to terrestrial environments. Inland running waters play an obvious role in this cycle by being conduits of P from weathering terrestrial materials to coastal oceans. Wetlands, lakes, and more recently, reservoirs can add a substantial time-lag to this delivery as a substantial part of the P being weathered can be stored temporarily in sediments of these inland waters rather than being delivered to coastal waters. The P carried by rivers to the oceans consists of both dissolved and particulate forms. The particulate forms make up the majority of the P being transported, and much of this is of very low reactivity and may never enter the biotic cycle before being buried in deltaic and coastal sediments.

Humans have altered the global cycle of P in several ways. First and perhaps foremost they directly mine P from rocks to make fertilizers and other compounds. Second humans have altered the transfer of P. Agricultural practices generally increase erosion of soils and the transport of phosphorus from land to the ocean via rivers. Further, sewering and animal production can increase the input of highly available P in rivers. Reservoirs on the other hand trap particulates and can decrease the transfer of P from land to sea. At present on the global scale the net balance of these reactions appears to be a higher transfer of P in rivers than occurred pre-industrially. The human impact on phosphate and other highly available forms of P is most apparent.

Nested within the global cycle of P are ecosystem- and landscape-scale cycles. For relatively pristine forested and grassland systems the inputs of P include weathering and atmospheric inputs but also P transferred by animal movement and migration. Dissolved P in soil solution travels with ground water to running waters and standing waters. Particulate forms of P are transferred by stream bank erosion and overland flow of water. Further organisms feeding in the terrestrial environment or in the coastal oceans can transfer P directly to inland waters. For some lakes geese have been shown to be important in the transfer of P from grasslands to aquatic systems and salmon are an often-cited example of transfer from coastal oceans and estuaries to rivers and lakes.

The amount of P transferred to aquatic systems from pristine systems depends on a number of factors including the soil texture and chemistry, the slope of the watershed and the amount and pattern of precipitation and water runoff in the watershed. Human-caused weather change can alter P outputs through their impact on water runoff from watersheds. Additionally, atmospheric pollution may alter these transfers. In particular, acid precipitation is thought to decrease soil pH and increase binding of P to soils and, hence, decrease the P transfer from terrestrial to aquatic systems. This decrease in P transfer, coupled with changes in P cycling within lakes, has been thought to be responsible for decreases in primary production in some poorly buffered lakes in Europe and North America whose watersheds are subject to high inputs of acid in precipitation. This decline in production has been termed oligotrophication. On the other hand for well-buffered lakes it is possible that watershed inputs of sulfuric acid (a component of acid precipitation originating from fossil fuel combustion) may increase production (**Figure 3**).

For cropland systems the transfers of P to aquatic systems from the watershed are often accelerated compared to transfers from intact forests or grasslands. When annual crops are grown erosion can increase markedly. This increase in erosion is primarily associated with the periods when soils are not vegetated. Additionally, fertilizer P is often applied in excess of that removed in crops and over decades can lead to increased P in soils that can leach into surface waters. In many areas these relatively diffuse (nonpoint) P inputs are sufficient to cause extreme eutrophication of streams, lakes and wetlands located within agricultural watersheds. In order to allay this increased P transfer best management practices can be applied. These measures include decreased fertilizer application, altering the timing of fertilizer and manure application and construction of riparian buffer strips that can capture P leaving croplands before reaching streams.

The food requirements of urban areas are supplied from agricultural areas that may be thousands of miles away. The import of food is generally the primary P input to urban areas although P in cleaning products can also be significant. The destination of both these P imports is generally the sewage treatment plants of cities. In the 1960s essentially all of the P in sewage

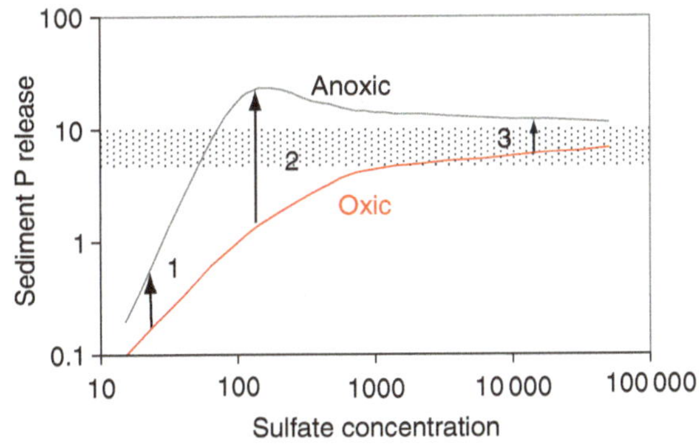

Figure 3 P release from sediments to the water column for systems with a large range of sulfate concentrations (in μM). P release is expressed relative to carbon release (P:C ratio in atoms times 100) and the general range of P:C for phytoplankton are shown by the stippled areas. General trends in relative P release are shown for both oxic (red line) and anoxic (grey line) conditions across the range of sulfate concentrations found among inland waters and three inland systems are labeled as examples. Systems 1 and 2, respectively, represent low and moderately high sulfate concentration freshwater systems. System 3 has sulfate that represents a saline (estuary or high sulfate saline lake) or highly sulfur polluted system (e.g., mine drain influenced). Notice for all three systems P release is higher under anoxic conditions but the among system release is greater for the higher sulfur system. Modified from Caraco *et al.* (1998).

was pumped into rivers, lakes and estuaries. Today the extent of P removal in these plants varies markedly; in general developed countries have more P removal than do developing countries. For developed countries the implementation of P removal from sewage has been associated with clear declines in dissolved available P and even total P in a number of lakes and rivers. For example, P concentrations in a number of rivers including the Thames and the Rhine have decreased to less than 50% of the maximum values observed in the 1970s coincident with changes in sewage treatment in the watershed. The fact that concentrations are still high compared with most pristine watersheds is related to high nonpoint loads that persist.

P Cycling and Retention in Inland Waters

Inland waters are active sites of storage and transformation of the P that enters these systems from point and nonpoint sources. There are three ultimate fates of all elements including P that enters a water body. The P can be stored in sediments (buried), exported down stream or lost to the atmosphere. For P there are very few examples where atmospheric loss is reported to be substantial. The two primary fates of P are hydrologic export and burial (**Figure 1**). The proportion of P that is lost to sediment burial and hence not transported down stream varies substantially between and within types of inland waters. Lakes can have some of the highest retentions (approaching 100%) while for streams and river stretches retention is generally the smallest. The between- and within-system variation in nutrient retention for inland waters is related to several factors. Perhaps the most important factor is residence time of the aquatic system. Systems with long residence times bury a higher percent of the P that enters the system and export a lower fraction of the P that enters. Other factors that can influence burial include a large number of physical, chemical and biotic variables. Some of these factors impact the delivery of P to sediments and other determine weather it remains buried in sediment or is returned to the water column.

Much of the P delivered to sediments is form of sinking particles. These particles include organic matter originating from phytoplankton production or from production in the watershed and inorganic particles (e.g., iron oxides). For either organic particles or inorganic particles the size and density of particles strongly influences sinking rates and hence the likelihood of particles reaching sediments. Phytoplankton which can be an important component of P delivered to sediments vary substantially in both size and density and also vary in mobility adding to great variation in sinking rate. Diatoms are often dense due to a silica shell and relatively large and sink rapidly compared to low-density cyanobacteria with gas vacuoles or small and mobile groups of plankton. The delivery of P in phytoplankton or other particulate matter can be enhanced by zooplankton grazing which can package small particles into larger relatively rapid-sinking fecal pellets. Just as phytoplankton vary in their capacity to sink so does the capacity of zooplankton vary in their capacity to package grazed and egested carbon into rapidly sinking packets. While pelagic grazers can alter the capacity of particles to sink to sediments benthic grazers such as bivalves can increase P delivery more directly. These organisms filter and remove particles from the water column some of the P is incorporated into their biomass and some is egested on the sediment surface. In the past decade the importance of P delivery to sediments by observing system changes following the introduction and establishment of the zebra mussel. In these systems grazing related P removal from the water column and delivery to sediments has contributed to a simultaneous decline in pelagic production by phytoplankton and increase of production by benthic algae.

Particulate matter that reaches sediments can return to the water column in dissolved form or by the re-suspension of particles. Re-suspension is likely in high-energy environments and in running waters can be a major fate of previously deposited P during floods. In lakes wave action generated by winds or in some cases human activity such as boating or dredging can be a major cause of re-suspension. Benthic organisms can act to stabilized sediments and prevent re-suspension or can increase re-suspension. Microbial and algal films can act as stabilizing elements as can macrophytes and bivalves that increase structural resistance at the benthic boundary and decrease hydrologic flow. On the other hand benthic-feeding fish such as carp can increase greatly the re-suspension of sedimented particles. Humans can also increase suspension in shallows directly through the use of motorized vehicles and indirectly by eliminating benthic plants and animals that previously stabilized sediments.

The particulate P that is not re-suspended but remains on sediments is subject to decomposition and chemical reaction associated with pH and oxidation-reduction differences within sediments. These reactions result in changes in the form of P but only sometimes result in an actual release of P to overlying waters. A classic example of these interactions takes phosphorus from particulate organic P to dissolved inorganic in porewater to P bound to particulate iron oxides at the sediment surface and finally to dissolved P released to bottom waters. More recently it has been demonstrated that the role of oxygen is mediated by a number of other factors including sulfate concentrations in the water (**Figure 3**). This interaction between sulfur and P may in part explain relatively high P concentrations found in many brackish to saline estuaries.

The link between oxygen concentrations in bottom waters and P release from sediments of many inland systems has the potential to generate a positive feed back between P loads and production. This positive feedback has been termed accelerated eutrophication and has been hypothesized to have occurred in both lakes and estuaries including the Baltic Sea. Briefly, P loads lead to greater primary production, which leads to greater sedimentation of organic matter and decomposition of this material resulting in lower oxygen in bottom waters. This lower oxygen in turn can lead to higher release of P from sediments. The link between P release and oxygen in bottom waters and the potential for a positive feedback accounts for the fact that bottom water aeration has been used to try to stop further eutrophication of water bodies or reverse ongoing eutrophication.

At the same time that P is entering and leaving the water column of inland aquatic systems there is an active cycle within the water column. Transfers of P occur within different parts of the water column. For many lakes, estuaries and reservoirs the most important transfer is between bottom waters and surface waters. As bottom waters and particularly anoxic bottom waters are often enriched in P compared with surface waters, inclusion of bottom waters in the surface zone can result in significant increases in P of surface waters. For north temperate lakes and reservoirs these inputs are generally largest in early fall and can result in fall phytoplankton blooms. In addition to locational transfers of P within the water column is the active cycling between forms that has been particularly well studied in well-lit surface waters of lakes. Dissolved inorganic P taken up by phytoplankton or bacteria is eventually released from the particulate organic fraction by the action of grazers or cell lysis that can be associated viral infection. The P released as dissolved organic P can be directly accessible to phytoplankton owing to the action of phosphatases or other enzymes or can be taken up by bacteria in organic form and used. What is notable about this food web study is the rapid rate of turnover of many of the P pools in surface waters. Phytoplankton and bacterial pools can turn over in less than a day and dissolved inorganic P can have turnovers of minutes to hours.

Continuing Research on the P Cycle

A large number of studies over the past 50 years have increased our understanding of P cycling within inland waters and the role of inland waters in landscape and global scale transfers of P. This research has shed light on the important role of P in inland waters, which resulted in the implementation of a several management practices to control P concentrations in both terrestrial systems and within inland waters themselves. Some of these practices have been very successful, and for a substantial number of lakes, reservoirs, estuaries and rivers, P concentrations have declined since the 1970s and along with these declines come a decrease in eutrophication. Problems of eutrophication persist, however, and in some cases we have seen new problems arising. Some of these changes may be associated with recent global climate change or species introductions. To date it is difficult to evaluate the relative importance of climate and species changes alone or their interaction with ongoing changes in land use and P management. The complex interactions that drive gradual or sudden changes in the P cycle of inland waters will continue to be studied in upcoming decades.

Glossary

Anoxic – Lacking oxygen.

Benthic – Sediment.

Benthos – Organisms that live on sediments of aquatic systems.

Eutrophic – High production generally associated with high nutrients.

Hypoxic – Low oxygen levels.

Macrophyte – Aquatic vascular plant.

Oligotrophic – Low production.

Phosphine – Common name for PH_3, a gaseous P compound.

Positive Feedback – A mechanism that enhances output.

Phytoplankton – Autotrophic organisms that inhabit the water of inland waters and oceans.

Redox – Short for oxidation–reduction, reactions where electrons are transferred between compounds.

See also: Nitrogen; Nitrogen Fixation.

Further Reading

Caraco NF, Cole JJ, and Likens GE (1989) Evidence for sulphate-controlled phosphorus release from sediments of aquatic systems. *Nature* 341: 316–318.

Carpenter SR, *et al.* (1998) Nonpoint pollution of surface waters with phosphorus and nitrogen. *Ecological Applications* 8: 559–568.

Emsley J (2001) *The Shocking History of Phosphorus: A Biography of the Devil's Element.* London: Pan Books.

Howarth RH, Jensen HS, Marion R, and Postma H (1995) Transport to and processing of P in near shore and oceanic waters. In: Tiessen H (ed.) *Phosphorus in the Global Environment.* SCOPE, Wiley.

Schindler DW (1977) Evolution of phosphorus limitation in lakes. *Science* 195: 260–262.

Scheffer M, *et al.* (1993) Alternate equilibria in shallow lakes. *Trends in Ecology & Evolution* 8: 275–279.

Vanni MJ (2002) Nutrient cycling by animals in freshwater ecosystems. *Annual Review of Ecology and Systematics* 33: 341–370.

Silica

D J Conley and E Struyf, Lund University, Lund, Sweden

Si Cycling at the Ecosystem Scale

Due to the importance of Si for structural support and the overall health of plant communities, it is considered an essential element for growth. Plants in terrestrial ecosystems take up DSi through their roots and deposit it as BSi in structures known as phytoliths. Through litterfall or when plants die, the BSi in plants is deposited in the upper organic soil layers. ASi can partly dissolve again into DSi and be subsequently recycled through the uptake of DSi by plants. DSi also reacts with Al to form insoluble hydroxyaluminosilicates or can inorganically precipitate as ASi in the soil horizon. Although most of the recyclable Si in terrestrial ecosystems is in the form of phytoliths, there is a variety of ASi compounds of various reactivity in soils. Mineral silicates dissolve very slowly compared to this ASi and can be considered unavailable for the growth of plants on biological timescales from minutes to years. It is, therefore, ASi recycling that drives the biological Si cycling in terrestrial ecosystems. It has been recently shown that the export of DSi from terrestrial environments to aquatic ecosystems is from the internal biogeochemical Si cycle in terrestrial ecosystems, rather than direct mineral silicate weathering, as was previously believed. Although all Si eventually derives from mineral Si weathering, mineral weathering replenishes the small amount of DSi leached from the biological ASi pool.

DSi enters aquatic ecosystems from terrestrial ecosystems through ground water, shallow soil water or surface run-off. Run-off from urban areas and domestic wastewater contain DSi concentrations in the range of 30–100 μM. This is low compared to the average DSi global riverine concentration of 150 μM. Current knowledge permits us to estimate that around 10% of the annual river load of DSi is derived from human inputs, either through urban runoff, industry, wastewater or through fertilizer application on specialized crops such as sugar cane using sodium meta-silicate. Direct inputs of DSi from human activity have not had as great an impact on DSi concentrations as N and P inputs from human activity have had on N and P concentrations.

ASi can enter natural waters from terrestrial ecosystems through overland runoff and the erosion of soils, which can contain high concentrations of ASi in phytoliths, especially in surface soils. ASi can also be produced within natural waters through the growth of organisms that contain Si, such as diatoms, chyrsophytes and sponges. ASi concentrations in river waters are often smaller than DSi concentrations. However, high ASi concentrations up to 100 μM have been observed in eutrophic waters and ASi can exceed DSi concentrations in natural waters through diatom uptake of DSi and deposition as ASi. Very high concentrations of ASi from biological origin (250–400 μM) have been found in tropical rivers.

Wetland ecosystems located along the land–ocean continuum also impact the transport of Si in natural waters. Wetland sedges and grasses have been identified as important accumulators of Si. *Oryza sativa* (rice) accumulates ASi in quantities up to 10% of the total dry weight. In commercial rice fields, this large accumulation can lead to DSi depletion of the soil, necessitating application of Si fertilizers such as sodium meta-silicate. *Phragmites australis* (the common reed) is probably the most ubiquitous natural wetland grass, with a typical biomass of around 1000 g dry weight per square meter. It accumulates BSi (80–120 g m^{-2} in tidal freshwater ecosystems) in amounts that are around the same order of magnitude as terrestrial ecosystems found in Amazon rain forests (the most productive terrestrial ecosystem in the world), although biomasses in rain forests exceed those of reed stands by several orders of magnitude. As such, wetland systems can exert a strong influence on Si transport in natural waters, through retention, burial and remineralization of ASi. Burial of plant phytoliths in sediments and soils can result in permanent removal of Si from the ecosystem, but wetland soil waters rich in DSi can also buffer DSi concentrations in adjacent rivers. The impact of wetland ecosystems on aquatic cycling of Si has received most attention in tidal marshes, probably because estuaries are often confronted with DSi limitation of diatoms. In the tidal freshwater marshes of the Scheldt River, Belgium, over 90% of the ASi pool was contained in *Phragmites* vegetation. The tidal exchange with the nearby eutrophic river waters results in a net sedimentation of suspended matter. Along with suspended matter, large amounts of diatom ASi are imported into the marsh ecosystem. Around 40% of the imported diatom ASi is permanently buried in the marsh sediments. The remaining 60% is recycled, as well as over 90% of the *Phragmites* reed BSi, resulting in enrichment of marsh pore waters with DSi. This enrichment of tidal freshwater marshes with DSi was found to play an important role in the buffering of the availability of DSi in periods of DSi depletion by diatoms. High export of DSi from the marshes was observed during summer when the lowest ambient DSi concentrations occurred in the estuary.

Silica Cycling in Natural Waters

Dissolved silica (DSi) concentrations in natural waters are strongly dependent on temperature and uptake by diatom. This results in seasonal concentration patterns of DSi (**Figure 1**). Diatoms occur in the upper surface layer where light penetration enables photosynthesis. In spring and summer, diatoms take up DSi and deposit it, forming an outer shell known as a frustule in the form of ASi. This annual growth of diatoms reduces DSi to low levels during summer. After a diatom dies, DSi is released again, along with other nutrients, through remineralization. However, N and P dissolve much faster as compared to Si, leading to a higher relative export of diatom ASi to deep waters, with a significant part of the diatom ASi settling on the bottom sediments. The highest DSi concentrations often occur in natural waters in the temperate zone during winter, when diatom production is low.

When diatom ASi settles at the bottom, it dissolves, increasing the pore water concentrations of DSi and reducing the dissolution of the settled diatom ASi. The export of DSi from the sediment is strongly dependent on ambient temperature and chemical processes such as inclusion of Al in the diatom frustule (which reduces the dissolution rate of the frustule)

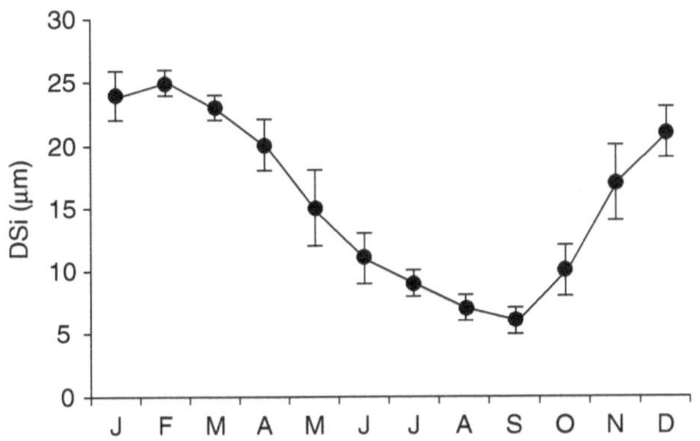

Figure 1 The annual cycle of dissolved silica in Lake Michigan. The highest DSi concentrations occur in the winter when biological production is the lowest, declining concentrations occur as diatoms reduce DSi concentrations during the spring diatom bloom with the lowest DSi concentrations during summer. Concentrations increase again when diatom production is at its minimum and remineralization of ASi to DSi continues. Modified from Rousar DC (1973) Seasonal and spatial changes in primary production and nutrients in Lake Michigan. *Water, Soil and Air Pollution* 2: 497–514.

and diffusion gradients. In addition, bioirrigation of the sediments by benthic organisms such as chironomids, e.g., the circulation of water in the upper sediments when organisms are irrigating their burrow structures, increases the export of DSi and enhances ASi dissolution. Although dissolution rates of ASi in aquatic sediments are in orders of magnitude below those observed in the water column, most of the ASi produced (up to 95%) is eventually recycled and returned to the water column as DSi. Still, 5% or more is permanently buried in lake sediments.

In shallow lakes, frequent mixing and resuspension of material deposited on the sediment–water interface causes higher recycling rates of DSi. A proportionally higher amount of matter located in deeper water with permanent sedimentation allows for a higher accumulation of ASi. Therefore, the amount of BSi which is permanently buried is also dependent on the morphological characteristics of a lake. The permanent loss of ASi in river ecosystems is generally considered small as most rivers do not have many permanent deposition areas. However, in rivers flowing through watersheds with many lakes, DSi concentrations are significantly lower due to uptake by diatoms and permanent ASi accumulation in sediments.

DSi Limitation

DSi is used and depleted from the upper lighted portion of the water column in natural waters by diatoms. When low concentrations of DSi are reached, the growth of diatoms is reduced, eventually leading to DSi concentrations that are limiting for their growth. The diatoms will then be replaced by other algae, such as green algae or cyanobacteria, that do not require DSi. The occurrence of diatom DSi limitation has been greatly enhanced by human eutrophication due to N and P inputs. In pristine ecosystems, N and P were usually available in limiting amounts compared to Si. Diatoms take up N, Si and P in a molar ratio of about 16:16:1 (N:P:Si). If N concentrations are below Si, and DSi concentrations exceed P concentrations by a factor 16, either N or P will be the limiting nutrient. This was usually the case before the anthropogenic perturbations of the N and P cycle. However, with increasing N and P inputs due to human activities, the occurrence of DSi limitation of diatoms has become more frequent. The consequent dominance of cyanobacteria and nonsiliceous algae is associated with a change in food quality at the bottom trophic levels and has significant negative effects on food webs and upper trophic levels. The increase in harmful algal blooms in estuaries over the past few decades may be due in part to lower Si availability. In addition, changes in the Si:P or Si:N ratios can have significant effects on the species composition of diatom communities through resource competition along resource gradients according to their relative abilities to grow under limitation by Si, N, P and light (**Figure 2**).

Human Influence on Si Cycling

Human activities have altered the biogeochemical Si cycle, both in terrestrial and aquatic systems. Fundamental changes have occurred in the landscape with man's activities. In addition, eutrophication of aquatic ecosystems has greatly altered the biogeochemical Si cycle in natural waters. Alteration by man of the aquatic biogeochemical Si cycle has been well documented through the effects of eutrophication and the enhanced accumulation of BSi in sediments, and therefore, lower concentrations in surface waters. The increased delivery of the essential plant nutrients N and P stimulates the growth of algae in aquatic ecosystems. In many cases, diatom growth is also stimulated. Subsequent sedimentation and burial of diatom frustules in sediments can remove large quantities of DSi from the water column in natural waters, causing a long-term decline in water column DSi concentrations, especially in systems with long residence times. Although first described in the North American Great Lakes (**Figure 3**), the generality of this mechanism can be applied to all aquatic ecosystems, both fresh water and marine.

The reduction of DSi transport by rivers with the construction of dams and impoundments is increasingly appreciated and has become known as the 'artificial lake effect.' Lakes act as nutrient traps and naturally retain some portion of nutrients transported along the aquatic continuum. The mechanism is simply that as one increases the residence time of water, the growth of diatoms is increased with subsequent sedimentation of diatoms behind the impoundment. Burial of diatoms and phytoliths in accumulating dam sediments removes them from the active biogeochemical Si cycle and thus contributes to the decline of DSi in rivers. Enhanced trapping of BSi behind the numerous dams that have been built along many rivers (**Figure 4**) and the associated long-term decline in water column DSi add to the effect of enhanced N and P input and increase vulnerability to DSi limitation. Construction of reservoirs is usually accomplished at the cost of flooding wetlands, reducing the potential of silica rich wetlands to buffer riverine DSi concentrations. Disconnecting the river from the riparian zone also alters the rates of weathering and recycling of Si and leads to reductions in DSi transport by rivers.

The effect of human activity on terrestrial Si transport has received too little scientific attention. We have only recently begun to understand that land use changes have changed Si fluxes from terrestrial ecosystems to natural waters. Enhanced losses of DSi from terrestrial organic matter reservoirs have

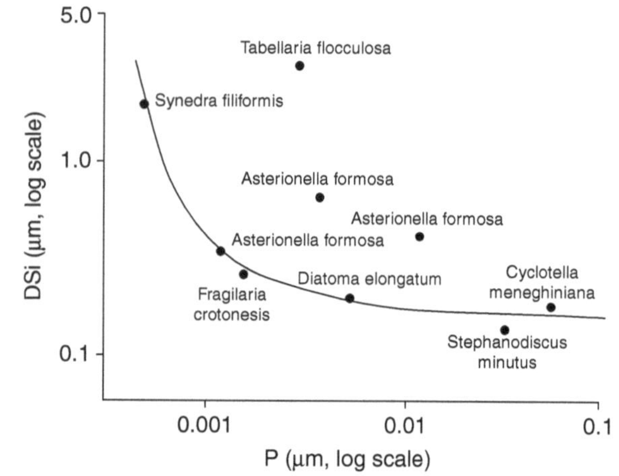

Figure 2 Interspecific tradeoffs in diatom competitive abilities for DSi and ortho-phosphate along resource gradients. Modified from Tilman D, Kilham SS, and Kilham P (1982) Phytoplankton community ecology: The role of limiting nutrients. *Annual Review of Ecology & Systematics* 13: 49–72.

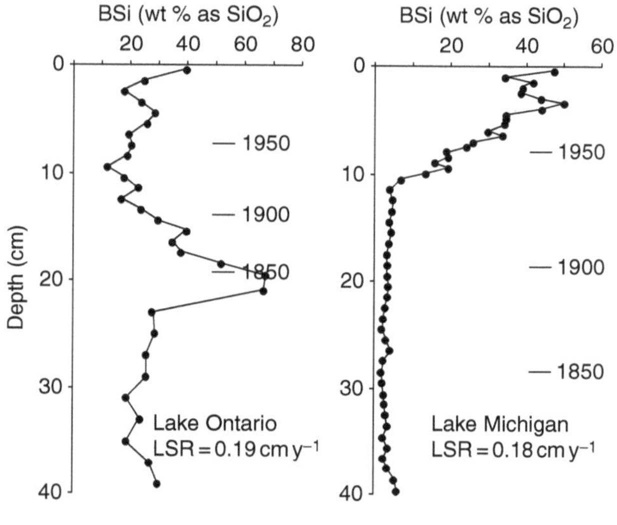

Figure 3 The sediment accumulation of biogenic silica (BSi) in two of the North American Great Lakes, Lake Michigan and Lake Ontario, with eutrophication. The enhanced accumulation of BSi occurred in Lake Ontario with widespread forest clearance and conversion to agriculture peaking in the period 1840–1850. Due to a smaller watershed and differences in soil types Lake Michigan did not respond as greatly to forest clearance, but rather large increases in BSi accumulation occurred with input of sewage and agricultural derived P. Modified from Schelske CL, Stoermer EF, Conley DJ, Robbins JA, and Glover R (1983) Early eutrophication in the lower Great Lakes: new evidence from biogenic silica in sediments. *Science* 222: 320–322.

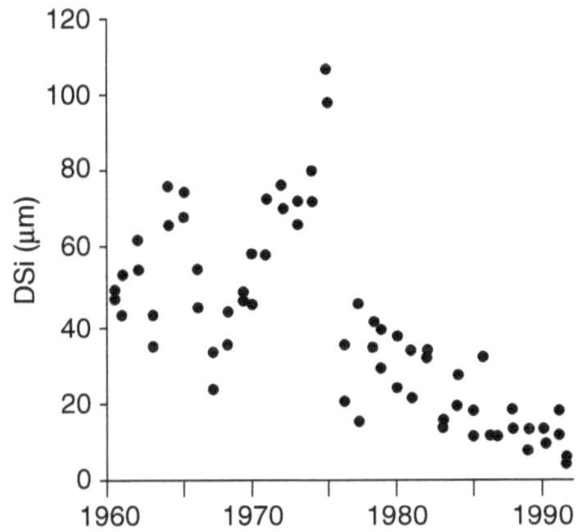

Figure 4 Reductions in dissolved silica (DSi) concentrations in the coastal zone of the Black Sea from reductions in the Danube River DSi concentrations following the construction of dams on the Danube River in the 1970s. Modified from Humborg C, Ittekkot V, Cociasu A, and Bodungen BV (1997) Effect of Danube River dam on Black Sea biogeochemistry and ecosystem structure. *Nature* 386: 385–388.

occurred through deforestation and conversion of terrestrial ecosystems to agriculture. Changing the vegetation will also change the cycling of Si in terrestrial ecosystems. Examination of runoff patterns following deforestation reveals that increases in DSi occur following forest clearance with the enhanced DSi transport continuing over a period of many decades. Disturbance of the soil with agriculture alters the biological, chemical and mineralogical properties of the soil and changes the compartmentalization of Si. This implies that fundamental changes have occurred in the terrestrial cycle of BSi with man's alteration of the landscape.

Continuing Research on the Si Cycle

We have only a limited understanding of the biogeochemical cycle of Si, especially in comparison with the other nutrient biogeochemical cycles. Rapid advances are presently occurring in our understanding with the recognition that there is an active terrestrial biogeochemical Si cycle with large stores of ASi in soils. The terrestrial biogeochemical Si cycle controls the inputs of both DSi and ASi to aquatic ecosystems, thus regulating the rates and quantities cycled in natural waters. Many basic questions remain to be answered regarding the cycling and inputs of Si to natural waters and the role of different ecosystems in regulating the flux of Si. Only recently has it become known that human modification of the terrestrial Si cycle is important, with large repercussions on long-term trends in Si cycling in natural waters.

Glossary

Amorphous silica (ASi) – Noncrystalline forms of hydrated silica containing Si and O bonds. Biogenic silica (BSi) is a type of ASi, but ASi also occurs as secondary weathering products in soils.

Biogenic silica (BSi) – An amorphous form of Si biogenically precipitated by a variety of aquatic organisms including chrysophytes, sponges, radiolarians, but most importantly by the diatoms. In addition, BSi is formed in plants.

Dissolved silicate (DSi) – Orthosilicic acid with the chemical formula of H_4SiO_4. DSi originates from the weathering of mineral silicates and is the chemical form used by organisms during biological uptake.

Mineral silicates – Highly ordered Si and O containing structures which react slowly on biological time scales. Si is the second most abundant element on Earth comprising 28% of the Earth's crust.

See also: Nitrogen; Phosphorus.

Further Reading

Anton SR (1983) *Silicon Geochemistry and Biogeochemistry*, 248pp. Academic Press.

Conley DJ, Schelske CL, and Stoermer EF (1993) Modification of silica biogeochemistry with eutrophication in aquatic systems. *Marine Ecololgy Progress Series* 101: 179–192.

Conley DJ (2002) Terrestrial ecosystems and the global biogeochemical silica cycle. *Global Biogeochemical Cycles* 16: 1121, doi: 10.1029/2002GB001894.

Humborg C, Ittekkot V, Cociasu A, and Bodungen BV (1997) Effect of Danube River dam on Black Sea biogeochemistry and ecosystem structure. *Nature* 386: 385–388.

Humborg C, Blomqvist S, Avsan E, Bergensund Y, and Smedberg E (2002) Hydrological alterations with river damming in northern Sweden: Implications for weathering and river chemistry. *Global Biogeochemical Cycles* 16: 1039, doi: 10.1029/2002GB001369.

Kilham SS, Theriot EC, and Fritz SC (1996) Linking planktonic diatoms and climate in the large lakes of the Yellowstone ecosystem using resource theory. *Limnology and Oceanography* 41: 1052–1062.

Rousar DC (1973) Seasonal and spatial changes in primary production and nutrients in Lake Michigan. *Water, Soil and Air Pollution* 2: 497–514.

Schelske CL, Stoermer EF, Conley DJ, Robbins JA, and Glover R (1983) Early eutrophication in the lower Great Lakes: new evidence from biogenic silica in sediments. *Science* 222: 320–322.

Struyf E and Conley DJ (2008) Silica: An essential nutrient in wetland biogeochemistry. *Frontiers in Ecology & Environment* In press.

Tilman D, Kilham SS, and Kilham P (1982) Phytoplankton community ecology: The role of limiting nutrients. *Annual Review of Ecology & Systematics* 13: 49–72.

Salinity

G Harris, Lancaster University, UK

The hydrological cycle in watersheds includes both the 'green' water – rainfall, evaporation, soil moisture and ground waters – as well and the 'blue' water in streams, rivers, and lakes (Falkenmark, 1998). From hill slope to the sea, water falling as rain flows across and through the landscape carrying dissolved inorganic and organic matter and modifying both its own composition and that of the landscape as it does so. Water is both a conveyor of materials in a watershed as well as a reactant in the weathering cycle (Stumm and Morgan, 1996). The water in lakes, rivers, and streams is in intimate contact with the rocks, vegetation, and soils of the watersheds in which they lie. The total dissolved matter in water is therefore a function of the geology of the watershed and the rate of water movement through surface runoff and groundwater flow. Waters flowing from heavily weathered tropical watersheds, which arise in regions of high seasonal rainfall (and hence high runoff; e.g., the Amazon river in South America) are usually quite dilute and may reflect the ionic composition of the rainfall. On the other hand waters flowing in arid or semiarid watersheds (where the runoff is low because of low rainfall and high evaporation rates; e.g., the Colorado river in the USA or the Murray Darling river system in Australia) may be quite concentrated by evaporation and modified by the geology of the watershed.

Saline inland waters differ markedly from the freshwater systems that are the traditional domain of limnology and freshwater biology (Williams, 2000). Saline waters are common in arid areas on all continents, even Antarctica (Burton, 1981; Laybourn-Parry *et al.*, 2002). Dryland areas occupy about 50% of the continents and are home to upwards of one billion people. Dry regions have some of the World's largest lakes and reservoirs (Williams, 2000), some of which may be highly saline: for example the salinity of the 'Big Sea' of the Aral Sea is $48\,\mathrm{g\,l^{-1}}$ – about one and a half times that of seawater (Létolle and Cherterikoff, 2000). Saline lakes and reservoirs are frequently monomictic or meromictic and possess a characteristic depauperate biota.

Salinity is defined as the weight in grams of the dissolved inorganic matter in one kilogram of water (Stumm and Morgan, 1996). It is therefore expressed as S‰ – in parts per thousand. Seawater has a remarkably constant salinity between 33‰ and 37‰. Salinity is also commonly measured by the electrical conductivity of the water, or more correctly its specific conductance. The units commonly used are either Siemens.$\mathrm{cm^{-1}}$ or mhos.$\mathrm{cm^{-1}}$. Distilled water has a very low conductivity – about $1\,\mu\mathrm{S\,cm^{-1}}$ or $10^{-6}\,\mathrm{Ohm^{-1}\,cm^{-1}}$. ($1\,\mu\mathrm{S\,cm^{-1}}$ is equivalent to 1 electrical conductivity unit [EC unit].) There is no precise conversion factor from specific conductance to total dissolved salts (TDS), if the total dissolved salts are expressed as mg $\mathrm{l^{-1}}$ or ppm then the conversion factor from EC units to TDS lies between 0.55 and 0.9 depending on the precise ionic composition of the water in question. Once the ionic composition of the water is known then the conductivity is proportional to concentration.

Gibbs (1970) examined the major factors controlling world water chemistry. He identified three major factors: first, precipitation dominance in dilute, low salinity waters where Na^+ and Cl^- dominated the ionic composition. This tends to be found in tropical regions in watersheds dominated by igneous rocks where the TDS composition was similar to that of seawater and the concentration was of the order of 10 ppm (e.g., the Rio Negro). Second, rock dominance in medium salinity waters where the waters were in equilibrium with the underlying geology and soils and where Ca^{2+} and $HCO3^-$ dominated. Here TDS concentrations were of the order of 200 ppm (e.g., the Columbia, Nile, Rhine and Indus Rivers). Last, high salinity waters with high total dissolved salts concentration (2000 ppm to seawater concentrations of 35 000 ppm, or higher) where a process of evaporation and precipitation once again led to Na^+ and Cl^- dominance. Gibbs noted that there were regional differences in salinity across the continents. In the high rainfall tropical regions of the old weathered continents (e.g., South America, Africa, and Australia) Na^+ and Cl^- dominated the low salinity river and surface waters. Similarly, Na^+ and Cl^- dominated the high salinity surface waters in arid regions of the subtropics, where evaporation rates are high. In regions where the surface rocks and soils were dominated by sedimentary rocks and limestones and evaporation rates were low (e.g., Europe and North America) Ca^{2+} and $HCO3^-$ dominated the ionic composition.

Radke *et al.* (2002) have explained the sequence of events that occurs in the evaporation of inland waters in terms of the Eugster–Jones–Hardie model of evaporation and precipitation (Drever, 1982) so that initial waters dominated by divalent cations (e.g., Ca^{2+}) may, through a stepwise process of concentration and precipitation, become to be monovalent cation (Na^+ and Cl^-) dominated. Branch points occur depending on the stoichiometeric ratios of major ions when calcite/aragonite, gypsum and dolomite/Mg-silicate are precipitated.

Ecological Responses

Although it is a simple matter to measure the EC of a water sample, interpreting what it means ecologically is a more complex task. The interactions of changing salinity with the biota are quite complex. The salinity of seawater is quite constant and most marine organisms are termed stenohaline – they have a narrow tolerance range for variability in salinity around 35‰. Similarly freshwater organisms are also stenohaline – they have a narrow tolerance range at low salinity levels. Estuarine organisms, and some of those that live in saline inland waters, are what is termed euryhaline and can tolerate quite large and rapid changes in salinity. Fresh waters are usually defined as waters with salinity below 3‰ (3000 ppm) although as a result of evaporation and saline inflows values as high as 60–100‰ (60 000 to 100 000 ppm) may be found in inland waters in arid regions; up to three times more saline than seawater. Stenohaline freshwater organisms are adapted to salinities in the region from 0–3‰ (Nielsen *et al.*, 2003). A survey of the biota of 230 wetlands of varying salinity by Pinder *et al.* (2005) showed little effect of salinity below 4.1‰ followed by a rapid decline in biodiversity at higher concentrations. If the euryhaline halophytic species (which occurred in the salinity range from 3–10‰) were removed from the analysis then biodiversity of the stenohaline freshwater invertebrates began to decline at 2.6 ‰. The biodiversity of highly saline inland waters is quite low and is usually dominated by a small number of stenohaline salt tolerant species, most notably species of cyanobacteria (e.g., *Arthrospira (Spirulina)* sp., *Nodularia* sp.), algae (e.g., *Dunalliella* sp.), small crustacea (e.g., *Artemia* sp.) and a few higher plants (e.g., *Ruppia* sp.). Radke *et al.* (2003) showed that as the salinity of inland waters rose owing to evaporation and precipitation (following the Eugster–Jones–Hardie model) then the occurrence of ostracods followed specific pathways of the model. Pathways of concentration and precipitation follow the ratios of Na^+/H^+, Na^+/Ca^{2+} and alkalinity/Cl^- and divided the ostracods into freshwater, transitional, halophile, and halobiont (stenohaline) groupings.

Hammer *et al.* (1990) found a steep decline in the biodiversity of benthic organisms as salinity increased in prairie lakes in Canada (see also Hammer, 1986). Williams *et al.* (1990) examined the response of organisms to increasing salinity in 79 lakes in western Victoria, Australia, and showed that the number of species declined less steeply than in Canada, particularly at higher salinities >10‰. While stenohaline freshwater organisms declined sharply at low salinities, there was a group of euryhaline species that spanned a wide range of habitats and salinities (Williams *et al.*, 1990). Clearly for euryhaline species salinity is not the primary determinant of success. In accord with this conclusion, Wood and Talling (1988) found that salinity was not the primary determinant of occurrence for a group of phytoplankton species in a series of Ethiopian inland waters varying in salinity.

More recent work at the high salinity end of the spectrum (from 10‰ to >35‰) has uncovered a range of specialized salt tolerant species that may be regionally diversified so that biodiversity in saline lakes is a matter of scale and evolutionary history (Williams *et al.*, 1990, Timms, 2007). Biodiversity may fall sharply with salinity in a subset of lakes – and in some cases many species may be absent (Seaman *et al.*, 1991) – but if biodiversity is studied in a wide geographical range of lakes then biogeographical factors and the evolutionary history ensures that the decline in species number is slower and regionally specific. In this way the history of isolation and evolution has a significant role to play in the ecological response of inland waters.

So deep time plays a role in what we see today. Nielsen *et al.* (2003) and Timms (2007) reviewed the effects of changing salinity on organisms in inland waters in Australia and concluded that the response of major groups depended on their evolutionary history. Many stenohaline freshwater phytoplankton, higher plants, and micro-invertebrates were quite sensitive to increased salinity because of a long evolutionary history in fresh waters. Biodiversity was reduced at salinities above 2–3‰. Australian macro-invertebrate and fish populations on the other hand, many of which have a more recent marine evolutionary history, are euryhaline, tolerant of increased salinity and regionally diversified. Macro-invertebrates and fish species have the ability to tolerate salinities in the 9–10‰ range or even higher, although Nielsen *et al.* (2003) note that some species have particularly sensitive life history stages. Adult fish may tolerate salinities up to 30‰ while their eggs fail to develop at salinities over 4‰. Desiccation is also a key factor in arid regions: while some species may tolerate high salinities they are unable to tolerate drying out completely (Timms, 2007). In arid inland regions survival of many euryhaline species may be very dependent on the persistence of small refuges during dry periods and may also reflect biogeographical distributions that arose during wetter periods in the distant past (Timms, 2007). Water birds, because of their migratory abilities are not so regionally differentiated, but they do show continental scale biogeographical patterns of distribution. Water birds migrate very long distances to find water depending on rainfall patterns in any particular year and saline lakes and wetlands may be important refigures on migratory flyways (Andrei *et al.*, 2006).

As noted earlier, increased TDS in waters in arid subtropical regions may be associated with dominance by monovalent cations (Na^+ and K^+) together with a preponderance of Cl^- resulting from long periods of rainfall, evaporation, and salt deposition. Early work by Pearsall (1932) on the effects on phytoplankton of changing the monovalent to divalent cation ratios in inland waters showed that this ratio influenced the major groups of plankton present. Diatoms appear to be most common in waters dominated by divalent ions, whereas desmids seem to occur most commonly in waters dominated by monovalent ions. Zooplankton appear to be equally sensitive to fluctuations in the ratio of $(Na^+ + K^+)/(Ca^{2+} + Mg^{2+})$ in inland waters (see e.g., Radke *et al.*, 2002).

In addition to the physiological effects of the changing ratio of $(Na^+ + K^+)/(Ca^{2+} + Mg^{2+})$ there are also effects on the coagulation of clays in the water. In dilute Na^+ dominated waters clay particles are slow to coagulate and settle, thus leaving very turbid waters in which there are sufficient particles in suspension to significantly influence light penetration and the availability of phosphorus and other elements (Nielsen *et al.*, 2003). The salinity of inland waters is rarely constant; levels may decrease during periods of high flows when salinity is diluted by runoff and increase during low flow periods as a result of evaporation and groundwater intrusions. Saline groundwater intrusions may be sufficiently concentrated as to induce density stratification in the water column, restricting the mixing depth and tending to lead to anoxic bottom waters in eutrophic situations. Organisms must therefore cope with fluctuations in salinity over space and time.

The salinity of inland waters, and particularly the salinity of saline ground water inflows into bottom waters, strongly influences the exchange of minor elements and nutrients between the sediments and the water column. Not only do microbial populations change along salinity gradients (Bouvier and del Giorgio, 2002, Del Giorgio and Bouvier, 2002) but the exchange of key nutrients such as phosphorus is critically determined by the concentrations of Fe and SO_4^{2-} in sediment pore waters and in bottom waters (Roden and Edmonds, 1997). The biogeochemistry of fresh waters and high SO_4^{2-} marine and saline waters is quite different because of stoichiometric influences on the major groups of organisms and on nutrient cycling (Harris, 1999; Sterner and Elser, 2002). A further feature of saline inland waters is the widespread occurrence of benthic bacterial and cyanobacterial mats (Bauld, 1981), which may be present because of a relative paucity of herbivores in these saline environments. These are the inland water equivalents of the stromatolites known from hypersaline coastal waters and from the fossil record. Hepatotoxins and neurotoxins produced by these cyanobacterial mats have been implicated in the deaths of Flamingos feeding in the saline lakes of East Africa (Krienitz *et al.*, 2003).

Recent work on the microbiology of saline inland waters has revealed a highly unusual group of bacteria and archaea that are polyextremophiles. Species of *Natranaerobius* and *Halonatronum* are able to grow in high temperature, alkaline pH, high NaCl concentration, and anoxic environments. Genomic and proteomic studies will reveal how these unusual organisms evolved to cope with the stresses of living in multiple extreme conditions (Mesbah and Wiegel, 2008).

Salt as a Conservative Tracer in Watersheds

Of all the major ions in inland waters chloride shows the most conservative behavior, so cyclical salts – derived from seawater carried inland and deposited as rainfall – may be used as a tracer of watershed function and to study the links between 'green' and 'blue' water flows (Simpson and Herczeg, 1994). Neal and his co-workers have long studied the movement of sea salt in the Plynlimon catchment in Wales using Cl^- as a conservative tracer of water movement. The results are a series of paradoxical conclusions that are set to revolutionize our understanding of watershed hydrology (Neal, 1997). Kirchner's analysis of Neal's data (Kirchner *et al.*, 2000) showed that the travel times of 'green' water exhibited fractal properties and turned the essentially 'white' noise spectrum of the Cl^- signal in the rainfall (equal amounts of variability across all input frequencies) into a 'pink' spectrum in the 'blue' water with variability inversely proportional to the frequency in the streams draining the watershed. There were delayed pathways of water flow in the ground waters of the watershed and, on occasion, rainfall displaced 'old' water from the watershed through the interaction of rainfall, past wetting histories and groundwater discharges. The signal of Cl^- in the stream draining the watershed showed very complex variability over a range of frequencies resulting from mixtures of delayed 'old' water and fresh 'new' water draining from the watershed by flow paths that varied in time and space (Kirchner, 2003; Feng *et al.*, 2004; Bishop *et al.*, 2004). The ecological significance of this is that the organisms in the streams and rivers see a spectrum of variability in salinity and water quality over time (Nielsen *et al.*, 2003), transformed from the rainfall input spectrum by watershed properties (Harris and Heathwaite, 2005).

New techniques of collecting high frequency data (e.g., 10 min sampling intervals) using water quality electrodes, wireless connectivity and web based data analysis systems are set to revolutionize our knowledge of the behavior of 'green' and 'blue' water in watersheds. We face a future high frequency data wave as new technologies for monitoring water quality come on line (Kirchner *et al.*, 2004). These data reveal hitherto unexpected aspects of pattern and process in watersheds – aspects certainly not resolved by the usual method of collecting data weekly or at less frequent intervals (Harris, 2007).

Dryland Salinity and Salinization of Inland Waters

Salinity in inland waters may be defined either as primary salinity – where the salinity is natural due either to sources of salt in watersheds or to high evaporation rates, or both – or secondary salinity where the salinity has arisen or increased due to human activity in the watershed (Williams, 1999). One simple mechanism which causes the salinity of inland waters to rise is diversion of inflowing streams and rivers for urban and irrigation water supplies. Classic examples of increased salinity due to river diversion include the Dead Sea (Middle East), Aral Sea (Asia), Pyramid Lake (Nevada, USA) and Mono Lake (California, USA). In all cases salinity rose markedly over a period of decades due to water extraction and the disturbance of the hydrological regime (Williams, 1999). The only remedy is to cease or reduce water extraction so as to (partially) restore the original hydrological balance. Many naturally saline water bodies occur in endorheic watersheds – where there is no outflow and evaporation is the primary means of water loss (e.g., Lake Eyre, Australia). In these lakes and streams salinity is reduced by dilution during periods of rainfall and inflow followed by increased salinity as the water evaporates.

Perhaps the most pressing applied issue concerning the secondary salinity of inland waters is that arising from human modification of the vegetation in watersheds in arid zones (Williams, 1999). Many arid regions – where the potential evaporation rate exceeds the rainfall – have, over long time periods, amassed large stores of salt in their soils. These salts, dominated by NaCl, are what are called cyclical salts because they arise from the deposition of marine salts by rainwater (Herczeg *et al.*, 2001). Over millennia the water evaporates and leaves the salt stored in the soil profiles. There is good evidence that over evolutionary time an equilibrium is established between rainfall and the evapotranspiration of the natural perennial vegetation leading to highly efficient 'green' water use and low infiltration (Harris, 2007). Naturally diverse perennial vegetation consists of many species with differing and complementary growth and rooting strategies which combine to make the most effective use of the scarce water (see e.g., Pate and Bell, 1999). Human action to clear vegetation for agriculture replaces the native perennial vegetation with annual or perennial crops with lower water use efficiencies which causes an increase in water infiltration. Irrigation of these annual or perennial crops (e.g., fruit and citrus trees) also further increases the infiltration of water into the soil profile. The result is mobilization of the salt in the soil profile, the salinization of ground waters and increasing salinity in streams, rivers and lakes (Simpson and Herczeg, 1994). This is now a widespread problem in arid regions of Asia, southern Africa, the Americas and Australia. Rivers such as the Colorado and the Murray Darling, and lakes such as the Aral Sea now show increasing salinity as cyclical salts are fed into the rivers by groundwater movement. The salinity of rivers feeding the Aral Sea has risen seven-fold in the last century since agriculture was developed in the watershed (Létolle and Chesterikoff, 1999). Infiltration and groundwater recharge is sporadic in time when rainfall occurs and, as noted above, the travel times of ground waters in the watershed modify the output of salt. The effects of saline ground waters are often most pronounced during low flow periods when surface runoff ceases and the rivers are largely fed by saline ground waters.

Over time the salt is slowly moved out of the soil profile and returned to the sea from whence it came (hence the term cyclical salts) and a new equilibrium is reached. Times to equilibrium can be long however: 300–400 years at least in watersheds with low relief and low rainfall to as little as 50 years in steeper watersheds. Thus, many watersheds which have been colonized by western agriculture in the last 200 years are still salinizing and the rivers are slowly responding to the new hydrological balance (Jolly *et al.*, 2001).

Increasing salinity in these rivers leads to the slow elimination of freshwater biodiversity, including the death of trees and wetland plants, the elimination of fish, and stimulation of toxic cyanobacterial blooms. One good sign of a dying wetland is the occurrence of dead or dying trees and wetland plants at the margins (Williams, 1999). In the lower reaches of the Murray Darling system in Australia an entire cohort of 250-year-old River Red Gum trees has been killed in the last decade by a combination of excessive water extraction and prolonged drought.

In many arid parts of the world the river systems are flood plain rivers where flow variability is large and the entire system of river channels, wetlands and floodplains depends on irregular flooding events to maintain the health and biodiversity of the system. River regulation and water extraction for stock, domestic, and irrigation water supplies reduces both the natural flow variability of the river systems and the number of over bank floods. Floodplain ecosystems that are not frequently inundated and flushed with fresh water tend to become salinized and the health of the vegetation declines over time. Remediation of these inland waters is either through reductions in water extractions – thus restoring a more natural flow regime – or by artificial environmental watering regimes where engineering structures employ smaller amounts of water to ensure flooding of flood plain ecosystems (see, for example, http://thelivingmurray.mdbc.gov.au/). Also, large-scale catchment revegetation programs are underway to replace perennial vegetation (trees and crops such as Lucerne) in an effort to reduce groundwater recharge and salt movement. Such programs are supported by computer modeling in an effort to optimize plantings and achieve the maximum return on investment (see e.g., Tuteja *et al.*, 2003).

Further Reading

Andrei AE, Smith LM, Haukos DA, and Surles JG (2006) Community composition and migration chronology of shorebirds using the saline lakes of the Great Plains, USA. *Journal of Field Ornithology* 77: 372–383.

Bauld J (1981) Occurrence of benthic microbial mats in saline lakes. *Hydrobiologia* 81: 87–111.

Bishop K, Seibert J, Kohler S, and Laudon H (2004) Resolving the double paradox of rapidly mobilized old water with highly variable responses in runoff chemistry. *Hydrological Processes* 18: 185–189.

Bouvier TC and del Giorgio P (2002) Compositional changes in free-living bacterial communities along a salinity gradient in two temperate estuaries. *Limnology and Oceanography* 47: 453–470.

Burton HR (1981) Chemistry, physics and evolution of Antarctic saline lakes. *Hydrobiologia* 82: 339–362.

Del Giorgio P and Bouvier TC (2002) Linking the physiologic and phylogenetic successions in free-living bacterial communities along an estuarine salinity gradient. *Limnology and Oceanography* 47: 471–486.

Drever JJ (1982) *The Geochemistry of Natural Waters*. 2nd edn. New Jersey: Prentice Hall.

Falkenmark M (1998) Dilemma when entering the 21st century: Rapid change but lack of sense of urgency. *Water Policy* 1: 421–426.

Feng X, Kirchner JW, and Neal C (2004) Measuring catchment-scale retardation using spectral analysis of reactive and passive tracer time series. *Journal of Hydrology* 292: 296–307.

Gibbs RJ (1970) Mechanisms controlling world water chemistry. *Science* 170: 1088–1090.

Hammer UT (1986) *Saline Lake Ecosystems of the World*. Dordrecht: Junk.

Hammer UT, Sheard JS, and Kranabetter J (1990) Distribution and abundance of littoral benthic fauna in Canadian Prairie saline lakes. *Hydrobiologia* 197: 173–192.

Harris GP (1999) Comparison of the biogeochemistry of lakes and estuaries: Ecosystem processes, functional groups, hysteresis effects and interactions between macro- and microbiology. *Marine and Freshwater Research* 50: 791–811.

Harris GP (2007) *Seeking Sustainability in an Age of Complexity*. Cambridge, UK: Cambridge University Press.

Harris GP and Heathwaite AL (2005) Inadmissible evidence: Knowledge and prediction in land and waterscapes. *Journal of Hydrology* 304: 3–19.

Herczeg AL, Dogramaci SS, and Leaney FWJ (2001) Origin of dissolved salts in a large, semi-arid groundwater system: Murray Basin, Australia. *Marine and Freshwater Research* 52: 41–52.

Jolly ID, Williamson DR, Gilfedder M, *et al.* (2001) Historical stream salinity trends and catchment salt balances in the Murray-Darling basin, Australia. *Marine and Freshwater Research* 52: 53–64.

Kirchner JW (2003) A double paradox in catchment hydrology and geochemistry. *Hydrological Processes* 17: 871–874.

Kirchner JW, Feng X, and Neal C (2000) Fractal stream chemistry and its implications for contaminant transport in catchments. *Nature* 403: 524–527.

Kirchner JW, Feng X, Neal C, and Robson AJ (2004) The fine structure of water quality dynamics: the (high frequency) wave of the future. *Hydrological Processes* 18: 1353–1360.

Krienitz L, Ballot A, Kotut K, Wiegand C, Pütz S, Metcalf JS, Codd GA, and Pflugmacher S (2003) Contribution of hot spring cyanobacteria to the mysterious deaths of Lesser Flamingos at Lake Bogoria, Kenya. *FEMS Microbiology Ecology* 43: 141–148.

Laybourn-Parry J, Quayle W, and Henshaw T (2002) The biology and evolution of Antarctic saline lakes in relation to salinity and trophy. *Polar Biology* 25: 542–552.

Létolle R and Chesterikoff A (1999) Salinity of surface waters in the Aral Sea region. *International Journal of Salt Lake Research* 8: 293–306.

Mesbah NM and Wiegel J (2008) Life at extreme limits. The anaerobic halophilic alkalithermophiles. *Annals of the New York Academy of Sciences* 1125: 44–57, doi: 10.1196/annals.1419.028.

Neal C (1997) A view of water quality from the Plynlimon watershed. *Hydrology and Earth System Sciences* 1: 743–753.

Nielsen DL, Brock MA, Rees GN, and Baldwin DS (2003) Effects of increasing salinity on freshwater ecosystems in Australia. *Australian Journal of Botany* 51: 655–665.

Pate JS and Bell TL (1999) Application of the ecosystem mimic concept to species-rich Banksia woodlands of Western Australia. *Agroforestry Systems* 45: 303–341.

Pearsall WH (1932) Phytoplankton in the English Lakes. II. The composition of the phytoplankton in relation to dissolved substances. *Journal of Ecology* 20: 241–262.

Pinder AM, Halse SA, McRae JM, and Shiel RJ (2005) Occurrence of aquatic invertebrates of the wheatbelt region of Western Australia in relation to salinity. *Hydrobiologia* 543: 1–24.

Radke LC, Howard KWF, and Gell PA (2002) Chemical diversity in south-eastern Australian saline lakes I: geochemical causes. *Marine and Freshwater Research* 53: 941–959.

Radke LC, Juggins S, Halse SA, De Dekker P, and Finston T (2003) Chemical diversity in south-eastern Australian saline lakes II: Biotic implications. *Marine and Freshwater Research* 54: 895–912.

Roden EE and Edmonds JW (1997) Phosphate mobilisation in iron-rich anaerobic sediments: microbial Fe(III) oxide reduction

versus iron-sulphide formation. *Archiv fur Hydrobiologie* 139: 347–378.
Seaman MT, Ashton PJ, and Williams WD (1991) Inland salt waters of southern Africa. *Hydrobiologia* 210: 75–91.
Simpson HJ and Herczeg AL (1994) Delivery of marine chloride in precipitation and removal by rivers in the Murray-Darling Basin, Australia. *Journal of Hydrology* 154: 323–350.
Sterner RW and Elser JJ (2002) *Ecological Stoichiometry: The Biology of Elements from Molecules to the Biosphere.* Princeton, NJ: Princeton University Press.
Stumm W and Morgan JJ (1996) *Aquatic Chemistry: Chemical Equilibria and Rates in Natural Waters.* New York: Wiley.
Timms BV (2007) The biology of saline lakes of central and eastern Australia: A review with special reference to their biogeographical affinities. *Hydrobiologia* 576: 27–37.
Tuteja NK, Beale G, Dawes W, *et al.* (2003) Predicting the effects of land use change on water and salt balance – a case study of a catchment affected by dryland salinity in NSW, Australia. *Journal of Hydrology* 283: 67–90.
Williams WD (1999) Salinisation: a major threat to water resources in the arid and semi-arid regions of the world. *Lakes and Reservoirs: Research and Management* 4: 85–91.
Williams WD (2000) Dryland lakes. *Lakes and Reservoirs: Research and Management* 5: 207–212.
Williams WD, Boulton AJ, and Taaffe RG (1990) Salinity as a determinant of salt lake fauna: A question of scale. *Hydrobiologia* 197: 257–266.
Wood RB and Talling JF (1988) Chemical and algal relationships is a salinity series of Ethiopian inland waters. *Hydrobiologia* 158: 29–67.

ORGANIC COMPOUNDS: CYCLES AND DYNAMICS

Contents

Allelochemical Reactions
Carbohydrates
Carbon, Unifying Currency
Dissolved Humic Substances: Interactions with Organisms
Interactions of Dissolved Organic Matter and Humic Substances
Lipids
Methane
Natural Organic Matter
Organic Nitrogen
Nutrient Stoichiometry in Aquatic Ecosystems
Redox Potential

Allelochemical Reactions

E M Gross, University of Konstanz, Konstanz, Germany

Introduction

Allelochemical reactions are of major significance in the adaptation of species and organization of communities. Allelochemicals are mostly so-called 'secondary metabolites' produced by organisms such as plants, animals, or microorganisms, and which are not needed for basic (primary) metabolism. Rather, they have ecological functions to counterbalance abiotic and biotic stressors. Allelochemicals affect other organisms, either in their physiology, growth, and behavior or life history. Effects range from stimulation to regulation and inhibition. Some allelochemicals are released by one organism and taken up or recognized by another organism. Other allelochemicals are located within the producing organism (i.e., a plant or algae) and affect grazers, herbivores, or pathogens upon damage. Allelochemical interactions should only encompass effects caused by allelochemicals, and should exclude any resource interference or trophic interactions. While evidence for the involvement of allelochemicals under controlled (laboratory) conditions is possible, often in situ proof remains difficult.

A significant difference between allelochemical reactions in terrestrial and aquatic systems lies in the medium in which active compounds have to be transported and have a sufficient stability. Many terrestrial allelochemicals are transported through air, thus are usually of low molecular weight and volatile. Such compounds can easily be identified by gas chromatography combined with mass spectrometry. More difficult, not only in aquatic but also in terrestrial systems, is the identification of hydro- or amphiphilic allelochemicals transported by water or in soil, especially when phenolic compounds exhibiting a high reactivity are involved. Analytical tools to identify hydro- or amphiphilic allelochemicals are much less developed, and also have not received the same intensity of research as lipophilic, low molecular compounds. As a consequence, most well-described aquatic allelochemicals are still not identified in their structure, despite intensive research yielding at least a preliminary characterization of their chemical properties.

Terminology

Allelochemical reactions between organisms are manifold and evoked a plethora of terms and definitions.

1. Kairomones are compounds produced by one organism and sensed by and beneficial to another organism. Allomones are compounds that are beneficial to the producing organism. Synomones evoke responses favorable to both organisms.
2. Enemy-avoidance kairomones warn potential prey for predators. Foraging kairomones enable predators, grazers, and herbivores to locate food.
3. Feeding deterrents occur in plants, algae, cyanobacteria, or other organisms and cause a behavioral reaction of grazers and herbivores leading to reduced consumption or complete avoidance of this food.

4. Inhibitors or toxins affect physiological processes in consumers ranging from sublethal to mortal effects.
5. Allelopathically active compounds comprise plant, algal, or cyanobacterial secondary metabolites negatively (or positively) affecting other primary producers.
6. Allelochemicals comprising information for the receiving organism are also termed infochemicals or chemical cues.

Experimental Approaches Studying Allelochemical Reactions

The investigation of allelochemical reactions not only in aquatic systems requires an interdisciplinary approach. Prerequisites are sound observations of phenomena in the field and successful transfer of the biotic interactions to a laboratory bioassay. The identification of allelochemicals requires sophisticated chemical analytical techniques (extraction procedures, solid phase extraction (SPE), chromatography (high-performance liquid chromatography – HPLC, gas chromatography/mass spectrometry – GC/MS, nuclear magnetic resonance spectroscopy – NMR) and excellent working bioassays. Apart from a few well-studied systems such as fish kairomone(s) and cyanobacterial toxins, many reports on aquatic allelochemical interactions are more or less anecdotal. This and an in part inconsequent use of terms make it difficult to find all known allelochemical reactions in common literature databases.

Range of Allelochemical Reactions

Allelochemical reactions occur between almost all groups of organisms and affect all biotic interactions such as grazing, competition, facilitation, or decomposition. They act on individuals and populations and may affect trophic interactions at the community or even ecosystem level. Allelochemical effects on a certain group of organisms might have further consequences for other organisms through direct and indirect mechanisms.

Allelochemical reactions occur in the following interactions:

1. Animal–animal interactions (predator–prey systems, mate and kin recognition);
2. Plant–animal interactions (grazer–algae interactions, plant–herbivore interactions, feeding deterrents, foraging kairomones);
3. Plant–plant interactions (allelopathy, autotoxicity);
4. Plant–microbe interactions (allelopathy, plant–pathogen interactions (**Figure 1**)).

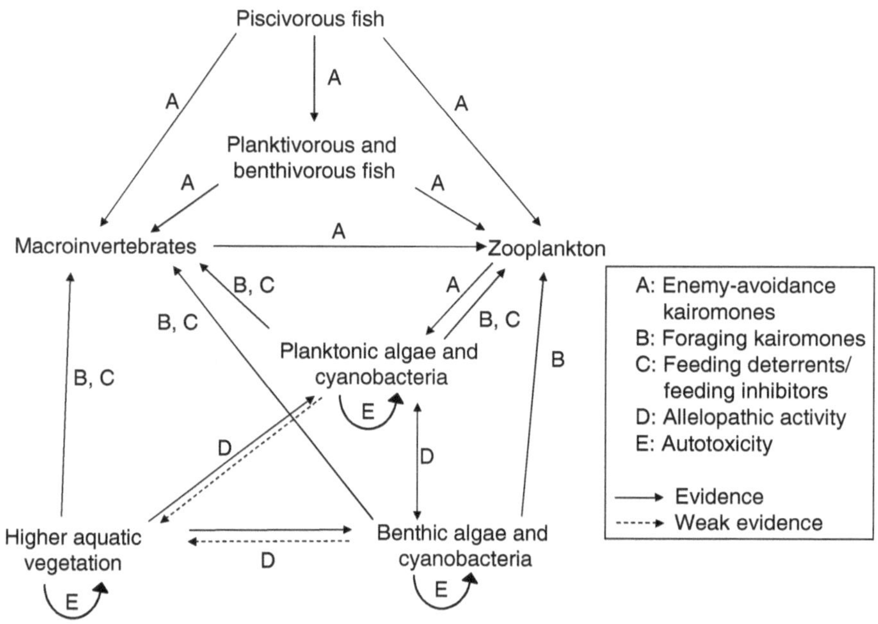

Figure 1 Possible aquatic allelochemical reactions. Different organisms interact by various allelochemical principles. Wide-spread at higher trophic levels are enemy-avoidance kairomones (A), which may affect prey behavior to minimize predation. The search for suitable prey or food is guided by foraging kairomones (B). Several primary producers possess feeding inhibitors (C) to deter grazers. Allelopathic interactions (D) exist by definition only among primary producers or between primary producers and microorganisms and involve both inhibitory and stimulatory effects. Autotoxicity (E) is a mechanism for density control in several primary producers.

Animal–Animal Interactions

Animals may interact intra- and interspecifically by allelochemicals. Animals can chemically sense mates, relatives, conspecifics, or predators by specific chemical cues.

Predator–prey interactions in pelagic and benthic aquatic environments may be controlled by allelochemical reactions. Prey can smell the presence of predators by kairomones released from them or by alarm cues liberated from injured conspecifics. Prey reacts towards these infochemicals by performing adaptive reactions to decrease mortality and increase fitness. None of the chemicals involved in these interactions has been identified so far, yet multiple interactions between fish and macroinvertebrates have been investigated. Some examples are provided in **Table 1**.

Fish Kairomones

One of the most investigated aquatic allelochemical reaction involves kairomones that are released by fish and evoke a response of the prey. Chemical cues originating from many different fish, e.g., sticklebacks (**Figure 2**), influence a wide range of potential prey in pelagic and benthic habitats, in still and running waters. Both vertebrates (planktivorous fish) and invertebrates (crustaceae, gammarids, insects, molluscs) react towards fish kairomone. Prey reactions range from behavioral adaptations (changes in

Table 1 Animal–animal allelochemical reactions

Producing organism	*Target organism*	*References*
Fish	Fish	[1]
Fish	Zooplankton	[2–5]
Fish	Insect larvae	[6]
Fish	Crayfish, gammarids	[7, 8]
Fish	Snails	[9, 10]
Chaoborus	Zooplankton	[11–13]
Bythothrepes, Leptodora, Mysis	Zooplankton	[14–16]

Selected examples of different allelochemical interactions between and among vertebrates and invertebrates. Fish are important predators in aquatic systems and various chemical cues have evolved in predator–prey interactions with other fish or pelagic and benthic macroinvertebrates. Predator kairomones also warn zooplankton of large insect larvae and large crustaceans, which also exert a strong predation pressure.

Sources

1. Bronmark C and Miner JG (1992) Predator-induced phenotypical change in body morphology in crucian carp. *Science* 258: 1348–1350.
2. Lass S, Vos M, Wolinska J, and Spaak P (2005) Hatching with the enemy: *Daphnia* diapausing eggs hatch in the presence of fish kairomones. *Chemoecology* 15: 7–12.
3. Laforsch C, Beccara L, and Tollrian R (2006) Inducible defenses: The relevance of chemical alarm cues in *Daphnia*. *Limnology and Oceanography* 51: 1466–1472.
4. Loose CJ, von Elert E, and Dawidowicz P (1993) Chemically-induced diel vertical migration in *Daphnia*—a new bioassay for kairomones exuded by fish. *Archiv für Hydrobiologie* 126: 329–337.
5. Ringelberg J (1999) The photobehaviour of *Daphnia* spp. as a model to explain diel vertical migration in zooplankton. *Biological Reviews of the Cambridge Philosophical Society* 74: 397–423.
6. Cowan CA and Peckarsky BL (1994) Diel feeding and positioning periodicity of a grazing mayfly in a trout stream and a fishless stream. *Canadian Journal of Fisheries & Aquatic Sciences* 51: 450–459.
7. Abjornsson K, Hansson LA, and Bronmark C (2004) Responses of prey from habitats with different predator regimes: Local adaptation and heritability. *Ecology* 85: 1859–1866.
8. Abjornsson K, Bronmark C, and Hansson LA (2002) The relative importance of lethal and non-lethal effects of fish on insect colonisation of ponds. *Freshwater Biology* 47: 1489–1495.
9. Turner AM, Bernot RJ, and Boes CM (2000) Chemical cues modify species interactions: The ecological consequences of predator avoidance by freshwater snails. *Oikos* 88: 148–158.
10. Turner AM and Montgomery SL (2003) Spatial and temporal scales of predator avoidance: Experiments with fish and snails. *Ecology* 84: 616–622.
11. Dodson SI, Hanazato T, and Gorski PR (1995) Behavioral responses of *Daphnia pulex* exposed to carbaryl and *Chaoborus* kairomone. *Environmental Toxicology and Chemistry* 14: 43–50.
12. Tollrian R and von Elert E (1994) Enrichment and purification of *Chaoborus* kairomone from water: Further steps toward its chemical characterization. *Limnology & Oceanography* 39: 788–796.
13. Pareijko K and Dodson S (1990) Progress towards characterization of a predator–prey kairomone: *Dapnia pulex* and *Chaoborus americanus*. *Hydrobiologia* 198: 51–59.
14. Effert EL and Pederson CL (2006) Failure of cyclomorphic features to deter size-dependent predation by *Leptodora kindti* on *Daphnia lumholtzi*. *Journal of Freshwater Ecology* 21: 457–466.
15. Peacor SD, Pangle KL, and Vanderploeg HA (2005) Behavioral response of Lake Michigan *Daphnia mendotae* to *Mysis relicta*. *Journal of Great Lakes Research* 31: 144–154.
16. Bungartz B and Branstrator DK (2003) Morphological changes in *Daphnia mendotae* in the chemical presence of *Bythotrephes longimanus*. *Archiv für Hydrobiologie* 158: 97–108.

swimming speed and direction, changes in location [DVM – diel vertical migration, in deeper, mostly stratified lakes; DHM – diel horizontal migration, in shallow lakes with extended submerged macrophyte vegetation]) to morphological changes (helmet or neck teeth formation, changes in outer shape) or adaptations of life cycle (size and age at first reproduction, growth rate, diapause). The kairomone affecting *Daphnia* apparently is the same in several piscivorous and planktivorous fish, but it is not clear yet if it is only one compound and how its exact chemical structure looks like. Trimethylamine has been proposed as active cue but is not present in high enough concentrations in fish incubation water, and also resulted in different reactions by some prey. Most likely bacteria are involved in the formation or activation of fish kairomone.

Individual daphnids react fast toward fish cues, indicating phenotypic plasticity. Different genotypes (clones) can exhibit distinct responses (e.g., magnitude of DVM) toward fish kairomone. Environmental factors may influence the effect of kairomones on prey. *Daphnia* change their kairomone-induced reactions towards fish, depending on light intensity or whether they are parasitized. Kairomones released by predators apparently evoke stronger response in

Figure 2 Three-spined stickleback (*Gasterosteus aculeatus*) among submerged vegetation. Sticklebacks, perch, roach, rudd, and several other planktivorous fish are known to release fish kairomones affecting behavior, morphology, and life history of zooplankton. Photograph © Martin Mörtl, Konstanz, Germany.

zooplankton than does alarm cues (explained later) by crushed conspecifics.

Several other invertebrates such as gammarids respond to fish and crayfish kairomones. Fish kairomones reduced the shredding activity of *Gammarus pulex*. Chironomid larvae react to the presence of fish kairomones by an increased digging activity. Snails respond to fish kairomones by moving to protective cover. Fish cues caused snails to prefer *Dreissena* covered surfaces over plain surfaces and to prefer interstices over exposed mussel shells.

Crucian carp develop a deeper body in the presence of fish kairomones minimizing attacks by gape-limited piscivores.

Invertebrate Kairomones

Small, grazing zooplankton may sense invertebrate predators by kairomones. Chemical cues originating from larvae of the phantom midge (*Chaoborus* spp.), *Notonecta*, *Bythotrephes*, *Leptodora*, and other invertebrate predators affect daphnids and other zooplankton. Reactions of zooplankton to invertebrate kairomones differ from that to fish kairomones and cause an adaptive response, i.e., helmet formation in *Daphnia cucullata* (**Figure 3**), counteracting the gape-limitation of invertebrate predators.

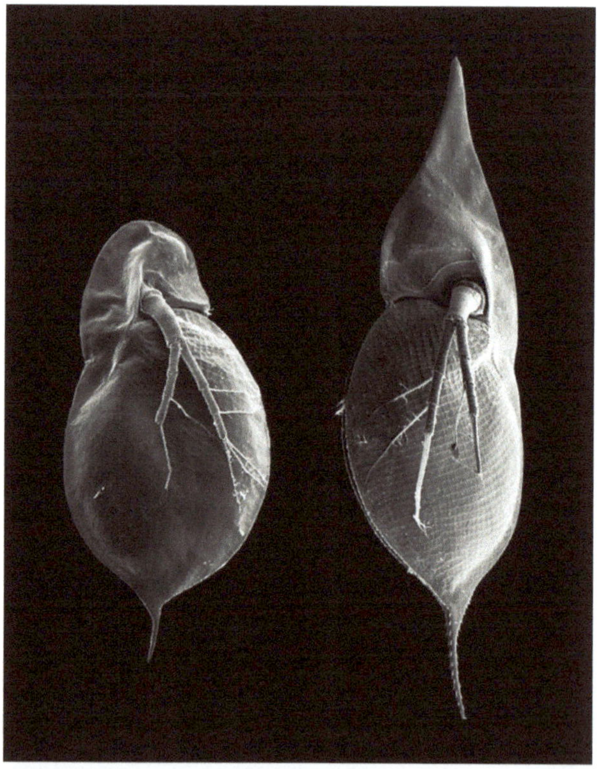

Figure 3 Cyclomorphosis in *Daphnia cucullata*. Fish kairomone induces helmet formation. Photograph © Christian Laforsch/Ralph Tollrian, Munich/Bochum, Germany.

Alarm Cues

Alarm cues are released when prey is damaged by predators. Crushed organisms are often used in experimental set-ups to create alarm cues, but the response of prey to these cues might be less than towards alarm cues created by predation. It seems unclear whether predator-specific additional signals (due to saliva or something else) add to the information and if prey can distinguish alarm cues originating from prey damaged by different predators. Prey reaction may depend on the combination of alarm cues and predator kairomones.

When injured, some fish release a cue alarming conspecifics and even other fish species about predators. This alarm cue is released by club cells in the skin. It is also called Schreckstoff, based on its first detection by Karl von Frisch in 1938 when studying minnows (*Phoxinus phoxinus*). Not all schooling fish release Schreckstoff; it has been described in the Ostariophysi and several Scorpaeniformes and Perciformes. The chemical nature is not fully solved. Some active compounds have been tested in their interaction with olfactory receptor cells in fish.

Also crushed invertebrates may provoke alarm reactions in conspecifics. Waterborne cues from macerated individuals alter morphology in several *Daphnia* species. Snails react to alarm cues originating from crushed conspecifics and change their habitat use in the presence of predators. When fish were present, snails moved under cover but selected the water surface in the presence of crayfish. Freshwater pulmonate snails exhibit different inter- and intraspecific behavioral responses to alarm cues from crushed conspecifics. Thin-shelled organisms moved to the water surface or out of the water while thick-shelled organisms moved under cover.

Partner Finding and Kin Recognition

Relatively little is known about pheromones involved in mate finding in freshwater organisms. A trihydroxylated progestin steroid stimulated oocyte maturation and pheromone release in Eurasian ruffe. A sex pheromone released during the breeding season by mature females of *Pacifastacus leniusculus* stimulates courtship and mating behavior in males. A specific surface glycoprotein might function as a contact mate recognition pheromone for freshwater rotifers. Chemical cues might affect nestling recognition and parental care by male bluegills. Kin recognition in freshwater systems is another field awaiting progress.

Plant–Animal Interactions

Plants will be used as term in the following to include also macro- and microalgae as well as cyanobacteria. Allelochemical reactions may occur both ways, plants affecting animals and animals affecting plants. The major interaction of plants and animals in aquatic systems is exploitative. Chemical cues attracting animal pollinators do not occur under water. Traditionally, feeding on phytoplankton or epiphytes is considered grazing while the consumption of higher plants is termed herbivory, but both terms are also used in the extended meaning of consumption of primary producers. Plant–herbivore or algal–grazer interactions are basically another form of predator–prey interactions.

Comparable to terrestrial systems, several herbivore deterrents have been elucidated from aquatic higher plants. Many toxic compounds against mammals and invertebrates are formed by cyanobacteria. While several invertebrates are attracted to suitable algae or cyanobacteria by volatile chemical cues, infochemicals released by grazing zooplankton can affect morphology of phytoplankton species. Examples of these allelochemical interactions are listed in **Table 2**.

Impact of Plants on Grazing and Herbivory

Several but not all higher plants, algae, and cyanobacteria contain compounds deleterious to grazing or herbivorous organisms. Feeding deterrents result in low or no consumption but in general have no further negative impact on the consumer. Inhibitors or toxins affect specific, in part essential, metabolic processes in the consumer and may even cause mortality. Negative impacts on consumers require sensitivity of these organisms, either in their capacity to detect and recognize feeding deterrents, or in their susceptibility to inhibitors or toxins. The impact of certain allelochemicals on different grazers or herbivores may vary strongly.

Cyanobacterial toxins Cyanobacteria produce numerous and potent environmental toxins, some of them affecting the feeding and performance of zooplankton. Microcystins and other hepatotoxic cyclic peptides are produced by species of *Microcystis*, *Planktothrix*, *Anabaena*, and *Nodularia*. Cyclic peptides inhibit protein phosphatases and can be fatal to mammals because of liver damage. Another group of toxins, cyanobacterial cyclic depsipeptides affect digestive enzymes and act as trypsin inhibitors in zooplankton. The major digestive proteases of *Daphnia magna* are trypsins and chymotrypsins and are different to digestive proteases

Table 2 Plant–animal allelochemical reactions

Producing organisms	*Target organism*	*Active compound(s)*	*References*
Feeding deterrents in macrophytes			
Nasturtium officinale	Macroinvertebrates (Trichoptera, Gammarids, Snails)	Glucosinolates	[1]
S. cernuus	Crayfish	Lignans	[2, 3]
Habenaria repens	Crayfish	Habenariol	[4]
M. spicatum	*A. ephemerella*, snails	Polyphenols	[5–7]
Toxins			
Cyanobacteria	Zooplankton	Cyclic depsipeptides	[8–11]
Cyanobacteria	Mammals	Cyclic peptides,	[12, 13]
Foraging kairomones			
Cyanobacteria	Nematodes	VOCs	[14]
Cyanobacteria, chlorophytes	Snails	VOCs	[15]
Miscellaneous			
Daphnia sp.	*Scenedesmus*	unknown	[16–21]
Zooplankton	Phytoplankton	unknown	[22, 23]

1. Newman RM, Kerfoot WC, and Hanscom ZI (1996) Watercress allelochemical defends high-nitrogen foliage against consumption: Effects on freshwater invertebrate herbivores. *Ecology* 77: 2312–2323.
2. Kubanek J, Fenical W, Hay ME, Brown PJ, and Lindquist N (2000) Two antifeedant lignans from the freshwater macrophyte *Saururus cernuus. Phytochemistry* 54: 281–287.
3. Kubanek J, Hay ME, Brown PJ, Lindquist N, and Fenical W (2001) Lignoid chemical defenses in the freshwater macrophyte *Saururus cernuus. Chemoecology* 11: 1–8.
4. Wilson DM, Fenical W, Hay M, Lindquist N, and Bolser R (1999) Habenariol, a freshwater feeding deterrent from the aquatic orchid *Habenaria repens* (Orchidaceae). *Phytochemistry* 50: 1333–1336.
5. Li YK, Yu D, and Yan X (2004) Are polyphenolics valuable in anti-herbivory strategies of submersed freshwater macrophytes? *Archiv für Hydrobiologie* 161: 391–402.
6. Choi C, Bareiss C, Walenciak O, and Gross EM (2002) Impact of polyphenols on the growth of the aquatic herbivore *Acentria ephemerella* (Lepidoptera: Pyralidae). *Journal of Chemical Ecology* 28: 2223–2235.
7. Walenciak O, Zwisler W, and Gross EM (2002) Influence of *Myriophyllum spicatum*-derived tannins on gut microbiota of its herbivore *Acentria ephemerella. Journal of Chemical Ecology* 28: 2045–2056.
8. Blom JF, Baumann HI, Codd GA, and Jüttner F (2006) Sensitivity and adaptation of aquatic organisms to oscillapeptin J and [D-Asp(3),(E)-Dhb(7)] microcystin-RR. *Archiv für Hydrobiologie* 167: 547–559.
9. Agrawal MK, Bagchi D, and Bagchi SN (2005) Cysteine and serine protease-mediated proteolysis in body homogenate of a zooplankter, *Moina macrocopa,* is inhibited by the toxic cyanobacterium, *Microcystis aeruginosa* PCC7806. *Comparative Biochemistry & Physiology Part B Biochemistry & Molecular Biology* 141: 33–41.
10. Weckesser J, Martin C, and Jakobi C (1996) Cyanopeptolins, depsipeptides from cyanobacteria. *Systematic & Applied Microbiology* 19: 133–138.
11. Rohrlack T, Christoffersen K, Hansen PE, *et al.* (2003) Isolation, characterization, and quantitative analysis of microviridin J, a new *Microcystis* metabolite toxic to *Daphnia. Journal of Chemical Ecology* 29: 1757–1770.
12. Dierstein R, Kaiser I, and Weckesser J (1989) Inhibition of prodigiosin formation in *Serratia marcescens* by extracts of toxic cyanobacteria. *Systematic and Applied Microbiology* 12: 244–248.
13. Carmichael WW (1994) Toxins of cyanobacteria. *Scientific American* 270: 78–86.
14. Höckelmann C, Moens T, and Jüttner F (2004) Odor compounds from cyanobacterial biofilms acting as attractants and repellents for free-living nematodes. *Limnology and Oceanography* 49: 1809–1819.
15. Fink P, von Elert E, and Jüttner F (2006) Volatile foraging kairomones in the littoral zone: Attraction of an herbivorous freshwater gastropod to algal odors. *Journal of Chemical Ecology* V32: 1867–1881.
16. Lampert W, Rothhaupt KO, and von Elert E (1994) Chemical induction of colony formation in a green-alga (*Scenedesmus acutus*) by grazers (*Daphnia*). *Limnology and Oceanography* 39: 1543–1550.
17. Lürling M and Van Donk E (1997) Morphological changes in *Scenedesmus* induced by infochemicals released in situ from zooplankton grazers. *Limnology and Oceanography* 42: 783–788.
18. von Elert E and Franck A (1999) Colony formation in *Scenedesmus*: grazer-mediated release and chemical features of the infochemical. *Journal of Plankton Research* 21: 789–804.
19. Lürling M and Beekman W (2002) Extractable substances (anionic surfactants) from membrane filters induce morphological changes in the green alga *Scenedesmus obliquus* (Chlorophyceae). *Environmental Toxicology and Chemistry* 21: 1213–1218.
20. Yasumoto K, Nishigami A, Yasumoto M, *et al.* (2005) Aliphatic sulfates released from *Daphnia* induce morphological defense of phytoplankton: isolation and synthesis of kairomones. *Tetrahedron Letters* 46: 4765–4767.
21. Yasumoto K, Nishigami A, Kasai F, Kusumi T, and Ooi T (2006) Isolation and absolute configuration determination of aliphatic sulfates as the *Daphnia* kairomones inducing morphological defense of a phytoplankton. *Chemical and Pharmaceutical Bulletin* 54: 271–274.
22. Lürling M (2003) Phenotypic plasticity in the green algae *Desmodesmus* and *Scenedesmus* with special reference to the induction of defensive morphology. *Annales de Limnologie* 39: 85–101.
23. Wiltshire K, Boersma M, and Meyer B (2003) Grazer-induced changes in the desmid *Staurastrum. Hydrobiologia* 491: 255–260.

Selected examples of allelochemicals found in primary producers and affecting herbivores or compounds released by animals affecting primary producers.

Microcystin LR [1]

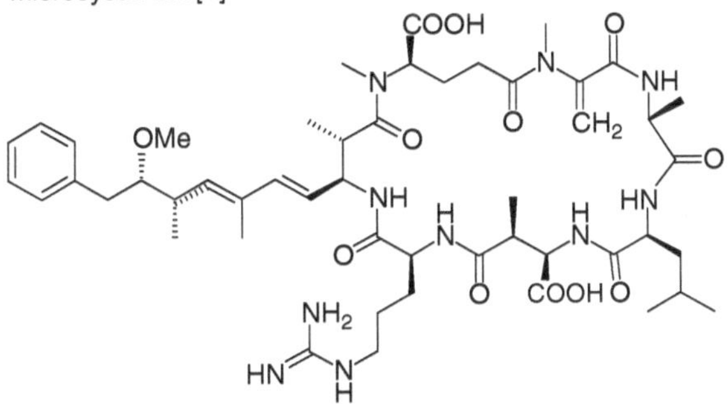

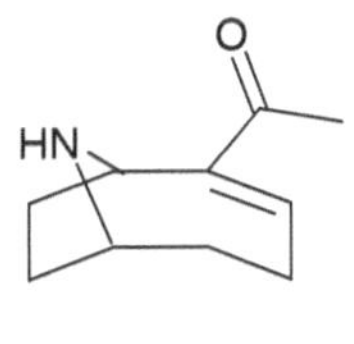

Anatoxin-A [2]

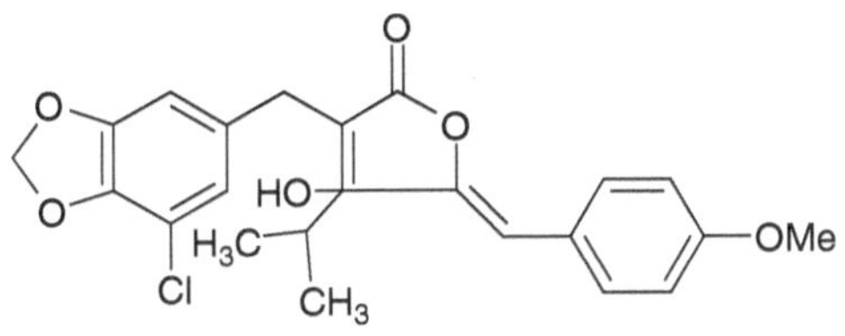

Cyanobacterin [3]

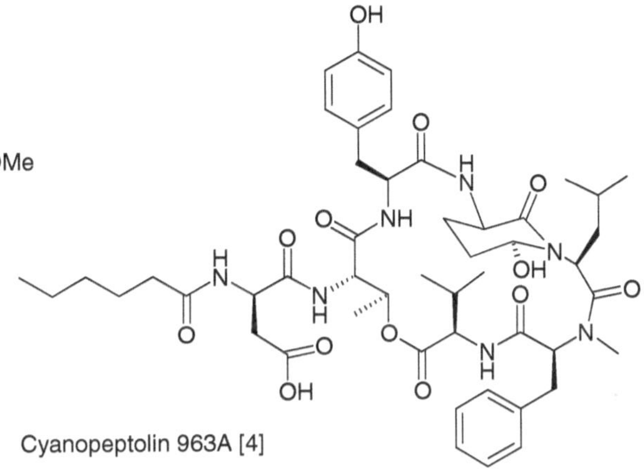

Cyanopeptolin 963A [4]

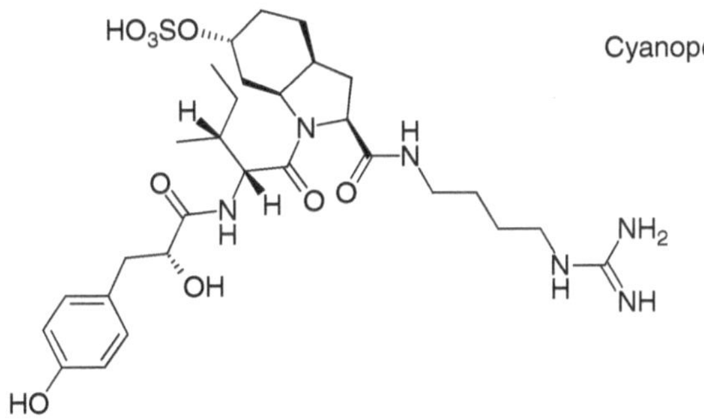

Aeruginosin 98-B [5]

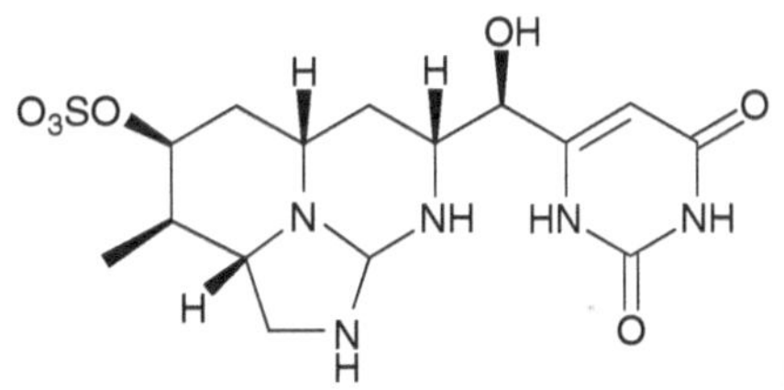

Cylindrospermopsin [6]

Figure 4 Some allelochemicals found in cyanobacteria. For references see **Tables 2** and **3**. Beresovsky D, Hadas O, Livne A, *et al.* (2006) Toxins and biologically active secondary metabolites of *Microcystis* sp. isolated from Lake Kinneret. *Israel Journal of Chemistry* 46: 79–87. Carmichael WW (1994) Toxins of cyanobacteria. *Scientific American* 270: 78–86. Gleason FK and Paulson JL (1984) Site of action of the natural algicide, cyanobacterin in the blue green alga, *Synechococcus* sp. *Archives of Microbiology* 138: 273–277; Bister B, Keller S, Baumann HI, *et al.* (2004) Cyanopeptolin 963A, a chymotrypsin inhibitor of *Microcystis* PCC7806. *Journal of Natural Products* 67: 1755–1757; Ishida K, Okita Y, Matsuda H, Okino T, and Murakami M (1999) Aeruginosins, protease inhibitors from the cyanobacterium *Microcystis aeruginosa, Tetrahedron* 55: 10971–10988; Harada K-I, Ohtani I, Iwamoto K, *et al.* (1994) Isolation of cylindrospermopsin from a cyanobacterium *Umezakia natans* and its screening method. *Toxicon* 32: 73–84.

found in mammals. Cyanopeptolins, a special class of cyclic depsipeptides exhibit protease-inhibitory and/or cytotoxic activity. Further protease inhibitors are oscillapeptins, found for example in *Planktothrix rubescence* and *Oscillatoria agardhii.*

Anabaena flos-aquae produces the alkaloid neurotoxin anatoxin-*a*. This allelochemical affects growth and performance of zooplankton, both daphnids and smaller zooplankton such as rotifers. *A. flos-aquae* decreased the lifespan, fecundity, and population growth rate in *Brachionus calyciflorus*. Effects of anatoxin-*a*-producing cyanobacteria on the population dynamics of susceptible zooplankton taxa in natural communities might be influenced by environmental factors such as food availability and temperature.

Herbivore deterrents Similar to terrestrial plants, several higher aquatic plants contain feeding deterrents against herbivores (**Figure 5**). Aquatic insects, snails, fish, and birds are major herbivores of aquatic vegetation. True herbivores only feed on living plant tissue while omnivores include both live and dead plant material and other food items in their diet. Gammarids and crayfish are omnivores feeding also

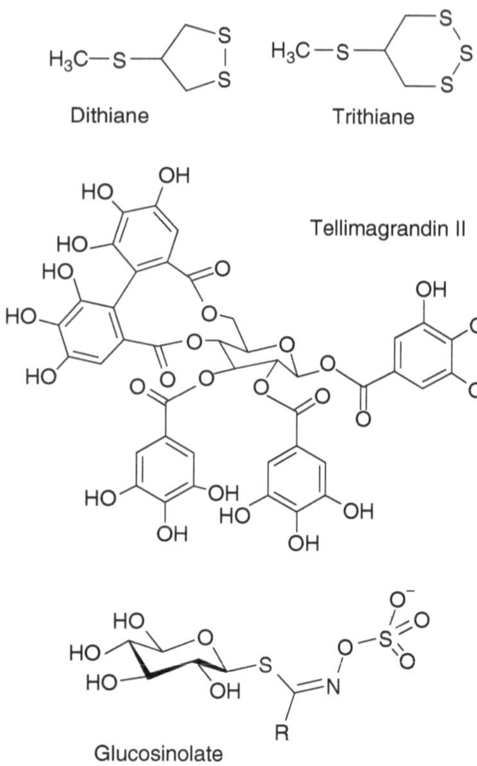

Figure 5 Some feeding deterrents and allelopathically active compounds found in aquatic macrophytes.

on higher aquatic plants. Glucosinolates present in watercress deter various herbivores, gammarids, trichoptera, and snails. Lignoid chemical defenses known from Lizard's-tail (*Saururus cernuus*) and habenariol isolated from the emersed growing water-spider orchid *Habenaria repens* affect crayfish feeding. High concentrations of tannins (hydrolysable polyphenols) in *Myriophyllum spicatum* influence aquatic invertebrate herbivores such as the aquatic moth larvae *Acentria ephemerella* and *Radix swinhoei.* Elucidation of feeding deterrents against true herbivores suffers so far from a lack of suitable feeding assays to perform bioassay guided fractions of active plant extracts.

Foraging kairomones Consumers may recognize suitable plants by specific chemical cues that are either released into the environment or attached to the plant surface. Not many foraging kairomones are known from aquatic systems so far. The aquatic weevil *Euhrychiopsis lecontei* is attracted by higher concentrations of glycerol and uracil to *M. spicatum* and seems to use these signals to distinguish between *M. spicatum* and other milfoil species. Snails respond to volatile foraging kairomones released from disintegrating algal cells. The response requires a certain mixture of volatile organic compounds (VOCs) comparable to organismic interactions known from terrestrial systems. Aquatic nematodes are attracted to cyanobacterial biofilms using multiple odor compounds.

Impact of Grazers on Plants

Grazers or herbivores may provoke inducible defenses in their plant prey. Plants may response by changes in their allocation to growth or reproduction, by changing the content of defensive compounds or the formation of defensive structures such as spines. The induction of such changes requires either direct feeding of grazers or herbivores on the plant or the action of specific allelochemical cues released by the grazer.

One of the most investigated systems is colony (coenobia) formation in the green alga *Scenedesmus* induced by chemicals released from *Daphnia.* The structure of the allelochemical is not known. Urea and aliphatic sulfates excreted by daphnids have been shown to induce colony formation. Whether these are the true kairomones affecting *Scenedesmus* remains an open question. SPE treatment of *Daphnia* water and testing the colony formation in a bioassay suggested that the active compound(s) may contain an olefinic carboxylic acid.

Plant–Plant Interactions

Plants experience inter- and intraspecific competition for resources such as light, nutrients, or space. Access to resources depends also on the environment they are living in (benthos, pelagial), and in aquatic habitats may either be from the water or from the substrate they grow on. In contrast to exploitative competition, where plants compete for resources, interference competition relates to allelopathy or other nonresource-based interactions. If a plant manages to limit the growth of adjacently growing primary producers, access to nutrients and light will increase. Allelopathically active compounds have been isolated from higher aquatic plants, macro- and microalgae and cyanobacteria (**Table 3**). Different classes of secondary metabolites are involved in these interactions, and leading to multiple modes of actions against target species. The inhibition of photosynthesis is a prominent mode of action for many allelopathically active compounds.

Allelopathy

Biochemical interference among plants and between plants and microorganisms is termed allelopathy *sensu strictu*. The term is, however, also used sometimes for biochemical interference in the animal

Table 3 Plant–plant allelochemical reactions

Producing organism	*Target organism*	*Allelopathically active compound(s)*	*Mode of action*	*References*
Macrophytes				
Berula erecta (*Sium erectum*)	Diatoms	Polyacetylenes (Falcarindiol, Falcarinol)	Growth	[1]
Ceratophyllum demersum	Diatoms (*Nitzschia*), Cyanobacteria	Unknown, elemental sulfur?	Growth, community composition, PS	[2–5]
Chara aspera	Cyanobacteria	Unknown	Growth	[6]
Chara globularis	Diatoms (Nitzschia), Chlorophytes, natural phytoplankton samples	Cyclic sulfur compounds (Dithiane, Trithiane), elemental sulfur	Growth, PS	[7–10] [11]
	Bacteria	Unknown		
		Charamin		
Eleocharis spp.	Submerged macrophytes	Dihydroactinidiolide, Trihydroxycyclopentenyl-fatty acid	Growth	[12, 13]
Juncus effusus	Autotoxic	Unknown	Seedling germination	[14]
Myriophyllum brasiliense	Cyanobacteria	Eugeniin (= Tellimagrandin II)	Growth	[15]
M. spicatum	Cyanobacteria, diatoms, chlorophytes	Hydrolysable polyphenols (Tellimagrandin II)	Growth, PS	[3, 16–19]
Typha latifolia	Autotoxic ?	Unknown	Germination	[20, 21]
Eukaryotic algae				
Chlamydomonas	Other algae, cyanobacteria (*A. flos-aquae*, *Microcystis*)	Unknown, fatty acid derivative ?	Growth, toxin production, heterocyst spacing	[22–24]
Peridinium gatunensis	*Microcystis*	Unknown	Lysis, growth	[25]
Cyanobacteria				
Scytonema hoffmanii	Cyanobacteria, higher plants	Cyanobacterin	Growth, PS	[26, 27]
Fischerella spp.	Cyanobacteria, Chlorophytes, Diatoms	Fischerellins A & B	Growth, PS	[28–30]
Nostoc punctiforme	Autotoxic	Unknown	Growth	[31]
Microcystis spp.	*Peridinium gatunensis*	Unknown	Growth, PS (internal carbonic anhydrase)	[32]
Anabaena flos-aquae	Chlamydomonas	Unknown	Settling	[33]

1. Wium-Andersen S, *et al.* (1987) Algal growth inhibitors in *Sium erectum* HUDS. *Archives of Hydrobiology* 111: 317–320.
2. Jasser I (1995) The influence of macrophytes on a phytoplankton community in experimental conditions. *Hydrobiologia* 306: 21–32.
3. Körner S and Nicklisch A (2002) Allelopathic growth inhibition of selected phytoplankton species by submerged macrophytes. *Journal of Phycology* 38: 862–871.
4. Kogan SI and Chinnova GA (1972) Relations between *Ceratophyllum demersum* (L.) and some blue-green algae. *Hydrobiological Journal* (*Ghidrobiol. Zh.*) 8: 14–19 (21–27).

5. Wium-Andersen S, Anthoni U, and Houen G (1983) Elemental sulphur, a possible allelopathic compound from *Ceratophyllum demersum*. *Phytochemistry* 22: 2613.
6. Berger J and Schagerl M (2003) Allelopathic activity of *Chara aspera*. *Hydrobiologia* 501: 109–115.
7. Mulderij G, Van Donk E, and Roelofs JGM (2003) Differential sensitivity of green algae to allelopathic substances from *Chara*. *Hydrobiologia* 491: 261–271.
8. Lürling M, van Geest G, and Scheffer M (2006) Importance of nutrient competition and allelopathic effects in suppression of the green alga *Scenedesmus* obliquus by the macrophytes *Chara, Elodea* and *Myriophyllum*. *Hydrobiologia* 556: 209–220.
9. Anthoni U, *et al*. (1980) Biologically active sulphur compounds from the green alga *Chara globularis*. *Phytochemistry* 19: 1228–1229.
10. Wium-Andersen S, Anthoni U, and Christophersen C (1982) Allelopathic effects on phytoplankton by substances isolated from aquatic macrophytes (Charales). *Oikos* 39: 187–190.
11. Anthoni U, *et al*. (1987) Charamin, a quaternery ammonium ion antibiotic from the green alga *Chara globularis*. *Journal of Organic Chemistry* 52: 694–695.
12. Stevens KL and Merril GB (1980) Growth inhibitors from Spikerush. *Journal of Agriculture and Food Chemistry* 28: 644–646.
13. Yeo RR and Thurston JR (1984) The effect of dwarf spikerush (*Eleocharis coloradoensis*) on several submersed aquatic weeds. *Journal of Aquatic Plant Management* 22: 52–56.
14. Ervin GN and Wetzel RG (2000) Allelochemical autotoxicity in the emergent wetland macrophyte *Juncus effusus* (Juncaceae). *American Journal of Botany* 87: 853–860.
15. Saito K, *et al*. (1989) Inhibitory substances from *Myriophyllum brasiliense* on growth of blue-green algae. *Journal of Natural Products* 52: 1221–1226.
16. Planas D, *et al*. (1981) Ecological significance of phenolic compounds of *Myriophyllum spicatum*. *Verhandlungen der Internationalen Vereinigung für Limnologie* 21: 1492–1496.
17. Nakai S, *et al*. (1999) Growth inhibition of blue-green algae by allelopathic effects of macrophytes. *Water Science & Technology* 39: 47–53.
18. Leu E, *et al*. (2002) Polyphenolic allelochemicals from the aquatic angiosperm *Myriophyllum spicatum* L. inhibit photosystem II. *Plant Physiology* 130: 2011–2018.
19. Gross EM, Meyer H, and Schilling G (1996) Release and ecological impact of algicidal hydrolysable polyphenols in *Myriophyllum spicatum*. *Phytochemistry* 41: 133–138.
20. McNaughton SJ (1968) Autotoxic feedback in relation to germination and seedling growth in *Typha latifolia*. *Ecology* 49: 367–369.
21. Grace JB (1983) Autotoxic inhibition of seed germination by *Typha latifolia*: an evaluation. *Oecologia* 59: 366–369.
22. Kearns KD and Hunter MD (2000) Green algal extracellular products regulate antialgal toxin production in a cyanobacterium. *Environmental Microbiology* 2: 291–297.
23. McCracken MD, Middaugh RE, and Middaugh RS (1980) A chemical characterization of an algal inhibitor obtained from *Chlamydomonas*. *Hydrobiologia* 70: 271–276.
24. Kearns KD and Hunter MD (2002) Algal extracellular products suppress *Anabaena flos-aquae* heterocyst spacing. *Microbial Ecology* 43: 174–180.
25. Vardi A, *et al*. (2002) Dinoflagellate-cyanobacterium communication may determine the composition of phytoplankton assemblage in a mesotrophic lake. *Current Biology* 12: 1767–1772.
26. Gleason FK and Case DE (1986) Activity of the natural algicide cyanobacterin on angiosperms. *Plant Physiology* 80: 834–838.
27. Gleason FK and Baxa CA (1986) Activity of the natural algicide, cyanobacterin, on eucaryotic microorganisms. *FEMS Microbiology Letters* 33: 85–88.
28. Papke U, Gross EM, and Francke W (1997) Isolation, identification and determination of the absolute configuration of fischerellin B. A new algicide from the freshwater cyanobacterium *Fischerella muscicola* (Thuret.). *Tetrahedron Letters* 38: 379–382.
29. Hagmann L and Jüttner F (1996) Fischerellin A, a novel photosystem-II inhibiting allelochemical of the cyanobacterium *Fischerella muscicola* with antifungal and herbicidal activity. *Tetrahedron Letters* 37: 6539–6542.
30. Gross EM, Wolk CP, and Jüttner F (1991) Fischerellin, a new allelochemical from the freshwater cyanobacterium *Fischerella muscicola*. *Journal of Phycology* 27: 686–692.
31. Harder R (1917) Ernährungsphysiologische Untersuchungen an Cyanophyceen, hauptsächlich dem endophytischen *Nostoc punctiforme*. *Zeitschrift für Botanik* IX: 145–242.
32. Sukenik A, *et al*. (2002) Inhibition of growth and photosynthesis of the dinoflagellate *Peridinium gatunense* by *Microcystis* sp. (cyanobacteria): A novel allelopathic mechanism. *Limnology and Oceanography* 47: 1656–1663.
33. Kearns KD and Hunter MD (2001) Toxin-producing *Anabaena flos-aquae* induces settling of *Chlamydomonas reinhardtii*, a competing motile alga. *Microbial Ecology* 42: 80–86.

Selected examples for different interactions between higher aquatic vegetation, algae and cyanobacteria both in benthic and planktonic habitats. PS – photosynthesis.

kingdom. Stimulatory or inhibitory effects due to, mostly secondary, plant metabolites occur. In a few cases a switch from stimulatory to inhibitory activity with increasing concentrations was observed. Especially inhibitory compounds enable the producing organism to limit growth of competitors and thus facilitate access to resources.

Allelopathic interference has been described for all types of macrophytes (emergent, floating-leaved, floating, submerged), macro- and microalgae as well as for cyanobacteria. Compounds are either released into the environment or transferred by direct cell–cell contact, e.g. for benthic organisms. The benthic cyanobacteria *Fischerella* spp. produce various fischerellins (**Figure 6**). Among them, fischerellin A strongly inhibits other cyanobacteria at concentrations as low as 14 nM. Some, but not all charophytes produce cyclic sulfur compounds, dithiane and trithiane, allelopathically active against diatoms and other phytoplankton species (see **Figure 5**). Tannins, especially tellimagrandin II, is the major allelopathically active compound in Eurasian watermilfoil (*M. spicatum*) (**Figure 7**). Several inhibitory compounds have been isolated and tested under laboratory conditions. Experimental proof for in situ activity is difficult and only indirect evidence for the role of allelopathic interference in the competition of primary producers exists. To distinguish allelopathic interference clearly from resource competition or other interactions (herbivory, pathogens), proof for the production and release of active compounds by the producing organism, proof for sufficient transport and uptake by the target organism and proof for a specific, mostly deleterious, effect on the target organism is needed. Since allelopathy affects other primary producers, inhibition of photosynthesis is a widespread and efficient mode of action of allelopathically active compounds. Several benthic cyanobacteria produce allelopathically active compounds, which interfere with photosystem II.

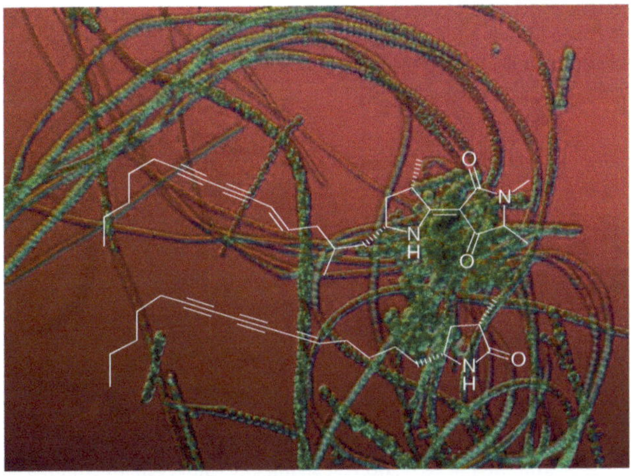

Figure 6 *Fischerella muscicola* and the allelopathically active compounds fischerellin A and B [1–4]. Adapted from Papke U, Gross EM, and Francke, W (1997) Isolation, identification and determination of the absolute configuration of fischerellin B. A new algicide from the freshwater cyanobacterium *Fischerella muscicola* (Thuret.). *Tetrahedron Letters* 38: 379–382; Hagmann L and Jüttner F (1996) Fischerellin A, a novel photosystem-II inhibiting allelochemical of the cyanobacterium *Fischerella muscicola* with antifungal and herbicidal activity. *Tetrahedron Letters* 37: 6539–6542; Gross EM, von Elert E, and Jüttner F (1994) Production of allelochemicals in *Fischerella muscicola* under different environmental conditions. *Verhandlungen der Internationalen Vereinigung für Limnologie* 25: 2231–2233; Gross EM, Wolk CP, and Jüttner F (1991) Fischerellin, a new allelochemical from the freshwater cyanobacterium *Fischerella muscicola*. *Journal of Phycology* 27: 686–692. Photograph copyright Elisabeth Gross, Konstanz, Germany.

Allelopathic interactions can influence phytoplankton succession in lakes. In a freshwater lake, dominant blue-green algae inhibited the growth of diatoms isolated from the same lake. Allelopathy, especially in benthic cyanobacteria, can influence the composition of benthic algal mats and may even lead to almost single species biofilms.

Only few reports suggest allelopathic effects of algae on higher aquatic plants. Evidence exists for allelopathic effects of especially submerged macrophytes on cyanobacteria and eukaryotic algae. Epiphytic cyanobacteria and algae appear to be less affected by plant allelochemicals than planktonic species. In situ allelopathic effects are difficult to assess due to the large dilution of released active compounds especially in phytoplankton.

Autotoxicity

Autotoxicity might be considered as a special form of allelopathic interference. Intraspecific density-regulation in plant stands is possible through the inhibition of germination, sprouting or other reproductive means by compounds released either into the sediment or water. Autotoxicity has been described for emergent macrophytes (*Typha*, *Juncus*). Inhibition of own seedlings in otherwise vegetatively spreading plants may benefit temporal dispersal of seedling establishment and population regeneration after disturbance. Autolytic compounds have also been found in some cyanobacteria (*Phormidium*, *Nostoc*) but none of the autolytic compounds has been isolated or identified.

Algal Pheromones

Several freshwater diatoms release volatile compounds that originate from eicosapentaenoic acid. They may act as pheromones. *Asterionella formosa* produces fucoserratene and *Gomphonema parvulum* a hormosirene and a dictyopterene.

Figure 7 *M. spicatum* and its major allelochemical tellimagrandin II. This and other hydrolysable polyphenols (also named tannins) have multiple functions in this macrophyte. These allelochemicals are allelopathically active, interfere with the growth of herbivorous aquatic moth larvae, exhibit antibacterial activity, but might also be degraded by specialized bacteria associated with this macrophyte. Abiotic factors also affect their concentration and composition. Adapted from Gross E M, Meyer, H and Schilling G. (1996) Release and ecological impact of algicidal hydrolysable polyphenols in *Myriophyllum spicatum*. *Phytochemistry* 41: 133–138; Leu E, Krieger-Liszkay A, Goussias C, and Gross EM (2002) Polyphenolic allelochemicals from the aquatic angiosperm *Myriophyllum spicatum* L. inhibit photosystem II. *Plant Physiology* 130: 2011–2018; Choi C, Bareiss C, Walenciak O, and Gross EM (2002) Impact of polyphenols on the growth of the aquatic herbivore *Acentria ephemerella* (Lepidoptera: Pyralidae). *Journal of Chemical Ecology* 28: 2223–2235; Walenciak O, Zwisler W, and Gross EM (2002) Influence of *Myriophyllum spicatum*-derived tannins on gut microbiota of its herbivore *Acentria ephemerella*. *Journal of Chemical Ecology* 28: 2045–2056; Müller N, Hempel M, Philipp B, and Gross EM (2007) Degradation of gallic acid and hydrolysable polyphenols is constitutively activated in the freshwater plant-associated bacterium *Matsuebacter* sp. FB25. *Aquatic Microbial Ecology* 47: 83–90; Gross EM (2003) Differential response of tellimagrandin II and total bioactive hydrolysable tannins in an aquatic angiosperm to changes in light and nitrogen. *Oikos* 103: 497–504. Photograph copyright Martin Mörtl/Elisabeth Gross, Konstanz, Germany.

Allelochemical Reactions Involving Heterotrophic Microorganisms

Several aquatic plants and animals produce antimicrobial compounds inhibiting fungi or heterotrophic bacteria. How important these compounds are for the fate of the producing organisms is unclear. It is also unknown if these compounds are actively excreted.

Besides being directly involved, bacteria often seem to affect allelochemical reactions. They can metabolize allelopathically active compounds. Fish mucus bacteria seem to be crucial for the production of the kairomone involved in fish–zooplankton interactions.

See also: Silica.

Further Reading

Bronmark C and Hansson LA (2000) Chemical communication in aquatic systems: An introduction. *Oikos* 88: 103–109.

Burks RL and Lodge DM (2002) Cued in: Advances and opportunities in freshwater chemical ecology. *Journal of Chemical Ecology* 28: 1901–1917.

Gross EM (2003) Allelopathy of aquatic autotrophs. *Critical Reviews in Plant Sciences* 22: 313–339.

Lass S and Spaak P (2003) Chemically induced anti-predator defences in plankton: A review. *Hydrobiologia* 491: 221–239.

McClure JW (1970) Secondary constituents of aquatic angiosperms. In: Harborne JB (ed.) *Phytochemical Phylogeny*, pp. 234–268. Academic Press: London.

Ruther J, Meiners T, and Steidle JLM (2002) Rich in phenomena-lacking in terms. A classification of kairomones. *Chemoecology* 12: 161–167.

Whittaker RH and Feeny PP (1971) Allelochemics: Chemical interactions between species. *Science* 171: 757–770.

Carbohydrates

N O G Jørgensen, University of Copenhagen, Frederiksberg, Denmark

Introduction

Carbohydrates are probably the most common organic substances in nature. Carbohydrates are found in all organisms and are involved in many vital functions. In living cells, the carbohydrate ribose is an essential part of nucleic acids (RNA and DNA), but carbohydrates are also elements in many other molecules such as coenzymes, antibiotics and toxins. In plants and algae, carbohydrates are among the first organic compounds formed during photosynthesis. Carbohydrates can be used as an immediate source of energy by largely all cells, but in many photosynthetic organisms a major portion of the carbohydrates will be used for production of structural compounds, e.g., cellulose in cells walls, or for synthesis of storage products such as starch. Living and dead cells can release carbohydrates to the ambient water, as discussed later, and this contributes to the pool of dissolved organic matter in the water. In lakes and streams, carbohydrates serve as important nutrient sources to microorganisms that hereby convert the dissolved carbohydrates into microbial cells that can enter the food webs and feed other organisms.

Carbohydrates and their Building Blocks

Carbohydrates share the same empirical formula $(CH_2O)_n$ and are divided into four major groups: monosaccharides, disaccharides, oligosaccharides, and polysaccharides.

Monosaccharides are the building blocks of di-, oligo-, and polysaccharides. Most monosaccharides contain 5 or 6 carbon atoms and are named pentoses ($C_5H_{10}O_5$) and hexoses ($C_6H_{12}O_6$), but monosaccharides with up to nine carbon atoms are found. The most common pentose is ribose and its derivative deoxyribose. Typical hexoses are glucose, fructose, and galactose. Examples of different saccharides are shown in **Figure 1**.

Disaccharides consist of two monosaccharide molecules and include sucrose (common sugar; one glucose and one fructose molecule), lactose (milk sugar; one glucose and one galactose molecule), and maltose (malt sugar, e.g., from degradation of starch; only glucose molecules).

Oligosaccharides include from 3 and up to 20 monosaccharide units and are less common carbohydrates than the other groups of polymer saccharides. Examples of oligosachharides are the trisaccharide raffinose (consists of galactose, glucose, and fructose; found in plants) and maltotriose (three glucose units; formed during degradation of starch and occur in blood of some arthropods).

Polysaccharides may contain up to 10 000 saccharide units that are linked together and often form widely branched structures. Most polysaccharides serve as energy reserves like starch, glycogen, and cellulose. The monomer units may be identical as in starch and glycogen (glucose units), but polysaccharides may also consists of arabinose or xylose. Some polymer compounds include sugar derivatives, e.g., aminosugars (in chitin of insects and crustaceans and in bacterial cell walls) and galacturonic acid (derivative of galactose) in pectin in plant material. Today more than a hundred of different sugar species and sugar derivatives are known.

Carbohydrates in Aquatic Environments

Carbohydrates in aquatic environments can either occur as free monosaccharides such as glucose, fructose, and galactose, or be combined in larger molecules and make up cellulose and starch. When dissolved in water, free and combined carbohydrates are often abbreviated DFCHO and DCCHO (dissolved free and combined carbohydrates), respectively.

Analytical Problems Measuring Carbohydrates in Natural Waters

Before presenting concentrations on DFCHO and DCCHO in different freshwater environments, it is necessary to discuss the analytical procedures that have been used for measurements of carbohydrates and their shortcomings.

Colorimetric Methods

Carbohydrates have often been measured using colorimetric procedures, in which the sugars react with various chemical compounds to form a colored substance that can be measured by spectrophotometry (light absorption). The weakness by most of the colorimetric procedures is that their sensitivity is too low to measure natural concentrations of free saccharides (DFCHO). This requires a concentration step.

The colorimetric techniques only react with free monosaccharides and this means that concentrations

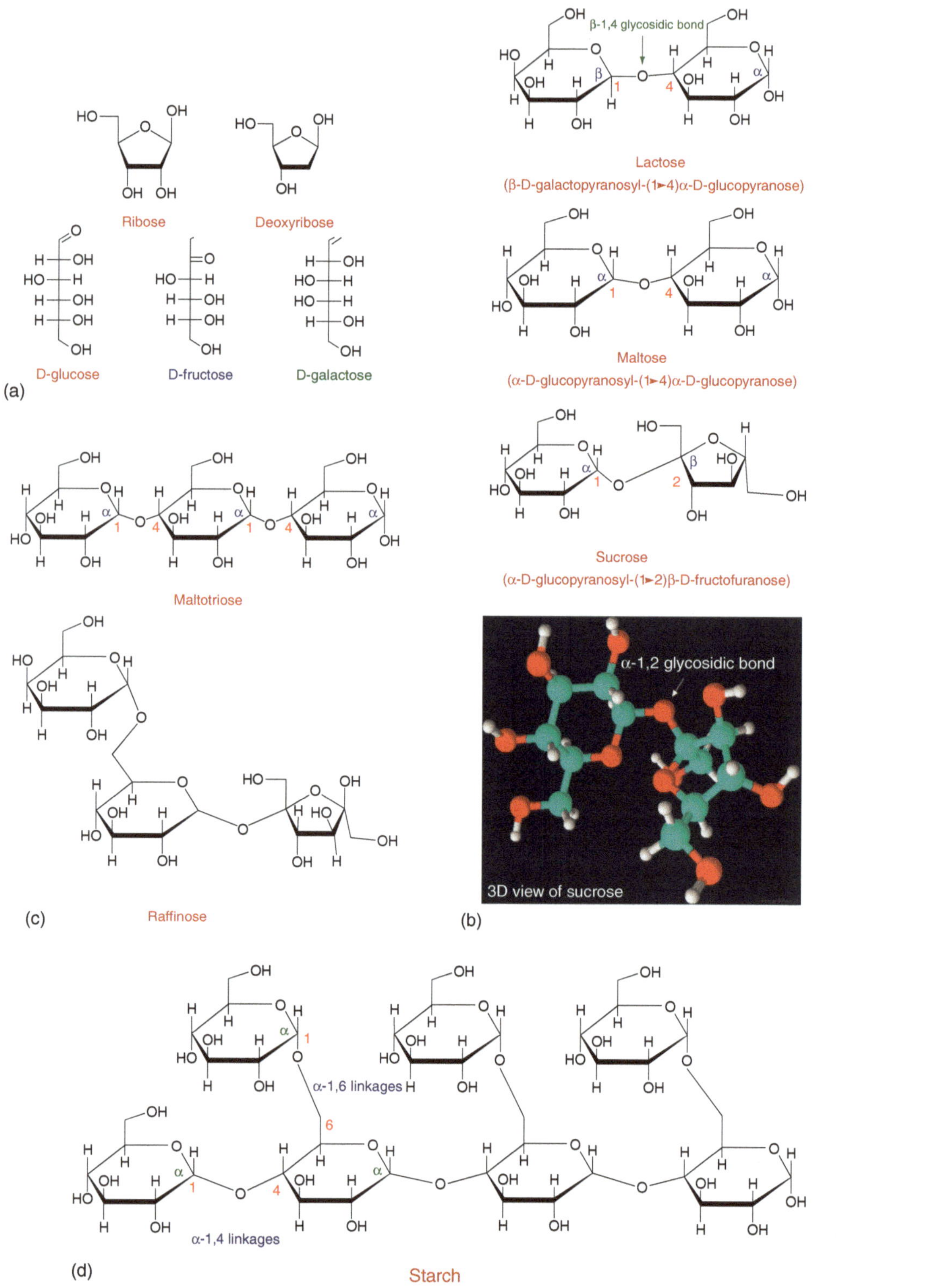

Figure 1 Examples of monosaccharides and carbohydrates in living organisms. (a) Monosaccharides: Pentoses (ribose and deoxyribose) and hexoses (glucose, fructose, and galactose). The most common enantiomer forms (D isomers) are shown. (b) Disaccharides: Lactose (milk sugar), maltose (malt sugar), and sucrose (common sugar). Note that both α-1,4 and β-1,4 bonds occur between the two monosaccharide molecules. (c) Trisaccharides: Maltotriose (consists of three glucose molecules linked by α-1,4 bonds) and raffinose (consists of galactose, glucoses, and fructose linked by β-1,4 bonds). (d) The polysaccharide starch, consisting of glucose molecules connected by both α-1,4 and α-1,6 linkages. Starch molecules can contain several millions of glucose residues. See text for more details.

of polysaccharides and complex carbohydrates (DCCHO) must be hydrolyzed to their monomer substances to be quantified. The hydrolytic step, e.g., treatment with dilute HCl at 100 °C, has been shown to degrade or convert some sugars. Furthermore, some carbohydrates, e.g., in humic matter, are not converted to free monosaccharides by hydrolysis. Also, the colorimetric methods do not produce an identical color formation by reaction with all monosaccharides. These circumstances introduce serious biases when analyzing natural concentrations of DFCHO and DCCHO, and they imply that concentrations of carbohydrates measured with different colorimetric methods cannot be directly compared.

Chromatographic Methods

Analysis of single monosaccharides (with or without hydrolysis) has been performed using gas chromatography (GC) or high-performance liquid chromatography (HPLC). The GC detection requires several derivatization steps, while the HPLC procedure in combination with pulsed amperometric detection (PAD) can produce a relative fast and direct quantification of mono- and disaccharides in natural water samples. The GC and HPLC techniques both involve some chemical manipulation of the water sample and may introduce minor analytical artifacts.

Comparison of Methods

When comparing amounts of carbohydrates in a natural water sample, determined by colorimetric approaches and chromatographic techniques, e.g., GC and HPLC, the colorimetric approaches in most cases indicate higher concentrations than the chromatographic techniques. As an example, concentrations of total dissolved carbohydrates in the Japanese Lake Nakanuma was about twofold higher when determined with a colorimetric method than with a GC method (**Table 1**). One reason for this discrepancy may be that many sugars in natural water samples have not yet been characterized and therefore they are not detected by the chromatographic techniques. The different analytical circumstances and occurrence of unknown saccharides in natural waters make it difficult to compare concentrations of DFCHO and DCCHO obtained by different methods.

Concentrations of Dissolved Free and Combined Carbohydrates

Total Amounts of Carbohydrates

Concentrations of total dissolved carbohydrates (TDCHO, consisting of DFCHO and DCCHO) range between 1 and 10 μM in most lakes and rivers, but large variations occur (**Table 1**). The lowest concentrations are typically found in environments with a low biological activity such as alpine and humic lakes, while higher concentrations usually are found in eutrophic environments with a high algal production. The very high concentrations (>20 μM) reported in some cases may represent specific biological events that release carbohydrates to the water, e.g., an intense phytoplankton bloom with a high algal biomass, but physical factors such as tidal changes in estuarine environments may also cause a significant discharge of organic matter as observed in the Indian Mangrove (**Table 1**).

Carbohydrates Relative to DOC

Carbohydrates make up an important portion of dissolved organic matter in fresh water. Relative to the total amount of dissolved organic carbon (DOC), TDCHO typically makes up from 5 to 20% of the DOC (**Table 1**). In highly productive lakes, carbohydrates can constitute about half or more of DOC, as observed in the German Lake Plußsee and the Antarctic lakes. The sources of carbohydrates in these two lakes were an intense algal production because of high input of inorganic nutrients from the surrounding environment (Lake Plußsee) and fur seal remains (Antarctic lakes). In contrast to these lakes, carbohydrates are of quantitative less importance in humic lakes, since the humic matter itself contains a large portion of DOC. In running waters, carbohydrates make up a similar proportion of DOC as in lakes, but the carbohydrates may originate from flooding of land rather than release by living organisms in the water.

Free Monosaccharides

Most carbohydrates in natural waters consist of polysaccharides or are part of large molecules such as humic compounds, and only a minor portion occurs as free mono- or disaccharides. In many fresh waters, DFCHO make up between 1 and 10% of the dissolved carbohydrates (**Table 1**). The reason for the lower abundance of DFCHO, relative to DCCHO, is that free saccharides are readily taken up by microorganisms, while DCCHO are more resistant to microbial utilization and must be degraded to monomers by enzymes before uptake. An example of lake water with rather high abundance of DFCHO is the humic Lake Mekkojärvi in Finland, where up to half of the carbohydrates were free saccharides. This appears not to be a general characteristic of humic lakes, as DFCHO only made up 3% of all sugars in the humic Lake Skärhultsjön in Sweden.

Table 1 Concentrations of total dissolved carbohydrates (TDCHO), dissolved combined carbohydrates (DCCHO) and dissolved free carbohydrates (DFCHO) in various freshwater environments

Location/environment	*TDCHO*		*DCCHO (μM)*	*DFCHO (μM)*	*Analytical method*	*Source*
	Conc. (μM)	*% of DOC*				
Lakes						
Lake Klintsjön (clearwater lake, Sweden)	10.6	26		0.100	PAD-HPLC	1
Sep Reservoir (oligo-mesotrophic lake, Massif Central, France)	1.5–31.6		0.56–12.4	0.97–19.2	Colorimetric	2
Lake Constance (mesotrophic lake, Germany)	1.2–3.6	6		0.020–0.440	PAD-HPLC	2–4
Lake Plußsee (eutrophic lake, Germany)	3.5–24.4	3–44			GC	5
Lake Nakanuma (eutrophic lake, Japan)	0.5–3.3	1–5			GC Colorimetric	6
	3.3–6.7	5–10				
Lake Au (eutrophic lake, Switzerland)				0.001–0.048	MS-HPLC	7
Lake Bagsværd (eutrophic lake, Denmark)			1.2–2.2	0.067–0.224	PAD-HPLC	8
Cima Lake systems (eutrophic lakes, South Brazil)	11.1–23.9	11–18	–	–	Colorimetric	9
Antarctic Lakes (eutrophic lakes, Signy Island)	1–11	7–65			PAD-HPLC	10
Lake Diktar-Erik (oligotrophic and boreal lake, Sweden)		0.1–0.4 (DFCHO only)	ND	0.034–0.150	PAD-HPLC	11
Lake Skärshultsjön (humic lake, Sweden)	6.6	4.5	6.4	0.190	PAD-HPLC	1
Lake Mekkojärvi (humic lake, Finland)	0.06–4.4	0.3	0.05–0.95	0.008–0.466	GC	5
Lake Brienz (oligotrophic and alpine lake, Switzerland)	0.6	14			Colorimetric	12
Running waters						
White Clay Creek (variable nutrient status, woodland area, PA, USA)	–	5.7–6.1	0.06–1.31	–	PAD-HPLC	13
	0.7–8.2	3.0–7.3	0.64–8.1 (stormflow: 12.7)	0.048–0.074 (stormflow: 0.380)	PAD-HPLC	14
Lena River (arctic drainage basin, Russia)	1.6	21			Colorimetric	15
Mangrove (Cochin estuary, southwest India)	22–315	–	–	–	Colorimetric	16

Danube River Delta (Sulina, Romania)	3 (spring) 9.2 (winter)	8 (spring) 25 (winter)	–	–	Colorimetric	17

The proportion of TDCHO in the pool of dissolved organic carbon (DOC) is indicated. For conversion of carbohydrate concentrations into molar concentrations from amounts published in weight per volume or content of carbon, a molecular weight of 180 g mol^{-1} or six carbon atoms per molecule has been assumed. ND: no data.

PAD-HPLC and MS-HPLC: High performance liquid chromatography in combination with pulsed amperometric detection (PAD) or mass spectrometry (MS); GC: Gas chromatography, including various derivatization steps.

Sources

1. Tranvik LJ and Jørgensen NOG (1995) Colloidal and dissolved organic matter in lake water: Carbohydrate and amino acid composition, and ability to support bacterial growth. *Biogeochemistry* 30: 77–97.
2. Jugnia LB, Richardot M, Debroas D, and Devaux J (2006) Bacterial production in the recently flooded Sep reservoir: Diel changes in relation to dissolved carbohydrates and combined amino acids. *Hydrobiologia* 563: 421–430.
3. Bunte C and Simon M (1999) Bacterioplankton turnover of dissolved free monosaccharides in a mesotrophic lake. *Limnology and Oceanography* 44: 1862–1870.
4. Weiss M and Simon M (1999) Consumption of labile dissolved organic matter by limnetic bacterioplankton: The relative significance of amino acids and carbohydrates. *Aquatic Microbial Ecology* 17: 1–12.
5. Münster U (1993) Concentrations and fluxes of organic-carbon substrates in the aquatic environment. *Antonie Van Leeuwenhoek International Journal of General and Molecular Microbiology* 63: 243–274.
6. Ochiai M and Ukiya T (1981) Seasonal variations of dissolved organic constituents in Lake Nakanuma during March and November 1979. *Verhandlungen Internationale Vereinigung für Theoretische und Angewandte Limnologie* 21: 682–687.
7. Meon B and Juttner F (1999) Concentrations and dynamics of free mono- and oligosaccharides in a shallow eutrophic lake measured by thermospray mass spectrometry. *Aquatic Microbial Ecology* 16: 281–293.
8. Jørgensen NOG and Jensen RE (1994) Microbial fluxes of free monosaccharides and total carbohydrates in fresh water determined by PAD-HPLC. *FEMS Microbiology Ecology* 14: 79–94.
9. Pedrosa P, Calasans CVC, and Rezende CE (1999) Particulate and dissolved phases as indicators of limnological and ecophysiological spatial variation in Cima Lake system, Brazil: a case study. *Hydrobiologia* 411: 89–101.
10. Quayle WC and Convey P (2006) Concentration, molecular weight distribution and neutral sugar composition of DOC in maritime antarctic lakes of differing trophic status. *Aquatic Geochemistry* 12: 161–178.
11. Jonsson A, Ström L, and Åberg J (2007) Composition and variations in the occurrence of dissolved free simple organic compounds of an unproductive lake ecosystem in northern Sweden. *Biogeochemistry* 82: 153–163.
12. Chanudet V and Filella M (2006) The application of the MBTH method for carbohydrate determination in fresh waters revisited. *International Journal of Environmental Analytical Chemistry* 86: 693–712.
13. Cheng XH and Kaplan LA (2001) Improved analysis of dissolved carbohydrates in stream water with HPLC-PAD. *Analytical Chemistry* 73: 458–461.
14. Gremm TJ and Kaplan LA (1997) Dissolved carbohydrates in streamwater determined by HPLC and pulsed amperometric detection. *Limnology and Oceanography* 42: 385–393.
15. Kattner G, Lobbes JM, Fitznar HP, Engbrodt R, Nothig E-M, and Lara RJ (1999) Tracing dissolved organic substances and nutrients from the Lena River through Laptev Sea (Arctic). *Marine Chemistry* 65: 25–39.
16. Geetha R, Chandramohanakumar N, and Mathews L (2006) Seasonal and tidal impact on the organic compounds and nutrients distribution in tropical mangroves, Kerala, India. *Chemistry and Ecology* 22: 29–45.
17. Saliot A, Derieux S, Sadouni N, Bouloubassi I, Fillaux J, Dagaut J, Momzikoff A, Gondry G, Guillou C, Breas O, Cauwet G, and Deliat G (2002) Winter and spring characterization of particulate and dissolved organic matter in the Danube-Black Sea mixing zone. *Estuarine Coastal and Shelf Science* 54: 355–367.

Sources of Carbohydrates in Water

Release by Phytoplankton

The main source of dissolved carbohydrates in aquatic environments is photosynthetic organisms like microalgae and aquatic plants, although other organisms may also contribute carbohydrates to the water. Studies of microscopic plankton algae (microalgae) have shown that the cells release a large variety of carbohydrates during photosynthesis in the light, while the release is lower in the dark. It is not known whether this release is intentional or accidental, but possibly the cells during the daytime produce more carbohydrates than that are needed in biosynthetic pathways or for energy purposes. The released carbohydrates can make up more than 15% of carbohydrates inside the cells and be a considerable loss of cell material.

Plankton algae depend on inorganic nutrients such as nitrate and phosphate for growth. However, when the nutrients are depleted as happens after periods with warm and sunny weather, the phytoplankton cells cannot survive and begin to disintegrate. During this breakdown, the cells release large amounts of carbohydrates. This means that phytoplankton blooms in many cases will be accompanied by a peak of both free and combined carbohydrates. It seems that especially diatom blooms can lead to high concentrations of carbohydrates in lakes.

Release by Colony-Forming Algae and Plants

Mucopolysaccharides produced by colony-forming algae are another source of carbohydrates in lakes and rivers. These carbohydrates consist of many different monosaccharides and help the cells adhere to each other and form colonies. When the colonies disintegrate, e.g., due to senescence, the carbohydrates are released to the ambient water. Other carbohydrates produced by algae include the intracellular storage product β-1,3-glucan, which mainly consists of glucose monomers. Both mucopolysaccharides and glucan are degraded to smaller saccharides by microorganisms that produce hydrolytic enzymes (see later).

In shallow waters, aquatic plants may be another major source of carbohydrates because carbohydrates can constitute about 80% of the organic matter in plant tissue. Following the decay of plants, variable amounts of carbohydrates can be released to the water.

Other Sources of Carbohydrates

In addition to direct release by algae and plants, dissolved carbohydrates in natural waters can also originate from (1) bacterial cell walls that contain several different saccharides, (2) fungi and insects in which the cell walls and external skeleton (exoskeleton) contain amino sugars that can be converted to free ammonium and glucose, and (3) zooplankton which during feeding on phytoplankton often loose dissolved organic compounds from the algal cells, a process known as sloppy feeding. Finally, all living cells include ribose in the DNA (as deoxyribose) and RNA and this sugar is also released to the water when the organisms die.

Composition of Carbohydrates in Water

Carbohydrates in Algae and Dissolved in Water

Since phytoplankton are the main sources of dissolved carbohydrates in many aquatic environments, it might be assumed that composition of carbohydrates in water and the phytoplankton cells is similar. This is only partly true, as can be seen in **Figure 2**. The composition of carbohydrates in phytoplankton (from a marine location but most likely identical to freshwater phytoplankton) is dominated by glucose, mannose + xylose (the two monosaccharides cannot always be separated during analysis), galactose, and arabinose. Other sugars, such as rhamnose, fucose, and ribose, were less common. When composition of DCCHO and DFCHO in lake water is compared (the Danish Lake Furesøen), it can be seen that DCCHO were dominated by glucose and galactose, while DFCHO had a rather equal abundance of mannose + xylose, galactose, glucose, and fructose. The change in major carbohydrates between algae, DCCHO and DFCHO most likely reflects that some monosaccharides are more easily degraded by microorganisms and this causes a change in the carbohydrate spectrum.

Natural Saccharides in Fresh Water

Glucose is the most universal building block in carbohydrates, but glucose is also easily assimilated by most living cells and used as source of energy or for production of new cells. This versatile role of glucose may explain the reduction of glucose in the pool of free saccharides in the lake water in **Figure 2**, as compared to abundance of glucose in phytoplankton and DCCHO. Among other carbohydrates identified in phytoplankton and water, are arabinose (pentose; found in hemicellulose and pectin), cellobiose (two glucose molecules; subunit of cellulose), fucose (pentose; found on some cell surfaces and can be linked to nitrogen), rhamnose (hexose; found in some plants), fructose (hexose; found in many plants and being sweeter than sucrose), and melibiose (disaccharide of galactose and glucose; found in some plants).

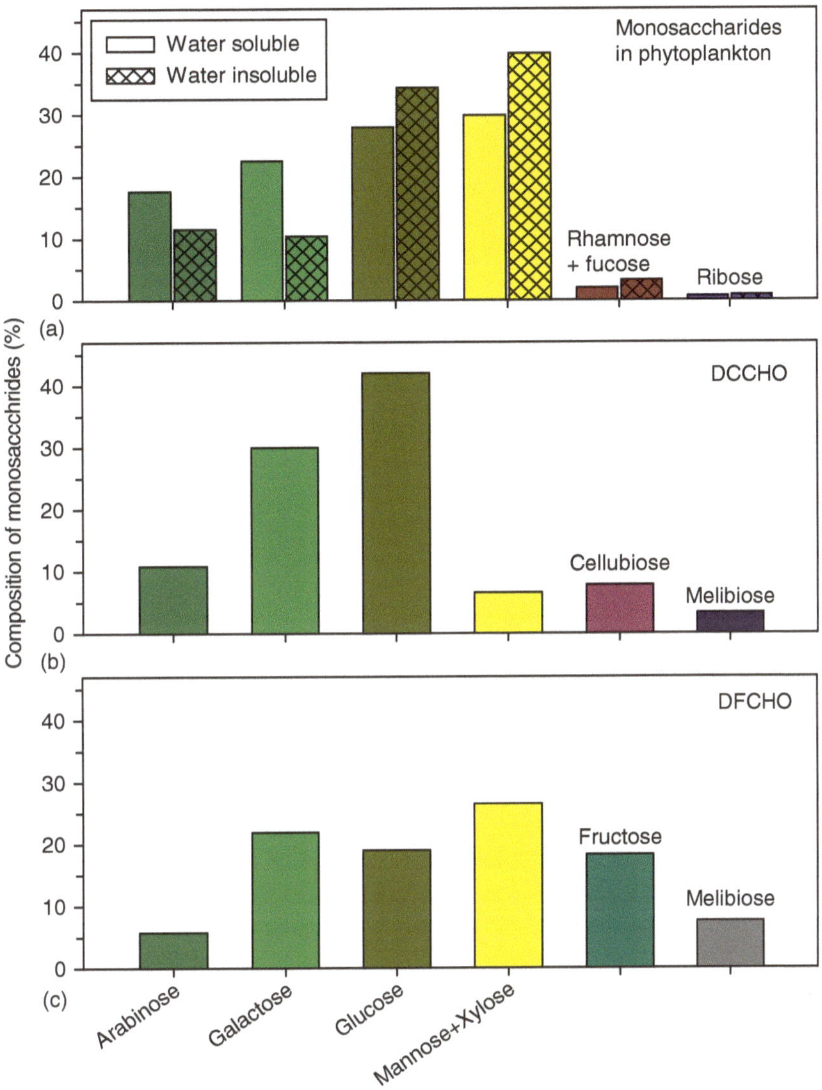

Figure 2 Composition of different monosaccharides in (a) phytoplankton from Akkeshi Bay, Japan (both water soluble and water insoluble saccharides are shown) and (b, c) in pools of DCCHO and DFCHO in water of Lake Furesø, Denmark. Part (a) reproduced from Goes JI, Nobuhiko H, Satoru T, Takeo H, and Hiroaki S (1996) Metabolism of neutral monosaccharide constituents of storage and structural carbohydrates in natural assemblages of marine phytoplankton exposed to ultraviolet radiation. *Limnology and Oceanography* 41: 1478–1489, with permission from The American Society of Limnology and Oceanography; Parts (b, c) reproduced from Jørgensen NOG and Jensen RE (1994) Microbial fluxes of free monosaccharides and total carbohydrates in freshwater determined by PAD-HPLC. *FEMS Microbiology Ecology* 14: 79–94, with permission from Elsevier.

Dynamic Concentrations of Dissolved Carbohydrates

Annual Variations in Carbohydrate Pools

Since carbohydrates are released by many organisms but also utilized by a large variety of organisms, it is not surprising that concentrations of dissolved carbohydrates are variable in natural waters. Studies of carbohydrates in lakes indicate that concentrations of free mono- and disaccharides (make up most of the DFCHO pool) typically fluctuate more than concentrations of all carbohydrates in the water. This can be illustrated for the large German Lake Constance which has been focus of several studies on occurrence and cycling of carbohydrates. In this lake, the total concentration of dissolved carbohydrates (TDCHO, consisting of DCCHO and DFCHO) was measured from spring to end of summer in 1994 and was found to vary between 2 and 5 μM (**Figure 3(a)**).

Free dissolved saccharides (DFCHO) were not measured in the lake in 1994, but during the same period in the following year, DFCHO varied between

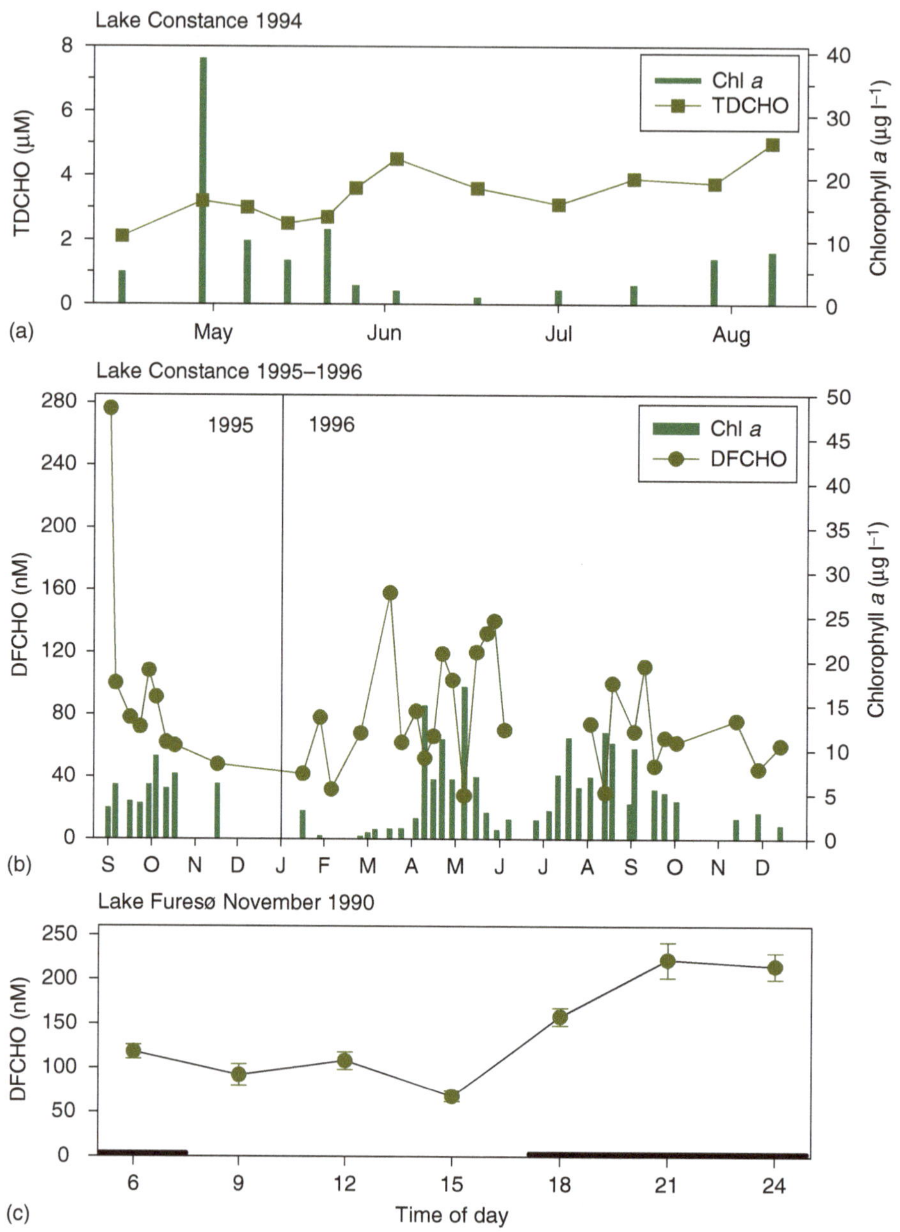

Figure 3 Seasonal and day–night changes of carbohydrates and monosaccharides in lakes. (a, b) Concentrations of total dissolved carbohydrates (TDCHO) and total dissolved free carbohydrates (DFCHO), and concentrations of chlorophyll *a* in Lake Constance, Germany. (c) Concentration of DFCHO during a diel cycle in Lake Furesø, Denmark. Error bars indicate standard deviation of the DFCHO concentrations: Light-dark periods are shown. Part (a) reproduced from Hanisch K, Schweitzer B, and Simon M (1996) Use of dissolved carbohydrates by planktonic bacteria in a mesotrophic lake. *Microbial Ecology* 31: 41–55, with permission from Springer-Verlag; Part (b) reproduced from Bunte C and Simon M (1999) Bacterioplankton turnover of dissolved free monosaccharides in a mesotrophic lake. *Limnology and Oceanography* 44: 1862–1870, with permission from The American Society of Limnology and Oceanography; Part (c) reproduced from Jørgensen NOG and Jensen RE (1994) Microbial fluxes of free monosaccharides and total carbohydrates in freshwater determined by PAD-HPLC. *FEMS Microbiology Ecology* 14: 79–94, with permission from Elsevier.

28 and 160 nM, and this variation was twofold higher than observed for TDCHO in the previous year (**Figure 3(b)**). If concentrations of DFCHO in the studied periods in 1994 and 1996 are considered, variations of DFCHO were even higher (up to 16-fold), emphasizing that free saccharides can be highly dynamic compounds in lake water.

Daily Variations in Carbohydrate Pools

Even on a shorter time scale, such as one-day, significant variations of DFCHO can occur. In the Danish Lake Furesø, the concentration of DFCHO was found to vary more than threefold (from 68 to 222 nM) during one day (**Figure 3(c)**). The reason for the higher concentrations in the night was not determined, but 'sloppy feeding' by zooplankton when grazing on phytoplankton may have been important in release of the free carbohydrates, as many zooplankton organisms are mainly active in the dark.

Carbohydrates and Algae in Running Waters

In running waters, the water flow means that concentrations of carbohydrates and densities of algae in the water rarely will vary simultaneously, despite algae in most cases will be the dominant source of the carbohydrates (see above). This is illustrated by concentrations of DCCHO and DFCHO and the content of chlorophyll *a* (indicator of algal biomass) in a reservoir in the River Sep and Riaux system in central France. The concentrations of dissolved carbohydrates were highest in early June, with concentrations of 75 μM (DCCHO) and 39 μM (DFCHO), but at that time very low concentrations of chlorophyll *a* occurred (**Figure 5(a)**). The specific source of carbohydrates was not determined in this study, but input of allochthonous organic matter originating from decomposition of flooded terrestrial organic matter might have produced the dissolved carbohydrates.

Relations Between Carbohydrates and Phytoplankton in a Large Lake

Even in lakes without a water flow, concentrations of carbohydrates may not coincide with the density of phytoplankton or follow the decay of phytoplankton cells. In the large Lake Constance, the amount of chlorophyll *a* did not coincide with concentrations of neither DCCHO nor DFCHO, although a positive relationship occasionally was found (**Figure 3**). Among reasons for this discrepancy can be (1) that few carbohydrates were released by living and senescent algal cells, (2) that release of carbohydrates during 'sloppy feeding' by zooplankton was independent of the algal chlorophyll *a* content, and (3) that microorganisms took up the carbohydrates as soon as they were released and hereby controlled the amount of DFCHO in the water. It should be mentioned that a more intense sampling, e.g., at daily intervals, might have given a more accurate illustration of relations between carbohydrates and phytoplankton in the lake, but such intense sampling is not always realistic.

The frequently higher variations in concentrations of DFCHO than of DCCHO most likely reflect a fluctuating input of free saccharides from phytoplankton during photosynthesis (at variable rates in day and night) or during grazing and decay of the cells, and by active microorganisms that immediately assimilate the released mono- and disaccharides. In contrast, DCCHO often consists of carbohydrates that are more recalcitrant (difficult for the microorganisms to degrade), such as cellulose and other structural carbohydrates from algae and plants. However, after an intense phytoplankton bloom, dead algal cells may release a substantial amount of polymer saccharides that are easily degraded by bacteria, e.g., β-1,3-glucan and mucopolysaccharides, and this will lead to a temporal increase of the DCCHO pool.

Assimilation of Carbohydrates

Uptake of Carbohydrates and Bacterial Growth

Most eukaryotic and prokaryotic cells can utilize different monosaccharides for production of energy and synthesis of new cell material. Bacteria are especially efficient in assimilating monosaccharides, even at very low concentrations, and this significantly influences the abundance of free carbohydrates in lakes and rivers. The ability by bacteria to assimilate glucose has previously been used as an indicator of bacterial activity in natural waters. In the 1970s, it was suggested that the maximum uptake rate of glucose by bacteria, V_{max}, reflected the potential bacterial growth in a lake or river and was named 'heterotrophic potential.' The approach was successfully used for comparison of different environments, although it did not indicate the actual bacterial growth rate.

Uptake of Glucose and Fructose

With the introduction of PAD-HPLC for measurements of concentrations of individual DFCHO in the early 1990s (see above) and in combination with radioactively labeled tracers of monosaccharides, bacterial assimilation rates of different monosaccharides have been determined in many aquatic environments. The measured assimilation rates were further related to the bacterial production to determine the nutritional importance of carbohydrates to the bacterial metabolism. As an example, assimilation of glucose has been found to vary from 0.4 to 75 nmol l^{-1} h^{-1} and sustain from 1% to 46% of the bacterial carbon demand (**Table 2**). Similar studies for fructose indicate a lower uptake and a corresponding lower importance to bacteria. When comparing the amount of glucose and fructose to the uptake rate, it was found that the time needed to turn over the pools of these two monosaccharides ranged from few hours to several days in the water.

Table 2 Microbial uptake of monosaccharides and carbohydrates in lakes

Compound	*Location*	*Concentration*[a]	*Uptake rate (nmol* l^{-1} h^{-1}*)*[b]	*Importance of uptake to bacterial production*	*Method*	*Comment*	*Source*
Glucose	Lake Furesøen (mesotrophic, Denmark)	9–49 nM	0.4–3.1	Glucose-C met 10% (diel study) to 21% (118 h study) of the bacterial C demand	Uptake of ^{14}C-glucose	Diel and laboratory study. November; 6 °C	1
Glucose	Lake Skärshultsjön (humic, Sweden)	110 nM	0.5 (mean value)	Glucose-C met 1.3% of the bacterial C demand	Uptake of ^{14}C-glucose	Laboratory incubation of size-separated water. June; 20 °C	2
Glucose	Lake Furesøen (eutrophic, Denmark)	46–67 nM	75 (mean value)	Glucose-C met 46% of the bacterial C demand	Uptake of ^{14}C-glucose	Surface water. May 5; 14 °C	3
Fructose	Lake Furesøen (eutrophic, Denmark)	6.5–17 nM	5.9 (mean value)	Fructose-C met 3.6% of the bacterial C demand	Uptake of ^{14}C-fructose	Surface water. May 5; 14 °C	3
Fructose	Lake Furesøen (mesotrophic, Denmark)	9–57 nM	0.6–2.1	Fructose-C met 4.7% (diel study) to 11% (118 h study) of the bacterial C demand	Uptake of ^{14}C-fructose	Diel and laboratory study. November; 6 °C	1
DFCHO	Lake Diktar-Erik (oligotrophic and boreal, Sweden)	34–150 nM	ND	DFCHO-C and organic acid-C together met ~30% of bacterial C demand	Mass balance estimate of gross values (include respiration)	Boreal lake with ice-free period from June to October	4
DFCHO	Lake Constance (mesotrophic, Germany)	>22–440 nM	0.1–3.7	DFCHO-C met 22–71% of bacterial C demand	Uptake of several ^{14}C-labeled monosaccharides	The highest uptake rates occurred in the upper 10 m of the 250 m deep lake	5
Total DCCHO	Lake Constance (mesotrophic, Germany)	1700–5500 nM	6.5–48	TCCHO-C met 17–305% of bacterial C demand	Uptake of TCCHO in lake water by particles <1.0 μm (assumed to be bacteria) over 24–48 h	Water was collected at 3 m depth	6

ND: no data

Sources

1. Jørgensen NOG and Jensen RE (1994) Microbial fluxes of free monosaccharides and total carbohydrates in freshwater determined by PAD-HPLC. *FEMS Microbiology Ecology* 14: 79–94.
2. Tranvik LJ and Jørgensen NOG (1995) Colloidal and dissolved organic matter in lake water: Carbohydrate and amino acid composition, and ability to support bacterial growth. *Biogeochemistry* 30: 77–97.
3. Jørgensen NOG (1990) Assimilation of free monosaccharides and amino acids relative to bacterial production in eutrophic lake water. *Archiv für Hydrobiologie* 34: 99–110.
4. Jonsson A, Ström L, and Åberg J (2007) Composition and variations in the occurrence of dissolved free simple organic compounds of an unproductive lake ecosystem in northern Sweden. *Biogeochemistry* 82: 153–163.
5. Bunte C and Simon M (1999) Bacterioplankton turnover of dissolved free monosaccharides in a mesotrophic lake. *Limnology and Oceanography* 44: 1862–1870.
6. Hanisch K, Schweitzer B, and Simon M (1996) Use of dissolved carbohydrates by planktonic bacteria in a mesotrophic lake. *Microbial Ecology* 31: 41–55.

[a]Some concentrations were determined in units of glucose, weight of carbohydrates or content of carbon in the carbohydrates. For comparison, all concentrations have been converted to molar concentrations, assuming a molecular weight of 180 g mol^{-1} or six carbon atoms per molecule saccharide. Concentrations were measured by PAD-HPLC, except for the study of total TDCHO in Lake Constance where a colorimetric technique was used.

[b]Except for the study of TDCHO in Lake Constance, all rates represent incorporation of carbohydrates into bacteria cell biomass (net uptake) and do not include respiration of carbohydrates. In the study of TDCHO, uptake rates were measured from changes in concentrations in the water and hereby include respiration.

The variable assimilation of free monosaccharides is illustrated for uptake of glucose in Lake Furesø (**Figure 4(a)**). Between morning and midnight, the assimilation demonstrated a fivefold variation while the bacterial production only varied 2.5-fold. This indicates that the bacteria had a variable uptake of glucose during the day. The similarity in concentration and uptake of glucose suggest that the availability of glucose probably controlled the uptake rate. Some studies indicate that bacterial uptake of glucose in aquatic environments is controlled by availability of nitrogen, since access to both a carbon source (carbohydrates) and nitrogen makes it possible for bacteria to synthesize proteins and other cell constituents that require both C and N. However, in most lakes nitrogen is rarely a limiting compound, suggesting that other mechanisms probably control the bacterial utilization of glucose and fructose.

Measurements of Uptake of Free Saccharides

In most studies, uptake of free saccharides by microorganisms has been determined from incorporation of a radioisotope of the saccharide such as ^{3}H-glucose or ^{14}C-fructose. The radioisotope is added at a trace amount and does not change the natural concentration of the saccharide in the sample. The amount of isotope incorporated by the microorganisms per time (minutes or hours) is used to determine the turnover time of the saccharide in the sample. The approach assumes that the radioisotope and the natural saccharide are incorporated by the microorganisms at a similar rate. As an example, if microorganisms in the sample incorporate 10% of an added ^{14}C-glucose in 1 h, the time needed to incorporate the entire glucose pool would be 10 h. If the concentration of glucose is known, the actual amount of glucose taken up by the microorganisms per hour can be determined.

The radioisotope approach was used to measure incorporation (or uptake) rates of glucose and fructose in Lake Furesø (see above) but has also been used to determine assimilation of all DFCHO in the German Lake Constance. In Lake Constance, a selection of different ^{14}C-labeled mono- and disaccharides were added to the water and their uptake by microorganisms was measured. The concentration of the saccharides was measured simultaneously. The researchers found that DFCHO sustained from 22% to 71% of the amount of carbon required by the bacteria for growth in the lake (**Table 2**). This shows that DFCHO are – or can be – among the most important carbon sources to the bacteria.

Uptake of DFCHO by bacteria in lakes has also been determined by a different approach. In the boreal and semi-arctic Lake Diktar-Erik in northern Sweden, changes in concentrations of DFCHO between inlet and outlet of the lake, relative to the bacterial production, indicated that free saccharides met about 30% of the bacterial carbon demand (**Table 2**).

Uptake of all Carbohydrates and Bacterial Growth

Little is known about turnover of the total pool of dissolved carbohydrates (TDCHO), but this was attempted determined during the spring and summer season in Lake Constance. In the laboratory, organisms other than bacteria were removed from the water and then concentrations of TDCHO (includes both DFCHO and DCCHO) were followed for about 2 days. A decline in the TDCHO concentration was assumed to reflect uptake by bacteria. The uptake rates were found rather variable during the studied period, ranging from 5.5 to 47 nmol l^{-1} h^{-1}, and they did not follow seasonal changes in concentration of TDCHO or bacterial numbers (**Figure 4(b)** and **4(c)**). When the cell-specific uptake rates were determined, the variation in uptake was even larger, varying from 1.4 to 11 amol $cell^{-1}$ l^{-1} h^{-1} ($a = atto = 10^{-18}$) during the period. The large fluctuations in uptake rates show that carbohydrates had a changing importance as nutrients to the microorganisms, either because other organic compounds were available, or that composition of the total carbohydrate pool was unattractive or required special enzymes to be taken up, or that different bacterial species dominated the microbial populations and had a changing preference for carbohydrates as nutrients.

When related to the amount of carbon incorporated into bacterial biomass during growth in the lake, it was determined that the TDCHO uptake could sustain from 17% to >300% of carbon needed by the bacteria. A large span in the importance of carbohydrates as nutrients for bacteria was expected due to the observed, variable uptake rates, but the apparent excess uptake relative to the carbon demand (>100%) was not expected. The amount of TDCHO was determined with a colorimetric approach and this may have biased and overestimated the actual amounts of carbohydrates taken up by the bacteria (see above). Alternatively, some TDCHO were released as organic matter by bacteria, instead of being incorporated into cell biomass.

Utilization of Polysaccharides Requires Enzymes

Before microorganisms can utilize polysaccharides, the saccharides must be converted (degraded) to mono- and disaccharides that can be transported across the cell membrane. Bacteria produce a large number of

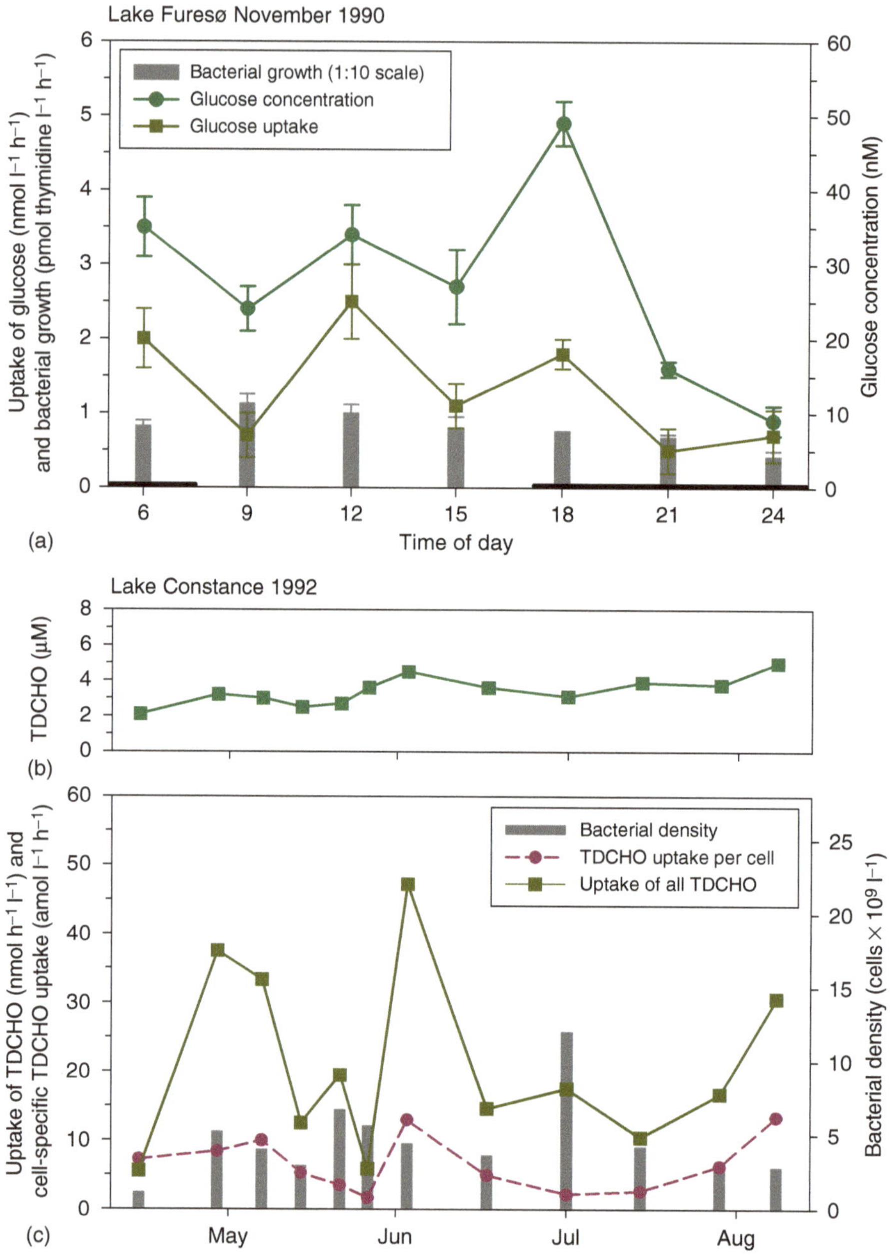

Figure 4 Uptake of glucose and total dissolved carbohydrates in lakes. (a) Day–night changes in uptake and concentrations of glucose, and of bacterial growth (shown at 1:10 scale) in Lake Furesø, Denmark. Error bars indicate standard deviation. For more information on application of the compound thymidine for measuring bacterial growth. (b, c) Concentration and uptake (per volume water and per bacterial cell) of total dissolved carbohydrates (TDCHO) in Lake Constance, Germany. The bacterial density is also indicated. Part (a) reproduced from Jørgensen NOG and Jensen RE (1994) Microbial fluxes of free monosaccharides and total carbohydrates in freshwater determined by PAD-HPLC. *FEMS Microbiology Ecology* 14: 79–94, with permission from Elsevier; Parts (b, c) reproduced from Hanisch K, Schweitzer B, and Simon M (1996) Use of dissolved carbohydrates by planktonic bacteria in a mesotrophic lake. *Microbial Ecology* 31: 41–55, with permission from Springer-Verlag.

enzymes for breaking down polysaccharides and other polymer compounds into their monomers. The enzymes are released to the water surrounding the microorganisms and are consequently named extracellular enzymes. If the enzymes encounter a substrate with chemical bonds that can be cut by the enzyme, the enzyme will release monomers from the substrate, and these monomers are taken up by the bacteria. In lakes and streams, the released monomers will be dissolved in the water and allow organisms, other than those that produced the enzymes, to utilize the monomers. This phenomenon is named 'cross-feeding.' The risk of cross-feeding will be lower if the bacteria are attached to the substrate, e.g., cellulose fibers in wood, as the organisms are closer to the substrate source. Studies have shown that there is often a tight coupling between activity of extracellular enzymes and uptake of the released monomers, as exemplified by an observed simultaneous change in glucosidase activity (releases free glucose, see below) and uptake of glucose in some freshwater environments.

Enzymatic Activity in a French River Reservoir

Among the most common extracellular enzymes for degradation of polysaccharides are α- and β-glucosidases. Glucosidases are produced by many microorganisms and are used to cleave α-1,4 bonds between glucose units in starch and glycogen, and β-1,4 bonds in cellulose. The presence of these two enzymes was examined in a reservoir in the French Sep River system from March to August. The activity of β-glucosidase peaked in June and August, while the activity of α-glucosidase was more variable during the period (**Figure 5(b)**). There were no correlations between enzyme activity, densities of bacteria (increased during the studied period), concentrations of DCCHO and DFCHO, or algal biomass (chlorophyll *a*) in the reservoir. When the bacteria were separated in different size groups, it was found that most of the enzyme activity belonged to bacteria that were attached to particle surfaces (particles from 2 to 100 μm size). Most likely, this indicates that these bacteria were fastened to algal and plant material that was rich in polysaccharides that could be degraded to α- and β-glucose monomers by the enzymes and

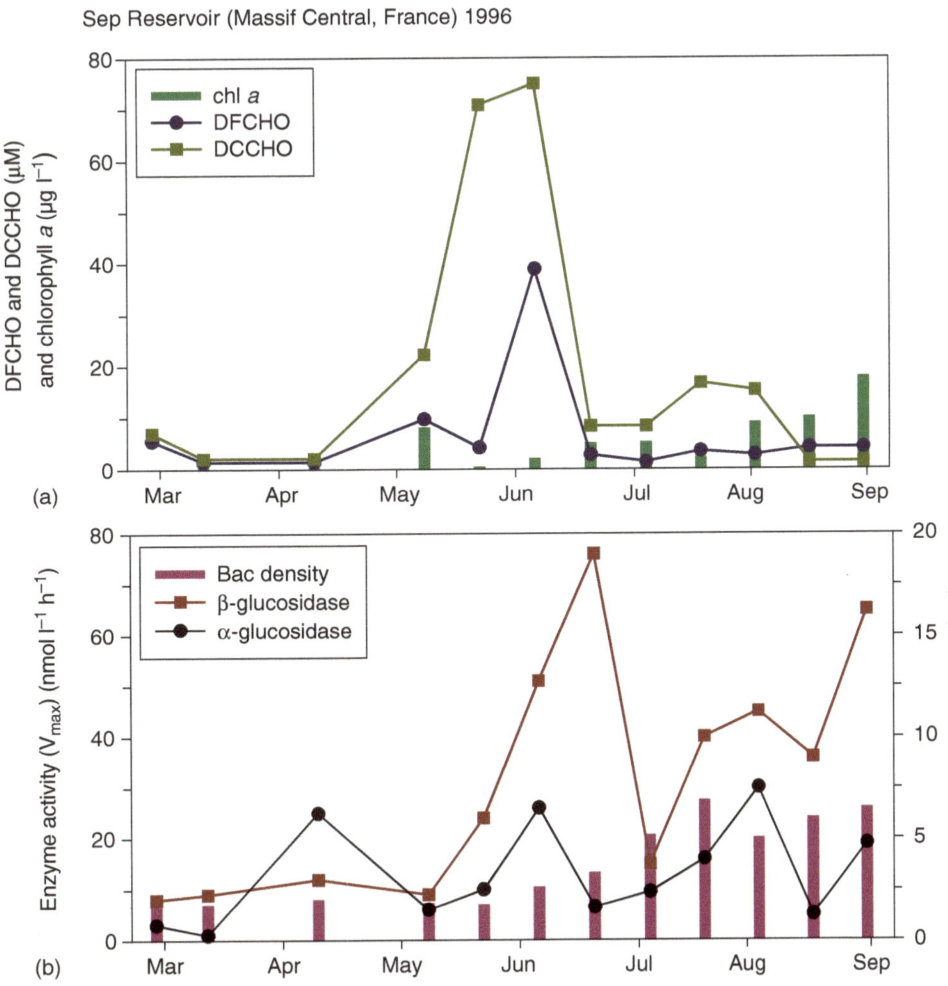

Figure 5 Seasonal variations in dissolved carbohydrates, algal biomass, and activity of carbohydrate-degrading enzymes in the Sep River Reservoir, France. (a) Concentration of dissolved free and combined carbohydrates (DCCHO and DFCHO) and chlorophyll *a*. (b) Activity of the enzymes α-glucosidase and β-glucosidase and abundance of bacteria in the water. Enzyme activity is shown as the calculated maximum rate for cleavage of α or β bonds. Reproduced from Richardot M, Debroas D, Thouvenot A, Romagoux JC, Berthon JL, and Devaux J (1999) Proteolytic and glycolytic activities in size-fractionated surface water samples from an oligotrophic reservoir in relation to plankton communities. *Aquatic Sciences* 61: 279–292, with permission from Birkhäuser Verlag.

subsequently taken up by the bacteria. The study in the Sep reservoir demonstrated that bacteria attached to particles also had high production of other enzymes involved in degradation of polymer compounds such as proteins. This indicates that the attached bacterial populations were well adapted to living on surfaces of particles.

Special Polysaccharides: TEP and Alginates

Transparent Exopolymer Particles

Aquatic researchers have recently found a high abundance of gel-like particles in the water column in lakes and the sea. The particles have been named transparent exopolymer particles or TEPs. TEPs are larger than bacteria and consist mainly of polysaccharides. The actual mechanisms forming TEPs are not known, but polysaccharides released by phytoplankton and bacteria as well as some macroalgae are believed to be responsible for producing the polysaccharides in TEPs.

TEPs act like glue in the water and aggregate organic matter to larger particles. The largest particles formed by TEPs are known as 'lake snow' that can be seen as flakes in the water. TEPs have several important functions. The particle formation allows organic matter to sink to deeper waters and the bottom, where it can feed on benthic organisms. TEPs also create microenvironments or refugia for microorganisms and hereby protect them from their predators. In addition, the high content of organic matter in TEP is an excellent food source to many microorganisms, and TEPs typically have a much higher microbial activity per volume than the ambient water. Research in the ecological impact of TEPs is still in its early stage, but occurrence of TEPs emphasize that carbohydrates play an important role in cycling of organic matter in aquatic food webs.

Alginates

The structural polysaccharide alginate is mainly produced by brown seaweeds but it is also formed by several bacterial species in freshwater environments. Building blocks of alginate are the two sugar acids α-L-guluronic acid and β-D-mannuronic acid. Alginates have a large capacity to absorb water and are of large industrial importance. In the industry, alginates are isolated from natural weeds and laboratory-grown bacteria and are used as stabilizers (gelling agents) in foods like ice cream and juices, in cosmetics, in paper production, and in environmental applications, e.g., for production of slow-releasing fertilizers.

Carbohydrates in Humic Matter

A large portion of dissolved organic matter in natural waters consists of chemically undefined compounds that are resistant to degradation by most organisms and is categorized as humic substances. In lakes, humic substances are formed by chemical and biological conversion of dissolved organic compounds in the water, or compounds released by plants, animals, and microorganisms. In rivers and streams, leaching of humic substances from soils may be an additional source of humic matter. Carbohydrates often make up a small portion of humic substances in fresh water and they are not utilized by microorganisms as the saccharides form complex and recalcitrant molecule structures with other organic substances.

Changes of environmental conditions have been shown to modify the composition of humic matter and also alter the abundance of carbohydrates in lakes. When rain with a low pH (acid rain) acidifies soft-water lakes, the low pH can increase the amount of carbohydrates (DCCHO) of the fulvic acid fraction (fulvic acids are water soluble compounds in humic matter). The mechanisms are not completely understood, but some carbohydrate-degrading enzymes in the water (originating from algae and bacteria) may be inhibited at low pH, leading to accumulation of carbohydrates in the fulvic acid fraction in the water.

Sunlight has also been shown to modify humic substances in lakes by releasing monosaccharides and hereby stimulate the growth of microorganisms. Most likely, abundance and composition of carbohydrates in fresh water are influenced by many other environmental effects that have not yet been determined. Composition and cycling of humic substances are treated elsewhere in the encyclopedia.

Carbohydrates and Stable Isotopes

Application of Stable Isotopes in Ecology

How can the impact of carbohydrates in an ecosystem be traced? One clever way of studying cycling of carbohydrates in food webs has been to take advantage of the natural abundance of different carbon (C) isotopes in carbohydrates of various groups of organisms. In nature, the dominant C isotope is ^{12}C, followed by ^{13}C (about 1.1% of C in nature; ^{13}C is a stable atom with one extra neutron) and ^{14}C (trace substance only; unstable and radioactive atom with two extra neutrons). Biochemical reactions discriminate between the three isotopes and will typically incorporate the heaviest isotope at the lowest rate in new organic molecules.

Tracing of Carbohydrates in a Food Web

American researchers have examined the abundance of ^{12}C and ^{13}C in carbohydrates of different marine organisms and cedar wood relative to the composition of ^{12}C and ^{13}C in the total cell biomass. As can be seen in **Figure 6**, there is a significant difference between the fraction of ^{13}C in glucose and in the biomass of different organisms, and this can be used to trace the food source of the organisms. As examples, similarity in $\Delta^{13}C$ in copepods and phytoplankton suggests that copepods feed on glucose in phytoplankton, while zooplankton in general has a lower $\Delta^{13}C$ value, which may indicate that, e.g., bacteria were part of their food. The negative $\Delta^{13}C$ value of cyanobacteria (the filamentous species *Anabaena*) and cedar wood suggests that their glucose content does not serve as food source for neither bacteria nor zooplankton. Similar isotope fractionation studies have not been made in freshwater environments, but the isotope approach is a powerful tool for analysis of food webs in natural ecosystems.

Carbohydrates – From Molecule to Trees

Carbohydrates have been part of the biosphere for more than 3 billion years and made it possible to form solid cell walls in organisms ranging from

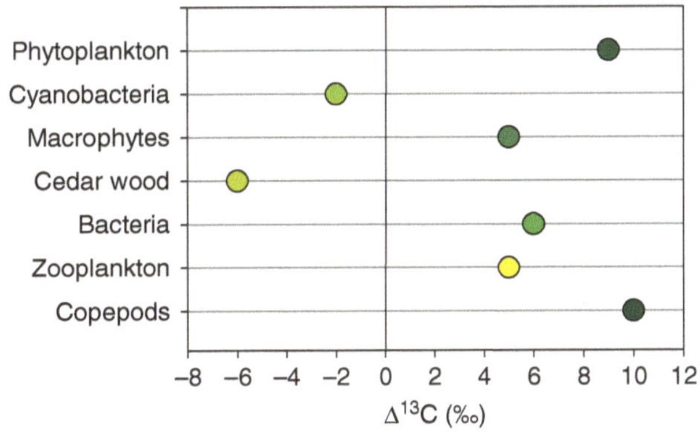

Figure 6 The stable carbon isotope composition of glucose relative to the total biomass (as $\Delta^{13}C$ ‰) of different aquatic organisms and wood. The $\Delta^{13}C$ ‰ value reflects content of ^{12}C and ^{13}C relative to a $^{12}C/^{13}C$ standard (the Pee Dee rock formation in South Carolina, USA). Reproduced from Teece MA and Fogel ML (2007) Stable isotope biogeochemistry of monosaccharides in aquatic organisms and terrestrial plants. *Organic Geochemistry* 38: 458–473, with permission from Elsevier.

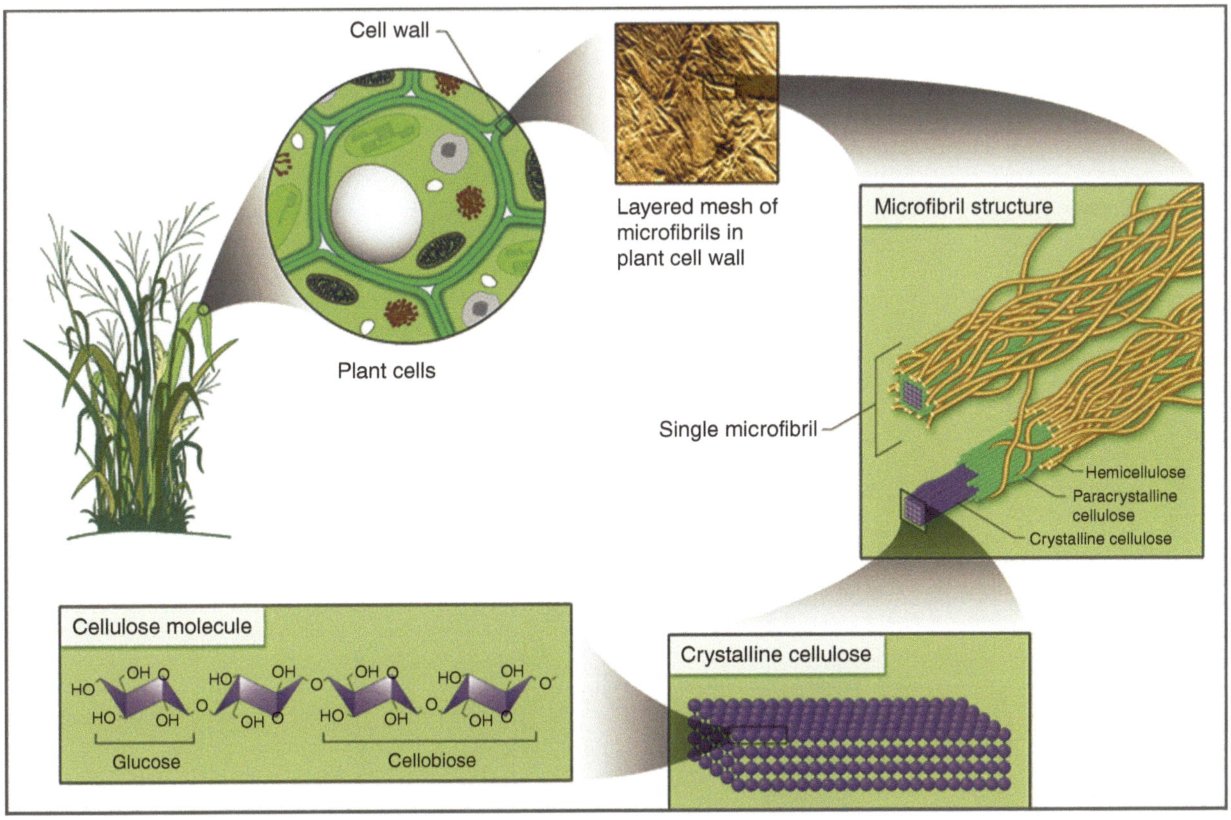

Figure 7 Cellulose is the most common structural carbohydrate in nature. Glucose molecules (arranged in pairs to form cellobiose) make up a crystalline structure that is used in the synthesis microfibrils that create rigid cell walls in plant cells. For more information on structure and composition of cellulose and other biologically important carbohydrates, see e.g., http://genomics.energy.gov/gallery/gtl/originals/420.jpg or http://www.lsbu.ac.uk/water/hycel.html.

microscopic algae and cyanobacteria to majestic trees. The building blocks in the cells are glucose molecules that are combined to the structural carbohydrate cellulose (**Figure** 7). Despite its firm structure, cellulose can be degraded to its building blocks by many bacteria and fungi and hereby enter a new biological cycle.

See also: Dissolved Humic Substances: Interactions with Organisms; Organic Nitrogen.

Further Reading

Bunte C and Simon M (1999) Bacterioplankton turnover of dissolved free monosaccharides in a mesotrophic lake. *Limnology and Oceanography* 44: 1862–1870.

Chanudet V and Filella M (2006) The application of the MBTH method for carbohydrate determination in freshwaters revisited. *International Journal of Environmental Analytical Chemistry* 86: 693–712.

Cheng XH and Kaplan LA (2001) Improved analysis of dissolved carbohydrates in stream water with HPLC-PAD. *Analytical Chemistry* 73: 458–461.

Jonsson A, Ström L, and Åberg J (2007) Composition and variations in the occurrence of dissolved free simple organic compounds of an unproductive lake ecosystem in northern Sweden. *Biogeochemistry* 82: 153–163.

Jørgensen NOG and Jensen RE (1994) Microbial fluxes of free monosaccharides and total carbohydrates in freshwater determined by PAD-HPLC. *FEMS Microbiology Ecology* 14: 79–94.

Jugnia LB, Richardot M, Debroas D, and Devaus J (2006) Bacterial production in the recently flooded Sep reservoir: Diel changes in relation to dissolved carbohydrates and combined amino acids. *Hydrobiologia* 563: 421–430.

Kattner G, Lobbes JM, Fitznar HP, Engbrodt R, Nothig E-M, and Lara RJ (1999) Tracing dissolved organic substances and nutrients from the Lena River through Laptev Sea (Arctic). *Marine Chemistry* 65: 25–39.

Quayle WC and Convey P (2006) Concentration, molecular weight distribution and neutral sugar composition of DOC in maritime antarctic lakes of differing trophic status. *Aquatic Geochemistry* 12: 161–178.

Richardot M, Debroas D, Thouvenot A, Romagoux JC, Berthon JL, and Devaux J (1999) Proteolytic and glycolytic activities in size-fractionated surface water samples from an oligotrophic reservoir in relation to plankton communities. *Aquatic Sciences* 61: 279–292.

Teece MA and Fogel ML (2007) Stable isotope biogeochemistry of monosaccharides in aquatic organisms and terrestrial plants. *Organic Geochemistry* 38: 458–473.

Tranvik LJ and Jørgensen NOG (1995) Colloidal and dissolved organic matter in lake water: Carbohydrate and amino acid composition, and ability to support bacterial growth. *Biogeochemistry* 30: 77–97.

Weiss M and Simon M (1999) Consumption of labile dissolved organic matter by limnetic bacterioplankton: The relative significance of amino acids and carbohydrates. *Aquatic Microbial Ecology* 17: 1–12.

Carbon, Unifying Currency

Y T Prairie, Université du Québec à Montréal, Montréal, QC, Canada
J J Cole, Cary Institute of Ecosystem Studies, Millbrook, NY, USA

Introduction

A unifying feature of all life on Earth is that it is based on a central element: carbon. The transformation of C from one form to another, either from inorganic to organic as in photosynthesis or from organic to inorganic as in respiratory or decomposition processes, basically defines the actions of living organisms and thus plays a unique and fundamental role. For this reason, carbon is often used as the unifying currency with which to quantify the rates of basic metabolic or ecological processes, from individual cells to whole ecosystems. The transformations of carbon in its various forms also serve to establish relationships among ecosystems, for example linkages between lakes and their surrounding land.

Historically, the examination of the functioning of aquatic ecosystems as a system of interconnected flows of material and/or energy was strongly influenced by the pioneering work of H.T. Odum. It also served as the conceptual framework of the International Biological Programme of the 1960s. This view emerged in contrast to a more species-based prior approach, in which interactions between species are central to the understanding of lake ecological dynamics. Instead, a carbon flow approach generally considers only the major biological compartments (i.e., bacteria, phytoplankton, zooplankton, etc.), without special regard to the identity of the species composing the particular assemblages present. By aggregating individual species into coarse functional groups, the approach subsumes variability among taxa to the benefit of greater generality and predictability.

The aim of this chapter is to describe briefly the functioning of freshwater ecosystems as systems of interconnected carbon pools and carbon flow processes, and to examine the main drivers determining them. While the emphasis is on lakes, the same general principles apply to other types of inland aquatic systems, such as small streams and rivers.

Carbon Pools

As a chemical constituent of living matter, carbon typically represents about half of the mass of organic material of aquatic organisms. Despite this constancy however, carbon pools are very unevenly divided among lake constituents and generally follow an inverse relationship with body size: in any freshwater ecosystem, there is much less carbon contained in fish than in bacteria.

In lake waters, the largest reservoirs of carbon are generally found as dissolved constituents, both organic (DOC) and inorganic (DIC). DOC is a largely uncharacterized amalgam of hundreds of individual compounds with molecular weights varying from simple carbohydrates to highly complex molecules of different aromaticity (cite other chapters, McKnight?). The concentrations of DOC in fresh waters vary enormously from <0.5 mg C l^{-1} in freshwater springs to over 300 mg C l^{-1} in hypersaline ponds of endorheic areas. However, for the vast majority of lakes (>85%), DOC concentrations vary within a much narrower range, between 1 and 20 mg C l^{-1}. Dissolved inorganic carbon concentrations in freshwater lakes varies within a similar range, typically between 0.5 and 15 mg C l^{-1}. In inland saline waters however, it can reach much higher concentrations.

The amount of carbon contained in the living biomass of water column organisms is generally small when compared with those of the other pools and will depend on the biological productivity of the systems. For the pelagia, particularly in oligotrophic waters, heterotrophic bacteria will comprise the largest living organic carbon pool, followed sequentially by phytoplankton, zooplankton, and fish. For benthic organisms, macrophytes can accumulate large quantities of carbon over the growing season. The approximate turnover time of several taxonomic groups are listed in **Table 1**.

Although not operating on the same time scale as the dissolved or biological carbon pools, the largest carbon reservoir of lakes is preserved within the accumulated bottom sediments. Depending on the age of the lake ecosystems (a few thousand to several million years), lakes can permanently store enormous quantities of carbon (e.g., L. Tanganyika alone contains several hundred gigatons of carbon in its sediments, i.e., within the same order of magnitude as the biomass of terrestrial plants in the entire biosphere). Even in lakes that were created following the last glaciation, lake sediments often contain as much carbon as the soils of their watersheds.

A Simplified Carbon Cycle for Inland Waters

Carbon enters inland aquatic ecosystems from several pathways and in different forms (**Figure 1**). It can enter from the atmosphere as carbon dioxide by simple diffusion across the air–water interface. There will be a net transfer from the atmosphere to the water when the partial pressure of CO_2 in the water is lower than that of the atmosphere. In general, this occurs only in eutrophic systems where the photosynthetic demand for CO_2 is large. In most oligo and mesotrophic systems, their surface waters tend to be oversaturated with CO_2 with respect to the atmosphere and therefore act as net sources. Other carbon sources entering through the atmosphere are volatile organic carbon compounds (VOCs), such as terpenes and other isoprene derivatives, as well as other organic material scavenged by dust particles. Their biological significance is largely unknown.

The major pathway for carbon to enter lakes is through stream and groundwater inflows as both inorganic (DIC: carbonate and bicarbonate ions plus dissolved CO_2) and organic fractions (particulate (POC) and dissolved (DOC)). In most aquatic systems, bicarbonate (HCO_3^-) and carbonate (CO_3^-) ions are the main source of alkalinity and are generated through weathering reactions of the dissolved CO_2, derived from the decomposition of soil organic matter, with various minerals contained in the parent rock material. Organic carbon inputs in streams largely originates from the incomplete decomposition of soil and terrestrial plant material into soluble compounds (DOC). Over large spatial scales, the amount of DOC exported from catchments to lakes will depend on the productivity of the terrestrial landscape as well as on the local hydrological flow path. Thus, although chemically very different, both the inorganic and organic fractions have ultimately the same origin: terrestrial primary production.

Table 1 Approximate turnover time for different taxonomic groups

Taxonomic group	*Turnover time*
Bacterioplankton	Hours to days
Phytoplankton	Days
Zooplankton	Days to weeks
Benthic invertebrates	Days to weeks
Fish	Months to years

Once it has reached the lake, only a portion of the carbon inputs will undergo transformation before it leaves the system through its outflow. For example, carbonate alkalinity (CO_3^- and HCO_3^- ions) is largely conservative in lakes except in cases where some primary producers (e.g., macrophytes) can use bicarbonate ions as inorganic substrates when dissolved CO_2 is scarce, or when carbonates precipitate as in marl lakes. Organic carbon inputs however have a much different fate. Most (typically 90%) of the organic carbon enters lakes as dissolved, while the particulate remainder is composed mostly of detrital terrestrial plant and soil material. Although the dissolved organic matter can be quite old and therefore suggestive of a high degree of recalcitrance, the conditions prevailing in aquatic systems are more conducive to rapid transformation. Thus, a significant portion of the DOC received by lakes will be transformed in situ by the combined action of photochemical reactions and microbial utilization. Although some of the DOC used by microbes can make its way into bacterial biomass and subsequent assimilation into organisms of higher trophic levels, most of

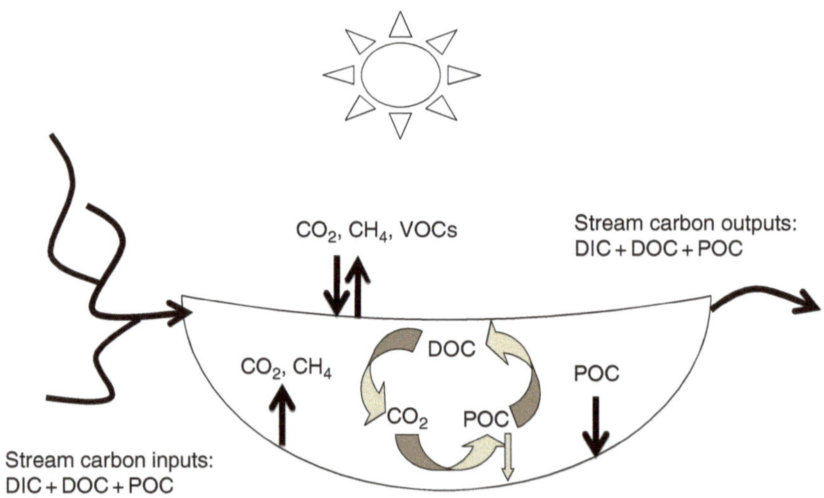

Figure 1 A simplified carbon cycle for inland aquatic ecosystems.

it will simply be respired and released as carbon dioxide. Depending on the direction of the pCO_2 gradient at the air–water interface, this CO_2 will either evade to the atmosphere or be available for photosynthetic assimilation by the phytoplankton and benthic plants. In anaerobic conditions of the sediments or the hypolimnion, organic matter degradation can lead to methanogenesis, thereby increasing the CH_4 partial pressure relative to that of the atmosphere. In many instances, methane can be oxidized to CO_2 before reaching the air–water interface but in strongly anoxic conditions, carbon can evade lakes as methane gas. In general however, methane flux is small (less than 1%) compared with the flux of carbon dioxide.

The photosynthetic assimilation of CO_2 also constitutes a major pathway of carbon transformation in lakes. The stoichiometry of carbon fixation by aquatic photoautotrophs generally follows the empirical reaction equation:

$$106\ CO_2 + 106\ H_2O + 16\ NH_3 + H_3PO_4 \leftrightarrow (CH_2O)_{106}(NH_3)_{16}(H_3PO_4) + 106\ O_2$$

in accordance with the so-called Redfield ratio (C:N:P = 106:16:1), which refers to the average elemental composition of aquatic biomass. The reverse reaction (i.e., to the left) corresponds to aerobic respiration. When oxygen is depleted, such as in sediments or in the hypolimnion of certain lakes, decomposition processes are achieved through alternate reactions, such as sulfate, nitrate, iron, or manganese reduction, as well as through methanogenesis.

In the broadest sense, all biological carbon transformations can be characterized as either a production or a respiration process, both of which apportioned among autotrophs and heterotrophs alike. Depending on the level of aggregation of the aquatic compartments under study, production and respiration processes in aquatic systems can be quantified at any scale: from individual cells to entire ecosystems, or for broad single compartments such as bacteria, phytoplankton, benthic plants, zooplankton, benthos, or fish.

The Predictability of Carbon-Based Aquatic Processes

In aquatic science as in other fields, the utility of a scientific construct is ultimately measured by the degree to which successful predictive theories can be built around it. In this regard, the description of aquatic processes in terms of carbon flow patterns has been remarkably reproducible.

Patterns of Organic Matter Production

Primary production by lake phytoplankton is largely determined by the concentration of its primary limiting nutrient, phosphorus. Except in hypereutrophic systems (total phosphorus concentration >150 μg P l^{-1}) where light can become limiting, the positive relationship between phytoplankton production and phosphorus is approximately proportional, implying that a doubling in phosphorus normally gives rise to a doubling in phytoplankton production. In contrast, benthic primary production is negatively related to phosphorus concentration because the increase in phytoplankton abundance decreases light penetration to the surface of the sediments. In general, the contribution of benthic primary producers to the overall production of the lake ecosystem declines with increasing lake trophic status. Conversely, it can represent most of the lake's productivity, particularly in shallow oligotrophic systems.

While secondary production (that is the production of consumers) is also ultimately regulated by nutrients, several factors modulate the ecological efficiency with which autochthonous carbon production is converted to secondary production. Thus, empirical relationships between secondary production and phosphorus are generally weak. In addition, recent evidence suggest that organic carbon of terrestrial origin, particularly particulate organic carbon, plays some role in sustaining the growth of higher trophic level organisms. Nevertheless, and in spite of this terrestrial load-secondary production shunt, the ecological efficiency between trophic levels is generally low in lakes (<10%). Existing models of invertebrate production in lakes are based largely on temperature and on the biomass of individuals.

Patterns of Organic Matter Oxidation

In lakes that can be considered in a long term equilibrium with the nutrients they receive, carbon accumulation will be in the form of sediment accumulation. Because sedimentary carbon burial is small relative to in situ primary production, it implies that most of it is respired either in the water column or at the sediment surface. Thus, respiration will be largely a function of external organic C input and in-system primary production and of the factors controlling production itself, nutrients. Although autotrophic respiration accounts for a sizeable fraction (about 30%), most of the respiration will be heterotrophic and, within the heterotrophic component, 80% is microbial. On average, benthic respiration represents only about 15% of the total respiration of the 'typical' lake ecosystem but can be much larger in shallow unstratified lakes. However, this proportion is not constant along the trophic gradient, with the contribution of benthic processes lower in more eutrophic systems.

It is now well established that for the water column compartment, respiration generally exceeds gross primary production, a condition termed net heterotrophy. For this situation to be sustained, it requires an external subsidy of organic carbon that is metabolized in situ. This pattern is found mostly in oligotrophic to mesotrophic environments. In eutrophic systems, the autochthonous net primary production is usually sufficient to counterbalance the excess respiration generated by the metabolized exogenous organic carbon.

Glossary

Autochthonous – Organic matter that was produced in the system of interest. Contrast it to allochthonous, which refers to organic matter that was produced elsewhere (for example in the watershed) and imported into the system of interest.

Autotrophs – An organism that produces all, or nearly all of its complex organic molecules from simple inorganic compounds and energy from either light (photoautotrophs) or by exploiting chemical reactions (chemoautotroph).

Heterotrophs – An organism that requires complex organic molecules from other sources. Consumer organisms are all heterotrophs.

International biological programme – A decade long (1964–1974) large scale program to coordinate ecological studies of the world's major biomes.

Lake Tanganyka – A very large lake in Central Africa; L. Tanganyka is the second largest lake in the world by volume.

Partial pressure – The total gas pressure of a mixture of gases is the sum of the partial pressures exerted by each individual gas.

Redfield ratio – The ratios (or compositional stoichiometry) in planktontic marine organisms. Coined in honor of A. Redfield who wrote extensively on this subject.

Stoichiometry – The quantitative measure of the ratios of chemical reactants and products in a chemical reaction. Compositional stoichiometry refers to the ratios of elements in a given substance or an organism.

See also: Alkalinity; Dissolved CO_2.

Further Reading

del Giorgio PA and Peters RH (1994) Patterns in planktonic *p–r* ratios in lakes – Influence of lake trophy and dissolved organic-carbon. *Limnology and Oceanography* 39: 772–787.

Pace ML and Prairie YT (2005) Respiration in lakes. In: del Giorgio PA and Williams PJ leB (eds.) *Respiration in Aquatic systems*, pp. 103–121. Oxford, UK: Oxford University Press.

Redfield AC (1958) The biological control of chemical factors in the environment. *American Scientist* 64: 205–221.

Sobek S, *et al.* (2007) Patterns and regulation of dissolved organic carbon: An analysis of 7,500 widely distributed lakes. *Limnology and Oceanography* 52: 1208–1219.

Smith VH (1979) Nutrient dependence of primary productivity in lakes. *Limnology and Oceanography* 24: 1051–1064.

Vadeboncoeur Y, *et al.* (2003) From Greenland to green lakes: Cultural eutrophication and the loss of benthic pathways in lakes. *Limnology and Oceanography* 48: 1408–1418.

Dissolved Humic Substances: Interactions with Organisms

C E W Steinberg, Humboldt Universität zu Berlin, Berlin, Germany
M A Timofeyev, Baikalean Research Centre, Irkutsk, Russia
R Menzel, Humboldt Universität zu Berlin, Berlin, Germany

Published by Elsevier Inc.

Introduction

On the global scale with huge geological formations, such as the bedrock shields in the Northern Hemisphere or the nutrient-poor regions in the tropics, the non-calcareous, fresh waters rich in humic substances (HSs) appear to predominate over calcareous, HS-poor ones. In freshwater systems, HSs derive from peat and dominantly from terrestrial plant debris, where lignin building blocks, tannins, and terpenoids are the main source material. HSs comprise the majority of the organic carbon in any freshwater type, including all organisms. This even applies to non-eutrophicated fresh waters, which do not have visible brownish colors. Yet, the knowledge of how HSs control freshwater life is comparably scarce. Many HS-rich waters seem to have reduced biodiversity, which applies to zooplankton, zoobenthos, and fish, phytoplankton, and macrophytes. It has been suggested that one explanation for the low diversity of the primary producers is the poor underwater light climate. From field observations, it is well understood that in HS-rich boreal as well as in tropical lakes, nympheids and helophytes dominate. Yet, there are also direct HS-mediated effects on freshwater organisms.

Many studies deal with the interaction of natural organic water constituents (dissolved organic carbon, HS, etc.) with xenobiotic chemicals and/or heavy metals and show a natural attenuation/quenching of the potentially adverse chemicals. But what is the 'regular' function of these water constituents in non-eutrophicated freshwater ecosystems? The quantitative aspect of these substances is well known: HSs exceed the organic carbon in all living organisms by roughly one order of magnitude; the concentrations lie between 1 and 100 mg l^{-1} DOC with extremes of up to >240 mg l^{-1} DOC in Brazilian coastal lagoons. In contrast to this quantitative significance of HSs in freshwater systems, the knowledge on their ecological function (qualitative significance) is still less. Only their function as an indirect external energy and carbon source for microbial growth has intensely been studied during the last two decades after the pioneering energy budget calculation of a Finnish humic lake and subsequent mechanistic laboratory studies. Although the microbiological papers were the major opening of the ecological understanding of dead organic matter in fresh waters, it also seems persistent to consider HSs as natural environmental chemicals. The presence of functional groups such as carboxylic, phenolic, ketonic, aromatic, aliphatic etc. groups, enable HSs to interact with both living and nonliving matter. Furthermore, the lipophilic moieties facilitate an interaction with lipids of the biomembranes.

In 1985, two papers appeared hypothesizing direct effects of HS with freshwater organisms. From studies of the electromobility of HSs, their ability to interact with biomembranes was postulated. This was also concluded by analogy to detergents: it was observed how low HS concentrations could strongly stimulate, and high concentrations inhibit microorganism development – a so-called hormetic effect. Meanwhile, it is accepted that HSs are taken up and induce a variety of response reactions in the organisms. Once taken up, HSs can exert specific as well as nonspecific effects. So far, specific effects comprise reduction of the photosynthetic oxygen production, estrogenicity, or chemical attraction. Nonspecific effects are physical and chemical membrane irritation, induction and modulation of biotransformation enzymes, induction of stress defense proteins (chaperons), or oxidative stress defense. Also, apoptotic (programmed cell death equivalent to anticancer response) reactions are induced as observed with human cells. Furthermore, HSs appear to have antiviral, anticoagulant, profibrinolytic, and antiinflammatory properties.

Interference within Photosynthetic Oxygen Production

From recent environmental and microbial studies, it is evident that HSs have the potential to act as external electron acceptors for microbial respiration. One may hypothesize that internalized HSs should interfere within all electron transport reactions in organisms or cells. This hypothesis has been supported by exposing several freshwater plants and cyanobacteria to different HS sources and measuring the photosynthetic oxygen release, and quinones, a major building block of HS which is measured as spin density, statistically count almost completely for this effect (**Figure 1**). HSs directly quench electrons or bind to the bioquinones in photosystem II and thereby block the electron transfer.

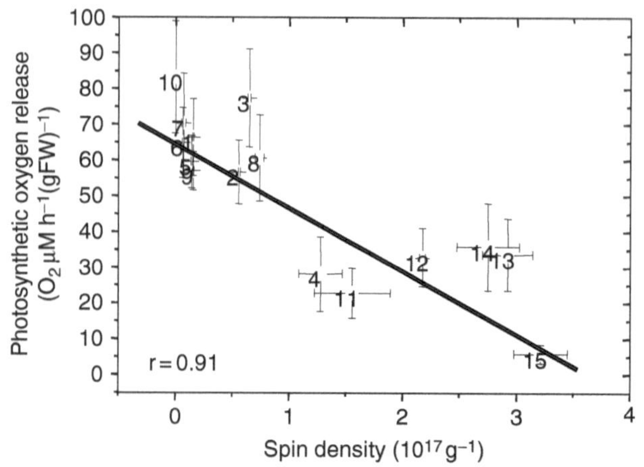

Figure 1 Quantitative structure–effect relationship relating photosynthetic oxygen release of the macrophyte *Ceratophyllum demersum* L. to spin densities of the HSs, equivalent to quinoid structures. (1) Suwannee River NOM, (2) Suwannee River FA, (3) Suwannee River HA, (4) HS1500, (5) Hellerudmyra NOM, (6) Svartberget NOM, (7) Valkea Kotinen NOM, (8) Hietajärvi NOM, (9) Nordic Standard NOM, (10) Birkenes NOM, (11) Soil-FA II, (12) Peat-R-HA, (13) Sum-HA, (14) Leon-HA, (15) Peat-S-HA. Paul A, Pflugmacher S, and Steinberg CEW (2003) Correlation of spin concentration in humic substances with inhibitory effects on photosynthesis of aquatic macrophytes. *Akademie gemeinnützige Wissenschaften zu Erfurt, Sitzungsberichte der Mathematischen-Naturwissenschaftlichen Klasse* 12: 205–217 (in German).

Another mode of action also occurs: development of internal oxidative stress and reduction of photosynthesis to avoid access production of reactive oxygen species (ROS).

The response of phototrophs upon HSs exposure is rather variable. Yet, as a general rule, cyanobacteria are more sensitive than eukaryotic algae, probably because of their simpler cell structure. As one extreme of response upon exposure to HSs, not any effect in the exposed algae may be detected as shown with the coccal green alga, *Pseudokirchneriella subcapitata*. During the logarithmic growth phase, the growth rates of this algal species were identical and independent of any HS exposure (**Figure 2(a)**). The other extreme is covered by almost complete growth inhibition with the coccal cyanobacteria *Chroococcus minutus* at rather low HS concentrations (**Figure 2(c)**). Another coccal green alga, *Desmodesmus communis* occupies an intermediate position between *P. subcapitata* and *C. minutus* (**Figure 2(b)**).

Empirically, the fact that cyanobacteria are more sensitive than eukaryotic phototrophs has already been the basis for combating cyanobacterial growth and blooms by adding cheap natural phenolic compounds, such as leachates from straw and leaf litter to lakes. The high sensitivity of cyanobacteria is also in good agreement with field observations, which show that they are not able to develop in eutrophicated brown water lakes. Some preliminary evidence is emerging that also shows that heterotrophic microorganisms may suffer from HS exposure.

Hormone-Like Effects

Hormone-like effects of HSs have recently been recorded with the nematode *Caenorhabditis elegans* (Maupas), the ornamental fish *Xiphophorus helleri* Heckel, and the clawed frog *Xenopus laevis* (Daudin). In *C. elegans*, exposure to HS often, but not always, led to increased numbers of offspring (**Figure 3**). The contrasting effects could not yet be related to any building blocks.

With the clawed frog, the estrogenic mode of action has been identified so far. Consistently, gene expression profiles of *C. elegans* elucidated that the exposure to any HS induced vitellogenin genes.

Another target organ of the hormone-like effect is the thyroid system. Preliminary results showed that also the thyroid system has been affected whereby the complexation of iodine by HS and subsequent reduced bioavailability may be one likely mode of action. Similar conclusions were also provided by epidemiological studies on drinking water in Denmark. Another, more direct, mode of action is the influence of HSs on thyroid hormone metabolism by inhibiting hepatic thyroxin 5′-monodeiodinase activity with the formation of goiters.

Chemical Attraction and Gene Regulation in *C. elegans*

HS induce several receptors in the compost-dwelling nematode *C. elegans*. Applying functional ecological genomics to this animal (whole genome microarrays), notable transcriptional upregulations were identified in chemosensors and olfactory receptors. HSs from different sources have the potential to act as environmental signals: adult worms migrated towards the humic material. This appears to be a genetically fixed property of *C. elegans*, since the animals, for countless generations, have not been in contact with HS.

Besides the chemoreceptors, transporters were also induced, which indicates that HSs might have actively been taken up. Furthermore, a limited number of genes coding for enzymes involved in biotransformation were found to be differentially expressed. Since HSs induce oxidative stress and other stress symptoms, the question arises as to why *C. elegans* actively

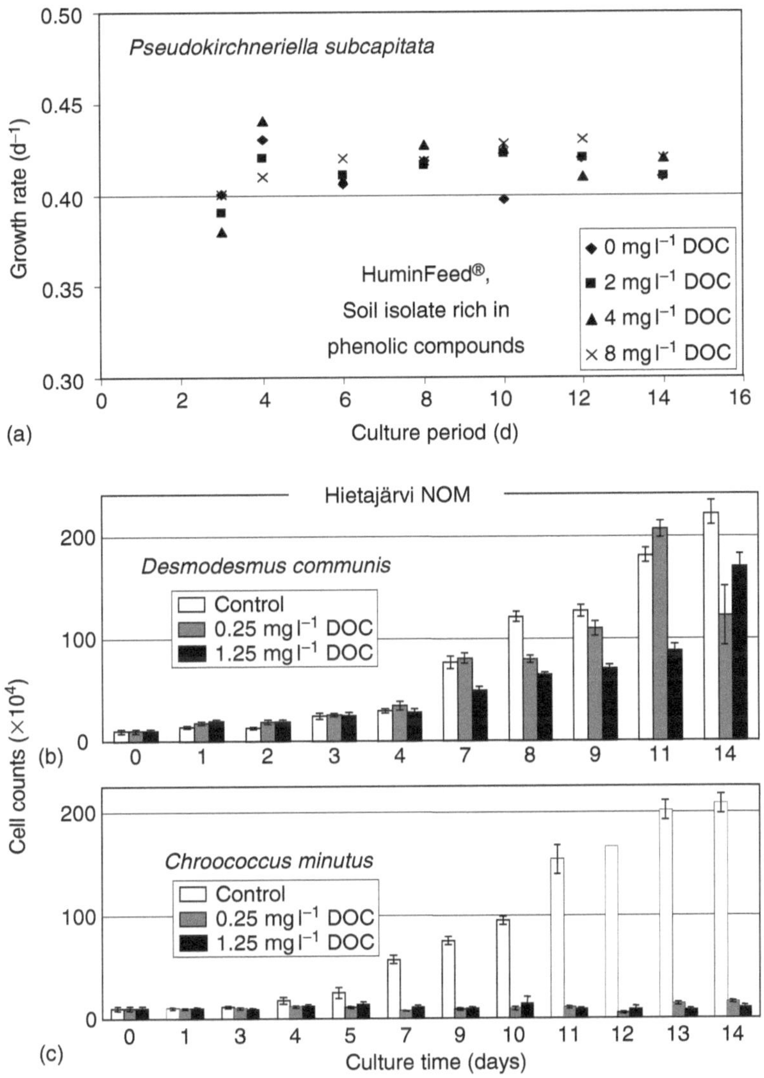

Figure 2 Growth rates and cell counts of two coccal green algae and one cyanobacterial species exposed to humic materials. (a) Growth rates of *P. subcapitata* (Korshikov) Hindak during the logarithmic growth phase in semi-continuous cultures in the absence and presence of increasing concentrations of the commercial HuminFeed (Claußen & Steinberg, unpublished data). (b), (c) Impact of Svartberget NOM (in mg l^{-1} DOC) on the cell yield of the coccal green alga *Desmodesmus communis* (E.H. Hegew.) (formerly *Scenedesmus quadricauda*) and the coccal cyanobacteria *Chroococcus minutus* (Kütz.) Näg. Prokhotskaya LY and Steinberg CEW (2007) Differential sensitivity of a coccal green alga and a cyanobacterial species to dissolved natural organic matter. *Environmental Science and Pollution Research* 14 *Special Issue* 1: 11–18.

looks for stressful environments? Does it only tolerate this stress, which can be considered adverse, or has the stress had a positive effects on the worms? This aspect will be discussed in more detail after completing the display of stress phenomena.

Interaction with Membranes

Exposed to HSs, biomembranes are physically as well as chemically irritated. To identify the physical irritation, the resting potential of giant cells of the charophyte *Nitellopsis obtusa* (Groves) was tested. This potential gets depolarized upon exposure to HSs. Chemical irritation occurs as membrane (lipid) peroxidation as shown with red blood cells or amphipods. The major underlying mechanisms may be the internal production of reactive oxygen and nitrogen species, ROS. Internal HSs are processed with ROS as byproducts, because any activation and subsequent reduction of O_2 automatically produces ROS. It is interesting to note that in the exposed *Gammarus lacustris* Sars, lipid peroxidation starts clearly before free H_2O_2 accumulates in the tissue, indicating that this process might be part of the antioxidative stress defense (**Figure 4**).

Chemical Stress Defense

As one clear response to HS exposure, defense proteins, better known as heat-shock proteins (Hsp) were induced in freshwater animals and algae in environmentally realistic HS concentrations. Beside the most common Hsp70, small crystalline Hsps have also been found as transcription and protein products. Crystalline Hsps are known to be fundamental in lifespan extension, as shown with the fruit fly *Drosophila melanogaster* Meigen.

All organisms have means to rid chemical burdens which originate, for instance, in animals, cyclic production and degradation of hormones, exotic food chemicals, xenobiotics and so on. They have developed so-called biotransformation pathways. Also, HSs behave like chemical clues in these pathways, and enzymes of both phases have been induced upon HS exposure. The induction of biotransformation phase I enzymes is indicated by the expression of the aryl hydrocarbon receptor (AhR), which is comparably unspecific and combines with a great variety of organic compounds. This is shown in comparison with tetraclorodibenzo-*p*-dioxin (TCDD), which binds strongly to the AhR (**Figure 5**).

In addition to the biotransformation pathways, another mechanism for handling toxins and xenobiotic chemicals is active; the so-called multixenobiotic resistance transporter (MXR). This is a membrane-bound group of P-glycoproteins, which act as an exporter against xenobiotics and/or their metabolites. The MXR pump can be inhibited by chemosensitizers. Chemosensitization of the MXR defense system could cause increases in intracellular accumulation and subsequent toxic effects of xenobiotic chemicals and heavy metals otherwise effluxed by the MXR system. HSs have the potential to block the MXR pump, which leads to increased concentrations of internalized xenobiotics and heavy metals.

All physiological responses to HSs require energetic costs, which have to be compensated on the expense of the energy resources of the individuals. As shown

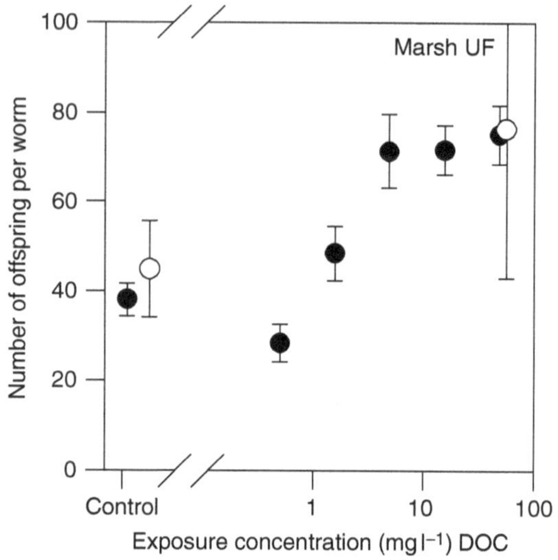

Figure 3 Number of offspring per worm in two bioassays with *C. elegans* after 72 h exposure to DOM of a marsh in southern Ontario, Canada. The numbers of offspring increase up to 8 mg l^{-1} and remain on that level. Höss S, Begtold M, Haitzer M, Traunspurger W, and Steinberg CEW (2001) Refractory dissolved organic matter can influence the reproduction of *Caenorhabditis elegans* (Nematoda) *Freshwater Biology* 46: 1–10.

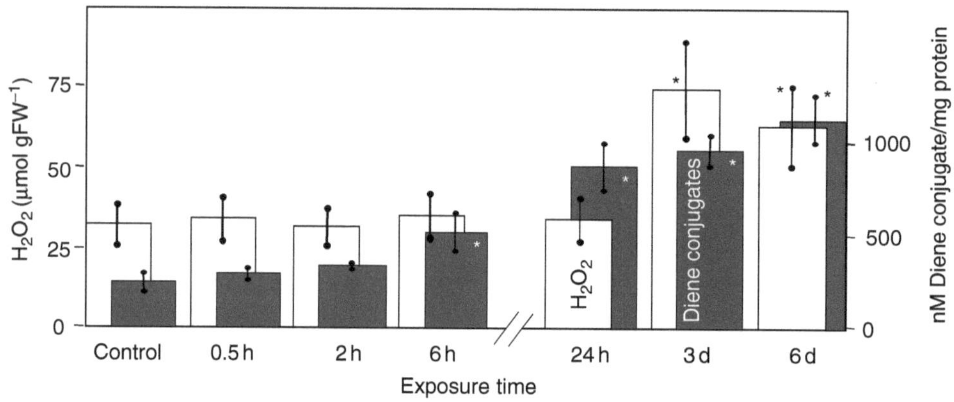

Figure 4 Development of internal H_2O_2 and lipidperoxidation in *Gammarus lacustris* exposed to 14 mg l^{-1} DOC of Lake Schwarzer See, Brandenburg State, Germany. Modified from Timofeyev MA, Shatilina ZM, Kolesnichenko AV, Kolesnichenko VV, Pflugmacher S, and Steinberg CEW (2006) Natural organic matter (NOM) induces oxidative stress in freshwater amphipods *Gammarus lacustris* Sars and *Gammarus tigrinus* Sexton. *Science of the Total Environment* 366: 673–681. Diene conjugates indicate lipidperoxidation, asterick (*) denotes statistically significant differences from controls. Note: Lipidperoxidation starts before free H_2O_2 accumulates in the tissues.

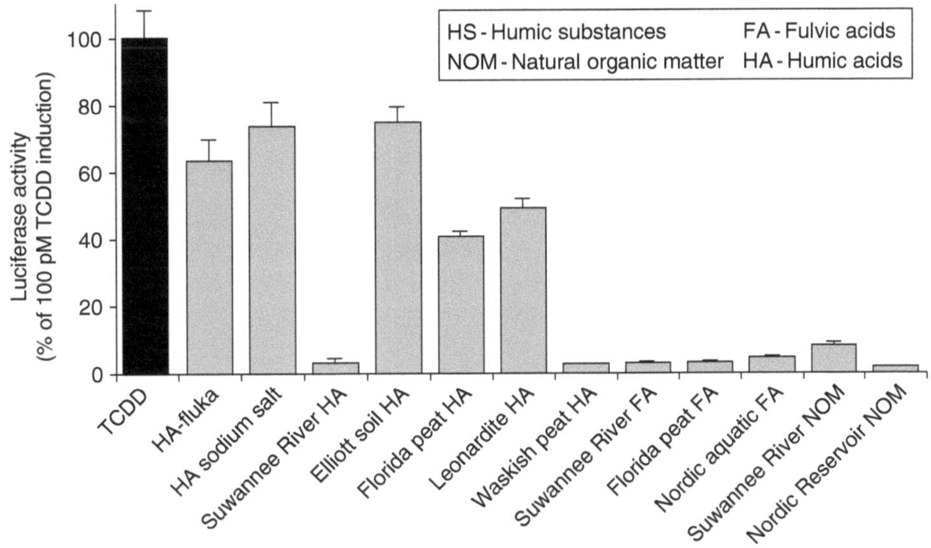

Figure 5 AhR-mediated activities of 12 HS in comparison to tetraclorodibenzo-*p*-dioxin (TCDD) as a well-known strong AhR agonist. The activity is indicated by a genetically modified cell line with the luciferase system under AhR control: when the AhR is activated, the luciferase system also starts to work. TCDD concentration was 3.2×10^{-5} mg l^{-1}, the HS concentration 150 mg l^{-1} DOC. Janošek J, Bittner M, Hilscherová K, Bláha L, Giesy JP, and Holoubek I (2007) AhR-mediated and antiestrogenic activity of humic substances. *Chemosphere* 67: 1096–1101.

in zebra fish embryos, this energy consumption affects their normal development and leads to birth defects. These findings are in agreement with a study on Finnish and Russian lakes, where a significant negative correlation was found between total organic carbon content of the lake water and the body length of several fish species: The higher the HS content of the lake, the smaller the individuals of a given year class.

Albeit the adverse energetic effects in the fish, there is increasing awareness in the research community of aging that mild chemical stresses (perhaps milder than the above exposure with the zebra fish embryos) is even beneficial. In general, mild chemical stress to individuals means training of the chemical defense system (biotransformation enzymes, antioxidant enzymes, stress proteins) with interesting consequences for the individuals.

Acquisition of Multiple Stress Resistance

It is worthwhile to note that one plausible theory of aging claims that the biotransformation and the antioxidant defense systems have to be trained, and this training leads to individual longevity. Recently, key molecules involved in longevity have been identified. Pro-longevity genes include some that encode antioxidant enzymes, biotransformation enzymes, and others encoding Hsp, particularly of low molecular weight (Hsp22). Genes of all enzyme families and Hsps have been found to be upregulated in *C. elegans* after exposure to natural and artificial HS. This is consistent with findings in gammarids, where the production of small heat proteins increased upon exposure to HS. Furthermore, more recent studies with this nematode have revealed significant lifespan extensions and upregulation of several antistress genes, including two genes encoding for longevity. The retardation of the aging process is often found to be combined with the acquisition of a multiple stress resistance. On the phenomenological basis, increased stress resistance was also found in fish with improved individual constitution after strong mechanical stress as exemplified with an ornamental fish (**Figure 6**). As a very recent quasi field validation, it has been shown that logging-induced increases in dissolved organic carbon improved the recruitment of yellow perch (*Perca flavescens* Mitchell) in Canadian Shield lakes. Intrigued by the *C. elegans* result, the potential of HS-mediated lifespan extension has been tested with other animals, for instance the water flea, *Daphnia magna* Straus. Upon exposure to HS, the usually parthenogentic females started to produce males. Subsequently, males and females were separately exposed. Surprisingly, the lifespans of males, but not of females, were extended. In fact, with females, the lifespan decreased upon HS exposure. To date, it is uncertain what this means in ecological terms.

The knowledge of potential benefits of mild stresses to plants is less extensive than that of animals. Since most plants do not age in the strict

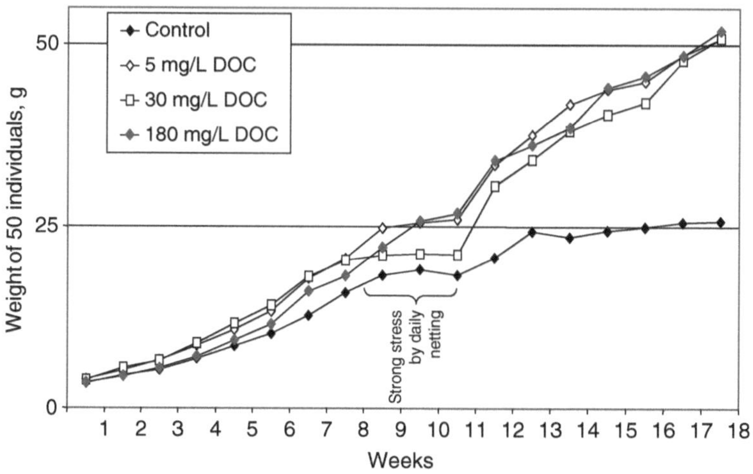

Figure 6 Increases of the survival rate of embryos and juveniles of the swordtail, *Xiphophorus helleri* Heckel, exposed to HSs. Modified from Meinelt T, Schrekenbach K, Knopf K, Wienke A, Stüber A, and Steinberg CEW (2004) Humic substances affect physiological condition and sex ratio of swordtail (*Xiphophorus helleri* Heckel). *Aquatic Sciences* 66: 239–245. Newly born juvenile swordtails were exposed to increasing concentration of the synthetic HS1500 and the development of juveniles was monitored till sexual differentiation of the individuals. Any addition of HS, even as low as 5 mg l^{-1} DOC, led to a weight gain by the young animals and also to a multiple stress resistance: strong physical stresses by daily netting caused only a small delay of the growth, whereas the control animals did not really recover from the stressful 14 days. In the meantime, it has been shown with carp that symptoms of density stress (e.g., cortisol levels) can be reduced if the fish are exposed to HS.

gerontological sense, it can be expected that mild stresses do not necessarily lead to lifespan extension of the complete plants, but of certain organs, which means that these organs have developed a stress resistance. With respect to HSs, there are first indications that HS-exposed wheat and apricot seedlings develop a water deficiency and salt resistance. It might be assumed that this acquisition is a more widespread phenomenon, for instance, in coastal lagoons where salt water intrusions occur frequently.

Conclusion

Because of their ubiquity and their variety of functional groups, dissolved HSs have the potential to interfere within almost any biotic structure and biochemical pathway in freshwater organisms. Because of the low molecular masses of their building blocks, they are capable to easily pass biomembranes. The eco-genomic study with *C. elegans* shows that they have the chance to be actively taken up. Inside the organisms, HSs are metabolized like xenobiotic chemicals with ROS as byproducts. They induce a variety of nonspecific and specific reactions in the organisms, starting from stress response, such as Hsp induction, to herbicide-like mode of action towards primary producers. They modulate the activity of membrane-bound pumps for xenobiotic chemicals and, in turn, should change bioconcentration of xenobiotics and heavy metals. Strong chemical stress leads to teratogenic effects and losses in lipid storages. However, mild chemical stress appears to be actively looked for, since the ecogenomic study with *C. elegans* revealed that also chemosensory and olfactory genes are switched on, and the nematode actively migrated into the stressful environment. As a consequence, the individual lifespan can be prolonged and a multiple stress tolerance can be acquired.

In summary, these rudimentary results show that HSs directly interact with aquatic organisms and modify both the individual fitness and the abiotic environment. Hence, dissolved HSs have to be considered an abiotic ecological driving force – more cryptic than light, nutrients, or temperature.

Glossary

AhR – Aryl hydrocarbon receptor, a cytosolic transcription factor that is normally inactive. Upon ligand binding to chemicals such as TCDD, the receptor is activated leading to changes in gene transcription.

Biotransformation – Reactions which transform lipophilic endobiotic and xenobiotic chemicals into hydrophilic products. The biotransformation comprises two main phases: Phase I – a polar functional group is applied to a lipophilic substrate by oxidation, reduction, or hydrolysis, so that it becomes more hydrophilic. Phase II – the product of Phase I is conjugated to a hydrophilic endobiotic substance in order to make it more hydrophilic.

Bioconcentration – Diffuse uptake from the environment and internal storage of xenobiotic chemicals and/or heavy metals by aquatic organisms.

DOC, TOC – Dissolved organic carbon, total organic carbon.

FA – Fulvic acid, soluble in the whole pH-range.

HA – Humic acid, soluble in alkaline, but not in acidic solutions.

Hormesis – *From (eco)-toxicology:* dose response phenomenon characterized by low dose stimulation, high dose inhibition of processes/development.

HS – Humic substances, include HA and FA.

Hsp – Heat shock protein, nowadays called stress protein, a chaperon induced by a variety of physical and chemical stresses. The figure behind HSP indicates the molecular mass in kDa; for instance, Hsp70 has a molecular mass of 70 000 Da.

HS1500 – Synthetic humic substance with a mean molecular mass of 1.5 kDa; autoxidation product of polyphenols with alkyl bridges; similar to lignin building blocks.

HuminFeed® – A product derived from leonardite by alkaline extraction and used as a food additive.

Molecular weight – Must be more stringently called molecular mass, in Da.

MXR – Multiple xenobiotic resistance, a membrane-bound P-glycoprotein which pumps chemicals out of cells.

NOM – Natural organic matter, isolated by reverse osmosis (or ultrafiltration).

ROS – Reactive oxygen and nitrogen species, such as singlet oxygen, hydrogen super oxide, super oxide anion radicals, nitric oxide, or peroxynitrite.

Xenobiotic chemicals – Organic chemicals, alien to biological systems, often halogenated, produced by man; however, consideration of the natural halogen chemistry suggests that the separation between natural and xenobiotic potentially toxic chemicals is somewhat fictitious.

Further Reading

Bittner M, Janošek J, Hilscherová K, Giesy J, Holoubek I, and Bláha L (2006) Activation of Ah receptor by pure humic acids. *Environmental Toxicology* 21: 338–342.

Ghabbour EA and Davies G (eds.) (2004) Humic Substances: *Nature's Most Versatile Materials*. New York: Taylor and Francis.

Hessen DO and Tranvik LJ (eds.) (1998) *Aquatic Humic Substances – Ecology and Biogeochemistry*. Berlin: Springer.

Keskitalo J and Eloranta P (eds.) (1998) *Limnology of Humic Waters*. Leiden: Backhuys.

Matsuo AYO, Woodin BR, Reddy CM, Val AL, and Stegeman JJ (2006) Humic substances and crude oil induce CYP1A expression in the Amazonian fish species *Colossoma macropomum* (tambaqui). *Environmental Science and Technology* 40: 2851–2858.

Münster U (1985) Investigation about structure, distribution and dynamics of different organic substrates in the DOM of lake Plusssee. *Archiv für Hydrobiologie, Supplement* 70: 429–480.

Steinberg CEW (2003) *Ecology of Humic Substances in Freshwaters*. Berlin: Springer.

Steinberg CEW, Kamara S, Prokhotskaya VYU, *et al.* (2006) Dissolved humic substances – Ecological driving forces from the individual to the ecosystem level? *Freshwater Biology* 51: 1189–1210.

Steinberg CEW, Saul N, Pietsch K, Meinelt T, Rienau S, and Menzel R (2007) Dissolved humic substances facilitate fish life in extreme aquatic environments and have the potential to extend lifespan of *Caenorhabditis elegans*. *Annals of Environmental Sciences* 1: 81–90.

Visser SA (1985) Physiological action of humic substances on microbial cells. *Soil Biology and Biochemistry* 17: 457–462.

Ziechmann W (1994) *Humic Substances*. Mannheim: BI-Wissenschaftsverlag.

Interactions of Dissolved Organic Matter and Humic Substances

L J Tranvik and E von Wachenfeldt, Uppsala University, Uppsala, Sweden

Introduction

In inland waters generallly, a minor fraction of the organic matter is found in living biomass. Hence, most of it is in different stages of degradation of plants and animals. It also exhibits a range in size, from small organic molecules such as sugars and amino acids, to macroscopic detrital particles. Most of the organic matter in the water column is typically dissolved or colloidal, i.e., it does not settle downwards to the sediment by gravitation. Hence, it stays in the water as long as the hydraulic retention time of the water body allows, or until it is mineralized or transformed into a particulate form that may sink out of the water column. Organic matter in the water is operationally divided into dissolved and particulate phases, based on filtration. The fraction that passes a filter, typically of 0.2–0.45 μm pore size, is classified as dissolved organic matter (DOM), while those components that are retained by the filter are considered particulate organic matter (POM). The 'dissolved' phase includes not only truly dissolved molecules, but also a range of colloids. For example, most viruses and a substantial number of bacteria pass the filters used to separate dissolved and particulate material (**Figure 1**). DOM typically constitutes in the order of 90% of the total organic carbon.

The chemical composition of DOM is highly complex and only a minor fraction can be identified as specific molecules. Much progress in the characterization of DOM comes from studies of average properties, such as apparent molecular weight distribution, elemental composition, and the relative importance of different chemical moieties such as aromatic and aliphatic structures. The bulk of the DOM can also be successfully divided into a number of major groups, including humic substances, which are operationally defined as the fraction of the DOM that is adsorbed into a chromatographic resin (originally XAD-8) at pH 2, and can be eluted again from the resin at high pH. The humic substances can be further divided into humic and fulvic acids based on solubility at different pH. The humic matter is often the major component of lake and river DOM. It can be formed within lakes from organic matter originating from indigenous primary production by phytoplankton and littoral plants, but most of the humic matter is imported from the watershed, being products of degradation of organic matter in soils and wetlands. Accordingly, humic substances are complex mixtures of heterogeneous organic compounds of biotic origin that have undergone extensive transformations since they were first produced by plants.

Considering that the operational definition is based on general properties of hydrophobicity, it is not surprising that the humic substances are a widely heterogeneous group of compounds. Moreover, since humic substances also make up a large proportion of the total DOM, the average characteristics of humic substances and total DOM are often very similar. Accordingly, this chapter addresses the interactions of physical processes with DOM in general, rather than humic substances specifically.

Important properties of humic substances, with implications for their interactions with physical processes, include an abundance of aromatic structures and the presence of ionic structures, most importantly carboxylic and phenolic groups. The aromatic structures absorb UV and short wavelength visible radiation, which results in photochemical reactions. Due to the acidic functional groups, humic substances have strong effects on the acid–base chemistry of inland waters. These functional groups are also involved in the complexation of metals.

The size distribution of organic matter in the water, with most of it being dissolved or colloidal, also means that the interface between organic matter and water with its other dissolved compounds, e.g., inorganic ions, is very large. This results in a very large potential for biological, physical, and chemical interactions. Some of these interactions (notably, utilization of DOM by bacteria and photochemical processes), are described in other chapters (cf. 'see also' list at the end of this chapter). This chapter focuses on two important types of interaction with physicochemical processes i.e., (1) processes that affect the size distribution of DOM, including the build-up of particles from dissolved compounds and colloids, resulting in sedimentation, and (2) interactions of DOM and humic substances with metals and nutrients.

The Physical Appearance of Humic Substances in Water

The chemical composition of DOM and humic substances are described in other chapters of this encyclopedia. The focus here is on their physical appearance, including the size of molecules and aggregates, and shape. Properties of humic substances

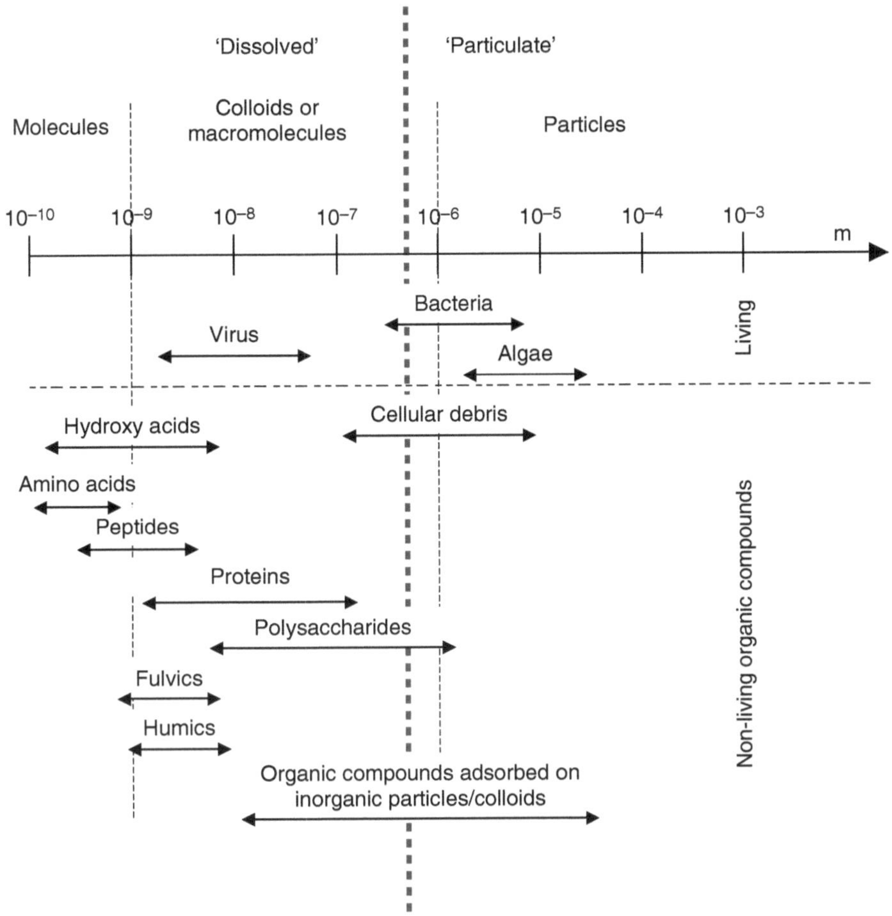

Figure 1 Size distribution of dissolved, colloidal, and particulate organic matter in natural waters. The gray dashed line indicates the common operational division between dissolved and particulate matter based on filtration. Developed from Stumm W and Morgan JJ (1996) *Aquatic Chemistry: Chemical Equilibria and Rates in Natural Waters*. New York: Wiley.

in water that are obvious at the macroscopic scale include color and surfactant characteristics.

Perhaps the most obvious trait of humic substances is the brown color and strong absorption of short wavelength light. The molecular structures responsible for the absorption are termed chromophores, which are associated with aromatic rings and conjugated double bonds. Light that is absorbed results in warming as well as photochemical reactions, both with far-reaching ramifications for aquatic ecosystems.

Humic substances act as surfactants with hydrophilic regions on the molecules, due to their hydrophobic character. The hydrophilic end of the humic molecule becomes faced towards the aqueous phase, while the hydrophobic end gets oriented towards hydrophobic parts of other humic molecules. Humic surfactants formed in this way produce foam, e.g., in the rapids of streams or when waves are breaking, which frequently accumulates at the shoreline or in backwaters (**Figure 2**). Due to the amphiphilic traits of humic substances, they form micelles or structures similar to biological membranes. The solubility of hydrophobic compounds, including pesticides and other anthropogenic substances, is enhanced by the presence of humic substances, which is possibly due to solubilization of hydrophobic molecules in the hydrophobic interior of micelles.

Regular biopolymers such as nucleic acids and proteins are highly complex, but their polymerization is strongly controlled, resulting in highly constrained sequences of monomers, but also a predictable secondary and tertiary structure. In contrast, the formation of humic substances is highly random, resulting in very heterogeneous and irregular molecules. Still, they exhibit structures analogous to the secondary and tertiary structures of biopolymers. At low concentrations, humic substances appear largely as dissolved ions. At higher concentrations, and under the influence of a range of conditions described later on,

Figure 2 A spectacular example of humic substances acting as surfactants. In a section of a stream where the elevation drops rapidly, foam is formed by surface active humic matter in the turbulent water under the ice. The foam is pressed up through a hole in the ice downstream, where it freezes around the edges of the hole as it gets exposed to cold air. Due to the much lower density of foam compared to water, a surprisingly high chimney (2.4 m, upper panel) is built up. The inside of the chimney is protected from freezing by isolation and by a continuous flow of foam (lower panel, cross section of 'humic foam chimney,' the direction of the arrow is from the base to the top of the chimney). Photos by Allan Rodhe.

they aggregate into larger assemblages of molecules and colloids. Electron micrographs reveal both fibrillar and globular appearances, as well as irregular networks.

Colloidal Organic Matter

A colloid is an entity large enough to have supramolecular structure and properties (e.g., a possibility of conformational changes for organic colloids or an electrical surface field for inorganic colloids) but small enough to remain in suspension in the absence of aggregation. Colloidal organic carbon (1.5 nm–0.45 μm) may comprise a substantial part of the DOM. Common organic aquatic compounds that possess colloidal properties include rigid biopolymers, humic substances and flexible biopolymers. The rigid biopolymers include fibrillar polysaccharides and peptidoglycans released from plankton as exudates or cell wall components. Humic substances appear to stabilize inorganic colloids through a modification of the surface potential. On the contrary, polysaccharides tend to favor the increase of coagulation and consequently the sedimentation of inorganic colloids.

Common methods to assess the size distribution of the components of DOM include size exclusion chromatography and ultrafiltration. The results from these methods reveal the apparent molecular weight distribution, rather than the size spectrum of individual molecules. Hence, as long as aggregates are not formed or broken up by the method, the results reflect the size of physical units, including aggregates of molecules as well as individual molecules. Often, and particularly in waters rich in humic substances, much of the DOM appear in sizes (apparent molecular weights larger than a few thousand Dalton) that show clear colloidal properties.

Aggregation

Aggregation of organic matter is a two-step process in which particles must first collide and then stick together. The collision may lead to aggregation depending on the strength of the energy barrier of their interaction, including an attractive energy due to van der Waals force and a repulsive due to negative electric charges. Coagulation and sedimentation of colloidal matter depend both on physical properties (e.g., size, density, compactness, rigidity) and chemical properties of the surfaces. The size of particles susceptible to flocculation processes may be initially small (colloidal range), ranging up to a size of around 1 mm. Polymers adsorbed to particles can both promote flocculation by bridging mechanisms and can also be antagonistic as a repulsive force.

The hydrophilicity and hydrophobicity of the colloids is important for the stability of formed aggregates. Hydrophilic molecules are water-soluble (e.g., polysaccharides, proteins) and because of their solubility they are thermodynamically stable, but they can be induced to aggregate or precipitate by changing the solvency conditions (by changing temperature or ionic concentrations). In contrast, the hydrophobic compounds, including humic substances and hence most often the majority of the dissolved organic matter, are not soluble in water. The hydrophobic compounds are thermodynamically unstable but can be kinetically stable by repulsion, preventing them from forming aggregates.

Mechanisms of Collisions and Formation of Colloids and Larger Particles

Aggregation of molecules and colloids may continue until larger flocs are formed, which are prone to sink out of the water column. This has recently been shown to be an important mechanism of sediment formation in unproductive lakes, where the indigenous primary production is low but DOM imported from the terrestrial environment is abundant. Several factors, separately as well as interactively, affect the formation of colloids and larger flocs of organic matter, as summarized below.

Size and Concentration

The smallest colloids (<0.1 μm) coagulate quickly into stable aggregates (0.1–1.0 μm) and larger colloids (>10 μm) are removed from suspension by sedimentation. Thus particles in the range of 0.1–10 μm are the most stable colloidal material in aquatic systems. Accordingly, aggregation is most rapid for the smallest colloids and the initial phases of aggregation. When the organic macromolecules are much smaller than inorganic colloids that occur in the same suspension or flexible enough to form coils, they may be adsorbed as a thin layer on the inorganic colloids, and modify the energy barrier. This results in a change of coagulation rate, for example when clay particles are coated with organic matter. When the organic macromolecules are much larger than the inorganic colloids they may bind to several colloids and induce aggregation by bridging mechanisms. Consequently loose aggregates (flocs) are formed, whose structure, size, density, and rate of formation depend on the flexibility of the macromolecules, their relative length compared to colloidal size, and the relative concentration of colloids and macromolecules.

Brownian Motion (Thermal Effects)

Particles in the colloid size range (less than few micrometers) experience a random multidirectional motion due to the impact of solute molecules. The motion has an effect on particles in the colloidal size range (**Figure 3(a)**).

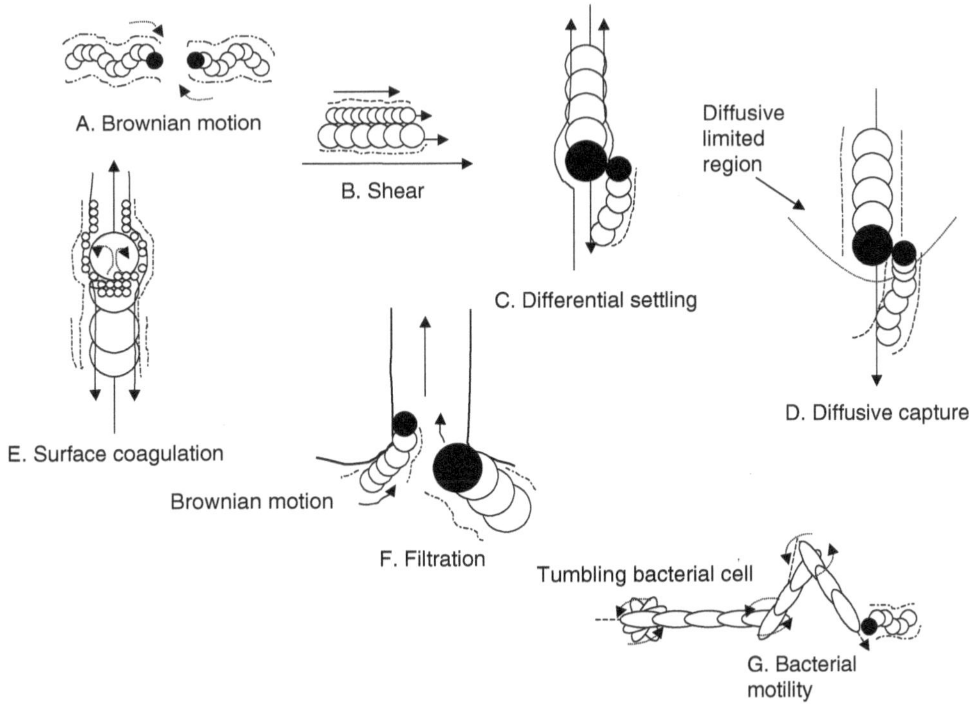

Figure 3 Collision mechanisms contributing to aggregation of DOM. Redrawn from Kepkay PE (1994) Particle aggregation and the biological reactivity of colloids. *Marine Ecology Progress Series* 109: 293–304.

Shear (Flow Effect) and Turbulence

Particles follow streamlines and/or eddies which cause collision between particles. This type of collision is particularly important for particles larger than a few micrometers in diameter. Turbulence is a strong controlling factor for flocculation where large shear stress in a turbulent environment promotes collision of colloids and particles. High turbulence can also have the opposite effect, leading to a breakdown of the flocs (**Figure 3(b)**).

Differential Settling (Gravity Effects)

Settling of particles at different velocities can lead to interceptions of particles. The density of the particles and the density of the surrounding water influence the mechanisms for collisions and settling velocities for the particles (**Figure 3(c)**).

Diffusive Capture

A diffusion-limited boundary layer is established around the surface of a settling particle and outside this layer the main transport mechanism of particle transport is advection. Within this layer, diffusion is of major importance, where colloids can be captured (**Figure 3(d)**).

Surface Coagulation and Bubbling

Molecules aggregate into colloids at the air–water interface and especially on rising air bubbles. When bubbles burst at the air–water interface, their organic coating collapses into colloids. This mode of formation of colloids has been shown to be prominent in the rapids of humic-rich running waters (**Figure 3(e)**).

Filtration

Macroaggregates possess a porous structure with the capability to capture settling particles through interception or size exclusion (**Figure 3(f)**).

Microbial Activity and TEP

Swimming microbes collide with colloids, inducing aggregation. In addition, microbes may release extracellular compounds (e.g., polysaccharides) that promote flocculation. Transparent exopolymer particles (TEP) are a chemically diverse and heterogeneous group of particles consisting mainly of acidic polysaccharides. The TEPs are sticky, flexible, and gel-like. The TEPs can glue particles together and therefore promote flocculation of other molecules and colloids, such as humic substances. The formation of polysaccharide particles can be a pathway for the conversion of DOM into POM during phytoplankton blooms in oceans, and probably also in lakes (**Figure 3(g)**).

Cations, pH, and Adsorption to Particles

Low pH and high concentrations of cations promote the flocculation of DOM. The solubility of humic substances increases with pH, and they are easily adsorbed to surfaces at low pH. This is utilized in the isolation of humic substances from environmental samples, but it also has implications for the solubility of humic substances in the natural environment. At a high ionic strength of water, the negative charge of humic substances (due largely to phenolic and carboxylic groups) is screened, and aggregation can occur of colloids and molecules that are stable at lower ionic strength. Ion bridging between divalent cations such as Ca^{2+} and Mg^{2+} promotes the formation of polysaccharide gels as well as humic substances. The low salt concentration in fresh water allows a wider electrochemical double layer around colloids and fine particles, which enhances the repulsive forces between the negatively humic substances. These repulsive forces are weakened in marine waters with high concentrations of salts. Thus larger flocs will potentially be formed in marine environments. In this way, riverine humic substances may aggregate in estuaries, where fresh water is mixed with saline marine water. Humic substances can be adsorbed to e.g., clay minerals and undergo precipitation. Lower pH or an increase in ionic strength augments the precipitation.

Sunlight

Flocculation has been observed during exposure of DOM to UV and short wavelength visible light, suggesting a photochemical stimulating effect on flocculation processes. Irradiation leads to cleavage of the DOM into a variety of photoproducts such as dissolved inorganic carbon, carboxylic acids, and a range of other carbonyl compounds, with implications for the bacterial utilization of DOM. Accordingly, photochemically produced carboxylic acids have been demonstrated to account for most of the carbon demand of bacteria in the epilimnion of a humic lake in summer. The background for the effect of light is unclear, but it is possible that the photochemical conditioning of DOM renders it increasingly hydrophobic, thereby increasing the potential for aggregation.

The Role of Iron for Aggregation

Iron oxides are common in natural waters, not the least in dystrophic waters, where large amounts of both iron and humic substances are imported from the watershed. Iron interacts with various aqueous dissolved compounds, and humic substances may coprecipitate with iron-oxide particles. Iron can catalyze the photodegradation of humic substances and colored dissolved organic matter (CDOM). In addition, light and iron interactively promote particle formation from dissolved humic substances. This is suggested to be due to products of photolysis, including reactive oxygen species, which transform DOM into less soluble forms. Moreover, oxic water iron (III) forms a complex with DOM which stays in solution. When the Fe(III)–DOM complex is exposed to UV light, the complex is broken, resulting in colloids of DOC and Fe(II). Fe(II) is oxidized into Fe(III) by photochemically produced H_2O_2 and the redox cycling (Fenton's reaction) is thereby complete. Possibly, humic colloids are stabilized by Fe(III), but are more prone to aggregation when the Fe(III)–DOM complex is broken. An indication that iron is of importance for sedimentation of organic matter in lakes is the relationship between sediment carbon stock burial and total iron concentration in the lake water.

Interactions of DOM and Humic Substances with Metals and Nutrients

Interactions of organic matter with iron are important for particle generation and stabilization of colloids, as explained in the section on colloid formation above. Metals exhibit a range of additional interactions with organic matter, and organic matter also interacts with nutrients in the water column. Some of the most important interactions are summarized below.

Metals

The mobility of metals in the aquatic environment is regulated strongly by humic substances and DOM. DOM can bind trace metals and toxic metals, affecting their bioavailability. Aluminum complexation is an important example of how humic substances can regulate the toxicity of metals. At low pH, complexation of Al^{3+} is weak, resulting in detrimental effects on biota. As pH rises, Al is increasingly complexed with DOM, decreasing its toxicity. Accordingly, Al toxicity is a major problem in anthropogenically acidified waters. Iron displays important interactions with implications for nutrient acquisition by organisms, described in a separate section below.

DOM, Iron, and Inorganic Phosphorus

Humic substance and iron both originate in the watershed. Their concentrations across lakes are correlated, as both tend to increase with increasing drainage of substances from the watershed. It has been demonstrated that Fe may be a limiting nutrient for phytoplankton growth in lakes with very low concentrations of DOM and Fe, such as Lake Superior and remote clearwater lakes in the Scandinavian mountains. DOM may act as a ligand of Fe, preventing the formation of iron oxides that may sink out of the water column.

Humic substances, Fe(III) and inorganic phosphorus (orthophosphate) form stable colloids. In humic-rich lakes a major fraction of the orthophosphate occurs in this colloidal state, which largely withdraws P from uptake by organisms. The association of P with humic substances is dependent on the presence of Fe(III), and on a low pH where humic substances are expected to be protonated and coiled, rather than at higher pH when molecules become more stretched. The photochemical redox cycle of iron described above also affects the P cycle of the water column, as reduction of Fe(III) to Fe(II) breaks the DOM–Fe (III)–P complex.

DOM and Nitrogen

Nitrogen is a structural element of DOM, including humic substances, with the order of 1% of DOM being nitrogen by weight. Some of this organic nitrogen can be released as ammonium upon exposure to solar UV radiation. In this way, photochemical ammonification can make previously inaccessible nitrogen available for producers. In addition, ammonium ions can be strongly bound to DOM. It has been hypothesized that this ammonium is unavailable to uptake by organisms, but can be released by ion exchange when DOM is transferred to brackish or marine waters, where other cations (sodium) compete more strongly than ammonium for binding sites on DOM. This is analogous to standard methodology in soil science, where ammonium is removed from organic matter by addition of potassium chloride, whereby ammonium is replaced by potassium by ion exchange.

Humic Substances and Hydrophobic Contaminants

Anthropogenic hydrophobic compounds, such as chlorinated biphenyls (PCB), can be bound to DOM. A possible mechanism is that they become associated with micelles of humic substances, as outlined above. The humic-associated contaminants are not available for uptake into organisms, resulting in less bioaccumulation in food webs. At lower pH in the water, the humic aggregates become less water soluble (more hydrophobic), and increasingly able to bind hydrophobic contaminants. Hence, there is a complex interaction between the acidity of inland waters and the bioaccumulation of hydrophobic contaminants, mediated by humic substances.

Concluding Remarks

Dissolved organic matter (DOM) is the dominant form of organic matter in the water column of inland waters, and it is chemically very heterogeneous and complex. As shown here, the physical appearance is also heterogeneous, including a range of molecules, aggregates, and colloids of different sizes and shapes. The humic substances have abundant functional groups, mostly carboxylic and phenolic, which interact with other compounds in the water. Many of these interactions are important regulators of ecosystem function, e.g., binding of metals and nutrients. Interactions between organic molecules, aggregates, and colloids determine the size distribution of organic matter, including the formation of entities large enough to sink out of the water column by gravity. Considering that DOM constitutes the majority of the organic matter in inland waters, the resulting translocation of organic matter from water to sediments may be of great importance for the carbon cycle of inland waters.

Glossary

Colloid – A macromolecule, collection of molecules or small particle (e.g., clay or virus particle) large enough to possess supramolecular properties (e.g., ability for conformational change or electrical surface field), but too small to settle by gravity, allowing it to stay suspended in the water column indefinitely. Colloids typically range is size from a few nanometers to a few micrometers.

Flocculation – The aggregation of molecules and colloids into larger particles (μm size range or larger) subject to sinking by gravity.

See also: Carbon, Unifying Currency; Light, Photolytic Reactivity and Chemical Products; Natural Organic Matter; Nutrient Stoichiometry in Aquatic Ecosystems; Organic Nitrogen.

Further Reading

Droppo IG, Leppard GG, Liss SN and Milligan TG (eds.). (2004) *Flocculation in Natural and Engineered Environmental Systems*. Boca Raton: CRC Press.

Findlay SEG and Sinsabaugh RL (eds.) (2003) *Aquatic Ecosystems: Interactivity of Dissolved Organic Matter*. Amsterdam: Academic Press.

Hessen DO and Tranvik LJ (eds.) (1998) *Aquatic Humic Substances: Ecology and Biogeochemistry*. Berlin: Springer.

Kepkay PE (1994) Particle aggregation and the biological reactivity of colloids. *Marine Ecology Progress Series* 109: 293–304.

Steinberg CEW (2003) *Ecology of Humic Substances in Freshwaters*. Berlin: Springer.

Stumm W and Morgan JJ (1996) *Aquatic Chemistry: Chemical Equilibria and Rates in Natural Waters*. New York: Wiley.

Thurman EM (1985) *Organic Geochemistry of Natural Waters*. Boston: Nijhoff/Junk.

Lipids

Y Olsen, Norwegian University of Science and Technology, Trondheim, Norway

Introduction

Lipids have a negative perception among people in the western world where being overweight is a severe and increasing health problem. This view does not appreciate that some lipid components are the main building blocks of animal's most sophisticated organ, their brain and nervous system. It is moreover a human's current understanding that it is our brain and mental capacities that make us superior to all other living creatures.

It was first in the 1960s that the family of n-6 fatty acids (explained later) were found to be essential nutrients for humans (essential fatty acids, EFAs). The family of n-3 fatty acids was proved to be essential later in the century. This resulted in comprehensive research activity on human health aspects of dietary EFAs. Unlike protein biosynthesis, the biosynthesis of lipids in our tissues is not strictly controlled by genes, but rather by a complex interaction between genes and the dietary lipid composition. This means that our selection of food items affect our health conditions quite broadly. A main current view is that humans, and other animals, tend to be evolutionarily adapted to a certain proportion of n-6 and n-3 fatty acids in their diets and that breeding of agricultural plants, intensive food production, and the food processing industries of our time have tended to reduce the dietary contents of n-3 fatty acids relative to n-6 fatty acids as compared with ancient times.

It has become apparent that an increased intake of specific long-chain fatty acids of the n-3 family may protect against cardiovascular disease by lowering of the blood pressure, reduction of the plasma triglyceride, and many related mechanisms. It is a more recent issue that our dietary fatty acid composition may also affect mental capacities and cause disorders, including depression and hyperactivity. Some of these fatty acids, in particular DHA and ARA (see structures below), are main components of the brain. A balanced n-3 to n-6 fatty acids ratio of 1:1 in the diet has been recommended to secure a healthy brain function, the ratio 20:1 has been common in red meat eating communities.

The most important food sources of long-chain n-3 fatty acids such as DHA and EPA is seafood, but apparently less known; it is important that most freshwater fishes are also valuable sources of these fatty acids. Freshwater resources are key food sources for many communities, and it is important to recognize how important these lipid resources are and will continue to be for human nutrition.

Also aquatic organisms depend on sufficient nutrition, and studies of essential nutrients have become an issue of ecology. Species of cold-blooded animal show different abilities to synthesize long-chain EFA (20 and 22 carbon chains) based on short-chain moieties (18 carbon chains). The availability of long-chain EFA in the diet of the animals is therefore likely an important driver of biodiversity.

Essential Fatty Acids

Fatty acids can be named by their number of carbon atoms, number of double bonds, and the position of the first double bond from the methyl end of the fatty acid, e.g., 18:3 n-3, which means that the fatty acid has 18 carbon atoms and 3 double bonds, with the first double bond between carbon number 3 and 4 from the methyl end. The most common fatty acids have trivial names, 18:3 n-3 is for example mentioned as alpha-linolenic acid or ALA (**Figure 1**).

Fatty acids are commonly grouped in saturated fatty acids (SAFA), mono-unsaturated fatty acids (MUFA), and poly-unsaturated fatty acids (PUFA). SAFA have no double bonds and show common chain lengths of 14–24 carbons. MUFA have one double bond and common chain length of 14–24 carbon. PUFA contain two or more double bonds, with common chain lengths of 16–22 carbon and 2–6 double bonds. Highly unsaturated fatty acids (HUFA) are PUFA with 20 or more carbon atoms and 3 or more double bonds.

All animals are capable of biosynthesis of 16:0 and 18:0 fatty acids and to modify these precursors into 16:1 n-7 and 18:1 n-9. It is very important for understanding lipid nutrition to note that animals cannot synthesize double bonds in positions closer than between carbon 7 and 8 from the methyl end (n-7) although such fatty acids are essential for their growth and development. Only plants and bacteria are capable of doing that, and these fatty acids must be supplied in the food for animals and are known as EFAs. EFAs are grouped in two families; the linolenic acid (n-3) and the linoleic acid (n-6) families, with the 18 carbon ALA and LA as the basic respective moieties (**Figure 1**). Some animals are capable of further elongation and desaturation of ALA and LA to form

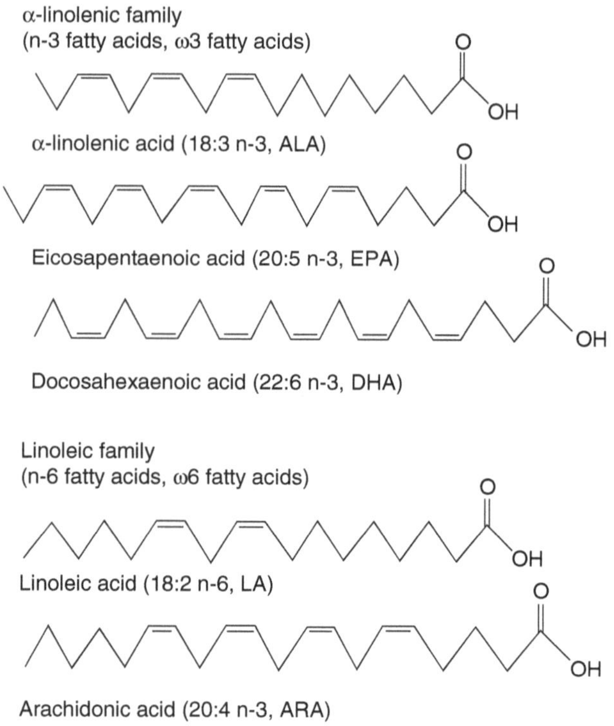

Figure 1 Some common n-3 and n-6 fatty acids in aquatic organisms.

n-3 and n-6 HUFA through successive steps (**Figure 2**). These abilities are, however, poor or apparently lost for many species.

The question of essentiality of given n-3 and n-6 fatty acids are complex, and may seem confusing. The confusion originates in the variable species specific ability to metabolically upgrade 18 carbon EFA moieties to 20 and 22 carbon homologues. The efficiency of this process is important for the actual EFA requirements of a species. The long-chain n-3 fatty acids will not be strictly essential for species that can synthesize these from the 18 carbon fatty acids, for example humans (vegetarians survive well). Other species that lack the ability to upgrade the 18 carbon EFA will need to have the long-chain EFA supplied in the food.

It may, however, still be argued that dietary supply of ARA, EPA, and DHA can counteract specific disorders and thereby affect health of humans and other animals. At the end, it becomes a question whether *essential* is defined as 'essential for life' or 'essential for more specific health aspects.' For example, there is accumulating evidence that DHA, supplied directly from the mother and in breast milk, is essential in early developmental stages for humans and mammals in general in order to build a well-functional brain and mental capacity.

Another aspect is that many carnivorous fish species appears to have little or no metabolic capacity to elongate and desaturate the shorter 18 carbon precursors to form ARA, EPA, and DHA. These fatty acids are paramount and needed for neural tissues, cell membrane functioning, and many regulatory functions. A major question of research has been whether carnivorous fish species lack the genes of the enzymes needed to synthesize DHA from ALA or whether the genes of these enzyme systems are simply not expressed? Recent research has indicated that the genetic information is present, but that evolution in some way has turned off gene expression in some species.

Important Lipid Groups and Functions

Lipids are, depending on their polarity, grouped as neutral or polar lipids (**Figure 3**). Triacylglycerides (TAG) and wax esters (WE) are neutral lipids and particularly abundant storage lipids providing metabolic energy through oxidative catabolism. TAG is the dominant energy and carbon storage product in higher animals. The fatty acid configuration of TAG is highly diverse and dependent of the dietary fatty acid composition.

Some marine calanoid copepods and euphausiids populating deep parts of the oceans or high latitudes store their surplus energy as WE, which may constitute >50% of dry matter at the end of the growth season. The WE molecules are formed by a single fatty alcohol moiety esterified to a single fatty acid moiety. WEs have a lower melting point and higher energy content per unit weight than do TAG.

The basic unit of phospholipids (PL) is a diacylglycerides molecule with a phosphate group bound to an organic group in a terminal position of the glycerol (**Figure 3**). Aquatic cold-blooded animals are characterized by having PL containing a high fraction of PUFA. Common organic groups esterified to the phosphate group of PL are ethanolamine, choline, serine, and inositol, forming the major phosphorglycerides of animals, denoted phospatidylethanolamine, phospatidylcholine, phospatidylserine, and phospatidylinositol (**Figure 3**). These PL, together with cholesterol and sphingolipids, are ubiquitous constituents of cell membranes and therefore both structurally and functionally important.

Phospholipids are main components of cell membranes, with a major functional role for cellular transport processes of ions and molecules, and metabolism in general. PL is synthesized by enzyme systems that exhibit higher affinity for PUFA than for other fatty acids, with the highest affinity for HUFA. This

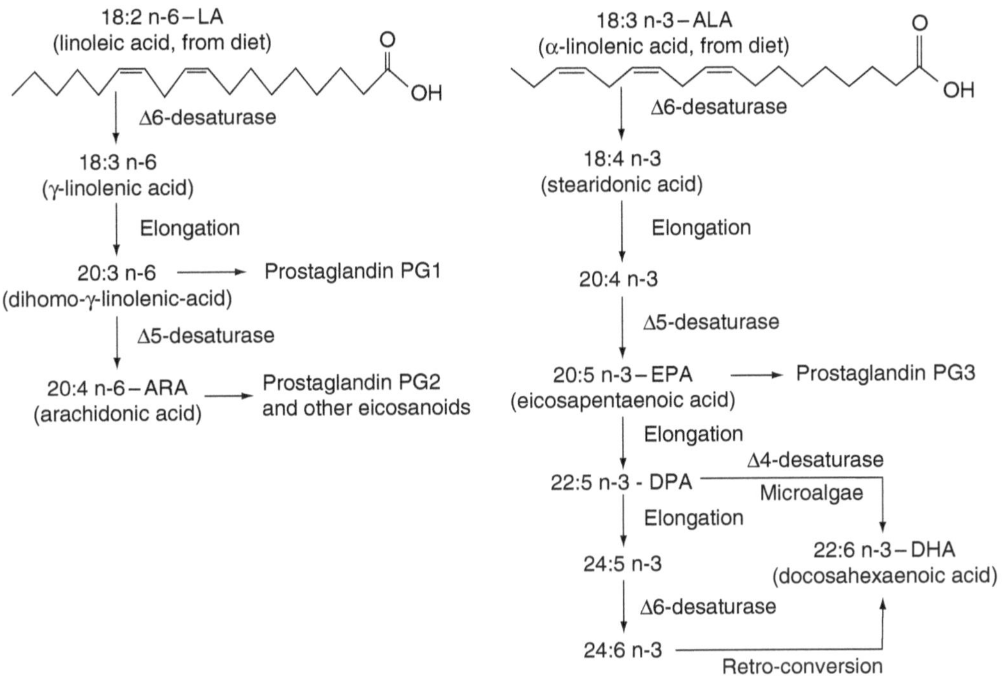

Figure 2 Pathways of biosynthesis of n-3 and n-6 HUFA from the respective C18 precursors ALA and LA. Biosynthesis is taking part through successive desaturation and elongation steps. Specific enzymes are required, and Δ indicate the position of the synthesized double bond from the acid side of the fatty acid chain where the enzyme can introduce a new double bond.

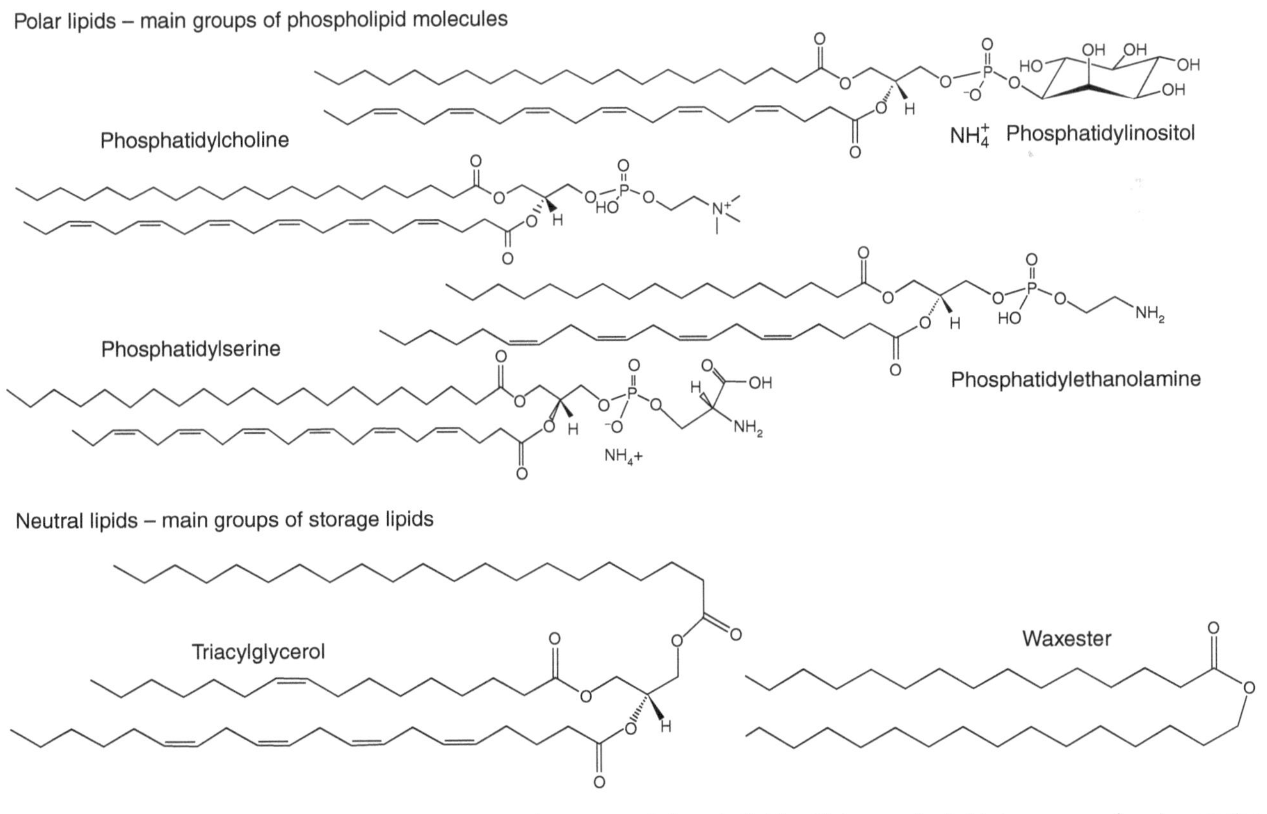

Figure 3 Examples of molecules belonging to the 4 major groups of phospholipids which are polar lipids (upper panel) and examples of a tri-acylglycerol and a wax ester molecule which belong to the neutral lipids (lower panel).

secures some degree of a genetic control of PL synthesis and accordingly of the fatty acid composition of cell membranes, which is very critical from a functional point of view. Enhanced levels of MUFA may be incorporated in membrane PL under conditions of PUFA deficiency. This will at some point, which is species-specific, result in a gradually reduced physiological capacity and state of health.

Neural tissues contain very high amounts of DHA (e.g., 25–30% of total fatty acids in brain PL of carp), suggesting that DHA is particularly important for reproduction and developing larvae. From an evolutionary point of view, we may assume that essential components that are found in high quantities in the eggs are nutritionally important, and eggs normally contain very high DHA levels. It is easy to imagine that an inadequate brain function and vision, as demonstrated for DHA-deficient herring larvae, will be fatal for all higher aquatic animals.

The 20 carbon fatty acids EPA and ARA of the membrane PL are precursors in prostaglandin synthesis. Prostaglandins are precursors for a number of regulating compounds known as tissue hormones. Prostaglandin G_3 is synthesized from EPA and acts as a regulatory antagonist to prostaglandin G_2 which is synthesized from ARA. The prostaglandin G_3:G_2 ratio, which is believed to modify many cellular processes, will depend on the EPA:ARA ratio in membranes. This implies that the fatty acid composition of the food will directly affect the regulatory hormonal processes of animals, and accordingly their state of health.

Fatty Acid Transfer and Modification in the Freshwater Food Web

Fatty acids are transferred from lower to higher levels in the freshwater food web, while metabolic modifications of individual n-6 and n-3 fatty acids from phytoplankton can take place in the animals during that process. The transport and accumulation of these fatty acids are therefore also dependent on the structure and function of the food web. **Figure 4** illustrates that metazoans and fish will mainly consume and retain EFAs originating from eukaryotic algae and the food of insects, which may be of both terrestrial and aquatic origin. Larger protozoan may occasionally contribute as feed for younger stages of fish, but a majority of the energy and material is normally lost during the extra trophic transfers from cyanobacteria, bacteria, and small eukaryotic algae through metazoans. The microbial food web may contribute more to the fatty acid flow in ultra-oliotrophic lakes.

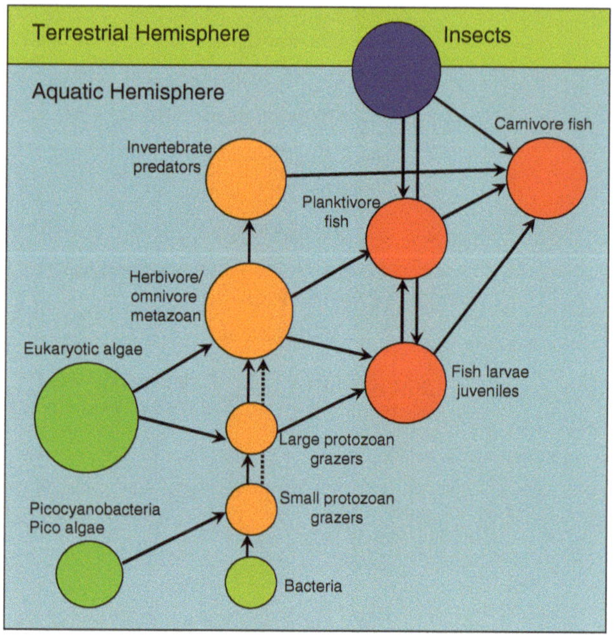

Figure 4 Schematic food web structure for fresh water with main trophic linkages included. Fish can feed on both aquatic and terrestrial food items.

Eukaryotic Algae

Planktonic algae respond to environmental changes by modifying their biochemical composition. Some algal species tend to store carbon in the form of lipids, more specifically TAG, in cases of nutrient limitation, and the stores tend to increase with increasing extent of nutrient limitation. This is because photosynthesis, generating carbon stores, is less sensitive to nutrient limitation than cell division, which directly requires nutrients. Some other algal species tend to store carbon in the form of carbohydrates, while keeping their lipid contents constant. Lipids in algae normally range from 5% to 70% of dry weight, with 15–30% and 8–12% as common ranges for species that tend to store lipids and carbohydrates, respectively. The highest extreme values are representative for severely nutrient-limited lipid-storing species. The fraction of fatty acids of total lipids is normally in the range of 30–50% in algae.

With the dynamics of total lipids (and total fatty acids; 30–50% of total lipids) in mind, there is still a strong genetic control of fatty acid composition of planktonic algae, be it marine or fresh water. The fatty acid profile, contents expressed in terms of percent of total fatty acids, is relatively constant for any given species, and there is a general trend in composition for the different main taxonomic groups. The percent distribution of fatty acids is only slightly affected by the nutritional state of the alga and the temperature. A severely nutrient-limited population contains a lower percent of PUFA than a fast growing, nutrient-sufficient population. This means that the variability of fatty acids in planktonic algae is driven by environmental conditions, but only within the range of variability allowed by genetic properties. Species or taxonomic group is the main criteria for fatty acid composition.

Compiled contents of the most important n-3 and n-6 fatty acids of some freshwater planktonic algal groups are shown in **Figure 5**. The corresponding level of total lipids (TL, not shown) is 86 ± 22, 283 ± 62, and 279 ± 66 mg g DW^{-1} in the taxonomic groups of Cyanobacteria, Chlorophyceae, and Flagellates, respectively, indicating that the majority of the species of green algae and flagellates store lipids whereas cyanobacteria store polysaccharides. Green algae exhibited the highest contents of n-3 and n-6 fatty acids, both expressed in quantitative and relative terms, whereas only species within the group of flagellates contain significant amounts of EPA and DHA. The content of ARA is throughout at trace levels, which is quite common also for marine planktonic algae.

Figure 5 also suggests a common general pattern of variation of fatty acid composition in taxonomic groups of marine and freshwater algae. This difference will also be reflected in the respective animals on higher trophic levels of those systems. Large cyanobacteria and green algae are uncommon in marine waters, and fatty acids of picocyanobacteria are not an important food source for metazoans and higher trophic levels (**Figure 4**). Marine phytoplankton is dominated by the groups of diatoms, dinoflagellates, and a broad taxonomic group of small brown/yellow flagellates, which are all relatively HUFA rich. Marine diatoms typically contain high levels of EPA, dinoflagellates are rich in DHA, and small flagellates are very diverse, but normally characterized by high levels of a mixture of n-3 HUFA. ARA and short-chain n-6 fatty acids are found in low amounts, only. This might suggest that marine zooplankton generally will have access to higher EPA and DHA levels in their food than that of freshwater zooplankton.

Direct measurements of fatty acid composition made for natural lakes reveals, however, that HUFA are available in the seston fraction in significant amounts through most of the vegetation period. On the other hand, sedimenting material forming food for sediment living animals is very poor in HUFA. The high nutritional value of seston must imply that diatoms and a diverse group of flagellates constitute a significant fraction of the phytoplankton biomass. The potential contribution by protozoan grazers is unknown.

In conclusion, it cannot be out ruled that freshwater animals may experience HUFA limitation if their capacity of upgrading short-chain 18 carbon fatty acids like LA and ALA to HUFA is low. The alternative of undertaking upgrading is to have a low physiological requirement of HUFA. This will be reflected by a sustained, low percent of HUFA in PL. Sediment living animals must unconditionally be capable of upgrading short-chain PUFA.

EFA in Freshwater Fish

Aquatic ecologists will normally find major variability in lipid and fatty acid composition in zooplankton and fish, be it in marine or fresh water. Measurements of essential lipids can for sure reflect state of health of animals, but it is not straightforward how such measurements can be used for this purpose. Such use will require a physiological understanding of lipid metabolism and natural variability.

Animals that have experienced high food availability, or severe limitation of a specific essential nutrient, will tend be fatter than starved animals. The fatty acid composition of the TAG storage lipids will in all events to great extent reflect the fatty acid composition of the recently consumed food. The fatty acid composition of PL is less variable, thanks to the higher degree of genetic control of the composition of this component. The fatty acid composition of PL will therefore better reflect EFA requirements of the species. Physiological disorders can be signified when the fatty acid composition of membrane PL deviates from the normal for the species.

It is important that the content of PL will be relatively constant per DW in animals, although dependent of species and developmental stage. In contrast, the content of TAG (or WE) is highly variable per DW. This means that the percent PL of total lipids decrease with increasing total lipids, whereas the quantitative amount per DW stays relatively constant. It follows that the fatty acid composition of a lean animal, or muscle tissue of species that do not store lipids in the muscle, to great extent reflects the composition of its PL, whereas the composition of a fat animal reflects the fatty acid compositions of TAG.

Figure 6 illustrates the above relationships for data of Chinese freshwater fish, compiled from literature. The fish was purchased in the market; the nutritional history was accordingly unknown. PL contents per DW remain relatively constant, whereas neutral lipids (NL) per DW, comparable with TAG, increases slightly with increasing TL of the fish meat. Meanwhile, the percent PL of TL decreases significantly. Data derived from natural populations of zooplankton and fish should be treated with these conceptual relations in mind.

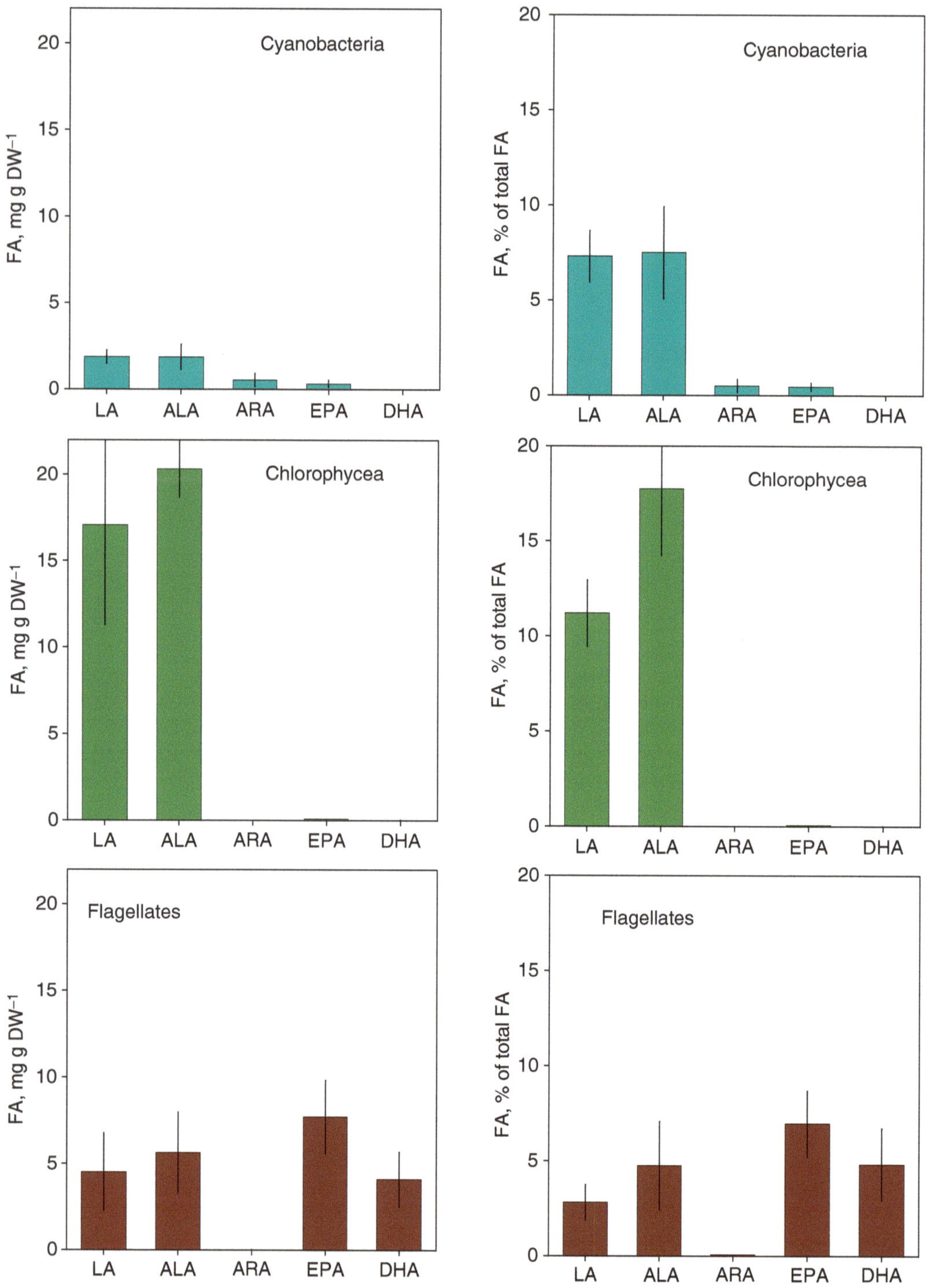

Figure 5 Essential and potentially essential n-3 and n-6 fatty acid contents (mg gDW^{-1} to the left, percent of totals fatty acids to the right) of some important taxonomic groups of freshwater planktonic algae. Bars indicate 1SE of the mean ($n = 12$ for cyanobacteria, $n = 6$ for chlorophyceae, and $n = 6$ for flagellates). For the abbreviations of fatty acids, see glossary.

A main general aim of ecologists and fish biologists is to use lipid and fatty acid proxies as measures for nutritional state of natural populations of zooplankton and fish. This requires a fundamental road map for the normal, healthy situation, including knowledge on EFA requirement of the species. Such information must include the EFA composition of PL, because TAG and TL composition to a great extent reflect the dietary fatty acid composition. To derive exact information on species requirements for EFA require

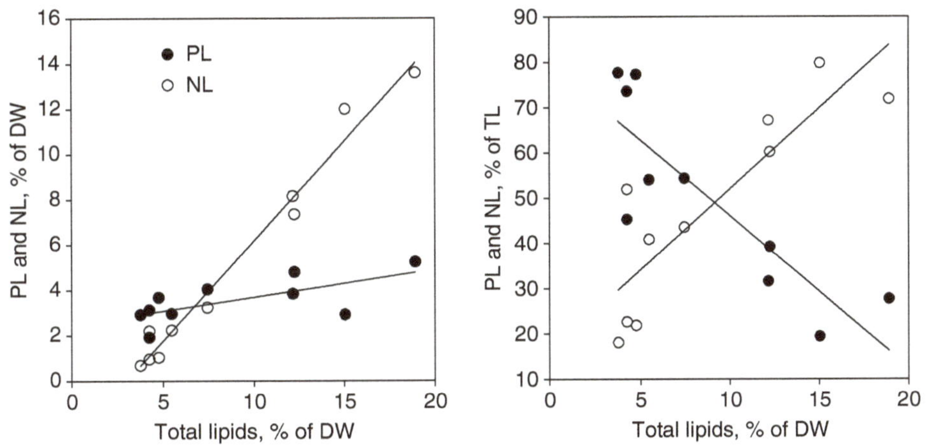

Figure 6 The contents of phospholipids (PL) and neutral lipids (NL, mainly tri-acylglycerols) as a function of total lipid contents (% of DW) of the fish meat in a selection of Chinese freshwater fish species ($n = 10$). Left panel expresses PL and NL contents as percent of dry weight. Right panel expresses PL and NL contents as percent of total lipids.

considerable research efforts, but compilations of data for several species or a species of different nutritional states can help to derive that road map.

A compilation of mean fatty acid composition for the 10 Chinese species of freshwater fish (mentioned earlier) revealed an accumulation of DHA and ARA, and to some extent EPA, in PL relative to NL and TL (**Figure 7**). NL and TL showed high levels of 18:1 n-9, which is a main energy source of the fish. DHA and ARA ranged 9–32% and 4.4–19% of total fatty acids, respectively, in PL. This shows that both fatty acids are important components of membranes in all species. Both fatty acids are found in far less amounts in NL.

There are low contents and no accumulation of LA and ALA in PL. These are the potential precursors for HUFA synthesis in the fish as well as in zooplankton. Many of the species involved have capabilities of desaturation and elongation of these short C18 moieties to form the longer HUFA, which may contribute to explain the low level.

The general picture in **Figure 7** is comparable with that of marine fish except for on one main point; the ARA content is far higher in freshwater fish than in marine fish (~1%), thereby coming closer to the composition of PL of land living animals and neural tissues. It follows that the EPA:ARA ratio is just below 1, and therefore far lower than in marine fish. The lower n-3/n-6 ratio of freshwater fish meat as compared with marine fish meat seems to be a general feature. This can be related to an evolutionary adaptation to terrestrial food (or lipid) sources containing more n-6 fatty acids in freshwater fish.

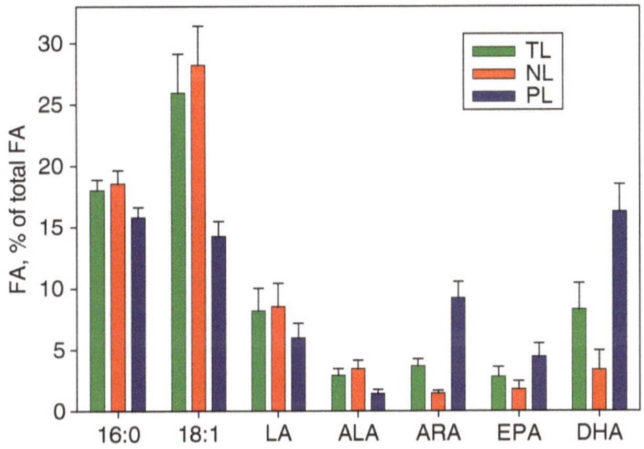

Figure 7 Percent fatty acid distribution of essential n-3 and n-6 fatty acid contents and some other abundant fatty acids in total lipids (TL), NL, and PL in of the fish meat in a selection of Chinese freshwater fish species. Bars indicate 1SE of the mean ($n = 10$).

Characteristics of Food for Freshwater Fish

A brief consideration of some common feed of freshwater fish (**Figure 4**) reveals fundamental differences with marine fish food, in particular, for planktivorous species. Water living insects contain 10–22% (mean 15%) EPA of total fatty acids, only traces of DHA and <1% ARA. *Daphnia* sp. is an important food source for planktivorous freshwater fish. Many species of daphnids tend to catabolize DHA selectively, and EPA and ARA appear to be the important HUFA of *Daphnia* sp. The concentration of EPA has been suggested as a limiting factor for growth in many lakes. Also, Gammerids show a relatively high ARA content (4–12%) and only traces of DHA. The content of EPA is variable, ranging from 0% to 13%. Freshwater copepods appear to have a similar DHA and EPA content as marine copepods, thus representing a food item with significant content of DHA (20–40%). Finally, LA and ALA are abundant in most food items.

The general picture reveals the higher availability of ARA and n-6 fatty acids in the food of freshwater fish than for marine fish. It also confirms a generally low availability of DHA and surplus of EPA. The short-chain LA and ALA, which are potential HUFA precursors, are available throughout in all food items, although in variable amounts.

Concluding Remarks

The main taxonomic groups of freshwater algae exhibit high EFA contents, but only the diverse group of flagellates, as defined here, contain HUFA in significant amounts. Green algae were poor in HUFA, but show high contents of the precursors for HUFA synthesis in animals, LA and ALA. Zooplankton and other food items for fish are poor in DHA, but their high ARA contents most likely reflect a metabolic upgrading of LA. An upgrading of ALA to form EPA and DHA is not that apparent for these food items; this is on the other hand most likely undertaken quite efficiently by freshwater fish.

The general road map above will not necessarily be representative for species or groups that catabolize DHA selectively and accumulate EPA in their membranes, which can be the case for daphnids. The freshwater food web is for sure more rich in n-6 fatty acids than the marine food web, including the physiologically very important ARA. There are, however, generally great similarities for given taxonomic groups. The higher abundance of calanoid copepods, dinoflagellates, and small flagellates in marine ecosystems reflects the higher DHA contents of the marine food web as compared with the freshwater food web.

Glossary

Arachidonic acid (ARA, 20:4 n-6) – Long-chain, highly unsaturated fatty acid of the n-6 family of fatty acids. ARA can be synthesized from LA by successive chain elongation and desaturation by some animals.

Docosahexaenoic acid (DHA, 22:6 n-3) – Long-chain, highly unsaturated fatty acid of the n-3 family of fatty acids. DHA can be synthesized from ALA by successive chain elongation and desaturation by some animals.

Eicosanoids – Biologically active lipid mediators of the 20 carbon fatty acids ARA and EPA and their metabolites, including prostaglandins, thromoxanes, leukotrines, and other oxygen derivatives with main regulatory functions in cells.

Eicosapentaenoic acid (EPA, 20:5 n-3) – Long-chain, highly unsaturated fatty acid of the n-3 family of fatty acids. EPA can be synthesized from ALA by successive chain elongation and desaturation by some animals.

Essential fatty acids (EFA) – Fatty acids which cannot be synthesized by a given organism and therefore needs to be supplied in the diet.

Highly unsaturated fatty acids (HUFA) – PUFA with 20 or more carbon atoms and 3 or more double bonds.

Linoleic acid (LA, 18:2 n-6) – Short-chain essential fatty acid of the n-6 family of fatty acids, can only be synthesized by plants.

α-Linolenic acid (ALA, 18:3 n-3) – Short-chain essential fatty acid of the n-3 family of fatty acids, can only be synthesized by plants.

Mono-unsaturated fatty acid (MUFA) – Fatty acid with one double bond, normally in positions n-7, n-9, and n-11, and with most common chain lengths of 16–24 carbon in aquatic organisms.

n-3 fatty acids – Fatty acids with 18–24 carbon chains, 2–6 double bonds, and the first double bond in between carbon numbers 3 and 4 from the methyl end of the fatty acid. The n-3 family of fatty acids constitutes all n-3 fatty acids.

n-6 fatty acids – Fatty acids with 18–22 carbon chains, 2–5 double bonds, and the first double bond in between carbon numbers 6 and 7 from the methyl end of the fatty acid. The n-6 family of fatty acids constitutes all n-6 fatty acids.

Phospholipid (PL) – A diacylglyceride molecule with a phosphate group, esterified to a specific organic group, esterified to the glycerol molecule in a terminal position of the diacylglycerol instead of a fatty acid in TAG. PLs are polar lipids, which are functional components of all cell membranes.

Poly-unsaturated fatty acids (PUFA) – Fatty acid with two or more double bonds and chain lengths of 16 (with 2–4 double bonds, in algae), 18 (with 2–5 double bonds), 20 (with 2–5 double bonds), and 22 carbon (with 2–6 double bonds). All n-3 and n-6 fatty acids are PUFA.

Saturated fatty acid (SAFA) – Fatty acid with no double bonds, showing most common chain lengths of 14–24 carbon in aquatic organisms.

Triacylglycerol (TAG) – A glycerol molecule esterified to 3 fatty acids, in positions sn-1, sn-2 and sn-3 of the glycerol. TAG is a common neutral lipid for storage of energy.

Wax ester (WE) – A single fatty acid moiety esterified to a single fatty alcohol moiety. WE are common neutral lipids for storage of energy, particularly in calanoid copepods and krill in cold environments.

Further Reading

Ahlgren G, Gustavsson IB, and Boberg M (1992) Fatty acid content and chemical composition of freshwater algae. *Journal of Phycology* 28: 37–50.

Ahlgren G, Goedkoop W, Markensten H, Sonesten L, and Boberg M (1997) Seasonal variations in food quality for pelagic and benthic invertebrates in Lake Erken – The role of fatty acids. *Freshwater Biology* 38(3): 555–570.

Allport S (2006) *The Queen of Fats: Why omega-3s were removed from Western Diet and What We Can do to Replace Them.* California: University of California Press.

Arts MT and Wainman BC (eds.) (1999) *Lipids in Freshwater Ecosystems.* New York: Springer-Verlag.

Brett MT, Muller-Navarra DC, Ballantyne AP, Ravet JL, and Goldman CR (2006) Daphnia fatty acid composition reflects that of their diet. *Limnology and Oceanography* 51(5): 2428–2437.

Bell JG, Ghioni C, and Sargent JR (1994) Fatty acid composition of 10 freshwater invertebrates which are natural food organisms of Atlantic salmon patt (*Salmon salar*): A comparison with commercial diets. *Aquaculture* 128: 301–313.

Broadhurst CL, Cunnane SC, and Crawford MA (1998) Rift Valley lake fish and shellfish provided brain-specific nutrition for early Homo. *British Journal of Nutrition* 79(1): 3–21.

Crawford MA, Bloom M, Broadhurst CL, *et al.* (1999) Evidence for the unique function of docosahexaenoic acid during the evolution of the modern hominid brain. *Lipids* 34: 39–47.

Evjemo JO, Reitan KI, and Olsen Y (2003) Copepods as live food organisms in the larval rearing of halibut larvae (*Hippoglossus hippoglossus* L.) with special emphasis on the nutritional value. *Aquaculture* 227(1–4): 191–210.

Henderson RJ and Tocher D (1987) The lipid composition and biochemistry of freshwater fish. *Progress in Lipid Research* 26: 281–347.

Kaneniwa M, Miao S, Yuan C, Iida H, and Fakuda Y (2000) Lipid components and enzymatic hydrolysis of lipids in muscles of Chinese freshwater fish. *JAOCS* 77(8): 825–831.

Sargent JR and Henderson RJ (1986) Lipids. In: Corner EDS and O'Hara SCM (eds.) *The Biological Chemistry of Marine Copepods.* Oxford: Clarendon Press.

Sargent JR, Tocher DR, and Bell JG (2002) The lipids. In: Halver JE and Hardy RW (eds.) *Fish Nutrition.* San Diego, CA: Academic Press.

Tocher DR (2003) Metabolism and functions of lipids and fatty acids in teleost fish. *Reviews in Fisheries Science* 11(2): 107–184.

Wang Y, Crawford MA, Chen J, *et al.* (2003) Fish consumption, blood docosahexaenoic acid and chronic disease in Chinese rural populations. *Comparative Biochemistry and Physiology A* 136: 127–140.

Relevant Websites

www.lipidlibrary.co.uk.
www.cyberlipid.org/.
http://lipidbank.jp/.
http://www.lipidmaps.org/.
http://www.dhaomega3.org/.
http://www.cytochemistry.net/Cell-biology/.

Methane

D Bastviken, Stockholm University, Stockholm, Sweden

Introduction

Methane (CH_4) is a major product of organic matter (OM) decomposition in freshwater environments. CH_4 is both produced and consumed in aquatic environments. This results in local within-system carbon cycling, providing a link between anoxic and oxic processes or habitats. It has been hypothesized that CH_4 produced in anoxic sediments can represent a significant carbon source for aquatic food webs in oxic habitats. The local consumption of CH_4 is not complete and CH_4 emissions from inland waters represent a substantial greenhouse gas contribution to the atmosphere. The aim of this chapter is to review various aspects of CH_4 cycling in inland waters, including

1. the fundamentals of CH_4 formation and oxidation in aquatic environments,
2. the contribution of CH_4 dynamics to overall carbon cycling in aquatic systems,
3. the potential contribution of CH_4 carbon to aquatic food webs,
4. the role of CH_4 emissions in linking aquatic carbon cycling, global atmospheric chemistry, and climate regulation.

CH_4 is an odorless organic trace gas melting at $-182.5\,°C$ and boiling at $-161.5\,°C$. It is combustible at concentrations of 5–15% in air, and it represents the major component of natural gas, biogas, and marsh gas. CH_4 has a Henry's law constant of 1.29×10^{-3} M atm^{-1} (1.27×10^{-5} mol m^{-3} Pa^{-1}) at 25 °C, expressing a low water solubility. CH_4 absorbs infrared radiation, with strong adsorption bands at 7–13 μm. The greenhouse warming potential for CH_4, a function of atmospheric residence time and radiative properties, is ~23 times greater than for carbon dioxide (CO_2) on a molecular basis over a 100-year timescale. The atmospheric mixing ratio of methane has doubled over the past 250 years and, between 1977 and 2002, has increased from 1.5 to ~1.8 ppmv. It has been estimated that CH_4 accounts for ~20% of the increase in the greenhouse gas effect observed since the mid 1700s.

CH_4 is constantly being degraded in the atmosphere by photooxidative and hydroxyl radical associated processes. The estimated lifetime of CH_4 in the atmosphere is 8–12 years. Hence, steady-state or increasing atmospheric concentrations require large, continuous emissions from various sources.

CH_4 Formation

Biochemical and Microbiological Aspects of Methanogenesis

Most of the CH_4 in the atmosphere is of biogenic origin, as determined by its ^{14}C dating. Biological production of CH_4, methanogenesis, is performed by methanogenic archaebacteria. All known methanogens share some characteristics, including high intracellular concentrations of coenzyme F_{420} (resulting in autofluorescence after stimulation under a fluorescence microscope), the presence of the enzyme methyl coenzyme M methylreductase, and the presence of unique phospholipid membrane components.

CH_4 production is a strictly anaerobic process relying on a limited number of low molecular weight substrates. The two principal pathways of production are acetotrophic (acetate dependent) methanogenesis and hydrogenotrophic (H_2 dependent) methanogenesis. In the former case, acetate (CH_3COO) is cleaved into CH_4 and carbon dioxide (CO_2). During hydrogenotrophic methanogenesis, H_2 reacts with CO_2 forming CH_4 and H_2O. Other low molecular weight organic molecules, such as formate, methanol, dimethyl sulfide (DMS), tri-, di-, and monomethylamines, and ethylamine, can also be used for methanogenesis, but acetate or H_2 and CO_2 are believed to be the quantitatively most important substrates.

Microbial methanogenesis represents a terminal step of anoxic OM degradation (**Table 1**). Before the terminal anoxic degradation can occur, complex organic substrates have to undergo several degradation steps (**Figure 1**). Complex OM is fermented to intermediate fermentation products, including acetate, H_2, and CO_2, which can then be used in methanogenesis. Typically, this requires the activity of several different bacterial groups, which organize in syntrophic consortia facilitating the transfer of the reaction products between cells. In the presence of oxidized compounds that can function as electron acceptors, e.g., nitrate (NO_3^-), manganese (IV) (Mn^{4+}), iron (III) (Fe^{3+}), and sulfate (SO_4^{2-}), terminal degradation steps like denitrification, iron reduction, manganese reduction, and sulfate reduction may yield more energy than does methanogenesis (**Table 1**). Hence, since bacteria capable of utilizing alternative terminal electron acceptors often have a superior energy gain, and sometimes also more efficient

Table 1 Examples of redox reactions used in the microbial degradation of OM[a]

Process	*Substrates*		*Final products*		$-\Delta G^0$ *(kJ/mole* e^-*, pH = 7)*
	e^- *donor*	e^- *acceptor*			
Aerobic respiration	OM	O_2	CO_2	H_2O	125
Denitrification	OM	NO_3^-	CO_2	N_2	112
Manganese reduction	OM	Mn(IV)	CO_2	Mn(III), Mn(II)	95
Iron reduction	OM	Fe(III)	CO_2	Fe(II)	24
Fermentation	OM	OM	CO_2	LMWOM, H_2	5–60
Sulfate reduction	LMWOM	SO_4^{2-}	CO_2	S^{2-}	18
Methanogenesis	LMWOM, H_2	LMWOM, CO_2	CO_2	CH_4	14–28

[a]The general principle of these metabolic processes is that organisms gain energy by transferring electrons from an electron (e^-) donor to an e^- acceptor. The e^- donor has to be more reduced than the e^- acceptor to allow energy gain. The difference in degree of reduction or oxidation between the e^- donor and acceptor is roughly proportional to the energy gain. $-\Delta G^0$ indicates the energy yielded from the listed processes. Values of $-\Delta G^0$ are related to specific reactions under standard conditions in Zehnder AJB and Stumm W (1988) Geochemistry and biogeochemistry of anaerobic habitats. In Zehnder AJB (ed.) *Biology of Anaerobic Microorganisms*, pp. 1–38, New York: John Wiley & Sons, and should only be used for comparison between the different processes in the table. OM and LMWOM denote OM and low molecular weight OM (i.e., fermentation products), respectively.

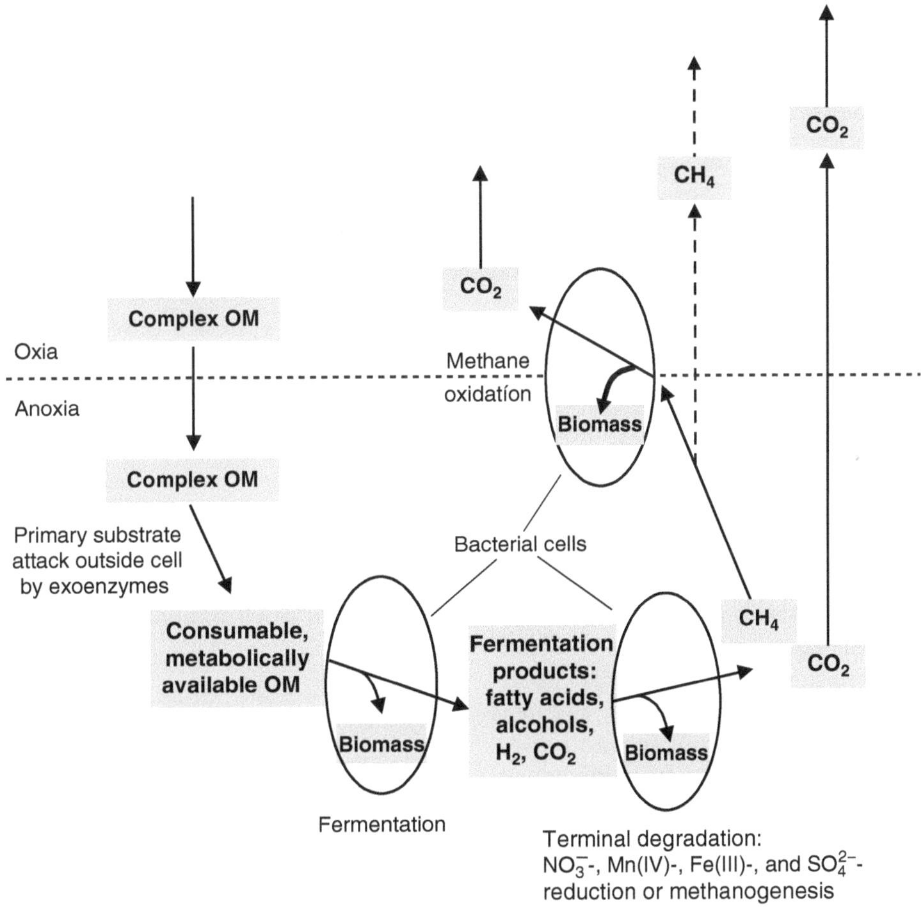

Figure 1 Simplified illustration of anoxic degradation of OM and of CH_4 oxidation in and aquatic sediment. Complex polymeric OM reaching the sediment is subjected to hydrolysis and fermentation to low molecular weight intermediary products. Production of CH_4 represents one of several alternative terminal anoxic degradation processes. CH_4 oxidation, primarily occurring at the oxic–anoxic interface, results in the formation of CO_2 and biomass from CH_4. Some CH_4 may escape oxidation and leave the system. See text for details.

substrate uptake systems (greater substrate affinity), they are likely to outcompete methanogenic bacteria. Thereby, the presence of alternate electron acceptors inhibits methanogenesis (**Table 2**).

Environmental Factors Affecting Methanogenesis

The effects of various environmental factors are summarized in **Table 2**. Previously, methanogenic

Table 2 Effects of some environmental factors on CH_4 production rates in aquatic systems[a]

Environmental factor	*Effect on CH_4 production (+/−)*	*Notes*	*Sources*[b]
O_2	−		1, 4
Alternative electron acceptors (NO_3^-, Fe^{3+}, Mn^{4+}, SO_4^{2-})	−		2, 4, 5
Temperature	+	Q_{10} 4.1 ± 0.4 ($n = 1046$; data from various wetland systems)	1, 2, 4, 6, 7
pH	Variable	Community-dependent pH optimum	4, 7
Roots	+ by OM leakage and root decay − by O_2 leakage	+ probably dominate	2, 4
Availability of recently produced OM	+	Yields indirect relation between system primary productivity and CH_4 production	4, 7, 8
Physical disturbance	+ at low intensity − at high to moderate intensity	Effects may differ among environments and communities	3, 4
Sulfide	− above threshold conc.	Thresholds of 0.1–1 mM found	4

Sources

1. Boon PI and Mitchell A (1995) Methanogenesis in the sediments of an Australian freshwater wetland: Comparison with aerobic decay, and factors controlling methanogenesis. *FEMS Microbioogy Ecology* 18: 175–190.
2. Conrad R (2002) Control of microbial methane production in wetland rice fields. *Nutrient Cycling in Agroecosystems* 64: 59–69.
3. Dannenberg S, Wudler J, and Conrad R (1997) Agitation of anoxic paddy soil slurries affects the performance of the methanogenic microbial community. *FEMS Microbiology Ecology* 22: 257–263.
4. Segers R (1998) Methane production and methane consumption: A review of processes underlying wetland methane fluxes. *Biogeochemistry* 41: 23–51.
5. van Bodegom PM, Scholten JCM, and Stams AJM (2004) Direct inhibition of methanogenesis by ferric iron. *FEMS Microbiology Ecology* 49: 261–268.
6. van Hulsen J, Segers R, van Bodegom P, and Leffelaar P (1999) Temperature effects on soil methane production: An explanation for observed variability. *Soil Biology and Biochemistry* 31: 1919–1929.
7. Whalen SC (2005) Biogeochemistry of methane exchange between natural wetlands and the atmosphere. *Environmental Engineering Science* 22: 73–94.
8. Yavitt JB, Williams CJ, and Wieder RK (2000) Controls on microbial production of methane and carbon dioxide in three Sphagnum-dominated peatland ecosystems as revealed by a reciprocal field peat transplant experiment. *Geomicrobiology Journal* 17: 61–88.

[a]Positive and negative effects are denoted by + or −, respectively. OM denotes organic matter.
[b]Only examples of sources given. Sources 2, 4, and 7 are reviews representing many studies on the regulation of methanogenesis.

bacteria were considered strict anaerobes unable to survive at elevated concentrations of O_2. However, it is now established that many methanogens can tolerate substantial O_2 exposure even though this inhibits methanogenesis. Likewise, the inhibition by sulfate is not strict, but primarily results from substrate competition. The temperature optimum of methanogenesis is usually well above in situ temperatures, and the potential CH_4 production rates increase about fourfold if the temperature increases by 10 °C (see Q_{10} value in **Table 2**). pH within the natural variation found in aquatic habitats does not seem to limit methanogenesis directly, but can affect what substrates are available and used. Under low pH, acetotrophic methanogenesis seem to be favored, while higher pH favors H_2-dependent methanogenesis. Often both acetotrophic and hydrogenotrophic methanogenesis can occur simultaneously with either process contributing between 20% and 80% of the overall CH_4 production in sediments.

Spatial Distribution of Methanogenesis in Freshwater Environments

CH_4 formation is restricted to environments with very low concentrations of alternate electron acceptors (e.g., O_2, NO_3^-, Mn^{4+}, Fe^{3+}, and SO_4^{2-}), but at the same time having a high enough concentration of the substrates is needed. Such conditions are common in freshwater sediments. If the sediments are overlain with oxic water, or in the proximity of roots of aquatic plants from which O_2 diffuses into the sediment, there is often a redox gradient over a few millimeters to centimeters in the sediment. In this gradient, the organic remineralization processes using alternate electron acceptors often occur in the energy yield sequence according to the order in **Table 1**. In these cases, methanogenesis dominate terminal OM degradation in subsurface sediments or at longer distance from the roots, where the other electron acceptors are depleted. If the sediment is overlain with anoxic water and lacks roots contributing O_2, methanogenesis may dominate terminal OM degradation in the whole sediment column. In wetlands, the depth distribution of methanogenesis often depends on the position of the groundwater table, which regulates the availability of O_2 and other electron acceptors.

So far, there is little evidence of significant rates of methanogenesis in anoxic water columns, possibly due to low concentration of substrate relative to that found in the sediment. It is also possible that the establishment of stable syntrophic consortia is more difficult in the water column due to higher turbulence than in the sediments, and this may also restrict methanogenesis. Anoxic microenvironments such as digestive tracts of animals or the interior of suspended detrital particles are sites of methanogenesis that may be of importance for the depth distribution of CH_4 in the oceans, but in fresh waters the sediments appear to be the dominant source.

CH_4 Oxidation

Aerobic Methane Oxidation

The fact that CH_4 is the most highly reduced organic compound means that energy can be gained by the oxidation of CH_4 in the presence of O_2 or other potent oxidants. Aerobic methane-oxidizing bacteria (MOB) take advantage of this by using CH_4 as energy and carbon source and O_2 as electron acceptor. MOB oxidize CH_4 sequentially to methanol, formaldehyde, formate, and finally to CO_2. The intermediate product, formaldehyde (CHO) is a primary carbon source for growth.

MOB can be divided into three different groups (type I, type II, and type X). Studies of isolated strains from these groups have revealed differences in morphology, membrane structures, resting stages, and metabolic pathways. Some phospholipid fatty acids represent biomarkers unique for MOB and can be used for identification and biomass estimation. The key enzyme responsible for the CH_4 oxidation is methane monooxygenase (MMO). There are both particulate and soluble types of MMO. Particulate MMO appears to be common to all MOBs while soluble MMO is not. Some MOBs are obligately methanotrophic, while others are methylotrophic and can use methyl groups on various small organic molecules. A chemically diverse catalytic capacity has been found for soluble MMO, including the oxidation of many aromatic and alicyclic organic compounds, and oxidative dehalogenation of halomethanes, haloalkanes, and haloalkenes. Ammonia monooxygenase, used by ammonia-oxidizing bacteria, is very similar to MMO, and many MOBs may be able to perform ammonia oxidation as well. Another link to nitrogen cycling, although not specifically associated with MMO, is that many MOB can fix molecular nitrogen (N_2) at low O_2 concentrations.

Environmental Factors Affecting CH_4 Oxidation

Aerobic CH_4 oxidation depends on the presence of both CH_4 and O_2 (**Table 3**). CH_4 oxidation rates appear less sensitive to temperature than does methanogenesis (**Tables 2** and **3**). There is no clear pattern regarding how pH affects CH_4 oxidation rates, possibly due to local adaptations to various pH ranges by the microbial communities. Likewise, the effects of ammonium and nitrate are contradictory and seem to differ between environments (**Table 3**). High salinity inhibits CH_4 oxidation as does high light intensity. Light can also indirectly favor CH_4 oxidation by enhancing photosynthesis and thereby oxygenation of sediments or water columns.

Anaerobic CH_4 Oxidation

It can be thermodynamically favorable to oxidize CH_4 using any electron acceptor that gives a higher energy yield than does methanogenesis (see **Table 1**). Although initially controversial, the anaerobic oxidation of CH_4 by syntrophic interactions between bacteria performing sulfate reduction and reverse methanogenesis, respectively, in marine environments is now recognized. In inland waters, anaerobic CH_4 oxidation with SO_4^{2-} has been indicated at SO_4^{2-} concentrations greater than 0.5 mM. Observations of CH_4 oxidation coupled to NO_3^- reduction (denitrification) in anoxic ground water and sludge have been presented. Anaerobic oxidation by Fe^{3+} may also occur, but in general very little is known about anaerobic CH_4 oxidation with electron acceptors other than SO_4^{2-}. Anaerobic CH_4 oxidation can be considerable in saline environments, but in fresh waters, anaerobic CH_4 oxidation appears to be of less importance than does aerobic CH_4 oxidation.

Spatial Distribution of Aquatic CH_4 Oxidation

Aerobic CH_4 oxidation is most extensive at the oxic–anoxic interface where both O_2 and CH_4 are available. The oxic–anoxic interface can be situated in the surface sediment, in deeper sediments close to roots releasing O_2, or in the water column (**Figure 2**). The zone with extensive CH_4 oxidation is usually confined to a layer only a few millimeters thick in

Table 3 Effects of some environmental factors on aerobic CH_4 oxidation rates (MOX) in aquatic systems[a]

Environmental factor	*Effect on CH_4 oxidation (+/−)*	*Notes*	*Sources[b]*
CH_4	+ if suitable electron acceptor is available (e.g., O_2)		9, 10
O_2	+ if enough CH_4 is present		4, 5, 9, 10
Temperature	+	Q_{10} 1.9±0.4 (n = 328; data from various wetland systems)	8, 11
pH	Variable effect	pH optimum may vary among communities	4, 8, 11
Salinity	−	Reduction from 40 mM, total inhibition from 9%	8
Light	− (direct effects)	Indirect effects related to photosynthesis and O_2 dynamics.	2, 3, 6
	+ (indirect effects)		
Nitrogen (NH_4^+ or NO_3^-)	Debated	Both + and − effects found	1, 4, 5, 7
Roots	+	Through O_2 release	8
Season		Highest MOX during summer in sediments	8

Sources
1. Bodelier PLE and Laanbroek HJ (2004) Nitrogen as a regulatory factor of methane oxidation in soils and sediments. *FEMS Microbiology Ecology* 47: 265–277.
2. Dumestre JF, Guézennec J, Galy-Lacaux C, *et al.* (1999) Influence of light in petit saut reservoir, French Guiana. *Applied and Environmental Microbiology* 65: 534–539.
3. King GM (1990) Regulation by light of methane emissions from a wetland. *Nature* 345: 513–515.
4. King GM (1992) Ecological aspects of methane oxidation, a key determinant of global methane dynamics. In Marshall KC (Ed.) *Advances in Microbial Ecology*, pp. 431–468, New York: Plenum Press.
5. Liikanen A and Martikainen PJ (2003) Effect of ammonium and oxygen on methane and nitrous oxide fluxes across sediment–water interface in a eutrophic lake. *Chemosphere* 52: 1287–1293.
6. Murase J and Sugimoto A (2005) Inhibitory effect of light on methane oxidation in the pelagic water column of a mesotrophic lake (Lake Biwa, Japan). *Limnology and Oceanography* 50: 1339–1343.
7. Rudd JWF, Furutani A, Flett RJ, and Hamilton RD (1976) Factors controlling methane oxidation in shield lakes: The role of nitrogen fixation and oxygen concenteration. *Limnology and Oceanography* 21: 357–364.
8. Segers R (1998) Methane production and methane consumption: a review of processes underlying wetland methane fluxes. *Biogeochemistry* 41: 23–51.
9. Sundh I, Bastviken D, and Tranvik L (2005) Abundance, activity, and community structure of pelagic methane-oxidizing bacteria in temperate lakes. *Applied Environmental Microbiology* 71: 6746–6752.
10. Sundh I, Mikkela C, Nilsson M, and Svensson BH (1995) Potential Aerobic Methane Oxidation in a Sphagnum-Dominated Peatland—Controlling Factors and Relation to Methane Emission. *Soil Biology and Biochemistry* 27: 829–837.
11. Whalen SC (2005) Biogeochemistry of methane exchange between natural wetlands and the atmosphere. *Environmental Engineering Science* 22: 73–94.

[a]Positive and negative effects are denoted by + or −, respectively.

[b]Only examples of sources are given. Source 4, 8, and 11 are review papers representing many studies.

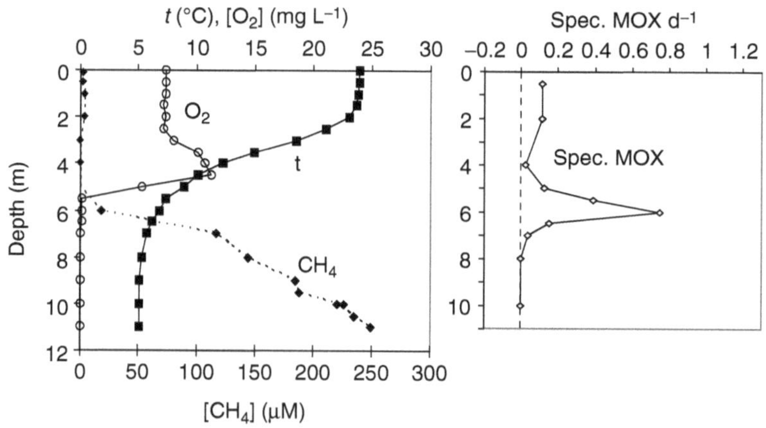

Figure 2 Example of water column profiles of temperature (*t*), O_2 concentration ($[O_2]$), CH_4 concentration ($[CH_4]$), and specific CH_4 oxidation rates (Spec. MOX; the fraction of the present CH_4 being oxidized each day), in a temperate lake during summer stratification. Data from Paul Lake, Northern Michigan, USA, July 2001.

the sediment, but can be extended over decimeter or meter scales in the water column. The location of this zone varies with time, depending on how the oxygenation of the water varies temporally. For example, in a lake that develops an anoxic hypolimnion during summer stratification, the most extensive CH_4 oxidation is expected at the oxic–anoxic interfaces in the water column and in the shallow sediment at depths above the oxycline. During the rest of the year, most of the CH_4 oxidation would be expected to occur in the sediment surface if the water column is oxygenated. In wetlands, the oxic–anoxic interface, and thereby the zone of CH_4 oxidation, follows the movements of the groundwater table.

Little is known about the spatial distribution of anaerobic CH_4 oxidation in freshwater ecosystems, but presumably, anaerobic CH_4 oxidation may also occur where both substrate (CH_4) and electron acceptor (e.g., SO_4^{2-}) are found. If anaerobic CH_4 oxidation depends on syntrophic relationships, the process may be sensitive to physical disruption such as turbulence, breaking microbial associations, and this may limit the distribution in the water column.

At marine hydrocarbon seeps, CH_4 oxidation can occur in association with animals (e.g., endosymbiotic MOB in mussels), but no such endosymbiotic distribution of MOB has yet been shown in inland waters.

Aquatic CH_4 Dynamics

CH_4 Concentrations in Aquatic Environments

CH_4 concentrations in sediments or water columns are the consequence of formation rates, import rates, CH_4 oxidation rates, and rates of export to the water column or the atmosphere. High pore water concentrations, at which CH_4 is saturated at the prevailing pressure, is typical for deep sediment layers, and any excess CH_4 that is produced will then form bubbles in the sediment. If surface sediments are oxygenated, CH_4 oxidation will keep CH_4 concentrations low, down to the micromole range, at the sediment surface. If the sediment is overlain with anoxic water and CH_4 oxidation in the sediment is low, concentrations will be high in the whole sediment. CH_4 concentrations in anoxic parts of stratified water columns can exceed 1000 μM. Consequently, steep concentration gradients influenced by physical transport patterns and CH_4 oxidation are frequently found in stratified waters (**Figure 2**). In strongly stratified systems, lack of advective transport may efficiently trap CH_4 below the thermocline. In such cases, most of the surface-water CH_4 may be derived from shallow sediments above the thermocline. CH_4 concentrations in well-mixed oxygenated parts of water columns are generally low, typically ranging from 0.01 to 10 μM.

Rates of CH_4 Formation and Oxidation

Most available assessments of CH_4 production and oxidation rates in sediments are based on incubations of sediment slurries. Slurries are made to ensure well-mixed conditions and to avoid the development of gradients that may affect CH_4 production. It is well recognized that such measurements, although providing good potential production rates under experimental conditions, may not reflect actual in situ rates. In situ rates are more difficult to measure and different approaches have been tried, including studies of intact sediment cores and whole system CH_4 budgets. **Table 4** provides an overview of CH_4 production, CH_4 oxidation, and CH_4 release rates, acquired by different methods, in some lake sediments. CH_4 production rates range from almost 0 to 68 mmol m^{-2} day^{-1}. In oxic surface sediment, 50–95% of the produced CH_4 was oxidized. CH_4 release, representing the net transport of CH_4 to the water, may to a large extent depend on CH_4 oxidation rates and regulation by the O_2 penetration depth has been suggested. The release of CH_4 to the water column in previously studied lakes was 0.04–46 mmol m^{-2} day^{-1} (**Table 4**). Production or release values greater than 2.9 mmol m^{-2} day^{-1} only occurred in the more nutrient-rich systems (mesotrophic or eutrophic), indicating some influence of nutrient levels, probably indirectly by affecting sedimentation rates and quality of the sediment OM as well as the frequency of anoxia in bottom waters. However, assuming that **Table 4** is representative of available lake data, there is a lack of studies in oligotrophic and dystrophic lakes, which constitute a large share of all lakes globally.

The oxidation of CH_4 in the water column is highly dependent on the mixing regime and the depth of lakes. Hence, reported water column oxidation rates are highly variable ranging from zero in deep anoxic water layers to very high rates in the oxycline where O_2 and CH_4 concentration gradients meet (**Figure 2**) or during lake overturn periods when concentrations of both CH_4 and O_2 are elevated throughout the entire water column. Hence, rates can vary from 0 to 20 mmol m^{-3} day^{-1} at individual depths (**Table 5**). Areal CH_4 oxidation rates, taking the average depth into account, range from 0 to 31 mmol m^{-2} day^{-1}. Apparently, a large proportion of the CH_4 entering the water column will be oxidized (45–100%; **Table 5**). Hence, considering both surface sediments and water

Table 4 Examples of CH_4 production and oxidation rates in sediment, or net diffusive flux from sediment to water in lakes[a]

Lake	*Trophic state*	*Method*	*Sediment origin*	*O_2 regime in water above sediment in situ*	*Sediment CH_4 production (mmol m^{-2} day^{-1})*	*CH_4 oxidation in surface sediment (% of production)*	*Net CH_4 flux from sediment to water (mmol m^{-2} day^{-1})*	*Source*
Fuchskuhle	Acid	Slurry incubation	Profundal		0.4–1.2			3
Dagow	Eu	Slurry incubation	Profundal		0.7–13			3
Kevätön	Eu	Slurry incubation	Profundal		27–68			11
Ontario	Meso	Slurry incubation	Prof + Litt		0–42.5			15
Stechlin	Oligo	Slurry incubation	Profundal		0.3–1.2			3
Two Siberian lakes	Dys	Radioisotopes	Prof + Litt		0.006	67%		6
Washington	Meso	Benthic chamber	Profundal	Oxic	0.6	50%	0.2–0.3	10
Tuusulanjärvi	Eu	Conc. gradient calc.	Profundal	Oxic	4.5			7
Postilampi	Eu	Conc. gradient calc.	Profundal	Anoxic	6.6			7
Soiviojärvi	Eu	Conc. gradient calc.	Profundal	Oxic	0.5			7
Takajärvi	Eu	Conc. gradient calc.	Profundal	Oxic	0.3			7
Ranuanjärvi	Eu	Conc. gradient calc.	Profundal	Oxic	4.8			7
Lokka	Eu	Conc. gradient calc.	Profundal	Oxic	0.03			7
Luiminkajärvi	Meso	Conc. gradient calc.	Profundal	Oxic	1.7			7
Porttipahta	Meso	Conc. gradient calc.	Profundal	Oxic	1.6			7
Big Soda Lake	Saline	Conc. gradient calc.	Profundal	Anoxic			2.9	8
Mono Lake	Saline	Conc. gradient calc.	Profundal	Anoxic			1.3	14
Kevätön	Eu	Intact cores	Profundal	Oxic	1.9–9.4	66–95%	0.1–3.2	11
Illersjön	Eu	Intact cores	Profundal	Anoxic	1.1–2.4			1
Mårn	Eu and Dys	Intact cores	Profundal	Anoxic	1.1			1
Constance	Meso	Intact cores	Profundal	Oxic	0.5	93%	0.04	5
Biwa	Meso	Intact cores	Profundal	Oxic	0.002–1.2	90%		13
Biwa	Meso	Intact cores	Littoral	Oxic	1–8.8	90%		13
Lillsjön	Dys	Intact cores	Profundal	Anoxic	0.6			1
Frain's Lake	Eu	Hypolimnion budget	Profundal	Anoxic			3.8–9.8	9

Wintergreen	Eu	Hypolimnion budget	Profundal	Anoxic	10–46	17
St George	Meso	Hypolimnion budget	Profundal	Anoxic	5–14.8	2
Third Sister	Meso	Hypolimnion budget	Profundal	Anoxic	3.2–4.5	9
Onodaga	Meso	Hypolimnion budget	Profundal	Anoxic	5.6–11.9	12
Mendota	Eu	Whole lake budget	Profundal	Anoxic	35.8	4
Lake 227	Eu	Whole lake budget	Profundal	Anoxic	10.8	16
Lake 227	Eu	Whole lake budget	Littoral	Oxic	0.8	16

Sources
1. Bastviken D, Ejlertsson J, and Tranvik L (2002) Measurement of methane oxidation in lakes-a comparison of methods. *Environmental Science and Technology* 36: 3354–3361.
2. Bédard C and Knowles R (1991) Hypolimnetic O_2 consumption, denitrification, and methanogenesis in a thermally stratified lake. *Canadian Journal of Fisheries and Aquatic Sciences* 48: 1048–1054.
3. Casper P (1992) Methane production in lakes of different trophic state. *Archiv für Hy drobiologie Beiheft Ergebnisse der Limnologie* 37: 149–154.
4. Fallon RD, Harris S, Hanson RS, and Brock TD (1980) The role of methane in internal carbon cycling in Lake Mendota during summer stratification. *Limnology and Oceanography* 25: 357–360.
5. Frenzel P, Thebrath B, and Conrad R (1990) Oxidation of methane in the oxic surface layer of a deep lake sediment (Lake Constance). *FEMS Microbiology Ecology* 73: 149–158.
6. Gal'chenko VF, Dulov LE, Cramer B, Konova NI, and Barysheva SV (2001) Biogeochemical processes of methane cycle in the soils, bogs, and lakes of western Siberia. *Microbiology* 70: 175–185.
7. Huttunen JT, Vaisanen TS, Hellsten SK, and Martikainen PJ (2006) Methane fluxes at the sediment–water interface in some boreal lakes and reservoirs. *Boreal Environment Research* 11: 27–34.
8. Iversen N, Oremland RS, and Klug MJ (1987) Big Soda Lake (Nevada). 3. Pelagic methanogenesis and anaerobic methane oxidation. *Limnology and Oceanography* 32: 804–814.
9. Kelly CA and Chynoweth DP (1980) Comparison of in situ and in vitro rates of methane release in freshwater sediments. *Applied and Environmental Microbiology* 40: 287–293.
10. Kuivila KM, Murray JW, Devol AH, Lidstrom ME, and Reimers CE (1988) Methane cycling in the sediments of Lake Washington. *Limnology and Oceanography* 33: 571–581.
11. Liikanen A, Huttunen JT, Valli K, and Martikainen PJ (2002) Methane cycling in the sediment and water column of mid-boreal hyper-eutrophic Lake Kevaton, Finland. *Archiv fur Hydrobiologie* 154: 585–603.
12. Matthews DA, Effler SW, and Matthews CM (2005) Long-term trends in methane flux from the sediments of Onondaga Lake, NY: Sediment diagenesis and impacts on dissolved oxygen resources. *Archiv fur Hydrobiologie* 163: 435–462.
13. Murase J, Sakai Y, Kametani A, and Sugimoto A (2005) Dynamics of methane in mesotrophic Lake Biwa, Japan. *Ecological Research* 20: 377–385.
14. Oremland RS, Miller LG, and Whiticar MJ (1987) Sources and flux of natural gases from Mono Lake, California. *Geochimica et Cosmochimica Acta* 51: 2915–2929.
15. Roy R, Legendre P, Knowles R, and Charlton M (1994) Denitrification and methane production in sediment of Hamilton Harbour (Canada). *Microbial Ecology* 27: 123–141.
16. Rudd JWM and Hamilton RD (1978) Methane cycling in a eutrophic shield lake and its effects on whole lake metabolism. *Limnology and Oceanography* 23: 337–348.
17. Strayer RG and Tiedje JM (1978) In situ methane production in a small, hypereutrophic, hardwater lake: Loss of methane from sediments by vertical diffusion and ebullition. *Limnology and Oceanography* 23: 1201–1206.

[a]Lake trophic states denoted are acidotrophic (Acid), eutrophic (Eu), mesotrophic (Meso), oligotrophic (Oligo), and dystrophic (Dys). Profundal (Prof) or littoral (Litt) sediment is represented. The studies have been grouped based on the approach and we refer to original references for method details. Mean values or ranges are given. CH_4 leaving the sediment by ebullition is not accounted for by sediment CH_4 production values, meaning that total sediment CH_4 production may be greater.

Table 5 Examples of methane oxidation rates in lake water columns

Lake	Lake information	CH_4 oxidation			% CH_4 oxidized	Sources
		mmol m⁻² day⁻¹	*day⁻¹*	*mmol m⁻² day⁻¹*		
Lake Kevätön, Finland	Hyper eutrophic	0–27		15–31		8
Lake 227, Canada	Eutrophic, summer stratification	0–9.6		1.2	100	10, 11
Lake 227, Canada	Eutrophic, fall overturn	1–17.3		2.9	60	11
Lake Mendota, USA	Eutrophic, summer stratification			10.6	45	2
Ilersjön, Sweden	Eutrophic, summer and winter stratification	0.004–3.3	0.01–0.21	0.6–4.3	57–90	1
Mårn, Sweden	Eutrofic, humic, summer and winter stratification	0.001–2.3	0.01–0.8	0.7–1.2	64–100	1
Lake Kasumigaura, Japan	Eutrophic, shallow, 5 whole years studied	0.02–1.1	0.01–6.5		74	13
Lake Nojiri, Japan	Mesotrophic, fall overturn	0–19.8	0.04–1	1	94	14
Williams Lake, USA	Whole year study			6.6		12
Shingobee Lake,USA	Whole year study			20		12
Lillsjön, Sweden	Dystrophic, summer and winter stratification	0.001–0.3	0–0.1	0.4–0.7	69–100	1
Lake Mekkojärvi, Finland	Dystrophic, fall overturn			3.3–4.2		7
Valkea-Kotinen, Finland	Dystrophic, whole year study			1.2	80	6
Lake Tanganyika, East Africa	Meromictic			0–9.6		9
Lake Kivu, Central Africa	Meromictic	0.04–0.9				4
Big Soda Lake, USA	Saline, anaerobic CH_4 oxidation	$2 \times 10^{-3} - 85 \times 10^{-3}$	$5.6 \times 10^{-4} - 8.1 \times 10^{-4}$	1.4		3
Mono Lake, USA	Alkaline, saline, primarily anaerobic CH_4 oxidation	$4 \times 10^{-5} - 4.8 \times 10^{-2}$	$1.3 \times 10^{-4} - 7 \times 10^{-3}$	0.07–0.7		5

Sources

1. Bastviken D, Ejlertsson J, and Tranvik L (2002) Measurement of methane oxidation in lakes—A comparison of methods. *Environmental Science and Technology* 36: 3354–3361.
2. Fallon RD, Harris S, Hanson RS, and Brock TD (1980) The role of methane in internal carbon cycling in Lake Mendota during summer stratification. *Limnology and Oceanography* 25: 357–360.
3. Iversen N, Oremland RS, and Klug MJ (1987) Big Soda Lake (Nevada). 3. Pelagic methanogenesis and anaerobic methane oxidation. *Limnology and Oceanography* 32: 804–814.
4. Jannasch HW (1975) Methane oxidation in Lake Kivu (Central Africa). *Limnology and Oceanography* 22: 814–832.
5. Joye SB, Connell TL, Miller LG, Oremland RS, and Jellison RS (1999) Oxidation of ammonia and methane in an alkaline, saline lake. *Limnology and Oceanography* 44: 178–188.
6. Kankaala P, Huotari J, Peltomaa E, Saloranta T, and Ojala A (2006) Methanotrophic activity in relation to methane efflux and total heterotrophic bacterial production in a stratified, humic, boreal lake. *Limnology and Oceanography* 51: 1195–1204.
7. Kankaala P, Taipale S, Grey J, *et al.* (2006) Experimental delta C-13 evidence for a contribution of methane to pelagic food webs in lakes. *Limnology and Oceanography* 51: 2821–2827.
8. Liikanen A, Huttunen JT, Valli K, and Martikainen PJ (2002) Methane cycling in the sediment and water column of mid-boreal hyper-eutrophic Lake Kevaton, Finland. *Archiv fur Hydrobiologie* 154: 585–603.
9. Rudd JWM (1980) Methane oxidation in Lake Tanganyika (East Africa). *Limnology and Oceanography* 25: 958–963.
10. Rudd JWM and Hamilton RD (1975) Factors controlling rates of methane oxidation and the distribution of the methane oxidizers in a small stratified lake. *Archiv fur Hydrobiologie* 75: 522–538.
11. Rudd JWM and Hamilton RD (1978) Methane cycling in a eutrophic shield lake and its effects on whole lake metabolism. *Limnology and Oceanography* 23: 337–348.
12. Striegl RG and Michmerhuisen CM (1998) Hydrologic influence on methane and carbon dioxide dynamics at two north-central Minnesota lakes. *Limnology and Oceanography* 43: 1519–1529.
13. Utsumi M, Nojiri Y, Nakamura T, *et al.* (1998) Oxidation of dissolved methane in an eutrophic, shallow lake: Lake Kasumigaura, Japan. *Limnology and Oceanography* 43: 471–480.
14. Utsumi M, Nojiri Y, Nakamura T, *et al.* (1998) Dynamics of dissolved methane and methane oxidation in dimictic Lake Nojiri during winter. *Limnology and Oceanography* 43: 10–17.

[a]Various methods were used including monitoring CH_4 concentration decline over time in closed vessels, changes in $\delta^{13}CH_4$ values, or ^{14}C-based approaches. Studies attempting to quantify actual in situ rates were selected (i.e., potential rate measurements with substantial additions of CH_4 were not included). Averages or ranges are reported.

columns, CH_4 oxidation substantially reduces the amount of CH_4 that can be emitted from aquatic environments by diffusion.

CH_4 and Aquatic Food Webs

Zooplanktons and chironomids are frequently depleted in ^{13}C relative to assumed food sources such as algae and bulk particulate OM, indicating that they may partly rely on more ^{13}C-depleted carbon sources. Biogenic CH_4, and consequently, the biomass of MOB are much depleted in ^{13}C. During the last decade, there has been a growing interest regarding to what extent CH_4 carbon transformed to MOB biomass can support higher food web levels (**Figure 3**). Carbon mass balances support the idea that this happens. CH_4 oxidation in aquatic environments is extensive and the high MOB growth efficiency ranging from 0.15 to 0.8 means that a high proportion of the consumed CH_4 is transferred to the biomass. The production of MOB biomass can be of the same order of magnitude as the total heterotrophic bacterial production and even comparable to the primary production in some lakes (**Table 6**). MOB biomass could represent a substantial food resource in many aquatic environments. Until now, several studies have indicated that CH_4 carbon indeed can support zooplankton and chironomid populations, at least in some lakes, during some seasons (e.g., when CH_4 oxidation can be extensive compared to algal production), or some bacterivore populations (e.g., zooplankton preferentially feeding at the oxic–anoxic interface) (**Table 7**). However, this idea of a methanotrophic microbial loop is still controversial, and the use of $\delta^{13}C$ as an indicator has been criticized because of difficulties to exclude the contribution of ^{13}C-depleted algal carbon formed from respired CO_2. While it is likely that CH_4 carbon can substantially support aquatic food webs in some cases, future studies are needed about under what conditions this is important.

CH_4 Emissions from Inland Waters

Flux Pathways

Emission of CH_4 from freshwater ecosystems to the atmosphere can occur in at least four different ways (**Figure 4**):

1. Ebullition of bubbles from the sediment that rapidly pass through the water column. Most of the released CH_4 thereby escapes CH_4 oxidation. The ebullition rate is related to CH_4 formation rates in the sediment and also to the hydrostatic pressure that has to be overcome to allow release of the bubbles. Thereby, ebullition is depth dependent and most intensive from shallow sediments where the hydrostatic pressure is low (**Figure 5**). The relation to hydrostatic pressure allows weather to influence the timing of bubble release, and a higher occurrence of ebullition has been observed coincident with drops in the air pressure due to storms or frontal passages (**Table 8**).
2. Diffusive flux from the water surface. The transport from the sediment through the water column occurs primarily through advective processes of eddy diffusion (turbulence-enhanced diffusion). Very low diffusion rates similar to molecular diffusion may occur across stably stratified water layers at the sediment–water interface,

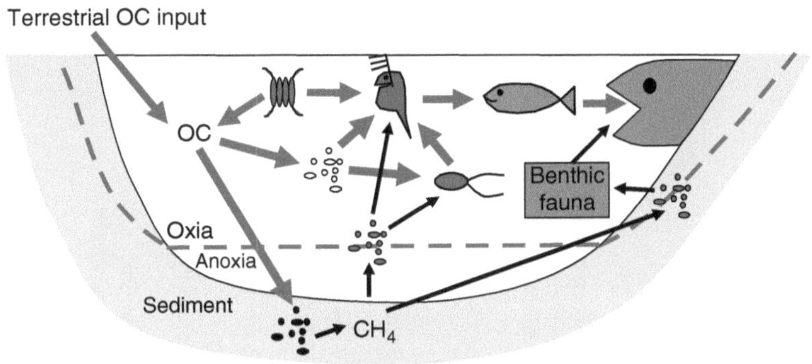

Figure 3 Schematic illustration of how CH_4 carbon can be incorporated into aquatic food webs. Organic carbon (OC) from terrestrial sources or aquatic primary production (symbolized by Scenedesmus algae) reach anoxic sediment where methanogenic bacteria (black dots) produce CH_4. At the oxic–anoxic interface in the water column or in sediment, CH_4-oxidizing bacteria (grey dots) transform CH_4 carbon to bacterial biomass that can be grazed upon by pelagic or benthic bacterivores. Narrow black arrows illustrate the potential routes for CH_4 carbon to various food web levels. The large grey arrows symbolize the classical food web carbon transfer including the microbial loop. White dots represent heterotrophic bacteria relying on OC, and being grazed upon by both large zooplankton (symbolized by Daphnia) and by protozoans (symbolized by flagellates). (The arrows were made different to make the figure easier to read, and size does not represent quantities or rates).

Table 6 Comparisons between parts of the CH_4 cycling and other parts of the carbon cycling in some freshwater environments

Lake	*System/study information*	*Comparison*	*Sources*
Methanogenesis (MG)			
7 North American lakes		MG accounted for 19–87% of the anaerobic or 20–25% of the total C mineralization	4, 13, 14
Ryans I Billabong, Australia	Small shallow oxbow lake	MG accounted for 30–60% of the sediment carbon flux or at least one-third of total wetland respiratory carbon losses	3, 7
Pantanal, Brazil	Various waters on floodplain	MG represented 20% of total heterotrophic metabolism in water and sediment	9
Lake St George, Ontario, Canada	Mesotrophic	MG accounted for 56% of organic carbon mineralization during lake stratification	2
Lake 227, Canada	Eutrophic, dimictic	MG equal to 55% of C input from runoff and atmosphere, and 13% of primary production	17
Methane oxidation (MOX)			
Lake Tanganyika, East Africa	Meromictic	MOX $\geq$ primary productivity on annual basis	16
Mekkojärvi, Finland	Dystrophic	Water column MOX rates during fall overturn similar to rates of primary production during July	12
Lake Sebasticook, Maine, USA	Hypereutrophic	MOX accounted for 62–70% of hypolimnetic O_2 demand	15
Lake 227, Canada	Eutrophic, dimictic	MOX consumed all available O_2 under ice during winter resulting in water column anoxia and widespread suffocation of fauna	17
Lake Constance	Mesotrophic	MOX corresponded to 9% of the sediment O_2 demand	8
Methane-oxidizing bacteria (MOB)			
Lake Plußsee, Germany	Eutrophic, hypolimnion	MOB biomass 1.4–3% of total bacterial biomass	6
Three Swedish lakes	Two eutrophic lakes and one dystrophic lake	MOB growth rates 0.5–119% of heterotrophic bacterial production, or 0.3–7% of primary production. MOB biomass 5.4–41% of total bacterial biomass at different depths in the water column	1, 19
Lake Nojiri, Japan	Mesotrophic	Production of MOB biomass during fall overturn similar to primary production	20
Lake Kjelåsputten, Norway	Dystrophic	Production of MOB biomass similar to all C fixation by heterotrophic bacteria	10

Valkea-Kotinen, Finland	Dystrophic	Production of MOB biomass equivalent to 23–81% of heterotrophic bacterial production and 5–10% of primary production	11
CH_4 emission			
Wintergreen Lake, Michigan, USA	Small, shallow, hypereutrophic	Summer CH_4 emissions corresponded to 24–37% of summer productivity	18
Everglades, Florida, USA	Plant-mediated wetland flux	CH_4 emissions corresponded to 3–14% of net ecosystem production	5

Sources

1. Bastviken D, Ejlertsson J, Sundh I, and Tranvik L (2003) Methane as a source of carbon and energy for lake pelagic food webs. *Ecology* 84: 969–981.
2. Bédard C and Knowles R (1991) Hypolimnetic O_2 consumption, denitrification, and methanogenesis in a thermally stratified lake. *Canadian Journal of Fisheries and Aquatic Sciences* 48: 1048–1054.
3. Boon PI and Mitchell A (1995) Methanogenesis in the sediments of an Australian freshwater wetland: Comparison with aerobic decay, and factors controlling methanogenesis. *FEMS Microbiology Ecology* 18: 175–190.
4. Capone DG and Kiene RP (1988) Comparison of microbial dynamics in marine and freshwater sediments: Contrasts in anaerobic carbon catabolism. *Limnology and Oceanography* 33: 725–749.
5. Chanton JP, Whiting GJ, Happell JD, and Gerard G (1993) Contrasting rates and diurnal patterns of methane emissions from emergent aquatic macrophytes. *Aquatic Botany* 46: 111–128.
6. Eller G, Kanel LK, and Kruger M (2005) Cooccurrence of aerobic and anaerobic methane oxidation in the water column of Lake Plußsee. *Applied and Environmental Microbiology* 71: 8925–8928.
7. Ford PW, Boon PI, and Lee K (2002) Methane and oxygen dynamics in a shallow floodplain lake: the significance of periodic stratification. *Hydrobiologia* 485: 97–110.
8. Frenzel P, Thebrath B, and Conrad R (1990) Oxidation of methane in the oxic surface layer of a deep lake sediment (Lake Constance). *FEMS Microbiology Ecology* 73: 149–158.
9. Hamilton SK, Sippel SJ, and Melack JM (1995) Oxygen depletion and carbon dioxide and methane production in waters of the Pantanal wetland of Brazil. *Biogeochemistry* 30: 115–141.
10. Hessen D and Nygaard K (1992) Bacterial transfer of methane and detritus; implications for the pelagic carbon budget and gaseous release. *Archiv für Hydrobiologie Beiheft Ergebnisse der Limnologie* 37: 139–148.
11. Kankaala P, Huotari J, Peltomaa E, Saloranta T, and Ojala A (2006) Methanotrophic activity in relation to methane efflux and total heterotrophic bacterial production in a stratified, humic, boreal lake. *Limnology and Oceanography* 51: 1195–1204.
12. Kankaala P, Taipale S, Grey J, *et al.* (2006) Experimental delta C-13 evidence for a contribution of methane to pelagic food webs in lakes. *Limnology and Oceanography* 51: 2821–2827.
13. Kuivila KM, Murray JW, Devol AH, Lidstrom ME, and Reimers CE (1988) Methane cycling in the sediments of Lake Washington. *Limnology and Oceanography* 33: 571–581.
14. Mattson MD and Likens GE (1993) Redox reactions of organic matter decomposition in a soft water lake. *Biogeochemistry* 19: 149–172.
15. Mayer L, Liotta F, and Norton S (1982) Hypolimnetic redox and phosphorous cycling in hypereutrophic Lake Sebasticook, Maine. *Water Research* 16: 1189–1196.
16. Rudd JWM (1980) Methane oxidation in Lake Tanganyika (East Africa). *Limnology and Oceanography* 25: 958–963.
17. Rudd JWM and Taylor CD (1980) Methane cycling in aquatic environments. In Droop MR and Jannasch HW (eds.) *Advances in Aquatic Microbiology*, pp. 77–150, London: Academic Press.
18. Strayer RG and Tiedje JM (1978) In situ methane production in a small, hypereutrophic, hardwater lake: Loss of methane from sediments by vertical diffusion and ebullition. *Limnology and Oceanography* 23: 1201–1206.
19. Sundh I, Bastviken D, and Tranvik L (2005) Abundance, activity, and community structure of pelagic methane-oxidizing bacteria in temperate lakes. *Applied and Environmental Microbiology* 71: 6746–6752.
20. Utsumi M, Nojiri Y, Nakamura T, *et al.* (1998) Dynamics of dissolved methane and methane oxidation in dimictic Lake Nojiri during winter. *Limnology and Oceanography* 43: 10–17.

MG, MOX, and MOB denote methanogenesis, methane oxidation, and methane-oxidizing bacteria, respectively.

Table 7 Examples of studies suggesting utilization of CH_4 carbon by different parts of the food web at least in some of the studied lakes or taxa, or during some of the studied seasons

Food web compartment	*System studied*	*Methods (indicator of CH_4-carbon, and type of study)*	*Sources*
Zooplankton	Billabongs, Australia	^{13}C, field observations	2
Zooplankton	Humic lakes	^{13}C, field observations	8
Zooplankton	Eutrophic and humic lakes, Sweden	^{13}C, carbon mass balance, field observation	1
Zooplankton	Polyhumic lake, Finland	^{13}C, carbon mass balance, laboratory and field experiments	9
Chironomids	Large mesotrophic Lake Biwa, Japan	^{13}C, field observations	11
Chironomids	Eutrophic lakes, Germany	^{13}C, field observations	5
Chironomids	Forest lakes, Finland	^{13}C, field observations	7
Chironomids	Large mesotrophic Lake Biwa, Japan	^{13}C, fatty acid compostion, field observations	10
Chironomids, several other invertebrates, fish (*Curimatidae*)	Pantanal, Brazil	^{13}C, field observations	3
Chironomids	Eutrophic lake, Germany	^{13}C, microbial community analyses, field observations	4
Off shore chironomids and oligochetes	Small oligotrophic arctic lakes, Canada	^{13}C, field observations	6
Several macroinvertebrates	Small stream, Japan	^{13}C, ^{15}N	12

Sources
1. Bastviken D, Ejlertsson J, Sundh I, and Tranvik L (2003) Methane as a source of carbon and energy for lake pelagic food webs. *Ecology* 84: 969–981.
2. Bunn SE and Boon PI (1993) What sources of organic carbon drive food webs in billabongs? A study based on stable isotope analysis. *Oecologia* 96: 85–94.
3. Calheiros DF (2003) *Influência do pulso de inundação na composição isotópica ($\delta^{13}C$ e $\delta^{15}N$) das fontes primárias de energia na planície de inundação do rio Paraguai (Pantanal-MS).* Piracicaba-SP: Universidade de São Paulo.
4. Eller G, Deines P, Grey J, Richnow HH, and Kruger M (2005) Methane cycling in lake sediments and its influence on chironomid larval partial derivative C-13. *FEMS Microbiology Ecology* 54: 339–350.
5. Grey J, Kelly A, Ward S, Sommerwerk N, and Jones RI (2004) Seasonal changes in the stable isotope values of lake-dwelling chironomid larvae in relation to feeding and life cycle variability. *Freshwater Biology* 49: 681–689.
6. Hershey AE, Beaty S, Fortino K, *et al.* (2006) Stable isotope signatures of benthic invertebrates in arctic lakes indicate limited coupling to pelagic production. *Limnology and Oceanography* 51: 177–188.
7. Jones RI and Grey J (2004) Stable isotope analysis of chironomid larvae from some Finnish forest lakes indicates dietary contribution from biogenic methane. *Boreal Environment Research* 9: 17–23.
8. Jones RI, Grey J, Sleep D, and Arvola L (1999) Stable isotope analysis of zooplankton carbon nutrition in humic lakes. *Oikos* 86: 97–104.
9. Kankaala P, Taipale S, Grey J, *et al.* (2006) Experimental delta C-13 evidence for a contribution of methane to pelagic food webs in lakes. *Limnology and Oceanography* 51: 2821–2827.
10. Kiyashko SI, Imbs AB, Narita T, Svetashev VI, and Wada E (2004) Fatty acid composition of aquatic insect larvae *Stictochironomus pictulus* (Diptera: Chironomidae): evidence of feeding upon methanotrophic bacteria. *Comparative Biochemistry and Physiology B: Biochemistry & Molecular Biology* 139: 705–711.
11. Kiyashko SI, Narita T, and Wada E (2001) Contribution of methanotrophs to freshwater macroinvertebrates: evidence from stable isotopes. *Aquatic Microbial Ecology* 24: 203–207.
12. Kohzu A, Kato C, Iwata T, *et al.* (2004) Stream food web fueled by methane-derived carbon. *Aquatic Microbial Ecology* 36: 189–194.

in the water column, or at the water–atmosphere interface. Because the diffusive transport of dissolved CH_4 is much slower than transport by ebullition, a large portion of dissolved CH_4 is oxidized in oxic sediments or water layers (**Tables 4** and **5**), resulting in low CH_4 concentrations in surface waters. However, due to the poor solubility of CH_4, surface fresh water is still nearly always oversaturated (3–3000 times oversaturation given a concentration interval of 0.01–10 M) with respect to atmospheric CH_4 (1.8 ppmv). The diffusive flux is driven by the concentration difference between the water and the air, and by factors affecting the physical transport rate across the water–air interface. Accordingly, the diffusive flux is enhanced by high surface-water concentrations or by high turbulence in the surface water making more water coming in contact with air (**Table 8**).

3. Release of CH_4 stored in hypolimnia upon water column mixing. This storage flux is a particular case of diffusive flux. It occurs when deep anoxic

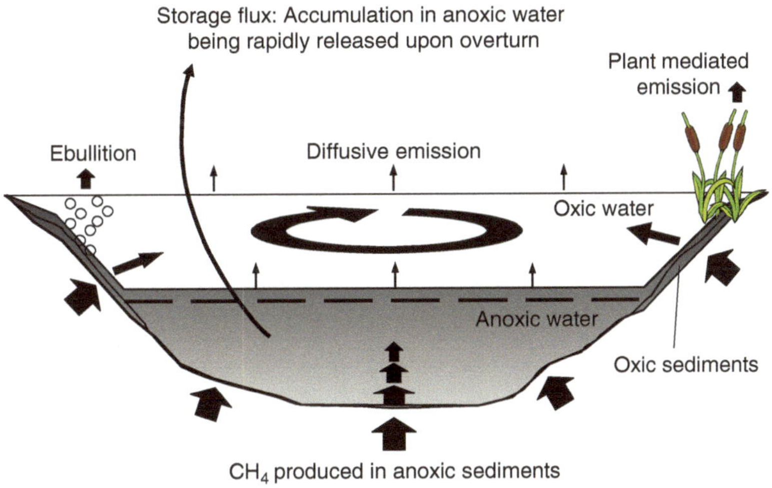

Figure 4 Illustration of CH_4 emission pathways and transport patterns in a stratified lake with anoxic bottom water. CH_4 is primarily produced in anoxic sediments and transported to the water column. The gray area represent the zone where CH_4 oxidation may be extensive given availability of suitable electron acceptors. When possible anoxic electron acceptors have been depleted CH_4 will start to accumulate in the bottom water, while most of the CH_4 oxidation occur at the oxic–anoxic interface in the water column or in the surface of the shallow sediment. See text for details. Adapted from Bastviken D, Cole J, Pace M, and Tranvik L (2004) Methane emissions from lakes: Dependence of lake characteristics, two regional assessments, and a global estimate. *Global Biogeochemical Cycles* 18, GB4009, doi: 10.1029/2004/GB002238 with permission from American Geophysical Union.

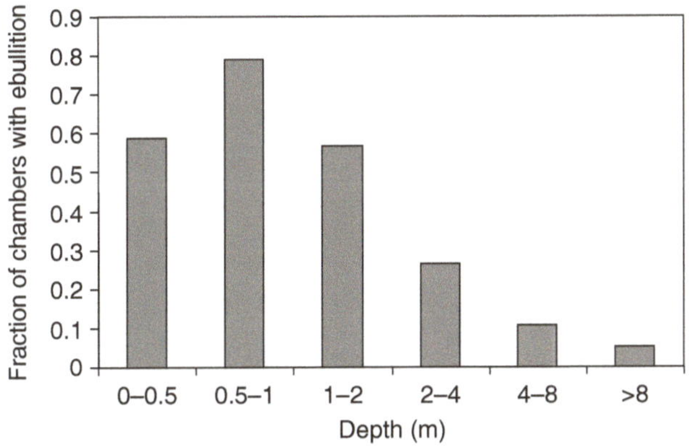

Figure 5 Fraction of floating chambers receiving ebullition over a 24-h period at different depths. Data are based on 242 measurements in 11 lakes. Adapted from Bastviken D, Cole J, Pace M, and Tranvik L (2004) Methane emissions from lakes: Dependence of lake characteristics, two regional assessments, and a global estimate. *Global Biogeochemical Cycles* 18, doi: 10.1029/2004/GB002238 with permission from American Geophysical Union.

and CH_4-rich water in stratified water bodies is mixed up to the surface and come in contact with the atmosphere resulting in a rapid diffusive release. Much of the CH_4 stored in the anoxic water can be vented to the atmosphere over a timescale of days. It is still debated to what extent enhanced water column CH_4 oxidation during lake overturn can reduce the storage flux, and estimates of how much of the stored CH_4 was oxidized instead of emitted range between >50% and 5%. In any event, the storage flux represents a significant flux pathway in many fresh waters (**Table 9, Figure 6**).

4. Flux through rooted emergent plants. Many aquatic plants transport air to the roots to supply them with O_2. Such gas transport systems also transfer gases such as CH_4 from the roots to the leaves and the atmosphere. These transport systems can be pressurized (e.g., *Typha* spp., *Phragmites* spp., *Nuphar* spp.) or passive (e.g., *Carex* spp., *Peltandra* spp., *Eriophorum* spp.) depending on the plant species. These plants function as conduits for CH_4 that enters the roots allowing rapid transport to the atmosphere bypassing the zones of CH_4 oxidation. At the same time, fewer bubbles are noted where emergent plants are growing. For some species, diurnal patterns in the transport as an effect of light and solar heating on the plant physiology has been shown (**Table 8**).

The Contribution of Different Flux Types and their Regulation

Plant-mediated emission and ebullition dominates from most lakes and wetlands. Ebullition accounts for 50% or more of the open water emission (**Figure 6, Table 9**). However, it should be emphasized that ebullition, being episodic and having a heterogeneous spatial distribution, is the most difficult of the emission pathways to quantify. At present, large-scale

Table 8 Environmental variables being related with different types of CH_4 emission from lakes and wetlands

Frameworks of study	*Ebullition*	*Diffusive flux*	*Storage flux*	*Plant flux*	*Sources*
Within-system studies					
Temperate lake	Air pressure				16
Lake Vesijärvi, Finland Littoral transect				Sediment temperature, plant biomass	13
Peat	Air pressure				18
Danish wetland		Light–O_2–MOX[a]			14
Typha and Cladium stands in the Everglades				Diel variability for Typha due to solar illumination driving active gas transport. No diurnal pattern for Cladium having diffusive (passive) gas transport. Spatial variability due to biomass and production	2
Boreal lake winter emissions	Total flux related to late summer water level, early winter precipitation, and summer biomass production				15
Sallie's Fen, New Hampshire	Total flux correlated with climate and weather related variables such as peat temperature, precipitation, net radiation, and soil moisture profiles				8
Boreal Mire	Total flux related to depth of water table and presence of vascular plants				9
Among systems studies					
Boreal and temperate lakes[b]	Lake area, Water depth, Tot P, CH_{4surf}	Lake area, CH_{4surf} Areal storage	Lake area Tot P, DOC AVF, CH_{4surf}		1
Temperate lakes: Potential spring emission[b]			Lake area Sed OM[b]		17
Boreal littoral zones with *Equisetum fluviatile, Schoenoplectus lacustris* and *Phragmites australis*				Species composition, plant biomass, plant growth dynamics, sediment temperature and type of sediment	11, 12
Freshwater marshes in Sanjiang Plain, China				Species composition, water depth; diel variability related to redox potential	5–7

Global wetland emissions	Total flux related with net ecosystem exchange or net primary production	19
Northern wetlands (45–70° N)	Emissions depend on soil moisture and temperature	10
Subarctic wetlands, Sweden	Total flux related with temperature, hydrology, plant community, and alterations in the permafrost distribution	4
Wetlands in Greenland, Iceland, Scandinavia, and Siberia	Total flux related with temperature and microbial substrate availability	3

Sources
1. Bastviken D, Cole J, Pace M, and Tranvik L (2004) Methane emissions from lakes: Dependence of lake characteristics, two regional assessments, and a global estimate. *Global Biogeochemical Cycles* 18, doi: 10.1029/2004/GB002238.
2. Chanton JP, Whiting GJ, Happell JD, and Gerard G (1993) Contrasting rates and diurnal patterns of methane emissions from emergent aquatic macrophytes. *Aquatic Botany* 46: 111–128.
3. Christensen TR, Ekberg A, Strom L, *et al.* (2003) Factors controlling large scale variations in methane emissions from wetlands. *Geophysical Research Letters* 30: Art. No. 1414.
4. Christensen TR, Johansson TR, Akerman HJ, *et al.* (2004) Thawing sub-arctic permafrost: Effects on vegetation and methane emissions. *Geophysical Research Letters* 31. Art. No. L04501.
5. Ding WX, Cai ZC, and Tsuruta H (2004) Diel variation in methane emissions from the stands of *Carex lasiocarpa* and *Deyeuxia angustifolia* in a cool temperate freshwater marsh. *Atmospheric Environment* 38: 181–188.
6. Ding WX, Cai ZC, and Tsuruta H (2005) Plant species effects on methane emissions from freshwater marshes. *Atmospheric Environment* 39: 3199–3207.
7. Ding WX, Cai ZC, Tsuruta H, and Li XP (2002) Effect of standing water depth on methane emissions from freshwater marshes in northeast China. *Atmospheric Environment* 36: 5149–5157.
8. Frolking S and Crill P (1994) Climate controls on temporal variability of methane flux from a poor fen in southeastern New Hampshire—Measurement and modeling. *Global Biogeochemical Cycles* 8: 385–397.
9. Granberg G, Sundh I, Svensson BH, and Nilsson M (2001) Effects of temperature, and nitrogen and sulfur deposition, on methane emission from a boreal mire. *Ecology* 82: 1982–1998.
10. Harriss R, Bartlett K, Frolking S, and Crill P (1993) Methane emissions from northern high-lattitude wetlands. In: Oremland, RS (ed.) *Biogeochemistry of Global Change: Radiatively Active Trace Gases*, pp. 449–486, New York: Chapman & Hall.
11. Kankaala P, Kaki T, Makela S, *et al.* (2005) Methane efflux in relation to plant biomass and sediment characteristics in stands of three common emergent macrophytes in boreal mesoeutrophic lakes. *Global Change Biology* 11: 145–153.
12. Kankaala PM, äkelä S, Bergström I, *et al*. (2003) Midsummer spatial variation in methane efflux from stands of littoral vegetation in a boreal meso–eutrophic lake. *Freshwater Biology* 48: 1617–1629.
13. Kankaala P, Ojala A, and Kaki T (2004) Temporal and spatial variation in methane emissions from a flooded transgression shore of a boreal lake. *Biogeochemistry* 68: 297–311.
14. King GM (1990) Regulation by light of methane emissions from a wetland. *Nature* 345: 513–515.
15. Larmola T, Alm J, Juutinen S, *et al*. (2004) Contribution of vegetated littoral zone to winter fluxes of carbon dioxide and methane from boreal lakes. *Journal of Geophysical Research-Atmospheres* 109: Art. No. D19102.
16. Mattson MD and Likens GE (1990) Air pressure and methane fluxes. *Nature* 347: 718–719.
17. Michmerhuizen CM, Striegl RG, and McDonald ME (1996) Potential methane emission from north-temperate lakes following ice melt. *Limnology and Oceanography* 41: 985–991.
18.Tokida T, Miyazaki T, and Mizoguchi M (2005) Ebullition of methane from peat with falling atmospheric pressure. *Geophysical Research Letters* 32: Art. No. L13823.
19. Whiting GJ and Chanton JP (1993) Primary production control of methane emission from wetlands. *Nature* 364: 794–795.

[a]Indirect effect by light on diffusive emissions since light affect O_2 concentrations which in turn affect CH_4 oxidation rates and thereby water CH_4 concentrations.

[b]Tot P and CH_{4surf} refer to total phosphorous and surface-water CH_4 concentrations, respectively. DOC is concentrations of dissolved OM. Areal storage means the CH_4 storage in the water column per square meter. AVF represents the anoxic volume fraction of the lake. Sed OM is sediment OM content.

Table 9 Open water emissions of CH_4 from lakes in different climate regions[a]

Region/lake type	*Ebullition (mmol m^{-2} yr^{-1})*	*Diffusive flux (mmol m^{-2} yr^{-1})*	*Storage flux (mmol m^{-2} yr^{-1})*	*Total open water emission (mmol m^{-2} yr^{-1})*	*Sources*
Arctic lakes (>50° N)					
Average	1478	50		1556	14, 28
Range		7–93			
n	1	5		1	
Boreal lakes (45–60° N)					
Average	322	68	57	311	2, 3, 13
Range	82–563	8–313	0–301	22–1000	
n	2	20	10	4	
Temperate freshwater lakes (20–45° N)					
Average	744	234	175	1110	2, 5, 6, 10–12, 15–22, 24–27
Range	8–4526	9–3660	1–1029	42–4926	
n	15	30	36	14	
Tropical freshwater lakes (20° S-20° N)					
Average	999	404	2281	2023	1, 4, 7–9, 23
Range	38–2635	151–1129		188–3764	
n	6	6	1	5	
Saline lakes					
Average		147			17
Range		15–375			
n		3			

Source

1. Bartlett KB, Crill PM, Sebacher DI, *et al.* (1988) Methane flux from the central Amazon floodplain. *Journal of Geophysical Research* 93: 1571–1582.
2. Bastviken D, Cole J, Pace M, and Tranvik L (2004) Methane emissions from lakes: Dependence of lake characteristics, two regional assessments, and a global estimate. *Global Biogeochemical Cycles* 18, doi: 10.1029/2004/GB002238.
3. Bastviken D, Ejlertsson J, and Tranvik L (2002) Measurement of methane oxidation in lakes—A comparison of methods. *Environmental Science and Technology* 36: 3354–3361.
4. Boon PI and Mitchell A (1995) Methanogenesis in the sediments of an Australian freshwater wetland: Comparison with aerobic decay, and factors controlling methanogenesis. *FEMS Microbiology Ecology* 18: 175–190.

5. Casper P, Maberly SC, Hall GH, and Finlay BJ (2000) Fluxes of methane and carbon dioxide from a small productive lake to the atmosphere. *Biogeochemistry* 49: 1–19.
6. Chau YK, Snodgrass WJ, and Wong PTS (1977) A sampler for collecting evolved gases from sediment. *Water Research* 11: 807–809.
7. Crill PM, Bartlett KB, Wilson JO, *et al*. (1988) Tropospheric methane from an Amazonian floodplain lake. *Journal of Geophysical Research* 93: 1564–1570.
8. Devol AH, Richey JE, Clark WA, and King SL (1988) Methane emissions to the troposphere from the Amazon floodplain. *Journal of Geophysical Research* 93: 1583–1592.
9. Engle D and Melack JM (2000) Methane emissions from an Amazon floodplain lake: Enhanced release during episodic mixing and during falling water. *Biogeochemistry* 51: 71–90.
10. Fallon RD, Harris S, Hanson RS, and Brock TD (1980) The role of methane in internal carbon cycling in Lake Mendota during summer stratification. *Limnology and Oceanography* 25: 357–360.
11. Fendinger NJ, Adams DD, and Glotfelty DE (1992) The role of gas ebullition in the transport of organic contaminants from sediments. *The Science of the Total Environment* 112: 189–201.
12. Howard DL, Frea JI, and Pfister RM (1971) In *Proceedings of the 14th Conference on Great Lakes Research*, pp. 236–240, Ann Arbor, Michigan, Toronto, Canada.
13. Huttunen JT, Alm J, Liikanen A, *et al*. (2003) Fluxes of methane, carbon dioxide and nitrous oxide in boreal lakes and potential anthropogenic effects on the aquatic greenhouse gas emissions. *Chemosphere* 52: 609–621.
14. Kling GW, Kipphut GW, and Miller MC (1992) The flux of CO_2 and CH_4 from lakes and rivers in arctic Alaska. *Hydrobiologia* 240: 23–36.
15. Mattson MD and Likens GE (1993) Redox reactions of organic matter decomposition in a soft water lake. *Biogeochemistry* 19: 149–172.
16. Michmerhuizen CM, Striegl RG, and McDonald ME (1996) Potential methane emmision from north-temperate lakes following ice melt. *Limnology and Oceanography* 41: 985–991.
17. Miller LG and Oremland RS (1988) Methane efflux from the pelagic regions of four lakes. *Global Biogeochemical Cycles* 2: 269–277.
18. Miyajima T, Yamada Y, Wada E, *et al*. (1997) Distribution of greenhouse gases, nitrite, and $\delta^{13}C$ of dissolved inorganic carbon in Lake Biwa: Implications for hypolimnetic metabolism. *Biogeochemistry* 36: 205–211.
19. Riera JL, Shindler JE, and Kratz TK (1999) Seasonal dynamics of carbon dioxide and methane in two clear-water lakes and two bog lakes in northern Wisconsin, U.S.A. *Canadian Journal of Fisheries and Aquatic Sciences* 56: 265–274.
20. Rudd JWM and Hamilton RD (1978) Methane cycling in a eutrophic shield lake and its effects on whole lake metabolism. *Limnology and Oceanography* 23: 337–348.
21. Schultz M, Faber E, Hollerbach A, Schröder HG, and Güde H (2001) The methane cycling in the epilimnion of Lake Constance. *Archiv Fur Hydrobiologie* 151: 157–176.
22. Smith LK and Lewis WM (1992) Seasonality of methane emissions from five lakes and associated wetlands of the Colorado Rockies. *Global Biogeochemical Cycles* 6: 323–338.
23. Smith LK, Lewis WMJ, Chanton JP, Cronin G, and Hamilton SK (2000) Methane emissions from the Orinoco River floodplain, Venezuela. *Biogeochemistry* 51: 113–140.
24. Strayer RG and Tiedje JM (1978) In situ methane production in a small, hypereutrophic, hardwater lake: Loss of methane from sediments by vertical diffusion and ebullition. *Limnology and Oceanography* 23: 1201–1206.
25. Striegl RG and Michmerhuisen CM (1998) Hydrologic influence on methane and carbon dioxide dynamics at two north-central Minnesota lakes. *Limnology and Oceanography* 43: 1519–1529.
26. Utsumi M, Nojiri Y, Nakamura T, *et al.* (1998) Oxidation of dissolved methane in a eutrophic, shallow lake: Lake Kasumigaura, Japan. *Limnology and Oceanography* 43: 471–480.
27. Utsumi M, Nojiri Y, Nakamura T, *et al.* (1998) Dynamics of dissolved methane and methane oxidation in dimictic Lake Nojiri during winter. *Limnology and Oceanography* 43: 10–17.
28. Walter KM, Zimov SA, Chanton JP, Verbyla D, and Chapin FS (2006) Methane bubbling from Siberian thaw lakes as a positive feedback to climate warming. *Nature* 443: 71–75.

[a]Note the small number of samples for some flux types or in some regions. The unit mmol m^{-2} yr^{-1} was chosen to allow easy comparisons between different flux types, since storage flux is best expressed on a yearly basis. "*n*" denote number of systems studied.

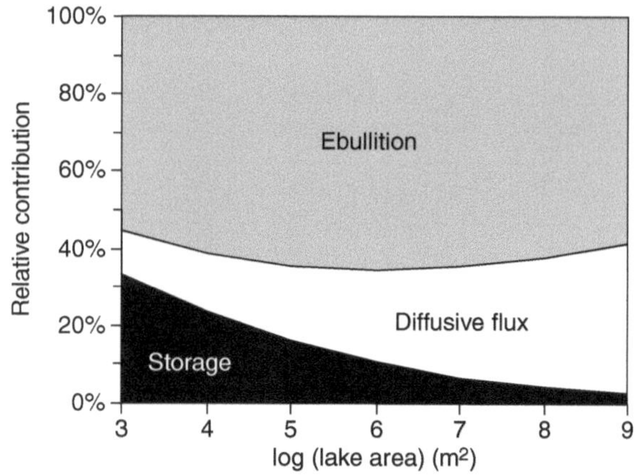

Figure 6 Relative contribution of different open water methane flux pathways versus lake area. Based on data from 73 boreal and temperate lakes. Reproduced from Bastviken D, Cole J, Pace M, and Tranvik L (2004) Methane emissions from lakes: Dependence of lake characteristics, two regional assessments, and a global estimate. *Global Biogeochemical Cycles* 18, doi: 10.1029/2004/GB002238 with permission from American Geophysical Union.

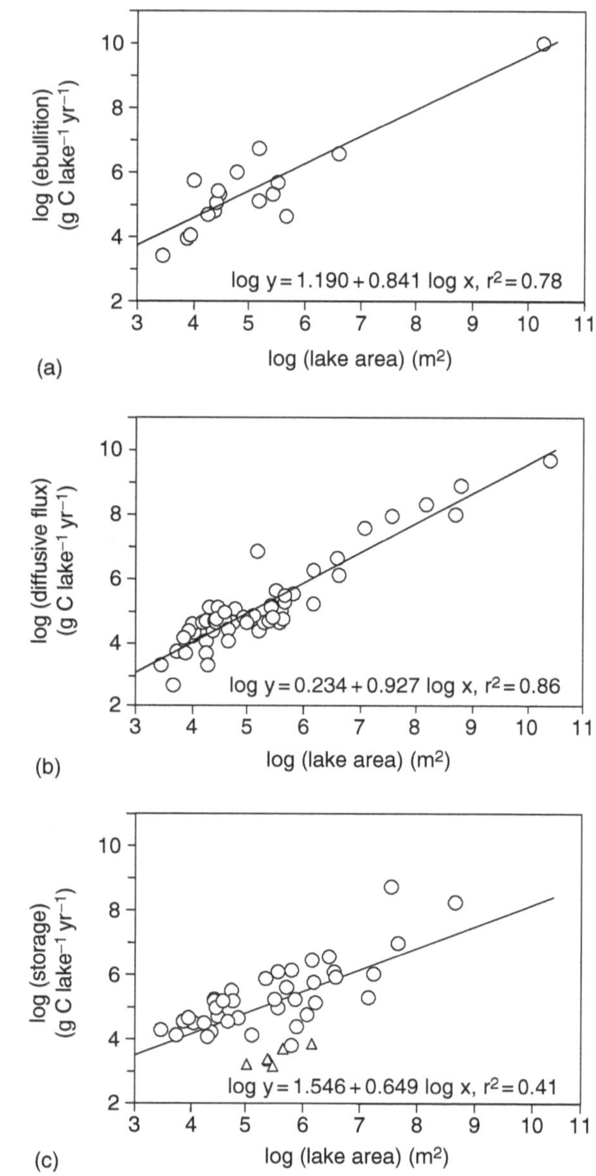

Figure 7 Ebullition (a), diffusive flux (b), and water column methane storage during stratification periods (c) for lakes of different size. The triangles in panel (c) represent lakes without anoxic parts of the water column. Linear regression lines are given in the graphs. Adapted from Bastviken D, Cole J, Pace M, and Tranvik L (2004). Methane emissions from lakes: Dependence of lake characteristics, two regional assessments, and a global estimate. *Global Biogeochemical Cycles* 18, doi: 10.1029/2004/GB002238 with permission from American Geophysical Union.

estimates of ebullition are likely to be underestimates because of the limited amount of data and the predominance of short-term and infrequent measurements (**Table 9**).

At a small-scale level ($<1\,m^2$), emission rates are regulated by a combination of CH_4 production and oxidation rates, and transport. In turn, this small-scale regulation depends on larger-scale environmental factors as summarized in Table 8. In general, open water emissions from lakes (ebullition, diffusive flux, and storage flux) seem related to lake area (**Figure 7**), lake morphometry, and the amount of OM reaching the sediments. Wetland emissions, dominated by plant-mediated flux, depend on sediment or peat temperature, the depth of the water table, plant community composition and biomass, and net primary production. In northern areas, alterations in the permafrost distribution are important because they affect the surface hydrology, inundation, and plant community structure.

Open water emissions from lakes range between 0.06 and 13.5 mmol m^{-2} day^{-1} (**Table 9**), while the range for wetland- and plant-mediated emissions is 0.01–36.9 mmol $m^{-2}\,day^{-1}$ (**Table 10**). Rivers also emit CH_4, seemingly at similar rates as lakes, but data are fewer (**Table 11**). Global CH_4 emissions from natural wetlands range between 100 and 237 Tg CH_4 $year^{-1}$ ($6.3 \times 10^{12} - 14.8 \times 10^{12}$ mol $year^{-1}$) corresponding to ~75% of all natural emissions. Lakes are believed to contribute 8–48 Tg $CH_4\,year^{-1}$ ($0.5 \times 10^{12} - 3 \times 10^{12}$ mol $year^{-1}$) or 6–16% of the natural emissions. Given total natural and anthropogenic global emissions estimates of 410–600 Tg CH_4 $year^{-1}$, natural inland waters account for more than 25% of the atmospheric CH_4. Reservoirs represent a special case. Old reservoirs emit CH_4 at rates similar to lakes. However, recently established reservoirs can emit much higher amounts of CH_4 because of the flooding of soils and rapid transformation of soil OM to CH_4. Consequently, reservoir fluxes are

Table 10 CH_4 emissions from littoral zones or wetlands (averages or ranges are given)[a]

Location	*CH_4 emission ($mmol\ m^{-2}\ d^{-1}$)*	*Dominating plants/Notes*	*Sources*
Arctic wetlands (>50° N)	7.8		4
Manitoba wetlands (northern study area of Boreas)	0.6–22		2
Tundra wetlands, northeast Siberia	0–17.6		10
Boreal wetlands (45–60° N)	13.8		4
Swedish mires	0.5–14.8		11
Lake Ekojärvi littoral, Finland		Mesotrophic–Eutrophic boreal lake	9
	27.8	*P. australis*	9
	2.2	*S. lacustris*	9
	1.7	*E. fluviatile*	9
	1.0	*N. lutea*	9
	1.0	*S. gramineum*	9
	0.5	*P. natans*	9
Lake Kevätön littoral, Finland	4.4	Hypereutrophic, mixed plant community	8
Lake Mekrijärvi littoral, Finland	3.0	Meostrophic, polyhumic, mixed plant community	8
Temperate wetlands (20–45° N)	5.8		4
Colorado wetlands			12
Dillon	0.01	Mixed plant communities	12
Long	0.9	Mixed plant communities	12
Pass	0.1	Mixed plant communities	12
Red Rock	5.5	Mixed plant communities	12
Red Rock	12.6	*Nuphar lutea* stand	12
Rainbow	2.4	Mixed plant communities	12
Temperate Sphagnum bog, Japan	28.1	Marsh trefoil (*Menyanthes trifoliata*)	14
	18.1	*P. australis*	14
	4.4	*Sphagnum* spp.	
Natural wetlands in China	3.2		6
Wuliangsu Lake, western China	3.2	*P. pectinatus*	7
	22.3	*P. australis*. Marked seasonal and diurnal variation	7
Everglades, Florida	8.9	Typha stands; 90% through plants	3
	2.8	Cladium stands	3
Tropical wetlands (~ 3° S)	11.7		4
Orinoco River floodplain, Venezuela			
Flooded forest	6.8	63% emitted as ebullition	13

Continued

Table 10 Continued

Location	*CH_4 emission ($mmol\ m^{-2}\ d^{-1}$)*	*Dominating plants/Notes*	*Sources*
Floating macrophytes	1.6	52% emitted as ebullition	13
Amazon floodplain			
Wetlands	24.4		5
Macrophytes	36.9		5
Flooded forests	6.9		5
Grass mats	24.4	64% emitted as ebullition	1
Flooded forests	13.7	54% emitted as ebullition	1

Source

1. Bartlett KB, Crill PM, Sebacher DI, *et al.* (1988) Methane flux from the central Amazon floodplain. *Journal of Geophysical Research* 93: 1571–1582.
2. Bubier J, Moore T, Savage K, and Crill P (2005) A comparison of methane flux in a boreal landscape between a dry and a wet year. *Global Biogeochemical Cycles* 19: Art. No. GB1023.
3. Chanton JP, Whiting GJ, Happell JD, and Gerard G (1993) Contrasting rates and diurnal patterns of methane emissions from emergent aquatic macrophytes. *Aquatic Botany* 46: 111–128.
4. Crill P (1996) Latitudinal differences in methane fluxes from natural wetlands. *Mit. Internat. Verein. Limnol* 25: 163–171.
5. Devol AH, Richey JE, Clark WA, and King SL (1988) Methane emissions to the troposphere from the Amazon floodplain. *Journal of Geophysical Research* 93: 1583–1592.
6. Ding WX, Cai ZC, and Wang DX (2004) Preliminary budget of methane emissions from natural wetlands in China. *Atmospheric Environment* 38: 751–759.
7. Duan XN, Wang XK, Mu YJ, and Ouyang ZY (2005) Seasonal and diurnal variations in methane emissions from Wuliangsu Lake in arid regions of China. *Atmospheric Environment* 39: 4479–4487.
8. Juutinen S, Alm J, Larmola T, *et al.* (2003) Methane (CH4) release from littoral wetlands of Boreal lakes during an extended flooding period. *Global Change Biology* 9: 413–424.
9. Kankaala P, Mäkelä S, Bergström I, *et al.* (2003) Midsummer spatial variation in methane efflux from stands of littoral vegetation in a boreal meso-eutrophic lake. *Freshwater biology* 48: 1671–1629.
10. Nakano T, Kuniyoshi S, and Fukuda M (2000) Temporal variation in methane emission from tundra wetlands in a permafrost area, northeastern Siberia. *Atmospheric Environment* 34: 1205–1213.
11. Nilsson M, Mikkela C, Sundh I, *et al.* (2001) Methane emission from Swedish mires: National and regional budgets and dependence on mire vegetation. *Journal of Geophysical Research-Atmospheres* 106: 20847–20860.
12. Smith LK and Lewis WM (1992) Seasonality of methane emissions from five lakes and associated wetlands of the Colorado Rickies. *Global Biogeochemical Cycles* 6: 323–338.
13. Smith LK, Lewis WMJ, Chanton JP, Cronin G, and KHS (2000) Methane emissions from the Orinoco River floodplain, Venezuela. *Biogeochemistry* 51: 113–140.
14. Sugimoto A and Fujita N (1997) Characteristics of methane emissions from different vegetations on a wetland. *Tellus* 49B, 382–392.

[a]Only examples from the numerous published wetland studies are given. For data from more wetlands see e.g., Bartlett K and Harriss R (1993) Review and assessment of methane emissions from wetlands. *Chemosphere* 26: 261–320; Harriss R, Bartlett K, Frolking S, and Crill P (1993) Methane emissions from northern high-lattitude wetlands. In (Oremland RS Ed.) *Biogeochemistry of Global Change: Radiatively Active Trace Gases*, pp. 449–486, New York: Chapman & Hall; Wetzel RG (2001) *Limnology-Lake and River Ecosystems*. San Diego: Academic Press; Whalen SC (2005) Biogeochemistry of methane exchange between natural wetlands and the atmosphere. *Environmental Engineering Science* 22: 73–94; and Yavitt JB, Williams CJ, and Wieder RK (1997) Production of methane and carbon dioxide in peatland ecosystems across North America: Effects of temperature, aeration, and organic chemistry of peat. *Geomicrobiology Journal* 14: 299–314.

Table 11 Average CH_4 emissions from rivers, streams, and reservoirs

System studied	*CH_4 emission (mmol m^{-2} yr^{-1})*		*n*	*Sources*
	Average	*Range*		
Temperate and boreal rivers and streams	530	3–3650	12	1–4
Temperate and boreal reservoirs	456	228–1825	17	5
Tropical reservoirs	6844	456–34 219	5	5

Source

1. Hope D, Palmer SM, Billett MF, and Dawson JJC (2001) Carbon dioxide and methane evasion from a temperate peatland stream. *Limnology and Oceanography* 46: 847–857.
2. Jones JB and Mulholland PJ (1998) Methane input and evasion in a hardwood forest stream: Effects of subsurface flow from shallow and deep pathways. *Limnology and Oceanography* 43: 1243–1250.
3. Kling GW, Kipphut GW, and Miller MC (1992) The flux of CO_2 and CH_4 from lakes and rivers in arctic Alaska. *Hydrobiologia* 240: 23–36.
4. Lilley M, De Angelis MA, and Olson E (1996) Methane concentrations and estimated fluxes from Pacific Northwest rivers. *Mit. Internat. Verein. Limnol* 25: 187–196.
5. St. Louis VL, Kelly CA, Duchemin É, Rudd JWM, and Rosenberg DM (2000) Resevior surfaces as sources of greenhouse gases to the atmosphere: A global estimate. *BioScience* 50: 766–775.

highly variable (**Table 11**). It has been suggested that reservoirs contribute 70 Tg CH_4 $year^{-1}$ globally, corresponding to 18% of total anthropogenic CH_4 emissions or 7% of the anthropogenic emissions of CO_2 equivalents.

CH_4 and Whole-System Carbon Cycling

Methanogenesis is the dominating terminal OM degradation process in anoxic freshwater environment. Methanogenesis has been reported to account for 20–56% of the total carbon mineralization in lakes (**Table 6**). CH_4 oxidation can be comparable to primary production, and has the potential to rapidly consume large amounts of O_2. In isolated water bodies, CH_4 oxidation has caused hypoxia or anoxia, resulting in fish kill events. CH_4 emissions may correspond to up to 37% of the primary productivity (**Table 6**). Altogether, the existing comparisons indicate that CH_4 dynamics can represent a substantial proportion of the total carbon cycling in aquatic systems. This means that lakes and wetlands represent environments where CO_2, fixed by terrestrial and aquatic primary production, is converted to CH_4 with a relatively high efficiency.

Knowledge Gaps

The knowledge about CH_4 in lakes and rivers relies on few studies of individual systems usually during single seasons or single years. Hence, there is a general lack of systematic comparisons across systems, and of temporal variability. Within such frameworks further attempts to relate CH_4 cycling to other aspects of aquatic ecology and biogeochemistry would be valuable, including e.g., CH_4 versus overall carbon cycling, the potential of CH_4 to support food web carbon demand, possible connections between the nitrogen fixing capacity of MOB and nitrogen cycling in nitrogen poor environments, and CH_4 oxidation versus O_2 consumption.

In spite of considerable efforts to assess wetland emissions, there is still a lack of understanding of the large spatial and temporal variation observed. Regarding emissions in general, large-scale estimates of ebullition- and plant-mediated flux represent the most important uncertainties. Both suffer from a lack of data and require future attention in all types of inland waters. Large-scale emission estimates also suffer from uncertainties in the areal estimates of different aquatic environments.

An important challenge is to estimate how CH_4 emissions from inland waters will be affected by climate change. Emissions from wetlands are considered particularly sensitive to climate, and there is a risk of increased CH_4 emissions in a warmer future. Such emission feedback responses represent important links between aquatic methane cycling and the future climate.

Acknowledgments

I wish to thank Patrick Crill and Nguyen Duc for very valuable comments on the manuscript.

See also: The Benthic Boundary Layer (in Rivers, Lakes, and Reservoirs); Carbon, Unifying Currency; Density Stratification and Stability; Distribution and Abundance of Aquatic Plants – Human Impacts; Effects of Climate Change on Lakes; Gas Exchange at the Air-Water Interface; Hydrodynamical Modeling; Natural Organic Matter; Nitrogen; Nitrogen Fixation; Pressure; Redox Potential; Small-Scale Turbulence and Mixing: Energy Fluxes in Stratified Lakes.

Further Reading

Bartlett K and Harriss R (1993) Review and assessment of methane emissions from wetlands. *Chemosphere* 26: 261–320.

Bastviken D, Cole J, Pace M, and Tranvik L (2004) Methane emissions from lakes: Dependence of lake characteristics, two regional assessments, and a global estimate. *Global Biogeochemical Cycles* 18, doi: 10.1029/2004/GB002238.

Bastviken D, Ejlertsson J, Sundh I, and Tranvik L (2003) Methane as a source of carbon and energy for lake pelagic food webs. *Ecology* 84: 969–981.

Capone DG and Kiene RP (1988) Comparison of microbial dynamics in marine and freshwater sediments: Contrasts in anaerobic carbon catabolism. *Limnology and Oceanography* 33: 725–749.

Conrad R (2002) Control of microbial methane production in wetland rice fields. *Nutrient Cycling in Agroecosystems* 64: 59–69.

Grey J, Kelly A, Ward S, Sommerwerk N, and Jones RI (2004) Seasonal changes in the stable isotope values of lake-dwelling chironomid larvae in relation to feeding and life cycle variability. *Freshwater Biology* 49: 681–689.

Harriss R, Bartlett K, Frolking S, and Crill P (1993) Methane emissions from northern high-lattitude wetlands. In: Oremland RS (ed.) *Biogeochemistry of Global Change: Radiatively Active Trace Gases*, pp. 449–486. New York: Chapman & Hall.

Kankaala P, Taipale S, Grey J, *et al.* (2006) Experimental delta C-13 evidence for a contribution of methane to pelagic food webs in lakes. *Limnology and Oceanography* 51: 2821–2827.

King GM (1992) Ecological aspects of methane oxidation, a key determinant of global methane dynamics. In: Marshall KC (ed.) *Advances in Microbial Ecology*, pp. 431–468. New York: Plenum Press.

Liikanen A, Huttunen JT, Valli K, and Martikainen PJ (2002) Methane cycling in the sediment and water column of mid-boreal hyper-eutrophic Lake Kevaton, Finland. *Archiv fur Hydrobiologie* 154: 585–603.

Reeburgh WS (2003) Global methane biogeochemistry. In: Keeling R (ed.) *Treatise on Geochemistry, Volume 4: The Atmosphere*, pp. 65–89. Amsterdam: Elsevier.

Rudd JWM and Taylor CD (1980) Methane cycling in aquatic environments. In: Droop MR and Jannasch HW (eds.) *Advances in Aquatic Microbiology*, pp. 77–150. London: Academic Press.

Segers R (1998) Methane production and methane consumption: A review of processes underlying wetland methane fluxes. *Biogeochemistry* 41: 23–51.

StLouis VL, Kelly CA, Duchemin E′, Rudd JWM, and Rosenberg DM (2000) Resevior surfaces as sources of greenhouse gases to the atmosphere: A global estimate. *BioScience* 50: 766–775.

Whalen SC (2005) Biogeochemistry of methane exchange between natural wetlands and the atmosphere. *Environmental Engineering Science* 22: 73–94.

Natural Organic Matter

E Michael Perdue, Georgia Institute of Technology, Atlanta, GA, USA

Introduction

Primary production in the Earth's system provides a continuously renewed supply of organic compounds to the Earth's soils and surface waters. The principal components of primary production are the building blocks of living organisms: lipids, proteins, carbohydrates, lignin, etc. It is the fate of these biomolecules to be converted back to carbon dioxide through heterotrophic activity and photochemical reactions. During this recycling process, a very small percentage (<1%) of the photosynthetically fixed organic carbon accumulates for relatively long periods of time in the Earth's soils, sediments, and natural waters.

The more resistant (or refractory) organic matter that temporarily avoids its ultimate fate of oxidation to CO_2 is definitely not a mixture of undegraded lipids, proteins, carbohydrates, and lignin. Moreover, scientists agree that this refractory organic matter is not the result of anthropogenic activities that occasionally introduce significant quantities of organic compounds into soils, sediments, and natural waters. Instead, it is believed to be formed naturally – either through biochemical reactions occurring in living organisms or through abiotic reactions occurring after the senescence and death of living cells. To emphasize its natural origin, this refractory organic matter in soils, sediments, and natural waters is commonly called natural organic matter (NOM).

This article explores the inventory of NOM in inland waters and global fluxes to the oceans, the methods by which NOM is isolated from natural waters for advanced chemical characterization, and the most commonly reported chemical properties of NOM. The goal is to provide a statistically robust description of NOM that distinguishes it clearly from the major classes of biomolecules from which it is derived.

Definitions

Natural organic matter in inland waters is a mixture of unsurpassed complexity. To deal with such complexity, scientists have historically subdivided NOM into a few fractions according to its physical state (dissolved molecules, colloids, and particles), it point of origin (within or outside the water body), its polarity (hydrophobic or hydrophilic), its acidity (acidic, neutral, or basic), etc. Such distinctions have guided the study of NOM for the past 30–40 years, and much of the current knowledge concerning NOM is built upon the study of these various fractions. Even though the science is evolving rapidly, it is useful to introduce these historically important terms:

- Dissolved organic matter – that fraction of NOM which passes through a 0.45 μm filter
- Particulate organic matter – that fraction of NOM which is removed by a 0.45 μm filter
- Autochthonous NOM – that fraction of NOM which is formed directly in a water body
- Allochthonous NOM – that fraction of NOM which is formed elsewhere in the catchment and transported into a water body
- Hydrophobic NOM – that fraction of NOM which is adsorbed on XAD-8 (or equivalent) resin at pH 2
- Hydrophilic NOM – that fraction of NOM which is not adsorbed on XAD-8 (or equivalent) resin at pH 2
- Acid, neutral, and basic NOM – fractions of NOM according to the effect of pH on their retention by XAD-8 (or equivalent) resin and anion exchange resins

Because modern research on NOM grew out of earlier studies of organic matter in soils, the terms humic acid (HA) and fulvic acid (FA) are very commonly used to describe fractions of NOM in inland waters. Both are classified as hydrophobic acids, and they differ only in the greater solubility of FA at very low pH. Specifically,

- Humic acid – that fraction of NOM which is adsorbed on XAD-8 (or equivalent) resin at pH 2, eluted from the resin at high pH, and precipitated from solution at pH 1
- Fulvic acid – that fraction of NOM which is adsorbed on XAD-8 (or equivalent) resin at pH 2, eluted from the resin at high pH, and not precipitated from solution at pH 1

In the remainder of this article, the characterization of NOM will draw upon published research dealing with any or all of these fractions as well as studies of unfractionated NOM. Several datasets and statistical summaries used in this article are based on a recent, comprehensive review by Perdue and Ritchie (See Further Reading).

Inventories and Fluxes of NOM

As noted earlier, the NOM in inland waters is a remarkably complex mixture, whose concentration is viewed rigorously as the sum of the concentrations of many thousands of individual organic compounds. The practical approach to measuring the concentration of NOM is to measure the concentration of CO_2 that is produced when NOM is completely oxidized in a carbon analyzer, using either high-temperature combustion or UV photooxidation in the presence of a strong oxidizing agent to convert all forms of organic carbon completely to CO_2. Accordingly, the concentration of NOM in inland waters is expressed in terms of carbon. The concentration of NOM in an unfiltered sample is expressed in terms of total organic carbon (TOC). When samples are filtered through a 0.45 μm filter prior to the analysis, the concentration of NOM is expressed in terms of dissolved organic carbon (DOC). The difference between TOC and DOC is the concentration of particulate organic carbon (POC). The first standard method for measuring TOC was introduced as a tentative method in 1971.

At least 5120 papers dealing with yellow organic acids, aquatic humus, dissolved organic matter, natural organic matter, total organic carbon, and dissolved organic carbon were published between January, 1900 and November, 2002; however, only 70 of those papers were published prior to 1970. It is thus understandable that nearly all estimates of either the annual flux of TOC from the continents to the oceans or the average concentration of TOC in rivers have been published since 1970.

Several published estimates of the fluxes of water and TOC from the continents to the oceans and the corresponding discharge-weighted average concentration of TOC in rivers are given in **Table 1**. It is noteworthy that the first (and earliest) estimate of 820 $\mu mol\,L^{-1}$ in **Table 1** (which was cited in the original paper as a 'personal communication' from another scientist) was published in 1971, at a time which pre-dates the availability of a significant number of measurements of the concentration of TOC in rivers.

Over the past thirty years, a robust dataset of TOC concentrations has become available, and several more-refined estimates of fluxes of water and TOC have been published. Using only the data from **Table 1** that were published after 1980, the annual flux of TOC from the continents to the oceans is $31.1 \pm 1.9\,Tmol\,yr^{-1}$. When this average flux is combined with the corresponding average annual discharge of water from the continents to the oceans ($38\,100 \pm 2400\,km^3\,yr^{-1}$), the corresponding average TOC concentration in rivers is 816 $\mu mol\,L^{-1}$. It is

Table 1 Selected estimates of global fluxes of organic carbon from the continents to the oceans

Land area (10^6 km^2)	*Discharge (km^{3}yr^{-1})*	*Flux (Tmol yr^{-1})*			*Conc. (μmol L^{-1})*			*Source*
		TOC	*DOC*	*POC*	*TOC*	*DOC*	*POC*	
	32 500	26.6			820			3
	36 000	16.5	10.7	5.8	458	297	161	4
	37 700	15.7	10.2	5.5	417	271	146	5
	42 000	30.8			733			9
113.0		34.2			814			9
99.9	37 400	32.8	17.9	14.9	877	479	398	7, 8
99.9	37 400	31.5			842			7, 8
125.9	35 319	27.9			790			2
106.3	38 170	31.5	17.1	14.4	825	448	377	6
129.0		30.1						1
129.0		30.2						1

Sources

1. Aitkenhead JA and McDowell WH (2000) Soil C:N ratio as a predictor of annual riverine DOC flux at local and global scales. *Global Biogeochemical Cycles* 14: 127–138.
2. Degens ET, Kempe S, and Richey JE (1991) Summary: biogeochemistry of major world rivers. In *SCOPE 42, Biogeochemistry of Major World Rivers* (eds. ET Degens, S Kempe, and JE Richey), pp. 323–347. New York: Wiley.
3. Garrels RM and Mackenzie FT (1971) *Evolution of Sedimentary Rocks*. Norton and Co., New York, pp. 101–109.
4. Garrels RM, MacKenzie FT, and Hunt C (1975) *Chemical Cycles and the Global Environment*, pp. 73–80. Kaufmann, Los Altos, CA.
5. Kempe S (1979) Carbon in the freshwater cycle. In *SCOPE 13, The Global Carbon Cycle* (eds. B Bolin, ET Degens, S Kempe, and P Ketner), pp. 317–342. New York: Wiley.
6. Ludwig W, Probst J-L, and Kempe S (1996) Predicting the oceanic input of organic carbon by continental erosion. *Global Biogeochemical Cycles* 10: 23–41.
7. Meybeck M (1981) River transport of organic carbon to the ocean. In *Flux of Organic Carbon by Rivers to the Ocean: CONF 8009140 UC-11*, pp. 219–269. Washington DC: U.S. Department of Energy.
8. Meybeck M (1982) Carbon, nitrogen, and phosphorus transport by world rivers. *American Journal of Science* 282: 401–450.
9. Schlesinger WH and Melack JM (1981) Transport of organic carbon in the world's rivers. *Tellus* 33: 172–187.

rather striking that the average result from the last thirty years of research is not significantly different from the initial estimate made in 1971.

Two of the references in **Table 1** that were published between 1980 and the present provide independent estimates of the discharge-weighted average concentrations of DOC and POC in rivers. If these estimates are averaged, the overall average concentrations of DOC and POC in rivers are 464 and 388 $\mu mol\,L^{-1}$, respectively. The corresponding TOC concentration (852 $\mu mol\,L^{-1}$) is somewhat greater than the average TOC concentration that was calculated from the more extensive set of estimated fluxes of water and TOC. It is very well established in the scientific literature that most of the TOC in rivers is generally found to be DOC, so the average proportion of POC on a global scale is surprisingly high. Global estimates of particulate matter (organic or inorganic) in rivers are often strongly weighted by the rather high particulate loads of some Asian rivers.

The average global inventories and fluxes of organic carbon in **Table 1** are relatively consistent and conceal the inherent temporal and spatial variations observed among inland waters. The original literature cited in **Table 1** clearly demonstrates that there are wide variations in both the concentrations and fluxes of TOC among individual rivers. A widely published set of estimated median TOC concentrations in a variety of fresh waters is given in **Table 2**. TOC concentrations in fresh waters range over almost two orders of magnitude. The estimated median TOC concentration for rivers in **Table 2** is lower than the global discharge-weighted estimated TOC concentration from **Table 1**.

Isolation of NOM

Chemical characterization of the NOM in inland waters is generally preceded by some process(es) by which the NOM is separated from the inorganic constituents of natural waters and concentrated by removal of water. The available methods differ significantly in yield of NOM, chemical conditions to which the NOM is exposed during the processes of desalination and concentration, and the rate at which water samples can be processed. Although many properties of NOM can be studied using concentrated liquid solutions, other important properties such as elemental composition can only be measured on dry samples. In addition, NOM is better preserved and stored as dry samples rather than in concentrated aqueous solutions. Therefore, the desalination and concentration of NOM are generally followed by lyophilization to obtain dry samples.

Table 2 Typical concentrations of organic carbon in fresh waters

Water type	*TOC (μmol L⁻¹)*
Ground water	60
River	580
Oligotrophic lake	180
Eutrophic lake	1000
Marsh	1420
Bog	2750

Source
Thurman EM (1985) *Organic Geochemistry of Natural Waters*. Dordrecht: Martinus Nijhoff/Dr. W Junk Publishers.

Solid Phase Extractions of NOM

XAD Resin Method. XAD resins are commonly used to isolate humic substances from natural waters. Even though recoveries of NOM using XAD resins are not especially high, much that is known (or thought to be known) about the role of NOM in many environmental processes has been derived from studies on NOM samples that were isolated and/or fractionated using XAD resins. This method was developed initially to isolate NOM, and it was later converted into a comprehensive analytical fractionation scheme for classification of organic solutes in water. In the analytical fractionation scheme, XAD-2 and XAD-8 resins, cation- and anion exchange resins, and aqueous HCl and NaOH solutions were employed to separate NOM into hydrophilic and hydrophobic fractions, each of which was further separated into acidic, basic, and neutral fractions. NOM could thus be quantified according to the relative proportions of these six fractions, but significant quantities of isolated samples were not generated by this analytical scheme.

In the early 1980s, the original classification scheme was extended to the preparative scale. That preparative-scale method employs XAD-8 resin to adsorb humic substances from acidified ($pH < 2$) water samples. The aquatic humic substances are then back-eluted with NaOH, and fractionated into aquatic HA (insoluble at pH 1) and aquatic FA (soluble at pH 1). The two fractions are further purified to remove inorganic solutes and freeze-dried separately. Many authors have used this approach to isolate HA and FA from fresh waters.

Almost half of the NOM in river waters is not adsorbed by XAD-8 resin, so the method has been modified to collect an additional fraction of NOM by passing the effluent from the XAD-8 column through a column of XAD-4 resin. About half of the NOM adsorbed by the XAD-4 resin cannot be subsequently desorbed with dilute base solutions and recovered.

Accordingly, an additional 7–25% of NOM in fresh waters is usually recovered by this step. The new fraction has been termed hydrophilic acids because it is not adsorbed by XAD-8 resin; however, the XAD-4 resin to which it does adsorb is actually much more hydrophobic than XAD-8 resin. Other names have been suggested for this fraction, including XAD-4 acids, transphilic acids, and amphiphilic acids.

For 58 samples of river water, the yield of hydrophobic acids (humic and fulvic acids) ranged from 19 to 90%, with a mean of 54% and a standard deviation of 14%. If an additional 10% of initial NOM that can potentially be recovered by processing the eluent from the XAD-8 process with XAD-4 resin is also considered, then the average total yield of isolated NOM by the XAD method is around 64%.

C_{18} Method. Another method of isolation of NOM from water is adsorption of NOM on either a cartridge or a disk filled with particles that are coated with C_{18} alkyl groups chemically bonded to a silica support. As in the XAD methods, samples are initially acidified to pH 2–2.8, and then passed through a filter or cartridge of C_{18} that has been thoroughly washed sequentially with methanol and acidified water. The adsorbed NOM is then eluted with methanol. Relative to the XAD method(s), the C_{18} method is less expensive, much faster, and less prone to contamination by components of the adsorbent. The trade-off is lower yields of NOM and fractionation of NOM. Recoveries of NOM are substantially lower than with XAD resins, and the C_{18} adsorbent has a rather high selectivity for adsorption of aromatic constituents of NOM. Even the task of isolating lignin-derived phenols from fresh waters has been found to be quite dependent on operational parameters of the method.

Membrane Isolation of NOM

Both organic and inorganic solutes can be concentrated from natural waters using ultrafiltration (UF) or reverse osmosis (RO) membranes. In both methods, a feed solution consisting of water and aqueous solutes is placed under pressure and passed across a semipermeable membrane, where the feed solution is separated into a permeate solution (relatively lower concentrations of solutes) and a retentate solution (relatively higher concentrations of solutes). As the feed solution is processed, the retentate solution is recycled back to the sample reservoir and the permeate solution is discarded. As more feed solution is added either continuously or discontinuously to the sample reservoir and processed, the concentrations of all solutes that are well rejected by the membrane gradually increase in the sample reservoir. The major practical difference between UF and RO membranes is their respective abilities to 'reject' solutes. Rejection is generally defined as

$$R = 1 - C_P/C_F \qquad [1]$$

where C_P and C_F are the concentrations of solute in the permeate and feed solutions, respectively. The rejection of any given solute by RO membranes is greater than its rejection by UF membranes. It is also generally true that rejection of solutes usually follows the basic order of polyvalent ions > monovalent ions > neutral molecules.

The fundamental problem that arises if membranes are used to concentrate NOM from inland waters is the co-concentration of inorganic solutes. The ideal membrane would allow all inorganic solutes to pass through the membrane ($R = 0$) while retaining all of the NOM ($R = 1$). Unfortunately, membranes that have relatively low R values for inorganic solutes also have relatively low R values for NOM. The best that can be achieved with a single type of membrane is relatively good removal of inorganic solutes and relatively good recovery of NOM.

The second major problem with membrane-based separations is that NOM is a complex mixture whose constituents certainly have a range of R values for any given type of membrane. Consequently, as a sample is concentrated, the retained NOM becomes selectively enriched in those constituents having greater R values.

Because R values are generally lower when UF membranes are used, the retention of larger and more highly charged organic constituents of NOM is significantly greater than the retention of inorganic solutes. NOM samples concentrated by UF are therefore also partially desalted. After the NOM has been concentrated to the desired extent, most of the remaining inorganic solutes can be removed by diafiltration – a process in which the concentrated sample of NOM and inorganic solutes is washed continuously with pure water until the inorganic solutes have been removed by dialysis.

UF membranes are available in a variety of pore sizes (nominal molecular weight cut-offs – MWCO). Summary statistics for isolation of NOM by UF membranes are given in **Table 3**. As might be expected, the yield of NOM increases with decreasing MWCO. Only when a MWCO of 1 kDa is used does the yield of NOM exceed the mean value of around 54% that is attainable using the XAD-8 method.

When used to concentrate NOM from fresh waters, the products obtained by UF are often relatively clean, i.e., most inorganic solutes have been removed. The greatest advantages of the UF method are its speed and chemical mildness. It should be

Table 3 Percent recovery of NOM using ultrafiltration membranes of varying molecular weight cut-off (MWCO)

MWCO (kDa)	*Obs.*	*Mean %*	*Std. Dev. %*
100	34	31	20
10	53	48	28
1	68	66	24

Source
Perdue EM and Ritchie JD (2003) Dissolved organic matter in freshwaters. In Drever JI, Holland HD, and Turekian KK (eds.) *Surface and Ground Water, Weathering, and Soils, Vol. 5, Treatise on Geochemistry*, pp. 273–318. Oxford: Elsevier–Pergamon.

remembered, however, that solutes having higher molecular weights and higher charge-to-mass ratios are selectively concentrated by this method.

When RO membranes are used, both NOM and inorganic solutes are very well retained. This poses some significant problems. First of all, most inland waters are not far from equilibrium with $CaCO_3(s)$ and atmospheric $CO_2(g)$. Consequently, as water is removed by RO, the ion activity product of $CaCO_3(s)$ or some other inorganic carbonate or hydroxide is quickly exceeded, causing one or more insoluble minerals to precipitate on the RO membrane, effectively clogging its pores and preventing further removal of water. This potential problem is avoided by 'softening' the water sample using cation exchange resin in either the Na^+ or H^+ form. The H^+ form has a distinct advantage, because the displaced H^+ reacts with HCO_3^- to form $CO_2(g)$ until all metallic cations and inorganic carbon are removed from the sample. In a typical application, the final concentrated solution contains NOM, inorganic acids (HCl, HNO_3, and H_2SO_4), and dissolved silica (H_4SiO_4).

Both HCl and HNO_3 are relatively volatile and can be removed along with water when the concentrated sample of NOM is lyophilized. $H_4SiO_4(aq)$ is converted into amorphous $SiO_2(s)$, which remains in the dry sample, and H_2SO_4 remains in the NOM sample as a concentrated, strong acid.

Even though some inorganic solutes are co-concentrated along with NOM, RO has been used on a wide variety of fresh waters. The yield of NOM ranges from 57 to 104%, with a mean yield of 88% and a standard deviation of 10%. This average yield is far better than that obtained by any other method. Furthermore, the RO method is very fast and chemically mild. Its only drawback is the co-concentration of H_4SiO_4 and H_2SO_4. When the concentrations of H_4SiO_4 and H_2SO_4 are comparable to the concentration of NOM in a water sample, it is inadvisable to lyophilize the sample unless the H_2SO_4 is neutralized prior to lyophilization by addition of an equivalent concentration of NaOH to the sample. The final dry sample will then contain NOM, $SiO_2(s)$, and $Na_2SO_4(s)$. If the addition of NaOH is omitted, the final sample will contain concentrated H_2SO_4, which will react vigorously with many components of NOM and cause undesirable alteration of the NOM.

Coupled Reverse Osmosis and Electrodialysis

Several recent studies have coupled RO with electrodialysis (ED) – a membrane-based method for desalting aqueous samples. In this method, water is removed from an aqueous sample by RO and inorganic salts are removed from the sample by electrodialysis. The two methods are highly complementary, because RO is most efficient when osmotic pressure is very low, and ED is most efficient when electrical conductivity is very high. As water is removed from a sample by RO, salinity, osmotic pressure, and electrical conductivity of the sample all increase. As discussed earlier, it is also very likely that sparingly soluble inorganic salts will precipitate on the RO membrane, further decreasing the efficiency of the RO process. With these changes, the efficiency of the RO process gradually decreases and the efficiency of the ED process gradually increases.

In practice, RO and ED can be coupled so that the two processes are used simultaneously or sequentially in any order. This flexibility makes it possible, in principle, to concentrate and desalt NOM samples from both fresh and saline natural waters. Although the number of observations is very limited, an average recovery of 84% of NOM has been obtained from several river water samples. The ability of the coupled RO/ED method to remove H_2SO_4 and H_4SiO_4 while retaining NOM in fresh water samples has also been demonstrated. Although it is outside the scope of this review, the most exciting development is the use of the coupled RO/ED method to recover an average of 75% of NOM from 16 seawater samples – a yield that roughly triples the state-of-the-science. Much more widespread use of the coupled RO/ED method in both fresh and saline waters is anticipated.

Molecular Weights of NOM

A complex mixture such as NOM contains many organic compounds, so the 'molecular weight' of NOM is actually a distribution of molecular weights. The average molecular weight of such a distribution is usually measured or calculated as either a number-average (M_n) or weight-average (M_w) molecular weight.

These average molecular weights are defined as:

number-average molecular weight:

$$M_n = \frac{\sum_i w_i}{\sum_i n_i} = \frac{\sum_i n_i M_i}{\sum_i n_i} \qquad [2]$$

weight-average molecular weight:

$$M_w = \frac{\sum_i w_i M_i}{\sum_i w_i} = \frac{\sum_i n_i M_i^2}{\sum_i n_i M_i} \qquad [3]$$

where n_i, M_i, and w_i are the number of moles, molecular weight, and mass of the *i*th component of a mixture (note that $w_i = n_i M_i$). For any mixture containing molecules having more than one molecular weight, it is readily shown that $M_w > M_n$. The ratio M_w/M_n is known as the polydispersity of the mixture.

Experimental methods for measurement of average molecular weights or distributions of molecular weight may lead naturally to either M_n or M_w, so average molecular weights that are measured on different samples using different methods may not always be comparable. The methods by which molecular weight distributions and average molecular weights are determined for mixtures can be classified as either non-colligative or colligative methods.

Non-Colligative Methods

Non-colligative methods are based directly on the properties of solute molecules – their sizes, shapes, densities, diffusion coefficients, charge densities, etc. Most non-colligative methods yield a distribution of apparent molecular weights, from which M_n and M_w values can be extracted. Examples of non-colligative methods for determination of distributions of molecular weight are size exclusion chromatography, flow field-flow fractionation, diffusivimetry, UV scanning ultracentrifugation, and mass spectrometry.

When raw data from non-colligative methods are converted into molecular weight distributions, it is generally necessary to make one or more simplifying assumptions about the physical properties of the constituents of NOM. Most generally, it is assumed that all components of NOM have some properties in common (e.g., specific UV absorptivity, partial specific volume, or shape) or that the relationship between molecular weight and some measured property for the components of NOM is identical to the corresponding relationship in a set of known organic solutes that are used to construct calibration curves. In the case of mass spectrometry, where only ions can be observed, the measured distribution of *m/z* peaks is strongly affected by the ionization efficiencies of various classes of organic compounds. Consequently, the properties of the observed distribution of peaks may differ from the actual distribution of molecular weights in the original NOM sample.

Colligative Methods

Colligative methods are based on the variation of a property of a solvent with its mole fraction in a solution (e.g., melting point, boiling point, vapor pressure, and osmotic pressure). The mole fraction of solvent depends directly on the mole fraction of solute, but not on the nature of the solute.

Colligative methods do not yield a distribution of molecular weights. In fact, only the M_n of a mixture can be determined by such methods. Unlike M_n values that can be obtained from non-colligative methods, however, the M_n values obtained by colligative methods are independent of the nature of the components of the mixture. There is, however, one complication. The dissociation of acidic functional groups in NOM in aqueous solution causes directly measured M_n values to be too low because the H^+ ions from the dissociation reactions become new solute species with a very low formula weight. Fortunately, if the concentration of released H^+ is also measured, there are straightforward methods to obtain the correct M_n values.

Statistical distributions of measured M_n and M_w values for NOM, freshwater FA, and freshwater HA depend to some extent on the methodology; however, the most striking aspect to the results in the published literature is their relative consistency. Summary statistics are given in **Table 4**. The statistical summary in **Table 4** does not include estimates of M_n and M_w that are sometimes calculated from mass spectroscopic data, because the results are not considered to be quantitative. Some trends can be identified in **Table 4**. Median and mean M_n values for any type of sample are greater when measured by non-colligative methods than when measured by colligative methods, but the difference is less than a factor of two. It is likely that this difference is largely due to the reliance on UV absorbance as a surrogate for concentrations of organic compounds in many of the non-colligative methods. Considering that the colligative methods are independent of the physical properties of NOM and the non-colligative methods exploit collectively several different physical properties of NOM, the results in **Table 4** are considered to be remarkably consistent.

Elemental Composition of NOM

The discussion of the elemental composition of NOM in inland waters is facilitated by comparison with

Table 4 Statistical summary of the molecular weight distributions of NOM, fulvic acids, and humic acids

Method/Sample[a]	*Obs.*	*Minimum*	*Maximum*	*Median*	*Mean*	*Std. Dev.*
Number-average (M_n)						
NOM (C)	1	614	614	614	614	n.a.
NOM (NC)	38	400	2700	1101	1112	447
FA (C)	14	540	900	633	678	118
FA (NC)	18	639	1790	1170	1174	334
HA (C)	1	1220	1220	1220	1220	n.a.
HA (NC)	6	1320	2374	1750	1837	402
Weight average (M_w)						
NOM (NC)	57	700	4900	1590	1767	790
FA (NC)	24	950	2800	1805	1796	546
HA (NC)	12	2090	6590	3310	3629	1160

[a]NOM, FA, and HA are natural organic matter, fulvic acid, and humic acid, respectively. C and NC are colligative and non-colligative methods, respectively. Molecular weights are in units of g mol^{-1}.

Source

Perdue EM and Ritchie JD (2003) Dissolved organic matter in freshwaters. In Drever JI, Holland HD, and Turekian KK (eds.) *Surface and Ground Water, Weathering, and Soils, Vol. 5, Treatise on Geochemistry*, pp. 273–318. Oxford: Elsevier–Pergamon.

the elemental compositions of some of the major building blocks of biomass. For this purpose, the assumed average empirical formula of lignin, which is found only in terrestrial biomass, is $C_{139}H_{148}O_{52}$. Lipids, proteins, and sugars, which occur in all living matter, are assigned average empirical formulas of $C_{18}H_{34}O_2$, $C_{106}H_{168}O_{34}N_{28}S_1$, and $C_6H_{10}O_5$, respectively.

A statistical distribution of the elemental composition of organic matter in fresh waters is given in **Table 5** for 57 samples of NOM, 117 samples of FA, and 107 samples of HA. Median elemental compositions in **Table 5** and average empirical formulae for major building blocks of biomass have been used to calculate average atomic ratios of H/C and O/C, and the results are plotted in a van Krevelen diagram in **Figure 1**. Lipids, lignins, and sugars define the *biomass triangle*, within which the chemical composition of any kind of biomass is expected to plot. The smaller triangle defines the compositional boundaries for aquatic biomass, which does not contain lignin. It is evident that the median atomic ratios of NOM, FA, and HA from fresh waters lie outside these triangles.

The average compositions of NOM, FA, and HA lie approximately parallel to the lignin–sugar mixing line, but their locations are shifted down and/or to the right (lower H/C and higher O/C). Any combination of chemical processes through which the H/C ratio of a sample decreases and its O/C ratio increases will shift average chemical compositions in the correct direction. A simple example of such a process is:

$$R-CH_2OH + O_2 \rightarrow R-COOH + H_2O \qquad [4]$$

The transformation of a primary alcohol to a carboxylic acid, which occurs, for example, in the oxidation of sugars to uronic acids, is the sort of chemical change that could account for observed differences between the composition of biomass and the composition of NOM, FA, and HA. It is important to note, however, that oxidation and dehydration do not have to occur simultaneously, or even in the same molecule, for the bulk chemistry of the mixture to be shifted in the correct direction. It is only possible to conclude that the bulk chemical compositions of FA, HA, and NOM are consistent with chemical changes due to oxidation and dehydration.

Redfield Biomass

The atomic ratios of C:N:P in aquatic biomass are approximately 106:16:1 (known as the Redfield ratios). When these ratios are combined with a highly simplistic chemical equation for photosynthesis ($CO_2 + H_2O = CH_2O(s) + O_2$), a more sophisticated chemical equation for photosynthesis in aquatic environments is obtained:

$$106\ CO_2 + 106\ H_2O + 16\ NH_3 + H_3PO_4 = (CH_2O)_{106}(NH_3)_{16}(H_3PO_4)(s) + 106\ O_2 \qquad [5]$$

The empirical formula for aquatic biomass in eqn. [5] is ubiquitous in the scientific literature, often transformed to the chemically equivalent formula $C_{106}H_{263}O_{110}N_{16}P(s)$. It is insightful to include the corresponding atomic H/C and O/C ratios of Redfield biomass in **Figure 1**. This composition plots impossibly far from the biomass triangle. The dashed line in **Figure 1** is the path along which that chemical composition evolves as H_2O is subtracted from the formula of Redfield biomass. In all likelihood the average chemical composition of aquatic biomass lies along the segment of the dashed line that lies between the lipid–sugar mixing line and the protein–sugar mixing line.

Table 5 Statistical summary of the elemental compositions (weight percent) of NOM, fulvic acids, and humic acids

Sample	*Obs.*	*Minimum*	*Maximum*	*Median*	*Mean*	*Std. Dev.*
Fulvic acid						
C	117	41.40	62.70	52.31	52.00	4.16
H	117	2.50	6.60	4.32	4.37	0.63
O	117	27.50	52.10	41.80	41.48	4.90
N	117	0.20	9.20	0.99	1.31	1.01
S	43	0.20	4.31	0.75	1.20	1.00
Humic acid						
C	107	38.69	62.70	53.50	53.49	3.90
H	107	2.60	6.90	4.30	4.34	0.77
O	107	23.50	47.15	39.20	38.64	4.83
N	107	0.60	9.80	1.90	2.41	1.68
S	36	0.28	3.20	0.93	1.23	0.79
Natural organic matter						
C	57	42.30	57.20	49.60	49.53	3.34
H	57	3.60	6.30	4.70	4.82	0.74
O	57	34.30	52.57	43.49	43.04	4.12
N	57	0.39	5.37	1.47	1.69	1.00
S	8	0.50	4.68	1.84	1.94	1.32

Source
Perdue EM and Ritchie JD (2003) Dissolved organic matter in freshwaters. In Drever JI, Holland HD, and Turekian KK (eds.) *Surface and Ground Water, Weathering, and Soils, Vol. 5, Treatise on Geochemistry*, pp. 273–318. Oxford: Elsevier–Pergamon.

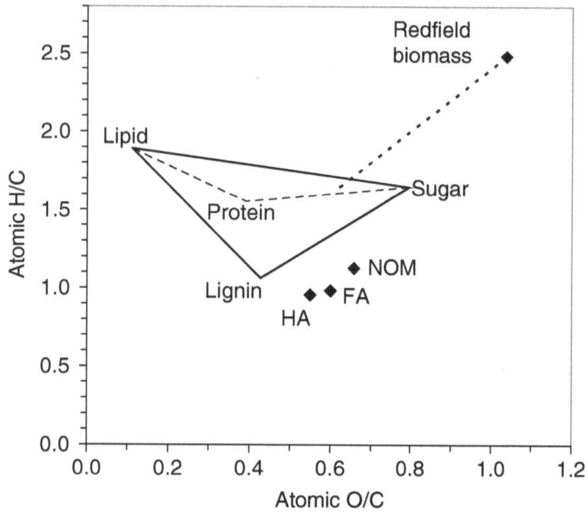

Figure 1 Atomic ratios of NOM, fulvic acid, humic acid, and building blocks of biomass.

The excessive amount of H_2O in the empirical formula of Redfield biomass is partially a consequence of the simplified representation of biomass as a mixture of CH_2O, NH_3, and H_3PO_4. The actual building blocks of biomass contain many hemiacetal, amide, and ester bonds, all of which are formed by condensation, i.e., the loss of H_2O. As an example, when n moles of glucose ($C_6H_{12}O_6$) are polymerized to starch or cellulose ($H(C_6H_{10}O_5)_nOH$), $n-1$ moles of H_2O are formed. This water of condensation is still included in the conventional empirical formula of Redfield biomass.

If H_2O is removed from Redfield biomass, its resulting composition will evolve along the dashed line in **Figure 1** until it falls within the biomass triangle. A limited range of more likely H/C and O/C can then be determined. Combining the equations for the loss of H_2O from Redfield biomass, the lipid–sugar mixing line, and the protein–sugar mixing line, the O/C ratio in Redfield biomass should be in the range of 0.613–0.657, corresponding to 65–70 O atoms in the empirical formula. Taking the lower limit, which lies on the protein–sugar mixing line, the photosynthetic production of algal biomass is much better described by:

$$106\ CO_2 + 61\ H_2O + 16\ NH_3 + H_3PO_4 = C_{106}H_{173}O_{65}N_{16}P(s) + 106\ O_2 \quad [6]$$

A second factor that could affect the validity of the Redfield formula is the assumption that the oxidation states of all elements other than carbon are not modified during photosynthesis and that the oxidation state of carbon is changed from +4 in CO_2 to 0 in Redfield biomass. The latter assumption, in particular, is not consistent with the chemical compositions of the major building blocks of biomass.

Oxidation State of Organic Carbon

The average oxidation state of carbon (Z_C) is readily calculated from elemental composition by the formula:

$$Z_C = 2(O/C) + 3(N/C) - (H/C) - 5(P/C) \quad [7]$$

where the contribution of sulfur is neglected. The dataset used previously to calculate the average elemental compositions which are given in **Table 5** have also been used to calculate Z_C for NOM, FA, and HA in fresh waters, and those values, together with the average Z_C for the major building blocks of biomass, are plotted in **Figure 2**. This plot shows clearly that the average oxidation state of carbon in FA, HA, and NOM is much more positive than in the major building blocks of biomass. **Figure 2** shows further that all building blocks of biomass have $Z_C \leq 0$, so any mixture of two or more of those building blocks would necessarily have a negative oxidation state (which contradicts the implicit assumptions in the Redfield formula and equation for photosysthesis).

It is clearly appropriate to view early diagenesis of biomass as an oxidative process. NOM, FA, and HA are utilized as both a source of carbon, nitrogen, and energy by microbial communities in inland waters. Chemical models of such respiration, whether aerobic or anaerobic, should perhaps be formulated in a manner that is consistent with the average oxidation state of the material being oxidized. Average Z_C values of +0.291, +0.275, and +0.247 for NOM, FA, and HA in inland waters can be calculated by eqn. [7] using the median elemental compositions in **Table 5**. Altogether, a weighted-average Z_C of +0.27 is characteristic of these materials.

Because organic matter in inland waters is more highly oxidized than Redfield biomass or any of the major components of biomass, correspondingly less O_2 will be required for aerobic respiration. As a simple example, consider the oxidation of a mixture whose average empirical formula is $C_CH_HO_ON_N$ and for which the average oxidation state of carbon is Z_C. The balanced chemical equation for oxidation of 1/C moles of this mixture can be expressed as:

$$\left(\frac{1}{C}\right)C_CH_HO_ON_N + \left(\frac{4-Z_C}{4}\right)O_2 \rightarrow CO_2 + \left(\frac{N}{C}\right)NH_3 + \frac{1}{2}\left(\left(\frac{H}{C}\right) - 3\left(\frac{N}{C}\right)\right)H_2O \qquad [8]$$

It is evident that a mixture for which Z_C is positive will require less than one mole of O_2 to oxidize a mole of carbon to CO_2.

The organic matter in inland waters has a positive oxidation state because it has already been *partially* oxidized by microbial processing. It may be anticipated that the further microbial oxidation of this organic matter can lead not only to CO_2 but also to more highly oxidized organic matter (an even more positive average Z_C). In such instances, the consumption of O_2 may be significantly greater than the CO_2 produced by the microbial processing.

Unsaturation

Another important descriptive characteristic of organic matter that can be calculated from its elemental composition is its unsaturation. The number of rings and/or pi-bonds per gram of molecules in a complex mixture can be calculated in the same manner that this calculation is performed for pure compounds. When the elemental composition is expressed in units of mmol g^{-1}, the total unsaturation (U_{total}) is:

$$U_{total} = C_{total} + \frac{N_{total}}{2} - \frac{H_{total}}{2} + \frac{1000}{M_n} \qquad [9]$$

where M_n is the number-average molecular weight of the complex mixture. When elemental data for NOM, FA, and HA, as well as average compositions of major components of biomass, are used in eqn. [9] (assuming $M_n = 1000$ Da if an average value is not available), the distribution of unsaturation shown in **Figure 3** is

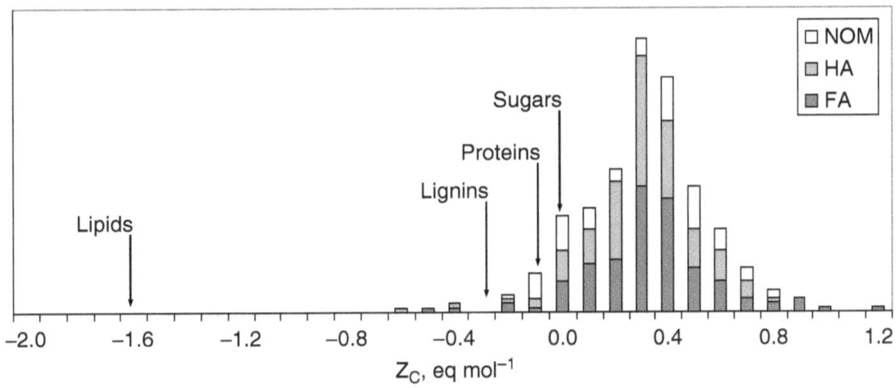

Figure 2 Average oxidation state (Z_C) in NOM, fulvic acids, and humic acids, compared with the average Z_C for major components of biomass.

obtained. This plot shows clearly that the average unsaturation in FA, HA, and NOM is much greater than in lipids and sugars, and in the range of the unsaturation of proteins and lignins.

The comparison of these fundamental properties of NOM, FA, and HA with those of major components of biomass leads to conflicting conclusions. How can organic matter in inland waters simultaneously be more highly oxidized than the most highly oxidized biopolymer (carbohydrates) and have as much unsaturation as lignin? The answer – carboxylic acids, and lots of them! The COOH group is highly oxidized and, at the same time, it has the highest U/C ratio of any structural moiety that is likely to be found in organic matter.

Acidic Functional Groups of NOM

As stated earlier, the acidic functional groups of NOM (mainly carboxyl and phenolic groups) are largely responsible for its positive oxidation state of carbon and high level of unsaturation. These functional groups not only determine the impact of NOM on the acid–base chemistry of inland waters but they also regulate its interactions with dissolved metal cations and its adsorption to some mineral surfaces. Methodologically, the concentrations of carboxyl and phenolic groups in NOM are measured using two basic approaches – indirect titrations and direct titrations.

Indirect titrations have the advantage of simplicity – titration of a filtered reaction mixture to a fixed pH end point following a 24 h equilibration with a weak or strong base. Total acidity is determined using the strong base $Ba(OH)_2$, and carboxyl content is determined using the weak base $Ca(CH_3CO_2)_2$. Phenolic contents are calculated in indirect methods as the difference between total acidity and carboxyl content. Since the early 1980s, direct titrations have been much more often used than indirect titrations to quantify the acidic functional groups of NOM, even though there is no standardized protocol. Experimental conditions for conducting direct titrations (i.e., titer, length of titration, ionic strength) are left to the discretion of the researcher.

Because pH is monitored continuously as increments of a strongly basic titrant (e.g., NaOH) are added, direct titrations provide much more detail about the acidic strengths of proton binding sites in NOM than do indirect methods. The lack of distinct inflection points in the titration curves of NOM is a result of the broad spectrum of acidic strengths of its functional groups. A problem in direct titrations is slow side reactions that generate additional acidity during the titration, causing noticeable downward drifts in pH under mildly alkaline conditions. These reactions are often invoked to explain the strong hysteretic effect that may be observed when titrated samples are subsequently re-acidified. Such reactions may also occur in the reaction mixtures used in indirect titrations, but the experimental protocol precludes their detection.

A statistical summary of more than 130 measurements of the acidic functional groups of NOM, FA, and HA in fresh waters is given in **Table 6**. Both indirect and direct titrations detect a higher average concentration of COOH groups in FA and NOM than in HA. Indirect and direct titrations yield essentially the same average concentration of COOH groups in HA. It is interesting to note, however, that indirect titrations give higher average COOH contents for both FA and NOM. In the $Ca(CH_3CO_2)_2$ reaction, COOH groups are quantified by titrating the acetic acid that is formed from the COOH groups of the sample, after the sample has been removed by filtration. When the method is applied to NOM and fulvic acids, a portion of the sample is not removed by

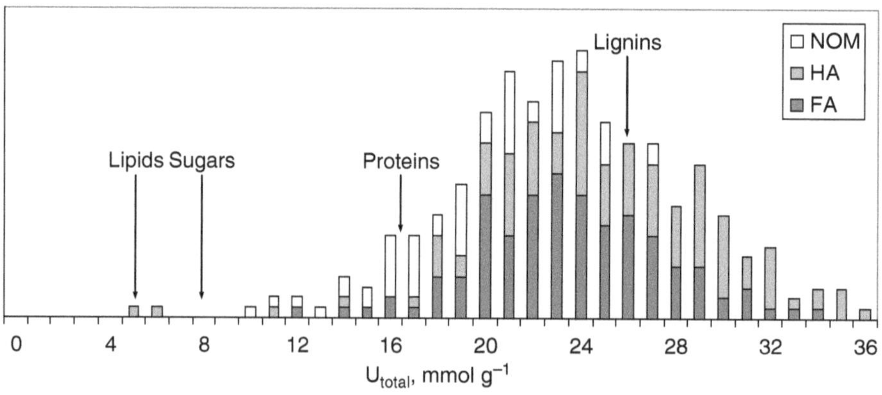

Figure 3 Average unsaturation in NOM, fulvic acids, and humic acids, compared with the average unsaturation in major components of biomass.

Table 6 Carboxyl and phenolic contents of freshwater fulvic acids (FA), humic acids (HA), and NOM[a]

Group	*Sample*	*Obs.*	*Mean*	*SD*
COOH (I)	FA	8	5.7	2.0
	HA	8	3.9	1.3
	NOM	11	8.7	1.4
ArOH (I)	FA	8	2.4	0.9
	HA	8	2.6	0.6
	NOM	11	3.6	2.0
COOH (D)	FA	73	5.1	1.2
	HA	22	4.0	1.0
	NOM	11	5.2	2.0
ArOH (D)	FA	65	1.7	0.8
	HA	22	1.8	0.5
	NOM	6	1.9	0.9

[a]COOH and ArOH are carboxyl and phenolic groups, respectively, and their concentrations (in meq g^{-1}) are measured by either indirect (I) or direct (D) methods. SD is standard deviation.

Source

Perdue EM and Ritchie JD (2003) Dissolved organic matter in freshwaters. In Drever JI, Holland HD, and Turekian KK (eds.) *Surface and Ground Water, Weathering, and Soils, Vol. 5, Treatise on Geochemistry*, pp. 273–318. Oxford: Elsevier–Pergamon.

the filtration step, and any acidic functional groups in that material which react with added NaOH up to the fixed endpoint of pH 9.8 are incorporated into the measurement and thus cause the COOH content to be overestimated.

The data in **Table 6** also show that indirect titrations yield higher estimates of phenolic content (ArOH) than direct titrations for FA, HA, and NOM. In the $Ba(OH)_2$ method that is used to measure total acidity, from which ArOH is calculated by subtraction of COOH, samples are reacted at pH $\approx$ 13 for 24 h before back-titration with strong acid. Interestingly, the method also requires complete removal of the sample from the reaction mixture prior to the titration to pH 8.4 with a strong acid such as HCl. Any residual organic mater in the filtered solution causes total acidity, and thus ArOH, to be underestimated. This is clearly not the trend in the average data in **Table 6**.

There are at least two possible explanations for the general trend. The first is simply pH. Direct titrations seldom achieve a pH that is greater than 11, so acids with $pK_a > 11$ are not detected at all. Those weaker acids are readily detected, however, in the $Ba(OH)_2$ method, for which the sample is equilibrated at pH $\approx$ 13. A second factor is the slow chemical reactions that generate additional acidity when NOM is directly titrated to alkaline pH. Such reactions should logically occur in the $Ba(OH)_2$ reaction mixture, perhaps to an even greater extent than is observed in direct titrations. All things considered, it appears that the indirect titrations lead to overestimation of ArOH content.

Using data from **Table 6** for direct titrations, the average ratio of COOH:ArOH in NOM and related materials is 2.7 ± 0.4. The WHAM series of models that are widely used to describe metal complexation by NOM generally fix this ratio at 2.0, thus significantly overestimating the role of ArOH in both the acid–base and metal complexation reactions of NOM.

From the mean elemental compositions of NOM, FA, and HA in **Table 5**, the overall weighted average concentrations of carbon and oxygen are calculated to be 43.35 and 25.44 mmol g^{-1}, respectively. The weighted average COOH and ArOH contents (from direct titrations) in **Table 6** are 4.89 and 1.75 meq g^{-1}, respectively. Together these results indicate that, on average, COOH groups account for 11.3% of carbon and 38.4% of oxygen in organic matter in inland waters. ArOH groups account for an additional 6.9% of oxygen. Without knowing the average number of ArOH groups per benzene ring, the percentage of carbon in ArOH groups is less certain. It is nonetheless quite clear from this simple analysis that more than 50% of the oxygen in NOM is in non-acidic functional groups, among which esters, alcohols, and ethers are likely contributors. The distribution of non-acidic oxygen-containing functional groups in NOM is best revealed by ^{13}C NMR spectra – the topic to be discussed next.

^{13}C NMR Spectroscopy of NOM

^{13}C NMR spectrometry is one of the most powerful tools for studying functional groups and structural subunits in NOM. The method is applicable to aqueous and non-aqueous solutions of NOM, and to solid samples. The chemical shift range of ^{13}C in organic compounds lies in the range of approximately 0–220 ppm relative to $Si(CH_3)_4$. The proportions of the major structural subunits of NOM are estimated by integrating spectra between user-specified limits of chemical shift. The number of structural subunits and the chemical shift range of each structural subunit vary widely from one study to another, depending on the interests of the author and the appearance of the spectra.

A statistical summary of ^{13}C NMR spectra of freshwater FA, HA, and NOM is given in **Table 7**. Original integrated peak areas have been re-interpreted, if necessary, to consistently subdivide carbon into the following five classes:

- carbonyl (aldehydes and ketones) – 190 to 220 ppm
- carboxyl (carboxylic acids, esters, and amides) – 160 to 190 ppm

Table 7 Overall percent distribution of carbon in freshwater fulvic acids, humic acids, and NOM

Carbon type	*Obs.*	*Average*	*Std. Dev.*
Fulvic acids			
Alkyl	31	30.0	10.0
Alkoxy	31	22.7	5.4
Aromatic	31	25.8	7.4
Carboxyl	37	17.3	2.4
Carbonyl	31	4.2	1.8
Humic acids			
Alkyl	32	26.1	9.5
Alkoxy	32	25.4	7.2
Aromatic	32	30.3	9.8
Carboxyl	38	14.2	2.7
Carbonyl	33	4.0	1.9
NOM			
Alkyl	18	24.1	9.0
Alkoxy	18	24.9	4.6
Aromatic	18	26.8	10.8
Carboxyl	19	19.0	2.0
Carbonyl	18	5.2	2.3

Source
Perdue EM and Ritchie JD (2003) Dissolved organic matter in freshwaters. In Drever JI, Holland HD, and Turekian KK (eds.) *Surface and Ground Water, Weathering, and Soils, Vol. 5, Treatise on Geochemistry*, pp. 273–318. Oxford: Elsevier–Pergamon.

Table 8 Integrated chemical description of natural organic matter in inland waters

Property	*Units*	*Average*
M_n (all methods)	g mol^{-1}	1099
M_w (all methods)	g mol^{-1}	2015
C	mmol g^{-1}	43.4
H	mmol g^{-1}	44.2
O	mmol g^{-1}	25.4
N	mmol g^{-1}	1.3
S	mmol g^{-1}	0.4
COOH (direct)	mmol g^{-1}	4.89
ArOH (direct)	mmol g^{-1}	1.75
C-alkyl (NMR)	mmol g^{-1}	11.8
C-alkoxy (NMR)	mmol g^{-1}	10.5
C-aromatic (NMR)	mmol g^{-1}	12.1
C-carboxyl (NMR)	mmol g^{-1}	7.1
C-carbonyl (NMR)	mmol g^{-1}	1.9
Z_C	eq mol^{-1}	0.26
U_{total}	mmol g^{-1}	22.9

Source
Perdue EM and Ritchie JD (2003) Dissolved organic matter in freshwaters. In Drever JI, Holland HD, and Turekian KK (eds.) *Surface and Ground Water, Weathering, and Soils, Vol. 5, Treatise on Geochemistry*, pp. 273–318. Oxford: Elsevier–Pergamon.

- aromatic (unsubstituted and substituted) – 110 to 160 ppm
- alkoxy (alcohols, hemiacetals, ethers, etc.) – 60 to 110 ppm
- alkyl (methyl, methylene, and methine) – 0 to 60 ppm

Because some structural moieties were reported in the literature more often than others, the direct calculated average distributions of carbon have been normalized to 100% to obtain the results in **Table 7**.

Alkyl, alkoxy, and aromatic moieties each account for around 24–30% of carbon, carboxyl-containing moieties account for 14–19% of carbon, and carbonyl groups account for the remaining 4–5% of carbon. The differences between FA, HA, and NOM are less evident than the similarities. In fact, no statistically significant differences are observed.

The results in **Table 7** include those obtained on solid samples and those obtained on dissolved samples. Some studies were conducted under more rigorous, quantitative conditions than others. Fortunately, information obtained in ^{13}C NMR spectra can be used to predict other compositional characteristics of HA, FA, and NOM, and this type of analysis is presented in the next section.

Integrated Description of NOM

Natural organic matter in inland waters is probably best represented chemically by well-characterized NOM samples. Around 80% of available data, however, are for aquatic FA and HA, which are major fractions of NOM (but not identical to NOM). So this wealth of information can be incorporated into an integrated statistical description of NOM; all datasets used earlier in this paper for NOM, FA, and HA from fresh waters have been merged here to obtain weighted-average chemical properties. In all cases, the weighting factor for each chemical descriptor is the number of observations (see **Tables 5–7**). The results are given in **Table 8**.

The tabulated results in **Table 8** include chemical measurements and spectroscopic measurements. It is appropriate to seek internal checks on the consistency of these two methods of characterization of NOM. It is immediately evident that the carboxylic acid content (COOH) from direct titrations is smaller than the 'C-carboxyl' content from ^{13}C NMR spectroscopy. This discrepancy is anticipated, however, because the NMR estimate of C-carboxyl includes carboxylic acids, esters, and amides. Making the reasonable assumption that virtually all N occurs as amides, which is strongly supported by many ^{15}N NMR measurements, the weighted-average NOM contains approximately 4.9, 0.9, and 1.3 mmol g^{-1} each of carboxylic acids, esters, and amides. When the phenolic (ArOH) content is combined with C-aromatic, assuming all aromatic carbon is in benzene rings, then the average aromatic ring contains 0.87 ArOH groups, which is also a realistic value.

A more stringent test can be conducted by using the ^{13}C NMR estimate of the distribution of carbon to calculate the corresponding distribution of oxygen in weighted-average NOM. The atomic O/C ratio of each carbon class in **Table 8** is rather tightly constrained. Each carbonyl carbon is bonded to one oxygen, thus accounting for 1.9 mmol g^{-1} (7.5%) of oxygen. Each carboxyl carbon is bonded to either two oxygens (carboxylic acid, ester) or one (amide) oxygen, thus accounting for 12.9 mmol g^{-1} (50.8%) of oxygen. An alkoxy carbon is bonded to one oxygen in esters, ethers, and alcohols; however, sugars contain slightly less than one oxygen per carbon (e.g., $H(C_6H_{10}O_5)_nOH$). The upper limit for alkoxy oxygen is 10.5 mmol g^{-1} (41.3%); however, if all alkoxy carbon is in sugars, this value could be as low as 8.75 mmol g^{-1} (34.4%). Depending on these constraints, the amount of oxygen that is attached to aromatic rings must fall in the range of 0.1–1.85 mmol g^{-1} (0.4–7.3%). Note that the estimated upper limit is very similar to the ArOH content, as determined by direct titrations. On the basis of this analysis, the chemical and spectroscopic data seem to be in good agreement.

A final direct test of internal consistency between direct chemical and spectroscopic data is provided by unsaturation. The total unsaturation (U_{total} = 22.9 mmol g^{-1}) must be distributed among various structural components of NOM. Both carbonyl and carboxyl carbons, for example, have one pi-bond per carbon, so they account for 1.9 mmol g^{-1} (8.3%) and 7.1 mmol g^{-1} (31.0%) of unsaturation, respectively. Aromatic carbon is assumed to occur primarily in benzene rings, which have 4 moles of unsaturation (one ring and three pi-bonds) per six carbons. Accordingly, an additional 8.1 mmol g^{-1} (35.4%) of unsaturation can be attributed to aromatic carbon. At this point, the remaining unsaturation (5.8 mmol g^{-1}, 25.3%) simply cannot be accommodated by the remaining alkoxy and alkyl carbon, both of which are sp^3-hybridized and can only be unsaturated if they form rings. Approximately one ring per four carbon atoms would be required, and this is unreasonably high. The observed amount of unsaturation (from chemical data) cannot be reconciled with NMR data in **Table 8**.

Solid-state ^{13}C NMR spectra are often non-quantitative, especially spectra obtained using cross polarization with magic angle spinning (CPMAS). The most commonly cited factor is the proximity of ^{13}C to H, which is lowest for unsaturated forms of carbon. A common outcome is the overestimation of sp^3 carbon and underestimation of sp^2 carbon. The chemical and spectroscopic estimates of unsaturation for weighted-average NOM can be reconciled if the NMR estimates of unsaturated forms of carbon are increased slightly and the estimates of saturated carbon are decreased proportionately. Because oxygen is very abundant in both unsaturated and saturated forms of carbon, such an adjustment can be made without compromising the excellent balance between chemical and spectroscopic estimates of oxygen in weighted-average NOM.

The weighted-average chemical properties of NOM in inland waters are those of a material that is already headed down the path to mineralization. Its relatively high oxidation state of carbon and its high unsaturation are both indicative of substantial diagenetic alteration relative to common major constituents of biomass. The build-up of COOH groups in NOM accounts for both of these trends, and endows NOM with its well-known ability to impart acidity to natural waters and to complex metal cations. The increased unsaturation leads to a statistically increased probability of conjugated, unsaturated structural moieties that strongly attenuate UV and visible light in surface waters. Finally, the increased oxidation state of carbon in NOM diminishes its energetic value as a food source to heterotrophic organisms and thus renders it less bioavailable than biomass.

See also: Carbon, Unifying Currency; Interactions of Dissolved Organic Matter and Humic Substances; Light, Photolytic Reactivity and Chemical Products; Lipids; Nutrient Stoichiometry in Aquatic Ecosystems; Organic Nitrogen; Redox Potential.

Further Reading

Aiken GR, McKnight DM, Wershaw RL, and MacCarthy P (eds.) (1985) *Humic Substances in Soil, Sediment, and Water.* New York: Wiley-Interscience.

Averett RC, Leenheer JA, McKnight DM, and Thorn KA (eds.) (1989) *Humic substances in the Suwannee River, Georgia: Interactions, properties, and proposed structures.* U.S. Geological Survey Open-File Report, Denver, CO.

Christman RF and Gjessing ET (eds.) (1983) *Aquatic and Terrestrial Humic Materials.* Ann Arbor: Ann Arbor Science Publisher.

Findlay SEC and Sinsabaugh RL (eds.) (2003) *Aquatic Ecosystems – Interactivity of Dissolved Organic Matter.* San Diego, CA: Academic Press.

Frimmel FH, Abbt-Braun G, Heumann KG, Hock B, Lüdemann HD, and Spiteller M (eds.) (2002) *Refractory Organic Substances in the Environment.* Weinheim: Wiley-VCH.

Frimmel FH and Christman RF (eds.) (1988) *Humic Substances and their Role in the Environment.* Chichester, UK: Wiley-Interscience.

Hessen DO and Tranvik LJ (eds.) (1998) *Aquatic Humic Substances – Ecology and Biogeochemistry.* Berlin: Springer.

Perdue EM and Gjessing ET (eds.) (1990) *Organic Acids in Aquatic Ecosystems.* Chichester, UK: Wiley-Interscience.

Perdue EM and Ritchie JD (2003) Dissolved organic matter in freshwaters. In: Drever JI, Holland HD, and Turekian KK (eds.) *Surface and Ground Water, Weathering, and Soils, Vol. 5, Treatise on Geochemistry*, pp. 273–318. Oxford: Elsevier–Pergamon.

Steinberg CEW (2003) *Ecology of Humic Substances in Freshwaters*. Heidelberg: Springer.

Thurman EM (1985) *Organic Geochemistry of Natural Waters*. Dordrecht: Martinus Nijhoff/Dr. W. Junk Publishers.

Wetzel RG (1975) *Limnology*. Philadelphia: Saunders.

Wilson MA (1987) *NMR Techniques and Applications in Geochemistry and Soil Chemistry*. Oxford: Pergamon Press.

Zepp RG and Sonntag Ch (eds.) (1995) *The Role of Nonliving Organic Matter in the Earth's Carbon Cycle*. Chichester: Wiley.

Organic Nitrogen

N O G Jørgensen, University of Copenhagen, Fredericksberg, Denmark

Introduction

Nitrogen is a major element in all organisms. Nitrogen is found in proteins, being structural tissues or enzymes, in nucleic acids (RNA and DNA), and in cell wall components. The content of nitrogen in most organisms is about 15% of the dry weight. This prevalence of nitrogen in living material means that large amounts of organically-bound nitrogen are returned to the environment, when the organisms die. The organic nitrogen compounds are degraded by microorganisms to simple nitrogen compounds. Because most of these nitrogen compounds are inorganic, the process has been named mineralization. The nitrogen compounds formed by the mineralization can be utilized for production of new cell material by many organisms. In this way nitrogen is recycled within the environment. In lakes and rivers, input of new nitrogen may come from fixation of atmospheric N_2, by leaching from farmland and woods, from atmospheric precipitation and deposition, and from discharge of sewage water. In this article, composition, sources and cycling of dissolved organic nitrogen will be presented and discussed.

Dissolved Nitrogen in Aquatic Environments

Definition of Nitrogen Pools

Organic nitrogen in aquatic environments consists of truly dissolved organic nitrogen (DON) and particulate organic nitrogen (PON). DON is defined as material that can pass a 0.2-μm filter, while PON is the material that is retained on the filter. This means that PON includes both dead organic matter and living organisms that are larger than 0.2 μm. The definition of DON as material being smaller than 0.2 μm is a practical, operational term rather than a biologically or chemically well-defined term. DON includes a variety of organic molecules and compounds, ranging from small molecules like urea and amino acids, to peptides and proteins, but does also include viruses and small bacteria.

Another important nitrogen pool in all aquatic environments is dissolved inorganic nitrogen (DIN), which consists of ammonium, nitrate, and nitrite. Thus, the content of total dissolved nitrogen (TDN) is made up by DIN and DON.

Concentrations of DON in rivers and lakes have been found to vary from 3 μM N in low-productive, oligotrophic waters to about 100 μM N in eutrophic rivers and lakes. Higher concentrations (>300 μM N) have been measured in waters receiving a high input of nitrogen such as tropical fish ponds. A large portion of DON in lakes and rivers typically originates from release by algae and phytoplankton. Since the biomass of both algae and phytoplankton varies with season and depth, DON concentrations will often show seasonal variations and will also vary with depth in lakes. Further, DON can be transported by rivers and streams into lakes and introduce large local variations in concentrations.

DON vs. DIN

Which nitrogen pool is largest: DON or DIN? At winter seasons in temperate and boreal zones, when the primary production is low and there can be a high input of DIN from drainage areas, the amount of DIN will typically be higher than the amount of DON. In contrast, there will often be more DON than DIN during summer periods when there is a high phytoplankton production, because DIN is being used for new primary production. This decline in DIN and a frequent release of DON by algae and macrophytes during photosynthesis leads to a relatively higher DON than DIN pool. In addition, dead phytoplankton and zooplankton organisms also contribute DON to the water as discussed later.

An example of annual changes in DON and DIN is illustrated for the large, mesotrophic Lake Kinneret in Israel (**Figure 1**). Before onset of the algal growth, rather similar amounts of DIN (ammonium, nitrite, and nitrate) and DON occur in the lake, but late in summer, most DIN has been consumed by the phytoplankton, and release of organic nitrogen has increased the DON pool. The amount of nitrogen bound in living and dead organic matter (PON) in the water was similar in March and September, but the composition of living and dead matter in PON may be different, e.g., due to zooplankton grazing on phytoplankton.

An interesting relation between DIN, DON, and PON can be found in some stratified lakes, where DON has an important function in transporting nitrogen from the surface to deeper layers. In the large, meromictic Lake Biwa in Japan, summer stratification leads to accumulation of PON and DON

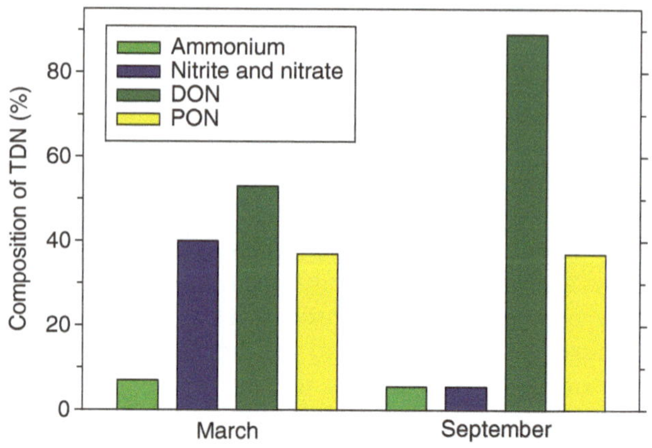

Figure 1 Composition of inorganic nitrogen (ammonium, nitrite and nitrate), dissolved organic nitrogen (DON) and particulate organic nitrogen (PON) in the water column of Lake Kinneret in Israel in March and September. Average values from 1975 to 1994 are shown. Reproduced from **Figure 2** in Berman T and Bronk DA (2003) Dissolved organic nitrogen: A dynamic participant in aquatic ecosystems. *Aquatic Microbial Ecology* 31: 279–305, with permission from Inter-Research.

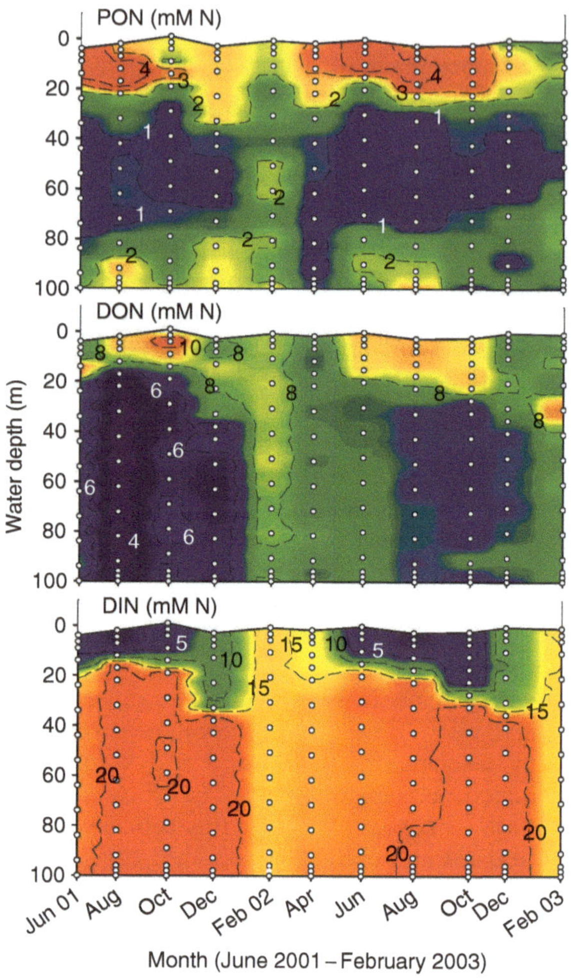

Figure 2 Depth profiles of PON, DON and DIN in Lake Biwa in Japan from June 2001 to February 2003. See text for details. Reproduced from **Figure 2** in Kim C, Nishimura Y, and Nagata T (2006) Role of dissolved organic matter in hypolimnetic mineralization of carbon and nitrogen in a large, monomictic lake. *Limnology & Oceanography* 51: 70–78, with permission from the American Society of Limnology and Oceanography.

and, as expected, depletion of DIN in the epilimnion (**Figure 2**). In the winter, convective mixing transports DON to the bottom layers where it is degraded by bacteria to DIN. In Lake Biwa, mineralization of DON increased by about 30% in the subsurface layers in the autumn. The produced DIN increased the ambient DIN pool and became evenly distributed in the entire water column. During the next summer stratification, the events are repeated. In this way, DON can have an important function as a 'nitrogen pump' in large, meromictic lakes. A similar winter deep-water mineralization was also found for the pool of dissolved organic carbon (DOC) in the lake.

Natural DON Compounds

Composition and Detection

Since most DON compounds originate from release by living organisms or from degradation of dead organisms, input of DON to lakes and rivers does to a large extent reflect the chemical composition of biological matter. Thus, common DON compounds include free and protein-bound amino acids, amino sugars (from cell walls), and nucleotides (from RNA and DNA), but metabolites like urea and methylamines are also among frequently found DON compounds in fresh water. Other DON compounds may possibly be important in lakes and rivers, but have escaped detection because of low concentration. Many DON compounds are quickly utilized by microorganisms and this lowers their natural concentration to levels that are difficult to quantify by current analytical procedures. As an illustration, sensitive fluorescent techniques for analysis of amino acids have been available from around 1980 and have led to numerous data on amino acid concentrations in natural waters. In contrast, natural concentrations of, e.g., purine and pyrimidine monomers of nucleotides such as adenine and uracil are largely unknown due to lack of sensitive and reliable techniques for analysis of these compounds.

Specific DON Compounds

A list of the most commonly found DON compounds in rivers and lakes are given in **Table 1**. All the shown DON compounds are actively involved in biological

Table 1 Dissolved organic nitrogen compounds in lakes and rivers

Compound	*Location*	*Concentration (μM N)*	*Source*
Amino acids			
Dissolved free amino acids (DFAA). Building blocks of proteins and metabolic products. Released to the water by living and dead organisms, e.g., during enzymatic degradation of proteins. Osmotic active compounds in some animals.	Oligotrophic rivers and lakes	0.01–0.2	5, 13, 21
	Meso- and eutrophic rivers and lakes	0.1–1.8	6, 12, 17, 18
	Mono Lake (soda lake); USA	0.250–3.0	8
	Organic Lake, Antarctica.	1–290	3
Dissolved combined amino acids (DCAA). Amino acids bound in proteins (muscles, enzymes, cell membranes and structural tissues), cell walls or adsorbed to humic material. Released to the water by living and dead organisms.	Oligotrophic rivers and lakes	0.3–2.5	20, 21
	Meso- and eutrophic rivers and lakes	1.0–5.0	11, 12, 18
	Humic lakes	5–25	14, 22
	Mono Lake (soda lake); USA	5–15	8
Nucleotides			
DNA (deoxyribonucleic acids). DNA occurs in chromosomes and plasmids. Released by living and dead organisms, and during exchange of genetic information among bacteria.	Lakes; Poland	0.006–0.75	19
	Lakes and rivers; USA mainland & Hawaii	0.018–0.95	10
RNA (ribonucleic acids). RNA occurs in ribosomal, messenger or transfer RNA. Released to the water by living and dead organisms.	Ponds; Hawaii, USA	0.25–9.30	10
Purines			
Adenine (free compound). Base building block in RNA and DNA	Mesotrophic lake; Denmark	0.06	6
Glucosamine			
Free glucosamine. Release product from degradation of bacteria, fungi, insects and crustaceans	Mesotrophic lake; Germany	0.005–0.03	1
Combined glucosamine. Building block in the polysaccharide chitin and in cells walls of bacteria and fungi. Chitin also makes up the outer skeletons of insects and crustaceans.	Mono Lake (soda lake); USA	0.005–0.05	8
	Lakes and ponds	0.04–2.1	15
	Mono Lake (soda lake); USA	2–4.5	8

Continued

Table 1 Continued

Compound	*Location*	*Concentration (μM N)*	*Source*
Urea			
Metabolic product in degradation of organic nitrogen, but can also be a degradation product of purines	Streams and lakes	0.2–5	4, 21
	Hypertrophic pond; Korea	2.6–18	16
Methylamines			
Degradation product by bacteria in anoxic environments	Flax Pond; USA	0.05–0.37	9
	East Lake; China	2.0	2

Sources

1. Bunte C and Simon M (1999) Bacterioplankton turnover of dissolved free monosaccharides in a mesotrophic lake. *Limnology & Oceanography* 44: 1862–1870.
2. Cai L, Zhao Y, Gong S, Dong L, and Wu C (2003) Use of a novel sol–gel dibenzo-18-crown-6 solid-phase microextraction fiber and a new derivatizing reagent for determination of aliphatic amines in lake water and human urine. *Chromatographia* 58: 615–621.
3. Gibson JAE, Qiang XL, Franzmann PD, Garrick RC, and Burton HR (1994) Volatile fatty and dissolved free amino adds in organic lake, Vestfold Hills, East Antarctica. *Polar Biology* 14: 545–550.
4. Glibert PM, Trice TM, Michael B, and Lane L (2005) Urea in the tributaries of the Chesapeake and coastal bays of Maryland. *Water, Air, & Soil Pollution* 160: 229–243.
5. Jonsson A, Ström L, and Åberg J (2007) Composition and variations in the occurrence of dissolved free simple organic compounds of an unproductive lake ecosystem in northern Sweden. *Biogeochemistry* 82: 153–163.
6. Jørgensen NOG, Engel P, Jellison R, and Hollibaugh JT (2008) Contribution of bacterial cell wall components to DOM in alkaline, hypersaline Mono Lake, California. *Geomicrobiology Journal* 25: 38–55.
7. Jørgensen NOG (1987) Free amino acids in lakes. Concentrations and assimilation rates in relation to phytoplankton and bacterial production. *Limnology & Oceanography* 32: 97–111.
8. Jørgensen NOG, Engel P, Jellison R, and Hollibaugh JT (2007) Contribution of bacterial cell wall components to DOM in alkaline hypersaline Mono Lake, California. *Geomicrobiology Journal*. Submitted.
9. Jørgensen NOG, Kroer N, Coffin RB, Yang XH, and Lee C (1993) Dissolved free amino acids, combined amino acids, and DNA as sources of carbon and nitrogen to marine bacteria. *Marine Ecology Progress Series* 98: 135–148.
10. Karl DM and Bailiff MD (1989) The measurement and distribution of dissolved nucleic acids in aquatic environments. *Limnology & Oceanography* 34: 543–558.
11. Keil RG and Kirchman DL (1991) Dissolved combined amino acids in marine waters as determined by a vapor-phase hydrolysis method. *Marine Chemistry* 33: 243–259.
12. Kroer N, Jørgensen NOG, and Coffin RB (1994) Utilization of dissolved nitrogen by heterotrophic bacterioplankton: A comparison of three ecosystems. *Applied Environmental Microbiology* 60: 4116–4123.
13. Langenheder S, Kisand V, Lindström ES, Wikner J, and Tranvik L (2004) Growth dynamics within bacterial communities in riverine and estuarine batch cultures. *Aquatic Microbial Ecology* 37: 137–148.
14. Münster U, Einiö P, Nurminen J, and Overbeck J (1992) Extracellular enzymes in a polyhumic lake: important regulators in detritus processing. *Hydrobiologia* 229: 225–238.
15. Nedoma J, Vrba J, Hejzlar J, Simek K, and Straskrabová V (1994) N-acetylglucosamine dynamics in freshwater environments: Concentration of amino sugars, extracellular enzyme activities, and microbial uptake. *Limnology & Oceanography* 39: 1088–1100.
16. Park MG, Shim JH, and Cho BC (1997) Urea decomposition activities in an ammonium enriched freshwater pond. *Aquatic Microbial Ecology* 13: 303–311.
17. Robarts RD, Wicks RJ, and Gehr R (1990) Seasonal changes in the dissolved free amino acid and DOC concentrations in a hypertrophic African reservoir and its inflowing rivers. *Hydrobiologia* 199: 201–216.
18. Rosenstock B and Simon M (1993) Use of dissolved combined and free amino acids by planktonic bacteria in Lake Constance. *Limnology & Oceanography* 38: 1521–1531.
19. Siuda W, Chróst RJ, and Gude H (1998) Distribution and origin of dissolved DNA in lakes of different trophic states. *Aquatic Microbial Ecology* 15: 89–96.
20. Stepanaukas R, Jørgensen NOG, Eigaard OR, Zvikas A, Tranvik LJ, and Leonardson L (2002) Summer inputs of riverine nutrients to the Baltic Sea: Bioavailability and eutrophication relevance. *Ecological Monographs* 72: 579–597.
21. Stepanaukas R, Laudon H, and Jørgensen NOG (2000) High DON bioavailability in boreal streams during a spring flood. *Limnology & Oceanography* 45: 1298–1307.
22. Tranvik LJ and Jørgensen NOG (1995) Colloidal and dissolved organic matter in lake water: Carbohydrate and amino acid composition, and ability to support bacterial growth. *Biogeochemistry* 30: 77–97.

processes and many of them are turned over at fast rates as discussed later. The molecular structure of selected DON compounds are shown in **Figure 3**.

Dissolved Free and Combined Amino Acids

Amino acids are the most studied DON compounds in fresh water. The amino acid pool consists of truly dissolved free amino acids (DFAA), and amino acids bound in peptides, proteins, or adsorbed to inorganic and organic matter, named dissolved combined amino acids (DCAA). Concentrations of DFAA and DCAA in lakes and streams reflect a balance between input and uptake. Concentrations of DCAA are usually higher than concentrations of DFAA, as DCAA must first be degraded by enzymes to free amino acids (DFAA) before being used by microorganisms. This means that microbial production of enzymes often will limit the utilization of DCAA.

Concentrations of amino acids in lakes and rivers will typically reflect the trophic status of the environment. The lowest concentrations (0.01–0.2 μM N for DFAA and 0.3–2.5 μM N for DCAA) occur in oligotrophic waters, while higher concentrations (up to 1.8 μM DFAA-N and 5.0 μM DCAA-N) usually are found in eutrophic lakes and rivers. Highest concentrations within the shown ranges are found during seasons with a high biological production, while significantly lower amounts of both DFAA and DCAA occur during winter.

The highest concentrations of DFAA have been measured in extreme lakes such as the hypersaline Antarctic Organic Lake (up to 290 μM DFAA-N in the anoxic hypolimnion) and the alkaline soda lake Mono Lake (up to 3 μM DFAA-N). Organic Lake receives large amounts of nitrogen and phosphorus from penguins and these nutrients sustain dense populations of the algae *Dunaliella* in the lake. Mono Lake is a hypersaline, alkaline, and phosphorus-rich lake in the Sierra Nevada Mountains, California, with specialized algal populations.

Like DFAA, some of the highest concentrations of DCAA are found in extreme lakes (up to 13 μM N in Mono Lake; no data on DCAA in Organic Lake are available), but in humic lakes even higher amounts of DCAA have been measured. In humic, Finnish lakes, the content of DCAA can vary from 5 to 25 μM N. A portion of the humic-bound amino acids is available to microorganisms, but most of the amino acids in humic matter are resistant to bacterial degradation.

Amino Acids in a Stratified Lake

Variations in the biological activity can easily modify natural pools of DCAA and DFAA, e.g., if the release by phytoplankton or the uptake by bacteria change. Similarly, in deep lakes variations in biological processes with depth will also lead to changes of the amino acid pools. As an example, concentrations of DFAA and DCAA in a depth profile of the stratified Mono Lake in mid May in 2005 are shown in **Figure 4(a)**. The main producers of amino acids in the lake are planktonic algae that occur at high densities to at least 30 m depth. The pool of DFAA decreased from 3.5 μM N at the surface to about 1 μM N between 5 and 35 m depth. The higher concentration in the 0–5 m layer may have been caused by a reduced bacterial uptake of DFAA (the bacterial density increased below 5 m), but a higher input (release) of DFAA by the phytoplankton may also have occurred. Concentrations of DCAA were significantly higher than DFAA, ranging from 10 to 15 μM N, except for a decrease to about 5 μM N in the oxic–anoxic thermocline. This decline probably reflects a higher microbial utilization of DCAA in the thermocline, since this layer allows bacteria with different metabolic and

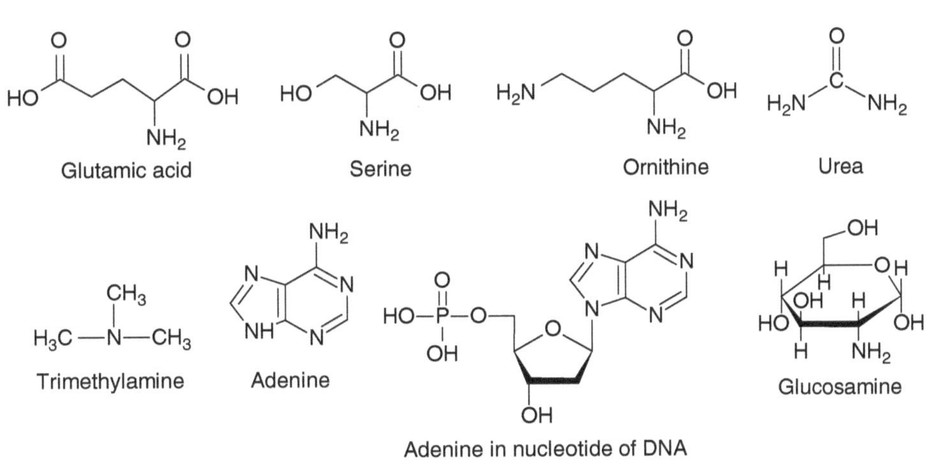

Figure 3 Molecular structures of important DON compounds.

oxygen preferences to have optimum growth here. The high DCAA concentrations in the lake represent a large pool of nutrients, corresponding to about 150 and 400 μg N and C l^{-1}, respectively, and may reflect both a high release as well as a low uptake, especially in the anoxic hypolimnion.

Composition of DFAA

Commonly found DFAA in lakes and rivers include the protein amino acids serine (Ser), glycine (Gly), alanine (Ala), aspartic acid (Asp), and glutamic acid (Glu). Other commonly found DFAA are nonprotein amino acids such as ornithine (Orn; intermediate in nitrogen metabolism in cells), taurine (Tau; osmotic active amino acid in many animals), and diaminopimelic acid (DAPA; component of bacterial cell walls). The dominance of the individual amino acids may vary and depend on the origin of the DFAA, e.g., a specific phytoplankton population, but the capacity for uptake of individual amino acids by bacteria is also important. As examples, the frequent abundance of Ser among DFAA in natural waters can be caused by the lack of uptake mechanisms for this amino acid as observed by some bacteria, while low concentrations of Glu may reflect a fast uptake, since this amino acid is an intermediate in the synthesis of other amino acids in bacteria.

As an example of fluctuations in DFAA composition, the depth profile of major DFAA in Mono Lake is shown in **Figure 4(b)**. The change in composition from 1 m depth (dominance by Asp and Ser) to the thermocline stratum and the anoxic bottom water (dominance by Orn, Ala, Gly, Ser, and Glu) clearly indicate that different biological processes occurred in the surface and in deeper water. The specific processes that caused changes of the DFAA spectra are not easily determined, unless the amino acid has a specific source. As an example, ornithine (Orn) is a typical free amino acid in many crustaceans such as copepods, and the abundance of this amino acid in deeper waters in Mono Lake may indicate that the abundant brine shrimp, *Artemia monica*, was the main source of this amino acid.

Composition of DCAA

The composition of DCAA is typically less variable than that of DFAA. In studies of various types of lakes and rivers, the DCAA composition was typically dominated by Gly, Ala, Glu, Val (valine), and Leu (leucine), but nonprotein amino acids such as β-Ala and derivatives of aminobutyric acid have also been identified among DCAA. In Mono Lake, the DCAA were dominated by Ala, Gly, Glu, and Asp in the entire 35 m water column.

The reduced variation in composition of DCAA appears to mirror the amino acid composition of particulate organic (dead and living) matter, at least in some environments. Supporting this, the DCAA composition in different European lakes as well as in Mono Lake had an amino acid composition almost similar to that found in particulate organic matter from

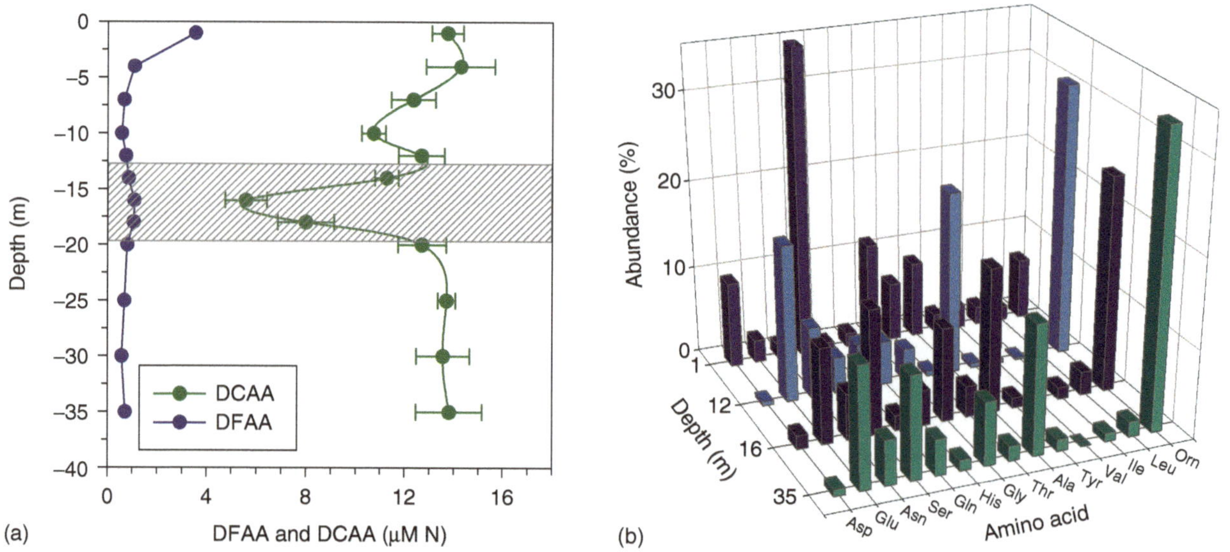

Figure 4 Concentrations of DFAA and DCAA (a) and composition of DFAA (b) in depth profiles of the stratified, alkaline and saline Mono Lake, California, USA, in May 2005. In (a), the dashed area indicates the stratified zone from 50% O_2 saturation to complete anoxic condition. Reproduced from Figures 3 and 4 in Jørgensen NOG, Engel P, Jellison R, and Hollibaugh JT (2008) Contribution of bacterial cell wall components to DOM in alkaline, hypersaline Mono Lake, California. *Geomicrobiology Journal* 25: 38–55, with permission from Taylor & Francis. In (b), the shown depths represent the upper oxic zone (1 m), the middle and lower stratified zone (12 and 16 m), and the anoxic bottom layer (35 m). Reproduced from **Figure 5** in Jørgensen NOG, Engel P, Jellison R, and Hollibaugh JT (2008) Contribution of bacterial cell wall components to DOM in alkaline, hypersaline Mono Lake, California. *Geomicrobiology Journal* 25: 38–55, with permission from Taylor & Francis.

geographically different locations, such as North America, Europe, and Japan. This similarity between particulate organic matter and DCAA strongly suggests that DCAA are released from particulate matter and that the composition has not been significantly changed by diagenetic transformations. This is in contrast to the more variable composition of DFAA.

Nonprotein Amino Acids

Amino acids form optical isomers and exist as L or D isomers or enantiomers. Proteins and enzymes consist of L isomers, but bacteria also produce D amino acid isomers that are vital components in peptides of bacterial peptidoglycan. Peptidoglycan makes up the cell wall in bacteria and consists of polymers of the two amino compounds, *N*-acetylglucosamine and *N*-acetylmuramic acid, together named glycan. The polymers are cross-linked by short peptides, made from L and D isomers of amino acids (**Figure 5(a)**). The D isomers include Asp, Glu, Ser, and Ala. Except from some algae, only bacteria produce D amino acid isomers, and therefore these isomers have been used as markers or tracers of bacteria. The most prominent study of bacterial D isomers was the examination of a Martian meteorite which turned out not to contain extraterrestrial D amino acids and thus, did not indicate life on this planet.

In aquatic studies, abundance of D isomers has been used to determine the importance of bacterial cell wall material in formation and preservation of organic material. In a study of 50 rivers entering the Baltic Sea, the contribution of D amino acids among DCAA was determined relative to the total content of DON. As can be seen in **Figure 6(a)**, the amount of D isomers increased with the DON content in the rivers and made up about 0.5% of DON. In Mono Lake, D amino acids constituted about 0.3% of the DON. In this lake, the D isomers made up between 30% and 40% of Asp, Glu, Ser, and Ala in the DCAA pool (**Figure 5(b)**). Since D isomers only make up 1–7% of all amino acids in bacteria, this enhanced proportion of D isomers among DCAA shows that bacterial cell wall material accumulates or has a long turnover time in the lake.

Other DON Compounds

Glucosamine and Other Amino Sugars

A common DON substance in fresh water is chitin, that consists of the monomer amino sugar glucosamine. Chitin is a major component in skeletons of insects and crustaceans and this makes chitin one of the most abundant organic compound in the biosphere. Other important sources of glucosamine are fungal and bacterial cell walls. In fresh water, concentrations of free glucosamine have been found to be lower than concentrations of DFAA, probably due to a fast uptake by bacteria, but the amount of dissolved, combined glucosamine is comparable to DCAA.

Occurrence of dissolved, combined glucosamine in the rivers entering the Baltic Sea showed that the glucosamine concentration increased with the DON

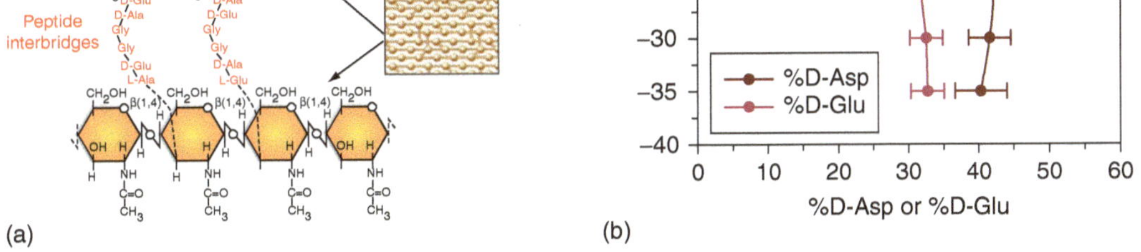

Figure 5 Schematic outline of the composition of the cell wall of a Gram-positive bacterium (a) and the proportion of D and L isomers of aspartic acid and glutamic acid in the DCAA pool in a depth profile of Mono Lake in May 2005 (b). In (a), L and D amino acids in peptide interbridges between the glycan strings in peptidoglycan are shown. NAG = *N*-acetylglucosamine and NAM = *N*-acetylmuramic acid. The cell wall structure at right is reproduced from Figure 4.31 in *BROCK Biology of Microorganisms*, 11th Edition, with permission from Prentice Hall. In (b), the dashed area indicates the stratified zone from 50% O_2 saturation to complete anoxic condition. Figure (b) is reproduced from Figure 6 in Jørgensen NOG, Engel P, Jellison R, and Hollibaugh JT (2008) Contribution of bacterial cell wall components to DOM in alkaline hypersaline Mono Lake, California. *Geomicrobiology Journal* 25: 38–55, with permission from Taylor & Francis.

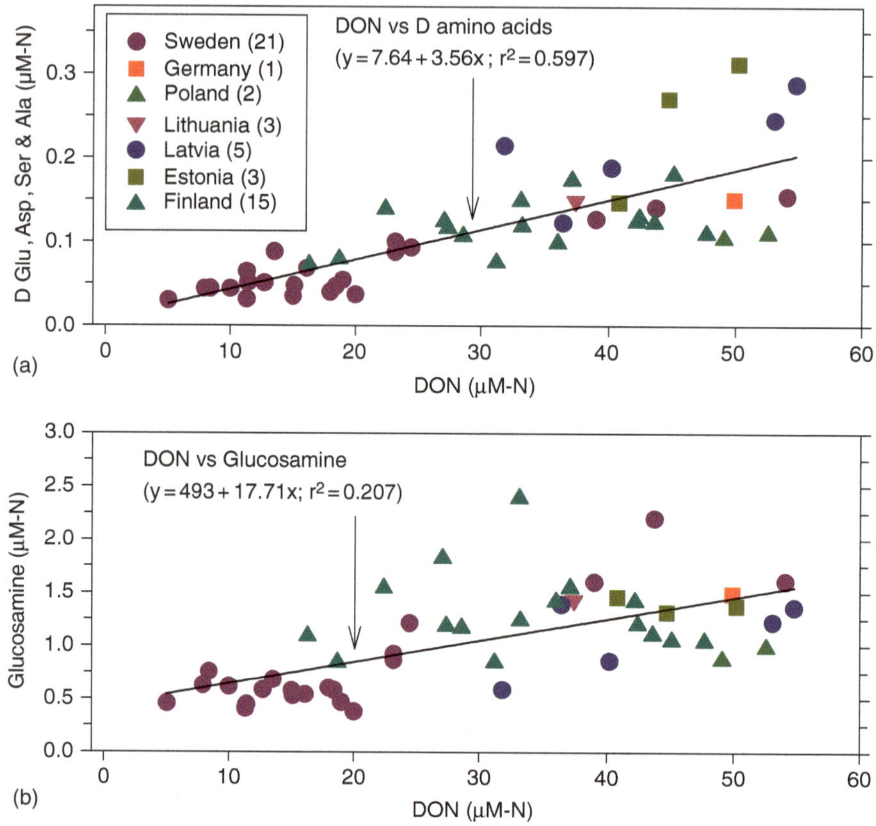

Figure 6 Concentrations of DON in 50 rivers entering the Baltic Sea relative to concentration of D isomers of four amino acids (glutamic acid, aspartic acid, serine, and alanine) (a) and glucosamine (b) in summer 1999. Countries in which the rivers are located and linear regressions between concentrations of DON and D isomers and glucosamine are shown. Reproduced from **Figure 1** in Jørgensen NOG, Stepanaukas R, Pedersen AGU, Hansen M, Nybroe O (2003) Occurrence and degradation of peptidoglycan in aquatic environments. *FEMS Microbiology Ecology* 46: 269–280, with permission from Elsevier.

content, as found for D amino acids, and that it constituted from 0.3% to 11% of DON (**Figure 6(b)**). The reason for the increased abundance of glucosamine in some areas, e.g., Finnish rivers, is unknown, but drainage of water from soils with a dense fungal biomass might explain the higher concentrations. The observation of the rivers supports that glucosamine is a major component of DON.

Other naturally occurring amino sugars include galactosamine and mannosamine. Both compounds are constituents of bacteria, algae, and crustaceans (not mannosamine), but they are less abundant than glucosamine. A few studies of bacterial turnover of amino sugars in lakes indicate that amino sugars in general are consumed by bacteria, although at a slower rate than glucose is consumed.

Dissolved RNA and DNA

Concentrations of dissolved RNA and DNA as well as their molecular constituents in fresh water have only been studied in a few cases. Dissolved DNA has been found to occur at concentrations from <0.01 to 1 μM N. The main source of this DNA, at least in lakes, appears to be phytoplankton. Analysis of the composition of dissolved DNA in marine waters has shown that viruses may constitute about half of the DNA, while the remaining DNA includes a continuum from small (<100 bp) to large (36 000 bp) DNA fragments. This means that dissolved DNA does also include plasmid-sized molecules. Dissolved DNA is probably degraded by bacteria (DNase is a ubiquitous enzyme in bacteria) and used as a source of energy and nutrients, but dissolved DNA may also participate in exchange of genetic information among bacteria as discussed later. RNA has been less studied than DNA, but it seems that – rather surprisingly – concentrations of dissolved RNA are similar or higher than concentrations of DNA, despite RNA is a rather instable molecule.

Occurrence of the base constituents of DNA and RNA (purines and pyrimidines) is relatively unknown in fresh water, but in one study the concentration of the purine adenine was determined to 0.06 μM N. The low concentration most likely indicates that adenine like other purines and pyrimidine are degraded fast by microorganisms as been observed in marine environments.

Urea

Urea is a microbial degradation product of purines and is also a common nitrogen waste product of many aquatic organisms. Urea typically constitutes up to 3.5 μM N, but significantly higher concentrations have been measured in special environments such as a eutrophic Korean pond (**Table 1**). Urea is an important nitrogen source to phytoplankton and is also used as supplementary nitrogen source by bacteria.

Methylamines

Methylamines are anaerobic metabolic products produced by many bacteria and are quantitatively important as DFAA. Methylamines include mono-, di-, and trimethylamine that all have a characteristic fish-like smell. Methylamines are typically degraded by bacteria to ammonium and formaldehyde.

Humic Matter

Humic matter constitutes another source of DON in fresh water (see **Interactions of Dissolved Organic Matter and Humic Substances**). Nitrogen makes up only a small portion, typically <5%, of the molecular composition of humic compounds. Autochthonous humic matter (produced within the lake or river) appears to have a higher nitrogen content than the humic matter being released from soil. Addition of humic matter to fresh water has shown that humic nitrogen can sustain growth of both bacteria and algae, and that even amino acids can be released from the humic compounds. This indicates that although humic matter is considered as low-labile organic matter, at least a portion of the nitrogen content is biologically available.

Sources of DON

All living organisms contribute organic nitrogen compounds to the water. A major input of DON occurs when organisms die and cell substances are released to the ambient water. This means that virtually all soluble cell constituents at some stage can be found dissolved in the ambient water. Other major sources of DON may be deliberate release of nitrogen compounds as metabolic waste products, or release of DON following stress or adjustment to new environmental conditions such as changes in temperature and salinity.

Examples of DON compounds produced by different organisms are mentioned in the earlier section, but here various groups of DON-producing organisms and other sources of DON shall briefly be mentioned.

Phytoplankton

Since phytoplankton cells are main producers of organic matter and hereby also major consumers of nitrogen in aquatic environments, there has been a large interest in characterizing both utilization and loss of nitrogen by phytoplankton. Release of DON by actively growing phytoplankton in natural environments and in laboratory cultures has been documented in several studies, although mainly by marine species. Typically the nitrogen loss has been measured as production of $DO^{15}N$ after uptake of an inorganic nitrogen source such as $^{15}NH_4^+$ (ammonium) or $^{15}NO_3^-$ (nitrate). The results have shown that 18–30% of the inorganic nitrogen taken up typically was released as DON. This is a surprisingly large release considering that nitrogen in many environments is a growth-limiting nutrient. From an evolutionary view point, it also appears to be a misplaced strategy to loose a large portion of the nitrogen after it has been taken up. The mechanisms causing the loss of nitrogen are still being discussed, and it has been argued that the methods used for studying release of DON actually may have increased the nitrogen loss.

Most likely the released DON consists of a large spectrum of organic nitrogen compounds, but in laboratory studies DCAA were found to be dominant DON compounds, while DFAA made up about 20%. In intact lake water samples, release of urea and ammonium by phytoplankton has also been found. Generally, it will often be difficult to determine the actual composition of released DON as bacteria in the experimental media (if present) will rapidly take up the organic substances, especially small molecules. This implies that some of the observed DON release rates may be underestimates, if a portion of the DON compounds were already assimilated by bacteria.

Bacteria

Most bacteria are known to be consumers rather than producers of organic matter, including DON, but bacteria may also release organic nitrogen to the environment, although actual release rates of DON by bacteria in aquatic environments remain to be studied.

Infection of bacteria by viruses has been shown to remove up to about 20% of the bacterial production in lakes. Following infection, new viruses are released from the bacteria, but in addition, several DON compounds are also released to the ambient water. The released compounds are dominated by DCAA, but DFAA, glucosamine, dissolved DNA, and cell wall fragments have also been found. Since bacterial production constitutes between 20% and 50% of the phytoplankton production, viral infection can be a significant source of DON in natural waters.

Bacteria also release various amino acids and their derivatives, such as glycine betaine (triggered by rapid changes of external osmotic pressure or temperature) and homoserine lactones (for communication between some bacteria, e.g., during biofilm formation). Further, extracellular enzymes (proteins) are released and used for degradation of dissolved and particulate organic matter and will be part of the DCAA pool in the water.

Other compounds released by bacteria include metabolic excretion products, especially from degradation of purine and pyrimidine bases in DNA and RNA. The main excretion product is urea, but hypoxanthine, β-alanine, and β-aminoisobutyric acid among others, have also been identified as degradation products. Finally, some natural bacterial mutants have an overproduction (and release) of certain amino acids, e.g., aspartate. This mutant function has inspired the medical and nutritional industry to generate new mutants that can produce other amino acids such as arginine, lysine, and tryptophane in excess amounts, which can be used for commercial production.

Protists, Zooplankton, and Fish

The main nitrogen waste product by protists (flagellates and ciliates), aquatic invertebrates, and vertebrates is ammonium, but a large variety of DON compounds can be released by these organisms.

Protists have been shown to produce urea, methylamines, and purines (guanine and hypoxanthine) as nitrogen waste products. When protists feed on bacteria, dissolved DNA, bacterial cell wall fragments, and proteins have also been found undigested in the water.

Zooplankton produce urea as the main nitrogen waste product, but release of free and protein amino acids has also been observed during feeding on phytoplankton. A portion of these amino acids probably originate from rupture of algal cells rather than metabolic processes. Supporting this, a high release rate of amino acids has been measured in the night, coinciding with the highest feeding rate by the zooplankton, and causing a stimulation of the bacterial activity due to uptake of amino acids. Another nitrogen excretion product by zooplankton is the important osmoregulatory amino acid taurine.

Fish produce urea as a major nitrogen waste product, but purines such as allantoin, nucleic acids, amino acids and trimethylamine are also common excretion products. In general, nitrogen compounds from fish are not only produced in the kidneys, but may also be released from gill and skin surface. Also, skin and scales in fish have a high content of the purine guanine that will be released to the ambient water from both living and dead fish. Finally, it should be mentioned that uric acid is the dominant nitrogenous waste product by some aquatic animals, e.g., snails.

Input of DON from Atmosphere, Soil, and Sewage Discharge

Atmosphere Consumption of fossil fuels for industrialization, heating, and transport has increased the atmospheric nitrogen content during the last 150 years. DON has been found to make up about 25% of the atmospheric nitrogen content, corresponding to 5–90 μM DON in rain and fog water. Between 1990 and 2000, the typical content of DON in rain above continental areas was 25–40 μM DON-N. The composition of DON in rain is only partly known, but (in increasing order) aliphatic amines, DFAA, DCAA, and urea are typically identified, while methyl cyanides, nitrogenous polyaromatic hydrocarbons, and other heterocyclic compounds occur occasionally in rain.

The significance of DON in precipitation to the nitrogen pool in lakes and rivers depends on the ambient DON pool in the water and the extent of precipitation. As an example, the ambient DON concentration in a mesotrophic river or lake will increase by 10%, if assuming a water depth of 5 m, 25 μM DON in rain and river or lake water, and an annual precipitation of 500 mm m^{-2}. In eutrophic waters the importance of DON will be minor, but in oligotrophic waters DON in precipitation can be a major source of nitrogen.

Soil Release of nitrogen from soils has mainly focussed on DIN (typically nitrate) rather than DON. However, recent studies show that concentrations of nitrate and DON can be similar in arable soils, while DON in natural grassland can be up to 10-fold more abundant than nitrate. When nitrogen is leaching from soils to surface water and ground water, DON can constitute up to 20% of total nitrogen leachate from arable soils. Leaching of DON is not restricted to arable soils, but does also occur from uncultivated land. As an example, a significant discharge of DON was observed during spring flood at a boreal site in northern Sweden during melting of ice and snow.

In leachates from the arable soils, DON has been found to be dominated by DCAA (20–50% of DON-N), while DFAA, urea, free amino sugars, and bacterial cell wall fragments made up <15% of DON. A similar proportion of DFAA, urea, and amino acids were found in the Swedish boreal spring flood water, but most DON was chemically unidentified. This agrees with observations from arable soils, where only about half of DON leachate has been unidentified.

Sewage The dominant DON compound in raw sewage is urea, making up about 80% of DON, but urea is converted to ammonium during the biological sewage treatment. In secondary treated wastewater (ready for discharge into rivers and lakes), DFAA and DCAA have been found to make up 0.05–3% and 0.6–13% of DON, corresponding to about 0.4 and 1–12 μM N, respectively. The remaining DON in treated wastewater consists of various amines (e.g., methylamines), chelating agents from detergents such as EDTA (ethylenediamine tetraacetic acid), pharmaceuticals, and a wide spectrum of organic nitrogen compounds at low concentrations, while about 10% of DON discharged in lakes and rivers typically is unidentified.

Cycling of DON

Largely all biological activity in aquatic environments leads to production of DON, as can be seen earlier. Yet, although a temporary increase in DON concentrations may occur, the amount of DON in lakes and rivers does not tend to accumulate over time. This is due to an efficient utilization by microorganisms in the water. Bacteria take up the DON compounds and utilize both carbon and nitrogen of DON for production of new cells. If specific DON compounds are rich in nitrogen relative to carbon, a portion of the nitrogen in DON may be released as DIN, typically ammonium. Some algae can also use DON substances as a nitrogen source, but usually they do not incorporate the carbon moiety. However, some photosynthetic microorganisms can use both nitrogen and carbon of DON for growth in the dark (explained later).

Cycling of Specific DON Compounds

A large spectrum of DON compounds has been identified in natural waters, but biological production and cycling of many of these substances have not yet been determined. Studies on turnover of DON compounds have focussed on free and combined amino acids, dissolved DNA and urea, but cycling of total DON pools has also been examined. Various methods have been used for analysis of cycling of specific DON compounds. Some of these methods are presented in the Appendix.

Uptake of DFAA by Bacteria

Free amino acids (DFAA) are rapidly taken up by most heterotrophic, planktonic microorganisms and serve as sources of energy or nitrogen, or both. In many environments, DFAA will be the single most important nitrogen source to bacteria, followed by ammonium. Examples of uptake rates and contribution of DFAA to the bacterial nitrogen demand is shown in **Table 2**.

The frequent dominance of DFAA as a major nitrogen source to bacteria means that uptake rates of DFAA and bacterial growth rates often will covary, and that concentrations of DFAA to some extent can control the bacterial production. This significance of DFAA to the bacterial production can be exemplified by the DFAA uptake in the arctic and oligotrophic Russian rivers and the eutrophic Thames River. At both locations, DFAA sustained about half of the bacterial nitrogen demand, despite concentrations and uptake rates of DFAA were significantly lower in the arctic rivers than in Thames River (**Table 2**). In lakes, uptake of DFAA relative to the bacterial production may be more variable due to shifts in composition of DON, e.g., during phytoplankton blooms. Yet, DFAA constitute an important nitrogen source to bacteria in lakes, as can be seen from the uptake rates in two mesotrophic lakes and a humic lake in **Table 2**.

The ecosystem significance of DFAA turnover relative to nitrogen cycling by phytoplankton and bacteria was studied in the Danish Lake Hylke. During a monthly period, concentrations of DFAA tended to decline (except for an initial increase) and varied from 330 to 1650 nM N (**Figure 7(a)**). Bacterial uptake of DFAA and total bacterial growth fluctuated during the period and did not covary (**Figure 7(b)**). Bacterial production and primary production also fluctuated, but had similar changes during the last half of the studied period. Calculation of nitrogen fluxes indicate that DFAA-N met on average 45% of the bacterial N demand (range 15–174%) and corresponded to about 13% of the nitrogen incorporation by the phytoplankton (range 6–31%). The bacterial production corresponded to about 29% (range 9–81%) of the primary production, based on nitrogen incorporation. The sustenance by DFAA to about half of the bacterial nitrogen demand emphasises the nutritional values of DFAA and their importance to the total nitrogen cycling in the lake.

In addition to nitrogen, DFAA also provide valuable carbon to the bacteria, as amino acids contain about threefold more carbon than nitrogen. After uptake by bacteria, carbon of DFAA is typically divided into equal portions that each are being used

Table 2 Uptake of DON components

Compound	*Location*	*Concentration of DON compound*	*Uptake rate (ng N l^{-1} h^{-1})[a,b]*	*Importance of DON uptake*	*Method*	*Source & comments*
DFAA	Arctic rivers Ob and Yenisei, Russia	12–43 nM N	24–50	DFAA-N met 33–54% of the bacterial N demand	Uptake of ^{3}H amino acids	Season: Summer at 9–14 °C water temperature. Turnover time of DFAA was about 12 h [9]
DFAA	Thames river, England	200 nM [^{15}N enrichment: 100 nM]	350 (mean)	DFAA-N met about 50% of the bacterial N demand	Uptake of ^{15}N amino acids	Season: February at 5–7 °C water temperature. Turnover time of DFAA ranged from 12 h to 5 days [10]
DFAA	Lake Mossø (mesotrophic), Denmark	400–1000 nM	75–470	DFAA-N met about 15% of the bacterial N demand	Uptake of ^{14}C amino acids	Season: April at 10 °C water temperature. Turnover time of DFAA ranged from 30 to 100 h [4]
DFAA	Lake Skärhultsjön (humic), Sweden	900 nM	250 (mean)	DFAA-N met 20–100% of the bacterial N demand	Uptake of ^{14}C amino acids	Season: June at 21 °C water temperature. Uptake by particles <1.3 μm (presumably mainly bacteria). Turnover time of DFAA of 20–80 h [7]
DFAA	Lake Constance (Bodensee), Germany	25–500 nM	3–95	DFAA-N met <1–80% of the bacterial N demand	Uptake of ^{3}H amino acids	Period: May-November. Turnover time of DFAA ranged from 12 to 920 h. Natural or 1.0 μm filtered lake water used
DCAA	Lake Constance (Bodensee; mesotrophic), Germany	2000–3200 nM	100–450	DFAA-N met 5–200% of the bacterial N demand	Concentration changes and ^{3}H amino acid dilution experiments	Period: May-November. Turnover time of DCAA ranged from 12 to 920 h. Natural or 1.0 μm filtered lake water used. DFAA were most important to the bacteria before a diatom bloom in July, after that DCAA were most important [13]
DCAA	Eleven Mile Creek, Florida, USA	3000 nM	525–1050	DCAA-N met about 70% of the bacterial N demand	Changes in concentration	Season: May at 25 °C water temperature. Uptake by particles <1.0 μm (presumably mainly bacteria) was studied [8]
DCAA	Lake Skärhultsjön (humic), Sweden	3100 nM	−16 (release) to 18	DCAA-N met from −30% (release) to 24% of the bacterial N demand	Changes in concentration	Season: June at 21 °C water temperature. Uptake by particles <1.3 μm (presumably mainly bacteria). Release of DCAA in some experiments [7]
N-acetyl-glycosamine	Lake Constance (Bodensee; mesotrophic), Germany	4–20 nM	30–50 (0–10 m) 2–15 (20 m) 1–10 (50 m)	N-acetyl-glycosamine-N met <5% of the bacterial N demand (estimated)	Uptake of ^{14}C N-acetylglucosamine	Season: 17 month period. Turnover time ranged from 56 to 118 h (between 3 and 10 m depth) but were >600 h at 50 m depth. Highest concentrations in autumn, lowest during spring algal bloom [2]
Urea	Lake Biwa, Japan (1)	0.2–2.5 μM N [^{15}N enrichment: ND]	150–600 53–172	Urea uptake was 14–22% of the total nitrate + ammonium + urea uptake by bacteria and algae	Uptake of ^{15}N urea	Seasonal study. Algal and bacterial uptake in natural lake water [11]
	Lake Biwa, Japan (2)	Ambient: ND [^{15}N enrichment: 0.7–7 μM]		Urea uptake was 4–58% of the total nitrate + ammonium + urea uptake by bacteria		Seasonal study. Uptake by particles <1.3 μm (presumably mainly bacteria) was studied [18]

Urea	Lake Kinneret, Israel	0.25–0.75 μM [^{15}N enrichment: >0.025 μM]	14–350	Urea uptake was 3–23% of total nitrate + ammonium + urea uptake by bacteria and algae	Uptake of ^{15}N urea	Seasonal study. Algal and bacterial uptake in untreated lake water. Lowest rates were measured in the winter [1]
Urea	Smith Lake, Alaska, USA	0.2–3.0 μM [^{15}N enrichment: 0.01–8 μM]	10–7590 (mean 1710)	Urea uptake was 4–31% of the total nitrate + ammonium + urea uptake by bacteria and algae	Uptake of ^{15}N urea	Season: May-June at 10–21 °C water temperature and during a bloom of cyanobacteria. Natural, untreated lake water was used [3]
Urea	Shallow pond, Korea	1.3–9.1 μM	80–1135	Bacterial and algae uptake of urea was 47% and 53%, respectively	Enzymatic hydrolysis of ^{14}C urea to $^{14}CO_2$	Seasonal study. Highest rates during summer. Bacterial fraction defined as particles <1.3 μm [12]
Urea	Roskilde Fjord estuary, Denmark	0.2–2.4 μM	2–183	Met 1–44% (mean 20%) of the bacterial N demand	Enzymatic hydrolysis of ^{14}C urea to $^{14}CO_2$	Seasonal study. Uptake by particles <1.3 μm (presumably mainly bacteria) was studied [5]
Dissolved DNA	Eleven Mile Creek, Florida, USA Santa Rosa Sound estuary, Florida, USA	180 μg DNA l^{-1} 6 μg DNA l^{-1}	(700) 38–50	Met 76% (?) of bacterial N demand Met 5% of bacterial N demand	^{3}H *E. coli* DNA used as tracer to estimate the turnover time of ambient D-DNA.	Season: May at 25 °C water temperature. In Eleven Mile Creek, the natural DNA concentration was probably overestimated due to interference from humic matter. This implies that the uptake rate was also overestimated [8]
Dissolved DNA	Lake Constance, Germany	3 mg DNA l^{-1} (added)	200–600	ND	Calf thymus DNA added to lake water	Season: June-August. The lake water was 1.0 μm prefiltered, implying that only bacteria remained in the water [15]
Dissolved DNA	Roskilde Fjord estuary, Denmark	2–6 μg DNA l^{-1}	5–150	Met 8% of bacterial N demand	^{3}H labelled DNA at sizes of 100–569 bp	Season: September at 16–19 °C water temperature. DNA was incorporated into non-protein molecules in bacteria [6]
High molecular (>1000 Da) DON	Various creeks, Sweden	2–40 μM N	ND	ND	Bacterial biomass production from uptake of DON > 1000 Da	Seasonal study. Size separation of DON from three creeks. From 0.1% to 5.9% of DON was bioavailable. Highest bioavailability in the autumn [17]
Total DON	50 rivers entering the Baltic Sea	3–55 (mean 35) μM N	ND	ND	Bacterial biomass production from uptake of DON	Season: Summer at water temperatures from 17 to 23 °C. About 30% of the total DON was utilized by the bacteria. This bioavailable fraction often corresponded to the content of DFAA, DCAA and urea [16]
Total DON	Various freshwater sources (urban runoff, pasture and forest), East USA	2.5–260 μM N	ND	ND	Uptake of DON by estuarine plankton, determined from changes in DON concentrations	Season: July-August at 18–30 °C water temperature. From 0% to 73% of the DON was bioavailable. Lowest bioavailability (23–30%) in forest and pasture runoff; highest in urban and suburban runoff [14]

Continued

Table 2 Continued

Compound	*Location*	*Concentration of DON compound*	*Uptake rate ($ng\ N\ l^{-1}\ h^{-1}$)[a,b]*	*Importance of DON uptake*	*Method*	*Source & comments*
Total DON	Nine rivers in Eastern USA	1–35 μM N	ND	DON met 43% (mean value) of the bacterial N demand.	Uptake of DON by bacterioplankton, determined from changes in DON concentrations	Season: June-August at water temperatures of 18–29 °C. On average 23% of DON was bioavailable; most of this could be urea [19]

Sources

1. Berman T, Bechemin C, and Maestrini SY (1999) Release of ammonium and urea from dissolved organic nitrogen in aquatic ecosystems. *Aquatic Microbial Ecology* 16: 295–302.
2. Bunte C and Simon M (1999) Bacterioplankton turnover of dissolved free monosaccharides in a mesotrophic lake. *Limnology & Oceanography* 44: 1862–1870.
3. Gu B and Alexander V (1993) Dissolved nitrogen uptake by a cyanobacterial bloom (*Anabaena flos-aquae*) in a subarctic lake. *Applied Environmental Microbiology* 59: 422–430.
4. Jørgensen NOG (1986) Fluxes of free amino acids in three Danish lakes. *Freshwater Biology* 16: 255–268.
5. Jørgensen NOG (2005) Uptake of urea by estuarine bacteria. Aquat Microb Ecol 42: 227–243.
6. Jørgensen NOG and Jacobsen CS (1996) Bacterial uptake and utilization of dissolved DNA. *Aquatic Microbial Ecology* 11: 263–270.
7. Jørgensen NOG, Tranvik L, Edling H, Graneli W, and Lindell M (1998) Effects of sunlight on occurrence and bacterial turnover of specific carbon and nitrogen compounds in lake water. *FEMS Microbiology Ecology* 25: 217–227.
8. Kroer N, Jørgensen NOG, and Coffin RB (1994) Utilization of dissolved nitrogen by heterotrophic bacterioplankton: A comparison of three ecosystems. *Applied Environmental Microbiology* 60: 4116–4123.
9. Meon B and Amon RMW (2004) Heterotrophic bacterial activity and fluxes of dissolved free amino acids and glucose in the Arctic rivers Ob, Yenisei and the adjacent Kara Sea. *Aquatic Microbial Ecology* 37: 121–135.
10. Middelburg JJ and Nieuwenhuize J (2000) Nitrogen uptake by heterotrophic bacteria and phytoplankton in the nitrate-rich Thames estuary. *Marine Ecology Progress Series* 203: 13–21.
11. Mitamura O and Matsumoto K (1981) Uptake rate of urea nitrogen and decomposition rate of urea carbon at the surface microlayer in Lake Biwa. *Verhandlungen der Internationalen Vereinigung für Theoretische und Angewandte Limnologie* 21: 556–564.
12. Park MG, Shim JH, and Cho BC (1997) Urea decomposition activities in an ammonium enriched freshwater pond. *Aquatic Microbial Ecology* 13: 303–311.
13. Rosenstock B and Simon M (1993) Use of dissolved combined and free amino acids by planktonic bacteria in Lake Constance. *Limnology & Oceanography* 38: 1521–1531.
14. Seitzinger SP, Sanders RW, and Styles, R (2002) Bioavailability of DON from natural and anthropogenic sources to estuarine plankton. *Limnology & Oceanography* 47: 353–366.
15. Siuda W, Chróst RJ, and Gude H (1998) Distribution and origin of dissolved DNA in lakes of different trophic states. *Aquatic Microbial Ecology* 15: 89–96.
16. Stepanaukas R, Jørgensen NOG, Eigaard OR, Zvikas A, Tranvik LJ, and Leonardson L (2002) Summer inputs of riverine nutrients to the Baltic Sea: Bioavailability and eutrophication relevance. *Ecological Monographs* 72: 579–597.
17. Stepanaukas R, Leonardson L, and Tranvik LJ (1999) Bioavailability of wetland-derived DON to freshwater and marine bacterioplankton. *Limnology & Oceanography* 44: 1477–1485.
18. Takahashi M, Hama T, Matsunaga K, and Handa N (1995) Nitrogenous nutrient uptake by phytoplankton and ammonium regeneration by microbial assemblages in Lake Biwa. *Journal of Plankton Research* 17: 1027–1037.
19. Wiegner TN, Seitzinger SP, Glibert PM, and Bronk DA (2006) Bioavailability of dissolved organic nitrogen and carbon from nine rivers in the eastern United States. *Aquatic Microbial Ecology* 43: 277–287.

[a] Where uptake rates were measured in units of carbon, the rates were converted to units of nitrogen using nitrogen content of the organic compounds and published carbon-to-nitrogen conversion factors for bacteria and algae.

[b] In some studies the uptake was measured as incorporation of substance into biomass. In most organisms, a portion of the substance taken up will be lost during metabolic processes and not incorporated into new biomass. This means that the amount of substance taken up was higher than the amount being incorporated into biomass. This difference between uptake and incorporation may also influence the turnover time of the substance, since this is based on the total amount of substance taken up over time relative to the concentration in the water.

ND: No data.

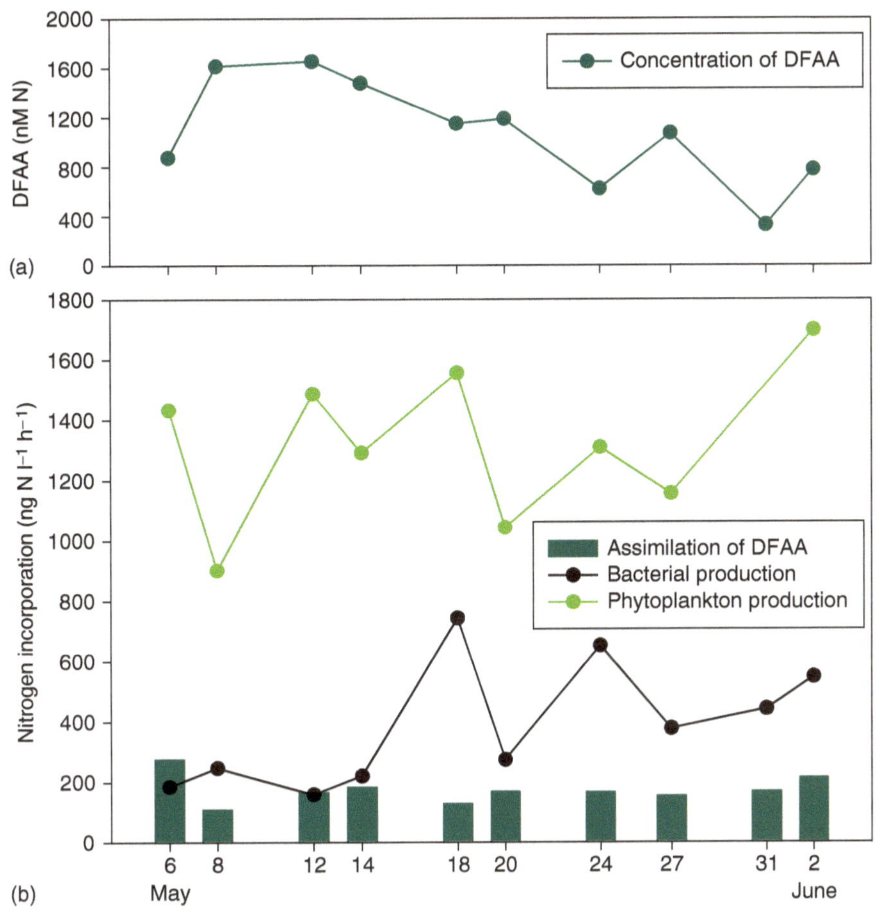

Figure 7 Concentrations of DFAA (a) and incorporation of DFAA and production of bacteria and phytoplankton (b) in the eutrophic Lake Hylke, Denmark, in summer 1984. The original rates were measured in units of carbon but were converted to units of nitrogen using published conversion factors. Reproduced from **Figure 2** in Jørgensen, N. O. G. (1987). Free amino acids in lakes: Concentrations and assimilation rates in relation to phytoplankton and bacterial production. *Limnology & Oceanography* 32: 97–111, with permission from the American Society of Limnology and Oceanography.

for synthesis of new cell material and production of energy through respiration. If nitrogen content of the DFAA exceeds the nitrogen demand by the bacteria, all DFAA carbon may be used by the bacteria while a portion of the amino nitrogen is being released as ammonium. Release of ammonium probably also explains the occasional higher uptake of DFAA-N than required by the bacterial production in Lake Hylke.

Uptake of DFAA by Microalgae

Some microalgae, being planktonic or benthic, are also reported to use DFAA as a supplementary nitrogen source during photosynthesis. Interestingly, instead of taking up the entire amino acid molecules, certain algae may use cell-surface enzymes that catalyze the oxidation of amino acids to extracellular ammonium and other products, but only ammonium is taken up. Recently, it has also been shown that some cyanobacteria can grow at very low light intensities when DFAA are present. These cyanobacteria probably use DFAA as a source of energy rather than nitrogen.

Uptake of Other Amino Compounds

Free amino sugars, such as glucosamine and galactosamine, are commonly found in natural waters, although at lower concentrations than DFAA. Bacterial utilization of amino sugars has not been widely studied, but most bacteria possess enzymes for uptake of at least the simplest amino sugar, glucosamine. In Lake Constance this amino sugar (as *N*-acetylglucosamine) was found to be assimilated at rates corresponding to single free amino acids and contribute up to 5% to the bacterial N demand (**Table 2**). Possibly, amino sugars are more important as carbon than nitrogen sources to bacteria due to the higher content of carbon.

Uptake of DCAA

Dissolved combined amino acids are utilized by many bacteria but before uptake, the DCAA must be broken down to single amino acids or small (less than five amino acid) peptides. Bacteria produce and release extracellular peptidases and proteases that hydrolyze peptide bonds in peptides and proteins to produce free amino acids. The DCAA pool in lakes and rivers does not only consist of peptides and proteins, but does also include amino acids in cell wall fragments and absorbed to organic and mineral surfaces. This complex nature of DCAA may mean that bacterial enzymes cannot always produce (release) free amino acids from DCAA, despite a sizable amount of DCAA is present.

Bacterial utilization of DCAA in natural waters has been often studied by using the substrate analogue compound leucine–MCA (see Appendix). As an example, the bacterial capacity for utilization of DCAA was estimated in the Finnish Lake Mekkojärvi (**Figure 8(a)**). Changes in DCAA concentrations and activity of the peptidase enzyme were similar during the studied summer season. This shows that a proportion of the DCAA must have been peptides with a terminal leucine molecule. Similarly, a close resemblance between leucine–MCA activity and incorporation of free leucine in the water was found in a depth profile of the Norwegian Lake Skjervatjern (**Figure 8(b)**). Incorporation of leucine into bacterial proteins has been used a reliable technique for measuring bacterial protein synthesis and with this, the bacterial production. Thus, the substrate analogue approach can be applied as a useful proxy for estimating enzymatic activity and hereby also often for a general characterization of the metabolic activity of the bacteria.

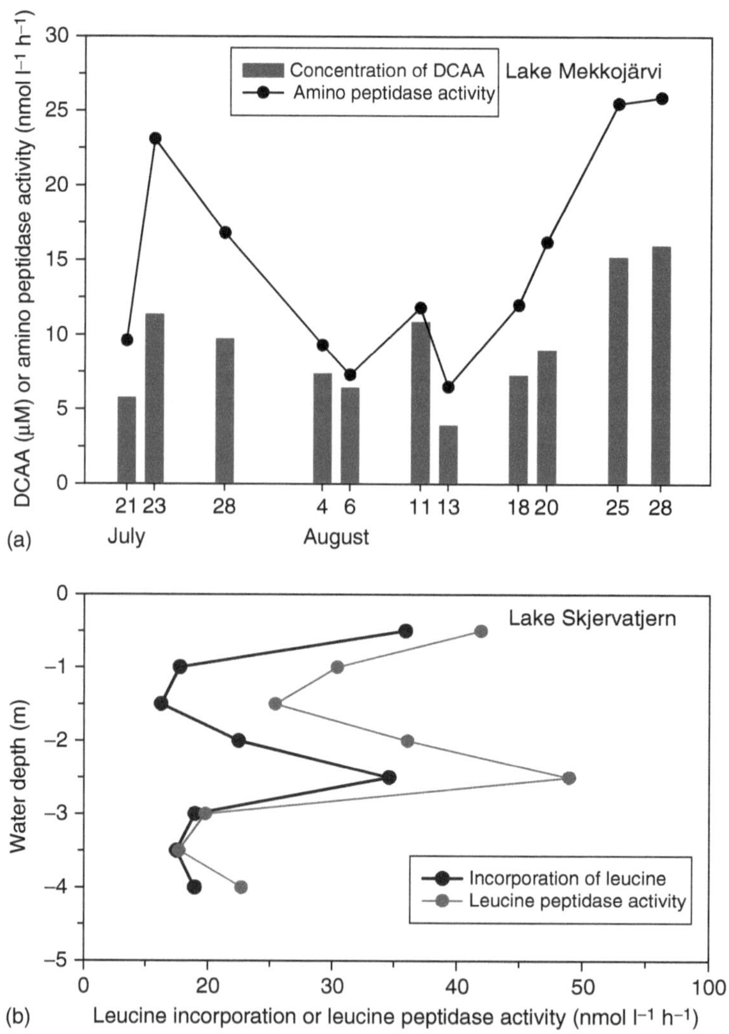

Figure 8 Aminopeptidase activity (measured with the leucine–MCA approach, see Appendix) relative to concentrations of DCAA at 0–1 m depth in Lake Mekkojärvi, Finland, in summer 1986 (a). Incorporation of the amino acid leucine and leucine–MCA peptidase activity in a depth profile of Lake Skjervatjern, Norway, in summer 1992 (b). Reproduced with permission from Dr. Uwe Münster, University of Tampere, Finland.

Bacteria may only produce a limited number of different extracellular enzymes, also named ectoenzymes. It may be speculated if bacteria that produce a specific ectoenzyme, e.g., a proteinase, actually also benefit other bacteria that cannot produce this ectoenzyme but live in the same microenvironment. This has been examined in laboratory cultures of a proteolytic (protein-degrading) and a nonproteolytic bacterial strain. The results clearly showed that although the nonproteolytic strain benefited from the enzyme production, the proteolytic strain was more competitive due to its enzyme production.

A more direct approach for measuring utilization of DCAA by bacteria in a lake or river is to study actual concentration changes of DCAA in the water over a period of time, e.g., 12 or 24 h. The disadvantages by this method are that organisms other bacteria must be removed in the experiment to eliminate input (or release) of DCAA, and that bacteria themselves may also release DCAA to the water, e.g., from dead bacteria. Examples of measured DCAA uptake rates using this approach are given in **Table 2**. They show that DCAA may sustain from <1% and up to 70–80% of the bacterial nitrogen demand. Thus, DCAA can be a major source of nitrogen to bacteria in some cases. In Lake Constance, uptake of DCAA apparently exceeded the bacterial nitrogen production, but this may indicate that there was a simultaneous, but not registered, release of other nitrogen compounds such as ammonium or urea by the bacteria. An example of release of DCAA by bacteria, the experiment in the Swedish lake indicated that up to 30% of the bacterial nitrogen demand was lost as DCAA, but the reason for this was not determined (**Table 2**).

Bacteria can also use D isomers of DCAA in bacterial peptidoglycan, but specific enzymes may be required to allow peptidases to hydrolyze D amino acids in the peptide interbridges (see **Figure 6**). Laboratory analyses demonstrate that amino acid isomers in peptidoglycan are degraded at rates similar to protein amino acids, while another amino component in peptidoglycan, *N*-acetylglucosamine, is rather resistant to degradation in intact peptidoglycan.

Uptake of Dissolved DNA

DNA constitutes a potentially valuable source of carbon, nitrogen, and phosphorus to bacteria, but DNA may also be involved in transfer of genetic information among bacteria. Most research on bacterial uptake of dissolved DNA has been performed in marine locations and only a few observations from fresh water exist. In general, dissolved DNA appears to be cycled fast by bacteria, with turnover time of the ambient pool of less than 1 day. In terms of nitrogen, uptake of dissolved DNA has been estimated to sustain between 5% and 8% of the bacterial nitrogen demand, if ignoring the high rates found in the humus-rich creek (**Table 2**). Possibly, dissolved DNA is more important as a source of phosphorus than nitrogen, especially in oligotrophic waters. Still, the actual fate of DNA moieties in bacteria remains to be more carefully examined.

Dissolved DNA has also been shown to be involved in horizontal gene transfer by transformation into competent (receptive) bacterial cells. Transformation may be an important mechanism for transfer of DNA, e.g., in plasmids, among bacteria. The steps involved in the uptake appears to include a rapid binding of DNA to the cell surface, followed by cleavage by cell-bound and extracellular nucleases (including a degradation of double-stranded DNA into single stranded DNA), and a final transport into the cells and incorporation into nucleic acids.

Bacterial utilization of the bases of DNA and RNA, purines and pyrimidines, has not been examined in fresh water. The ultimate degradation product of purines is urea and this pathway is assumed to be a dominant source of urea in natural environments. Pyrimidines can also be degraded to urea, but other nitrogen degradation products are also found.

Uptake of Urea

Urea has typically been considered as primarily a nitrogen source to photosynthetic organisms such as algae and plants, but urea can also be a supplementary nitrogen source to bacteria. Before urea can be used for synthesis of new cell material, it is hydrolyzed by urease to two ammonium molecules. Urease is widely found in eukaryotic organisms, but seems only to occur in some bacteria.

Methodologically it is difficult to separate uptake of urea by phytoplankton and bacteria, e.g., by size-separation procedures, as both groups of organisms take up urea and often overlap in cell sizes. Consequently, most observations of planktonic uptake of urea include both algae and bacteria. As can be seen in **Table 2**, the highest urea uptake rates (up to 7600 ng N l^{-1} h^{-1}) have been measured in untreated water samples with algae and bacteria present. Significantly lower rates were found in water when only bacteria were present (rates <200 ng N l^{-1} h^{-1}).

The results in **Table 2** also show that urea is a less abundant nitrogen source than ammonium and nitrate to phytoplankton. Among reasons for this may be that natural concentrations of urea often are below concentrations of ammonium and that ammonium is being produced at a higher rate than urea, e.g., from degradation of organic matter. It should be mentioned that some of the rates in **Table 2** may be overestimates due to enrichment of the natural urea concentration by addition of ^{15}N urea.

The quantitative significance of urea uptake by bacteria has not been determined in fresh water, but in a seasonal study of the Danish Roskilde Fjord, the average contribution of urea-N to the bacterial N production was determined to be 20%. The relatively sparse utilization by bacteria has been suggested to be caused by a small number of bacterial species being capable of taking up urea, a relatively slow uptake system, and a general bacterial nitrogen control uptake that favors ammonium to urea.

Uptake of Total DON

Utilization of total, unidentified DON by microorganisms has been studied in several rivers that drain a large spectrum of environments, ranging from boreal peat bogs to urban areas. The standard experimental approach has been addition of bacterioplankton or phytoplankton or both to the water, followed by measurements of changes in the DON concentration. Reduction in DON in such bioassays is assumed to reflect availability of DON. The results show that rivers draining boreal forest and bogs have the lowest proportion of bioavailable DON, typically $<20\%$, while rivers receiving urban runoff can have $>70\%$ bioavailable DON (**Table 2**).

Analyses of specific nitrogen compounds that are used by microorganisms in such bioassays show that DCAA, DFAA, and urea often are the predominantly assimilated DON compounds. Apparently, nonspecific nitrogen compounds, such as humic substances, appear to have a low bioavailability and be of minor importance to the microbial cycling of DON in rivers. Supporting that humic substances are rather recalcitrant, size-fractionation of DON from Swedish rivers indicated that high-molecular DON (>1000 Da) was less labile than the total DON pool, and that only 0.1–5.9% of the high-molecular DON was bioavailable. It is questionable whether there is always such a low uptake of high-molecular DON. Thus, experiments with DOC from the Mississippi river have shown that about 50% of the high-molecular weight fraction (>1000 Da) was degradable by bacteria, while only about 10% of the low-molecular DOC was utilized by bacteria. The nitrogen content of this DOM was not determined, but the high-molecular DOC could not support the observed bacterial growth if no nitrogen was present. Biological degradation of humic substances and other DOC compounds are treated further in **Dissolved Humic Substances: Interactions with Organisms.**

Bacterial DON Dynamics

Bacterial activity has a central function in cycling of nitrogen in lakes and rivers. When bacteria take up DON, the dissolved nitrogen is transformed into bacterial biomass (particulate nitrogen) and can be utilized by higher organisms within the pelagic food web. Although some photosynthetic microorganisms can utilize DON compounds, most DON in natural waters is assimilated by bacteria. In addition to producing new biomass, bacteria may also release inorganic nitrogen (DIN) during consumption of DON, as mentioned earlier, and hereby provide nitrogen for phytosynthetic organisms. On the other hand, if bacteria assimilate DOM with a low nitrogen content, they will often use DIN, typically ammonium, to obtain nitrogen for synthesis of amino acids and nucleotides. In this way, bacteria and algae may compete for nitrogen in some situations.

The significance of different DON sources to heterotrophic bacteria is illustrated in **Figure 9**. Amino acids, being both DFAA and DCAA, are the main nitrogen sources to many bacteria, followed by ammonium. The nutritional importance of other DON compounds such as urea, nitrate, DNA and humic matter will depend on their concentrations and availability in the environment and on the ability by the different microorganisms to take up these substances and incorporate them into biomass.

The ecosystem importance of bacteria in the cycling of dissolved nitrogen is indicated in **Figure 9**. As can be seen, when bacteria become infected by viruses and disintegrate and when protists (flagellates and ciliates) feed on the bacteria, DON and DIN compounds will typically be released and may sustain growth by other bacteria and algae. The ability of bacteria to assimilate DON and make it available to higher organisms in the food web is an important step in the microbial loop.

Appendix: Methods for Studying Cycling of DON Compounds

Uptake of Single DON Compounds

Uptake of DON by microorganisms is ideally studied by measuring a reduction in concentration during a time period. But since microorganisms themselves produce a wide spectrum of DON compounds during

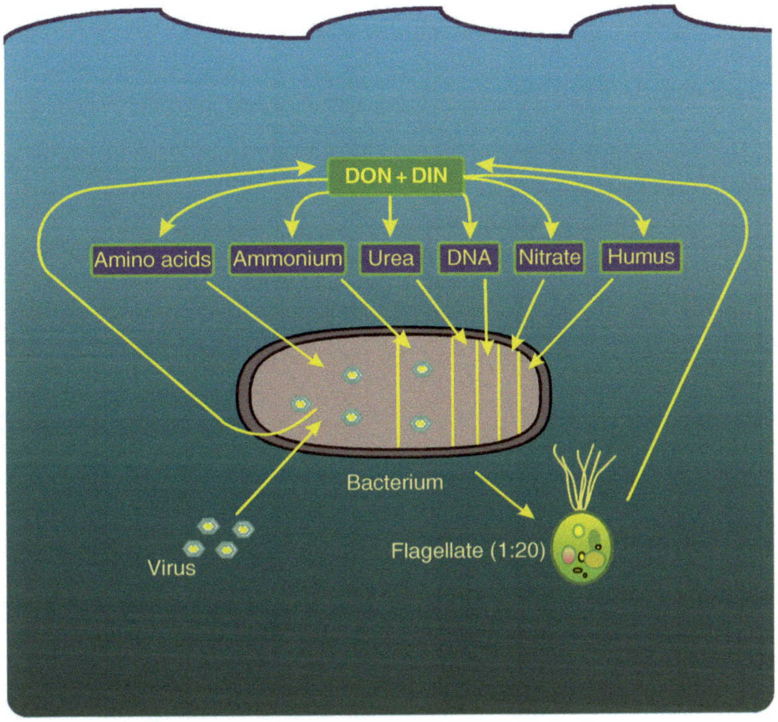

Figure 9 Significance of different DON sources to a typical aquatic, heterotrophic bacterium, based on own and published data. Compartments inside the bacterium represent the dominance of the various DON and DIN compounds. The arrows illustrate that DON and DIN may also be lost by bacteria during viral infection and grazing of bacteria by flagellates and ciliates.

growth, a net uptake of DON may not always be detected. Instead uptake of specific DON compounds can be followed by using ^{15}N-labeled molecules, e.g., ^{15}N-glycine, as tracer of the ambient amount of glycine. ^{15}N is a stable nitrogen isotope and can be detected in a mass spectrometer. After incorporation into bacterial biomass, the cells are separated from the media and their content of ^{15}N is determined. As an example, if 10% of the ^{15}N-glycine is taken up in 2 h by bacteria in a lake, turnover time of the glycine pool will be 20 h. When the natural glycine pool is determined, the glycine uptake rate by bacteria in the water can be quantified.

Unfortunately, the analytical sensitivity of the ^{15}N technique is low and requires addition of a relatively high concentration of the ^{15}N label to measure uptake of a DON compound. This means that the natural pool, e.g., of glycine, in most cases is enriched above the natural level by the ^{15}N addition and thus, it is no longer the natural glycine pool that is being studied. This problem can be met by application of radioactive isotopes of atoms that occur in DON, e.g., carbon (C) or hydrogen (H). Typically used radioactive isotopes are ^{14}C (carbon-14) and ^{3}H (tritium) that both have a high analytical sensitivity (require only addition at very low amounts) and can be used as tracers of many low-molecular weight DON compounds at naturally low concentrations. Based on incorporation of ^{14}C or ^{3}H into the bacterial biomass and molecular composition of the studied compounds, the bacterial nitrogen uptake can be estimated. The radioactive isotope analysis can be supplemented by measurement of DIN compounds in the ambient media to control that the nitrogen moiety of the studied substances actually was incorporated into bacteria and not excreted.

Uptake of Polymer DON Compounds

Bacterial utilization of polymer DON compounds such as proteins and nucleic acids is also difficult to study from concentration changes by chemical techniques as there can be a simultaneous release of the same compounds by the bacteria. In some cases commercially available, radioactively labeled polymers are available, e.g., ^{14}C-proteins and ^{3}H-DNA, and they can be used to measure of uptake these compounds.

Alternatively, bacterial degradation of polymer compounds can be measured by using so-called substrate analogue compounds. This approach exploits that bacteria release ectoenzymes for degradation of polymer organic substances into monomers, e.g., cleavage of chitin into free *N*-acetylglucosamine molecules. The free monomers are subsequently assimilated by bacteria. Substrate analogue compounds typically consist of organic carrier molecules that have a monomer such as a monosaccharide or an amino acid attached. If bacterial ectoenzymes are present in the water, e.g., peptidase (for cleavage of amino acids from peptides), the amino acid attached to the carrier compound will also be cut-off. When the amino acid is removed, the carrier compound becomes fluorescent and can relatively easily be detected. With this technique the bacterial potential for enzymatic degradation of a many organic polymers can rather easily be measured.

For analysis of bacterial utilization of DCAA in natural waters, the compound leucine–MCA (leucine-methoxycoumarin-4-acetyl) is among the most frequently used substance. If a leucine-removing peptidase is present in the water, the fluorescent ACM compound is produced from leucine–MCA in proportion to the enzymatic activity. This means that fluorescence of the ACM compound can be used to quantify the peptidase activity or, to be more specific, the release of leucine from peptides.

See also: Chemical Properties of Water; Ground Water and Surface Water Interaction; Groundwater Chemistry; Hydrological Cycle and Water Budgets.

Further Reading

Antia NJ, Harrison PJ, and Oliveira L (1991) The role of dissolved organic nitrogen in phytoplankton nutrition, cell biology and ecology. *Phycologia* 30: 1–89.

Berman T and Bronk DA (2003) Dissolved organic nitrogen: a dynamic participant in aquatic ecosystems. *Aquatic Microbial Ecology* 31: 279–305.

Cornell S, Rendell A, and Jickells T (1995) Atmospheric inputs of dissolved organic nitrogen to the oceans. *Nature* 376: 243–246.

Jørgensen NOG (2006) Uptake of urea by estuarine bacteria. *Aquatic Microbial Ecology* 42: 227–243.

Jørgensen NOG, Stepanaukas R, Pedersen AGU, Hansen M, and Nybroe O (2003) Occurrence and degradation of peptidoglycan in aquatic environments. *FEMS Microbiology Ecology* 46: 269–280.

Kim C, Nishimura Y, and Nagata T (2006) Role of dissolved organic matter in hypolimnetic mineralization of carbon and nitrogen in a large, monomictic lake. *Limnology & Oceanography* 51: 70–78.

Meon B and Amon RMW (2004) Heterotrophic bacterial activity and fluxes of dissolved free amino acids and glucose in the Arctic rivers Ob, Yenisei and the adjacent Kara Sea. *Aquatic Microbial Ecology* 37: 121–135.

Murphy DV, Macdonald AJ, Stockdale EA, *et al.* (2000) Soluble organic nitrogen in agricultural soils. *Biology and Fertility of Soils* 30: 374–387.

Rosenstock B and Simon M (2003) Consumption of dissolved amino acids and carbohydrates by limnetic bacterioplankton according to molecular weight fractions and proportions bound to humic matter. *Microbial Ecology* 45: 433–443.

Stepanaukas R, Jørgensen NOG, Eigaard OR, Zvikas A, Tranvik LJ, and Leonardson L (2002) Summer inputs of riverine nutrients to the Baltic Sea: Bioavailability and eutrophication relevance. *Ecological Monographs* 72: 579–597.

Relevant Websites

http://www.monolake.org – Mono Lake Web Site.

http://en.wikipedia.org/wiki/Lake_Constance – Geography on the German Lake Constance.

Nutrient Stoichiometry in Aquatic Ecosystems

R W Sterner, University of Minnesota, St. Paul, MN, USA

Introduction

In chemistry, the term 'stoichiometry' refers to the number of atoms of elements on both sides of a reaction. Stoichiometry tells you how many different molecules of each type of reactant you need to generate a specific product or set of products. Ecologists and limnologists can use these same principles of stoichiometry to understand aquatic ecosystems because individual species, like molecules, have defined chemical composition. The elemental formula of an individual species is normally not as strictly defined as the kinds of molecules we are most accustomed to thinking about, those with specific, unvarying composition. However, stoichiometry can still be a very useful approximation. Stoichiometry has much to say about the linkage of cycling of different elements and about patterns of nutrient limitation in primary producers and other parts of food webs. It bridges studies on individual species with studies dealing with the flow of matter and energy in ecosystems.

In this article, we first consider some of the patterns in elemental content of particular aquatic organisms and then we consider the ecological consequences of those patterns.

Stoichiometric Homeostasis

Homeostasis is one of the hallmarks of life. Homeostasis is a resistance to change. For example, homeothermic organisms maintain a constant body temperature in spite of fluctuating environmental temperature. There are many other examples of homeostasis in biology, like the pH of blood and the concentration of ATP in a cell. Homeostasis results from negative feedbacks. In ecological stoichiometry, we use this same term to refer to the tendency of an organism to maintain its chemical content in spite of variation in the chemical content of its resources (**Figure 1**). The figure illustrates the general principle. Mathematically, stoichiometric homeostasis (H) is defined as

$$y = cx^{1/\mathrm{H}}$$

where y = organismal elemental content (e.g., moles P per dry mass of animal), c = a constant, x = elemental content of resources (measured in the same units as 'y'), and H (the Greek letter 'eta') is the measure of homeostasis. If homeostasis is so strong that organismal content is constant no matter what the chemical content of the organism's food is, the stoichiometric coefficient H approaches infinity and we refer to that as a 'strict homeostasis'.

Not all species maintain stoichiometric homeostasis to the same degree, and within any given species not all resources are regulated equally. These differences will be elaborated upon later, but in terms of nutrients such as nitrogen and phosphorus, primary producers generally have more variable chemical content than animals. Algae and macrophytes, like other 'plants' and many microorganisms, can store nutrients by assimilating them at rates much larger than immediate requirements. Later, these organisms can then tap into these reserves for future growth. Animals have more limited capacity for nutrient storage. Differences in mean chemical content between plants and animals have also been suggested recently by analysis of selected proteomes (all of the proteins produced by a species) of these groups; animals had about 7% higher number of N atoms per amino acid side chain than did plants.

Within an animal taxon, variation in element content is lowest for macronutrients, intermediate for essential micronutrients, and highest in nonessential metals (**Table 1**). Animals apparently regulate macroelements most strictly and they regulate elements with definite biological function more than those elements that are incorporated into cells even though they are not essential for growth or survival.

There has been progress recently in defining patterns of homeostasis across elements and across organisms, but we are still a long way from a comprehensive understanding of all the potentially important patterns and contrasts. Homeostasis is an important concept in ecological stoichiometry because the stricter the homeostasis the closer the more an organism's chemical composition resembles a molecule with defined composition. Maintaining a given chemical content in spite of the composition of resources has some detrimental effects on fitness. As we will see later, homeostasis makes organism growth sensitive to the composition of the food that is eaten. Without homeostasis, organisms would 'be what they eat' and there would be no slowing of growth when an individual element becomes scarce. There must be substantial benefits to homeostasis to make up for the easily observed costs, but much less is known about the nature of these benefits.

Stoichiometry at the Organism Level

Let us now examine some of the stoichiometric patterns for particular aquatic organisms. Organisms vary in nutrient content for many reasons because there are many functions of different elements in biology. Several fundamental causes for organism-level variation in stoichiometry are phylogeny (closely related species are more likely to resemble each other in terms of stoichiometry as well as many other phenotypic traits), structure (especially for large organisms or those with major investments in support or defense), allometry (effects of body size), and life histories (high rates of growth have a unique stoichiometric signature). In considering these factors, we will start at the base of the food chain, the primary producers.

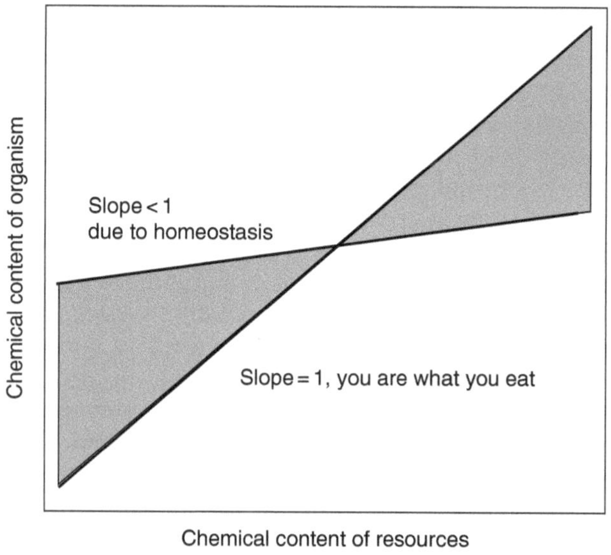

Figure 1 Stoichiometric homeostasis deals with the question of whether organisms 'are what they eat'. Organisms have varying strengths of regulating their own content of different elements. If the chemical content of organisms passively reflects the resources they take in, their own chemistry would match the chemistry of their resources, and observations would lie along the 1:1 line marked 'you are what you eat'. This would be an absence of stoichiometric homeostasis. Stoichiometric homeostasis reduces the slope of the relationship below the 1:1 line (shaded regions). Different organisms and different chemical resources produce different degrees of homeostasis (see for example **Table 1**).

Primary Producers

As is apparent from many examples given throughout this article, stoichiometric variability is the hallmark of primary producers. This variability arises because nutrient uptake and biomass gain can be decoupled. For example, if one exposes a nutrient-limited algal cell suddenly to high concentrations of nutrients, within minutes it can replenish cellular nutrient stores. It then puts those stores to work over time scales of hours to days as the cell gains organic carbon via photosynthesis and eventually divides by fission or mitosis. Cells also may take up nutrients for more than any immediate growth needs, resulting in higher nutrient content than seen in long-term, high growth-rate situations. These cellular nutrient stores sometimes are in the form of specialized storage compounds such as poly phosphate (for P) or cyanophycin (for N). The autotroph cell can also divert resources among different functional pools, for example photosynthetic pigments are N-rich and often cells growing under low light and excess N will be very rich in these pigments. If that cell moves to a high light but low N environment, those pigments can be broken down and the N moved into different pools. In this way, photosynthetic pigments serve as an N reservoir, even though they are not storage compounds per se.

The general model used to describe the linkage between cellular stores and cell growth was proposed years ago by M.R. Droop. Droop's model applies

Table 1 Coefficients of variation (CV) of content of different elements (log_{10}-transformed) measured over a season in a single pond

Organism	*Content of elements*									
Amphipod	C	N	P	Cu	Zn	Se	As	Pb	Hg	Cd
	0.004	0.005	0.007	0.009	0.018	0.019	0.021	0.059	0.079	0.124
Damselflies	C	N	P	Cu	Se	Zn	As	Hb	Cd	Pb
	0.004	0.004	0.016	0.024	0.025	0.030	0.035	0.05	0.053	0.074
Snails	C	N	P	Se	Zn	As	Cu	Cd	Hb	Pb
	0.002	0.005	0.007	0.018	0.018	0.018	0.019	0.040	0.045	0.085
Zooplankton	C	P	N	Zn	Se	Pb	As	Cu	Cd	Hg
	0.003	0.003	0.006	0.007	0.017	0.02	0.026	0.035	0.041	0.059

The higher the CV, the more variable is the element in the given organism relative to the element's mean concentration. Elements are arranged left to right by increasing CV. Macronutrients are unshaded. Essential micronutrients are lightly shaded. Non-essential trace elements are darkly shaded.

Adapted from **Figure 2** in Karimi R and Folt CL (2006) Beyond macronutrients: Element variability and multielement stoichiometry in freshwater invertebrates. *Ecology Letters* 9: 1273–1283. and [permission being sought].

under conditions of a single limiting chemical resource. Droop performed experiments with algae in continuous culture and suggested that growth rate was negatively and linearly related to ratio of biomass to nutrients within the cells (**Figure 2**). The Droop relationship ties a stoichiometric quantity (C:P) to a dynamic variable (growth rate).

Aquatic primary producers come from deeply branched evolutionary lineages. Some of the stoichiometric variability across diverse algal forms was recently described. Two of the major lineages, the 'red superfamily' and the 'green superfamily' can be separated on the basis of their patterns in trace metal composition. Members of the red superfamily (e.g., diatoms) tend to be higher in Cd, Co, and Mn whereas members of the green superfamily (e.g., green algae) tend to be higher in Cu, Zn, and Fe. Quite possibly these differences reflect major geological events such as the relative concentration of ocean oxygen during the evolutionarily origin of these groups.

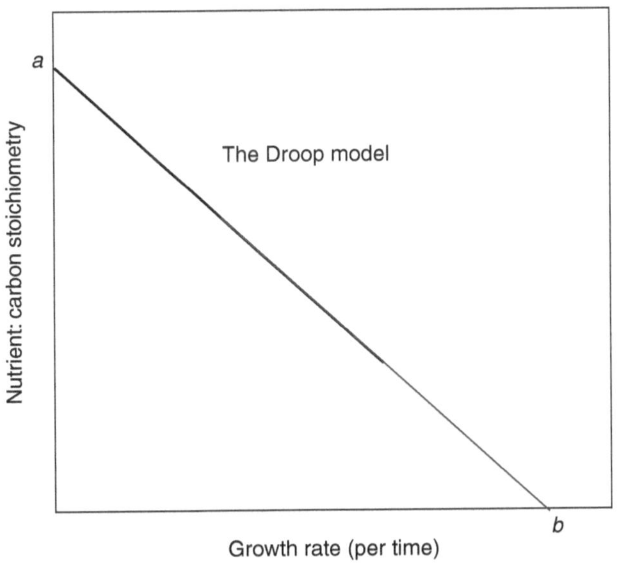

Figure 2 The functional coupling between organismal stoichiometry (e.g., C:P or C:N, vertical axis) and growth (per capita rate, time^{-1}) is often linear when one resource is limiting. This straight-line model is normally written algebraically as

$$\mu = \mu'(1 - (Q_{min}/Q))$$

where μ = cellular growth rate (time^{-1}), Q_{min} = the cell quota at zero growth, and Q is cell quota (in this case, P:C or N:C). The term 'quota' refers to the nutrients per cell or per biomass in the cell, e.g., P:C or N:C. In the Figure, *a* is $1/Q_{min}$ and *b* is μ'. The parameter μ' is a theoretical (high) growth rate that would be reached at infinite quota. The thick part of the line represents a range where one would expect to see actual observations. The narrow part of the line would be growth rates that were never observed.

Freshwater algae attached to substrates include microscopic forms ('periphyton') living in close association with other microbes and nonliving organic materials, all collectively forming a biofilm. These biofilms are the main source of autochthonous production in streams and sometimes in nearshore lake habitats as well. Living algae are often a minor contributor to the total carbon within such biofilms. Especially in streams, the stoichiometry of the entire biofilm is made up by living and nonliving components. In order of decreasing N and P content, those include wood, leaf litter, green leaves, and both periphyton and fine particulate organic matter (**Figure 3**). Stoichiometric variability at the base of stream food webs is probably more dependent on the relative contributions of these different components than it is on variation within any one component.

Zooplankton

Moving up the food chain from primary producers, we encounter a variety of herbivores. In water columns, these herbivores are members of the zooplankton. The freshwater zooplankton is not a particularly taxonomically diverse group; protozoans, rotifers, cladocerans and copepods as well as juvenile insects are common members. Cladocerans and copepods both belong to the Crustacea, and they plus the insects are all Arthropods. Though species from any number of these several major taxonomic groups may coexist, may have overlapping diets, and may share many other aspects of their ecology such as predators, they nevertheless can have divergent stoichiometries. In particular, some species, such as *Bosmina*, have low P content, whereas others, such as *Scapholeberis* or *Daphnia* have P content several times higher. All of these species, diverse from the standpoint of their elemental content, are members of the same taxonomic order. Homeostatic regulation of element content is highly variable, even within members of a single genus of zooplankton. Thus, stoichiometric patterns include not just average nutrient content but also the degree of variation within the taxon. We have made much progress in understanding the importance of average nutrient content, which will be apparent later in this article, but we have a very poor understanding of the ecological relevance of differences in homeostatic regulatory ability. Why some species are very homeostatic in element content and others are less tightly regulated is not at all understood yet.

Cross-species variability in average phosphorus content has been linked to life-history differences because biochemicals responsible for growth have a

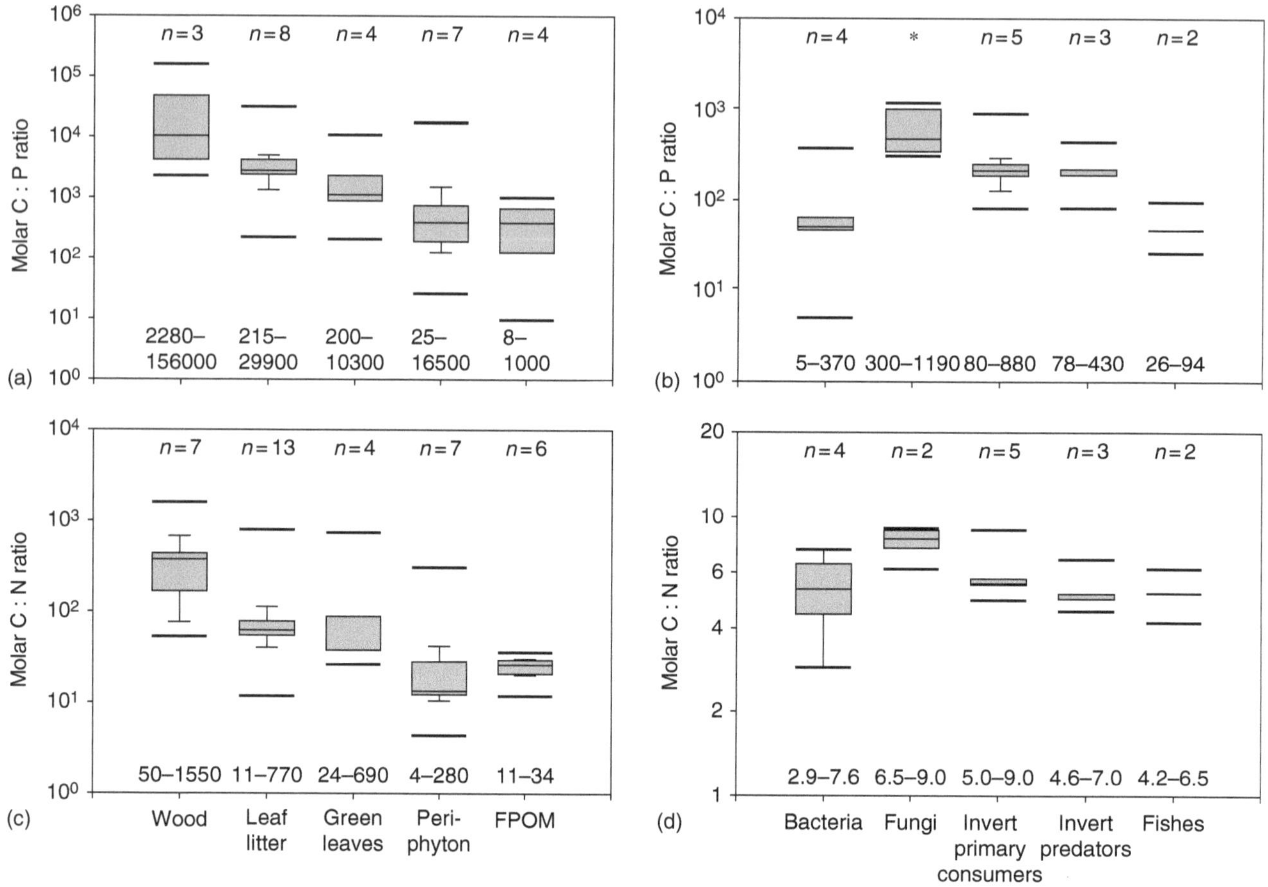

Figure 3 Means and variability of C:N:P stoichiometry in common freshwater benthic resources and consumers. FPOM: fine particulate organic matter. Boxes and whiskers give the median, 10th, 25th, 7th, and 90th percentiles of the data within each category. Original sources of data include 49 literature studies. Ranges are given in text below the boxes. Reproduced from **Figure 1** in, Cross WF, Benstead JP, Frost PC, and Thomas SA (2005) Ecological stoichiometry in freshwater benthic systems: Recent progress and perspectives. *Freshwater Biology* 50: 1895–1912, with permission from Blackwell Publishing.

distinct elemental signature. Species with a potential for rapid biomass growth rates must contain large complements of P-rich biosynthetic machinery. This machinery is dominated by ribosomes, subcellular organelles involved in protein synthesis and which have distinctly high P content. The Growth Rate Hypothesis (GRH) proposes that elevated P demands caused by increased allocation to P-rich ribosomal RNA under rapid growth drives variation in whole-organism P content and thus whole-organism C:P and N:P ratios. Through the GRH, there may be very strong couplings between life histories and stoichiometry and therefore nutrient and energy flow. One example of couplings between growth and P content is given in **Figure 4**. Note that the GRH coupling exists when overall food concentration is high. When food concentration is low, animals are likely mainly energy limited and couplings to nutrient content are muted or absent. Linkages among growth rate and stoichiometry seem to depend upon the nature of growth limitation, what resource is limiting at any given time.

Other Aquatic Invertebrates

The stoichiometry of stream and lake benthic invertebrates has been explored, and wide chemical variation among the diverse members of this group has been observed. In one such study, %P ranged from ~0.2% to 1.5% and the N:P ratio ranged from <20 to >100. Trichoptera and Ephermeroptera were lower in %P and %N than Diptera, Odonata, and Plectoptera. When examined as a function of feeding group, predators were found to have higher P and N than other feeding groups. This is an impressive amount of stoichiometric variation. Not many studies have examined the consequences of this stoichiometric variation in streams, but perhaps some of the same processes we describe later relative to planktonic systems hold as well in streams.

Bacteria and Protozoa

Heterotrophic microorgansms include bacteria and Protozoa. Not much work has been done yet on the stoichiometry of freshwater Protozoa. Freshwater bacteria are often found to have relatively high N and P content (low C:N and C:P) compared with other aquatic organisms. Studies suggest that bacterial N content is very homeostatic but P content varies with the availability of P for uptake as well as the growth rate of the bacteria. One study on *E. coli* that varied the chemistry of the growth substrate and the growth rate itself found a range in C:P from 40 to 75, a range in N:P from 11 to 18 and a very constant C:N of 4. At high growth rate, bacteria had low C:P and elevated RNA content. In comparing the results for *E. coli* with literature findings, this same study suggested that individual bacterial strains were strongly homeostatic whereas in nature bulk bacterial communities varied much more widely, suggesting that varying dominance of individual strains may be responsible for much of the variation in bacterial C:N:P composition in the field. In this respect, stoichiometric patterns in the bacteria have a resemblance to other organisms, including freshwater zooplankton.

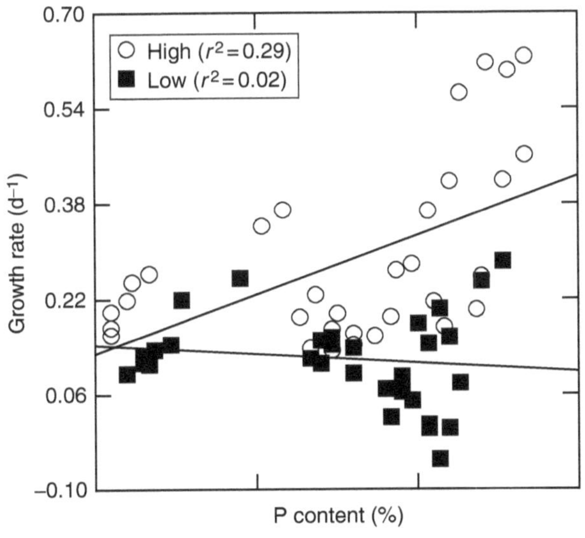

Figure 4 In a laboratory bioassay study involving four species from three genera of cladocerans, growth rate and P content were positively related when food concentration was high (open circles) but not when it was low (closed squares). At low food concentration, animals were energy limited, but at high food concentration they were likely P limited so that growth was linked to P levels in the animals. Reproduced from **Figure 2** in Ferrão-Filho AS, Tessier AJ, and DeMott WR (2007) Sensitivity of herbivorous zooplankton to phosphorus-deficient diets: Testing stoichiometric theory and the growth rate hypothesis. *Limnnology and Oceanography* 52: 407–415, with permission from American Society of Limnology and Oceanography.

Fish

Fish are often the ecologically dominant vertebrates in aquatic ecosystems. Fish can be herbivores as well as carnivores or even detritivores or omnivores. Fish have numerous effects on aquatic ecosystems and at times they can be so abundant that they channel great quantities of nutrients in and out of their own biomass through feeding, excreting, etc.

This group too is characterized by wide stoichiometric variation and the ecological consequences of such a wide stoichiometric variability are just beginning to be understood. Phosphorus content of fish ranges from less than one to over four percent. Low-P fish include minnows in the Cyprinidae family. High-P fish occur in both temperate and tropical ecosystems; they are bony and are typical of the Percidae, Centrachidae and Loricariidae families. In fish, evolutionary patterns associated with structure (boniness) account for most of the stoichiometric variation. We see this in a very strong correlation between fish Ca and fish P (**Figure 5**). Fish skeletons can account for from one to more than five percent of body mass. Bony skeletal material is made of a calcium phosphorus mineral called apatite with chemical formula $Ca_5F(PO_4)_3$ where other anions may replace F. This large

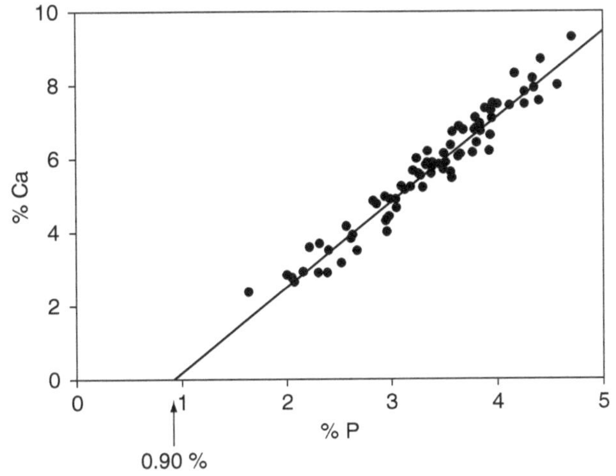

Figure 5 Ca vs. P for 18 species of fish. The line has equation $y = 2.31x - 2.09$ ($r^2 = 0.95$). Reproduced from **Figure 4** in Hendrixson HA, Sterner RW, and Kay AD (2007) Elemental stoichiometry of freshwater fish in relation to phylogeny, allometry and ecology. *Journal of Fish Biology* 70: 121–140, with permission from Blackwell Publishing.

quantity of mineral has a great effect on the stoichiometry of individual species of fish – the slope of the line in **Figure 5** is almost exactly as predicted from the stoichiometry of apatite mineral.

We will now consider the ecological ramifications on these stoichiometric properties of species.

Food Quality

Consumers rely on organic food resources that may or may not closely match their own chemical requirements for survival and growth. The idea of a 'balanced' diet is familiar to everyone from the standpoint of human nutrition – it means that the chemical composition of food matches closely an ideal composition. Stoichiometry helps us understand how balanced vs. imbalanced a particular diet may be for any given aquatic consumer. Natural ecosystems often present highly imbalanced diets to consumers. Effects of many different nutritive substances such as vitamins, amino acids and fatty acids are known. Here, we will focus solely on individual chemical elements.

To help understand how different elements may limit consumer growth, we use a tool called a 'Threshold Element Ratio' (TER). A TER is the ratio of elements in an organism's resources where that organism is equally limited by both elements. When the ratio of elements (e.g., C:P) in the food exceeds the TER, the element in the denominator (P) limits consumer growth. Likewise, when the ratio is lower than the TER, the element in the numerator (C) limits growth. TERs are determined by two main things: the nutrient content of the consumer organism, which we considered earlier, and growth efficiency, which we will consider next.

Growth efficiency is defined as the amount an organism gains divided by the mass it ingests. Growth efficiencies based on carbon for aquatic consumers average around 0.3 with large variation around that average. Stoichiometry helps explain some of that variation. The growth efficiency of a strongly homeostatic organism must adjust to the composition of the food, otherwise the consumer would not be able to maintain its homeostasis. There are many well studied examples where decreased growth efficiency on poor quality (low nutrient) foods has been observed in aquatic systems. A good example is heterotrophic bacteria where growth efficiency is typically well above 50% for growth on substrates with C:N < 5, but declines to 10% or less on substrates with C:N ratios > 30. Because bacterial processing of organic carbon is often a very big part of the organic carbon cycle in lakes, these stoichiometric modulations of C shunted into cell growth versus loss as CO_2 or organic C can play a very large role in the biogeochemical cycling of carbon.

TERs have been calculated for many aquatic organisms (**Figure 6(a)**). Fish have low C:P TERs (typically about 100–200), a reflection of their P-rich skeleton. The highest C:P TERs (about 500) are found for aquatic insects, which of course do not have bony skeletons but still require P for making RNA and other biomolecules. In fact, within the invertebrates, there is a statistically significant negative relationship between maximal potential growth rate and the C:P TER (**Figure 6(b)**, open symbols). Here we see the elemental requirements for the machinery for growth affecting whole-organism ecology, consistent with the Growth Rate Hypothesis described earlier.

There have been numerous studies on stoichiometric dimensions to food quality in zooplankton.

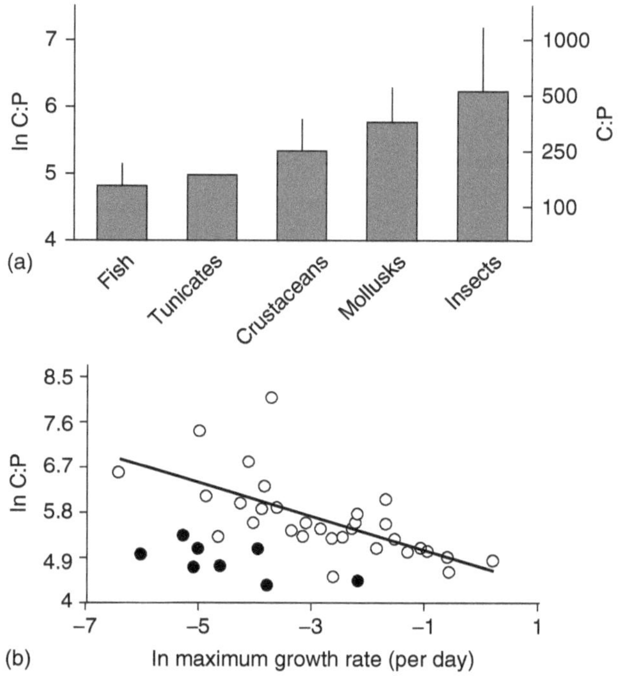

Figure 6 Ratios of C:P (moles C per moles P), where different kinds of aquatic animals switch between growth limitation by carbon vs. growth limitation by phosphorus. When food C:P exceeds an animal's TER, the animal is limited by P and when food C:P is less than an animals's TER, carbon limits growth. (a) Their higher TERs (mean ± SD) indicate that less food P is required per unit food carbon for mollusks and insects relative to fish, tunicates, and crustaceans. (b) A higher growth rate depresses the C:P TER in invertebrates (open circles), a signal consistent with the growth rate hypothesis. Further, at any given growth rate, high P content depresses fish C:P TER (closed circles) below that observed for invertebrates. Adapted from **Figures 2** and **4** in Frost PC, Benstead JP, Cross WF, Hillebrand H, Larson GL, Xenopoulos MA, and Yoshida T (2006) Threshold elemental ratios of carbon and phosphorus in aquatic consumers. *Ecology Letters* 9: 774–779, with permission from Blackwell Publishing.

One example comes from the study of a model organism in ecological stoichiometry, *Daphnia*, the waterflea. *Daphnia* are keystone herbivores in planktonic systems; they merit special attention because of the role they can play in decreasing algal biomass. Above, we discussed how *Daphnia* are generally high in P content. As one would therefore expect, *Daphnia* growth is sensitive to the P content of their food (**Figure 7**). Algae with high C:P ratio (and thus low P content) dramatically lowers *Daphnia* growth rate, in this case lengthening by three-fold the time to double biomass. Other studies on zooplankton stoichiometry have shown: (a) low P zooplankton such as *Bosmina* are much less affected by the C:P of algal food than are high P zooplankton like *Daphnia*; and (b) particular life stages of some zooplankton may be particularly sensitive to C:P of food; thus, an overall effect on the animal may be determined most at specific points in its life cycle One recent study showed that hybrids of two *Daphnia* species performed better on high C:P food than the two parental species, suggesting that success in stoichiometrically imbalanced systems might be responsible for maintenance of hybrids in the ecosystem.

Stoichiometry predicts that consumers of high element content should be rare in environments where that element is scarce. A large Norwegian study showed a strong relationship between the abundance of *Daphnia* and seston C:P ratios (**Figure 8**). In this study, all sites with high *Daphnia* abundance had seston C:P well below this organism's C:P TER. Seston C:P ratios were the most important variable in explaining the abundance of *Daphnia*. In general, physiological studies of TERs, laboratory growth studies, and field studies all point to a consistent conclusion, that high P herbivores like *Daphnia* are affected by stoichiometric imbalances between food and consumer.

There are some other very high imbalances between consumer and resource stoichiometry in streams and lake benthos. Certain consumers in those habitats subsist largely on nutrient-poor detritus. An example of stoichiometry affecting a stream invertebrate is given in **Figure 9**. In this study, snails were given two quantities of periphyton food that had been grown under different nutrient treatments. Periphyton grown with added P (treatments +P and +N +P) had C:P of about 200 whereas C:P was 400 or more when P was not added (treatments Ambient and +N). The snails were strongly homeostatic and they showed pronounced stoichiometric effects on growth at low but not high periphyton quantity.

Although most studies of food quality have emphasized what happens when food is deficient in some element, recent studies have also shown that food may be imbalanced when it is too high in a particular element. Whether reduced food quality at high nutrient content is due to overt toxicity or because very high content of some element indicates some other chemical imbalance is not yet understood.

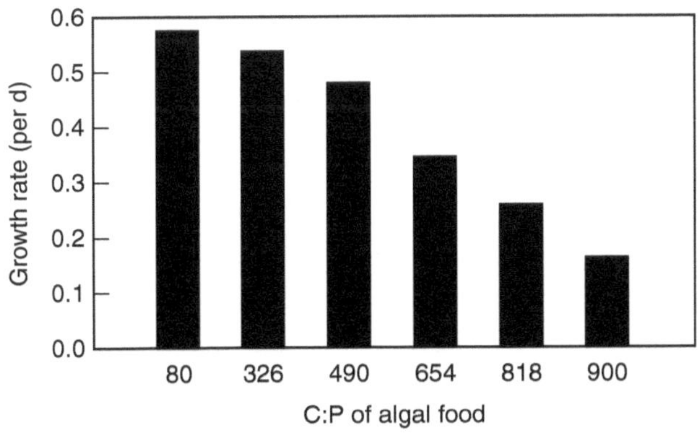

Figure 7 The planktonic herbivore, *Daphnia magna*, has much reduced growth rate when it feeds on algae of high C:P ratio (low P content). All algal food treatments in this study were the same species of algae, *Scenedesmus acutus*, grown under different conditions to get a stoichiometric range from 80 to 900. Converting from growth rate units one can find that at low C:P, these *Daphnia* double their biomass in 1.2 days, whereas at high C:P it takes 3.8 days, or more than three times as long. Adapted from DeMott WR, Gulati RD, and Siewertsen K (1998) Effects of phosphorus-deficient diets on the carbon and phosphorus balance of *Daphnia magna*. *Limnology and Oceanography* 43: 1147–1161, with permission from American Society of Limnology and Oceanography.

Nutrient Recycling

Next we consider stoichiometric aspects of nutrient cycling. In particular, we consider processes that deliver nutrients to the dissolved pool. Nutrients enter the dissolved pool in inland ecosystems from the air, the surrounding land, the sediments or substrate, and also from the organisms living in the water, which release nutrients as waste products. There are stoichiometric aspects to all of the above, but here we will focus on biological resupply, or nutrient recycling by consumers. Consumers here can range from Protozoa to fish.

The stoichiometry of nutrient recycling is related to food quantity, considered above. When an organism's food is deficient enough in a particular element for it to be limited by that element, a great premium is placed on not wasting that element. In order to maintain a stoichiometric homeostasis, organisms must efficiently retain elements in short supply in their food relative to stoichiometric requirements. For the same reason, they must either not assimilate or they must release upon assimilation, elements that are in

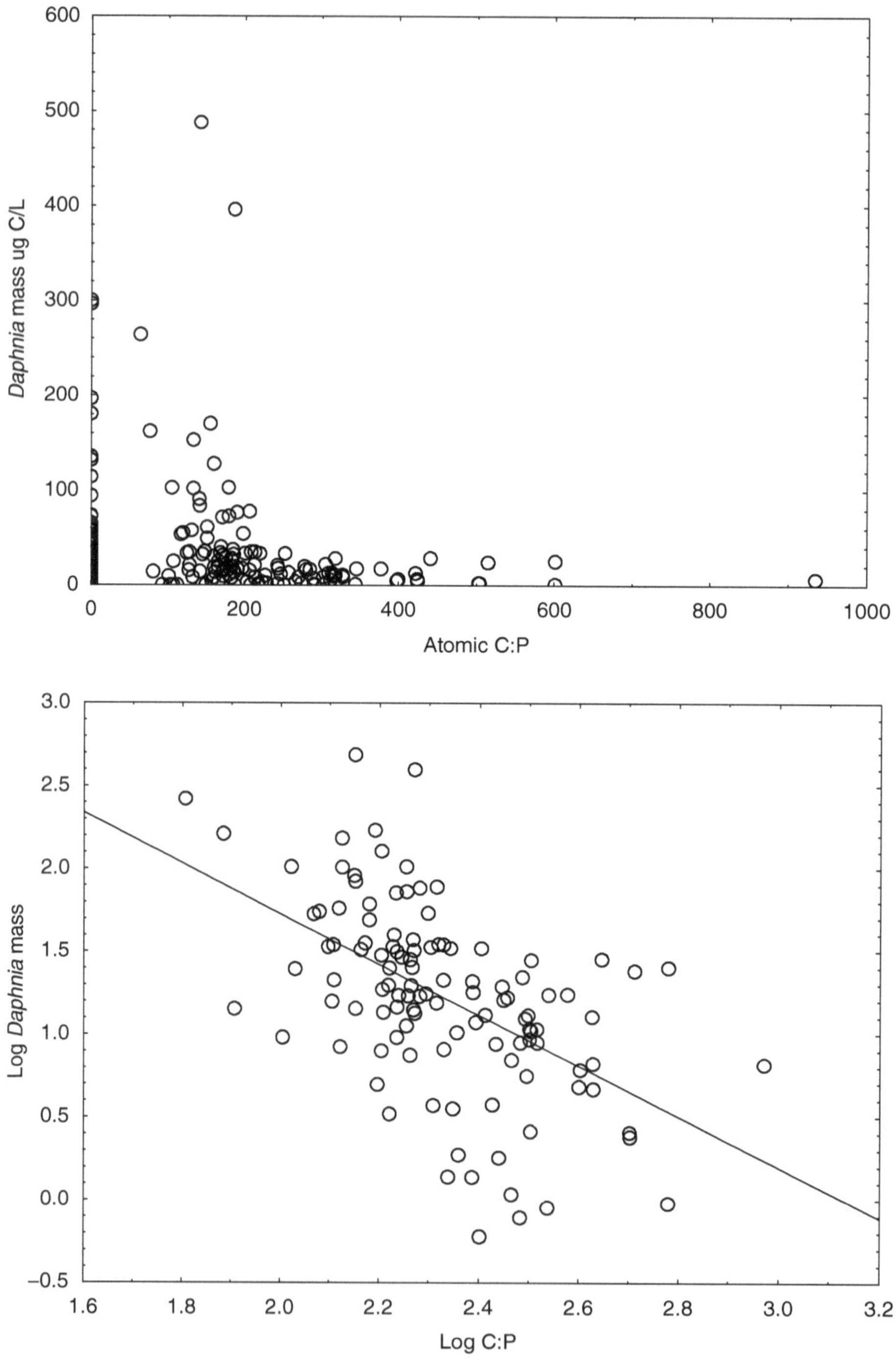

Figure 8 Consistent with predictions of stoichiometric theory, lakes with high C:P in their seston have lower abundance of the high-P, keystone genus, *Daphnia*. Data are from a survey of 400 Norwegian lakes. The upper panel shows the data plotted on linear axes and emphasizes how large populations build up only at C:P below the *Daphnia* TER, which typically is about 300. The lower panel shows the same data plotted on logarithmic axes and shows a strong consistently negative slope. Replotted from data published in Hessen DO (2006) Determinants of seston C:P ratio in lakes. *Freshwater Biology* 51:1560–1569. [permission being sought].

surplus to their own requirements. Stoichiometric homeostasis thus has a highly constraining effect on patterns of nutrient recycling by consumers.

Many studies on many organisms have considered the N:P ratio of release by individual aquatic organisms. In **Figure 10** we see data gathered from freshwater and marine zooplankton. Other studies with similar results have been conducted on bacteria, Protozoa, and vertebrates including fish. **Figure 10** illustrates the two principal stoichiometric effects on

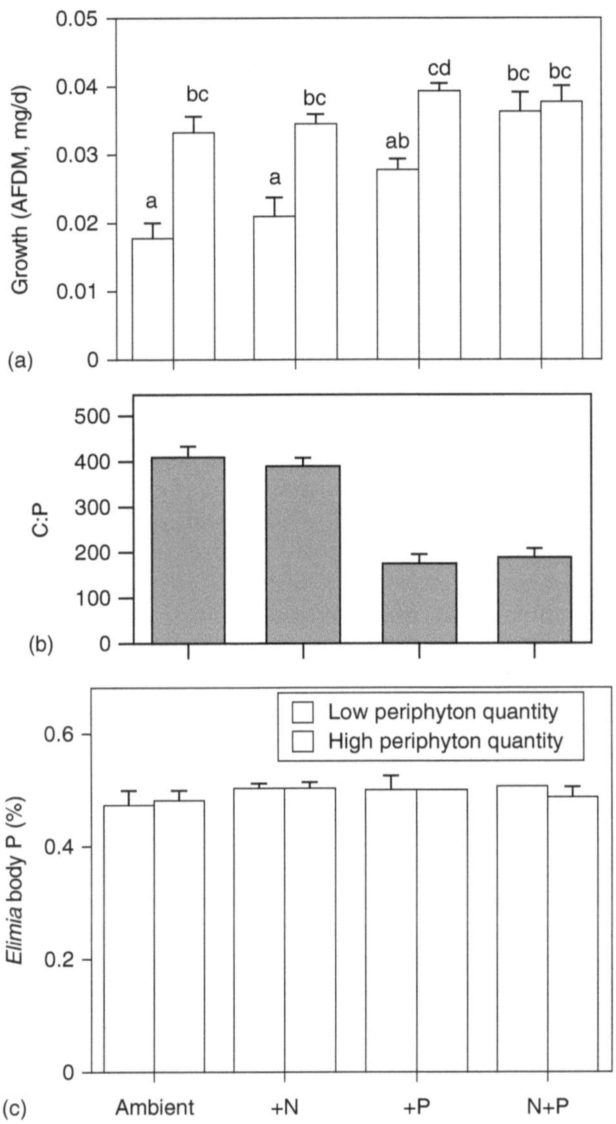

Figure 9 Effect of nutrient addition on periphyton and snail stoichiometry. Snails (*Elimia livescens*) were reared in experimental streams (flumes) and fed periphyton on clay tiles. An approximately similar periphyton biomass (low or high) was given to snails in all treatments. The experiment lasted 34 days. (a) Snails growth varied with nutrient treatments at low periphyton quantity but not high periphyton quantity. At low periphyton quantity, additions of P stimulated snail growth. (b) The C:P of the periphyton under different nutrient treatments. (c) The %P of the snails under different nutrient treatments. Note that the periphyton have more flexible nutrient content than the snails. Adapted from Stelzer RS and Lamberti GA (2002) Ecological stoichiometry in running waters: Periphyton chemical composition and snail growth. *Ecology* 83:1039–1051, with permission from Ecological Society of America (ESA).

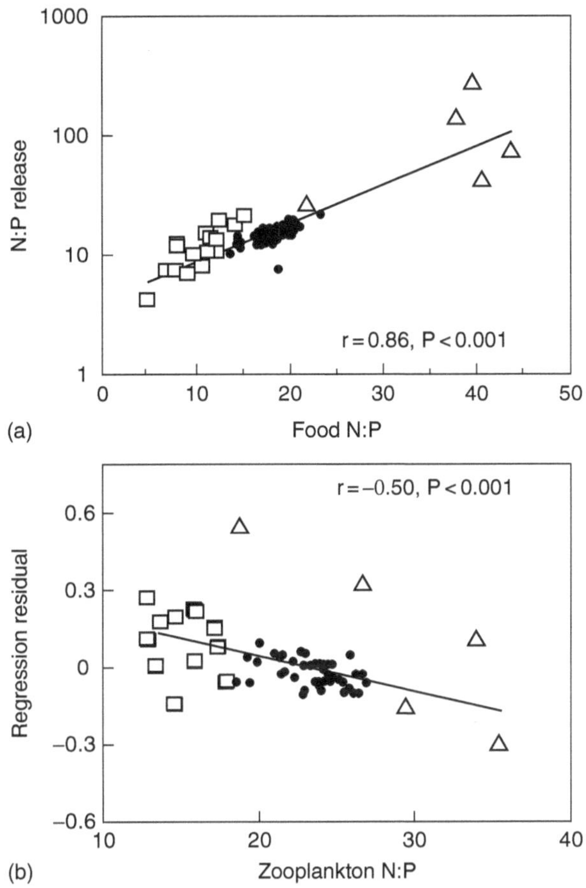

Figure 10 Two aspects of stoichiometry altering nutrient fluxes through organisms. (a) The N:P released by zooplankton is positively related to the N:P of the food they are eating. (b) After accounting for the trend in (a), the consumers' own N:P ratio is negatively related to their N:P release (b). Reproduced from **Figure 4** in Elser JJ and Urabe J (1999) The stoichiometry of consumer-driven nutrient recycling: Theory, observations, and consequences. *Ecology* 80:735–751, with permission from Ecological Society of America (ESA).

nutrient recycling. First, the N:P recycled is positively related to the N:P in the food ingested. Note that because of the log scale in this Figure, the N:P released is not directly proportional to the N:P eaten. Instead, as N:P in the food increases slowly, N:P released increases very quickly. This upward trajectory was predicted mathematically, and it has been shown repeatedly in different nutrient cycling studies. When feeding on food of low N:P, a homeostatic consumer must keep the N for itself but recycle excess P, so that N:P released is very low. Likewise, when feeding on food of high N:P a homeostatic consumer must release excess N but keep relatively scarce P, making N:P recycled high. The second principle stoichiometric effect on nutrient recycling comes about because of the chemical variability among consumers. Consumers that are low in N:P will recycle at high N:P ratios and organisms that are high in N:P will recycle at low N:P. Given that many studies point to the critical importance of N:P loading as determining algal community structure, this example illustrates how ecological stoichiometry integrates species-level information with ecosystem-level processes. It is remarkable that something as potentially complicated as the rates of nutrient release by living animals relates so strongly to simple measures such as nutrient content.

Community Dynamics

The separate effects of stoichiometry on animal growth and nutrient cycling considered above combine to influence community dynamics. One long-term culture study found two markedly different outcomes. In one, cultures had high abundance of algae but low abundance of grazers (a 'green' world). This first outcome was seen in the cultures with high algal C:P ratio. In the other outcome, cultures had low abundance of algae but high abundance of grazers (a 'clear' world). This second outcome was seen where algal C:P ratios were low. There was a sharp threshold between the two outcomes near the threshold, small differences in growth conditions promoting different stoichiometric ratios had large effect on community dynamics. Stoichiometry played a role in separating the green and the clear world. A second interesting aspect of this study was that high grazing promoted high food quality so that zooplankton had a positive demographic effect on their own populations – more zooplankton meant better food and therefore higher zooplankton growth. This kind of positive feedback between population size and population growth contrasts markedly with classical ideas about population regulation.

Mathematical, theoretical analyses have uncovered other surprising aspects of community dynamics associated with stoichiometry. Stoichiometrically explicit models of grazers, algae and nutrients suggest the existence of complex dynamics, including steady population changes through time, irregular, even chaotic, nonrepeating cycling of populations, and, in some versions of models, deterministic grazer extinction.

These different community dynamics emerge quite easily and naturally from models that explicitly incorporate stoichiometry. They suggest that classical, nonstoichiometric, ideas about population regulation, dynamics, and competition may be far off base when stoichiometric effects are important.

Whole-System Scale

These many stoichiometric patterns and processes that we have considered so far operate simultaneously to form aggregate stoichiometric properties of entire ecosystems. A touchstone for understanding ecosystem-level stoichiometry is the Redfield Ratio, a concept put forward by the oceanographer A.C. Redfield in the 1930s–1950s. Redfield noticed a correspondence between the C:N:P ratio in particulate matter of the surface ocean and dissolved nutrients in the deep ocean. The atomic ratio he observed was C_{106}:N_{16}:P_1, which is now called the Redfield Ratio. From this similarity he deduced that the entire global ocean had its nutrient content determined mainly by the biota. His argument was that both C and N come into stoichiometric balance with P, as set mainly by the demands of the oceanic biota. If either C or N are in short supply, the biogeochemical processes of primary production or N fixation can tap into atmospheric sources of these elements. Likewise, excesses of C or N can be vented to the atmosphere by respiration or denitrification. Phosphorus lacks an important gaseous phase, and its supply to the oceans is set by geological (or, now, human) action on the continents, which controls the rate of runoff from major river systems. The Redfield Ratio is similar to the ratios that many phytoplankton obtain when growing at either high growth rate or light limitation. The variation among species is such that if one takes an average of many different species, a C:N:P of 106:16:1 emerges. The Redfield ratio thus describes an overall mean of many species growing without strong single-element nutrient limitation.

The same kinds of measurements can be made for inland waters, but lakes, ponds, streams and rivers exhibit a greater degree of variability in their stoichiometric C:N:P ratios than do the oceans. This variability across sites can be due to many factors. Inland ecosystems are subject to widely varying inputs of elements from the atmosphere and from their watershed. They flush relatively rapidly compared with the time scales necessary for Redfield balancing mechanisms. Individual lakes, ponds, streams or rivers may be more homogeneous in terms of the many variable factors that determine biogeochemical potentials than the entire world's oceans; for example, highly unproductive lakes may lack anoxic zones that would support high levels of denitrification. They may be dominated by a small number of species during bloom conditions. There is a potentially large but variable detrital pool, poor in N or P, which may be found. Finally, they may be strongly limited by a single element, whether it be P or N or perhaps even some other nutrient. Given this large number of factors that can perturb C:N:P ratios away from Redfieldian balance, it is not surprising to see that freshwater systems are generally more stoichiometrically variable at the ecosystem level than are oceans (**Figure 11**). This great stoichiometric variability is the primary force that sets into motion many of the processes already described in this article, such as food quality constraints or stoichiometric determination of nutrient recycling patterns. Let us therefore now further analyze some of this great stoichiometric variability.

One explanation for this variation is the Light: Nutrient hypothesis. Light stimulates photosynthesis and thus is related to the input of carbon into particulate matter in inland ecosystems. As we discussed earlier, rates of carbon and nutrient acquisition in autotrophs are not tightly linked. When light is plentiful but nutrients are scarce, photosynthesis can be high compared to nutrient uptake and thus there develops a high C:P or C:N ratio in autotroph biomass. Characteristics of high light environments are clear water, shallow mixing or, in the case of benthos, shallow depth. The Light:Nutrient hypothesis suggests that stoichiometric variation a the base of food chains results from the relative balance of energy (light) and nutrients (N and P) available for primary producers.

Support for the Light:Nutrient hypothesis is accumulating. C:P ratios in 65 European lakes generally declined with increasing mixing depth (and zooplankton biomass correspondingly increased, as one would predict from food quality constraints). An elegant mathematical model of how mixing depth should affect lakes by way of stoichiometric shifts corresponded to the same patterns in the data. In another study, experimentally incubating phytoplankton from three Ohio lakes at different light levels and different nutrient levels caused seston C:N:P to shift in ways predicted by the Light:Nutrient hypothesis. C:P and C:N went up at higher light and down with increased nutrients. Third, seston in 112 Norwegian lakes covering a C:P from 24 to 1842 was related to measures of nutrient availability and it showed several different effects of decreased light owing to shading by algae in more productive water columns. Seston C:P in these Norwegian lakes declined strongly with increasing phosphorus (**Figure 12**). No effects of lake area, season or latitude on C:P were detected. Particulate

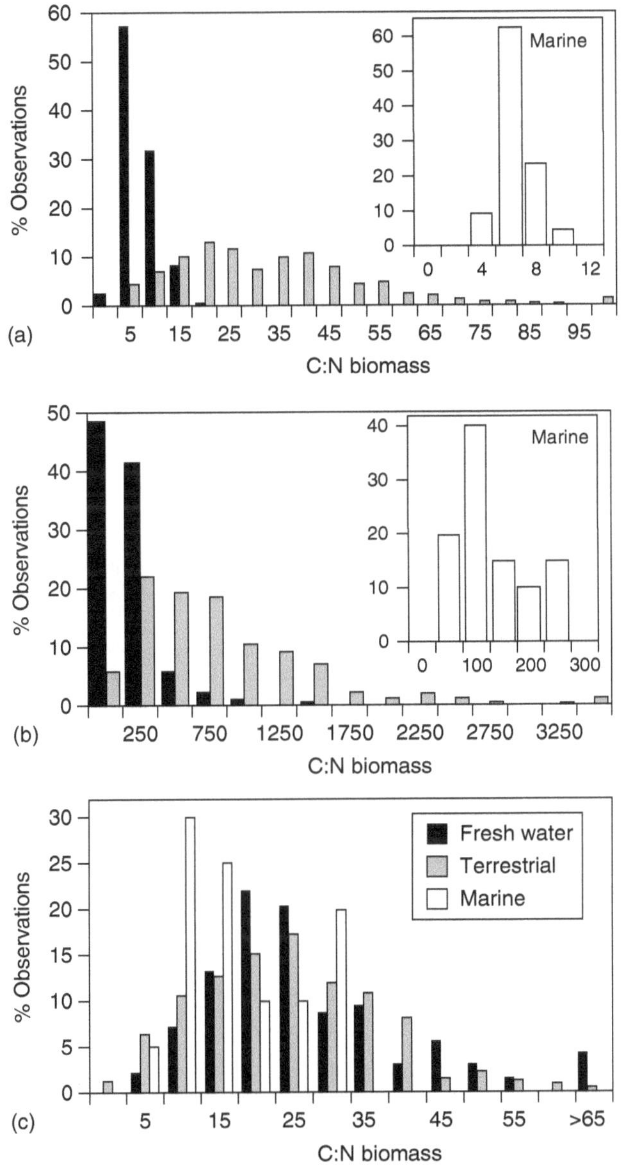

Figure 11 Variations in C:N:P stoichiometry at the base of marine (white), freshwater (black), and (for reference) terrestrial (gray) food webs. Data for lakes and oceans represent values of seston, including all particles living and dead, whereas data for terrestrial habitats reflect values for individual leaves. Note that freshwater systems have more variable C:N, C:P, and N:P ratios than do marine systems. This means the 'raw material' for stoichiometric determination of growth rate, nutrient cycling, etc. is greater in freshwater than marine systems. Though outside the scope of this article, this Figure also shows that terrestrial systems too have great stoichiometric variability at the base of their food chains. Reproduced from Figure 3.13 in Sterner RW and Elser JJ (2002) *Ecological Stoichiometry: The Biology of Elements from Molecules to the Biosphere*. Princeton, NJ: Princeton University Press. [permission being sought]

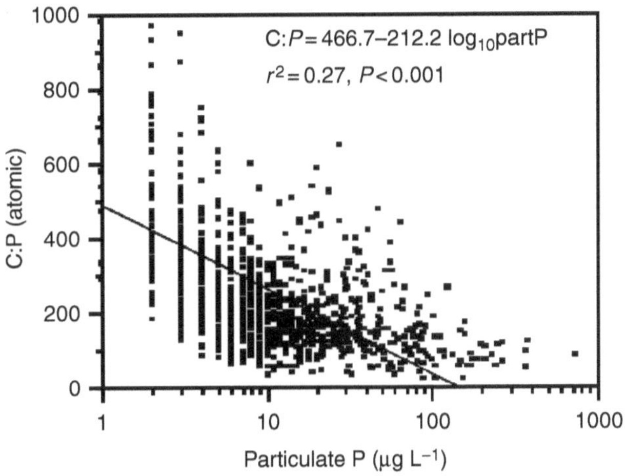

Figure 12 Seston C:P ratios decline with increasing particulate P and therefore with overall productivity. Reproduced from Figure Hessen DO (2006) Determinants of seston C:P ratio in lakes. *Freshwater Biology* 51:1560–1569. [permission being sought]

C did not correlate with C:P. The highest C:P ratios occurred in low P, therefore unproductive, lakes. C:P increased with increased values of the ratio of Secchi depth:total P and similarly C:P increased with the POC:chlorophyll ratio (representing the relative proportion of particulate detritus; **Figure 5**). As we saw earlier (**Figure 8**), high P *Daphnia* in these lakes were less abundant when seston C:P was high. There is mounting evidence that inland aquatic ecosystems do respond to relative availability of light vs. nutrients and thus these abiotic factors influence ecosystem dynamics in a way that is strongly mediated by stoichiometry.

One final stoichiometric aspect at the whole ecosystem scale relates to carbon fluxes in ecosystems. The fate and cycling of carbon in ecosystems is a topic of increasing concern in today's world. The human economy is run in part on cycling of carbon and nutrients for energy production, transportation, agriculture, etc. Fossil fuel combustion combined with land use changes have resulted in excess CO_2 building up in the atmosphere. Lakes often are carbon sources (not sinks) relative to global carbon cycling due to decomposition of terrestrially derived organic matter. However, lakes, and especially reservoirs, do build up significant reservoirs of carbon in their sediments. One early study suggested that lakes and reservoirs globally accumulated organic carbon at a rate of about 200 Tg year^{-1}, more even than the world's oceans even though the oceans cover a much greater fraction of Earth's surface. The fate of carbon cycling through ecosystems is closely related to stoichiometric ratios. It has been noted that ecosystems that are high in C:P or C:N show more carbon cycled through detrital food chain pathways, whereas those ecosystems with lower C:P or C:N ratios show more carbon channelized through herbivore/grazing pathways. We can expect that lakes and reservoirs that are higher in C:P and C:N likely bury a larger fraction of their organic carbon than do lakes at the opposite end of the stoichiometric gradient. Thus, ecological stoichiometry can be an important aspect of how inland ecosystems fit into global carbon cycles.

Further Reading

Andersen T, Elser JJ, and Hesson DO (2004) Stoichiometry and population dynamics. *Ecology Letters* 7: 884–900.

Boersma M and Elser JJ (2006) Too much of a good thing: On stoichiometrically balanced diets and maximal growth. *Ecology* 87: 1325–1330.

Cross WF, Benstead JP, Rosemond AD, and Wallace JB (2003) Consumer-resource stoichiometry in detritus-based streams. *Ecology Letters* 6: 721–732.

Elser JJ, Acharya K, Kyle M, *et al.* (2003) Growth rate–stoichiometry couplings in diverse biota. *Ecology Letters* 6: 936–943.

Frost PC, Benstead JP, Cross WF, *et al.* (2006) Threshold elemental ratios of carbon and phosphorus in aquatic consumers. *Ecology Letters* 9: 774–779.

Frost PC, Hillebrand H, and Kahlert M (2005) Low algal carbon content and its effect on the C:P stoichiometry of periphyton. *Freshwater Biology* 50: 1800–1807.

Hecky RE, Campbell P, and Hendzel LL (1993) The stoichiometry of carbon, nitrogen, and phosphorus in particulate matter of lakes and oceans. *Limnology and Oceanography* 38: 709–724.

Hendrixson HA, Sterner RW, and Kay AD (2007) Elemental stoichiometry of freshwater fish in relation to phylogeny, allometry, and ecology. *Journal of Fish Biology* 70: 121–140.

Hessen DO (2006) Determinants of seston C:P ratio in lakes. *Freshwater Biology* 51: 1560–1569.

Karimi R and Folt CL (2006) Beyond macronutrients: Element variability and multielement stoichiometry in freshwater invertebrates. *Ecology Letters* 9: 1273–1283.

Quigg A, Finkel ZV, Irwin AJ, *et al.* (2003) The evolutionary inheritance of elemental stoichiometry in marine phytoplankton. *Nature* 425: 291–294.

Sterner RW (1995) Elemental stoichiometry of species in ecosystems. In: Jones C and Lawton J (eds.) *Linking Species and Ecosystems*, pp. 240–252. New York: Chapman and Hall.

Sterner RW and Elser JJ (2002) *Ecological Stoichiometry: The Biology of Elements from Molecules to the Biosphere*. Princeton, NJ: Princeton University Press.

Sterner RW and Hessen DO (1994) Algal nutrient limitation and the nutrition of aquatic herbivores. *Annual Review of Ecology and Systematics* 25(1–29).

Vanni MJ, Flecker AS, *et al.* (2002) Stoichiometry of nutrient recycling by vertebrates in a tropical stream: Linking species identity and ecosystem processes. *Ecology Letters* 5(2): 285–293.

Relevant Website

www.wikipedia.com.

Redox Potential

M Søndergaard, University of Aarhus, Denmark

Definitions and General Aspects

Redox potential is a quantitative measure of the ability of a substance or a biogeochemical system to receive or donate electrons. In environmental studies, it is usually used to characterize the reducing or oxidizing capacity of a system, typically in an anoxic or close to anoxic environment as, for example, the surface sediments of freshwater, marine, or estuarine systems. Here, a steep gradient in redox potential is found around the sediment–water interface because of the consumption of oxygen and other electron acceptors when organic material is being decomposed and the supply of oxygen from the water column is limited. The redox potential gives a general description of many redox reactions and can be viewed as a measure of a system's overall reducing or oxidizing capacity.

Redox potential is measured in millivolts (mV), and is defined relative to a standard hydrogen electrode, generally reported as E_h. The standard electrode is arbitrarily given a potential of 0 mV. All systems donating electrons to the hydrogen electrode have a negative potential and all systems taking up electrons from a hydrogen electrode have a positive redox potential. A low and negative redox potential indicates a strongly reducing environment, such as that found in anoxic sediments, while a high and positive redox potential is indicative of oxic conditions such as those found in water saturated with oxygen.

Measurements of the redox potential in environmental studies have been used widely since the early 1940s. First, it was used to describe redox changes in stratified lakes and at the sediment surfaces, and later also in rivers and wetlands in many different types of studies on redox reactions with important implications for organic carbon cycling and nutrient dynamics. Most frequently, redox potential measurements have been used to describe aspects related to nutrient dynamics in lakes.

Measuring the Redox Potential

Redox potential is most commonly measured by an inert metal electrode in conjunction with a saturated calomel electrode or an Ag/AgCl electrode as reference electrode. These electrodes have well-defined redox potentials in relation to the standard hydrogen electrode. The correction factor for a calomel electrode relative to the standard hydrogen electrode is +247 mV. The inert metal electrode is usually made of platinum. The platinum electrode is inserted into the sediment or the media to be measured and the redox potential is read on a millivolt meter.

In sediments, the electrode may be inserted into intact cores either laterally through ports in acrylic cores or vertically down through the sediment–water interface. Then redox profiles can be generated through a stepwise insertion of the electrode. The electrode equilibrium time for each measurement is normally a few minutes when the drift is minimal, and a whole-sediment profile can usually be finished within less than 1 h depending on how detailed the profile should be described. If the electrode is left for too long in reducing sediments, it may be poisoned with sulfide coating, which will lead to erroneous negative redox values.

Although the redox potential is quick and very simple to measure, the results often vary considerably depending on the sampling and measurement techniques used. Therefore, redox measurements are often used in relation to other environmental parameters and often regarded more as a relative rather than an exact measure. When measuring in natural systems, the redox potential cannot describe a specific chemical redox couple, but is rather a measure of the overall redox reaction potential, including numerous redox reactions. These are usually not in thermodynamic equilibrium because some redox reactions take a long time to reach equilibrium and because a number of biological and chemical processes continuously take place.

The redox potential is influenced by the pH of the solution in which it is measured. Therefore, the redox potential is often corrected to pH 7 by adding 58 mV for every pH unit above 7 and by subtracting 58 mV for every pH unit below 7. The redox potential also depends marginally on temperature, but decreases only by 60 mV from 0 to 30 °C.

Oxygen has a strong effect on the redox potential because redox potential is influenced by the fourth root of the partial pressure of oxygen, and in oxygenated water the redox potential will remain above 300–500 mV and stay high as long as the water is not near anoxia. Microbial-mediated redox processes can decrease the redox potential to a level as low as −300 mV.

Redox Reactions

Redox reactions or processes are chemical reactions that involve the transfer of electrons from one element to another. The donor loses an electron and changes from the reduced to the oxidized form, while the acceptor receives an electron and changes from the oxidized to the reduced form. This process is accompanied by a change in free energy, equivalent to the tendency to donate or accept electrons. The oxidized and reduced forms of an element are said to form a redox couple.

Redox processes in natural waters involve a number of elements among which carbon (C), oxygen (O), nitrogen (N), sulfur (S), iron (Fe), and manganese (Mn) are the most prominent. Photosynthesizing plants harvest light energy and produce reduced states of compounds, particularly carbon, which are eventually oxidized back to an equilibrium state, and the energy is again released through energy-yielding redox reactions and used by respiratory and fermentative organisms. Thus, organisms can be regarded as catalysts and mediators of the transfer of electrons in the reduction and oxidation processes of C and other compounds.

When a transfer takes place from oxic to more reduced conditions, the main biological and chemical reactions are often ordered along a typical redox (and often depth) gradient. Aerobic reactions dominate in the oxidized water column, but different redox processes become more important in the turnover of organic matter in anaerobic systems. The rate and types of chemical compounds produced in the mineralization of organic matter at increasingly reduced conditions depend on the level of productivity (eutrophication; **Table 1**). In lakes, primary production is normally limited by nutrient availability and here phosphorus (P) is particularly important.

One of the most important redox reactions for the exchange of P between sediment and water is the Fe^{2+}/Fe^{3+} system: the $Fe^{3+} + e^- \leftrightarrow Fe^{2+}$, since iron is often found in relatively high quantities in the sediment and because iron (II) is easily oxidized to iron (III) ions and iron (III) easily reduces to iron (II) ions. The oxidation–reduction of iron takes place at a redox potential below 200 mV and thus under conditions typically prevailing at the sediment–water interface in most freshwater systems.

Redox Potential and Lakes

Stratified Lakes

In deep, temperate lakes, thermal stratification is usually established during late spring or early summer due to decreasing density of water at temperatures above 4 °C. This divides the water column into a warm surface layer (epilimnion) and a cold bottom layer (hypolimnion) that do not mix because of the different density. Between the epilimnion and hypolimnion, a metalimnion with distinct gradients in temperature, oxygen, and redox conditions may develop during the summer. In nutrient poor lakes, the stratification will not significantly affect the redox conditions as the water remains more or less saturated with oxygen throughout the water column because of the low respiration rate (**Table 1**).

However, in more eutrophic lakes with a higher primary production and also higher oxygen consumption in the lake water and the sediment, a redox gradient develops when oxygen concentrations are reduced to low levels (**Figure 1**). In the metalimnion and parallel to the oxygen gradient, a redox gradient develops, particularly when oxygen concentrations drop below 1 mg O_2 l^{-1}. An oxygen concentration of about 0.1 mg O_2 l^{-1} corresponds to a redox potential of about 200 mV. At this stage and at further reduced conditions, increased concentrations of reduced compounds such as hydrogen sulfide, ammonia, and iron may also develop.

As the bacterial decomposition proceeds in the hypolimnion and in the sediment, the concentrations of reduced Mn, then Fe, and eventually sulfide increase. The order of accumulation reflects their different ability to act as electron acceptors and thus the changing redox conditions. How far lake hypolimnions proceed in the progressive accumulation of reduced elements depends on the mineralization rate and the reduction potential of the water column.

Table 1 Changes in oxygen (O_2), redox potentials (E_h), iron (Fe^{2+}), hydrogen sulfide (H_2S), and phosphate (PO_4^{3-}) in the hypolimnia of stratified lakes with increasing productivity

Lake status	O_2	E_h *(mV)*	Fe^{2+}	H_2S	PO_4^{3-}
Oligotrophic	High	400–500	Absent	Absent	Very low
Mesotrophic	Much reduced	400–500	Absent	Absent	Very low
Eutrophic	Much reduced	~250	High	Absent	High
Hypertrophic	Much reduced	$<$100	Decreasing	High	Very high

Source (in a slightly modified form) Wetzel RG (2001) *Limnology*, 3rd edn. Florida: Academic Press.

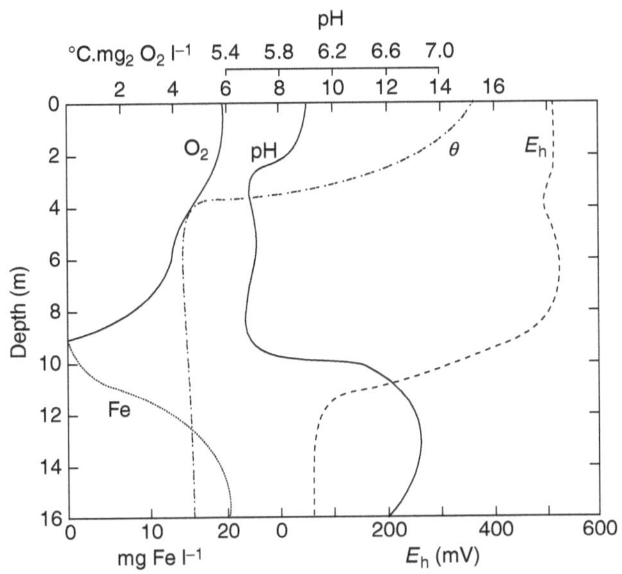

Figure 1 Vertical distribution of temperature, oxygen, pH, total iron, and redox potential in the permanently meromictic Lake Skiennungen, Norway. Reproduced from Wetzel RG (2001) *Limnology*, 3th edn. Florida: Academic Press, with permission from Academic Press.

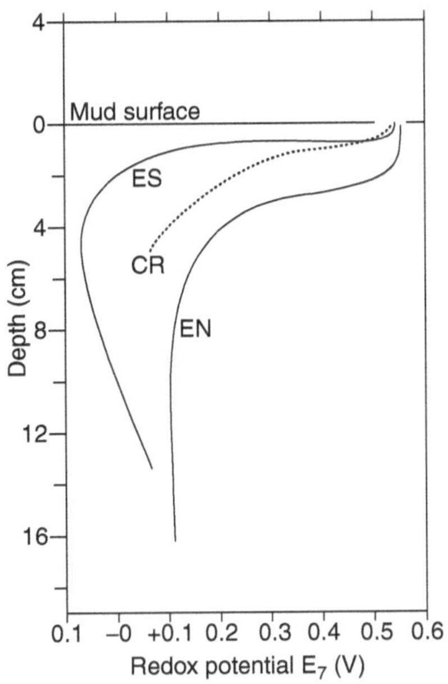

Figure 2 The redox potential in the surface sediment of a eutrophic lake (Eastwaite Water, ES) and in more oligotrophic lakes (Crummock Water (CR) and Ennerdale Water (EN)). Reproduced from Gorham E (1958) (after Mortimer, 1942). Observations on the formation and breakdown of the oxidized microzone at the mud surface in lakes. *Limnology and Oceanography* 3, 291–298, with permission from American Society of Liminology and Oceanography.

Although the theoretical sequence of redox processes is identical for all lakes and redox couples, the decrease rate of the redox potential at anoxia is not the same, because the type and concentrations of the redox couples involved varies from lake to lake.

At water circulation in autumn and introduction of oxygenated water into the hypolimnion, precipitation of ferric hydroxides and manganese oxide can take place as the redox potential increases and Fe^{2+} and Mn^{2+} oxidize. This may influence the nutrient availability for phytoplankton production because of a scavenger effect of phosphorus on ferric precipitates. In polymictic lakes with no permanent stratification, large fluctuations of oxygen concentrations in time and space may create large variability in the sedimentation and redissolution of redox-dependent elements. A coupling between the redox cycle of iron and manganese may induce the precipitation of ferric hydroxide even under anoxic conditions.

Redox Potential around the Sediment–Water Interface

During stratification in deep and nutrient-rich lakes, the gradient of the redox potential develops in water strata where oxygen concentrations become low. However, in shallow and well-mixed lakes or sections of lakes where the oxygen concentrations and redox potential in the water column remain high throughout the season, the redox potential gradient develops around the sediment–water interface. The shape and position of the gradient relative to the sediment surface depend on the amount and types of organic matter in the sediment, and the access and transport of electron donors from the water column. Thus, the redox potential in eutrophic lakes will be considerably lower at the sediment surface than in oligotrophic lake sediments (**Figure 2**).

A typical zonation within the sediment of eutrophic lakes would include (1) a brown top layer (depth 0–1 cm) with the presence of oxygen and nitrate and with redox potentials above 100 mV; (2) a green-brown layer (depth 1–5 cm) with a high iron reduction activity and negative redox potential; (3) a black-colored layer (5–10 cm) indicating the presence of iron sulfide and a zone with high sulfate reduction activity where the redox potential will be below −100 mV; (4) and a grayish-colored layer (>10 cm) acting as the zone of methane production with a very low redox potential (**Table 2**).

When nitrate is present in the overlying water it diffuses into the sediment and serves as an alternative electron acceptor in the absence of oxygen. As demonstrated in **Figure 3**, the presence of nitrate is followed by a considerably higher redox potential. Introduction of nitrate to reduced sediment can be

Table 2 Main respiration reactions listed in order of decreasing redox conditions and with approximate redox range when they are most important

Reaction	*Redox potential (mV)*
Aerobic decomposition	>250
Oxygen–nitrate	100–250
Iron–manganese (IV)	0–100
Sulfate	−200–0
Methane–hydrogen	<−200

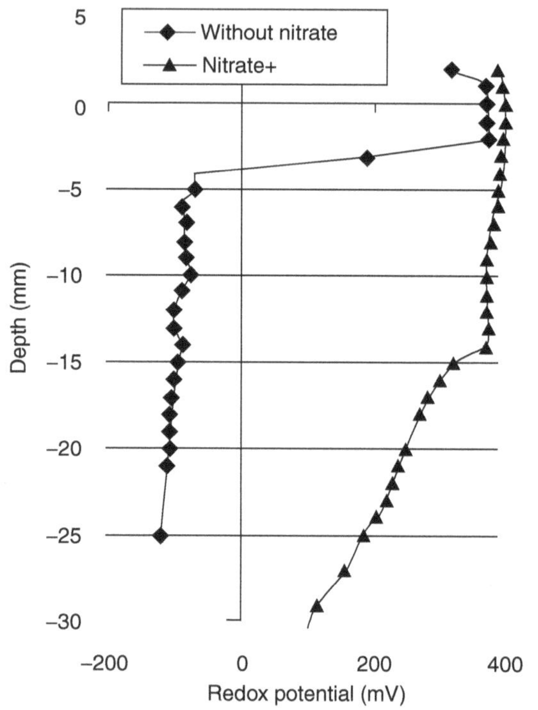

Figure 3 The redox potential in the sediment from a fish pond without nitrate addition and after incubation of 30 days with 15 ppm N-NO_3 in the overlaying water. Reproduced from Meijer LE, Avnimelech Y (1999) On the use of micro-electrodes in fish pond sediments. *Aquacultural Engineering* 21: 71–83, with permission from Elsevier.

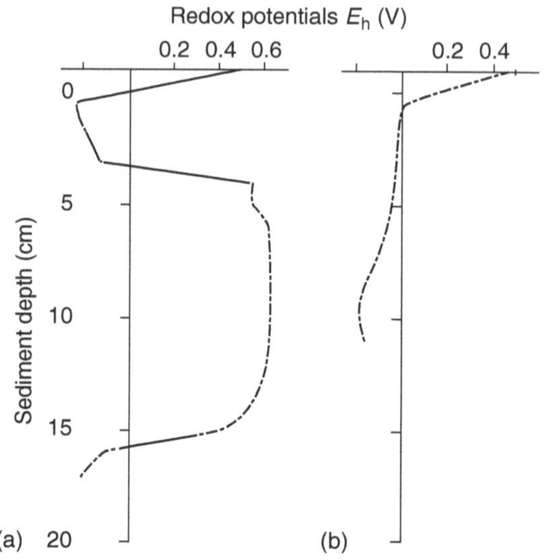

Figure 4 Redox potential in Lake Kalgaard, Denmark, in a vegetation covered sediment (a) and sediment without vegetation (b). Reproduced from Wium-Andersen S and Andersen JM (1972) The influence of vegetation on the redox profile of the sediment of Grane Langsø, a Danish Lobelia lake. *Limnology and Oceanography* 17: 948–952, with permission from American Society of Limnology and Oceanography.

recognized as a shift from black to brown color as the black iron sulfide is oxidized.

Submerged macrophytes, in particular species with well-developed roots such as isoetid plants, not only grow mainly in oxidative sediments of oligotrophic lakes, but also establish a high redox potential deep into the sediment. As shown in **Figure 4**, the presence of an oxidized rhizosphere around *Littorella* in an oligotrophic lake creates a much higher redox potential compared to areas without *Littorella*. The redox potential is low because of the presence of a layer of decaying plant material in the upper few centimeters, but markedly increased in the rhizosphere extending about 15 cm into the sediment. Other macrophyte species with a less-developed root system and less radial oxygen losses from their roots have a less marked impact on the sediment redox potential. The high redox potential in the presence of macrophytes with well-developed root systems has significant implications for the concentrations of dissolved iron and phosphate. Thus, by oxidizing the sediment isoetid species immobilize iron and phosphate and prevent accumulation of mobile phosphorus in the sediment and water layers. Also, the presence of chironomid larvae, tubificids, and other invertebrates in the surface sediment can influence the redox conditions. Diffusion and water transport are facilitated by burrows, and oxidized sediment is often formed along the wall of burrows.

As primary production and decomposition rates change during the season, seasonal fluctuations occur in the redox potential. Typically, the redox stratification in the surface sediment of eutrophic lakes becomes more compressed in spring and summer because of the higher oxygen consumption, while less reducing conditions are found near the sediment–water interface during fall and winter (**Figure 5**).

Redox Potential and Nutrient Cycling

Iron is a very abundant element and the most prevalent redox-active metal. In oxic environments, only

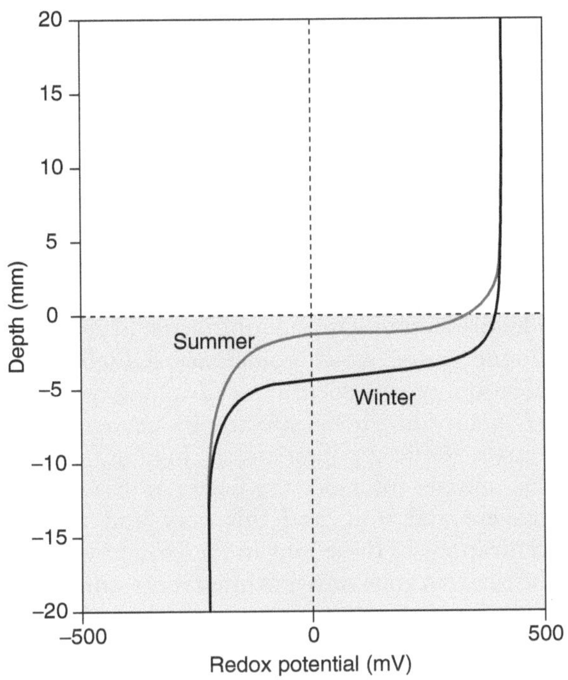

Figure 5 Changes in general redox potential from winter to summer in a eutrophic lake sediment.

the ferric state is energetically stable and comprises mainly insoluble forms, whereas the reduced ferrous minerals, apart from ferrous sulfide (FeS) and pyrite (FeS_2), are relatively soluble. At the oxic–anoxic interface close to the sediment surface, ferrous iron (Fe^{2+}) is produced biologically by iron-reducing bacteria or chemically with sulfide, and consumed again by the rapid chemical or biological reoxidation of ferrous iron with oxygen or nitrate to ferric iron (Fe^{3+}). This cycling between the oxic and the reduced form of iron can be repeated multiple times before the iron eventually is buried in deeper parts of the sediment as ferrous iron.

The high affinity of oxidized iron (Fe^{3+}) hydroxides to bind phosphorus makes the oxidized microzone, which is usually present in the surface sediment in lakes, very important for the internal cycling of phosphorus and thereby also for the overall lake water quality. This is because phosphorus often is the limiting nutrient for primary production. Phosphate becomes adsorbed onto iron hydroxides under oxic conditions, but mobilized again if iron is reduced, and thereby the iron phosphate precipitation may control the behavior of phosphorus under oxic conditions. Phosphorus is not only released from stratified lakes with an anoxic hypolimnion, but substantial amounts may also be released under aerobic conditions in the lake water, as is often seen in eutrophic shallow lakes after a reduction of the external phosphorus loading. This release takes place in spite of oxic conditions in the lake water because anoxic conditions prevail a few centimeters into the sediment.

In some lakes and particularly in brackish waters with high sulfate concentrations or freshwater systems with increased loading of sulfate, sulfate-reducing bacteria become a more important pathway for the decomposition of organic matter. Sulfate reduction leads to higher production of sulfide which reduces iron: $2Fe^{3+} + S^{2-} \rightarrow 2Fe^{2+} + S$, and in a next step binds available iron into iron-sulfide minerals. This diminishes the pool of iron hydroxides and reduces the capacity for the adsorption of phosphorus, thus increasing the risk of eutrophication because of the higher phosphorus availability for phytoplankton production. Sulfate-mediated eutrophication is common in aquatic systems.

Redox Potential and Lake Restoration

Lake restoration has been widely used in recent decades to combat the effects of eutrophication caused by present or former high loading of nutrients, in particular, phosphorus. As one of the most important inorganic binding mechanisms of phosphorus in lakes is the redox-sensible binding to iron minerals, lake restoration measures influencing the redox-sensible sorption of phosphorus have been in focus.

In deep stratified lakes, restoration has been conducted by introducing oxygen into the hypolimnetic waters during summer with the aim to increase the redox potential and to reduce the accumulation of phosphorus released from reduced iron compounds. Oxygen is usually added through diffusors placed on the bottom of the lake at the deepest points.

The ability of nitrate to oxidize the sediment surface in the absence of oxygen has been implemented in chemical lake restoration techniques to reduce the internal loading of phosphorus. By injecting a concentrated nitrate solution into the surface sediment or the water immediately above the sediment the redox potential will be increased to a level maintaining iron in its oxidized form. This increases the phosphorus retention of the sediment and reduces the overall phosphorus availability for phytoplankton in the lake. Nitrate may, however, also contribute to increased phosphorus release because of the increased mobilization of organically bound phosphorus.

To increase the sorption capacity of phosphorus, iron has been added to the sediments. However, the overall binding capacity of phosphorus depends on a high redox potential to maintain iron in its oxidized form. Alum has also been used to precipitate phosphorus and increase the sorption capacity of the sediment.

Contrary to iron, its sorption capacity does not depend on the redox conditions.

Redox Potential and Rivers

Compared to lakes, rivers and streams generally have a more steady input of oxygen because of the turbulence of water and continuous atmospheric reaeration. Generally, this also creates a more stable and higher redox potential in the whole water column without any clear stratification. There are, however, marked variations in sediment redox conditions according to river type and the position on the river system, i.e., between the upper fast flowing and lower slowly floating stretches. Redox conditions similar to those recorded in lakes may develop at the sediment–water interface and, particularly, in sediment parts accumulated in more slowly floating sections of the river.

Seasonal changes in the redox potential due to changes in water flow and temperature influence the amount of soluble reactive phosphorus in a similar manner as in lake sediments. During summer, the higher temperatures enhance the microbial activity and lower the redox potential in the sediment. This may cause dissolution of manganese and iron hydroxides, increasing phosphorus concentrations also in the ground water close to the riverbank. Redox changes may also influence the retention and release of phosphorus from the sediment and of the water in the river. In Polish River Narew, considerably higher phosphate concentrations occurred when the redox potential was reduced below 200 mV (**Figure 6**). A low redox potential was recorded during summer under low-flow conditions and high temperatures. As in lakes, the presence of nitrate is able to increase the redox-sensitive sorption of phosphorus and, with it, the amount of phosphorus in the river water.

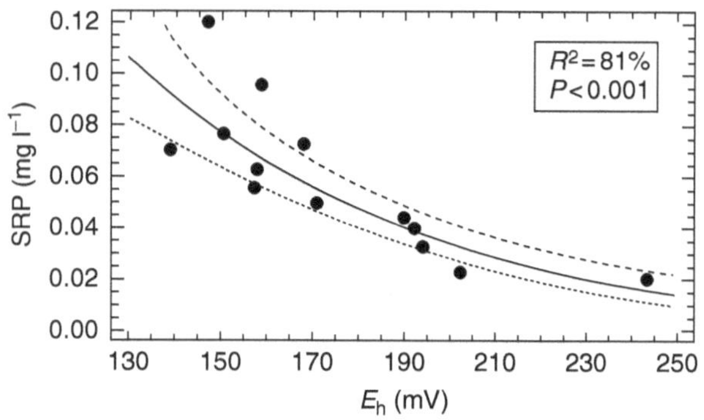

Figure 6 Relationship between the concentration of orthophosphate (SRP) and redox potential (E_h) in Polish river Narew during low-flow conditions. Reproduced from Banaszuk P and Wysocka-Czubaszek A (2005) Phosphorus dynamics and fluxes in a lowland river: The Narew anastomosing river system, NE Poland. *Ecological Engineering* 25 and, 429–441, with permission from Elsevier.

In the hyporheic zone of the river bed, i.e., the active ecotone between the surface stream and the groundwater containing water originating from both the river and the neighboring aquifers, small-scale dynamic gradients may be found in the redox potential. These may have significant implications for the chemically and biologically mediated nutrient transformation, including, for example, nitrification and denitrification. During infiltration of river water to ground water, which sometimes is used also for purification or production of drinking water, the water quality might be affected by changing redox potentials. Redox potentials as low as −100 mV during summer influence the bacterial dissolution of manganese and iron, and this may lead to higher concentrations of these ions in the ground water.

The riparian zone of rivers often represents areas of high species macrophyte richness, this being sustained by the seasonal and interannual flow variability. When soils become waterlogged oxygen is rapidly depleted and the redox potential reduced, this having considerable influence on plant growth, although the critical levels are species specific. In wet riparian meadows, anaerobic conditions with low redox potential (<300 mV) prevail throughout the summer, whereas in drier meadows, soil conditions are predominantly aerobic throughout the year with high redox potentials.

Redox Potential and Wetlands

Wetlands characterized by more or less permanently waterlogged soils are temporarily and spatially very heterogeneous with alternating cycles of oxidation and reduction of sulfur, iron, nitrogen, and carbon. The electron acceptors used by microorganisms depend on the redox potential, pH, and availability of electron acceptors and comprise oxygen, nitrate, manganese, iron, sulfate, and carbon dioxide in a sequential chain. In mesoscale systems, there may be an overlap of different redox processes.

Especially water level fluctuations may have large implications for the redox potential as illustrated by a soil profile along a hydrological gradient (**Figure 7**). Soil oxygen concentrations generally show an inverse linear relation to the redox potential with a redox potential of about 250 mV at 10% oxygen saturation and ~0 mV at 1–2% oxygen saturation. The change from aerobic to anaerobic metabolism occurs at oxygen concentrations below 1%. As in lake sediments, the depth of the oxic surface layer depends on the consumption rate of oxygen by the microflora and the supply rates of oxygen from above. In wetlands, redox potentials vary considerably during the season in areas with changing water table depth. For example, in riparian meadows the dynamics of the soil redox potential follows the seasonal fluctuations in stream stage and water table. This may also have important implications for the diversity of plants.

In relation to nutrient retention and release, wetlands on noncalcareous soils retain phosphorus under aerobic conditions by oxidized iron, manganese, and aluminum minerals. Under anaerobic conditions, phosphate is released to the interstitial water by the same mechanisms as in lake sediments. Iron and manganese are reduced to more soluble forms when the redox potential is reduced. The critical redox potential depends on pH and is around 300 mV at pH 5, 100–300 mV at pH 6–7, and −100 mV at pH 8. In calcareous systems, the apatite equilibrium, which is controlled by pH rather than the redox potential, is more important for phosphorus solubility.

In some parts of the world it has become common to establish new wetlands in order to increase recreational areas, wildlife abundance or the retention of nutrients. Water table fluctuations and variable hydraulic loading can alter soil redox conditions and thereby the phosphorus availability in these wetlands because of the direct relation between the redox potential and the reduction of ferric phosphate. Wetlands constructed on lands used for agricultural

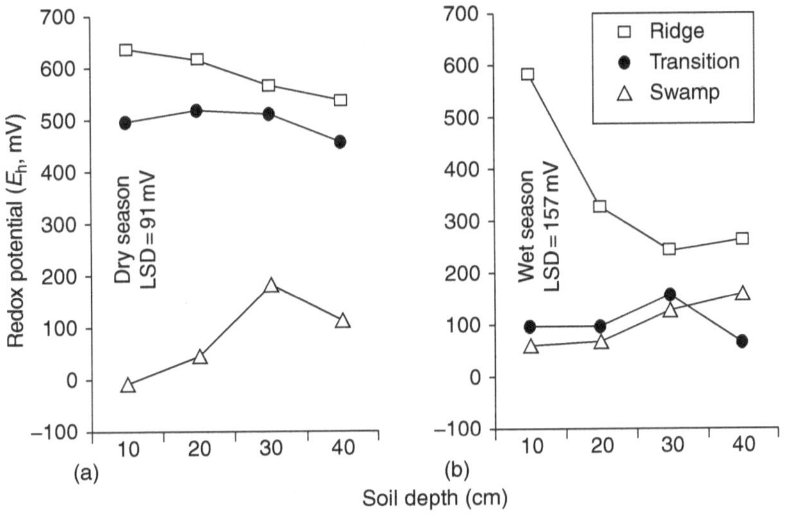

Figure 7 Soil profiles of redox potential in different hydrological seasons and at different distances to wetlands. Reproduced from Yu KW, Faulkner SP, and Patrick WH (2006) Redox potential characterization and soil greenhouse gas concentration across a hydrological gradient in a Gulf coast forest. *Chemosphere* 62: 905–914, with permission from Elsevier.

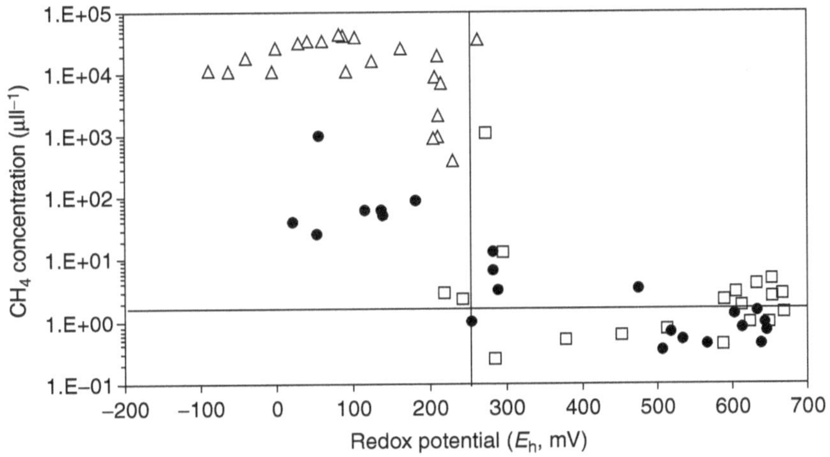

Figure 8 Concentration of methane in relation to redox potential at three locations across a hydrological gradient in a Gulf coast forest (Louisiana). Reproduced from Yu KW, Faulkner SP, and Patrick WH (2006) Redox potential characterization and soil greenhouse gas concentration across a hydrological gradient in a Gulf coast forest. *Chemosphere* 62: 905–914, with permission from Elsevier.

purposes for a number of years resulting in accumulation of phosphorus may be ineffective in reducing phosphorus levels in effluents.

In relation to global warming and the emission of greenhouse gasses such as methane (CH_4) from wetlands, the redox potential may be a good indicator of expected changes in the water table and increasing soil saturation and flooding periods. Thus, the soil redox potential is a good quantitative indicator of anoxic conditions, which is important for the production and uptake of methane. Significant methane production can usually only occur under strongly reducing conditions (redox potential ≤ 150 mV). Correspondingly, swamp areas and wetlands will have considerably higher methane concentrations and production than drier areas have (**Figure 8**). The major part of the diffusive flux of methane from deeper strata may, however, be immobilized or oxidized to CO_2 in surface layers of freshwater wetlands and will therefore not necessarily escape to the atmosphere.

Further Reading

Boström B, Jansson M, and Forsberg C (1982) Phosphorus release from lake sediments. *Archiv für Hydrobiologie Beiheft Ergebnisse der Limnologie* 18: 5–59.

Cooke DG, Welch EB, Peterson SA, and Nichols SA (2005) *Restoration and Management of Lakes and Reservoirs,* 3rd edn. Boca Raton: CRC Press, Taylor & Francis.

Fenchel T and Blackburn TH (1979) *Bacteria and Mineral Cycling.* London: Academic Press.

Mortimer CH (1941) The exchange of dissolved substances between mud and water in lakes (parts I and II). *Journal of Ecology* 29: 280–329.

Mortimer CH (1942) The exchange of dissolved substances between mud and water in lakes (part III and IV). *Journal of Ecology* 30: 147–201.

Schüring J, Schulz HD, Fischer WR, Böttcher J, and Duijnisveld WHM (eds.) (2000) *Redox–Fundamentals, Processes and Applications.* 251pp. Berlin: Springer-Verlag.

Søndergaard M, Wolter KD, and Ripl W (2002) Chemical treatment of water and sediments with special reference to lakes. In: Perrow MR and Davy AJ (eds.) *Handbook of Ecological Restoration: Principles of Restoration,* vol. 1, pp. 184–205. Cambridge: Cambridge University Press.

Wetzel RG (2001) *Limnology,* 3rd edn. Florida: Academic Press.

POLLUTION AND REMEDIATION

Contents

Acidification
Aquatic Ecosystems and Human Health
Bioassessment of Aquatic Ecosystems
Deforestation and Nutrient Loading to Fresh Waters
Distribution and Abundance of Aquatic Plants – Human Impacts
Effects of Climate Change on Lakes
Eutrophication
Fires
Floods
Mercury Pollution in Remote Fresh Waters
Pollution of Aquatic Ecosystems I
Pollution of Aquatic Ecosystems II: Hydrocarbons, Synthetic Organics, Radionuclides, Heavy Metals, Acids, and Thermal Pollution
Vector-Borne Diseases of Freshwater Habitats

Acidification

W Geller and M Schultze, UFZ – Helmholtz Center for Environmental Research, Magdeburg, Germany

Introduction

Natural rain is weakly acidic by atmospheric carbon dioxide, reacting to carbonic acid in water. Therefore, the chemical characteristic of natural waters is dominated by dissolved carbonates, since these are the only minerals in rock and sediments that can easily be dissolved by rain water. The composition of fresh waters, in both soft and hard waters, is differing by the concentrations of the components of the system (eqns. [5c–e]), which together are buffering standard fresh waters to about pH 7.

Acidification is the additional input of acids to waters – from natural and anthropogenic sources (**Plate 1**) – shifting the pH to lower values and usually eliminating the carbonate buffering system. The two most important mineral acids in this respect are sulfuric and nitric acid. Industrial acid emissions have altered lakes and rivers because of wind transport away from the source areas and lack of neutralizing materials in the geology of the receiving regions, e.g., in North America and Scandinavia.

Accompanying the acidification, further problems arise beyond low pH and aciditiy:

- rising concentrations of dissolved solids, particularly of sulfate in the drainage waters,
- mobilization of potentially toxic metals, and
- loss of biologic diversity.

The adverse factors prohibit the use of the impacted water for irrigation, fishery and aquacultures, and its use as drinking water.

Following, the sources, distribution pathways, reactive modifications, effects on soils and biota, and the importance of the different kinds of acidification are described. The approaches how to mitigate or to remediate acidifications are presented elsewhere within this encyclopedia.

Chemistry of Acidified Waters and Buffering Mechanisms

Carbonic Acid and Fresh Water

Natural fresh waters contain mainly carbonates, HCO_3^- and CO_3^{2-}, and most fresh waters show nearly identical chemical compositions with pH 7 and the carbonate buffering system dominating. Carbonate-poor geological regions have soft waters with low mineral content, and carbonate-rich areas show higher mineralized hard waters.

Definitions and Dimensions

The capacity to neutralize additions of acids or bases depends on the respective concentrations of base

Plate 1 Anthropogenic and natural acid waters: Upper left panel – Acid Sulfate Soil: Bottle Bend Lagoon, Australia, acidified after lowering of water level by drought, with aeration of sulfidic sediments. (Image by courtesy of CSIRO 2004, Land and water Technical Report 28/04: http://www.clw.csiro.au/scientific_reports.html). Upper right – Acid Rio Tinto draining the Pyrite Belt in Southern Spain (Photograph: M. Koschorreck). Lower panels – Acid lignite pit lake (pH 2.3) in Lusatia, East Germany (Photograph: G. Packroff). See also view of acid volcanic Lake Voui – http://www.altitude-photo.com/fiche-photo.php?id photo=11994/, http://www.volcanodiscovery.com/volcano-tours/typo3temp/pics/15e7421124.jpg

cations, of strong acid anions, and the concentrations of two- or three-valent weak acids that function as buffering systems by stepwise dissociation of their protons. The relationship between base cations and strong acid anions determine whether the water is alkaline, neutral, or acidic.

Alkalinity and acidity The term alkalinity is equivalent to the acid-neutralizing capacity (ANC). Negative values of alkalinity are termed acidity, equivalent to the base-neutralizing capacity (BNC).

$$\begin{aligned}\mathrm{ANC} = {} & (\text{sum of base cation concentrations}: \\ & Ca^{2+},\ Mg^{2+},\ Na^{+},\ K^{+},\ NH_4^{+} \\ & - (\text{sum of strong acid anion concentrations}: \\ & SO_4^{2-},\ NO_3^{-},\ Cl^{-})[\text{concentrations in} \\ & \text{milliequivalents per litre}: \text{meq } L^{-1}]\end{aligned} \quad [1]$$

ANC = –BNC thus.

The BNC of acid waters can be measured directly by titration, usually with a solution 0.01N sodium hydroxide till pH 8.2, resulting in a measure of (mmol L^{-1}), or (meq L^{-1}). The acidity of acid water also can be calculated from the pH-value and from the major weakly acid components of water (Fe, Al, Mn; metal concentrations given in mg L^{-1}) by eqn. [2] (if the concentrations of carbon dioxide, dissolved silicate, humic acids, and other heavy metals are small compared to the sum of iron, aluminum, and manganese):

$$\begin{aligned}\mathrm{Acid_{calc}}[\mathrm{meqL^{-1}}] = {} & \frac{2Fe^{2+}}{56} + \frac{3Fe^{3+}}{56} + \frac{3Al}{27} + \frac{2Mn}{55} \\ & + 1000 \times 10^{-\mathrm{pH}}\end{aligned} \quad [2]$$

The pH-value is a bulk measure comprising the strong mineral acids, which are completely dissociated into their anions and protons, and the dissociated part of the weak acids. The non-dissociated protons of the weak acids are set free stepwise during the titration process, thereby functioning as buffering systems.

According to the US standards, acidity is measured by the amount of $CaCO_3$ that has to be added to reach neutrality. The dimension (mg $CaCO_3$ L^{-1}) can be converted to (meq L^{-1}) following eqn. [3]:

$$\mathrm{Acidity(meqL^{-1})} = (\mathrm{mgCaCO_3L^{-1}})/50 \quad [3]$$

Buffering systems: the weak acids of carbon, aluminum, and iron Acids may enter aquatic systems directly or after passage of soil and deeper underground. The contacts of the solid phase of soil and underground to the passing acids result in buffering, i.e., proton-consuming reactions. Thereby, the dissolution of minerals is the main source of metal mobilization among the buffering processes. The relevant processes are:

- Buffering by carbonates and carbonic acid eqns. [5a–d]

  Carbonic acid system:

$$CaCO_3 + 2H^+ \Leftrightarrow Ca^{2+} + CO_2 + H_2O \quad [5a]$$

$$CaCO_3 + H^+ \Leftrightarrow Ca^{2+} + HCO_3^- \quad [5b]$$

$$CO_3^{2-} + H^+ \Leftrightarrow HCO_3^- \quad [5c]$$

$$HCO_3^- + H^+ \Leftrightarrow H_2CO_3 \quad [5d]$$

$$H_2CO_3 \Leftrightarrow CO_2 + H_2O \quad [5e]$$

- Buffering by silicates, especially aluminosilicates and aluminum silicates (eqns. [6a–d] give examples). The liberated aluminum and the formed gibbsite behave as described in eqns. [7a–c].

  Feldspar dissolution:

$$KAlSi_3O_8 + 7H_2O + H^+ \Leftrightarrow Al(OH)_3 + 3H_4SiO_4 + K^+ \quad [6a]$$

$$2KAlSi_3O_8 + 9H_2O + 2H^+ \Leftrightarrow Al_2Si_2O_5(OH)_4 + 4H_4SiO_4 + 2K^+ \quad [6b]$$

$$KAlSi_3O_8 + 4H_2O + 4H^+ \Leftrightarrow Al^{3+} + 3H_4SiO_4 + K^+ \quad [6c]$$

  Clay dissolution:

$$Al_2Si_2O_5(OH)_4 + 6H^+ \Leftrightarrow 2Al^{3+} + 2H_4SiO_4 + H_2O \quad [6d]$$

- Buffering by pedogenic oxides and (oxi)hydroxides (eqns. [7a–c] – dissolution of gibbsite which forms the so-called aluminum buffering system and eqns. [8a–c] – dissolution of ferrihydrite which forms the so-called iron buffering system).

  Aluminum system:

$$Al(OH)_3 + H^+ \Leftrightarrow Al(OH)_2^+ + H_2O \quad [7a]$$

$$Al(OH)_2^+ + H^+ \Leftrightarrow AlOH^{2+} + H_2O \quad [7b]$$

$$AlOH_2^+ + H^+ \Leftrightarrow Al^{3+} + H_2O \quad [7c]$$

  Iron system:

$$Fe(OH)_3 + H^+ \Leftrightarrow Fe(OH)_2^+ + H_2O \quad [8a]$$

$$Fe(OH)_2^+ + H^+ \Leftrightarrow FeOH_2^+ + H_2O \quad [8b]$$

$$FeOH^{2+} + H^+ \Leftrightarrow Fe^{3+} + H_2O \quad [8c]$$

- Buffering by ion exchange (eqn. [9] gives an example; X – active site at the surface of particles or solid material).

  Sites of ion exchange:

$$X - K + H^+ \Leftrightarrow X - H + K^+ \quad [9]$$

- Buffering by protonation of humic substances

  Humic acids:

$$R - COO^- + H^+ \Leftrightarrow RCOOH \quad [10]$$

When acids enter aquatic systems directly, the relevant buffering systems are in the liquid phase: the carbonate system eqns. [5b–d], the aluminum buffer eqns. [7a–c], the iron buffer eqns. [8a–c], and reactions of further weak acids (e.g., humic/fulvic acids, silicate). The chemical equilibrium between sulfate and hydrogensulfate comes into play if pH drops below 2.5 (eqn. [11]).

Formation of hydrogensulfate:

$$SO_4^{2-} + H^+ \Leftrightarrow HSO_4^- \quad [11]$$

The action of the different buffering systems results in typical distributions of pH-values obtained from regional surveys in lakes which are partly acidified. **Figure 1** shows the pH-values of German pit lakes resulting from lignite mining forming a three-modal distribution. Surveys of lakes which are partly impacted by atmospheric deposition usually result in a two-modal pH-distribution formed by the aluminum buffer and by the carbonate buffer.

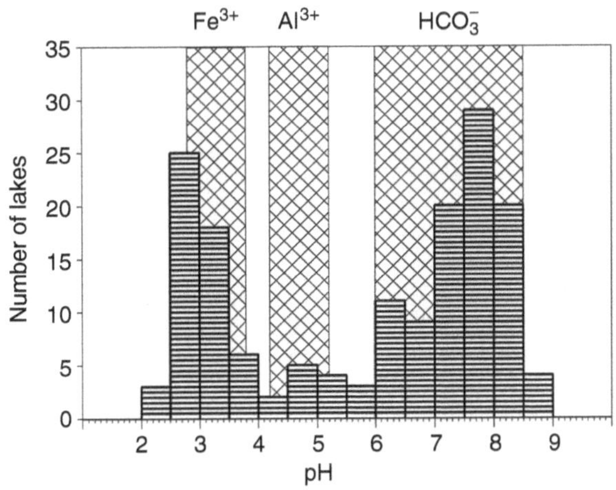

Figure 1 Frequency distribution of the pH values of 159 mining lakes in Germany (data from Nixdorf *et al.* (2001) *Tagebauseen in Deutschland*. Umweltbundesamt, Berlin. UBA-Texte 01/35 http://www.umweltdaten.de/publikationen/fpdf-l/1996.pdf). The hatched areas indicate the pH ranges of the buffering system of ferric iron (Fe^{3+}), aluminium (Al^{3+}), and bicarbonate (HCO_3^-).

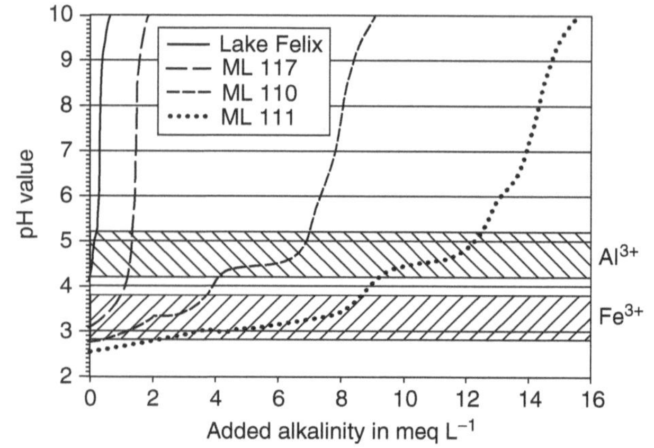

Figure 2 Titration curves of four different mining lakes in the Lusatian lignite mining district, Germany (data by courtesy of O. Totsche). The shaded areas indicate the pH ranges of the buffering systems of ferric iron (Fe^{3+}) and aluminum (Al^{3+}). Depending on the concentrations of ferric iron and aluminum and the formed minerals in the particular lakes, the extent of the according plateaus differs in the titration curves. Below pH 2.5, hydrogensulfate (HSO_4^-) may act as buffer. Ion exchange with particles, silicates, and the carbonate buffering system are relevant at pH above 5, depending on the availability of suspended particles and (bi-)carbonate.

The buffering systems govern not only the decrease of pH as the result of impact of acids. The progress of neutralization is shown by typical titration curves of water from acid German pit lakes in **Figure 2**. The plateaus indicate the action of the iron and the aluminum buffering systems, respectively. The extent of the plateaus depends on the concentration of the respective metals in the lake water.

Pyrite weathering Acids of geogenic origin can be set free by natural weathering or by processes induced by mining or by agriculture. The most important process is the oxidation of pyrite or similar sulfides (eqns. [12–15]). These oxidation processes are accelerated by sulfur and iron oxidizing bacteria. The overall products of the multi-step reaction are sulfuric acid and ferric iron hydroxide.

$$FeS_2 + 7/2O_2 + H_2O \Leftrightarrow Fe^{2+} + 2SO_4^{2-} + 2^{H+} \quad [12]$$

$$Fe^{2+} + 1/4O_2 + H^+ \Leftrightarrow Fe^{3+} + 1/2H_2O \quad [13]$$

$$Fe^{3+} + 3H_2O \Leftrightarrow Fe(OH)_3 + 3H^+ \quad [14]$$

$$FeS_2 + 14Fe^{3+} + 8H_2O \Leftrightarrow 15Fe^{2+} + 2SO_4^{2-} + 16H^+ \quad [15]$$

In the case of mining-induced pyrite oxidation, dewatering operations bring pyrite in contact with air. Before, the pyrite had been stable in the water-saturated anoxic underground. Once in contact with air, the oxidation goes on rapidly and forms acid mine drainage (AMD). Pyrite is not only one of the most important iron ores but also occurs as an accompanying mineral in many sulfidic ores of other metals, in lignite and coal deposits, in shales, or in marine sediments, such as mined clay deposits.

Outcrops of natural pyrite deposits provide natural conditions for pyrite oxidation. Such natural acidic waters are known from the Iberian Pyrite Belt in southern Spain, mountainous regions in New Mexico, or the lakes of the Tyrell Basin and the Yilgarn Block in southern Australia.

Concentrations of dissolved substances in acid waters The concentrations of total dissolved solids in acid waters increase from about 50 mg L^{-1} at pH 5 in soft waters, which result from atmospheric deposition, to 100 g L^{-1} at pH 0 in extremely acidic brines of volcanic springs. The spectrum of dissolved elements originates from the involved mineral acids and from the composition of the thereby dissolved minerals in soils and rocks.

We show in **Figure 3**, the concentration of some constituents of acid waters resulting from the described types of acidification. For all types, data from ground water, springs, streams and lakes are included. However, these data comprise only waters with pH < 6.

The occurring concentrations of sulfate and metals result from sulfide oxidation as well as from buffering processes in soil and water (eqns. [5–9,12,15]). The pH-values of these acid waters range from 6 to values below zero under extreme conditions in abandoned mine workings in California. We show in **Figure 4**, the pH-values of the waters comprised in **Figure 3**.

Types and Extent of Acid Waters

Atmospheric Deposition and Acid Rain

The atmospheric pathway distributes acidic gaseous SO_2, NH_3, and NO_x over large areas of land. The sources of SO_2 and NO_X are combustion of coal, lignite, and oil in power plants and in households, waste gases of motor cars, and various industrial waste gases, e.g., resulting from roasting of sulfidic ores. NH_3 emissions often result from stock farming, especially when done in large-scale units. The acidity of NH_3 emissions results from oxidation/nitrification during atmospheric transport or after deposition in the top soil.

Although the term acid rain is commonly used, this kind of acidification usually includes the

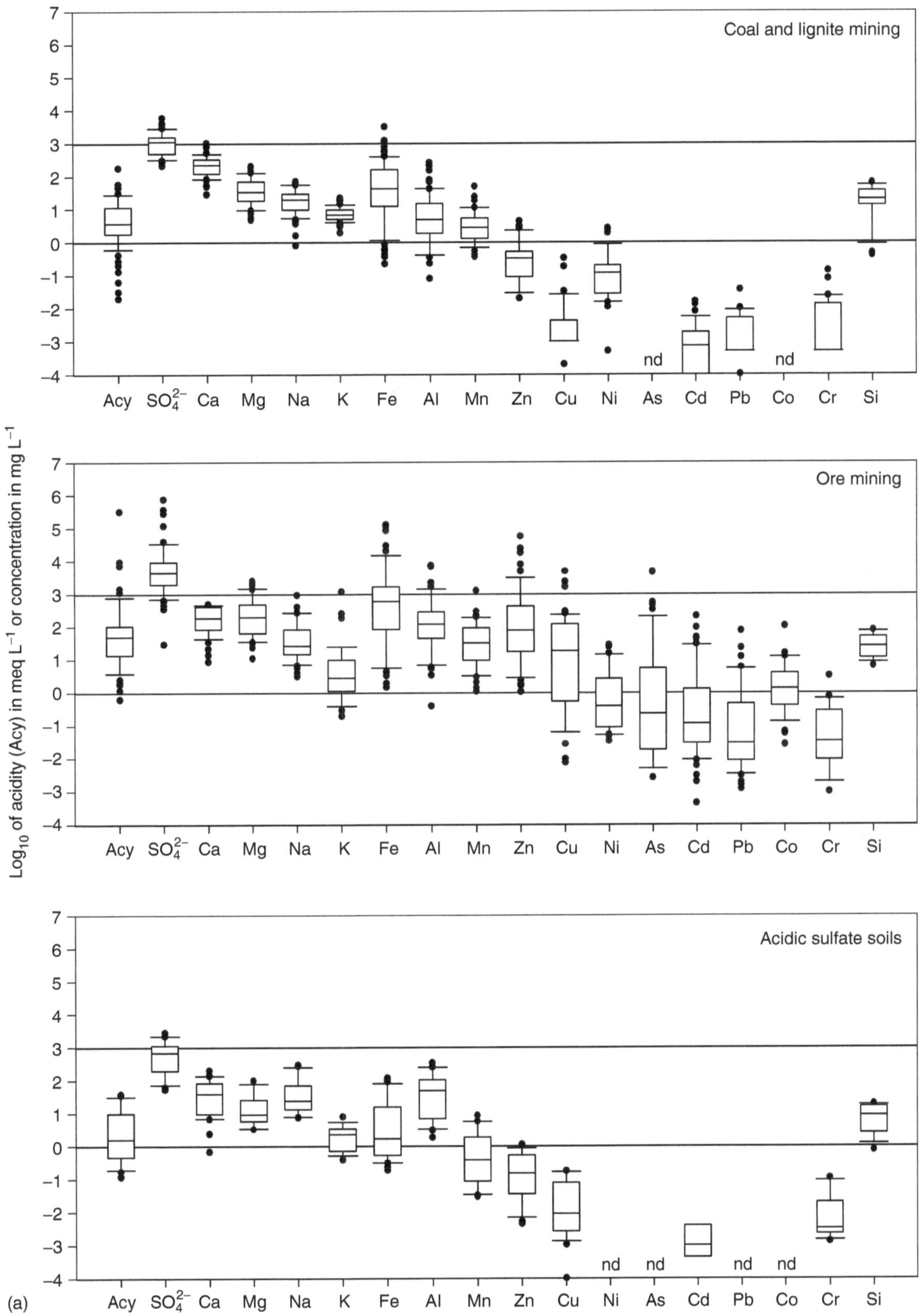

Figure 3 (Continued)

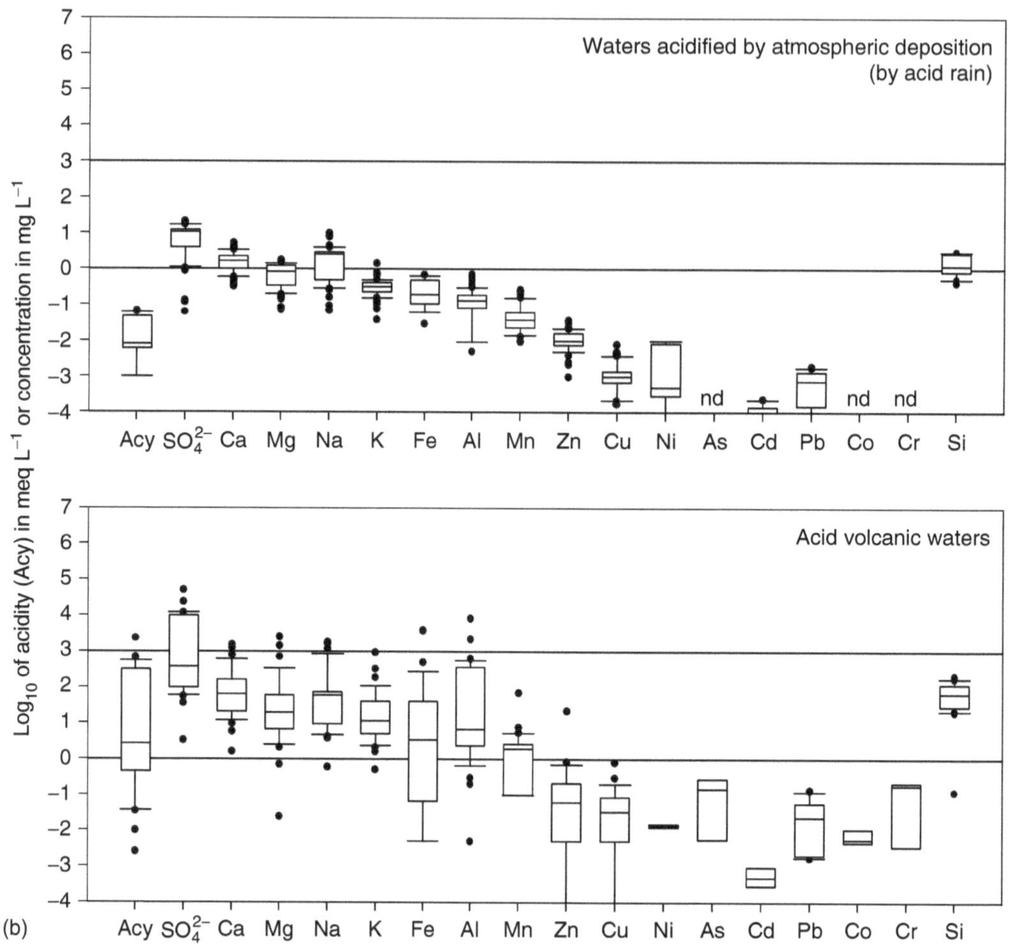

Figure 3 Characteristics of acid drainage by the concentrations of chemical constituents. Lines within the boxes are median values, box limits are 25 and 75 percentiles, and whiskers show the 10- and 90-percentile values. Single dots indicate data below 10 percentile or above 90 percentile, respectively (nd – not detected).

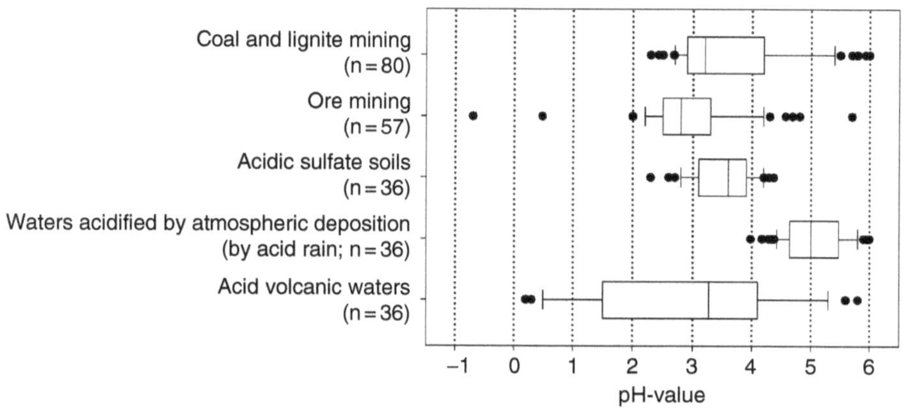

Figure 4 pH values of the waters comprised in **Figure 3**. Lines within the boxes are median values, box limits are 25 and 75 percentiles, and whiskers show the 10- and 90-percentile values. Single dots indicate data below 10 percentile or above 90 percentile, respectively (nd – not detected).

deposition of aerosols, of fog, the dry deposition of the mentioned gases, and of dust. In a two-step process, the soils first become acidic after complete loss of carbonate minerals, then, acidic waters seeping through the soils dissolve further minerals, especially Al-silicates. These waters cause the acidification of ground water and surface waters.

Beyond the critical limits, the interaction between soil and trickling water leads to accumulation of acidity (N and S content) in soils. The carbonate and hydrogen carbonate anions in soil and surface waters are replaced by sulfate, and Ca^{2+} and Mg^{2+} cations by H^+ and Al^{3+}. After a large-scale process of titration, the impacted soils and waters are carbonate-free, and the affected waters are enriched in sulfate and aluminum, with pH-values in a range between 4.5 and 5.5.

Under certain conditions, this kind of acidification occurs episodically. If the reaction time between introduced acidity and soil is long enough the mentioned buffering processes are able to neutralize the trickling water. Elevated precipitation rates or the sudden liberation of accumulated acidity, e.g., from melting snow covers, however, may cause temporarily elevated acidity.

Soft waters of carbonate-poor geological regions are more affected by acidic inputs because of the low contents of ions and low buffering capacities. They are sensitive to acid depositions. The critical load limits are reached where the percentage of the base cations Ca^{2+}, Mg^{2+}, and K^+ is lower than 20% of the total cation exchange capacity, or the acid neutralization capacity (ANC) is below 0–25 $\mu eq\ L^{-1}$. Estimations of the critical loads in geologically sensitive areas showed a range of annual sulfur depositions between 300 and 800 $mg\ m^{-2}$.

After the emissions of sulfur dioxide and nitrogen oxides, transport times of 15–21 h were observed from the states in the midwestern United States, where the emissions occurred, to the northeastern states in New England and Canada, where the acid rain was precipitating. Similar transport times and distances are known from England and from central Europe to southern Scandinavia. The percentages of acidified lakes were 27% in Norway. In Sweden, 17 000 (20%) of 85 000 surveyed lakes were affected until 1992. The trend of atmospheric acidification was reversed during the 1980s and 1990s by countermeasures in the western industrial countries. The present global problem areas with risk of surface water acidification are in mid- and northeastern North America, in eastern Canada, in western, central and northern Europe, in the eastern parts of China, in India, and in northwestern Russia.

Acid Mine Drainage (AMD)

Acidification of inland waters by acid mine drainage is common in many mining areas where we find acidic open pit lakes, acidic streams or acid mine drainages from shafts, adits, waste rock or overburden dumps, and tailings. If the acid drainage occurs after cessation of ore-mining operations or if it originates from the naturally outcropping deposits, the term acid rock drainage (ARD) is used.

The impact of acid mine drainage on rivers was estimated to 19 300 km in the United States. Avoca River in southeastern Ireland is affected from an abandoned copper and sulfur mine area by AMD of pH 2.7. The annual load amounts to 300 tons of metals: 108 tons Zn, 276 tons Fe, 6 tons Cu, 0.3 tons Cd. In Germany, about 500 lakes result from open cast lignite mining. Surface coal mining left hundreds of pit voids and lakes in the Appalachians and the US Midwest during the first half of the last century. In this area, new post-mining lakes will appear from 86 major ore mining plants, and in Nevada 30 pit lakes will emerge within 20 years. Many surface-mine voids will become future lakes, largely acidic and contaminated with toxic metals, from 19 metal mines in Canada, 74 in Australia, 37 in Chile, 75 in Kazakhstan, and in several other countries.

In southern Spain, the rivers Rio Tinto and Rio Odiel are natural drainages of the Iberian Pyrite Belt, the world's largest deposit of pyrite and other sulfidic ores. Rio Tinto is a naturally extreme environment with mean pH-values of 2.2 and high concentrations of heavy metals (Fe 2.3 g/L, Zn 0.22 g/L, Cu 0.11 g/L). The metals are deposited in estuarine and near-shore marine sediments containing 11.2% Fe, 0.93 g/kg Cu, 1.15 g/kg Zn, 0.73 g/kg Pb, and 0.66 g/kg Ba. These loads are caused by natural processes since 300 000 years and, additionally, by mining activities since about 4000 years.

Drainage from Acid Sulfate Soils

The acidification of sulfate soils originates from sedimentary deposits where sulfate-rich waters previously were mixed with organic loads. The decay of organic matter reduced sulfate and iron, resulting in pyrite as final product. Where these layers are disturbed, for example, by agriculture, or by lowered water tables after drought or drainage, the contact with air gives the same results as in oxidized sulfidic mining areas, and the sulfate soils turn to acid sulfate soils (ASS).

Acid sulfate soils are distributed worldwide across an area estimated at 170 000 km^2. Sulfidic deposits can form in sedimentary basins where organic loads meet with sulfate waters, both in coastal areas, where the sulfate source is marine water, and in river basins with non-marine saline loads. Examples are coastal mangrove forests, coastal lagoons, estuaries, flood-plains of inland rivers in semi-arid regions, and agricultural land after long-term irrigation with gypsiferous water. Reports are given from the Carribean region, Guyana, Surinam, Trinidad, Venezuela, Africa, South-East

Asia, Thailand, Vietnam, The Netherlands, England, Wales, Scotland, Russia, Finland, and from Australia.

The case of western Finland The coastal area of western Finland between Helsinki in the south and Oulu in the north previously was covered by the sea and, after post-glacial isostatic land uplift, is now up to 100 m a.s.l. across a total area of 3360 km^2. The sediments containing metal sulfides, usually pyrite, were deposited during the Litorina period (6000–7000 years BP) and emerged above the sea about 4000 years ago. Fifty to eighty years ago these areas were cultivated and drained. The artificial draining lowered the groundwater table and gave access to atmospheric oxygen. After oxygenation the drain water became highly acidic, contaminating the receiving rivers.

The case of Australia Acid soils are found in Australia in both coastal areas and river floodplains, and are a general problem after draining the land for agriculture. The unprecedented drought of 2007 led to an additional drop in inland water levels, and more sulfidic sediment may become exposed along the affected rivers, e.g., Murray, Darling. The total area extent in Australia is estimated to be 40 000 km^2 of ASS. A national strategy for the management of coastal and inland acid sulfate soils was developed to identify suitable countermeasures. These, however, are limited to only a few options, such as tidal flushing of coastal areas, better control of drainage, reforestation, or liming campaigns.

Volcanic Waters and Crater Lakes

Volcanic activities are a natural geogenic source of acidity. Volatile mineral acids are thermally set free as gaseous SO_2, HCl, and HF. After mixing with meteoric waters, strong acids are formed that dissolve the volcanic rock, causing leaching and weathering. The outflowing water is extremely acidic, emerging as highly mineralized brines in geothermal hot springs at flanks and top of the volcanoes, or collect in crater lakes.

Rivers which receive acidic and toxic volcanic inputs are heavily affected. The crater lake of the volcano Kawah Ijen, East Java, contains 32 Mio m^3 of hot and acidic brine water (pH<0.4, TDS > 100 g/kg, SO_4^{2-} 70 g/kg, Cl^- 21 g/kg, F^- 1.5 g/kg). The outflowing brine is contaminating the Banyupahit River. Since the acidic inflow is only incompletely neutralized, the river biota has disappeared. Downstream rice fields are irrigated with 4 m^3/s of the river water, containing daily-loads of 150 tons SO_4^{2-}, 2.8 tons F^-, 50 tons Cl^-, 10 tons Al, 35 kg Ti, and 4 kg Cu. The crater lake and its environmental impacts are estimated to be more than 200 years old. A similar situation is found at the Patuha volcano in West Java, where an acidic crater lake, Kawah Putih, and springs of acid brines with pH < 1 drain into the Citarum River. The river is contaminated by the toxic elements and, also, is used for irrigation.

Lago Copahue is an acid crater lake in Argentina at Copahue Volcano, supplying the 13 km-long Rio Agrio, an extremely acidic river with pH 0.6–1.6. The river discharges into the glacial Lake Caviahue, diluting the water to pH 2.5. The outflow of the lake, Lower Rio Agrio, is further diluted in its course reaching neutrality after ca. 50 km in downstream stretches.

Biological Effects of Acidification

With increasing acidity and decreasing pH inland waters lose all species of sensitive groups, fish, mollusks, and cyanobacteria. Species richness and diversity of phyto- and zooplankton, and of benthic invertebrates decrease to low levels. Herbivorous insects decrease, whereas carnivorous groups increase in numbers. Some groups become more abundant, the dinoflagellates become the dominant phytoplankton, fungi increase in number and diversity, and sulfate reducing bacteria in anaerobic zones. In the littoral zone *Sphagnum* moss or mats of filamentous green algae grow to masses. The food webs in rain-acidic lakes became simpler. The rates of system metabolism and productivity are reduced to low levels, and, thereby, detritus and dead woody debris can accumulate.

With lower pH, an increase is observed of acidity, toxic species of aluminum, and more intensive and deeper reaching UV irradiation. Most fish species disappeared at pH below 5.7 in the acidic lakes of the La Cloche Mountains (Canada), and no species survived <pH 4.3. The area of Norway was affected to 25%. By 1975 in southern Norway, 50% of the populations of brown trout were lost. Finally, 9630 fish populations disappeared up until 1990, and 5405 populations did not reproduce. In several thousand acidified lakes in southern Norway the macroinvertebrates disappeared with pH below 4.8. In a survey of neutral and rain-acidified streams, and in acidic lignite mining lakes in Germany, the number of benthic macroinvertebrates decreased from about 50 species at pH 8 to zero at pH 2 (**Figure 5**).

In geogenically acidic drainages, often toxic heavy metals are found as contaminants. Although these constituents are diluted, neutralized, and minerals are precipitated in the receiving rivers, the acids and toxic metals damage the freshwater biota. In Europe, a typical case is the above-mentioned Avoca River in southeastern Ireland which is affected by continuous AMD from an abandoned copper and sulfur mining

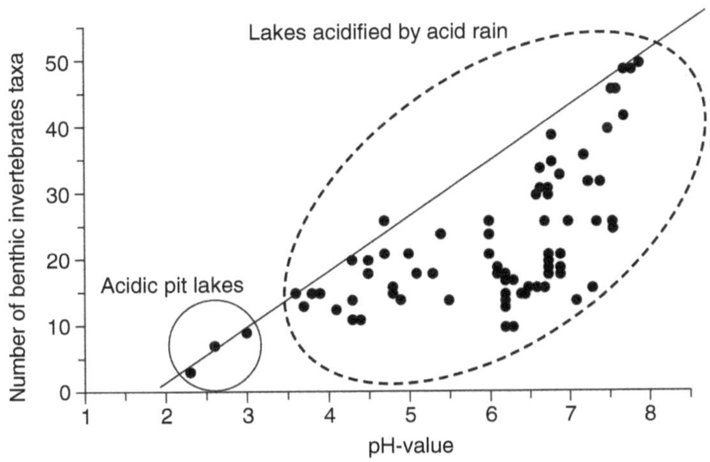

Figure 5 Macrozoobenthos in acidic lakes. (By courtesy of G. Rodrigues, Diss. Techn. Univ. Braunschweig 2001).

area. In the contaminated river, the pH of which is 5.8, macrophytes and fish are eliminated, and macroinvertebrates survive only for short periods. The damage to the indigenous biota is due to the combination of metal toxicity, sedimentation, acidity, and salinization. In permanently acidic and metal-rich waters, unexpectedly diverse communities of extremophiles can develop.

Success of Countermeasures and Long-Term Developments

Rain-Acidified Waters

The control of the atmospheric acidification of soils, lakes, and rivers was achieved in two steps:

1. the emissions of acidic smoke could be reduced with the result of less acidic atmospheric deposition in these regions. In North America and in Europe, the emissions were reduced during the last two decades by legal regulation and technical improvements, as documented by monitoring networks and by long-term observations at single sites (**Figure 6**). In Europe, the former eastern bloc countries followed with a delay of one decade; the area of former East Germany could be identified by the atmospheric SO_2 content till 1992. In other countries (China, India, Russia, South-America) the trend shows still increasing emissions, and improvements are in delay.
2. The remediation of already acidic waters and soils was successfully reached in a few countries with campaigns of liming rivers, lakes, and soils of the catchment areas. In southern Sweden, the world-wide largest liming program was conducted over 20 years by spreading annually 200 000 tons of suspended limestone. Thereby, 6000 lakes and half of the acidified area could be treated. Remediation appears as a two-step process, chemical remediation preceding biological response. Biological recovery is emerging along the given generation lengths, zooplankton and macroinvertebrates need 3–10 years, fish populations will follow. The chemistry of surface waters is fast responding to lime treatment, but the recovery of soils across the catchment areas is generally a slow, centennial process during which the soil system still remains sensitive. Therefore, continuing programs of liming surface waters and forest are needed to consolidate the reached state of neutralization and to increase the number of remediated watersheds.

Acid Mine Drainage and Acid Sulfate Soils

In areas exposed to geogenic acidification from acid mine drainage, a broad spectrum of countermeasures is available and under development. The acidification of lakes and rivers by acid drainages can be prevented by preclusive measures and direct treatment of the polluting waters as (a) restrictions on running mining activities, (b) neutralizing the waters by active treatment of the AMD by addition of alkaline chemicals, (c) passive treatment using natural geochemical and biological reactions to reduce acidity, and ion concentrations, and (d) the in situ treatment of acidic lakes.

The different approaches for remediation and treatment are described in a separate article within this encyclopedia.

The problems of acid sulfate soils are complex, and the approaches for countermeasures are limited. The primary option appears as reducing and controlling the artificial drainage, thereby declining the access of oxygen to the sulfidic layers. Acid

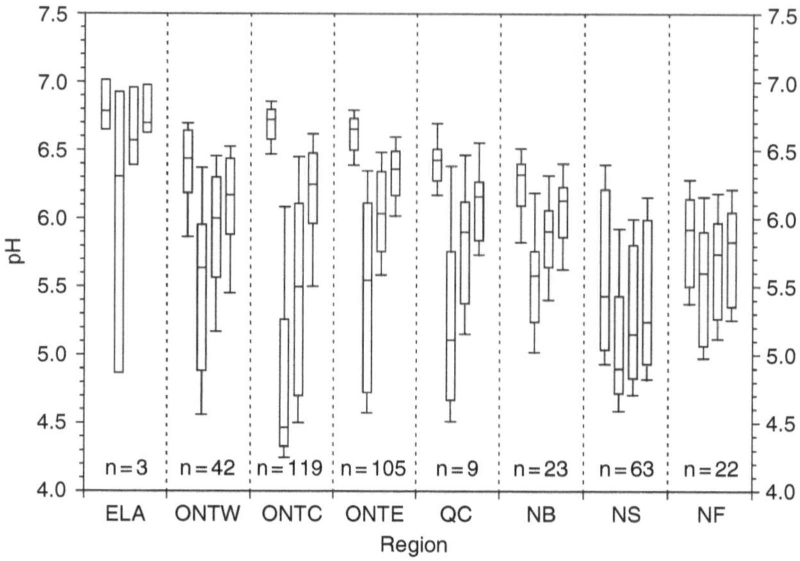

Figure 6 Distribution of modeled and measured pH values of lakes from different regions of Eastern Canada (ELA – Experimental Lake Area, western Ontario; ONTW – region east of Lake Superior, Ontario; ONTC – region north of the central part of Lake Huron, Ontario; ONTE – region east of Georgian Bay of Lake Huron, Ontario; QC – southern Quebec; NB – southern New Brunswick; NS – Nova Scotia; NF – Newfoundland). Within each regional block, the first box shows the distribution of modeled pre-acidification values, the second shows 1975 (worst case) distribution, the third shows year 2000 values, while the fourth box describes expected values under further agreed-to Canadian and currently proposed US emission reductions. The number of sites in each region is given. Lines within the boxes are median values, box limits are 25 and 75 percentiles, and whiskers show the 10 and 90 percentile values. (By courtesy of © Environment Canada 2007). Source: Clair *et al.* (2007) Past and future changes to acidified eastern Canadian lakes: A geochemical modeling approach. *Applied Geochemistry* 22: 1189–1195.

drainage water from sulfate soils can be treated with the same methods as acid mine drainage. The soil treatment with lime is more difficult, since the alkaline substances must be introduced into the subsurface layers in depths of about 1–1.5 m, where the sulfidic metals had been accumulated and, after drainage, were subsequently oxidized in the past. Coastal sites with acid sulfate soils can be treated with tidal flushing. National strategies for the remediation of acid sulfate soils, at coastal and at inland sites, presently are under discussion mainly in Australia and Finland.

Lake Orta: Acidified by Industrial Waste and Remediated by Liming

The prealpine Lake Orta in northern Italy is 143 m deep, with a water volume of 1.3 km^3. The water was heavily polluted from 1926 to 1982 by industrial wastewater (ammonium sulfate, heavy metals Cu, Cr, Ni, and Zn). Between 1960 and 1980, the lake became acidic (pH 3.9–4.5) by oxidation of the load of 3000 tons of ammonium-N per year. After stopping the input of industrial wastewaters, the lake was treated 1989–1990 with 18 000 tons of powdered limestone and could be neutralized by this chemical treatment. To date, this was the worldwide biggest liming application at a single lake.

Conclusions

Acidification of inland waters affects ground water, streams, rivers and lakes. The reasons may be natural ones, such as volcanism or oxidation at natural outcrops of pyrite deposits, or artificial ones, such as atmospheric deposition of acidity (acid rain) or pyrite oxidation caused by mining or agriculture. The anthropogenically induced acidification requires both the mitigation of the symptoms and the remediation of the causes. The recovery of acidified waters in northern America and in northern Europe was only possible after removal of the original causes, the major sources of acid waste gases. Usually, the effort for mitigation is high, but smaller than the costs of loosing aquatic ecosystems by acidification. Treatment of drainage from natural sources may also be necessary where the impacted waters are used for irrigation or drinking water supply.

Generally, acidification appears – after eutrophication – as the most important threat and stress factor for continental water resources and the ecosystems of lakes and rivers. In many impacted countries the current regulations are not sufficient to reach the remediation targets.

Glossary

Acid drainage (AD) – Drainages with acidic water both from anthropogenic and from natural acid sources, as from mining of coal, sulfidic ores, from aerated sulfide-containing soils, and from volcanic sources.

Acid mine drainage (AMD) – Mine drainage water acidified after oxidation of pyrite from mining of coal and sulfidic ores.

Acid rain – Rain, snow, sleet, and dry deposition, which are acidified by waste gases from industrial and other anthropogenic sources, often used as simplifying synonym for atmospheric deposition of acidic substances.

Acid sulfate soil (ASS) – Soils with sulfide content that release acidic and sulfate containing water after aeration by agriculture and/or lowered ground water.

Acidification – Change of water chemistry after input of acids by acid rain or acids from geogenic sources.

Acid-neutralization capacity (ANC) – The capacity of waters to neutralize added acids.

Atmospheric acidification – Impact on soils and waters by atmospheric transport of acid waste gases, aerosols, and dust from the anthropogenic sources to the affected areas which are impacted by wet and dry deposition of acid substances.

Atmospheric deposition – Sum of wet and dry deposition.

Buffering system – The two- or three-valent weak acids which keep a system within a certain pH range by their ability to change their dissociation state, i.e., to remove protons from the system (protonation of the weak acids' anions) or to liberate protons into the system (by de-protonation of the weak acids).

Dry deposition – Amount of substances deposited as gases, aerosols and dust from the atmosphere onto the earth surface.

Geogenic acidification – Acidification by natural or anthropogenically induced acid impacts from geological sources via ground water.

Liming – Treament of acid waters or soils by powdered limestone or suspended lime to mitigate the symptoms of acidification.

Remediation – Measures to restore acidified systems by eliminating the cause and the source of acidification.

Restoration – Treaments and measures to restore acidified waters and ecosystems from the disturbed state to reach again the original, natural state.

Volcanic acidification – Acid impact on soils and waters from active volcanoes via air and volcanic water.

Wet deposition – Amount of substances deposited as fog, rain, snow and other kinds of wet precipitation from the atmosphere onto the Earth's surface.

Further Reading

Bouwman AF, van Vuuren DP, Derwent RG, and Posch M (2002) A global analysis of acidification and eutrophication of terrestrial ecosystems. *Water, Air and Soil Pollution* 141: 349–382.

Brodin YW (ed.) (1992) Critical loads in nordic countries. *Ambio* Special issue, 21.

Brown M, Barley B, and Wood H (2002) *Minewater Treatment: Technology, Application and Policy.* London: IWA Publishing.

Driscoll CT, Lawrence GB, Bulger AJ, Butler TJ, Cronan CS, Eagar C, Lambert KF, Likens GE, Stoddard JL, and Weathers KC (2001) Acidic deposition in the northeastern United States: Sources and inputs, ecosystem effects, and management strategies. *BioScience* 51: 180–198.

Eloranta P (ed.) (2004) *Inland and Coastal Waters of Finland.* Helsinki: Publ. Univ, ISBN 952-10-1141-6.

EPA – U.S. Environmental Protection Agency (2006) Management and Treatment of Water from Hard Rock Mines, EPA/625/R-06/014. http://www.epa.gov.

ERMITE consortium, Younger PL and Wolkersdorfer C (eds.) (2004) Mining impacts on the freshwater environment: technical and managerial guidelines for catchment scale management. *Mine Water and the Environment* 23: 1–80.

Fleischer S and Kessler E (eds.) (1993) Acidifcation of surface waters in Sweden – Effects and countermeasures. *Ambio* 22(5): 257–337.

Fölster J and Wilander A (2002) Recovery from acidification in Swedish forest streams. *Environmental Pollution* 117: 379–389.

Geller W, Klapper H, and Salomons W (eds.) (1998) *Acidic Mining Lakes.* Berlin, Heidelberg, New York: Springer.

Hem JD (1992) Study and interpretation of the chemical characteristics of natural water. *U.S. Geological Survey Water-Supply Paper 2254*, 264 pp. Washington: U.S. Geological Survey.

Henrikson L and Brodin YW (eds.) (1995) *Liming of acidified surface waters.* Berlin, Heidelberg: Springer.

Johnson DB and Hallberg KB (2005) Acid mine drainage remediation options: A review. *Science of the Total Environment* 338: 3–14.

Olem H (1991) *Liming of Surface Waters,* 331pp. Chelsea: Levis Publishing.

Persson G (2008) Zooplankton response to long-term liming: Comparison of 15 limed and 15 reference lakes in Sweden. *Limnologica* 38: 1–13.

PIRAMID Consortium (2003) Engineering guidelines for the passive remediation of acidic and/or metalliferous mine drainage and similar wastewaters. European Commission 5th Framework RTD Project no. EVK1-CT-1999-000021: Passive in-situ remediation of acidic mine/industrial drainage (PIRAMID). University of Newcastle Upon Tyne. http://www.ncl.ac.uk/piramid.

Schindler DW (1988) Effects of acid rain on freshwater ecosystems. *Science* 239: 149–157.

Sheoran AS and Sheoran V (2006) Heavy metal removal mechanism of acid mine drainage in wetlands: A critical review. *Minerals Engineering* 19: 105–116.

Varekamp JC and Rowe GL Jr. (eds.) (2000) Crater Lakes *J Volcanology and geothermal Research*. Special Issue, vol. 97.

Weathers KC, Likens GE, Butler TJ, and Elliott A (2006) Acid rain. In: Rom W (ed.) *Environmental and Occupational Medicine,* 4th edn., pp. 1549–1561. Philadelphia: Lippincott-Raven Publishers.

Younger PL, Banwart SA, and Hedin RS (2002) *Mine Water: Hydrology, Pollution, Remediation.* Dordrecht: Kluwer Academic Publishers.

Relevant Websites

http://www.cciw.ca/gems – Canada Centre of Inland Waters (CCIW).

http://www.ce.cmu.edu/~acidmine/resources.html – Carnegie Mellon University, Pittsburgh, AMD Resources.

http://www.mines.edu/fs_home – Colorado School of Mines, AMD-homepage.

http://www.emep.int – Co-operative Programme on Long-range Transboundary Air Pollution.

http://www.claire.co.uk/costar.php – CoSTaR research facility, Hydrogeochemical Engineering Research & Outreach (HERO), Group at Newcastle University, the Coal Authority, Durham and Northumberland Councils and CLAIRE.

http://themes.eea.eu.int – European Environment Agency (EEA).

http://europa.eu.int – European Union (EU).

http://www.apps.fao.org – Food and Agricultural Organisation of the United Nations (FAO).

http://www.gemswater.org – Global Environmental Monitoring System on inland water quality.

http://www.bafg.de/grdc.htm – Global Runoff Data Centre.

http://www.gwpforum.org – Global Water Partnership.

http://www.inap.com.au – International Network for Acid Prevention (INAP). Includes clear overview of topics and reports on INAP-funded research.

http://www.clw.csito.au/scientific-reports.html.

http://www.mdbc.gov.au – Murray_Darling Basin Commission.

http://www.nrcan.gc.ca/mms/canmet-mtb/mmsl-lmsm/mend – Mine Environment Neutral Drainage (MEND) Program, Canadian mining companies and provincial/territorial and federal departments.

http://nadp.sws.uiuc.edu – National Atmospheric Deposition Program. Illinois State Water Survey.

http://wvwri.nrcce.wvu.edu – National Mine and Reclamation Center, West Virginia University.

http://www.dpie.gov.au/dpie/armcanz/pubsinfo/ASS/ASS.html – National Strategy for the Management of Coastal Acid Sulfate Soils, Australia, New Zealand.

http://www.oecd.org – Organisation for Economic Co-operation and Development (OECD).

http://www.nmnh.si.edu/gvp/volcano – Smithsonian Institution. Global volcanism program.

http://www.acidrain.org – Swedish NGO Secretariat on Acid Rain.

http://www.uba.de – Umweltbundesamt, Berlin.

http://toxics.usgs.gov/topics/minelands.html – U.S. Geological Survey Toxic Substances Hydrology Program section on Hard-Rock Mining Contamination.

http://www.epa.gov – US Environmental Protection Agency (EPA).

http://www.wmo.ch – World Hydrological Cycle Observing System.

Aquatic Ecosystems and Human Health

R M McNinch, J B Rose, and E A Dreelin, Michigan State University, East Lansing, MI, USA

Introduction

Aquatic ecosystems provide recreational opportunities and essential services to humans. However, there are a number of threats to human health, including physical and biological threats, from inland waters. Inland waters are commonly managed by humans for recreation and to maintain wildlife systems. As a result, these waters are often tourist and/or recreational areas. Humans may be exposed to health risks via recreational activities such as swimming or boating, or via ingestion of contaminated water. This chapter discusses major threats to human health in inland waters, the sources of those threats and measures to protect human health.

Physical Threats to Human Health in Inland Waters

Drowning is the most serious physical danger to a person recreating in inland waters. US data on fatalities due to non-boating related drowning show that children aged over 5 years are more likely to drown in natural fresh waters, such as rivers and lakes, than in other types of waterbodies. Flowing and deep waters are of greatest concern as large waves generated during storm events can submerse a swimmer and swift currents in rivers can overwhelm boaters or swimmers. Activities such as boating, water skiing, jet skiing, and rafting allow people access to more dangerous waters where they are at an increased risk of drowning. Drowning may result in death or nonfatal brain damage ranging from memory loss and learning disabilities to a permanent vegetative state. Cuts, broken bones, and head trauma are among other physical injuries that can occur at inland swim areas as a result of running and falling in areas containing debris, or diving in shallow or rocky areas. Exposure to the sun is also one of the physical risks while recreating on the water; it can cause sunburns, with long-term exposure being associated with cancer. Heat stress and heat stroke may also occur during extreme heat.

In order to prevent physical injuries when visiting or using various types of waters, people should feel comfortable in the environment they are in, know how to swim before entering any waterbody and never swim alone. They should only recreate in designated areas and follow all applicable guidelines, rules, or laws. The use of personal floatation devices (also known as a PFD, lifejacket, or life preserver) may be used for safety, if approved by an appropriate authority (e.g., the US Coastguard, Canadian Coast Guard or Transport Canada) and in some places are required by law when participating in rafting, boating, tubing, or water and jet skiing activities. Lifeguards, park managers, and water patrol officers are sometimes on duty at recreational waters and can be contacted in times of emergency or to simply provide information. In addition, sunscreen can protect against sunburn and drinking fluids can help to prevent dehydration.

Waterborne Disease Associated with Inland Waters

The main biological dangers associated with inland waters come from exposure to harmful pathogens and toxins that cause disease in humans. Most of these microorganisms are what are known as fecal–oral pathogens and enter the waters from fecal pollution from animals and humans (see section below for more information on sources). Humans become infected when they swim and ingest small amounts of contaminated water or via exposure to the nasal passages or directly to the skin. Common human pathogenic microorganisms found in water include viruses such as Adenovirus and Norovirus and bacteria such as *Shigella*. Pathogens associated with both animal (zoonotic pathogens) and human waste include *Cryptosporidium* and *Giardia*, both of which are protozoan parasites, and the bacteria *Escherichia coli* O157:H7 and *Campylobacter* spp. Leptospirosis, caused by a bacterium called *Leptospira*, which is found in wild and domestic animal urine, is another disease to which humans may be exposed.

Table 1 describes common types of waterborne bacteria, parasites and viruses and their associated diseases. The timing of the start of the illnesses after exposure ranges from minutes to hours to days depending on the pathogen. Skin, wound, ear, eye, respiratory, neurological and gastrointestinal infections may result from exposure to pathogens found in contaminated waters. Enteric illnesses, those associated with the gastrointestinal tract (stomach and intestines), are the most frequent adverse health outcome. These illnesses typically cause diarrhea, nausea and stomach cramps. The World Health

Table 1 Common bacteria, parasites and viruses which may cause recreational water illnesses

	Disease	*Incubation time*
Bacteria		
E. coli (O157:H7) from cattle and human feces	Gastrointestinal	12 h to 3 days
Campylobacter spp. from animal, bird, cattle, and human feces	Gastrointestinal	2–5 days
Cyanobacteria, known as blue green algae, naturally occurring associated with nutrients	Gastrointestinal, respiratory, and nervous system	30 min to 2 days
Leptospira from animal urine	Fever, headache, vomiting	2 days to 4 weeks
Salmonella spp. from humans and animal feces	Gastrointestinal	1–3 days
Shigella spp. from human feces only	Bacillary dysentery	2–4 days
Parasites		
C. parvum from animal and human feces and *C. hominus* (oocysts) from human feces only	Diarrhea	2–10 days
Entamoeba histolytica (cysts) from human feces only	Amoebic dysentery	1–7 weeks
Giardia (cysts) from animal and human feces	Diarrhea	1–2 weeks
N. fowleri a free-living amoeba	Pan Amoebic Meningoencephalitis	1–3 days
Schistosomes from bird feces and found in snails.	Swimmers Itch, cercarial dermatitis	Minutes
Viruses (from human feces only)		
Adenovirus	Respiratory	<10 days
Hepatitis A	Diarrhea Hepatitis	15–50 days
Norovirus	Diarrhea, vomiting	1–2 days
Rotavirus	Diarrhea, vomiting	3–8 days

Organization (WHO) estimates that approximately 2 million people die due to diarrheal disease each year, the majority of them children under 5 years of age. Disease outbreaks, which occur when two or more people are infected with the same disease due to the same causal agent, are usually enteric in nature when the source of infection is a water source. Young children, elderly and immuno-compromised people should take particular caution when recreating at inland waters, as they are most at risk for recreational water illnesses.

Schistosome dermatitis or swimmer's itch is not associated with fecal contamination but is a common problem for swimmers. It is caused when cercaria, a larval parasite released by snails, penetrate human skin instead of a natural host (usually waterfowl). The cercaria, unable to enter the vascular system and complete its life cycle in humans, dies in the skin. The resulting skin irritation is caused by an allergic reaction to the cecaria. Symptoms of swimmer's itch usually involve a tingling sensation shortly after wading or swimming, followed by hives or a skin rash that usually subsides within an hour but may leave red patches that can develop into papules (small solid bumps) or pustules (those containing pus). Symptoms usually peak in 48–72 h and subside within 4–7 days. The rash is usually treated topically with antihistamines. There are three species of Schistosoma flatworms that complete their life cycle in humans and cause schistosomiasis. There are over 600 million people at risk of schistosomiasis and over 200 million infected worldwide. The species that cause schistosomiasis are found mainly in Africa, the Middle East, and eastern Asia. Signs of infection are initially similar to swimmer's itch but progress to fever, cough, chills and muscle aches 2 months after infection. Chronic infections can cause liver, kidney and bladder damage.

Algal blooms, large amounts of algae in one area, may also pose a threat to human health. About 50–70% of algal blooms are harmful algal blooms (HABs); meaning that they produce toxins that can harm humans and wildlife and/or cause other negative impacts such as decreased oxygen levels in waterbodies. Excessive inputs of nutrients are related to algal blooms. Some species of Cyanobacteria or blue-green algae found in fresh water have the ability to produce toxins, such as microcystins that are harmful to human and animal health. Currently, there are 65 known microcystin toxins that attack the liver, nervous system, or cause skin irritations of animals and humans who contact or ingest the toxin. Symptoms include itchy eyes, irritated skin, vomiting, diarrhea, allergic responses and liver damage. Algal toxins, including microcystin, have been correlated with illness and off-tastes in fish and shellfish. This is also a concern for pets and other animals that may drink from these waters. Removal of the toxin from water is difficult and management approaches for controlling the blooms, include use of copper sulfate. This can actually worsen conditions by increasing the amount of toxins released into the water as cells lyse and die. Algal blooms, even those not composed of harmful algae, may also act as a suitable environment for the survival and accumulation of fecal bacteria and parasites. Green algae have been found to contain high concentrations of fecal indicator pathogens which may pose a threat to human health upon exposure.

Naegleria fowleri is a free-living protozoan parasite that lives in lakes and ponds, particularly in warm waters. Humans, commonly young children, are exposed during swimming, diving, and other water sports. The parasite enters the body through the nose and then migrates into the brain causing Primary Amoebic Meningoencephalitis (PAM) resulting in a fever and then death. It is difficult to identify and many waters may contain *Naegleria* spp. but not necessarily the type that is associated with disease.

In the US, the Center for Disease Control compiles data from states on recreational waterborne outbreaks. From 1985 through 2004 in natural fresh recreational waters, there were 154 outbreaks, with 7123 cases of illness. Gastrointestinal illness caused by bacteria was one of the major causes of these outbreaks (27%) and an additional 23% were diarrhea-associated but the pathogen in the outbreak was not identified. There were 25 outbreaks caused by *Shigella* and 12 by *E.coli* 0157:H7, both of which cause very serious illness. Interestingly there were 32 individual cases of infections associated with *Naegleria*, which results in death. The source of the outbreaks was predominately lakes (117 outbreaks) as compared with rivers or streams (9 outbreaks).

Sources of Pathogens and Toxins: Pollution in Inland Waters

Biological, chemical or physical substances can reduce actual or potential benefits of inland water and pose threats to the humans, wildlife, or other organisms that use it as a resource. Sources of pollutants are categorized as either point or non-point sources. Point source pollution can be traced to a specific location, such as a pipe that directly discharges into a waterbody. Non-point source pollution is diffuse and cannot be traced to one specific site. Non-point source pollution is caused by precipitation; as rain or snowmelt moves over the ground, it picks up and carries pathogens and other pollutants to waterbodies.

Once pathogens and toxins enter inland waters, their fate is determined by their surrounding environment and the nature of the pathogen. Sediments provide a protective barrier from solar irradiation, extreme temperatures, and saline conditions, allowing some microorganisms to survive or even re-grow. Thus, sediments may have higher abundances of bacterial populations and lower die-off rates than water. In addition, microbes adsorb to sediments which provide a means of transporting microorganisms across spatial distances via stream flow and runoff. Chemical pollutants such as lead and mercury can also accumulate in the sediment and sand where they can be re-suspended in the water column, causing elevated levels of pollution to which humans can be exposed via recreation. Humans may also be exposed to pathogens directly from ingestion of contaminated beach sand (e.g., sand that has been contaminated with waterfowl feces or sewage). Some common sources of pollution that harbor pathogens and toxins in inland waters are listed below and highlighted in **Figure 1** along with the pathways by which humans may be exposed.

Fecal Pollution

Human and animal feces are one of the most common sources of waterborne pathogens. Some microbes excreted in feces are able to survive for a period of time outside a host and in the environment. People can then ingest these microbes by swallowing contaminated waters, breathing in aerosols, or touching contaminated surfaces (e.g., sand) and transferring microbes to self and others via their hands. Some common sources of fecal pollution are described below.

Animals (wildlife, farm animals, pets) All animals, including pets and wildlife, excrete feces that may be transported to inland waters. Waterfowl often congregate near inland waters, particularly public beaches, in high numbers and contribute significantly to fecal pollution of water and beach sand. Gulls, geese, and ducks excrete the fecal bacteria *E. coli*, enterococci and *Clostridium perfringens*, in addition to the pathogens *Campylobacter*, *Salmonella*, and parasites that cause schistosomiasis. Wildlife, deer, and other ruminants (e.g., cows, sheep, and goats) can also harbor zoonotic bacteria including *Campylobacter*, *E. coli* O157:H7, *Salmonella* spp., and *Streptococcus* spp., pathogenic protozoa including *Giardia lamblia* and *Cryptosporidium parvum* and a number of host-specific viruses. Domesticated animals such as dogs and cats can harbor these same pathogens. Cats in particular can harbor *Toxoplasma*. Dog feces can contribute to fecal loading of inland waters through runoff from yards and direct defecation on beaches or in the water, when allowed access. Feral cats have also been known to contaminate waterbodies. Similarly, horse, cow, and other animal manure is a source of fecal pollution in inland waters via runoff and direct deposition to inland

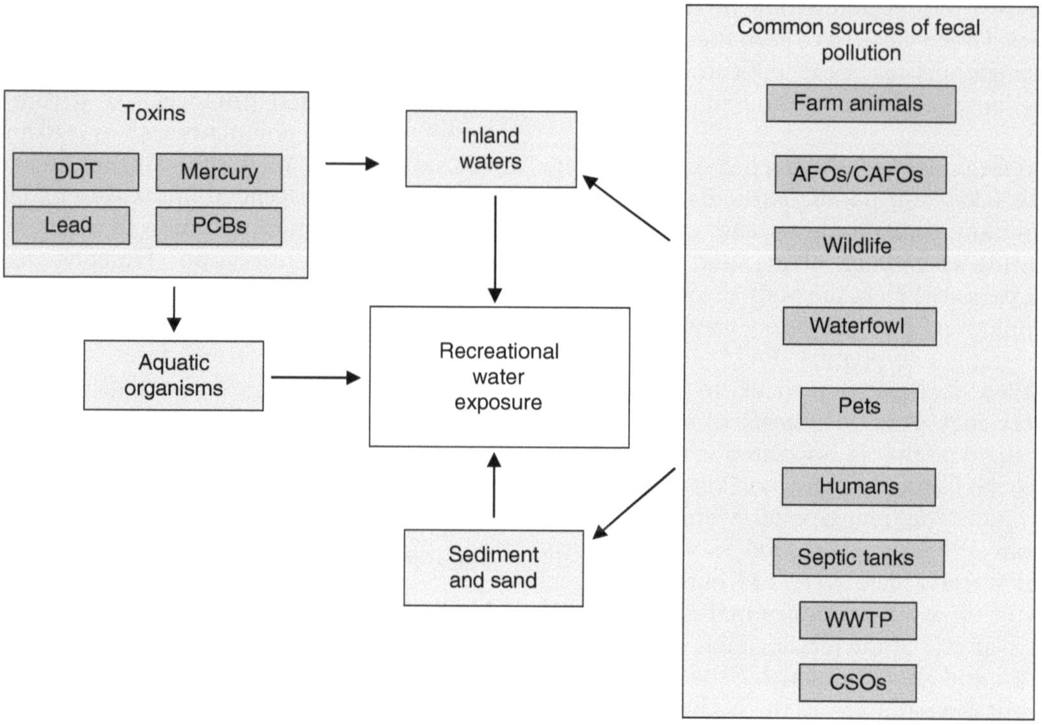

Figure 1 Common sources of pollution and the pathways by which humans may have recreational water exposure (AFO, Animal Feeding Operations; CAFO, Concentrated Animal Feeding Operations; WWTP, Wastewater Treatment Plants; CSOs, Combined Sewer Overflows).

waters when cattle are allowed access to waterbodies for drinking. To prevent animal feces from entering waterways, animal waste should be disposed of properly and never left in or near aquatic ecosystems.

Animal feeding operations Concentrated Animal Feeding Operations (CAFOs) are defined in the US as facilities in which animals have been, are or will be stabled, or confined for a total of at least 45 days in any 12-month period. CAFOs commonly use lagoon systems for animal waste management. These lagoons typically consist of open air pits filled with both urine and feces from the animals. Lagoons are susceptible to leaks and ruptures, which may contaminate nearby waterbodies. Waste from CAFOs can also enter neighboring waterways through surface runoff or groundwater filtration.

Humans Humans can directly contribute to fecal contamination of inland waters via improperly discarded diapers, accidental fecal releases, and through small amounts of feces rinsing off their bodies upon entering the water. If a person has diarrhea at the time that they are swimming, their feces are likely to contain pathogens that can cause recreational water illnesses. Human waste may also enter aquatic ecosystems via failures in waste management practices as described in the following sections. To help prevent fecal contamination at public waters, people should shower before entering the water and not swim when they have diarrhea. Dirty diapers should also be disposed of properly.

Septic tank systems Septic tanks are often used in rural areas, campgrounds, and picnic areas in place of sewer systems to treat human waste and separate solids and liquids in wastewater. The liquid portion of the waste is disposed of through a drain field where natural filtering takes place in the soil. In areas where there is a high concentration of septic tanks, it is possible for pathogenic organisms to enter shallow ground waters or nearby surface waters. In sandy and limestone soil areas including Florida, islands in the Caribbean, the Mediterranean and Asia, bacteria from waste are transferred to nearby recreational sites easily because the soil does not allow the necessary extent of filtering. One study in Florida (US) demonstrated that viruses reached surface waterways in as little as 2 h after being flushed down a toilet. The US Environmental Protection Agency (EPA) reviewed literature regarding failure rates of septic tanks and found cited failure rates ranging from 10 to 20%.

Wastewater treatment plants Sewered communities send their waste to a treatment plant where the sewage is treated to remove suspended solids and convert ammonia to nitrate (nitrification). Following treatment, wastewater is discharged to neighboring waterbodies through pipes and outfalls. In the US, the Clean Water Act mandates secondary treatment for all sewage treatment facilities, meaning that oxygen demanding substances and nitrification is required. Disinfection to kill the microorganisms may be required if the discharge is thought to impact recreational or drinking water supplies. The European Union (EU) also now often requires secondary disinfection. The most common form of disinfection used is treatment with chlorine to reduce fecal coliform bacteria. This process is less expensive than other methods of disinfection, including the use of ultraviolet radiation; however, it may negatively affect fish if not used properly and does not necessarily destroy parasites and viruses.

Many facilities use partial treatment or do not disinfect prior to discharging wastewater into inland waters. This potentially allows fecal pollution, toxic chemicals, and pharmaceuticals to enter aquatic ecosystems. The New River, bordering the US and Mexico, holds 76–95 million liters of raw sewage daily. In developing countries, approximately 90–95% of all domestic sewage and 75% of all industrial waste is discharged into surface waters without any treatment. Lack of access to safe water supplies and basic sanitation is a major cause of the diarrheal diseases reported each year.

Combined sewer overflows (CSOs) and sanitary sewer overflows (SSOs) Communities with combined sewer systems collect both stormwater runoff and domestic sewage in the same piping system. During times of heavy runoff, caused by rainfall or snowmelt, the capacity of the combined system can be exceeded and untreated sewage and storm water is released into nearby waterways. This discharge from CSOs and failures in separate sanitary sewage systems (i.e., SSOs) can contain high levels of pathogens. These systems must now be addressed in the US by separating the two systems or increasing retention of the CSO and then treatment.

Chemical Toxins and Fish Consumption

Heavy metals and other toxins can accumulate in the fatty tissue of fish, shellfish, and other aquatic organisms, posing a health risk to humans who consume them. Top predators in particular have a higher level of toxins in their body than organisms in lower trophic levels because of bioaccumulation and biomagnification. Fish consumption advisories are a commonly used approach for preventing excess human exposure to chemical contaminants in the fatty tissue of aquatic organisms. These advisories inform people what levels of fish are safe for them to consume based on geographic location and species of fish being consumed as well as the age, sex and pregnancy status of the consumer. Pregnant women should be particularly cautious about consuming fish, as toxins in fish tissue can have adverse affects on their unborn child. Some common sources of toxins to inland waters and fish are described below.

Heavy metals Mercury is a neurotoxin, meaning it affects the nervous system of exposed individuals. Mercury poisoning can cause personality changes, nervousness, trembling, and dementia. The largest sources of mercury pollution are chlorine chemical plants and coal-fired power plants that release mercury, which occurs naturally in coal, into the air through smokestacks. The mercury then contaminates natural waters through atmospheric deposition. Other sources of mercury include facilities that recycle auto scrap, thermometers, thermostats, other switches, and light bulbs. The US EPA issued a Clean Air Mercury Rule in 2005 to reduce and cap mercury emissions from coal-fired power plants, which emit approximately 50 tons of mercury pollution annually. In 2005, there were 9 mercury based chlorine plants in the US and 50 in Europe. As of 2008 two of the US plants had closed, two converted to mercury-free methods and another had committed to do so.

Lead poisoning symptoms can include nausea, sluggishness, vomiting, painful gastrointestinal irritation, diarrhea, loss of appetite, colic, weakness, and dehydration. It can result in developmental disorders, sterility, and abortion and in more severe cases produce convulsions, external limb paralysis, coma, or death. Lead is found in high levels in mining, older paints, industrial emissions and some petroleum products, and in lower levels in some fertilizers, building materials, batteries and bullets. Leaded petroleum is no longer used in the US or Canada, but its emissions can still be found in high levels in soils. Lead and other heavy metals (copper and zinc) that are mined using a smelting process can settle in natural waters where they are taken up by minerals and made bioavailable to fish and other organisms. These toxins can remain in river and lake sediments for decades, even after their use has been discontinued. One case of copper mining in Montana (US) produced contaminants found over 100 years later, 300 miles down stream of a river with its headwaters in the mining area.

Polychlorinated biphenyls (PCBs) Exposure to PCBs can cause liver, stomach, and thyroid gland injuries and impaired reproduction, immune, and behavioral responses. PCBs are mixtures of up to 209 individual chlorinated compounds; they do not occur naturally in the environment. Because PCBs do not burn easily and act as good insulators, they have been used as coolants and lubricants in transformers, capacitors and other electrical equipment, as well as in fluorescent lighting fixtures, electrical devices, microscopes and hydraulic oils. The production of PCBs was banned in the US in 1977; however, PCBs are long-lived compounds and persist in the environment.

Dichlorodiphenyltrichloroethane (DDT) DDT is an organochlorine insecticide that was used around the world after World War II. It can cause cancer, birth defects, reproductive effects, and have chronic effects on the nervous system, immune system, liver, and kidneys. Most uses of DDT were banned in the US in 1972 following the controversy that its effects were not fully known. It was banned for agricultural use worldwide shortly thereafter but is still allowed for limited use in disease vector control in certain parts of the world, an issue that remains controversial.

Protecting Human Health

At the individual level, people can help control pollution by disposing hazardous waste products properly, keeping pollutants out of drains and toilets, minimizing the use of fertilizers and pesticides, monitoring septic tanks and throwing away trash and recyclables in their proper receptacles. People can also help minimize polluted runoff by not overwatering yards, lawns, and crop fields, which may be polluted. However, because not all pollution can be eliminated, management and remediation techniques have been established, including beach monitoring guidelines to help reduce human exposure to potentially contaminated waters.

Water Quality Standards: Monitoring Fresh Waters

A standard is a legally enforceable criterion that has been set by a governing body, based on the evidence regarding appropriate use of water for recreating. The WHO designed Guidelines for Safe Recreational Water Environments to be used as the basis of recreational water quality standards that are imposed at most public recreational swimming areas today. In the case that a beach fails to meet the water quality standard, a beach closure or advisory may be issued. The WHO, US EPA, and EU guidelines are listed in **Table 2**.

A similar indicator system is used to monitor the safety of both inland and marine recreational waters. Most coastal beaches are monitored regularly whereas many inland waters, unless designated specifically as a public beach, are not monitored. An indicator system typically uses fecal coliform bacteria, *E. coli*, enterococci, or *C. perfringens* as a basis of measurement for fecal contamination. Coliforms live in the digestive tract of warm-blooded animals and are excreted in their feces naturally; hence, they are used to indicate the presence of feces. Indicators are generally presumed to be present in elevated concentrations when other, more harmful pathogens, are also present. This has been demonstrated in studies where there was an increased disease risk (caused by bacteria, viruses, and parasites) when high levels of these bacteria were present in the tested water. Given this relationship, elevated concentrations of fecal

Table 2 WHO, US EPA, and EU guidelines for fresh recreational waters

Indicator	*Levels per 100 ml*	*Interpretation*	*Agency*
Enterococci	≤ 40	Very Good, risk <1%	WHO
	41–200	Good, risk 1–5%	
	201–500	Fair, risk 5–10%	
	>500	Poor, risk >10%	
Enterococci	<33 Geometric mean	Risks <0.8%	US EPA
	61 Single sample max		
E.coli	<126 Geometric mean	Risks <0.8%	US EPA
	<235 Single sample max		
Enterococci	200 (95 percentile)	Excellent	EU
	400 (95 percentile)	Good	
	330 (90 percentile)	Sufficient	
E. coli	500 (95 percentile)	Excellent	EU
	1000 (95 percentile)	Good	
	900 (90 percentile)	Sufficient	

indicator bacteria may therefore be indicative of the presence of other harmful pathogens and that the water is unsafe for human exposure.

To test for these indicator bacteria, water samples are collected and analyzed for the level of the live indicator bacteria that can be cultured in the laboratory. Results take 18–24 h to be obtained using this method. This poses a problem because people may have come into contact with water contaminated with harmful pathogens but not known until 24 h later. Currently, no indicator test is available that can be done on location and provides immediate results.

In addition to the time lag of the current testing methods, indicator testing does not identify the full extent of pathogen contamination or the sources of contamination. Pathogen testing is an alternative method of monitoring water quality but there are numerous pathogens and testing for them all on a regular basis is currently impractical for time and cost reasons. Predictive models are another method being used at some beaches to predict beach water quality prior to human exposure. These models, such as Now-cast and SwimCast in the Great Lakes region, take into account factors such as temperature, wind and current patterns, pH, turbidity, rainfall and historical data.

Water Purification and Outdoor Recreation

Fresh waters are likely to be consumed by hikers, boaters, and other people recreating outdoors in or near inland waters. Although natural waters often give the appearance of being pristine, they can contain harmful microbes that are not readily visible to the human eye. All water consumed from inland lakes and rivers should therefore be treated or disinfected before consumption. Boiling water is the simplest and most effective way to purify water but it does not remove any particulate matter or chemical contaminants and is not effective in removing algal toxins such as microcystin. Filters are another method of effectively removing most biological contaminants, particulate matter, and chemical contaminants from water for immediate consumption. Chemical treatments such as chlorine, iodine, and oxidizing agents, are also very effective at treating viruses and bacteria. Since no one method can remove all toxins, the best method of water purification is to use a combination of treatments.

General Safety

The best way for people to prevent biological threats when recreating at inland waters is to follow good hygiene practices. This includes always washing hands after exposure to recreational waters or sand, especially before eating or drinking. People should avoid swallowing water when swimming and, in the case of small children, ingesting sand. In areas where swimmer's itch is a problem, people can reduce the risk of acquiring swimmer's itch by avoiding shallow areas, avoiding areas with snails, and toweling the skin or showering immediately after leaving the water. Areas near storm drains and small pools of water on beaches may have elevated levels of contaminants and should be avoided along with areas of visible fecal pollution or algal blooms. People may also want to familiarize themselves with the quality of the water in which they are recreating by talking with beach or park staff members, checking with the local health department, or looking on websites to obtain a list of water quality testing results.

Conclusion

Humans use inland waters for a number of recreational activities from swimming to fishing and although these natural waters may look pristine they can pose both physical and biological threats to their users. Much has been learned about the source of pollutants and progress made to stop and/or remediate the effects they have on the environment and its users. However, damage has already been done in the case of chemical pollutants which can persist in the environment for years to come and progress is still needed in terms of finding an accurate, on-site test for monitoring recreational water quality. People can protect themselves from waterborne threats by practicing good hygiene and learning safe swimming and boating practices.

Glossary

Algae – Microscopic photosynthetic organisms that are usually aquatic and lack true stems, roots, and leaves.

Algal blooms – An accumulation of algal cells of high enough concentration to be visible to the naked eye.

Bioaccumulation – Net accumulation of pollutants over time within an organism due to environmental factors, occurs within a trophic level.

Biomagnification – Accumulation of toxins which is amplified up trophic levels. The result is that top predators generally have high concentrations due to eating prey that have accumulated toxins.

Cercaria – The parasitic larva of a trematode worm.

Combined sewer overflows (CSOs) – Occur when the storage system containing storm water and untreated sewage is released into a water body during periods of heavy rainfall.

Cyanobacteria – Aquatic and photosynthetic bacteria, can have unicellular, filamentous, or colonial morphology; mostly planktonic, also known as blue-green algae.

Dichlorodiphenyltrichloroethane (DDT) – An organochlorine insecticide used after World War II.

***Escherichia coli* (*E. coli*)** – One of the main species of bacteria living in the lower intestines of mammals, known as gut flora. When located in the large intestine it actually assists with waste processing, vitamin K production, and food absorption.

Enteric bacteria – A large group of rod shaped, gram negative bacteria that typically multiply in the gastrointestinal tract of humans and animals and are excreted in feces.

Guideline – A suggested level for concentration of contaminants which is not to be surpassed but is not legally enforceable.

Indicator organism – Microbes whose presence in water signals the presence of fecal matter and potentially pathogens (e.g., *E. coli*, Fecal coliforms).

Non-point source – Source of pollution from diffuse sources such as surface run-off.

Pathogenic – Microorganisms that cause disease.

Polychlorinated Biphenyls (PCBs) – A class of organic compounds with 1–10 chlorine atoms attached to biphenyl. Its chemical formula is $C_{12}H_{10-x}Cl_x$.

Point source – Source of pollution originating from a known source such a discharge pipe combined sewer overflow.

Pollution – Introduction of substances into the environment resulting in negative effects on ecosystems.

Septic tank – On-site waste disposal system used for sewage storage and discharge.

Sanitary sewer overflow – Discharges of municipal raw untreated domestic waste. Occurs when there is heavy rainfall or when a blockage of the sewer occurs. Differs from Combined Sewer Overflow as it does not contain storm water.

Standard – A legally enforceable limit of contamination after which action needs to be taken to ensure public safety (e.g., beach closings, discharge permitting).

Trophic level – The feeding position of an organism in a food chain (i.e., primary producer, herbivore, primary carnivore).

See also: Ground Water and Surface Water Interaction.

Further Reading

Brenner RA, Tumble AC, Smith GS, Kessler EP, and Overpeck MD (2001) Where Children Drown, United States, 1995. *Pediatrics* 108(1): 85–89.

Dufour AP (1984) *Health Effects Criteria for Fresh Recreational Waters.* (EPA-600/1–84–004) Washington, DC: USEPA. Office of Research and Development.

Kalff J (2002) *Limnology: Inland Water Ecosystems.* Upper Saddle River, NJ: Prentice Hall. 41–52 pp.

Pond K (2005) *Water Recreation and Disease. Plausibility of Associated Infections: Acute Effects. Sequelae and Mortality.* World Health Organization (WHO). http://www.who.int/water_sanitation_health/bathing/recreadis/en/index.html.

Prescott LM, Harley JP, and Klein DA (1999) *Microbiology.* 4th edn. WCB/McGraw-Hill.

Schlossberg D. *Infections of Leisure.* 2nd edn. Copyright American Society for Microbiology. 1999.

United States Environmental Protection Agency (USEPA) (1986) *Ambient Water Quality Criteria for Bacteria.* (EPA-440/5–84/002) Washington, DC: Office of Water Regulations and Standards, Criteria and Standards Division.

United States Environmental Protection Agency (USEPA) (2002) *Implementation Guidance for Ambient Water Quality Criteria for Bacteria.* (EPA-823-B-02–003) Washington, DC: Office of Water.

United States Environmental Protection Agency (USEPA) (2003) *Bacterial Water Quality Standards for Recreational Waters (Freshwater and Marine Waters).* (EPA-823-R-03–008) Washington, DC: Office of Water.

World Health Organization (WHO) (1999) *Health-Based Monitoring of Recreational Water: The Feasibility of a New Approach (The 'Annapolis Protocol').* (WHO/SDE/WSH/99.1) World Health Organization – Sustainable Development and Healthy Environments.

World Health Organization (WHO) (2003) *Guidelines for safe recreational water environment Volume 1 – Coastal and fresh waters.* Retrieved July, 2007 from. http://whqlibdoc.who.int/publications/2003/9241545801_contents.pdf*http://www.who.int/water_sanitation_health/bathing/srwe1/en/

Relevant Websites

http://www.cdc.gov – Centers for Disease Control and Prevention (CDC).
http://www.epa.gov – EPA.
http://europa.eu – EU.
http://www.worldwater.org – Pacific Institute, The World's Water.
http://www.nrdc.org – Natural Resources Defense Council (NRDC).
http://www.fsis.usda.gov – United States Department of Agriculture (USDA).
http://water.usgs.gov/ – United States Geological Survey (USGS).
http://www.who.int – WHO.

Bioassessment of Aquatic Ecosystems

R H Norris, University of Canberra, Canberra, ACT, Australia
M T Barbour, Tetra Tech, Owings Mills, MD, USA

Why Biological Assessment and What is It?

The restoration and maintenance of 'healthy' aquatic ecosystems have become common goals for their sustainable well-being. These goals recognize the importance of the biological integrity of aquatic ecosystems for providing the goods and services for consumptive and nonconsumptive uses of fresh water on which humans rely. Several developments such as the US Clean Waters Act and the European Water Framework Directive (and similar initiatives in other countries) have served to heighten the importance of aquatic biota and ecological values. The ultimate effect of human activities in catchments and directly on water bodies is to change the biota and biological processes. Thus, information needed for the protection of inland waters has shifted from mainly physical and chemical measures (on the assumption that acceptable river condition would be achieved if these were met) to the use of biological indicators.

Bioassessments are implemented where information is needed on the ecological health of a waterbody or catchment. Normally, an environmental or water quality agency conducts the bioassessment within an established monitoring and assessment program, but citizen programs are also gaining momentum in several countries. The evolution of bioassessment and the associated scientific foundation has been fostered by agencies for providing ecological information for management decisions regarding the status, protection, and restoration of waterbodies. Regulatory statutes in many countries provide the impetus for gathering and interpreting biological data to address the inherent issues in the water laws and regulations.

Biota represent the health of the aquatic ecosystems – both good and bad – showing the effects from chemical pollution, organic effluents, agricultural runoff, habitat change, damming, water extraction, draining, and diversions and other uses. Biological assessment (bioassessment) is an evaluation of the condition of a waterbody using biological surveys and other direct measurements of the resident biota in surface water (**Figure 1**). Sites assessed as 'good' will usually have a diverse range of organisms, including many environmentally sensitive ones. Sites assessed as 'fair' will have lost several of the least tolerant organisms and sites in 'poor' biological condition will have mostly characteristic environmentally tolerant organisms such as several species of bloodworms (midges) and true worms. The more tolerant worms and midges are often dominant in places with low dissolved oxygen concentrations and high organic loads that are common in degraded sites. Bioassessment may be focused on either single (e.g., fish) or multiple (e.g., fish, invertebrates, algae) assemblages or biological processes (e.g., decomposition rates), and incorporates measurements of the surrounding habitat to enable an accurate evaluation of the health and ecological status of the aquatic community. Bioassesments may include toxicity testing for direct assessment and for recommending limits on chemical concentrations, use of biomarkers (physiological, biochemical, histopathological responses), and several methods using community structure. Fish, invertebrates, and algae are the most commonly used organisms for bioassessment and each will respond slightly different to environmental stressors. When the stressor is known, selecting the type of organism to use for assessment may be straightforward. However, if the stressor is not known, or there are multiple stressors, using more than one type of organism may provide better information.

Bioassessments are an application of basic ecological principles, and are designed to facilitate a better understanding of the influences of human activities on waterbodies. The central premise of bioassessments is that the ecological function of waterbodies is of interest and concern, and that uses of waterbodies for human use and consumption may confound or compromise the natural ecology.

Choices in Approach to Bioassessment

There are three main types of bioassessment:

1. indication of the presence of known or expected pollutants through bioaccumulation or other suborganism tests;
2. toxicity studies and bioassays to set limits to protect ecosystems, or test direct effects of pollutants on biota;
3. assessment of the impact of pollutants or land uses on ecosystems.

Aquatic biota can be used to directly measure the bioavailable portion of a pollutant input as in item 1 in the list. Such tests may be particularly important for protecting consumptive uses where

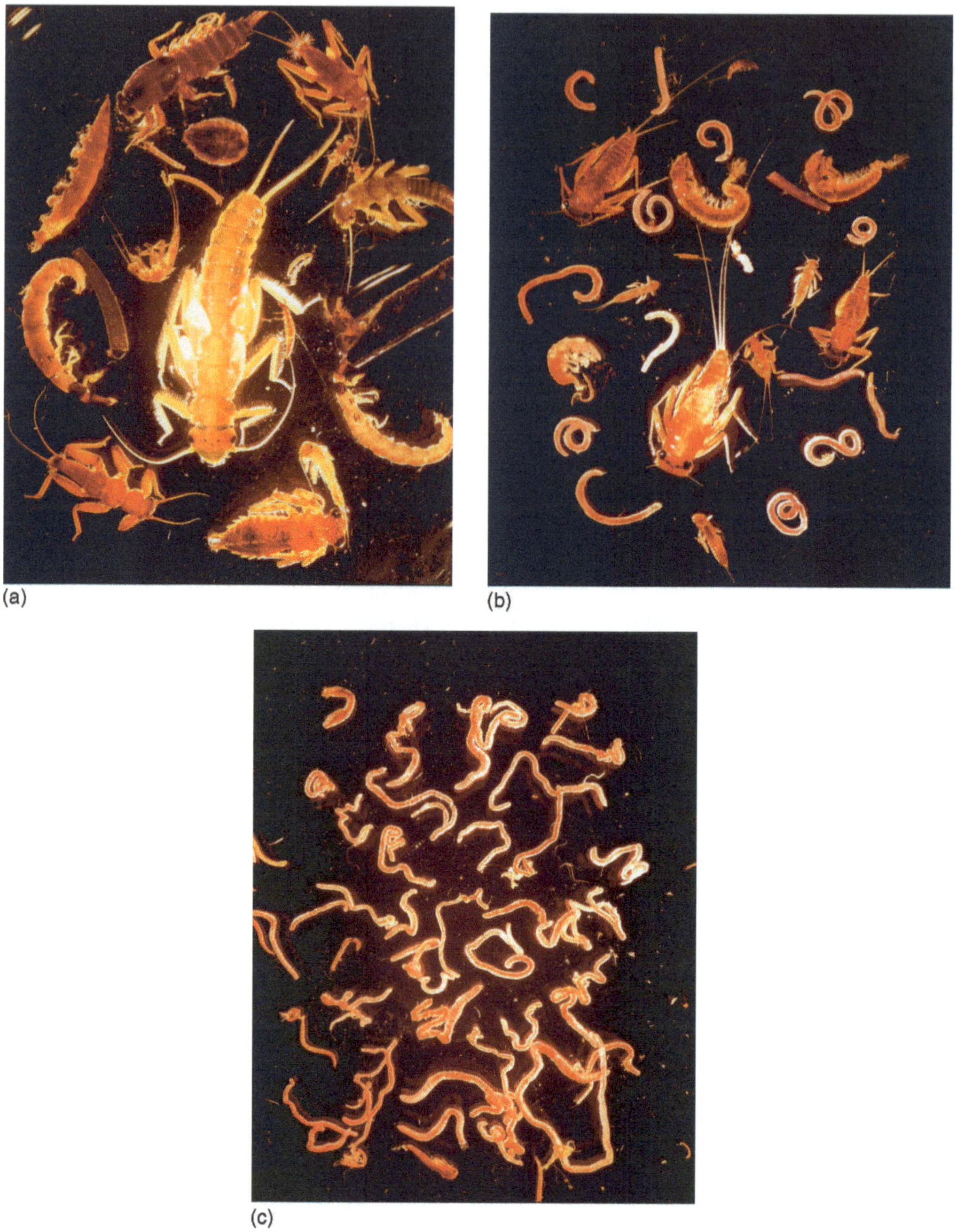

Figure 1 Evaluation of the benthic macroinvertebrate assemblage to assess biological condition of streams.

information on the potential level of pollutant exposure to humans or animals is needed. However, this kind of testing suffers from several limitations. The pollutant must be prejudged to be measured and also requires adequate physiological knowledge of the uptake and bioaccumulation, and period of retention in the test organisms. Some organisms will regulate metal levels, and others will breakdown or eliminate organic compounds over different periods. Uptake and retention will also be dependent on chemical speciation, that is, whether the pollutant is dissolved or adsorbed to particles and may be a function of other chemicals, which can have a synergistic or antagonistic effect on uptake and retention.

Where uptake by organisms is used as an assessment method, the natural variability in the effects of pollutant levels attributed to physiological factors (seasonal growth, sex, metabolism, etc.) must also be known to separate random variation from significant uptake. To be of local relevance, such testing is best done on local organisms, meaning that much work on local organisms may be needed to use this approach. Further complicating the use of organisms to indicate the presence of known or expected pollutants is the rapidity at which new chemicals are being created and discharged to the environment.

Toxicity studies and bioassays are important for providing information about environmental tolerances and the consequences of adaptation to pollutants (e.g., behavioral changes) for management strategies to be formulated. Such tests allow the effects of known pollutants on biota to be assessed and standards (or criteria) to be set. Synergistic or antagonistic effects of components in effluents on biota can also be detected where the whole effluents are tested. The main drawback of toxicity testing is that the results of studies with a few species are often extrapolated to complex systems. Sensitivity of species to pollutants also varies. These problems may be overcome by multispecies toxicity testing, or the testing of whole or enclosed systems, in field studies. Unfortunately, despite increasing sophistication in laboratory testing, there is often disagreement between test results and the concentrations found to be ecologically damaging in the field.

The effects of known, unexpected, or multiple stressors on aquatic ecosystems can be assessed by changes in the numbers of indicators species, community structure, or change in biological function (e.g., denitrification). The basis of the indicator species concept, incorporated into biotic indices, is that changes in physical and chemical conditions will favor some species, have little effect on others and be detrimental to the remainder. The appearance or disappearance of a particular species may be an indicator of some specific change in physical or chemical conditions of environment. The ecological risk assessment of threatened and endangered species is a common use of indicator species. Often, this approach focuses on megafauna, such as endangered fish species or aquatic plants; however, sometimes distribution ranges of key macroinvertebrates are evaluated with this approach. The use of indicator species requires accurate taxonomic work at the species level because even species from the same genus may respond differently to the same environmental stress. Autecological work will generally be needed for local biota to underpin biotic indices.

The intensive local ecological understanding needed to develop biotic indices has led to measurements of community structure becoming widely used, globally. The ecological consequences of cumulative and multiple stressors are best studied at higher levels of biological organization and population or community levels. Such studies have therefore been widely adopted to evaluate the ecological condition of inland waters, and coupled with a growing knowledge of the ecological requirements of the aquatic biota enable interpretation of the factors causing damage. In addition, methods that provide early warning and predict impending damage before it occurs have been developed, but these are generally in the early phases of testing. This approach is termed ecological risk assessment, which integrates multiple lines of evidence to provide a prediction capability.

Management goals for inland waters are often framed in terms of processes (e.g., maintaining river health), yet most programs detect changes in composition, or pattern. This is fine if the linkage between pattern and process is known to be close; that is, change in pattern implies a change in process. However, changes in community structure could occur in the absence of a change in process (e.g., species replacement by others from functionally similar groups). It may also be possible for ecological processes to alter without affecting community structure (e.g., a decrease in growth rate may reduce productivity but not change the taxonomic composition). In the last few years, methods have been developed for evaluating changes in ecological processes such as photosynthesis to respiration ratios and stable isotope analysis to trace the flow of nutrients and energy through ecosystems.

History of Bioassessment

Use of ambient biological communities, assemblages, and populations to provide information for the protection, management, and even exploitation of water resources has been around in some fashion for over 150 years. Early researchers in Great Britain documented the relationship among human illness and poor sanitary and drinking water conditions in the mid-1800s, which led to the development of the first set of national legislation addressing water pollution. The evolution of biological assessment of freshwater ecosystems took root in the twentieth century. It was in the early 1900s that researchers codified the study of microbiota into a system called the 'Saprobien System' that could be used to gauge the severity of organic pollution. This system became the basis from which many types of bioassessment indices have been developed over the past few decades. Bioassessments are generally indicator species based or community based in principle, although multiple and complementary approaches are becoming more sought after in the twenty-first century (**Figure 2**).

1900–1950

Surface water quality assessment based on biological indicators of pollution began in Europe in the early twentieth century as rapidly increasing human populations and industrial development began to severely affect aquatic ecosystems. The concept of biological

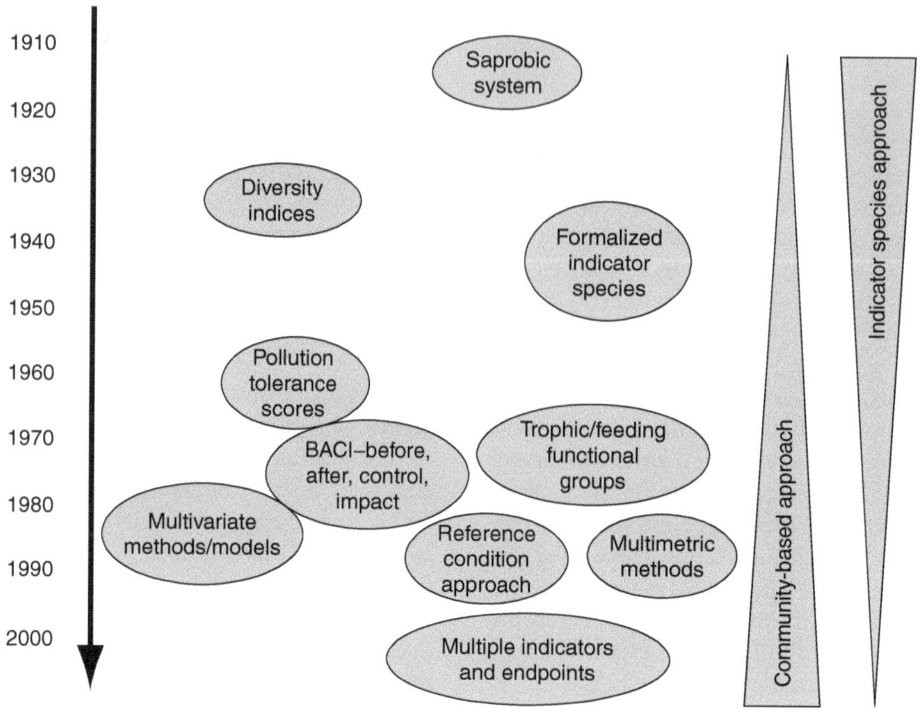

Figure 2 Historical development of bioassessment approaches for aquatic ecosystems.

indicators of aquatic conditions originated with the development of the Saprobien System for rivers, which is based on the presence of microorganisms of the plankton and periphyton community in response to sewage contamination, and the resulting decrease in oxygen. The Saprobien System was developed in Germany and has been extensively revised by European scientists and is still the basis of some systems currently in use.

1950–1970

The development of environment agencies and the establishment of environmental mandates generated a market for formal approaches to bioassessment. This market was part of the impetus for the development of the Trent Biotic Index in 1964, which was one of the earliest forms of biotic index. It was restricted to six groups of key organisms, and assigned scores to individual taxa based on their responses to certain pollutants. The index was the precursor of many that followed and still currently used, including the widely used Hilsenhoff index in North America, the Biological Monitoring Working Party (BMWP) score system used in the UK and Spain, the Index Biotique (France) and the Belgian Biotic Index.

1970–2002

1970–2002 has been a period of proliferation of bioassessment literature and research activity. Noteworthy is the detailed development of (1) parametric statistical analysis (based largely on Before, After, Control, Impact or BACI study designs); (2) rapid sampling protocols, which include the formalized integration of many indices in the multimetric approach, which was initially established in the United States; and (3) multivariate methods and predictive models, now largely available because of the enhanced power of personal computers.

The BACI design, promoted strongly in the late 1970s, was an important framework in designing bioassessment approaches. Unfortunately, resource constraints (money, time, political will) force most bioassessments to use the crudest of these designs, so all the refinements of the BACI design developed over the last 25 years have resulted in the formalization of rapid bioassessment protocols for running waters in the United States that provided much of the impetus for the proliferation of such programs in the late twentieth century. Compared with protocols and procedures for running waters, technical guidance for the bioassessment of lakes has made more modest progress in the United States, although extensive national programs have been undertaken in Europe, especially Sweden.

The multimetric approach defines an array of measures or metrics that individually provide information on diverse biological attributes and, when integrated, provide an overall indication of biological condition. A metric is a calculated term or enumeration representing some aspect of biological assemblage structure, function, or other measurable characteristic that changes in a predictable way with increased human influence. For a broad range of human impacts on aquatic ecosystems, a comprehensive, multiple metric approach has been found to be an effective assessment tool in the United States.

Predictive models dependent on multivariate statistical methods were first developed for national biological monitoring programs in the United Kingdom beginning in 1977 for classifying unpolluted running-water sites based on benthic macroinvertebrates, and to determine whether the biota expected at unstressed sites could be predicted from physical and chemical features of the site. From this, the River Invertebrate Prediction and Classification System (RIVPACS) approach was recommended to regulatory authorities as a general classification scheme and bioassessment for waters of the United Kingdom. The predictive models were in their third generation by 1995, and were the basis of a national stream assessment in a second, quinquennial, national river quality survey in the United Kingdom. A slight modification of this approach has also been adopted for Australia's National River Health Program that first reported nationally on the condition of flowing waters in 2001.

In the early 1990s, the Reference Condition Approach (RCA) was developed to add power to the bioassessment strengthening the benchmark from which to assess the biological condition. The RCA advocates the use of reference sites, or those best able to represent the expected biological condition. It is most commonly applied for broad-scale assessments where comparisons are made between suspect sites and groups of sites that, by their physical and chemical features, are unlikely to be affected by human activities and are expected to harbor similar biota. The RCA differs fundamentally from other approaches commonly used for assessing water bodies (e.g., BACI designs) in that sites, rather than multiple collections within sites, serve as replicates. The RCA has been successfully implemented in Canadian streams of the Fraser River basin, British Columbia, American streams on a statewide scale, and the nearshore, fine sediments of the Great Lakes. It is also the basis for many broad-scale bioassessment programs in the United Kingdom, Europe, the United States, and Australia, and has been used for most types of water bodies, including rivers, lakes, and wetlands in wide array of environments.

Types of Bioassessment Studies

- *Ecological status and condition assessments.* This type of bioassessment is to address the questions: What is the condition or quality of my waters? What percentage of my waters is in 'good,' or 'poor' condition? Sampling designs are structured to enable extrapolation of bioassessment results to unassessed waters, and, thereby, provide estimates of ecological condition to the entire water resource.
- *Trends analysis and monitoring over time.* Monitoring programs generally want to know, is ecological condition getting better or getting worse? Bioassessments designed to be conducted on a periodic basis are used to evaluate this question. This type of bioassessment is one of the more common uses of biological information.
- *Problem characterization, stressor identification, and diagnosis.* Management questions relating to impairment include the following: Is there a problem? If so, how bad is the problem? What is causing the problem? How do we mitigate the problem? Biological indicators, sometimes referred to as biological response signatures, are inherent in many bioassessments to provide information on identifying the causal agent to impairment. These bioassessment tools are very much still in developmental stage in many parts of the world, and depend on the knowledge of sensitivity of different biota to specific stressors.
- *Ecological risk assessments and predictive ecology.* Bioassessments are beginning to be used more in evaluating the risk to an aquatic ecosystem from a human disturbance activity. These risk assessments depend on an understanding of the biological condition gradient relationship to a continuum of the magnitude of a stressor or multiple stressors, and the knowledge of an expected community structure given no disturbance at all. Several countries have developed indices and techniques for providing an ecological risk assessment using biological data.
- *Restoration and ecosystem recovery monitoring.* The physical and chemical restoration of specific waterbodies requires some level of monitoring to verify whether ecological recovery actually occurs. Bioassessments are used to determine whether and when recovery occurs. As in ecological risk assessment, an understanding of reference condition (expected community structure) and the biological condition gradient is important to assess the progress and attainment of recovery.

Biota Used in Bioassessments

Bioassessment of an aquatic community consists of surveying and analyzing various assemblages. An assemblage is an association of interacting

Table 1 The aquatic assemblages that are used in bioassessments in different waterbody types, where lotic systems are flowing waters (streams and rivers) and lentic systems are lakes and reservoirs

Community assemblage	*Lotic systems*	*Lentic systems*	*Wetlands and marshes*
Benthic macroinvertebrates	✓	✓	✓
Fish	✓	✓	✓
Algae (periphyton)	✓		✓
Algae (phytoplankton)		✓	
Aquatic vegetation	✓	✓	✓
Amphibians	✓		✓
Reptiles			✓
Zooplankton		✓	

populations of organisms in a given waterbody – for example, the fish assemblage or benthic macroinvertebrate assemblage. Ideally, a bioassessment program would have the capability of using multiple assemblages. However, many bioassessments rely on a single assemblage for evaluating ecological condition. Those assemblages used most commonly in the various waterbody types are listed in **Table 1**.

Examples of Bioassessment Programs – See above in History

Bioassessments can be conducted at different scales, depending on the needs and management questions. National and multinational scale assessments are generally conducted to estimate ecological status and condition as a basis for broad-scale decision making. State and local bioassessments are directed toward multiple questions and involve all types described earlier. Watershed assessments typically focus on targeted problems and restoration goals. Examples of bioassessment programs are as follows:

1. National scale
   a. United States – A study of the ecological condition of small streams and tributaries was conducted throughout the United States of America from 2004 to 2006. The study was designed as an opinion poll, that is, 1400 stream sites were selected at random (probabilistic sampling design) to represent the condition of all streams in regions that share similar ecological characteristics. A biological assessment was the core of this program and provided a basis for reporting on the ecological condition of the streams in the USA, and to note where the major problems were with the water resource quality. As such, an additional 600 sites that were considered to be in reference condition (minimally damaged) were incorporated in the design to include a RCA benchmark for different regions of the country. Two basic questions were asked of this program: (1) What percentage of the Nation's stream resource is in good condition? (2) What is the nature and relative importance of stressors in different regions of the country?
   b. Great Britain – Great Britain has implemented the third version of their River Invertebrate Prediction and Classification system (RIVPACS). This version includes 614 reference sites throughout Great Britain and 70 sites in Northern Ireland. The major use of RIVPACS is for prediction of the macroinvertebrate fauna to be expected at a site in the absence of environmental stress. The RIVPACS models may be used to make predictions at different taxonomic levels and for different seasons, according to the needs of users. Procedures are available for comparing the fauna observed at a site with the expected fauna, based on presence/absence data or log categories of abundance. RIVPACS has been used for several national assessments at five yearly intervals.
   c. Australia – Biological assessment of freshwater ecosystems has developed markedly during the last decade as a result from concern about ecological values that has led to an emphasis on aquatic biota for assessing conditions directly. A National River Health Program (NRHP) was set up to, among other things, assess and monitor the health of Australian rivers. The NRHP adopted a modified version of the RIVPACS program from the United Kingdom. This required a standard, working platform (called the Australian River Assessment System, AUSRIVAS) that would allow water managers to compare the condition of streams and rivers both within and among all Australian states and territories. In 1994–1995 macroinvertebrate were sampled from almost 2000 sites throughout Australia, in spring and autumn, so that the variability of sites in Reference Condition could be assessed, and so that models could be built to

predict communities in Reference Condition, by matching reference sites with test sites using environmental variables. In 1997, Australia embarked on the First National Assessment of River Health, in which about 6000 sites are sampled in spring and autumn each year. This is probably the first continent-wide assessment of river condition.

2. Multinational
   a. European Union AQEM – The European Union published the Water Framework Directive (WFD) in 2002, which aims to harmonize many aspects of European water policy, and has a major focus on ecological assessment to determine future restoration measures and catchment management plans. One research project organized in response to the mandates of the WFD is AQEM, which refers to the development and testing of an integrated assessment system for the ecological quality of streams and rivers throughout Europe using benthic macroinvertebrates. AQEM addresses the various stream types throughout the countries of Austria, The Czech Republic, Germany, Greece, Italy, The Netherlands, Portugal, and Sweden. The multimetric AQEM assessment system is used to classify a stream stretch into an Ecological Quality Class ranging from 5 (high quality) to 1 (bad quality) and often provides information on the possible causes of degradation.
3. State or Provincial
   a. Ohio, USA – The State of Ohio is an example of having a well-developed bioassessment program, particularly for streams and rivers. The Ohio Environmental Protection Agency (OEPA) has been conducting long-term bioassessment since the late 1970s. Both fish and benthic macroinvertebrate assemblages are surveyed to address various management decision needs relevant to three central issues: (1) assessing ecological condition of catchments, (2) locating and diagnosing impairment, and (3) monitoring improvements after mitigation activities are imposed. One of the best proven uses for bioassessment is for spatial and trend analysis. OEPA has been able to demonstrate the longitudinal condition along a river and showed that over time after improvements were made to discharges, the ecological condition also improved.
4. Community programs
   a. Community-based programs generally revolve around an environmental interest group that may or may not include professional biologists and other scientists. However, the primary impetus of community programs is a profound interest in protecting or restoring local water resources, oftentimes, a watershed that is vital to the community. Community programs can be found in many countries, and they can operate at different scales, i.e., regional, provincial, or local. Examples of community programs in the United States include River Watch Networks that maintain monitoring programs in key river basins or watersheds; Save Our Streams (SOS) operates in several American states and conducts training of community volunteers in conducting bioassessments of streams and rivers to identify impacts from suspected causes; and many other entities that foster organizations to assist communities in establishing viable programs that educate, train, and maintain the program for the general populace. These community programs have become instrumental in influencing management decisions regarding the protection and restoration of waterbodies.

See also: Biogeochemistry of Trace Metals and Mettaloids; Distribution and Abundance of Aquatic Plants – Human Impacts; Pollution of Aquatic Ecosystems I; Pollution of Aquatic Ecosystems II: Hydrocarbons, Synthetic Organics, Radionuclides, Heavy Metals, Acids, and Thermal Pollution; Rivers.

Further Reading

Alexander G (1876) *The Law as to the Pollution of Rivers: The Rivers Pollution Prevention Act, 1876.* London: Knight and Co., Local Government Publishers.

Bailey RC, Norris RH, and Reynoldson TB (2004) *Bioassessment of Freshwater Ecosystems. Why and How.* New York: Kluwer.

Barbour MT, Gerritsen J, Snyder BD, and Stribling JB (1999) *Rapid Bioassessment Protocols for Use in Streams and Wadeable Rivers: Periphyton, Benthic Macroinvertebrates and Fish*, 2nd edn. Washington, D.C.: U.S. EPA, Office of Water. EPA/841-B-99–002.

Barbour MT and Yoder CO (2000) The multimetric approach to bioassessment, as used in the United States. In: Wright JF, Sutcliffe DW, and Furse MT (eds.) *Assessing the Biological Quality of Freshwaters: RIVPACS and other Techniques*, pp. 281–292. Ambleside, UK: Freshwater Biological Association.

Barbour MT (1997) The re-invention of biological assessment in the U.S. *Human and Ecological Risk Assessment* 3: 933–940.

Davis WS (1995) Biological assessment and criteria: building on the past. In: Davis W and Simon T (eds.) *Biological Assessment and Criteria: Tools for Water Resource Planning and Decision making*, pp. 15–29. Ann Arbor, Michigan: Lewis Publishers.

De Pauw N and Van Hooren B (1983) Method for biological quality assessment of water courses in Belgium. *Hydrobiologia* 100: 153–168.

Green RH (1979) *Sampling Design and Statistical Methods for Environmental Biologists.* New York: Wiley.

Hilsenhoff WL (1987) An improved biotic index of organic stream pollution. *Great Lakes Entomologist* 20: 31–39.

Hynes HBN (1994) Historical perspective and future direction of biological monitoring of aquatic systems. In: Loeb SL and Spacie A (eds.) *Biological Monitoring of Aquatic Systems*, pp. 11–21. Ann Arbor, Michigan: Lewis Publishers.

International, Organization for Standardization (ISO) (1979) *Assessment of the Biological Quality of Rivers by Macroinvertebrate "score"*, p. 18. ISO/TC 147/SC 5/WG 6 N 5.

Johnson RK and Goedkoop W (2000) The 1995 national survey of Swedish lakes and steams: Assessment of ecological status using macroinvertebrates. In: Wright JF, Sutcliffe DW, and Furse MT (eds.) *Assessing the Biological Quality of Freshwaters. RIVPACS and Other Techniques*, pp. 229–240. Ambleside, UK: Freshwater Biological Association.

Karr JR and Chu EW (1999) *Restoring Life in Running Waters. Better Biological Monitoring.* pp. 206, Washington, D.C: Island Press.

Kolkwitz R and Marsson M (1909) Okologie der tierischen Saprobien. Beitrage zur lehre von des biologischen Gewasserbeurteilung. *Internationale Revue der Gesamten Hydrobiologie und Hydrographie* 2: 126–152.

Lenat DR and Barbour MT (1994) Using benthic macroinvertebrate community structure for rapid, cost-efficient, water quality monitoring: Rapid bioassessment. In: Loeb SL and Spacie A (eds.) *Biological Monitoring of Aquatic Systems*, pp. 187–215. Ann Arbor, MI: Lewis Publishers.

Plafkin JL, Barbour MT, Porter KD, Gross SK, and Hughes RM (1989) *Rapid Bioassessment Protocols for Use in Streams and Rivers. Benthic Macroinvertebrates and Fish.* Washington D.C: Office of Water Regulations and Standards, U.S. Environmental Protection Agency. EPA/440/4-89/001.

Simpson JC and Norris RH (2000) Biological assessment of water quality: development of AUSRIVAS models and outputs. In: Wright JF, Sutcliffe DW, and Furse MT (eds.) *Assessing the Biological Quality of Freshwaters: RIVPACS and Similar Techniques*, pp. 125–142. Ambleside: Freshwater Biological Association.

Tuffery G and Verneaux J (1968) *Method de determination de la qualite biologique des eaux courantes.* (Paris: Centre National d'etudes techniques et de Recherches technologiques por l'Agriculture, les Forets et l'Equipment rural. Section Peche et Pisiculture) pp. 28.

US EPA (1998) *Guidelines for Ecological Risk Assessment.* Washington, DC: US Environmental Protection Agency. EPA/630/R-95/002F.

Woodiwiss FS (1964) The biological system of stream classification used by the Trent River Board. *Chemistry and Industry* 11: 443–447.

Wright JF, Moss D, Armitage PD, and Furse MT (1984) A preliminary classification of running-water sites in Great Britain based on macroinvertebrate species and the prediction of community type using environmental data. *Freshwater Biology* 14: 221–256.

Wright JF (1995) 'Development and use of a system for predicting the macroinvertebrate fauna in flowing waters' *Australian Journal of Ecology* 20: 181–197.

Deforestation and Nutrient Loading to Fresh Waters

M C Feller, University of British Columbia, Vancouver, BC, Canada

Introduction

Nutrient cycling within forest ecosystems involves nutrient uptake and retention by biota, which retards nutrient movement to fresh waters. Deforestation, or killing of forest vegetation, initially disrupts this uptake and retention resulting in altered nutrient fluxes to fresh waters. These fluxes are in both dissolved and particulate form. Dissolved fluxes dominate for N, S, C, K, Na, Mg, and Ca but particulate fluxes tend to dominate for P, and can also be important for N (**Table 1**). Fresh water nutrient fluxes have decreased or increased to variable extents, or have remained unaffected by deforestation (**Table 2**). Their response to deforestation depends on the nutrient and the effects of deforestation on the factors controlling freshwater chemistry and particulates.

Factors Controlling Nutrient Loading in Fresh Waters

Dissolved Nutrients

The following factors are the most important determinants of chemistry:

1. geological weathering;
2. atmosphere precipitation and climate, including (1) precipitation chemistry, (2) stream discharge, and (3) temperature;
3. terrestrial biological processes, including (1) nutrient uptake, (2) nutrient transformations, and (3) production of soluble chemicals;
4. physical–chemical reactions in the soil; and
5. physical, chemical, and biological processes within aquatic ecosystems, including the physical–chemical processes of (1) ion exchange, (2) oxidation–reduction, (3) evaporation–crystallization processes, (4) pH-induced chemical transformations, and the biological processes of (5) chemical uptake and (6) microbial transformations.

Due to the complexity of these factors, nutrient loading in streams draining nearby, apparently similar watersheds can vary from one watershed to the next as a result of small changes in geology, soil, streambed materials, or stream shading, for example, none of which might be visually obvious. Freshwater nutrient fluxes may also vary temporally from one year to the next as a result of changes in weather or precipitation chemistry, for example. Consequently, accurate determination of the effect of deforestation on freshwater nutrient loading requires sampling of a control watershed which remains undisturbed, as well as a deforested watershed prior to as well as after deforestation, as occurred in the studies used in **Table 2**. Sampling of a watershed only after deforestation has occurred and comparing its nutrient loading to that of an undisturbed watershed or sampling of only a deforested watershed without a control, will not allow a conclusive assessment of the effects of deforestation on freshwater nutrients.

Temporal variations in stream nutrient concentrations, which vary with discharge, require frequent sampling, at least once every 1–2 weeks, to quantify the effects of deforestation on nutrient loading. Such frequent sampling is less necessary for lakes, whose chemistry changes less rapidly than that of streams.

Nutrients in Particulate Form

The most important factors controlling particulates in fresh water are:

1. watershed topography, including (1) slope and (2) roughness or uniformity,
2. soil erodibility, as determined by soil physical properties,
3. precipitation characteristics,
4. watershed susceptibility to mass wasting as determined by topography and bedrock,
5. stream channel characteristics, including presence and stability of debris jams, and the nature of materials lining the streambed and streambanks,
6. proximity of vegetation to surface water (which controls the extent to which plant litter is transferred to water),
7. extent of roading, particularly new unsealed roads, in a watershed.

The two concerns noted above for solute sampling also apply to particulate sampling, even more so in the case of sampling frequency, because particulate transport increases greatly while stream discharge is increasing in response to rain or snowmelt.

Table 1 Dissolved nutrient load as a percentage of the total nutrient load carried by streams flowing from essentially undisturbed watersheds, and the influence of disturbance on this percentage

Area	*Watershed size (ha)*	*N*	*P*	*S*	*C*	*K*	*Na*	*Mg*	*Ca*	*Source*
A. Undisturbed streams										
North America										
Wilson and Blossom Rivers, SE Alaska	18 100–29 400	–	–	–	>90	–	–	–	–	20
Beaver Creek, Quebec	1 800	82	–	–	–	–	–	–	–	16
Lake Memphremagog area, Quebec	200	–	38	–	–	–	–	–	–	17,18
Experimental Lakes area, Ontario	<1–7 (4 watersheds)	82–88	44–52	–	94–97	–	–	–	–	1
	12–170 (3 watersheds)	84–90	64–72	–	–	–	–	–	–	3
H.J. Andrews Experimental Forest, Oregon	10	78	–	–	–	–	–	–	–	21
	10	–	–	–	–	99	–	99	98	8
	21	71	81	–	–	–	–	–	–	12
Green Lakes Area, Colorado	8	88	–	–	–	–	–	–	–	23
Ward Creek, California	526	–	16–32	–	–	–	–	–	–	9
Nineteen streams in United States	1 800–85 000	75 (range 34–99)	–	–	–	–	–	–	–	11
Hubbard Brook, New Hampshire	13	96	43	99	–	74	96	93	98	5
	22.5	94	26	100	71	67	91	88	96	4
	130	–	11		–	–	–	–	–	15
Coweeta, North Carolina	13–14 (2 watersheds)	–	–	–	–	100	–	–	98	22
Lexington, Tennessee	<1–1 (8 watersheds)	70	31	–	–	–	–	–	–	14
Coffeeville, Mississippi	2–3 (5 watersheds)	53 (range 50–59)	29 (range 24–36)	–	–	–	–	–	–	7,19
Europe										
Rivers throughout Finland	?									
	15 rivers	90	40	–	94	–	–	–	–	13
Rest of World										
Caura R., Venezuela	4 750 000	64	49	–	–	–	–	–	–	10
BuKit Tarek, Malaysia	3	66	80	–	–	86	99	29	36	24
Ballance Stream, New Zealand	9	75	48	–	–	–	–	–	–	2
Purukohukohu basin, New Zealand	2 watersheds	97–100	14–38	–	–	–	–	–	–	6

B. Effects of disturbance										
Hubbard Brook, New Hampshire										
Undisturbed	13.2	96	43	99	–	74	96	93	98	5
After cutting + herbicide		100	10	97		88	87	91	98	
Hubbard Brook, New Hampshire										
Undisturbed	22.5	94	26	100	71	67	91	88	96	4
After debris jam removal		74	5	99	34	17	62	53	80	
Experimental Lakes area, Ontario				–	–	–	–	–	–	3
Undisturbed	12–170	84–87	64–65							
After windstorm	(2 watersheds)	74–87	48–63							

Sources

1. Allan CJ, Roulet NT, and Hill AR (1993) The biogeochemistry of pristine, headwater Precambrian shield watersheds: an analysis of material transport within a heterogeneous landscape. *Biogeochemistry* 22: 37–79.
2. Bargh BJ (1977) Output of water, suspended sediment and phosphorus and nitrogen forms from a small forested catchment. *New Zealand Journal of Forest Science* 7: 162–171.
3. Bayley SE, Schindler DW, Beaty KG, Parker BR, and Stainton MP (1992) Effects of multiple fires on nutrient yields from streams draining boreal forest and fen watersheds: nitrogen and phosphorus. *Canadian Journal of Fisheries and Aquatic Sciences* 49: 584–596.
4. Bilby RE (1981) Role of organic debris jams in regulating the export of dissolved and particulate matter from a forested watershed. *Ecology* 62: 1234–1243.
5. Bormann FH, Likens GE, Siccama TG, Pierce RS, and Eaton JS (1974) The export of nutrients and recovery of stable conditions following deforestation at Hubbard Brook. *Ecological Monographs* 44, 255–277.
6. Cooper AB and Thomsen CE (1988) Nitrogen and phosphorus in streamwaters from adjacent pasture, pine and native forest catchments. *New Zealand Journal of Marine and Freshwater Research* 22: 279–291.
7. Duffy PD, Schreiber JD, McClurkin DC, and McDowell LL (1978) Aqueous- and sediment-phase phosphorus yields from five southern pine watersheds. *Journal of Environmental Quality* 7: 45–50.
8. Fredriksen RL (1971) Comparative chemical water quality – natural and disturbed streams following logging and slashburning. In Krygier JT, and Hall JD (eds.) *Forest Land Uses and Stream Environment*, pp. 125–138. Corvallis: Oregon State University.
9. Leonard RL, Kaplan LA, Elder JF, Coats RN, and Goldman CR (1979) Nutrient transport in surface runoff from a subalpine watershed, Lake Tahoe basin, California, *Ecological Monographs* 49: 28–310.
10. Lewis WM, Jr. (1986) Nitrogen and phosphorus runoff losses from a nutrient poor tropical moist forest. *Ecology* 67: 1275–1282.
11. Lewis WM, Jr. (2002) Yield of nitrogen from minimally disturbed watersheds of the United States. *Biogeochemistry* 57/58: 375–385.
12. Martin CW and Harr RD (1989) Logging of mature Douglas-fir in western Oregon has little effect on nutrient output budgets. *Canadian Journal of Forest Research* 19: 35–43.
13. Mattsson T, Kortelainen P, and Räike A (2005) Export of DOM from boreal catchments: impacts of land use cover and climate. *Biogeochemistry* 76: 373–394.
14. McClurkin DC, Duffy PD, Ursic SJ, and Nelson NS (1985) Water quality effects of clearcutting upper coastal plain loblolly pine plantations. *Journal of Environmental Quality* 14: 329–332.
15. Meyer JL and Likens GE (1979) Transport and transformation of phosphorus in a forest stream ecosystem. *Ecology* 60: 1255–1269.
16. Naiman RJ and Melillo JM (1984) Nitrogen budget of a subarctic stream altered by beaver (*Castor canadensis*). *Oecologia* 62: 150–155.
17. Prairie YT and Kalff J (1988) Dissolved phosphorus dynamics in headwater streams. *Canadian Journal of Fisheries and Aquatic Sciences* 45: 200–209.
18. Prairie YT and Kalff J (1988) Particulate phosphorus dynamics in headwater streams. *Canadian Journal of Fisheries and Aquatic Sciences* 45: 210–215.
19. Schreiber JD, Duffy PD, and McClurkin DC (1980) Aqueous- and sediment-phase nitrogen yields from five southern pine watersheds. *Soil Science Society of America Journal* 44: 401–407.
20. Sugai SF and Burrell DC (1984) Transport of dissolved organic carbon, nutrients, and trace metals from the Wilson and Blossom rivers to Smeaton Bay, southeast Alaska. *Canadian Journal of Fisheries and Aquatic Sciences* 41: 180–190.
21. Triska FJ, Sedell JR, Cromack K, Jr., Gregory SV, and McCorison FM (1984) Nitrogen budget for a small coniferous forest stream. *Ecological Monographs* 54: 119–40.
22. Webster JR and Patten BC (1979) Effects of watershed perturbation on stream potassium and calcium dynamics, *Ecological Monographs* 49: 51–72.
23. Williams MW, Hood E, and Caine N (2001) Role of organic nitrogen in the nitrogen cycle of a high-elevation catchment, Colorado Front Range. *Water Resources Research* 37: 2569–2581.
24. Yusop Z, Douglas I, and Nik AR (2006) Export of dissolved and undissolved nutrients from forested catchments in Peninsula Malaysia. *Forest Ecology and Management* 224: 26–44.

Table 2 Estimated effect of deforestation on the flux (kg ha^{-1} $year^{-1}$) of nutrients dissolved in streamwater using studies which involve control and deforested watersheds for periods both before and after deforestation

Area	*Vegetation*[a]	*Percent*[b] *of vegetation removed*	*Type of disturbance*[c]	*Watershed size*[d] *(ha)*	*Post disturbance period (years)*	*Annual precipitation (mm)*	*Soil texture*[e]	*Bedrock*[f]	*Org – C*	*NO_3 – N*	*NH_4 – N*	*DIN*	*Org – N*	*TDN*	*PO_4 – P*	*Org – P*	*TDP*	*SO_4 – S*	*K*	*Na*	*Mg*	*Ca*	*Cl*	*Source*
North America																								
Experimental Lakes	C (H)	50–100	W	12	1	820	S	G	–	0.1	0.0	–	0.3	–	–	–	0.0	–	0.9	–	–	–	–	4
area, Ontario	C (H)	70–100	W	170	1	820			–	0.0	0.0	–	0.2	–	–	–	0.0	–	1.2	–	–	–	–	32
Nashwaak,	C-H	91 (B)	C	391	3	1320		G	–	1.5	–	–	–	–	–	–	–	–	–	–	–	–	–	16
New Brunswick	C-H	91 (B)	C	391	6	1320			–	2.3	0.9	–	–	–	–	–	–	–	5.1	–	0.3	3.3	–	39
Turkey Lakes,	H	100	C	5	3	1230	GSiL	B,G	0.1	18.4	–	–	–	–	–	–	–	–	–	–	–	20.5	–	11
Ontario	H	50	PC	68	3	1230			0.1	4.4	–	–	–	–	–	–	–	–	–	–	–	0.6	–	11
	H	33	PC	24	3	1230			3.8	3.1	–	–	–	–	–	–	–	–	–	–	–	7.9	–	11
UBC Research Forest	C	61	C	23	5	2220	SL	G	–	1.5	0.0	–	–	–	0.0	–	–	–0.1	3.0	–2.4	0.1	–0.7	–2.4	10
British Columbia	C	19	C+S	68	5	2430	SL		–	0.9	0.0	–	–	–	0.0	–	–	–1.5	0.3	–6.9	–2.1	–7.4	–7.5	10
	C	43	H	23	5	2220	SL		–	4.1	0.0	–	–	–	0.0	–	–	–1.7	1.4	1.6	0.6	3.2	2.8	9
Hubbard Brook,	H	100	F+H	16	3^{g}	1300	SL	G,M	–	116.5	–	–	–	~107^{h}	–	–	<0.1^{h}	–3.9	29.0	9.3	12^{h}	64.2	–1.7	2,6,15, 19–22,25
New Hampshire					3^{g}	1300			–	42.3	–	–	–	–	–	–	–	2.3	15.5	2.7	–	23.8	2.0	2,15, 19–22,25
	H	100 (B)	PC	36	6	1300			–	5.3	–	–	–	–	–	–	–	0.4	2.9	2.0	–	4.7	0.1	2,15, 19–22,25
	H	100	C	12	6	1300			–	11.8	–	–	–		–	–	–	0.3	6.0	0.7	–	7.5	–0.3	
White Mountains, New Hampshire	H (C)	100	C	2–24(9)	4	?	SL	G,M	–	–	–	14.2	–	–	–	–	–	–	3.7	–	–	15.2	–	26
Catskill Mountains, New York	H	23	PC	10	1	1530	Med	?	–	6.3	–0.2	–	–	–	–	–	–	–	0.8	0.7	–0.1	–0.4	–	36
Isaac Creek, North	C	98	C	91	3	1430	L+	?	–	–	–	0.7	1.5	–	0.2	0.1	–	–	–	–	–	–	–	18
Carolina	C	82	C	109	3		Org		–	–	–	0.9	1.2	–	0.3	0.1	–	–	–	–	–	–	–	18
	C	25	C	359	3				–	–	–	0.5	1.6	–	0.0	0.1	–	–	–	–	–	–	–	18
Clemson Experimental Forest, South Carolina	C	100	C	1–2(3)	3	1440	C	G,M	–	0.0	0.0	–	–	–	0.0	–	–	–	0.2	0.1	0.2	0.4	–	35
Grant Memorial Forest, Georgia	C-H	<100(B)	C+M	33	6	1310	?	?	–	0.1	–	–	–	–	–	–	0.6	–	0.9	1.1	2.4	4.1	–	13
Starke, Florida	C	100	C+M	64	2	1400	S	?	–	0.0	–0.1	–	0.9	–	0.0	–	0.0	–	0.4	–	0.6	–	–	30
	C	100	C+M+B	48	2	1400			–	0.1	0.1	–	1.3	–	0.0	–	0.0	–	4.1	–	0.9	–	–	30
Cherokee County,	C-H	<100(B)	C+M+B	3	5	1070	SiL	S	–	0.1	0.0	–	0.2	–	0.0	–	0.1	–	1.6	1.5	0.6	0.2	–	5
Texas	C-H		C+M+B	3	5	1070	above C			0.0	0.0	–	0.2	–	0.0	–	0.0	–	0.5	0.4	0.2	–0.1	–	5
Fraser Experimental	C	40	C	41	2	580	GS	M	–	0.0	–	–	–	–	–	–	–	0.5	0.7	0.6	0.5	5.1	–	33
Forest, Colorado	C	30	C	78	2	580			–	0.2	–	–	–	–	–	–	–	1.8	1.5	4.7	9.6	–	–	33
Blue Mountains,	C	50	PC+B	30	2	1430	?	B	–	0.0	–	–	–0.2	–	–	–	<0.1	–	–	–	–	–	–	34
Oregon	C	43	C+B	24	2	1430			–	0.5	–	–	–0.4	–	–	–	<0.1	–	–	–	–	–	–	34
		22	C+B	118	2	1430				0.2	–	–	0.1	–	–	–	<0.1	–	–	–	–	–	–	34
Alsea watershed,	C	100	C+B	71	2	2540	GL	S	–	12.1	–	–	–	–	0.0	–	–	–	–	–	–	–	–	7
Oregon	C	25	C+B	303	2	2540			–	2.1	–	–	–	–	0.0	–	–	–	–	–	–	–	–	7
H.J. Andrews	C	100	C+B	13	7–10	2190	SL-SiL	V	–	0.3	–	–	0.2	–	0.0	0.1	–	–	1.8	3.5	3.0	4.4	–	24
Experimental Forest,	C	60	PC+B	15	7–10	2190			–	0.0	–	–	0.0	–	0.0	0.0	–	–	1.6	2.9	2.7	5.9	–	24
Oregon	C	100	C+B	96	1–2	2390	L-SiL			1.3	–	–	–	–	0.3	–	–	–	0.8	4.4	3.9	13.6	–	12
Europe																								
Hälsingland, Sweden	C	95	C	40	9	?	SSi	?	–	0.7	0.4	–	1.2	–	–	–	–	6.0	2.7	4.0	1.3	5.1	2.5	31
	C	50	C	200	9				–	0.2	0.1	–	0.7	–	–	–	–	5.4	2.3	2.2	1.1	4.6	1.5	31

Tegernsee Alps, Germany	C-H	40	PC	4	1	1990	C	S	1.3	8.1	2.7	–	–	–	0.0	–	–	5.1	8.6	10.3	28.8	134.6	8.1	3
Kershope Forest, England	C	100	C	2	2	1440	C	?	–	28.0	4.5	–	–	–	–	–	–	–	16.4	–11.8	1.1	1.1	–	1
Beddgelert, Wales	C	62	C	1	5	2600	CL-	S	–	16.8	–	–	–	–	–	–	–	–	12.4	–	–	–	–	29
	C	28	C	6	5	2600	SCL		–	3.2	–	–	–	–	–	–	–	–	3.2	–	–	–	–	29
Plynlimon, Wales	C	47	C	340	4	2400	SiCL-SiL	S	5.2	8.2	0.0	–	–	–	0.0	–	–	9.9	1.3	2.1	3.8	5.0	3.5	8
Karpenissi, Greece	C	15	PC	147	2	1400	L-CL	S	–	0.0	–0.3	–	–	–	–	–	–	–23.4	0.6	0.6	–2.7	7.2	5.0	27
Rest of World																								
Collie River basin, W Australia	DE	100	C	94	5	1120	GS	G	–	–	–	–	–	–	–	–	–	–	–	–	–	–	447.2	38
	DE	54	C	344	5	820	above C		–	–	–	–	–	–	–	–	–	–	–	–	–	–	4.6	38
	DE	38	C	350	5	800			–	–	–	–	–	–	–	–	–	–	–	–	–	–	2.4	38
Cropper Creek, SE Australia	ME	76(B)	C	46	2	1410	L	S	–	–	–	–	–	–	–	–	–	–	2.1	5.0	2.9	2.0	9.0	14
Coranderrk, SE Australia	ME	86(B)	C	53	5	1100	C-	D	–	0.1		–	–	–	–	–	–	–0.7	0.7	12.0	–6.8	3.9	30.2	17
		42(B)	C	65	5	1270	CL		–	0.0		–	–	–	–	–	–	–1.9	0.2	6.7	0.5	2.0	8.0	17
Maimai, New Zealand	HP	100	C+B	5	1	1310	G	?	–	1.0	0.8	–	–	5.7	0.6	–	1.3	–	126.8	31.6	21.7	46.8	–	28
Sabah, Malaysia	TF	100	C+B	6	2.75	3220	C+LS	S	47.2	1.1	0.8	–	4.4	–	0.4	0.3	–	–8.5	30.5	–3.4	–1.9	10.3	0.3	23
	TF	100	C+M	3	2.75	3490	C+LS		72.1	2.5	1.1	–	6.2	–	0.1	0.2	–	3.6	38.5	0.4	2.8	9.1	13.9	23
	TF	100	C+B	10	2.75	3490	C+LS		61.0	4.6	1.2	–	8.7	–	0.2	0.3		4.1	68.9	3.6	5.6	9.7	17.4	23
Central Amazon basin, Brazil	TF	9	C+B	23	1	2750	SCL-LS	?	–	0.2	0.1	–	2.6	–	0.0	0.1	–	0.1	4.4	7.4	0.9	3.1	3.0	37

Fluxes were calculated usually by assuming that the post-deforestation ratio of flux in the deforested stream to that in the control stream would have been the same as the pre-deforestation ratio, had deforestation not occurred.

[a]Vegetation is C, coniferous; DE, dry eucalyptus; C(H), mainly coniferous with some deciduous hardwood; C-H, mixed conifer and deciduous hardwood; H, deciduous hardwood; HP, hardwood-podocarp; ME, moist eucalyptus; TF, tropical forest.

[b]B indicates a buffer strip was left beside the stream.

[c]Type of disturbance is H, herbicide; C, clearcutting; F, tree felling; M, mechanical site preparation; PC, partial cutting; S, slashburning; W, windstorm.

[d]For multiwatershed studies, the number of watersheds is in parentheses.

[e]Soil texture is C, clay; G, gravelly; L, loamy; S, sandy; Si, silty; Org, organic soil; Med, medium.

[f]Bedrock is B, basalt; D, dacite; G, granitic; M, metamorphic (mainly gneiss and schist); S, sedimentary (mainly sandstone, siltstone, mudstone, and shale); V, mixed volcanic.

[g]The first 3-year period was during deforestation; the second 3-year period was the first 3 years after deforestation.

[h]Estimated for a 4-year period during, and for the first year following, deforestation [Bormann, Likens, Siccama, Pierce, and Eaton (1974)].

Sources

1. Adamson JK, Hornung M, Pyatt DG, and Anderson AR (1987) Changes in solute chemistry of drainage waters following the clearfelling of a sitka spruce plantation. *Forestry* 60: 165–177.
2. Bailey SW, Buso DC, and Likens GE (2003) Implications of sodium mass balance for interpreting the calcium cycle of a forested ecosystem. *Ecology* 84: 471–484.
3. Bäumler R and Zech W (1999) Effects of forest thinning on the streamwater chemistry of two forest watersheds in the Bavarian Alps. *Forest Ecology and Management* 116: 119–128.
4. Bayley SE, Schindler DW, Beaty KG, Parker BR, and Stainton MP (1992) Effects of multiple fires on nutrient yields from streams draining boreal forest and fen watersheds: nitrogen and phosphorus. *Canadian Journal of Fisheries and Aquatic Sciences* 49: 584–596.
5. Blackburn WH and Wood JC (1990) Nutrient export in stormflow following forest harvesting and site preparation in east Texas. *Journal of Environmental Quality* 19: 402–408.
6. Bormann FH, Likens GE, Siccama TG, Pierce RS, and Eaton JS (1974) The export of nutrients and recovery of stable conditions following deforestation at Hubbard Brook. *Ecological Monographs* 44: 255–277.
7. Brown GW, Gahler AR, and Marston RB (1973) Nutrient losses after clear-cut logging and slash burning in the Oregon Coast Range. *Water Resources Research* 9: 1450–1453.
8. Durand P, Neal C, Jeffery HA, Ryland GP, and Neal M (1994) Major, minor and trace element budgets in the Plynlimon afforested catchments (Wales): general trends, and effects of felling and climate variations. *Journal of Hydrology* 157: 139–156.
9. Feller MC (1989) Effects of forest herbicide applications on streamwater chemistry in southwestern British Columbia. *Water Resources Bulletin* 25: 607–616.
10. Feller MC and Kimmins JP (1984) Effects of clearcutting and slashburning on streamwater chemistry and watershed nutrient loss in southwestern British Columbia. *Water Resources Research* 20: 29–40.
11. Foster NW, Beall FD, and Kreutzweiser DP (2005) The role of forests in regulating water: the Turkey Lakes watershed case study. *Forestry Chronicle* 81: 142–148.
12. Fredriksen RL (1971) Comparitive chemical water quality – natural and disturbed streams following logging and slashburning. In Krygier JT, and Hall JD (eds.). *Forest Land Uses and Stream Environment*, pp. 125–138. Corvallis: Oregon State University.
13. Hewlett JD, Post HE, and Doss R (1984) Effect of clear-cut silviculture on dissolved ion export and water yield in the Piedmont. *Water Resources Research* 20: 1030–1038.
14. Hopmans P, Flinn DW, and Farrell PW (1987) Nutrient dynamics of forested catchments in southeastern Australia and changes in water quality and nutrient exports following clearing. Forest Ecology and Management 20: 209–231.
15. Hornbeck JW, Martin CW, Pierce RS, Bormann FH, Likens GE, and Eaton JS (1986) Clearcutting northern hardwoods: effects on hydrologic and nutrient ion budgets. *Forest Science* 32: 667–686.
16. Krause HH (1982) Nitrate formation and movement before and after clear-cutting of a monitored watershed in central New Brunswick, Canada. *Canadian Journal of Forest Research* 12: 922–930.

17. Langford KJ, and O'Shaughnessy PJ (1980) *Second progress report. Coranderrk.* Melbourne and Metropolitan Board of Works Report No. MMBW-W-0010. Melbourne, Australia.
18. Lebo ME and Herrmann RB (1998) Harvest impacts on forest outflow in coastal North Carolina. *Journal of Environmental Quality* 27: 1382–1395.
19. Likens GE, Bormann FH, Pierce RS, and Reiners WA (1978) Recovery of a deforested ecosystem. *Science* 199: 492–496.
20. Likens GE, Driscoll CT, Buso DC, Mitchell MF, Lovett GM, Bailey SW, Siccama TG, Reiners WA, and Alewell C (2002) The biogeochemistry of sulfur at Hubbard Brook. *Biogeochemistry* 60: 235–316.
21. Likens GE, Driscoll CT, Buso DC, Siccama TG, Johnson CE, Lovett GM, Ryan DF, Fahey T, and Reiners WA (1994) The biogeochemistry of potassium at Hubbard Brook. *Biogeochemistry* 25: 61–125.
22. Lovett GM, Likens GE, Buso DC, Driscoll CT, and Bailey SW (2005) The biogeochemistry of chlorine at Hubbard Brook, New Hampshire, USA. *Biogeochemistry* 72: 191–232.
23. Malmer A (1996) Hydrological effects and nutrient losses of forest plantation establishment on tropical rainforest land in Sabah, Malaysia. *Journal of Hydrology* 174: 129–148.
24. Martin CW, and Harr RD (1989) Logging of mature Douglas-fir in western Oregon has little effect on nutrient output budgets. *Canadian Journal of Forest Research* 19: 35–43.
25. Martin CW, Driscoll CT, and Fahey TJ (2000) Changes in streamwater chemistry after 20 years from forested watersheds in New Hampshire, USA. *Canadian Journal of Forest Research* 30: 1206–1213.
26. Martin CW, Pierce RS, Likens GE, and Bormann FH (1986) Clearcutting affects stream chemistry in the White Mountains of New Hampshire. USDA Forest Service Research Paper NE-579. Northeastern Forest Experiment Station.
27. Nakos G and Vouzaras A (1988) Budgets of selected cations and anions in two forested experimental watersheds in central Greece. *Forest Ecology and Management* 24: 85–95.
28. Neary DG, Pearce AJ, O'Loughlin CL, and Rowe LK (1978) Management impacts on nutrient fluxes in beech-podocarp-hardwood forests. *New Zealand Journal of Ecology* 1: 19–26.
29. Reynolds B, Stevens PA, Hughes S, Parkinson JA, and Weatherley NS (1995) Stream chemistry impacts of conifer harvesting in Welsh catchments. *Water, Air and Soil Pollution* 79: 147–170.
30. Riekerk H (1983) Impacts of silviculture on Flatwoods runoff, water quality, and nutrient budgets. *Water Resources Bulletin* 19: 73–79.
31. Rosén K, Aronson J-A, and Eriksson HM (1996) Effects of clear-cutting on streamwater quality in forest catchments in central Sweden. *Forest Ecology and Management* 83: 237–244.
32. Schindler DW, Newbury RW, Beaty KG, Prokopowich J, Ruszczynski T, and Dalton JA (1980) Effects of a windstorm and forest fire on chemical losses from forested watersheds and on the quality of receiving streams. *Canadian Journal of Fisheries and Aquatic Sciences* 37: 328–334.
33. Stottlemyer R (1987) Natural and anthropic factors as determinants of long-term stream chemistry. In Troendle CA, Kaufmann MR, Hamre RH, and Winokur RP (tech. cords.) *Management of subalpine forests: building on 50 years of research*. USDA Forest Service General Technical Report RM-149. Rocky Mountain Research Station.
34. Tiedemann AR, Quigley TM, and Anderson TD (1988) Effects of timber harvest on stream chemistry and dissolved nutrient losses in northeast Oregon. *Forest Science* 34: 344–358.
35. Van Lear DH, Douglas JE, Cox SK, and Augspurger MK (1985) Sediment and nutrient export in runoff from burned and harvested pine watersheds in the South Carolina Piedmont. *Journal of Environmental Quality* 14: 169–174.
36. Wang X, Burns DA, Yanai RD, Briggs RD, and Germain RH (2006) Changes in stream chemistry and nutritional export following a partial harvest in the Catskill Mountains, New York, USA. *Forest Ecology and Management* 223: 103–112.
37. Williams MR, and Melack JM (1997) Solute export from forested and partially deforested catchments in the central Amazon. *Biogeochemistry* 38: 67–102.
38. Williamson DR, Stokes RA, and Ruprecht JR (1987) Response of input and output of water and chloride to clearing for agriculture. *Journal of Hydrology* 94: 1–28.
39. Zhu A, Arp PA, Mazumber A, Meng F, Bourque CP-A, and Foster NW (2005) Modeling stream water nutrient concentrations and loadings in response to weather condition and forest harvesting. *Ecological Modeling* 185: 231–243.

Effects of Deforestation on the Factors Controlling Freshwater Nutrient Loads

Dissolved Nutrients

Effects on geological weathering By exposing the land surface to greater temperature extremes, and greater amounts of often more acidic water leaching through the systems, deforestation can increase weathering rates. Increased nitrification following forest harvesting can lead to increased acid leaching which increases weathering rates. Quantitative estimates of the impacts of harvesting on weathering rates, however, appear to be generally unavailable, with the exception of the Hubbard Brook Experimental Forest in northeastern United States.

Effects on atmospheric precipitation/climate

Effects on precipitation chemistry and acidity Deforestation has affected the acidity of water passing through the soil as well as the acidity of streamwater (pH decreases of 0.2–0.5 pH units). This occurs due to release of organic acids from decomposing logging debris or to enhanced nitrification which produces HNO_3. Such enhanced acid production has had similar effects to those of acid rain, in terms of flushing nutrient cations from soil into streams. Enhanced weathering release and dissolution of chemicals, such as Ca oxalate, in soil can contribute to a sustained increased loss of Ca to streamwater for 10 years or more.

Deforestation can also increase H^+ production if it results in a greater net biomass increment (excess of biomass gain from growing vegetation over biomass loss from decomposition), which is likely to occur when slow growing or steady state forests are converted to more rapidly growing younger ones. Enhanced acid production within watersheds would appear to be a general response to deforestation, although the intensity of this response varies widely, depending on the extent of nitric acid generation and the degree to which net biomass increment is affected.

The extent of HNO_3 leaching through a watershed after deforestation depends primarily on (1) the extent of N uptake by regrowing vegetation, greater N uptake leaving less N available for nitrification, (2) N immobilization in the soil, greater N immobilization (indicated by higher soil C/N ratios) also leaving less N available for nitrification, (3) lags in nitrification whereby nitrifying organisms are suppressed or exist only in low populations, and (4) soil N availability prior to deforestation, for which the greater the availability, the greater the amount of N for nitrification.

Deforestation generally increases nitrification to a greater extent in the northern hemisphere temperate deciduous hardwood than in conifer forests. Thus, for the studies in **Table 2**, the mean increase in streamwater NO_3-N loading during the first few years following forest harvesting was 13.1 kg ha^{-1} year^{-1} for seven studies in pure hardwood forests (average deforestation of 72%) compared to 3.5 zkg ha^{-1} year^{-1} for 22 studies in pure coniferous forests (average deforestation of 61%). However, considerable variation within both hardwood and conifer forest datasets (coefficients of variation of both mean values are >100%) indicates that deforestation in a hardwood forest will not always increase streamwater NO_3-N loading to a greater extent than in a conifer forest.

Effects on hydrology Deforestation causes an initial increase in streamflow, the extent of which depends on the degree of forest removal. Streamflow decreases to preharvesting levels over a period of years and may decrease below preharvesting levels if the regenerating forest has a higher evapotranspiration or is less able to capture cloud water than the pre-harvesting forest, as is the case for wetter Australian eucalypt forests.

Deforestation can also increase soil water, allowing rises in water tables. If the rising water table contacts nutrient rich soil material, these nutrients can be taken into freshwater bodies, considerably enriching them. Transport of salts from salt-rich soil parent material to streams via rising water tables following deforestation has been notable in southern Australia (**Figure 1**).

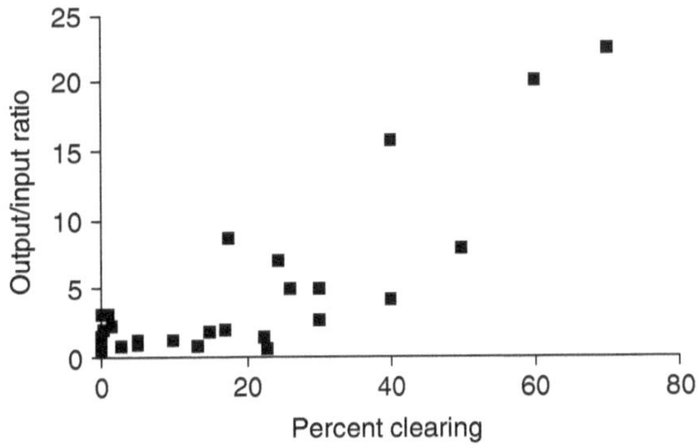

Figure 1 Watershed chloride output/input ratios as a function of the percentage of a watershed deforested by forest harvesting and clearing for agriculture in southwestern Australia. Redrawn from **Figure 5** in Trotman CH (ed.) (1974) *The Influence of Land Use on Stream Salinity in the Manjimup Area, Western Australia*. W.A. Department of Agriculture Technical Bulletin no. 27. Perth, Australia, with the kind permission of the Department of Agriculture and Food of the Government of Western Australia.

Deforestation has had variable effects on snowpacks, depending on the size and orientation of the deforested area. Greater snowpacks may result in higher streamwater nutrient loadings during the early snowmelt period as a result of preferential elution of ions from the snowpack, but this is unquantified. Thus, deforestation can alter water movement through watersheds, but the influence on nutrient loading of this water increment from deforestation has not been precisely quantified.

Effects on temperature Deforestation has increased summer soil surface temperatures as well as streamwater temperatures, depending primarily on the extent to which the tree canopy is removed. Higher temperatures during summer may accelerate weathering and decomposition as well as chemical and microbial reactions, which may then alter nutrient movement through soil into fresh water, although this, too, has not been quantified.

Effects on terrestrial biological processes

Chemical uptake Deforestation initially reduces nutrient uptake by terrestrial vegetation leading to enhanced nutrient flow from terrestrial to aquatic ecosystems, for periods of usually 1–7 years. If plant uptake was a dominant process influencing freshwater nutrient loading, then this loading would respond to deforestation as shown in **Figure 2**. Thus, essential and growth limiting nutrients would exhibit highest loadings at time 0, shortly after deforestation, then decline as plant growth increased, but limiting nutrients to a greater extent, before increasing again

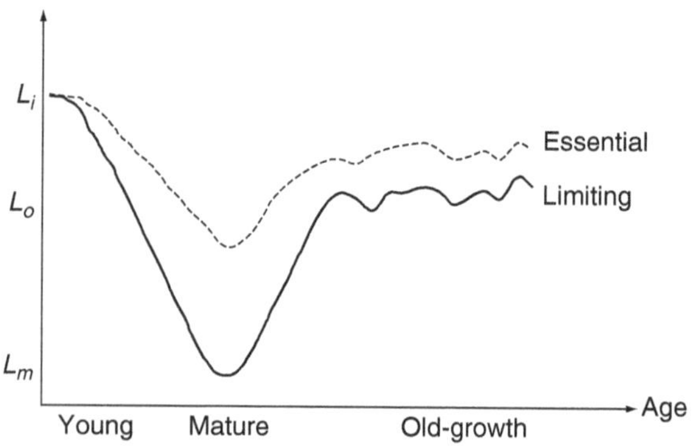

Figure 2 Loading of essential and limiting nutrients in streamwater, expressed as percentage of the maximum, as a function of the age of vegetation in the surrounding watershed. Adapted from Feller MC (2005) Forest harvesting and streamwater inorganic chemistry in western North America: A review. *Journal of the American Water Resources Association* 41: 785–811, and Vitousek PM and Reiners WA (1976) Ecosystem development and the biological control of stream chemistry. In: Nriagu JO (ed.) *Environmental Biogeochemistry*, pp 665–680. Ann Arbor, MI: Ann Arbor Science.

in old-growth forests where mortality reduces uptake. A consequence of this temporal pattern in nutrient loading is that deforestation of mature forests has the potential to cause greater increases in essential and limiting streamwater nutrient concentrations (L_i–L_m for a limiting nutrient), than does deforestation of old growth forests (L_i–L_o for a limiting nutrient) (**Figure 2**).

Dissolved inorganic N can be immobilized in the soil by mycorrhizal assimilation. Deforestation has at least temporarily adversely affected soil mycorrhizae, which might lead to significant initial increases in nitrification and NO_3^- leaching through soil into streamwater. The extent to which deforestation affects mycorrhizae depends on the severity of the disturbance, inherent soil ectomycorrhizal diversity, dryness of the climate, and the abundance of hosts, but the influence of deforestation-induced changes in mycorrhizal populations on NO_3^- leaching is unknown.

Chemical transformations Deforestation usually increases nitrification. This increases NO_3^- flow through soil into streams. This flow may be enhanced if N fixation also increases after deforestation, as it can if the amount of more well-decayed coarse woody debris (CWD) on the ground surface increases and if the pre-existing CWD is not greatly fragmented during deforestation. N fixation may also be increased if the cover of N-fixing plant species, such as *Alnus* spp., increases after deforestation. NO_3^- flow can be retarded if enhanced denitrification offsets enhanced nitrification. Denitrification has been enhanced by deforestation as a result of increased soil moisture (due to decreased evapotranspiration), soil NO_3^- content (due to increased nitrification), and soil dissolved organic C (DOC) (due to enhanced decomposition of organic material). Nitrification may be decreased, with subsequent reduced freshwater NO_3 loading, if deforestation removes *Alnus* trees. The balance between these different N transformations will help determine the extent of NO_3^- movement to streamwater.

Deforestation has generally lead to declines in streamwater loading more often for SO_4 than for any other nutrient ion (**Table 2**). This decline may be due to (1) decreased activity of S-oxidizing bacteria, perhaps due to toxic effects of high NO_3^- concentrations, (2) increased SO_4^{2-} reduction resulting from increased anaerobic conditions in soil, (3) decreased inputs of dry deposition resulting from loss of foliage area, (4) enhanced microbial immobilization of SO_4^{2-} or reduced organic S mineralization resulting from increased forest floor decomposition, (5) dilution from increased water flow, (6) precipitation of aluminum hydroxy-sulfate minerals, or (7) increased SO_4^{2-} adsorption in the mineral soil as a result of an acid-induced increase in the SO_4^{2-} adsorption capacity of the soil. Studies in the northeastern United States have suggested that soil adsorption (process number 7) was the most important process.

Production of soluble chemicals Deforestation which causes an initial decrease in litter-producing riparian vegetation would result in less organic matter input into streams immediately postharvesting, and an initial decline in litter-originating soluble chemicals in streamwater, although this has not been quantified. Deforestation leaves variable amounts of organic material on the ground surface. This material produces easily soluble chemicals which are leached away as the organic material decomposes (**Figure 3**), although this may not always contribute greatly to enhanced soil N availability postharvesting.

Deforestation has had variable effects on litter decomposition rates, due to differences in (1) the type of material being decomposed, (2) climate, (3) impacts of deforestation on soil organisms, and (4) the degree of mixing of organic matter with soil materials. Thus, deforestation may, or may not, lead to increased organic matter decomposition rates and hence increased production of soluble, easily leached chemicals.

Effects on physical–chemical reactions in the soil Direct effects of deforestation on physical–chemical reactions in the soil have not been reported. Indirect effects, however, include an increase in the anion exchange capacity of a soil as pH decreases. Thus, if deforestation enhances nitrification, increased nitric acid generation can increase the anion exchange capacity of the soil, helping to retard movement of ions through soil into streams, partly explaining the decline in SO_4^{2-} loading in streamwater found after deforestation, as mentioned above. Retardation of P movement through the soil as pH decreases, due to increased fixation by hydrous oxides of Fe and Al, might also help to explain the minimal response of streamwater P loading to deforestation (**Table 2**).

Effects on processes within aquatic ecosystems Deforestation can influence most processes within aquatic ecosystems, but the extent of this influence will depend strongly on the amount of organic debris, fine sediment, solar radiation, and acid reaching the aquatic ecosystem. Ion exchange and chemical redox reactions and microbial transformations all increase with the surface area of the streambed substrate, so these processes are likely to be enhanced if deforestation increases the quantity of colloidal and fine particulate organic material in a stream. Although many studies have investigated nutrient cycling processes within undisturbed forested streams, the direct effects of deforestation on these individual processes have not been studied.

Figure 3 Large amounts of woody debris close to a stream channel following clearcutting of *Eucalyptus delegatensis* forest across a stream in eastern Victoria, Australia, 1987. Photograph by Michael Feller.

Primary production in a stream ecosystem increases with solar radiation and, to a lesser extent, with temperature, so when these parameters are increased by deforestation, as often happens, enhanced primary production occurs. Increased primary production increases removal of nutrients from water, and hence decreases nutrient loading.

If deforestation increases acid generation within a watershed, and if some of that acid reaches a stream, then ion exchange and pH-dependent chemical redox reactions within the stream will be affected, as will be pH-induced microbial transformations. Additions of acid to streams have resulted in increased streamwater Ca^{2+}, Mg^{2+}, K^{+}, and micronutrient loadings, but not NO_3^-, and NH_4^+ loadings. Increased acid levels in streamwater may also enhance primary production, and hence uptake by aquatic primary producers, as the acid may adversely affect organisms that graze on the primary producers.

The impacts of deforestation on these different processes will affect freshwater nutrient loading in different ways, depending on the nutrient and the process (**Table 3**). Thus, deforestation has the potential to increase, decrease, or have no effect on the loading of every nutrient considered, as has been found. Each arrow in **Table 3** should not be considered to have the same weight for a given nutrient. Thus, Na, Mg, and Ca all have more than twice as many upward as downward arrows but deforestation has sometimes decreased fluxes of these nutrients (**Table 2**).

Other factors contributing to the variable effects of deforestation on nutrient loading Freshwater nutrient loading depends not only on the response of the preceding factors to deforestation, but also on characteristics of the deforestation itself, soil characteristics, and rate of revegetation following deforestation. These additional factors add further complexity so that deforestation-induced changes in nutrient fluxes can vary widely. These changes have ranged over four orders of magnitude for NO_3-N, for example (**Table 2**). These other factors include:

1. *Extent of the watershed deforested*: In general, the greater the percentage of the watershed deforested the greater will be the impacts on freshwater nutrient loading in that watershed (**Figure 1**). For a wide range of environmental conditions (given in **Table 2**), deforestation of northern hemisphere temperate forests is likely to cause a maximum initial increase in NO_3-N, K^+, and Ca^{2+} loading of approximately 3, 2, and 5 kg ha^{-1} year^{-1}, respectively, for each 10% of a watershed deforested (**Figure 4**).
2. *Presence of buffer strips between freshwater and harvested areas*: Undisturbed buffer strips adjacent to a water body can filter out increased amounts of nutrients flowing from a deforested area to that water body. The efficiency of filtering by buffer strips increases with buffer strip width, with 100% efficiency suggested to occur for widths of approximately 100 m, although the effectiveness of buffer

Table 3 Initial trends in dissolved nutrient chemical loading in fresh waters resulting from deforestation impacts on the different factors controlling freshwater nutrient loading

Factor	*C*	*Ca*	*Cl*	*Fe*	*H*	*K*	*Mg*	*Mn*	*N*	*Na*	*P*	*Si*	*SO$_4$*	*Trace metals*
Geological weathering	↑	↑	–	↑	↓	↑	↑	↑	↑–	↑	↑	↑	↑	↑
Atmospheric precipitation/climate														
Precipitation chemistry	↑↓	↑–	–	↑–	↑–	↑–	↑–	↑–	↑–	↑–	↑–	↑–	↑–	↑–
Hydrologic influences	↑↓	↑↓	–	↑↓	↑↓	↑↓	↑↓	↑↓	–	↑↓	↑↓	↑↓	↑↓	↑↓
Temperature	↑	↑	–↑	↑	↑↓	↑	↑	↑	↑	↑	↑	↑	↑	↑–
Terrestrial biological processes														
Chemical uptake	–	↑	↑	↑	↑	↑	↑	↑	↑	↑	↑	–	↑	–
Chemical transformations	↑↓	–	–	–	↑↓	–	–	–	↑↓	–	–	–	↑↓	–
Production of soluble chemicals	↑↓	↑–	↑	↑–	↑–	↑–	↑–	↑–	↑–	↑–	↑–	–	↑–	↑–
Physical–chemical reactions in the soil	↓–	–	↓–	–	↓–	–	–	–	↓–	–	↓–	–	↓–	–
Processes within aquatic ecosystems														
Ion exchange reactions	–	↑↓	↑↓	↑↓	↑↓	↑↓	↑↓	↑↓	↑↓	↑↓	↑↓	–	↑↓	↑↓
Chemical redox reactions	–	–	–	↑↓	–	–	–	↑↓	↑↓	–	–	–	↑↓	↑↓
Evaporation–crystallization	↑–	↑–	–	–	–	–	–	–	–	–	–	–	–	–
pH-induced transformations	↑↓	↑↓	↑↓	↑↓	↓↑	↑↓	↑↓	↑↓	↓↑	↑↓	↓↑	↑↓	↑↓	↓↑
Uptake by primary producers	–	↓–	–	↓–	–	↓–	↓–	↓–	↓–	↓–	↓–	↓–	↓–	↓–
Microbial transformations	–	–	–	–	–	–	–	–	↑↓	–	–	–	↑↓	–

↑ Effects lead to an increase in loading.

↓ Effects lead to a decrease in loading.

– Effects have little to no impact on loading.

Adapted from **Table 3** in Feller MC (2005) Forest harvesting and streamwater inorganic chemistry in western North America: A review. *Journal of the American Water Resources Association* 41: 785–811, with the kind permission of the American Water Resources Association.

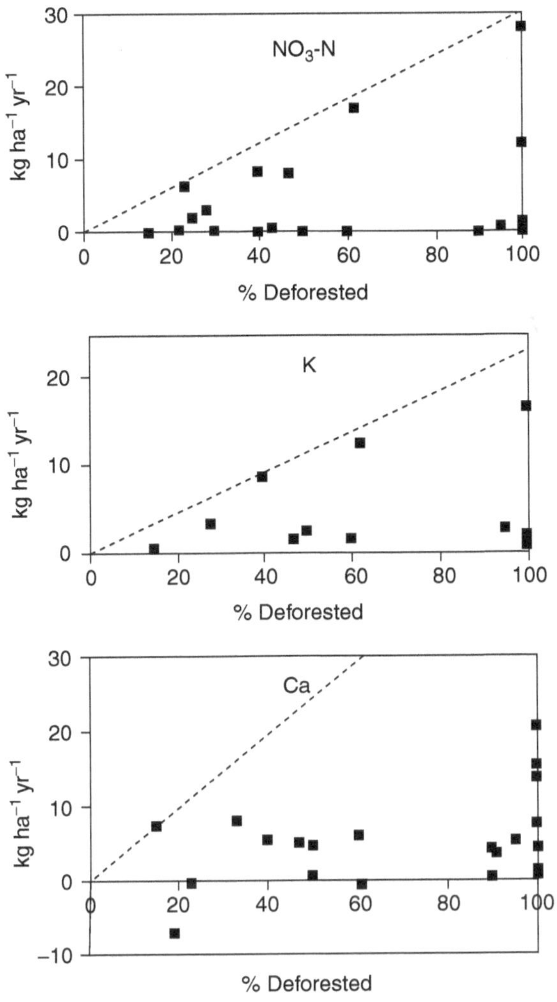

Figure 4 Deforestation-induced changes in streamwater nitrate – nitrogen, potassium, and calcium loads as a function of the extent of deforestation in northern hemisphere temperate forests during the first few years after nonherbicide deforestation. Data were obtained from **Table 2**. Dashed lines represent the maximum likely increase in stream nutrient loading due to deforestation. Calcium data exclude one extremely high outlier.

strips also depends on other factors, such as soil properties, slope angle, presence or absence of small ephemeral channels running through the buffer strips to the water body, and the type of vegetation present (**Figure 5**).

3. *Nature of the treatment given to a watershed following deforestation*: Site preparation treatments, such as mechanical scarification, slashburning, and herbicide application, can potentially enhance the effects of deforestation on streamwater chemistry. Thus, clearcutting plus slashburning causes greater increases in streamwater nutrient loading than just clearcutting alone. When deforestation is caused by herbicide application, nutrient loadings in streams appear to increase to a greater extent than from other causes of deforestation (**Figure 6**). This may be explained by greater nitrification following herbicide-induced deposition of lower C/N ratio litter. Removal of woody debris after deforestation can decrease the magnitude of nutrient loading increases, unless this removal causes substantial mechanical disturbance to the soil, in which case nutrient loading increases can be enhanced.
4. *Rate of revegetation following deforestation*: The more rapidly an area revegetates after harvesting, the more rapidly will streamwater nutrient loading return to predeforestation levels. If vigorous revegetation results in greater nutrient immobilization rates than occurred prior to deforestation, nutrient fluxes to fresh water can decline to below predeforestation levels (**Figure 6**).
5. *Nutrient content of the soil (soil fertility) prior to deforestation*: Nutrient loss to fresh water after deforestation increases with soil nutrient content. Consequently, N loading in fresh water following deforestation is likely to be greater as the degree of N saturation of a watershed increases.
6. *Buffering capacity of the soil*: The greater the ability of soil to retain nutrients, the lesser the amount of nutrients that will be washed through soil into a water body, and hence the lesser the increase in freshwater nutrient loading after deforestation.
7. *Abundance of large water bodies in a watershed*: Large nutrient and water storage areas, such as lakes, peatlands, and swamps, can trap nutrients flushed from the land following deforestation. Consequently, the greater the abundance or the larger the size of such nutrient storages in a watershed, the lesser will be the nutrient loading in the stream leaving the watershed (**Figure 7**). If the flush of nutrients into lakes following deforestation does not exceed their storage capacity, then nutrient loading in outlet streams may be little affected by deforestation.
8. *Timing of deforestation*: Deforestation which occurs before a heavy rainfall period at a time when plant nutrient uptake is relatively low is likely to result in greater movement of nutrients into surface waters than a deforestation that occurs during a period of low rainfall and high plant nutrient uptake.

Nutrients in Particulate Form

Effects on Watershed Topography, Soil Erodibility, Precipitation Characteristics, and Watershed Susceptibility to Mass Wasting

Deforestation generally has little effect on the first four factors controlling particulates in fresh water.

Figure 5 Small buffer strips of undisturbed vegetation adjacent to major, but not ephemeral, streams in extensively cleared *Eucalyptus* forest prior to planting *Pinus radiata* in the Acheron valley, central Victoria, Australia, 1980. Photograph by Michael Feller.

Deforestation may increase ground surface roughness by adding coarse woody debris which can prevent materials moved by surface erosion from entering fresh waters. Roads constructed during deforestation activities can concentrate water flows and facilitate movement of runoff into fresh waters (**Figure 8**), but the effects of this on freshwater nutrient loading have not been well quantified.

In general, however, these four factors simply help explain the variable effects of deforestation on freshwater particulate loads. Deforestation is more likely to increase freshwater particulate loads if a watershed (1) has steep, uniform slopes, (2) has erodible soils, such as those with high silt contents, (3) is in an area with intense and heavy precipitation, and (4) is prone to mass wasting by having moderately steep slopes, depressions where soil water can accumulate, or bedrock with fracture planes parallel to the ground surface.

Effects on Stream-Channel Characteristics

Deforestation is most likely to affect stream-channel characteristics if it occurs close to a stream channel. Debris jams tend to trap particulates, decreasing their downstream flux. Creation of stable debris jams is therefore likely to decrease stream particulate fluxes. Destabilization of debris jams will have the opposite effect. If deforestation removes trees immediately adjacent to water bodies, particularly if soil disturbance is involved, bank erosion is likely to enhance particulate loading in the water bodies.

Effects on Proximity of Vegetation to Surface Water

Deforestation which removes vegetation over or close to freshwater surfaces will decrease, at least temporarily, plant litter input into the water, and hence organic particulate loading. If the deforestation kills vegetation without removing it (e.g., herbicides or wind storms), the opposite occurs and organic particulate loads temporarily increase.

Effects on Extent of Roading

Road construction has often lead to greater increases in freshwater particulates than other human activities, such as tree felling. Consequently, deforestation which involves road construction can increase freshwater particulates to an extent dependent on the care taken with road layout and construction (**Figure 9**).

Deforestation has had variable effects on freshwater particulate loads, depending on the above factors. It has affected the particulate load/dissolved load ratio (**Table 1**) although there have been too few studies to generalize. Many studies have quantified

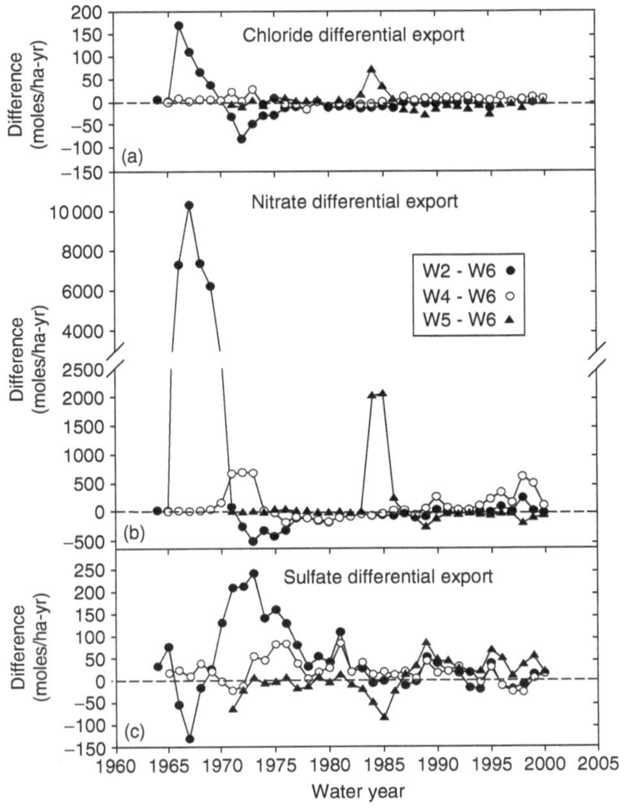

Figure 6 Difference between streamwater ion export in deforested and undisturbed (W6) deciduous hardwood watersheds in the Hubbard Brook Experimental Forest, NE. U.S. A. Deforestation involved tree cutting in 1965–66 followed by herbicide application in 1966–68 for W2, forest harvesting in 25-m wide strips covering one-third of the watershed in each of 1970, 1972, and 1974, for W4, and clearcutting in 1983–84 for W5, leaving a buffer strip adjacent to the stream. Reproduced with the kind permission of Springer Science and Business Media from Figure 19 in Lovett GM, Likens GE, Buso DC, Driscoll CT, and Bailey SW (2005) The biogeochemistry of chlorine at Hubbard Brook, New Hampshire, U.S.A. *Biogeochemistry* 72: 191–232.

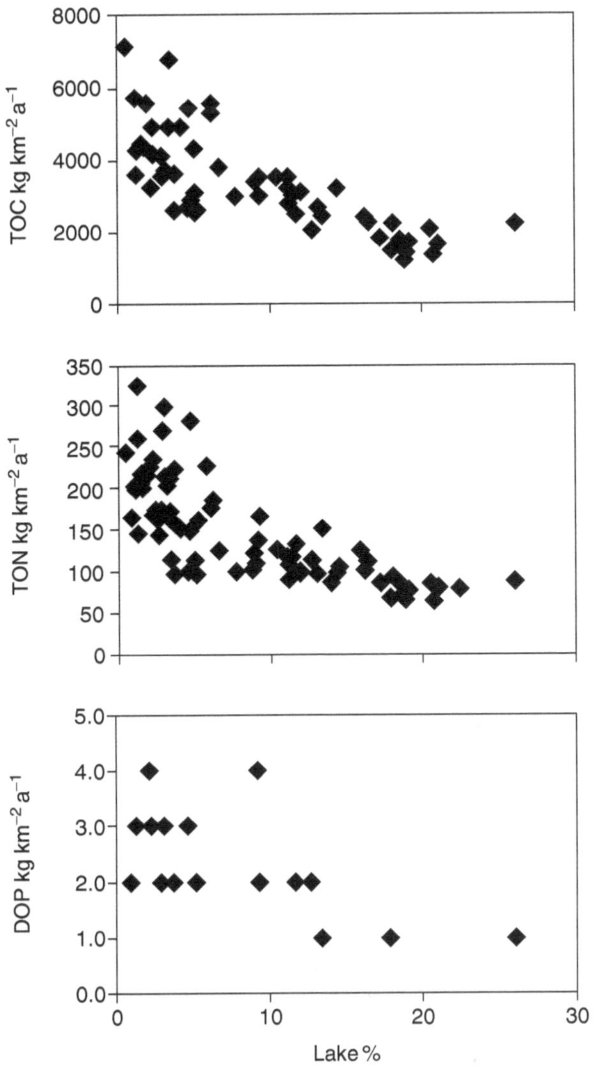

Figure 7 Relationships between export of total organic C (TOC), total organic N (TON), and dissolved organic P (DOP) and the proportion of lakes in coniferous forest watersheds in Finland. Watersheds contained <1–44% agricultural land and 0–6.5% urban land. Reproduced with the kind permission of Springer Science and Business Media from **Figure 5** in Mattsson T, Kortelainen P, and Räike A (2005) Export of DOM from boreal catchments: impacts of land use cover and climate. *Biogeochemistry* 76: 373–394.

the effect of deforestation on freshwater sediment and particulate loads, but very few have quantified the effect on nutrient loads in particulate form. Increases in sediment and particulate loads, however, will usually mean increases in nutrient loads in particulate form.

Conclusions

Dissolved nutrient loading in fresh water has been explained by five major factors, while particulate nutrient loading has been explained by seven major factors, many of which interact. Thus, nutrient loading is characterized by complexity and variability, both temporal and geographic, and even within one stream at the same time. This presents a challenge to sampling programs and to the development of models which predict nutrient loading.

Variable effects of deforestation on freshwater nutrient loading can be explained by the variability in deforestation effects on the major processes controlling this loading. When this level of variability is added to the variability inherent in undisturbed freshwater bodies, some major knowledge gaps are inevitable. Some of the more important knowledge gaps are as follows.

Figure 8 Dense network of logging roads on relatively steep slopes in clearcut *Eucalyptus delegatensis* forest in north eastern Victoria, Australia, 1980. Photograph by Michael Feller.

Figure 9 Road construction on moderately steep slopes adjacent to a stream channel, with no buffer strip of undisturbed vegetation in south western British Columbia, Canada, 1999. Photograph by Michael Feller.

Knowledge Gaps

1. *Detailed freshwater nutrient budgets are scarce*:
   Very few studies have quantified the different inputs and outputs, whose balance determines freshwater nutrient loading.
2. *Studies of longitudinal variation in stream nutrient loading are scarce*:
   Few studies have quantified nutrient flux trends along extended lengths of streams. Such studies would be important in determining how far downstream the impacts of deforestation might extend.
3. *Studies of deforestation impacts on processes within freshwater ecosystems are scarce*:
   The one published study that could be found on the impacts of deforestation on streamwater nutrient fluxes resulting from within-stream processes suggested that partial deforestation by an ice storm at the Hubbard Brook Experimental Forest increased within-stream retention and transformation of NO_3^-, reducing the potential increase in streamwater NO_3^- loading. Greater quantification of the impacts of deforestation on within-stream nutrient cycling processes is necessary to accurately determine the quantity of nutrients moving into fresh water after deforestation. Nearly all studies which have purported to determine this, have made the tacit assumption, of questionable validity, that deforestation has had no significant effect on processes within the freshwater ecosystem. The true deforestation-induced loss of nutrients to fresh water remains uncertain. Nutrient movement from land to streamwater, for example, is likely to include an amount exported by the stream as well as an amount removed from the stream (**Figure 10**), so that the deforestation-induced loss of nutrients to fresh water may be greater than that inferred from streamwater fluxes alone.
4. *The influence of deforestation on freshwater particulate nutrient loads has been poorly quantified*:
   The effects of deforestation on freshwater sediment and particulate loads has been well studied but not the chemical composition of these particulates or sediment.
5. *The influence of deforestation on geological weathering release of chemicals has been poorly quantified*:
   As geological weathering is a dominant process influencing freshwater nutrient loading, it is desirable to understand its quantitative response to deforestation. This would help in understanding the origin of solutes in fresh water after deforestation and, hence, in determining methods of reducing solute loading, if considered necessary.

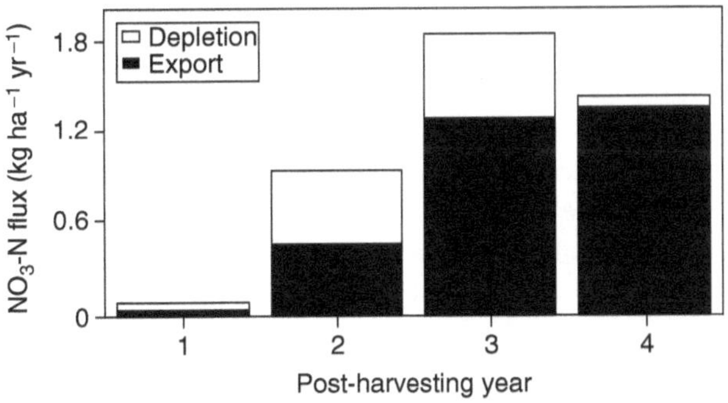

Figure 10 Annual NO_3-N fluxes measured in a stream flowing out of a 59 ha deciduous hardwood-conifer watershed at Coweeta, North Carolina, USA for the first 4 years following forest harvesting. Export is the streamwater dissolved load while depletion is the amount estimated to have been removed from the stream by in-stream processes. Produced from data presented in Swank WT and Caskey WH (1982) Nitrate depletion in a second-order mountain stream. *Journal of Environmental Quality* 11: 581–584, with the kind permission of the American Society of Agronomy.

6. *The influence of deforestation on the snowmelt flux of nutrients into streams has been poorly quantified*:
   The effect of deforestation on snowpacks and snowmelt has been reasonably well documented but the influence of altered snow hydrology on freshwater nutrient loading during the snowmelt runoff period has not been studied.
7. *The influence on freshwater nutrients of deforestation-induced changes in soil and stream temperatures has been poorly quantified*:
   Changes in soil and stream temperature regimes can affect geological weathering, terrestrial biological processes, physical–chemical reactions in the soil, and processes within aquatic ecosystems. None of these effects seems well quantified.
8. *The effects of deforestation on nutrient loading in drier environments, tropical regions, and the southern hemisphere have been poorly quantified*:
   Although many studies have quantified deforestation effects on freshwater nutrient concentrations, from which relative trends in loadings can sometimes be inferred, most studies that have more accurately quantified deforestation effects on freshwater nutrient loading have been conducted in areas receiving >1000 mm of precipitation, and in northern hemisphere temperate deciduous hardwood and coniferous forests (**Table 2**).

General Effects of Deforestation on Freshwater Nutrient Loading

Despite the complexity and variability in freshwater nutrient loading, some important generalizations can be made. These are:

1. The usual increase in nitrification following deforestation is one of the most important determinants of not only NO_3 loading in fresh water, but also the loading of other nutrients whose fluxes can be affected by the nitric acid produced by nitrification.
2. The type of vegetation influences its response to deforestation with temperate deciduous hardwood forests likely to exhibit greater nutrient flux changes than temperate coniferous forests in the northern hemisphere.
3. Nutrient flux changes occurring when herbicides cause or accompany deforestation are likely to be greater than those following deforestation without herbicides.
4. Deforestation-caused changes in freshwater nutrient loading are usually short lived, particularly for nitrogen – generally up to 7 years, but usually considerably less. However, base cation changes may occur for more than 10 years.

See also: Acidification; Alkalinity; Carbon, Unifying Currency; Chemical Fluxes and Dynamics in River and Stream Ecosystems; Chemical Properties of Water; Chloride; Dissolved CO_2; Eutrophication; Fires; Fluvial Transport of Suspended Solids; Ground Water and Surface Water Interaction; Groundwater Chemistry; Hydrological Cycle and Water Budgets; Iron and Manganese; Major Cations (Ca, Mg, Na, K, Al); Micronutrient Elements (Co, Mo, Mn, Zn, Cu); Natural Organic Matter; Nitrogen; Nitrogen Fixation; Nutrient Stoichiometry in Aquatic Ecosystems; Organic Nitrogen; Phosphorus; Rivers; Salinity; Silica.

Further Reading

Bernhardt ES, Likens GE, Buso DC, and Driscoll CT (2003) In-stream uptake dampens effects of major forest disturbances on watershed nitrogen export. *Proceedings of the National Academy of Sciences* 100(18): 10304–10308.

Bernhardt ES, Likens GE, Hall RO Jr, *et al.* (2005) Can't see the forest for the stream? In-stream processing and terrestrial nitrogen exports. *BioScience* 55: 219–230.

Binkley D and Brown TC (1993) Forest practices as nonpoint sources of pollution in North America. *Water Resources Bulletin* 29: 729–740.

Bormann FH and Likens GE (1979) *Pattern and Process in a Forested Ecosystem.* New York, NY: Springer-Verlag.

Brown GW (1985) *Forestry and Water Quality.* Corvallis, OR: Oregon State University Press.

Feller MC (2005) Forest harvesting and streamwater inorganic chemistry in western North America: A review. *Journal of the American Water Resources Association* 41: 785–811.

Harris GP (2001) Biogeochemistry of nitrogen and phosphorus in Australian catchments, rivers and estuaries: effects of land use and flow regulation and comparisons with global patterns. *Marine and Freshwater Research* 52: 139–149.

Hornbeck JW, Martin CW, Pierce RS, *et al.* (1986) Clearcutting northern hardwoods: effects on hydrologic and nutrient ion budgets. *Forest Science* 32: 667–686.

Malmer A (1993) *Dynamics of Hydrology and Nutrient Losses as Response to Establishment of Forest Plantation. A Case Study on Tropical Rainforest Land in Sabah, Malaysia.* Umeå, Sweden: Dissertation.

Mann LK, Johnson DW, West DC, *et al.* (1988) Effects of whole tree and stem-only clearcutting on postharvest hydrologic losses, nutrient capital, and regrowth. *Forest Science* 34: 412–428.

Martin CW, Noel DS, and Federer CA (1984) Effects of forest clearcutting in New England on stream chemistry. *Journal of Environmental Quality* 13: 204–210.

Naiman RJ, Décamps H, and McClain ME (2005) *Riparia. Ecology, Conservation and Management of Streamside Communities.* Elsevier Academic Press.

Relevant Websites

http://www.hubbardbrook.org – Hubbard Brook.
http://www.fsl.orst.edu/lterhome.html/ – H.J. Andrews.
http://coweeta.ecology.uga.edu/webdocs/1/index.htm – Coweeta.
http://www.umanitoba.ca/institutes/fisheries/ – Experimental Lakes Area.
http://www.tlws.ca/index2.shtml – Turkey Lakes.
http://bangor.ceh.ac.uk/plynlimon – Plynlimon.

Distribution and Abundance of Aquatic Plants – Human Impacts

S Knight, University of Wisconsin Trout Lake Station and Wisconsin Department of Natural Resources, Boulder Junction, WI, USA
J Hauxwell, DNR Science Operations Center, Madison, WI, USA

Introduction

Most aquatic plants inhabit the shallow water or littoral zone of lakes and streams. Aquatic plants growing along a lake's edge are both a protective and nourishing component of the lake ecosystem. From a human viewpoint, aquatic plants are often seen as a hindrance to human recreation, but many people also recognize the importance of macrophytes for healthy lakes. The aquatic plant community is a critical habitat and nursery for fish, a source of oxygen for all organisms, a refuge for prey as well as a foraging area for predators, a buffer against erosion and sediment resuspension from both waves and shoreline inputs, and can significantly contribute to overall lake primary productivity (**Figure 1**).

Over the past several decades, losses of or changes in assemblages of native submersed aquatic vegetation has been a reoccurring phenomenon in shallow freshwater systems worldwide, due to a relatively limited number of factors. Repeatedly, human-induced disturbances as a result of anthropogenic alterations of landscapes and atmospheric conditions have increasingly affected the ecology of adjacent aquatic systems, including aquatic plant communities. In addition, in-lake aquatic plant management activities have increased due to the increasing spread of invasive exotic plants.

Balancing the abundance and distribution of aquatic plants is often the formidable task of an aquatic plant manager. If plants are so abundant that they impede fishing, boating, or swimming (for people or fish), then there may be 'too many' plants. Some lakes may naturally have 'too few' plants because of inhospitable bottom sediments, physical barriers such as wind or waves, or turbid water preventing plant growth. Pollution, especially increased nutrient inputs causing eutrophication, can paradoxically lead to either 'too many' or, eventually, 'too few' plants. Acid rain, industrial wastes, changing global temperatures, and carbon dioxide and UV levels all contribute to more recent changes in macrophyte community composition, diversity, and growth rates. Invasive species introductions of both exotic plants and/or animals in addition to management activities are another mechanism by which people can dramatically alter aquatic plant communities. Together, these largely human-caused changes create a complex set of variables influencing macrophytes in the lake ecosystem.

To understand the various controls of aquatic plant ecology, one must first learn (1) the basic abiotic and biotic controls on plant distribution and abundance, including nutrients, light, and competition; (2) the way people inadvertently affect plant distributions by altering these abiotic and biotic factors via eutrophication, pollution, UV, acid rain, increased atmospheric carbon dioxide concentrations, and invasive species introductions; and (3) the effects of various purposeful aquatic plant management techniques. In the following sections, we explore the variety of ways people, through our activities on land and water, both inadvertently and purposefully, affect the distribution and abundance of aquatic plants.

Eutrophication

Light, nutrients, temperature, carbon dioxide, and oxygen interact to regulate aquatic plant growth. Eutrophication is largely a human-caused change in the nutrient status of a lake. Nutrients from sewage, agricultural and lawn fertilizers transported by rivers, runoff, ground water, or direct atmospheric deposition are early suspects when previously clear waters turn green. In a classic experiment in Canada begun in 1973 conducted by D.W. Schindler, nitrogen, carbon, and phosphorus were added to one side of a partitioned lake and just nitrogen and carbon were added to the other. The phosphorus (P) side developed tremendous algal growth and was clear evidence that excess P, more than any other nutrient, was responsible for explosive algal growth. This experiment led to wide recognition of the problems associated with P inputs into lakes and then to strong regulations against phosphates in laundry products.

Effects of Increased Nutrients on Primary Producers

At the beginning of the eutrophication process, biomasses of plants and algae (both in the water column and living on the plants directly) tend to increase. However, under high nutrient conditions in the water column, algae, rather than macrophytes, will proliferate and prevent light from reaching far into

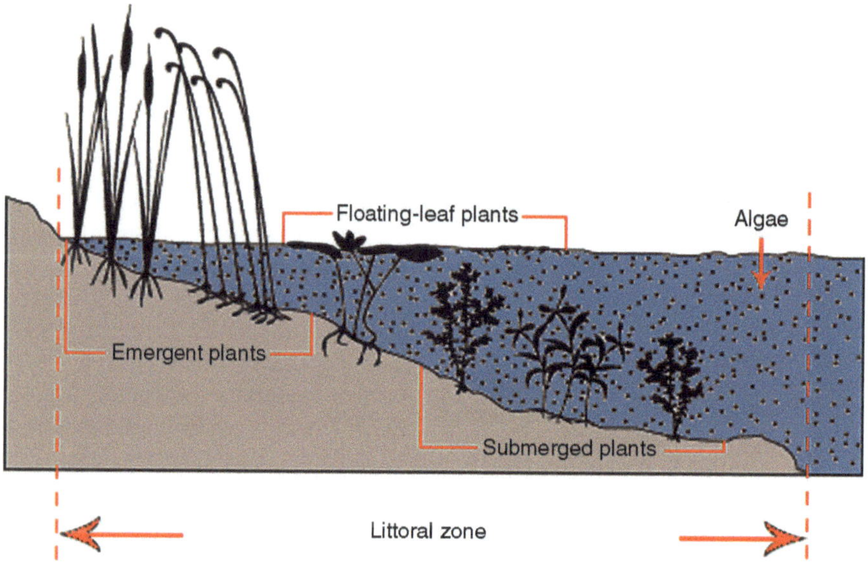

Figure 1 Zonation of emergent, floating and submersed plants in the littoral zone (http://www.extension.umn.edu/distribution/horticulture/images/6955f02.gif).

the water. Under these conditions, rooted macrophytes may become light-limited so that their distribution decreases as the maximum depth of plant growth moves shoreward becoming restricted to just the shallowest rim of the lake. As more biomass from primary producers lives and then dies, oxygen demand from bacterial decomposition increases. This can lead to anoxic conditions at the bottom of the lake, augmenting P internal loading favored by anoxic sediment conditions. This positive feedback is difficult and sometimes impossible to counteract even if the external nutrient inputs are halted. Under severe conditions, the lake, especially if shallow, may 'flip' (sometimes called a 'stable state' change) from a macrophyte-dominated littoral zone to being phytoplankton-dominated. This change in state is often irreversible, resulting in a cascade of consequences associated with no plants and low light (**Figure 2**).

Light and nutrients are intricately coupled in lakes and, in addition to distribution and the maximum depth of plant growth, have important effects on macrophyte species' diversity, growth form, and biomass. Macrophytes largely take up nutrients, such as phosphorus and nitrogen, from the sediments through their roots, rather than from the water column through their leaves and stems. Some lakes naturally have nutrient-rich sediments and may have luxuriant macrophyte growth, depending on light conditions. Under natural lake conditions, there is greater species diversity with increased light, regardless of sediment richness. As site fertility increased, there tends to be an increased proportion of plant

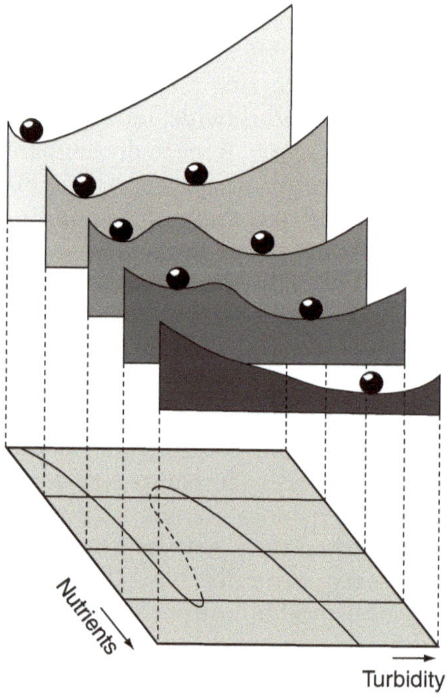

Figure 2 A shallow lake's littoral zone may experience a 'stable state' shift and change from being macrophyte-dominated to algal-dominated. Scheffer M, Hosper SH, Meijer M-L, Moss B, and Jeppesen E (1993) Alternative equilibria in shallow lakes. *Trends in Ecology and Evolution* 8(8): 275–279.

biomass in canopy-forming plant species such Eurasian water-milfoil (*Myriophyllum spicatum*), which can negatively affect the light regime for understory plants. In addition, some plants, such as the submersed, rootless *Ceratophyllum demersum* (coontail) and free

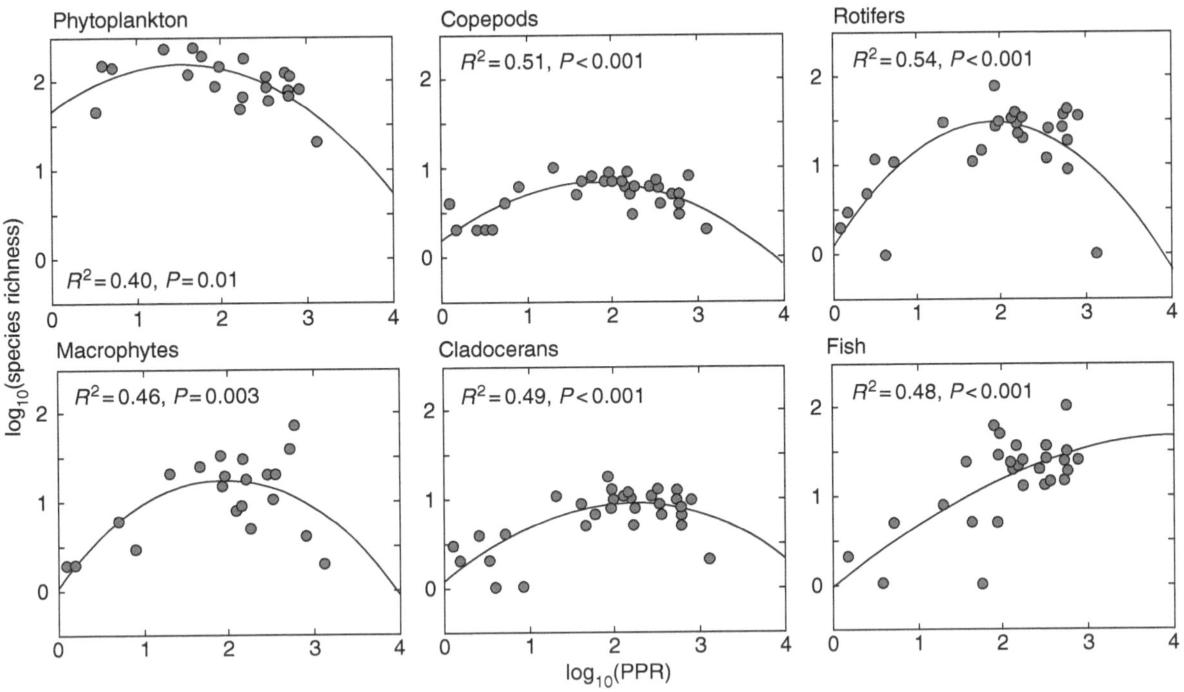

Figure 3 Species richness is related to primary productivity in a 'hump-shaped' pattern. Dodson SI, Arnott SE, and Cottingham KL (2000) The relationship in lake communities between primary productivity and species richness. *Ecology* 81(10): 2662–2679, Per Figure.

floating species such as *Lemna* (duckweed) do rely on water column nutrients and may proliferate in a nutrient-rich water column.

Regression analyses indicate that species richness is related to primary productivity in a unimodal, 'hump-shaped' pattern for many groups of aquatic organisms, including macrophytes (**Figure 3**). The shape of this relationship may be understood by looking at both ends of the 'hump'. Thermodynamics may explain an increase in species richness with productivity. There is more energy available to a lake and therefore more potential to support additional species and trophic levels. While phytoplankton experience their highest diversity at relatively low levels of productivity, macrophyte diversity is highest at higher levels of production. This may be because more photosynthate must go to structural components in plants, as compared to phytoplankton. At the high end of productivity, relatively few species have the 'winning' growth form to withstand shading and competition at decreasing light concomitant with increased productivity, but there may be other explanations.

Macrophytes display a wide variety of growth forms, and these are consistent with their growth patterns in different ranges of nutrients and light levels. Under constant light, *Potamogeton praelongus*, growing near to the surface, achieves its greatest height and biomass under high nutrient conditions. In experimental conditions, plants of shorter stature, such as *Potamogeton robbinsii*, do not display much variation in biomass with different sediment nutrient conditions, indicating it was most affected by light.

Feedbacks between Plants and Algae and Trophic Interactions

An abundant plant community has a positive feedback effect on water clarity. Water clarity tends to be greater in lakes with macrophytes than those without, even with comparable total P levels. In one study, chlorophyll *a* concentration was inversely related to the percent of a lake's total volume containing aquatic macrophytes, probably due to many factors. Healthy vegetation diminishes turbulence near the sediments so that suspended materials have a greater chance to settle and limits resuspension of nutrients from the bottom sediments. Zooplankton, taking refuge from planktivorous fish among the vegetation, may consume vast quantities of phytoplankton, also decreasing shading of plants. In experimental removal of zooplanktivorous fish in a small, shallow eutrophic lake, there was a marked increase in aquatic plant growth despite no lessening of high nutrient inputs. Another experiment performed a food web manipulation in a hypertrophic lake with no macrophytes and abundant phytoplankton. This lake experienced explosive population growths, or blooms, of *Microcystis*, a blue green algae (also known as cyanobacteria) in summer. The planktivorous and benthivorous fish species were completely removed and a new fish community of pike and rudd was introduced. This change in fish community led to much lower phytoplankton densities and an increase in the abundance of macrophytes. Both of these studies indicate the potential influence of trophic interactions on the light regime for plants.

Most plants have some periphyton, that is, algae and other microorganisms resting on the macrophyte leaf surfaces. In *Lobelia dortmanna*, dense periphyton growth in the spring slows growth of the macrophyte, until summer when periphyton growth is diminished due to N and P limitation. While periphyton will shade plants and decrease a plant's ability to photosynthesize, periphyton can absorb water-column nutrients, sequestering it from the phytoplankton. Plants may tolerate some epiphytic growth, which may be minimized by grazing invertebrates and thus eliminate the shading effects of phytoplankton.

Effects of Watershed and Shoreline Development on Aquatic Plant Communites

Development of watersheds and shoreline areas and subsequent increases in nutrient loading affect aquatic plant communities (**Figure 4**). At the watershed scale, transformation of land parcels from natural conditions toward agriculture or urban development result in increased nutrient loads to adjoining water bodies. Decreases in species richness in Ontario lakes have been linked to urban development (measured by road density). In addition, in Midwestern lakes, the abundance of submersed, floating-leaf, and emergent macrophytes decreased with urban development of watersheds. Human impacts associated at the lakeshore scale include nutrient effects associated with septic leachate and lawn fertilizer runoff. In addition, effects may be magnified due to shoreline interface modifications and the direct removal of plants. Light availability for macrophytes is reduced under docks or piers. Seawalls and beaches prevent the growth of near-shore species such as emergents and floating-leaf macrophytes. In Minnesota, there was a 66% decline in floating-leaf and emergent vegetation lakewide due to lakeshore residential development.

Figure 4 Removal of natural vegetation along shore by property owners can affect lake water quality and habitat (photograph by Robert Korth).

Increased Ultraviolet Radiation

With the recent thinning of the stratospheric ozone resulting in increased ultraviolet solar radiation (UVR) penetration to Earth, there has been considerable interest in UVR effects on organisms and natural systems. Ultraviolet A (UVAR, wavelength 320–400 nm) and ultraviolet B (UVBR, wavelength 290–320 nm) have somewhat different effects on living organisms. It is clear that aquatic macrophytes respond to different levels of photosynthetically active radiation (PAR) and that some will grow deeper in the water depending on their tolerance to lower light levels, but it is less clear what the effects of increased UVR would be. There is a tremendous range over which UVR can penetrate aquatic systems, with UVR penetrating less in lakes with increased dissolved organic carbon (DOC) concentrations. In experiments conducted in a deep, clear water lake in New Zealand (where UVR penetrates more than 5 m), the shallow species *Potamogeton cheesemanii* experienced little growth difference if UVR was present or filtered from the light source and slightly deeper-growing *Isoetes alpinus* experienced slightly different growth. However, very deep-growing *Chara fibrosa* and *Chara corallina* experienced marked, and sometimes irreversible declines in growth rates when experimentally brought up to the shallows and exposed to UVR. Other aquatic plants, such as aquatic mosses and liverworts also demonstrate a UV-induced stress response. Under experimental situations, UVBR negatively affected growth of *Chara aspera* and caused DNA damage, but did not naturally penetrate as far as the natural *Chara* beds. The degree of UVR inhibition seems to be related to the upper depth limits to which these plants grow. Others have found that removal of UVBR improves the photosynthetic activity of marine angiosperms in shallow, sandy areas.

Acid Rain

Lake acidification is largely the result of precipitation of sulfur compounds from coal-burning power plants. In low pH waters, dissolved inorganic carbon (DIC) is largely present as carbon dioxide, while at higher alkalinities and higher pH values, DIC is largely available as bicarbonate (HCO_3^-). Many macrophytes can use bicarbonate as their source of carbon, but at an energetic cost to the plant. Despite these costs, many of these plants that can use HCO_3^- may not be viable at lower pH sites where CO_2 is the only source of carbon. However, since higher pH lakes tend to be fairly well buffered, these lakes and their macrophytes will be less affected by acid rain. Poorly buffered lakes are especially vulnerable to the pH-lowering effect of acid rain, and acidified lakes may see a change in macrophyte species.

Many macrophytes, commonly known as isoetids, are common in poorly buffered, often acidic, oligotrophic lakes. These isoetids are usually quite small, have tough leaves and many take up CO_2 through their roots. In Adirondack lakes, pH-related factors were the most important factors influencing species assortment. If acid rain changes the alkalinity and pH of lakes, it is likely there will be a shift toward more isoetid-type plants and fewer of the leafier and taller elodeid-type plants as represented by *Elodea* (water weed) and many *Potamogeton* (pondweeds). Further, low densities of macrophytes in Adirondack lakes may be due to increased acidification, lower concentrations of base cations, and increased concentrations of trace metals. In a survey of some Adirondack lakes, species richness and diversity were lower in lakes with pH less than 5.5. Acid-intolerant species, such as slender naiad (*Najs flexilis*), stonewort (*Nitella*) and small pondweed (*Potamogeton pusillus*) and others may be displaced when lakes become acidified. Experiments and observations reveal there is often a shift from vascular macrophytes to mosses associated with lake acidification. In other experiments, ammonium sulfate was experimentally applied to macrophytes. Ammonium sulfate has a eutrophying effect because it adds a source of nitrogen, but it also has an acidifying effect because of the biologically mediated nitrification of ammonium, lowering the pH from 5.6 to 3.5. The ammonium sulfate had a long-lasting impact on plant species that are sensitive to both acidification and eutrophication.

Chemical Pollutants

Plants are sometimes exposed to toxic chemicals from industrial wastes. Heavy metals, such as mercuric chloride, lead acetate, cadmium chloride, and cupric sulphate cause senescence in aquatic plants. These chemicals affect respiration, photosynthesis, and other metabolic pathways. Heavy metals and halogenated hydrocarbons are also toxic to aquatic plants and decimated the species diversity in an alluvial floodplain of the Elbe Estuary.

Increased Carbon Dioxide

Since the partial pressure of CO_2 is the same in the water as in the air at equilibrium, there will be increased CO_2 in the water as global atmospheric CO_2 levels rise. Increased CO_2 will tend to lower the pH, so many of the issues pertaining to increased CO_2 will also be the same issues raised by increasing acidification. Some lakes are naturally supersaturated with CO_2, and these lakes may experience increased growth rates of macrophytes when the lakes are supersaturated, but may not be affected by increased global carbon dioxide increases. Complicating this further, lakes may be supersaturated with CO_2 part of the year or at some depths and under-saturated at other times, so the effects of increased CO_2 levels are difficult to predict.

In experiments augmenting the DIC (as CO_2) concentration, two species that rely on carbon in the water column, *Elodea canadensis* and *Utricularia vulgaris*, experienced increased growth. A third species, *Eriocaulon aquaticum* takes up CO_2 through its roots and is more common in softer, lower pH lakes and did not show increased growth with increased DIC. In other experiments, wild celery (*Vallisneria americana*) had increased growth rate with increased DIC, when grown at a relatively low pH 5. This suggests that increased CO_2 levels may have a 'fertilizing' effect on the growth rates of many macrophytes. In modeling exercises where P was not limiting, it appears that macrophytes restricted to using CO_2 will increase growth rates threefold, while plants that can also use bicarbonate may double their growth rate. The model further predicts that there will be a strong effect of increased CO_2 levels under eutrophic conditions with high plant biomass, where CO_2 concentrations may fall to very low levels and the pH may rise.

Complex Interactions - Increased CO_2, Acidification and Eutrophication

Softwater lakes are probably the most vulnerable to these global, atmosphere-related and nutrient-related stressors. They are an important component of the worldwide community diversity resource base in northern North America and northern Europe. There could be changes at the population level, species level, and in functional diversity as the combined stresses of acidification, increased CO_2 levels, and eutrophication develop. Shallow and oligotrophic moorland pools in The Netherlands, are subject to and extremely vulnerable to sulfate and ammonium acid deposition and agricultural nutrient inputs. Since acidification may cause some macrophyte species changes and increased CO_2 concentrations may increase macrophyte growth rates, combined effects of these atmospheric changes may lead to complex interactions and unpredictable outcomes for macrophytes.

Complex Interactions – Plants, Algae, Nutrients, Carbon and Consumers

In simple experiments, it is clear that there is an increase in periphyton with increased nutrients and a strong negative effect of periphyton on plants. Additionally, there are complex interactions among plants, periphyton, DIC, grazers, and grazer predators. Lakes vary in their concentrations of DIC, and in experimental manipulations, it is clear that plants and periphyton compete for DIC and light. In this study, periphyton had a negative effect on plants, but the periphyton biomass was not directly controlled by nutrients but rather by the number of grazers. The grazers, in turn, were controlled by the number of grazer-eating fish. Increased nutrient loading led, not to increased biomass of periphyton, but to increased numbers of invertebrates on the plants, leading to less periphyton on the plants in an analogous process to zooplankton feeding on algae. DIC may be an important factor in the response of plants to eutrophic conditions, but the success of plants will also be influenced by complex interactions of changing water chemistry, grazers, and fish. Another study confirmed less periphyton on macrophytes when fish were excluded, allowing snail grazers to scrape the plants clean. In addition, plant biomass was often much higher in fish-free enclosures and in one case, a macrophyte species was only found in fish-free enclosures, suggesting that the plant–periphyton–snail–fish food web can control macrophyte species diversity.

Invasive Species Introductions

As humans increase in both number and geographic mobility, we increasingly become a vector for the spread of nonindigenous aquatic species, including plants and other taxa that may directly or indirectly affect plants. It is estimated that in the Great Lakes alone, 183 non-native species have established largely via ballast water discharge from ocean-going ships. This initial vector then provides the opportunity for potential migration of these taxa into abundant Midwestern inland lakes. Invasives are a global problem for lake ecosystems, with major modifications of plant habitat reported across latitudes and from East to West.

Invasive aquatic plant species can affect the distribution and abundance of natives dramatically. This interaction is often mediated via competition, with invasives often better competitors for light, nutrients, carbon, or space. For example, infestations of non-native macrophytes, including Eurasian watermilfoil (*M. spicatum* L.) and hydrilla (*Hydrilla verticillata* (L.f.) Royle) competitively exclude native freshwater macrophytes. Because these exotics initiate rapid growth early in the season, they are able to concentrate a large percentage of their biomass toward the water's surface, shading out later-growing light-limited taxa. Hydrilla, for example, limits light penetration by 95% within the first 30 cm of the water column. Additionally, allelopathic interactions can result in exclusion of a native plant over an exotic.

Other invasive taxa can also affect aquatic plants either directly or indirectly. Non-native herbivores (e.g., grass carp (*Ctenopharyngodon idella* Val.) or the rusty crayfish (*Orconectes rusticus*)) may severely affect aquatic vegetation through consumption or simple destructive habitat modification. Zebra mussels can become so numerous that they can improve the light regime for aquatic plants by filtering phytoplankton. In Long Point Bay, Lake Erie and Saginaw Bay, Lake Huron, for example, aquatic plants expanded in coverage, most likely as a result of improved water clarity. In Oneida Lake, New York, an invasion of zebra mussels also led to an increase in the maximum depth of plants and an increase in macrophyte species richness.

Additional physical or biological disturbances to lake ecosystems can affect aquatic plants and competitive interactions between native and invasive species. Eutrophic waters are often a result of disturbance to a watershed and may experience more instances of invasion by exotic, invasive species. Gap occurrence in native beds of vegetation via dredging or plant management activities can also benefit the colonization and dominance of invasive plant taxa.

Aquatic Plant Management Activities

When humans perceive that macrophyte growth reaches nuisance levels, they may enlist a number of strategies to reduce the biomass or kill the plants. Often, the problem is excessive growth of an invasive exotic such as Eurasian water-milfoil (*M. spicatum*), curly-leaf pondweed (*Potamogeton crispus*), water chestnut (*Trapa natans*), hydrilla (*H. verticillata*), or many other alien species that can come to dominate the macrophyte community. These species tend to grow very quickly and form a surface canopy that impedes water recreation such as swimming, fishing, and boating. Native plant species, less likely to reach the surface, are often eliminated by shading from these aggressive exotics.

Fortunately, most people recognize the value of macrophytes to lake ecosystems and do not necessarily aim for a weed-free lake. Many states have laws requiring riparian owners to apply for a permit to manage aquatic plants. There are many factors to consider when choosing a plant management strategy, including the situation and size of the project, available methods, best timing for management, the short-term and long-term ecological effects and costs of various strategies, and environmental risks. A comprehensive review of techniques will also bring an awareness of hidden costs for maintaining equipment, repeating treatments, and restoring habitat. Management actions fall under four general categories: mechanical control, physical control, biological control, and chemical control. There are advantages and disadvantages associated with all techniques, reviewed extensively in **Table 1**. Below, we describe management activities that can significantly affect plant distributions at the largest of spatial scales (tens of hectares up to whole-lake scale).

Mechanical Control

Mechanical control involves the direct removal or cutting of plant material. Mechanical harvesting, for example, is like underwater lawn-mowing, cutting the plants about 2 m below the surface, collecting them on a boat with a conveyor belt, and then transporting them to a suitable off-loading site (**Figure 5**). Harvesting is not recommended for small or dispersed populations of invasive species. The harvester does not collect every cut plant fragment and these pieces could become established elsewhere, further spreading the plant. Mechanical harvesting is recommended in lakes with very heavy native or invasive species communities that prohibit swimming, water skiing, fishing, or other recreational activities. Harvesting will remove plants from the lake surface, but like grass on your lawn, the plants will grow back and will probably have to be cut more than once per season. Since the plants are removed from the lake, there is a net loss of nutrients from the lake. However, it may take many years of harvesting before there is a significant loss of nutrients available for plant growth, especially if the lake has an inlet. Seepage lakes or lakes with only one outlet may see one effect more quickly.

Physical Control

Physical management techniques, such as dredging and draw-down manipulate the plant environment. Dredging is most often effective when soft, nutrient-rich sediments overlay a hard lakebed. By deepening the lake, the bottom may be too deep for plants to grow. However, if the water body is without a firm bottom, dredging may increase algae or aquatic plants by exposing and suspending bottom sediments that contain more available nutrients. Dredging is often cost prohibitive and can have detrimental effects on a lake's ecosystem. Drawdowns can be done on waters that have a dam controlling the water level. Water level is lowered during the drawdown, and aquatic plants are controlled through sediment drying, compacting, or freezing. Drawdowns can be an effective aquatic plant management technique on larger reservoirs. Most reservoirs have little fluctuation of water levels, although a varying water level is a more natural condition. Many reservoir plants, including the beneficial near-shore plants, require water level fluctuation to survive. Mimicking this fluctuation may increase the desirable species while decreasing the undesirable species. Drawdowns can have a significant impact on populations of amphibians and fish especially if they are done in the winter.

Biological Control

Biological control, or biocontrol, uses living organisms (such as herbivores or fungi) to control unwanted species. Often, invasive species arrive in new waters unaccompanied by their natural enemies that would have kept their populations in check in their native land. One approach to managing these nuisance plants is to find local, or import the invasive species' natural, enemies that will keep them under control. Unlike chemical or mechanical controls, biological controls can be self-sustaining as the organism may overwinter and then resume eating its host the next year. Generally, biocontrol is unlikely to eliminate its host plant, but it may be effective at lowering the density and restoring a healthy, diverse native plant community. Biocontrol is not without its own risks as well. It is important to remember that the biocontrol agent itself may be invasive and potentially could cause a worse nuisance than the aquatic plant you are trying to manipulate.

Chemical Control

Many aquatic herbicides are applied to lakes every year all over the USA (**Figure 6**). Aquatic herbicide chemicals fall into two general categories. Systemic herbicides are absorbed and transported through the plant's vascular system, killing the entire plant. Contact herbicides kill the part of the plant in contact with the chemical but the roots may survive and the

Table 1 General management approaches for aquatic plant control, including area affected, how each works, and associated advantages and disadvantages (specific outcomes highly dependent on specific techniques chosen within these general categories)

Approach	*Area*	*How it works*	*Advantages*	*Disadvantages*
Mechanical	m^2 – Whole lake	Plants reduced by mechanical means	Immediate results	Must be repeated, often more than once per season
		Wide range of techniques, from manual to highly mechanized	Flexible control	Can resuspend sediments and increase turbidity and nutrient release
		Techniques include hand-pulling, rototilling, hydroraking, cutting, harvesting	Can balance habitat and recreational needs	Requires plant disposal
			Usually minimal impact to lake ecology	Fragments of vegetation can re-root
Physical	m^2 – Whole lake	Plants are reduced by altering variables that affect growth, such as water depth or light levels	Variable benefits depending on specific technique	Variable drawbacks depending on specific technique
		Techniques include: bottom barriers, drawdown, dredging, dyes, nutrient inactivation		
Biological	Whole lake	Living organisms (e.g., herbivores or fungi) eat or infect plants	Possibly self-sustaining; organism may over-winter, resume eating its host the next year	Effectiveness will vary as control agent's population fluctuates
		Techniques include: stocking of weevils, carp, crayfish, pathogens, allelopathic plants, native plantings	Lowers density of problem plant to allow growth of natives	Provides moderate control – complete control unlikely
				Control response may be slow
				Must have enough control agent to be effective
				May cause unintended consequences, including invasivity itself
Chemical	m^2 – Whole lake	Granules or liquid chemicals kill plants or cease plant growth; some chemicals used primarily for algae	Effective plant control from small to large scale	Possible toxicity to aquatic animals or humans, especially applicators
		Chemicals must be used in accordance with label guidelines and restrictions	Flexibility for different situations	Often affect desirable plant species that are important to lake ecology and compete with invasive species
		Techniques include: contact herbicides (endothall, diquat) and systemic herbicides (2,4-D, fluridone, glyphosate, triclopyr, copper compounds)	Some can be selective if applied correctly	Treatment set-back requirements from potable water sources and/or drinking water use restrictions after application, usually based on concentration
			Can be applied toward restoration activities, if timing and existing plant community considered	May cause severe drop in dissolved oxygen causing fish kill – depends on plant biomass killed, temperatures, and lake size and shape
			Results within days to months of treatment	Nutrient release from decaying vegetation can cause increased algae and bacteria and be a nuisance to riparian property owners

Continued

Table 1 Continued

Approach	*Area*	*How it works*	*Advantages*	*Disadvantages*
				Repeat treatments usually needed, within season or across seasons, dependent on herbicide
				Often controversial
No management	Whole lake	Do not actively manage plants	Minimizing disturbance can protect native species that provide habitat for aquatic fauna, reduce shoreline erosion, may improve water clarity, and may limit spread of invasive species	May allow small population of invasive plants to become larger, more difficult to control later
			No immediate financial cost	Excessive plant growth can hamper navigation and recreational lake use
			No system disturbance	May require modification of lake users' behavior and perception
			No unintended effects of other control options	

Figure 5 Aquatic plant harvester.

Figure 6 Aquatic plant herbicide applicator.

plant may regrow. Broad spectrum (non-selective) herbicides will kill most plants they encounter. Selective herbicides will affect only some plants, often broad-leaf or dicot plants such as Eurasian water milfoil. Although herbicides are screened to minimize harm to other aquatic organisms at recommended application rates, it is important to remember that the effect of herbicide breakdown products, herbicide interactions in the environment, and impacts of herbicides on all life stages of all aquatic organisms are not well studied. Also, aquatic organisms living in the vegetation may be affected simply by habitat loss.

Conclusion

Human impacts on aquatic plant communities are numerous and involve many mechanisms, all of which may interact. Protection of these important environments requires an understanding of the factors that influence plant distribution and abundance and how human activities affect with these factors.

Further Reading

Bronmark C and Hanson L-A (1998) *The Biology of Lakes and Ponds*. Oxford, UK: Oxford Universtiy Press.

Carpenter SR and Lodge DM (1986) Effects of submersed macrophytes on ecosystem processes. *Aquatic Botany* 26: 341–370.

Cooke GD, Welch EB, Peterson SA, and Nichols SA (2005) *Restoration and Management of Lakes and Reservoirs*. 3rd edn. Boca Raton, FL: CRC Press.

Dodson SI, Arnott SE, and Cottingham KL (2000) The relationship in lake communities between primary productivity and species richness. *Ecology* 81(10): 2662–2679.

Jeppesen E, Sondergaard M, Sondergaard M, and Christoffersen K (eds.) (1998) *The Structuring Role of Submerged Macrophytes in Lakes*. New York: Springer.

Scheffer M, Hosper SH, Meijer ML, Moss B, and Jeppesen E (1993) Alternative equilibria in shallow lakes. *Trends in Ecology and Evolution* 8(8): 275–279.

Schindler DW (1974) Eutrophication and recovery in experimental lakes: implications for lake management. *Science* 184: 897–899.

Wetzel RG (2001) *Limnology: Lake and River Ecosystems*. San Diego: Academic Press.

Effects of Climate Change on Lakes

W F Vincent, Laval University, Quebec City, QC, Canada

Introduction

Lake ecosystems are vital resources for aquatic wildlife and human needs, and any alteration of their environmental quality and water renewal rates has wide-ranging ecological and societal implications. The increasing accumulation of greenhouse gases in the atmosphere as a result of human activities has begun to affect the structure, functioning, and stability of lake ecosystems throughout the world, and much greater impacts are likely in the future. Current global circulation models predict an increase in air temperatures of several degrees by the end of the twenty-first century, combined with large changes in the regional distribution and intensity of rainfall. These changes will also be accompanied by massive disruption of the cryosphere, the ensemble of ice-containing environments on Earth. These shifts in climate forcing appear to have already begun, and the onset of changes in the physical, chemical, and biological attributes of lakes is affecting their ability to maintain the present-day communities of aquatic plants, animals, and microbes, and their capacity to provide ecosystem services such as safe drinking water and inland fisheries (**Figure 1**).

Lakes have always been subject to the impacts of climate change, and natural climate variations in the past have been one of the main reasons that lakes are ephemeral features of the landscape. Most of today's lakes are the result of climate amelioration and the retreat of the Pleistocene glaciers some 10 000 years ago, and so most present-day lakes are relatively young. A powerful approach toward understanding the potential impacts of future climate change on lakes is the application of paleolimnological methods in which lake sediment cores are dated and analyzed to infer climate impacts in the past. Most such analyses have been restricted to the time period of the last few 1000 years; however, detailed records of greater than 100 000 years are becoming available from studies of ancient lakes. Studies of present-day lakes of different ages (chronosequences) and across latitudinal gradients are also providing valuable insights into the consequences of climate change. Additional knowledge about climate impacts is coming from modeling and experiments, combined with multi-decade, regional analyses of lakes that are currently experiencing shifts in temperature and precipitation.

Some of the most striking examples of climate impacts to date are from limnological and paleolimnological studies in the polar regions. Global circulation models predict that the fastest and most pronounced warming will be at the highest latitudes because of a variety of feedback processes that amplify warming in these regions. These include the capacity of warm air to store more water vapor, itself a powerful greenhouse gas, and the reduced albedo (reflection of sunlight) as a result of the melting of snow and ice, leaving more solar energy to be available for heating. Some of the immediate impacts of climate change on high-latitude lakes include loss of perennial ice cover, increasing duration of open water conditions, increasing water temperatures, stronger water column stratification and shifts in water balance, in some cases leading to complete drainage or drying up of the waterbodies.

Changes in air temperature and precipitation have direct effects on the physical, chemical, and biological characteristics of lakes, and they also operate on lakes indirectly via modifications in the surrounding watershed, e.g., through shifts in hydrological flow pathways, landscape weathering, catchment erosion, soil properties, and vegetation. Of particular interest to limnologists (lake and river scientists) are the interactions between variables, the feedback effects that accelerate or dampen environmental change, and threshold effects by which lakes may abruptly shift from one environmental state to another.

Physical Effects of Climate Change

Basin Integrity

Changes in the precipitation regime that accompany climate change have the potential to cause shifts in the connectivity of lakes (with biological implications, e.g., for migratory fish species), as well as in erosion rates that could affect the inflow and outflow dynamics of lakes. The latter effects are particularly conspicuous in the tundra permafrost. Many thaw ponds (also called thermokarst ponds) on the Siberian and Alaska tundra are draining as a result of increased melting of the permafrost, while certain thaw ponds further south in discontinuous, subarctic permafrost have begun to expand as a result of the differential melting of such

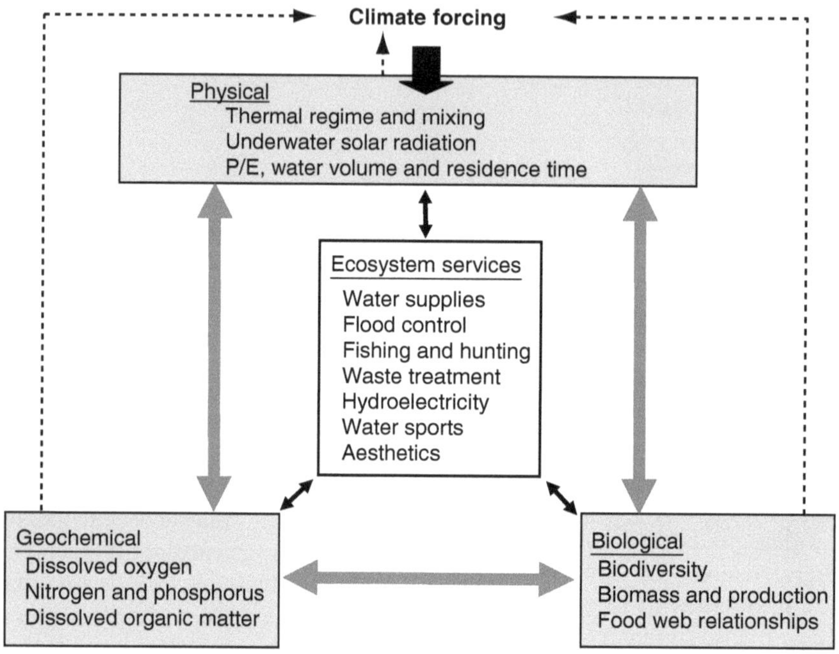

Figure 1 Changes in climate forcing affect the physical environment of lake ecosystems and thereby alter their chemical and biological properties. These changes affect the capacity of lakes to provide ecosystem services. P/E, precipitation to evaporation ratio. Dotted lines indicate positive feedback effects, e.g., via decreased ice cover or the release of greenhouse gases from lakes into the atmosphere.

soils. Ice-dammed lakes, e.g., those behind thick ice shelves (epishelf lakes), ice-jams on rivers, glacial dams, and behind coastal sea ice barriers (stamukhi lakes) are prone to complete loss because of climate-dependent ice fracturing and collapse.

Water Balance

Shifts in precipitation relative to evaporation (the P/E ratio) cause changes in the water budget and hydraulic residence time of lakes, as well as in their depth and areal extent. Ponds and wetlands are especially vulnerable to changes in P/E because of their shallow depths and large surface to volume ratio. For example, some rock basin ponds in the Canadian High Arctic have been drying up as a result of climate warming in the region, perhaps for the first time in millennia. Even large, shallow lakes can undergo major climate-driven fluctuations and loss. Lake Chad in north-western Africa is less than 7-m deep and has responded greatly to changes in climate over the last few decades. From the 1960s onward it experienced a precipitous decline in rainfall, and this has led to a major contraction in lake area, from 25 000 km^2 in the early 1960s to 1350 km^2 in the 1990s. This climate effect has been compounded by the increasing need to irrigate farmland during these drought conditions, and human water use appears to account for about half of the decrease in lake surface area.

Glaciers in many parts of the world are undergoing accelerated retreat, and glacial fed lakes and reservoirs are therefore particularly vulnerable to changes in their inflow regime. For example, the Chinese Academy of Sciences estimates that by 2050, up to 64% of China's glaciers may have disappeared, with serious consequences for the estimated 300 million people who live in China's arid west and who depend upon this vital water supply. At Lake Biwa, Japan, the reduced flow from snowpack in the surrounding mountains is thought to be a mechanism reducing cold underflows that play an important role in recharging the bottom waters of the lake with oxygen.

At the other extreme, global change may affect the frequency of extreme storm events and flooding in some regions. There is evidence that warming of the sea surface is resulting in an increasing intensity and duration of storms, and certain coastal lakes may therefore be exposed to increased flooding and wave erosion of their shorelines, as well as increased hydraulic flushing and storm-induced mixing.

Ice Cover

Many lakes in the north temperate zone are covered by ice and snow through much of the year. Climate warming is already showing an effect on such lakes by causing earlier dates of ice break-up, and later dates of freeze-up. The 2007 Intergovernmental Panel on Climate Change report noted that for all available lake and river ice data over the last 150 years, the break-up date had become (mean ± 2SD) 6.5 ± 1.2 days earlier per century, while the freeze-up date had occurred 5.8 ± 1.6 days later per century. At polar latitudes, the earlier occurrence of rain instead of snow is likely to hasten the ice melting process, which is also affected by changes in the wind field. Loss of snow and ice cover results in a major shift in the underwater irradiance regime, with increased solar energy for radiative heating of the water column, increased availability of photosynthetically available radiation (PAR) for primary production, and increased exposure to ultraviolet radiation (UVR). The latter will accelerate photobiological and photochemical processes, with a variety of positive and negative effects on ecosystem structure and productivity. Modeling analyses show that loss of ice cover can have orders of magnitude larger effects on underwater UVR exposure than does moderate ozone depletion in the stratosphere. The prolonged loss of ice cover also allows a longer period of wind-induced mixing of the surface waters of lakes, and may change the transport pathways and fate of inflows that during ice conditions are retained immediately beneath the ice.

Temperature and Stratification

Warmer air temperatures result in warmer surface waters via conduction, although this may be dampened in part by increased evaporation rates. For north temperate lakes, the increased duration of ice-free conditions will allow increased warming of the surface waters by sensible heat transfer (conduction) and by radiative transfer (penetration of solar radiation into the water column). These effects in turn cause changes in the density structure of lakes, with a layer of warmer, lighter water at the surface that prevents the transport of heat via turbulent mixing to deeper parts of the lake. This retention of heat in the upper waters acts as a positive feedback mechanism, and with increasing heating the thermal stratification becomes stronger and more resistant to wind-induced mixing. By trapping more heat in the surface-mixed layer, less is available for heating the lower column, and deep waters can become cooler as a result of climate warming. An example of the strong effects of climate change on water temperature is the current trend in Lake Superior, Canada–USA. The summer surface water temperatures of this lake increased by 2.5 °C over the period 1979–2006, well in excess of the air temperature increase over the same period. This has been primarily attributed to earlier ice break-up and the earlier onset of stratification and warming each year. This differential effect is not apparent in tropical lakes that lack winter ice cover, and such waters appear to be more closely tracking the regional trend in air temperatures.

At high latitudes, increased warming may lead to a complete change in the stratification and mixing regime. Deep polar lakes are usually cold monomictic; that is, they remain stratified under the ice through most of the year, and then mix at temperatures less than 3.98 °C (the temperature of the maximum density of water) throughout their summer period of open water conditions. However, if the water temperatures warm above 3.98 °C in summer, the lake will become stratified, and therefore dimictic, with periods of mixing before and after this summer period. This shift from cold monomixis to dimixis is a critical threshold that allows the retention of heat in the surface waters and it may lead to a higher biological productivity. Paleolimnological analyses have suggested that such transitions occurred in certain Finnish lakes over the last 200 years, and that the resultant increased productivity caused biological shifts toward a cladoceran zooplankton community.

Chemical Effects of Climate Change

In-Lake Effects

The hydraulic residence time of a lake (the time required to completely replace all water in a lake by its river, groundwater and rainfall inputs) affects the chemical composition of lake waters by controlling the time available for biogeochemical and photochemical processes to operate, the extent of accumulation and loss of dissolved and particulate materials, and the duration of biogeochemical interactions with the lake sediments and littoral zone. In lakes that experience anoxic bottom water conditions and nutrient release from the sediments, a prolonged residence time caused by reduced precipitation and inflows can result in increased phosphorus accumulation (internal phosphorus loading) and eutrophication. Conversely, in regions that experience increased precipitation and water flow, the increased flushing of nutrients and phytoplankton may result in reduced algal production. This latter effect may be offset by increased catchment erosion and transport of nutrient-rich soils into the lake if storm events become more frequent and severe.

The stronger and more prolonged stratification of lake waters is also likely to lead to more pronounced gradients in chemical variables down through the water column. In enriched lakes, this may trigger the shift from oxygenated to anoxic conditions in the bottom waters, and crossing this threshold is typically accompanied by major upshifts in the concentrations of algal nutrients, including available phosphorus, and in various toxic compounds, notably hydrogen sulfide (H_2S). Anoxia can also lead to increased nitrogen losses from the ecosystem by denitrification, the bacterial process that converts nitrate to nitrogen gas.

Catchment Effects

Climate affects many landscape properties, including vegetation and soils, which in turn exert a strong influence on water quality and quantity in the receiving lake basin. Increased temperatures and precipitation may lead to increased rock weathering, thereby affecting solute composition and concentrations in the runoff. Long-term studies at Toolik Lake Alaska have detected a doubling of alkalinity over time, and this may be the result of climate-dependent increases in active layer depth in the permafrost that have exposed new soil material to weathering. Changes in precipitation may affect the magnitude of overland versus subterranean flow, thereby changing the nature and quantity of nutrient, major ion, and dissolved organic carbon (DOC) inputs from different soil horizons to lakes. DOC loading is particularly responsive to climate change, through the leaching and erosion of soils (compounded by permafrost melting in the Polar Regions), changes in microbial transformation rates of particulate and dissolved organics, and through changes in catchment vegetation. The effects of the latter are apparent from high-latitude paleolimnological studies, which indicate an increase in DOC concentrations associated with colonization of the catchments by woody shrubs and trees during periods of climate warming. The increased transport of the colored fraction of this DOC (colored dissolved organic matter, CDOM) can have both positive and negative effects on aquatic photosynthesis, by shading cells against damaging UVR and by absorbing PAR that would otherwise be available for photosynthesis.

Biological Effects of Climate Change

The biological responses to the physical and chemical changes noted here involve multiple interactions, feedbacks and complex nonlinear responses that are not possible to fully predict on the basis of current knowledge. However, a great variety of direct and indirect effects can be identified as some of the likely impact pathways of ongoing climate change. These operate at multiple scales, from changes in whole ecosystem structure and dynamics, down to physiological and molecular responses at the individual and cellular level that in turn feed back into overall ecosystem dynamics.

Ecosystem Integrity

At the broadest scale, climate change has the potential to radically alter the physical structure of lake ecosystems, and thereby cause the extinction or alteration of aquatic biota. The melting of the Siberian permafrost is one of the most striking examples, where 11% of thaw ponds in a large study area have recently drained and been lost, thereby eliminating the habitat for aquatic biological communities. The contraction of large shallow lakes also results in major habitat loss, and even deep lakes may lose important ecological features as a result of relatively small fluctuations in water level. For example, the coastal wetlands of the North American Great Lakes are important for migratory birds, and the majority of species of fish in the region depend on these littoral and supralittoral environments for reproduction. Small changes in lake levels as a result of shifting water balance make these environments especially vulnerable to climate change. Changes in the connectivity between aquatic habitats as a result of rising or falling water levels will also influence the species composition of aquatic biota, especially fish communities.

Vertical Habitat Structure

Changes in ice cover, stratification, and mixing, and thereby the vertical gradients in lake properties, will have far reaching effects on pelagic communities and production. The associated changes in light and nutrients will affect phytoplankton production, and therefore the availability of food to high trophic levels. For some high-latitude systems, decreased ice cover and stronger stratification may allow for increased irradiance supply and rates of primary production. For example, the strong climate warming at Signy Island in the maritime Antarctic has been accompanied by increased overwintering phytoplankton biomass. This was attributed to reductions in lake snow and ice cover and improved light conditions for summer growth, as well as increased nutrient inputs following deglaciation of the catchments. In contrast to this response, simulation analyses for the North American Great Lakes have shown negative effects of projected climate change. Earlier summer stratification results in a decreased period of mixing of nutrients from the sediments and from deeper waters into the euphotic zone, and therefore increased nutrient limitation of algal growth. These effects are compounded by projected increases in cloud cover, which would reduce the availability of light for photosynthesis.

In Lake Tanganyika, the present trend of climate warming, decreased wind speeds, and increased water column stability appears to have resulted in decreased nutrient supply to the euphotic zone by mixing, resulting in a 20% decrease in phytoplankton production. The consequences of this effect are serious given that a large portion of animal protein available to the surrounding human populations is derived from the pelagic fishery, and this decrease in algal photosynthetic rates is estimated to have caused a 30% drop in fish yield.

A prolonged ice-free period in northern, temperate lakes, and therefore a decreased duration of winter ice cover, may prove to be advantageous for some fish species. In certain locations (e.g., small, enriched lakes in southern Canada and northern United States), prolonged winter stratification leads to complete oxygen depletion of the hypolimnion, resulting in the "winter kill" of fish stocks, and such effects may lessen as the period of ice cover contracts.

Photosynthetic Communities

Changes in light and nutrient availability that accompany climate change will also have a qualitative effect on species composition and diversity at the primary producer level, which in turn may impact on higher trophic levels. Paleolimnological studies have shown that the diatom composition of the periphyton and phytoplankton in ponds at many sites in the circumpolar Arctic has changed markedly over the last 150 years, consistent with a strong response to decreased ice cover and possibly increased nutrient supply. In ancient Lake Baikal, there is evidence that endemic, cold-dwelling diatom taxa that support the current food-web will be replaced by faster growing, nondiatom species, including picophytoplankton. Changes in lake level are likely to affect the species composition of aquatic macrophyte communities in the littoral zone of lakes, with implications for the birds, fish, and other biota that depend on these plants.

A major climate change concern at temperate latitudes is the prospect of a shift in phytoplankton species composition toward dominance by species of cyanobacteria that form noxious blooms. These organisms create many water quality problems in lakes and reservoirs, including the release of taste and odor compounds, the production of various toxins, and the overproduction of biomass that clogs water filters, disrupts zooplankton feeding, and causes oxygen depletion. Bloom-forming cyanobacteria are likely to be favored in a warming climate by several mechanisms. First, their temperature for optimum growth tends to be high (25 °C or above), and warmer conditions will favor their more rapid accumulation and dominance. Second, these species can regulate their position in stable water columns by way of gas vacuoles that allow them to sink or rise to the optimal depths for photosynthesis or nutrient uptake. As noted above, future climate change scenarios predict increased stratification and water column stability that would favor this strategy. Third, a warmer climate may lead to increased anoxia and internal phosphorus loading (perhaps also in external loading in some circumstances), and bloom-forming cyanobacteria tend to become increasingly prevalent with increasing degree of phosphorus enrichment. Finally, a highly toxic species from tropical and subtropical regions, *Cylindrospermopsis raciborskii*, has been increasingly observed in temperate lakes. This apparent invasion has been ascribed to climate change, with warmer temperatures favoring the early germination of its resting spores (akinetes).

Fish Communities and Migration

Many fish species are sensitive to even small changes in the temperature of their surroundings, and the warming trend associated with climate change is likely to cause a shift in the geographic distribution of many taxa. Cold water species such as walleye and trout may be driven to extinction in some lake waters because their thermal tolerance is exceeded, while other species may be able to colonize new habitats as a result of warmer conditions. In Toolik Lake, Alaska, a bioenergetics analysis showed that a 3 °C rise in summer lake water temperature could cause the young-of-the-year lake trout to consume eight times more food that is currently necessary to maintain their condition, which greatly exceeds the present food availability. In Ontario lakes, the northern range of smallmouth bass is currently limited by temperature because the length of the growing season determines the winter survival of the young-of-the-year. Under a climate change scenario, this species could invade some additional 25 000 northern lakes, and because of its strongly negative effects on other fish, such invasions would likely cause the extirpation from these lakes of four native cyprinid species.

Climate change may affect the migratory behavior of some fish species. Arctic char for example migrate from lakes to the sea where the higher food availability allows them to grow to a large size. If arctic lakes become more productive as a result of climate change, then char migrations are projected to decrease, and the resultant lake char would be of smaller size and less suitable as a food resource for northern aboriginal communities.

Microbial Processes

Many lakes are net heterotrophic, with the respiration (R) of organic materials derived from autochthonous (within the lake) and allochthonous (outside the lake) sources exceeding photosynthesis (P). Heterotrophic microbial activity is likely to be enhanced in a warmer climate because of the direct physiological effects of temperature, and in some environments by the increased availability of organic carbon and nutrients, e.g., from increased catchment runoff. Photosynthesis on the other hand is strongly regulated by light availability and may be less responsive to the temperature increase, or even negatively affected via the mechanisms noted earlier. This combination of effects would exacerbate the negative P/R balance in lakes, and may result in lakes becoming even greater net sources of CO_2 to the atmosphere. Production of the more powerful greenhouse gas methane will be favored by prolonged anoxia in some environments, and in high northern latitudes by its release from permafrost surrounding thaw ponds. The production and release of greenhouse gases from arctic thaw lakes has been identified as a potentially important feedback mechanism that will accelerate climate warming.

See also: Eutrophication.

Further Reading

Austin JA and Colman SM (2007) Lake Superior summer water temperatures are increasing more rapidly than regional air temperatures: A positive ice-albedo feedback. *Geophysical Research Letters* 34: L06604, doi:10.1029/2006GL029021.

Brooks AS and Zastrow JC (2002) The potential influence of climate change on offshore primary production in Lake Michigan. *Journal of Great Lakes Research* 28: 597–607.

Coe MT and Foley JA (2001) Human and natural impacts on the water resources of the Lake Chad basin. *Journal of Geophysical Research Atmospheres* 106(D4): 3349–3356.

Jackson DA and Mandrak NE (2002) Changing fish biodiversity: Predicting the loss of cyprinid biodiversity due to global climate change. In: McGinn NA (ed.) *Fisheries in a Changing Climate*, pp. 89–98. Bethesda, Maryland, USA: American Fisheries Society.

Korhola A, Sorvari S, Rautio M, *et al.* (2002) A multi-proxy analysis of climate impacts on the recent development of subarctic Lake Sannajaärvi in Finnish Lapland. *Journal of Paleolimnology* 28: 59–77.

Lemke P, Ren J, Alley RB, *et al.* (2007) Observations: Changes in snow, ice and frozen ground. In: Solomon S, Qin D, Manning M, *et al.* (eds.) *Climate Change 2007: The Physical Science Basis. Contribution of Working Group I to the Fourth Assessment Report of the Intergovernmental Panel on Climate Change.* Cambridge, UK: Cambridge University Press.

Mackay AW, Ryves DB, Morley DW, Jewson DH, and Rioual P (2006) Assessing the vulnerability of endemic diatom species in Lake Baikal to predicted future climate change: a multivariate approach. *Global Change Biology* 12: 2297–2315. doi: 10.1111/j.1365-2486.2006.01270.x.

McDonald ME, Herschey AE, and Miller MC (1996) Global warming impacts on lake trout in Arctic lakes. *Limnology and Oceanography* 41: 1102–1108.

McKnight D, Brakke DF, and Mulholland PJ (eds.) (1996) Freshwater ecosystems and climate change in North America. *Limnology and Oceanography* 41(special issue): 815–1149.

Mueller DR, Vincent WF, and Jeffries MO (2003) Break-up of the largest Arctic ice shelf and associated loss of an epishelf lake. *Geophysical Research Letters* 30: 2031 doi:10.1029/2003GL017931.

O'Reilly CM, Alin SR, Plisnier PD, Cohen AS, and McKee BA (2003) Climate change decreases aquatic ecosystem productivity of Lake Tanganyika, Africa. *Nature* 424: 766–768.

Prowse TD, Wrona FJ, Reist JD, Hobbie JE, Lévesque LMJ, and Vincent WF (2006) Climate change effects on hydroecology of arctic freshwater ecosystems. *Ambio* 35: 347–358.

Quayle WC, Peck LS, Peat H, Ellis-Evans JC, and Harrigan PR (2002) Extreme responses to climate change in Antarctic lakes. *Science* 295: 645.

Quesada A, Vincent WF, Kaup E, *et al.* (2006) Landscape control of high latitude lakes in a changing climate. In: Bergstrom D, Convey P, and Huiskes A (eds.) *Trends in Antarctic Terrestrial and Limnetic Ecosystems*, pp. 221–252. Dordrecht: Springer.

Schindler DW and Smol JP (2006) Cumulative effects of climate warming and other human activities on freshwaters of arctic and subarctic North America. *Ambio* 35: 160–168.

Smol JP and Douglas MSV (2007) Crossing the final ecological threshold in high arctic ponds. *Proceedings of the National Academy of Sciences USA* 104: 12395–12397.

Smith LC, Sheng Y, MacDonald GM, and Hinzman LD (2005) Disappearing Arctic lakes. *Science* 308: 1429. doi: 10.1126/science.1108142.

Vincent WF, Rautio M, and Pienitz R (2007) Climate control of underwater UV exposure in polar and alpine aquatic ecosystems. In: Orbaek JB, Kallenborn R, Tombre I, Hegseth E, Falk-Petersen A, and Hoel AH (eds.) *Arctic Alpine Ecosystems and People in a Changing Environment*, pp. 227–249. Berlin: Springer.

Walter KM, Zimov SA, Chanton JP, Verbyla D, and Chapin FS (2006) Methane bubbling from Siberian thaw lakes as a positive feedback to climate warming. *Nature* 443: 71–75.

Wiedner C, Rücker J, Brüggemann R, and Nixdorf B (2007) Climate change affects timing and size of populations of an invasive cyanobacterium in temperate regions. *Oecologia* 152: 473–484. doi: 10.1007/s00442-007-0683-5.

Eutrophication

V H Smith, University of Kansas, Lawrence, KS, USA

Introduction

Fluxes of nutrients from the landscape into receiving waters vary greatly over both time and space, depending upon regional soil composition and texture, geomorphology, vegetation type and cover, and land use. These nutrient fluxes have been greatly enhanced by human activities, as a consequence of rapid population growth and major technological achievements occurring since the Industrial Revolution. Eutrophication, or overenrichment with nutrients, is an environmental issue of concern for wetlands, streams, rivers, lakes, and reservoirs worldwide. Excess nutrient inputs have been found to cause a variety of effects, both desirable and undesirable, on the structure and function of surface waters in every continent of the globe.

Origins and Evolution of the Eutrophication Concept

More than 160 years ago, pioneering research by the German agricultural chemist Justus von Liebig established that the growth of agricultural crops was proportional to the supply rates of key nutrients, including nitrogen (N) and phosphorus (P). Liebig's Law of the Minimum states that the local yield of terrestrial plants should be limited by the nutrient that is present in the environment in the least quantity relative to its demands for plant growth, and this statement has been confirmed worldwide. This Law also provides a conceptual foundation for the response of aquatic ecosystems to nutrient enrichment: surface waters receiving low inputs of nitrogen and phosphorus are typically very unproductive, whereas highly nutrient-enriched waters typically exhibit abundant growth of aquatic plants.

A remarkably consistent response of inland waters to nutrient enrichment has been observed worldwide. On the basis of pioneering research on European bogs, German and Swedish scientists in the early 20th Century began to categorize water bodies by their relative fertilities. Their careful analyses of diverse European lake districts revealed that phytoplankton production was determined primarily by the availability of nitrogen and phosphorus; that regional variations in lake productivity correlated well with the underlying geology of the lakes' watersheds; and that the lake's productivity level also influenced the biology of the lake as a whole. These analyses ultimately led to global acceptance of the Trophic State Concept, which places surface waters into a series of categories on the basis of their relative supplies of nutrients. These trophic states range from oligotrophic (poorly fed, or poorly nourished) to eutrophic (well-fed, or well-nourished) systems. Systems receiving intermediate levels of nutrients are termed mesotrophic, and eutrophication is the process by which receiving waters become more eutrophic through increases in their overall rate of nutrient supply from all internal and external sources. Although these trophic state categories have most frequently been used to describe freshwater lakes, they can also be applied to reservoirs and flowing waters (**Tables 1–3**).

Current Understanding and Management of Eutrophication

Our knowledge of eutrophication, its consequences and its management has expanded markedly (**Table 2**): (1) the principal sources of nutrient loading to receiving waters have been identified and quantified; (2) conceptual and empirical models have been developed that can be used to predict nutrient concentrations in receiving waters from nutrient inputs and basin-specific hydrology; (3) a firm mechanistic understanding of the relative roles of N and P as growth-limiting nutrients has evolved; (4) a sophisticated set of models has been developed that provides quantitative linkages between nutrient availability and critical aspects of water quality recognized by the public as being valuable and worthy of protection; (5) a sophisticated understanding has emerged of the physical, chemical, and biological factors that moderate the local expression of eutrophication; and (6) an extensive set of tools has been developed that can be used to help manage or restore many of the consequences of eutrophication, once they occur. Our knowledge base is most advanced for lakes and reservoirs; however, the development of eutrophication science for fluvial and wetland systems is advancing rapidly as well.

Sources of Nutrients to Inland Surface Waters

Nutrient supplies to inland waters are typically derived from the following primary sources:

1. geological weathering, which generates natural exports of nutrients from the landscape in which the water body is embedded;

Table 1 Water quality characteristics of four trophic state categories in lakes

Trophic state category	*Total nitrogen* ($\mu g\ l^{-1}$)	*Total phosphorus* ($\mu g\ l^{-1}$)	*Chlorophyll* a ($\mu g\ l^{-1}$)	*Secchi disk transparency (m)*
Oligotrophic	<350	<10	<3.5	>4
Mesotrophic	350–650	10–30	3.5–9	2–4
Eutrophic	650–1200	30–100	9–25	1–2
Hypertrophic	>1200	>100	>25	<1

The terms oligotrophic, mesotrophic, and eutrophic correspond to systems receiving low, intermediate, and high nutrient inputs, respectively. Hypertrophic is the term used for systems receiving greatly excessive nutrient inputs.

Source

1. Nürnberg G (1996) Trophic state of clear and colored, soft- and hard-water lakes with special consideration of nutrients, anoxia, phytoplankton and fish. *Lake and Reservoir Management* 12: 432–447.

Table 2 Water quality characteristics of three trophic state categories in temperate streams and rivers

Trophic state category	*Total Nitrogen* ($\mu g\ l^{-1}$)	*Total Phosphorus* ($\mu g\ l^{-1}$)	*Mean benthic Chlorophyll* a ($mg\ m^{-2}$)	*Maximum benthic Chlorophyll* a ($mg\ m^{-2}$)	*Mean suspended Chlorophyll* a ($\mu g\ l^{-1}$)
Oligotrophic	<700	<25	<20	<60	<10
Mesotrophic	700–1500	25–75	20–70	60–200	10–30
Eutrophic	>1500	>75	>70	>200	>30

Source

1. Dodds WK, Jones JR, and Welch EB (1998) Suggested criteria for stream trophic state: Distributions of temperate stream types by chlorophyll, total nitrogen and phosphorus. *Water Research* 32: 1455–1462.

Table 3 Timeline of key events and advances in the eutrophication of inland waters

Event	*Source*
Development of the flush toilet	1
Liebig's Law of the Minimum	2
Initial links between plant nutrient availability and aquatic production	3,4
First nutrient mass balance for an aquatic ecosystem	4
Origins of trophic state classification systems for European fresh waters	5
Establishment of critical N and P concentrations for algal blooms in lakes	6
Development and use of phosphate-based laundry detergents	7
Origins of stoichiometric views of biological chemistry in aquatic systems	8
Empirical evidence for interactions between N and P as limiting nutrients	9
First OECD synthesis report on eutrophication	10
U.S. National Academy of Sciences report on eutrophication	11
First special ASLO symposium on eutrophication	12
Confirmation of the pivotal role of P in freshwater eutrophication	13
Development and refinement of empirical eutrophication models	14,15,16,17,18
Development of quantitative trophic state indices	19
Evidence of the existence of alternative stable states	20
Evidence that food web structure alters responses to eutrophication	21,22,23
Second OECD symposium on eutrophication	24
OECD summary report	25
Second special ASLO symposium on eutrophication	26

Sources

1. Horn LJ (1996) *The porcelain god: A social history of the toilet*. London: Robson Books.
2. von Liebig J (1855) Principles of agricultural chemistry with special reference to the late researches made in England. In Pomeroy LR (ed.), *Cycles of essential elements. Benchmark papers in Ecology*, vol. 1, pp 11–28. London: Dowden, Hutchinson, and Ross.
3. Brandt K (1901) Life in the ocean. In *Annual report of the Board of Regents of the Smithsonian Institution for the year ending June 30, 1900*, pp. 493–506. Washington, DC: Smithsonian.
4. Johnstone J (1908) *Conditions of life at sea*. Cambridge, UK: Cambridge University Press (Reprinted in 1977 by Arno Press, New York).
5. Carlson RE and Simpson JT (1995) Trophic state. In *A coordinator's guide to volunteer lake monitoring methods*, pp. 73–92. Madison, WI: North American Lake Management Society.
6. Sawyer CN (1947) Fertilization of lakes by urban drainage. *Journal of the New England Water Works Association* 61: 109–127.
7. Lief A (1958) *It Floats: The story of Procter and Gamble*. New York: Rinehart.
8. Redfield AC (1958) The biological control of chemical factors in the environment. *American Scientist* 46: 205–222.

9. Sakamoto M (1966) Primary production by phytoplankton community in some Japanese lakes and its dependence on lake depth. *Archiv für Hydrobiologie* 62: 1–28.
10. Vollenweider RA (1968) *The scientific fundamentals of lake and stream eutrophication, with particular reference to phosphorus and nitrogen as eutrophication factors*. Tech. Rep. DAS/DSI/68.27. Paris: Organisation for Economic and Cooperative Development.
11. National Academy of Sciences (1969) *Eutrophication: Causes, Consequences, Correctives*. Washington, DC: U.S. National Academy of Sciences.
12. Likens GE (ed.) (1972) *Nutrients and Eutrophication*. American Society of Limnology and Oceanography Special Symposium 1. Lawrence, KS: Allen Press.
13. Schindler DW, Brunskill GJ, Emerson E, *et al.* (1972) Atmospheric carbon dioxide: Its role in maintaining phytoplankton standing crops. *Science* 177: 1192–1194.
14. Dillon PJ and Rigler FH (1974) The phosphorus–chlorophyll relationship in lakes. *Limnology and Oceanography* 19: 767–773.
15. Dillon PJ and Rigler FH (1975) A simple method for predicting the capacity of a lake for development based on lake trophic state. *Journal of the Fisheries Research Board of Canada* 31: 1518–1531.
16. Jones JR and Bachmann RW (1976) Prediction of phosphorus and chlorophyll levels in lakes. *Journal of the Water Pollution Control Federation* 48: 2176–2182.
17. Vollenweider RA (1976) Advances in defining critical loading levels for phosphorus in lake eutrophication. *Memorie dell'Istituto Italiano di Idrobiologia* 33: 53–83.
18. Nürnberg GK (1984) The prediction of internal phosphorus load in lakes with anoxic hypolimnia. *Limnology and Oceanography* 29: 111–124.
19. Carlson RE (1977) A trophic state index for lakes. *Limnology and Oceanography* 22: 361–369.
20. Holling CS (1973) Resilience and stability of ecological systems. *Annual Review of Ecology and Systematics* 4: 1–23.
21. Hrbáček J, Dvořáková M, Kořínek M, and Procházková L (1961) Demonstration of the effect of the fish stock on the species composition of zooplankton and the intensity of metabolism of the whole plankton association. *Verhandlungen der Internationale Vereinigung für Theoretische und Angewandte Limnologie* 14: 192–195.
22. Shapiro J (1979) The need for more biology in lake restoration. In *Lake Restoration*, US EPA 440/5-79-001, pp. 161–167. Washington, DC: Environmental Protection Agency.
23. Carpenter SR, Kitchell JF, and Hodgson JR (1985) Cascading trophic interactions and lake productivity. *BioScience* 35: 634–639.
24. U.S. EPA. 1981. *Restoration of inland lakes and waters*. EPA 440/5–81–010. Washington, DC: U.S. Environmental Protection Agency.
25. OECD 1982. *Eutrophication of Waters: Monitoring, Assessment and Control*. Paris: Organisation for Economic and Cooperative Development.
26. Joye SB, Smith VH, Howarth RW, *et al.* (eds.) (2006) Eutrophication of freshwater and marine ecosystems. *Limnology and Oceanography* 51: (1, part 2), 351–800.

2. enhanced nonpoint exports of nutrients from the landscape, which stem from human activities, including land clearing, manipulation of vegetation cover, farming, and the manuring or application of commercial fertilizers to terrestrial vegetation;
3. point source exports of nutrients from human sources, which include raw sewage, treated wastewater effluent, and treated or untreated industrial effluents;
4. transport of these nutrients directly via overland and subsurface water flows into wetlands, stream channels, or lake or reservoir basins;
5. delivery of additional nutrients directly to the surface of the water body from both particulate and gaseous atmospheric sources;
6. internal regeneration of nutrients (= internal loading) through the biological activities of microbes and other living organisms; and
7. internal regeneration of nutrients (internal loading), which is caused by shifts in the oxidation–reduction potential of the underlying soils and sediments within the water body.

The relative magnitude of each of these nutrient sources varies over both time and space. As a result, there is remarkable heterogeneity in the fertility of surface waters located across the globe. For example, wetlands, rivers, and lakes embedded in a local geology dominated by granitic rock will naturally be oligotrophic; examples of these include waters located in the Canadian Shield of North America, and Lake Baikal (Siberia). Conversely, water bodies embedded in landscapes dominated by nutrient-rich marine and fluvial sediments may become naturally eutrophic even in the absence of human-associated inputs; examples include some of the shallow, fertile water bodies found in central and western Canada, and in the Netherlands.

However, even in geologically nutrient poor landscapes, N and P exports derived directly or indirectly from human activities can cause rapid and sometimes extreme eutrophication. Similarly, naturally eutrophic lakes can be fertilized further by human-derived point and nonpoint nutrient sources. This acceleration of nutrient inputs to surface waters is termed cultural eutrophication.

Cultural eutrophication most likely began to occur with increasing frequency and intensity worldwide as humans moved from nomadic to more agricultural behavior, especially once manure and other fertilizers began to be applied to the landscape. Evidence from the paleolimnological literature suggests significant human disturbances of local surface water quality occurred even in Neolithic and Bronze age times. However, the pace of cultural eutrophication was accelerated further by the advent of the flush toilet, when human wastes began to be routed directly into receiving waters. Although its invention has been generally attributed to John Harrington (*c.* 1596, in Great Britain), archaeological evidence suggests that sitting toilets incorporating flowing water may have been used more than 2000 years ago in both ancient Egypt and China. Modern-day examples of cultural eutrophication include Lake Washington, USA (see **Figure 1**) and the lower Laurentian Great Lakes (Canada/USA) in the 1970s.

Nitrogen and Phosphorus are Essential Growth-Limiting Nutrients

Although a brief but sharp controversy arose in the 1970s about the relative roles of carbon, nitrogen, and phosphorus as drivers of the eutrophication process, experimental research on Canadian lakes demonstrated unequivocally that N and P are the primary growth-limiting nutrients for algae and vascular plants (**Table 3**). Both these mineral elements are essential components of plant tissues; and, consistent with Liebig's Law of the Minimum, both N and P are typically present in low supplies relative to the metabolic demands of aquatic primary producers. In the 1960s, careful comparative studies of Japanese lakes revealed significant interactions between these two nutrients (**Table 3**); however nitrogen limitation of summer algal biomass was found to be most likely when the total nitrogen:total phosphorus (TN:TP) ratio in the water column of a lake was less than 10:1 by weight. Algal biomass was limited by phosphorus when the TN:TP ratio was greater than 17 by mass, but either N or P could limit algal growth at intermediate TN:TP ratios. Although debates about the exact values of these critical ratios may continue for some time to come, the assessment of nutrient limitation based upon N:P ratios has been broadly accepted by investigators worldwide.

Of these two key limiting nutrients, phosphorus does not have a significant gaseous phase; in contrast, 78% of the Earth's atmosphere by volume is dinitrogen gas (N_2). Under conditions of nitrogen limitation, dissolved N_2 is available in almost limitless supply to support the growth of phytoplankton species that are able to fix molecular nitrogen (cyanobacteria), which can lead to toxic blooms and other cyanobacteria-associated water quality problems in many P-enriched surface waters (see later). The careful management and control of P loading thus has been adopted worldwide as the keystone of eutrophication management in inland waters.

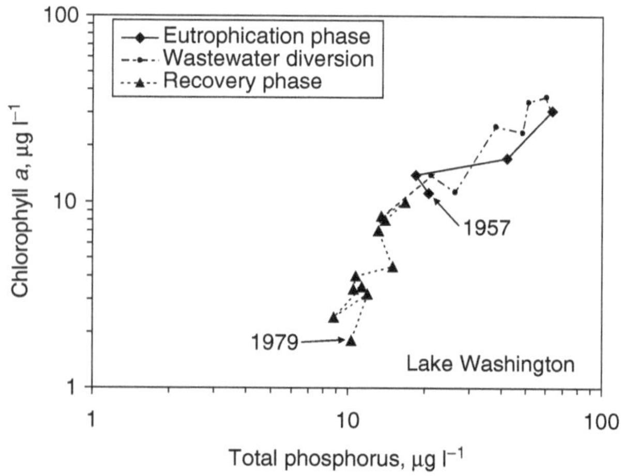

Figure 1 Time trends in the mean concentrations of total phosphorus and phytoplankton biomass (estimated as the photosynthetic pigment chlorophyll *a*) during the eutrophication and subsequent restoration of Lake Washington (USA). Lake Washington received wastewater effluents from Seattle, Washington, for many decades until these nutrient inputs were diverted in a stepwise series of stages during 1963–1968. Data from Edmondson WT and Lehman JT (1981) The effect of changes in the nutrient income on the condition of Lake Washington. *Limnology and Oceanography* 26:1–29.

Nutrient Loading Models

The inputs of biologically available nitrogen and phosphorus to most inland waters are derived primarily from atmospheric deposition, fluvial inputs, and groundwater inflows. The sum of these three external inputs is termed the external load. In addition to the external load, biologically-driven internal sources (see *Sources of Nutrients to Inland Surface Waters* section) may contribute significantly to nutrient fluxes to inland waters. However, both the absolute and the relative loads of N and P often differ significantly among water bodies, and the loadings of N and P to individual water bodies also may change over time as a result of natural climatic variability and changes in local or regional human activities.

To forecast the effects of eutrophication on receiving waters, it is first necessary to be able to predict how water column nutrient concentrations will change when the magnitudes of N and P inputs are altered. This major advance in eutrophication modeling (**Table 2**) stemmed from the successful merger of chemical mass-balance concepts with the fundamental principles of hydrology. Central to these predictive models is a calculation of quantitative input–output budgets for both nitrogen and phosphorus, and the calculation of a similar budget for water. In lakes and reservoirs, nutrient losses to outflowing water, and nutrient losses due to gravitational sedimentation from the water column, both correlate strongly with the hydraulic retention time of the water body; these nutrient loss terms are typically estimated empirically using details from the lake's water budget.

Effects of Eutrophication on Water Quality

Just as fertilizing land typically stimulates vascular plant growth, the most common effect of increasing the N and P supplies to aquatic ecosystems is an increase in the abundance of algae and aquatic vascular plants. For example, the biomass of phytoplankton algae depends strongly upon the concentration of total phosphorus in lakes and reservoirs located on every major continent of the world (**Figure 2**); a similar nutrient stimulation of algal biomass and primary productivity has been found in wetlands, streams, and rivers.

However, the consequences of eutrophication are much diverse and troublesome than an overgrowth of aquatic primary producers. Eutrophication can cause a wide array of symptoms and environmental costs that are either directly or indirectly related to the nuisance growth of aquatic plants (**Table 4**).

Many of the symptoms of eutrophication are highly undesirable for both recreational and commercial users of the affected water body. For example, consumers of drinking water from eutrophic supply reservoirs are much more likely to incur high water treatment costs. In addition, eutrophic drinking water supplies may lead to a greater likelihood of consumer complaints owing to taste and odor problems (**Figure 3**); similarly, eutrophic lakes may exhibit high concentrations of biologically produced toxins in the water column (**Figure 4**). Both these major water quality problems are frequently associated with the eutrophication-enhanced proliferation of nuisance cyanobacteria (blue-green algae), which can occur in fluvial systems as well as in lakes and reservoirs.

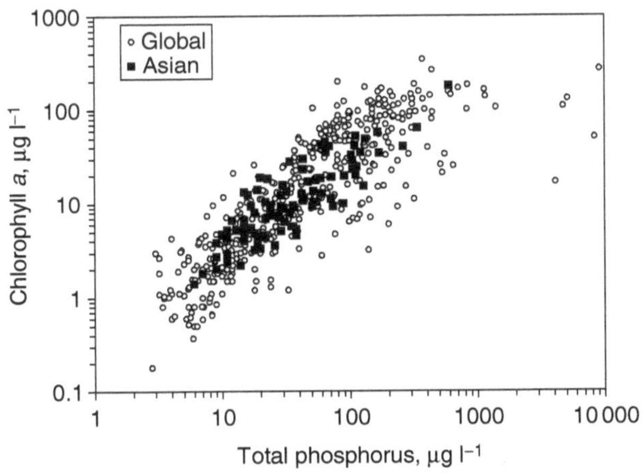

Figure 2 Positive relationship between mean phytoplankton biomass and mean total phosphorus in more than 500 lakes and reservoirs located worldwide. Global non-Asian data (open circles) are from Smith (1982) *Limnology and Oceanography* 27:1101–1112, and from Prairie YT, Duarte CM, and Kalff J (1989) Unifying nutrient–chlorophyll relationships in lakes. *Canadian Journal of Fisheries and Aquatic Sciences* 46:1176–1182. Asian data (closed squares) are from Jones JR, Perkins, BD, Witt A Jr., *et al.* (2000) Limnological characteristics of some reservoirs in Thailand. *Verhandlungen der Internationale Vereinigung für Theoretische und Angewandte Limnologie* 27: 2158–2166; Jones JR, Knowlton MF, and An KG (2003) Trophic state, seasonal patterns and empirical models in South Korean reservoirs. *Lake and Reservoir Management* 19: 64–78; and Kim B, Park, J-H, Hwang G, *et al.* (2001) Eutrophication of reservoirs in South Korea. *Limnology* 2: 223–229.

Factors that Modify the Local Expression of Eutrophication

It has been known for more than a century that shallow water bodies often tend to be much more productive than deeper waters that may be located nearby; this variability in productivity often reflects the relative volume of water that is available within the lake basin to dilute incoming nutrients. However, a diverse set of additional physical, chemical, and biological factors can potentially modify the local expression of nutrient loading (**Table 5**).

Effects of salinity Even among lakes of similar size and fertility, the presence of high salinity levels often tends to reduce phytoplankton production relative to the values expected from total phosphorus concentrations in the water column, causing deviations from the general relationships that have been observed in most freshwater lakes and reservoirs (**Figures 1** and **2**).

Effects of inorganic turbidity and dissolved organic color Much of our understanding of the relationships between nutrients and primary production has been derived from studies of clear natural lakes. However, highly erodable landscapes are distributed worldwide, and high concentrations of suspended soils often accompany nutrients exported in terrestrial runoff from these systems. Similarly, many landscapes export high concentrations of plant-derived dissolved organic matter (including humic and fulvic acids), which can impart a very strong tea-colored stain to receiving waters. Inorganic turbidity or such coloration of waters can greatly restrict the amount of subsurface light available to support plant growth in rivers, lakes, and reservoirs. For example, in waters with high mineral turbidity such as those found in some regions of Australia, the Americas, Europe, and Asia, phytoplankton biomass may be depressed as much as 90% from the values expected from measured nutrient concentrations. Although perhaps lower in absolute magnitude, reductions in primary production can also sometimes be seen in highly colored surface waters worldwide.

Table 4 Effects of eutrophication on inland waters

Effect	*Source*
Increases in the productivity and biomass of phytoplankton and suspended algae in lakes, reservoirs, and rivers	1,2,3
Shifts in algal composition, often to nuisance bloom-forming species, many of which may be toxic or which may not be consumed effectively by aquatic grazers	4,5
Increases in the productivity and biomass of attached microalgae (periphyton)	6,7
Shifts in periphyton composition, often to less desirable species	7,8
Changes in the productivity and biomass of aquatic vascular plants	4,9,10
Increases in the productivity and biomass of aquatic consumers	11,12
Shifts in both invertebrate and vertebrate consumer species composition, often to less desirable species	4,13
Increases in the probability of summer and winter fish kills	14,15
Decreases in water transparency	1,2,16
Increases in the probability and the intensity of taste, odor, and water filtration problems in surface water supplies	17
Depletion of deep water oxygen	1,4,18
Accumulation of dissolved nutrients in anoxic deep waters	19,20
Decreases in the perceived aesthetic value of the water body	21
Negative economic impacts, including decreased property values and reduced recreational uses	22, 23
Changes in the processing and fates of nonnutrient contaminants	24
Possible increases in disease risk and intensity	25

Sources

1. OECD (1982) *Eutrophication of Waters: Monitoring, Assessment, and Control*. Paris: Organisation for Economic Cooperation and Development.
2. Jones JR and Knowlton MF (1993) Limnology of Missouri reservoirs: An analysis of regional patterns. *Lake and Reservoir Management* 8: 17–30.
3. Van Nieuwenhuyse EE and Jones JR (1996) Phosphorus–chlorophyll relationship in temperate streams and its variation with stream catchment area. *Canadian Journal of Fisheries and Aquatic Sciences* 53: 99–105.
4. Smol JP (2002) *Pollution of Lakes and Rivers: A Paleoenvironmental Perspective*. London: Arnold.
5. Paerl HW, Fulton RS, III, Moisander PH, and Dyble J (2001) Harmful freshwater algal blooms, with an emphasis on Cyanobacteria. *The Scientific World JOURNAL* 1: 76–113.
6. Biggs BJF (2000) Eutrophication of streams and rivers: Dissolved nutrient–chlorophyll relationships for benthic algae. *Journal of the North American Benthological Society* 19: 17–31.
7. Dodds WK and Welch EB (2000) Establishing nutrient criteria in streams. *Journal of the North American Benthological Society* 19: 186–196.
8. DeNicola DM, de Eyto E, Wimaere A, and Irvine K (2004) Using epilithic algal communities to assess trophic status in Irish lakes. *Journal of Phycology* 40: 481–495.
9. Sand-Jensen K, Riis T, Vestergaard O, and Larsen SE (2000) Macrophyte decline in Danish lakes and streams over the last 100 years. *Journal of Ecology* 88: 1030–1040.
10. Hilt S, Gross EM, Hupfer M, *et al.* (2006) Restoration of submerged vegetation in shallow eutrophic lakes—A guideline and state of the art in Germany. *Limnologica* 36: 155–171.
11. McCauley E and Kalff J (1981) Empirical relationships between phytoplankton and zooplankton biomass in lakes. *Canadian Journal of Fisheries and Aquatic Sciences* 38: 458–463.
12. Bachmann RW, Jones BL, Fox DD, *et al.* (1996) Relations between trophic state indicators and fish in Florida (U.S.A.) lakes. *Canadian Journal of Fisheries and Aquatic Sciences* 53: 842–855.
13. Tammi J, Lappalainen A, Mannio J, *et al.* (1999) Effects of eutrophication on fish and fisheries in Finnish lakes: A survey based on random sampling. *Fisheries Management and Ecology* 6: 173–186.
14. Barica J (1975) Summer kill risk in prairie ponds and possibilities of its production. *Journal of the Fisheries Research Board of Canada* 32: 1283–1288.
15. Fang X and Stefan H (2000) Projected climate change effects on winterkill in shallow lakes in the northern United States. *Environmental Management* 25: 291–304.
16. Edmondson WT (1972) Nutrients and phytoplankton in Lake Washington. American Society of Limnology and Oceanography Special Symposium 1, pp. 172–193. Lawrence, KS: Allen Press.
17. Watson S (2004) Aquatic taste and odor: A primary signal of drinking-water integrity. *Journal of Toxicology and Environmental Health Part A* 67: 1779–1795.
18. Cornett RJ and Rigler FH (1979) Hypolimnetic oxygen deficits: Their prediction and interpretation. *Science* 205: 580–581.
19. Nürnberg G (1997) Coping with water quality problems due to hypolimnetic anoxia in Central Ontario Lakes. *Water Quality Research Journal of Canada* 32: 391–405.
20. Pettersson K (1998) Mechanisms for internal loading of phosphorus in lakes. *Hydrobiologia* 373/374: 21–25.
21. Smeltzer E and Heiskary SA (1990) Analysis and applications of lake user survey data. *Lake and Reservoir Management* 6: 109–118.
22. Pretty JN, Mason CF, Nedwell DB, *et al.* (2003) Environmental costs of freshwater eutrophication in England and Wales. *Environmental Science and Technology* 37: 201–208.
23. Carpenter SR, Bolgrien D, Lathrop RC, *et al.* (1998) Ecological and economic analysis of lake eutrophication by nonpoint pollution. *Australian Journal of Ecology* 23: 68–79.
24. Berglund O (2003) Periphyton density influences organochlorine accumulation in rivers. *Limnology and Oceanography* 48: 2106–2116.
25. Johnson PTJ and Carpenter SR (2007) Influence of eutrophication on disease in aquatic ecosystems: patterns, processes, and predictions. In Ostfeld R, Keesing F, and Eviner V (eds.), *Reciprocal Interactions between Ecosystems and Disease*. Princeton, NJ: Princeton University Press (in press).

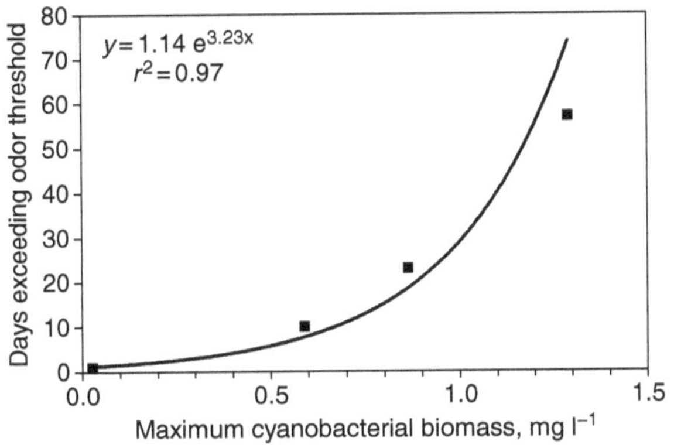

Figure 3 Relationship between annual maximum cyanobacterial biomass and the number of days that the water odor exceeded thresholds for objectionable drinking water quality in Saginaw Bay, Lake Huron (USA/Canada). Each point represents one year of record during the period 1974–1980. Data from Bierman VJ Jr., Dolan DM, Kasprzyk R, and Clark JL (1984) Retrospective analysis of the response of Saginaw Bay, Lake Huron, to reductions in phosphorus loadings. *Environmental Science and Technology* 18: 23–31.

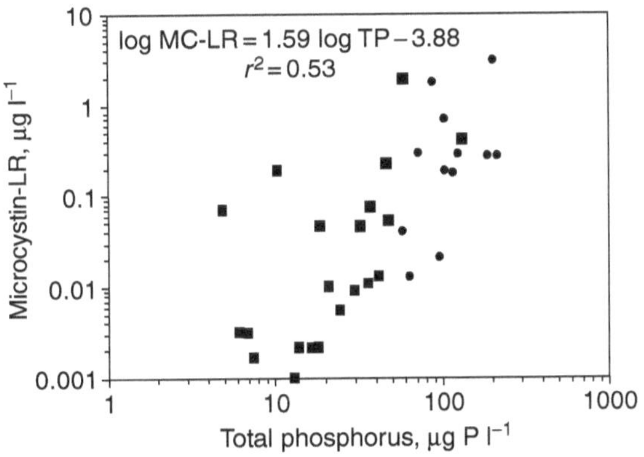

Figure 4 Relationship between summer mean concentrations of the cyanobacterial toxin Microcystin-LR and total phosphorus in 35 Canadian lakes. Data are from Kotak BG, Lam AK-Y, Prepas EE, and Hrudey SE (2000) Role of chemical and physical variables in regulating microcystin-LR concentration in phytoplankton of eutrophic lakes. *Canadian Journal of Fisheries and Aquatic Sciences* 57: 1584–1593 (closed circles); and from Giani A, Bird DF, Prairie YT, and Lawrence JF (2005) Empirical study of cyanobacterial toxicity along a trophic gradient of lakes. *Canadian Journal of Fisheries and Aquatic Sciences* 62:2100–2109 (closed squares).

Effects of hydrology The hydraulic residence time of surface waters varies tremendously. Fluvial systems, for example, are characterized by parcels of water that are in constant motion. Although the growth of algae suspended and transported within these parcels may continue as the water moves downstream, the biomass of suspended and attached algae at any fixed point in the stream channel can exhibit enormous variation over time; this temporal variance is tremendously elevated following runoff events, when base flow conditions can sometimes be replaced by torrential scouring events.

Although there is typically much less variability, some natural lake basins and artificial impoundments may exhibit water residence times shorter than 1 week; such temporal variations in hydraulic flushing can be particularly pronounced in run-of-the-river reservoirs. Strong increases in flushing can impose unusually high washout loss rates on suspended algae in rapidly flushed systems, and can potentially result in significant reductions in algal standing crop following major runoff events.

Effects of vascular plants (macrophyte vegetation) Although they frequently produce highly visible phytoplankton blooms, shallow lakes and deep lakes with extensive littoral zones can exhibit abundant growths of vascular plants during the normal growing season. Enriched, slow-moving rivers may also produce dense aquatic plant growth. This macrophyte vegetation can be present in one or more of three major growth forms: emergent plants, such as cattails (*Typha* sp.); submerged plants, such as coontail (*Ceratophyllum demersum*) and the invasive species *Hydrilla verticillata*; and floating-leaved plants, such as duckweed, water lilies (*Nyphea, Nuphar*), and the invasive water hyacinth (*Eichhornia crassipes*).

Aquatic vascular vegetation provides an essential habitat and primary forage base for many resident aquatic animals, as well as for many migratory waterfowl. However, the nutrient-stimulated overgrowth of macrophytes can create problems for human recreational and commercial use if their biomass attains nuisance levels. Moreover, catastrophic shifts from algae-dominated to macrophyte-dominated alternative stable states may occur in some shallow eutrophic systems (see later).

Effects of food web structure Beginning with pioneering work performed on Czechoslovakian ponds in the 1950s, a rapidly expanding body of research on food web structure has made it clear that changes in the density and species composition of consumer populations can have profound effects on aquatic ecosystem structure and function. For example, high populations of benthivorous fishes can strongly enhance nutrient fluxes from bottom sediments into the water column, thereby contributing significantly to internal nutrient loading. In addition, the presence of high population densities of zooplanktivorous fish tends to maximize the standing crop of phytoplankton; this enhancement of algal production is typically associated with reductions in grazing by large-bodied

Table 5 Key factors that modify the expression of nutrient enrichment in inland waters

Factor	*Source*
Salinity	1,2
Dissolved organic color	3,4
Suspended soils (inorganic suspended solids)	4,5,6
Hydrology (water flow regimes)	7,8,9,10
Aquatic macrophyte vegetation	11,12
Food web structure (aquatic consumer biomass and composition)	13,14,15,16
Alternative stable states	12,17,18,19

Sources

1. Campbell E and Prepas EE (1986) Evaluation of factors related to the unusually low chlorophyll levels in prairie saline lakes. *Canadian Journal of Fisheries and Aquatic Sciences* 43: 846–854.
2. Jeppesen E, Søndergaard M, Kanstrup E, *et al.* (1994) Does the impact of nutrients on the biological structure and function of brackish and freshwater lakes differ? *Hydrobiologia* 275/276: 15–30.
3. Havens KE and Nürnberg G (2004) The phosphorus–chlorophyll relationship in lakes: Potential influences of color and mixing regime. *Lake and Reservoir Management* 20: 188–196.
4. Carlson RE and Havens KE (2005) Simple graphical methods for the interpretation of relationships between trophic state variables. *Lake and Reservoir Management* 21: 107–118.
5. Hoyer MV and Jones JR (1983) Factors affecting the relation between phosphorus and chlorophyll a in midwestern reservoirs. *Canadian Journal of Fisheries and Aquatic Sciences* 40: 192–199.
6. Descy J-P, Reynolds CS, and Padisák J (1994) Phytoplankton in turbid environments: Rivers and shallow lakes. *Hydrobiologia* 289(1–3): 1–214.
7. Reynolds CS (1997) *Vegetative Processes in the pelagic: A Model for Ecosystem Study*. Oldendorf/Luhe, Germany: Ecology Institute.
8. Jones JR, Thompson RA, Seong CN, *et al.* (2006) Monsoon influences on the limnology of Juam Lake, South Korea. *Verhandlungen der Internationale Vereinigung für Theoretische und Angewandte Limnologie* 29: 1215–1222.
9. Van Nieuwenhuyse EE and Jones JR (1996) Phosphorus–chlorophyll relationship in temperate streams and its variation with stream catchment area. *Canadian Journal of Fisheries and Aquatic Sciences* 53: 99–105.
10. Biggs BJF (2000) Eutrophication of streams and rivers: dissolved nutrient–chlorophyll relationships for benthic algae. *Journal of the North American Benthological Society* 19: 17–31.
11. Bachmann RW, Horsburgh CA, Hoyer MV, *et al.* (2002) Relations between trophic state indicators and plant biomass in Florida lakes. *Hydrobiologia* 470: 219–234.
12. Scheffer M, Szabó S, Gragnani A, *et al.* (2003) Floating plant dominance as a stable state. *Proceedings of the National Academy of Sciences* 100: 4040–4045.
13. Carpenter SR, Christensen DL, Cole JJ, *et al.* (1995) Biological control of eutrophication in lakes. *Environmental Science and Technology* 29: 784–786.
14. Hansson L-A, Annadotter H, Bergman E, *et al.* (1998) Biomanipulation as an application of food-chain theory: Constraints, synthesis, and recommendations for temperate lakes. *Ecosystems* 1: 558–574.
15. Moss B (2004) Fish and nutrient effects in shallow lakes: A pan-European mesocosm experiment. *Freshwater Biology* 49: (Special issue), 1517–1656.
16. Caraco NF, Cole JJ, and Strayer DL (2006) Top-down control from the bottom: Regulation of eutrophication in a large river by benthic grazing. *Limnology and Oceanography* 51(1, part 2): 664–670.
17. Blindow I, Andersson G, Hargeby A, and Johansson S (1993) Long-term pattern of alternative stable states in two shallow eutrophic lakes. *Freshwater Biology* 30: 159–167.
18. Scheffer M, Carpenter S, Foley JA, *et al.* (2001) Catastrophic shifts in ecosystems. *Nature* 413: 591–596.
19. Carpenter SR (2005) Eutrophication of aquatic ecosystems: Bistability and soil phosphorus. *Proceedings of the National Academy of Sciences* 102: 10002–10005.

cladocerans in the genus *Daphnia*, which are released from size-selective fish predation. Conversely, the presence (or deliberate introduction) of large populations of piscivorous fish often tends to result in sharp reductions in the algal standing crop, unless the phytoplankton community is dominated by large, inedible taxa. Similar food web interactions have been found to govern the biomass and community structure of algae in fluvial systems as well.

Alternative stable states The trophic state of almost any inland water body can be strongly influenced by changes in its external nutrient loading. In some cases, a eutrophic system will rapidly return to a more oligotrophic (and more publicly desirable) condition following the successful implementation of external nutrient controls. However, eutrophic ecosystems typically do not return to exactly the same physical, chemical, and ecological conditions that existed prior to the onset of nutrient enrichment. For example, some degree of hysteresis was evident during the eutrophication and wastewater diversion phases of the restoration of Lake Washington (**Figure 1**).

An emerging body of literature on the temporal dynamics of ecological systems predicts that although the trajectory of some ecological systems may be relatively consistent and unchanged over time (states 1–4 in **Figure 5(a)**), it is possible for other systems to exhibit a series of alternative conditions (deep 'valleys' 2 and 4 in **Figure 5(b)**). For example, such alternative stable states have been particularly evident in shallow eutrophic lakes, which may shift

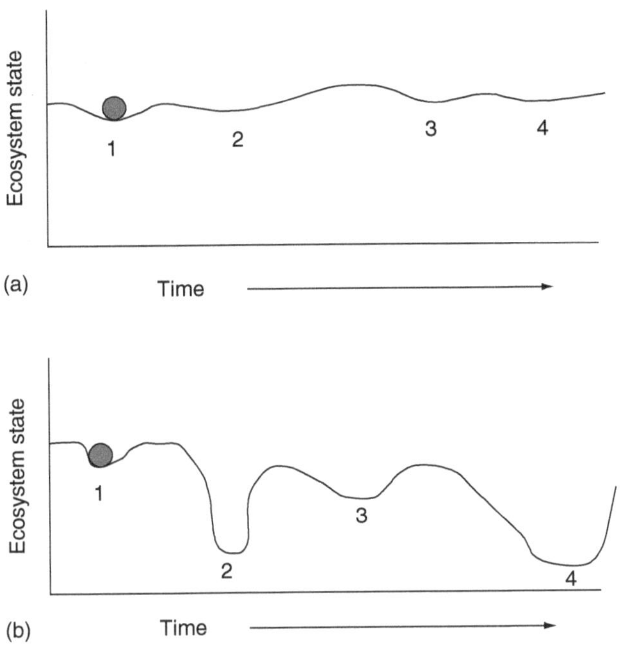

Figure 5 Graphical illustration of alternative stable states in aquatic ecosystems: (a) An ecosystem exhibiting relatively little change in moving from state 1 to any of states 2–4. (b) An aquatic ecosystem that could change from state 1 to three markedly different ecological states. Once it moves into alternative stable states 2 or 4, very strong ecological or environmental forcing is needed to change this system to a different condition.

from turbid, algae-dominated to clear, macrophyte-infested conditions. Recent research in North America and Europe suggests that eutrophication is reversible in principle, but from the perspective of human lifetimes, it can appear to be permanent unless there are substantial changes in watershed and lake basin management.

Correcting the problem: Eutrophication control, restoration, and management Public and government reactions to eutrophication led aquatic scientists and civil engineers worldwide to develop management frameworks for nutrient enrichment in inland waters. Restricting external nutrient loading is the centerpiece of eutrophication management. During the past three decades, reductions or diversions of external nutrient inputs have been used effectively to restore many nutrient-impaired water bodies. The outcome has been particularly positive for lakes in which eutrophication was caused primarily by highly controllable point source inputs, such as wastewater treatment plant effluents (e.g., Lake Washington, USA; **Figure 1**).

However, such lake restoration efforts have not been universally successful (**Table 6**). In some cases, the restriction of point source nutrient inputs alone was found to be insufficient, because nonpoint inputs continued to contribute sufficient N and P to maintain eutrophic conditions. In other cases, a chemical 'memory' of past nutrient loading to shallow systems became stored within the lake sediments, and post-restoration fluxes of regenerated N and P from these enriched sediments were sufficiently high to slow (or even prevent) recovery. For lakes embedded in naturally eutrophic landscapes, restoration may not be economically feasible or physically possible.

As a result, a diverse set of methods has been developed to help correct eutrophication problems in systems that are resistant to external nutrient regulation alone (**Table 7**). However, each of these methods has its own unique limitations and success rate. For example, restoration efforts in shallow eutrophic lakes involving both nutrient controls and biomanipulation (through deliberate modifications of food web structure) have not been universally successful, and their outcomes have been found to vary with lake size and climate. Other potential restoration techniques, such as the use of toxic chemicals and biological control agents, may have associated risks, which must be carefully considered as well.

Conclusions

Humans now strongly influence almost every major aquatic ecosystem, and their activities have markedly altered the flow of growth-limiting nutrients from the landscape into receiving waters. A remarkable unity has been observed in the response of almost all inland waters to nutrient overenrichment, and reducing the symptoms of eutrophication to publicly acceptable levels is a feature common to all successful eutrophication control efforts. Moreover, quantitative frameworks for the assessment and control of eutrophication have been developed and implemented successfully worldwide. However, several key questions remain.

Knowledge Gaps

Causes and Composition of Harmful Algal Blooms

Increases in the frequency and intensity of cyanobacterial blooms are one of the most visible and most undesirable symptoms of eutrophication in inland waters. However, despite decades of research, it is clear that nutrient enrichment does not always guarantee cyanobacterial dominance, and the factors that regulate the ecological success of nuisance species of cyanobacteria remain incompletely understood. Moreover, cyanobacteria are not the sole source of phytoplankton-related water quality problems. Some eukaryotic algae, including the invasive chrysophyte *Gonyostomum semens* and the chlorophyte *Uroglena american*, can cause significant taste and odor problems in fresh waters;

Table 6 Global efforts to restore eutrophication via external nutrient loading control, and their outcomes in inland waters

Inland water	*Outcome*	*Source*
Lake Washington, USA	Improved	1
Laurentian Great Lakes	Improved	2
North Temperate lakes	Variable	3
Global lakes	Variable	4
Swedish lakes	Variable	5
Swedish Great Lakes	Variable	6
Western European lakes	Variable	7
Shallow Dutch lakes	Variable	8
Irish rivers and lakes	Variable	9,10
British rivers	Variable	11,12
North American and Western European lakes	Variable	13,14
Danish streams and lakes	Variable	15
Belarus lakes	Improved	16
Meretta Lake, Canada	Improved	17
Lake Maggiore, Italy	Improved	18
Lake Nemi, Italy	Improved	19
Lake Balaton, Hungary	Improved	20,21
Baldeggersee, Switzerland	Improved	22
Lake Lucerne, Switzerland	Improved	23
Piberger See, Austria	Improved	24
Lake Albufera, Spain	Improved	25
Lake Suwa, Japan	Improved	26
Lake McElwaine, Africa	Improved	27
Lake Eymir, Turkey	Improved	28
Lake Arresø, Denmark	Improved	28
Lake Paranoá, Brazil	Improved	29
Bow River, Canada	Variable	30
Kuparuk River, USA	Improved	31

Sources

1. Edmondson WT and Lehman JT (1981) The effect of changes in the nutrient income on the condition of Lake Washington. *Limnology and Oceanography* 26: 1–29.
2. Beeton AM (2002) Large freshwater lakes: Present state, trends, and future. *Environmental Conservation* 29: 21–38.
3. Smith VH and Shapiro J (1981) Chlorophyll–phosphorus relations in individual lakes: Their importance to lake restoration strategies. *Environmental Science and Technology* 15: 444–451.
4. Cullen P and Forsberg C (1988) Experiences with reducing point sources of phosphorus to lakes. *Hydrobiologia* 170: 321–336.
5. Forsberg C and Ryding S-O (1980) Eutrophication parameters and trophic state indices in 30 Swedish waste-receiving lakes. *Archiv für Hydrobiologie* 89: 189–207.
6. Wilander A and Persson G (2001) Recovery from eutrophication: Experiences of reduced phosphorus input to the four largest lakes of Sweden. *Ambio* 30: 475–485.
7. Sas H (ed.) (1989) *Lake Restoration by Reduction of Nutrient Loading; Expectations, Experiences, Extrapolations*. St. Augustin: Academia-Verlag Richarz.
8. Van Liere L and Gulati RD (1992) Restoration and recovery of shallow eutrophic lake ecosystems in The Netherlands: epilogue. *Hydrobiologia* 233: 283–287.
9. Foy RH, Champ WST, and Gibson CE (1996) The effectiveness of restricting phosphorus loadings from sewage treatment works as a means of controlling eutrophication in Irish lakes. In Giller PS and Myers AA (eds.) *Disturbance and Recovery in Ecological Systems*, pp. 136–154. Dublin: Royal Irish Academy.
10. Foy RH and O'Connor WCK (2002) Managing the effects of agriculture on water quality in Northern Ireland. In Haygarth PM and Jarvis SC (eds.), *Agriculture, Hydrology, and Water Quality*, pp. 417–440. Wallingford, Oxfordshire: CAB International.
11. Jarvi HP, Lycett E, Neal C, and Love A (2002) Patterns in nutrient concentrations and biological quality indices across the upper Thames river basin, UK. *Science of the Total Environment* 23: 282–283.
12. Jarvie HP, Neal C, and Williams RJ (2004) Assessing changes in phosphorus concentrations in relation to in-stream plant ecology in lowland permeable catchments: Bringing ecosystem functioning into water quality monitoring. *Water Air and Soil Pollution* 4(2–3), 641–655.
13. Anderson NJ, Jeppesen E, and Søndergaard M (2005) Ecological effects of reduced nutrient loading (oligotrophication) on lakes: An introduction. *Freshwater Biology* 50: 1589–1593.
14. Jeppesen E, Søndergaard M, Jensen JP, *et al.* (2005) Lake responses to reduced nutrient loading – an analysis of contemporary long-term data from 35 case studies. *Freshwater Biology* 50: 1747–1771.
15. Kronvang B, Jeppesen E, Conley DJ, *et al.* (2005) Nutrient pressures and ecological responses to nutrient loading reductions in Danish streams, lakes and coastal waters. *Journal of Hydrology* 304(1–4): 274–288.
16. Parparov A, Hambright KD, Hakanson L, and Ostapenia A (2006) Water quality quantification: basics and implementation. *Hydrobiologia* 560: 227–237.
17. Michelutti N, Douglas MSV, and Smol JP (2002) Tracking recent recovery from eutrophication in a high arctic lake (Meretta Lake, Cornwallis Island, Nunavut, Canada) using fossil diatom assemblages. *Journal of Paleolimnology* 28: 377–381.
18. de Bernardi R, Calderoni A, and Mosello R (1996) Environmental problems in Italian lakes, and lakes Maggiore and Orta as successful examples of correct management leading to restoration. *Verhandlungen der Internationale Vereinigung für Theoretische und Angewandte Limnologie* 26: 123–138.
19. Margaritora FG, Fumanti B, Alfinito S, *et al.* (2005) Trophic condition of the volcanic Lake Nemi (Central Italy): environmental factors and planktonic communities in a changing environment. *Journal of Limnology* 64: 119–128.
20. Istvanovics V, and Somlyódy L (2001) Factors influencing lake recovery from eutrophication—The case of basin 1 of Lake Balaton. *Water Research* 35: 729–735.
21. Istvanovics V, Clement A, Somlyódy L, *et al.* (2007) Updating water quality targets for shallow Lake Balaton (Hungary), recovering from eutrophication. *Hydrobiologia* 581: 305–318.
22. Lotter AF (1998) The recent eutrophication of Baldeggersee (Switzerland) as assessed by fossil diatom assemblages. *Holocene* 8: 395–405.
23. Bührer H and Ambühl H (2001) Lake Lucerne, Switzerland, a long term study of 1961–1992. *Aquatic Sciences* 63: 1–25.
24. Tolotti M and Thies H (2002) Phytoplankton community and limnochemistry of Piburger See (Tyrol, Austria) 28 years after lake restoration. *Journal of Limnology* 61: 77–88.
25. Romo S, Villena M-J, Sauquillo M, *et al.* (2005) Response of a shallow Mediterranean lake to nutrient diversion: Does it follow similar patterns as in northern shallow lakes? *Freshwater Biology* 50: 1706–1717.
26. Okino T and Kato K (1987) Lake Suwa—Eutrophication and its partial recent recovery. *GeoJournal* 14: 373–375.
27. Thornton JA (ed.). (1982) Lake McIlwaine: The eutrophication and recovery of a tropical African human-made lake. *Monographiae Biologicae*, vol. 49. The Hague: Junk.
28. Beklioglu M, Ince O, and Tuzun I (2003) Restoration of the eutrophic Lake Eymir, Turkey, by biomanipulation after a major external nutrient control. I. *Hydrobiologia* 489: 93–105.
29. Jeppesen E, Meerhoff M, Jacobsen BA, *et al.* (2007) Restoration of shallow lakes by nutrient control and biomanipulation—The successful strategy varies with lake size and climate. *Hydrobiologia* 581: 269–285.
30. Sosiak A (2002) Long-term response of periphyton and macrophytes to reduced municipal nutrient loading to the Bow River (Alberta, Canada). *Canadian Journal of Fisheries and Aquatic Sciences* 59: 987–1001.
31. Benstead JP, Green AC, Deegan LA, *et al.* (2007) Recovery of three arctic stream reaches from experimental nutrient enrichment. *Freshwater Biology* (in press; online doi:10.1111/j.1365-2427.2007.01723.x).

Table 7 In-lake eutrophication management methods

Method	*Source*
Aeration and artificial destratification	1,2,3
Increased flushing (enhanced water through-flow)	1,4,5
Hypolimnetic withdrawals	1,5,6
Nutrient inactivation and sediment oxidation	1,7,8
Sediment removal (dredging)	1,9,10
Manipulation of food web structure (biomanipulation)	1,11,12
Macrophyte dredging and harvesting	1,13,14
Additions of toxicants/poisons	1,15,16
Additions of biological control agents	1,17,18

Sources

1. Cooke GD, Welch EB, Peterson S, and Nichols SA (2005) *Restoration and Management of Lakes and Reservoirs*, 3rd edn. Boca Raton: CRC Press.
2. Antenucci JP, Ghadouani A, Burford M, and Romero JR (2005) The long-term effect of artificial destratification on phytoplankton species composition in a subtropical reservoir. *Freshwater Biology* 50: 1081–1093.
3. Lepistö L, Sammalkorpi I, Jokipii R, and Niemelä M (2003) Effects of destratification and fish removal on the genus *Microcystis* in a shallow hypertrophic lake. *Archiv für Hydrobiologie* Suppl. 148(1, parts 1/4): 375–386.
4. Jagtman E, Van der Molen DT, and Vermij S (1992) The influence of flushing on nutrient dynamics, composition and densities of algae and transparency in Veluwemeer, The Netherlands. *Hydrobiologia* 233(1–3): 187–196.
5. Klapper H (2003) Technologies for lake restoration. *Journal of Limnology* 62(Suppl. 1), 73–90.
6. Nurnberg GK (1987) Hypolimnetic withdrawal as lake restoration technique. *Journal of Environmental Engineering* 113: 1006–1017.
7. Welch EB and Cooke GD (1999) Effectiveness and longevity of phosphorus inactivation with alum. *Lake and Reservoir Management* 15: 5–27.
8. Prepas EE, Pinel-Alloul B, Chambers PA, *et al.* (2001) Lime treatment and its effects on the chemistry and biota of hard water eutrophic lakes. *Freshwater Biology* 46: 1049–1060.
9. Cronberg G, Gelin C, and Larsson K (1975) Lake Trummen restoration project II. Bacteria, phytoplankton and phytoplankton productivity. *Verhandlungen der Internationale Vereinigung für Theoretische und Angewandte Limnologie* 19: 1088–1096.
10. Van der Does J, Verstraelen P, Boers P, *et al.* (1992) Lake restoration with and without dredging of phosphorus-enriched upper sediment layers. *Hydrobiologia* 233(1–3): 197–210.
11. Meijer M-L, Boois I, Scheffer M, *et al.* (1999) Biomanipulation in shallow lakes in The Netherlands: An evaluation of 18 case studies. *Hydrobiologia* 408/409: 13–30.
12. Kasprzak P, Benndorf J, Mehner T, and Koschel R (2002) Biomanipulation of lake ecosystems: An introduction. *Freshwater Biology* 47: 2277–2281.
13. Collett LC, Collins AJ, Gibbs PJ, and West RJ (1981) Shallow dredging as a strategy for the control of sublittoral macrophytes: A case study in Tuggerah Lakes, New South Wales. *Australian Journal of Marine and Freshwater. Research* 32: 563–571.
14. Boylen CW, Eichler LW, and Sutherland JW (1996) Physical control of Eurasian watermilfoil in an oligotrophic lake. *Hydrobiologia* 340(1–3): 213–218.
15. Van Hullebusch E, Deluchat V, Chazal PM, and Baudu M (2002) Environmental impact of two successive chemical treatments in a small shallow eutrophied lake: Part II. Case of copper sulfate. *Environmental Pollution* 120: 627–634.
16. Madsen JD (2006) Techniques for managing invasive aquatic plants in Mississippi water resources. *Proceedings of the 36th Annual Mississippi Water Resources Conference* 36: 42–51.
17. Pieterse AH, Kettunen M, Diouf S, *et al.* (2003) Effective biological control of *Salvinia molesta* in the Senegal River by means of the weevil *Cyrtobagous salviniae*. *Ambio* 32: 458–462.
18. Cofrancesco AF (2000) Factors to consider when using native biological control organisms to manage exotic plants. *Journal of Aquatic Plant Management* 38: 117–120.

however, the environmental factors that lead to the proliferation of these and other nuisance phytoplankton species are not known, and require further research.

Health Risks

In some cases eutrophic systems may be associated with increasing concentrations of biological toxins (**Figure 4**), but the subsequent health risk to humans and other animals exposed to these waters is still not fully known or quantified. In addition, eutrophic surface waters potentially provide a rich growth medium for animal and plant pathogens delivered by point or nonpoint sources. The interactions between eutrophication, pathogen loads, disease transmission, and disease outcomes remain very poorly understood.

Interactions with other Pollutants and Environmental Stressors

In addition to pathogens, nutrient-enriched waters experience a wide array of additional environmental stressors. For example, the runoff and effluents that transport nitrogen and phosphorus to receiving waters also typically contain a wide variety of organic and inorganic pollutants. However, empirical and theoretical frameworks have not yet been developed to predict with confidence the effects of eutrophication on the biochemical transformations and fates of other water-borne pollutants. In addition, local climatic conditions may change markedly throughout much of the world during the next 100 years, and we still have only a rudimentary understanding of the interactions that may exist between eutrophication and global climate change.

General Effects of Eutrophication

The responses of inland waters to nutrient enrichment can be highly variable over space and time. However, several important generalizations can be made:

1. Increases in the supplies of nitrogen (N) and phosphorus (P) to inland waters are typically accompanied by increases in both primary and secondary productivity.

2. These responses to nutrient enrichment are consistent with the large body of knowledge on growth-limiting factors and ecological interactions that has emerged during the past 150 years.
3. Most physical, chemical, and biological expressions or symptoms of eutrophication are often site-specific.
4. Several physical, chemical, and ecological factors can potentially modify the expression of eutrophication; the effects of these factors are often site-specific.
5. Logical, objective, quantitative methods have been developed to manage and control eutrophication, and these frameworks are now in accepted use worldwide.

Glossary

Benthivorous fish – Fish that feed upon food items located in or on sediments, e.g., carp, roach.

Biomanipulation – The deliberate manipulations of food web structure in aquatic ecosystems in order to obtain improved water quality in eutrophic systems.

Cyanobacteria – Photosynthetic, prokaryotic microorganisms that can dominate the phytoplankton or periphyton of eutrophic water bodies; also called blue-green algae.

Hydraulic residence time – The average time in which a parcel of water resides within a water body, once it enters; it is equal to quotient between the total volume of the water body itself, and the rate of water flow through it.

Periphyton – The community of attached photosynthetic organisms that inhabit submerged surfaces within an aquatic ecosystem.

Phytoplankton – The community of free-floating photosynthetic microorganisms that inhabit the water column of inland waters (primary producers).

Piscivorous fish – Fish that feed upon other fish.

Trophic cascade – Controls of energy flows in aquatic ecosystem that propagate downwards in the food chain, from the top (= upper consumer levels) to the bottom (= primary producers).

Zooplanktivorous fish – Fish that feed upon zooplankton.

Zooplankton – The community of microscopic animal consumers, both herbivorous and carnivorous, that inhabit the water column of inland waters.

See also: Aquatic Ecosystems and Human Health; Carbon, Unifying Currency; Chemical Fluxes and Dynamics in River and Stream Ecosystems; Chemical Properties of Water; Deforestation and Nutrient Loading to Fresh Waters; Fluvial Transport of Suspended Solids; Ground Water and Surface Water Interaction; Groundwater Chemistry; Hydrological Cycle and Water Budgets; Natural Organic Matter; Nitrogen; Nitrogen Fixation; Nutrient Stoichiometry in Aquatic Ecosystems; Organic Nitrogen; Phosphorus; Pollution of Aquatic Ecosystems I; Pollution of Aquatic Ecosystems II: Hydrocarbons, Synthetic Organics, Radionuclides, Heavy Metals, Acids, and Thermal Pollution; Salinity; Streams.

Further Reading

Bachmann RW, Jones BL, Fox DD, Hoyer M, *et al.* (1995) Relationship between trophic state indicators and fish in Florida (U.S.A.) lakes. *Canadian Journal of Fisheries and Aquatic Sciences* 53: 842–855.

Biggs BJF, Francoeur SN, Huryn AD, *et al.* (2000) Trophic cascades in streams: Effects of nutrient enrichment on autotrophic and consumer benthic communities under two different fish predation regimes. *Canadian Journal of Fisheries and Aquatic Sciences* 57: 1380–1394.

Carpenter SR, Caraco NF, Correll DL, *et al.* (1998) Nonpoint pollution of surface waters with phosphorus and nitrogen. *Ecological Applications* 8: 559–568.

Codd CA, Morrison LR, and Metcalf JS (2005) Cyanobacterial toxins: Risk management for health protection. *Toxicology and Applied Pharmacology* 203: 264–267.

Correll DL (1998) The role of phosphorus in the eutrophication of receiving waters: A review. *Journal of Environmental Quality* 27: 261–266.

Drenner RW and Hambright KD (2002) Piscivores, trophic cascades, and lake management. *The Scientific World JOURNAL* 2: 284–307.

Ekdahl EJ, Teranes JL, Guilderson TP, *et al.* (2004) Prehistorical record of cultural eutrophication from Crawford Lake, Canada. *Geology* 32: 745–748.

Jarvie HP, Neal C, and Withers PJ (2006) Sewage-effluent phosphorus: A greater risk to river eutrophication than agricultural phosphorus? *Science of the Total Environment* 360(1–3): 246–253.

O'Sullivan PE and Reynolds CS (eds.) (2004) *The Lakes Handbook, vol. 2: Lake Restoration and Rehabilitation.* London: Blackwell Publishing.

Prepas EE and Charette T (2003) Worldwide eutrophication of water bodies: Causes, concerns, controls. In: Lollar BS (ed.) *Treatise on Geochemistry*, vol. 9, pp. 311–331. London: Elsevier.

Sharpley AN, Daniel T, Sims T, *et al.* (2003) *Agricultural Phosphorus and Eutrophication.* 2nd edn. Agricultural Research Service Publ. ARS–149. Washington, DC: United States Dept. of Agriculture.

Smith RA, Alexander RB, and Schwarz GE (2003) Natural background concentrations of nutrients in streams and rivers of the conterminous United States. *Environmental Science and Technology* 37: 3039–3047.

Smith VH (2003) Eutrophication of freshwater and marine ecosystems: A global problem. *Environmental Science and Pollution Research* 10: 126–139.

Relevant Websites

http://www.bio.tcu.edu/drenner/lakebib2979.html – Effects of Fish on Lake Systems.

http://www.cdc.gov/hab/cyanobacteria/facts.htm – Harmful Algal Blooms: Cyanobacteria.

http://www.aquatics.org/pubs/madsen2.htm – Advantages and Disadvantages of Aquatic Plant Management Techniques.

http://lepo.it.da.ut.ee/~olli/eutr/html/htmlBook_0.html – Drainage Basin Nutrient Inputs and Eutrophication.

http://www.unep.or.jp/Ietc/Publications/techpublications/TechPub-11/index.asp#1 – Planning and Management of Lakes and Reservoirs.

Fires

E Prepas, N Serediak, G Putz, and D W Smith, Lakehead University, Thunder Bay, ON, Canada

Introduction

Fire is a periodic occurrence in vegetated landscapes, with variable impacts dependent on both biotic and abiotic factors. It has a distinct impact on freshwater systems compared with other landscape disturbances. Stable watersheds exhibit chemical (mass concentration) and hydrological (discharge and velocity) fluxes within a predictable range of variability. This can be altered dramatically by wildfire, which is essentially an oxidative chemical reaction of varying duration and intensity. The heat generated by wildfires of sufficient size can shift nutrient composition, redistribute cations, and alter pH balances in burned watersheds and subsequently in the waters that drain them. Fire effects can also induce soil hydrophobicity, decrease infiltration, and enhance lateral subsurface flow, thus increasing runoff and the risk of sedimentation and landslip.

Fire is a natural disturbance in most instances, and some forests have been managed in attempts to emulate the physical changes fire exerts within watersheds. In contrast to a disturbance like forest harvesting, however, wildfire is first a chemical process, which is then compounded by mechanical disruption. This is a fundamental difference from commercial harvesting, where the mechanical disturbance (i.e., tree removal, soil disruption, and compaction) initiates the resulting chemical response. This is a central factor in the effects of wildfire on freshwater systems.

The Structure of Wildfire

To understand the basic effects of fire on aquatic systems, it is helpful to first qualify what constitutes a wildfire and to differentiate between wildfire and prescribed burns. Although both types of fires can be caused by human activity, the components of a prescribed burn are held under tight control to achieve a desired land management objective. Generally, this outcome is to reduce fine fuel loading (accumulation of shed needles, leaves, bark, and other organic materials), manipulate the understorey structure, control susceptible tree pathogens, and condition sites for tree regeneration. In contrast to prescribed burns, wildfires behave randomly and have varying degrees of impact relative to the particular physical descriptors that constitute a watershed, and to the size and structure of the fire itself.

Fire originates where an available fuel source coincides with a successful ignition event. Ignition can be anthropogenic (accidental or prescribed) or natural (lightening strike). In general, fires can be classified as one of three types: surface (lowest impact), intermittent crown (moderate impact), and crown fire (highest impact). Several countries now use the Fire Weather Index (FWI) to provide a numerical assessment of fire type (**Figure 1**).

Wildfires generally move in an elliptical pattern from the ignition source, driven by prevailing winds. In forested regions, combustible understorey often provides the initial fuel, after which fires may develop progressively into the overstorey or crown. Crown fires can produce frontal fire intensities sufficient to ignite vegetation too humid to serve as an initial fuel source. Although wildfires often have a single ignition source, wind transport of hot ash and embers, especially from crown fires, can result in multiple ignition events attributable to the same fire. After successful ignition, fires are self-propagating until the fuel source is exhausted, oxygen is consumed, or until fuel is rendered incombustible (e.g., through wetting from rainfall or active suppression).

The characteristics of wildfires can be divided into four primary components: intensity, severity, size, and frequency.

Fire Intensity

The intensity of an active fire is a measure of vertical heat transfer above ground, and the degree to which vegetation mortality has occurred. Although temperature extremes are an integral part of any fire, temperature itself is not a good indicator of fire intensity. Woody fuels ignite at roughly 350 °C and maximum temperatures in forest fire flames can reach 1000 °C. However, flames 5 cm high have a markedly different measure of intensity relative to flames 5 m high, despite having the same core temperature. Therefore, standardized measures of fire intensity have been developed. One example is the Canadian Forest Fire Behaviour Prediction System, which quantifies fire intensity as energy released per unit length of fire front in kilowatts per meter ($kW\,m^{-1}$) (**Table 1**). Fire intensities become enhanced as flammable vapors are released and themselves ignited as fuel combustion progresses.

Fire Severity

The vertical transfer of heat from surface to below ground, as well as general biological impacts, are a

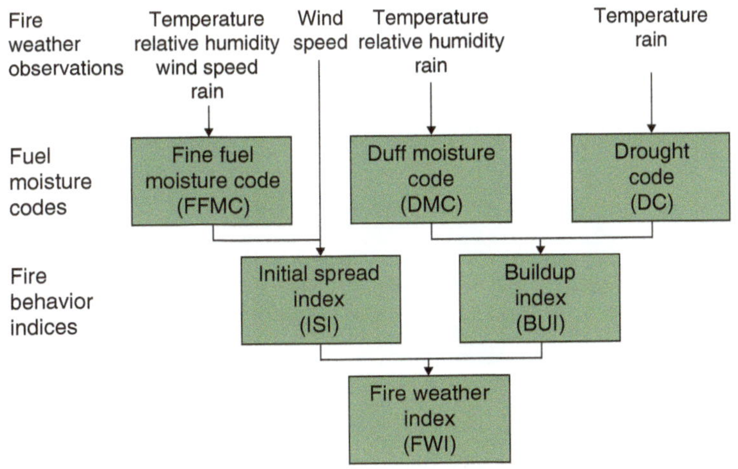

Figure 1 Variables involved in quantifying the FWI rating. Reproduced with permission from Canadian Forest Service, Natural Resources Canada.

Table 1 Ranges of fire severity and intensity with associated fire types

	Severity (°C)	*Intensity (kW h^{-1})*	*Associated fire type*
Low	<180	100–2000	Surface
Moderate	180–300	2000–10 000	Intermittent crown
High	>300	>10 000	Continuous crown

measure of fire severity. Crown fires are often far more severe than grass fires, even though they typically move at slower speeds. For example, fast moving grass fires have typical spread rates ranging from 25 to 80 m min^{-1} and have been clocked at 300 m min^{-1}. In comparison, crown fires advance at a rate of 25–50 m min^{-1} and reach 100 m min^{-1} at top speed. Forested watersheds can experience intense downward heat transfer during crown fires, enough to completely ash fine fuels and organic layers down to mineral soils (**Figure 2**). Extreme heat during wildfire can sterilize the soil microbial community, volatilize soluble inorganic and organic compounds in plants, soil and water, and liberate labile mono- and divalent dominant ions.

Fire Size

The area and volume of materials consumed by wildfire constitute the overall fire size. While human establishment in many areas has increased the number of fires, this may not have had as large an impact on total area burned per year as once thought, because fire size is often affected by the ignition source. Lightening-induced fires that occur in remote areas can develop into large wildfires before detection. Human-caused fires often occur in or near populated areas where they are reported quickly and their effects are minimized before sizeable areas are burned. These generalizations do not take into account global regions where traditional burning for agricultural purposes is more frequent and can occasionally develop into uncontrolled bush or crown fires.

Fire Frequency

Naturally occurring wildfires are now seen as more of an integral component rather than strictly a disruption of watershed processes, thus the practice of fire suppression has begun to shift focus from a mandate of complete elimination to one of controlling progression (where feasible). More effort has been exerted in determining the periodicity of wildfires in presettlement forests through pollen counts, dendrochronolgy, and carbon deposition in lacustrine systems, and comparing these results with what occurs when fire suppression becomes a part of watershed management. There is a growing belief that climate has the greatest impact on the frequency of large, severe fires. It is the large fires that cause most of the damage associated with wildfire (**Figure 3**). Periodicity of naturally occurring wildfire ranges from 5 to 20 years in vulnerable areas, up to estimates of 500 years in less fire-prone regions.

Wildfires as an Ecological Impact on Aquatic Systems

The effects of wildfire are felt differently within and among watersheds, and are dependent on many factors. Watersheds with a predominance of dry, resinous, and easily ignitable plant matter have a higher probability of impact from fire, as well as those with broad tracts of uninterrupted vegetation with few

Figure 2 Examples of fire intensity in a burning jack pine forest based on the FWI rating system: (a) low intensity surface fire (FWI 9), (b) moderate intensity intermittent crown fire (FWI 20), (c) high intensity continuous crown fire (FWI 34). Reproduced from Alexander, MJ and De Groot, WJ (1988) Fire behavior in jack pine stands as related to the Canadian Forest Fire Weather Index (FWI) System, with permission from the Canadian Forestry Service, Northern Forestry Centre, Edmonton, Alberta.

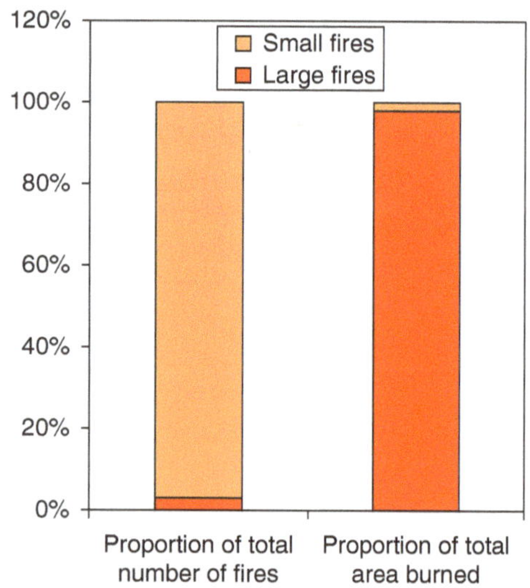

Figure 3 Average fire size relative to number of reported fire starts (from North American estimates). Large fires are defined as burning over 200 ha of land area.

naturally occurring firebreaks, which serve to limit spread (firebreaks include features such as ridges, surface waters, and areas with little or poor quality fuel). Fires burn more readily uphill because of the inherent upward movement of hot air, which results in differential burn patterns dependent on topography. The presence of a tinder source is key: watersheds where sufficient fine fuel sources exist with concurrent low moisture levels are prone to more frequent and severe fires than humid watersheds or areas where the litter layer remains thin and soil organic content is low. Seasonal timing plays an important role in ignition, intensity, and severity. Wildfires that occur in early spring before adequate precipitation and prior to leaf-out can be much larger in size compared with late-spring fires that ignite after plants are foliated and the watershed is sufficiently wet.

Globally, the effects of wildfire are more pronounced in regions where (1) hot, dry conditions occur regularly, (2) watersheds are primarily closed canopy systems, and (3) fuel sources are readily combustible (**Figure 4**). Fire-prone regions possess many factors both pre- and post-fire that contribute to cumulative watershed impacts.

Pre-Fire Considerations

Conditions that exist prior to wildfire events have a direct impact on the structure of fires passing through watersheds. These components affect fire movement, intensity, severity, and duration, and post-fire responses within the watershed.

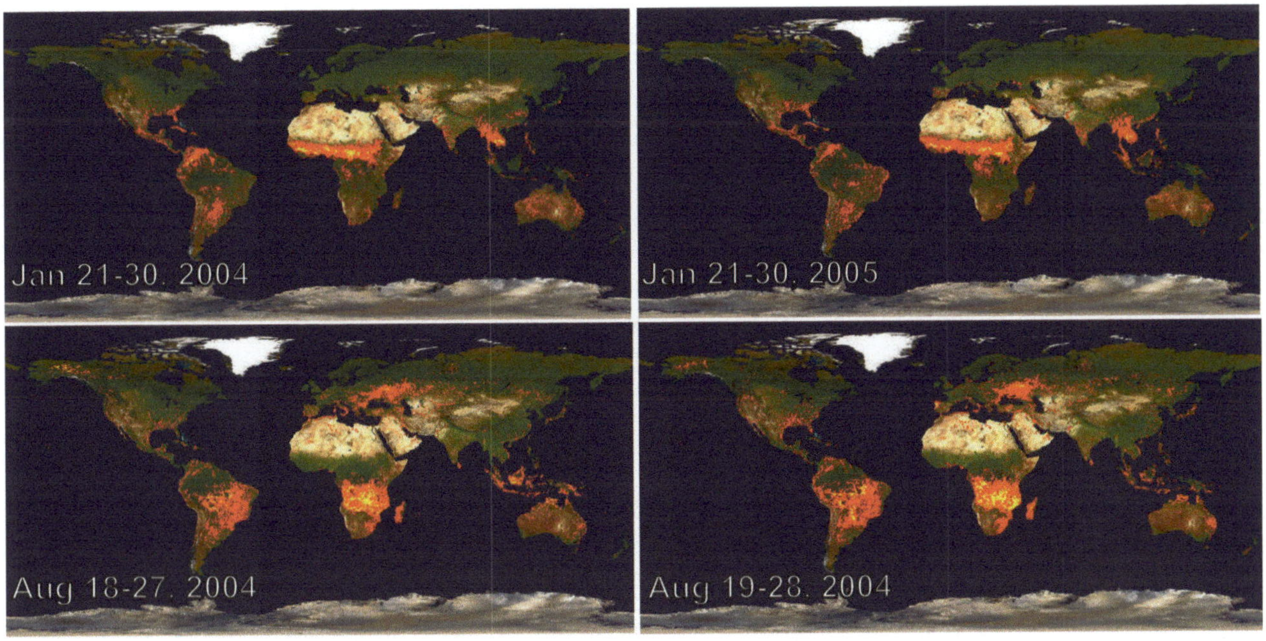

Figure 4 Four 10 day averages of global fire occurrences showing seasonal differences in fire locations over 2 years. Fire intensities are represented by red (lowest), orange (moderate), and yellow (highest). Image courtesy of MODIS Rapid Response Project at NASA/GSFC.

Watershed Slope

Steeper, leeward slopes have a higher probability of burning than windward areas of lower relief. Steep slopes are at risk for landslide after severe fires, especially if the fire is followed by intense precipitation. Also, runoff and sediment loads are higher from steep slopes following fire.

Catchment Size

Burned watersheds of first- and second-order stream size generally have a larger, faster response than those associated with streams of increasing size, where responses may be dampened by proportionally more unburned areas.

System Morphology

Watersheds with a high proportion of low-order streams draining high relief areas display more measurable shifts in chemistry post-fire in comparison to regions of low topographic relief, where fire-related chemical changes may be ameliorated by bogs, fens, swamps, marshes, or open-water areas. The overall percentage of surface water in a drainage basin will affect burn area, and the size of individual water features will also affect fire movement (e.g., the ability of the fire to cross a first-order *versus* a fifth-order stream).

Soil Characteristics

The structure, porosity, existing hydrophobicity, and type of soil underlying the drainage basin will affect chemical and infiltration responses post-fire. The condition and depth of the litter layer (alternately known as duff or LFH horizon) contributes to fuel loading pre-fire, surface spread during fire, and protection of mineral horizons post-fire.

Baseline Water Chemistry

The chemical composition of surface waters pre-fire can have a direct effect on post-fire response. Streams with inherently low alkalinity are at increased risk for pH reduction from influxes of soil-bound, soluble acidic anions (e.g., sulfate, chloride) released during precipitation events following wildfire.

Weather Patterns

Ecoregions exist in concert with long-term weather patterns and there is a growing realization that periodic wildfires do as well. Areas with prolonged seasonal drought are often at increased risk for shorter fire cycles that can include very large wildfires. In addition, global phenomena, such as El Nino and La Nina events, can produce localized drying cycles that create conditions for wildfire outside of expected regional normals, or exacerbate those which usually occur.

Timing

Wildfires that occur after very dry winters, in early spring before significant precipitation, or after drying events, are more severe than those that occur when vegetation is actively growing. The importance of timing can be diurnal as well as seasonal. Fires that burn over several days tend to cycle, causing less damage at night, as available moisture condenses out of the cooler air. This reduces the severity of the burn and aids in creating unburned patches.

Fuel Load and Forest Age

Combustion for naturally occurring wildfires is dependent on the amount of fine fuel available for either ignition or propagation. In early successional forests, fine fuel inputs are primarily from the canopy in the form of shed needles, leaves, and bark. This gradually progresses to a composition including lower growing vegetation (e.g., mosses, lichens, and short grasses) as the forest matures. Successful ignition is less likely in younger forests or recently burned areas where fuel loads have yet to reach an adequate density or quality. In harvested forests, fire risk may be high if sufficient fine fuel remains as an exposed ignition source in proximity to unharvested areas.

Canopy Type

In general, the greater the fire intensity, the greater is the severity of impact. Frontal fire intensities reach their highest levels in closed canopy systems where wildfires have the opportunity to burn upwards as well as outwards. Watersheds consisting of open canopy systems are more likely to experience lower intensity surface fires that have decreased post-fire impacts. Alternately, in grassland ecosystems where urban settlement has created a shift in plant composition approaching a closed canopy, crown fires may occur where they historically had not.

Post-Fire Impacts

Watershed characteristics affect the movement, pattern, and behavior of wildfire. Subsequently, fire impacts watershed functions with immediate, intermediate, and long-term effects (**Figure 5**).

Immediate Effects: 0–5 Years

What remains following wildfire is a fire legacy, the most visible impact of which is consumption of aboveground biomass (**Figure 6**). In severe fires, much of the over- and understorey vegetation is reduced to standing dead and ash. If enough of the aboveground portion of trees and shrubs are destroyed, root systems may die as well. Post-fire impacts are influenced by the patchiness of the burn (burned *versus* unburned areas within the limits of the fire's spread), mean distance to undisturbed areas, depth of ash, coarse woody debris inputs, and surviving cover and root systems.

Nutrient cycling Nutrient losses are variable post-fire depending on the ecosystem and where the majority of nutrients were distributed before the fire (i.e., soil or vegetation). In general, combustion of overstorey plants releases a more homogenous mixture of inorganic nutrients into the ash layer relative to what existed prior to fire. Nutrient loss is felt differently depending on location: tropical forests tend to be phosphorus (P) limited, while northern boreal forests tend to be nitrogen (N) limited. Tropical regrowth will be negatively affected if heavy rains follow severe burns and remove forest floor P fractions already in short supply. Northern forests may have a greater lag time before regrowth occurs if volatilized N losses are high, followed by P-leaching precipitation events (**Table 2**).

Heat-induced volatilization greatly affects N stores. Organic N released during combustion is oxidized to gaseous forms (NO, NH_3), which is then lost to the atmosphere. Inorganic N (primarily NH_4) remains in the ash layer and may lead to a stimulation of nitrification in the surviving microbial community, resulting in high levels of nitrate (NO_3) that can persist for several years. With little appreciable atmospheric loss other than particulate transport, P remains as an ash constituent along with base cations (calcium, magnesium, potassium) and acidic anions (sulfate, chloride). Carbon exists post-fire in soluble (dissolved organic carbon) and insoluble forms (charcoal). The presence of charcoal as a strong adsorbent is thought to inhibit certain fire-liberated compounds from joining the post-fire soil nutrient cycle and from moving into surface waters. These include both particulate P, and secondary metabolites that can inhibit plant growth.

Water movement following fire results in leaching and transport of soluble materials. Receiving waters may experience multiyear increases in nitrates, dissolved P, potassium, sulfate, and chloride. Less mobile divalent calcium and magnesium tend to remain in soil, temporarily increasing soil pH until regeneration begins to consume stores. Particulate inputs into surface waters also increase post-fire, most notably particulate P following heavy rain events. The movement of previously soil-bound acidic anions can decrease

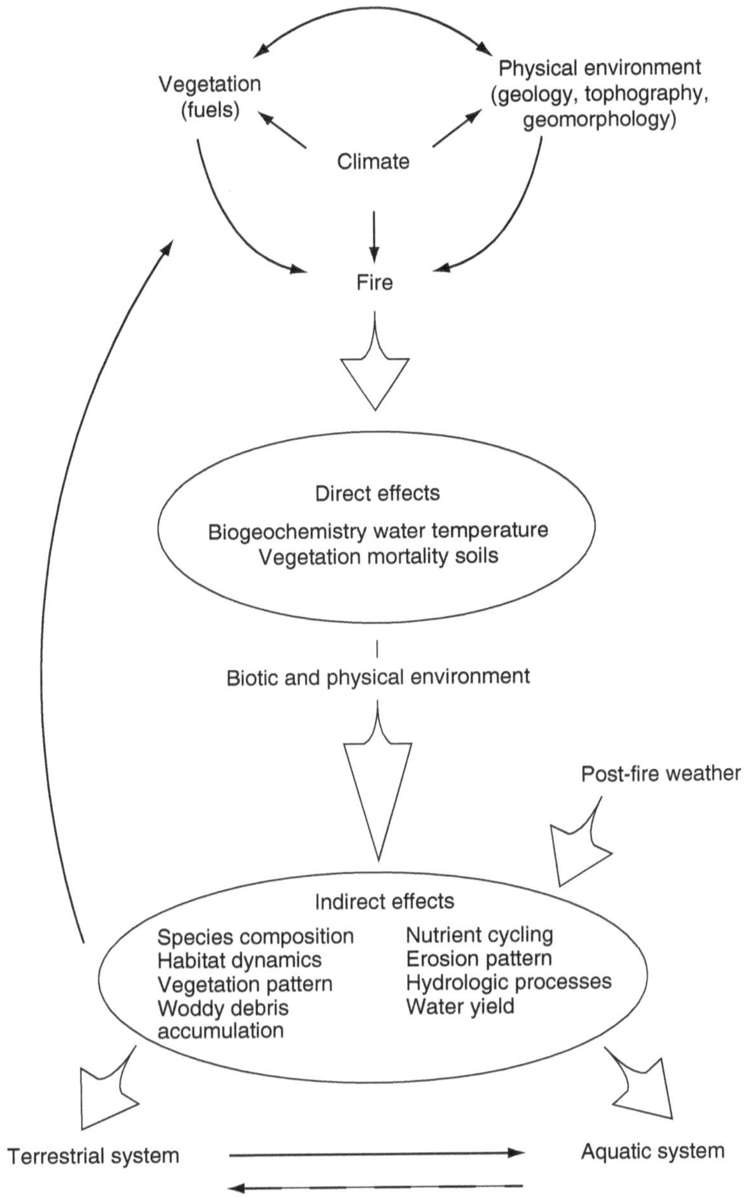

Figure 5 The interaction of fire, biota, and the physical environment. Adapted from **Figure 1** in Gresswell RE (1999) Fire and aquatic ecosystems in forested biomes of North America. *Transactions of the American Fisheries Society* 128: 193–221, with permission from the American Fisheries Society.

pH levels in stream waters, which is exacerbated in catchments draining regions with a high percentage of peatlands or with a history of acid deposition (i.e., low pre-fire alkalinity). As well, rapidly developing regrowth is more likely to uptake basic cations over acidic anions, creating an additional relative increase in acidic export. Post-fire revegetation is often from NO_2-tolerant plants, such as fireweed (*Epilobium angustifolium*) and raspberry (*Rubus* spp.) in boreal forest regions, which are gradually replaced by later successional plants as ammonification increases relative to nitrification as the primary mechanism for inorganic (bioavailable) N formation.

Infiltration and runoff The heat generated by fire results in water repellent soil conditions of varying degrees: ground fires in grass or chaparral can destroy naturally occurring surface hydrophobicity but create hydrophobic layers at depth, and intense crown fires are capable of producing a highly water repellent subsurface layer in forest soils. The burning of evergreen plants with naturally high levels of waxes, resins, or oils (e.g., eucalyptus, pine and fir) is prone to inducing or increasing hydrophobicity in soils after fire. Combustion of overstorey vegetation liberates large numbers of long-chain organic compounds. Flammable portions of these organic compounds are

Figure 6 Standing dead trees and heavy ash layer on scorched soil following a 2001 continuous crown fire near Delorme Lake, Alberta, Canada. Photograph by Mark Serediak.

volatilized during the active fire, and the remaining fractions migrate into the soil to a depth governed by the temperature gradient. Condensation occurs subsurface and the hydrophobic compounds are deposited in a much more concentrated and uniform layer than would normally occur (**Figure 7**). Hydrophobic conditions are also thought to increase with increasing soil organic matter and litter depth.

The establishment of post-fire hydrophobicity has a direct impact on water infiltration. Water movement can shift from primarily downward migration to increased lateral flow over the hydrophobic soil layer. Induced hydrophobicity degrades following fire, attenuating in influence anywhere from directly after the first substantial rain event up to 6 years post-burn. Although temporary, water export can be dramatically enhanced, even if hydrophobic conditions exist for only one episode of peak flow. Increased discharge has been noted in many watersheds following fire, with corresponding rises in coarse woody debris inputs to streams and element export from watersheds. This is especially true for particulate fractions, as post-fire runoff moves laterally through unconsolidated ash and burned soil layers uninhibited by vegetation.

Sedimentation and erosion In burned watersheds, sediment movement is most noticeable in precipitation events immediately following fire, but may be sustained for several years, especially during seasonal peak flows. The post-fire shift from infiltration-dominated to overland flow-dominated processes tends to proportionally increase the sediment carrying capacity. Many watershed studies have found substantial rises in water export and associated sediment loading after fire. This can substantially increase the level of suspended solids in receiving waters, which in turn can affect light penetration and productivity. The results of sediment relocation include two major factors: increased deposition of upland sediments in streams and ultimately lakes, and increased erosion and bank incision as soil layers normally covered are exposed and peak flows increase after fire.

Burned soil horizons are more vulnerable to detachment from rainsplash and other erosion processes if hydrophobic conditions exist, the ash layer is thin, vegetation is absent, or relief is high. Any or all of these situations result in a decrease of a watershed's ability to reduce sediment transport to surface waters. The movement of ash and silt into receiving waters can leave mineral layers unprotected and open to further erosive impacts. If relief is high, sediment export can reach landslide extremes either through surficial debris flows or subsurface failures resulting in soil slip. In-stream effects range from an increase in fine-fraction sediments to massive shifts in channel morphology.

Biota Relative to terrestrial plants, animals, and bacteria, aquatic organisms are largely protected from fire by virtue of location. However, intense fires can produce enough smoke and ash to cause immediate chemical toxicity, even in aquatic systems. In areas with established fire regimes, animals such as salmonid fishes exhibit adaptive strategies to surviving fires. Migration can follow heat-induced stress in lower-order streams where fires are capable of raising water temperatures to over 20 °C. However, mobility and other response mechanisms may fail for local populations when fires are extreme and escape routes are blocked. Phytoplankton, zooplankton, and benthic macroinvertebrates are more likely to be affected by in situ hydrological and chemical effects post-fire than by physical effects of the fire itself (e.g., high temperature). These effects can also have large impacts on the re-establishment of fish populations after fire, delaying colonization until fluvial processes have stabilized.

Lake zooplankton and phytoplankton community composition can also shift in response to fire. Many factors affect lake biotic response post-fire, including lake position in the landscape: shallow headwater

Table 2 Selected studies where response to fire was investigated in burned watersheds

Fire location	*Year of fire*	*Area burned*	*Sampling focus*	*Dominant cover*	*Fire source*	*Fire type*	*Time since fire*	*Effects of fire*	*Source*
Maine, USA	1947	>4000 ha	Permanent streams	Birch (*Betula papyrifera*), striped maple (*Acer pennsylvanicum*)	Wildfire	Continuous crown	50 years	Lower C and N concentrations in forest floor litter but higher C and N concentrations in upper mineral soil layers.	12
NE Minnesota, USA	1971	6000 ha	Watershed streams and lakes	Black spruce (*Picea mariana*), *Pinus* spp., maple (*Acer rubrum*), aspen (*Populus tremuloides*), birch (*B. papyrifera*)	Wildfire	NA	0–3 years	Concentrations of P in runoff elevated for 2 years following fire and decreased in third year. No detectable effects on P concentrations in receiving waters.	9
South Carolina, USA	1973	NA	Runoff	*Pinus* species	Experimental	Surface	0–1 month	Solubility increased following artificial leaching for Ca (20×), Mg (10×), Na, and K (both 2.3×). No change in biologically available N and P in litter.	8
W Montana, USA	Mid-1970s	60 plots of 0.3 ha	Soil water	Douglas fir (*Pseudotsuga menziesii*), larch (*Larix occidentalis*)	Experimental	Surface	0–2 years	Net losses of Ca and Mg below the root zone when soil surface temperatures > 300 °C. Highest nutrient release from ash followed first precipitation event, tapering with successive wettings. If ash was initially hydrophobic, nutrient release was delayed.	16
South Carolina, USA	1978–1979	60% of watershed over 2.5 years	Ground water and streams	*Pinus* species	Experimental	Surface	0–3.5 years	Low severity fires consumed less than 30% of forest floor matter, resulting in low ash production. No significant nutrient effects for ground water or stream water.	14

Continued

Table 2 Continued

Fire location	*Year of fire*	*Area burned*	*Sampling focus*	*Dominant cover*	*Fire source*	*Fire type*	*Time since fire*	*Effects of fire*	*Source*
Experimental Lakes Area, NW Ontario, Canada	1980	Most over- and understorey in watershed	Lake catchment	Jack pine (*Pinus banksiana*), black spruce (*P. mariana*), birch (*B. papyrifera*), aspen (*P. tremuloides*)	Wildfire	Continuous crown	0–3+ years	ANC decreased 20%, corresponding increase in stream acidity. Reduced ANC linked to increase in SO_4 produced by organic matter oxidation. Recovery generally within 3 years.	2
NW Montana, USA	1988	15 000 ha	Third-order streams	Lodgepole pine (*Pinus contorta*)	Wildfire	Continuous crown	Fire actively burning	5–60× increase in NH_4–N and Ortho-P, with ≤90% of total nutrient pool as soluble portions. Return to baseline levels in 2 weeks for P and up to 6 weeks for NP increase attributed to ash falling directly into the water and N increase attributed to diffusion from smoke.	15
N Australia	1990–1994	670 ha	Intermittent streams	*Eucalyptus*	Experimental	Intermittent crown	1–4 years	Increases in N, P, and TSS ≤10× only after storm runoff in most heavily burned catchment.	17
NW Quebec, Canada	1995	>50 000 ha	Forest floor	Boreal forest species	Wildfire	Variable	1 year	Forest floor dry weight reduced. Total and exchangeable Ca concentrations increased. Exchangeable K increased in 0–10 cm mineral layer. Magnitudes of pH changes proportional to fire severity. Nutrient concentrations reduced only in severely burned areas.	3

Quebec, Canada	1995	Three separate fires, 30 000–50 000 ha	Lake catchment	Transition of Boreal Mixedwood and Boreal Coniferous forests	Wildfire	Variable	0–3 years	Increased K (3–8×), TN (2–3×), TP (2×), Mg (2–3×), and SO_4 (3–4×) export rates. Element export highest in the first year and declined overall in the remaining two years. Mobile monovalent ions and SO_4 rapidly flushed, declining 50% by the third year.	5, 7
N Alberta, Canada	1995	50–100% tree cover killed per watershed	Headwater lakes	Black spruce (*P. mariana*) dominated peatlands, aspen (*P. tremuloides*) in upland areas	Wildfire	Variable	2 years and 20–40 years	Recently impacted lakes had >2× increases in P, >1.2× increases in N and 1.5× increases in dissolved organic carbon. Mean pH 9% lower in burnt catchments, including lakes with longer times since fire explained 74% of the variance in P, using time since disturbance and percent disturbance combined.	10
Thunder Bay, Ontario, Canada	1996	10 ha blocks	Forest floor	Aspen (*P. tremuloides*), spruce (*Picea* spp.), post-infestation standing dead balsam fir (*Abies balsamifera*)	Experimental	Variable	24 h to 3 months	Fire severity, as manipulated through fuel loading, reduced depth of forest floor and was negatively associated with nutrient availability for regenerating plants.	6
NW Alberta, Canada	1998	>150 000 ha	Lake catchment	*Pinus* species, spruce (*Picea* spp.), balsam fir (*A. balsamifera*), poplars (*Populus* spp.)	Wildfire	Variable	2 years	Dissolved organic carbon concentrations increased 1.4-fold. Color increased proportional to catchment area burned divided by lake volume.	1

Continued

Table 2 Continued

Fire location	*Year of fire*	*Area burned*	*Sampling focus*	*Dominant cover*	*Fire source*	*Fire type*	*Time since fire*	*Effects of fire*	*Source*
NW Alberta, Canada	1998	84–89% of watersheds burned	Thrid- and fourth-order watersheds	White spruce (*Picea glauca*), lodgepole pine (*P. contorta*), poplars (*Populus* spp.)	Wildfire	Variable	0–4 years	Runoff and particulate P fractions higher for 4 years, with average increases of 1.6× and 3.7×, respectively. Discharge, with enhanced responses to summer storm events. Increased P exports attributed to peak flow periods.	4, 13
Nevada, USA	1999	NA	Soil water	Native sagebrush (*Artemesia tridentata*) v. native and invasive annuals	Wildfire	Surficial grass	2 years	Conversion of cover to annual species may stimulate water loss immediately following fires.	11

Sources

1. Allen EW, Prepas EE, Gabos S, Strachan W, and Chen W (2003) Surface water chemistry of burned and undisturbed watersheds on the Boreal Plain: an ecoregion approach. *Journal of Environmental Engineering and Science* 2: S73–S86.
2. Bayley SE, Schindler DW, Parker BR, Stainton MP, and Beaty KG (1992) Effects of forest-fire and drought on acidity of a base-poor boreal forest stream: similarities between climatic warming and acidic precipitation. *Biogeochemistry* 17: 191–204.
3. Brais S, Par D, and Ouimet R (2000) Impacts of wild fire severity and salvage harvesting on the nutrient balance of jack pine and black spruce boreal stands. *Forest Ecology and Management* 137: 231–243.
4. Burke JM, Prepas EE, and Pinder S (2005) Runoff and phosphorus export patterns in large forested watersheds on the western Canadian Boreal Plain before and for 4 years after wildfire. *Journal of Environmental Engineering and Science* 4: 319–325.
5. Carignan R, D'Arcy P, and Lamontagne S (2000) Comparative impacts of fire and forest harvesting on water quality in Boreal Shield lakes. *Canadian Journal of Fisheries and Aquatic Sciences* 57(Supp.2), 105–117.
6. Johnston M and Elliott J (1998) The effect of fire severity on ash, and plant and soil nutrient levels following experimental burning in a boreal mixedwood stand. *Canadian Journal of Soil Science* 78: 35–44.
7. Lamontagne S, Carignan R, D'Arcy P, Prairie YT, and Par, D (2000) Element export in runoff from eastern Canadian Boreal Shield drainage basins following forest harvesting and wildfires. *Canadian Journal of Fisheries and Aquatic Sciences* 57(Suppl. 2), 118–128.
8. Lewis Jr, WM (1974) Effects of fire on nutrient movement in a South Carolina pine forest. *Ecology* 55: 1120–1127.
9. McColl JG and Grigal DF (1975) Forest fire: effects on phosphorus movement to lakes. *Science* 188: 1109–1111.
10. McEachern P, Prepas EE, Gibson JJ, and Dinsmore WP (2000) Forest fire induced impacts on phosphorus, nitrogen, and chlorophyll a concentrations in boreal subarctic lakes of northern Alberta. *Canadian Journal of Fisheries and Aquatic Sciences* 57(Suppl. 2), 73–81.
11. Obrist D, Delucia EH, and Arnone III JA (2003) Consequences of wildfire on ecosystem CO_2 and water vapor fluxes in the Great Basin. *Global Change Biology* 9: 563–574.
12. Parker JL, Fernandez IJ, Rustad LE, and Norton SA (2001) Effects of nitrogen enrichment, wildfire, and harvesting on forest-soil carbon and nitrogen. *Soil Science Society of America Journal* 65: 1248–1255.
13. Prepas EE, Burke JM, Chanasyk DS *et al.* (2003) Impact of wildfire on discharge and phosphorus export from the Sakwatamau watershed in the Swan Hills, Alberta, during the first two years. *Journal of Environmental Engineering and Science* 2: S63-S72.
14. Richter DD, Ralston CW, and Harms WR (1982) Prescribed fire: effects on water quality and forest nutrient cycling. *Science* 215: 661–663.
15. Spencer CN and Hauer FR (1991) Phosphorus and nitrogen dynamics in streams during a wildfire. *Journal of the North American Benthological Society* 10: 24–30.
16. Stark NM (1977) Fire and nutrient cycling in a Douglas fir/larch forest. *Ecology* 58: 16–30.
17. Townsend SA and Douglas MM (2000) The effect of three fire regimes on stream water quality, water yield and export coefficients in a tropical savanna (northern Australia). *Journal of Hydrology* 229, 118–137.

ANC – acid neutralizing capacity, TN – total nitrogen, TP – total phosphorus, TSS – total suspended solids.

Figure 7 Schematic representation of fire-induced hydrophobicity in soil layers. Subsurface water repellency develops after organic fractions condense in soil, enhancing lateral flow. Surface runoff increases as the wettable zone above the hydrophobic layer becomes saturated.

lakes have experienced increased cyanobacteria levels, whereas deeper lakes in large watersheds have experienced increased diatom biomass. Lake effects seem to be linked more to area burned within the watershed relative to lake surface area than to other indicators.

Intermediate Effects: up to 10 Years

Nutrient cycling Following the initial pulse after fire, chemical components in soil and receiving waters begin to stabilize. The levels of readily available nutrients begin to decline as plant regeneration starts to occur and litter depth accumulates and recovers. Nitrogen losses are ameliorated through inputs from precipitation and dry deposition, as well as N fixation from the recovering soil microbial community. Particulate P losses are reduced as runoff rates approach pre-fire conditions. Nitrate and acidic anion exports from watersheds decrease and stream pH recovers proportionally.

Infiltration and runoff Fire-induced hydrophobicity is gradually degraded and downward flow is re-established. Initial water losses can cause soil compression, producing a runoff effect more persistent than that caused solely by hydrophobic soil. However, the death of root systems can increase available flowpaths to soil layers at depth, reducing lateral flow. In addition, as revegetation occurs, compacted soil layers are loosened by new root growth, which also serves to stabilize watershed hillslopes, stream banks, and lakeshores. Coarse woody debris inputs associated with runoff can continue for several years following fire as standing dead vegetation topples to ground level.

Sedimentation Accumulation of in-stream sediment gradually decreases following fire. Soils that are not rapidly revegetated are especially prone to erosion and bank incision, which can alter mid- to long-term channel morphology. Channel profiles gradually return to pre-fire conditions, although widening and deepening may persist for several years in areas of geomorphic sensitivity (i.e., higher relief, coarse textured soils).

Biota Recovery of aboveground vegetation continues as the microbial community is re-established and nitrate-tolerant plants are succeeded. Most post-fire monitoring for other biotic components (e.g., fish, benthic macroinvertebrates, phytoplankton, zooplankton) has occurred over the short-term, and intermediate to long-term responses across a variety of fire-impacted watersheds are not known or poorly understood. There is some evidence that functional feeding groups among stream macroinvertebrates shift in dominance from allochthonous (shredder and collector) to autochthonous (scraper and filter feeder) communities in accordance with stream productivity for several years following fire (**Figure 8**).

Long-Term Effects: 10± Years Post-Fire

Most wildfire effects are immediate and short-lived, with a rapid return to pre-fire conditions for many ecosystem components. However, some long-term effects have been noted. The decomposition of coarse

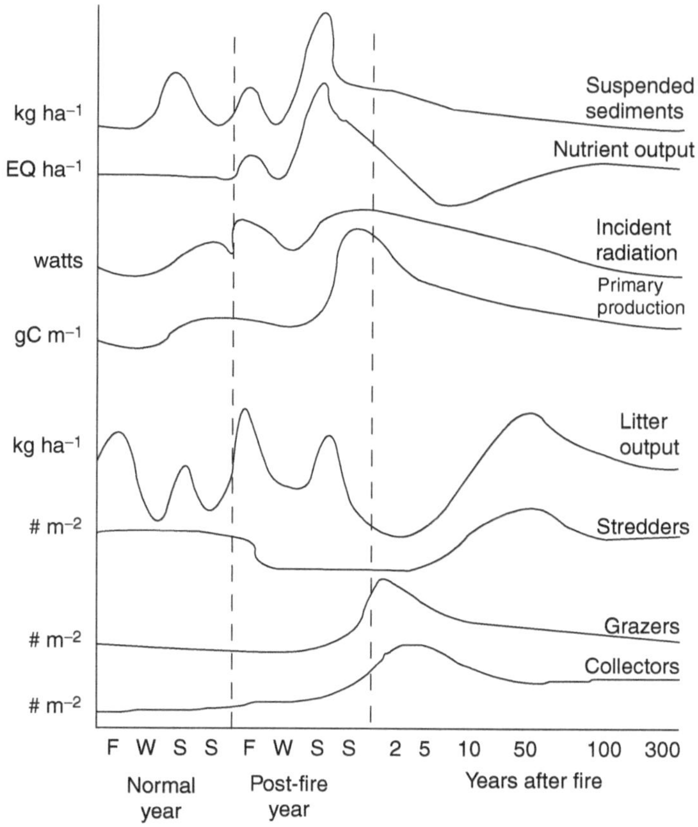

Figure 8 Watershed component response relative to time since fire. Adapted from Minshall GW, Brock JT, and Varley JD (1989) Wildfires and Yellowstone's stream ecosystems. *BioScience* 39: 707–715, with permission from the American Institute of Biological Sciences.

woody debris both in-stream and within the watershed can provide a persistent source of nutrient inputs for many years. Coarse woody debris loading can also affect stream flow and water chemistry, creating impounded areas that did not exist prior to wildfire. There can be cumulative effects on channel morphology related to initial increased discharge post-fire, and to sediment pulses that move through the watershed. Wildfires often reset the successional stage of an ecosystem, the effects of which are felt for many years following fire. Long-term monitoring is an area of increasing research, as the complexities of living in ecosystems governed by periodic fires become better understood.

Summary

The effects of wildfire on freshwater systems are shaped not only by the fire but also by characteristics of the system itself. The impacts of fire can be generalized with the following major points:

1. Fire regimes are established through the interaction of physical, chemical, and biological elements within a watershed in combination with climate.
2. Climate affects fire in both the short-term (e.g., localized drought) and the long-term (e.g., seasons, global weather patterns), and is a primary driver for both fire frequency and occurrence.
3. The impacts of fire are governed both by intensity (vertical height) and severity (downward heat transfer).
4. Intense, severe fires are more likely to occur in closed canopy ecosystems where crown fires are possible.
5. Most ecosystem components recover rapidly after fire, although persistent effects can be present for many years afterwards.
6. Fire is a naturally occurring, periodic component of functioning watersheds.

What is known about wildfire is sure to be amended as research continues into its effects on aquatic systems. While much is already understood, there are still many areas for increased focus. These include:

1. long-term monitoring at the watershed scale;
2. response and recovery of lacustrine systems;

3. modeling the effects of fire;
4. quantifying watershed risk predictors.

Continued interest in wildfire and its impacts will only strengthen the understanding of aquatic systems and their capacity to respond to large-scale perturbations.

Glossary

Continuous crown fire – Also referred to as a crown fire, bush fire, or brush fire.

Hydrophobicity – A condition of water repellency, either permanent or transient.

Infiltration – The movement of water through pores, gaps, or interstices, generally from surface to deeper soil layers.

$kW\,h^{-1}$ – Unit of fire intensity equal to energy release per unit length of fire front.

Pyrolysis – Chemical change brought about by the action of heat.

Surface fire – Often interchanged with grass fire, although a surface fire can occur under any cover type.

Volatilization – Causing to pass off in vapor.

See also: Acidification; Alkalinity; Carbon, Unifying Currency; Chemical Fluxes and Dynamics in River and Stream Ecosystems; Chemical Properties of Water; Chloride; Fluvial Transport of Suspended Solids; Ground Water and Surface Water Interaction; Groundwater Chemistry; Major Cations (Ca, Mg, Na, K, Al); Natural Organic Matter; Nitrogen; Nitrogen Fixation; Nutrient Stoichiometry in Aquatic Ecosystems; Organic Nitrogen; Phosphorus; Rivers.

Further Reading

Clark JS (1997) *Sediment Records of Biomass Burning and Global Change.* Berlin: Springer.

Doerr SH, Shakesby RA, and Walsh RPD (2000) Soil water repellency: its causes, characteristics and hydro-geomorphological significance. *Earth-Sciences Reviews* 51: 33–65.

Goldammer JG and Jenkins MJ (Des) (1990) *Fire in Ecosystem Dynamics: Mediterranean and Northern Perspectives.* The Hague: SPB Academic Publishers.

Gresswell RE (1999) Fire and aquatic ecosystems in forested biomes of North America. *Transactions of the American Fisheries Society* 128: 193–221.

Kasischke ES and Stocks BJ (2000) *Fire, Climate Change, and Carbon Cycling in the Boreal Forest.* New York: Springer.

Kozlowski TT and Ahlgren CE (1974) *Fire and Ecosystems.* New York: Academic Press.

Likens GE (1989) *Long-Term Studies in Ecology: Approaches and Alternatives.* New York: Springer-Verlag.

Mooney HA, Bonnicksen TM, Christensen NT, Lotan JE, and Reiners WA (eds.) (1981) *Fire Regimes and Ecosystem Properties.* U.S. Forest Service General Technical Report WO-26.

Pickett STA and White PS (eds.) (1985) *The Ecology of Natural Disturbance and Patch Dynamics.* San Diego: Academic Press.

Sala M and Rubio JL (1994) *Soil Erosion and Degradation as a Consequence of Forest Fires.* Logrono, Spain: Geoforma Ediciones.

Strauss D, Bednar L, and Mees R (1989) Do one percent of forest fires cause ninety-nine percent of the damage? *Forest Science* 35: 319–328.

Swanson FJ (1981) Fire and geomorphic processes. U.S. Forest Service General Technical Report WO-26, pp. 401–420.

Turner MG and Romme WH (1994) Landscape dynamics in crown fire ecosystems. *Landscape Ecology* 9: 59–77.

Wein RW and MacLean D (1983) *The Role of Fire in Northern, Circumpolar Ecosystems.* New York: Wiley.

Wright HA and Bailey AW (1982) *Fire Ecology.* New York: Wiley.

Relevant Websites

http://www.fire.uni-freiburg.de/welcome.html – Global Fire Monitoring Centre.

http://rapidfire.sci.gsfc.nasa.gov/ – MODIS Rapid Response System.

http://www.idrc.ca/imfn/ – International Model Forest Network.

http://www.ciffc.ca/ – Canadian Interagency Forest Fire Centre.

Floods

J A Hubbart and J R Jones, University of Missouri, Columbia, MO, USA

Definition of Flood

A flood is loosely defined as river discharge exceeding bankfull limitations. It is also considered a temporary rise of the water level, as in a river or lake or along a seacoast or wetland, resulting in its spilling over and out of its natural or artificial confines onto land that is normally dry. Floods are sometimes described according to their statistical occurrence. For example, a 50-year flood is a flood having a magnitude that is reached in a particular location on average once every 50 years. This is often referred to as a return interval (Tr), and is calculated as follows:

$$\mathrm{Tr} = (n + 1)/m$$

where n is the total number of events, and m is the specific event number in question. With this calculation, probability (p) of event occurrence is calculated by

$$p = 1/\mathrm{Tr}$$

A helpful way to visualize the difference in flow between a flood and normal runoff is to visualize the flood channel width, which is the floodplain portion that will discharge the 50-year flood (**Figure 1**). In this example, the flood zone is centered over the main channel, an unusual situation in natural systems where the flood zone can be offset or split into several zones depending on the topography.

Forecasting annual flows and the magnitude and frequency of flood events is a challenge. Hydrologic data from unaltered, wildland systems show great variation in annual floods (coefficients of variation ≥ 1.0), making it difficult to predict when the floodplain will be inundated. Human land use changes further complicate flood forecasting because impermeable surfaces reduce infiltration and accelerate runoff. Ultimately, both natural and anthropogenic processes can result in watershed morphological changes that may modify flow. It is noteworthy that even minor changes in flood magnitude, duration or frequency, although statistically undetectable, should not be misinterpreted as ecologically or culturally benign.

Causes and Effects of Floods

All rivers and streams are subject to fluctuations in flow. During a rainstorm, the amount, intensity, duration, area, and path of the storm all influence the runoff reaching the stream. Multiple land form and use factors affect the ability of land to absorb precipitation and therefore affect the rate of runoff. Area and path of the storm relate to the area of the basin receiving rainfall, which in turn, represents the area contributing runoff. Area and the runoff rate determine the volume of water that will pass a given point downstream.

Modification of runoff rates occurs by variations in topographic relief, soil infiltration processes, vegetative cover, and surface retention (e.g., ponding) within a given catchment or watershed. The key physical mechanism controlling runoff and flooding processes is precipitation. River flow is largely determined by the precipitation regime (i.e., amount and type of precipitation). Precipitation type is determined by factors such as elevation (rain, snow) and orographic uplift, and whether it is on the leeward or windward side of mountains. Orographic uplift often results in heavier precipitation on the crests and windward slopes of mountain ranges. It also accounts for much heavier precipitation than in surrounding lowlands. The same process also causes rain shadow effects on leeward sides of mountains. The nature and condition of the drainage basin, and variable climates also affect streamflow, and therefore flood potential. Vegetative cover also affects the rate at which surface water flows to a main channel by slowing and spreading out runoff. The passage of water is similarly ameliorated in basins with natural storage areas, including lakes and wetlands. Consequently, smaller peak flows are produced in basins with dense vegetation and lakes, reservoirs and wetlands than those without.

Snowmelt: In high elevations and northern latitudes, most precipitation is snow. During snowmelt, large quantities of water are released. Snowmelt induced runoff floods are the most common type of flooding in these areas and generally occur in the spring but also occur during sudden winter thaws and as a consequence of rain-on-snow events. Heavy runoff results from rapid melting of snow under the combined effect of sunlight, winds, warmer temperatures, and rain. When there is an above average snow depth, a sudden thaw, or both, the potential for high volumes of runoff and subsequent flooding increases. This process is made more severe when the rising snowmelt runoff is compounded by runoff from heavy rainfall. Because climatic factors influencing the rate of snowmelt are often widespread, snowmelt runoff flooding conditions can exist over vast areas

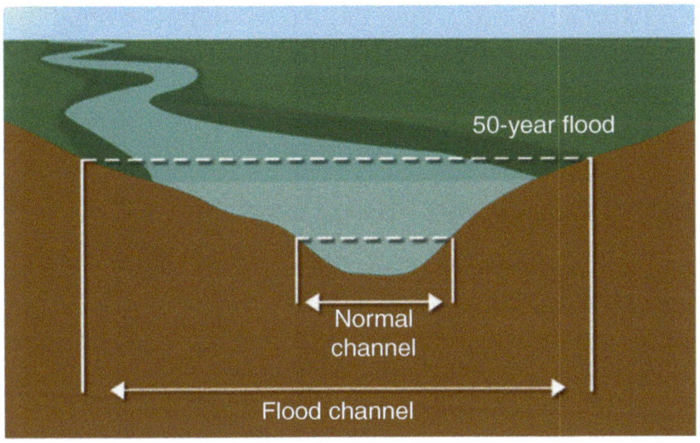

Figure 1 Example 50-year flood event.

and mobilize and transport a great deal of debris and sediment.

Rainfall: Heavy rainfall can result in flash flooding. Although thousands of hectares are frequently flooded as a result of flash flooding (for example, northern Queensland, Australia, and the upper Mississippi River drainage, USA), flash floods usually occur in small watersheds as a result of large rain events and are characterized by peak flow within six hours of the onset of rainfall. Flood conditions develop rapidly because heavy rainfall surpasses the infiltration capacity of the soil, resulting in a very high runoff rate. These types of events are generally locally intense and damage is usually restricted to a limited area. Large rivers generally remain unaffected, while smaller streams can overtop their banks, even in a drought year.

Seasonal and inter-annual variability in rainfall and flooding are often a result of El Niño-Southern Oscillation (ENSO). ENSO is a global system of ocean–atmosphere climate fluctuations arising from warmer ocean currents coupled to higher air temperatures. The result is markedly increased evaporation and large-scale interaction between the ocean and the atmosphere. Effects of ENSO are observed by changes in the distribution of rainfall, causing floods in some areas and drought in others. This process leads to drastic alterations to normal weather patterns, including heavy rains and catastrophic flooding in the United States, Asia, and other parts of South and Central America. For example, large areas of Asia receive more than 80% of annual precipitation during the seasonal monsoon season. Conversely, regions in Australia, Indonesia, and India may undergo severe drought because moisture normally dispersed around the world is evaporating too quickly and staying within the Eastern Pacific Ocean. ENSO is the most prominent known source of inter-annual variability in weather and climate around the world (range: 3–8 years).

Ice jams: Ice jams are a major cause of flooding in northern latitudes. In fact, for most northern rivers, the annual peak water levels are due to ice jams resulting from the accumulation of ice fragments, which build up and restrict the flow of water. A rise in water levels may result from spring snowmelt, or a sudden midwinter thaw accompanied by substantial rainfall, resulting in a rapid increase in water levels and severe ice jams. Ice jams can lead to flooding because of two main features. First, ice jam thickness can be considerable, amounting to several meters. Second, the underside of the ice cover is usually very rough. Therefore, the flow depth has to be much greater than for open water, resulting in relatively high water levels with relatively small discharges. This condition leads to a great deal of water and pressure that when released can lead to substantial flooding.

Outburst floods: Outburst floods are also common in northern latitudes, and have some similarities to ice jams. Lakes dammed by glaciers or moraines suddenly drain and large quantities of water, mud, and debris are released. An outburst flood typically occurs when the water level becomes high enough to actually float the ice or when a small channel forms under the ice and causes rapid melting, thus expanding the channel. Another common mechanism is overtopping of the ice dam and the rapid thermal and mechanical erosion of a channel, leading to sudden large scale drainage. The release of water is often sudden and catastrophic.

Coastal storms: Humans living along the shores of major lakes or along ocean coasts are occasionally subject to flooding as a result of high wind and wave action, or the interaction between high estuarine flows and tides. Shoreline flooding may be caused by storm surges often occurring simultaneously with high waves. Surges are caused by sudden changes in atmospheric pressure and by wind stress accompanying moving storm systems. In certain coastal regions, maximum storm surges are produced by severe tropical hurricanes. Along the coasts, severe storms can produce surges of up to 2 m, but in some areas of the world, for example, Bangladesh, severe storms can produce surges in excess of 8 m. Generally, surges in lakes and reservoirs are less, though they have been recorded as high as 2.5 m on Lake Erie. Specific types of coastal flooding events include tsunamis, cyclones, and hurricanes.

Urban runoff: Urban stormwater runoff can flood local rivers as well as the urban area itself. Urbanization drastically alters the drainage characteristics of natural catchments by increased impervious surface area and thus volume and rate of surface runoff (**Figure 2**). Other effects of increased impervious

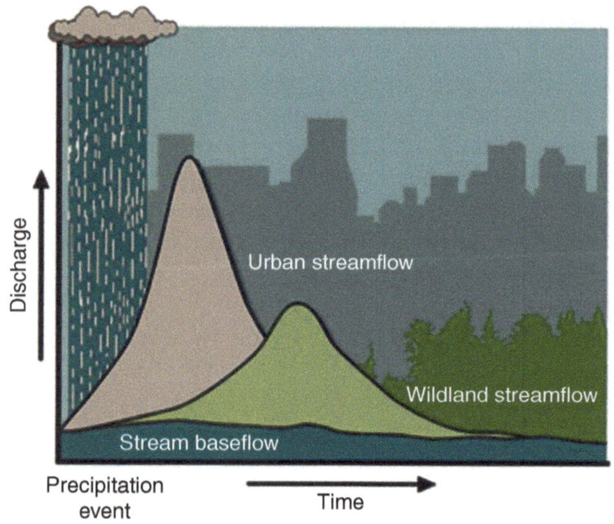

Figure 2 Example of difference between wildland (unaltered) and urban (human modified) event based runoff regimes.

surface area can include decreased water quality, changes to microclimates, habitat degradation and destruction, and diminished stream and landscape aesthetics. Although the impact of urban runoff on major river systems may be minimal, the carrying capacity of small streams may be quickly exceeded, causing flooding and channel erosion problems. Runoff from intense rainfall can exceed the carrying capacity of sewer systems, creating a backup and thus, flooding. In urban settings, streamflow-gauging stations are often used to provide continuous flow records that can be used in the design of new urban infrastructure, including roads, bridges, culverts, channels, and detention structures.

Tropical flooding: Tropical floods are usually caused by cyclones, otherwise known as hurricanes, typhoons, or tropical storms. Cyclones often result in large quantities of rain falling in a short time and can result in a great deal of flooding and sometimes human loss of life. There are two basic types of flood caused by tropical cyclones. Flash flooding occurring in streams and urban areas almost immediately following rainfall and rising water can reach depths of multiple meters. River flooding generally occurs from heavy rains coupled with recent cyclonic activity and can persist for weeks. The impact of tropical floods is locally variable. Water levels that exceed flood stage can constitute minor, moderate, and major flooding over relatively short geographic distances due to topographic variability and ability for the terrain to attenuate flood flows.

Dam failure: Flooding also results from the failure of dams or other hydraulic structures. The suddenness and magnitude of these events often have disastrous results. The failure of dams formed by beaver (*Castor canadensis*) can also result in an outburst flood of impounded water.

Flood Costs and Mitigation

Floods can be disastrous. Often, personal property, businesses, industries, crops, and roads are damaged and human lives can be lost. Floods cost humans many millions of dollars every year in property damage, lost production, lost wages, and lost business. Floods however, are also a natural phenomenon and are often necessary for ecosystem health. The concept of 'environmental flows' was developed in this case to determine how much water needs to be left in a river in order to maintain its ecological health. These flows are critical (especially in dry regions) to provide water for floral and faunal communities, as well as security for human use and socio-economic stability.

Economic development in concert with an increasing population has brought pressure altering the flow regime of surface water systems. Human efforts to constrict the active zone of floodplains that attenuate floods include the construction of dams, dykes, and concrete diversion channels, channel dredging and realignment, and drainage of wetlands. These measures, although perhaps beneficial in the interests of economic development, have resulted in the decline of fish and wildlife habitats, and the disruption of entire ecosystems. Many humans have a false sense of security owing to the size and proclaimed strength of levees protecting their livelihood. Hence, levees may increase the potential for floods by a reduction in flood preparedness and by creating incentives to build structures in areas subject to flooding.

Flood amelioration practices can include materials that are put in place to increase bank stabilization, reduce bank erosion, and stop the meandering of streams and rivers. Common strategies include dikes, loose rock paving, and establishment of plant communities on banks. Often, these practices are successful in reducing bank erosion, but hinder the ability of the channel to widen naturally through flooding, leading to increased flooding by reducing the length of the stream or river, and increasing flow velocity resulting in channel incision. The amount of sediment transported depends mainly on flow volume (**Figure 3**), which also depends on size of the drainage basin, and rate and volume of precipitation. Increases in suspended sediment concentrations above natural levels often have a detrimental impact on fish and invertebrate habitat in streams.

High levels of suspended sediment can reduce the effectiveness of drinking water treatment processes

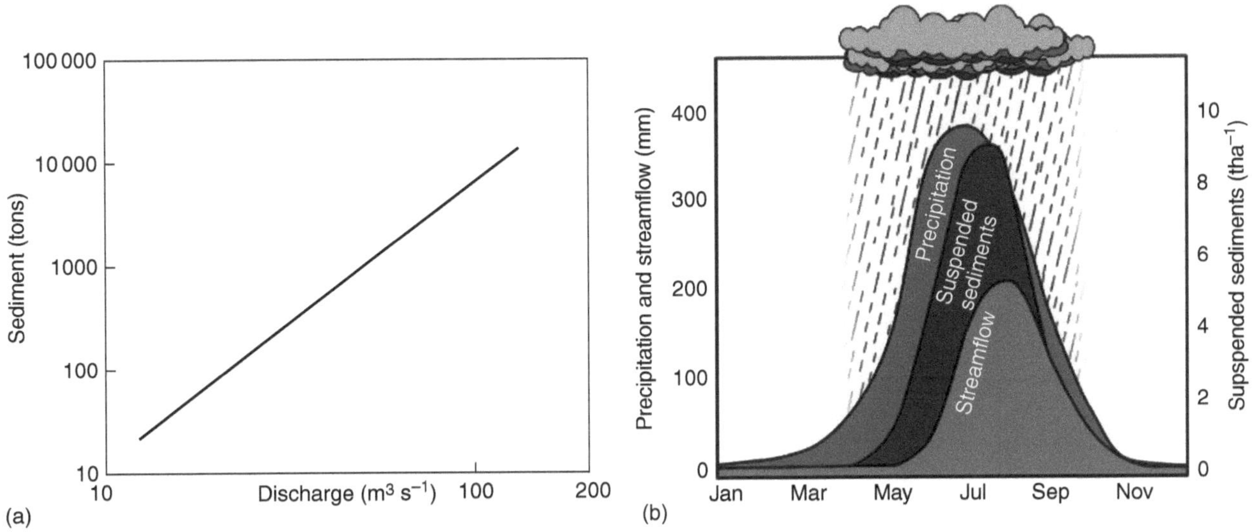

Figure 3 Relationships between (a) sediment flux and stream or river discharge, and (b) precipitation, streamflow and suspended sediment.

and may increase maintenance costs by clogging or reducing the capacity of filtration systems. Suspended sediment and other particulates are aesthetically undesirable for domestic use and can be associated with higher bacterial concentrations. Suspended sediment carried by flood waters can reduce light penetration and temporarily decrease plant productivity in lakes and streams. The same flood waters can also transport nutrients such as phosphorus with the sediment often resulting in an infusion of otherwise nutrient limited waters.

Flood waters are a natural driving force in maintaining the productivity of rivers and floodplain systems. Floods inundate adjacent floodplains and connect river channels with streamside soils and vegetation that result in chemical and faunal exchanges that influence both communities. Peak flows that move or abrade stream substrate can scour attached algae and aquatic macrophytes and reduce or change the community structure of benthic invertebrates and fish populations. Often recovery of these communities to pre-flood conditions is quite rapid. Floods are major contributor to annual inputs of nutrients to lakes and reservoirs. Dissolved nitrogen and phosphorus, delivered with flood inflow, drive aquatic productivity and generally lake systems are more productive during wet years than during periods of drought.

Because of the intricate nature of river ecosystems, water quality sampling and analysis programs are necessary to provide data essential to the understanding and behavior of natural systems and influences of human activity. Thus, simultaneous and continuous sampling of suspended sediment is critical for accurate understanding of pollutant loading of the watershed which is (a) governed by hydrological processes, and (b) closely correlated to chemical pollutant concentrations. These processes are controlled to a large degree by local microclimates.

Ultimately, flooding is a necessary natural process that only has negative effects when humans are involved. Human inhabitation of land that naturally floods causes great losses of life and property. Anthropogenic alterations to these flood-based systems have only worsened the problem in most cases.

See also: Atmospheric Water and Precipitation; Fluvial Export; Fluvial Transport of Suspended Solids; Ground Water; Ground Water and Surface Water Interaction; Streams; Vadose Water.

Further Reading

Junk WJ (2005) Flood pulsing and the linkages between terrestrial, aquatic and wetland systems. *Verh. Internat. Verein. Limnol.* 29: 11–38.

Mays LW (1996) *Water Resources Handbook*. NY: McGraw-Hill.

Novotny V and Olem H (1994) *Water Quality. Prevention, Identification, and Management of Diffuse Pollution*. New York: Van Nostrand Reinhold.

Wanielista M, Kersten R, and Eaglin R (1997) *Hydrology: Water Quantity and Quality Control*. 2nd edn. NJ: Wiley.

Ward AD and Elliot WJ (1995) *Environmental Hydrology*. Boca Raton, FL: CRC.

Mercury Pollution in Remote Fresh Waters

C J Watras, Wisconsin Department of Natural Resources, Madison, WI, USA

Introduction

Mercury pollution in fresh waters emerged as a widespread environmental issue during the 1980s. Prior environmental concern had focused on geographically isolated incidents involving direct mercury discharges into surface waters or the contamination of large, long-lived marine fish like swordfish and tuna. However, during the 1980s, reports of mercury-contaminated fish from remote fresh waters across eastern North America and northern Europe rose dramatically.

In Sweden, roughly 10 000 lakes were 'black-listed' due to unacceptably high mercury levels in pike. In the Great Lakes region of North America, the number of lakes with public health advisories cautioning people against the consumption of native fish increased more than 10-fold. By 1994, 34 US states had mercury advisories, rising to 41 states in 2000. Many of the affected fresh waters were otherwise pristine ecosystems.

Concern about widespread fishery contamination prompted biogeochemical investigations to identify the sources and fates of mercury in remote lakes and small watersheds. However, early field studies were hampered by sampling artifacts (due to inadvertent sample contamination) and by high analytical detection limits. To overcome these obstacles, 'clean' sample handling protocols were adopted, and new analytical procedures were developed to detect aqueous mercury at picomolar concentrations. By the late 1980s, it was possible to determine waterborne concentrations of mercury and methylmercury accurately in most fresh waters. By the early 1990s, credible mass balances for mercury and methylmercury in fresh waters began to emerge.

The Distribution of Mercury in Fresh Waters

Lake Water and Sediments

In the surface waters of relatively pristine lakes, concentrations of mercury and methylmercury (meHg) typically range from about 0.1–5 ng L^{-1} (**Table 1**). Over this range, there is often a direct relationship with dissolved organic carbon (DOC). DOC of terrestrial origin (i.e., humic matter) is a strong ligand for Hg, meHg, and other trace metals. Equilibrium calculations indicate that most of the dissolved Hg_T in remote fresh waters is bound to DOC. Organic complexation increases metal residence time in the water column, thus, partly explaining the direct relationship. Mercury and DOC are also co-transported to lakes from terrestrial watersheds (especially from wetlands), so there is a common-source factor as well. Concentrations of waterborne Hg and meHg may co-vary with pH, but the relationship tends to be non-linear. The non-linearity may reflect the pH-dependence of several relevant processes, such as Hg-DOC binding, DOC coagulation and settling, and the photochemical reduction of $Hg(OH)_2$ to volatile Hg^0.

In contaminated lakes that received historical discharges of industrial or geologic mercury, Hg can reach much higher concentrations in surface waters and sediments. However, meHg concentrations often remain disproportionately low (**Table 1**). Relatively low concentrations of meHg in contaminated lakes may reflect the biocidal effect of $Hg^{(II)}$ on methylating bacteria or the stimulation of inducible demethylation pathways, as described below. Perhaps the most striking difference between remote and contaminated lakes is observed in sediments, where Hg_T concentrations may differ by several orders of magnitude. This observation reflects the rapid downward transport and burial of Hg via settling particles and, consequently, the short residence time for Hg in the water-column of many fresh waters ($\leq$ 1 year). Although the remobilization of some sedimentary Hg undoubtedly occurs, estimates are poorly constrained and vary widely. The conservative behavior of Hg in many sediment cores suggests that once below the surface, Hg is effectively buried. The ubiquitously low concentrations of meHg in sediments relative to source material (i.e., pelagic biota and settling seston) indicate that demethylation is the dominant reaction within sediments over annual time-scales.

Hypolimnetic Enrichment

In anoxic hypolimnetic waters, concentrations of waterborne Hg_T and meHg are often much higher than in surface waters. In remote lakes, hypolimnetic enrichment may approach 10-fold for Hg_T and 100-fold for meHg (**Figure 1**). Hypolimnetic enrichment has not been observed in lakes with oxic hypolimnia. Concentrations of waterborne Hg_T and meHg are not consistently higher in contaminated than in

Table 1 Concentration of mercury (Hg_T, total mercury) and methylmercury ($meHg_T$) surface waters of freshwater lakes and reservoirs (na, data not available)

Category	*Water body*	*Surface water*		*Surficial sediments*		*Source*
		Hg_T (ng L^{-1})	*$meHg_T$ (ng L^{-1})*	*Hg_T (ng g^{-1}, dwt)*	*$meHg_T$ (ng g^{-1}, dwt)*	
Remote	Swedish forest lakes	0.8–10	0.04–0.8	50–300	na	1
	Northern Wisconsin lakes, USA	0.5–4.4	0.04–0.8	63–289	0.5–7.4	2
	Adirondack lakes, NY, USA	0.8–5.3	0.07–0.6	na	na	3
	Rocky Mountain lakes, USA	0.4–2.9	0.01–0.1	90–100	<1–5	4
	Mountain lakes, western USA	0.3–3.3	0.01–0.4	na	na	5
	Finnish lakes	1.3–7.2	0.12–2.2	134–277	0.1–8	6
	Lake Baikal	0.25	0.01	40	na	7
	Lake Superior	0.2–1	0.003–0.2	16–209	0.1–0.5	8
	Voyageurs National Park, MN	0.4–3.3	<0.04–0.3	102–364	na	9
Polluted	Onondaga Lake, NY, USA	3–9	0.2–1.4	1 000–50 000	4–11	10
	Clear Lake, CA, USA (Oaks Arm)	10–105	0.02–0.25	30 000–150 000	3–13	11
	Clay Lake, Ontario, Canada	5–80	0.2–4.1	3000–14 000	7–22	12
	Lahontan Reservoir, Nevada, USA	4–2100	0.3–7.2	25 000–30 000	25–35	13

Sources:

1. Lindqvist, O. (ed.) (1991) *Water, Air, Soil Pollut* 55: 1–261.
2. Watras CJ, *et al.* (1998) *Sci Tot Environ* 219: 183–208.
3. Driscoll CT (1995) *Environ Sci Technol* 28: 136A–143A.
4. Watras CJ, *et al.* (1995) *Can. J Fish Aquat Sci* 52: 1220–1228.
5. Krabbenhoft DP, *et al.* (2002) *Water, Air, Soil Pollut Focus.* 2(2): 233–249
6. Verta M, *et al.* (1994) In Watras CJ and Huckabee JW, (eds). *Mercury Pollution: Integration and Synthesis*. Boca Raton, FL: Lewis Publishers. pp. 119–136; Verta, M and Matilainen T. (1995) *Water, Air, Soil Pollut* 80: 585–588.
7. Leermakers M, *et al.* (1996) In Baeyens W, *et al.* (eds). Global and Regional Mercury Cycles: Sources, Fluxes and Mass Balances. *NATO ASI Series 2 – Environment*. Vol. 21, pp. 303–315.
8. Rolfus K, *et al.* (2003) *Env Sci Technol* 37: 865–872.
9. Wiener JG, *et al.* (2006) *Env Sci Technol* 40: 6261–6268.
10. Henry EA, *et al.* (1995) *Water, Air, Soil Pollut* 80: 509–518; Jacobs LA, *et al.* (1995) *Water, Air, Soil Pollut* 80: 553–562; Gbondo-Tugbawa S and Driscoll CT. (1998) *Water, Air, Soil Pollut* 105: 417–426.
11. Bale AE (2000) *J Environ Eng* 126: 153–163.
12. Parks JW, *et al.* (1989) *Can J Fish Aquat Sci* 46: 2184–2202; Jackson TA, Environment Canada, personal communication.
13. Bonzongo JC, *et al.* (1996) *Environ Pollut* 92: 193–201; Diamond M, *et al.* (2000). *Water, Air, Soil Pollut* 117: 133–156.

remote anoxic hypolimnia. These observations may reflect the dominant role that sulfide plays in controlling the aqueous speciation of mercury in anoxic waters. Both $Hg^{(II)}$ and meHg have a strong affinity for sulfide. In anoxic hypolimnia, dissolved sulfide can strip $Hg^{(II)}$ and meHg from settling particles and shunt them back into the dissolved phase. Equilibrium speciation models indicate that mercury-sulfide complexes dominate the dissolved phase in sulfidic waters (i.e., HgS^o, $Hg(SH)_2$, CH_3HgSH^o, and polysulfides), providing a geochemical explanation for the observed hypolimnetic build-up.

Interactions with sulfide may also set the stage for meHg production in anoxic waters. Laboratory studies indicate that neutrally-charged Hg-S complexes, such as HgS^o and $Hg(SH)_2$, are passively transported across the cell membrane of microbes that inhabit sulfidic hypolimnia. Certain of these microbes, particularly sulfate reducing bacteria (SRB), are capable of methylating this $Hg^{(II)}$. Thus, a biogeochemical scenario explaining the hypolimnetic build-up of meHg may proceed in four steps: (1) downward transport of $Hg^{(II)}$ on settling particles, (2) stripping by S^{2-} to yield dissolved HgS^o and $Hg(SH)_2$, (3) uptake by SRB and production of meHg, and (4) accumulation of CH_3HgSH^o in the deep water. Measurements of methylation rates using stable and radio isotopes of $Hg^{(II)}$ have shown that meHg is produced at sufficiently high rates to account for the hypolimnetic build-up.

Alternative explanations for hypolimnetic build-up include: (1) the reductive dissolution of Hg-laden and meHg-laden oxyhydroxides of Fe and Mn and (2) the remineralization of Hg and meHg from settling detrital seston.

Seasonality

The seasonal cycle of mercury in the waters of temperate/boreal lakes often involves a period of increase during spring and a period of decline during fall

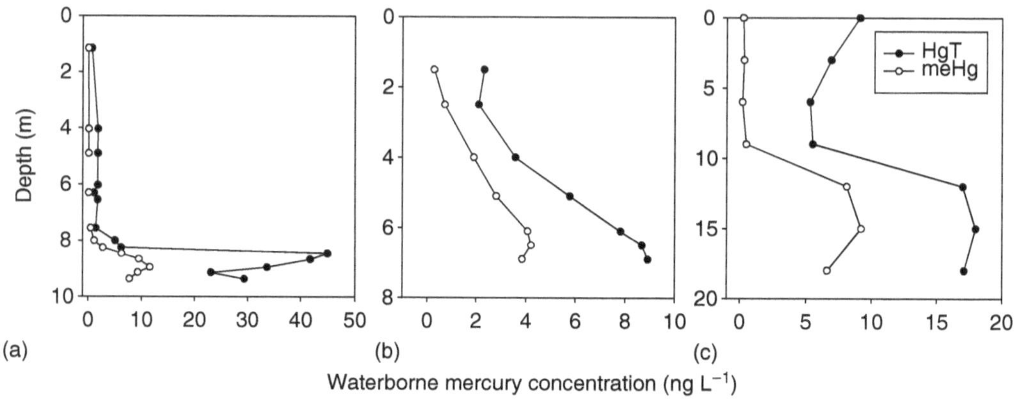

Figure 1 Hypolimnetic enrichment of waterborne Hg_T and meHg in the deep, anoxic waters of three lakes. (A) Little Rock Lake – a small, precipitation-dominated, clear-water seepage lake in northern Wisconsin; (B) Devils Lake, a small, wetland-dominated, dark-water drainage lake in northern Wisconsin; (C) Onondaga Lake, a large, Hg-polluted drainage lake in Syracuse, New York. Data from Watras *et al.* (1994) in Watras CJ and Huckabee JW (eds) *Mercury Pollution: Integration and Synthesis*. Boca Raton, FL: Lewis Publishers. pp. 153–177; Watras *et al.*, unpublished data; Jacobs LA *et al.* (1995) *Water, Air, Soil Pollut* 80: 53–562.

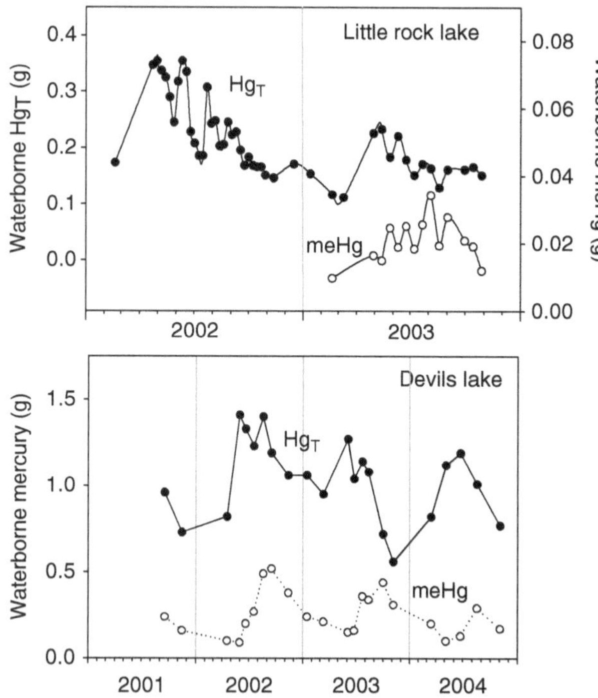

Figure 2 Annual cycles of waterborne Hg_T and meHg in Little Rock Lake and Devils Lake in northern Wisconsin. Data from Watras *et al. Can J Fish Aquat Sci* 65: 100–116.

(**Figure 2**). In northern Wisconsin, seasonal changes in waterborne mercury concentrations reflect the annual cycle of atmospheric $Hg^{(II)}$ deposition and the annual cycle of microbial meHg production. During spring, melting snow and ice can combine with heavy rain to maximize the loading of $Hg^{(II)}$ to remote Wisconsin lakes. As a result, waterborne concentrations of $Hg^{(II)}$ in surface waters rise. During summer, settling seston can remove $Hg^{(II)}$ from the epilimnion faster than it is re-supplied by the atmosphere. Under these conditions, waterborne Hg_T declines. Simultaneously, the activity of methylating microbes can increase in deep anoxic regions of the lake, leading to a peak in waterborne meHg production.

Thus, the peak in waterborne MeHg tends to occur several weeks after the peak in waterborne $Hg^{(II)}$ has begun to subside. MeHg production in wetland areas of contributing watersheds also peaks during summer, providing an additional source of meHg to receiving waters. During winter, the lakes are sealed by ice from atmospheric inputs of $Hg^{(II)}$ and microbial activity presumably subsides. Concentrations of waterborne Hg_T and meHg then decrease to minimum values.

Biota

Mercury is one of the few toxic elements that is known to biomagnify in aquatic foodwebs. Analysis of pelagic organisms from both remote and contaminated lakes indicates that meHg, rather than $Hg^{(II)}$, is preferentially transported up the food chain. In sediments, in water, and at the base of the aquatic foodchain, meHg constitutes a small but increasing fraction of the Hg_T pool. However, at high trophic levels, almost all of the Hg_T is in the methylated form (**Figure 3**). This behavior suggests that meHg turns over more slowly than carbon or $Hg^{(II)}$ in higher organisms.

Within simple aquatic foodchains, the major step in meHg bioaccumulation occurs at low trophic levels. In phytoplankton, the concentration of meHg is typically 10^4-fold higher than in water. Between successive trophic levels, biomagnification factors typically vary from 2-fold to 10-fold.

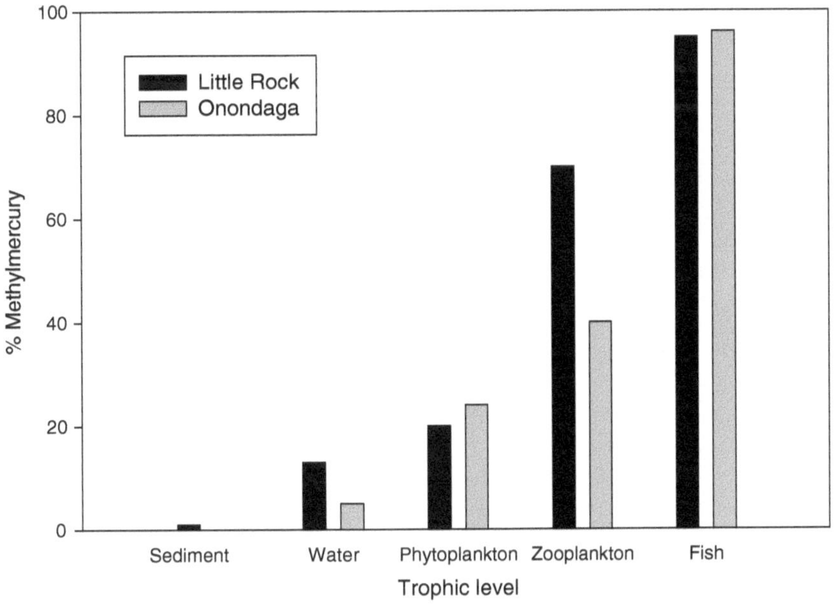

Figure 3 Biomagnification of meHg in the pelagic foodwebs of Little Rock Lake (remote) and Onondaga Lake (contaminated). Data from Watras CJ *et al.* (1994) in Watras CJ and Huckabee JW (eds). *Mercury Pollution: Integration and Synthesis*. Boca Raton, FL: Lewis Publishers. pp. 153–177; Watras *et al.* (1996) in Baeyens W *et al.* (eds). Global and regional mercury cycles: sources, fluxes and mass balances. *NATO ASI Series 2 – Environment*. Vol. 21, pp 329–358; Henry EA *et al.* (1995) *Water, Air, Soil Pollut.* 80: 509–518.

The concentration of meHg in fish tissue can be 10^7-fold higher than that in water. Unlike other organic contaminants, such as DDT and PCBs, meHg is stored primarily in the muscle tissues of fish rather than in fatty deposits. Within muscle tissue, meHg appears to bind preferentially to the sulfur-containing amino acids, cysteine and methionine, reflecting its strong affinity for reduced sulfur.

Within fish communities, concentrations of meHg increase with trophic position, fish size and pH. High concentrations of meHg are typically found in large, predatory fish from low pH lakes, while lower concentrations are generally found in small, forage fish from high pH lakes (**Figure 4**). The effects of trophic position and body size reflect the importance of diet in the bioaccumulation of meHg. However, the negative relationship between fish meHg and pH or ANC is likely the result of factors that co-vary with lake acidification rather than the direct effect of pH *per se*. Several studies indicate that such factors affect the net production of meHg and its availability to the base of aquatic food chains.

Growth rates can affect the concentration of meHg in aquatic biota, a phenomenon known as growth dilution. The simplest and most general form of the expression relating biotic meHg concentration and growth can be written as

$$C_{\text{biota}} = V_{\text{net}}/u,$$

where C_{biota} is the concentration of meHg in an organism, V_{net} is the net uptake rate from all sources, and u is the growth rate or dilution term. More complex models expand the terms on the right-hand side of this expression to reflect specific physiological and biogeochemical processes. Biodilution is a variant of growth dilution. It occurs when meHg is diluted by algal biomass so that herbivores are exposed to lower carbon-specific meHg concentrations in their diet.

For humans and wildlife, meHg in fish constitutes the health risk of concern. Survey data from Wisconsin indicate a direct relationship between rates of fish consumption and body burdens of meHg in humans (**Figure 5**). Of the 2000 survey respondents, 29% of the men and 13% of the women had body burdens above the level of concern (1 $\mu g\ g^{-1}$ in hair). Respondents who consumed sportfish had higher meHg burdens than other respondents for the same level of fish consumption. Fetuses and young children constitute the human population at the greatest risk from meHg, due to effects on the developing nervous system.

Laboratory and field studies of fish, birds, and mammals indicate that dietary meHg can lead to deleterious changes in behavior and reproduction at concentrations observed in natural environments. Some piscivorous wildlife have detoxification mechanisms that reduce the effects of meHg exposure. In loons and other aquatic birds that feed on fish, meHg is partially shunted to feathers which, being proteinaceous, contain cysteine and methionine. When feathers are shed during the molting season, they take their meHg burden with them.

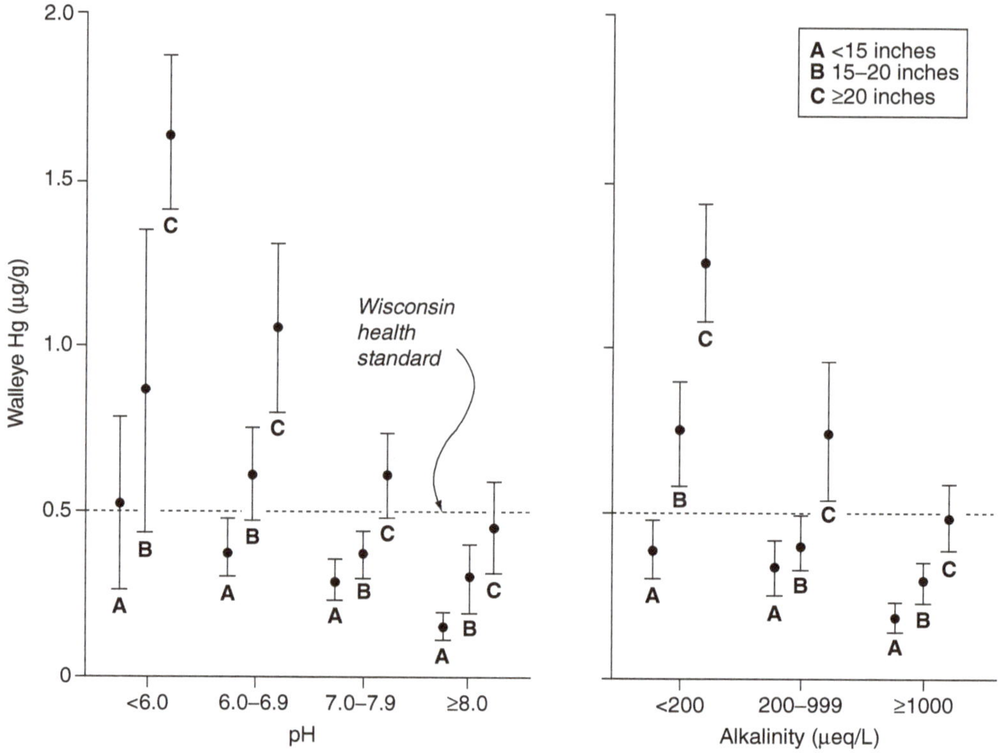

Figure 4 Mercury concentrations in Wisconsin walleye as a function of fish size, lake water pH and alkalinity. Data From Lathrop RC *et al.* (1989) *Technical Bull 163*. Wisconsin Department Natural Resources, with permission.

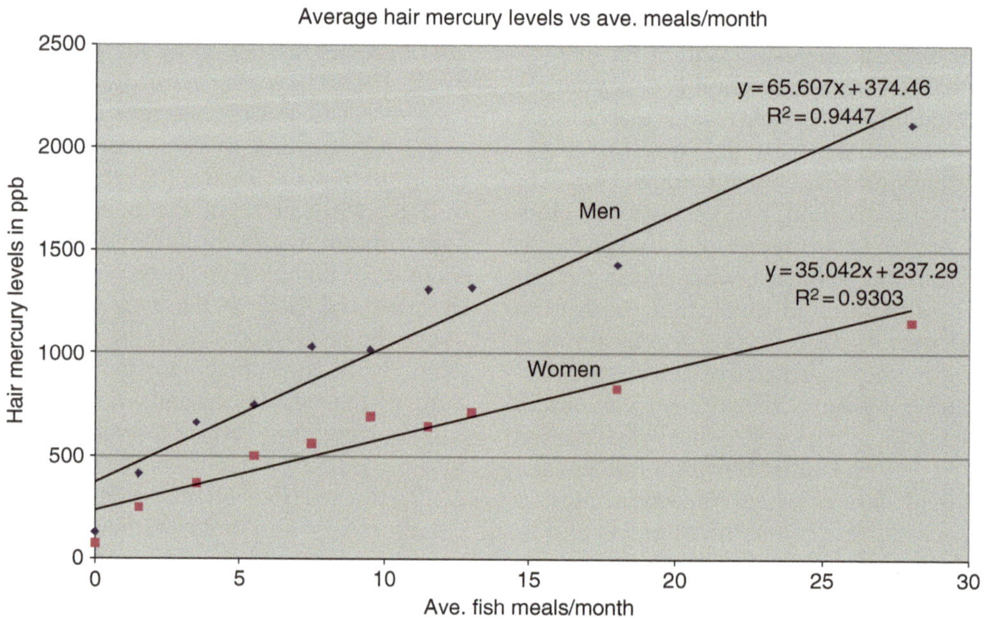

Figure 5 Relationship between rates of fish consumption and body burdens of meHg in Wisconsin residents. Data from Knobeloch L *et al.* (2007) *Environ Res* 103: 205–210. (Graphic courtesy of Lynda Knobeloch.)

Mass Balances for Mercury

Annual Budgets for Total Mercury

Annual budgets for Hg_T indicate that atmospheric deposition is the dominant source to remote lakes, whether delivered directly to the lake surface or indirectly delivered via runoff from the contributing watershed (**Table 2**). Wet deposition constitutes 50–70% of the direct atmospheric load. Runoff contributes 0–60% of the Hg_T load to lakes with water

Table 2 Annual Budgets for Hg_T in large and small freshwater lakes (units = $\mu g\ m^{-2}\ y^{-1}$, normalized to lake surface area, unless indicated otherwise)

Budget Term	*Lake*							
	Little Rock[a]	*Devils*[b]	*Pallette*[c]	*Champlain*[d]	*Michigan*[e]	*Superior*[f]	*Baikal*[g]	*Onondaga*[h]
Inputs								
Atmospheric deposition	7.8 ± 1.9	9.8 ± 2.4	10.3	15.1	16.6	9.0	5.5	36.7[k]
Wet	na	na	6.8	7.9	na		3.8	33.2[k]
Dry	na	na	3.5	7.2	na		1.7	3.5
Streamflow in[i]	0	15.1 ± 10.7	0	24.7	3.4	3.4	1.9	1,133
Groundwater in	0	0	1	na	0.2	0.04	na	1.7
Total	7.8 ± 1.9	25.5 ± 13.7	11.3	39.8	20.2	12.4	7.4	1171.4
Outputs								
Streamflow out	0	5.7 ± 1.7	0	1.3	0.3	0.4	0.5	233
Groundwater out	0.03 ± 0.01	0.6 ± 0.3	2.1	na	na	na	na	na
Evasion	7.7 ± 2.0	19.3 ± 9.6	4.3	29.5	9	8.8	na	1.3
Sedimentation			4.9	21.4	10.9	3.3	6.9	941
Total	8.0 ± 2.0	25.6 ± 11.4	11.3	52.2	20.2	12.5	7.4	1175
Storage change	−0.2 ± 0.2	0 ± 3.8	na	na	Na	na	na	na
Residence time (Hg_T, years)	0.3 ± 0.03	0.37 ± 0.13	0.73	na	1.3	5.9	20–65[j]	na
Residence time (H_2O, years)	4	1		na	33	107	330	0.2–0.4
Surface area (km^2)	0.08	0.12	0.7	1179	58,016	82,362	31,500	12
Volume (km^3)	0.00025	0.00035	0.007	25.8	4916	12,080	23,000	0.14

[a]Mean ± range for south basin, 2002–03; Watras CJ, *et al.* (2006) *Limnol. Oceanogr.* 51: 257–270; Watras CJ, *et al. Can J Fish Aquat Sci* (in press).

[b]Mean ± SD for 2002–2004; Watras et al. *Can J Fish Aquat Sci* 65: 100–116.

[c]Watras CJ, *et al.* (1994) In Watras CJ and Huckabee JW. (eds). *Mercury Pollution: Integration and Synthesis*. Boca Raton, FL: Lewis Publishers. pp. 153–177; Watras CJ, *et al.* (1996) In Baeyens W, *et al.* (eds). *Global and Regional Mercury Cycles: Sources, Fluxes and Mass Balances. NATO ASI Series 2 – Environment. Vol. 21*, pp. 329–358. cf. Krabbenhoft DP and Babiarz CL. (1992) *Wat Res Res* 28: 3119–3128.

[d]Gao N *et al.* (2006) *Environ Sci Technol* 40: 82–89.

[e]Mason RP and Sullivan KA (1997) *Environ Sci Technol* 31: 942–947.

[f]Rolfus KR, *et al.* (2003) *Environ Sci Technol* 37: 865–872.

[g]Leermakers M *et al.* (1996) In Baeyens W, *et al.* (eds). *Global and Regional Mercury Cycles: Sources, Fluxes and Mass Balances. NATO ASI Series 2 – Environment. Vol. 21*, pp. 303–315.

[h]Henry EA, *et al.* (1995) *Water, Air, Soil Pollut* 80: 509–518; cf. Gbondo-Tugbawa S, and Driscoll CT (1998) *Water, Air, Soil Pollut* 105: 417–426.

[i]Includes runoff and wastewater outfall where data available.

[j]Ignores potentially large evasive efflux.

[k]Anomolous value.

Hg-polluted Onondaga Lake in New York shown for comparison with more remote fresh waters (na, data not available).

residence times >1 year. For lakes with short water residence times and for lakes receiving runoff from contaminated watersheds, fluvial inputs of Hg_T can be much more important. Groundwater inputs of Hg_T are generally a minor part of the annual budget.

Sedimentation and evasion are generally the major loss terms for Hg_T. Since Hg_T inputs and outputs roughly balance over the course of a year, the magnitude of these two loss terms explains the short residence time of waterborne mercury relative to water residence time. Annual rates of Hg^0 evasion are poorly constrained for all lakes; but in most budgets sedimentation is the dominant loss term – reflecting the high particle reactivity of $Hg^{(II)}$. There is some evidence that Hg^0 evasion rates are highest during summer following large inputs of dissolved $Hg^{(II)}$ as natural rainfall or as experimental additions. The rapid photo-reduction of $Hg(OH)_2$ near the water surface could explain these high evasion rates.

Seasonal Budgets for Methylmercury

Since the production and destruction of meHg in lakes and watersheds often balance over an annual timescale, annual budgets can distort the relative importance of external inputs and outputs. As a result, seasonal budgets can provide a more complete inventory of meHg sources and sinks. **Table 3** shows three seasonal budgets constructed during the summer period when meHg accumulates in the water column.

The three seasonal budgets indicate that direct atmospheric deposition is a significant external source of meHg, but runoff can be the dominant external source when wetlands occupy the

Table 3 Seasonal budgets for waterborne meHg in three remote northern lakes

Budget term	*Lake*			
	Little Rock Lake		*Devils Lake*	*Lake 240*
	Acidified (1990–96)	*De-acidified (1998–03)*	*(2002–04)*	
Δ Storage	1010 ± 707	612 ± 337	1564 ± 1455	500
Atmospheric deposition	71 ± 21	51 ± 20	76.9 ± 25.4	29.5
Inflow	0	0	299 ± 152	
Methylation (net)	(990 ± 693)	(592 ± 337)	(1872 ± 1365)	600
Outflow	10 ± 10	10 ± 10	573 ± 137	6.8
Net sedimentation	30 ± 20	30 ± 20	137 ± 85	na
Demethylation	na	na	na	122.7
Surface area (ha)	9.8		11.7	44.1
Hydrology	seepage		drainage	drainage
Wetland influence	low		high	moderate

Mass balances were constructed for the period spring through summer when meHg accumulates in the anoxic, hypolimnetic water column. Data are pg meHg m^{-2} (mean ± SD) from Watras CJ, *et al.* (2006) *Limnol. Oceanogr.* 51: 257–270; 2006. Watras, CJ, *et al. Can J Fish Aquat Sci* 65: 100–116. Sellers, P. (2001). *Limnol Oceanogr* 46: 623–631.

contributing watershed (**Table 3**). For wetland-dominated Devils Lake in northern Wisconsin, runoff constitutes 60% of the annual hydrologic budget and the export of meHg from the wetland to the lake is 4-fold higher than direct depositional inputs, on average. Across the three lakes given in **Table 3**, the relative importance of runoff as a meHg source increases with the amount of wetland in the watershed. Field data for watersheds in eastern North America and northern Europe indicate that the annual meHg output from wetlands to surface waters varies from about 0.1 to 0.5 μg meHg m^{-2} wetland y^{-1}.

Despite the importance of wetlands as an external source of meHg, the seasonal budgets shown in **Table 3** indicate that internal production can be a much larger meHg source. In these three lakes, internal meHg production constitutes ~80% to >95% of the meHg that accumulates in the water column during summer. However, the export of $Hg^{(II)}$ and microbial nutrients from wetlands to lakes can be high, potentially fueling in-lake methylation in wetland-dominated lakes.

Experiments under field and laboratory conditions indicate that factors affecting the supply of $Hg^{(II)}$ and substrates governing microbial activity interact synergistically to control the seasonal production of meHg in lakes and wetlands. The synergistic interaction of $Hg^{(II)}$ and SO_4 has been described theoretically by the expression

$$\mathrm{MMR} = \mathrm{MMR}_{\max}\left(\frac{[\mathrm{Hg}^{(\mathrm{II})}]}{k_{\mathrm{Hg}^{(\mathrm{II})}} + [\mathrm{Hg}^{(\mathrm{II})}]}\right)\left(\frac{[\mathrm{SO}_4^{2-}]}{k_{\mathrm{SO}_4} + [\mathrm{SO}_4^{2-}]}\right),$$

where MMR is the observed mercury methylation rate, MMR_{max} is the maximum rate possible for a given microbial community, and k represents the half-saturation constants for $Hg^{(II)}$ and SO_4 uptake. The dependence on SO_4 implies that SRB are the dominant methylators. The expression can be expanded to include other substrates that may limit SRB activity, such as organic carbon, nitrogen or phosphorous.

In addition to synergism, the saturation model implies that relatively large effects occur at low substrate concentrations. This implication has relevance for remote, relatively pristine lakes where the concentration of key substrates is near limiting levels. In this type of ecosystem, one might expect to see a more dramatic response to modest increases or decreases in key substrates than one would observe in a heavily contaminated ecosystem.

Methylation Pathways

MeHg production in lakes is the net result of two processes, methylation and demethylation. It is not governed by a single forward–backward chemical reaction, but rather results from separate reaction pathways. In many instances, the pathways are catalyzed by microbial activity.

Assuming first-order kinetics, the net rate of meHg production can be expressed as the difference between rates of methylation and demethylation:

$$\mathrm{d[meHg]/dt} = K_{\mathrm{m}}[\mathrm{Hg}^{(\mathrm{II})}] - K_{\mathrm{dm}}[\mathrm{meHg}])$$

where K_m and K_{dm} are the substrate-specific rate constants. Under simulated field conditions for anoxic sediments and hypolimnetic waters, rates of methylation approaching 10% per day have been estimated using trace additions of isotopically labeled $Hg^{(II)}$ in incubation experiments. Such high methylation rates can be reconciled with observed waterborne concentrations only if the pool of available $Hg^{(II)}$ is assumed to be low or if the specific rate of demethylation is high. Since concentrations of meHg are typically much lower than concentrations of $Hg^{(II)}$ under natural conditions, relatively high specific rate constants for demethylation are not unreasonable at steady-state (i.e., $K_m/K_{dm}) = (meHg/Hg^{(II)})$.

$Hg^{(II)}$ Methylation

Although no specific $Hg^{(II)}$ methylation gene has been identified in SRB or other microbes, the methylation process appears to be under enzymatic control, albeit as an inadvertent by-product of normal microbial metabolism. $Hg^{(II)}$ methylation requires the transfer of a carbanion from a methyl donor, such as methylcobalamine (vitamin B_{12}). There are three known cobalamine-containing enzymes that could be the methyl donor: methane synthase, methionine synthase, and the corrinoid of the acetyl-CoA pathway. As yet, there is no strong evidence implicating any one of these enzymes as the key methylator in SRB or any other anaerobic microbe. However, strong biogeochemical links between the aquatic cycles of meHg, C, S, and Co are implied.

MeHg Demethylation

MeHg demethylation occurs by at least three pathways, two of which involve microbes. Inducible genes, known as the Mer-A and Mer-B operons, govern one microbial pathway. The mer-A operon encodes for organomercurial lyase, which cleaves the C–Hg bond, yielding CH_4 and $Hg^{(II)}$. The mer-B operon encodes for mercury reductase, which reduces $Hg^{(II)}$ to Hg^0, which may then evade from the system. The low meHg concentrations observed in highly contaminated lakes might be due to induction of the Mer operons. A second microbial demethylation pathway involves oxidative demethylation, producing CO_2 as an end product rather than CH_4. It is not clear whether this pathway is inducible.

Incubation experiments in several lakes indicate that meHg can also be destroyed in near-surface waters by sunlight. Experimentally determined rates of photo-demethylation were used to construct the seasonal budget for meHg in Lake 240 in the Canadian ELA. As shown in **Table 3**, photo-demethylation was estimated to be the major sink for meHg in Lake 240 during summer.

The Aquatic Mercury Cycle

An overview of the aquatic mercury cycle in a remote freshwater lake is shown in **Figure 6**. Mercury is emitted to the atmosphere by natural and anthropogenic processes and it is transported into the airshed of a lake along with other pollutants, such as SO_2. The chemical speciation and physical form of atmospheric Hg can change during transport with interconversions between oxidized, reduced, and

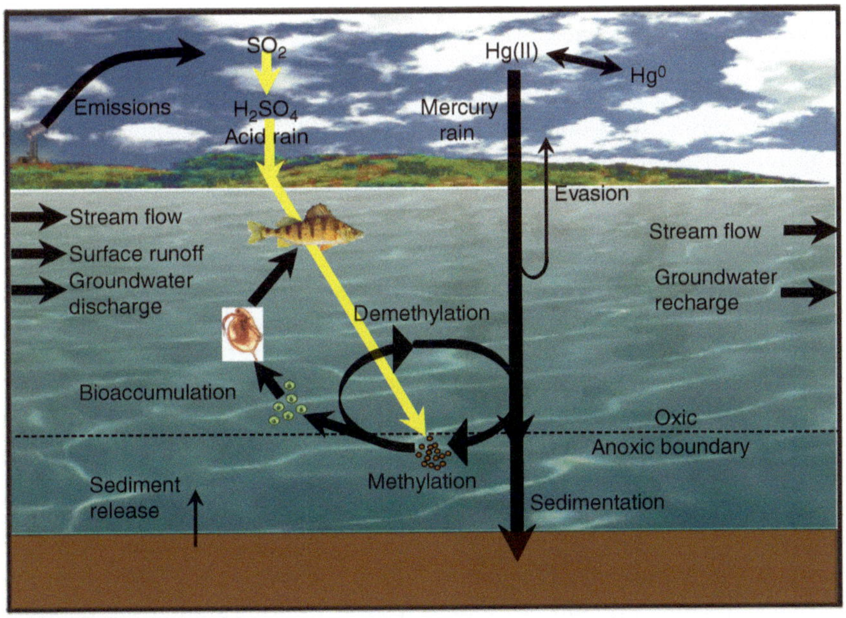

Figure 6 The aquatic mercury cycle in remote fresh waters, showing the interactive effects of acid rain and mercury rain. Acid rain stimulates the meHg subcycle, yielding a synergistic effect on levels of fish contamination.

particulate forms. The nature and kinetics of these atmospheric reactions remain uncertain, but they govern the fate of Hg emissions. Wet and dry depositional processes remove the oxidized and particulate forms of Hg from the atmosphere most readily. As indicated in **Table 2,** wet deposition appears to be the predominant pathway from the atmosphere to most lakes.

Once deposited on the lake surface, most of the Hg is eventually scavenged by particulate matter and transported toward sediments. A variable fraction of the atmospherically deposited Hg can be reduced and returned to the atmosphere by evasional efflux. During transport toward eventual sediment burial, some Hg enters the methylation/demethylation subcycle. Methylated Hg (meHg) enters the base of the aquatic food chain and is biomagnified through trophic transfer leading to elevated concentrations in fish.

The methylation pathway can be further fueled by inputs of SO_4, which enter the lake during acid deposition events. Following emission to the atmosphere, SO_2 is converted to sulfuric acid in reactions involving atmospheric aerosols. The resultant acid rain stimulates the activity of SRB in anoxic zones of the lake. SRB are generally considered the principal methylators of Hg in aquatic ecosystems. As noted above, inputs of Hg and SO_4 can interact synergistically to yield larger amounts of meHg than either substrate would yield individually.

Additional sources of Hg to lakes include runoff and groundwater inseepage, but the ultimate source to streams and ground water is atmospheric deposition – except in mercuriferous regions or industrially polluted sites. Additional Hg sinks include outflow via a distributary stream and groundwater outseepage.

Atmospheric processes similarly dominate the global Hg budget. Current global budgets indicate that the atmospheric pool contains 5000–6000 tonnes of Hg. Under assumptions of steady-state with an average atmospheric Hg residence time of $\sim$1 year, this pool implies global emission and deposition rates of 5000–6000 tonnes/year. The simple picture is complicated by evidence of rapid cycling and poorly understood transformation process in the atmosphere and at the earth's surface.

Humans have had a large impact on the global Hg cycle. Estimates of current anthropogenic emissions to the atmosphere are in the range of 2000–2400 tonnes/year. This emission rate is several orders of magnitude smaller than anthropogenic emissions of many other atmospheric pollutants, like SO_4 and CO_2. However, given the biomagnification potential of meHg and its high toxicity, current emissions are sufficient to yield unhealthy concentrations in the fish of many remote lakes.

Estimated oceanic emission rates for Hg are similar to current anthropogenic emissions, but much of it is the re-emission of Hg attributable to past human activity. Human activities that have perturbed the global cycle include fossil fuel combustion, gold-mining, ore refining, manufacturing, and the disposal of Hg-containing products. The reduction of Hg emissions from all these sectors is possible with current technology and proper handling practices.

Remediation Strategies

Although atmospheric deposition is the major source of Hg to remote lakes, and although anthropogenic emissions constitute a major source of Hg to the atmosphere, there is no consensus about how long it would take emission controls to have a measurable effect on levels of fish contamination. The lack of consensus stems largely from two uncertainties. Firstly, there is uncertainty about the relative effect of regional and global Hg sources on local Hg deposition. If global sources predominate, then emission reductions in particular regions may have little impact on local deposition. Secondly, there is uncertainty about the responsiveness of lakes to depositional change. Over the past century, a large pool of Hg has accumulated in soils and sediments. The remobilization of this historically deposited Hg could delay recovery for decades or even centuries.

With respect to the first uncertainty, data for northern Europe and the Great Lakes region of North America suggest that regional emission reductions can have a measurable effect on local Hg deposition. In Sweden and Germany, Hg in air and/or rain declined along with declining Hg emissions from downwind sources in the former Soviet Union during the early 1990s. In northern Wisconsin, there have been declines in wet Hg deposition since 1994 – a downward trend that follows regional reductions in the commercial and industrial use of Hg and the closure of a large smelter. Throughout the Great Lakes region, there has been evidence of declining Hg accumulation in recent lake sediments.

With respect to the second uncertainty, long-term data for a precipitation-dominated lake in northern Wisconsin suggest that the response to depositional change can be rapid. In Little Rock Lake (LRL), seasonal changes in waterborne Hg closely followed the annual cycle of atmospheric Hg deposition for more than ten consecutive years. During this decadal time period, there also were gradual downward trends in both annual Hg deposition and lake-water Hg_T. Between two consecutive years when atmospheric Hg deposition declined by 30%, waterborne

Hg in the lake declined by a similar amount (−27%). In LRL, downward trends in fish have followed the downward trends in lake-water and rain. Across hundreds of other lakes in northern Wisconsin, long-term data indicate a similar region-wide decline in fish contamination during the past two decades.

Although these findings indicate that certain lakes can respond rapidly to depositional change, the response time for lakes that are influenced more by their terrestrial watershed remains uncertain. For example, in the wetland-dominated Devils Lake in northern Wisconsin, waterborne Hg declined by only 13% during the two years when atmospheric Hg deposition declined by 30%. However, the modest response was not due to buffering by fluvial inputs of soil-derived Hg from the watershed. Instead, fluvial Hg inputs were lost from the lake more rapidly during the year when inputs were high, thereby diminishing the apparent effect of input reductions.

There remain large gaps in our understanding of the atmospheric reactions governing Hg transport and deposition. Reliable atmospheric models for predicting the impact of emission reduction require these gaps to be filled. Nevertheless, long-term monitoring data strongly suggest that regional emission reductions can have a positive effect on sensitive water resources. However, it is also important to reduce anthropogenic Hg emissions at a global scale. Global processes apparently drive Hg inputs to the world's oceans, and much of this Hg evades back into the atmosphere for eventual transport to continental regions. Even in continental regions where improvements in air, water and/or fish have been documented over recent years, fish remain contaminated above acceptable levels. Reaching a safe consumption threshold will likely require emission reductions at local, regional and global scales.

An alternative strategy to ameliorate the health effects of Hg contamination relies on the increased issuance of fish consumption advisories along with increased publicity to promote broader compliance. Unfortunately, this strategy has no effect on non-human populations; and it removes an important source of protein from the diet of many humans.

See also: Pollution of Aquatic Ecosystems I.

Further Reading

Baeyens W, *et al.* (eds.) (1996) *Global and Regional Mercury Cycles: Sources, Fluxes and Mass Balances, NATO ASI Series 2 – Environment. Vol. 21.* Dordrecht: Kluwer. 563pp.

Ebinghaus R, *et al.* (eds.) (1999) *Mercury Contaminated Sites.* Berlin: Springer-Verlag.

Lindqvist O, *et al.* (eds.) (1991) Mercury as an Environmental Pollutant. *Water, Air, Soil Pollut* 56: 1–847.

Lindqvist O (ed.) (1991) Mercury in the Swedish Environment. *Water, Air, Soil Pollut* 55(1–2): 1–261.

Morel FMM, *et al.* (1998) The chemical cycle and bioaccumulation of mercury. *Ann Rev Ecol Syst* 29: 571–576.

Munthe J, *et al.* (2007) *Ambio* 36(1): 33–44.

Porcella DB, *et al.* (eds.) (1995) Mercury as a global pollutant. *Water, Air, Soil Pollut* 80(1–4): 1–1334.

Ullrich SM, *et al.* (2001) Mercury in the aquatic environment: a review of factors affecting methylation. *Crit Rev in Environ Sci Technol* 31: 241–293.

Watras CJ and Huckabee JW (eds.) (1994) *Mercury Pollution: Integration and Synthesis.* Boca Raton, FL: Lewis Publishers, 727pp.

Weiner JM, *et al.* (2003) Ecotoxicology of mercury. In Hoffman DJ, *et al.* (eds.) *Handbook of Ecotoxicology.* Boca Raton, FL: Lewis Publishers. pp. 409–463.

Pollution of Aquatic Ecosystems I

F M Wilhelm, University of Idaho, Moscow, ID, USA

Introduction

The pollution of aquatic ecosystems occurs worldwide, includes a variety of sources and impacts, and is increasing. Water pollution has been variously defined (**Table 1**), but generally involves degradation resulting from human activities that cause water to become unusable for an intended purpose. It is important to distinguish the effects of human activities from natural phenomena, e.g., mud slides, volcanic eruptions, etc. that impact water because natural phenomena represent forces which fundamentally shape the evolution of natural ecosystems in the absence of human activity. Water pollution associated with anthropogenic activities is typically rapid, persistent, and outstrips the evolutionary potential of ecosystems, leaving them in a depauperate state.

Water pollution threatens the existence of life on this planet and efforts to eliminate sources of pollution and restore impacted systems must become a priority internationally. Although over 70% of the planet is inundated with water, less than 3% is fresh water and of that, less than 1% is available to support life (**Table 2**). Exponential population growth, urbanization, (by 2030, 60% of the human population is expected to live in an urban setting), industrialization, and expanding food production increase the demand for water and further reduce the finite quantity available. A 'water crisis' is not far off, or is already upon us, depending on where you live. This is, and will continue to be, exacerbated by the uneven global distribution of water, the expanding human population, and climate change. The UN Secretary-General recently informed the General Assembly of its potential to fuel or intensify conflicts around the world in the future, while an article in 'Fortune' magazine states, 'Water promises to be to the twenty-first century what oil was to the twentieth century: the precious commodity that determines the wealth of nations.' To be sure, there will be conflicts over water. Resolving issues ranging from the allocation to the protection and restoration of the resource will require a unified effort and will touch all sectors of society.

The Evolution of Human Societies and Water Pollution

Human physiology requires water and thus civilizations tended to emerge around or along easily accessible sources. General patterns in the evolution of human societies include a progression from hunter-gathers to sedentary agrarians, culminating in industrialization. Reliance on fixed water sources and their use and pollution increase along this evolutionary path. Early Egyptian farmers extensively irrigated via canals, as did the Mayan civilization in the new world. Romans engineered the mass transfer of water via aqueducts from remote sources to supply their cities and industrial sites. Industrialization has led to the largest increase in use and pollution of water resources along with the recognition that water pollution has negative consequences and recovery efforts must be undertaken. However, civilizations generally pollute first, then succumb or attempt to clean up. Well known examples of this cycle include formerly sewage laden rivers such as the Thames and the Potomac in the capitals of England and the United States, respectively. These river's stink during hot summers caused politicians to flee the cities. After much effort, the Thames once again has salmon, while the Potomac still needs further work. Neither is free of encumbrances. The money spent recovering aquatic ecosystems far outpaces any perceived or real costs to avoid pollution in the first place. Given our collective knowledge and available technology, we must find the will to protect our well being and avoid repeating past cycles. This is especially crucial as developing nations are encouraged to strive for the affluent standards of living enjoyed in developed nations.

Identifying Sources of Pollution

Generally, the sources of aquatic pollution can be classified as point source (PS), and nonpoint source (NPS). PS pollution is also referred to as end-of-pipe pollution and includes sources that are typically easy to identify and highly localized, such as the termination of a pipe or culvert from industry or a waste water treatment plant (WWTP) that delivers some effluent to a receiving water body. NPS pollution on the other hand comes from one or more diffuse source(s) that cannot be readily identified. Its source is water from rainfall or snowmelt that moves over or through the landscape as run-off. As it interacts with the landscape, it picks up pollutants, eventually depositing them in lakes, rivers, coastal waters, or even ground water. Because of its diffuse nature, NPS pollution is much more difficult to control and remediate than PS pollution.

Point Source Pollution

Because of the highly localized nature of PS pollution, it is relatively easy to sample, analyze, and determine if it poses a threat to a receiving water body. In general, much progress has been made in the control of PS pollution in developed countries. For example, the use of flush toilets that delivered waste into the Thames, resulting in its degradation and offensive odor, was identified and harnessed into specially built sewers for delivery into a sewage farm to prevent direct entry into the river. Preventing PS pollution has greatly improved the quality of receiving water bodies around the world. Most developed nations now have laws and permitting systems such as the Clean Water Act of 1972 in the United States to manage PS pollution. Although the results of implementing such a law and controlling PS pollution are obvious, its development and enforcement require the availability of funds and an infrastructure to deal with effluents. Unfortunately, these necessities are often lacking in developing nations, where other more pressing societal needs top national agendas. As a result, areas of the globe currently experiencing some of the highest population growth and scarce water resources are least able to address PS pollution directly. It is this lack of funds and infrastructure that contributes to the estimated 2 million tons of waste disposed of daily in water and the approximately 14 000 waterborne disease-related deaths.

Although WWTPs generally improve the quality of the water released compared to that received, the degree to which effluent may still be a pollutant in receiving waters depends on the level of treatment (**Table 3**). In general, all WWTPs perform primary treatment with some degree of efficiency. This includes the removal of easily separated materials including solids, grit (sand and rocks), fats, oils, grease, and other floating matter. It is typically accomplished by mechanical means such as screens, rakes, and settling areas. Secondary treatment involves the substantial breakdown of organic matter contained in the effluent to reduce its biological oxygen demand (BOD), and the consumption of oxygen during the respiration of organic matter, on receiving waters. Removal of carbon from sewage liquor can be achieved via several means, including fixed-film or suspended growth systems. In fixed-film systems, microbial films form on a medium which is then contacted by sewage. Trickle filter systems consist of a media bed made of rocks, foam, peat moss, or plastic, which receives a constant stream of sewage. In rotating biological contactors, the

Table 1 Operational definitions of water pollution

Definition	*Source*
The presence in water of harmful and objectionable material – obtained from sewers, industrial wastes and rainwater run-off – in sufficient concentrations to make it unfit for use.	1
Any human caused contamination of water that reduces its usefulness to humans and other organisms in nature.	2
A polluted waterway is one that is ecologically out of balance as a result of contamination.	3
Addition of harmful or objectionable material in concentrations or sufficient quantities to adversely affect its usefulness or quality	4

[1]Glossary of Environment Statistics, Studies in Methods, Series F, No. 67, United Nations, New York, 1997.
[2]City of Dallas Storm Water Management. http://www.dallascityhall.com/
[3]Greyson JC (1990) *Carbon, Nitrogen, and Sulfur Pollutants and Their Determination in Air and Water*, 416 pp. CRC Press.
[4]Washington Department of Natural Resources, Treatment Division. http://www.dnrNR/metrokc.gov

Table 2 Availability of water resources

Pool	*Percentage*
Oceans	97.61
Ice, glaciers	2.08
Soil moisture, atmospheric vapor, and saline lakes	0.014
Rivers, lakes, and accessible ground water	0.304

Source: Wetzel RG (2001) *Limnology, Lake and River Ecosystems*, 3rd edn. New York: Academic Press.

Table 3 Treatment levels, materials removed from waste stream and methods used to achieve them in waster water treatment plants

Treatment	*Material removed*	*Methods*
Primary	Solids, floating matter, fats, oils, and grease	Mechanical, screens, filters, settling areas
Secondary	Organic matter	i Fixed film, sludge passed over substrate on which bacteria grow that degrade organic matter-trickle filter, or rotating biological contactors ii Suspended media-Activated sludge, surface aerated basins,
Tertiary	Disinfection, nitrogen, phosphorus	Sand filtration, lagoons, *Daphnia*, rotifers, bacteria, constructed wetlands, UV light, ozonation, chlorine

medium with attached biofilm is passed through the pool of sewage. In suspended growth systems, e.g., activated sludge, the microbial biomass is mixed with the sewage. The objective of both systems is to remove carbon from sewage by conversion to biomass that can be easily handled by the plant. Tertiary treatment can involve one or several additional steps and may remove nutrients such as nitrogen and phosphorus that can stimulate excessive algal growth. It may also involve disinfection with ozone, chlorine, or ultraviolet light. Estimates released in early 2008 from the US Environmental Protection Agency (USEPA) indicating needs for waste water treatment facilities in the United States are in excess of USD 202.5 billion. This includes facility upgrades to provide treatment, manage sewer infrastructure, and storm water. The number of individuals receiving secondary treatment in the United States has increased from 84.1 million in the early 1970s to 205 millions in 2004 and is expected to increase to 285 million by 2024 if the above costs are met. This compares favorably with African and Asian countries, where sewage treatment is nearly unavailable and raw sewage is discharged directly into the environment. In some middle eastern countries, raw sewage is injected into ground water that serves as metropolitan drinking water supplies.

In developed countries, concerns are surfacing about the inability of WWTPs to process trace contaminants such as hormones from animal farming, residues from human hormonal contraception, and synthetic compounds that mimic hormones, very low concentrations of which can negatively affect biota in receiving waters. Widespread testing of natural waters in North America and Europe has shown the presence of some or all of the above compounds of concern in a surprising number of water bodies. Caffeine is rapidly becoming a marker of choice, given its specificity to indicate human contamination and the potential presence of drugs, contraceptive products, and intestinal pathogens. How these compounds will be dealt with in effluent in the future remains to be seen.

Nonpoint Source Pollution

The diffuse nature of NPS pollution, unlike PS pollution, makes it difficult to identify and thus eliminate it. Generally, reducing or eliminating NPS pollution requires a concerted sampling effort over large geographic areas to identify contributing sources (e.g., direct rainfall that is contaminated versus contamination after contact with soil). Once contaminant sources have been identified, their elimination requires further effort targeting the identified pools. For example, agriculture, especially the application of nitrogen-based fertilizers to grow corn in the US Midwest, has been identified as one of the major sources contributing to the large dead zone (area that lacks oxygen) in the Gulf of Mexico. Excess fertilizer not taken up by plants during the growing season enters waterways via run-off and is eventually delivered into the Mississippi River and ultimately the Gulf. Because primary production in marine environments tends to be limited by nitrogen, its arrival stimulates primary production. Once algal cells die and sink, their decomposition decreases dissolved oxygen concentrations, leading to hypoxic and anoxic conditions. Identifying and implementing actions to control the application of fertilizer, which may limit farm income, on farms far removed from where pollution problems are manifested will be difficult. Generally, the restoration of aquatic ecosystems suffering NPS pollution requires a multipronged approach, including best management practices, e.g., leaving or reestablishing native vegetation (buffer strips) near water bodies; reducing pollution sources, e.g., acid rain producing emissions; and direct in-system management, e.g., application of alum to immobilize phosphorus or removal of contaminated sediment. All these approaches are expensive and require considerable coordination at a watershed scale to ensure success. Remediation processes are also hampered by the long-distance transport of pollutants, which may make the problem international and require nations to cooperate. Because developed countries have made considerable progress in harnessing PS pollution, NPS pollution is receiving attention and funding. However, in developing countries, NPS pollution will not be dealt with for a significant time to come. Can we afford to wait as the population continues to expand and nations become affluent?

Specific Sources of Aquatic Pollution

Sources of aquatic pollution vary from individuals to large industries and encompass a wide range of contaminants that have broad effects (**Table 4**). Here, I will address the first three sources (see Further Reading list at the end for other chapters dealing with the other contaminants). Although traditionally not considered water pollution, the loss of, or reduction in, water quantity eliminates its use for any other purpose and thus meets the basic definition of water pollution (**Table 1**). The reduction of inland seas to dust bowls (**Figure 1**) and rivers to trickles requires our immediate attention. We must attempt to balance our need for water versus the function of ecosystems and concern for populations from which we wish to remove it. This has consumed, and will continue to consume significant amounts of time and money, as

Table 4 Major sources of freshwater pollution, their effects and contaminants of concern

Pollution type	*Primary sources*	*Effect(s)*	*Constituents of concern*
1. Water loss	Abstraction	Reduces available water mass, salinization, increased temperature	Water quantity
2. Organic matter	Industrial wastewater and domestic sewage.	Depletes oxygen from the water column as it decomposes. Stresses or kills aquatic life.	BOD, Dissolved Organic carbon (DOC), dissolved oxygen (DO)
3. Nutrients	Run-off from agricultural lands and urban areas (lawns, golf courses), also from industrial discharge, e.g., food processing industries	Over-stimulates growth of algae that results in loss of oxygen when blooms die thereby harming aquatic life. High levels of nitrate in drinking water lead to illness in humans, especially infants	Total N (organic + inorganic), total P (eutrophication) which then decomposes, For eutrophication: (Dissolved Oxygen, Individual N species Orthophosphate)
4. Pathogens and microbial contaminants	Human, livestock, and natural sources	Contaminated drinking water leads to spread of infectious diseases (e.g., diarrheal disease and intestinal parasites), increased childhood mortality in developing countries	*Shigella, Salmonella, Cryptosporidium*, Fecal coliform (Coliform), *Escherichia coli* (mammal feces – *E. coli*)
5. Salinization	Leached from soils by over irrigation or by over-pumping coastal aquifers resulting in saltwater intrusion or rise of water table dissolving salts	Salt build-up in soils reduces yield or kills crops. Renders fresh water supplies undrinkable	Electrical conductivity, chloride (followed, by full suite of major cations (Ca, Mg), anions
6. Acidification	Oxides of sulfur and nitrogen associated with particulates from electric power generation, industrial stack and auto/truck emissions (wet and dry deposition). Acid mine drainage from tailings as well as mines	Acidifies lakes and streams which leaches heavy metals such as aluminum from soils into water bodies	pH
7. Heavy metals	Industries and mining	Persist in freshwater environments, especially sediments for long periods. Can be remobilized by change in redox. Can accumulate in fish and shellfish tissue and be toxic to aquatic organisms and humans who consume them	Pb, Cd, Zn, Cu, Ni, Cr, Hg, As (particularly ground water)
8. Toxic organic compounds and microorganic pollutants.[a]	Wide variety of sources from industrial sites, automobiles, farmers, home gardeners, municipal waste waters	Range of toxic effects in aquatic fauna and humans from mild immune suppression to acute poisoning or reproductive failure	PAHs, PCBs, pesticides (lindane, DDT, PCP, Aldrin, Dieldrin, Endrin, Isodrin, hexachlorobenzene)
9. Thermal	Rivers fragmented by dams and reservoirs that slow it allowing it to warm. Industry from cooling towers and other end-of-pipe above-ambient temperature discharges	Change in oxygen levels and decomposition rate of organic matter in the water column. May shift the species composition of the receiving water body	Temperature
10. Silt and suspended particles	Soil erosion, agriculture, road building, deforestation, construction and other land use changes	Reduces water quality for drinking and recreation and degrades aquatic habitat by smothering with silt, disrupting spawning and interfering with feeding	Total suspended solids, turbidity
11. Exotic species	Commercial shipping, anglers, public	Alters ecosystem functions by replacing, removing key species or altering processes	Any nonnative species, plant or animal, that disrupts aquatic ecosystems (e.g., zebra mussels in North America Great Lakes, Nile perch in Lake victoria, Africa

Other pollutants include radioactivity, fluoride, selenium.

Source: UNESCO (2006) *Water a Shared Responsibility*. The United Nations World Water Development Report 2. United Nations Educational, Scientific and Cultural Organization (UNESCO), Paris France and Bergahn Books, NY USA.

[a]Microorganic pollutant list now includes a suite of endocrin disrupters, antioxidants, plasticizers, fire retardants, insect repellents, solvents, insecticides, herbicides, fragrances, food additives, prescription drugs and pharmaceuticals (e.g., birth control, antibiotics, etc.), nonprescription drugs (e.g., caffeine, nicotine and derivatives, stimulants).

demonstrated by the current ongoing legal disputes to manage the flow of western US rivers for irrigation and generation of electricity, as well as to ensure adequate flows for salmon stocks listed as endangered under the US federal endangered species act. Reducing the quantity of water may seem relatively harmless and may not seem to be much of a pollution concern given that water does remain in the system. However, abstraction or diversion of water from endorheic systems can lead to the salinization of the remaining water to the point where it can no longer support biota or becomes too saline for human consumption (**Figure 1**).

Oxygen Demand in Aquatic Ecosystems

Oxygen demand in aquatic ecosystems results from the oxidation of organic matter or chemical reactions involving compounds such as nitrogen. The introduction of organic matter into the aquatic environment can lead to hypoxic or anoxic conditions during its break down. Because oxygen is the electron acceptor during aerobic cellular respiration, organisms exhibiting this metabolic pathway consume oxygen from the environment. Not only classic decomposers such as bacteria and fungi, but also vertebrates and invertebrates break down organic material that enters aquatic environments. For example, frugivorous fish

Figure 1 continued

in the Amazon consume nuts and fruit that fall into the water, while mayflies and amphipods may be invertebrates involved in shredding leaf material that enters a stream. Bacteria and fungi serve to break down cellulose and other refractory organic material that cannot be easily broken down by other organisms.

Standardized tests have been developed to examine the concentration and composition of organic matter in water supplies, waste effluent, and receiving water bodies. Chemical oxygen demand (COD) quantifies the oxygen used to fully oxidize all organic compounds to carbon dioxide in the presence of a strong oxidizing compound and acidic conditions. Gross fractions of the organic matter, such as the material susceptible to microbial degradation, can be identified with a BOD test.

Chemical Oxygen Demand

Tests for COD measure all organic matter in a sample if oxidized to carbon dioxide. However, it does not provide a time frame or the degree to which this material is actually oxidized in situ, and thus may overestimate the actual amount of oxygen consumed. The test uses a known excess of a strong oxidizing agent such as potassium dichromate ($K_2Cr_2O_7$) which is reduced by the organic material in the sample under acidic conditions. After digestion of the

Figure 1 LANDSAT image showing extent of loss of Aral Sea between 1973 and 2004. Image courtesy of UNEP GRID-Sioux Falls.

sample, the remaining $K_2Cr_2O_7$ is titrated with ferrous ammonium sulphate to determine the $K_2Cr_2O_7$ consumed. The oxidizable organic matter is then calculated in oxygen equivalents. The test does not include the oxygen demand due to nitrification because dichromate does not oxidize ammonia to nitrate. However, oxygen consumption due to nitrification can be significant in aquatic environments and can be calculated. Currently, COD is not used in a regulatory manner by the EPA in the USA; rather, BOD is regulated. However, other countries do use it as a regulatory limit, for example, in Switzerland the COD for wastewater and industrial waste must be less than 200 and 1000 $mg\,l^{-1}$, respectively, before it can be returned to the environment. It is possible to empirically relate COD to BOD, but it is specific to individual samples.

Biological Oxygen Demand

BOD is the oxygen consumed by microorganisms as they break down organic material. Although similar to COD in that it measures the amount of organic compounds, it differs in that it only measures the biologically active organic matter. The standard test for BOD is the five-day (BOD_5) test carried out in the dark at 20 °C. The five-day time interval originated from the estimated time it took water to travel from source to estuary in the United Kingdom where the test was developed at the turn of the last century. To undertake the test, water samples are collected and seeded with microorganisms obtained from activated sludge diluted with deionized water. Initial oxygen concentrations are obtained before sealing the samples and incubation is done in the dark at 20 °C. Incubation in the dark ensures that no additional oxygen is produced in the samples from photosynthesis. After 5 days, the dissolved oxygen is measured again. The difference between the initial and final oxygen concentrations corrected for BOD of the seed and dilution factor is the BOD. Results can be influenced by the oxidation of ammonia, but allylthiourea can be added to prevent it. This BOD is then solely the result of carbonaceous oxidation and is identified as CBOD. BOD values range widely; generally, pristine waters have a value below 1 $mg\,l^{-1}$, moderately polluted waters 2–8 $mg\,l^{-1}$, and treated municipal sewage 20 $mg\,l^{-1}$. Standards for the discharge of effluent from WWTPs have the range 20–30 $mg\,l^{-1}$ and require a minimum flow in receiving waters to ensure sufficient dilution.

Factors Affecting Oxygen Demand

Oxygen demand varies as a function of temperature, the nature of the organic material, and the quantity of material entering the aquatic environment. Because many aquatic organisms are ecotherms, their metabolic rate depends on the temperature of the environment. The rate approximately doubles for every 10 °C rise in temperature and is known as van't Hoff's rule. Thus metabolic rates, oxygen consumption, and oxygen demand at 20 °C will be much greater than at 4 °C. This nonlinear increase has important consequences when considering restoration options (see below). The introduction of large amounts of organic material will have a greater oxygen demand than small quantities. Additionally, material containing simple sugar, protein and lipid molecules will be broken down more rapidly and thus have a higher oxygen demand than material containing complex molecules such as wood fibers high in cellulose and lignin. This explains the higher impact of sewage compared to a comparable quantity of woody material. Refractory material such as wood must generally be exposed first to fungi and bacteria before its energy becomes available to other organisms.

Restoration Techniques to Avoid Low Oxygen

Significant advances have been made over the last several decades to avoid hypoxic or anoxic conditions in aquatic environments. Chief among these is the advancement of secondary and tertiary treatment at WWTPs to reduce the release of high COD material. Significant advances in in situ methods have also occurred to increase the mass of oxygen in water. For example, increased understanding of hydrodynamic and geomorphologic processes in lotic environments has resulted in new in-stream designs that minimize habitat disturbance to fauna, yet maximize the amount of oxygen introduced into degraded systems. For lentic systems, a plethora of techniques has been developed to increase the mass of oxygen. This is likely due to the constraints imposed on natural circulation and mixing by direct stratification, which isolates the hypolimnion in lentic systems. Methods to increase oxygen supply in lentic systems range from the installation of compressor operated aerators, impellers/mixers, and specialized cones to supersaturate water, to the injection of molecular oxygen.

Typical components for aerators and injectors include some form of compressor or pump to move air or oxygen gas to the bottom of the lake where it is released via a diffuser. Compressors must be sufficiently large to overcome the hydrostatic pressure exerted by the water column to produce bubbles. Much effort has gone into designing optimum diffusers and bubble plumes to achieve or prevent the destratification of the water column. In cases where the entire water column is destratified, most of the oxygenation occurs at the air–water interface. To maintain stratification, small bubbles maximize the surface area:volume ratio and facilitate the diffusion and dissolution of oxygen into the water. Often, pure oxygen gas instead of air is injected to increase the diffusion gradient. Alternative approaches involve removing the water from the hypolimnion by pumping it to the surface where it is allowed contact with the atmosphere. Once oxygenated, it is then pumped back into the hypolimnion before warming significantly. Speece cone installations also pump water from the hypolimnion, but it is super- or hypersaturated with oxygen gas in situ before release along the lake bed via a diffuser pipe. A final approach involves the installation of impellers surrounded by specially designed cones/funnels which deflect and accelerate water raised from depth across the surface of the lake, where oxygenation occurs.

Capital investment costs of each system vary and can range from several thousand dollars for a small aerator installation to millions for an oxygen injection system. Added to this are annual operating costs which can be significant for oxygen injection systems. While small aerator installations typically require access only to electricity, oxygen injection systems require an on-site facility to receive and store liquid oxygen, as well as infrastructure to move it into the lake. Installation of any of these systems is a commitment to continuous management.

All the above methods have been successful but failures too have occurred. Failures typically occur due to equipment failure, failure to identify or correctly quantify oxygen demands, and installing an inappropriately sized system. Determining requirements is an involved process, and although general guidelines can be found in the literature, any installation should be based on empirical data obtained from the particular system under consideration. For example, an aerator used to destratify and oxygenate a lake must be sufficiently large to overcome the increase in oxygen demand from increasing the metabolic rate of microbes in the sediment. Because destratifying the water column generally results in overall warming, the metabolic rate of the sediment microbes, and thus overall oxygen consumption, will also increase (see metabolic rate discussion above). It may be possible for this increase in oxygen consumption to completely consume any oxygen transferred at the surface, resulting in failure. Hence the caution to use system-specific empirical data when sizing a system. Although these methods offer solutions to increase oxygen concentrations in water, they represent treatment of symptoms rather than the cause. Additional efforts are needed to return ecosystem functions so that dependence on these energy intensive methods can be reduced.

Nutrients in Aquatic Ecosystems

Nutrients are essential to the function of aquatic ecosystems. Primary producers synthesize these elements into macromolecules and biomass via photosynthesis to fuel higher trophic levels and food webs. Algal production, similar to the production of terrestrial plants, is limited by the essential element present in the smallest quantity, which is known as Liebig's Law of the minimum. In fresh waters, much attention has been focused on the elements nitrogen (N) and phosphorus (P), because they individually or together limit algal productivity. Temperate lakes are generally limited by P as demonstrated repeatedly by the experimental whole-lake addition of nutrients in different parts of the world e.g., Lake 226 at the Experimental Lakes Area in northwestern Ontario, Canada. Nitrogen limitation occurs frequently but most often occurs in mountain lakes, tropical and subtropical systems, coastal waters, and in some temperate lakes in late summer. Specific algal groups may be seasonally limited by other elements. For example, diatom growth can be limited by the availability of silica. In rare cases, other elements limit primary productivity; but these are minor in comparison to N and P.

Eutrophication, Supply Ratios, and Species Composition

Nutrients supplied in excess stimulate algal growth and result in eutrophication. It is one of the most widespread forms of aquatic pollution. Cultural eutrophication differs from natural eutrophication in that the latter occurs naturally in the evolution of aquatic ecosystems from open water to solid land, while the former is the result of anthropogenic activities such as the discharge of sewage, land use changes, and deforestation. Consequences of excessive inputs of nutrients include unsightly blooms of algae that can result in taste and odor problems in drinking water, the production of toxic compounds such as microcystin, and loss of oxygen, which further exacerbates eutrophication via a positive feedback.

The ratio at which N and P are supplied to aquatic systems can influence the composition of the algal community. A ratio of total nitrogen:total phosphorus (TN:TP) of 7:1 (by mass) is considered a balanced supply for algal communities. Although individual species depart significantly from this ratio, at the community level it appears to be highly conserved across a wide range of ecosystems. The ratio is often operationally defined as 10:1, given relatively small differences in productivity. Low ratios indicate nitrogen limitation, while high ratios indicate phosphorus limitation. Nitrogen limitation favors cyanobacteria to dominate algal communities because species of this group can fix atmospheric nitrogen and thus overcome the limitation. The dominance of cyanobacteria can lead to taste and odor problems in drinking water supplies, as well as the production of highly toxic hepato- and neurotoxins that result in the death of livestock, pets, and humans. Higher than average rates of liver cancer have also been documented from geographic regions in China which have chronic blooms of cyanobacteria. The number of aquatic systems susceptible to harmful algal blooms is expected to increase in the future as eutrophication continues and climatic warming provides an ideal thermal regime. Finally, the development of tests to rapidly detect the presence of toxins associated with harmful algal blooms is an area of intense research.

Other Consequences of Eutrophication

Algal blooms resulting from high concentrations of nutrients also lead to large fluctuations in the concentration of dissolved oxygen. During blooms, high concentrations of dissolved oxygen occur in surface waters during the day as a result of photosynthesis. However, at night, respiration of algal cells may result in hypoxia or anoxia and can lead to the death of fish. Typically, wind mixing and diffusion across the air–water interface prevent such low oxygen events from lasting for significant periods of time. When algal blooms decline, the break down of cells consumes oxygen (see above) and can contribute to hypoxic or anoxic conditions in the hypolimnion of lakes. This reduces the volume of water available to biota, increasing encounter rates and alters predator–prey relationships or community composition. Furthermore, reduced conditions at the sediment–water interface promote the remobilization of elements, or internal loading, in soluble ionic form, e.g., orthophosphate. Internal loading can be a significant fraction of the annual nutrient budget in lakes. High internal loading of nutrients also creates a positive feedback if metalimnetic entrainment occurs. The focus is on maintaining oxic conditions in the hypolimnion (see above) and reducing the input of nutrients (see below).

Reducing Nutrients in Aquatic Ecosystems

Ameliorating the effects of N and P in aquatic ecosystems requires multiple strategies. As mentioned below, the best strategy is to prevent their accumulation in the first place. Because N occurs in many different forms (e.g., ammonia, amonium ion, nitrate, nitrite, urea and N_2 gas) and processes such as nitrification and denitrification occur under oxic and anoxic conditions, respectively, treatment strategies for it may differ from those for P, which occurs primarily as a variant of phosphate. Due to the strong negative valence of phosphate, it is highly hydrophobic and under oxidized conditions seldom occurs in ionic form, but is adsorbed to small particles such as clay. Hence its strong loading correlation with total suspended solids (TSS). Denitrification is the removal of nitrogen as a gas when nitrate is reduced via nitrite by bacteria under anoxic conditions. However, anoxic conditions promote the internal loading of P (see above), meaning that the removal of or immobilization of one element can enhance the mobilization of the other.

Dredging of sediments is one strategy used to remediate aquatic ecosystems in which a significant mass of N and P has accumulated. Although dredging is similar to wiping the slate clean and starting over, unless inputs are reduced, the effectiveness will be limited. In addition, dredging is expensive, limiting the strategy to small areas or lakes. Removed sediments must be dewatered and disposed securely to prevent their re-entry into the system. Because many other contaminants including metals and synthesized organics such as polychlorinated biphenyls (PCB) also accumulate in sediments, the disturbance of sediments may release such materials back into the water, where they may pose a risk to biota. Other strategies include immobilization through the addition of chemicals or polymers. These methods are also costly and treatment of large areas or entire lakes is rarely undertaken. Currently these strategies are limited to aquatic ecosystems that serve multiple purposes such as drinking water supply and recreation.

Restoring Aquatic Ecosystems

The simplest and most cost-effective strategy to deal with all aquatic pollution is to avoid it in the first place. Time and again it has been shown that clean-up and restoration costs far out-distance any perceived negative economic impacts or costs associated with preventing pollution. However, because we do not plan on a long-term basis (beyond the time-frame of the polluting event) and we have not paid proper regard to things such as a healthy environment in economic models, we are unable to accurately determine the value of an unpolluted ecosystem before we begin its destruction. In rare cases in which a healthy ecosystem has been valued, we are all only too ready to greedily give it up for short-term gains.

Conclusions

The existence of life on earth depends entirely on access to and availability of clean, fresh water, which is of a finite quantity that is dwindling rapidly as the human population continues to expand and alter earth's ecosystems in profound ways. Although many individuals are acutely aware of the importance of a clean water resource, too many others are oblivious to the tenuous thread by which life clings to this planet. Generally, we do not give a second thought to bottled drinking water; having running water at the tap; depositing the weekly garbage at the curb; eating fruit grown in a commercial operation far away; or filling up the family SUV with fuel containing corn-based ethanol. In some way, each of these affects our water resources negatively. Part of our responsibility as scientists is to share our understanding and raise awareness of the need for clean freshwater resources. Given that it is much easier to protect than to attempt recovery, I encourage each reader of this article to share its contents with at least one other person with the request to do the same. We can no longer afford to be ignorant about our impending demise.

Glossary

Aerobic cellular respiration – The process of harnessing energy from chemical bonds in fuel molecules using oxygen as the electron acceptor.

Anoxia – Total absence of dissolved oxygen from water.

Endorheic – A water body with no outflow. Water loss occurs predominantly via evaporation or abstraction, although some leakage to ground may occur.

Epilimnion – The upper stratum of warm, low density and uniformly circulating water in a lake.

Hypoxia – Reduction of dissolved oxygen to a low concentration that excludes aerobic organisms, often 2–3 $mg\,l^{-1}$.

Hypolimnion – A stratum of cold, dense water underlying warmer, less dense water in a lake.

See also: Alkalinity; The Benthic Boundary Layer (in Rivers, Lakes, and Reservoirs); Chemical Fluxes and Dynamics in River and Stream Ecosystems; Chemical Properties of Water; Chloride; Deforestation and Nutrient Loading to Fresh Waters; Dissolved CO_2; Eutrophication; Major Cations (Ca, Mg, Na, K, Al); Mercury Pollution in Remote Fresh Waters; Methane; Micronutrient Elements (Co, Mo, Mn, Zn, Cu); Nitrogen; Nitrogen Fixation; Organic Nitrogen; Phosphorus; Pollution of Aquatic Ecosystems II: Hydrocarbons, Synthetic Organics, Radionuclides, Heavy Metals, Acids, and Thermal Pollution; Redox Potential; Salinity; Vector-Borne Diseases of Freshwater Habitats.

Further Reading

Cooke DG, Welch EB, Peterson S, and Nichols SA (eds.) (2005) *Restoration and Management of Lakes and Reservoirs*, 3rd edn. Boca Raton: CRC Press.

Fast AW and Lorenzen MW (1976) Synoptic survey of hypolimnetic aeration. *Journal of the Environmental Engineering Division American Society of Civil Engineers* 102: 1161–1173.

National Academy of Sciences (2002) *Estimating Water Use in the United States: A New Paradigm for the National Water-Use Information Program.* Washington, DC USA: National Academy Press.

Newbury RW and Bates D (2006) Hydraulics: dynamics of flow. In: Hauer FH and Lamberti GA (eds.) *Methods in Stream Ecology*, 2nd edn., pp. 79–102. New York: Academic Press.

Reddy MV (2005) *Restoration and Management of Tropical Eutrophic Lakes.* USA: Science Publisher, Inc.

Schindler DW (1972) Evolution of phosphorus limitation in lakes. *Science* 195: 260–262.

Schindler DW (2006) Recent advances in the understanding and management of eutrophication. *Limnology and Oceanography* 51: 356–363.

Schindler DW and Donahue WF (2006) An impending water crisis in Canada's western prairie provinces. *Proceedings of the National Academy of Sciences USA* 103: 7210–7216.

Scrimgeour GJ and Chambers PA (1997) Development and application of a nutrient diffusing bioassay for large rivers. *Freshwater Biology* 38: 221–231.

Tek Bahadur G (2007) Restoration of small lakes through cooperative management: A suitable strategy for poverty-laden areas in developing countries? *Lakes & Reservoirs: Research & Management* 12: 237–246.

UNESCO (2006) *Water a Shared Responsibility.* The United Nations World Water Development Report 2. United Nations Educational, Scientific and Cultural Organization (UNESCO), Paris France and Berghahn Books, NY, USA.

Zhang DD, Brecke P, Lee HF, He Y-Q, and Zhang J (2007) Global climate change, war and population decline in recent human history. *Proceedings of the National Academy of Sciences* 104: 19214–19219.

Relevant Websites

UN World Water Assessment Program – http://www.unesco.org/water/wwap/.

USEPA (2008) The Clean Water Act – United States Environmental Protection Agency – http://www.epa.gov/r5water/cwa.htm.

USEPA (2008) Water. United States Environmental Protection Agency – http://www.epa.gov/ebtpages/water.html.

Water on the web – http://waterontheweb.org/.

Pollution of Aquatic Ecosystems II: Hydrocarbons, Synthetic Organics, Radionuclides, Heavy Metals, Acids, and Thermal Pollution

K R Echols, J C Meadows, and C E Orazio, US Geological Survey, Columbia, MO, USA

Published by Elsevier Inc.

Introduction

Inland waters can be negatively impacted by transportation, agriculture, mining, power generation, nuclear energy, weapons production, and other activities. Although some aquatic pollution occurs naturally, most aquatic pollution by hydrocarbons, synthetic organics, radionuclides, and heavy metals is the result of human use, misuse, and improper disposal of chemicals. Some anthropogenic activities acidify or thermally pollute inland waters and lower the habitat quality. Production and use of petroleum products has resulted in inland waters polluted with hydrocarbons. Some of the myriad of synthetic organic chemicals (and their associated toxic byproducts) used by modern society can become aquatic pollutants. Certain pollutant chemicals that are persistent and toxic, such as halogenated organic chemicals, are of great concern. The heavy metals lead, cadmium, and mercury, and other toxic elements emitted by smelters, chloralkali plants, mining, agricultural irrigation drains, power production plants, and industrial processes can negatively impact fish and wildlife. Mining operations leave behind metal sulfides that result in acid drainage that creates toxic aquatic conditions. Radionuclides in the aquatic environment from accidental releases from nuclear power plants and plutonium production can reach harmful levels. Thermal pollution of aquatic systems from warm industrial effluents lowers the quality of the habitat for aquatic organisms. The impact of these various stressors of inland waters depends on their magnitude, relative toxicity, the nature of the aquatic system, and sensitivity of the organisms.

Hydrocarbons

Hydrocarbons are a simple class of organic chemicals that are obtained from the earth's two large reservoirs of organic material: petroleum and coal. A wide range of aliphatic hydrocarbons are produced by cracking and fractional distillation of petroleum (crude oil). These range from light petroleum products that contain small molecules such as pentane, hexane, cyclohexane, benzene, and toluene, up to heavy viscous long-chained oils and waxes. Coal coking, liquefaction, and distillation also produce various hydrocarbons. Organic solvents, fuels, polycyclic aromatic hydrocarbons (PAHs), and synthetic organics are obtained from coal and petroleum.

Organic Solvents

Organic solvents are used in numerous industrial processes and in everyday products (paints, oils, varnishes, and dry cleaning liquids to name a few) and may be released to the environment in industrial effluents or through improper disposal by users. Many solvents are classified as volatile, contaminating the air as well as water. Common solvent chemicals found in surface and ground water include chlorinated solvents such as trichloroethane, chloroform, dichloromethane, dichloroethylene, trichloroethylene, and *p*-dichlorobenzene. Gasoline components, such as methyl *tert*-butyl ether, benzene, toluene, and xylenes, are also polluters of inland waters. Most of these compounds have been shown to be more toxic to mammals than to aquatic organisms; however, some are toxic to fish when chronically exposed. Although organic solvents generally volatilize and degrade in surface water relatively quickly, there are cases where elevated levels persist. For example, an environmental disaster in November 2005 dumped 100 metric tons of benzene, aniline, and nitrobenzene into the Songhua River after an explosion at a chemical plant in the Jilin Province, China. The cold Siberian conditions in the area reduced the volatility of these chemicals and allowed them to move downstream and threaten drinking water supplies. Organic solvents can also mobilize less soluble toxic organic chemicals, resulting in increased organism exposure.

Petroleum Hydrocarbon Oils

Crude oil is predominantly a mixture of straight and branched chain aliphatic hydrocarbons, ranging from gaseous methane to viscous tar-like compounds with up to 80 carbons. Crude oil also contains aromatic hydrocarbons as well as sulfur- and nitrogen-containing hydrocarbons. Leaks of petroleum crude oil from pipelines, tanker trucks and ships, and oil wells have polluted rivers and lakes. The toxicity of the bulk oil to an aquatic system depends both on the organisms involved and on the type of the oil. The petroleum industry has four classifications: Class A includes volatile, low molecular weight, typically clear, low density oils. In water, these oils disperse rapidly, coating solid surfaces as well. Because of their low density, they float on the water, forming a layer at the water–air interface which allows their volatile components to evaporate quickly. In some spill scenarios, igniting this flammable oil may be either an option or a hazard. Their low viscosity allows the oils to penetrate sediment where they may persist. These light oils can be highly toxic to humans, fish, and other biota. Class B oils have a waxy or oily consistency and are less toxic and less mobile than the lighter fraction oils, but can persist longer in sediment. The black-brown colored oils of Class C are so heavy and sticky that removal from an aquatic system can be difficult. Although Class C oils lack the more toxic compounds found in lighter oils, wildlife can be smothered or drowned when contaminated. Residual fuel oils and medium-heavy crude oils are in this category. Class D oils are high molecular weight, heavy crude oils, heavy paraffins, and residual oils. Although they are less toxic than lighter oils, they are very difficult to clean up and will remain coated to rocks and sediments.

Polycyclic Aromatic Hydrocarbons

Complex mixtures of polycyclic aromatic hydrocarbons (PAHs) are ubiquitous environmental contaminants that result from both natural processes and anthropogenic activities. The molecular structure of the PAH class of chemicals is a series of fused aromatic rings (**Figure 1**), with potential for alkyl group substitution on the rings. Generally, petroleum-derived PAHs have a higher percentage of alkylated PAHs relative to the nonsubstituted (parent) PAHs. Several PAHs are of concern because they are mutagenic and carcinogenic, while others fall into the category of being acutely toxic to aquatic life. Benzo [*a*]pyrene is a well-known mammalian carcinogen that also has carcinogenic effects on fish. The bioavailability of PAHs in aquatic environments is a factor of the water solubility and the octanol–water partition coefficient (K_{OW}) of the particular PAH. A large fraction of the larger ringed PAHs in an

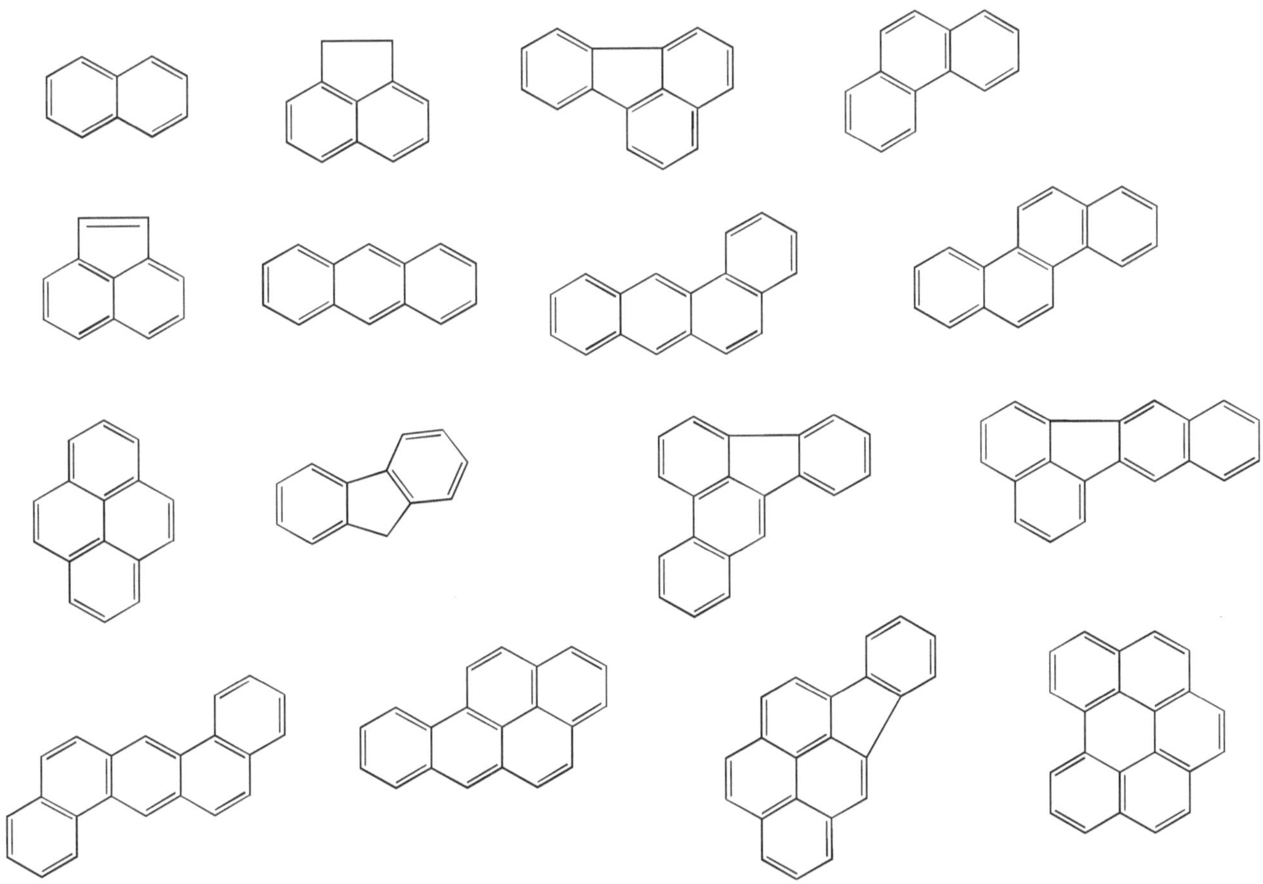

Figure 1 Structures of USEPA priority pollutant PAHs.

aquatic system is adsorbed by bed and suspended sediments. Those PAHs with a bay region structure, such as found with benzo[*a*]pyrene, are more toxic; these PAHs also break down more rapidly in ultraviolet light. Rates of abiotic and microbial degradation vary for each PAH. This differential weathering of a PAH mixture selectively increases the relative concentration of the more stable PAHs.

Natural sources of PAHs in the environment are seeps of oil, bitumen (natural rocks that contain hydrocarbons), forest fires, and volcanic activity. However, the major sources of aquatic PAH pollutants are due to incomplete combustion of fossil fuels from gasoline and diesel powered engines; combustion of coal from power plants; and open burning of plant-based material. Other sources of aquatic PAH pollution include wastes from coal gassification plants and storm water runoff over fossil fuel-based materials such as asphalt and spills of oil and gas. Coal tar contaminated soils of former manufactured gas plants and coking plants can be long-term inland water pollution sources of PAHs. An example of a PAH-polluted inland water is the Indiana Harbor/Canal-Grand Calumet River System of Grand Calumet, Indiana, USA. This area of the Great Lakes has a high level of PAHs in waters and sediments. Activities associated with intensive industries such as coking plants, steel mills, and petroleum refineries release PAHs, resulting in degradation of the habitat. Other sources of PAHs in industrial areas are automobile exhaust, runoff from pavement, and sewage outfall. The burning of wood and fossil fuels has been implicated in the PAH contamination of Lake Michigan and other inland waters. Other releases of PAHs are related to creosote wood preservative impregnation of railroad ties, telephone poles, and other treated woods. Recently, PAH pollution of urban streams has been traced to certain types of parking lot sealants.

Synthetic Organic Chemicals

Polychlorinated Biphenyls

Polychlorinated biphenyls (PCBs) are industrial chemicals that were produced as various mixtures in the United States from 1930 to1977 and sold under the trade name Aroclor®. They were used, and are still in use in many applications, as dielectric fluids for capacitors and transformers, in heat transfer applications, in hydraulic fluids, and as additives in the manufacture of caulk, paint, paper, pesticides, adhesives, and plastics. The properties that made them useful in all these different applications – stability, low flammability, low water solubility, low vapor pressure, and solubility in organic solvents – also made them problematic in the environment. The PCB chemical structure consists of biphenyl molecule with 1–10 chlorine atoms attached (**Figure 2**). Although there are 209 theoretically possible isomers (congeners) of PCBs, approximately 150 congeners are prominent in the commercial mixtures and are likely to be found in the environment.

PCBs released into the environment are now persistent decades after their production was banned by the US Environmental Protection Agency. Because their use was so widespread, PCBs have been found in aquatic environments all over the world. They are found in virtually all compartments of the biotic environment, including humans, and including areas remote from their initial use. PCBs and many other high K_{OW} compounds discussed in this article may become associated with fine particulate matter and globally distributed to pristine environments via high altitude atmospheric circulation. Because of their high lipid solubility, they tend to bioaccumulate in species (such as piscivorous fish and piscivorous birds like ospreys, cormorants, gulls, and bald eagles) at the top of food webs. Certain PCBs have been shown to cause cancer, endocrine disruptions, and compromise immune, reproductive, and neurological systems. Those PCBs which do not have chlorines substituted on the carbon adjacent to the biphenyl linkage (ortho position) have effects similar to that of dioxin. These are the coplanar or non-ortho, dioxin-like PCBs. Serious cases of PCB contamination in aquatic ecosystems have been associated with industrial sites that synthesized PCB mixtures, used PCBs for manufacturing electrical transformers and capacitors, or used PCBs as hydraulic fluids in die casting aluminum such as in automobile manufacturing. Examples of heavily PCB-contaminated sites include the Hudson River in New York, the Fox River in Wisconsin, the Saginaw River in Michigan, and the Coosa River in Alabama.

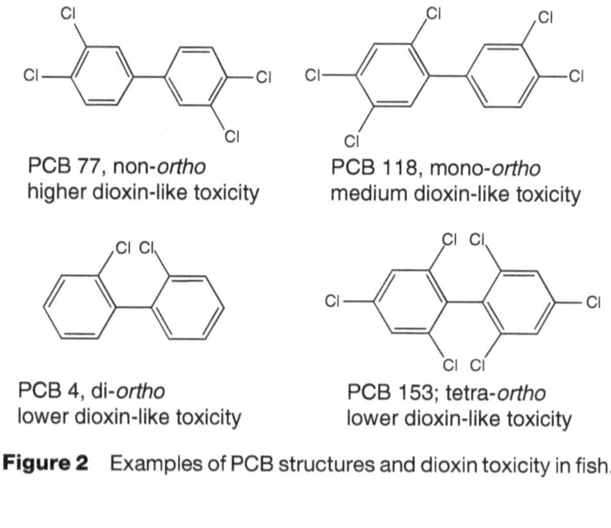

Figure 2 Examples of PCB structures and dioxin toxicity in fish.

Polybrominated Diphenyl Ethers

Polybrominated diphenyl ethers (PBDEs) are industrial chemicals whose concentrations are rising exponentially in the aquatic environment. They are similar to PCBs, with a basic diphenyl ether structure substituted with up to 209 possible combinations of bromine atoms on both rings (**Figure 3**). The sources of these persistent chemicals are diffuse, with a major source thought to be PBDE flame retardants leaching from everyday products such as textiles, computers, furniture foam, plastics, electronics, automobiles, and construction materials. The polymeric materials in these products can contain 5–25% PBDE by weight. Despite very recent reduction in the manufacture of PBDEs, plastic products containing these flame retardants will continue to leach PBDEs into the environment for years to come.

The biological hazards posed by PBDEs and their metabolites include reduction in thyroid function, and impaired neurological development and behavioral performance. Although levels in the environment worldwide are surpassing PCBs, chlordane, and DDT, the rise is most marked in North America where PBDE levels in aquatic biota are projected to continue to rise under current levels of manufacturing and use, doubling every 5 years (reported in 2004). Because of the lipophilic and bioavailable nature of PBDEs, they readily accumulate in organisms in the same manner as PCBs. The pentabrominated diphenylether, congener number 47 (PBDE-47), typically comprises 40–70% of the total PBDE concentration in fish tissue. In Europe, frogs from pristine areas of Scandinavia have been found to contain residues of PBDEs. Aquatic organisms are exposed to PBDEs primarily through the food chain. A persistent pollutant of such widespread distribution as PBDE is of concern because of its increasing environmental concentration and potential toxicity. In spite of these concerns with PBDE pollution, a paucity of information exists on the toxicity of these compounds to aquatic organisms. Amphibians and fish currently serve as model organisms to understand the environmental impact of PBDEs. Adverse impacts are expected because PBDEs have been shown to interfere with thyroid function in some vertebrates, which is critical for development and metamorphic changes. In frogs, increases in the length of the larval period and decreases in the mass at metamorphosis have been linked to declines in adult fitness. PBDEs have the potential to negatively impact amphibian populations via alterations in the life history characteristics that are closely linked with fitness. In addition, impairments to neurobehavioral development and performance would also significantly limit survival, and could create a shift in life history characteristics that would limit population growth.

Pesticides

Pesticides are chemicals used to control pests whether they are insects, plant, fungi, or other animals. The structures of the major classes of pesticides are given in **Figure 4**. Natural products were used for centuries but beginning in the early twentieth century, chemicals were specifically synthesized for pest control. The majority of the chemicals that were developed were organochlorine pesticides, such as aldrin, benzene hexachlorocylcohexanes (HCHs), chlordanes, dichlorodiphenyltrichloroethanes (DDTs), dimethyl tetrachloroterephthalate (DCPA), diazinon, dieldrin, endosulfans, endrin, hexachlorobenzene (HCB), methoxychlor, and mirex. Most of these chemicals persisted in the environment long after they were used,

Figure 3 Computer scrap is a potential source of PBDE pollution.

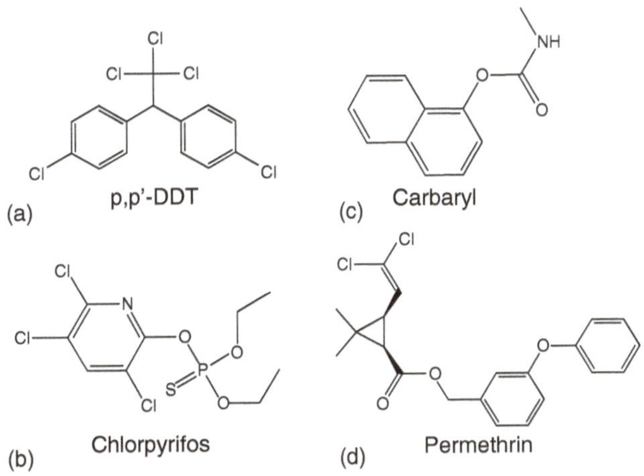

Figure 4 Examples of pesticide classes: (a) organochlorine, (b) organophosphate, (c) carbamate, and (d) pyrethroid.

accumulated in lipid tissues of a variety of biota, and increase in concentration at the higher levels of food webs.

In the late 1940s, a new class of insecticides developed as a corollary to chemical warfare research with organophosphorus compounds. Some members of this class of compounds are nervous system agents that inhibit acetylcholinesterase and causes excessive stimulation of cholinergic nerves. Some of these agents are among the most potent acute toxins ever produced. From the study of these compounds arose the family of organophosphate insecticides. Common organophosphate insecticides were malathion, chlorpyrifos, diazinon, and parathion. This class of pesticides has higher toxicity to mammals than do the organochlorine pesticides.

Carbamates are pesticides with ester and amide functionality. They are polar compounds which inhibit acetylcholinesterase as much as the organophosphates do. Carbamates have limited environmental persistence. The ester group is easily hydrolyzed when the molecule is exposed to water, so they are not particularly persistent in the aquatic environment. Because of the polarity and reactivity of carbamates, they break down within a span of days to weeks in water. As such they do not bioaccumulate significantly in organisms. They have similar half-lives in both soil and water.

Another class of pesticides is the pyrethrins/pyrethroids. Pyrethrin is a natural product of the chrysanthemum plant, which has been used for thousands of years as a pesticide. The synthetic versions of pyrethrins are called pyrethroids. These pesticides are neurotoxic, affecting the sodium channel in nerve cells, eventually causing paralysis. The first synthetic version of a pyrethroid was developed in the mid-1950s with the synthesis of allethrin. Numerous other pyrethroids continue to be manufactured and are used in household pesticide products for mosquito, flea, and tick control as well as for agricultural uses. In the aquatic environment, pyrethroids are extremely toxic to fish; however, they are unstable in sunlight and air, typically being degraded in 1–2 days. They bind strongly to soil or sediment so have relatively low mobility in the environment.

Unwanted Waste Byproducts

Many chemicals that are found as contaminants in aquatic environments are unintentional byproducts of commercial manufacturing or other industrial processes or degradation products of specific-use chemicals. Combustion, which produces PAHs, can also release or produce other unwanted chemicals. The poster child of chemical byproducts is dioxin, which has been labeled the most toxic human-made chemical. The term dioxin typically refers to polychlorinated dibenzo(*p*)dioxins (PCDDs) and polychlorinated dibenzofurans (PCDFs). These chemicals are created unintentionally as byproducts of paper production by chlorination of wood pulp, as byproducts of chemical production of chlorinated phenols, and are released from incineration of organic material that contains chlorine (whether organochlorine or chloride). There have been many inadvertent releases of these compounds. During the Vietnam conflict, massive amounts of the defoliant Agent Orange, a mixture of 2,4-D and 2,4,5-T, were produced and released in the zones of operation. It was later found that percent levels of dioxins were produced in the manufacture of the defoliant and were released as well. Entire communities have been literally shut down by these compounds. In 1972, a contractor sprayed waste oil containing dioxins on roads and arenas in the community of Times Beach, Missouri. When this was discovered, the entire community was closed and the occupants relocated.

One isomer of the dioxin family, the 2,3,7,8-tetrachlorodibenzo(*p*)dioxin, is the most toxic isomer of all the PCDDs and PCDFs. Parts per trillion concentrations have negative impacts on aquatic fish and wildlife. Because dioxins are lipophilic compounds, they have a high propensity to bioaccumulate up aquatic food webs. Dioxins and dioxin-like compounds in the aquatic environment have been shown to cause reproductive and developmental effects in fish and fish-eating birds (**Figure 5**), egg mortality, yolk sac edema, and embryo deformities, which reduce fry and chick survival.

Chemicals of Emerging Concern

An emerging pollution concern for inland waters comes from chemicals used in pharmaceuticals and personal care products. Some of these chemicals or their metabolites/degradation products have been shown to pass through waste water treatment plants (**Figure 6**). These are chemicals used as antimicrobials in soaps, as fragrances in personal care products, as components or metabolites of commonly used medicines, and so on. The synthetic polycyclic musk fragrances have been found in surface water, waste water, sediments, and have been shown to accumulate in aquatic biota. Several of these fragrant compounds also have weak estrogenic activity. The antimicrobial compound triclosan has also been found in water, sediment, and fish as its methylated metabolite. Nonyl phenols and nonyl phenol ethoxylates are a group of chemicals that are used as nonionic surfactants (i.e., designed to reduce the surface tension of water) in a variety of products, as inert ingredients in pesticide formulations, as spermicides, and as

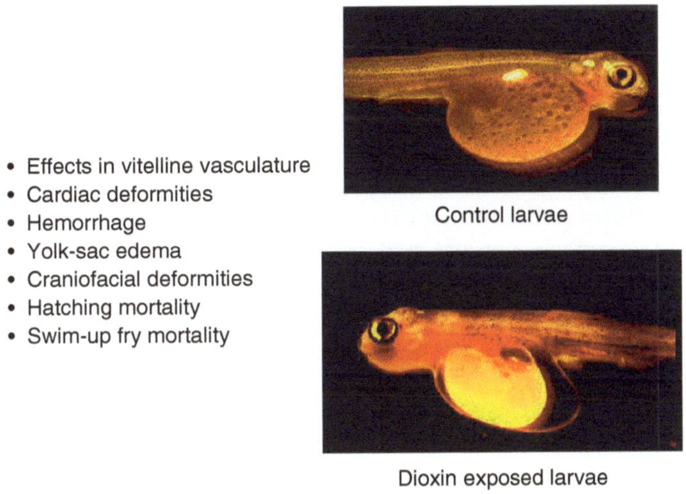

Figure 5 Lake trout fry exposed to dioxin. Deformities and egg sac edema. Reproduced from Wright PJ and Tillitt DE (1999) Embryotoxicity of Great Lakes lake trout extracts to developing rainbow trout. *Aquatic Toxicology* 47(2): 77–92.

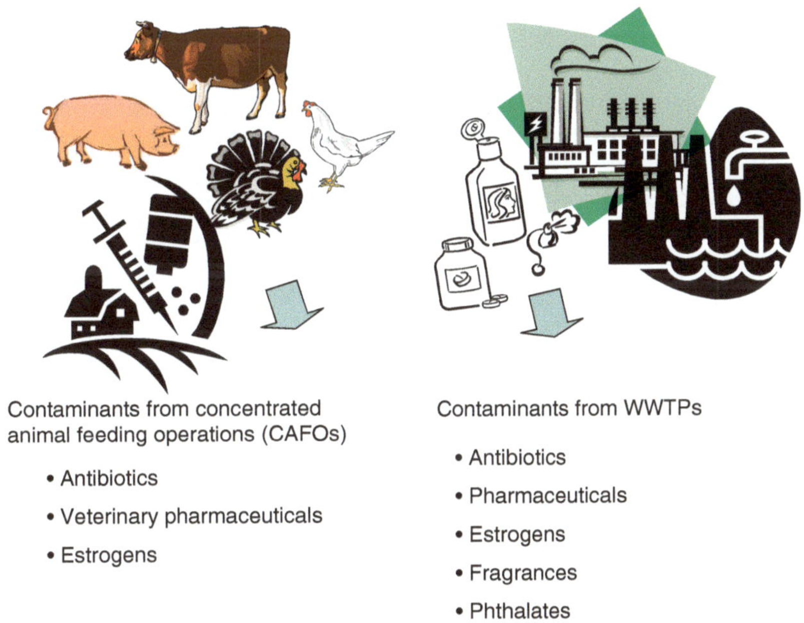

Figure 6 Examples of emerging contaminants and their sources.

components in certain plastics such as PVC (polyvinyl chloride). They are toxic to a variety of aquatic organisms and have been studied for endocrine disrupting effects. Several phthalates or other additives that are added to plastics are found in aquatic systems. Bisphenol-A has been determined to have estrogenic activity, as has butyl benzyl phthalate. Other compounds that are being evaluated in aquatic environments are pharmaceutical compounds such as antibiotics, analgesics, antidepressants, estrogens (such as birth control pills or the growth hormone diethylstilbestrol), and their metabolites, which may still have pharmacological or hormonal activity. Many of these types of compounds survive treatment and are released in water from waste water treatment plants. Concentrated animal feeding operations, where sustained feeding of antibiotics is practiced, are a significant source of veterinary pharmaceuticals as well. Many of these chemicals fall into the general class of chlorinated hydrocarbons and while many of the compounds break down quickly in the environment, the results are unpredictable. Only now are studies showing if they or their breakdown products have any impacts on aquatic ecosystems.

Heavy Metals

Contamination of an aquatic environment by toxic elements can result from metal mining operations, industrial processes, coal-fired power plants, and other activities. The term heavy metal is not well defined and often refers to trace elements or trace metals. The US EPA has a priority pollutant list that includes so-called heavy metals such as mercury, cadmium, chromium, and lead, and the metals copper, nickel, and zinc, and the metalloid elements arsenic and selenium. Other elements can be of concern environmentally if their concentrations reach levels that are beyond the narrow range needed for growth and maintenance of organism health.

Mercury

Mercury pollution of inland waters is a global problem. Among the sources of mercury pollution are coal-fired power plants, mining wastes, chloralkali plants, and municipal waste incinerators. Mercury is released and cycled in the environment as either elemental mercury, or various inorganic and organic forms (**Figure 7**). In aquatic ecosystems, the inorganic forms of divalent mercury (Hg^{2+}) in sediment are converted to the more toxic organic form, methylmercury. Typically, 90% of the mercury found in fish is the methylmercury form. Methylmercury is produced inside sulfate-reducing bacteria. The amount of methylmercury varies among ecosystems and is not directly related to the amount of atmospheric mercury deposition, as the ability of the aquatic system to methylate mercury and to bioaccumulate depends on environmental conditions. Fish and other species in several inland waters of the United States have excess methylmercury levels, which is a concern for consumers (both human and wildlife). Certain aquatic systems are prone to mercury contamination; for example, in the United States, the Everglades in Florida, the lower Great Lakes, and the Ohio Valley have among the highest deposition rates.

Until recently, the US Food and Drug Administration regulations set a limit of 1 ppm mercury as being safe for human consumption. In 2002, these action limits were lowered to 0.3 ppm to reflect the newly discovered health risks associated with the consumption of fish containing elevated Hg concentrations. As a result, the US EPA and US Department of Health (USDOH) posted a consumer advisory about the risk of mercury in fish, advising pregnant women, and women of childbearing age who may become pregnant, and young children not to eat identified large fish that accumulate the highest levels of MeHg. In the United States, there are more fish consumption

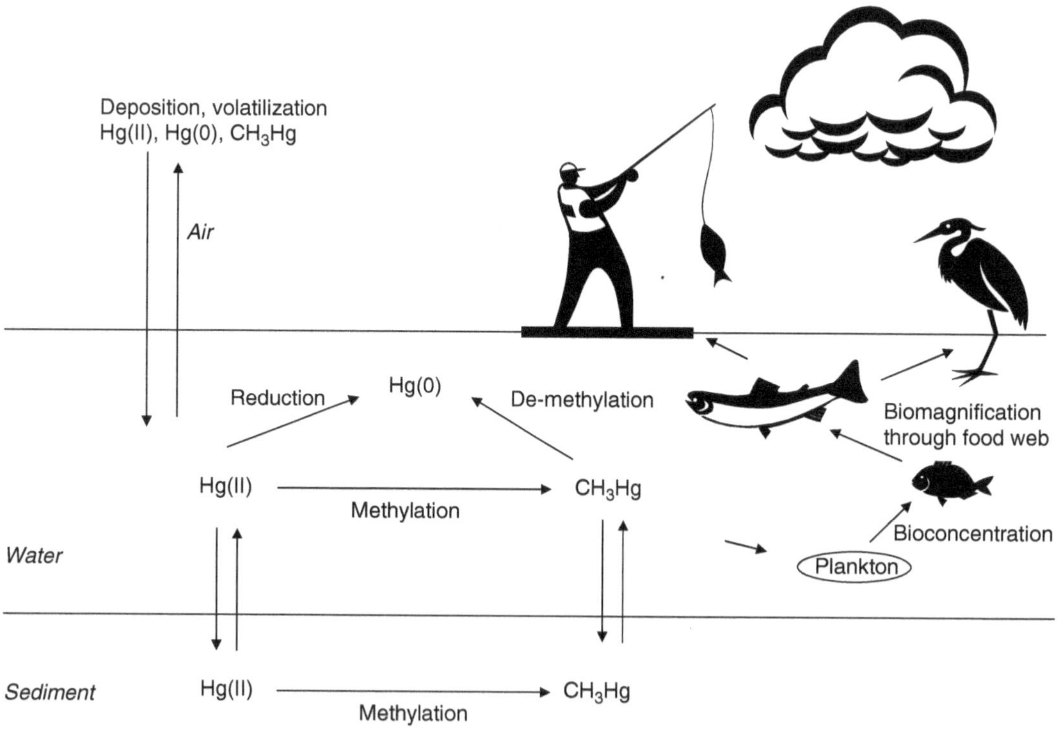

Figure 7 Mercury cycling in the aquatic environment.

advisories and bans posted for Hg than any of the 46 chemical contaminants that appear on the advisory. Advisories are used to alert sensitive populations to the risk of potential exposure to specific chemical contaminates, specific species and size, and geographic location (i.e., river, lake, stream).

Aquatic organisms obtain methylmercury from water, sediment, and food. It biomagnifies up the food chain and can be found in fish at two orders of magnitude higher than that of surface water they swim in. Inorganic mercury does not easily transfer into food webs and does not biomagnify. Methylmercury is a neurotoxin that can cause incoordination, poor appetite or feeding behavior, lessened swimming activity, starvation, and eventually mortality. Fish-eating birds are adversely affected by methylmercury although birds have better mechanisms to eliminate the body burden of methylmercury by excretion into growing feathers.

Other Toxic Elements

Lead is a heavy metal that has been used by humans for centuries because it is one of the easiest metals to mine. Starting in the industrialization of the nineteenth and through the twentieth century, lead was used in a variety of applications: as gasoline anti-knock additive tetraalkyl lead; pigment and paint additive; in storage batteries; in construction and building, such as lead pipes; and for the aquatic environment particularly lead shot used in hunting and shooting. As other uses have been phased out to protect human populations from exposure to lead, the mining, smelting, and recycling of lead batteries have become the major source of lead pollution. In aquatic ecosystems, lead has been shown to cause death in waterfowl due to the ingestion of lead shot.

Cadmium is an unnecessary trace element that is found typically as oxide in the environment. It is used in batteries, in pigments, metal coatings, and metal alloys. Cadmium is primarily released in the environment from mining activity and the burning of fossil fuels.

Chromium is a necessary trace element that is used in humans in insulin to regulate use of fat, sugar, and protein. The industrial uses of chromium are steel manufacturing, chrome plating, dyes and pigments, and leather tanning. When released in an aquatic environment, chromium usually associates with sediment and it is not accumulated at large levels in fish.

Arsenic is a toxic metalloid that is typically found in oxide forms (As_2O_3 and As_2O_5) that are soluble in water. Arsenic has been used as an insecticide in its arsenate form, and as a component of treated lumber as chromated copper arsenate. Arsenic-contaminated ground water is a problem in countries around the world, including parts of the United States. Arsenic is a carcinogen and can be acutely toxic to humans. Arsenic does not adversely affect aquatic organisms such as fish, but aquatic plants, fish, and shellfish are able to bioconcentrate arsenic and store it in an organic form, arsenobetaine, which is a relatively nontoxic form.

Selenium is a metalloid element that is an essential trace nutrient in very low amounts but becomes toxic at levels just slightly over the necessary levels. It is naturally occurring in inorganic – selenate, selenite, selenide, elemental selenium – and organic forms – selenomethionine, selenocysteine, dimethylselenide, and dimethyldiselenide. Selenate is the form found in mostly arid areas and in alkaline soils, but is soluble in water and is incorporated into plants, which then convert to organic forms. Selenium in dietary excess will substitute for sulfur in protein synthesis causing toxic problems in organisms. There have been documented effects on fish reproduction and poor survivability of larval or young fish as well as effects on older fish.

Radionuclides

Radionuclide pollution in aquatic ecosystems typically is a byproduct of nuclear power generation, nuclear weapons testing, nuclear waste disposal (low and high level), and natural sources. Once in the aquatic environment, radionuclides may cause negative effects on biota from radiation or by physical contact with the radionuclide, whether by ingestion or absorption. However, radiation is ultimately the cause of any damage to any organism, unless the chemical itself also has a toxic effect on the organism. Typical effects of radiation are cellular and genetic damage. The impact of radionuclides is dependent upon the type of radiation they produce and their radioactive half-life. Alpha radiation causes damage typically only when the source is taken internally, whether ingested, inhaled, or injected; beta radiation and gamma radiation (the most energetic) cannot be stopped by the top dead skin layer and causes thermal burns, genetic damage, and suppresses immune function. Some radionuclides have half-lives of a few hours to a few days (sodium-24, 15 h, phosphorus-32, 14 days); others have half-lives of many years, often hundreds to thousands of years (cesium-137, 27 years, carbon-14, 5568 years). Nuclides that have longer half-lives and produce higher radiation are of the most concern. Strontium-90 and cesium-137 are of the two most environmental concerns. Another consideration is whether a radionuclide can replace an important element in biological tissues. Several examples are radioactive iodine; strontium can substitute for calcium; cesium can substitute for potassium.

Internationally, accidents at the Chernobyl and the Mayak reactors in Russia have been a source of aquatic radionuclide pollution. Freshwater environments in the United States that have been impacted by release of radionuclides include: White Oak Lake and the Clinch River from Oak Ridge National Laboratory, Tennessee, operational from 1943 to the present; the Savannah River from the Savannah River Nuclear Plant, in South Carolina operational from 1951 to the present; and the Columbia River from the Hanford Nuclear Site in Washington where the US government produced plutonium for nuclear weapons from 1944 to 1971. Uranium fuel was partially transformed into plutonium inside the nuclear reactors along the Columbia River. This created other radioactive elements as well, including the radionuclides cerium, strontium, and ruthenium. At the Hanford site iodine-131, cerium-144, plutonium-239, ruthenium-103, ruthenium-106, and strontium-90 were released into the air environment, while arsenic-76, neptunium-239, phosphorus-32, sodium-24, and zinc-65 were released into the Columbia River. From Oak Ridge, the predominant isotopes released into aquatic systems were cesium-137, cobalt-60, strontium-90, and tritium. The Savannah River Plant released similar radionuclides, including cesium-137, and tritium, which have been monitored in effluent, sediment, and biota.

Thermal Pollution

Heat released into rivers, streams, lakes, or reservoirs by use of surface water as a coolant lowers the quality of the habitat, and is thus thermal pollution. Estimates are that 50% of the surface water in the United States is used as a coolant in power plants or other industrial uses. Water that is circulated back into the environment at a warmer temperature can cause problems for aquatic biota in the system. Warm water has lower dissolved oxygen content. In warmer water, the respiration rates of organisms increase, depleting oxygen faster. Populations of certain species of organisms may decline with increase in water temperature because they are acclimated to certain temperature ranges while other species may flourish, changing the dynamics of the ecosystem. Spawning may be affected or young fish may die.

Further Reading

Brooks BW, Chambliss CK, Stanley JK, *et al.* (2005) Determination of select antidepressants in fish from an effluent-dominated stream. *Environmental Toxicology and Chemistry* 24: 464–469.

Committee on EPA's Exposure and Human Health Reassessment of TCDD and Related Compounds, National Research Council. (2006) *Health Risks of Dioxins and Related Compounds: Evaluation of the EPA Reassessment.* Washington, DC: The National Academies Press.

Connell DW (1997) *Basic Concepts of Environmental Chemistry.* Boca Raton, FL: CRC Press.

Erickson MD (1997) *Analytical Chemistry of PCBs,* 2nd edn. Boca Raton, FL: CRC Press.

Forstner U and Wittmann GTW (1979) *Metal Pollution in the Aquatic Environment.* NewYork: Springer-Verlag.

Frontasyeva MV, Perelygin VP, and Vater P (eds.) (2001) Radionuclides and heavy metals in environment. Proceedings of the NATO Advanced Research Workshop on Monitoring and Human-Made Radionuclides and Heavy Metals Waste in Environment, Dubna, Russia, 3–6 October 2000, *Earth and Environmental Sciences,* vol. 5. *NATO Science Series,* Vol. 1V. Boston: Kluwer Academic Publishers.

Hites RA (2004) Polybrominated diphenyl ethers in the environment and in people: A meta-analysis of concentrations. *Environmental Science and Technology* 38: 945–956.

Hoffman DJ, Rattner BA, Burton GA Jr., and Cairns J Jr. (eds.) (2003) *Handbook of Ecotoxicology.* Boca Raton, FL: CRC Press.

Kolpin DW, Furlong ET, Meyer MT, *et al.* (2002) Pharmaceuticals, hormones, and other organic wastewater contaminants in U.S. streams, 1999–2000: a national reconnaissance. *Environmental Science and Technology* 36: 1200–1211.

Neff JM (1979) *Polycyclic Aromatic Hydrocarbons in the Aquatic Environment – Sources, Fate, and Biological Effects.* London: Applied Science Publishers.

Nowell LH, Capel PD, and Dileanis PD (1999) *Pesticides in Stream Sediment and Aquatic Biota: Distribution, Trends, and Governing Factors.* Boca Raton, FL: Lewis Publishers.

Peterle TJ (1991) *Wildlife Toxicology.* New York: Van Nostrand Reinhold.

Rand GM (ed.) (1995) *Fundamentals of Aquatic Toxicology: Effects, Environmental Fate and Risk Assessment.* Boca Raton, FL: CRC Press.

Rathburn RE (1998) *Transport, behavior, and fate of volatile organic compounds in streams. US Geological Survery professional paper 1589.*

Reemtsma T and Jekel M (eds.) (2006) *Organic Pollutants in the Water Cycle.* New York: Wiley.

Simonich SL, Begley WM, Baere G, and Eckhoff WS (2000) Trace analysis of fragrance materials in wastewater and treated wastewater. *Environmental Science and Technology* 34: 959–965.

Relevant Websites

http://www.atsdr.cdc.gov/ – The Agency for Toxic Substances and Disease Registry (ATSDR) is an agency of the US Department of Health and Human Services.

http://www.epa.gov – US Environmental Protection Agency website.

Vector-Borne Diseases of Freshwater Habitats

V H Resh, University of California, Berkeley, CA, USA

Introduction

Diseases transmitted by organisms living in fresh water have been a major force in shaping both the size of the human population and the development of our civilization. Certainly, far more people have been killed by diseases transmitted by freshwater vectors than those caused by all the wars in the history of the world. This is true even in recent times; during the Vietnam War, many times more casualties resulted from malaria than from combat.

Water-related human diseases currently kill more than 5 million people a year, and some 2.3 billion people suffer from diseases linked to water. Water-borne diseases can be classified into four major categories: (1) water-borne diseases that are spread through contaminated drinking water, such as cholera; (2) water-washed diseases arising from poor sanitation, such as typhoid; (3) diseases that are directly transmitted by vectors (usually biting flies) that depend on fresh water for one or more of the stages of their life cycle, such as malaria; and (4) diseases that are transmitted to humans that involve an aquatic animal that serves as an intermediate host, usually snails, such as schistosomiasis. Other articles in the Encyclopedia concentrate on the first two categories (e.g., coliforms); this article concentrates on the latter two categories.

Human Diseases and Freshwater Vectors

A variety of human diseases are transmitted by vectors that have life cycles associated with various types of water bodies, including lakes, ponds, rivers, streams, reservoirs, and irrigated fields. Of these diseases, malaria, schistosomiasis, lymphatic filariasis, onchocerciasis, Japanese encephalitis, and dengue are generally considered now to be the most important in terms of their infection rate, morbidity (i.e., either the incidence or to prevalence of a disease), or mortality (i.e., the number of deaths resulting from a particular disease).

These diseases have had and continue to have a devastating toll on human life. For example, malaria is endemic in more than 100 developing countries and more than 2 billion humans are at risk. 80–90% of the cases occur among the poorest countries of the world in Africa, but malaria is also a major problem in parts of South America, India, and Southeast Asia. Between 100 and 200 million people are estimated to be infected worldwide, and ~2 million deaths occur per year (even higher during epidemic years). An estimated cost of $1.7 billion in lost productivity and treatment costs applies in sub-Saharan Africa alone.

Malaria is caused by several species of *Plasmodium* protozoans: *P. vivax*, *P. falciparum*, *P. malariae*, and *P. ovale*. It is vectored by various species of *Anopheles* mosquitoes that often occur in high numbers in irrigation ditches and canals, lakes and ponds, riverine floodplains, wetland-rice cultivation areas, and human settlements. Current control measures involve personal protection (e.g., repellents, pesticide-impregnated bed nets, anti-malarial drugs), vector reduction through insecticide applications, and parasite elimination in infected humans through drug therapy.

Schistosomiasis occurs in most subtropical and tropical areas of the world, and 200 million people being affected with this debilitating disease; 20 million suffer consequent renal failure, bladder cancer, and liver fibrosis. Like malaria, the majority of the schistosomiasis cases occur in sub-saharan Africa. Children are particularly at risk; 88 million children under 15 are infected each year. The passing of blood in the urine of adolescent boys, a consequence of the disease, is so common in some parts of West Africa that it is considered normal and analogous to menarche in adolescent girls. Certain species of snails serve as the intermediate host of the blood fluke parasites in the genus *Schistosoma*; *S. mansoni*, *S. japonicum*, and *S. haematobium* are the most important agents of this disease. Control involves snail elimination through molluscicide applications or providing drug therapy to infected individuals. The latter is complicated by the frequency of reoccurrence of infections that occurs among individuals that have already been treated.

Lymphatic filariasis (or elephantiasis), which is transmitted by a few genera of culicine mosquitoes (culicine is a subfamily of the mosquito family Culicidae), is a global problem with high morbidity but low mortality. It is estimated that this disease affects some 120 million people living in 80 countries, and high infection rates were noted in Egypt, sub-Saharan Africa, India, the western Pacific Islands, and parts of the Caribbean and South America. The disease is caused by two species of filarial nematodes *Wucheria bancrofti* and *Brugia malayi*. It is especially prevalent in areas with irrigation systems and in weed-infested reservoirs. Control involves mosquito reduction and drug therapy, the latter either as a prophylactic or a treatment.

Japanese encephalitis (or brain fever) is also transmitted by some genera of culicine mosquitoes and is the result of a flavovirus infection. It occurs in South, Southeast, and East Asia, and in some Pacific Islands. During epidemic outbreaks, high mortality rates occur, especially among children. Control involves mosquito reduction through habitat modification or insecticidal applications, and through vaccination.

Dengue fever (or hemorrhagic fever) occurs mostly during and shortly after the rainy season in most tropical and subtropical areas of world. More than 100 million people suffer from it each year and the number of infections is rising annually. Likewise, its distribution has spread greatly over the past three decades. It is caused by four related forms of flavovirus and is vectored by *Aedes* mosquitoes, most commonly *Aedes aegypti*. This species of mosquito is also the infamous vector of yellow fever, a disease that has been eliminated in much of the world but which is now gaining renewed importance because of its transmission of dengue.

Adult females of many *Aedes* mosquitoes lay their eggs on moist soil that will later be flooded and from which the mosquitoes emerge as adults. Control is directed at mosquito reduction and biting prevention; no vaccine is available.

In contrast to diseases mentioned so far, in which the vector occurs in still or at least very slow-moving water, onchocerciasis is vectored by black flies whose immature stages occur attached to vegetation in fast-flowing streams. The parasite is the filarial nematode *Onchocerca volvulus*. Although most cases occur in Africa, it is also found in parts of South America and the Arabian Peninsula. It has been successfully controlled in West Africa through insecticide applications to reduce aquatic stages of the black fly vector and through the distribution of antihelminthic drugs (see later text) but it is still a problem in central and east Africa, and in parts of South America.

Insect Vectors of Human Disease

Biting flies (insects of the arthropod order Diptera) are the most important of the aquatic vectors of human disease. Of the flies, mosquitoes (the family Culicidae) are the most important vectors because of the high mortality and morbidity of the many diseases they transmit and because of the range of diseases for which they can serve as vectors (**Table 1**). For example, certain species of mosquitoes can transmit many types of nematodes, protozoans, or viruses. However, it should be remembered that while ~3000 described species of mosquitoes take blood for nourishment, most are nuisance biters (and some not even of humans) and do not transmit human diseases.

Table 1 Examples of human diseases vectored by mosquitoes

Disease vectored	*Parasite transmitted*
Lymphatic filariasis	*Brugia* and *Wucheria* nematodes
Malaria	*Plasmodium* protozoans
Yellow fever	Flavovirus
Dengue	Flavovirus
St. Louis encephalitis	Flavovirus
Japanese encephalitis	Flavovirus
Murray Valley encephalitis	Flavovirus
Russia spring-summer encephalitis	Flavovirus
Omsk Hemorrhagic fever	Flavovirus
West Nile fever	Flavovirus
Kyasanur forest disease	Flavovirus
Louping Ill	Flavovirus
California encephalitis	Bunyavirus
La Crosse encephalitis	Bunyavirus
Rift Valley fever	Bunyavirus
Eastern equine encephalitis	Togavirus
Western equine encephalitis	Togavirus
Venezuelan encephalitis	Togavirus
Ross River fever	Togavirus

As with terrestrial insect vectors, disease transmission involves two different blood meals. Only adult females take blood meals, using the blood of their victims to provide protein for their eggs to develop and pass through their larval and pupal stages in the oftentimes nutrient-poor aquatic habitats that they occupy. Using mosquitoes as an example, during the first blood meal a female injects saliva into the feeding wound; the saliva contains substances to reduce blood clotting. Using muscle pumps in her head, she then ingests blood from an infected human. Adult mosquitoes generally cannot undergo ovarian development until they have taken a blood meal (although some, referred to as being autogenous, can do this). Females tend to be attracted to their hosts by the heat, moisture, and carbon dioxide that they emit but certainly other factors (e.g., volatile chemicals emitted through the skin) are involved. Species can be selective in terms of hosts; some choose only birds, while some others chose only mammals. In species where both are selected, disease transmission may be enhanced when one of the hosts can serve as a disease reservoir (e.g., birds in West Nile Fever; dogs, cats, and monkeys in *Brugia* lymphatic filariasis).

Depending on the disease agent resulting in disease, there are three processes that can occur within the mosquito as the ingested pathogen migrates from her gut to her salivary glands. The pathogen can multiply but not undergo developmental changes (e.g., as in yellow fever), the pathogen can undergo developmental changes but not multiply (e.g., onchocerciasis), or the pathogen can do both (e.g., malaria). In all cases, when a mosquito takes a blood meal from another host, the disease agent may be transmitted through the injection of the anticoagulant saliva.

Other Diptera are important as well in disease transmission to humans and their animals. Black flies (the family Simuliidae) transmit onchocerciasis (river blindness), which is caused by a nematode (roundworm). This filarial worm lives in the human body for over a decade, and adult females produce millions of larvae that migrate throughout the body causing a series of progressive symptoms as the number of parasitic worms increases, including rashes and lesions, intense itching, loss of skin pigmentation, general debilitation, and eventually blindness.

Many species of biting flies that occur in moist or semi-aquatic habitats as larvae or that are common riparian-dwelling species as adults also transmit diseases. For example, some sand flies (Ceratopogonidae) can transmit the protozoan causing leishmaniasis, tsetse (Glossinidae) transmit the protozoan causing African sleeping sickness, and some deer flies (Tabanidae) can transmit *Loa loa*, the eye worm of humans.

Even biting flies that do not transmit disease can be major nuisances. In North America, horse flies and deer flies (both in the family Tabanidae) and black flies are irritating pests, oftentimes with painful bites that produce allergic reactions and infections that result from scratching the bites. There is also an economic cost associated with them in terms of lost tourist revenue and lower population densities in areas where they are a problem.

Other insects that are not important as direct transmitters of diseases (e.g., aquatic bugs, dragonfly larvae, beetle larvae, along with snails, small fish, and the even biofilm of aquatic plants) may serve as reservoirs of a water-associated disease, Buruli ulcer. Some naucorid bugs have even been shown experimentally to have the ability to transmit this disease.

Snails and Crustacenas as Intermediate Hosts of Human Disease

Freshwater snails are the intermediate hosts of a variety of trematodes (flukes) and some nematodes (roundworms) that cause many human diseases. The most important of these diseases is schistosomiasis (sometimes referred to as Bilharzia). This disease is caused by parasitic trematodes (blood flukes) that must find and invade a particular species of snail to continue their life cycle. The ecological requirements of these snails are a key determinant in the distribution and prevalence of this disease.

Schistosomiasis is a major public health problem in the world and blood flukes in general are important parasites of cattle and other large animals that humans depend on for survival. In humans, the blood flukes causing schistosomiasis lie in small veins in the lower abdomen; here males and females copulate and the females lay millions of eggs for 10 years or more. The eggs are passed into water supplies from feces and urine of infected people. After hatching the parasite actively swims and invades the body of a snail, usually by penetrating through the snail's foot. The parasite requires the presence of certain species of snails (e.g., *Oncomelania*, *Biomphalaria*, *Bulinus*) to transform (and multiply) itself into an infective stage, called cercaria. This life stage actively swims to the human host and penetrates the skin, which leads to infection.

A human's susceptibility to schistosomiasis depends on the species and strain of the invading blood fluke, the person's age, number of other parasites carried, nutritional status, and previous exposure to the parasite. Infection may cause incapacitation but the main effect is long-term damage to the intestine, bladder, and liver. Consequently, death may result from other infective agents attacking an already debilitated body.

Although schistosomiasis commonly occurs in developing countries, an infection related to this disease that occurs in developed countries is cercarial dermatitis or 'swimmer's itch'. This disease occurs worldwide and affects people who are swimming, wading, or working in littoral areas of both marine and freshwater habitats. Infection results when cercaria meant for birds or small mammals such as muskrats penetrate human skin, eliciting an immune response. The resulting skin rash goes away in about a week, the parasites degenerate, and the disease does not progress because humans are not the appropriate host for these animal schistsomes. Prevention involves brisk toweling after leaving the water (because penetration increases with drying and exposure length) and snail control.

Snails are vectors to several other human diseases (although much less common than schistosomiasis) as well, such as clonorchiasis and fasciolopsiasis. Like with schistosomiasis, the most medically important members of the Phylum Platyhelminthes (the flatworms) have definitive hosts that are vertebrates and intermediate hosts that are mollusks (mostly gastropods).

Another human disease that has been important in Africa and parts of the developing world is Guinea worm or dracunculiasis. The large nematode

(~1 m long) causing this disease releases larvae from an adult female worm (usually embedded in a human leg) through a skin lesion, when an infected human comes in contact with a pond or well. A larva is then ingested by a water flea (*Cyclops*), in which it develops and becomes infective. When a person drinks this infected water, the *Cyclops* is dissolved by the acidity in their stomach. The nematode larva is then activated, migrates through the subcutaneous tissue, and stays within the new host for about a year. It then emerges and starts the life cycle again. The treatment for guinea worm that has been used for thousands of years is to slowly wind the emerging worm around a stick over a several day period. This treatment is immortalized in the staff of Asclepius wound with a serpent, which is also the symbol of the modern-day healer or physician. An extensive Guinea worm eradication program has been highly successful and has eliminated the disease in most parts of the world and reducing human cases from 3.5 million to ~32 000.

Habitats of Human-Disease Vectors

Three important categories can be distinguished that provide habitat for vectors and intermediate stages of the causative agents of human diseases: (1) natural water bodies; (2) human-made water bodies, and (3) water bodies that form in human settlements and household environments. Freshwater vectors of human disease can occur in all of these habitat categories.

In terms of natural water bodies, streams and rivers are sources of the black flies that serve as vectors of onchocerciasis, and lakes and ponds provide habitats for snails that are intermediate hosts of schistosomiasis. Although mosquito vectors do occur in natural systems as eggs, larvae, and pupae, their densities tend to be far higher when they occur in human-made or settlement habitats, probably because of reductions in predation and competition compared to natural systems. Likewise, standing water in barrels and pots in human settlements, habitats where predation and competition are also lacking, are important habitats for mosquitoes.

The creation of human-made water bodies such as impoundments and irrigation ditches often result in hydrological changes that favor intensified vector breeding. This can result in increased *Anopheles* mosquito populations that may transmit malaria or in snail populations that may increase prevalence of schistosomiasis. Likewise, shifts in species composition may occur that reduce numbers of predators or competitors, and allow vectors to increase in number. Recent studies have also suggested that habitat disturbance may increase the prevalence of Buruli ulcer in ponds.

The creation of the Aswan High Dam on the Nile River in Egypt is an example of a project that resulted in an expansion of schistosomiasis. This disease was long prevalent in Lower Egypt, which is in the northern part of the country, such as in the Nile delta. Schistosomes have been found in mummies that are thousands of years old, and in the biblical story about the pharaoh's daughter finding Moses' floating cradle in the weeds, she certainly would have been attacked by cercariae when she went in to get him! In contrast, schistosomiasis was far less common in Upper Egypt (which is located in the south). However, the construction of the dam moved infected people from Lower Egypt to Upper Egypt and created favorable snail habitat because snails thrived in the slower-moving water conditions provided by the dam. Consequently, the disease became established there as well.

Potential increases in disease vectors are now generally a design consideration in development projects and this has been especially important when irrigation schemes are planned because their introduction may create new or more favorable habitats for disease vectors. A variety of hydraulic engineering approaches such as improvements in drainage, concrete lining or covering of canals, land leveling, filling in land depressions, controlling seepages, diking, and dewatering are useful additions. For example, it has been estimated that water improvements may result in disease reductions of 80% of Gambian sleeping sickness, 20% for onchocerciasis, and 10% for yellow fever. In some cases, however, pesticide applications to control insect vector and snail problems may be inevitable. Many of these habitat modifications and pesticide applications, however, may have a deleterious effect on the nontarget fauna that keeps vectors under control or is of economic or conservation value.

Tires are not a typical freshwater habitat but they are extremely important breeding sources for some mosquitoes. Tires are perfectly designed to hold water (there is no orientation from which water can drain from it), and many tree-hole occupying species use tires as breeding sites. The shipping of tires from Asia to North America for recapping and reselling (they do not recap tires in Asia) introduced the Asian tiger mosquito *Aedes albopictus* to North America. This mosquito is an effective transmitter of 23 human viruses, including those that cause St. Louis encephalitis, dengue, and yellow fever. Likewise, it has been speculated that reintroduction of yellow fever to Mexico was the result of shipments of

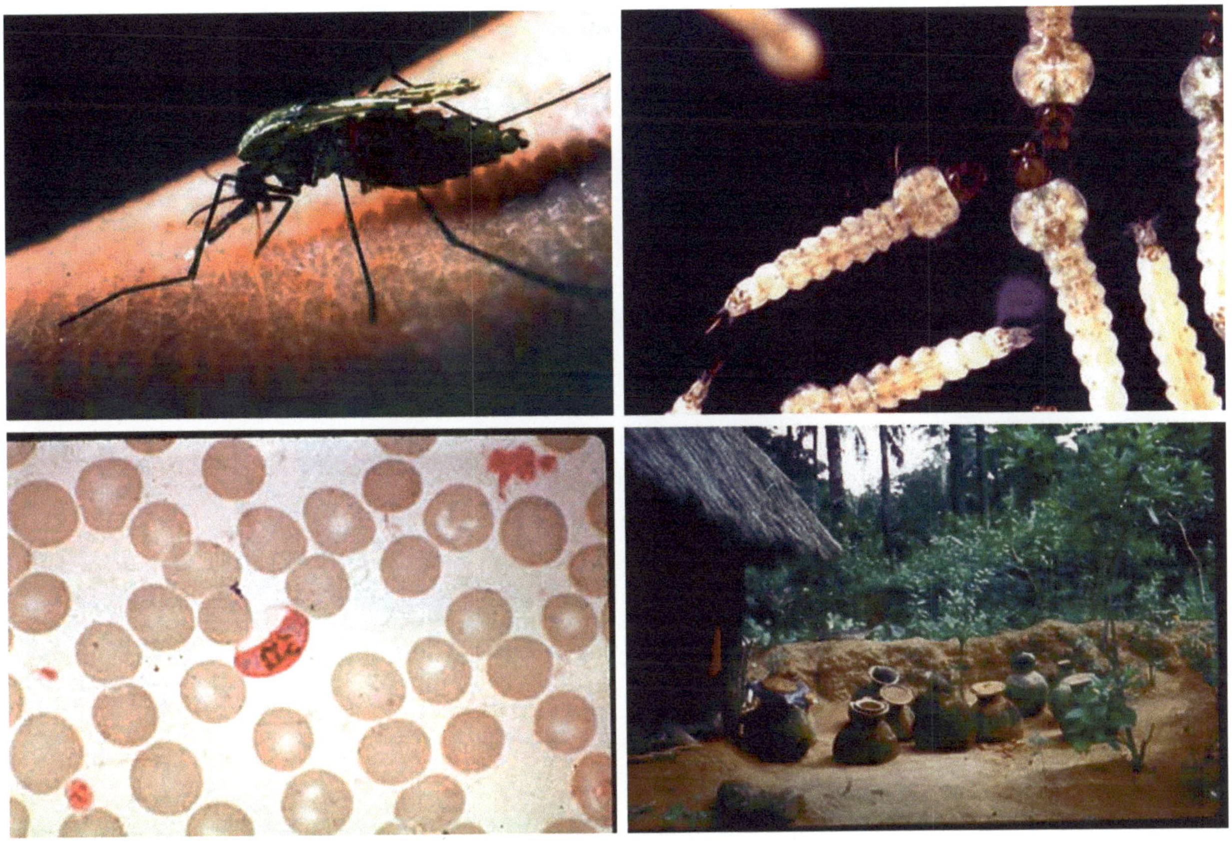

Plate 1 Top left. *Anopheles gambiae*: an adult female mosquito taking a blood meal and possibly vectoring malaria (WHO/TDR/Stammers). Top right. *Anophelese gambiae*: fourth-instar mosquito larva; all of the immature stages of mosquitoes occupy aquatic habitats (WHO/TDR/Stammers). Bottom left. *Plasmodium falciparum*: Mature female gametocyte stage of the malaria pathogen (WHO/MAP/TDR). Bottom right. Water-storage pots that *Anopheles gambiae* and other mosquitoes can use as breeding sites (WHO/TDR/Service).

discarded tires from Texas to Mexico to make huarache sandals because of a shortage of tires used in making them there!

Vector Control Strategies

In addition to advances in chemotherapy and vaccination strategies to control the disease agents, a great deal of attention has been applied to control strategies that can reduce populations of freshwater vectors of human diseases. The most commonly used approach is the application of toxicants to control the vectors. These can be done over large spatial scales (e.g., aerial applications of insecticides in 11 West African counties to control black fly vectors of onchocerciasis) or locally (e.g., the distribution of pesticide impregnated bed nets to control malaria vectors at the village level).

A second approach is the introduction of environmental modifications. These include draining standing or stagnant water, creating ditches for recirculation of water, and diverting flow so streams go dry for part of the year.

A third approach, the introduction of biological control agents such as predatory fish (e.g., *Gambusia*) and invertebrates, along with a variety of bacteria, fungi, and other organisms, has been widely used with mixed success and, like the above control approaches, sometimes with major undesired ecological consequences. In general, predators do not restrict themselves to mosquitoes, black flies, and other vectors as prey. However, the more specific the control agent (such as the bacterial toxins from *Bacillus thuringiensis israelensis*), the less likely the chance of such unwanted consequences.

Finally, the addition of competitive organisms that are not vectors (e.g., closely related snails to compete with schistosomiasis vectors) has worked in isolated areas. Swamping populations with sterilized males (as was effective in the screw worm control program) and the development of genetic modifications may have future value but have not thus far been found to be effective.

Plate 2 Top left. *Simulium damnosum:* An adult black fly female taking a blood meal and possibly vectoring river blindness or onchocerciasis (WHO/TDR/Stammers). Top right. *Simulium damnosum:* Larva of the black fly; all of the immature stages of black flies occupy aquatic habitats (WHO/TDR/Stammers). Bottom left. *Onchocerca volvulus*: The adult worms that causes onchocerciasis (WHO/TDR/OCP). Bottom right. A young girl leads her father, blind from onchocerciasis, through their village, a common scene in West Africa before this disease was controlled (WHO/TDR/Crump).

The large-scale control of aquatic vectors of human disease has had mixed success. Malaria, for which eradication through DDT and other pesticide applications was discussed as a possibility in the 1950s, soon re-emerged as a major source of human mortality. Malaria and other mosquito-vectored diseases continue to be an area of research in which both 'high' and 'low tech' solutions are being tested. Many of the former involve molecular biology and genetic engineering approaches. The latter involve widespread distribution of pesticide-impregnated bed nets to protect people during sleep.

Control of schistosomiasis has achieved somewhat better success than malaria and has involved breaking the life cycle of the parasite or interrupting the chain of infection. Clearly, if all snails are eliminated or all humans treated and remain free of infection, the chain of infection would be interrupted. Snail control is achieved through chemical control and, although biological control (e.g., with predatory snails or birds) has not been generally effective, habitat control has been very effective in China. There, removing snails from canals with chopsticks, sealing latrines, and preventing human wastes from reaching water have been very successful measures. Habitat modification for snail control, such as modifying habitats by reducing vegetation, and altering stream bed and channel conditions to increase water velocity are also effective but these approaches require continuous long-term effort.

Guinea worm control has revolved around providing clean water through filtration, treating contaminated water, and preventing persons who have active infections from contact with water supplies. This approach has been very successful in Africa. However, when political instability and migration of humans from infected to disease-free areas occurs, the threat of re-invasion of this and other diseases with freshwater vectors increases.

The true success story about the control of aquatic vectors of human disease, and really of any human disease, is the control of river blindness in

Plate 3 Top left. *Bulinus globosus*: The snail that is the intermediate host for *Schistosoma haematobium* (WHO/TDR/Stammers). Top right. *Schistosoma haematobium*: The adult worm that causes schistosomiasis (WHO/TDR/Stammers). Bottom left. *Schistosoma haematobium*: The cercaria that penetrates the skin of humans and leads to schistosomiasis (WHO/TDR/Stammers). Bottom right. A sample of normal urine and blood-containing urine from a child suffering from schistosomiasis (WHO/TDR/Lengeler).

West Africa. Although river blindness is known from 35 different countries, 99% of the cases occur in 26 countries of Africa. Prior to intensive control efforts, as many as 30% of adults living in streamside villages were blind, and blindness was viewed as an inevitable part of a person's life cycle if they lived near the fertile flood plains of West African rivers.

The Onchocerciasis Control Programme in West Africa (OCP) was begun in 1974 and continued through 2002. Covering an operational area of over 1.2 km^2, it included 30 million people who were vulnerable to the disease. OCP initially was based entirely on aerial application of insecticides to rivers to control black fly larvae. The extent of the program was enormous; over 50 000 km of streams received applications of insecticides at the peak of activity, and some rivers for as long as 20 years.

Beginning in the 1980s, ivermectin, a drug used in veterinary practice such as for treatment of dog heartworm, was distributed to humans to control the larval worms in their bodies and to complement the aerial application of insecticides in breaking the cycle of transmission of this disease.

OCP has been heralded as both a public health (the elimination of river blindness as a public health concern) and economic development (the opening of the river valleys free of onchocerciasis to settlement has enabled food to be grown for an estimated 17 million Africans) success. It has been expanded to 19 other African countries, primarily based on ivermectin distribution. Likewise, the success of the control program in West Africa prompted efforts to eliminate onchocerciasis from affected areas of Latin America.

A remarkable feature of OCP was the establishment of the first large-scale, long-term monitoring of the fish and benthic macroinvertebrate populations in the areas treated with insecticides. This, coupled with earlier experimental and taxonomic studies, has been the basis for much of what is known today about tropical African streams and rivers.

Human Culture, Freshwater Vectors and Disease

Cultural practices may make humans more susceptible to certain diseases with freshwater vectors.

For example, fish and other organisms can be hosts to parasites, some of which can be transmitted to humans through eating raw or partially cooked fish such as salmonids, pike, perch, and burbot, carp and other cyprinids. Jewish women have had higher rates of infection by the broad fish tapeworm (*Diphyllobothrium latum*) than women of other cultures. This is because they often ingested infective stages of this tapeworm while tasting raw fish during the preparation of a traditional dish, gefilte fish. The tendency of restaurants to cook fish 'rare' and the consumption of more raw fish may lead to an increase in the rate of other such parasites in humans.

Many Asian cultures use aquatic insects as part of their diet. Ingestion of uncooked caddis fly larvae while working in rice paddies is known to pass parasites on to humans. Likewise, eating uncooked freshwater prawns (*Macrobrachium*) may pass on parasites as well.

The creation of cities has certainly affected the prevalence of freshwater vectors and human diseases. For example, when early Africans were hunters and gatherers, malaria was likely to have been far less of a problem to the human population than it is today because of their continual movements and migrations to other areas to find food. However, when people settled in villages and cities with the onset of agriculture and animal domestication, the transmission of the protozoan and its cycling within the population increased greatly. Likewise, the introduction of root crops in agriculture created ideal habitat in which mosquitoes can propagate.

The modification of cultural practices may have disease consequences as well. For example, river blindness was likely less of a problem in West Africa before European colonization because people tended to live some distance (1 km) from the rivers and avoided going there during hours of peak black fly biting. However, when the European colonists and missionaries encouraged settlement close to the river, these long-standing cultural taboos disappeared and the infection rate greatly increased.

Finally, the enhancement of global biodiversity has become a cultural goal of many developed countries. The issue of biodiversity in human disease has been a major topic in the conservation biology literature; however, it is invariably about the loss of biodiversity from large-scale disease control programs (loss of fish in OCP and nontarget terrestrial and riparian fauna in tsetse control programs). However, biodiversity is also an important component in the control of freshwater vectors of human diseases as well. For example, all major malaria vectors consist of species complexes, and differences are evident among *Anopheles* strains in their vector competency, resting-site choice, biting behavior, and development of insecticide resistance. In cost-effective control of schistosomiasis, knowledge of snail diversity, distribution, and the epidemiology of parasite susceptibility by different snail species have been essential.

Onchocerciasis may provide the best example of how biodiversity relates to freshwater vectors and the diseases that they transmit. For example, cryptic species or strains, with varying competence in transmitting the disease or influencing its severity, are characteristic of both the black fly vector and the parasitic roundworms. Likewise, much of the success of OCP was also based on biodiversity. For example, the choice of insecticides used was based on maintenance of fish and nontarget invertebrate biodiversity. But even more unusual, the drug used to control onchocerciasis was also based on biodiversity. It was derived from a natural strain of *Streptomyces avermitilis*, a fungus discovered while screening a soil core from a golf course in Japan for potential nematocidal toxicity.

Glossary

Bilharzia – A human disease caused by various species of trematode worms that use snails as an intermediate host; also called schistosomiasis.

Black flies – Insects in the family Simuliidae of the order Diptera; some species are the vectors of the nematode worms that cause onchocerciasis.

Buruli ulcer – A human disease caused by infection with the bacteria *Mycobacterium ulcerans.*

Dengue fever – A human disease caused by a flavovirus that is transmitted by infected mosquitoes.

Dranunculiasis – A human disease caused by nematode worms, with the crustacean *Cyclops* as an intermediate host; also known as guinea worm.

Elephantiasis – A human disease caused by nematode worms that are transmitted by infected mosquitoes; also called lymphatic filariasis.

Filariasis – A group of human diseases caused by the presence of nematode worms.

Guinea worm – A human disease caused by nematode worms, with the crustacean *Cyclops* as an intermediate host; also known as dranunculiasis.

Intermediate host – A host that a human parasite uses as part of its life cycle.

Japanese encephalitis – A human disease caused by a flavovirus that is transmitted by infected mosquitoes.

Lymphatic filariasis – A human disease caused by nematode worms and transmitted by infected mosquitoes.

Malaria – A human disease caused by a protozoan that is transmitted by infected mosquitoes.

Morbidity – The incidence of a disease in a population.

Mortality – The death rate caused by a disease in a population.

Mosquitoes – Insects in the family Culicidae of the order Diptera that transmit many human and animal diseases.

Onchocerciasis – A human disease caused by nematode worms and transmitted by infected black flies; also known as river blindness.

River blindness – A human disease caused by nematode worms and transmitted by infected black flies; also known as onchocerciasis.

Schistosomiasis – A human disease caused by various species of trematode worms that use snails as an intermediate host; also called bilharzia.

Swimmer's itch – A mild form of schistosomiasis, which results in a skin rash, and is caused by trematode worms, that have snails as the intermediate host.

Vector of disease – An organism, such as a biting fly, that transmits an infectious disease.

Yellow fever – A human disease caused by a flavovirus that is transmitted by infected mosquitoes.

Further Reading

Desowitz RS (1991) *The Malaria Capers*. NY: W.W. Norton & Co.

Gratz NG (1999) Emerging and resurging vector-borne diseases. *Annual Review of Entomology* 44: 51–75.

Lane RP and Crossky RW (1993) *Medical Insects and Arachnids*. London: Chapman & Hall.

Mahmoud AAF (ed.) (2001) *Schistosomiasis*. London: Imperial College Press.

Marquardt WC, Demaree RS, and Grieve RB (2000) *Parasitology & Vector Biology*, 2nd edition, p. 702. San Diego: Academic Press.

Merritt RW, Benbow ME, and Small PLC (2005) Unravelling an emerging disease associated with disturbed aquatic environments: the case of Buruli ulcer. *Frontiers in Ecology and Environment* 3: 323–331.

Mullen G and Durden L (2002) *Medical and Veterinary Entomology*. San Diego: Academic Press.

Patz JA, Graczyk TK, Geller N, and Vittor AY (2000) Effects of environmental change on emerging parasitic diseases. *Australia Society for Parasitology* 30: 1395–1405.

Remme JHF (2004) Research for control: The onchocerciasis experience. *Tropical Medicine and International Health* 9: 243–254.

Resh VH, Lévêque C, and Statzner B (2004) Long-term, large-scale biomonitoring of the unknown: Assessing the effects of insecticides to control river blindness (onchocerciasis) in West Africa. *Annual Review of Entomology* 49: 115–139.

Ross AGP, Li YS, Sleigh AC, and McManus DP (1997) Schistosomiasis control in the People's Republic of China. *Parasitology Today* 13: 152–155.

Sakanari J, Moser M, and Deardoff TL (1995) *Fish Parasites and Human Health* (Rept. T-CSGP034). LaJolla: California Sea Grant.

Secor WE and Colley DG (eds.) (2005) *Schistosomiasis*. New York: Springer.

Spielman A and D'Antonio M (2001) *Mosquito, A Natural History of our Most Persistent and Deadly Foe*. NY: Hyperion.

Yaméogo L, Resh VH, and Molyneux DH (2004) Control of river blindness in West Africa: Case history of biodiversity in a disease control program. *EcoHealth* 1: 172–183.

Subject Index

Notes

Cross-reference terms in italics are general cross-references, or refer to subentry terms within the main entry (the main entry is not repeated to save space). Readers are also advised to refer to the end of each article for additional cross-references - not all of these cross-references have been included in the index cross-references.

The index is arranged in set-out style with a maximum of three levels of heading. Major discussion of a subject is indicated by bold page numbers. Page numbers suffixed by T and F refer to Tables and Figures respectively. vs. indicates a comparison.

This index is in **letter-by-letter** order, whereby hyphens and spaces within index headings are ignored in the alphabetization. For example, acid rain is alphabetized after acidity, not after acid(s). Prefixes and terms in parentheses are excluded from the initial alphabetization.

Where index subentries and sub-subentries pertaining to a subject have the same page number, they have been listed to indicate the comprehensiveness of the text.

Abbreviations

DCAA - dissolved combined amino acids

DFAA - dissolved free amino acids

DIN - dissolved inorganic nitrogen

DOC - dissolved organic carbon

DOM - dissolved organic matter

DON - dissolved organic nitrogen

TDS - total dissolved solids

TSS - total suspended solids

A

Abiotic processes, affecting dissolved elements in river/stream ecosystems 338
'Absolute' pressures 18–19
Accelerated loading, nutrients 8
Acetotrophic methanogenesis 480
Acetylene reduction assay 392
N-Acetylglucosamine, dissolved organic nitrogen (DON) 523
N-Acetylmuramic acid, dissolved organic nitrogen (DON) 523
Acid(s)
  production
    deforestation effects on dissolved nutrients 594
    see also Acidification; pH
  water pH and 345
  weak, acidified waters buffering mechanisms 558
  see also Acidity
Acid drainage
  characteristics
    acid sulfate soils 563
  definition 566
  see also Acidified waters
Acidification **557–568**
  accompanying problems 557
  biological effects 564, *566f*
  counter measures, and success of 565
  definition 557, 567
  deforestation effects on dissolved nutrients 594
  industrial waste in Lake Orta (Italy) 566
  volcanic *562f*, 564, 567
  see also Acid drainage; Acidified waters
Acidified waters
  atmospheric deposition 560
    see also Acid rain
  buffering mechanisms 557, *559f*
    aluminum 558
    iron 558
    pyrite weathering 560
    silicates 559
    weak acids 558
  chemistry 557
    acidity and alkalinity definitions 558
    carbonic acid and water 557
    definitions 557
    dimensions 557
    dissolved substances, concentrations 560, *562f*
    pyrite weathering and 560
  extent 560
  liming 566
  rain-acidified, remediation 565
  remediation 565
    acid mine drainage *see* Acid mine drainage (AMD)
    Lake Orta 566
    liming 566
  types 560
  volcanic and crater lakes *562f*, 564
  see also Acid drainage; Acidification
Acidity
  definition 558
  groundwater chemistry 114–115
  see also Acid(s); pH
Acid mine drainage (AMD) 560, 563
  avoidance/mitigation 565
  coal mining *562f*
  counter measures 565
  definition 567
  impacts 563
  iron 375, *375f*
  lignite mining *562f*
  ore mining *562f*
  see also Acidified waters
Acid-neutralizing capacity (ANC) 348, 558

Acid-neutralizing capacity (ANC) (*continued*)
  carbon dioxide–pH relationship 345–346, *346f*
  definition 360, 567
Acid rain 351, 405, 560, *562f*
  counter measures 565
  definition 567
  effects on aquatic plants 604–605
Acid sulfate soil (ASS) *562f*
  acid/water drainage 563
  Australian 564, *565f*
  counter measures 565
  definition 567
  Finland 564
Acid volcanic waters *562f*, 564
Adiabatic lapse rate 66
Adirondack lakes
  cation concentrations 359–360
  manganese 371
Advection, longitudinal 318
Advective processes, in density-driven flows 247
Advective transport 194–195
Adventitious roots
  definition 187
Aedes
  control strategies 678
  vectors of human diseases 678
Aedes aegypti 678
Aedes albopictus 680–681
Aerators 664–665
  cost 665
  failure 665
Aerenchyma
  definition 187
Aerobic respiration 455
Aeruginosin 98-B, structure *431f*
Age
  groundwater flow systems (time since recharge) 89, *89f*
  of water, groundwater chemistry 107, *108f*, *109t*
Agent Orange 672
Agriculture
  ammonia 379
  fertilizers, pollution 660
  impacts on/of
    ground water/surface water interaction 104
    hydrologic cycle 104–105
    rivers 127
    saline inland waters 421
  phosphorus transfer to aquatic systems 411
  silica cycle, effects 416–417
  water resources and *see* Water
Air–water interface
  gas exchange *see* Gas exchange at air–water interface
  rivers 149–150
Akkeshi Bay (Japan), monosaccharides *443f*
Alarm cues
  allelochemicals 429
  Daphnia 429
Alaska
  Peninsular lakes, micronutrient elements 378
  Toolik Lake *see* Toolik Lake (Alaska)
Algae
  benthic *see* Benthic organisms (benthos)
  biofilm component 335–336
  biomass
    photosynthetic production 510
  blooms *see* Algal blooms
  boundary layer exploitation 191–192
  buoyancy 193
  carbohydrates 442
    release 442
  cobalt requirements 378
  definition 575
  effects of increased nutrients 601–602
  eukaryotic, fatty acids 474
    composition *474f*, 475, *476f*
    genetic control 474–475
  green *see* Green algae
  manganese requirements 380
  metal-resistant mechanisms 388
  nutrient requirements 7–8, 128, 383
  nutrient stoichiometry 537, 539, *540f*
    carbon:phosphorus ratios 546
    limitation 538
  pheromones 435
  of river ecosystems *see* Algae, river ecosystems
  salt tolerance 419
  suitability to life in flowing waters 335
  turbulence effects 191
  uptake of metals 387
  see also Cyanobacteria; Diatom(s); Periphyton (attached microalgae)
Algae, river ecosystems 150–151
Algaecides, copper 381
Algal blooms
  causes 625
  composition 625
  definition 575
  harmful 570–571
    see also Cyanobacteria
  hydrodynamic modeling 321
  resulting from density currents generated under ice cover 254
Alginates 450
Alien species
  see also Exotic species
Alkalinity
  aquatic ecosystems **348–353**
  buffers 348, 351
  carbonate system, diagram for 348–349, *349f*
  carbon dioxide effect 349
  cation exchange effect 350
  chemical factors affecting 351
  chemistry of buffers 348
    chemical background 348
  definition 347, 348, 558
  evaporation effect 351
  measurement 348
    acid titration 348
      methods 348–349
    Gran titration 349
  natural buffering 351
  nitrate uptake effect 350
  origin 350
    assimilatory uptake 350
    cation exchange 350
    dissimilatory redox reactions 350
    evaporation and precipitation 351
    sources 350
    weathering 350
  photosynthesis effect 349–350
  precipitation effect 351
  relationship with nutrients 352
  use as tracer 348
  see also pH
Allelochemicals/allelochemical reactions **425–436**
  animal–animal interactions 426, 427, *427t*
    alarm cues 429
    fish kairomones 427
    invertebrate kairomones 428
    kin recognition 429
    partner finding 429
    predator–prey 427
    Schreckstoff 429
  definition 425
  heterotrophic microorganisms 426, 436
    antimicrobial compounds 436
  plant–animal interactions 426, 429, *430t*
    foraging kairomones 432
    grazing effects on plants 432
    herbivore deterrents 429, 431, *432f*
    impact of plants on grazing 429
  plant–plant interactions 426, 432, *433t*
    algal pheromones 435
    allelopathy 432
    autotoxicity 435
  range of 426
  terminology 425
  terrestrial *vs.* aquatic ecosystems 425
Allelopathy
  allelochemicals 432
Allomones, definition 425–426
Alluvial channels
  currents 239
  flow curvature effects 239

Alpha-linoleic acid (ALA) 472*f*
definition 478
Altamaha River (USA), photochemical reactivity, apparent quantum yield 41–42
Alternative electron acceptors, methanogenesis 480–481, 482*t*
Aluminosilicate minerals 354–355
Aluminum
acidified lakes 357
acidified water buffering mechanisms 558
Birkenes watershed (Norway) 357
functions 354
Gardsjon watershed (Sweden) 357
humic substance complexation 469
precipitation 355–356
valency 356
Aluminum hydroxides 354, 355–356, 356*f*
Amazon River Basin (South America)
floodplain
inundation near tributary 153*f*
fluxes 123*t*
nutrients 126*t*
rankings 124
water/sediment delivery 144*t*
water yields 143
Amictic lakes, in density stratification and stability 196–197
Amino acids in stratified lakes, dissolved organic nitrogen (DON) and 521, 522*f*
Ammonia (NH_3)/ammonium (NH_4^+) 400
agricultural applications 379
concentrations in rivers 126*t*
anthropogenic impacts 127–128
contaminant 104–105
emissions (atmospheric) 560
nitrogen fixation and 393
photochemical production 46*t*, 47
Ammonification, photochemical 469
Amorphous silica (ASi) 414
aquatic ecosystem entry 414
definition 417
Amphibians
lotic systems *see* Lotic ecosystems
parasitic diseases 406
polybrominated diphenyl esters and 671
Amphibites 99
Anabaena 394*f*
Anabaena flos-aquae, anatoxin-a production 431
Anadromous, definition 154
Anaerobic conditions
methanogenesis 480
wetlands hydrology 180
see also Anoxia; Anoxic water(s)
Anaerobic methane oxidation *see* Methane oxidation
Anaerobic oxidation of ammonium to nitrogen (ANAMOX) 401, 402*f*
Anatoxin-A
Anabaena flos-aquae 431
structure 431*f*
Animal(s)
feeding operations, pollution and 572, 673*f*
impact in riparian zones *see* Riparian zones
plant–animal interactions, foraging kairomones 432
Animal–animal interactions, allelochemicals *see* Allelochemicals/allelochemical reactions
Animas River (Colorado, USA) 375*f*
Anions 358
Annual changes/variability
carbohydrates, dynamic concentrations 443
dissolved inorganic nitrogen *vs.* dissolved organic nitrogen 517, 518*f*
wetland hydrographs 182
Anopheles 681*f*, 684
habitats 680
vectors of malaria 677
Anoxia
decomposition and reduced vertical mixing causing 6–7
Anoxic, definition 341, 398, 406, 413
Anoxic processes
organic material degradation, methanogenesis 480–481, 481*f*, 481*t*
river/stream ecosystems 340
biogeochemical changes and reversals 340–341, 340*t*
electron acceptors 337, 340
respiration 336*f*, 340
Anoxic water(s)
columns, methanogenesis 482–483
Antarctic lakes
carbohydrates 440*t*
Anthropogenic impacts
on aquatic plant distribution/abundance *see* Aquatic plant(s)
hydrologic cycle 104
nitrogen cycle 404
rivers 153–154
irrigation 127
mining 127
nutrient concentrations 127
nutrient ratios 128
sediments 121, 127
saline inland waters 421
trace element concentrations 384
water quality 148, 604*f*
see also Agriculture, impacts on/of; Human(s)
Antibiotics
pollution, emerging concerns and sources 672–673, 673*f*
Anticyclonic rotation
definition 285
see also Gravity waves
Antimicrobial compounds
allelochemicals 436
see also Antibiotics
Aphelenchoides
temperature tolerance 166–167, 167*t*
Appalachian Mountains
acid mine drainage 563
cation chemistry 356
Apparent quantum yield (AQY)
photochemical reactions 41
CDOM 41
variability 41, 42*f*
Apparent roughness, surface mixed layer dynamics 209–210
Aquatic ecosystems
biodiversity *see* Biodiversity, of aquatic ecosystems
biomanipulation *see* Biomanipulation of aquatic ecosystems
historical aspects 453
human health and *see* Aquatic ecosystems and human health
nutrient stoichiometry *see* Nutrient stoichiometry
restoration 666
vegetation *see* Vegetation
Aquatic ecosystems and human health 569–576
guidelines for recreational waters 574*t*, 576
human use of aquatic ecosystems 569
physical threats to humans 569
drowning, injury, exposure 569
prevention 569
pollution
chemical pollution 573
dichlorodiphenyltrichloroethane (DDT) 574, 576
heavy metals 573
polychlorinated biphenyls (PCBs) 574, 576
fecal 571
animal feeding operations 572, 673*f*
birds and animals 571
combined sewer overflows (CSOs) 573, 575
humans 572
sanitary sewer overflows (SSOs) 573, 576
septic tank systems 572, 576
wastewater treatment plants 573
protecting human health 574
general safety 575
individual level 574, 575
monitoring fresh waters 574
water purification 575
sources of pathogens and toxins 572*f*
pollution 571
point and non-point sources 571
role of sediment type 571
toxins 570–571
waterborne diseases 569, 570*t*
fecal-oral pathogens 569
incubation lengths and infection types 569–570
primary meningoencephalitis 571
statistics 571

Aquatic invertebrates *see* Invertebrate(s)
Aquatic macrophytes *see* Macrophyte(s)
Aquatic plant(s) 301, 305*f*
  alternative equilibria 602*f*
  human impact on distribution/abundance **601–610**
    acid rain 604
      buffers 604–605
      impact on species diversity 605
    carbon dioxide, increased 605
    chemical pollutants 605
      carbon-acidification-eutrophication 605
      plants-algae-nutrients-carbon-consumers 606
    eutrophication 601
      feedbacks and trophic interactions 603
      increased nutrients effect on primary producers 601
      stable state change 601–602
      watershed and shoreline development effects 604
    invasive species introduction 601, 606
    management activities 606, 608*t*
    ultraviolet radiation 604
  importance 601
  management
    aquatic herbicides 607–609, 609*f*
    biological control 607
    chemical control 607
    drawdowns 607
    dredging 607
    factors to consider 601, 607
    mechanical control 607
    mechanical harvesting 604–605, 609*f*
    physical control 607
  recent changes 601
    contributing factors 601
  relationship between primary productivity and diversity 603*f*
  ultraviolet radiation 604
    plant response 604
  see also Plant(s); Vegetation; Wetland plants
Aquifers 88
  confined (artesian) 88, 89*f*
  definition 174
  hydraulic head measurement 88
  recharge 175
    vadose water 86
  spring classification 158*t*
  unconfined 88, 89*f*
  water flow 155, 157*f*
  water removal 88
  see also Ground Water
Arachidonic acid (ARA) 471, 472*f*
  definition 478
  dietary supply 472
Aral Sea
  anthropogenic salinization 421
Archaea
  definition 174
  thermal springs 165–166
Archaebacteria, methanogenesis 480
Arid environments
  chloride concentrations 362
  other climate zones in Köppen–Geiger classification system *vs.* 362
  river discharge 143–144
Arkansas River (USA) 103*f*
Aroclor® 670
Aroostook River (North America), cation concentrations 360
Arsenic (As) 387, 388, 389*f*
  pollution 675
Artesian aquifers 88, 89*f*
'Artificial lake effect' 416
Aryl hydrocarbon receptor (AhR)
  definition 462
  tetraclorodibenzo-*p*-dioxin binding 460, 461*f*
Asia
  endorheic rivers *see* Endorheic rivers/waters
  rivers
    sediment loads 148
  springs 169*t*
Assimilatory nitrate reduction 400
Assimilatory nitrite reduction 400
Asterionella, mixed layer depth, effects 192–193
Aswan High Dam 680
Atlantic basins (South America), South American rivers draining to *see* South America
Atmosphere
  dissolved organic nitrogen 526
Atmospheric acidification, definition 567
Atmospheric component, hydrological cycles 51–52, 52*t*
Atmospheric deposition
  acidified waters 560
  definition 567
Atmospheric exchange
  dissolved carbon dioxide (CO_2) 343, 344
  rivers 149–150
Atmospheric precipitation
  effect on dissolved nutrient loads, deforestation effects 591
  see also Precipitation
Atmospheric pressure, changes 18–19
Atmospheric water **58–67**
  cloud formation and 60
  evaporation leading to 58, 59
    see also Evaporation
  water vapor concentration and flux 59
Atriplex portuloides, water flow modification 313–314
Aufwuchs, definition 187
  see also Periphyton
Australia
  acid sulfate soils 564, 565*f*
  ecosystem bioassessment 582–583
  springs 169*t*
Australian River Assessment System (AUSRIVAS) 582–583
Authogenic minerals 368–369
  definition 377
Autochthonous organic matter
  definition 456
Autotoxicity, allelochemicals 435
Autotroph(s)
  as biofilm component in river/stream ecosystems 335–336
  definition 398, 456
  suitability to life in flowing waters 335
Autotrophy, river/stream ecosystems 336*f*
Available potential energy (APE), internal stratification 258
Avoca River, acid mine drainage 563, 564–565
Avon River (Australia), seasonal patterns of precipitation/evapotranspiration 145*f*
Axolotls 172

B

BACI study, ecosystem bioassessment 580
Backwater(s)
  biogeochemistry 336*f*, 337
  definition 341
Bacteria
  bacterioplankton *see* Bacterioplankton
  benthic mats 420
  biofilm component 335–336
  carbohydrates 442
    growth and 447
    uptake 445
  cell wall material, dissolved organic nitrogen 523
  chemolithotrophic 387
  definition 174
  dissolved combined amino acid (DCAA) source 525–526
  dissolved free amino acid (DFAA) source 525–526
  dissolved free amino acids (DFAA) uptake *see* Dissolved free amino acids (DFAA)
  as dissolved organic nitrogen (DON) source 525
  dissolved RNA/DNA 525–526
    uptake 533
  dynamics, dissolved organic nitrogen (DON) *see* Dissolved organic nitrogen (DON)
  enteric 576
  extracellular enzymes 447–449
  growth 445, 447, 542
    efficiency 47
  metabolism, in absence of oxygen 337
  metal-resistant mechanisms 388
  nitrogen fixation *see* Nitrogen fixation
  nutrient stoichiometry 541
  organic matter
    bacteria as source 525

photochemical transformation 45–47
pathogenic 569, 570*t*, 571–572, 574–575
sulfate reducing *see* Sulfate-reducing bacteria (SRB)
thermal springs 166
trace elements 387
uptake of acid anions 350–351
Bacterial decomposers
in mineralization 339–340
suitability to life in flowing water 335
Bacterial decomposition
in stratified lakes 550–551
Bacterial mineralization 339–340
Bacterioplankton
carbon turnover 454*t*
Balaton, Lake (Hungary), eutrophication *see* Eutrophication
Balneology, definition 174
Baltic Sea, nitrogen fixation 396
Baroclinic flow 223*f*, 277, 278
definition 285
momentum equations 278
Barotropic flow 277, 278
definition 285
momentum equations 278
Base(s)
cations, definition 360
constituents of dissolved RNA/DNA 524–525
water pH and 345
Base flow, definition 174
Base neutralizing capacity (BNC) 558
Basin(s), river
Amazon River *see* Amazon River Basin (South America)
in flood potential 644
largest worldwide
hydrology/geomorphology 144
water/sediment delivery 144*t*
Mexico
see Mexico basins
Mississippi *see* Mississippi basin
size, stream flow characteristics and 139–140
South American *see* South America
see also Watershed(s)
Basin-scale internal wave degeneration 264, 265, 266*t*
damped linear waves 265
diapycnal mixing 264–265, 267*f*
isopycnals 264–265, 267*f*
Kelvin–Helmholtz billows *see* Kelvin–Helmholtz billows
mechanisms 265, 265*f*
solitary waves 267, 268*f*
supercritical flow 267
Froude numbers 267
internal hydraulic jumps 267, 273*f*
thermocline jet 267, 274*f*
turbulent benthic boundary layer and 264–265
see also Seiche(s)
Basin-scale standing wave motions *see* Seiche(s)
Bedload
stream flow characteristics 141
Bed material load, equilibrium transport of suspended solid 131
Bedrock
spring geology 158*t*
Beer's law, surface energy exchanges 211–212
Belgian Biotic Index, ecosystem bioassessment 580
'The bends' 26
Benthic, definition 398, 413
Benthic boundary layer (BBL) 191–192, **230–238**
bottom roughness effects 232
form drag 232–233
skin friction 232–233
smooth *vs.* rough flow 232
current velocity 230, 231*f*
definition 230
exploitation 191–192
oscillatory boundary layers 233
particle dynamics 237
quantification 238
resuspension–settling cycles 238
relevance 230
sediment direct effects 230
solute transport 234
diffusive sublayer 234
molecular diffusion 231*f*, 234–235, 235*f*
oxygen 235–236, 235*f*
porosity 235–236
reaction–diffusion models 235–236
tortuosity 235–236
nondiffusive fluxes 237
bioirrigation 237
bioturbation 237
convection 237
temperature 237
in situ flux measurements 237
oxygen 235*f*, 237
small-scale sediment topography effects 236, 236*f*
stratified 233
2-dimensional mixing 233
density stratification effects 233
induced turbulence 234, 235*f*
isotherms 234, 234*f*
parallel flow 234, 234*f*
solute diffusion 233
structure 230, 231*f*
turbulence production 225
turbulent *see* Turbulent benthic boundary layer (TBBL)
vertical structure 227
viscous sublayer 231*f*, 232, 232*f*
Benthic invertebrate fauna
carbon turnover 454*t*
rivers 151
wetlands *see* Wetland ecosystems
Benthic organisms (benthos)
algae
rivers
predominance in shallow, fast-flowing waters 150–151
see also Periphyton
cyanobacteria 392, 397
definition 413
invertebrates *see* Benthic invertebrate fauna
mats 420
nitrogen fixation 392, 397–398
response to hydrologic exchange 101*f*
response to increased salinity 419
rivers
predominance in shallow, fast-flowing waters 150–151
see also Periphyton
Benthic zone
primary production 409–410
Benthos *see* Benthic organisms (benthos)
Bernoulli Equation, dynamic pressure 23, 23*f*, 24
Bioaccumulation
definition 575
Bioassays, ecosystem bioassessment 579
Bioconcentration, definition 463
Biodilution 651
Biodiversity, of aquatic ecosystems
pH effect on 351, 352
salinity effect on 419, 421–422
Biofilm(s) 539
composition
in river/stream ecosystems
channel surface 335–336
hyporheic zone 337
definition 342
Biogenic silica (BSi) 414
definition 417
Biogeochemical cycling
rivers in 153
Biogeochemical zones, rivers/streams *see* River ecosystems; Stream ecosystems
Biogeochemistry
hyporheic zone 336–337, 336*f*
lake ecosystems *see* Lake ecosystem(s)
trace metals *see* Trace metals and metalloids
water quality, effect on 102
see also Water chemistry
Bioindicator(s)
fish, in rivers 152
see also Water quality
Bioirrigation, solute transport, benthic boundary layer 237
Biologically available photoproducts (BAPs) 39
food webs and 47
photochemical production 46*t*
Biological oxygen demand (BOD) 664

Biological–physical interactions **189–195**
mixed layer depth, nonmotile plankton maintenance 192
nutrient fluxes *see* Nutrient(s)
organism size 191
organism structure 191
turbulence *see* Turbulence
Biological Working Party (BMWP) score, ecosystem bioassessment 580
'Biologic zero'
definition 187
wetlands hydrology 180
Biomagnification 575
mercury 650
Biomanipulation of aquatic ecosystems
definition 628
Biomass
accumulation
backwaters 337
flowing waters 335
algal *see* Algae, biomass
aquatic plants *see* Aquatic plant(s)
decrease
phytoplankton 621
phytoplankton 621
Biomembranes, humic substance interactions 457
Biota
definition 174
ecosystem bioassessment 577–578, 581, 582*t*
fire impacts 636, 641
fluxes, modeling 320
littoral zone *see* Littoral zone, biota
pelagic *see* Pelagic biota
streams *see* Stream(s)
Biotic processes, affecting dissolved elements in river/stream ecosystems 338
Biotransformation
definition 462
humic substances 458–459, 460
Bioturbation
solute transport, benthic boundary layer 237
Birds *see* Waterbirds
Birkenes watershed (Norway), aluminum 357
Biting flies *see* Diptera (biting)
Biting midges *see* Ceratopogonidae
Black body radiation, surface energy exchanges 212
Black flies (Simuliidae) 682*f*
definition 684
habitats 680
larvae
boundary layer exploitation 191–192
vectors of onchocerciasis 678, 679
Blood flukes 679
life cycle in humans 679
see also Schistosomiasis
'Blue baby syndrome' 405–406
Blue-green algae *see* Cyanobacteria
Blue Ridge Mountains, cation chemistry 356
Bodensee
phytoplankton growth 192
wind-generated turbulence 190
Boiling point, water 15
Bottle Bend Lagoon (Australia), acidification 558*f*
Bottom roughness effects, benthic boundary layer (BBL) *see* Benthic boundary layer (BBL)
Boundary layer model, in gas exchange at air–water interface 29
Bouquet Spring 157*f*
Boussinesq approximation, pressure 25
Brackish water
in density stratification and stability 199
Broad fish tapeworm *(Diphyllobothrium latum)* 683–684
Brownian motion, colloid formation mechanisms 467, 467*f*
Brugia malayi 677–678
Brunt-Väisälä frequency 222–224
in continuous water-column stratification 256
in density stratification and stability 203
Bryophytes (mosses)
river flow resistance 308
Bubbles, in gas exchange at air–water interface 32, 35
Bubbling
colloid formation 467*f*, 468
Buffer strips, deforestation effects on dissolved nutrients 594–595, 596*f*
Bulk fluid structure models, water 1–2
Bunsen solubility coefficient 31, 31*t*
Buoyancy
algae 193
cyanobacteria *see* Cyanobacteria
fish *see* Fish
flux, turbulence 208
frequency, in continuous water-column stratification 256
Henry's Law 26
hydrostatic pressure and 22, 22*f*
nonmotile plankton 193
Physostomes 26
'relative' 23
'submerged weight' 22
Burger number
definition 277–278
gravity waves and 278, 280, 281, 281*f*, 282*f*, 284
Bur-reed *(Sparganium)*
hydrodynamic characteristics 302*t*
river flow resistance 308–309
Buruli ulcer 679, 680, 684

C

Cabbeling 251–252
Cadmium (Cd) 102
pollution 675
Caenorhabditis elegans
humic substances
chemical attraction 458
gene regulation 458
hormone-like effects 458, 460*f*
longevity genes 461
Calcium 124, 418
algal nutrient 8
freshwater concentrations 356
functions 354
magnesium, relationship in streams 356–357, 359*f*
nutrient loading 586*t*
deforestation effects 588*t*
environmental effects 594*t*
precipitation 355–356
valency 356
in water 7
Calcium nutrient loading, environmental effects 594*t*
Callitriche (starwort), river flow resistance 308–309
Campylobacter spp. 569
Canada
nonthermal springs, macroinvertebrates 164*t*
reservoirs *see* North America, reservoirs
Canadian Forest Fire Behavior Prediction System 630, 631*t*
Canopies, flow modification 314*f*
dense 316
exchanges 316*f*, 317, 317*f*
flexible 316, 316*f*
morphology 313
porosity 313
sparse 316
submerged 314–315, 314*f*, 315*f*
transitional length scales 318, 318*f*
unconfined 315–316
water renewal 316
water residence time 317, 318*f*
Capillary effect, vadose water 76
Capillary fringe, vadose water, hydrologic cycle 84
Capillary theory, vadose water unsaturated hydraulic conductivity 81
Carbamates, pollution and 672
Carbaryl 671*f*
Carbohydrates **437–452**
algae *see* Algae
aquatic environments 437
assimilation 445
rates 445, 446*t*
bacterial growth 447
building blocks 437
cellulose 451–452, 451*f*
composition 442
dissolved combined *see* Dissolved combined carbohydrates (DCCHO)
dissolved free *see* Dissolved free carbohydrates (DFCHO)
dissolved organic carbon *vs.* 439
dynamic concentrations 443
annual variations 443

daily variations 444
free monosaccharides 439
fungal 442
measurement
analytical problems 437
chromatographic methods 439
colorimetric methods 437
method comparisons 439
microbial uptake 446*t*
phytoplankton *see* Phytoplankton
running waters 445
sources 442
stable isotopes and 450
total dissolved 437, 440*t*
tracing 451, 451*f*
Carbon **453–456**
^{14}C
labeled uptake assays, dissolved organic nitrogen 534–535
aquatic processes based on, predictability 455
budgets
rivers 153–154
C_{18} method, natural organic matter isolation 506
classes 513–514
concentrations 126*t*, 127
cycle *see* Carbon cycle
dissolved inorganic *see* Dissolved inorganic carbon (DIC)
dissolved organic *see* Dissolved organic carbon (DOC)
flow 453
(DIC)
nutrient loading 586*t*
deforestation effects 588*t*
environmental effects 594*t*
nutrient pools, ground water–surface water 100
organic
dissolved *see* Dissolved organic carbon (DOC)
inputs 454–455
oxidation state 510, 510*t*, 511*f*
photochemistry 37
pools 100, 453
reduction, photosynthesis *see* Photosynthesis
transformation 453
Carbonates
acidified water buffering 558–559
alkaline buffers 348, 351
alkalinity, diagram 348–349, 349*f*
cation concentrations 359*t*
effect of dominance 418
Carbon cycle 48*f*, 453, 454*f*
gas exchange at air–water interface and 28
methane 490*t*, 501
Carbon dioxide (CO_2) 71–72
diffusion, turbulence and 193–194
dissolved **343–347**
atmospheric exchange 343, 344
calculation 347
in Hudson River 150*f*
lake ecosystems 344*f*
measurement 346, 346*f*
origin 343
pH regulation 345
regulation 345
variability 343
effect on alkalinity 349
fertilizing effect on macrophytes 605
gross primary production (GPP) 345
influence of levels on aquatic plants 605
natural organic matter oxidation 503–504
oxic layers 345
partial pressure 343–344, 345, 346, 453–454
air–water interface 454–455
photochemical production 46*t*
photosynthetic assimilation 455
springs 172
Carbonic acid (H_2CO_3) 343
acidified water buffering 558–559
fresh water and 557
Carboxylic acids, natural organic matter 512, 513, 515
Carex (sedges) 302*t*
Caribbean basins, South American rivers draining to *see* South America
Cartesian grids 327, 329*f*, 333–334
Cascade Mountains (USA), cations 358
Castle Lake (California, USA)
molybdenum deficiency 379
phytoplankton 381
Cation(s) **354–360**
analytical distinctions 356
anthropogenic influences 359
base, definition 360
biological sinks 355
colloid formation 467*f*, 468
environmental influences 359
exchange
natural organic matter isolation 507
fresh waters 356, 357*f*, 358*f*, 359*t*
functions 354
geochemical sinks 355
seasonal hydrologic changes 360
sources 354, 355*t*
atmospheric inputs 354
springs 155–156
Cellular transport 193
Cellulose 451–452, 451*f*
Central European Weather Radar Network (CERAD) 62
Central nervous system (CNS), chloride ions 361
Ceratium, turbulence effects 190–191
Ceratophyllum (hornwort) 302*t*
Ceratopogonidae 679
Cercaria
definition 575
Cesium-137 675–676
Chad, Lake, water balance 612
Chambers, for estimating gas exchange rates 33
Champlain Lake (North America), mercury pollution 653*t*
Changed water levels, wetland hydrographs 182
Channel(s)
formation, streams 134
river 146–147
stream *see* Stream channel
surface, biogeochemistry 335–336, 336*f*
Channel form, rivers 146–147
Charophyceae (stoneworts) 302*t*
Chelates, definition 381
Chemical composition, rivers/streams 340*t*, 341
Chemical conversion, rivers/streams 340, 340*t*
Chemical effects, lakes, climate change 613
Chemical fluxes, river/stream ecosystems **335–342**
Chemical oxygen demand (COD) 663
Chemical processes, ground water 53–54
Chemical removal, groundwater chemistry 113
Chemical transformations, deforestation effects on dissolved nutrients 592
Chemical uptake, deforestation effects on dissolved nutrients 592, 592*f*
Chemocline, in density stratification and stability 200
Chezy formula, river currents 240
Chitin
dissolved organic nitrogen (DON) 523
Chlorella
diffusion
carbon dioxide 193–194
phosphorus 194
Chlorella salina, cobalt 378–379
Chloride (Cl^-) **361–367**, 418, 420–421
ecological implications
concentration increases 364
intraannual fluctuations 364–365
research directions 365
physiological roles 361
retention effects on inland waters 365
sources 361
environmental factors 361–362
toxicity 361
Chloride–bicarbonate exchanger 361
Chlorine (Cl) 124
nutrient loading, environmental effects 594*t*
see also Chloride (Cl^-)
Chlorophyceae
fatty acid composition 476*f*
Chlorpyrifos 671*f*
Chromatography, carbohydrate measurements 439
Chromium (Cr) 387
pollution 675

Chromophoric dissolved organic matter (CDOM) 37
apparent quantum yield *see* Apparent quantum yield (AQY)
excited 37
photobleaching 37, 39
photochemical products 38*f*, 39
properties 39
see also Humic substances (HSs)
Chroococcus minutus, humic substances, responses 458
Cima Lake (Brazil), carbohydrates 440*t*
Circulation
direct 289
natural systems 297
patterns, in density stratification *see* Density stratification
residual 289
sources 293
stratified lake water, redox potential 551
vertical 293, 294*f*
Circulation period, in density stratification and stability 196
Cladocera
water viscosity effects 191
see also Daphnia
Clausius-Clapeyron equation 69–70
Clays
cations 354, 355
dissolution 559
iron 368
manganese 368
particle suspension 190–191
Clear Lake (Iowa, USA)
circulation patterns 298*t*
currents 298*t*
wind velocity 294*f*
Climate
effects on groundwater budgets *see* Groundwater budgets
fire and 642–643
warming *see* Climate change
Climate change
effects of 611, 612*f*
effects on cryosphere 611
effects on hydrologic cycle 70
effects on lakes *see* Climate change and lakes (below)
methane emissions 501
warming impact on density currents 255
see also Global warming
Climate change and lakes **611–616**
basin integrity 611
ice-dammed lakes 611–612
precipitation changes 611–612
biological effects 614
catchment effect 614
colored dissolved organic matter (CDOM) 614
dissolved organic carbon (DOC) 614
Toolik Lake (Alaska) 614
chemical effects 613
ecosystem integrity 614
species loss 614
fish communities 615
cold water species 615
migratory behavior 615
historical evidence 611
ice cover 612
photosynthetically active radiation (PAR) 612–613
ultraviolet radiation 612–613
in-lake chemical effects 613
gradient changes 613–614
hydraulic residence time 613
microbial processes 616
photosynthetic communities 615
phytoplankton species composition 615
physical effects 611
polar regions 611
studies 611
temperature/stratification 613
Lake Superior 613
mixing regimes 613
turbulent mixing 613
vertical habitat structure 614
ice-free periods 615
Lake Tanganyika 615
overwintering phytoplankton 614–615
pelagic communities 614–615
summer stratification 614–615
water balance 612
extreme storm events 612
glaciers 612
Lake Chad 612
Climatic circulation patterns, stream sources 136
Cloud(s)
formation 59, 60
particles and precipitation 60
Cloud condensation nuclei (CCN) 60
definition 66
Cloud resolving models (CRMs)
definition 66
precipitation modeling 64
Coal mining, acid drainage 562*f*
Coarse woody debris (in lakes/streams) 641–642
deforestation effect on dissolved nutrients 592
Coastal plain(s)
wetlands hydrology 177–178
Coastal storms, as cause of flooding 645
Cobalt 378
algal requirements 378
chemical properties 378
pH effects 378–379
salinity effects 378–379
sources 378
sulfide 378
Codex Leicester (da Vinci) 69
Cohesiveness, definition 14
Cold water species, lakes, climate change 615
Colligative properties, water *see* Water
Colloid(s)
definition 466, 470
formation mechanisms 467, 467*f*
adsorption 468
Brownian motion 467
bubbling 468
cations 468
concentration 467
differential settling 468
diffusive capture 468
filtration 468
flow effect 468
microbial activity 468
pH 468
size 467
sunlight 468
surface coagulation 468
transparent exopolymer particles 468
turbulence 468
properties 466
Colloidal organic matter 464, 466
see also Organic matter
Colorado River (South America) 103*f*
water/sediment delivery 144*t*
Colored dissolved organic matter (CDOM), catchment effect, climate change 614
Colorimetry, carbohydrate measurements 437
Columbia River (USA)
water/sediment delivery 144*t*
Comanche Springs (Texas, USA) 173*f*
Common reed *(Phragmites australis)*
biogenic silica accumulation 414–415
hydrodynamic characteristics 302*t*
Community-based programs, ecosystem bioassessment 582–583
Community organization/structure
bacteria *see* Bacteria
lake ecosystems *see* Lake ecosystem(s)
measurements, ecosystem bioassessment 579
Component assessments, water budgets 55
Composition effects, groundwater chemistry 108–109, 110*f*, 111*f*
Concentration change studies, bacterial uptake of dissolved combined amino acids 528*t*, 533
'Concentric shell' model, water solvation 5–6, 5*f*
Condensation, energy release 59
Condition assessment, ecosystem bioassessment 581
Confined aquifers 88, 89*f*
Confluences, river current structure *see* River currents
Congo River (Africa)
fluxes 123*t*
nutrients 126*t*

rankings 124
water/sediment delivery 144*t*
Conservative substance
definition 342
river/stream ecosystems 338
Conservative tracers, definition 365
Constricted rivers, *vs.* floodplain rivers and levees 144–145, 146*f*
Consumers, nitrogen:phosphorus ratios 544–546
Contact zones, definition 174
Contaminants *see* Pollutant(s)
Contamination
groundwater quality 92
see also Pollution
Continuous stratification
internal wave rays in 269, 270*f*, 271*f*, 272*f*
water-column 257*f*
Contributing areas, groundwater quality 92
Convection
definition 310–311
solute transport, benthic boundary layer 237
Convective available potential energy (CAPE), precipitation modeling 65–66
Convective clouds, formation 60
Converged grid, in hydrodynamic modeling 327
Coontail 302*t*
Coordination changes, water reactivity 6
Copepoda
fresh water
fatty acid composition 477
marine calanoid, wax esters 472
Copper (Cu) 381, 384, 573
algaecide function 381
pollution 381
Cordgrass *(Spartina)*, hydrodynamic characteristics 302*t*
Coriolis force(s) 291*f*
currents 287
surface mixed layer and 219
Coriolis parameter 278
definition 277
Costs
financial, floods 646
Coulombic forces, water structure 1–2
Courant-Lewy-Friedrichs (CFL) condition, hydrodynamic modeling 330, 331
Craik-Leibovich force 210–211
Crater lakes 564
Crenarchaeota 165
Crenobiology 157
Critical flow, river currents 239–240
Crooked Creek 157*f*
Croplands, phosphorus transfer 411
'Cross-feeding' 447–449
Crown fires 630–631
continuous, definition 643
Crude oil, pollution and 668–669
Crustacea 99
ground water–surface water exchange and 102–103
mesoplanktic, water viscosity effects 191
rivers 151
saline water biodiversity and 419
spring biota 157
see also Copepoda; Ostracods
Cryosphere, climate change 611
Cryptosporidium 569
Culicidae *see* Mosquitoes (Culicidae)
Cumulus clouds 60
Current(s) **286–300**
concepts 287
shallow layers, rotational effects 288
shallow water equations 287
deep-water 222
density *see* Density currents
effects of earth's rotation **277–285**
governing equations 278
gravity waves *see* Gravity waves
vorticity waves *see* Vorticity waves
homogenous basins *see* Homogenous basins (HB)
internal wave *see* Internal waves
lateral circulation 287, 289
meters, for measurement of
gravity waves 283, 284
vorticity waves 284–285
research directions 299
surface 222
unstratified water bodies **286–300**
see also Homogenous basins (HB)
velocity
benthic boundary layer (BBL) 230, 231*f*
Curvilinear grid 327–329, 329*f*, 333–334
Cyanobacteria 603–604
allelochemicals, structures 431*f*
benthic mats 420
buoyancy 25–26, 193
definition 398, 628
drinking water and 623*f*
fatty acid composition 476*f*
humic substances, responses 458
molybdenum
uptake 379
use and deficiency effect 379
nitrogen fixation 392, 394*f*, 401*t*, 410
benthic 397
coastal marine mats 397
controlling factors 394–395, 395*f*
energetic costs 394
epiphytic 397–398
freshwater lakes 393
rates 393–394, 397
saline waters *see* Saline waters
species associated 394
zooplankton grazing, effects *see* Zooplankton
nutrient changes favouring, in coastal zones 128
salt tolerance 419
toxins 429, 570–571
protease inhibitors 429–431
turbulence effects 191
see also Algae; Diatom(s)
Cyanobacterin, structure 431*f*
Cyanopeptolin 963A, structure 431*f*
Cycling **xv–xvi**
biogeochemical
rivers in 153
carbon, fossil fuel combustion 548
heavy metals 374–375
silica *see* Silica, cycling
Cyclones
as cause of flooding 646
Cyclonically rotating winds, anticyclonically *vs.* 290–291
Cyclonic rotation
definition 285
see also Gravity waves
Cyclops (water fleas) 679–680
Cylindrospermopsin, structure 431*f*

D

Daily variability
carbohydrate concentrations 444
Dalton, John 69
Dalton's law 69–70, 70–71
Dam(s) 127
effects on river hydrology 147–148
failure 646
in flood mitigation 646
storage capacity 147–148, 148*t*
Damming
effects
silica cycling 416, 417*f*
Damped linear waves, in internal wave degeneration 265
Damselflies
nutrient stoichiometry 538*t*
Danube River
Delta, carbohydrates 440*t*
water/sediment delivery 144*t*
Daphnia
fish kairomones 427–428
as food source 477
kairomones, alarm cues 429
nutrient stoichiometry 542–543, 543*f*, 544*f*, 547–548
water viscosity effects 191
Daphnia magna
humic substance-induced longevity 461

Darcy's law 73–74, 96
  groundwater movement/runoff 56, 88
  vadose water 80
Da Vinci, Leonardo, Codex Leicester 69
DCAA *see* Dissolved combined amino acids (DCAA)
DCCHO *see* Dissolved combined carbohydrates (DCCHO)
DDT *see* Dichloro-diphenyl-trichloroethane (DDT)
Dead Sea
  anthropogenic salinization 421
Debris *see* Coarse woody debris (in lakes/streams); Detritus
Debye forces, water structure 1–2
Deep-water
  currents 222
  renewal
    density currents and 247, 254
    episodic, in density stratification and stability 197–198
Deforestation
  nutrient load effects **585–600**
    dissolved nutrients *see* Dissolved nutrients, deforestation effects
    general effects 599
      herbicides 599–600
      nitrification 592, 599–600
      timescales 599–600
      vegetation type 599–600
    knowledge gaps 599
    particulate nutrients *see* Particulate nutrients, deforestation effects
Deicer, chloride concentrations 363, 365
Dendritic drainage, streams 134, 135*f*
Dengue fever 678, 684
Denitrification 401, 401*t*, 402–403, 403*f*
  chemosynthetic oxidation reactions 401, 402*f*
  definition 398, 406
  effect on alkalinity 350–351
  sediments 393
  see also Nitrogen cycle
Denmark
  nonthermal springs, macroinvertebrates 164*t*
Dense nonaqueous phase liquids (DNAPLs), groundwater chemistry 113
Density
  stratification *see* Density stratification
  stream drainage network organization 134
  water *see* Water
Density currents 247, 248*f*
  characteristics 247, 248*f*
  impact of environmental changes on 254
    climate warming 255
    enhanced river discharge 254–255
  intrusions 247–248, 248*f*
  in tropical and saline lakes 254
  see also Density plumes
Density-driven flows **247–255**
  advective processes 247
  currents *see* Density currents
  differential cooling 247, 251, 251*f*, 253
  differential heating 253
  plumes *see* Density plumes
  stratification 247, 253
  thermal bar mixing 247
  thermobaric instabilities 247
    Ekman transport 253–254
  thermocline 249–250
  turbulent mixing 247
  underwater channels 249–250
Density plumes 247
  generated by external inputs 249
    interbasin exchange 250, 250*f*
    river inflows 247, 249, 249*f*
    subsurface inflows 251
  generated by internal processes 251
    differential cooling 247, 251, 251*f*, 253
    under ice cover 254
    thermal bar 251, 252*f*
    thermal baricity 253
      mesothermal maximum 253, 253*f*
    turbidity currents 249–250, 254
  thermocline and 249–250
Density stratification 192–193, **196–206**
  basin scale 222
  circulation patterns 196
    amictic lakes 196–197
    circulation period 196
    holomictic lakes 196–197
    meromictic lakes 196, 199, 200
    monimolimnion 196–197, 199, 200*f*
    nutrients 196–197, 205
    overturning 196, 203–204
    temperature cycle 196, 197*f*
  density differences/formation of layers 197
    salinity stratification *see* Salinity, stratification
    temperature stratification *see* Temperature stratification
    thermobaric stratification *see* Thermobaric stratification
  effects, stratified benthic boundary layers 233
  episodic deep water renewal 197–198
  ground water inflows 200
  hydrostatic pressure 20, 20*f*
  internal waves and 196, 203
  potential energy 222
  processes forming gradients of dissolved substances 200
    diffusive 200
    ground water inflows 200
    mine lakes 200
    turbulent 200
  small scale 222
  stability, quantification 201, 202
    bulk quantities 204
    density 202, 203*t*
    differential quantities 203, 203*f*, 204*f*
    electrical conductivity/electrical conductance 201–202
    salinity 202
    temperature 201, 202*f*
  temperature effects 222, 223*f*
  thermoclines in 197, 198, 198*f*, 199
  turbulent transport 225
Depth
  equivalent, definition 277, 282–283, 285
  hydrostatic pressure (P_h) 19, 19*f*
  lakes *see* Lake(s)
  mixing, in density stratification and stability 197
  profiles
    dissolved free amino acids 522, 522*f*
  of saturation, wetlands hydrology 179
  variation, equilibrium transport of suspended solid 131–132
Desalting 507
Desiccation 419–420
Desmids, saline waters and 420
Destratification 664–665
Detritus
  definition 382
  microbial mineralization 355
  photochemical reactions 41
  woody *see* Coarse woody debris (in lakes/streams)
Developing nations, pollution and 659
Devil's Lake (Wisconsin, USA), mercury pollution 650*f*, 653–654, 653*t*, 654*t*, 657
DFAA *see* Dissolved free amino acids (DFAA)
Diafiltration, natural organic matter isolation 506
Diapycnal mixing, in internal wave degeneration 264–265, 267*f*
Diatom(s)
  ionic preference 420
  marine, fatty acid composition 475
  nitrogen, uptake 415–416
  nutrient requirements 128
  phosphorus
    sediment burial 412
    uptake 415–416
  planktic
    mixed layer depth effects 192–193
    sinking velocities 190–191
  silica cycling 415–416, 416*f*
DIC *see* Dissolved inorganic carbon (DIC)
Dichloro-diphenyl-trichloroethane (DDT) 671*f*
Differential cooling
  in density-driven flows 247, 251, 251*f*, 253
  in tropical and saline lakes 254
Differential heating, in density-driven flows 253
Diffuse attenuation coefficient
  surface energy exchanges 211–212
Diffuse recharge, groundwater recharge 90
Diffuse unsaturated flow, vadose water *see* Vadose water
Diffusion
  benthic boundary layer *see* Benthic boundary layer (BBL)
  carbon dioxide (CO_2), turbulence and 193–194

definition 310–311
Fick's law *see* Fick's law of diffusion
nutrient fluxes 193–194
phosphorus 194
Diffusion coefficient, gas exchange at air–water interface 31
Diffusive flux, methane emissions 486*t*, 488*t*, 489, 494*t*
Diffusive heat transport, in density stratification and stability 197
Diffusive sublayer, solute transport *see* Benthic boundary layer (BBL)
Diffusive transport
benthic boundary layer *see* Benthic boundary layer (BBL)
in gas exchange 28, 29*f*
Dikes
in flood mitigation 646
Dimictic lakes
density stratification and stability 196–197, 197*f*
DIN *see* Dissolved inorganic nitrogen (DIN)
Dineutes 14–15
see also Gyrinidae (whirligig beetles)
Dioxins
pollution and 672, 673*f*
toxicity 670, 670*f*
Dioxygen, photolytic reactivity 37
Diphyllobothrium latum (broad fish tapeworm) 683–684
Diptera (biting)
ecology
Suborder Nematocera
Simuliidae Family (black flies) 680
nuisance pests 679
vectors of human diseases 678
Direct circulation 289
Disaccharides 437, 438*f*
Discharge
definition 154
see also Flow
Discharge-weighted mean concentration 338
Disease(s)
nematodes causing 677–678, 679–680
parasitic *see* Parasitic diseases
Diseases (human) of freshwater habitats 677–685
classifications 677
freshwater vectors 677
habitats of human-disease vectors 680
categories 680
human-made water bodies 680
natural water bodies 680
temporary anthropogenic water bodies 680–681, 681*f*
hydraulic engineering considerations 680
human culture, freshwater vectors and disease 683
cultural practices 683–684
biodiversity 684
cities 684
ingestion of raw fish 683–684
ingestion of raw insects & crustaceans 684
modification of cultural practices 684
root crop agriculture 684
insect vectors 678
black flies 679
mosquitoes 678, 678*t*
pathogen life-cycle 678
other Diptera 679
other insects 679
transmission of disease 678
intermediate hosts 684
morbidity 685
mortality 685
most importance diseases 677
dengue fever 678
Japanese encephalitis 678
lymphatic filariasis (elephantiasis) 677–678, 684
malaria 677
onchocerciasis (river blindness) 678, 682*f*
schistosomiasis 677
scope of problem 677
snails and crustaceans as intermediate hosts 679
schistosomiasis 679
vector control strategies 681
application of toxicants 681
biological control agents 681
competitive organisms 681
environmental modifications 681
Guinea worm 682
malaria 682
onchocerciasis (river blindness) 682–683
schistosomiasis 682
vectors of disease 677, 685
waterborne diseases 569, 570*t*
fecal-oral pathogens 569
incubation lengths and infection types 569–570
primary meningoencephalitis 571
statistics 571
see also Aquatic ecosystems and human health
Dissimilar material effects, vadose water unsaturated hydraulic conductivity 81
Dissimilatory nitrate reduction 401
Dissimilatory nitrate reduction to ammonium (DNRA) 401, 402*f*
Dissolved carbon dioxide *see* Carbon dioxide (CO2)
Dissolved chemicals, groundwater chemistry 107
Dissolved combined amino acids (DCAA) 343, 521
bacterial uptake
concentration change studies 528*t*, 533
extracellular enzymes 532
Lake Skjervatjern studies 532, 532*f*
substrate analog studies 532, 532*f*
composition 522
reduced variability 522–523
concentrations 521
definition 521
sources
bacteria 525–526
phytoplankton 525
soils 526–527
uptake (bacterial) 532
Dissolved combined carbohydrates (DCCHO) 437, 439, 440*t*
concentrations 439
lake water 442
Dissolved DNA, as nitrogen source *see* Dissolved organic nitrogen (DON)
Dissolved free amino acids (DFAA) 521
bacterial uptake 527, 528*t*
galactosamine 531
glucosamine 531
growth rates 527, 528*t*
Lake Hylke studies 527, 531*f*
composition 522
depth profiles 522, 522*f*
fluctuations 522
specific amino acids 522
concentrations 521
definition 521
microalgal uptake 531
as nitrogen source 531
sources
bacteria 525–526
phytoplankton 525
soils 526–527
Dissolved free carbohydrates (DFCHO) 437, 439, 440*t*
concentrations 439
lake water 442
Dissolved gases
rivers 149, 150*f*
see also Water, dissolved gases
Dissolved inorganic carbon (DIC) 343, 453
chemical components 343
concentrations 343
Dissolved inorganic nitrogen (DIN)
definition 517
immobilization in soil 592
particulate organic nitrogen (PON) *vs.*, stratified lakes 517–518, 518*f*
sources, soils 526
Dissolved (solute) load, stream flow characteristics 141
Dissolved matter
mass transport 338, 339*t*
organic *see* Dissolved organic matter (DOM)
rivers 148
Dissolved nutrients *see* Nutrient loading
Dissolved nutrients, deforestation effects 591
atmospheric precipitation/climate 591
geological weathering 591
hydrology effects 591
snowpacks 592
soil water effects 591, 591*f*
streamflow 591
precipitation pH 591
nitric acid leaching 591

Dissolved nutrients, deforestation effects (*continued*)
nitrification 588*t*, 591
organic acid release 591
proton production 591
soil physical–chemical reactions 593
temperature 592
terrestrial biological processes 592
chemical transformations 592
chemical uptake 592, 592*f*
coarse woody material 592
litter decomposition 593, 593*f*
nitrification 592
soluble chemical production 593
water physical–chemical reactions *see* Water, physical–chemical reactions
Dissolved organic carbon (DOC) 453, 454
carbohydrates *vs.* 439
catchment effect, climate change 614
cations and 358
discharge-weighted average 505
mercury and 648
Dissolved organic matter (DOM) 464
chemical composition 464
colored, catchment effect, climate change 614
metal interactions 469
iron 469
nitrogen and 469
nutrient interactions 469
phosphorus interactions 469
photoproducts 38*f*, 468
size distribution 466
Dissolved organic nitrogen (DON) 400, 518, 519*t*, 523
amino acids in stratified lakes 521, 522*f*
bacterial dynamics 534
ecosystem importance 534, 535*f*
heterotrophic bacteria 534, 535*f*
chitin 523
composition 518
compounds 518
concentrations 517
cycling 527
definition 517
detection 518
dissolved inorganic nitrogen (DIN) *vs.* 517
annual changes 517, 518*f*
seasonality 517
stratified lakes 517–518, 518*f*
dissolved RNA/DNA 524
from bacteria 525–526
bacterial uptake 533
base constituents 524–525
fragment sizes 524
horizontal gene transfer 533
from phytoplankton 524
galactosamine 524
glucosamines 523, 524*f*
from bacteria 525–526
humic matter 525
mannosamine 524
methylamines 525
nonprotein amino acids 523
N-acetylglucosamine 523
N-acetylmuramic acid 523
bacterial cell wall material 523
glycans 523, 523*f*
particulate organic nitrogen (PON) *vs.*, stratified lakes 517–518, 518*f*
photochemistry 40
sources 525
atmosphere 526
bacteria 525
deliberate release 525, 526
fish 526
phytoplankton 525
protists 526
sewage 527
soil 526
zooplankton 526
uptake 528*t*, 534
polymer compound measurement 535
single compound measurement 534
study methods 534
urea 525
Dissolved oxygen
in Hudson River 150*f*
Dissolved silicate (DSi) 414
aquatic ecosystem entry 414
definition 417
limitation 415
seasonal concentrations 415, 415*f*
sedimentary export 415
time scale 414
Dissolved substances *see* Dissolved matter
Disturbance(s)
exotic species impact on native species 606
seasonal flow, river/stream ecosystems 336*f*, 337
Dithiane, structure 432*f*
DOC *see* Dissolved organic carbon (DOC)
Docosahexaenoic acid (DHA) 471, 472*f*
definition 478
dietary supply 472
neural tissues 474
Dolomite(s), definition 174
DOM *see* Dissolved organic matter (DOM)
DON *see* Dissolved organic nitrogen (DON)
Downwelling zones 99, 100, 101*f*, 102–103
Dracunculiasis 679–680, 682
definition 684
Drag 315, 316, 316*f*, 317
Drag coefficient, definition 310–311
Drainage
acid *see* Acid drainage
acid mine *see* Acid mine drainage (AMD)
dendritic, of streams 134, 135*f*
network organization, streams *see* Stream(s)
Drier environments *see* Arid environments; Dry environments
Drinking water
cyanobacteria and 623*f*
eutrophication, effects 621, 623*f*
nitrates 405–406
Droop, M R, nutrient stoichiometry 538–539, 539*f*
Drought
salt concentrations 364–365
Dry deposition, definition 567
Dry environments
deforestation effect on nutrient load 599
see also Arid environments
Dual tracers, for estimating gas exchange rates 33
Duration of wetness, wetlands hydrology 179
Dykes
in flood mitigation 646
Dynamic pressure 23, 23*f*, 24*f*
stagnation point 24
Dynamics
food-web, rivers *see* River ecosystems, food-web dynamics
particle, benthic boundary layer (BBL) *see* Benthic boundary layer (BBL)
river/stream ecosystems **335–342**

E

Ebullition
in gas exchange at air–water interface 35
methane emissions *see* Methane emissions
Ecological status, ecosystem bioassessment 581
Ecosystem(s)
aquatic *see* Aquatic ecosystems
bioassessment *see* Ecosystem bioassessment
hydrology *see* Ecosystem hydrology
importance of bacteria, dissolved organic nitrogen 534, 535*f*
integrity, climate change effects in lakes *see* Climate change and lakes
lake *see* Lake ecosystem(s)
river/stream *see* River ecosystems; Stream ecosystems
streams *see* Stream ecosystems
Ecosystem bioassessment **577–584**, 578*f*
approaches 577
bioassays 579
biota 577–578
community structure measurements 579
indicator species choice 579
management goals 579
organism uptake 578–579

pollutants 577
toxicity studies 577, 579
biota used 577–578, 581, 582*t*
definition 577
examples 582
Australia 582–583
Community-Based Programs 582–583
European Union 582–583
GB 582–583
national scale 582–583
USA 582–583
historical aspects 579, 580*f*
1900–1950 579
1950–1970 580
1970–2002 580
BACI study 580
Belgian Biotic Index 580
Biological Working Party (BMWP) score 580
Hilsenhoff Index 580
Index Biotique 580
metrics 580
predictive models 580–581
Reference Condition Approach (RCA) 580–581
River Invertebrate Prediction and Classification System (RIVPACS) 580–581, 582–583
Saprobien system 579–580
surface water quality assessment 579–580
Trent Biotic Index 580
implementation 577
study types 581
condition assessment 581
ecological status 581
predictive ecology 581
problem characterization 581
recovery monitoring 581
restoration 581
risk assessments 581
stressor identification 581
trends analysis 581
Ecosystem hydrology 95, 98*f*, 101*f*
lakes 104
rivers 102–103
streams 99
Eddies
meadows 309–310
turbulent, in gas exchange at air–water interface 28, 29–30
Eddy correlation methods 33
Eddy diffusivity 225
concept 224*f*
estimation 225, 226*t*, 227*f*, 228
in hydrodynamic modeling 331–332
surface mixed layer 209–210, 214*f*
Eddy viscosity
in hydrodynamic modeling 331–332
turbulence 207–208
wind on lakes 209–210
Effective porosity, groundwater movement 88–89
Eicosanoids, definition 478
Eicosapentaenoic acid (EPA) 472*f*
definition 478
dietary supply 472
Ekman layers 293–295
Ekman transport, thermobaric instabilities 253–254
Electrode types, redox potential measurement 549
Electrodialysis (ED)
reverse osmosis-coupled 507
Electron acceptors
anoxic energy production in river/stream ecosystems 337, 340
Electron flow tracers, dissolved gases 6–7
Electronic factors, water solvation 5
Electron transport, humic substances, effects 457–458
Element(s)
in flowing waters 339*t*
concentration, dependency on flow 338, 340*t*
rate of transformation 338
Elephantiasis (lymphatic filariasis) 677–678, 684
El Niño-Southern Oscillation (ENSO) 59
flooding and 645
Elodea (waterweed) 302*t*
Emergent plants
rooted, methane emissions 489, 494*t*
Empirical formulae, vadose water unsaturated hydraulic conductivity 81
Enclosures
for estimating gas exchange rates 33
End-of-pipe pollution *see* Point source (PS) pollution
Endorheic basins
saline lakes 199
see also Lake(s), endorheic
Endorheic lakes *see* Lake(s), endorheic
Endorheic rivers/waters 124–127
Energy
potential, water flow 239–240
solar, hydrological cycles 53, 54–55
surface *see* Surface energy exchanges
turbulent kinetic *see* Turbulent kinetic energy (TKE)
Energy balance, hydrological cycles 54
Energy line (EL), dynamic pressure 23
Energy resources, river ecosystems *see* River ecosystems
Engineering
structures, river currents 243
Enhanced river discharge, impact on density currents 254–255
Entrainment velocity, surface mixed layer 213–214
Environmental assessment, diatoms *see* Diatom(s)
Environmental change
impact on density currents 254
phytoplankton productivity *see* Phytoplankton productivity
see also Climate change
Environmental effects
cations 359
nutrient loading 594*t*
Environmental factors
aerobic methane oxidation 482*t*, 483, 484*t*
methanogenesis *see* Methanogenesis
Environmental flows 646
Environmental variability, phytoplankton productivity *see* Phytoplankton productivity
Enzymes, extracellular 447–449
Ephemeroptera (mayflies)
larvae 191–192
Epilimnion
definition 377
in density stratification and stability 197, 198*f*, 200
in water-column stratification 256
Epiphytes
cyanobacterial nitrogen fixation *see* Cyanobacteria
definition 310–311
Equilibrium transport, suspended solids *see* Suspended solids, fluvial transport
Equivalent depth, definition 277, 282–283, 285
Equivalent-medium measurement, vadose water preferential flow 82–83
Erosion 104–105
alluvial channels 239
deforestation effects, particulate nutrients 595
fire impact 636
nutrient flux 193
soils *see* Soil(s)
stream drainage network organization 135
trace element sources 383
Escherichia coli, nutrient stoichiometry 471, 541
Escherichia coli O157:H7 569, 576
Essential fatty acids (EFAs) 471, 472*f*
definition 478
fish 475, 477*f*
metazoans 474
protazoans 474
Esthwaite Water (England)
cobalt concentrations 378
Estrogens, pollution and 673*f*
Estuaries
distinguishing characteristics 144*t*
nitrogen fixation 396
nutrient input 403–404
rivers *vs.*, water movement 145
Estuarine flows, flooding and 645
Eugster-Jones-Hardie model of evaporation 418–419
Eulerian velocity, surface mixed layer 210
Euphausiids, wax esters 472
Eurasian watermilfoil *(Myriophyllum spicatum)* 606

Europe
  springs 169*t*
European Union, ecosystem bioassessment 582–583
Eutrophication 601, **617–629**, 665
  concept
    evolution 617
    origin 617
  consequences 617, 618*t*
  control 625
  definition 381, 398, 413
  effects 622*t*, 627
    dissolved organic color 621
    food-web structure 623
    hydrology 623
    inorganic turbidity 621
    modifying factors 621, 624*t*
    predictions 618*t*, 620
    salinity 621
    trophic states 624, 625*f*
    vascular plants 623
    water quality 621
  environmental stressors 627
  health risks 623*f*, 627
  historical aspects 618*t*
  human activity and 617–619
  management 617, 618*t*, 625, 627*t*
  nitrogen 405
  nutrients
    growth-limiting (N and P) 620
    loading models 620
    sources 617
    transport 617–619
  organic matter, photochemical transformation 47
  phosphorus 408
  pollutant interactions 627
  research directions 625
  restoration 625, 626*t*, 627*t*
  silica cycle 416
  see also Oligotrophication
Eutrophic lakes
  stratified lakes, redox potential 550, 551*f*
Eutrophic systems, chemical enhancement 31–32
Evaporation 99, 104, 418–419
  cycles 59
  defined 69
  in gas exchange at air–water interface 32
  land surfaces 59
  oceans 59
  rate 59
Evapotranspiration **69–75**
  continental-scale hydrologic cycle 70–71
  definition 69, 154
  evaporation and acceleration of global hydrologic cycle 69–70
  hypotheses to explain recent decrease 70–71, 72*t*
    land-use changes 72–73
      afforestation experiments 72–73, 73*f*
    solar dimming 70–71
      complementary hypothesis 71
      effect on pan evaporation 71
    stomatal conductance 71–72
  problem of quantification 73
  role in hydrologic cycle, historic studies 69
  salinization effects 362
  seasonal patterns 145*f*
  soil-plant-atmosphere system 73–74
  uncertainty of future trends 75
  vadose water 84
  water budgets *see* Water budgets
  wetlands hydrology 185
Exchange *see* Hydrologic exchange
Exotic species
  biological control 607
  impacts on native species 606
    effects of disturbances 606
    indirect effects 606
Export
  material from rivers 153–154
Extracellular enzymes 447–449
  bacterial uptake of dissolved combined amino acids 532
Extreme storm events, lakes, climate change 612

F

Falls Brook (New Hampshire, USA), cations 358, 359*f*
Farming *see* Agriculture
Fatty acids (FAs)
  algae, eukaryotic *see* Algae, eukaryotic
  biosynthesis 471–472, 473*f*
  dietary 471
  essential *see* Essential fatty acids (EFAs)
  food web 474*f*
    modification 474
    transfer 474
  groups 471
  highly unsaturated *see* Highly unsaturated fatty acids (HUFAs)
  mono-unsaturated 472–474, 478
  n-3 471
    definition 478
  n-6 471
    definition 478
  nomenclature 471
  polyunsaturated *see* Polyunsaturated fatty acids (PUFAs)
  saturated (SAFAs) 478
Fauna
  benthic invertebrate *see* Benthic invertebrate fauna
  invertebrate, ground water and 99, 102–103
  saline lakes *see* Saline lakes
  surface water, types 99
Fe^{2+}/Fe^{3+} system, redox reactions 550
Feeding
  animals, pollution and 572, 673*f*
  benthic invertebrates *see* Benthic invertebrate fauna
  fish *see* Fish
  modes, zooplankton *see* Zooplankton, in ecosystems
  zooplankton *see* Zooplankton, in ecosystems
Feeding deterrents, definition 425–426
Feldspars
  cations associated 354–355
  dissolution 559
Ferromanganese concretions 372*f*, 376*f*
Ferrous pyrite, redox potential 552–553
Ferrous sulfide, redox potential 552–553
'Ferrous wheel', iron cycling 373, 374*f*
Fertilizer(s)
  application, groundwater chemistry 111–112, 113*f*
  nitrogen cycle and 412–413
  pollution and 660
Fetch
  wind on lakes 208–209
Fick's law of diffusion 28, 193–194, 344
  wetland vegetation 306–307
Field capacity (water)
  vadose water, hydrologic cycle 84
  water budgets 56
Figeh spring, water supply for Damascus (Syria) 173
Filamentous cyanobacteria *see* Cyanobacteria, filamentous
Filariasis 684
  lymphatic 677–678, 684
Filtration, colloid formation 467*f*, 468
Financial costs, floods 646
Finite difference hydrodynamic models 330
Finite element hydrodynamic models 330
Finite volume hydrodynamic models 330
Finland
  acid sulfate soils 564
Fire(s) **630–643**
  classification 630
  frequency 631, 632*f*, 633*f*
  impacts 630, 634, 635*f*, 636*f*
    biota 636, 641
    erosion 636
    immediate effects 634, 636*f*
    infiltration 635, 641
    intermediate effects 641
    long-term effects 641
    nutrient cycling 634, 641
    runoff 635, 641
    sedimentation 636, 641
    tropical regrowth 634, 637*t*
    water movement 634–635
  intensity 630
  lightening-induced 631

pre-fire considerations 632
baseline water chemistry 633
canopy type 634
catchment size 633
forest age 634
fuel load 634
soil characteristics 633
system morphology 633
timing 634
watershed slope 633
weather patterns 633
seasonal timing 631–632
severity 630
size 631, 632*f*
surface, definition 643
see also Crown fires; Wildfires
Fire Weather Index (FWI) 630, 631*f*, 632*f*
'First-dissociation constant', dissolved carbon dioxide and 345
First law of thermodynamics, surface energy exchanges 212
Fischerella 435*f*
allelochemicals 435
Fischerellins 435
Fish
benthivorous 623–624, 628
buoyancy 26
carbon turnover 454*t*
carnivorous, essential fatty acids 472
communities in lakes, climate change effect *see* Climate change and lakes
contaminants accumulation 573
dissolved organic nitrogen and 526
fire impacts 636
fish consumption advisories 573
fresh water
essential fatty acids 475, 477*f*
food characteristics 474*f*, 477
kairomones 427, 428*f*
Daphnia, effects on 427–428
grammarids 428
phenotypic plasticity 428
mercury poisoning 390
multiple stress resistance 461, 462*f*
nutrient stoichiometry 541, 541*f*
threshold element ratio 542, 542*f*
nutritional state 476–477
phosphorus and 541–542
piscivorous 628
pollution effects
methylmercury 651, 674
polybrominated diphenyl esters 671
rivers 152
research knowledge/gaps 152
turbulence effects 191
zooplanktivorous 623–624, 628
Fisheries
mercury levels 648
Flagellates
fatty acid composition 476*f*
Flash flooding 644, 646
Flathead River (Montana, USA) 102
Flatwood, definition 187
Flies *see* Diptera (biting)
Floating chambers, for estimating gas exchange rates 33
Flocculation, definition 470
Flood(s) **644–647**
causes and effects 644
coastal storms 645
dam failure 646
ice jams 645
rainfall 645
snowmelt 644–645
urban runoff 645–646, 646*f*
costs and mitigation 646
definition 644
fifty-year 644, 645*f*
flash 644, 646
hydrographs, stream flow characteristics 139, 140*f*
outburst 645
return interval 644
tropical 646
zone 644
see also Flooding; Floodplain(s);
see also Floodwater
Flood-frequency curves, stream flow characteristics 140
Flooding
adaptation by riverine/riparian communities 152–153
definition 187
see also Flood(s)
Floodplain(s) 101–102, 105
floods costs and effects 646, 647
inundation 646, 647
sediment movement to streams 138–139
support of riverine communities 152–153, 153*f*
wetlands hydrology 177–178
see also River(s)
Floodplain rivers
vs. levees and constricted rivers 144–145, 146*f*
Flood-pulse concept (FPC) 152–153
Florida (USA)
streams, cation chemistry 357*f*, 358*f*, 359–360
Flow
density-driven *see* Density-driven flows
duration curves, stream flow 140
modification, submerged vegetation *see* Vegetation
paths *see* Flow paths
rivers *see* River(s), flow
sources *see* Stream(s), flow sources
streams *see* Stream(s), flow
systems 95, 96*t*, 97*f*
variation
concentration of dissolved substances and 338, 340*t*
influence on river/stream biogeochemistry 336*f*
see also Water, flow
Flow fields, lateral circulation 289
Flowing waters **143–154**
biogeochemically important elements within 339*t*
ranking by size 143
see also River(s); Stream(s)
Flow paths 95, 99, 100–101, 104–105
baseflow 101–102, 103*f*, 105
effects on groundwater chemistry *see* Groundwater chemistry
underflow 101–102, 103*f*
Flow sources, streams *see* Stream(s)
Flow systems 95, 96*t*, 97*f*
ground water *see* Ground Water
Fluid mechanics 17, 18*f*
normal stresses 17
shear stresses 17
see also Pressure
Fluid motion, dynamic pressure and 23–24
Flush toilets, pollution and 659
Fluvial systems
currents 239, 240, 240*f*, 241*f*
Fluxes **xv–xvi**
river *see* River(s), fluxes
Flux magnitudes, groundwater recharge 90
Food webs
eutrophication effects 623
fatty acids in *see* Fatty acids (FAs)
hot springs 166–167
lake ecosystems *see* Lake ecosystem(s)
methane 489
nitrogen 400
periphyton *see* Periphyton
rivers 146–147
benthic invertebrates 151
invertebrates, importance of 151
research/conceptual models 152
Foraging kairomones
definition 425–426
plant–animal interactions 432
Forest(s)
age, pre-fire considerations 634
deforestation *see* Deforestation
Form drag, benthic boundary layer (BBL) 232–233
Fossil fuel combustion
carbon cycling 548
nitrogen cycle and 412–413
Fractal objects, definition 66
Fractured-rock systems, groundwater flow systems 89
Fragrances, pollution and 673*f*
Free surface, wind set-up 256

Freezing point, water 15
Fresh water(s)
  carbon distribution 514*t*
  cation chemistry 356, 357*f*, 358*f*, 359*t*
  definition 419
  diseases from habitats with *see* Diseases (human) of freshwater habitats
  humic substances 457
  lakes, nitrogen fixation 393
  total organic carbon 505
Freshwater fish *see* Fish, fresh water
Frictional forces, river currents 239
Friction coefficients, river currents 240
Frogbit *(Hydrocharis morsus-ranae),* hydrodynamic characteristics 302*t*
Froude number *(Fr)*
  gas exchange at air–water interface 35
  river currents 239–240
  super critical flow 267
Fructose 438*f*
  uptake 445, 446*t*
Fugacity, definition 343
Fulvic acids (FA)
  atomic ratios 510*f*
  elemental composition 510*t*
  molecular weight 509*t*
  oxidation state 511*f*
Fungi
  carbohydrates 442
Funneled flow, vadose water preferential flow 82

G

Galactosamine
  bacterial uptake 531
  dissolved organic nitrogen (DON) 524
Galactose 438*f*
Gammarus lacustris, reactive oxygen species 460*f*
Gammerids, fatty acid composition 477
Ganges–Brahmaputra River basin, water/sediment delivery 144*t*
Gardsjon watershed (Sweden), aluminum 357
Gas(es)
  concentrations, groundwater chemistry 110–111
  dissolved, rivers 149, 150*f*
Gas chromatography (GC), carbohydrate measurements 439
Gas exchange
  surface slicks and 32
Gas exchange at air–water interface **28–36**
  basic principles 28
  bubble formation 32
  carbon cycles and 28
  chemical enhancement 29*f*, 31
    eutrophic systems 31–32
    stable isotope dynamics 31–32
  chemical factors 30
  diffusion coefficient 31
  diffusive transport 28, 29*f*
  ebullition in 35
  empirical models 34, 34*t*
  Ficks law of diffusion 28
  gas exchange coefficient *see* Gas exchange coefficient
  lotic systems 35
  macrophyte-mediated 35
  mechanistic models 29
    boundary layer model 29
    stagnant film model 29–30, 29*f*
    surface renewal model 29–30, 34–35
  meteorological conditions 32
  methods for estimating rates 33
    chambers and enclosures 33
    eddy correlation methods 33
    purposefully released tracers 33
  physical factors 28
    buoyancy flux 28
    molecular diffusivity 28
    Schmidt number *see* Schmidt number
    turbulent diffusivity 28
    turbulent eddies 28
  solubility of gases 31
    Bunsen coefficient 31, 31*t*
    Setschenow effect 31
  surfactants 32
Gas exchange coefficient 28, 34–35, 35–36
  boundary layer model 29
  Schmidt number and 30*f*, 31, 34*t*
Gas exchange velocity *(k)* 344–345
Gasoline 668
Gas transfer velocity *see* Gas exchange coefficient
Gaula River, cation chemistry 356
Geochemical models, groundwater chemistry *see* Groundwater chemistry
Geogenic acidification, definition 567
Geological weathering
  deforestation effects on dissolved nutrients 591
  eutrophication and 617–619
  see also Weathering
Geology
  groundwater chemistry 107, 108*f*
Geometric factors, water solvation 5
Germany
  mercury levels, emissions 656
Gerris (pond skaters) 14–15
Giardia 569
Glacial lakes
  climate change effect 612
  origin/formation
    ice dams *see* Ice dam lakes
Glaciers, hydrological cycles and 52*t*
Glaciers/snowfield
  lakes, climate change 612
    see also Glacial lakes
  outburst floods and 645
  wetlands hydrology 177–178
  see also Ice caps
Glaciolacustrine plain, definition 187
Glasswort *(Salicornia),* hydrodynamic characteristics 302*t*
Glatt River (Switzerland) 102
Global warming
  effect on hydrologic cycle 69–70
  see also Climate change
Global water balance, water budgets 57
Glossinidae (tsetse) 679
β-1,3-Glucan 442
Glucosamine 523
  bacterial uptake 531
  sources and use in chitin 523
  see also Dissolved organic nitrogen (DON)
Glucose 438*f*, 442, 443*f*
  uptake 445, 446*t*
Glucosinolates
  as herbivore deterrents 431–432
  structure 432*f*
Glycans 523, 523*f*
Gneiss, cation concentrations 359*t*
Gonyostomum semens, bloom effects 625–627
Gradients
  changes in lakes, climate change 613–614
  chemical/biological longitudinal in river/stream ecosystems 335
  redox potential 549
Graminoids, definition 187
Grammarids, fish kairomones 428
Granite, cation concentrations 359*t*
Gravel bars, effect in river channels 147*f*
Gravimetric lysimeter, water budgets 55–56
Gravimetric measurement, vadose water 76
Gravitational forces, river currents 239
Gravity, colloid formation 467*f*, 468
Gravity waves 278
  anticyclonic 281, 281*f*, 282*f*
  Burger number and 278, 280, 281, 281*f*, 282*f*
  current structure and measurement 282
  cyclonic 281, 281*f*, 282*f*
  definition 285
  internal *see* Internal waves
  measurement 282–283, 284
Grazer(s)
  flagellates *see* Flagellates
  phosphorus
    cycle 412
    limitation 410
Grazing
  bacteria *see* Bacteria
  effects on
    allelochemicals 429, 432

mixotrophic phytoplankton 47
process, zooplankton *see* Zooplankton, in ecosystems
zooplankton *see* Zooplankton, in ecosystems
Great Britain, ecosystem bioassessment 582–583
Great Lakes *see* Laurentian Great Lakes (USA); North American lakes
Green algae 101*f*, 570–571
fatty acids 475
humic substances, responses 458
Green superfamily, primary producers/production 539
Gross primary productivity (GPP)
carbon dioxide and 345
Ground water **88–94**
budgets *see* Groundwater budgets
chemical signature 102–103
chemistry *see* Ground Water chemistry
data and models 92, 93*t*
definition 88
discharge 95
hydrological cycles 53
exchange with lakes 103–104
flow, streams 136–137, 137*f*
flow paths 101–102
flow systems 89, 89*f*
age (time since recharge) 89, 89*f*
chemical composition changes 107
definition 89
fractured-rock systems 89
tracer techniques 89–90
hydrological cycles *see* Hydrological cycles
inflows
density currents and 251
wetlands hydrology 183–185
invertebrate fauna and 99, 102–103
movement
Darcy's law 56, 88
effective porosity 88–89
hydraulic conductivity 88–89
hydraulic gradient 88–89
outflow (recharge) *see* Ground Water, recharge
quality 92
contamination 92
contributing areas 92
receiving stream quality 92
saline water 92
water table contamination 92
recharge 90, 105
diffuse recharge 90
flux magnitudes 90
human activities 90
localized recharge 90
precipitation 90
seasonal fluctuation 90
urbanization effects 90
wetlands hydrology 185
runoff 56
see also Water budgets
springs and 168–170
storage, water budgets 56
surface water interaction *see* Ground Water and surface water interaction
temperatures 96
wetlands hydrology 179
see also Aquifers
Groundwater budgets 56, 90
climate effects 91
climate change 91–92
long-term droughts 91
long-term equilibrium 90–91
pumping effects 91
recharge balance 91, 91*f*
source changes 91, 92*f*
streamflow capture 91
Groundwater chemistry **107–117**
flow path effects 107
age of water 107, 108*f*, 109*t*
composition effects 108–109, 110*f*, 111*f*
dissolved chemicals 107
gas concentrations 110–111
geology 107, 108*f*
ion exchange 107–108
land use *vs.* 111, 112*f*
microbial activity 109
Morgan Creek (Maryland) study 111, 112*f*
nitrogen 110–111
noble gases 110–111
saturated water 107–108
sorption effects 107–108
stable isotope ratios 109–110
time 107, 108*f*
geochemical models 116
inverse modeling 116
NETPATH 116
PHREEQC 116
human activity effects 111
acidity 114–115
chemical removal 113
dense nonaqueous phase liquids (DNAPLs) 113
fertilizer application 111–112, 113*f*
nonaqueous phase liquids (NAPLs) 113, 115*f*
nonpoint source contamination 111–112, 113*f*
organic compounds 108*f*, 113–114
pharmaceuticals 112
plume formation 112–113, 114*f*
reactions 114–115
three-dimensionality 113, 114*f*
tracer studies 109*t*, 112
wastewater infiltration 112
surface water *vs.* 116
hyporheic zone 116
lakes 116
matrix interactions 116–117
wetlands 116
Ground Water and surface water interaction **95–106**
human impacts and future research 104
agriculture 104
future research 105
hydrosphere modifications 105
urbanization 105
interactions across the inland hydrosphere 99
hyporheic zones and transient storage modeling 100
lakes and wetlands 103
rivers and floodplains 101
streams 99
measuring 95
direct measures 96
heat tracer methods 96
mass-balance approaches 97
use of Darcy's law 96
scales and effects 95
Group velocity, internal wave rays 270
Growing season, wetlands hydrology *see* Wetlands hydrology
Growth dilution, definition 651
Growth efficiency, definition 542
Growth Rate Hypothesis (GRH), nutrient stoichiometry 539–540, 541*f*
Groyne fields 243
flow patterns 244*f*
Groynes (spur dykes) 243
Guinea worm 679–680, 682, 684
Gulf of Riga, normalized stream function 291*f*
Gyres 243, 293, 294*f*
topographic, homogenous basins 289
Gyrinidae (whirligig beetles)
Dineutes 14–15
Gyrinus 14–15
Gyrinus 14–15

H

Habitat(s)
archaea *see* Archaea
fungi *see* Fungi
stationary, biogeochemical influence on rivers/streams 335
Habitat-scale factors, algae in river ecosystems *see* Algae, river ecosystems
Hagen-Poiseulle's law 73–74
Halocline
in density stratification and stability 200
Halonatronum spp. 420
Harmful algal blooms *see* Algal blooms, harmful
Heat
storage/release 12
Heat flux, in surface energy exchanges *see* Surface energy exchanges

Heat of fusion (melting) 12
  definition 12
Heat of vaporization (condensation), water 13
Heat shock proteins (Hsp), humic substances, effects 460
Heavy metals
  cycling 374–375
  pollution 674
  see also Metal(s)
Henry's Law 343
  buoyancy 26
  definition 347
Herbicides
  deforestation, nutrient load effects 599–600
Herbivores
  deterrents, allelochemicals 429, 431, 432*f*
Heterocysts
  definition 398
  nitrogen fixation 394, 398
Heterotroph(s)
  allelochemicals effect *see* Allelochemicals/allelochemical reactions
  biofilm component 335–336
  definition 398
Heterotrophic bacteria
  dissolved organic nitrogen and 534, 535*f*
  see also Periphyton (attached microalgae)
Heterotrophs, definition 456
Hexameric ice, water structure 2–3, 2*f*, 3*f*
Hexose sugars 438*f*
Highly unsaturated fatty acids (HUFAs)
  biosynthesis 473*f*
  definition 478
  lakes, levels 475
High-performance liquid chromatography (HPLC)
  carbohydrate measurement 439
Hillslope characteristics, streams 137, 138*f*
Hilsenhoff Index, ecosystem bioassessment 580
Holmboe waves, in basin-scale internal waves degeneration 265
Holomictic lakes
  definition 196
  dimictic *see* Dimictic lakes
  monomictic 196–197
  oligomictic 196–197
  polymictic 196–197
Homeostasis
  definition 537
  stoichiometric 537, 538*f*
Homogenous basins (HB)
  currents **286–300**
    Ekman layers 293–295
    interaction with bottom and gyres 286
    wind setup responses 288–289
  future research 299
  topographic gyres 289
  topographic waves 290
Homogenous layers
  circulation generation 289
  lateral circulation 289
  vertical circulation 293
  wind setup 288
Horizontal density currents, generated under ice cover 254
Horizontal gene transfer, dissolved RNA/DNA 533
Horizontal mode seiches *see* Seiche(s)
Hormesis, definition 463
Hornwort *(Ceratophyllum)* 302*t*
Hot springs 156–157
  biology 172
  food webs 166–167
  human applications 173
  see also Thermal springs
Huang He River basin, water/sediment delivery 144*t*
Hubbard Brook Experimental Forest (USA) 599*f*, 599
  chloride concentrations 363, 365
  see also Falls Brook
Hudson River (USA)
  chloride concentrations 363
  dissolved gases 150*f*
Human(s)
  activity, effects on
    eutrophication 617–619
    groundwater chemistry *see* Groundwater chemistry
    groundwater recharge 90
    in river basins 153–154
    river/stream ecosystems 337
    water regulation 147, 148*t*
  diseases *see* Diseases (human) of freshwater habitats
  diving 26
  health *see* Aquatic ecosystems and human health; Diseases (human) of freshwater habitats
  impacts *see* Anthropogenic impacts
Humic acids (HA)
  atomic ratios 510*f*
  definition 503
  elemental composition 510*t*
  molecular weight 509*t*
  oxidation state 511*f*
Humic matter, dissolved organic nitrogen and 525
Humic substances (HSs)
  aggregation 466
  appearance 464
  carbohydrates 450
  dissolved organic nitrogen and 525
  dissolved oxygen content concentrations 457
  environmental conditions 450
  flocculation 466
  formation 464, 465–466
  functional groups 457
  HS-rich waters 457
  hydrophobic contaminants 469
  iron interactions 468, 469
  metal interactions 469
  nutrient interactions 469
  organism interactions **457–463**
    biomembranes 457, 459
    chemical stress defense 460
    hormone-like effects 458
    multiple stress resistance 461
    photosynthetic oxygen production, interference 457
    phototrophs 458
    strong *vs.* mild chemical stress 462
  photodegradation 468–469
  physical process interactions **464–470**
  properties 464
  protonation 559
  surfactant role 465, 466*f*
Humidity-based measurement, vadose water 77
HuminFeed®, definition 463
Humus
  cations 355
  photochemical formation 47
Hurricanes
  as cause of flooding 645
Hutchinson, George Evelyn
  trace elements 378
Hydraulic conductivity *(K)* 96
  definition 187
  groundwater movement 88–89
Hydraulic engineering projects, consideration of disease vectors 680
Hydraulic grade line (HGL), dynamic pressure 23
Hydraulic gradient, groundwater movement 88–89
Hydraulic head 96
  measurement, aquifers 88
Hydraulic radius, definition 310–311
Hydraulic residence times
  definition 628
  lakes, climate change 613
Hydric soils
  definition 187
Hydrilla verticillata (hydrilla) 606
Hydrocarbons, pollution and 668
  organic solvents *see* Organic solvents
  petroleum oils 668
  polycyclic aromatic *see* Polycyclic aromatic hydrocarbons (PAHs)
Hydrocharis morsus-ranae (frogbit) 302*t*
Hydrochemical facies, ground water 54, 54*f*
Hydrodynamic modeling **320–334**
  biota fluxes 320
  boundary and initial conditions 322
    'spin-up' time 322–324
  calibration 324, 333
  computational power 332
  dimensionality and capabilities 320

1-D models 320–321, 323*t*, 328*t*
2-D models 320–321, 325*t*, 328*t*
3-D models 320–321, 321–322, 326*t*, 328*t*
stratification in 320
future directions 334
grids 324
horizontal systems 327, 329*f*
Cartesian 327, 329*f*, 333–334
curvilinear 327–329, 329*f*, 333–334
unstructured 329, 329*f*, 333–334
size and convergence 327
vertical systems 329, 329*f*
isopycnal coordinate 329*f*, 330, 334
sigma coordinate 329–330, 329*f*, 334
z-level 329, 334
hydrostatic approximation 324
incompressible fluid Newtonian continuum mechanics 320, 322*f*
internal waves 331–332
lake *vs.* river 332
model errors 331
numerical diffusion of mass 331
numerical dispersion of waves 331
numerical dissipation of energy 331
numerical methods 330
finite difference 330
finite element 330
finite volume 330
nutrient fluxes 320
order of accuracy 331
pollutant fluxes 320
river *vs.* lake 332
spectral signal processing 320
temperature profiles 320, 321*f*, 332–333
thermocline 320, 321*f*, 331–332
time step 330
Courant-Lewy-Friedrichs (CFL) condition 330, 331
viscous limitation 330
tracers 332, 333*f*
turbulence and mixing 331
Hydrogen (H)
nutrient loading, environmental effects 594*t*
Hydrogen bonds, water structure 1–2
^{3}H-labeled uptake assays, dissolved organic nitrogen uptake 534–535
Hydrogenotrophic methanogenesis 480
Hydrogen sulfate, formation 559
Hydrographic change, river/stream ecosystems 337
Hydrographs
stream flow characteristics 139
wetlands hydrology *see* Wetlands hydrology
Hydrological cycles **51–57**, 52*f*, 58, 69, 95, 104
acceleration 69
evaporation and 69
atmospheric component 51–52, 52*t*
continental-scale, evapotranspiration and 70–71
definition 51, 69, 95, 174
description 51, 52*f*
energy balance 54
estimates of temperature effects on rainfall 69–70
evapotranspiration and *see* Evapotranspiration
ground water 51–52, 52*t*, 53, **95–106**
chemical processes 53–54
groundwater discharge 53
hydrochemical facies 54, 54*f*
see also Ground Water
ice caps/glaciers 52*t*
lithospheric components 52*t*, 53
ground water 53
soil water 53
surface water 53
multiyear climatic patterns 58
nutrient cycling 54
plants 54
sulfate 54
sulfide minerals 54
sulfur 54
oceans/seas 52*t*
precipitation and *see* Precipitation
snow/ice cover 68
soil water 52*t*, 53
movement 53
solar energy 53, 54–55
surface water 51–52, 52*t*, 53
ground water interaction **95–106**
see also Ground Water and surface water interaction
sediments 53
solutes 53
see also Surface water
temporal patterns 58
vadose water *see* Vadose water, hydrologic cycle
water budgets and **51–57**
wetlands *see* Wetlands
see also Hydrologic exchange; Water budgets
Hydrologic cycle *see* Hydrological cycles
Hydrologic exchange 95
direction 95, 101–102, 102–103, 104–105
effects on ecosystems 95, 98*f*, 99, 100, 102
fluxes 105
ground water and lakes 103–104
measuring 95–96, 98*f*
difficulties 95–96, 96–97
heat tracer methods 96
mass-balance approaches 97–99
piezometers 96, 98*f*
seepage meters 96, 98*f*
transient storage zone modeling 100, 101*f*
use of Darcy's law 96
metrics used to describe 102*t*
scales 99*t*
variations 98*f*, 99, 100*f*, 105
see also Ground Water and surface water interaction; Hydrological cycles
Hydrologic residence time, definition 347
Hydrology
definition 174, 187
deforestation effects on dissolved nutrients *see* Dissolved nutrients, deforestation effects
eutrophication and 623
Hydrophobicity, definition 643
Hydrophytes *see* Aquatic plant(s); Vegetation
Hydrophytic vegetation, definition 187
Hydrostatic approximation, in hydrodynamic modeling 324
Hydrostatic pressure (P_h) 17, 18*f*
'absolute' 18–19
atmospheric pressure changes 18–19
biological responses 25
buoyancy *see* Buoyancy
density variations 20, 20*f*, 25
depth 19, 19*f*
downward acceleration 19
forces 17
'gauge' 18–19
horizontal forces 19, 20*f*
solid-body rotation 19–20
static fluids 19
submerged surfaces 20, 21*f*
curved 21–22, 22*f*
water column weight 18–19
Hydrostatic profiles, vadose water, hydrologic cycle 83–84, 83*f*
Hydrothermal vents
in density current generation 251
Hydrothermal waters
iron and 376, 376*f*, 376*t*
manganese and 376, 376*f*, 376*t*
Hypertrophied lenticels, definition 187
Hypolimnion
definition 377
in density stratification and stability 197, 198*f*, 256
internal mode seiches 263–264
mercury pollution 648
methane emissions 489, 496*t*, 498*f*
oxygenation 664–665
surface mixed layer and 207, 208*f*, 216–217
vertical mode seiches 262–263
in water-column stratification 256
Hyporheic zone 99, 101–102
biogeochemistry 336–337, 336*f*
definition 342, 377
groundwater chemistry 116
iron cycling 373
manganese cycling 373
river redox potential 554
Hyporheos 99

Hyporheos (*continued*)
  definition 154
Hypoxia
  definition 154
Hypoxic environments, definition 406, 413
Hysteresis, vadose water 86

I

Ice
  cover
    density currents generated under 254
    lakes, climate change *see* Climate change and lakes
    see also Snow/ice cover
  hexameric, water structure 2–3, 2*f*, 3*f*
Ice caps
  hydrological cycles 52*t*
  see also Glaciers/snowfield
Ice dam lakes
  climate change 611–612
Ice-free periods, lakes, climate change 615
Ice jams
  as cause of flooding 645
'Ice-like' structures, water structure 3, 4*f*
Ice particles
  distribution 60
  growth 60
  radii 60
Impervious surfaces
  stream chloride concentrations 363–364, 363*f*
Impoundments
  effect on river hydrology 148*t*
  source of plankton for rivers 152*f*
  see also Reservoir(s)
Incoming/flood tides, wetlands hydrology 183–185
Incompressibility assumption, hydrostatic pressure 20
Index Biotique, ecosystem bioassessment 580
Indiana Harbor/Canal-Grand Calumet River System, PAH pollution 670
Indicators
  water quality *see* Water quality
  wetlands hydrology 185, 186*t*
  see also Bioindicator(s)
Indicator species choice, ecosystem bioassessment 579
Induced turbulence, stratified benthic boundary layers 234, 235*f*
Indus River (Asia)
  basin, water/sediment delivery 144*t*
Industrialization, pollution and 658
  see also Human(s), activity
Inertial period, definition 277
Infiltration
  definition 643
  vadose water 84, 84*f*
Inland seas, reduction 660–662
Inorganic carbon, dissolved *see* Dissolved inorganic carbon (DIC)
Inorganic phosphorus
  levels 128
  see also Phorphorus; Phosphorus (P)
Insect(s)
  carbohydrates 442
  downwelling zones 102–103
  fatty acid composition 477
  hyporheic zones 99
  rivers 151
  spring biota 157, 163*f*
Insecticide(s), pollution and 672
Instantaneous-profile method, vadose water unsaturated hydraulic conductivity 81
Interbasin exchange, density currents and 250, 250*f*
Internal current shear, in density stratification and stability 197
Internal hydraulic jumps 273*f*
  in internal wave degeneration 267
Internal mode seiche *see* Seiche(s)
Internal Rossby radius 277–278, 278*f*, 284
Internal seiches 258*f*
  in hydrodynamic modeling 324
  see also Seiche
Internal stratification, wind set-up 257
Internal waves, in lakes **256–276**, 288
  degeneration *see* Basin-scale internal wave degeneration
  in density stratification and stability 196, 203
  earth's rotational effects 278, 279
  forced 271, 273*f*
  geometry 256
  hydrodynamic modeling 331–332
  Lake Number 259–260, 263
  nonlinear *see* Nonlinear internal waves (NLIWs)
  rays, in continuous stratification 269, 270*f*, 271*f*, 272*f*
    dispersion relation 269–270
    group velocity 270
    turning depth 270
  resonant 271
  seasonal stratification 256
  seiche *see* Seiche(s)
  surface momentum transfer 256
  time series data analysis 271
  turbulent benthic boundary layer and 256, 273*f*
  water-column stratification 256, 257*f*
  Wedderburn number 259, 263
  wind set-up 256, 258*f*, 259*f*, 260*f*
    free surface 256
    internal stratification 257
  see also Gravity waves
International Biological Programme (IBP) 456
Inverse modeling, groundwater chemistry 116
Invertebrate(s)
  benthic *see* Benthic invertebrate fauna; Benthic organisms (benthos)
  hyporheic zones 99
  kairomones 428
  nutrient stoichiometry 540
  rivers 151
  river/stream ecosystems 335–336
Ion(s)
  major, in water 7
  rivers, presence in 149
  spring classification 158*t*
Ion bridging, colloid formation 468
Ion exchange
  acidified water buffering 559
  groundwater chemistry 107–108
Ionic composition
  biological responses 420
  effects on clays 420
  factors controlling 124
  fluxes 119
  measurements 124
Ionic structures, water solvation 4–5, 4*f*
Ion pairing, definition 347
Iribarren number, wave breaking 268–269
Iron (Fe) 350, 351, **368–377**, 386, 387
  acidified water buffering mechanisms 558
  acid mine drainage *see* Acid mine drainage (AMD)
  chemistry 368
    chemical forms 368, 369*t*
    photochemical reactions 40, 40*f*, 374
    redox reactions 368, 369*f*, 369*t*
    solubility 368
  clays 368
  environmental roles 368
  Fe^{2+}/Fe^{3+} system redox reactions 550
  humic substance interactions 468, 469
  hydrothermal waters and 376, 376*f*, 376*t*
  lake restoration and 553–554
  lakes 371, 372*f*
    cycling 371, 372*f*, 373
    inputs 371
    losses 371
  nitrogen fixation 395
  nutrient loading
    environmental effects 594*t*
    phosphorus levels and algal growth 8
  oxic layers 371
  in rivers/streams 369, 370*f*
    cycling 373, 374
    larger streams/rivers 373
    smaller streams/rivers 370
Iron minerals, lake restoration 553–554
Iron oxides
  adsorption-mediated chemical interactions 369
  phosphorus and 369, 372–373

trace metals and 374
Iron sulfides 371
Irrigation 104–105, 421
impact of 127
reservoirs *see* Reservoir(s)
Isoetes spp.
effects of acid rain 605
Isolated wetlands, definition 187
Isopycnal(s)
coordinate vertical grid systems 329*f*, 330, 334
internal seiche modes 263–264
internal wave degeneration 264–265
Isotes lacustris (Lake Quillwort) 302*t*
Isotherms, stratified benthic boundary layers 234, 234*f*
Isotopes
stable, carbohydrates and 450
water 13

J

Japanese encephalitis 678, 684
Juncus (rushes) 302*t*

K

Kairomones
definition 425–426
Karstic environments, subsurface inflows 251
Karst lakes
nitrogen limitation 379
Karst terrains
definition 174, 187
springs, flow 155
Kawah Ijen (Java), acidification 564
Keesom forces, water structure 1–2
Kelvin–Helmholtz billows
in basin-scale internal wave degeneration 265, 267*f*
Taylor-Goldstein equation 265–266
mixing and turbulence energetics 213–214, 215–216, 217*f*
Kelvin–Helmholtz instabilities 223*f*, 315, 315*f*
Kelvin waves 279–280, 280*f*, 281
Kentucky River (USA), seasonal patterns of precipitation/evapotranspiration 145*f*
Kinematic viscosity, definition 28
Kin recognition, allelochemicals 429
Korteweg-de-Vries equation, nonlinear internal waves (NLIWs) 268

L

Labile constituent(s)
definition 342
river/stream ecosystems 338, 339*t*
Lactose 438*f*
Lago Copahue, acidification 564
LaGrangian sampling 145–146
LaGrangian velocity, surface mixed layer 210
Lake(s)
alkaline and acid 351
highly alkaline lakes 353
low alkalinity acidic waters 352
Antarctica *see* Antarctic lakes
Australia *see* Australia, lakes
chemistry, groundwater chemistry *vs.* 116
climate change and *see* Climate change and lakes
coarse woody debris *see* Coarse woody debris (in lakes/streams)
crater 564
density stratification *see* Density stratification
dimictic *see* Dimictic lakes
distinguishing characteristics 144*t*
ecosystems *see* Lake ecosystem(s)
endorheic 421
endorheic basins 199
eutrophic *see* Eutrophic lakes
gas exchange at air–water interface *see* Gas exchange at air–water interface
glacial *see* Glacial lakes
holomictic *see* Holomictic lakes
hydrodynamic modeling 332
see also Hydrodynamic modeling
hydrologic exchange 103–104
ice cover *see* Ice; Ice dam lakes
ice dam *see* Ice dam lakes
internal waves *see* Internal waves
karst 379
layers, water density 11, 11*t*
meromictic *see* Meromictic lakes
methane emissions 496*t*, 498–501
mineral acids 352
minerals *see* Lake ecosystem(s)
monomictic *see* Monomictic lakes
North America *see* North American lakes
organic acids in 352
redox potential 550
see also Redox potential
restoration, redox potential *see* Redox potential
saline *see* Saline lakes
shallow lakes (and ponds)
silica cycling 415
softwater lakes, vulnerable to stresses 605–606
stratified *see* Stratified lakes/waters
surface mixed layer *see* Surface mixed layer (SML)
Sweden, mercury levels 648
water movement 145
wetlands hydrology 177–178
see also Lake ecosystem(s); Reservoir(s)
Lake 240 (North America), mercury pollution 654*t*
Lake Alpnach (Switzerland), vertical diffusivity 226*f*
Lake Au (Switzerland), carbohydrates 440*t*
Lake Bagsværd (Denmark), carbohydrates 440*t*
Lake Baikal (Siberia, Russia)
algaecides 381
mercury pollution 653*t*
turbulence 228
vertical diffusivities 227*f*
Lake Balaton (Hungary), eutrophication *see* Eutrophication
Lake basins
homogenous *see* Homogenous basins (HB)
integrity, climate change and *see* Climate change and lakes
Lake Belau (Europe), circulation patterns 296*t*
Lake Brienz (Switzerland), carbohydrates 440*t*
Lake Chad (Africa)
water balance 612
Lake Constance (Europe)
carbohydrates 440*t*, 443, 444*f*, 445, 446*t*
total dissolved, turnover 447, 448*f*
uptake measurement 447
circulation patterns 298*t*
gyres 293
wind forcing 295*f*
Lake Diktar-Erik (Sweden), carbohydrates 440*t*, 446*t*
uptake 447
Lake ecosystem(s)
acidified 357
see also Acidified waters
carbonate alkalinity 454–455
carbon pools 453
currents 287*f*
linear dynamics 286
nonlinearities 286
vorticity 290*f*
see also Current(s)
dissolved carbon dioxide 344*f*
dissolved combined carbohydrates 442
fatty acids 475
hydrodynamic processes 301
iron *see* Iron
manganese *see* Manganese
phosphorus
input 409–410, 412–413
sediments and 412
river/stream ecosystems *vs.* 335, 341
sediments
accumulation 455–456
see also Sediment(s); Sedimentation
shallow *see* Lake(s), shallow lakes (and ponds)
stratified *see* Stratified lakes/waters
sulfate reduction 373
vegetation 301
canopies 305
density 305
distribution 301

Lake ecosystem(s) (*continued*)
littoral zone 301, 305–306, 305*f*, 306*f*
pelagic zone 305–306
wind stress 301
Lake Eire, circulation patterns 296*t*
Lake Erie (Canada/USA)
chloride concentrations 363
Lake Furesø (Denmark), carbohydrates 443*f*, 444–445, 446*t*, 447, 448*f*
uptake measurement 447
Lake Geneva (Switzerland)
circulation patterns 298*t*
Lake Greifen (Switzerland) 102
Lake Huron (North America)
circulation patterns 296*t*
Lake Hylke (Denmark), bacterial dissolved free amino acid uptake 527, 531*f*
Lake Kinneret (Israel)
circulation patterns 298*t*
wind stress 292, 292*f*
Lake Klintsjön (Sweden), carbohydrates 440*t*
Lake Mekkojärvi (Finland) 440*t*
carbohydrates 439
Lake Michigan (USA)
circulation patterns 296*t*
mercury pollution 653*t*
silica cycle 415*f*, 416*f*
Lake Nakanuma (Japan), carbohydrates 439, 440*t*
Lake Number (L_N)
in density stratification and stability 205
internal waves 259–260, 263
surface mixed layer dynamics 216–217, 219, 219*f*
Lake Ontario (Canada/USA)
currents 291*f*
silica cycling 416*f*
Lake Orta (Italy)
acidification 566
liming 566
Lake Plußsee (Germany)
carbohydrates 440*t*
Lake Quillwort *(Isotes lacustris)* 302*t*
Lake Sammanmish (USA), cobalt concentrations 378
Lake Skärhultsjön (Sweden), carbohydrates 439, 440*t*
Lake Skjervatjern (Norway), bacterial uptake of dissolved combined amino acids 532, 532*f*
'Lake snow' 450
Lake Superior (Canada/USA)
climate change 613
copper 381
mercury pollution 653*t*
Lake Tahoe (USA)
currents 287*f*
nitrogen fixation 397
Lake Tanganyika (Africa)
circulation patterns 296*t*
climate change effect 615
definition 456
Lake Trichonis (Greece), circulation patterns 296*t*
Lake trout fry, development 673*f*
Lake Tuscaloosa reservoir (USA), photolytic reactions 48*f*
Lake Valkea-Kotinen (Finland), photochemical mineralization 41, 43–45, 44*f*
reaction rates 45
Lake Voui (Vanuatu) 558*f*
Lake Washington (USA)
eutrophication 624
Metallogenium bacteria 380
Lake Xochimilco (Mexico) 172
Laminar flow
stream flow characteristics 140–141, 141*f*
Lancangjiang (Mekong) River
basin, water/sediment delivery 144*t*
Land use
agriculture *see* Agriculture
changes in 147–148
groundwater chemistry *vs.* 111, 112*f*
Langmuir cells/circulations 209–210, 210*f*, 221
Large-Eddy Simulation (LES) methods, in hydrodynamic modeling 331–332
Latent heat transfer, surface energy exchanges 211
Laurentian Great Lakes (USA)
invasive species 606
mercury levels 648
silica cycling 416, 416*f*
see also Lake Erie, Lake Ontario, Lake Michigan, Lake Superior, North American lakes
Law of the Minimum, eutrophication 617
Law of the wall 209–210, 211, 226–227
turbulent benthic boundary layers 232
Layered media, vadose water 86
Lead (Pb)
poisoning (humans) 573
pollution 675
Lemna
hydrodynamic characteristics 302*t*
Lena River (Russia)
carbohydrates 440*t*
Lenticels, definition 187
Lentic environments
definition 174
oxygenation 664
Lentic littoral zone, definition 310–311
Leonardo da Vinci, Codex Leicester 69
Leptospira 569
Levees
effects on river hydrology 144–145, 147–148, 148*t*
vs. floodplain rivers and constricted rivers 144–145, 146*f*
Liebig, Justus von, eutrophication 617
Liebig's Law of the minimum 665
Ligands 385
definition 382
Light
attenuation
vertical 192
effects on diversity 602–603
harvesting, photosynthesis *see* Photosynthesis
interception, effects on vegetation 602–603
photolytic reactivity *see* Photolytic reactivity
relationship to nutrient levels 602–603
turbidity effects *see* Turbidity
UV *see* Ultraviolet (UV) radiation
vertical attenuation 192
'Light adaptation' 192
Light:Nutrient hypothesis 547
Lignin
empirical formula 508–509
Lignite mining, acid drainage 562*f*
Lignoids, as herbivore deterrents 431–432
Limestone
definition 174
weathering 354–355
Liming
definition 567
Lake Orta 566
soils 360
Limnology
water chemistry, trends/patterns 6
Linear wave equation, horizontal seiche modes 260, 262
Linoleic acid (LA)
definition 478
see also Alpha-linoleic acid (ALA)
Lipid(s) **471–479**
algal 474
biosynthesis 471
functions 472
groups 472
historical aspects 471
polarity 472, 473*f*
see also Fatty acids (FAs); Triacylglycerol (TAG)
Liquid-water cloud droplets 60
Lithospheric components, hydrological cycles 52*t*, 53
Litter, plant *see* Plant litter
Litter decomposition
deforestation effects on dissolved nutrients 593, 593*f*
see also Plant litter
Little Rock Lake (USA)
mercury pollution 650*f*, 653*t*, 654*t*, 656–657
Littoral zone
food-webs
vegetation role 301, 305–306, 305*f*, 306*f*
primary producers
zonation 602*f*

Littorella uniflora (shoreweed) 302*t*
Lobelia dortmanna 604
Local abiotic factors, algae in river ecosystems *see* Algae, river ecosystems
Localized recharge, groundwater recharge 90
'Lodging velocity' 309
London (dispersion) forces, water structure 1–2
Longitudinal gradients
  chemical/biological in river/stream ecosystems 335
Longitudinal movement, rivers/streams 341
Long-range phases, water structure 2–3
Long-term changes, wetland hydrographs 182
Long-term droughts, groundwater budgets 91
Long-term equilibrium, groundwater budgets 90–91
Long wave radiation, in gas exchange at air–water interface 32
Lotic ecosystems
  gas exchange in 35
Lotic environments
  definition 174
  oxygenation 664
Lotic macroalga, definition 310–311
Lymphatic filariasis (elephantiasis) 677–678, 685

M

Macaca fuscata 171*f*
Mackenzie River (Canada)
  basin, water/sediment delivery 144*t*
Macroalgae 301
Macroinvertebrates
  definition 174
  nonemergent, springs 162–163, 163*f*, 164*t*, 166*f*
  turbulence effects 191
Macrophyte(s)
  carbon dioxide effects 605
  definition 174, 310–311, 413
  in gas exchange at air–water interface 35
  growth forms 623
  nutrient stoichiometry 537
  Potamogeton spp. 603
  productivity, effects on diversity 603
  in rivers 150–151, 151*f*
  variety of growth forms 603
Macrophyte beds, flow **301–312**
Macropores, vadose water preferential flow 82, 82*f*
Magnesium (Mg) 124
  calcium relationship in streams 356–357, 359*f*
  functions 354
  nutrient loading 586*t*
    deforestation effects 588*t*
    environmental effects 594*t*
  valency 356
Major ions
  water 7
Malaria 406, 677, 682, 684
  definition 685
Maltose 438*f*
Maltotriose 438*f*
Man *see* Human(s)
Management
  aquatic plants *see* Aquatic plant(s)
  goals, ecosystem bioassessment 579
  reservoirs *see* Reservoir(s)
  wetlands *see* Wetland(s)
Manganese (Mn) 102, 351, **368–377**, 380, 387
  algal requirements 380
  chemistry 368, 380*f*
    chemical forms 368, 369*t*
    redox reactions 368, 369*f*, 369*t*
    solubility 368, 371
  clays 368
  environmental roles 368
  hydrothermal waters 376, 376*f*, 376*t*
  lakes 371, 372*f*
    cycling 371, 372*f*, 373
    inputs 371
    losses 371
  oxic layers 371
  rivers/streams 369
    cycling 373, 374
    larger streams/rivers 373
    smaller streams/rivers 370
  toxicity 380
Manganese oxides
  adsorption-mediated chemical interactions 369
  precipitation 371, 374*f*
  trace metals and 374
Mangroves
  water flow 310
Manning's formula, river flow hydraulics 308
Manning–Strickler formula, river currents 240
Mannosamine, dissolved organic nitrogen (DON) 524
Mass conservation, nonequilibrium transport of suspended solids 132–133
Mass movement, sediment movement to streams 138, 139*f*
Mass transport 337
  calculation 337
  definition 337
  dissolved substances 338, 339*t*
  particulate matter (solids) 338
  in river/streams, biogeochemical zones 336*f*
Mayflies *see* Ephemeroptera (mayflies)
Meadows, eddies 309–310
Mean annual air temperature (MAAT), springs 156–157, 158*t*
Mekong River *see* Lancangjiang (Mekong) River
Mer-A operon 655
Mer-B operon 655
Mercury (Hg)
  annual budgets 652, 653*t*
  atmospheric processes and 656
  biogeochemistry 388
  controls of methylation 390
  cycling 390*f*, 655, 655*f*, 674, 674*f*
  emissions control 390
    emissions sources 390–391
  methylation 654
    Hg(II) 655
    rate 654
    see also Methylmercury
  monomethyl mercury (MeHg) *see* Monomethyl mercury (MeHg)
  need for monitoring programs 391
  poisoning 573
  pollution *see* Mercury pollution
  transport 389, 390*f*
Mercury methylation rate (MMR) 654
Mercury pollution 674
  anthropogenic emissions 656
  biota 650
  concentrations 648, 649*t*, 650*f*
  distribution 648
  hypolimnetic enrichment 648, 650*f*
  lake waters 648
  mass balances 652
  oceanic emissions 656
  remediation strategies 656
  remote fresh waters **648–657**
  seasonality 649, 650*f*
  sediments 648, 649*t*
  sources 656, 674
  see also Methylmercury
Merian's formula 288–289
Meromictic lakes
  in density stratification and stability 196–197, 199, 201
Metabolism
  bacteria 337
  lake ecosystems *see* Lake ecosystem(s)
  stream ecosystems *see* Stream ecosystems
Metal(s)
  complexation, natural organic matter 513
  heavy *see* Heavy metals
  photochemical reactions 40
  see also Heavy metals
Metalimnion 198, 198*f*
  stratified lakes, redox potential 550
  turbulent velocities 190–191
  vertical seiche modes 262–263
  in water-column stratification 256
Metallogenium bacteria 382
  copper not taken up 381
  Lake Washington 380
  manganese source 380
  zinc preciptiation 381
Metapopulation, definition 174

Metazoans
  essential fatty acids 474
Meteoric water, definition 187
Meteorological conditions, in gas exchange at air–water interface 32
Methane **480–502**
  carbon cycling 490*t*, 501
  concentrations 485
  emission *see* Methane emissions
  food-webs 489
  formation *see* Methanogenesis
  future work 501
  as greenhouse gas 480
  oxidation *see* Methane oxidation
  physical properties 480
Methane emissions 489
  climate change 501
  diffusive flux 486*t*, 488*t*, 489, 494*t*
  ebullition 489, 493*f*, 494*t*
    amounts 493–498, 496*t*, 498*f*
  flux pathways 489, 493*f*
    small scales 498
  flux types and regulation 493
  from hypolimnia 489, 496*t*, 498*f*
  lakes 496*t*, 498–501
  rivers 498–501, 501*t*
  rooted emergent plants 489, 494*t*
  wetlands 498, 499*t*, 501
Methane monooxygenase, aerobic methane oxidation 483
Methane oxidation 480, 483
  aerobic 483
    environmental factors 482*t*, 483, 484*t*
    methane monooxygenase 483
    methane-oxidizing bacteria (MOB) 483
    spatial distribution 483–485, 484*f*
  anaerobic 483
    spatial distribution 485
  aquatic environments 485
  rates 485, 486*t*
    mixing dependence 484*f*, 485, 488*t*
    sediment slurry studies 485
  spatial distribution 483
Methane-oxidizing bacteria (MOB), aerobic methane oxidation 483
Methanogenesis 480
  biochemistry 480
    acetotrophic 480
    anaerobic conditions 480
    hydrogenotrophic 480
  environmental factors 481
    alternative electron acceptors 480–481, 482*t*
    oxygen 481–482
    pH 481–482
    sulfate 481–482
    temperature 481–482
  microbiology 480
    anoxic organic material degradation 480–481, 481*f*, 481*t*
    archaebacteria 480
  rates 485, 486*t*
    sediment slurry studies 485
  spatial distribution 482
    anoxic water columns 482–483
    sediments 482–483
  wetlands, redox potential 555*f*, 556
Methylamines, dissolved organic nitrogen (DON) 525
Methylmercury
  consumption 651, 652*f*
    detoxification mechanisms 651
  cycling 656
  demethylation 655
  pollution and 674
    bioaccumulation 650
    concentrations 648, 649*t*, 650*f*
    food chain transport 650, 651*f*
  seasonal budgets 653, 654*t*
  see also Mercury (Hg); Mercury pollution
Metrics, ecosystem bioassessment 580
Microbes
  spring biota 156
  see also Bacteria
Microbial activity
  colloid formation 467*f*, 468
  groundwater chemistry 109
Microbial communities
  salt effects 364
Microbial processes, lakes, climate change 616
Microcystin-LR
  phosphorus and 623*f*
  structure 431*f*
Microcystis
  turbulence effects 190–191
Microinvertebrates, definition 174
Micronutrient elements **378–382**
  concentrations 378, 379*t*
  deficiencies 378
Migration
  fire-induced 636
Migratory behavior, lakes, climate change 615
Miles-Howard criterion, Kelvin-Helmholtz billows 265–266
Mine lakes, in density stratification and stability 200
Mineralization
  by bacterial decomposers 339–340
Mineral silicates, definition 417
Mineral weathering, cations 354–355, 355*t*
Mining
  acid drainage *see* Acid mine drainage (AMD)
  pollution and 668
Mirror Lake (New Hampshire, USA), nitrogen fixation 397
Mississippi basin (USA)
  water/sediment delivery 144*t*
Mississippi River (USA) 123*t*, 126*t*, 128
  basin *see* Mississippi basin
Mixing (and small-scale turbulence) **222–229**
  basin scale 222
    turbulent transport 225
  dependence, methane oxidation 484*f*, 485, 488*t*
  depth, in density stratification and stability 197
  effects 223*f*
  hydrodynamic modeling 331
  rates 228
  small scale 222
  temperature effects 222, 223*f*, 225–226
  water density 10–11
  water movement/energy 145
Mixing regimes, lakes, climate change 613
Model(s), water structure 3
Modeling
  hydrodynamic *see* Hydrodynamic modeling
  inverse, groundwater chemistry 116
  nonhydrostatic 324
  transient storage zone *see* Transient storage zone modeling
Moisture dynamics, vadose water *see* Vadose water, hydrologic cycle
Moisture state, vadose water *see* Vadose water, hydrologic cycle
Molar charge units, definition 360
Molecular diffusion, solute transport, benthic boundary layer 231*f*, 234–235, 235*f*
Mollusca
  spring biota 157
Mollusks *see* Mollusca
Molybdate 379
  nitrogen fixation 395
Molybdenum 379, 379*t*
  availability 379–380
  cyanobacteria use 379
  nitrogen fixation 379, 395, 398
  solubility 379–380
Monimolimnion
  in density stratification and stability 196–197, 199, 200*f*
Monin Obukhov length 211
Mono Lake (California, USA) 421
Monomethyl mercury (MeHg) 388–389
  accumulation in food webs 390
  demethylation and mercury reduction 390
    processes 390
  microbial process 389
  production 389
  solving the problem 390
Monomictic lakes
  in density stratification and stability 196–197
Monosaccharides 437, 438*f*
  composition 443*f*
  free 439
Mono-unsaturated fatty acids (MUFAs) 472–474
  definition 478

Monsoon
flooding and 645
Morainal deposits, definition 187
Moraines, outburst floods and 645
Morgan Creek (Maryland) study, groundwater chemistry 111, 112*f*
Mortality
human, diseases of freshwater habitats 685
Mosquitoes (Culicidae) 677
control strategies 682
definition 685
habitats 680
vectors of human diseases 677–678, 678*t*
Mosses *see* Bryophytes (mosses)
Mountains/mountain ranges
wetlands hydrology 177–178, 178*f*
Mouth
placement, fish *see* Fish
Mucopolysaccharides 442
Multifractal objects, definition 66
Multixenobiotic resistance transporter (MXR) 460
definition 463
Murray-Darling river system (Australia) 125*t*, 126*t*, 421–422
Mycobacterium ulcerans 684
Myriophyllum (water milfoil) 302*t*
Myriophyllum spicatum (Eurasian watermilfoil) 606

N

Naegleria fowleri 571
National Oceanic and Atmospheric Administration (NOAA) 62
National parks
Yellowstone National Park (USA) 172
National River Health Program (Australia) 582–583
Natranaerobius spp. 420
Natural organic matter (NOM) **503–516**
acidic functional groups 512, 513*t*
carboxylic acids 512, 513, 515
measurement 512
metal complexation 513
phenolic content 512, 513
concentrations 503–504
definition 503
elemental composition 508, 510*t*
biomass triangle 509, 510*f*
chemical changes 509
organic carbon oxidation state *see* Carbon
redfield biomass *see* Redfield biomass
unsaturation 511, 512*f*, 515
fluxes 503, 504*t*
integrated statistical description 514, 514*t*
alkoxy carbon 515
aromatic carbon 515
carbonyl bonds 515
isolation 505
couple reverse osmosis/electrodialysis 507
membranes 506
solid phase extraction 505
measurement 503
molecular weights 507
colligative methods 508, 509*t*
non-colligative methods 508
^{13}C NMR spectroscopy 513, 514*t*
Navier–Stokes equations 73–74
water vapor flux 59
Navigation-induced river currents 244, 245*f*
Negro River (South America)
photochemical mineralization 45, 46*t*
Nematoda (and nematodes)
human diseases 677–678, 679–680
Nepheloid layer, definition 238
NETPATH, groundwater chemistry 116
Neural tissues, docosahexaenoic acid 474
Neutron logs, soil-water storage 56
Neutron scattering, vadose water 76
Newtonian fluids 17
Newton's Laws, currents 287
New Zealand
springs 169*t*
NEXRAD system, precipitation observations 62
Niger River (Africa)
water/sediment delivery 144*t*
Nile River (Africa) 125*t*, 127
basin, water/sediment delivery 144*t*
Nitellopsis obtusa, humic substances, effects 459–460
Nitrate (NO_3^-) 400
concentrations 126*t*, 127
contaminant 104–105
drinking water 405–406
effect on alkalinity 350
introduction, lake restoration 553
sediment–water interface redox potential 551–552, 552*f*
Nitrate reductase
definition 382
Nitric acid leaching, precipitation pH 591
Nitrification 400–401, 401*t*
definition 406
deforestation effects
dissolved nutrients 592
nutrient load effects 592, 599–600
nitrogen cycle *see* Nitrification
precipitation pH 588*t*, 591
Nitrogen (N) **400–407**, 665
^{15}N-labeled uptake assays, dissolved organic nitrogen uptake 534–535
biologically available 400
cellular stores 538
concentrations 126*t*, 127
contaminant 104–105
cycle *see* Nitrogen cycle
dissolved inorganic *see* Dissolved inorganic nitrogen (DIN)
dissolved organic *see* Dissolved organic nitrogen (DON)
dissolved organic matter and 469
fixation *see* Nitrogen fixation
food webs 400
forms 400
groundwater chemistry 110–111
health risks 406
impact of hydrologic exchange 100, 101*f*
inorganic 400
limitation of net primary production 403
molybdenum and 379
nutrient loading 586*t*
deforestation effects 588*t*
environmental effects 594*t*
organic *see* Organic nitrogen
phosphorus and 403, 409, 409*f*
photochemistry 38*f*, 40
stores, fire impacts 634
total dissolved (TDN), definition 517
transformation 400
Nitrogenase
definition 398
inhibition 393
synthesis 395–396
Nitrogen cycle
denitrification *see* Denitrification
ecosystem scale 402
uptake rates 402
human acceleration of 411
human impact 404
nitrification *see* Nitrification
Nitrogen fixation **392–399**, 401*t*
bacterial 400
energetic cost 393
heterotrophic 393, 397–398
benthic *see* Benthic organisms (benthos)
biological, in aquatic systems 392
rates 392–393
cyanobacterial *see* Cyanobacteria
definition 398, 406
freshwater lakes 393
heterocysts 394, 398
saline lakes 396, 412
Nitrogen oxides, emissions 563
Nitrous oxide (N_2O) 406
^{13}C NMR spectroscopy, natural organic matter 513, 514*t*
Noble gases, groundwater chemistry 110–111
Nonaqueous phase liquids (NAPLs), groundwater chemistry 113, 115*f*
Nondiffusive fluxes, solute transport in benthic boundary layer *see* Benthic boundary layer (BBL), solute transport

Nonequilibrium transport, suspended solids *see* Suspended solids, fluvial transport
Nonhydrostatic modeling 324
Nonlinear internal waves (NLIWs)
in internal wave degeneration 267, 267*f*, 275*f*
shoaling of 268, 269*f*, 273*f*, 274*f*
Nonpoint source (NPS) pollution 658, 660
definition 576
groundwater chemistry 111–112, 113*f*
restoration 660
Nonsuspended particles *see* Bedload
Nontidal wetlands 306
hydrodynamics 306
vegetation
dispersion control 306–307, 307*f*
functions 306
management 306
spatial distribution 306, 307*f*
Nonyl phenol(s), pollution and 672–673
Nonyl phenol ethoxylates, pollution and 672–673
North America
lakes *see* North American lakes
springs 169*t*
see also Canada; United States of America (USA)
North American lakes
Great Lakes region
effects of earth's rotation 281, 282*f*
see also Laurentian Great Lakes (USA)
North Atlantic, South American rivers draining *see* South America
Norwegian lakes
acidification 564
Seston carbon:phosphorus ratio 547*f*
Nuclear magnetic resonance (NMR) spectroscopy, natural organic matter 513, 514*t*
Nutrient(s) 665
agriculture and *see* Agriculture
concentrations 126*t*
dependence, deforestation effects on dissolved nutrients 594, 594*t*
eutrophication *see* Eutrophication
in flood water 646–647
fluxes 119, 128
microscale 193
modeling 320
physical influences 193
impact of hydrologic exchange 100, 102
limitation
definition 398, 406
wetland plants *see* Wetland plants
loading *see* Nutrient loading
recycling, zooplankton *see* Zooplankton, in ecosystems
reduction 666
relationship to light 602–603
release, wetlands, redox potential 555
retention, wetlands
redox potential 555
rivers 149
sources 193
species composition 665
spiraling *see* Nutrient spiraling
supply ratios 665
surface mixed layer and 207
water *see* Water
wetlands
redox potential 555
Nutrient loading 585
dissolved gases 7
dissolved nutrients 585
annual variations 585
models 620
external loading 620
internal loading 617–619
particulate nutrients 585
sampling 585
Nutrient spiraling
definition 145–146, 154, 342
in streams 341
Nutrient stoichiometry
algae *see* Algae, nutrient stoichiometry
aquatic ecosystems **537–548**
bacteria 541
carbon:nitrogen:phosphorus 540*f*
community dynamics 546
Droop model 539*f*
fish *see* Fish
food quality 542
growth rates and (Growth Rate Hypothesis) 539–540, 541*f*
Light:Nutrient hypothesis 547
organism level 538, 538*t*
periphyton 543, 545*f*
primary producers 538
protozoa 541
recycling 543
Redfield ratio 546, 547*f*
snails 538*t*, 543, 545*f*
whole-system scale 546
zooplankton 538*t*, 539, 542–543, 545*f*
Nymphaea (water lily) 302*t*

O

Oak Ridge, radionuclides 676
Oceans/seas
dissolved inorganic carbon concentrations 343
evaporation 59
hydrological cycles 52*t*
Oil leaks 668–669
Oligomictic lakes, in density stratification and stability 196–197
Oligosaccharides 437
Oligotrophication 411, 411*f*
definition 406, 413
nitrate/ammonium uptake 402
see also Eutrophication
Omidov scale 228
Onchocerca volvulus 678, 682*f*
Onchocerciasis (river blindness) 678, 682–683, 682*f*, 685
Onchocerciasis Control Programme 683
ecosystem monitoring 683
ivermectin use 683
success 683
vector relation to biodiversity 684
Onondaga Lake (New York/USA)
mercury pollution 650*f*, 653*t*
Ore mining, acid drainage 562*f*
Organic acid(s)
release, precipitation pH 591
Organic carbon
dissolved *see* Dissolved organic carbon (DOC)
dissolved gases varying with 7
particulate *see* Particulate organic carbon (POC)
total *see* Total organic carbon (TOC)
Organic compounds
groundwater chemistry 108*f*, 113–114
Organic matter 464
autochthonous *see* Autochthonous organic matter
bacteria *see* Bacteria, organic matter
chromophoric *see* Chromophoric dissolved organic matter (CDOM)
colloidal 464, 466
dissolved *see* Dissolved organic matter (DOM)
natural *see* Natural organic matter (NOM)
oxidation patterns 455
particulate *see* Particulate organic matter (POM)
photochemical reactions 40, 40*f*, 47
bacteria 45–47
biogeochemical contributions 47
food web responses 45
reactivity 41
regulators 42
production patterns 455
rivers/streams, concentration in 338
size distribution 465*f*
see also Natural organic matter (NOM)
Organic nitrogen **517–536**
definitions 517
dissolved *see* Dissolved organic nitrogen (DON)
particulate *see* Particulate organic nitrogen (PON)
Organic pollution 668
Organic solvents
classification 668
pollution and 668
Organism uptake, ecosystem bioassessment 578–579
Organophosphate insecticides 672

Orinoco River (South America)
basin, water/sediment delivery 144*t*
Orthophosphate *see* Phosphate (PO43-)
Oryza sativa (rice) 414–415
Oscillatory boundary layers, benthic boundary layer (BBL) 233
Osmotic pressure, water 15
Ostracods 419
Ostwald solubility coefficient 30–31
Outburst floods 645
Outgoing/ebb tides, wetlands hydrology 185
Overlying surface water, vadose water 86, 86*f*
Overturning, in density stratification and stability 196, 203–204
Overwintering phytoplankton, lakes, climate change 614–615
Oxic environments, definition 399, 406
Oxic layers
carbon dioxide concentrations 345
iron concentrations 371
manganese concentrations 371
Oxic processes
definition 342
in river/stream ecosystems 339, 339*t*
mineralization (by bacterial decomposers) 339–340
photosynthesis 339
respiration 336*f*, 339
Oxidation
effect on alkalinity 351
methane *see* Methane oxidation
patterns, organic matter 455
Oxidation/reduction reactions *see* Redox reactions
Oxygen (O)
concentration
redox potential measurement 549
demand 662
factors affecting 664
dissolved *see* Dissolved oxygen
gradients, stratified lakes, redox potential 550, 550*t*
inputs, river redox potential 554
iron oxidation 368
low levels
restoration techniques 664
soil, wetland plants *see* Wetland plants
manganese oxidation 368
methanogenesis 481–482
reactive species, formation 37
see also Reactive oxygen species (ROS)
solute transport, benthic boundary layer 235–236, 235*f*
spring classification 158*t*
status in flow systems 95, 96*t*
turbulent mixing and 224*f*, 225
Oxygenation, lake restoration 553
Oxygen-dependent organisms, river/stream ecosystems 335–336
Oxygen injectors 664–665
cost 665
failure 665
Ozmidov scale, shoaling of nonlinear internal waves 268–269

P

Pacific basins (South America) *see* South America
Pallette Lake (USA), mercury pollution 653*t*
Paludified landscape
definition 187
wetlands hydrology 177–178
Pan evaporation 70–71
relation to solar dimming 71
Parallel flow, stratified benthic boundary layers 234, 234*f*
Parasitic diseases
amphibians 406
Parker River (USA), photochemical reactivity 41–42
Partial pressure
carbon dioxide *see* Carbon dioxide (CO2)
definition 347, 456
Particle dynamics, benthic boundary layer (BBL) *see* Benthic boundary layer (BBL)
Particle electrical charges, turbidity *see* Turbidity
Particle-size distribution, vadose water retention 78
Particulate matter
mass transport 338
organic *see* Particulate organic matter (POM)
rivers 148
see also Sediment(s)
Particulate nutrients 585
deforestation effects 595
erosion 595
precipitation 595
roading effects 596, 598*f*
stream-channel characteristics 596
vegetation proximity to fresh water 596
watershed mass wasting 595
watershed topography 595
Particulate organic carbon (POC)
discharge-weighted average 505
Particulate organic matter (POM) 464
Particulate organic nitrogen (PON) 400
definition 517
Partner finding, allelochemicals 429
Passive imaging radiometers, precipitation observations 63–64
Passive scalar, definition 66
Pathogens in aquatic ecosystems 570*t*, 576
Adenovirus 569
Campylobacter spp. 569
coliforms 574–575
Cryptosporidium 569
Escherichia coli O157:H7 569, 576
Giardia 569
Leptospira 569
Norovirus 569
Schistosome dermatitis (swimmer's itch) and 570
Shigella spp. 569
in springs 173
Patuha volcano, acidification 564
Péclet numbers *(Pe)* 194–195
Pedogenic oxides, acidified water buffering 559
Pelagic biota
evolutionary ecology 189
invertebrates, rivers 151
Pelagic communities, lakes, climate change 614–615
Pelagic habitats, turbulence 192
Pelagic zone
definition 310–311
lake ecosystems *see* Lake ecosystem(s)
Pentose sugars 438*f*
Perching, vadose water 86
Periphyton (attached microalgae) 192, 604
definition 628
elevated nutrients, effects 606
nutrient stoichiometry 543, 545*f*
Permafrost
wetlands hydrology 177–178
Permethrin 671*f*
Personal care products, pollution and 672–673
Pesticides
pollution and 671, 671*f*
Petroleum oils
classification 668–669
pollution and 668
Petryk–Bosmajian equation, river flow resistance 308
pH 348, 351*see* Acid, acidity
aluminum and 357
colloid formation 468
effect on ecosystems 351, 352
methanogenesis 481–482
natural organic matter, acidic functional groups 512, 513
precipitation, deforestation effects *see* Dissolved nutrients, deforestation effects
redox potential measurement 549
water 343, 345, 346*f*
see also Acid, acidity
Pharmaceuticals
groundwater chemistry 112
pollution and 672–673
Phenotypic plasticity
fish kairomones 428
Pheromones, algae 435
Phosphate (PO_4^{3-}) 126*t*
Phosphatidylcholine, structure 473*f*
Phosphatidylethanolamine, structure 473*f*
Phosphatidylinositol, structure 473*f*
Phosphatidylserine, structure 473*f*
Phosphides 408
Phosphine 408

Phosphine (*continued*)
  definition 413
Phospholipids (PLs)
  definition 478
  freshwater fish 475
  functional roles 472–474
  structure 472, 473*f*
Phosphorus (P) 665
  cellular stores 538
  chemical properties 408
    oxidation states 408
  concentrations 126*t*, 127
  as contaminant 127–128
  cross-species variability 539–540
  cycle *see* Phosphorus cycle
  diffusion 194
  eutrophication 601
  fish 541–542
  historical aspects 408
  industrial sources 408
  inorganic
    levels 128
  iron oxide and 369, 372–373
  microcystin-LR and 623*f*
  nitrogen and 403, 409, 409*f*
  as nutrient 8
  nutrient loading 586*t*
    deforestation effects 588*t*, 593
    environmental effects 594*t*
  organic, levels 128
  oxygen concentrations and 412
  particulate forms 410–411, 412
  photochemistry 40
  phytoplankton productivity 455
  pollution 127–128
  production limitation 409
    decomposition effects 410
  reserves
    fire impacts 634
  retention 411
  roles 408
    in eutrophication 601
  water quality relationship 102
Phosphorus-binding, lake restoration 553
Phosphorus cycle **408–413**, 409*f*
  global regional transfer 410
  human impacts 410, 411
  hydrological export 411–412
  inland waters 411
  landscape transfer 410
  primary production 409
  research directions 413
  sediment burial 404, 411–412
  sewage 411
Photochemistry 37
  apparent quantum yield *see* Apparent quantum yield (AQY)
  detritus 41
  heterogeneous 40
  nitrogen 38*f*, 40
  organic carbon 37
  phosphorus 40
  solar radiation spectrum 45
Photolytic reactivity **37–49**
  apparent quantum yield *see* Apparent quantum yield (AQY)
  dioxygen 37
  rate predications 43, 45, 46*t*
  see also Photochemistry
Photon(s)
  energy delivery 37
  see also Light
Photophysics, primary 37
  reaction pathways 37
Photosynthesis
  dissolved gases 6–7
    carbon dioxide in lakes and 345
  manganese role 380
  oxic 339
  redox reactions 550
  see also Primary production
Photosynthetically active radiation (PAR)
  lakes, climate change 612–613
Photosynthetic communities, lakes, climate change *see* Climate change and lakes
Photosynthetic organisms
  thermal springs 166, 168*f*
  turbulence effects 190
  wind forcing and 192
Photosynthetic pigments
  photolysis 41
Phototrophs
  humic substance responses 458
PHREEQC, groundwater chemistry 116
Phthalates, pollution and 673*f*
Physical–biological interactions *see* Biological–physical interactions
Physical–chemical reactions, water *see* Water, physical–chemical reactions
Physical properties, water **10–16**
Physics, gas exchange at air–water interface *see* Gas exchange at air–water interface
Physoclists, buoyancy 26
Physostomes, buoyancy 26
Phytoplankton
  arsenic reduction 388
  biomass 621
    depression 621
  carbohydrates 442, 445
    release 442
  carbon turnover 454*t*
  Castle Lake (CA, USA) 381
  cations effect on 420
  community composition
    lakes, climate change 615
  definition 413, 628
  dissolved combined amino acid sources 525
  dissolved free amino acid sources 525
  dissolved organic nitrogen 525
  fire impacts 636–641
  marine, fatty acid composition 475
  methylmercury accumulation 650
  mixotrophic, grazing 47
  nitrogen cycle and 402
  nutrition
    nutrients 193
    phosphorus
      requirements 409, 410
  phosphorus delivered to sediments from 412
  production, phosphorus relationship 455, 617
  productivity *see* Phytoplankton productivity
  in rivers 150–151
  salt tolerance 419
  succession
    allelochemicals 435
  suitability to life in flowing waters 335
  turbidity effects *see* Turbidity
  turbulence and 192
Phytoplankton productivity
  effects on diversity 603
  see also Photosynthesis
Piezometers 96, 98*f*, 99
Piezometric head, dynamic pressure 23
Pipes/macropores, streams 136–137
Piston velocity 344–345
  definition 347
Pit-gauges 61
Pit lakes 558*f*
  neutralization 560, 560*f*
  pH values 559, 559*f*
Plains, coastal *see* Coastal plain(s)
Planetary Rossby waves 278, 283
Plankton
  definition 399
  nonmotile
    buoyancy 193
    maintenance 192
  sources to rivers 152*f*
  see also Phytoplankton; Zooplankton
Plant(s)
  aquatic *see* Aquatic plant(s)
  carbohydrate release 442
  emergent *see* Emergent plants
  hydrodynamics associated 301
  litter *see* Plant litter
  multiple stress resistance 461–462

nutrient cycling 54
silica cycling 414
transpiration 69
vascular
eutrophication and 623
see also Macrophyte(s)
vascular, nitrogen fixation 393
wetlands *see* Wetland plants
see also Vegetation
Plant–animal interactions, allelochemicals *see* Allelochemicals/allelochemical reactions
Plant litter
decomposition
deforestation, dissolved nutrients 593, 593*f*
Plant–plant interactions, allelochemicals *see* Allelochemicals/allelochemical reactions
Plasmodium spp. 677, 681*f*
'Platonic solids', water structure 3
Plume formation, groundwater chemistry 112–113, 114*f*
Poincaré waves 279, 279*f*, 280
Point sediment samples, equilibrium transport of suspended solid 131, 132*f*
Point source (PS) pollution 658, 659
definition 576
prevention 659
Polar regions
lakes, climate change 611
Pollutant(s)
agricultural 104–105, 127
bioaccumulative 390
copper (Cu) 573
dichlorodiphenyltrichloroethane (DDT) 574, 576
ecosystem bioassessment 577
fluxes, modeling 320
heavy metals 573, 605
lead (Pb) 573
mercury (Hg) 573
polychlorinated biphenyls (PCBs) 574, 576
rivers 150
surface mixed layer and 207
trace metals and metalloids 384, 388–389
urban 105
zinc (Zn) 573
see also Pollution
Pollution **658–667**, 659*t*, **668–676**
acidification *see* Acid drainage; Acidification; Acidified waters
anthropogenic 668
natural *vs.* 658
definition 576, 658, 659*t*
emerging concerns 672, 673*f*
fecal 571
animal feeding operations 572
animals 571–572
humans 572
septic tank systems 572, 576
sewer overflows 573
wastewater treatment plants 573
developing countries 573
human society evolution and 658
mercury *see* Mercury pollution
nitrogen 403
organic *see* Organic pollution
sources 658, 660, 661*t*
nonpoint *see* Nonpoint source (NPS) pollution
point *see* Point source (PS) pollution
thermal 676
trace metals and metalloids
atmospheric sources 384
nonpoint sources 384
point sources 384
see also Pollutant(s)
Polybrominated diphenyl ethers (PBDEs), pollution and 671, 671*f*
biological hazards 671
Polychlorinated biphenyls (PCBs)
persistence 670
pollution and 670
properties 670
structure 670, 670*f*
Polycyclic aromatic hydrocarbons (PAHs)
pollution and 669
sources 670
structure 669–670, 669*f*
Polymer compound measurement, dissolved organic nitrogen uptake 535
Polymictic lakes
in density stratification and stability 196–197
surface mixed layer 221
Polysaccharides 437
utilization 447
Polyunsaturated fatty acids (PUFAs)
definition 478
Ponding, definition 187
Pondweed *(Potamogeton)* 302*t*
Pore water, definition 238
Porosity
definition 187
nonequilibrium transport of suspended solids 132–133
solute transport, benthic boundary layer 235–236
Porous medium, vadose water 76, 77*f*
Positive feedback mechanisms, definition 413
Potamogeton (pondweed) 302*t*, 603
Potassium (K) 124
functions 354
nutrient loading 586*t*
deforestation effects 588*t*
environmental effects 594*t*
valency 356
Potassium dichromate, chemical oxygen demand tests 663–664
Potential energy
available (APE), internal stratification 258
convective available (CAPE), precipitation modeling 65–66
stratification, density 222
water flow 239–240
Potomac River (USA), pollution 658
Practical salinity units, in density stratification and stability 202
Prairie streams, salinity, fluctuation effects 364–365
Precipitation **58–67**, 100*f*, 101–102, 104, 105
changes, basin integrity 611–612
deforestation effects, particulate nutrients 595
effects, on stream flow characteristics 139–140
formation 59, 60
cloud 59, 60
groundwater recharge 90
measurement 58
scale-related problems 58
modeling 64
cascade models 64–65, 65*f*
difficulties 66
direct numerical models 64
global circulation 64
scale-invariant statistical models 64, 65*f*
observations 61
radar 61
rain gauges *see* Rain gauges
satellite remote sensing *see* Satellite remote sensing
pH, deforestation effects on dissolved nutrients *see* Dissolved nutrients, deforestation effects
processes associated 58
seasonal and nonseasonal influence on rivers/streams 336*f*, 337
seasonal patterns 145*f*
space–time distribution 58
wetlands hydrology 183–185
Predator–prey interactions
allelochemicals 427
Predictive ecology, ecosystem bioassessment 581
Predictive models, ecosystem bioassessment 580–581
Preferential flow, vadose water *see* Vadose water
Preferential flow zones, stream drainage network organization 134
Pressure **17–27**
'absolute' 18–19
atmospheric, changes 18–19
body forces 17
dynamic *see* Dynamic pressure
equations of motion 24
hydrostatic *see* Hydrostatic pressure
osmotic, water 15
reference surface gradients 25
surface forces 17
water vapor 15
Primary producers
benthic 409–410
changes, eutrophication *see* Eutrophication
eutrophication *see* Eutrophication
green superfamily 539

Primary producers (*continued*)
  littoral zone *see* Littoral zone, primary producers
  nutrients 665
    stoichiometry 538
  red superfamily 539
  rivers 150
Primary production 503
  benthic environments 409–410
  definition 406
  deforestation effects on dissolved nutrients 594
  lake ecosystems *see* Lake ecosystem(s)
  nutrients
    limitation 403
    stoichiometry 538
  pollution and 660
  reservoirs *see* Reservoir(s)
  see also Photosynthesis
Primary productivity
  surface mixed layer and 207
Problem characterization, ecosystem bioassessment 581
Prokaryotes, definition 399
Property-transfer models
  vadose water retention 79
  vadose water unsaturated hydraulic conductivity 81
Prostaglandins, fatty acids and 474
Protease inhibitors, cyanobacteria 429–431
Protistans, planktic, turbulence effects 191
Protists
  dissolved organic nitrogen (DON) 526
Proton hopping, water reactivity 5*f*, 6
Proton production, precipitation pH 591
Protozoa
  biofilm component 335–336
  essential fatty acids 474
  nutrient stoichiometry 541
Proximate flow paths 95, 99, 100–101, 104–105
Pseudokirchneriella subcapitata, humic substances, responses 458
Pumping effects, groundwater budgets *see* Groundwater budgets
Pycnocline
  definition 222
  in density stratification and stability 200
Pyramid Lake (Nevada, USA) 421
  nitrogen fixation 396
Pyrethrins/pyrethroids, pollution and 672
Pyrite 375–376, 375*f*
  oxidation 376
  weathering 560
Pyrolysis, definition 643

Q

Quantification
  density stratification *see* Density stratification
  soil moisture dynamic, vadose water retention 79
Quasi-geostrophic waves *see* Vorticity waves

R

Radar, precipitation observations 61, 63*f*
  dual-polarization 61–62
  Z–R relation 61–62
Radial drainage, streams 134
Radiation 675–676
  long wave, in gas exchange at air–water interface 32
  photosynthetically active *see* Photosynthetically active radiation (PAR)
  solar *see* Solar radiation
  UV *see* Ultraviolet (UV) radiation
Radiative cooling, surface mixed layer and 207, 208*f*
Radiative heating, surface mixed layer and 207, 208*f*
Radionuclides, pollution and 675
  accidents 676
Raffinose 438*f*
Rain
  acid *see* Acid rain
  continental, chloride sources 361–362
  see also Precipitation; Rainfall
Rainfall
  in flash flooding 645
Rain gauges 58, 61, 62*f*
  error sources 61
  limitations 61
Ranunculus (water crowfoot) 302*t*
Raupach–Thom model, water flow modification 314–315, 315*f*
Reach (segment) scale factors, algae in river ecosystems *see* Algae, river ecosystems
Reaction–diffusion models, solute transport, benthic boundary layer 235–236
Reactions, groundwater chemistry 114–115
Reactive oxygen species (ROS)
  membrane interactions 459–460
  production 37
Reactivity, water *see* Water
Reaeration, in rivers 149–150
Reaeration rate, gas exchange in lotic systems 35
Receiving stream quality, groundwater quality 92
Recharge, ground water *see* Ground Water
Recharge balance, groundwater budgets 91, 91*f*
Recovery monitoring, ecosystem bioassessment 581
Recreational waters, guidelines for, human health and 574*t*, 576
Rectangular drainage, streams 134
Rectilinear motion 285
Redfield biomass 509
  atomic ratios 509
  water removal effects 510
Redfield ratio 403
  definition 456
  nitrogen fixation 395–396
Red mangrove *(Rhizophora mangle)* 302*t*
Redox gradient 549
  redox reactions 550, 550*t*
Redox potential **549–556**
  definitions 549
  gradients 549
  lake restoration 553
    iron minerals 553–554
    nitrate introduction 553
    oxygenation 553
    phosphorus-binding 553
  lakes 550
  measurement 549
    electrode types 549
    oxygen concentration *vs.* 549
    pH *vs.* 549
    sediments 549
    units 549
    variability 549
  nutrient cycling 552
    ferrous pyrite 552–553
    ferrous sulfide 552–553
    ferrous *vs.* ferric iron 552–553
    sulfate-reducing bacteria 553
  rivers 554
    hyporheic zone 554
    oxygen inputs 554
    riparian zone effects 554
    seasonal changes 554, 554*f*
    temperature 554
  sediment–water interface 551, 551*f*
    nitrate 551–552, 552*f*
    seasonal fluctuations 552, 553*f*
    submerged macrophytes 552, 552*f*
    zonation formation 551, 552*t*
  stratified lakes 550
    bacterial decomposition 550–551
    eutrophic lakes 550, 551*f*
    metalimnion 550
    oxygen gradients 550, 550*t*
    thermal stratification 550
    water circulation 551
  wetlands 554
    aerobic *vs.* anaerobic metabolism 554–555
    establishment 555–556
    methane production 555*f*, 556
    nutrient retention/release 555
    water level fluctuations 554–555, 555*f*
    water table predictor 556
Redox reactions 550
  definition 377, 413
  Fe^{2+}/Fe^{3+} system 550
  photosynthesis 550

redox gradient 550, 550*t*
river/stream ecosystems 340, 340*t*
water reactivity 6
Red superfamily, primary producers/production 539
Reducing conditions, wetlands hydrology 180
Reference Condition Approach (RCA), ecosystem bioassessment 580–581
Refractory organic matter *see* Natural organic matter (NOM)
Regulation, water *see* Water regulation
Remediation 660
acidified waters *see* Acidified waters
definition 567
mercury pollution 656
saline lakes 422–422
Representative elementary volume (REV), diffuse unsaturated flow 79, 80*f*
Reservoir(s)
Asia *see* Asia
Australia *see* Australia
construction, effects 416
hydrodynamic modeling *see* Hydrodynamic modeling
mixing 225
New Zealand *see* New Zealand
North America *see* North America
phosphorus transfer 410
South America *see* South America
surface mixed layer *see* Surface mixed layer (SML)
turbulence 225
see also Impoundments
Residual circulation 289
Respiration
aerobic *see* Aerobic respiration
anoxic 336*f*, 340
carbon and 455–456
carbon dioxide and 345
organic carbon oxidation state 511
oxic 336*f*, 339
Restoration, ecosystem bioassessment 581
Resuspension–settling cycles, benthic boundary layer (BBL) 238
Retention curves, vadose water retention 78, 78*f*
Return currents, navigation-induced 245*f*
Revegetation, deforestation effects on dissolved nutrients 597*f*
Reverse osmosis (RO)
natural organic matter isolation 506
limitations 506, 507
Reynolds-Averaged Navier Stokes (RANS) equations, in hydrodynamic modeling 331–332
Reynolds averaging, turbulence 207
Reynold's number *(Re)*
definition 238, 310–311
in gas exchange at air–water interface 35
nontidal wetlands 307
Reynolds stresses, turbulence 207–208
Reynolds Transport Theorem 24
Rhizophora mangle (red mangrove) 302*t*
Rhizosphere
definition 406
Rhône river (France) 102–103
Riaux system, carbohydrates 445
Ribose 437, 438*f*
Rice *(Oryza sativa)*, biogenic silica accumulation 414–415
Richards equation 73–74
Richardson number
in density stratification and stability 203–204, 204*f*
Kelvin-Helmholtz billows 265–266
surface mixed layer dynamics 207
Rills, drainage network organization 134
Rio Agrio River (Argentina), acidification 564
Rio Calaveras (New Mexico, USA) 100*f*
Rio Negro (South America) *see* Negro River (South America)
Rio Odiel River (Spain), acid mine drainage 563
Rio Solimões *see* Amazon River basin (South America)
Rio Tinto River (Spain) 375*f*
acid mine drainage 558*f*, 563
Riparian zones
influence on biogeochemical processes in rivers/streams 337
river redox potential effects 554
vegetation 102
Risk assessments, ecosystem bioassessment 581
River(s) **143–154**
algae 150–151
alluvial deposition 102
anoxic processes *see* Anoxic processes
Asia *see* Asia, rivers
atmospheric exchange of gases 149–150
benthic invertebrates 151
biogeochemical cycling 153
biogeochemical zones 335, 336*f*
backwaters *see* Backwater(s)
channel surface 335–336, 336*f*
chemical composition of water and 341
flowing water of channel 335, 336*f*
hyporheic zone *see* Hyporheic zone
biology 150
carbon budget 153–154
channels
form/morphology 146–147
subsurface exchange zones 147*f*
chemical composition 340*t*, 341
chemical conversions 340, 340*t*
chemical fluxes and dynamics **335–342**
comparison of different-sized flowing waters 338
conservative substances 338, 342
discharge-weighted mean concentration 338
human influence 337
hydrographic change 337
labile constituents 338, 339*t*, 342
longitudinal gradients in 335
longitudinal movement 341
mass transport 336*f*, 337
constricted, *vs.* floodplain rivers and levees 144–145, 146*f*
crustaceans 151
currents *see* River currents
dams 147–148, 148*t*
as threat *see* River(s), threats
definition 143
discharge 143
arid/semi-arid regions 143–144
basins, largest worldwide 144*t*
enhanced, impact on density currents 254–255
seasonal variation 143–144
short-term events affecting 143
dissolved gases 149, 150*f*
CO_2 344*f*
dissolved organic carbon 505
dissolved substances 148, 338, 339*t*
distinguishing characteristics 144*t*
drainage areas
basins, largest worldwide 144*t*
see also Basin(s), river
ecosystems *see* River ecosystems
elements (chemical) in
biogeochemically important 339*t*
concentration dependent on flow 338, 340*t*
rate of transformation/lability 338, 339*t*
endorheic *see* Endorheic rivers/waters
energetics, trophic, research/conceptual models 152–153
estuaries *vs.*, water movement 145
export, material 153–154
fish 152
research knowledge/gaps 152
flooding 144
floodplains 101–102, 422–422
vs. levees and constricted rivers 144–145, 146*f*
see also Floodplain(s)
flow
element changes associated 338, 340*t*
seasonal, conditions 100*f*, 101–102
vegetation effects 307
blockage factor 309
sediment transport 310
spatial heterogeneity 309, 309*f*
turbulent field 309–310
see also Vegetation
see also River currents
fluvial export **118–130**
anthropogenic impacts, TSS and TDS 127
fluxes 118
anthropogenic impacts 118, 127
comparison of fluxes 118
definition 118
discrete sampling 119
durations 120
duration curves 120, 121*f*

River(s) (*continued*)
factors affecting 120
earth system relationship 118
estimation procedures 120
exorheic 123*t*, 125*t*, 127
factors affecting suspended matter 120–121
flow-material relationship 119
global river fluxes 119
major ions and total dissolved solids 124
measurements 118
rating curves 119
stations 119
variations 119
particulate fluxes and TSS levels 120, 122*t*
ranked for major rivers 124
river-borne material 118
river databases 118–119
components 118
sediments 127
total suspended solids (TSS) 118
trajectories 128
anthropogenic impacts 128–129, 128*f*
uncertainties in measuring 120
uses in science 118
variations
geographic 120
interannual 120
river size 120
food webs *see* River ecosystems, food-web dynamics
geomorphology 146
gravel-bed montane 102
human impacts on 153–154
hydrodynamical modeling 332
hydrologic exchange 101–102
hydrologic regimes 145*f*
hydrology 143
arid/semi-arid regions 143–144
large basins 144
hyporheic zones 101–102
inflow, density currents generated by 247, 249, 249*f*
insects 151
invertebrates 151
ionic presence 149
length 127
levees *see* Levees
macrophytes 150–151, 151*f*
mass transport 336*f*, 337
methane emissions 498–501, 501*t*
morphologies, floodplain *vs.* constricted 144–145, 146*f*
North America *see* North America
nutrients 149
oxic processes *see* Oxic processes
particulate matter 148
pelagic invertebrates 151
phytoplankton 150–151
plankton, sources of 152*f*
pollutants 150
primary producers 150
proximate flow paths 101–102
ranking variations 124
reaeration 149–150
redox potential *see* Redox potential
respiration 149–150
seasonal flow conditions 100*f*, 101–102
sediment load/yield
Asia 148
basins, largest worldwide 144*t*
sources of water 143
spatial scales 101–102
temperate *vs.* tropical 145*f*
temporal scales 101–102
threats 422–422
transit time 145–146
transparency, water 148
trophic energetics, research/conceptual models 152–153
water movement 145
estuaries *vs.* 145
water quality 148
water regulation 147, 148*t*
water sources 143
water transparency 148
water yields 143
basins, largest worldwide 144*t*
zooplankton 151
River basins *see* Basin(s), river
River blindness *see* Onchocerciasis (river blindness)
River Continuum Concept (RCC) 151, 152–153
River currents **239–246**
Chezy formula 240
classification 239
at confluences, structure 242, 242*f*
zones of maximum velocity 242–243, 243*f*
controlling factors 239
critical flow 239–240
engineering structures, influence 243
flow rates
effects 239
see also River(s), flow
fluvial channels 240
frictional forces 239
gravitational forces 239
Manning–Strickler formula 240
navigation-induced 244, 245*f*
nomenclature 245
origins 239
pattern, effects 239
secondary currents 241, 241*f*
classification 241–242
dynamical equation of flow 241–242, 242*f*
shear stresses 240
subcritical flow 239–240
supercritical flow 239–240
time–mean streamwise velocity 240*f*
River ecosystems
algae *see* Algae, river ecosystems
biotic/abiotic processes
affecting algae *see* Algae, river ecosystems
affecting dissolved elements 338
dissolved CO_2 344*f*
dissolved organic carbon 505
dissolved substances in 148, 338, 339*t*
flow affecting *see* River(s), flow
food webs *see* River ecosystems, food-web dynamics
iron *see* Iron
lake ecosystems *vs.* 335, 341
manganese *see* Manganese
nitrogen cycle 402, 404–405, 405*f*
oxic processes *see* Oxic processes
particulate organic carbon 505
phosphorus transport 410
precipitation influences 336*f*, 337
riparian zone/watershed influences 337
stationary habitats, biogeochemical influence 335
total organic carbon 504, 504*t*, 505
River ecosystems, food-web dynamics 146–147
research/conceptual models 152
Riverine fluxes *see* River(s), fluxes
Riverine Productivity Model (RPM) 152–153
River Invertebrate Prediction and Classification System (RIVPACS), ecosystem bioassessment 580–581, 582–583
Road de-icing, chloride sources 362, 363*f*, 365
Roading effects, deforestation effects, particulate nutrients 596, 598*f*
Rock fissures/fractures, definition 174
Rooted emergent plants, methane emissions 489, 494*t*
Root mean squared velocity, in gas exchange at air–water interface 30
Rossby radius of deformation 278, 278*f*, 279–280
currents and 288
definition 277
surface mixed layer dynamics 219
Rossby waves 278, 283
Rotary spectra of currents 283, 283*f*, 284
Roundworm *see* Nematoda
Rouse number, equilibrium transport of suspended solid 131–132, 132*f*
Running water
algae 445
carbohydrates 445
Runoff
continental-scale runoff 70
recent increases 70
definition 175
mercury and 652–653
streams 136–137

urban areas
as cause of flooding 645–646, 646*f*
dissolved silicate 414
Rushes *(Juncus)* 302*t*

S

Saccharides 442
uptake measurements 447
Salicornia (glasswort) 302*t*
Saline lakes
arid regions 353, 418
bird habitat 419–420
dryland salinity and stalinization 421
nutrients 420
salinity stratification 199
see also Salinity, stratification
threats 422–422
crops 421
remediation 422–422
water diversion 421
Saline waters
groundwater quality 92
nitrogen fixation 396, 412
water chemistry 396
Salinity **418–423**
anions associated 361
definition 399, 418
effects on biodiversity 419, 421–422
eutrophication effects 621
factors affecting 418
hydrostatic pressure and 20
inland waters 361, **418–423**
ecological responses 419
influences on circulation 420
variable 420
ionic composition 418
movement towards equilibrium 421
primary and secondary 421
sources 361–362
specific conductivity 418
proportional to ionic composition 418
stratification 199
brackish water 199
dissolved gases 200*f*
salt lakes 199
total dissolved substances 199
tolerances of organisms 419
biogeographical factors 419
evolutionary factors 419–420
total dissolved salts (TDS) 418
Salinization 361
definition 361
economic costs 362
land use effects 362
Salix spp. (willow)
river flow resistance 308–309
Salt lakes *see* Saline lakes
Saltmarshes
nitrogen cycle 402
nitrogen fixation 393
pioneer zones 310
sediment deposition 310
vegetation, canopy morphology 310
water flow 310
Salts 420, 421
see also Salinity
Sandstone, cation concentrations 359*t*
Saprobien System, ecosystem bioassessment 579–580
Satellite remote sensing, precipitation observations 62
microwave sensors 62–63
passive imaging radiometers 63–64
Saturated fatty acids (SAFAs), definition 478
Saturated water, groundwater chemistry 107–108
Scale invariance, definition 66
Scenedesmus
grazers, impact of 432
Scheldt River (Belgium), amorphous silica pools 414–415
Schindler, David, nitrogen fixation studies 396
Schistosoma 677
Schistosoma haematobium 683*f*
Schistosomiasis 677, 679
blood fluke life cycle in humans 679
control strategies 682, 684
definition/description 685
eutrophication leading to increase in snails 406
susceptibility factors 679
swimmer's itch 679, 685
vectors 683*f*
Schmidt number
definition 28
in gas transfer 29, 31, 32*t*, 33, 34
Schmidt stability, in density stratification and stability 204
Schreckstoff (alarm cue), allelochemicals 429
Seagrass beds, nitrogen fixation 393, 397
Seas *see* Oceans/seas
Seasonal changes/fluctuations
flow disturbances in rivers/streams 336*f*, 337
groundwater recharge 90
precipitation/evapotranspiration patterns 145*f*
river redox potential 554, 554*f*
sediment–water interface redox potential 552, 553*f*
wetland hydrographs 182
Seasonality
dissolved inorganic *vs.* organic nitrogen 517
Seawater
chloride composition 361
Secondary metabolites *see* Allelochemicals/allelochemical reactions
Second class waves *see* Vorticity waves
'Second-dissociation constant', dissolved CO_2 and 345
Sedges *(Carex)* 302*t*
Sediment(s)
delivery of world's largest river basins 144*t*
direct effects, benthic boundary layer (BBL) 230
iron and manganese cycling 371
lakes *see* Lake ecosystem(s)
loads
rivers of Asia 148
world's largest river basins 144*t*
mercury pollution 648, 649*t*
methanogenesis 482–483
movement to streams *see* Stream(s)
nitrogen fixation 393
redox potential measurement 549
river ecosystems *see* River ecosystems
silica cycling 415
surface water 53
suspended *see* Suspended sediments
transport
stream flow characteristics 141, 141*f*
vegetation effects 310–311
Sedimentation
anthropogenic impacts 104–105
nitrogen cycle and 402–403, 404
resuspension 305
Sediment flux per unit area, equilibrium transport of suspended solid 132
Sediment slurry studies
methane oxidation 485
methanogenesis 485
Seepage meters 96, 98*f*
Seepage springs 156
Seiche(s) 260
disruption *see* Basin-scale internal wave degeneration
horizontal modes 260, 261*f*, 263*t*
linear wave equation 260, 262
oscillatory currents 261–262, 262*f*
internal 258*f*
in hydrodynamic modeling 324
internal modes 263, 263*t*
isopycnals 263–264
stratification 263–264
wave function of 263–264, 264*f*
wind stress 258*f*, 259*f*, 263–264, 273*f*
vertical modes 262, 264*f*
Seiching 288–289
Selenium (Se) 387, 388
pollution 675
Sensible heat flux, surface energy exchanges 212–213
Sensible heat loss, in gas exchange at air–water interface 32
Sensible heat transfer, surface energy exchanges 211
Sep River reservoir (France)

Sep River reservoir (France) (*continued*)
  carbohydrates 440*t*, 445
  enzymatic activity 449, 449*f*
Setschenow effect 31
Sewage
  carbon removal 659–660
  dissolved organic nitrogen 527
  phosphorus cycle 411
Shale, cation concentrations 359*t*
Shallow lakes *see* Lake(s), shallow lakes (and ponds)
Shallow water currents, equations 287
Shark River (USA), photoreactivity 43
Sharks (Chondrichthyes)
  swim bladders 26
Shear production, turbulence 208
Shear stress
  colloid formation 467*f*, 468
  definition 238
  river currents 240
  turbulent benthic boundary layers 230–232
Shear velocity, surface mixed layer dynamics 209–210
Sherwood Number *(Sh)* 194–195
Shigella spp. 569
Shoots, wetland ecosystems *see* Wetland ecosystems
Shoreline irregularities, circulation and 293, 293*f*
Shoreweed *(Littorella uniflora)* 302*t*
Short-range phases, water structure 2–3
Sigma coordinate vertical grid systems 329–330, 329*f*, 334
Silica (SiO_2, silicon dioxide) 124, 127, 350, **414–417**
  amorphous *see* Amorphous silica (ASi)
  cycling
    ecosystem level 414
    human impacts 416
    natural waters 415
    research directions 417
  removal 414–415
Silicates
  acidified water buffering mechanisms 559
  dissolved *see* Dissolved silicate (DSi)
  dissolving, time scale 414
Silicon (Si)
  nutrient loading, environmental effects 594*t*
Silicon dioxide (SiO_2) *see* Silica (SiO_2, silicon dioxide)
Simuliidae *see* Black flies (Simuliidae)
Single compound measurement, dissolved organic nitrogen uptake 534
Sink, definition 187
Sinkhole, definition 175
'Sinters', definition 156
Skin friction, benthic boundary layer (BBL) 232–233
Sluice gate(s)
  dynamic pressure forces 24, 24*f*
Small-scale sediment topography effects, benthic boundary layer 236, 236*f*
Smelting, copper pollution 381
Snail(s)
  habitats 680
  kairomones, alarm cues 429
  nutrient stoichiometry 538*t*, 543, 545*f*
  population increases 406
  vectors of human diseases 683*f*
    control strategies 682
    other diseases 679
    schistosomiasis 677, 679, 682
Snow
  precipitation measurement 61
Snowfield *see* Glaciers/snowfield
Snow/ice cover **68**
  amounts 68
  hydrological cycle 68
  melting effects 68
  see also Ice, cover
Snowmelt, as cause of flooding 644–645
Snowpacks, deforestation effects on dissolved nutrients 592
Sodium (Na) 124, 418
  functions 354
  marine-derived 356–357, 359*f*
  nutrient loading 586*t*
    deforestation effects 588*t*, 593
    environmental effects 594*t*
  surface waters 359–360
  valency 356
Soft water(s) 343
  acid neutralization 563
  chloride concentrations 361–362, 362*f*
Soil(s)
  acidification 563
  acid sulfate *see* Acid sulfate soil (ASS)
  buffering capacity, deforestation effects 594–595
  chloride concentrations 361
  dissolved combined amino acid sources 526–527
  dissolved free amino acid sources 526–527
  dissolved inorganic nitrogen sources 526
  dissolved organic nitrogen 526
  erosion
    sediment movement to streams 138
  fires and 633, 636
  hydrophobicity 635–636
  liming 360
  nutrient content, deforestation effects 594–595
  phosphorus 410–411
  physical–chemical reactions, deforestation effects 593
Soil moisture dynamic quantification, vadose water retention 79
Soil water
  effects, deforestation effects 591, 591*f*
  hydrological cycles *see* Hydrological cycles
  storage, water budgets *see* Water budgets
Solar brightening 71
Solar dimming 70–71
Solar energy, hydrological cycles 53, 54–55
Solar ponds, salt lakes 199
Solar radiation
  absorption 37
  in density stratification and stability 197
  global changes 71*t*
  spectrum 45
  see also Sunlight; Ultraviolet (UV) radiation
Solids, mass transport 338
Solitary waves, in internal wave degeneration 267, 268*f*
Solubility of gases, in gas exchange at air–water interface 31, 31*t*
Solubility parameter (δ), water solvation 4, 5*f*
Soluble chemical production, deforestation effects 593
Solute(s)
  diffusion, stratified benthic boundary layers 233
  effects on water density 10
  surface water 53
  transport, benthic boundary layer *see* Benthic boundary layer (BBL), solute transport
Solvation, water *see* Water
Solvents, organic *see* Organic solvents
Songhua River (China), pollution 668
Sonic anemometer, in measurement of gas exchange 34
Sorption effects, groundwater chemistry 107–108
Source changes, groundwater budgets 91, 92*f*
South America
  Orinoco River *see* Orinoco River (South America)
  springs 169*t*
South Atlantic basins, South American rivers draining to *see* South America
Spain
  springs 167–168
    frequency 171*f*
    macroinvertebrates 164*t*
Sparganium 302*t*, 308–309
Spartina (cordgrass) 302*t*
Spartina alterniflora, water flow modification 313
Species
  loss, lakes, climate change 614
Specific yield, mass transport 338
Spectral signal processing, in hydrodynamic modeling 320
Sphagnum (moss) 352
Spiraling *see* Nutrient spiraling
Spring(s) **155–176**
  biology 157
    macroinvertebrates 162–163, 163*f*, 164*t*, 166*f*
    microbes 156
    nonthermal 157
    temperature limits 167*t*, 168*f*
    thermal *see* Hot springs; Thermal springs
  classification 155, 158*t*
  conservation 173, 173*f*
  conservation priorities 158*t*, 172
  definition 155
  dissolved gases 155–156
  ecology 158*t*

geographical locations 169*t*
geological origin 155, 156*f*
human uses 158*t*, 173
hydrological properties 156
importance 167
cultural 172
ecological 168, 171*f*
economic 173
geological 167, 169*t*
scientific 168
societal 172
island-like properties 172
seepage 156
temperature 155, 156–157, 158*t*, 170–172, 171*f*
water chemistry 155–156, 158*t*
Spring water age, definition 174
Spur dykes (groynes) 243
Sree River (Germany), currents 240*f*, 241*f*
St. Lawrence River basin (North America), water/sediment delivery 144*t*
Stability, density stratification *see* Density stratification
Stable isotopes
carbohydrates and 450
ratios, groundwater chemistry 109–110
Stagnant film model, in gas exchange at air–water interface 29, 29*f*
Stagnation period, in density stratification and stability 196
Starch 438*f*
Starwort *(Callitriche)*, river flow resistance 308–309
Stationary habitats, biogeochemical influence on rivers/streams 335
Stefan Boltzman constant, surface energy exchanges 212
Stereochemistry, definition 399
Stirring limit, surface mixed layer dynamics 214–215
Stoichiometry, definition 456, 537
Stokes drift velocity, surface mixed layer dynamics 210
Stoneworts *see* Charophyceae (stoneworts)
Storms
coastal, as cause of flooding 645
turbulence and 228
Storm surges 645
Stratification
continuous, internal wave rays in 269, 270*f*, 271*f*, 272*f*
density *see* Density stratification
hydrodynamic modeling and 320, 331–332
internal, wind set-up 257
maintenance 664–665
meromictic lakes
density 196–197, 199, 201
surface mixed layer and 207, 219*f*, 220*f*
thermal *see* Thermal stratification
turbulence and 207–208
water-column 256, 257*f*
Stratified lakes/waters
benthic boundary layer *see* Benthic boundary layer (BBL)
dissolved inorganic *vs.* organic nitrogen 517–518, 518*f*
iron concentrations 371–372, 373
manganese concentrations 371–372, 373
redox potential *see* Redox potential
thermal *see* Thermal stratification
turbulence 190, 225, 226*t*
in surface/bottom boundaries 225
Stratified mixed layers (SML), currents **286–300**
vertical circulation 293
Stratiform, definition 66
Stratiform clouds, formation 60
Stream(s) **134–142**
see also rivers and vice versa
benthic zone 99, 101*f*
biogeochemical zones 335, 336*f*
backwaters *see* Backwater(s)
channel surface 335–336, 336*f*
chemical composition of water and 341
flowing water of channel 335, 336*f*
channel *see* Stream channel
chemical composition 340*t*, 341
chemical conversions 340, 340*t*
chemical fluxes and dynamics **335–342**
chloride concentrations 363–364, 363*f*, 364*f*
coarse woody debris *see* Coarse woody debris (in lakes/streams)
definitions 99
discharge-weighted mean concentration 338
dissolved substances within 338, 339*t*
downwelling zones 99, 100, 101*f*
drainage network organization 134, 135*f*
channel formation 134
dendritic drainage 134, 135*f*
density 134
erosion 135
preferential flow zones 134
radial drainage 134
rectangular drainage 134
rills 134
subsurface flow 134–135
ecosystems *see* Stream ecosystems
elements (chemical)
biogeochemically important 339*t*
concentration in rivers/streams dependent on flow 338, 340*t*
rate of transformation/lability 338, 339*t*
flow
deforestation effects on dissolved nutrients 591
flow characteristics 139
basin size 139–140
bedload 141
dissolved (solute) load 141
duration curves 140
flood-frequency curves 140
flood hydrographs 139, 140*f*
flow duration curves 140
hydrographs 139
laminar flow 140–141, 141*f*
precipitation effects 139–140
sediment transport 141, 141*f*
stream hydraulics 140–141
stream hydrology 139
suspended load 141
suspensions 141
turbulent flow 140–141, 141*f*
washload 141
flow sources 135
climatic circulation patterns 136
cyclonic storms 136
monsoonal storms 136
rainfall 135
rain-on-snow 136
reservoir failure 136
snowmelt 135
gaining reaches 99, 100
hydraulics, stream flow characteristics 140–141
hydrographic change 337
hydrologic exchange 95
hydrology, stream flow characteristics 139
hyporheic zones 99, 100
iron *see* Iron (Fe)
labile constituents 338, 339*t*, 342
losing reaches 99, 100
manganese *see* Manganese (Mn)
sediment movement to 138
floodplains 138–139
mass movement 138, 139*f*
soil erosion 138
sources 138
stream bed/bank erosion 139
slope 99
upwelling zones 99, 100, 101*f*
water movement to 136
groundwater flow 136–137, 137*f*
hillslope characteristics 137, 138*f*
pipes/macropores 136–137
runoff 136–137
throughflow 137
see also River(s)
Stream bed/bank erosion 135
sediment movement to streams 139
Stream channel
characteristics, deforestation effects on particulate nutrients 596
formation 134
Stream ecosystems
anoxic processes *see* Anoxic processes
biotic/abiotic processes affecting algae *see* Algae, river ecosystems
biotic/abiotic processes affecting dissolved elements 338
comparison of different-sized flowing waters 338
conservative substances 338, 342
discharge-weighted mean concentration 338

Stream ecosystems (*continued*)
human influence 337
hydrographic change 337
hyporheic zone 99, 100
iron *see* Iron (Fe)
labile constituents 338, 339*t*, 342
lake ecosystems *vs.* 335, 341
longitudinal gradients in 335
longitudinal movement 341
manganese *see* Manganese (Mn)
mass transport *see* Mass transport
nitrogen cycle 402
nutrient spiraling *see* Nutrient spiraling
oxic processes *see* Oxic processes
precipitation influences 336*f*, 337
riparian zone/watershed influences 337
stationary habitats, biogeochemical influence 335
see also Stream(s)
Streamflow *see* Stream(s), flow
Streamflow capture, groundwater budgets 91
Stressors
algae in river ecosystems *see* Algae, river ecosystems
environmental
effect on eutrophication 627
identification, ecosystem bioassessment 581
Stromatolites 172
definition 175
Strontium-90 675–676
Stygobites 99
Stygophiles 99, 102–103
Subcritical flow, river currents 239–240
Sublimation, water 14
Submerged macrophytes
sediment–water interface redox potential 552, 552*f*
Substances, humic *see* Humic substances (HSs)
Substrate analog studies, bacterial uptake of dissolved combined amino acids 532, 532*f*
Subsurface exchange zones, importance to ecosystem function in river channels 147*f*
Subsurface flow, streams 134–135
Subsurface inflows, density plumes and 251
Succession
phytoplankton, allelochemicals and 435
Sucrose 438*f*
Sulfate ($SO_4$2-) 102–103, 124, 350–351
hydrological cycles 54
methanogenesis 481–482
nitrogen fixation 395, 396–397, 398
nutrient cycling 54
reduction *see* Sulfate reduction
Sulfate-reducing bacteria (SRB)
mercury methylation 649, 656
redox potential 553
Sulfate reduction
lakes 373
Sulfide (S_2-) 387
mercury speciation 648–649
Sulfide minerals
hydrological cycles 54
nutrient cycling 54
Sulfur (S) 350, 389
hydrological cycles 54
nutrient cycling 54
nutrient loading 586*t*
deforestation effects 588*t*, 593
environmental effects 594*t*
Sulfur dioxide, emissions 563
Sulphur *see* Sulfur (S)
Summer stratification, lakes, climate change 614–615
Sump, definition 187
Sunlight
colloid formation 468
see also Light; Ultraviolet (UV) radiation
Supercritical flow
in internal wave degeneration *see* Basin-scale internal wave degeneration
river currents 239–240
Surface coagulation, colloid formation 467*f*, 468
Surface energy exchanges 211
Beer's law 211–212
black body radiation 212
diffuse attenuation coefficient 211–212
first law of thermodynamics 212
heat flux 211, 213*f*, 214*f*
latent 212–213
sensible 212–213
turbulent 212
heat transfer
latent 211
sensible 211
Stefan Boltzman constant 212
temperature of maximum density 211–212
Surface fire, definition 643
Surface layers (SL), diffusivity 227
Surface mixed layer (SML) **207–221**, 208*f*
deepening and entrainment 213
mixing and turbulence energetics 213, 215*f*, 216*f*
synthesis 217
upwelling 216, 218*f*, 221
energy exchanges *see* Surface energy exchanges
motions/turbulence 207
basic ideas 207
winds on lakes 208
nutrients and 207
pollutants and 207
polymictic lakes 221
Surface momentum transfer, wind set-up and 256
Surface renewal model, in gas exchange at air–water interface 29–30, 34–35
Surface shear stress, turbulence and 225–226
Surface slicks, gas exchange and 32
Surface tension 14
definition 14
surfactant effects 14
Surface water
chemical signature 102–103
fauna 99
hydrological cycles *see* Hydrological cycles
inflow, wetlands hydrology 183–185
mercury levels 649*t*
outflow, wetlands hydrology 185
quality assessment, ecosystem bioassessment 579–580
sodium 359–360
temperatures 96
wetlands hydrology 179
see also Ground Water and surface water interaction
Surface waves, surface mixed layer dynamics 208–209, 209*f*
Surfactants
in gas exchange at air–water interface 32
Suspended loads
stream flow characteristics 141
see also Sediment(s); Sedimentation; Suspended sediments
Suspended sediments
flooding and 646, 647, 647*f*
Suspended solids, fluvial transport **131–133**
equilibrium transport 131
bed material load 131
calculations 131
depth variation 131–132
point sediment samples 131, 132*f*
Rouse number 131–132, 132*f*
sediment flux per unit area 132
wash load 131
nonequilibrium transport 132
calculations 133
examples 133
mass conservation 132–133
porosity 132–133
trap efficiency 132–133
urban areas 131
Suspensions
stream flow characteristics 141
see also Suspended loads; Suspended sediments
Sweden
mercury levels
emissions 656
lakes 648
Swim bladders, buoyancy 26
Switzerland
nonthermal springs, macroinvertebrates 164*t*
Sycamore Creek (Arizona, USA) 101*f*
Sycytonemin A, structure 431*f*
Synomones, definition 425–426
Synthetic organic chemicals, pollution and 670

T

Tabanidae (deer flies) 679
Tannins
  as allelochemicals 435, 436*f*
  as herbivore deterrents 431–432
'Taste-and-odor' algae, harmful algal blooms *see* Algal blooms, harmful
Taylor-Goldstein equation, Kelvin-Helmholtz billows 265–266
Tellimagrandin II, structure 432*f*
Telluric water, definition 188
Temperate zones
  nitrogen fixation 393, 405
  nutrient limitation 403
Temperature
  deforestation effects on dissolved nutrients 592
  effects on
    dissolved gases 7, 7*f*
    viscosity 15
    water density 10, 11*t*
  hydrostatic pressure and 20
  of maximum density
    differential cooling 251
    surface energy exchanges 211–212
  methanogenesis 481–482
  profiles, hydrodynamic modeling 320, 321*f*, 332–333
  river redox potential 554
  solute transport, benthic boundary layer 237
Temperature stratification 197
  chemocline 200
  diffusive heat transport 197
  epilimnion 197, 198*f*, 200
    see also Epilimnion
  halocline 200
  hypolimnion 197, 198*f*, 200
    see also Hypolimnion
  internal current shear 197
  lakes, climate change *see* Climate change and lakes
  metalimnion 198, 198*f*
    see also Metalimnion
  mixing depth 197
  nutrients 197–198, 205
  pycnocline 200
  solar radiation 197
  thermocline 197, 198, 198*f*
    see also Thermocline(s)
  turbulent heat transport 197
  wind stress 197
  see also Thermal stratification
Tensiometers, vadose water 76–77, 78*f*
  retention 78*f*, 79
Tension infiltrometer, vadose water unsaturated hydraulic conductivity 81
2,3,7,8-Tetrachlorodibenzo-*p*-dioxin (TCDD) 672
  aryl hydrocarbon receptor binding 460, 461*f*
Tetrahedral geometry, water structure 1
Thames River (UK), pollution 658
Thermal bar/baricity
  cabbeling 251–252
  density plumes 251, 252*f*, 253
    mesothermal maximum 253, 253*f*
  mixing 247
  tropical and saline lakes 254
Thermal expansion coefficient, in density stratification and stability 201
Thermal forcing, solar radiation-induced 310–311
Thermal pollution 676
Thermal springs 156–157
  biology 165, 166, 167*t*, 168*f*
  density 163*f*
  importance 172
  see also Hot springs
Thermal stratification
  redox potential in stratified lakes 550
  see also Temperature stratification
Thermistors, in measurement of gravity waves 283, 284
Thermobaric instabilities
  density currents and 247
  Ekman transport 253–254
Thermobaric stratification 198
  temperature of maximum density (T_{md}) 198–199, 199*f*
Thermocline(s)
  density plumes and 249–250
  in density stratification and stability 197, 198, 198*f*, 199
  internal waves
    nonlinear shoaling of 268–269, 273*f*
    resonant forced 272–273
    surface momentum transfer and wind set-up 257–258
  Kelvin-Helmholtz billows 265
  modeling 320, 321*f*, 331–332
  surface mixed layer and 207, 208*f*, 212, 216, 217*f*
  see also Temperature stratification
Thermocline jet, in internal wave degeneration 267, 274*f*
Thermodynamics, first law, surface energy exchanges 212
Thermus aquaticus 173
Thiessen's polygon area-averaging method, precipitation measurement 61
Thorpe scale 228
Three-dimensionality, groundwater chemistry 113, 114*f*
Threshold Element Ratio (TRE), nutrient stoichiometry 542, 542*f*
Throughflow
  definition 188
  streams 137
Tides 645
  incoming/flood, wetlands hydrology 183–185
  outgoing/ebb, wetlands hydrology 185
Time-domain reflectometry (TDR)
  soil-water storage 56
  vadose water 76
Timescales
  deforestation, nutrient load effects 599–600
Tires, habitat for human disease vectors 680–681
Titrations, natural organic matter, acidic functional group
    measurement 512
Toolik Lake (Alaska/USA)
  catchment effect, climate change 614
  iron 370*f*
  manganese 370*f*
Topographic waves *see* Vorticity waves
Topography
  wetlands hydrology 177–178
Total dissolved carbohydrates (TDCHO) 439, 440*t*
Total dissolved nitrogen (TDN), definition 517
Total dissolved solids (TDS)
  in density stratification and stability 199
  fluxes 127
  spring classification 158*t*
Total head, dynamic pressure 23
Total organic carbon (TOC) 503–504
  rivers 504, 504*t*
Total suspended solids (TSS)
  fluxes 118, 119, 127
    duration curve 120
    geographic variability 121, 122*t*
Toxicity studies, ecosystem bioassessment 577, 579
Toxins
  cyanobacterial 420
  see also Pollutant(s)
Trace elements
  concentration ranges
    earth's crust 383
    surface waters 383
  cycling and transport 387
  definition 383
  essential and nonessential 383
    essential 383
    nonessential 383
  sources 383
    anthropogenic 384
    geological 383
      erosion and dissolution 383
      geothermal 383
      human mediated 384
      transport in ground water 383
  unique properties 383
Trace metals and metalloids **383–391**
  atmospheric pollution 384
  bioavailability models
    Biotic Ligand Model (BLM) 387–388
    free metal ion activity model 387
    other models 388
  biogeochemistry in fresh water 384
    aqueous complexation 385
    complexing agents 386
    dissolved organic matter interactions 386
    equilibrium association constants 385

Trace metals and metalloids (*continued*)
calculation 385–386
ligand binding 385
major processes 385, 385*f*
natural organic matter interactions 386
oxidation, reduction, speciation 386
microbial 387
processes 386
particle adsorption 386
phase partitioning 386
phases 384
biological uptake, accumulation and transformation 387
metal resistance mechanisms 388
arsenic and selenium 388
mercury 388
toxic, biological protection pathways 387
water pollution 384
common contamination sources 384
Tracers
for estimating gas exchange rates 33
hydrodynamic modeling 332, 333*f*
measuring transit time in flowing waters 145–146
Tracer techniques
groundwater chemistry 109*t*, 112
groundwater flow systems 89–90
Transient storage zone modeling 100, 101*f*
method 100
Transit time, rivers 145–146
Transparency, water, rivers 148
Transparent exopolymer particles (TEP) 450
colloid formation 467*f*, 468
Transpiration 59, 69
see also Evapotranspiration
Trap efficiency, nonequilibrium transport of suspended solids 132–133
Trapped air, vadose water retention 78
Travertine springs, classification 156
Trends analysis, ecosystem bioassessment 581
Trent Biotic Index, ecosystem bioassessment 580
Triacylglycerol (TAG) 472
definition 478
eukaryotic algae 474
structure 473*f*
Triclosan, pollution and 672–673
Trithiane, structure 432*f*
Trophic cascades
definition 628
Trophic energetics, research/conceptual models 152–153
Trophic levels
definition 576
Trophic State Concept, eutrophication 617
Trophic structure, definition 175
Tropical floods 646
Tropical Rainfall Measuring Mission (TRMM) 62–63, 64*f*
Truckee River (USA), molybdenum 379
Tsetse (Glossinidae) 679
Tsunamis, as cause of flooding 645
Turbidity 420
currents, in density-driven flow 249–250, 254
Turbulence 190
carbon dioxide diffusion 193–194
colloid formation 467*f*, 468
density stratification 207–208
energy flux 228
in gas exchange in lotic systems 35
hydrodynamic modeling 320–321, 331
induced, stratified benthic boundary layers 234, 235*f*
intensity 190
internal wave rays in continuous stratification 270
internal waves 223*f*, 227
mechanisms 225–226
open 191–192
particle entrainment 190–191
pelagic habitats 192
small-scale 190, **222–229**
energy fluxes 223*f*
mixing *see* Mixing (and small-scale turbulence)
storms and 228
stratified lakes *see* Stratified lakes/waters
submerged vegetation 315
Turbulence coefficients, in hydrodynamic modeling 324
Turbulence velocity scale
equations 190
Turbulent benthic boundary layer (TBBL) 225, 230
internal wave degeneration and 264–265
internal waves and 256, 273*f*
'law of the wall' 232
shear stress 230–232
Turbulent eddies
definition 238
in gas exchange at air–water interface 28, 29–30
vertical eddy viscosity, hydrodynamic modeling 330
Turbulent flow, stream flow characteristics 140–141, 141*f*
Turbulent heat transport, in density stratification and stability 197
Turbulent kinetic energy (TKE)
definition 224–225, 238
dissipation rate, definition 238
lake ecosystems 304–305
production-dissipation balance 226–227
surface mixed layer 208, 209–210, 215*f*
Turbulent mixing
lakes, climate change 613
surface mixed layer 207
Turbulent stresses, canopy effects 313–314, 315*f*
Turbulent velocity scales (w_*)
in gas exchange at air–water interface 30
surface mixed layer dynamics 209–210
Turbulent vertical eddy viscosity, hydrodynamic modeling 330
Turning depth, internal wave rays 270
Turtosity, solute transport, benthic boundary layer 235–236
2-dimensional mixing, stratified benthic boundary layers 233

U

Ultrafiltration (UF), natural organic matter isolation 506
limitations 506
molecular weight cut-offs 506, 507*t*
Ultraviolet (UV) radiation 45
effects on aquatic plants 604
lakes, climate change 612–613
UV-A 45
UV-B 45
see also Solar radiation
Unconfined aquifers 88, 89*f*
Underwater channels, density plumes 249–250
UNESCO formula, in density stratification and stability 202
United Kingdom, nonthermal springs, macroinvertebrates 164*t*
United States Environmental Protection Agency (USEPA), wastewater treatment plants 659–660
United States of America (USA)
ecosystem bioassessment 582–583
lakes *see* North American lakes
nonthermal springs, macroinvertebrates 164*t*
water regulation effects of impounded rivers 148*t*
Unsaturated hydraulic conductivity, vadose water *see* Vadose water
Unsaturated hydrostatics, vadose water 76
Unsaturated zone *see* Vadose water
Unstable variations, vadose water preferential flow 82, 83*f*
Unsteady diffuse flow, vadose water 80
Unsteady flow measurement, vadose water unsaturated hydraulic conductivity 81
Unstratified water bodies, currents **286–300**
see also Homogenous basins (HB); Homogenous layers
Upper mixed layers *see* Stratified mixed layers (SML), currents
Upwelling
internal waves and 258, 262–263
surface mixed layer 216, 218*f*, 221
Upwelling zones 99, 100, 101*f*
Uranium 676
Urbanization
chloride concentrations 363
effects on groundwater recharge 90
effects on hydrologic cycle 105
Urban runoff, as cause of flooding 645–646, 646*f*
Urea
dissolved organic nitrogen (DON) 525
uptake 533
detection 528*t*, 533

Uroglena american, bloom effects 625–627
USA *see* United States of America (USA)
US Department of Health (USDOH), mercury
 pollution 674–675
US EPA National Stream Survey, cation chemistry 356, 357*f*
UV *see* Ultraviolet (UV) radiation

V

Vadose water **76–87**
 content 76
  gravimetric measurement 76
  neutron scattering 76
  time-domain reflectrometry 76
 Darcy's law 80
 definition 76
 diffuse unsaturated flow 79
  representative elementary volume 79, 80*f*
 hydrologic cycle *see* Vadose water
 porous medium 76, 77*f*
 preferential flow 81
  equivalent-medium measurement 82–83
  funneled flow 82
  macropores 82, 82*f*
  quantification 82
  types 81, 82*f*
  unstable variations 82, 83*f*
 processes 76
 retention 77
  examples 78
  formulae 79
  measurement 79
   property-transfer models 79
   tensiometers 78*f*, 79
  particle-size distribution 78
  practical significance 79
  retention curves 78, 78*f*
  soil moisture dynamic quantification 79
  trapped air 78
 unsaturated hydraulic conductivity 80, 80*f*
  capillary theory 81
  dissimilar material effects 81
  empirical formulae 81
  instantaneous-profile method 81
  measurement 80
  property-transfer models 81
  tension infiltrometer 81
  unsteady flow measurement 81
 unsaturated hydrostatics 76
 unsteady diffuse flow 80
 water pressure/energy 76, 77*f*
  capillary effect 76
  humidity-based measurement 77
  tensionmeter 76–77, 78*f*
Vadose water, hydrologic cycle 83
 aquifer recharge 86
  definition 86
 land surface interactions 84
 moisture dynamics 84
  evapotranspiration 84
  hysteresis 86
  infiltration 84, 84*f*
  layered media 86
  overlying surface water 86, 86*f*
  perching 86
  redistribution 85, 85*f*
 moisture state 83
  capillary fringe 84
  distribution effects 83–84, 83*f*
  field capacity 84
  hydrostatic profiles 83–84, 83*f*
Van der Waal forces, water structure 1–2
Van Krevelen diagram, biomass triangle 509, 510*f*
Van't Hoff's rule 664
Vapor pressure, water 15
Vegetation
 affecting river flow *see* River(s), flow
 aquatic ecosystems
  flexibility 309
  grassed channels 308
  implications 301
  shoots 308, 309
  turbulence fields 310–311
  see also Aquatic plant(s)
 hydrodynamic characteristics 302*t*
 hydrophytic, definition 187
 lake ecosystems *see* Lake ecosystem(s)
 littoral primary producers *see* Littoral zone, primary producers
 nontidal wetlands *see* Nontidal wetlands
 proximity to fresh water, deforestation effects on particulate nutrients 596
 riparian zones 102
 salt tolerance 419
 selenium uptake 388
 submerged, flow modification **313–319**
  distribution 313, 314*f*
  equations 314–315, 317
  stem scale 313, 315
  vertical profiles 314*f*
  see also River(s), flow; Wetland ecosystems
 turbulence effects 191
 types, deforestation effects on nutrient load 599–600
 wetlands *see* Wetland plants
 see also Plant(s)
Velocity, vertical, fluctuations 224–225
Velocity gradients, currents and 293*f*
Velocity profiles, currents and 295*f*
 basin depth 295–297, 295*f*
 model 294*f*, 295
Vernal pools
 definition 188
Vertebrates
 definition 175
Vertical differences, water density 10
Vertical habitat structure, lakes, climate change *see* Climate change and lakes
Vertical hydraulic gradient (VHG) 96, 98*f*, 100*f*
Vertical mode 285
Vertical mode seiches 262, 264*f*
Vertical velocity, fluctuations 224–225
Virus(es)
 thermal springs 165–166
Viscosity, water *see* Water
Viscous sublayer, benthic boundary layer (BBL) 231*f*, 232, 232*f*
Vivianite 373
Volatile tracers, for estimating gas exchange rates 33
Volatilization, definition 643
Volcanic acidification 562*f*, 564
 definition 567
Volcanic rock, cation concentrations 359*t*
Volcanic waters 564
 acid 562*f*, 564
Von Karman constant, surface mixed layer dynamics 209–210
Vortical modes *see* Vorticity waves
Vorticity (ζ) 289, 290*f*
 definition 285
 submerged vegetation and 315, 316–317, 318–319
Vorticity waves 278, 283
 definition 285
 measurement 284–285

W

W
Walleye, mercury concentrations 652*f*
Waquit Bay (off Massachusetts), Metallogenium bacteria 380
Warming, global *see* Global warming
Warm Spring (Huntingdon, USA) 156*f*
Wash load
 equilibrium transport of suspended solid 131
 stream flow characteristics 141
Waste byproducts, pollution and 672
Wastewater
 dissolved silicate 414
 infiltration, groundwater chemistry 112
Wastewater treatment plants (WWTPs) 659–660
 trace contaminants 660, 673*f*
 treatment levels 659*t*, 664
Water
 air–water interface *see* Air–water interface
 atmospheric *see* Atmospheric water
 balance

Water (*continued*)
global, water budgets and 57
lakes, climate change *see* Climate change and lakes
budgets *see* Water budgets
chemical properties **1–9**
see also Water chemistry
clarity 603–604
colligative properties 15
boiling point 15
freezing point 15
osmotic pressure 15
vapor pressure 15
column *see* Water column(s)
cycle *see* Water cycle
delivery to river basins, largest worldwide 144*t*
density 10
definition 10
lake layers 11, 11*t*
mixing 10–11
solute effects 10
temperature effects 10, 11*t*
vertical differences 10
discharge, definition 174
dissolved gases 6
anoxia 6–7
electron flow tracers 6–7
nutrient loading 7
organic carbon 7
photosynthesis 6–7
temperature effects 7, 7*f*
distribution 10
flow
potential energy 239–240
formation 10
heat capacity (specific heat) 11, 12*t*
consequences 12
definition 11–12
heat storage/release 12
heat of fusion (melting) 12
definition 12
heat of vaporization (condensation) 13
indicators of quality *see* Water quality
isotopes 13
level fluctuations
wetlands, redox potential 554–555, 555*f*
major ions 7
movement
scale concept 189
to streams *see* Stream(s)
nutrients 7
accelerated loading 8
algal requirements 7–8
calcium 8
iron 8
phosphorus 8
pH 343, 345, 346*f*
physical–chemical reactions, deforestation effects 593
acid production 594
amount of deforestation 594–595
buffer strips 594–595, 596*f*
deforestation timing 594–595
nutrient dependence 594, 594*t*
primary production 594
revegetation 597*f*
soil buffering capacity 594–595
soil nutrient content 594–595
treatment effects 594–595, 597*f*
water bodies 594–595, 597*f*
physical properties **10–16**
pollution *see* Pollution
pressure 15
see also Pressure
properties 189
purification 575
quality *see* Water quality
reactivity 6
coordination changes 6
proton hopping 5*f*, 6
redox reactions 6
regulation *see* Water regulation
removal, aquifers 88
see also Aquifers
resources, availability 659*t*
solvation 3
'concentric shell' model 5–6, 5*f*
electronic factors 5
geometric factors 5
ionic structures 4–5, 4*f*
solubility parameter (δ) 4, 5*f*
structure 1, 2*f*, 10, 11*f*
bulk fluid structure models 1–2
Coulombic forces 1–2
Debye forces 1–2
hexameric ice 2–3, 2*f*, 3*f*
hydrogen bonds 1–2
'ice-like' structures 3, 4*f*
Keesom forces 1–2
London (dispersion) forces 1–2
long-range phases 2–3
models 3
'Platonic solids' 3
short-range phases 2–3
tetrahedral geometry 1
van der Waal forces 1–2
sublimation 14
surface tension *see* Surface tension
table *see* Water table
viscosity 15, 189–190, 191
temperature effects 15
see also Flow
Waterbirds
saline lake habitat 419–420
Water budgets **51–57**
Californian example 56*t*, 57
component assessments 55
evapotranspiration 55
gravimetric lysimeter 55–56
potential evaporation *vs.* precipitation 55–56
framework 55
equations of 55
management 55
global water balance 57
groundwater runoff 56
Darcy's law 56
groundwater storage 56
soil-water storage 56
field water capacity 56
neutron logs 56
Time Domain Reflectometry 56
wetlands *see* Wetlands hydrology
see also Hydrological cycles
Water chemistry **1–9**
factors controlling 418
response to contaminants 104–105
response to ecological conditions 102
rivers 148, 149
Water circulation *see* Circulation
Water column(s)
carbon biomass 453, 454*t*
iron cycling 373
manganese cycling 373
nitrogen inputs 403
optical characteristics 45
stratification 256, 257*f*
turbulence 190
'Water crisis' 658
Water crowfoot *(Ranunculus)* 302*t*
Water cycle 58, 69
multiyear climatic patterns 58
temporal patterns 58
see also Hydrological cycles
Water fleas (*Cyclops* spp.) 679–680
Water Framework Directive (WFD) 582–583
Water lily *(Nymphaea)* 302*t*
Waterlogging, definition 188
Water milfoil *(Myriophyllum)* 302*t*
Water pollution *see* Pollution
Water quality
effect of biogeochemistry 102
effects of transient storage 100–101
guidelines 576
indicators 574–575

organisms 576
stream health *see* Stream(s)
see also Bioindicator(s)
monitoring 421
pathogen monitoring 574
indicator systems 574–575
indicator organisms 576
testing methods 575
predictive models 575
standards 574
rivers 148
standards 576
surveys 119–120
Water regulation
rivers 147, 148*t*
Water resources, availability 659*t*
Watershed(s)
see also catchment and vice versa
chloride retention effects 365
fire impacts 630–631, 632*f*
slope 633
studies 637*t*
tinder sources 631–632
influence on biogeochemical processes in rivers/streams 337
mass wasting, deforestation effects on particulate nutrients 595
topography, deforestation effects, particulate nutrients 595
see also catchment
Water starwort *(Callitriche)* 302*t*
Water table
contamination, groundwater quality 92
definition 155
predictor, wetlands, redox potential 556
Waterton Creek 375*f*
Water vapor
atmospheric 59
dynamic effects 59
concentration 59
cooling mechanisms 60
life cycle 58
transport 59
Waterweed *(Elodea)*, hydrodynamic characteristics 302*t*
Waves
gravity *see* Gravity waves
internal *see* Internal waves
topographic 290
vorticity *see* Vorticity waves
Wax esters (WEs) 472
definition 478
structure 473*f*
Weathering
definition 377
effect on alkalinity 350
geological *see* Geological weathering
limestone 354–355
mineral, cations 354–355, 355*t*
pyrite 560
Wedderburn numbers *(W)*
in density stratification and stability 204
internal waves 259, 263
surface mixed layer 215, 217*f*
Wet deposition, definition 567
Wetland(s)
chemistry, groundwater chemistry *vs.* 116
definitions 177, 188
ecosystems *see* Wetland ecosystems
flow and vegetation effects *see* Wetland ecosystems
hydrologic exchange 104
hydrology *see* Wetlands hydrology
methane emissions 498, 499*t*, 501
monomethyl mercury production 390
non-tidal *see* Nontidal wetlands
plants *see* Wetland plants
redox potential *see* Redox potential
see also Floodplain(s); Lake(s)
Wetland ecosystems
flow, vegetation effects **301–312**, 305*f*, 306*f*, 311*f*
lakes *see* Lake ecosystem(s)
mangroves 310
nontidal *see* Nontidal wetlands
rivers *see* River ecosystems
saltmarshes *see* Saltmarshes
shoots 308, 309
species diversity 310–311
tidal 306
nitrogen fixation 393
silica 414–415
wind stress 301
Wetland plants
invasive species
introduction 601, 606
Wetlands hydrology **177–188**
anaerobic conditions 180
contributing factors 177
coastal plains 177–178
floodplains 177–178
glaciers/snowfield 177–178
lakes 177–178
mountains 177–178, 178*f*
paludified landscapes 177–178
permafrost 177–178
topography 177–178
definition 177, 179, 188
depth of saturation 179
duration of wetness 179
frequency of prolonged wetness 179, 180
episodic 179
growing season 180
'biologic zero' 180
definition 180
specific plants 180
hydrographs 181, 183*f*, 184*f*
annual changes 182
changed water levels 182
long-term changes 182
seasonal changes 182
indicators 185, 186*t*
reducing conditions 180
regimes 181, 181*t*
water budget 182, 185*f*
annual 185, 186*t*
evapotranspiration 185
groundwater inflow 183–185
groundwater outflow (recharge) 185
incoming/flood tides 183–185
inputs 182–183
outgoing/ebb tides 185
outputs 185
precipitation 183–185
surface water inflow 183–185
surface water outflow 185
water sources 177
ground water 179
surface water 179
Whirligig beetles *see* Gyrinidae (whirligig beetles)
White Clay Creek (USA), carbohydrates 440*t*
WHO *see* World Health Organization (WHO)
Wildfires
ecological impacts 631
ignition 630
patterns 630
prescribed burn *vs.* 630
structure 630
Willow (*Salix* spp.)
river flow resistance 308–309
Wind(s)
cyclonically *vs.* anticyclonically rotating 290–291
fetch *see* Fetch
nutrient flux 193
speed, gas exchange at air–water interface 28, 30*f*, 34, 34*t*, 35–36
surface mixed layer motions/turbulence 208
turbulence and 190
surface shear stress 225–226
see also Turbulence
vertical mixing 222, 223*f*
Wind fetch *see* Fetch
Wind fields 292
effecting factors 292–293
Wind forcing 286, 295*f*
spatial variability 292
Wind setup 288
characterization 288–289
surface momentum transfer and 256

Wind stress
  in density stratification and stability 197
  internal waves and 256, 257, 258*f*, 259*f*, 260*f*
  lake ecosystems 301
  spatial variability 289–290
Woody debris *see* Coarse woody debris (in lakes/streams)
World Health Organization (WHO)
  Guidelines for Safe Recreational Water Environments 574, 574*t*
Wucheria bancrofti 677–678

X

XAD-4 resin, natural organic matter isolation 505–506
XAD-8 resin, natural organic matter isolation 505
XAD resin(s), natural organic matter isolation 505
Xenobiotics, definition 463
Xenopus laevis, humic substances, hormone-like effects 458

Y

Yellow fever 685
Yellow River basin (China), water/sediment delivery 144*t*
Yellowstone Lake (USA), ferromanganese concretions 376*f*
Yellowstone National Park (USA), biology 172
Yucatan Peninsula (Mexico), springs 172–173

Z

Zebra fish, embryonic development 460–461
Zebra mussel *(Dreissena polymorpha)* 606
Zinc (Zn) 380, 573
  soluble concentrations 381
  sources 381
Z-level vertical grid systems 329, 334
Zonation formation, sediment–water interface redox potential 551, 552*t*
Zone of aeration *see* Vadose water
Zooplankton
  carbohydrates 442
  carbon turnover 454*t*
  cations effect 420
  definition 628
  dissolved organic nitrogen effect 526
  ecosystems *see* Zooplankton, in ecosystems
  fire impacts 636–641
  grazing *see* Zooplankton, in ecosystems
  impact on phytoplankton populations 603–604
  nutrient stoichiometry 538*t*, 539, 542–543, 545*f*
  phosphorus, sediment burial 412
  rivers 151
  taxonomic diversity 539
  turbidity effects *see* Turbidity
  see also Copepoda
Zooplankton, in ecosystems
  grazing
    nitrogen fixation, effects 394–395, 397
Z–R relation, precipitation observations, radar 61–62

FSC
www.fsc.org
MIX
Papier | Fördert
gute Waldnutzung
FSC® C083411

Kolibri 360 GmbH
Ferdinand-Jühlke-Straße 7
99095 Erfurt, Deutschland
produktsicherheit@kolibri360.de